WASTEWATER REUSE AND WATERSHED MANAGEMENT

Engineering Implications for Agriculture,
Industry, and the Environment

WASTEWATER REUSE AND WATERSHED MANAGEMENT

Engineering Implications for Agriculture,
Industry, and the Environment

Edited by
Ajai Singh, PhD, FIE

Apple Academic Press Inc.
3333 Mistwell Crescent
Oakville, ON L6L 0A2
Canada

Apple Academic Press Inc.
1265 Goldenrod Circle NE
Palm Bay, Florida 32905
USA

© 2020 by Apple Academic Press, Inc.

First issued in paperback 2021

Exclusive worldwide distribution by CRC Press, a member of Taylor & Francis Group

No claim to original U.S. Government works

ISBN 13: 978-1-77463-431-8 (pbk)
ISBN 13: 978-1-77188-746-5 (hbk)

Library and Archives Canada Cataloguing in Publication

Title: Wastewater reuse and watershed management : engineering implications for agriculture, industry, and the environment / edited by Ajai Singh, PhD, FIE.

Names: Singh, Ajai, 1970- editor.

Description: Includes bibliographical references and index.

Identifiers: Canadiana (print) 20190071753 | Canadiana (ebook) 2019007177X | ISBN 9781771887465 (hardcover) | ISBN 9780429433986 (PDF)

Subjects: LCSH: Water reuse. | LCSH: Watershed management.

Classification: LCC TD429 .W37 2019 | DDC 628.1/62—dc23

Library of Congress Cataloging-in-Publication Data

Names: Singh, Ajai, 1970- editor.

Title: Wastewater reuse and watershed management : engineering implications for agriculture, industry, and the environment / editor: Ajai Singh.

Description: Palm Bay, Florida : Apple Academic Press, 2019. | Includes bibliographical references and index.

Identifiers: LCCN 2019006910 (print) | LCCN 2019008242 (ebook) | ISBN 9780429433986 (ebook) | ISBN 9781771887465 (hardcover : alk. paper)

Subjects: LCSH: Water reuse. | Watershed management.

Classification: LCC TD429 (ebook) | LCC TD429 .W353 2019 (print) | DDC 628.1/62--dc23

LC record available at https://lccn.loc.gov/2019006910

Apple Academic Press also publishes its books in a variety of electronic formats. Some content that appears in print may not be available in electronic format. For information about Apple Academic Press products, visit our website at **www.appleacademicpress.com** and the CRC Press website at **www.crcpress.com**

Printed in the United Kingdom
by Henry Ling Limited

ABOUT THE EDITOR

Ajai Singh, PhD, FIE

Ajai Singh, PhD, FIE, is an Associate Professor in the Department of Water Engineering and Management at the Central University of Jharkhand, Ranchi, India. He has more than 18 years of experience in teaching, research, and extension. He is a Fellow of the Institution of Engineers (India) and a life member of various societies. He has more than 30 research papers published in national and international journals of repute. Dr. Singh has authored one textbook, *Introduction to Drip Irrigation Systems*, which has proved to be very useful for graduate and postgraduate students of agricultural/civil engineering. He has been conferred the Distinguished Services Certificate (2012) by the Indian Society of Agricultural Engineers, New Delhi. He teaches subjects such as finite element methods, groundwater hydrology, micro irrigation, watershed management, and numerical methods.

CONTENTS

CONTRIBUTORS

Niva Bara
Head, Department of Agricultural Extension, Birsa Agricultural University, Kanke, Ranchi, Jharkhand, 834006, India

Amartya Kumar Bhattacharya
Chairman and Managing Director, MultiSpectra Consultants, 23, Biplabi Ambika Chakraborty Sarani, Kolkata – 700029, West Bengal, India, E-mail: dramartyakumar@gmail.com

Sabinaya Biswal
AGFE Department, Indian Institute of Technology, Kharagpur – 721302, West Bengal, India, E-mail: sabin9937@gmail.com

Nupur Bose
Associate Professor, Department of Geography, A.N. College, Patna – 23, Bihar, India

Anita Chakraborty
Senior Technical Officer, Centre for Environmental Management and Participatory Development, Salt Lake, Kolkata – 700106. E-mail: anitagcenator@gmail.com

Saswati Chakraborty
Department of Civil Engineering, Indian Institute of Technology Guwahati, Guwahati – 781039, India

Surendra Chandniha
Research Associate, National Institute of Technology, Roorkee, Uttrakhand, India

Ravish Chandra
Central Agricultural University, Pusa, Samastipur (Bihar), India, E-mail: ravish.cae@gmail.com

Suman Kumar Dey
Professor, Department of Geography, Ramsaday College, Amta, Howrah, West Bengal, India

Ashok Ghosh
Professor & Head, Research Section, Mahavir Cancer Sansthan & Research Centre, Patna – 801505, Bihar, India

Pranab Kumar Ghosh
Department of Civil Engineering, IIT Guwahati – 781039, India

A. K. Gosain
Professor, Indian Institute of Technology Delhi, Hauz Khas, New Delhi – 110016, India

S. Goswami
Professor, Department of Earth Science, Sambalpur University, Sambalpur, Odisha

Saurabh Kumar Gupta
Research Scholar, School of Natural Resource Management, Centre for Land Resource Management, Central University of Jharkhand, Brambe, Jharkhand, India

Subrata Halder
Executive Engineer (A-I), State Water Investigation Directorate, Government of West Bengal, India

V. K. Jain
Subject Matter Specialist, KVK, RVSKVV, Ashok Nagar, Madhya Pradesh, India

Kiran Jalem
Assistant Professor, Centre for Land Resource Management, Central University of Jharkhand, Ranchi, India, E-mail: jalemkiran@gmail.com

Madan Kumar Jha
Professor, AGFE Department, Indian Institute of Technology, Kharagpur – 721302, West Bengal, India

Garima Jhariya
Institute of Agricultural Sciences, Banaras Hindu University, Varanasi (U.P.) 221005, India, E-mail: garima2304@gmail.com

Mintu Job
Assistant Professor, Department of Agricultural Engineering, Birsa Agricultural University, Kanke, Ranchi, Jharkhand, 834006, India, E-mail: mintujob@rediffmail.com

D. R. Kaushal
Professor, Indian Institute of Technology Delhi, Hauz Khas, New Delhi – 110016, India

Anil Kumar
Associate Director, Water Business Group, CH2M HILL (India) Pvt. Ltd, Noida – 201301, India, E-mail: anil.kumar2@ch2m.com

Chandrajeet Kumar
Research Scholar, UGC-RGNF-SRF, Department of EWM, A.N. College, Patna 23, Bihar, India, E-mail: mail4chandrajeet@yahoo.com

Nishant Kumar
Technical Specialist, Drinking Water and Sanitation Department, Government of Jharkhand, Ranchi, India

Randhir Kumar
Research Scholar (PhD), Center for Water Engineering and Management, Central University of Jharkhand, Ranchi, India

Satish Kumar
PhD Scholar, Indian Institute of Technology Delhi, Hauz Khas, New Delhi, India – 110016, E-mail: satish.kumar140@gmail.com

Neeta Kumari
Department of Civil and Environmental Engineering, B. I. T. Mesra, Ranchi – 835215, India, E-mails: neetak@bitmesra.ac.in, neeta.sinha2k7@gmail.com

Priti Kumari
Assistant Professor, Department of Civil Engineering, GGSESTC, Bokaro, Jharkhnad. E-mail: priti.9407@gmail.com

A. K. Lohani
Scientist G, National Institute of Hydrology, Roorkee, Uttrakhand, India

B. K. Mandal
Meteorological Centre Ranchi, India Meteorological Department, B. M. Airport Road, Hinoo, Ranchi – 834002, India

Swati Maurya
Institute of Environment and Sustainable Development, Banaras Hindu University, Varanasi – 221005, India, E-mail: mauryaswati35@gmail.com

Brijesh Kumar Mishra
Department of Environmental Science and Engineering, Indian Institute of Technology
(Indian School of Mines), Dhanbad – 826004, India, E-mail: bkmishra3@rediffmail.com

K. L. Mishra
College of Agricultural Engineering, Jawaharlal Nehru Krishi Vishva Vidhyala Jabalpur (M.P.) – 480661,
India

D. Nandi
Assistant Professor, Department of RS & GIS, North Orissa University, Baripada, Odisha, India,
E-mail: debabrata.gis@gmail.com

Ipsita Nandi
Institute of Environment and Sustainable Development, Banaras Hindu University, Varanasi – 221005,
India, E-mail: ipsitabhu@gmail.com

Sadaf Nazneen
School of Environmental Sciences, Jawaharlal Nehru University, New Delhi – 110067, India,
E-mail: sadafnazneen01@gmail.com

Arvind Chandra Pandey
Professor, Centre for Land Resource Management, School of Natural Resource Management,
Central University of Jharkhand, Brambe, Jharkhand, India, E-mail: arvindchandrap@yahoo.com

H. K. Pandey
Department of Civil Engineering, MNNIT Allahabad, India

Varsha Pandey
Institute of Environment and Sustainable Development, Banaras Hindu University, Varanasi – 221005,
India, E-mail: varshu.pandey07@gmail.com

Prabeer Kumar Parhi
Assistant Professor, Centre for Water Engineering and Management, Central University of Jharkhand,
Brambe, Ranchi, India, E-mail: prabeer11@yahoo.co.in

Gopal Pathak
Department of Civil and Environmental Engineering, B.I.T. Mesra, Ranchi – 835215, India

Rajani K. Pradhan
Institute of Environment and Sustainable Development, Banaras Hindu University, Varanasi – 221005,
India, E-mail: rkpradhan462@gmail.com

N. Janardhana Raju
School of Environmental Sciences, Jawaharlal Nehru University, New Delhi – 110067, India

Rajeev Ranjan
Collage of Agricultural Engineering, Jawaharlal Nehru Krishi Vishva Vidhyala Jabalpur (M.P.) – 480661,
India

Priti Sagar
Research Scholar, Centre for Water Engineering and Management, Central University of Jharkhand,
Brambe, 835205, Jharkhand, India, Email: sagarpriti68@gmail.com

P. C. Sahu
Reader, Department of Geology, MPC Autonomous Colleges, Baripada, Odisha

Abhisek Santra
Department of Civil Engineering, Haldia Institute of Technology, Haldia, West Bengal

Kavita Shah
Institute of Environment and Sustainable Development, Banaras Hindu University, Varanasi, 221005, India

Arvind Kumar Shakya
Research Scholar, Department of Civil Engineering, IIT Guwahati, India – 781039,
E-mail: a.shakya@iitg.ernet.in

R. S. Sharma
Meteorological Centre Ranchi, India Meteorological Department, B. M. Airport Road, Hinoo,
Ranchi – 834002, India, E-mail: radheshyam84@rediffmail.com

Astha Singh
Department of Environmental Science and Engineering, Indian Institute of Technology
(Indian School of Mines), Dhanbad – 826004, India

C. S. Singh
Assistant Professor, Department of Agronomy, Birsa Agricultural University, Kanke, Ranchi,
Jharkhand, 834006, India

P. K. Singh
Professor, Department of Irrigation & Drainage Engineering, College of Technology,
G. B. Pant University of Agriculture and Technology, Pantnagar – 363145, Uttarakhand,
E-mail: singhpk68@gmail.com

Prashant K. Srivastava
Institute of Environment and Sustainable Development, Banaras Hindu University, Varanasi – 221005, India

Stuti
Research Scholar, School of Natural Resource Management, Centre for Land Resource Management,
Central University of Jharkhand, Brambe, Jharkhand, India, E-mail: stuti@cuj.ac.in

Sarika Suman
Banaras Hindu University, Varanasi, India

Sushmita
Research Scholar, Department of EWM, A.N. College, Patna 23, Bihar, India

Nity Tirkey
Research Scholar, Centre for Water Engineering and Management, Central University of Jharkhand,
Ranchi, India, E-mail: nitytirkey@gmail.com

A. K. Tiwari
Assistant Professor, Department of Horticulture, Birsa Agricultural University, Kanke, Ranchi,
Jharkhand, 834006, India

Vivek Tiwari
Banaras Hindu University, Varanasi, India, E-mail: viveektiwary@gmail.com

Sachin Kumar Tomar
Department of Civil Engineering, Indian Institute of Technology (Guwahati), Guwahati – 781039, India,
E-mail: sachintomar306@gmail.com

Vinod Kumar Tripathi
Department of Farm Engineering, Institute of Agricultural Sciences, Banaras Hindu University,
Varanasi, U.P., 221005, India, E-mail: tripathiwtcer@gmail.com

Utkarsh Upadhayay
Research Scholar, Centre for Water Engineering and Management, Central University of Jharkhand,
Jharkhand, Ranchi, India, E-mail: utkarsh.wem.cuj@gmail.com

Devendra Warwade
Assistant Professor, Department of Physics, Government College Sehore, Barktulla University, Bhopal, M.P., India

Pratibha Warwade
Assistant Professor, Center for Water Engineering and Management, Central University of Jharkhand, Brambe, Ranchi – 835205, India, E-mail: pratibhawarwade@gmail.com

Naval Kishor Yadav
Department of Civil Engineering, Haldia Institute of Technology, Haldia, West Bengal

ABBREVIATIONS

AC	activated carbon
AGR	aerobic granular reactor
AHP	analytical hierarchy process
ANN	artificial neural network
AOI	area of interest
APHA	American Public Health Association
ASCII	American Standard Code for Information Interchange
BC	biochar
BHPV	Bharat Heavy Plates & Vessels
BOD	biochemical oxygen demand
BSIL	Beekay Steel Industries Ltd.
BWDB	Bangladesh Water Development Board
CCI	climate change initiative
C_{eff}	coefficient of efficiency
CGWB	Central Ground Water Board
CNTS	carbon nanotubes
COD	chemical oxygen demand
CSI	cumulative score index
CVD	chemical vapor deposition
DAS	days after sowing
DDW	doubled distilled water
DEM	digital elevation model
DIC	dissolved inorganic carbon
DMICDC	Delhi Mumbai Industrial Corridor Development Corporation
DOC	dissolved organic carbon
DS NVPH	double span naturally ventilated polyhouse
DSSAT	decision support system for agrotechnology transfer
DVC	Damodar Valley Corporation
DW	dug well
EC	electrical conductivity
ECs	emerging contaminants
EDC	endocrine disrupting chemicals
EDS	environmental design solutions
EKW	East Kolkata Wetlands

EPA	Environmental Protection Agency
E_R	effective rainfall
ET_O	reference evapotranspiration
FESEM	field emission scanning electron microscope
gcm^{-3}	gram per centimeter cube
gl^{-1}	gram per liter
GLOF	glacial lake outburst floods
GPS	global positioning system
GSA	grain-size analysis
GSDC	grain-size distribution curves
GSV	granule settling velocity
ha	hectare
HMI	human-machine interface
HP	hand pump
HRT	hydraulic retention time
HSZ	hydrological sensitive zones
ICAR	Indian Council of Agricultural Research
ICASA	International Consortium for Agricultural Systems Applications
IMDMCC	Inter-Ministerial Disaster Management Co-Ordination Council
IPCC	Intergovernmental Panel on Climate Change
IR	net depth of irrigation
IWMED	Institute of Wetland Management and Ecological Design
IWRM	integrated water resource management
KRB	Kosi River Basin
LAI	leaf area index
LDR	linear decision rule
LST	land surface temperature
MBC	modified biochar
MBIR	Manesar Bawal Investment Region
MCDM	multi-criteria decision-making
MCL	maximum contaminant level
mgl^{-1}	milligram per liter
mh^{-1}	meter per hour
MIF	multi-influencing factor
MLD	million liter per day
MLVSS	mixed liquor volatile suspended solids
$mmday^{-1}$	millimeter per day
NASA	National Aeronautics and Space Administration
NDMC	National Disaster Management Council
NDVI	normalized difference vegetation index

NEO	NASA Earth Observation
NGA	National Geospatial-Intelligence Agency
NOAA	National Oceanic and Atmospheric Administration
OECD	Organization for Economic Co-Operation and Development
PCA	principal component analysis
POWER	prediction of worldwide energy resources
qha^{-1}	quintal per hectare
RIMCS	remote irrigation monitoring and control system
RLD	root length density
RMSE	root mean square error
RWSP	Rural Water Supply Scheme of Panchayati Raj
SBC	single board computer
SBR	sequencing batch reactor
SMC	soil moisture content
SMD	soil moisture deficit
SOP	standard operating policy
SRTM	shuttle radar topographic mission
SWMM	stormwater management model
TDS	total dissolved solids
TDSS	total dissolved and suspended solids
TEAP	terminal electron accepting processes
TSS	total suspended solids
TWI	topographic wetness index
UHI	urban heat island
VES	vertical electrical soundings
VSA	variable source area
VSS	volatile suspended solids
WHO	World Health Organization
WRC	water retention curve
WT NVPH	walking tunnel naturally ventilated polyhouse
WUE	water use efficiency

PREFACE

The Centre for Water Engineering and Management came into existence in July 2010 and is presently offering five-year integrated MTech, two-year MTech degrees, and a separate PhD program in Water Engineering and Management. With a mission to encourage innovative approaches of thinking, the center is in the process of developing highly skilled manpower in the field of water engineering. Water is at the core of sustainable development and is critical for socioeconomic development, healthy ecosystems, and for human survival itself.

The world's population is increasing and concentrating more and more in urban areas. This trend is particularly intense in developing countries, where an additional 2.1 billion people are expected to be living in cities by 2030. These cities produce billions of tons of waste every year, including sludge and wastewater. In India, the estimated sewage generation from Class I cities and Class II towns (representing 72% of urban population) is 38,524 million liters/day (MLD), of which there exists a treatment capacity of only 11,787 MLD (about 30%). How the treated wastewater is being used is something that needs to be looked into. India needs a national wastewater reuse policy to help address the urban and rural water demand by quantifying the targets and laying out legislative, regulatory, and financial measures to achieve those targets. It is a matter of great pride that the Government of Jharkhand has devised a Jharkhand Waste Water Policy 2017 to ensure increased use of recycled water for other purposes apart from drinking, through the provision of appropriate technologies for water recycling and protection of the environment.

The International Conference on Water and Wastewater Management and Modelling was held January 16–17, 2018 at Ranchi, India. The conference received a good number of research papers and review articles. The research papers were reviewed critically, and we are happy to have them collected in this volume. This will ensure larger distribution and circulation of the edited book in the research and teaching community across the world. The book is organized in chapters covering major themes of the conference; the chapters are divided into sections, and the sections into topical subsections.

The authors want to record their gratitude to all the contributing authors who participated in the conference. We take this opportunity to express our gratitude towards our Honorable Vice Chancellor, Prof. Nand Kumar Yadav 'Indu,' for his constant encouragement and support. I would like to thank the editorial staff, Sandy Jones Sickels, Vice President, and the production team at Apple Academic Press, Inc., for considering this book to publish when reuse of wastewater needs to be encouraged and streamlined in developing nations. Special thanks are due to the AAP production staff for bringing the quality production.

I request readers to offer their valuable and constructive suggestions that may help to improve future endeavors.

I express my deep admiration to my wife, Punam, and daughter Anushka, for their unconditional support and cooperation during the preparation of this book.

—Ajai Singh, PhD, FIE

PART I
Wastewater Management

VERMIFILTRATION OF ARSENIC CONTAMINATED WATER USING VERMIFILTRATION TECHNOLOGY: A NOVEL BIO-FILTER MODEL

CHANDRAJEET KUMAR[1], SUSHMITA[2], NUPUR BOSE[3], and ASHOK GHOSH[4]

[1]PhD Scholar, UGC-RGNF-SRF, Department of EWM, A.N. College, Patna 23, Bihar, India, E-mail: mail4chandrajeet@yahoo.com

[2]PhD Scholar, Department of EWM, A.N. College, Patna 23, Bihar, India

[3]Associate Professor, Department of Geography, A.N. College, Patna–23, Bihar, India

[4]Professor and Head, Research Section, Mahavir Cancer Sansthan and Research Centre, Patna 801505, Bihar, India

ABSTRACT

Arsenic toxicity has become a global concern owing to the increasing contamination of soil, groundwater, and crops in many regions of the world, and Bihar is one of the worst affected states of India where Arsenic concentration has been found up to 1861 ppb in groundwater. As Arsenic has a high magnitude of solubility, its removal from contaminated water and soil is very difficult. Vermifiltration of Chemically contaminated and sewage water using earthworms is a newly conceived novel technology with several advantages over the conventional water filtration systems. Certain species of earthworms (*Eisenia fetida* and *Eudrillus euginae*) have been found to purify the contaminated wastewater. Their bodywork as a 'bio-filter' having the capacity to bio-accumulate high concentrations of toxic chemicals in their tissues and the resulting purified water (termed as Vermiaqua) becomes almost chemical and pathogen-free. To make vermifiltration unit, a vermifilter bed was prepared using plastic drum which was in depth filled with a large, medium, small size stone chips followed by sand and humid soil at the top layer and 5000 earthworms were released in the moist soil top layer. Water controller knob was attached with a sample container which allows 50–60 drops per minute water input into the top layer of vermifilter bed which was also attached with Filtrate collector container through pipe for filtrate collection. Prepared arsenic trioxide solutions of 10,000 μgl^{-1} and 20,000 μgl^{-1} were poured in separate sample collection container and allowed to pass through layers of vermifilter bed for continuous three days and on the fourth day purified water was collected in nitric acid washed filtrate collection container and analyzed by atomic absorption spectrophotometer (AAS). According to the AAS analysis, the arsenic concentration of 10,000 μgl^{-1} and 20,000 μgl^{-1} values were decreased to 7.716 μgl^{-1} and 6.186 μgl^{-1}, respectively and accordingly in the body tissue of earthworms, 127.9 μgl^{-1} and 63.81 μgl^{-1} arsenic were found, respectively.

1.1 INTRODUCTION

Arsenic is a known carcinogen and a mutagen. Since the arsenic contaminated groundwater is leading to a host of health problems, often culminating in diseases with high fatality rates, like cancer. Tragically, most of the worst geogenic arsenic affected areas are located in developing economies of the world, and there is an ever-increasing demand for arsenic-free water in these heavily populated areas. The fluvial plains of South Asia have borne the brunt of incidences of arsenicosis, or chronic diseases due to arsenic poisoning, across countries of Bangladesh, India, Nepal, and Pakistan. This global health issue can be best controlled by halting direct and indirect ingestion of arsenic that is taking place through contaminated drinking water and arsenic contaminated agricultural produce, and simultaneously supplying clean water for drinking and irrigation purposes. Several proven filtration technologies are now globally in place, broadly categorized in Mandal et al. (2002). Earthworms have over 600 million years of history in waste and environmental management. Charles Darwin called them as the 'unheralded soldiers of mankind,' and the Greek philosopher Aristotle called them as the 'intestine of earth,' meaning digesting a wide variety of organic materials including the waste organics from earth (Darwin and Seward, 1903). Earthworms harbor millions of 'nitrogen-fixing' and 'decomposer microbes' in their gut. The distribution of earthworms in soil depends on factors like soil moisture, availability of organic matter, and pH of the soil. They occur in diverse habitats especially those which are dark and moist. Earthworms are generally absent or rare in soil with a very coarse texture and high clay content or soil with pH 4 (Gunathilagraj, 1996). In a study made by Kerr and Stewart (2006), it was opined that fetida can tolerate soils nearly half as salty as seawater. Earthworms can also tolerate toxic chemicals in the environment. *E. fetida* also survived 1.5% crude oil containing several toxic organic pollutants (OECD, 2000). Some species have been found to bio-accumulate up to 7600 mg of lead (Pb) per gm of the dry weight of their tissues (Ireland, 1983). They can tolerate a temperature range of 5 to 29°C. A temperature of 20–25°C and moisture of 60–75% are optimum for good worm function (Hand, 1988).

Vermifiltration of wastewater using waste eater earthworms is a newly conceived novel; an innovative technology developed by our research collaborator. Earthworms bodywork as a 'biofilter' and they have been found to remove the 5 days BOD by over 90%, COD by 80–90%, total dissolved solids (TDS) by 90–92% and the total suspended solids (TSS) by 90–95% from wastewater by the general mechanism of 'ingestion' and biodegradation of organic wastes, heavy metals and solids from wastewater and also by their 'absorption' through body walls (Sinha et al., 2015). Most successful species are the Tiger Worms (*Eisenia fetida*). Vermifiltration system is low energy dependent and has a distinct advantage over all the conventional biological wastewater treatment systems—the 'activated sludge process,' 'trickling filters,' and 'rotating biological contractor' which are highly energy intensive, costly to install and operate, and do not generate any income. This is also an odor free process. The most significant advantage is that there is 'no sludge formation' in the process as the earthworms eat the solids simultaneously and excrete them as vermicast. This plagues most municipal council in the world as the sludge being a biohazard requires additional expenditure on safe disposal in secured landfills. In the vermifilter process, there is 100% capture of organic & inorganic materials and any pathogen, and capital and operating costs are much lesser. Earthworm's bio-accumulate all toxic chemicals including the 'endocrine disrupting chemicals' (EDCs) from sewage which cannot be removed by the conventional systems. A pilot study on vermifiltration of sewage was made by Xing et al. (2005) at Shanghai Quyang Wastewater Treatment Facility in China. Taylor (2003) studied the treatment of domestic wastewater using vermifilter beds and concluded that worms could reduce BOD and COD loads as well as the TDSS (total dissolved and suspended solids) significantly by more than 70–80%. Hartenstein and Bisesi (1989) studied the use of earthworms for the management of effluents from intensively housed livestock which contains very heavy loads of BOD, TDSS, nutrients nitrogen (N) and phosphorus (P). The worms produced clean effluents and also nutrient-rich vermicompost. Bajsa et al. (2003) also studied the Vermifiltration of domestic wastewater using vermicomposting worms with significant results.

1.2 EXPERIMENTAL DESIGN

1.2.1 CONSTRUCTION AND INSTALLATION OF THE VERMIFILTRATION UNIT

To make vermifiltration unit, three plastic drums of 80-liter capacity were taken. Out of these three drums, two of them were prepared as vermifilter unit for arsenic filtration and remaining one was prepared as control (Figure 1.1). In the control unit, all other materials were organized in the same way as in vermifilter unit except earthworms. All three plastic drums were filled with different layers consisting of large size pebbles with 10" height, medium size pebbles (10"), small size pebbles (10") followed by sand (6–7"), and humid soil (20") at the top layer to prepare filter (Figure 1.2). Finally, 5000 number of earthworms (*Eisenia fetida*) weighing approximately 5 kg were released in the moist soil layer of two filter unit to prepare vermifiltration unit.

FIGURE 1.1   Construction and installation of the vermifiltration unit.

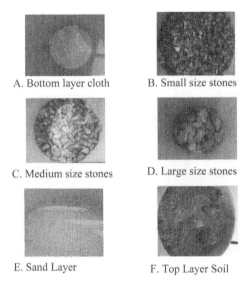

A. Bottom layer cloth   B. Small size stones

C. Medium size stones   D. Large size stones

E. Sand Layer   F. Top Layer Soil

FIGURE 1.2   Different layers of vermifiltration unit.

A movable iron stand of approximately 150 kg was prepared to have 4 columns and 3 rows to keep 3–4 filtration unit, sample containers, and filtrate collector containers. Upper row carries the sample container, middle row carries the plastic drum (vermifilter bed), and the lower row carries the filtrate collector container. Water controller knob was attached with a sample container which allows 50–60 drops per minute water input into the top layer of vermifilter bed (Figure 1.3). A base of plastic drums were joined with filtrate collector container through the pipe for filtrate collection. Filtrate collection containers were rinsed with nitric acid and tightly packed for prevention of arsenic oxidation.

FIGURE 1.3   Top layer of the vermifiltration Unit.

1.3   METHODOLOGY

1.3.1   PREPARATION OF DIFFERENT CONCENTRATION OF ARSENIC Trioxide Sample

About 1.320 gm of arsenic trioxide and 4.0 gm of NaOH were weighed and dissolved completely in a round bottom volumetric flask containing 100 ml distilled water and a total volume of 1 liter was maintained for preparing a stock solution of 1000 μgl^{-1}. From this stock, different concentrations of 10,000 μgl^{-1} and 20,000 μgl^{-1} were prepared.

1.3.2   FILTRATION OF ARSENIC THROUGH VERMIFILTRATION PROCESS

Prepared arsenic trioxide solutions of 10,000 μgl^{-1} and 20,000 μgl^{-1} were poured in separate sample collection container and allowed to pass through controlled water knob 50–60 drops per minute into top layer (soil + earthworm) of vermifiltration bed for continuous three days and on fourth day, vermifiltered water was collected in nitric acid washed filtrate collection container. In control unit (without earthworms), 10,000 μgl^{-1} arsenic trioxide solution was allowed to pass through controlled water knob 50–60 drops per minute into soil layer and the remaining process was the same.

1.3.3   ANALYSIS OF VERMIFILTERED ARSENIC TRIOXIDE SAMPLE BY ATOMIC ABSORPTION SPECTROPHOTOMETER (AAS)

About 50 ml of vermifiltered water was mixed with concentrated HNO_3 in Griffin beaker and covered with watch glass then placed on hotplate (90–120°C) to evaporate continuously till getting the final volume up to 5 ml. Again 5 ml concentrated HNO_3 was added in cooled 5 ml remaining sample and placed on hotplate (90–120°C) to evaporate continuously till getting the final volume to 3 ml. Griffin beaker wall and watch glass were washed with distilled water or 1% HNO_3. Finally, digested samples were filtered through What's Mann Filter Paper No. 1 to remove silicate, and other insoluble impurity and volume were adjusted to 50 ml with 1% HNO_3 and analyzed by AAS.

1.3.4   ANALYSIS OF EARTHWORMS' BODY TISSUE BY AAS

About 0.5 mg earthworms were weighed and washed with 1% HNO_3 for cleaning and treated with 5 ml concentrated HNO_3 and kept for overnight for digestion, then placed on hotplate (90–120°C) to evaporate continuously till getting the final volume 1 ml. Now, nitric acid and perchloric acid were added in the digested sample in the 6:1 ratio. Again, this was placed on hotplate (90–120°C) to evaporate continuously till getting the final volume to 1 ml. Finally, the digested sample was filtered through What's Mann Filter Paper No. 1 to remove silicate, and other insoluble impurity and volume were adjusted to 50 ml with 1% HNO_3 and analyzed by AAS.

1.4   RESULTS AND DISCUSSION

According to the AAS analysis, the arsenic concentration of 10,000 µgl⁻¹ and 20,000 µgl⁻¹ value were decreased to 7.716 µgl⁻¹ and 6.186 µgl⁻¹, respectively, but the control value decreased only to 80.780 µgl⁻¹ (Table 1.1). Accordingly, in the earthworm's body tissue, 127.9 µgl⁻¹ and 63.81 µgl⁻¹ arsenic were found in Test 1 and Test 2, respectively. In control, soil absorbs only 19.58 µg/l arsenic because of normal arsenic holding capacity and due to long exposure of air, oxidation, and precipitation of Arsenic.

TABLE 1.1   Arsenic Levels in Vermifiltered Solutions

	Control 10,000 µgl⁻¹	Test 1 10,000 µgl⁻¹	Test 2 20,000 µgl⁻¹
Vermifiltered Arsenic Water	80.780 µgl⁻¹	7.716 µgl⁻¹	6.186 µgl⁻¹
Earthworms' Body Tissue (0.5 mg)	No Earthworms were used	127.9 µgl⁻¹	63.81 µgl⁻¹
Soil Testing (0.5 mg)	19.58 µgl⁻¹	144.7 µgl⁻¹	92.37 µgl⁻¹

In Test 1, because of Earthworm's high arsenic holding capacity along with soil, 144.7 µgl⁻¹ Arsenic was absorbed. However, in Test 2, under twice the amount of arsenic concentration in the input solution, earthworms were found to absorb less arsenic in their body in comparison to those of Test 1. Likewise, the soil also held 92.37 µgl⁻¹ of arsenic which is comparatively lesser from the soil of Test 1.

1.5   CONCLUSIONS

Although this is indicative that Earthworms are highly tolerable to arsenic assimilation because of high absorbance capacity and thereby they are able to minimize the concentration of arsenic, certain research questions have been generated by this experiment:

1. The pathway of arsenic through the soil-earthworm-sand-small stone chips-medium stone chips large stone chips in both Test 1 and Test 2 environments need to be studied.
2. Reduced arsenic concentration levels in Test 2 phases need immediate research interventions. The results imply that with increasing doses of arsenic, the threshold values of arsenic uptake by soils and earthworms of filter beds decline significantly. If this be nature's behavior in limiting arsenic uptake by soil microbes and earthworms, this has tremendous potential in scientifically identifying the gene's responsible for this.
3. To identify the role of microbes, present in the gut of earthworms in arsenic conversion.

KEYWORDS

- arsenic
- bio-filter
- earthworms
- vermiaqua
- vermifiltration technology

REFERENCES

Bajsa, O., Nair, J., Mathew, K., & Ho, G. E., (2003). Vermiculture as a tool for domestic wastewater management, *Water Science and Technology*, IWA Publishing, *48*(11/12), 125–132.

Darwin, F., & Seward, A. C., (1903). *More Letters of Charles Darwin*. A record of his work in series of hitherto unpublished letters, John Murray, London, *2*, 508.

Gunathilagraj, K., (1996). *Earthworm: An Introduction*. Indian Council of Agricultural Research Training Program, Tamil Nadu Agriculture University, Coimbatore.

Hand, P., (1988). Earthworm biotechnology. In: *Green Shields Resources and Application of Biotechnology: The New Wave*. MacMillan Press Ltd., US.

Hartenstein, R., & Bisesi, M. S., (1989). Use of earthworm biotechnology for the management of effluents from intensively housed livestock, *Outlook Agriculture*, *18*(2), 72–76.

Ireland, M. P., (1983). *Heavy Metals Uptake in Earthworms, Earthworm Ecology* (pp. 247–265). Chapman & Hall, London.

Kerr, M., & Stewart, A. J., (2006). Tolerance test of *Eisenia fetida* for sodium chloride, US Department of Energy, *Journal of Undergraduate Research*. doi: https://www.osti.gov/servlets/purl/1051306.

Mandal, B. K., & Suzuki, K. T., (2002). Arsenic round the world: A review. *Talanta.*, *58*, 201–235.

OECD, (2000). Guidelines for testing organic chemicals. Proposal for new guidelines earthworms' reproduction tests (*E. fetida andrei*). *Organization for Economic Co-Operation and Development*. https://www.oecd.org/env/ehs/testing/Draft-Updated-Test-Guildeline-222-Earthworm-Reproduction-Test.pdf.

Sinha, R. K., Kumar, C., Hahn, G., Patel, U., & Soni, B. K., (2015). Embarking on second green revolution by vermiculture for production of chemical-free organic foods, production of crops and farm soils and elimination of deadly agrochemicals from earth: Meeting the challenges of food security of 21st century by earthworms. Nova Science Publishers, Inc. *Sir Charles Darwin's "Friend of Farmers" Agricultural Research Update*, *10*, 1–47.

Sinha, R. K., Misra, N. K., Singh, P. K., Ghosh, A., Patel, U., Kumar, J., et al., (2015). Vermiculture technology for recycling of solid wastes and wastewater by earthworms into valuable resources for their reuse in agriculture (organic farming) while saving water and fertilizer, In: Rajeev, P. S., & Abhijit, S., (eds.), *Waste Management: Challenges, Threats and Opportunities*, Nova Science Publications, USA. pp. 233–256, ISBN: 978-1-63482-150-6.

Taylor, (2003). The treatment of domestic wastewater using small-scale vermicompost filter beds. *Ecol. Eng.*, *21*, 197–203.

Xing, M., Yang, J., & Lu, Z., (2005). *Microorganism-Earthworm Integrated Biological Treatment Process: A Sewage Treatment Option for Rural Settlements*. ICID 21st European Regional Conference. http://www.zalf.de/icid/ICID_ERC2005/HTML/ERC2005PDF/Topic_1/Xing.pdf.

CHAPTER 2

PERFORMANCE OF AN AEROBIC GRANULAR REACTOR TREATING ORGANICS AND AMMONIA NITROGEN WITH TIME

SACHIN KUMAR TOMAR and SASWATI CHAKRABORTY

Department of Civil Engineering, Indian Institute of Technology (Guwahati), Guwahati – 781039, India,
E-mail: sachintomar306@gmail.com

ABSTRACT

Aerobic granules were developed in a sequencing batch reactor (SBR) for treating phenol (400 mgl⁻¹), thiocyanate (SCN⁻) (100 mgl⁻¹) and ammonia-nitrogen (NH_4^+-N) (100 mgl⁻¹). Reactor performance was analyzed in terms of pollutant degradation profile in one cycle of 6 h with respect to time. Mean biomass size and volatile suspended solids (VSS) were 1334.24 ± 30.56 μm and 4.90±0.40 gl⁻¹ in the reactor, respectively. Within initial 30 min, effluent phenol, SCN⁻, and COD (chemical oxygen demand) decreased to 1.93, 6.64 and 145.81 mgl⁻¹ from initial values of 400, 100 and 1015.71 ± 33.83 mgl⁻¹, respectively. Complete phenol and SCN⁻ degradations required almost 120 min. Maximum COD removal achieved was 90.37%. NH_4^+-N removal required a longer time than other pollutants. Kinetic analysis showed that 180 min was adequate for the degradation of all pollutants and to achieve nitrification by aerobic granules.

2.1 INTRODUCTION

The economic and efficient treatment of industrial wastewaters is a great challenge since they comprise of complex matter consists of several organic compounds (like aromatic compounds), inorganic compounds, and ammonia (Kim and Kim, 2003). Usually, physicochemical processes are practiced to treat industrial wastewaters in spite of having serious drawbacks such as high operational cost, incomplete degradation of aromatic compounds and production of other hazardous byproducts (Kim and Ihm, 2011). The drawbacks of physicochemical processes can be overcome by biological processes. However, the aromatic compounds can cause inhibition to the biological processes (Khalid and Naas, 2012). Phenol with a concentration of 110–487 mgl^{-1} is found with other inhibitory and toxic compounds like thiocyanate (SCN$^-$) and ammonia-nitrogen (NH$_4^+$-N) in wastewaters from coal liquefaction, coal gasification and synthetic fuel processing, etc. (Li et al., 2011; Zheng and Li, 2009). When phenol and (SCN$^-$) are present with ammonia, these compounds inhibit nitrification; therefore require to be removed before nitrification (Jeong and Chung, 2006; Kim et al., 2008). The rate-limiting step in wastewater treatment is the removal of ammonia by nitrification because of the very slow growth rate of nitrifying bacteria and susceptibility to environmental factors as well as other factors like inhibition by aromatic compounds. Therefore, growth and maintenance of sufficient nitrifying bacteria are very difficult in wastewater treatment systems based on suspended or fixed culture (Ochoa et al., 2002; Ramos et al., 2016).

Aerobic granule based biological treatment systems serve as an alternative approach for treating the complex industrial wastewaters (Gao et al., 2011). Aerobic granules are the microbial aggregates contain millions of organisms per gram of biomass with a regular round shape, a distinct outline and a compact structure formed via self-immobilization of microorganisms without any support/carrier under aerobic condition by applying controlled loading and operating conditions (Beun et al., 1999; Liu and Tay, 2007). Aerobic granulation has become an efficient and promising technology in biological wastewater treatment systems because of several advantages over conventional treatment systems such as excellent settling behavior, a compact structure, high metabolic activity, tolerance towards higher and shock loadings and ability to degrade toxic compounds like phenol (Adav et al., 2008; Corsino et al., 2015). That's why it has been extensively used in treating various wastewaters including industrial, nutrient-rich, and toxic wastewaters (Kishida et al., 2009; Val del Rio et al., 2012; Zhao et al., 2015). Aerobic granulation has a great potency for removing both organic and ammonium pollutants simultaneously in an economic manner (Singh and Srivastava, 2011). The works of literature are very limited for removing both phenols with ammonia nitrogen simultaneously by aerobic granular reactor (AGR) (Liu et al., 2005; Ramos et al., 2016). The limitation in the efficient operation of AGR for nitrification can be overcome by providing a selective approach to improve nitrifying granulation for effective ammonia conversion (Wu et al., 2017). Aerobic granule can tolerate the toxic effect of high concentration of toxic compounds due to a mass transfer shield of embedded cells provided by them (Ho et al., 2010). AGR is largely operated as a sequencing batch reactor (SBR). SBR is having a unique feature to be operated in a cyclic mode comprising of filling, aeration, settling, and withdrawal as compared to a continuous culture. To the best of authors' knowledge, the very limited study addressed the kinetic behavior of toxic pollutants like phenol and (SCN⁻) along with nitrification in one cycle. The present study aims to investigate the performance of AGR in removing toxic pollutants like phenol, (SCN⁻), and to achieve nitrification with time in one cycle.

2.2 MATERIALS AND METHODS

2.2.1 EXPERIMENTAL SET-UP

The present study was carried out in a laboratory scale reactor operated in sequential batch mode. The reactor had a working volume of 6 l. The working height, and an inner diameter (ID) of the reactor was 212 and 6 cm, respectively (H/D ratio of 35). Aeration was provided by an oil-free compressor at a rate of 2 l min⁻¹ by air stone kept at the bottom in the reactor. The influent synthetic wastewater was introduced from the bottom of the reactor with the help of a peristaltic pump. The up-flow liquid velocity in the reactor was maintained at a rate of 2 mh⁻¹. The reactor was maintained at room temperature (25–30°C).

2.2.2 CHARACTERISTICS OF SEED

The inoculum was collected from activated sludge unit of wastewater treatment plant of Indian Oil Corporation Limited (IOCL), Noonmati, Guwahati, Assam. The suspended and volatile solids in IOCL sludge were 2.42 ± 0.16 and 1.72 ± 0.15 gl⁻¹, respectively. The particle size of sludge was of 32.78 ± 0.01 μm. The sludge volume index (SVI_{30}) was 50.13 mlg⁻¹. In the reactor, 3 L sludge was used for working volume of 6 L.

2.2.3   FEED CHARACTERISTICS

The influent synthetic wastewater consisted of phenol of 400 mgl^{-1}; ammonia nitrogen (NH$_4^+$–N as NH$_4$Cl) of 100 mgl^{-1} and thiocyanate (SCN$^-$ as KSCN) of 100 mgl^{-1}. Feed pH was maintained between 7.5–8 by using sodium hydrogen carbonate and phosphate buffer (using 72.3 gl^{-1} of KH$_2$PO$_4$ and 104.5gl^{-1} of K$_2$HPO$_4$). This phosphate buffer also worked as a phosphorus source for microorganisms. Phosphate buffer of 1 mll^{-1} and trace metals solution of 1 mll^{-1} was added in the synthetic feed in the reactor. The stock trace metal composition was taken from previous literature (Sahariah and Chakraborty, 2011). The composition of stock trace metal solution was: MgSO$_4$.7H$_2$O: 10,000 mgl^{-1}, CaCl$_2$.2H$_2$O: 10,000 mgl^{-1}, FeCl$_3$.6H$_2$O: 5000 mgl^{-1}, CuCl$_2$: 1000 mgl^{-1}, ZnCl$_2$: 1000 mgl^{-1}, NiCl$_2$.6H$_2$O: 500 mgl^{-1}, CoCl$_2$: 500 mgl^{-1}.

2.2.4   OPERATIONAL STRATEGY

The operational schedule of the reactor is given in Table 2.1. The reactor was operated in the sequential batch mode with a volume exchange ratio of 50%; i.e., 50% of reactor working volume was decanted in each cycle, and the similar amount of fresh feed was added to the reactor. Settling time was kept constant for 5 min in the reactor throughout the study, except for the initial 15 days, when it was 15 min to prevent severe washout of biomass from the reactor. The reactor was operated at a cycle time of 6 h. Hydraulic retention time (HRT) was calculated using Eq. (1) and HRT value is given in Table 2.1.

$$\text{HRT (day)} = \frac{\text{Reactor volume (L)}}{\text{Volume decanted per cycle (L)} \times \text{No. of cycles per day}} \qquad (1)$$

Reactor was acclimatized for concentrations of phenol, (SCN$^-$), and ammonia up to 400, 100 and 100 mgl^{-1}, respectively. After 45 days of acclimatization period, the desired concentrations of pollutants were obtained. Then the reactor was operated for another 27 days with same feed and the same operating condition for steady state.

2.2.5   ANALYTICAL METHODS

The granule size was measured by a laser particle size analyzer (Mastersizer 2000, Malvern Instruments) and occasionally by field emission scanning electron microscope (FESEM) (Sigma, Zeiss). For FESEM analysis the

TABLE 2.1 Operational Schedule of Reactor

Operational time (day)	No of cycles/d	Cycle time (h)	Distribution of cycle time (min)				HRT (h)	OLR*	NLR**	ULV***
			Feeding	Reaction	Settling	Withdrawal				
1–75	4	6	30	320	5	5	12	2.13	0.20	2

*Organic loading rate (kg COD/m³.day);
**Nitrogen loading rate (kg NH_4^+-N/m³.day);
***Up flow liquid velocity (mh^{-1}).

granules were fixed with 2% glutaraldehyde overnight at 4°C after washing
with phosphate buffer (pH 7.0) and then was dehydrated with ethyl alcohol
and dried (Wang et al., 2007). Granule settling velocity (GSV) (mh⁻¹) was
determined by the free settling test as described by Yu et al. (2009). Analysis
for phenol, ammonia nitrogen, nitrite, (SCN⁻), chemical oxygen demand
(COD), suspended solids, volatile suspended solids (VSS), SVI_{30} was carried
out according to standard methods (APHA, 2005). Nitrate was analyzed by
ion chromatograph (792 Basic IC, Metrohm) using anion column (Metrosep
ASupp 5–250/4.0) and carbonate fluent.

2.3   POLLUTANTS DEGRADATION IN ONE CYCLE IN THE REACTOR

An analysis was carried out during a steady-state condition of the reactor
(70th–72nd day). At steady state VSS, average size and SVI_{30} were 4.90±0.40
gl⁻¹, 1334.24 ± 30.56 μm and 67.69±7.13 mlg⁻¹, respectively. GSV was
35.79 ± 2.20 m/h. FESEM images of granule are given in Figure 2.1.

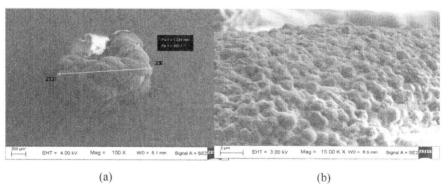

(a)   (b)

FIGURE 2.1   FESEM images of the granule.

In the present study, organic loading rate and nitrogen loading rate in
the reactor were 2.13 kg COD/m^3.day and 0.20 kg NH_4^+-N/m^3.day, respectively with influent COD: $NH4^+$-N ratio of 10.65. From Figure 2.2, it can be
observed that within the initial 30 min, effluent phenol and COD decreased
to 1.93 and 145.81 mgl⁻¹ from initial values of 400 and 1015.71 mgl⁻¹,
respectively. Complete phenol degradation required almost 120 min and
maximum COD removal achieved was 90.37% (effluent 111.97 mgl⁻¹) after
240 min. 93.36% SCN⁻ (concentration of 6.64 mgl⁻¹) was removed in initial
30 min and required almost 120 min to degrade completely (Figure 2.3).

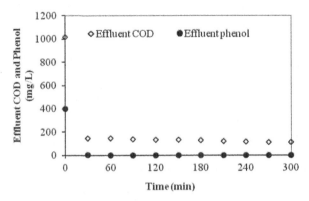

FIGURE 2.2 Effluent concentration profile of COD and phenol (mgl⁻¹) with time.

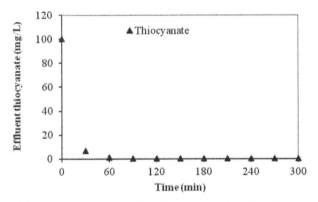

FIGURE 2.3 Effluent concentration profile of thiocyanate (mgl⁻¹) with time.

NH_4^+-N removal though started along with other pollutants, required a longer time than other pollutants. Only 40% removal was achieved in initial 30 min and complete removal required 180 min from an initial concentration of 100 mgl⁻¹ and 24 mgl⁻¹ NH_4^+-N generated from complete degradation of 100 mgl⁻¹ SCN^- (Figure 2.4). NO_3^--N profile (Figure 2.5) shows that at the initial time almost 70 mgl⁻¹ of NO_3^--N was already present in the reactor due to accumulation from the previous cycle. The concentration of NO_3^--N increased with time. At 300 min, NO_3^--N concentration was 128.79 mgl⁻¹ from the total initial NH_4^+-N of 124 mgl⁻¹ (Figure 2.5). From Figure 2.5, it was observed that during the first 150 min, NO_2^--N concentration increased up to 98.15 mgl⁻¹ and afterward it started to reduce. At 300 min, it was almost negligible, indicating the complete nitrification process. Deng et al. (2016) reported 93% COD and 41% nitrogen removals within the initial 30 min in acetate fed AGR from initial values of 700 and 37 mgl⁻¹, respectively. In the present study, COD removal was a little bit slower than this reported value. However, NH_4^+-N removal was faster. Wu et al., (2017) observed relatively long time period (>15 h) for achieving complete nitrification during the study of the impact of exogenous addition of nitrifying granular sludge cellular extract on nitrification efficiency, which was much higher as compared to the present study (6 h).

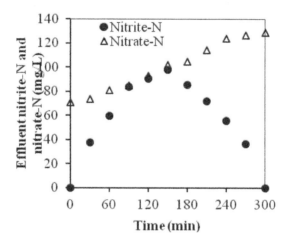

FIGURE 2.4   Effluent concentration profile of ammonia-N (mgl⁻¹) with time.

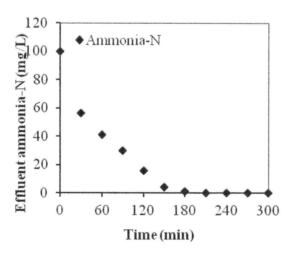

FIGURE 2.5   Effluent concentration profile of nitrite-N and nitrate-N (mgl⁻¹) with time.

2.4 CONCLUSIONS

Aerobic granules were successfully cultivated with phenol, (SCN⁻), and ammonia nitrogen at a cycle time of 6 h and kinetic analysis in one cycle was observed. Complete degradation of phenol and (SCN⁻) (99%) required 120 min. Maximum COD removal was observed after 240 min, and it was around 90%. Complete nitrification required a longer time than other pollutants and was achieved after 180 min. Kinetic analysis showed that 180 min was sufficient for removal of all pollutants, indicating that cycle time of 6 h can be decreased further to 3 h for better utilization of the reactor.

ACKNOWLEDGMENT

Authors would like to acknowledge their Institute providing instrument facility for FESEM images and to Indian Oil Corporation Limited (IOCL), Noonmati, Guwahati for providing sludge.

KEYWORDS

- **aerobic granules**
- **chemical oxygen demand**
- **nitrification**
- **sequencing batch reactor**
- **volatile suspended solids**

REFERENCES

Adav, S. S., Lee, D. J., Show, K. Y., & Tay, J. H., (2008). Aerobic granular sludge: Recent advances. *Biotechnology Advances, 26*(5), 411–423.

Al Khalid, T., & El Naas, M. H., (2012). Aerobic biodegradation of phenols: A comprehensive review. *Critical Reviews in Environmental Science and Technology, 42*(16), 1631–1690.

APHA, (2005). *Standard Methods for the Examination of Water and Wastewater* (21ˢᵗ edn.). APHA, AWWA, WPCF, American Public Health Association, Washington, DC, USA.

Beun, J. J., Hendriks, A., Van Loosdrecht, M. C. M., Morgenroth, E., Wilderer, P. A., & Heijnen, J. J., (1999). Aerobic granulation in a sequencing batch reactor. *Water Research, 33*(10), 2283–2290.

Corsino, S. F., Campo, R., Di Bella, G., Torregrossa, M., & Viviani, G., (2015). Cultivation of granular sludge with hypersaline oily wastewater. *International Biodeterioration and Biodegradation, 105*, 192–202.

Deng, S., Wang, L., & Su, H., (2016). Role and influence of extracellular polymeric substances on the preparation of aerobic granular sludge. *Journal of Environmental Management, 173*, 49–54.

Gao, D., Liu, L., Liang, H., & Wu, W. M., (2011). Aerobic granular sludge: Characterization, mechanism of granulation and application to wastewater treatment. *Critical Reviews in Biotechnology, 31*(2), 137–152.

Ho, K. L., Chen, Y. Y., Lin, B., & Lee, D. J., (2010). Degrading high-strength phenol using aerobic granular sludge. *Applied Microbiology and Biotechnology, 85*(6), 2009–2015.

Jeong, Y. S., & Chung, J. S., (2006). Simultaneous removal of COD, thiocyanate, cyanide, and nitrogen from coal process wastewater using fluidized biofilm process. *Process Biochemistry, 41*(5), 1141–1147.

Kim, K. H., & Ihm, S. K., (2011). Heterogeneous catalytic wet air oxidation of refractory organic pollutants in industrial wastewaters: A review. *Journal of Hazardous Materials, 186*(1), 16–34.

Kim, S. S., & Kim, H. J., (2003). Impact and threshold concentration of toxic materials in the stripped gas liquor on nitrification. *Korean Journal of Chemical Engineering, 20*(6), 1103–1110.

Kim, Y. M., Park, D., Jeon, C. O., Lee, D. S., & Park, J. M., (2008). Effect of HRT on the biological pre-denitrification process for the simultaneous removal of toxic pollutants from cokes wastewater. *Bioresource Technology, 99*(18), 8824–8832.

Kishida, N., Tsuneda, S., Kim, J., & Sudo, R., (2009). Simultaneous nitrogen and phosphorus removal from high-strength industrial wastewater using aerobic granular sludge. *Journal of Environmental Engineering, 135*(3), 153–158.

Li, H. Q., Han, H. J., Du, M. A., & Wang, W., (2011). Removal of phenols, thiocyanate, and ammonium from coal gasification wastewater using moving bed biofilm reactor. *Bioresource Technology, 102*(7), 4667–4673.

Liu, Y. Q., & Tay, J. H., (2007). Influence of cycle time on kinetic behaviors of steady-state aerobic granules in sequencing batch reactors. *Enzyme and Microbial Technology, 41*(4), 516–522.

Liu, Y. Q., Tay, J. H., Ivanov, V., Moy, B. Y. P., Yu, L., & Tay, S. T. L., (2005). Influence of phenol on nitrification by microbial granules. *Process Biochemistry, 40*(10), 3285–3289.

Ochoa, J., Colprim, J., Palacios, B., Paul, E., & Chatellier, P., (2002). Active heterotrophic and autotrophic biomass distribution between fixed and suspended systems in a hybrid biological reactor. *Water Science and Technology, 46*(1/2), 397–404.

Ramos, C., Suárez-Ojeda, M. E., & Carrera, J., (2016). Biodegradation of high-strength wastewater containing a mixture of ammonium, aromatic compounds, and salts with simultaneous nitritation in an aerobic granular reactor. *Process Biochemistry, 51*(3), 399–407.

Sahariah, B. P., & Chakraborty, S., (2011). Kinetic analysis of phenol, thiocyanate and ammonia-nitrogen removals in an anaerobic–anoxic–aerobic moving bed bioreactor system. *Journal of Hazardous Materials, 190*(1), 260–267.

Singh, M., & Srivastava, R., (2011). Sequencing batch reactor technology for biological wastewater treatment: A review. *Asia-Pacific Journal of Chemical Engineering, 6*(1), 3–13.

Val del Rio, A., Figueroa, M., Arrojo, B., Mosquera-Corral, A., Campos, J., García-Torriello, G., & Méndez, R., (2012). Aerobic granular SBR systems applied to the treatment of industrial effluents. *Journal of Environmental Management, 95*, S88–S92.

Wang, S. G., Liu, X. W., Gong, W. X., Gao, B. Y., Zhang, D. H., & Yu, H. Q., (2007). Aerobic granulation with brewery wastewater in a sequencing batch reactor. *Bioresource Technology, 98*(11), 2142–2147.

Wu, L. J., Li, A. J., Hou, B. L., & Li, M. X., (2017). Exogenous addition of cellular extract N-acyl-homoserine-lactones accelerated the granulation of autotrophic nitrifying sludge. *International Biodeterioration and Biodegradation, 118*, 119–125.

Yu, G. H., Juang, Y. C., Lee, D. J., He, P. J., & Shao, L. M., (2009). Enhanced aerobic granulation with extracellular polymeric substances (EPS)-free pellets. *Bioresource Technology, 100*(20), 4611–4615.

Zhao, X., Chen, Z., Wang, X., Li, J., Shen, J., & Xu, H., (2015). Remediation of pharmaceuticals and personal care products using an aerobic granular sludge sequencing bioreactor and microbial community profiling using Solexa sequencing technology analysis. *Bioresource Technology, 179*, 104–112.

Zheng, S., & Li, W., (2009). Effects of hydraulic loading and room temperature on the performance of the anaerobic/anoxic/aerobic system for ammonia-ridden and phenol-rich coking effluents. *Desalination, 247*(1–3), 362–369.

CHAPTER 3

EAST KOLKATA WETLANDS (EKW), INDIA: A UNIQUE EXAMPLE OF RESOURCE RECOVERY

ANITA CHAKRABORTY[1], SUBRATA HALDER[2], SADAF NAZNEEN[3], and SUMAN KUMAR DEY[4]

[1]Senior Technical Officer, Centre for Environmental Management and Participatory Development, Salt Lake, Kolkata – 700106, India, E-mail: anitagcenator@gmail.com

[2]Executive Engineer (A-I), State Water Investigation Directorate, Government of West Bengal, India

[3]School of Environmental Sciences, Jawaharlal Nehru University, New Delhi – 110067, India

[4]Professor, Department of Geography, Ramsaday College, Amta, Howrah, West Bengal, India

ABSTRACT

The age-old practice of utilizing wastewater into fishpond in the East Kolkata Wetlands (EKW), India is a unique example of resource recovery. The wetlands, providing a range of ecosystem services, form the base of ecological security of the entire region and livelihoods of the dependent communities. Being a dynamic ecosystem, the wetland is subject to influence from various natural as well as human factors. Integrated management of this ecosystem is crucial for maintaining the rich productivity of the wetland ecosystem as well as achieving the wise use of resources. The resources recovered from city sewage are used in three kinds of economic activities, i.e., wastewater fisheries, vegetable farming on a garbage substrate and paddy cultivation using pond effluent. In the entire recovery operation, the fishponds play a crucial and central role in the waste recycling process. The wetlands act as 'sink' for sewage and waste material from Kolkata that lacks any substitutable sewage treatment facility for its 4.5 million residents. About 3500 tons of municipal waste and 68 million liters of raw sewage drain into the wetland system on a daily basis. Sewage-fed fisheries that utilize such large volumes of sewage generated by the city started functioning as early as 1883. In 1940, about 4,682.22 ha of wetland area produced 0.14 t of fish per ha utilizing the municipal sewage. The profit-generating potential of the wetlands in terms of aquaculture aroused interest in local farmers as well as local landlords who leased out most of the ponds to commercial managers. The conspicuous result was employment generation through the production of fish and vegetables as well as a steady supply of food materials to the urban markets. There are 264 fish farms operating on a commercial basis. They cover a total area of about 2858.65 ha. The fish farms consist of units of various sizes from large holdings locally called bheries and relatively smaller ones called jheels due to their trench-like elongated shapes. But all these fish farms generally have similar types of produce, farming practice and distribution system. However, inappropriate understanding of the significance of these wetland practices has led to the gradual loss of system efficiency.

3.1   INTRODUCTION

The East Kolkata Wetland (EKW) sustains the World's largest and perhaps, the oldest integrated resource recovery (Figure 3.1) and practice based on a combination of agriculture and aquaculture, and provide livelihood support to a large, economically underprivileged population of around 20,000 families which depend upon the various wetland products, primarily fish and vegetables for sustenance. The wetland complex is located on the eastern fringes of Kolkata city as is one of the largest sewage fed fish ponds spread over an area of 12,500 ha. The Wetland Complex forms a part of the extensive inter-distributaries wetlands regimes formed by the Gangetic detla.

Based on its immense ecological and socio-cultural importance, the Government of India declared EKW as a wetland of International Importance under the Ramsar convention in the year 2003. At present wetland system produces over 15,000 MT per annum fish from its 264 functioning aquaculture pond, locally called *bheries*. Additionally, about 150 MT of vegetables are produced daily. Thus it is prudent to say that EKW servers as the backbone of food security of the Kolkata City (Figure 3.2).

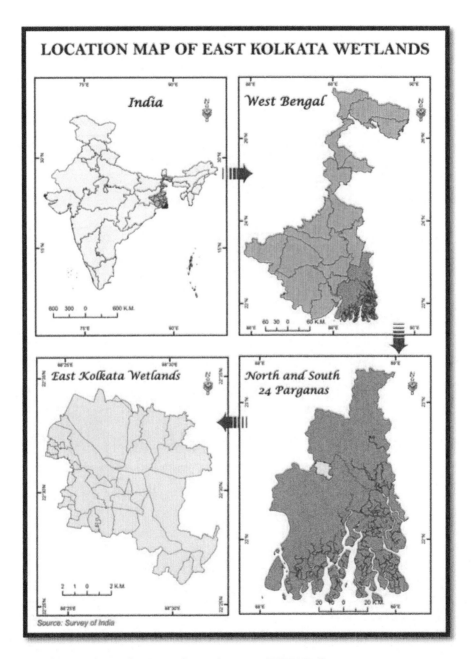

FIGURE 3.1 (See color insert.) Location map of EKW, Kolkata.

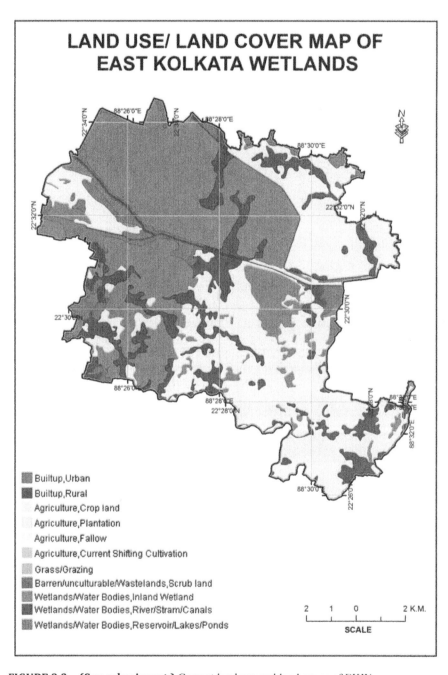

FIGURE 3.2   (See color insert.) Current land use and land cover of EKW.

EKW sets a conventional example of harnessing natural resource of wetland ecosystem for aquaculture and agriculture (including horticulture) ensuring community participation and their traditional knowledge. This wetland system also gives the best practice of augmentation of local communities into conservation and management practices. Thus this makes the wetland complex stands amongst one of the 17 case studies of wise use of wetlands by the Ramsar Convention (Bunting et al., 2011).

3.1.1 GENESIS OF EKW

The genesis of the East Kolkata Wetlands is strongly connected with the development of the city, changing courses of the River and waste management system of Kolkata. The city is located right on top of the mature Sundarbandelta which exceeds up to the northern fringe. Historically it was a part of mangrove and forestland of delta system. The Rivers in this deltaic region was characterized by improper drainage pattern and massive swamp formation with numerous wetlands (locally called as *bheries*). This complex network of river and its floodplain wetlands attract the East India Company as a navigational path to reach the hinterland of Bengal. Thus the city was established by Job Charnock in the year 1690 on the bank of River Hugli with a view to promote trade of goods. During this period the wetland complex boarding the eastern fringes of the city was nothing but was an insignificant jungle. These wetlands were called the salt lake. Presently this wetland complex is a part of EKW which also represents the remnant beds of mighty tidal River Bidyadhari and its spill areas. Initially brackish in character the wetlands were turned into freshwaters after the death of River Bidyadhari and subsequent loss of the connection with the other river system both upstream and downstream (Kundu and Chakraborty, 2017).

During the period of late eighteen century, a sudden annihilation in Rajmahal results in the shift of the Ganga to the Padma (Sengupta, 1980). This results in the disruption and gradually death of the connecting rivers, tributaries, estuaries, and network of channels between the Bidyadhari. River Damodar which principally contribute to the upland discharge to the Jamuna-Bidyadhari also changes its course in the due process. Thus in the absence of the upland water and sudden clogging of the interconnected channels, the Bidyadhari become depended solely on rainfall in its drainage basin. The situation further aggravated with an increase in tidal silt ingress and at the face of improper upland discharge to flush it down. Construction of bridges and channels further deep-rooted the situation and finally sealed the fate of the whole region. This event was perhaps marked as the landmark in the transformation of the ecosystem. Gradually the region converted into a large number of marsh and swamp area of varying size which later converted in ponds for both fish and paddy cultivation (Kundu and Chakraborty, 2017).

Gradually Kolkata also grew up into a large urban and trade center without any proper sewerage and solid management systems in place. The expansion of the city since its inception shows an adhocism rather than following a planned approach with respect to its spatial dimension and extension of the infrastructure, keeping its topology in mind. This unplanned approach often results in drainage congestion resulting in health impacts. The entire waste was initially dumped into the River Hugli, a practice which abandoned due to the frequent outbreak of malaria in the year 1757, 1762, 1770 affecting more than 76, 000 lives in Kolkata (Kundu et al., 2012). The practice was abandoned, and a committee was established to look for an alternate solution of the drainage problem. The committee recommended the transfer of all waste to salt lakes, as the city has a natural eastward slope. The wetlands were nearly 8.5 feet below the highest point of the city and thus recommendation strongly suggested for the construction of a series of sewers and pumping stations towards the salt lake. In 1864, a portion of the salt lakes was acquired for the dumping of solid waste. Though the first attempt of freshwater aquaculture was opted much later in 1918 (Kundu and Chakraborty, 2017).

Subsequent development of the wastewater channel in the city itself and the rapid growth in the settlements ensure more sewerage directly promoting and adoption of the waste fed aquaculture in the lakes. The wetland system presently has 264 functioning ponds (*bheries*). The solid waste dumping areas on the western fringe of the wetland complex were fully conversed to horticulture since 1876. Application of sewage was sequenced skillfully on the basis of detention time needed to improve the water quality appropriate for aquaculture activity.

The wetlands act as 'sink' for sewage and waste material from Kolkata that lacks any substitutable sewage treatment facility for its 4.5 million residents (Census of India, 2011). About 3500 tons of municipal waste and 68 million liters of raw sewage drain into the wetland system on a daily basis. Sewage-fed fisheries that utilize such large volumes of sewage generated by the city started functioning as early as 1883. In 1940, about 4,682.22 ha of wetland area produced 0.14 ton of fish per ha utilizing the municipal sewage. The profit-generating potential of the wetlands in terms of aquaculture aroused interest in local farmers as well as local landlords who leased out most of the ponds to commercial managers. The conspicuous result was employment generation through the production of fish and vegetables as well as a steady supply of food materials to the urban markets. The present manuscript is an effort to broadly represent the ecology-economic interface of wetland uses with special emphasis on the aquaculture practices in the wetland complex.

3.2   MATERIALS AND METHODS

Strategically observation and discussion were compiled on the data provided by the State Government Authorities of West Bengal responsible for the Conservation of EKW (Figure 3.3). Personal interviews with the stakeholders and the concerned authority depended on EKW was done to underpin the key issues faced by them. Also, a review on secondary literature for the survey authentication and ground proofing were done.

FIGURE 3.3   East Kolkata Wetlands, India.

3.3   OBSERVATION AND DISCUSSIONS

East Kolkata Wetland is the best existing example which provides a range of goods and services (Table 3.1) that contribute to human well-being and poverty alleviation. Communities living near wetlands are highly dependent on these services and are directly harmed by their degradation.

TABLE 3.1   Wetland Goods and Services Provided by the EKW

Goods	Services
• Fishery through aquaculture	• Flood control
• Agriculture	• Navigational path
• Horticulture	• Sewarage channels
• Animal husbandary	• Sewage treatment
• Water supply	• Garbage dumpyard
• Rugs from waste	• Nutrient and pesticides removal
• Recreation and scientific interest	• Heavy metal removal
	• Biodiversity

The economic activities that have mushroomed all over the have literally converted urban waste into wealth. Hence, this vast wetland has earned the title of Waste Recycling Region (WRR). These economic activities, of which agriculture, horticulture, and fisheries are the most important ones, have provided employment to thousands of people who dwell in the East Kolkata Wetland region and its outskirts.

The sewage of the entire Kolkata city enters the wetlands through a network of drainage channels which flows into the canal and ultimately falls in the fish ponds. During the entire process, the fishers believe the sunlight trigger the biochemical process/reaction which ultimately purifies the water entering the fish pond. For example, BOD (biochemical oxygen demand) is believed to be reduced through a symbiotic bio-chemical process between the algae and the bacteria present inside the sewage, where energy is drawn from algal photosynthesis. Each hectare of a shallow water body is capable of removing about 237 kg of BOD per day. This helps in the reduction of coliform bacteria prone to be pathogenic in nature, which even in a conventional mechanical sewage treatment plants may not be able to eliminate fully.

The effluents from the fish ponds are channelized to the southeastern region where abundant of paddy field was located. At present and still in the record there are 264 fish farms operating on a commercial basis. They cover a total area of about 2858.65 ha. The fish farms consist of units of various sizes from large holdings locally called *bheries* and relatively smaller ones called *jheels* due to their trench-like elongated shapes. But all these fish farms generally have similar types of produce, farming practice and distribution system.

A variety of sweet water fishes were cultured in the *bheries*. The main varieties in the existing practice of polyculture practice include:

1. Indian Major Carp – Rahu (*Labeo rohita*), Catla (*Catla catla*), Mrigal (*Cirrihinus mrigala*).
2. Indian Minor Carp – Bata – (*Labeo bata*).
3. Exotic Variety – Silver Carp (*Hypophthalmichthys molitrix*), Common Carp (*Cyprinuscarpio*), Grass Carp (*Tenopharyngodon idella*).
4. Tilapia–Nilotica (*Oreochromisnilotica*), Mosambica (*Tilapia mosambica*).

Apart from these cultured varieties (except Tilapia), some other varieties including forage fishes are also occasionally available in the *bheries*. These varieties are Punti (*Puntius japonica*), Sole (*Channa striatus*), Lata (*Channa punctatus*), Chyang (*Channa gachua*), Singi (*Heteropneustes fossillis*), Magur (*Clarias batrachus*), Fouli (*Notopterus notopterus*), Pungus (*Pangasius pangasius*), etc.

Despite of such a promising overview of fishery it has been observed that the average sizes of marketable fishes are not of the optimum weight (in grams). The prime reason behind the scenario is being that the fishes are normally netted/harvested much before they attain mature marketable growth and sizes. The main reasons for this are that the management/owners are compelled to create the maximum number of man-days possible in a year to provide employment to direct laborers viz. harvesters, carriers, etc. and the impending threat of poaching. Production and yield per hectare of fish varies among the *bheries* depending upon the conditions of production. Till today it has not been possible to study all possible variables that affect the production and yield of fish in this region. However, data which are available to give a reasonably sensible indication in this context require reconfirmation and updating.

The maximum numbers of *bheries* (125) fall within the area range of above two and below 10 hectares, followed by 76 *bheries* within the range up to two hectares, and 34 *bheries* fall within the range above 10 hectares to 20 hectares. More than 89% of *bheries* fall within the area range of up to 30 hectares. The maximum average yield (i.e., 6.48 MTha^{-1}) has been achieved in the *bheries* of more than 70 hectares in size. The yield is recorded to be increased with an increase in the size of *bheries*; thereafter, it tends to decrease as the size gets bigger and is lowest (i.e., 2.89 MT where *bheri* size is between 30 and 40 hectares). Yield per hectare again increases steadily with the increase in the *bheri* size and is the highest at 6.48 MT in fish farms where the effective area of the water body is above 70 hectares. Larger size *bheries*, though few, are more organized in terms of operations, planning through efficient management of production schedules, utilization of manpower, sewage, better procurement planning, monitoring water quality and fish health and efficient personnel management, etc. All these combine to provide the right synergy for achieving better production performance and per hectare yield.

In terms of annual production, 45 operating *bheries* reported 'increasing trend,' 61 *bheries* reported 'decreasing trend,' while in 76 *bheries* production was reported to be at an even level during the past three years. No proper indication of production trend was available in the remaining 82 *bheries*.

For a *bheri* to be operated efficiently, the following conditions are critical:

a) Maintenance of the required depth of water at all the three stages of the production process, e.g., at nursery pond, rearing pond and stocking pond with proper inlet-outlet management of sewage.

b) Availability of quality spawn/fry/fingerlings at required time and quantity.

c) Proper and efficient deployment of working personnel, ensuring satisfactory labor productivity and congenial labor relations.

d) Monitoring fish health.

The most important requisite to run a fishpond efficiently is an adequate and safe supply of wastewater. This is related to a number of other factors and deserves more elaborate discussion and research. Poor quality of sewage brings in lower quantity of nutrients and higher toxic load for the fish to feed upon. Low quality of sewage-borne nutrients requires supplementing with nutrients from outside. This entails more expenditure and increased operational costs that affect viability. In such situations, the *bheri* owners, with additional input costs, seek to add value to their produce (fish) to get higher returns and recover the additional expenditure incurred on fish nutrients/feed.

One way of countering this situation is by allowing the fishes to grow bigger. As this means lesser number of netting (harvesting) days resulting in the loss of man-days, the workers' union does not allow this to happen (Figure 3.4). This, in turn, gives birth to a situation of conflicting interests.

The following three types of ponds are required according to the stage of culture operation and ultimately production.

1. Nursery pond;
2. Rearing pond; and
3. Stocking pond.

Each of them is facilitate with proper inlet-outlet for the management of sewage inflow and waste outflow.

FIGURE 3.4   Fishponds of EKW.

Introduced stocks were raised in five major phases:

1. Pond preparation (mostly done in winters);
2. Primary fertilization (initial introductions of wastewater into the pond and allow to undergo natural purification, and stirring of the pond in order to reduce anaerobic conditions in the sediments);
3. Stocking of fish (initially stock a small number of fish for the test of acclimatization, growth, and water quality and subsequently introduction of seed in measured number);
4. Secondary fertilization (periodic introductions of wastewater into the ponds throughout the culture period), and finally; and
5. Fish harvesting (taken at different times according to species, growth, and market demand).

A flow diagram of pond preparation to harvesting of fish is shown in Figure 3.5.

During the entire operation, silt traps are pits at the edges of *bheries*, which trap the silt buildup in the ponds. Traps were periodically dredged, and the residues were used to strengthen dikes. Dredging and draining out of ponds were periodically done before the monsoons to release the nutrient locked in the bottom as well as deseeding of unwanted fish species and macrophytes. Farmers also believe that this step helps to kill the pathogenic parasites affecting the fish in culture period.

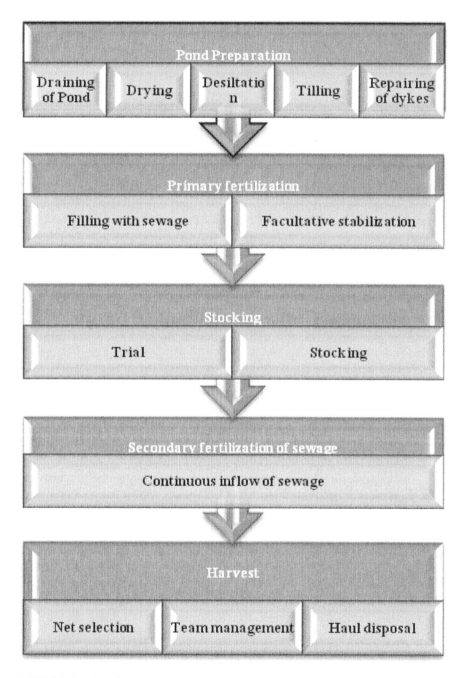

FIGURE 3.5   Flow diagram of pond preparation to harvest.

The stakeholders specifically involved for their livelihood in the wetland migrate at various part of the wetland complex with respect to the seasonal need. Usually, they get engaged in agriculture farming, garbage sorting, and aquaculture activities like trading, auctioneering, selling, making nets, maintaining drainage, and reinforcing banks. Apart from livelihood the stakeholders fully depend on the wetland for the fishes and vegetables required for their daily consumption need. During the entire process of self-purification of sewage, one macrophyte species (*Eichhornia crassipes*) commonly known as water hyacinth, play a role in the system's function. Some of the major supports it provides were:

1. Breaks the surface waves.
2. Stabilizes the bank erosion.
3. Used as replacement of expensive like stone or concrete to strengthen the banks.
4. Provide shade to fish during the month of summer.
5. The roots, well known to absorb metal ions, help to leach heavy metals out of the water.
6. Used as feed for cattle like buffalo usually decomposed and fed to carp.

Land Ownership: There are various types of land ownership of fishponds in EKW (Ghosh and Sen, 1987):

1. Owner-managed: but this is disappearing because of uncertainties of the wetlands' future.
2. Cooperatives: 13 Cooperatives with 300 fishers are present, these function well without any external assistance.
3. State Government Corporation: only two ponds are presently operational.

3.3.1   INCOME AND EXPENDITURE

This has not been studied extensively in any scientific study and approaches thus reliability on the presently available data is questionable (Ghosh, 1985a, 1985b; IWMED, 1986). Private fishers and farmers were also reluctant share to share the financial nitty-gritty, and thus relying on the State-managed fishponds is too few to represent the entire area fully.

3.4   CONCLUSIONS

Despite of the potential to provide multiple natural and social services, the cultural practice in the wetlands are still operated an informal way. As such, they need further study, effective maintenance, monitoring, and upgrading. Fast urban development and encroachment on eastward of the city is one of the major threats to the systems. Waste from the tanneries releasing untreated effluent directly into the wetlands, further threatening the water quality. Siltation of the ponds is a common phenomenon affecting the water depth in culture operation.

The presence of the Mafia in the control of fishponds had impacted the overall operation of the wetland complex. Impact of sewage and the threshold limit of nutrient (macro and micro) for the sustainable culture operation required to be scientifically established. Although researches confirmed the production of fish was not affected even by levels of ammonia-nitrogen concentration of 5.13 mgl^{-1} (the maximum limit is 0.1 mg^{-1}). But there still lies the possibility of transfer of the fatal micronutrient and heavy metals in the higher food chain.

Indian municipalities rely on the informal sector to supplement over-burdened waste management services; there remain questions about the system's effectiveness and more importantly, the environmental and moral issues of relying on so-called "ragpickers" to sort municipal garbage. First, these are the poorest members of society, typically women and children. The work is filthy and dangerous as it puts workers most of whom do not wear protective clothing or gloves into contact with broken glass and medical and other hazardous wastes. Many suffer from injuries and chronic skin diseases. Secondly, only garbage which can be sold is removed, which also means that some non-biodegradable wastes which have no commercial value, such as plastic and foil wrappings, remain in the soil where vegetables are grown. Third, the composting plants are taking biodegradable garbage (which is sold to tea gardens) from the vegetable farmers, is presently, adversely impacting the soil quality drastically. As a result, vegetable farmers have begun using chemical fertilizers to enhance soil fertility.

There is an absence of policies and strategies to guide coordinated actions within river basin linking coastal processes. Full ranges of ecosystem services of EKW were not integrated into the developmental plan. Water allocations are biased towards human uses ignoring ecological aspects. There is a huge lack of involvement of stakeholders, particularly marginalized communities. Lack of baseline information for planning and decision making is a major hindrance for the sustainable management of this wetland. Absence of effective institutional mechanisms is also a noteworthy issue to be tackled in the present and future. Management Planning Framework which is required for the sustainable conservation and wise use of wetland complex can be summarized as:

1. Management zoning identifying entire wetland area as core zone accommodating the ongoing practices and rationalizing proposed land use planning and direct basin as buffer zone.
2. Establishing a hierarchical and multi-scalar inventory of hydrological, ecological, socioeconomic, and institutional features to support management planning and decision making.
3. Ensuring hydrological connectivity of EKW with freshwater and coastal processes at basin level.
4. Environmental flows as a basis for water allocation for conservation and developmental activities.
5. Biodiversity conservation through habitat improvement of endangered and indigenous species.
6. Ecotourism development for enhancing awareness income generation and livelihood diversification.
7. Poverty reduction through sustainable use of land and other resource development and utilization.
8. Formation of multi-stakeholder groups for planning, implementation, and monitoring of MAP.
9. Strengthening EKWMA with adequate legal and administrative powers.
10. Capacity building at all levels for technical and managerial skills.
11. Result oriented monitoring and evaluation at activity, outcome, and impact levels.
12. Integrated management planning for EKW to achieve conservation and wise use of wetlands.
13. The Ramsar framework for wetland inventory assessment and monitoring which is a multi-scalar approach has been adopted for the purpose.
14. Interconnectivity in management planning.

ACKNOWLEDGMENT

The authors are thankful to East Kolkata Wetland Authority for providing all available information pertaining to East Kolkata Wetland. They are also thankful to Dr. Nitai Kundu, Senior Scientist, IESWM, Kolkata for helping in developing the manuscript.

KEYWORDS

- bheries
- East Kolkata Wetlands
- fish farms
- jheels
- sewage

REFERENCES

Bunting, S. W., Edward, P., & Kundu, N., (2011). *Environmental Management Manual: East Kolkata Wetlands.* CEMPD and Manak Publishers, New Delhi, pp. 1–156.

Ghosh, D., & Sen, S., (1987). Ecological history of Calcutta's wetland conversion. *Environmental Conservation, 14*(3), 219–226.

Ghosh, D., (1985). *Cleaner Rivers: The Least Cost Approach.* A village linked programme to recycle municipal sewage in fisheries and agriculture for food, employment, and sanitation. Government of West Bengal, India, 32.

Gosh, D., (1985). *Dhapa Report, From Disposal Ground to WAR (Waste-as-Resource) field.* Submitted to Calcutta Municipal Corporation (Preliminary Draft). Government of West Bengal, India, 32.

IWMED, (1986). *Growing Vegetables on Garbage, A Village Based Experience of City Waste Recycling.* Institute of Wetland Management and Ecological Design, Calcutta, India, 33.

Kundu, N., & Chakraborty, A., (2017). East Kolkata Wetlands: Dependence for ecosystem goods and services, In: Anjan Kumar Prusty, B., Rachna Chandra, & Azeez, P. A., (eds.), *Wetland Science, Perspective from South Asia* (pp. 381–406).

Kundu, N., Pal, M., & Saha, A., (2012). *East Kolkata Wetlands – Demographic & Livelihood Profile,* Manak Publishers, New Delhi.

Sengupta, B. K., (1980). *Kalikatar Pasei Laban Radh: Mahanagarer Sathi Ebom Bandhu Harader Cromo Bibartener Etihas Matsa Chasi Diibas Palan-o Seminar* (A salt marsh near Kolkata: a companion and friend, Fish farmers day celebration cum seminar).

PART II
Integrated Water Resources Management

CHAPTER 4

INTEGRATED WATER RESOURCE MANAGEMENT PLAN

ANIL KUMAR

Associate Director, Water Business Group, CH2M HILL (India) Pvt. Ltd, Noida – 201301, India, E-mail: anil.kumar2@ch2m.com

ABSTRACT

The freshwater shall be recycled and reused for the nonpotable and industrial process, as fresh water is a finite and vulnerable natural resource. Unique and innovative solutions are the demand for newly developed Indian cities to meet their water demands and thus looking at future water challenges, there is a need to search for alternate potential sources of water. For long-term development of a region, it is necessary to plan sustainable future water resource programme to meet demand during peak summers, which can be a potential concern for the industrial growth of the region. An IWRM plan involves across the board usage of available water resources in several combinations and manages sources in such a way that they are reliable and sustainable throughout the planning horizon. IWRM modeling together with detailed cost-benefit analysis was found to be economical than the traditional schemes to meet the demand of the MBIR region.

4.1 INTRODUCTION

Freshwater is a finite and vulnerable resource essential to sustain life, development, and the environment. With the continued growth of Indian cities and its periphery areas, the population is continuously increasing and thus the demand for water. Competing demands for water within the context of climate uncertainty and a growing population require unique and innovative solutions. Failure to recognize the economic value of water has led to wasteful and environmentally damaging uses of resources. Looking at the future water challenges, there is a need for an approach that leads to the development of potential alternative potential sources of water to minimize the overexploitation of freshwater sources.

Integrated water resource management (IWRM) plan is a systematic process for the sustainable development, allocation, and monitoring of water resource use in the context of social, economic, and environmental objectives to ensure effective, equitable, and sustainable water management. In simple terms, all the different uses of water resources are considered together for the sustainable water management approach to meet potable and non-potable demands of water. IWRM plan promotes development that coordinates management of water, land, and related resources so as to maximize the resultant economic and social welfare. One of the major aspects considered in the development of this approach includes the identification and development of an optimized solution that is based on a techno-economic evaluation of various available water sources or combinations thereof.

4.2   IWRM MODEL

It is observed that many water utilities around the world find it difficult to maintain desired water supply levels during the peak summer months and dry spells especially when only one source of water, which can be a potential concern for the industrial growth of the region. Several water utilities have implemented water reclamation (wastewater recycle and reuse) to augment the water supplies, especially for non-potable applications to meet the water scarcity where potable water demands are the demands that require drinking water quality for end-users that involve direct consumption or a high likelihood of direct consumption of the water by people. These demands include water for sinks, showers, and dishwashers and non-potable water demands are the demands for end-users that do not involve direct consumption by people can be met with non-potable water. For long-term development of a region, it is necessary to plan future water resource programmes in a rational and integrated for all users.

The above schematic diagram include potential system components such as rainfall, rivers, external supply and groundwater under water supply; residential, institutional, horticultural, and fire fighting under water demand; drinking water treatment plant, reuse/recycle treatment plant, and wastewater treatment plant for the complete water demand cycle of a region (Figure 4.1). Capital and operating costs for each step in the cycle are factored into the optimization analysis and balanced with other essential criteria to arrive at optimal decisions. The complete model can be developed to work with multiple data sets to reflect a range of future conditions. This allows a region to evaluate changes in system operations in response to climate change effects, or deviation from planning scenarios resulting in changes to supply and/or demand projections.

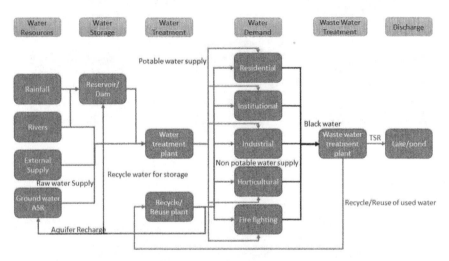

FIGURE 4.1   Schematic diagram of water resources and demand.

4.3   KEY COMPONENTS AND INTERRELATIONSHIP OF INTEGRATED WATER MANAGEMENT

IWRM can be developed by relating components of a region such as water supply, wastewater, water, etc. Figure 4.2 shows the interrelationship of all the key components of IWRM. The planning cycle of IWRM is given in Figure 4.3.

4.4   PLANNING OF IWRM

Typical issues that are considered and addressed while developing an IWRM Plan includes:

- Developing interfaces between macroeconomic and water resource decision making.

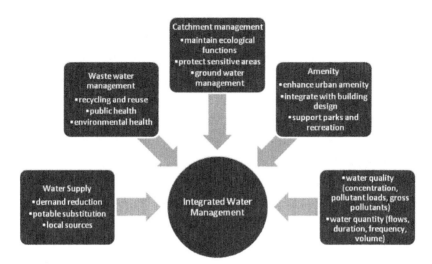

FIGURE 4.2   Interrelationship of key components of IWRM.

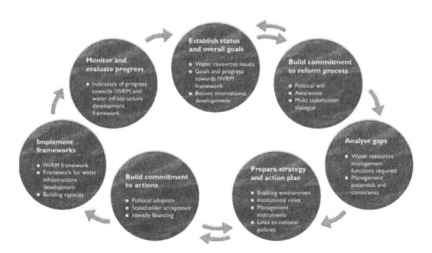

FIGURE 4.3   (See color insert.) IWRM planning cycle.

- Assessment of the efficiency of existing water infrastructure such as water and wastewater treatment plant, water supply infrastructure, stormwater drainage system, sewerage network system.
- Availability of non-conventional water resources and conservation technologies in the region, including an assessment of potential options for the development of these resources.

- Water quality and broader environmental issues related with existing conventional and nonconventional sources of water.
- Existing data collection system and access to information by users.
- Policy instruments and the legal and regulatory framework governing water resource management in the region.
- The role of the state in managing water resources in the region and potential for public-private partnership in any water resource development program.
- Processes for reconciling water quantity and quality needs of all water users.
- Mechanisms for consultation and public participation.
- Water allocation systems.
- Capacity building.
- Management agencies.
- Mechanisms to achieve financial sustainability.

4.5   RELIABILITY OF SOURCES IDENTIFIED IN THE IWRM PLAN

Reliability of sources is a very important aspect for the overall success of an IWRM plan. An IWRM plan involves across the board usage of available water resources in several combinations, apart from identifying the new and sustainable water sources such as the use of recycled water. While prioritizing the identification of existing or derived water source, such as recycled water, it is imperative that a comprehensive and detailed reliability analysis for a given source should be performed. It should be aligned with the national water policies, statutes governing the usage of water resources, and any proposed water infrastructure development project.

Overall feasibility of an identified option in an IWRM plan largely depends on its life cycle cost, which includes operational, maintenance, and any associated cost related with technology up gradation. While prioritizing the implementation plan, it should be kept in mind that resources optimizing the life cycle cost of the identified option are selected. The ultimate objectives of an IWRM plan are to develop and implement national/regional strategies, plans, and programs with regard to IWRM, and introduce measures to improve the efficiency of water infrastructure to reduce losses and increase recycling of water. A successful IWRM plan employs the full range of policy instruments, including regulation, monitoring, voluntary measures, market, and information-based tools, land-use management and cost recovery of water services.

4.6   DEMAND MANAGEMENT

Demand management should be a critical part of any IWRM Plan. As opposed to the supply side solutions, the cost for implementing demand management measures is modest to relatively low. In the long term, effective demand management would enable best practice management of overall water supply and infrastructure. There is sufficient scope to suggest ways to reduce the actual water consumption rate by using the world's best management water practices such as:

- Water-saving fixtures/devices.
- Behavior change and social awareness/education programmes.
- Leakage detection and repair.
- Minimization of non-revenue water losses.

The 150 lpcd is a guideline or aspirational design figure. In actual practice, water usage can be made significantly less and therefore reduce the amount of water resources to find, minimize the size of water infrastructure and reduce the capital cost and operating costs. It is likely that any new Indian city could realistically achieve a domestic water demand of at least 135 lpcd and may be able to achieve 120 lpcd with a little more effort aggressively and implementing the following measures:

- Higher plumbing and piping standards to prevent leaks.
- Strict construction standards, contract supervision, and leak testing.
- Leakage and non-revenue water monitoring and correction.
- Pricing signals to deter the waste of water and leakages.
- Strict controls and enforcement of water use rules.
- High levels of public awareness campaigns and education.

4.7   CASE STUDY

DMICDC is developing Manesar Bawal Investment Region (MBIR) in Haryana state as one of the industrial cities proposed to be developed in Phase 1 and CH$_2$M HILL International has developed IWRM plan for the region.

- The development of MBIR is projected to transform the land uses from predominately agriculture-based production to one that is dominated by residential, industrial, and open space land use with a projected population of 3.2 billion and estimated water demand of 1060 MLD for the year 2040.

- MBIR is located in within a water-stressed area of Haryana state that does not have any perennial river, and the state receives surface water from Yamuna, Sutlej, Ravi, and Beas Rivers under various interstate water-sharing agreements and Water to MBIR is provided under a Canal Rotation Program and Water to these canals is supplied in rotation, for an average of 15 days in a month.
- The area falls under the overexploited category for extraction of Groundwater and extensive over-exploration of shallow aquifer, and often uncontrolled groundwater extraction has been done at such a high rate that it has exceeded the rate of groundwater recharge.
- MBIR has limited availability of existing wastewater treatment infrastructure, and most of the sewage that is currently being generated is unutilized.

The recommended IWRM Plan requires the development of a range of surface water and recycled water resources. Surface water resources are predominately from the JLN Canal System. Recycled water is produced from the wastewater or sewage generated within the MBIR development area. Rainwater harvesting by the collection of water from rooftops is also recommended. The alternatives were subjected to a high-level screening exercise to shortlist potential alternatives. The primary criterion's adopted for recommending potential alternatives are:

- **Economic Criteria (35%)** is a "relative cost" to provide a high-level indication of the cost of the option.
- **Environmental Criteria (20%)** is dominated by residuals generation since this is expected to have the largest environmental impact.
- **Social Criteria (20%)** prioritized by health and safety risks along with the impact on users. Slightly less weight was given to organizational benefits and community acceptance.
- **Technical Criteria (25%)** all factors are equally weighted since each of these factors is essential in integrated water management.

IWRM modeling together with detailed cost-benefit analysis identified following combination to satisfy the requirements for robustness, reliability, and resilience as well as deliver the most optimal balance of environmental, social, and financial outcomes for the proposed MBIR development.

- JLN canal water freed up from agriculture.
- Use of water savings from WJC system relining and/or de-silting (between Munak to Khubru Head).
- Use of water savings from WJC system relining and/or de-silting (between Khubru and Loharu Head).
- Use water savings from WJC system relining and/or de-silting (between Loharu and JC III Pump House).
- Recycled water generated within MBIR (internal water recycling).
- Local rainwater harvesting (rooftop).

Thus, a combination of the option of only the relining and desilting of existing canals is able to save such a large amount of water and meet most of the demand of the MBIR region. Economically, IWRM scheme was found to be more economical as compared with the traditional schemes of meeting demands of MBIR region. For meeting the demand of 1060 MLD in MBIR region, master plan was developed presenting the block cost estimate to be 11,436 crores which included development of reservoir, irrigation canals, underground transmission and conveyance system for potable water only and no scheme for management of wastewater whereas IWRM plan of MBIR for meeting the same demand estimated the cost to be 4,998 crores including treatment plants, water distribution system, sewage collection system, sewage treatment plants and rising mains for potable and non potable water supplies.

From the above comparison, it can be concluded that IWRM projects can be implemented for prestigious regions like MBIR to meet potable and nonpotable water demands utilizing even the used water for nonpotable requirements.

KEYWORDS

- **demand management**
- **IWRM model**
- **IWRM plan**

EVALUATION OF GRAVITY-BASED DRIP IRRIGATION WITH PLASTIC MULCH ON RAISED BED CULTIVATION OF SUMMER OKRA AT FARMERS FIELD IN RANCHI DISTRICT

MINTU JOB[1], NIVA BARA[2], A. K. TIWARI[3], and C. S. SINGH[4]

[1]*Assistant Professor, Department of Agricultural Engineering, Birsa Agricultural University, Kanke, Ranchi, Jharkhand, 834006, India, E-mail: mintujob@rediffmail.com*

[2]*Head, Department of Agricultural Extension, Birsa Agricultural University, Kanke, Ranchi, Jharkhand, 834006, India*

[3]*Assistant Professor, Department of Horticulture, Birsa Agricultural University, Kanke, Ranchi, Jharkhand, 834006, India*

[4]*Assistant Professor, Department of Agronomy, Birsa Agricultural University, Kanke, Ranchi, Jharkhand, 834006, India*

ABSTRACT

Tank-based drip irrigation with plastic mulch is a viable option for farmers with small and fragmented land holdings and in places where the power supply is interrupted. Based on the survey in Chipra village, Nagri Block of Ranchi, a gravity-based low-pressure drip irrigation system was designed so that the system can be operated in low holding areas. Tank placed at a temporary platform of 2 to 3 meters was found to be sufficient for obtaining an acceptable uniformity of water distribution for a bed length of 15 meters with a water application rate of 1.2 lh^{-1}. Based on the field study with ten farmers on a unit size on 500 m^2 area, Okra production was successful with this system which enabled each farmer to get an average profit of Rs. 7209 from each plot for summer season okra. An average yield of 144 qha⁻¹ was obtained with this system of cultivation which was 56% more compared to the conventional method of cultivation. In addition, there was a water saving of 37% through this method. Use of Silver-black polyethylene mulch beside moderating the microclimate inside the bed also restricted weed growth with weed dry mass of 8 g/m^2 and 12 g/m^2 after 30 and 60 days after sowing. All other growth parameters of okra under drip irrigation and plastic mulch were superior to the conventional method of cultivation which resulted in increased yield, water use efficiency and economic benefits.

5.1   INTRODUCTION

"Don't earn enough to adopt TECHNOLOGY-Don't have the technology to earn ENOUGH." Farmers in Jharkhand are caught up in this vicious cycle for years. It is generally believed that the benefits of modern farm technology have been availed of only by large farmers. However, the fact is that even small farmers can utilize selected farm technologies for efficient farm operations. Drip irrigation along with a mulching system of cultivation is among one of the combinations which is technically feasible and economically viable for almost all orchards and vegetable crops. Cultivation of off seasonal crop using a method like drip irrigation, mulching, can give a higher income to the farmers which otherwise is not possible through conventional method. Okra is one of the most remunerative crops grown in this region and cultivated mostly in as a rainy season crop. Although some pockets of Ranchi district where there are perennial sources of water it is also grown as summer crops. Due to inclement climatic conditions like severe cold in January, its cultivation is delayed till the start of the early summer season; as a result, the available water sources get dried up and coupled with inefficient irrigation system the yield is much lower than its potential. Many researchers have reported higher application efficiency of drip irrigation system over the conventional irrigation methods. Sivanappan and Padmakumari (1980) compared drip and furrow irrigation systems and found that about 1/3rd to 1/5th of the normal quantity of water was enough for the drip-irrigated plots compared to the normal quantity of water applied to plots under surface irrigation in vegetable crops. Sivanappan et al. (1987) recommended drip system of irrigation in place of conventional furrow irrigation due to the economy in water utilization to the extent of 84.7% without any loss of yield. The response of okra to drip irrigation in terms of yield improvement was found to be different in different agroclimatic and soil conditions in India. The increase in the yield of okra to the tune of 40% was reported under drip irrigation (Patil, 1982). In another study, the yield was reported to be slightly lower to that of conventional furrow irrigation (Patil et al., 1993). Based on the study conducted at Rahuri, India, Khade (1987) reported 60.1% higher yield of okra with water saving of 39.5% under drip irrigation as compared to conventional furrow irrigation.

A gravity fed drip irrigation system is a cheap and effective way to apply water for a smaller sized crop area. This system does not require any electricity for pumps as the system uses the force of gravity to push water through the drip line. Most of the smallholder farmers in India rely on rainfed agriculture and frequently face dry spells and droughts that affect agricultural productivity. The gravity fed drip irrigation technologies are the solution to them as it bridges dry spells, mitigate against droughts and ensure food security. Thus, appropriate, affordable, accessible, gravity fed drip irrigation system can be the better alternative for small landholders. Thus a technically proven system consisting of raised bed cultivation of okra under plastic mulch and gravity based drip irrigation system was evaluated for small farm holding for its technical feasibility and economic viability.

5.2 MATERIALS AND METHODS

An effort was made in this direction through Farmers First Programme Under Directorate of Extension, Birsa Agricultural University, which worked with around 1000 farm families in villages Chipra and Kudlong under Nagri Block Ranchi. These farm families were marginalized by remoteness, inaccessibility, very small landholding, and traditional farming methods. Cash income was very low, and in many households, the main problem was still to achieve basic food security. The income was so poor that they had to struggle to make both ends meet. Out of compulsion most of the families including women were forced to work in a brick kiln as seasonal workers to bring cash flow. They had literally abandoned farming for quite some time as traditional farming with conventional crops in mostly rain-fed uplands gave very less dividends.

With the belief that technology interventions, which address land productivity of marginalized farmers, hold the key to usher effective means of addressing the issue of rural poverty alleviation. New interventions were pondered to bring back these families into cultivation for making a living. Now the first and the most difficult step was to bring back the confidence in these farmers who have left farming for quite some time and had adjusted themselves with being kiln workers. Initial Survey was conducted in Chipra and Kudlong villages by Farmer FIRST team of BAU. It was felt that intervention which is tangible and effect of which can be seen in a short time would be best to bring back the confidence in these farm families. After doing much exercise, plasticulture intervention was found most suited to start with. The difficulty was, as it always happens with new technologies, was to convince farmers about the concept. On the principle of seeing is believing, the group of farmers were given exposure visit to Birsa Agricultural University and ICAR for Eastern region where cultivation under drip irrigation (both pressurized and gravity fed). Vegetable crops under mulching in raided bed, cultivation, and nursery growing under low-cost polyhouse were shown. The same group was imparted multiple training at Birsa Agricultural University on different plasticulture application. Scientist from Birsa Agricultural University also imparted training though guest lecturer in multiple trainings organized by Farmers FIRST programme during Mahila Saptaah organized from March 6–12, 2017. After these training and seeing into the ground reality of their situation, farmers were inclined towards gravity based drip irrigation and black polyethylene mulching. 10 Farmers five each from Chipra and Kudlong were selected for this intervention.

Ten Farmers were chosen for this intervention and crop selected based on a survey done and seeing the seasonality and market condition okra (var-Osaka, F1 hybrid) was selected and planted in 500 m² area under gravity drip irrigation and mulching. Pressure for irrigation was made through an overhead tank placed at the height of 1.5 to 2 m. All technical aspects were looked into while designing the system and all agronomic measures followed.

5.2.1   LOCATION AND LAYOUT OF FIELD PLOT

The field experiment was conducted during February to May in 2017 at the Chipra and Kudlong villages of Nagri Block, Ranchi. The soil at the farmers' field were mostly sandy loam (18.4% clay, 22.6% silt, and 59.0% sand) having a bulk density of 1.39 g/cm³ with a basic infiltration rate of 1.8 cmh⁻¹. A field plot measuring 1000 m² was vertically divided into three equal parts (i.e., T1, T2 and T3) and 10 farmers from two villages viz. chipra and kudlong were the replications. The layout of the field with drip irrigation network is shown in Figure 5.1. Osaka-Eastern seeds (F_1 hybrid) variety of okra was selected and the seeds were sown at a spacing of 30 cm in the third week of Feb. Paired row planting was adopted with one lateral catering to the water needs of two rows of okra in raised bed plantation spaced at 0.5m in a bed with a top width of 80 cm. Standard agronomic practices such as fertilization and plant protection measures were applied during the crop period. The fertilizer doses of 100 kg N, 50 kg each of P and K along with 20 t of farmyard manure per hectare were applied to meet nutritional requirement of crop. In order to prevent fungal infection and attack of insects, Carbendazim with Dethane M 45 and Emida Cloropid were applied. The lateral lines were laid parallel to the crop rows and each lateral served two rows of crop. The laterals were provided with 'online' emitters of 2 lph discharge capacity at 0.6 m interval. In this arrangement there may be a possibility of overlapping of moisture front in the longitudinal direction under mulched condition partly due to lateral spread of moisture regime and mainly due to condensation of water vapor underneath the much so as it act as line source of irrigation. The treatments followed for the study were as stated below:

T₁   Control-farmers practice (Flood irrigation with 3 cm of water in furrows);

T₂   100% of irrigation requirement met through drip only;

T₃   100% of irrigation requirement met through drip with black plastic mulch.

Ten farmers were chosen for this intervention and crop selected based on a survey done and seeing the seasonality and market condition okra (var-Osaka, F1 hybrid) was selected and planted in 500 sq.m area under gravity drip irrigation and mulching. Pressure for irrigation was made through an overhead tank placed at the height of 1.5 to 2 m. All technical aspects were looked into while designing the system and all agronomic measures followed.

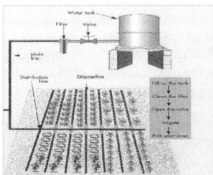

FIGURE 5.1   Drip layout with the overhead tank at farmers field.

5.2.2   ESTIMATION OF IRRIGATION WATER REQUIREMENT

The daily irrigation water requirement for the okra crop was estimated using the following relationship

$$IR = ETo \times Kc - ER \qquad (1)$$

where,

IR is net depth of irrigation (mmday^{-1});
ETo is the reference evapotranspiration (mmday^{-1});
Kc is the crop coefficient;
ER is the effective rainfall (mmday^{-1}).

The net volume of water required by the plant can be calculated by the relationship

$$V = IR \times \hat{A} \qquad (2)$$

where,

V is the net volume of water required by a plant (l day^{-1}).
$\hat{A}$ is the area under each plant (m^2).

That is, the spacing between rows (m) and spacing between plants (m). Since there was no effective rainfall during the crop growth period conducted at farmers field at Nagri, Block of Ranchi, rainfall occurring during these months can be taken as effective rainfall (Michael, 1981). In control plots, furrow irrigation was given at an interval of 10 days. The depth of irrigation water required for all the plants was roughly given up to 3 cm depth. Biometric observations were taken from selected plants in each treatment at 30 days interval. The operational and fixed costs of cultivation of were factored in a while calculating the cost of cultivation. The income from produce was estimated using the prevailing average market price @ Rs 18–20 per kg of produce. The net seasonal income from produce was estimated by subtracting the total seasonal cost from the income of the produce. The benefit-cost ratio, net income, and water-use efficiency were determined.

5.3   RESULTS AND DISCUSSION

The number of irrigation were divided into the four distinct crop growth stages, and the amount of irrigation was determined by climatological parameters based on the average monthly evapotranspiration of the area obtained from the Meteorological Department of the Birsa Agricultural University Ranchi. The number of irrigation at various stages of growth were 5, 21, 16 and 4 at an interval of 1 day for drip irrigation only and in the treatment drip irrigation with plastic mulching the number of irrigation provided at an interval of 2 days were 3, 14, 7, 7 at germination, vegetative, flowering, and harvest stages. In control (furrow irrigation) irrigation to the depth 3 cm was provided at an interval of 10 days. Total irrigation for furrow method of irrigation (control) was 27 cm while through drip irrigation without plastic mulch was 15.6 cm and for the treatment drip irrigation with plastic mulch was around 9 cm (Tables 5.1 and 5.2). Thus, it could be easily said that there is water saving of more than 80% through drip irrigation alone and if it is used in combination with plastic mulch, then there is further saving of water and irrigation frequency can also be reduced. Apart from this mulch also helps to arrest weed growth and moderate the soil temperature. This facilitates better growth and development of the crop.

The effect of drip irrigation on biometric parameters such as plant height, plant girth, fruits per plant, fruit length and fruit yield was compared with that of furrow irrigation treatments. The experimental results of these biometric observations for the year under consideration are presented in Table 5.3.

The results have shown increased yield attributing characters like plant height (56, 96 and 176 cm), plant girth (9.4, 13.9 and 21.0 mm) at 30, 60 and 90 DAS, respectively were highest for treatment T3 (Drip irrigation + mulching). Fruit per plant (15–23), fruit length (13.5 cm), fruit diameter (25.2 cm) and fruit yield (142.6 qha^{-1}) under plastic mulch and drip irrigation were also higher as compared to other treatments. The results corroborated the findings of Sivanappan and Padmakumari (1980) and Khade (1987). It can be seen that the plant growth and yield were greater in drip with mulch as compared to drip alone.

5.4   CONCLUSIONS

The drip irrigation is economical and cost-effective when compared with furrow irrigation. The use of drip either alone or in combination with mulch can increase the okra crop yield significantly over furrow irrigation to the tune of 35.5 to 51.2%. To irrigate 1 ha of okra crop with drip irrigation 150 l/m^2 water will be needed for this agroclimatic condition. The maximum duration of operation of drip irrigation is 42.3 min during peak demand of the crop with the emitter capacity of 2 lph. The net income could be increased

TABLE 5.1 Total Water Applied in Okra Crop During Entire Cropping Season (February 26, 2012 to June 10, 2012) Under Drip Irrigation

Month	Number of irrigations	Crop Stage	Frequency	Water/irrigation (l/m²)	Liters/ plant	Time/ irrigation (h-minutes)	Monthly water application
February	2	Germination	Daily	1.16	0.17	10.53	2.32
March	14	Germination(3)	Daily	1.29	0.19	11.61	3.87
		Vegetative(11)	Alternate day	1.80	0.27	16.20	19.8
April	15	Vegetative (10)	Alternate day	2.66	0.40	24	26.6
		Flowering(5)	Alternate day	3.64	0.55	32.82	18.2
May	15	Flowering (11)	Alternate day	4.70	0.70	42.30	51.7
		Harvesting(4)	Alternate	4.48	0.67	40.32	17.92
June	3	Harvesting	Alternate	3.40	0.51	30.60	10.2
Total	50						150.61

TABLE 5.2 Total Water Applied in Okra Crop During Entire Cropping Season (February 26, 2012 to June 10, 2012) Under Drip Irrigation and Plastic Mulch

Month	Number of irrigations	Crop Stage	Frequency	Water/irrigation (l/m²)	Liters/ plant	Time/ irrigation (h-minutes)	Monthly water application
February	1	Germination	Alternate days	1.16	0.17	10.53	1.16
March	10	Germination(2)	Alternate	1.29	0.19	11.61	2.58
		Vegetative(8)	3 days	1.80	0.27	16.20	14.4
April	9	Vegetative (6)	3 days	2.66	0.40	24	15.96
		Flowering(3)	3 days	3.64	0.55	32.82	10.92
May	10	Flowering (4)	3 days	4.70	0.70	42.30	14.1
		Harvesting(6)	3 days	4.48	0.67	40.32	26.88
June	1	Harvesting	3 days	3.40	0.51	30.60	3.4
Total	31					Total 89.4	

TABLE 5.3 Growth and Yield as Influenced by Different Treatments

Treatment	Plant Height (cm)			Plant girth (mm)			Fruits per plant	Fruit length (mm)	Fruit dia (mm)	Fruit yield qha⁻¹
	30	60	90	30	60	90				
Control (Flood Irrigation)	33	70	112	6.7	10.4	16.3	11–16	10.5	23.0	94.4
Irrigation with Drip System	49	88	148	8.1	12.6	19.2	14–21	12.8	24.5	128.2
Drip Irrigation + Plastic Mulch	59	96	176	9.4	13.9	21.0	15–23	13.5	25.2	142.6

by about 47.74% by adopting drip with plastic mulch and 29.21 for drip over furrow irrigation The benefit-cost ratio was found to be highest (1.64) for treatment under drip irrigation with mulch followed by (1.44) for drip without mulch (Table 5.4). The water use efficiency for the treatment drip with mulch was found out to be 1620 Kg/ha-cm, and that of treatment with only drip irrigation was 1440 kg/ha-cm while in farmers practice (control) it was 350 kg/ha-cm.

TABLE 5.4 Calculation of Benefit-Cost Ratio of Onion for First Year Under Different Treatments (Prices are Given in Indian Rupees)

S. No	Cost Economics	Treatments		
		T_0	T_1	T_2
1	Fixed cost	-	195000	244,049
	a) Depreciation	-	8404	10,532
	b) Interest	-	19500	12,202
	c) Repair and maintenance	-	1950	2,640
	Total (a+b+c)	-	25374	25374
2	Fixed cost total		207304	269423
3	Annual fixed cost (considering life of 10 years)	-	20730	20730
2	Cost of cultivation	72000	92730	92730
3	Fertilizer cost (Rs/ha)	3650	3650	3650
4	Mulch material (Rs/ha) (life 3 season)	-	-	9500
4	Irrigation charges	4896	3480	2088
5	Seasonal total cost (Rs/ha)	80546	99860	107968
6	Yield of produce (Kg/ha)	9440	12820	14260
7	Selling price (Rs/kg)	18	19	20
8	Income from produce (6×7), (Rs/ha)	169920	243580	285200
9	Net seasonal income (8–5), (Rs/ha)	89374	143720	177232
10	Benefit-Cost ratio (9/5)	1.11	1.44	1.64

ACKNOWLEDGMENT

Authors are thankful to Farmers First Programme of ICAR and Department of Extension Education, BAU, Ranchi for providing the financial assistance for conducting this work and for providing necessary facilities to carry out the programmes at farmers field.

KEYWORDS

- **drip irrigation**
- **plastics mulch**
- **uniformity**
- **water use efficiency**

REFERENCES

Khade, K. K., (1987). *Highlights of Research on Drip Irrigation* (pp. 20–21). Mahatma Phule Agricultural University. India, Pub. No. 55.

Michael, A. M., (1981). *Irrigation Theory and Practice* (Reprint 1st edn., pp. 539–542), Vikas Publishing House, New Delhi, India.

Sivanappan, R. K., & Padmakumari, O., (1980). *Drip Irrigation* (p. 15). Tamil Nadu Agricultural University, Coimbatore, India, SVNP Report.

Sivanappan, R. K., Padmakumari, O., & Kumar, V., (1987). *Drip Irrigation* (1st edn., pp. 75–80). Keerthi Publishing House, Coimbatore, India.

CHAPTER 6

ASSESSMENT OF AGRICULTURAL DROUGHT USING A CLIMATE CHANGE INITIATIVE (CCI) SOIL MOISTURE DERIVED/SOIL MOISTURE DEFICIT: CASE STUDY FROM BUNDELKHAND

VARSHA PANDEY, SWATI MAURYA, and PRASHANT K. SRIVASTAVA

Institute of Environment and Sustainable Development, Banaras Hindu University, Varanasi – 221005, India,
E-mail: varshu.pandey07@gmail.com

ABSTRACT

Soil moisture information is very important for agricultural drought monitoring, which is now possible to be measured using the Earth observation datasets. Drought conditions can be directly related to the Soil Moisture Deficit (SMD) variable by using the soil moisture data and soil physical properties. In this chapter, we have used Climate Change Initiative (CCI) Soil Moisture data in integration with soil hydraulic parameters for the derivation of SMD. The field capacity estimated from the simulated Water Retention Curve (WRC) through ROSETTA, which can be used for the transformation of CCI soil moisture data into SMD over Bundelkhand region. The analysis of results indicates that a large part of the Bundelkhand region is facing moderate to severe drought conditions derived from the SMD model developed in this study.

6.1 INTRODUCTION

Drought is more damaging and diverse in nature in comparison to other hazard types such as floods, tsunamis, etc. It emerges gradually over a large area and generally lasts for the long time span. Monitoring of drought is complex as it varies both spatially and temporally; therefore, it requires state-of-the-art tools and techniques. The extremity of agricultural drought can be best estimated by the soil moisture levels. The deficit in soil moisture due to rainfall shortage is highly correlated with drought phenomenon. Soil moisture is a key variable in hydrological modeling, meteorological modeling (Nandintsetseg and Shinoda, 2011) hazard event estimation such as flood and drought, as well as plant growth vegetation health monitoring. Therefore, precise monitoring and spatiotemporal estimation of soil moisture are important. Now, under the climate change initiative (CCI), a long-term record of soil moisture is available for hydrological modeling. On the other hand, the Soil Moisture Deficit/Depletion (SMD) is the important variable for flood and drought forecasting. The SMD provides an estimation of agricultural drought by representing the amount of water requirement to raise the soil moisture content of the plant root zone to field capacity (FC) (Srivastava, 2014). The prolonged condition of SMD in the soil may lead to drought, if the same situation persists for a long period (Srivastava, 2013). Therefore, in the purview of the above, the main focus of our study is to use CCI soil moisture for SMD evaluation during the Kharif season.

6.2   MATERIALS AND METHODOLOGY

6.2.1   STUDY AREA

The Bundelkhand region lies at the heart of India located below the Indo-Gangetic plain to the north with the undulating Vindhyan mountain range spread across the northwest to the south. The Uttar Pradesh region of Bundelkhand has worst affected drought area, lies between $24^{0}00'$ and $26^{0}05'$ N latitudes and $78^{0}00'$ and $82^{0}05'$ E longitudes (Figure 6.1). The main rivers are the Sindh, Betwa, Ken, Bagahin, Tons, Pahuj, Dhasan, and Chambal, and constitute the part of Ganga basin. The topography of the region is highly undulating, with rocky outcrops and boulder-strewn plains in a rugged landscape. The major soils include alluvial, medium black, and mixed red and black soil (Singh and Phadke, 2006).

6.2.2   DATASETS

The ESA's remote sensing CCI soil moisture (CCI-SM) combined product used in this study for the estimation of agricultural/soil moisture drought.

The CCI-SM combined product has a 25 km spatial and daily temporal resolution with its reference time at 0:00 UTC, having product version 02.2 in NetCDF–4 classic file format over a global scale. CCI-SM represents the values from the upper few millimeters to centimeters from the soil surface to estimation of the influence of soil depth on drought phenomenon. In this study, the blended product made by fusing active and passive is used (Nicolai, 2017). Soil maps can provide soil inputs such as soil texture, bulk density, organic carbon, infiltration, soil depth, and water holding capacity, etc., to models for predicting hydrological and climatic conditions. The Digital Soil Map of World (DSMW) used in this study and prepared by the Food and Agriculture Organization (FAO) of the United Nations is in vector form, at 1: 5,000,000 scale, in the geographic projection (Latitude-Longitude) (https://searchworks.stanford.edu/view/4059679). The main types of soil in this region are Chromic Luvisols (Sandy Clay loam soil), Eutric Cambisols (Loam soil), Orthic Luvisols (Sandy Loam soil), and Chromic Vertisols (Clay soil). In the study area, the dominant soil types that exist are sandy clay loam and sandy loam soil. These types of soils show very poor water holding capacity and high infiltration rate, which cause SMD in the area.

6.2.3   ROSETTA MODEL

There are number of PTFs based on linear and non-linear regression, and Artificial Neural Network (ANN) is used in the scientific study that relates the descriptive equations such as (Van Genuchten, 1980) with calculated soil properties. In this paper, Rosetta Lite (version 1.1) inbuilt in HYDRUS 1D (version 4.16.0110) model, based on ANN analysis is used for predicting soil hydraulic properties. The ROSETTA model executes five hierarchical input parameters for WRC prediction. WRC is the curve which represents the relationship between the soil moisture content and the soil water potential. In this study, for WRC estimation percentage of sand, silt, and clay + bulk density have been taken into account. The soil moisture content at different potential using Rosetta is based on algorithms described by Van Genuchten (1980) and Schaap (2001) represented in the following equation:

$$\theta(h) = \theta_r + \frac{\theta_s - \theta_r}{[1 + (\alpha h)^n]^m} \qquad (1)$$

where $\theta(h)$ represents the WRC in terms of soil moisture content, θ(cm^{3}cm^{-3}) as a function of the soil water potential h (cm), θr and θs (cm^{3}cm^{-3}) are the residual and saturated soil water contents respectively, n, and α (1/cm) are

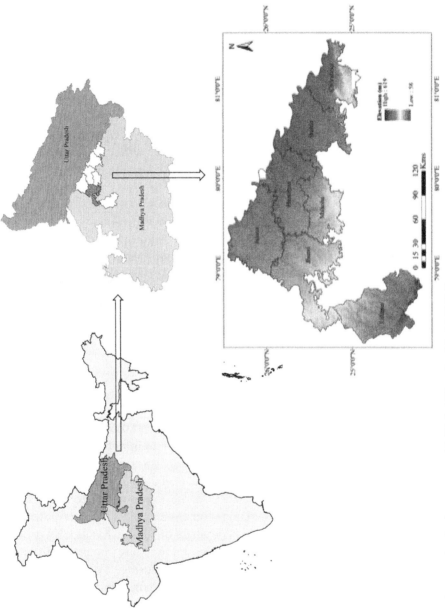

FIGURE 6.1   Study area map with Digital Elevation Model (DEM).

the curve shape parameters and m is empirical constant, which can be related to the n as:

$$m = 1 - \frac{1}{n}, \text{ For } n > 1 \tag{2}$$

6.2.4   SOIL MOISTURE DEFICIT (SMD)

SMD is calculated based on soil moisture content and FC. It is the difference between the moisture content available in the plant root zone and the amount of moisture content that the soil can hold against gravity. The primary thematic layers for SMD estimation were prepared using software Arc GIS (version 10.1) for SMD estimation. All the soil datasets were geo-referenced to the World Geographic System 84 (WGS84) coordinate system and then reclassified by assigning FC values estimated from the ROSETTA model. Then FC map is prepared by importing FC values generated by WRC (Figure 6.2), and soil moisture climatology map is also generated by CCI-SM datasets from 2002 to 2014 using Cell Statistic tool. The obtained layer was resampled for any geographical mismatch. The difference between FC and SMC (Soil Moisture Content) gives SMD. In hydrological modeling, FC is considered as the upper limit for soil moisture because extra water above FC cannot be held in the soil and will be drained away very quickly either as surface runoff or groundwater runoff. The SMD is calculated using a different hydrological model such as PDM by the following equation.

$$SMD = FC - SMC \tag{3}$$

6.3   RESULTS AND DISCUSSION

6.3.1   HYDRAULIC SOIL PARAMETERS

Despite soil hydraulic properties are typically measured in laboratories by in-situ soil samples, we used the Rosetta model, which is well tested at many locations, because of practical and economic constraints. The basic soil information (such as soil type, percents of sand, silt, and clay, bulk density) along with FC is shown in Table 6.1. Using the basic soil information, Rosetta depending upon Pedotransfer functions, measured soil hydraulic parameters for the formulation of water retention curve (WRC). WRC is defined as the amount of water retained in soil under a definite matric potential. It is based

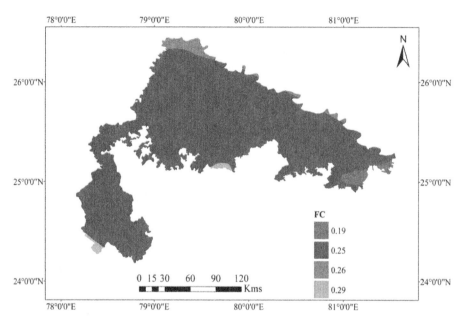

FIGURE 6.2   (See color insert.) Field capacity map of study area.

on soil texture, bulk density, organic matter, etc. and can be used for FC and Permanent Wilting Point (PWP) estimation of the soil.

TABLE 6.1   Soil Properties Derived From FAO-UNESCO for Bundelkhand-UP Region

Soil Type	% of Sand	% of Silt	% of Clay	Bulk Density (gcm^{-3})	Field Capacity (at 33 Kpa)
Sandy Clay Loam	59	11.2	29.8	1.6	0.25
Clay	20.8	23.5	55.7	1.7	0.29
Sandy Loam	71.9	8.9	19.2	1.5	0.19
Loamy	41.7	32.1	26.2	1.3	0.26

In the Bundelkhand-UP region, the dominant soil type is sandy clay loam and sandy loam as compared to the clay and loamy soil type. As the variation in soil types determines soil moisture retention capacity of the soil, ROSETTA based on soil type is used for estimation of the FC. The estimated WRCs for four soil type of the region is shown in Figure 6.3. After perusal of obtained result, we find that the FC of sandy loam (0.18 m^3m^{-3}) is found lower than among four soil type. Sandy clay loam (0.24 m^3m^{-3}), the dominant soil type of the area having moderate FC which indicates the poor water holding capacity and consequently fast infiltration the region. Therefore, soil water drained into the lower surface of soil layers and occurrence of SMD. Hence the region tends to drain out soil water and causes a higher chance of drought condition.

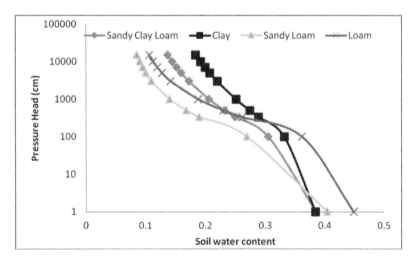

FIGURE 6.3 Water retention curve for different soils in Bundelkhand-up region.

6.3.2 SOIL MOISTURE DEFICIT (SMD)

The SMD is an important indicator for representing the degree of soil saturation; it is inversely related to the soil moisture in the soil layer. Therefore, a low (or negative) SMD indicates water surplus or little capacity for infiltration whereas, a high SMD (or positive) values means below FC and rain can infiltrate to the capacity of the SMD amount. In saturated soil, all of the available soil pores are full of water, but water will drain out of large pores under the force of gravity. The SMD represents the degree of soil saturation and is inversely proportional to the soil moisture content. The estimated SMD of the study area ranges from –0.04 to 0.07 m as shown in Figure 6.4. Negative SMD values observed less than 5% of the total study area indicating water surplus or lower infiltration capacity mostly seen in the parts of Lalitpur and Banda districts. Whereas, positive SMD values observed the rest of the area (>95%) found in all the districts indicating moisture content below FC or higher infiltration rate (Figure 6.4). We observed negative SMD in the southern regions at the high altitudes and positive values in the northern region at low altitudes.

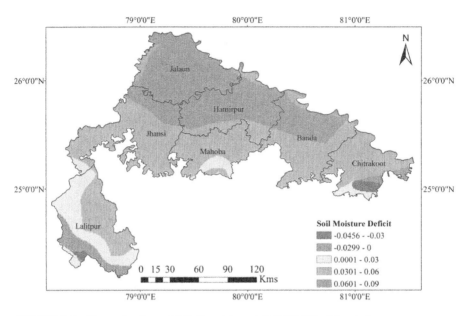

FIGURE 6.4 **(See color insert.)** Soil moisture deficit (SMD) of the study area.

6.4   CONCLUSIONS

Agricultural drought is one of the major natural hazards due to its impact at large spatial scale, gradually emergence over a long period, and long-lasting behavior. Agricultural drought management in the past is based on meteorological observations and water supply, but decision-makers have required drought response operations that should be available and accessible for use in a short period of time and need to be updated on the latest drought situation, especially soil moisture conditions. This study explores the satellite-based CCI soil moisture derived SMD for monitoring agricultural drought spatially. The development of a spatial map of SMD is based on soil moisture content and FC of the soil. The results depicted that more than 90% area is prone to the drought with 0.09 to 0.03 SMD value. These regions namely Jalaun, Hamirpur, and Jhansi are situated in the middle of the study region and require better irrigation scheduling for crop production. Further study will be continued to consider information from other sources to improve the spatial resolution of soil moisture estimation and better algorithms to utilize CCI-SM products as well as derivation of agricultural drought.

KEYWORDS

- **agricultural drought**
- **CCI soil moisture**
- **field capacity**
- **soil moisture deficit**

REFERENCES

Digital soil map of the world and derived soil properties. https://searchworks.stanford.edu/view/4059679 (accessed on 10/12/2018).

Nandintsetseg, B., & Shinoda, M., (2011). Seasonal change of soil moisture in Mongolia: Its climatology and modeling, *International Journal of Climatology, 31*, 1143–1152.

Nicolai-Shaw, N., Zscheischler, J., Hirschi, M., Gudmundsson, L., & Seneviratne, S. I., (2017). A drought event composite analysis using satellite remote-sensing based soil moisture. *Remote Sensing of Environment. 203*, 216–225.

Schaap, M. G., Leij, F. J., & Van Genuchten, M. T., (2001). ROSETTA: A computer program for estimating soil hydraulic parameters with hierarchical pedotransfer functions. *Journal of Hydrology, 251*, 163–176.

Singh, S., & Phadke, V., (2006). Assessing soil loss by water erosion in Jamni River Basin, Bundelkhand region, India adopting Universal Soil Loss Equation using GIS. *Current Science,* 1431–1435.

Srivastava, P. K., Han, D., Ramirez, M. A. R., & Islam, T., (2013). Appraisal of SMOS soil moisture at a catchment scale in a temperate maritime climate. *Journal of Hydrology, 498*, 292–304.

Srivastava, P. K., Han, D., Rico-Ramirez, M. A., O'Neill, P., Islam, T., & Gupta, M., (2014). Assessment of SMOS soil moisture retrieval parameters using tau–omega algorithms for soil moisture deficit estimation. *Journal of Hydrology, 519*, 574–587.

Van Genuchten, M. T., (1980). A closed-form equation for predicting the hydraulic conductivity of unsaturated soils. *Soil Science Society of America Journal, 44*, 892–898.

CHAPTER 7

EVALUATING THE USE OF "GOODNESS-OF-FIT" MEASURES IN A WATER MOVEMENT MODEL

VINOD KUMAR TRIPATHI

Department of Farm Engineering, Institute of Agricultural Sciences, Banaras Hindu University, Varanasi, U.P.–221005, India, Email: tripathiwtcer@gmail.com

ABSTRACT

To reduce hunger and eradicate poverty, then not only is achieving security for water, energy, and food for people critical, but also in doing so a far more integrated and cross-sector planning framework will be needed. The demand of wastewater (WW) for irrigation is gradually increasing due to escalating competition for freshwater by urban, industrial, and agricultural users. To sustain or increase agricultural production, there is a need to adopt highly efficient irrigation technologies such as surface or subsurface drip irrigation (SDI) systems. Studies related to water distribution under any irrigation system and water quality are important for efficient water and nutrients application. In the present study, the water dynamics under surface and subsurface drip irrigation was evaluated by taking cauliflower as a test crop on sandy loam soil. The calibrated model predicted all the parameters close to observed values with RMSE values ranging from 0.05 to 0.92. HYDRUS-2D model has the ability to predict water distribution with reasonably good accuracy in present crop and soil condition.

7.1 INTRODUCTION

Sources of freshwater at places in Africa, Asia, and South America are fast running out owing to accelerated net extraction for human use. Freshwater availability for irrigation in arid and semi-arid regions is a major concern around the World (Sikdar, 2007; Tripathi et al., 2016a). In the Millennium Development Goal formulated at the UN Millennium Development Summit in the year 2000, lack of access to safe water by the world's poor was pledged to be cut into half by 2015. There is plenty of water around us in the oceans, in terrestrial water bodies such as rivers and lakes, as ice and snow in the Polar Regions, on mountain tops, and in subsurface aquifers. But easily accessible fresh water is dwindling because of extensive agriculture, enhanced industrial activities and increasing domestic use. By all these anthropogenic activities polluted water is generated as byproduct called as wastewater (WW). Potentially the most efficient irrigation systems over traditional systems are often recommended to overcome this problem. Sustainability of water resources depends upon the magnitude of the overall productivity gain following the shift from traditional irrigation method to micro-irrigation system, the pattern of use of the saved water, and the type and a potential number of adopters (Namara et al., 2007).

To obtain the best possible delivery of water and solute under drip irrigation (DI) system decision for the optimum distance between emitters and depth of placement of lateral tube play an important role. It depends upon the dimensions of the wetted volume and the distribution of water and solute within the wetted volume. To control the groundwater contamination, subsurface drip irrigation (SDI) is the safest way of WW application (WHO, 2006; Tripathi et al., 2016b). It also leads to the reduction of weeds, evaporation from the soil surface and consequently to an increase in the availability of water for transpiration and overall water use efficiency in comparison to surface DI system (Romero et al., 2004). Uniform distribution of WW and nutrients in the crop root zone increases the efficacy of fertilizers and to maintain a dry soil surface to reduce water losses due to evaporation in case of SDI. Irrigation with WW through a subsurface drip system alleviates health hazards, odor, and runoff into surface water bodies due to no aerosol formation and produce does not come into direct contact with poor quality water. Longevity of emitters with a lateral tube also increases by subsurface placement. SDI has a special advantage of securing system safety against pilferage and damage by animals and during intercultural operations (Tripathi et al., 2014). Several empirical, analytical, and numerical models have been developed to simulate soil water content and wetting front dimensions for surface and SDI systems (Angelakis et al., 1993; Cook et al., 2003). Due to advances in computer speed, and the public availability of numerical models simulating water flow in soils, many researchers have become interested in using such models for evaluating water flow in soils with DI systems (Lazarovitch et al., 2007; Provenzano, 2007).

HYDRUS–2D (Simunek et al., 1999) is a well-known Windows-based computer software package used for simulating water, heat, and/or solute movement in two-dimensional, variably saturated porous media. This model's ability to simulate water movement for DI conditions has been assessed by many researchers (Simunek et al., 2008). Cote et al. (2003) used the HYDRUS–2D model to simulate soil water transport under SDI. They discussed that soil water and soil profile characteristics were often not adequately incorporated in the design and management of drip systems. Results obtained from simulation studies indicated that in highly permeable coarse-textured soils, water moved quickly downwards from the dripper.

Skaggs et al. (2004) compared HYDRUS–2D simulations of flow from an SDI line source with observed field data involving a sandy loam soil and an SDI system with a 6 cm installation depth of drip lateral and 3 discharge rates. They found very good agreement between simulated and observed soil moisture data. Ben-Gal et al. (2004) explained that one of the main problems with SDI systems is soil saturation near the emitter and its effects on emitter discharge resulting from the net pressure on the emitter outlet. To solve this problem, they installed the drip tube in a trench, and filled it with gravel to eliminate saturation, and net pressure around the emitter then simulated their conditions using HYDRUS–2D, and found good agreement between observed and simulated data. Lazarovitch et al. (2007) modified HYDRUS–2D further so that it could account for the effects of backpressure on the discharge reduction using the characteristic dripper function. Provenzano (2007) assessed the accuracy of HYDRUS–2D by comparing simulation results and experimental observations of matric potential for SDI systems in a sandy loam soil with a 10 cm installation depth in thoroughly mixed or repacked soils, and also found satisfactory agreement. However, our study was performed under field condition on undisturbed soil profiles.

Studies on the effect of depths of placement of drip laterals with WW do not appear to have caught researchers' attention so far. No measured data of soil water distribution in the root zone of drip irrigated with WW for cauliflower crop are available. Rahil and Antonopoulos (2007) using WANISIM, a 1-D model investigated the effects of irrigation on soil water and nitrogen dynamics with reclaimed WW using DI and application of nitrogen fertilizer for plant growth. The model simulated the temporal variation of soil water content with reasonable accuracy. However, an overestimation of the measured data was observed during the simulation period. Therefore, the present study was undertaken to understand the dynamics of municipal WW for simulating the water transport processes in the soil under the surface and SDI system. Such an understanding can help in identifying the best irrigation strategy for efficient use of WW. The simulation model Hydrus–2D (Simunek et al., 1999) was selected for the current study for simulation and modeling of soil water content under DI system for cauliflower crop. The simulated results were compared with field data involving placement of emitter lateral at surface and subsurface (15 cm depth).

7.2   MATERIALS AND METHODS

7.2.1   LOCATION AND SOIL OF EXPERIMENTAL SITE

The present study was conducted at Research Farm of Indian Agricultural Research Institute, New Delhi, India The soil of the experimental area was deep, well-drained sandy loam soil comprising 61% sand, 18% silt, and 21% clay. The bulk density of soil was 1.56 gcm^{-3}, field capacity was 0.16 percent, and saturated hydraulic conductivity was 1.13 cmh^{-1}.

7.2.2   CROP PRACTICE AND DESCRIPTION OF IRRIGATION SYSTEM

The cauliflower (cv: *Indame 9803*) seeds were sown in the seed tray (plug tray) under the poly house in the third week of September 2008 and 2009. Twenty-five days old cauliflower seedlings were transplanted at a plant to plant and row to row spacing of 40 cm x 100 cm, respectively. Daily irrigation was applied following the methodology formulated by Allen et al. (1998).

A DI system was designed for cauliflower crop in sandy loam soil using the standard design procedures. The control head of the system consisted of sand media filter, disk filter, flow control valve, pressure gauges, etc. Drip emitters with rated discharge 1.0×10^{-6} m^3s^{-1} at a pressure of 100 k Pa were placed on the lateral line at a spacing of 40 cm. The best treatment, i.e., WW filtered by the combination of gravel media and disk filter with the placement of lateral at surface and subsurface (15cm) was considered for the simulation study. The crop water demand for irrigation was estimated on the basis of Penman-Monteith's semi-empirical formula. The actual evapotranspiration was estimated by multiplying reference evapotranspiration with crop coefficient ($ET = ET_0 \times K_C$) for different crop growth stages. The crop coefficient during the crop season 2008–2009, and 2009–2010 was adopted as 0.70,

0.70, 1.05 and 0.95 at initial, developmental, middle, and maturity stages, respectively (Allen et al., 1998).

7.2.3 SOIL WATER CONTENT

The soil water contents were collected from the crop root zone along and across the DI lateral tube. It was collected from the surface (top visible layer within 2 cm), 2–15, 15–30, 30–45 cm layers of soil for the placement of lateral at surface and subsurface at 15 cm depth. Frequency Domain Reflectometry (FDR) was used for the determination of soil water content.

7.2.4 DESCRIPTION OF MODEL

HYDRUS–2D (Simunek et al., 1999) is a finite element model, which solves Richard's equation for variably saturated water flow and convection-dispersion type equations for heat transport. The flow equation includes a sink term to account for water uptake by plant roots. The model uses the convective-dispersive equation in the liquid phase and the diffusion equation in the gaseous phase to solve the solute transport problems. It can also handle nonlinear non-equilibrium reactions between the solid and liquid phases, linear equilibrium reactions between the liquid and gaseous phases, zero-order production, and two first-order degradation reactions: one which is independent of other solutes, and one which provides the coupling between solutes involved in sequential first-order decay reactions. The model can deal with prescribed head and flux boundaries, controlled by atmospheric conditions, as well as free drainage boundary conditions. The governing flow and transport equations are solved numerically using Galerkin-type linear finite element schemes.

7.2.5 ROOT WATER UPTAKE

The root uptake model (Feddes et al., 1978) assigns plant water uptake at each point in the root zone according to soil moisture potential. The total volume of the root distribution is responsible for 100% of the soil water extraction by the plant, as regulated by its transpiration demand. The maximum root water uptake distribution reflects the distribution in the root zone having roots that are actively involved in water uptake. The root zone having maximum root density was assigned the value of 1. Root distribution was assumed to be constant throughout the growing season. Maximum depth for simulation was taken as 60 cm.

7.2.6 INPUT PARAMETERS

There are two commonly used models describing soil moisture behavior, the Brooks-Corey model and the van Genuchten model. The van Genuchten model is most appropriate for soils near saturation (Smith et al., 2002). Soils within the root zone under DI system remains at near saturation throughout the crop season. Therefore, van Genuchten analytical model without hysteresis was used to represent the soil hydraulic properties. Sand, silt, and clay content of soil were taken as input and by Artificial Neural Network (ANN) prediction; the soil hydraulic parameters were obtained and are given in Table 7.1. A simulation was carried out applying irrigation from a line source as in the real case for each individual dripper.

TABLE 7.1   Estimated Soil Hydraulic Parameters

Soil layer	Soil depth (cm)	Qr (θ_r)	Qs(θ_s)	Alpha(α) (cm^{-1})	η	Ks (cmh^{-1})
1	0–15	0.0403	0.3740	0.0079	1.4203	1.09
2	15–30	0.0396	0.3748	0.0059	1.4737	0.7
3	30–45	0.0338	0.3607	0.0048	1.5253	1.39
4	45–60	0.0261	0.3682	0.0142	1.3875	1.22

Where θ_r and θ_s are the residual and saturated water contents, respectively; α is a constant related to the soil sorptive properties; η is a dimensionless parameter related to the shape of water retention curve, and K_s represent the saturated hydraulic conductivity.

7.2.7 INITIAL AND BOUNDARY CONDITIONS

Observed soil water in the soil profile was taken as initial water content. For all simulated scenarios, the bottom boundary was defined by a unit vertical hydraulic gradient, simulating free drainage from a relatively deep soil profile (Rassam and Littleboy, 2003). The no-flux boundary was used on the vertical side boundaries of the soil profile because the soil water movement will be symmetrical along these boundaries. The system was divided into four layers depending on the variability of the soil physical properties. To account the dripper discharge during irrigation, a flux type boundary condition with a constant volumetric application rate of dripper for irrigation duration was considered. During no irrigation period, flux was kept as zero. Time variable boundary condition was used in HYDRUS–2D simulations to manage the flux boundary depending on irrigation water requirement during irrigation and no irrigation period. In surface placement of drip lateral, the top boundary was considered as at atmospheric condition but a small part of the top boundary, around the dripper from where the water is applied to crop, was taken as time variable boundary condition. Under subsurface placement of drip lateral at 15 cm depth, the topsoil surface was considered at an atmospheric boundary condition. The atmospheric boundary is usually placed along the top of the soil surface to allow for interactions between the soil and the atmosphere. These interactions include rainfall, evaporation, and transpiration (root uptake) given in the time variable boundary conditions. The flux radius and subsequently fluxes per unit area, resulting from one meter of drip lateral was determined. No-flux boundary is impermeable and does not allow water into or out of the soil profile through it.

7.2.8   MODEL VALIDATION BY COMPARISON BETWEEN THE SIMULATED AND OBSERVED VALUES

To quantitatively compare the results of the simulations, observed and simulated values for water content was compared. The coefficient of efficiency (C_{eff}) and the root mean square error (RMSE) were the two statistical indices used to evaluate the predictions of the model quantitatively. The RMSE has also been widely used to evaluate the models (Skaggs et al., 2004).

7.3   RESULTS AND DISCUSSION

Distribution of water in the root zone soil profile with filtered WW is influenced by soil type, dripper discharge, depth of placement of drip lateral, quality of irrigation water and extraction of water by crop.

7.3.1   CALIBRATION OF MODEL

The HYDRUS–2D model was calibrated mainly for hydraulic conductivity values of the sandy loam soil. The model worked well with the measured hydraulic conductivity values. The model gives the spatial and temporal distribution of water content in simulated layers at pre-decided time steps. Field observations for water content in the soil were taken at 4 and 24 h after irrigation. Simulated and observed values of water at 4 and 24 h after irrigation were used to evaluate the performance of the model. RMSE values varied from 0.013 to 0.015. This indicates that Hydrus–2D can be used to simulate the water distribution with very good accuracy.

7.3.2   SOIL WATER DISTRIBUTION

Soil water content was determined using FDR by placing three access tubes at a distance of 0,15, and 30 cm away from lateral pipe up to a depth of 1.0 m. Observed soil water distribution at the initial, development, middle, and maturity stages are presented in Figure 7.1. During the initial growth stage (after 25 days of transplanting), when root length density (RLD) and leaf area index (LAI) was less than 1.0, 23% water content was observed within 10 cm of the radius. The downward movement of water was more than its lateral movement at all growth stages of crop due to gravity force playing a predominant role in comparison to the capillary force in an experimental plot. The higher values of water content near the drip lateral confirming the result obtained by Souza et al., (2003). Soil water content just below the dripper, i.e., 0.0 cm away from the lateral pipe was more throughout the crop season, almost at the level of field capacity, in all depths of placement of laterals. Soil water content at the surface at initial, developmental, middle, and maturity stages of the cauliflower was found to be 23.5, 24.1, 25.0, and 26.1%, respectively.

The soil surface appeared moist under subsurface placement of drip lateral at 15 cm depth in all growth stages of cauliflower. Soil water content above the dripper (at the surface) at initial, developmental, middle, and maturity stage of the cauliflower were found to be 17.1, 16.8, 16.3, and 15.6%, respectively under subsurface drip by placement of drip lateral at 15 cm depth (Figure 7.1). A significant difference (P<0.01) was observed in soil water contents of surface soil between placement of lateral at the surface and 15 cm depth.

Wetted soil bulb of 30 cm in width and 50 cm depth had more than 17% soil water content, which was very conducive for good growth of crop during development stage resulting in higher cauliflower yields at subsurface placed drip lateral (placement of drip lateral at 15 cm depth). At the initial and developmental stage of the crop, the active root was confined up to 15 cm soil depth. However, the placement of drip lateral at 15 cm soil depth, adequate soil water was found at 30, 45 and 60 cm soil depths (Figure 7.1).

Water that moved beyond the 40 cm soil depth was not available for plants at any stage.

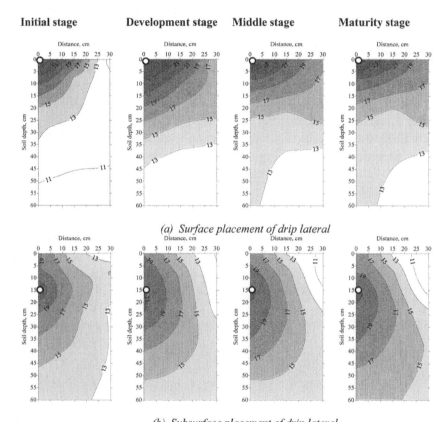

(b) Subsurface placement of drip lateral

FIGURE 7.1   (See color insert.) Observed soil water (volumetric in percent) distribution in the field.

Higher yield was achieved by maintaining relatively high water content in root zone conducive to good plant growth by placement of lateral at 15 cm depth under successive irrigation event. The high water content of the soil around the drippers facilitates better water transmission to the surrounding soil and keeps on replenishing the crop root zone (Segal et al., 2000). Therefore, keeping the drip lateral within the crop root zone and sufficiently below the soil surface replenishes the root zone effectively due to gravity flow in light soils and simultaneously reduces evaporation losses due to restricted upward capillary flow.

7.3.3   SIMULATION OF SOIL WATER DISTRIBUTION

The soil water content distribution from the model-simulated values are presented in Figure 7.2, and after comparison from observed values, statistical parameters are presented in Table 7.2. It shows good agreement between predicted and measured soil water content. The simulated values of water content at soil surface under surface placement of drip lateral were 24.2, 25.1, 25.8, and 25.9% at initial, developmental, middle, and maturity stage of the crop. Simulated soil water content above the dripper on soil surface at initial, developmental, middle, and maturity stage of the cauliflower were found 20.3, 18.6, 18.5, and 18.2%, respectively, under subsurface placement of drip lateral at 15 cm depth (Figure 7.2). The lower coefficient of efficiency and RMSE values were observed by subsurface placement of drip lateral at 15 cm depth.

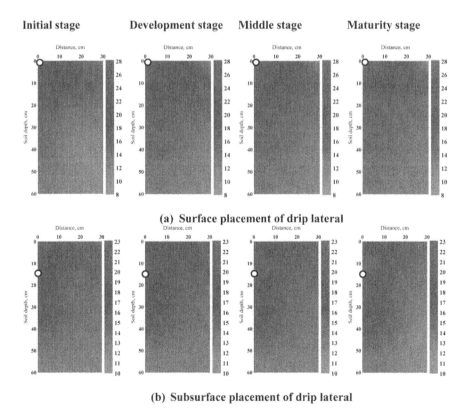

FIGURE 7.2   (See color insert.) Simulated soil water (volumetric in percent) distribution with the model.

TABLE 7.2 Statistical Parameters Indicative of Performance of Model for Water Content

Depth of placement of lateral	Statistical parameters	Crop growth stages			
		Initial	Developmental	Middle	Maturity
Surface	RMSE	0.05	0.87	0.68	0.92
	C_{eff}	−0.41	−0.01	−1.29	−2.25
Subsurface	RMSE	0.06	0.23	0.52	0.39
	C_{eff}	−2.36	−2.59	−1.98	−3.51

The input parameters for the simulation of HYDRUS–2D model were determined by detailed field experimentation. However, a few were taken from published literature matching to our soil and similar crop condition. It was found that the wetting patterns obtained during application of water generally consist of two zones, a saturated zone close to the dripper (5 cm around the dripper). The wetting pattern of elliptical shape was found under subsurface placement of drip lateral at 15 cm depth. Wetted depth was found larger than the surface-wetting radius resulting in more water below dripper under subsurface placement of drip lateral at 15 cm depth because of dominant nature of gravity force in comparison to capillary forces. The saturated radius was taken constantly throughout the crop season, from where flux entered.

The current version of Hydrus–2D has the limitation and does not calculate the time variable saturated radius. The difference observed between experimental and simulated soil water distribution may be attributed to the differences in saturated hydraulic conductivity of soil (observed and simulated by the model as an intermediate step). The root water uptake model was taken from the literature. Many researchers have tried this software and reported its usefulness for simulation and modeling of water distribution (Mailhol et al., 2001; Cote et al., 2003; Gardenas et al., 2005).

7.4 CONCLUSIONS

In this chapter, the software Hydrus–2D predicted the soil water content with high accuracy under DI. Adequate distribution of water content in the root zone of cauliflower is possible with the placement of drip lateral at 15 cm depth from the soil surface. The requirement of a large number of accurate parameters matching with the field condition is important for the simulation of soil water content in the root zone of the crop. Subsurface application of water increases water availability in the root zone of cauliflower crop enhanced the crop yield. The results of this study will be supportive for prediction of water availability in the root zone as well as a selection of irrigation method for different crops.

KEYWORDS

- **HYDRUS–2D**
- **root zone**
- **simulation**
- **wastewater**
- **water content**

REFERENCES

Allen, R. G., Pereira, L. S., Raes, D., & Smith, M., (1998). *Crop Evapotranspiration-Guidelines for Computing Crop Water Requirements* (p. 300). FAO Irrigation and Drainage Paper No., 56, FAO, Rome, Italy.

Angelakis, A. N., Rolston, D. E., Kadir, T. N., & Scott, V. N., (1993). Soil-water distribution under trickle source. *J. Irrig. Drain. Eng. ASCE, 119*, 484–500.

Ben-Gal Alon Lazorovitch, N., & Shani, U., (2004). Subsurface drip irrigation in gravel-filled cavities. *Vadose Zone Journal, 3*, 1407–1413.

Cook, F. J., Thorburn, P. J., Fitch, P., & Bristow, K. L., (2003). Wet Up: A software tool to display approximate wetting pattern from drippers. *Irrigation Science, 22*, 129–134.

Cote, C. M., Bristow, K. L., Charlesworth, P. B., Cook, F. J., & Thorburn, P. J., (2003). Analysis of soil wetting and solute transport in subsurface trickle irrigation. *Irrigation Science, 22*, 143–156.

Dukes, M. D., & Scholberg, J. M., (2005). Soil moisture controlled subsurface drip irrigation on sandy soils. *Applied Engineering Agriculture, 21*(1), 89–101.

Feddes, R. A., Kowalik, P. J., & Zaradny, H., (1978). *Simulation of Field Water Use and Crop Yield* (pp. 189). John Wiley and Sons, New York.

Lazarovitch, N., Warrick, A. W., Furman, A., & Simunek, J., (2007). Subsurface water distribution from drip irrigation described by moment analyses. *Vadose Zone J., 6*(1), 116–123.

Mailhol, J. C., Ruelle, P., & Nemeth, I., (2001). Impact of fertilization practices on nitrogen leaching under irrigation. *Irrig. Sci., 20*, 139–147.

Namara, R. E., Nagar, R. K., & Upadhyay, B., (2007). Economics, adoption determinants, and impact of micro-irrigation technologies: Empirical results from India. *Irrigation Science, 25*, 283–297.

Provenzano, G., (2007). Using HYDRUS–2D simulation model to evaluate wetted soil volume in subsurface drip irrigation systems. *J. Irrig. Drain. Engg., 133*(4), 342–349.

Rahil, M. H., & Antonopoulos, V. Z., (2007). Simulating soil water flow and nitrogen dynamics in a sunflower field irrigated with reclaimed wastewater. *Agricultural Water Management, 92*, 142–150.

Rassam, L., & Littleboy, M., (2003). Identifying vertical and lateral components of drainage flux in hill slopes. In: *Proceeding of MODSIM 2003*. Townsville, Queensland, Modeling and Simulation Society of Australia and New Zealand.

Romero, P., Botia, P., & Garcia, F., (2004). Effects regulated deficit irrigation under subsurface drip irrigation conditions on water relations of mature almond trees. *Plant Soil, 260*, 155–168.

Segal, E., Ben-Gal, A., & Shani, U., (2000). Water availability and yield response to high-frequency micro-irrigation in sunflowers. *Proceedings of the 6th Int. Micro-Irrigation Congress, Int. Council Irrigation Drainage*. Cape Town, South Africa.

Sikdar, S. K., (2007). Water; water everywhere, not a drop to drink? *Clean Technology Environmental Policy, 9*, 1–2.

Simunek, J., Sejna, M., & Van Genuchten, M. T. H., (1999). The HYDRUS–2D software package for simulating the two-dimensional movement of water, heat and multiple solutes in variably saturated media, version 2.0. *Rep. IGCWMC-TPS-53, Int. Ground Water Model. Cent., Colo. Sch. of Mines, Golden, CO*, 251.

Simunek, J., Van Genuchten, M. T. H., & Sejna, M., (2008). Development and applications of the HYDRUS and STANMOD software packages, and related codes. *Vadose Zone J., 7*(2), 587–600.

Skaggs, T. H., Trout, T. J., Simunek, J., & Shouse, P. J., (2004). Comparison of HYDRUS–2D simulations of drip irrigation with experimental observations. *Journal of Irrigation Drainage Engineering, 130*(4), 304–310.

Smith, R., Smettem, K. R. J., Broadbridge, P., & Woolhiser, D.A. (2002). *Infiltration Theory for Hydrologic Applications. American Geophysical Union Water Resources Monograph Series*, Vol. 15, 210 p.

Smith, R., Smettem, K. R. J., & Broadbridge, P., (2002). *Theory for Hydrological Applications*. American Geographical Union, USA.

Souza, C. F., & Matsura, E. E., (2003). Multi-wire time domain reflectometry (TDR) probe with electrical impedance discontinuities for measuring water content distribution. *Agricultural Water Management, 59*, 205–216.

Tripathi, V. K., Rajput, T. B. S., & Patel, N., (2014). Performance of different filter combinations with surface and subsurface drip irrigation systems for utilizing municipal wastewater. *Irrigation Science, 32*(5), 379–391.

Tripathi, V. K., Rajput, T. B. S., & Patel, N., (2016b). Biometric properties and selected chemical concentration of cauliflower influenced by wastewater applied through surface and subsurface drip irrigation system. *Journal of Cleaner Production, 139*, 396–406.

Tripathi, V. K., Rajput, T. B. S., Patel, N., & Kumar, P., (2016a). Effects on growth and yield of eggplant (*Solanumm elongema L.*) under placement of drip laterals and using municipal wastewater. *Irrigation and Drainage, 65*, 480–490.

WHO, (2006). *Guidelines for the Safe Use of Wastewater, Excreta, and Greywater* (Vol. II & IV), WHO Press, Geneva, Switzerland.

MOLLUSCS AS A TOOL FOR RIVER HEALTH ASSESSMENT: A CASE STUDY OF RIVER GANGA AT VARANASI

IPSITA NANDI and KAVITA SHAH

Institute of Environment and Sustainable Development, Banaras Hindu University, Varanasi, 221005, India, Email: ipsitabhu@gmail.com

ABSTRACT

Rivers are the main source of fresh water. Recent scenarios of anthropogenic activities are degrading the health of the river. Any river conservation or restoration programs require a proper insight into the extent of degradation towards the health of the river. Monitoring only the physicochemical properties does not give a proper picture of the health of the river. It requires a holistic approach wherein the health is assessed on certain well-defined parameters. This further requires some simple and robust tools for easy monitoring of river health. Bio-indicator serves as one such tool since their diversity changes with changing characteristics of river water quality. River Ganga at Varanasi is also home to a large variety of aquatic fauna. This paper utilizes this species richness for designing a tool taking mollusks as bio-indicator for health assessment of River Ganga at Varanasi. It also tries to define the reliability and accuracy for using mollusks as a robust and easy tool for river health assessment.

8.1 INTRODUCTION

Rivers in the current scenario are under permanent pressure of various forms of pollution causing anthropogenic activities. River ecosystems act as a sink for various kinds of toxicants emerging from different sources of pollution (Salánki et al., 2003). These toxic pollutants are severely degrading the health of the rivers (Cairns et al., 1993). Riverine biodiversity act as an indicator of the health of the river system (Boulton, 1999). Altered flow regimes and pollution are adversely affecting the biotic makeup of the river (Bunn and Arthington, 2002; Strayer and Dudgeon, 2010). Evaluating biotic components using bioindicator can reveal the real scenario of river health (Nandi et al., 2016). A bioindicator is an organism (or a part or community of an organism) that provides information about the quality of environment making changes in the morphological, histological or cellular structures, their metabolic processes, their population structure or in their behavioral pattern (Markert et al., 2003). Bioindicator response to the aquatic pollutants in two ways, i.e., either by accumulating the pollutants or by showing visible specific changes in the characteristics or population structure (Markert et al., 2003). Bioindicator are selected for pollution monitoring based on certain specific characteristics (Füreder and Reynolds, 2003) which includes: (1) clear taxonomy; (2) wide distribution; (3) occurrence in large numbers with appropriate visible body size; (4) restricted mobility; (5) site specificity; (6) narrow specific ecological demands and tolerance; (7) clear feeding structure; (8) long lifespan and generation time; (9) sensitivity to specific pollutants; (10) low genetic and ecological variability; (11) response to specific pollutant/substance should be representative to other taxas or even ecosystem; and (12) easy to sample, store, recognition, and robust during handling.

Degrading water quality in this era demands shift in the management strategies for pollution abatement towards a proper understanding of the extent of pollution. Simply measuring the physicochemical characteristics of water does not reveal the real scenario of river health (Nandi et al., 2016). Furthermore, these techniques are tedious and time-consuming as well as need expertise. The recent time now demands a shift in river health assessment program from traditional water quality monitoring towards assessment health using bioindicators. Mollusk belongs to the highly diverse group of macroinvertebrate (Patil, 2013). With a species diversity of 2,00,000, the mollusk stands second in species richness after arthropods (Patil, 2013; Strong et al., 2008). They also have diverse forms of ecological habitats with a long and sedentary form of living (Grabarkie-wicz and Davis, 2008; Patil, 2013). Slight changes in the physicochemical characteristics of the river water adversely affect species characteristics, richness, and species diversity of the mollusk (Pérez-Quintero, 2011; 2012). These features make mollusk a potent indicator for assessing water quality thereby health status of rivers.

In a preview of the above, the study is focused upon identifying certain species of mollusk from aquatic biodiversity which can be used as a tool for river health assessment studies.

8.2   MATERIALS AND METHOD

8.2.1   STUDY AREA

Varanasi located at 25.28° N latitude and 82.96° E longitude is the oldest living city in the world (Singh, 2011). Being situated at the bank of holy River Ganga, it is also considered as the spiritual capital of India (Singh, 2011). The city is situated in the eastern part of Uttar Pradesh along the left crescent-shaped bank of River Ganga (Mohanty, 1993). River Ganga of Varanasi serves as a home to a wide variety of aquatic biodiversity (Sarkar et al., 2012). The right bank usually consists of large stretches of sand bed (Nandi et al., 2017). These stretches of sand bed form the home to various types of macroinvertebrates (Sinha et al., 2007). Massive changes have been observed in the water quality of Varanasi which continues to degrade each day. Literature reveals the deterioration of water of Varanasi and thus is unfit for human consumption (Mishra et al., 2009; Namrata, 2010). These changes are also affecting the general characteristics of benthic macroinvertebrates (Sinha et al., 2007). This forms the aquatic biodiversity which can serve the role of bioindicator for easy monitoring of the health of River Ganga at Varanasi. The right bank from beyond Asi River to Teliyanala ghat (Figure 8.1) was selected as a study site for studying the mollusk species specific to this stretch which can be used as a bioindicator for river health.

8.2.2   COLLECTION OF SAMPLE

The mollusk was collected along the respective Ghats using handpick method (Fuller, 1978). Proper photographs were taken, and the collected mollusk was kept in a glass bottle in 10% formalin. Photographs of the natural site containing the mollusk are shown in Figure 8.2. Water samples were collected in polyethylene bottles and borosil bottles and taken to a lab for further analysis using the APHA method.

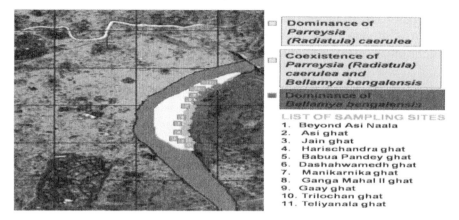

FIGURE 8.1   Study site.

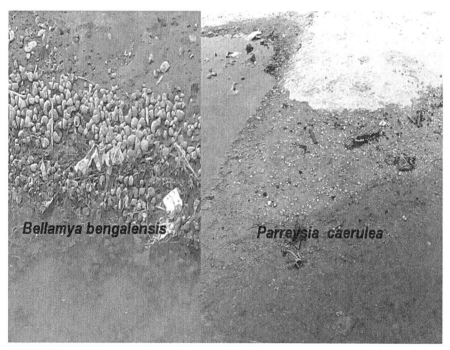

FIGURE 8.2   Molluscs species at their natural habitat

8.2.3 METHODOLOGY

The collected mollusk species were identified from Zoological Survey of India, Pune. The nomenclatures of these species are detailed in Table 8.1. Numbers of Mollusk obtained are listed in Table 8.2. The water samples were analyzed for various physicochemical parameters and heavy metals and the value of analysis thus obtained are listed in Table 8.3. A correlation matrix was also computed between water quality parameters and a number of individuals of mollusk obtained at each site using IBM SPSS statistical 20 software Table 8.4.

TABLE 8.1 Nomenclature of the Species Observed Along the Right Bank of River Ganga at Varanasi

Species Name	1. *Parreysia (Radiatula) caerulea*	Species 2. *Bellamya bengalensis*
Picture		
Phylum	Mollusca	Mollusca
Class	Bivalvia	Gastropods
Order	Trigoinoida	Mesogastropoda
Family	Unionidae	Viviparidae
Genus	*Parreysia (Radiatula)*	*Bellamya*
Species	*caerulea*	*bengalensis*

TABLE 8.2 Number of Mollusk Obtained in Each Representative Ghat

Sites	Gastropods (Bellamya)	Bivalvia (Parreysia)
R1	0	10
R2	0	10
R3	2	12
R4	2	10
R5	3	12
R6	11	2
R7	8	9
R8	10	6
R9	7	8
R10	11	0
R11	11	0

TABLE 8.3 Water Quality Analysis at the Representative Sites at the Right Bank of River Ganga

Sites	Calcium (mgl⁻¹)	COD (mgl⁻¹)	Nitrate (mgl⁻¹)	Dissolved Oxygen (mgl⁻¹)	Chloride (mgl⁻¹)	As (mgl⁻¹)	Fe (mgl⁻¹)	Ni (mgl⁻¹)
R1	30.602	30	3.057	9	34.253	0.2975	1.494	0.0291
R2	29.98	35	3.419	8	35.228	0.1102	1.143	0.0034
R3	30.247	37	3.753	8	35.264	0.0638	1.38	0.0018
R4	30.12	25	2.314	7.4	35.804	0.0469	0.561	0.0017
R5	30.31	37	3.311	7.4	35.708	0.0368	0.843	0.0077
R6	31.123	101	1.949	6.4	36.687	0.0341	0.905	0.0012
R7	27.314	85	1.867	6.8	37.425	0.0334	1.901	0.0014
R8	26.709	105	1.82	6.4	37.655	0.0295	0.575	0.0009
R9	26.79	117	1.365	6	37.783	0.0262	2.085	0.0018
R10	27.674	110	1.361	6	37.977	0.0331	0.488	0.0007
R11	23.769	135	1.307	6	38.992	0.0345	0.734	0.0008

8.3 RESULTS AND DISCUSSION

A change in dominance trend was observed along the entire stretch of the river bank. The dominance of *Parreysia (Radiatula) caerulea* (bivalvia) was observed in the stretch of R1 (beyond Asi River)-R5 (Babua Pandey ghat) after which there appeared a change in trend with rising in *Bellamya bengalensis* (gastropod) population. Codominance of the two species was observed in the stretch of R6 (Dashashwamedh ghat)-R9 (Gaay ghat). R10 (Trilochan ghat)-R11 (Teliyanala ghat) observed the dominance of *Bellamya bengalensis*. This change in dominance trend of the two species can be attributed to the topology of the river bank. The sites which are located on the extreme sides of the meander observe the dominance of a particular species while at the meander bend there is codominance of both the species (Figure 8.1). The correlation matrix reveals that both the species showed a significant correlation with dissolved oxygen, calcium, nitrate, and COD. *Parreysia* (bivalvia) showed positive correlation with DO, nitrate while negative correlation with COD and chloride. On the other hand, *Bellamya* (gastropod) showed negative correlation with dissolved oxygen, nitrate, and calcium but positive correlation with COD and chloride. Apart from these Bellamya showed negative correlation with arsenic (Table 8.4). The observation from correlation matrix highlights that *Parreysia* species are more adapted to cleaner environment

TABLE 8.4 Correlation Matrix of Mollusk Species and Water Quality

		Calcium	COD (mgl⁻¹)	Nitrate	Dissolved Oxygen	Chloride	As	Fe	Ni	Bellamya
Calcium	Pearson Correlation	1	-0.786**	0.725*	0.683*	-0.871**	0.396	0.006	0.380	-0.630*
	Sig0. (2-tailed)		0.004	0.012	0.021	0.000	0.228	0.987	0.248	0.038
	N	11	11	11	11	11	11	11	11	11
COD (mg/l)	Pearson Correlation	-0.786**	1	-0.882**	-0.901**	0.935**	-0.512	-0.036	-0.470	0.928**
	Sig0. (2-tailed)	0.004		0.000	0.000	0.000	0.108	0.917	0.145	0.000
	N	11	11	11	11	11	11	11	11	11
Nitrate	Pearson Correlation	0.725*	-0.882**	1	0.872**	-0.891**	0.460	0.095	0.403	-0.850**
	Sig0. (2-tailed)	0.012	0.000		0.000	0.000	0.154	0.780	0.219	0.001
	N	11	11	11	11	11	11	11	11	11
Dissolved Oxygen	Pearson Correlation	0.683*	-0.901**	0.872**	1	-0.946**	0.795**	0.192	0.720*	-0.906**
	Sig0. (2-tailed)	0.021	0.000	0.000		0.000	0.003	0.573	0.012	0.000
	N	11	11	11	11	11	11	11	11	11
Chloride	Pearson Correlation	-0.871**	0.935**	-0.891**	-0.946**	1	-0.681*	-0.144	-0.630*	0.895**
	Sig0. (2-tailed)	0.000	0.000	0.000	0.000		0.021	0.672	0.038	0.000
	N	11	11	11	11	11	11	11	11	11
As	Pearson Correlation	0.396	-0.512	0.460	0.795**	-0.681*	1	0.232	0.942**	-0.606*
	Sig0. (2-tailed)	0.228	0.108	0.154	0.003	0.021		0.492	0.000	0.048
	N	11	11	11	11	11	11	11	11	11
Fe	Pearson Correlation	0.006	-0.036	0.095	0.192	-0.144	0.232	1	0.231	-0.246
	Sig0. (2-tailed)	0.987	0.917	0.780	0.573	0.672	0.492		0.494	0.465
	N	11	11	11	11	11	11	11	11	11

TABLE 8.4 *(Continued)*

		Calcium	COD (mgl⁻¹)	Nitrate	Dissolved Oxygen	Chloride	As	Fe	Ni	Bellamya
Ni	Pearson Correlation	0.380	-0.470	0.403	0.720*	-0.630*	0.942**	0.231	1	-0.538
	Sig0. (2–tailed)	0.248	0.145	0.219	0.012	0.038	0.000	0.494		0.088
	N	11	11	11	11	11	11	11	11	11
Bellamya	Pearson Correlation	-0.630*	0.928**	-0.850**	-0.906**	0.895**	-0.606*	-0.246	-0.538	1
	Sig0. (2–tailed)	0.038	0.000	0.001	0.000	0.000	0.048	0.465	0.088	
	N	11	11	11	11	11	11	11	11	11
Parreysia	Pearson Correlation	0.543	-0.821**	0.775**	0.733*	-0.750**	0.318	0.431	0.338	-0.853**
	Sig0. (2–tailed)	0.084	0.002	0.005	0.010	0.008	0.341	0.186	0.310	0.001
	N	11	11	11	11	11	11	11	11	11

*Correlation is significant at the 0.05 level (2-tailed).

**Correlation is significant at the 0.01 level (2-tailed).

with higher DO content and less COD (Subba Rao, 1989; Patil, 2011). High chloride content indicate water polluted with human urine and feces (Piocos and de La Cruz, 2000). A positive correlation of *Bellamya* with chloride shows that they are more adapted in water contaminated with human urine and feces. Nitrate is known to cause oxidative stress in species of *Bellamya* hence there exist a negative correlation between *Bellamya* and nitrate concentration (Chinchore and Mahajan, 2013). Calcium plays a significant role in the growth and shell formation of *Bellamya* species (Baby et al., 2010) and hence it could be the reason for a strong negative correlation between population of *Bellamya* species and calcium content. Studies have reported sensitivity of *Bellamya bengalensis* towards arsenic (Ray, 2016; Ray et al., 2013); hence, this could probably be the reason for a negative correlation between arsenic concentration and population density of *Bellamya bengalensis*. These results highlights the role of *Bellamya* and *Parreysia* as bioindicator for water quality and pollution wherein *Parreysia* preferred more or less polluted habitat while *Bellamya* sustained better in polluted habitat. These species can be used as tool for assessing health of River Ganga at Varanasi.

8.4 CONCLUSIONS

Traditional monitoring of physicochemical properties of river water does not provide a real idea about the status of the health of the river. The time now demands a change towards a holistic approach involving biomonitoring of riverine diversity. The macroinvertebrate population provides the real scenario of the river health. Mollusc has highly rich diversity, and hence they provide the scope for application as bioindicator. The study focused on identifying mollusk diversity along the river stretch. Two mollusk varieties, i.e., *Parreysia (Radiatula) caerulea* (bivalvia) and *Bellamya bengalensis* (gastropod) were observed along the stretch with a change in dominance along the stretch. River water quality assessment conducted in parallel and the correlation matrix thus obtained reveals that *Parreysia* shows a positive correlation with DO while the negative correlation with COD and chloride whereas *Bellamya bengalensis* shows a negative correlation with DO and positive correlation with COD suggesting *Parreysia* as a better indicator of good water quality. Furthermore, *Bellamya* shows a positive correlation with chloride which substantiates the fact that it can be used as an indicator of human feces polluted water. *Bellamya* also showed significant correlation with heavy metals which further highlight its role as a bioindicator. These aspects of the study highlight the role of *Bellamya* and *Parreysia* as an indicator of water quality. These mollusk species can be used as bioindicator as simple tools for assessing river health. Further studies can be conducted in these lines to further strengthen the role of mollusk species as bioindicators for river health assessment.

KEYWORDS

- **bioindicator**
- **mollusk**
- **River Ganga**
- **river health assessment**
- **Varanasi**

REFERENCES

Baby, R., Hasan, I., Kabir, K., & Naser, M., (2010). Nutrient analysis of some commercially important mollusks of Bangladesh. *Journal of Scientific Research, 2*(2), 390–396.

Boulton, A. J., (1999). An overview of river health assessment: Philosophies, practice, problems, and prognosis. *Freshwater Biology, 41*(2), 469–479.

Bunn, S. E., & Arthington, A. H., (2002). Basic principles and ecological consequences of altered flow regimes for aquatic biodiversity. *Environmental Management, 30*(4), 492–507.

Cairns, J., McCormick, P. V., & Niederlehner, B., (1993). A proposed framework for developing indicators of ecosystem health. *Hydrobiologia, 263*(1), 1–44.

Chinchore, S., & Mahajan, P., (2013). Protective role of Coriandrum sativum (coriander) extracts on lead-induced alterations in the oxygen consumption of freshwater gastropod snail, Bellamya Bengalensis (Lamarck). *International Journal of Pharmaceutical Sciences and Research, 4*(7), 2789.

Fuller, S. L. (1978). *Fresh-Water Mussels (Mollusca: Bivalvia: Unionidae) of the Upper Mississippi River: Observations at Selected Sites within the 9-Foot Channel Navigation Project on Behalf of the US Army* (No. 78-33). Academy of Natural Sciences of Philadelphia PA Div of Limnology and Ecology.

Füreder, L., & Reynolds, J., (2003). Is Austropotamobius pallipes a good bioindicator? *Bulletin Français de la Pêche et de la Pisciculture, 370/371*, 157–163.

Grabarkiewicz, J., & Davis, W. (2008). *An Introduction to Freshwater Mussels as Biological Indicators (Including Accounts of Interior Basin,* Cumberlandian and Atlantic Slope Species) EPA-260-R-08-015. US Environmental Protection Agency, Office of Environment.

Markert, B. A., Breure, A. M., & Zechmeister, H. G., (2003). Definitions, strategies, and principles for bioindication/biomonitoring of the environment. *Trace Metals and Other Contaminants in the Environment, 6*, 3–39.

Mishra, A., Mukherjee, A., & Tripathi, B., (2009). Seasonal and temporal variations in physicochemical and bacteriological characteristics of River Ganga in Varanasi. *Int. J. Environ. Res*, *3*(3), 395–402.

Mohanty, B., (1993). *Urbanization in Developing Countries: Basic Services and Community Participation.* Concept Publishing Company.

Namrata, S., (2010). Physicochemical properties of polluted water of river Ganga at Varanasi. *International Journal of Energy and Environment*, *1*(5), 823–832.

Nandi, I., Srivastava, P. K., & Shah, K., (2017). Floodplain mapping through support vector machine and optical/infrared images from land sat 8 OLI/TIRS sensors: Case study from Varanasi. *Water Resources Management*, *31*(4), 1157–1171.

Nandi, I., Tewari, A., & Shah, K., (2016). Evolving human dimensions and the need for continuous health assessment of Indian rivers. *Curr. Sci., 111*, 263–271.

Patil, J. V., (2011). *Study of Selected Faunal Biodiversity of Toranmal Area.* Toranmal Reserve Forest.

Patil, S. R., (2013). A preliminary study of molluscan fauna of singhori wildlife sanctuary, raisen, Madhya Pradesh, India. *Indian Forester*, *139*(10), 932–935.

Pérez-Quintero, J. C., (2011). Freshwater mollusk biodiversity and conservation in two stressed Mediterranean basins. *Limnologica-Ecology and Management of Inland Waters*, *41*(3), 201–212.

Pérez-Quintero, J. C., (2012). Environmental determinants of freshwater mollusk biodiversity and identification of priority areas for conservation in Mediterranean watercourses. *Biodiversity and Conservation*, *21*(12), 3001–3016.

Piocos, E.A., & De la Cruz, A. A., (2000) *Solid Phase Extraction and High Performance Liquid Chromatography With Photodiode Array Detection of chemical indicators of Human Faecal contamination in water*, 23:8,1281-1291, DOI: 10.1081/JLC-100100414

Ray, M., Bhunia, A. S., Bhunia, N. S., & Ray, S., (2013). Density shift, morphological damage, lysosomal fragility and apoptosis of hemocytes of Indian mollusks exposed to pyrethroid pesticides. *Fish & Shellfish Immunology*, *35*(2), 499–512.

Ray, S., (2016). Levels of toxicity screening of environmental chemicals using aquatic invertebrates-a review. *Invertebrates-Experimental Models in Toxicity Screening: InTech.* DOI: 10.5772/61746

Salánki, J., Farkas, A., Kamardina, T., & Rózsa, K. S., (2003). Mollusks in biological monitoring of water quality. *Toxicology Letters*, *140*, 403–410.

Sarkar, U., Pathak, A., Sinha, R., Sivakumar, K., Pandian, A., Pandey, A., et al., (2012). Freshwater fish biodiversity in the River Ganga (India): Changing pattern, threats and conservation perspectives. *Reviews in Fish Biology and Fisheries*, *22*(1), 251–272.

Singh, R. P., (2011). Varanasi, India's cultural heritage city: contestation, conservation & planning. *Heritages Capes and Cultural Landscapes*, 205–254.

Sinha, R., Sinha, S. K., Kedia, D., Kumari, A., Rani, N., & Sharma, G., (2007). A holistic study on mercury pollution in the Ganga River system at Varanasi, India. *Current Science*, *92*(9), 1223–1228.

Strayer, D. L., & Dudgeon, D., (2010). Freshwater biodiversity conservation: Recent progress and future challenges. *Journal of the North American Benthological Society*, *29*(1), 344–358.

Strong, E. E., Gargominy, O., Ponder, W. F., & Bouchet, P., (2008). Global diversity of gastropods (Gastropoda, Mollusca) in freshwater. *Hydrobiologia*, *595*(1), 149–166.

Subba, R. N., (1989). *Handbook, Freshwater Mollusks of India.* Zoological Survey of India. Ed. Director, Zoological Survey of India. Calcutta, India.

CHAPTER 9

SPATIAL VARIABILITY IN THE WATER QUALITY OF CHILIKA LAGOON, EAST COAST OF INDIA

SADAF NAZNEEN and N. JANARDHANA RAJU

School of Environmental Sciences, Jawaharlal Nehru University, New Delhi–110067, India, E-mail: sadafnazneen01@gmail.com

ABSTRACT

Wetlands provide a range of ecological services which depends on the optimum health of the ecosystem. Chilika lagoon lying on the east coast of India is the largest lagoon in Asia and a biodiversity hotspot sustaining rich fishery resources. Freshwater inflows from the drainage basin along with saline water flow from the ocean results in a wide range of fresh, brackish, and saline water environments within the lagoon which gives rise to a very productive ecosystem. Various processes affecting the water quality of the lagoon system include agricultural drainage from the catchment area, municipal, and domestic waste and sewage intrusion and freshwater discharged into the lagoon from various rivers and rivulets. In the present study, an attempt has been made to understand the spatial variation in the water quality of Chilika lagoon. A total of 26 water samples were collected from different hydro-ecological regions of the lagoon during the 1st week of June 2013. The four regions remained significantly different from one another with respect to many parameters. Physicochemical properties like pH, salinity, EC (Electrical Conductivity), DO (Dissolved Oxygen), DIC (Dissolved Inorganic Carbon), DOC (Dissolved Organic Carbon), and nutrients like PO_4^{3-}, NO_3^-, $Si(OH)_4$ varied significantly between the sectors. A strong positive correlation between salinity, EC, Na^+, K^+, Ca^{2+}, and Mg^{2+} is a result of seawater dominance into the lagoon. Results of the Principal Component Analysis (PCA) suggest that all the sampling stations can be classified into three groups based on the spatial variations of salinity, EC, nutrients, and other physiochemical parameters. Lagoon's connections with the Bay of Bengal, residual fertilizers washed away from the agricultural fields, domestic sewage, biological activity and allochthonous materials brought by the river discharge play a significant role in governing the water quality of the lagoon. The non-conservative nature of the nutrients in the lagoon suggests that the lagoon is well mixed which gives rise to rich biodiversity.

9.1 INTRODUCTION

Lagoons comprise 13% of the earth's coastlines and are considered to be the most productive ecosystems in the biosphere with a very complex environment (Beltrame et al., 2009). Lagoon hydrodynamics is governed by the circulation and mixing caused by freshwater inputs via river discharge and inflow of saline waters through tidal flows (Statham, 2011). Coastal lagoons exhibit high variability in their physical, chemical, and biological characters, thus they tend to be typically unstable environments and the short and long-term variations that occur in these ecosystems are large in comparison to other saline environments. Lagoons usually exhibit estuarine characteristics and are regions of high primary productivity (Panigrahy et al., 2009). Salinity is an important factor, which is having direct and indirect control over the nutrient availability and their transformations in the saline aquatic environments. Tropical coastal lagoons are under huge anthropogenic pressures but have been studied to a lesser extent when compared to temperate estuaries (Burford et al., 2008). The rapid increase of human activities has increased nutrient transport from land to sea in the past decades, resulting in environmental deterioration and changes to biogeochemical processes. The nutrient-laden effluent discharged from shrimp farming is also a cause of eutrophication of coastal water, and its impacts have been a significant concern (Balasubramanian et al., 2004).

In Chilika, lagoon changes in land use, pattern, and agricultural practices in the catchment basin, as well as adjoining areas, has led to the addition of an enormous amount of nitrogen and phosphorous as residual fertilizers. Another important factor for the addition of nutrients is direct disposal of untreated wastewater from Bhubaneshwar City and the villages in and around Chilika. The discharge of fresh water into the lagoon is quite high during monsoon and post-monsoon while it is negligible or absent during other periods (Mohanty and Panda, 2009). This heavy discharge brings enormous amounts of nutrients into the lagoon. Water balance calculations indicate that an average 72% of the total freshwater inflow to the lake is due to runoff from the rivers, 15% from direct rainfall and 13% through drainage from the western catchment.

9.2 MATERIALS AND METHODS

9.2.1 STUDY AREA

Chilika lake is the largest tropical lagoon of Asia spread between latitude 19°28'–19°54'N and longitude 85°05'–85°38'E. It is a semi-enclosed, coastal lagoon on the east coast of India in the state of Odisha. The pear-shaped lagoon is about 65 kms long and varies in width from 18 km in the north to 5 km in the south (Figure 9.1). The lagoon is situated on the southern part of the Mahanadi delta-complex. The lagoon is cut off from the Bay of Bengal by a continuous sandy barrier-spit measuring 60 km in length and150 m in breadth, where backshore dunes are developed in the southern half of the spit in two or three parallel groups (Khandelwal et al., 2008). According to earlier estimates, the average water spread area of the lake is 906 km² in pre-monsoon and 1165 km² in the monsoon (Ghosh and Pattnaik, 2006). Present studies have estimated the lagoon area to be nearly 704 km² during summer which spreads to 1020 km² in monsoon (Gupta et al., 2008). The water depth in the lake varies from 0.9 to 2.6 m in the dry season and from 1.8 to 3.7 m in the rainy season. Chilika was designated a wetland of international importance under Ramsar convention in 1981. The lagoon is connected to the Bay of Bengal in the east. The barrier spit separates the lagoon from the Bay of Bengal and provides an inlet at its northeast extremity which remains open due to continuous flushing action brought by waves and currents. Changes in the position, shape, and breadth of the inlet have occurred from time to time depending on the interplay of available energy from land and sea and also due to the silt brought by the tributaries of Mahanadi Daya and Bhargavi rivers. At present, the lagoon is connected with the Bay of Bengal near Satapada (Sipakuda) by means of an artificial opening made in September 2000. The lagoon has several hydro-logical influences; the most important is its connection to the Bay of Bengal in the east and freshwater flow in the north due to the joining of tributaries of Mahanadi River. Many rivers and rivulets join the lagoon through its western catchment. Almost 52 small rivers and streams join the lagoon. The main tributaries of Mahanadi (such as Bhargavi, Daya, and Makara) account for almost 61% (850 m^3 s^{-1}) of the total freshwater.

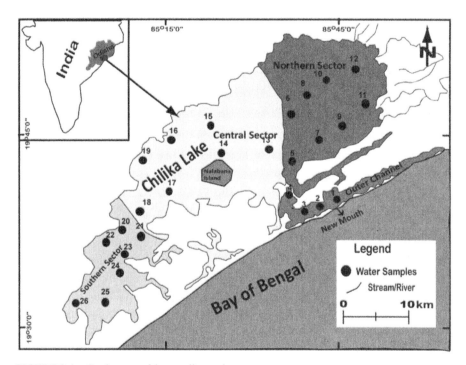

FIGURE 9.1   Study area with sampling points.

9.2.2   METHODOLOGY

A total of 26 surface water samples were collected from four hydro-ecological different sectors of the lagoon during pre and post-monsoon seasons (Figure 9.1). The pre-monsoon sampling was conducted in the 1st week of June 2013, and post-monsoon sampling was conducted in January 2014. The water samples were collected in 500 ml clean propylene bottles. Cleaning of all the plastic bottles was carried out by soaking in 5 % nitric acid (HNO$_3$ v/v) for 24 hours and then rinsing with doubled distilled water (DDW). After collection, the water samples were poisoned with HgCl$_2$ for nutrients and DIC analysis Another set of samples were collected in 100ml bottles for dissolved metals analysis. The water samples were filtered through 0.45µm nylon filters and acidified below pH < 2 by addition of concentrated HNO$_3$. The samples were transferred to the laboratory in the boxes and stored at 4°C until analysis. All the analysis was completed within 15 days of sampling. In every analysis, appropriate blanks were run as controls. pH, salinity, and EC in the water samples were analyzed on the field with portable multiparameter kit after calibrating with respective buffers for pH and EC. The parameters like DO, DIC, DOC, NO_3^-, PO_4^{3-}, H_4SiO_4 were analyzed by standards procedure for surface water.

- The Dissolved oxygen (DO) was measured by Wrinkler's method.
- HCO_3^- by titrimetric method.
- DOC was analyzed by TOC analyzer (Shimadzu TOC–5000 analyzer).
- NO_2^- by indophenol method.
- NO_3^- by cadmium reduction method PO_4^{3-} by the ascorbic acid method.
- H_4SiO_4 (dissolved silica) molybdosilicate method.
- SO_4^{2-} by the turbidimetric method.
- Na^+ and K^+ by Flame photometer (Elico Flame Photometer, CL–378).
- Ca^{2+} titration method and.
- Mg^{2+} by calculation method.
- The heavy metals were analyzed on Atomic Absorption Spectrophotometer (Thermoscientific M series).

9.3 RESULTS AND DISCUSSIONS

The distribution of nutrients and other physio-chemical parameters have been described through spatial diagrams.

9.3.1 pH

The lagoon water is alkaline in nature. The mean pH value in was 8.1 (7.5–8.6). Spatial distribution map shows that the highest pH values were observed in the southern sector of the lagoon (Figure 9.2). High pH values in southern sector are because of greater photosynthetic activity by algae and phytoplankton which remove dissolved CO_2 from the surface water and shift the equilibrium towards alkaline (Hennemann and Petrucio, 2011; Jayakumar et al., 2013). The pH value remained low in the northern sector due to large freshwater input into the lagoon. Rest of the sectors had saline water dominance; thus the pH is higher when compared to the northern sector.

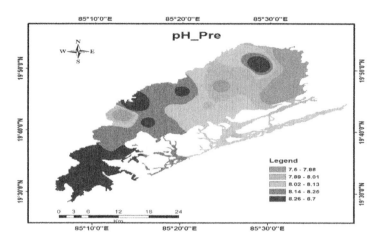

FIGURE 9.2   Spatial distribution of pH in pre-monsoon season.

9.3.2   ELECTRICAL CONDUCTIVITY (EC)

The mean EC value was 34750.4 µS/cm (2720–51,400 µS/cm). The conductivity of water is the ability to conduct electric current by means of the dissolved ions present in its milieu. Electrical conductivity is mainly governed by the ion concentration of the liquid. Since in the month of June saline water is dominant in the lagoon, so is the EC, as seawater contains much more ions when compared to the freshwater. Spatial distribution of EC showed the lowest values in the northern sector (Figure 9.3). This is due to the large freshwater influx in this region brought by the tributaries of Mahanadi River which discharge a significant amount of freshwater in the northern part of the lagoon throughout the year.

9.3.3   SALINITY (SAL)

Salinity is one of the most important factors which determine the concentration of other nutrients in a brackish water system. The mean salinity value was 21.44 PSU (1.3–32.8). The salinity value in the pre-monsoon period was similar to the previous studies in this lagoon (Panigrahy et al., 2007; Panigrahy et al., 2009; Muduli et al., 2013). The higher and comparatively stable salinity values observed in the southern sector could be a result of the input of saline water from Rushikulya estuary through the enclosed Palur canal throughout the year which maintains an even salinity value (Figure

9.4). Another reason for higher and stable salinity values in the southern sector could be less freshwater discharge. Moderate salinity values observed in the central sector is due to good circulation and mixing of both marine and freshwater in this region. This region is well mixed due to the freshwater discharge from small rivulets and streams joining the western catchment of the lagoon and entry of the seawater from the inlets present in the outer channel region.

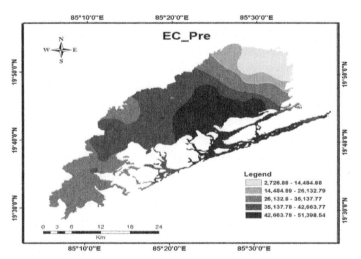

FIGURE 9.3   Spatial distribution of EC in pre-monsoon season.

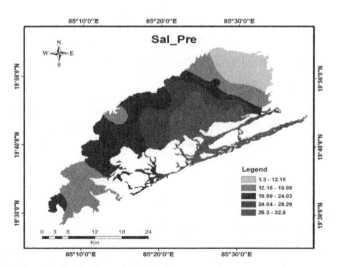

FIGURE 9.4   Spatial distribution of salinity in pre-monsoon season.

9.3.4 *DISSOLVED INORGANIC CARBON (DIC)*

DIC mean value during the pre-monsoon period was 170.4 mgl^{-1} (132–203.6 mgl^{-1}). In aquatic systems, DIC concentration is governed by chemical, physical, and biological processes. Spatial and temporal variation of DIC could be related to the catchment area characteristics and variability in hydrology (Figure 9.5). The spatial and temporal variability of DIC is mostly due to the weathering of the bedrock in the drainage basin. However, biological activity also has a large influence in controlling the DIC values in Chilika Lake (Gupta et al., 2008). Exchange of CO_2 with atmosphere, pH, salinity, carbonates, respiration, and photosynthesis are some of the factors that govern DIC concentration in a water body. Photosynthetic uptake of CO_2 and precipitation of calcareous material lowers DIC (Muduli et al., 2013). Moderate to high values of DIC has been observed in most parts of southern and some parts of the central sector during the pre-monsoon season as these regions do not have higher plants which make use of dissolved carbon for photosynthesis. Lower values of DIC observed could be a result of primary production and bacterial respiration of carbon which causes variations in DIC concentration (Muduli et al., 2012; Kanuri et al., 2013).

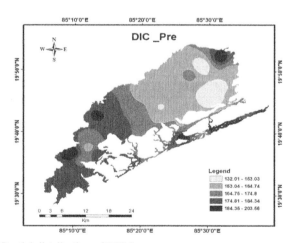

FIGURE 9.5   Spatial distribution of DIC in pre-monsoon season.

9.3.5 *DISSOLVED ORGANIC CARBON (DOC)*

The mean DOC value was 5.65 mgl^{-1} (3.87–7.89 mgl^{-1}). The chief sources of organic carbon to these environments are allochthonous materials exported from the land through rivers and autochthonous production of organic matter by algae through photosynthesis and intertidal vegetation (Kanuri et al., 2013). The combination of primary production of plant matter and decomposition rates controls the amount of DOC in water. Spatial distribution of DOC has been presented in Figure 9.6. DOC fluxes from sediments are also an important carbon source in many estuaries, and benthic remobilization is expected to be an internal source of DOC in shallow systems. Rivers draining into Chilika carry higher DOC concentration; thus Chilika has appreciable DOC load (Gupta et al., 2008).

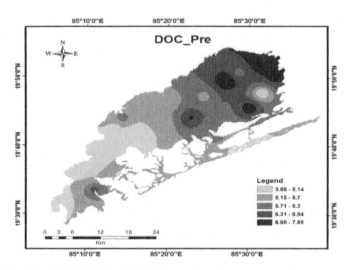

FIGURE 9.6   Spatial distribution of DOC in pre-monsoon season.

9.3.6   DISSOLVED OXYGEN (DO)

The mean DO value was 7.34 mgl^{-1} (2.34–13.43 mgl^{-1}). The spatial distribution of DO shows higher values in the northern sector when compared to the other sectors except near the Balugaon town in the central sector (Figure 9.7). Northern sector is dominated by a large number of freshwater macrophytes which engage in active photosynthesis. Oxygen is released as a by-product of photosynthesis. Therefore, the northern sector remains well oxygenated. Higher DO concentration near the Balugaon region could be a result of high concentrations of nutrients like NO_3^+ and PO_4^{3+} entering the lagoon from Balugaon township which gives rise thick algal mats. Presence of plants leads to photosynthetic activity and release of oxygen (Bose et al., 2012).

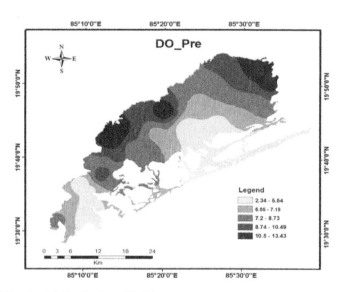

FIGURE 9.7   Spatial distribution of DO in pre-monsoon season.

9.3.7   NITRATE (NO_3^-)

Mean value for NO_3^- was 2.2 mgl^{-1} (0.2–7.5 mgl^{-1}). Nitrate in aquatic environments is mainly derived from terrestrial runoff. Concentration and rate of supply of nitrate are related to land use practices in the drainage basin since NO_3^- moves easily through soils and is rapidly lost from land through natural drainage systems. NO_3- is the most stable form of inorganic nitrogen in well-oxygenated waters. It is added through nitrogenous fertilizers, industrial effluents, human, and animal wastes through the biochemical activity of nitrifying bacteria, such as Nitrosomonas and Nitrobacter (Raju et al., 2009). The spatial distribution of NO^{3-} has been presented in Figure 9.8. These patches may be witnessing NO_3^- inputs due to agricultural runoff, untreated sewage discharge, aquaculture ponds, and other household wastes being directly dumped into the lagoon water. Organic nitrogen present in domestic sewage is bound to carbon-containing compounds as proteins ($R-NH_2$). Decomposition of organic nitrogen in the sewage soil by a variety of microorganisms slowly transforms the organic nitrogen to ammonia (NH_3) by a process called "mineralization" (Raju et al., 2012). Further, during the nitrification process NH_3 gets converted to NO_2^- first and then NO_3^-. NO_2^- is unstable therefore its gets oxidized to NO_3^-. Nitrate is quite soluble in water and being stable does not get adsorbed to soil particles as well. The majority of agricultural fertilizers contain nitrogen in the forms of ammonium and nitrate, often as ammonium nitrate (NH_4NO_3), which could be the prime source of nitrogen species in the coastal water (Newton and Mudge, 2005). High NO_3^- concentration in the post-monsoon season may be due to a large amount of nitrate brought about from the agricultural fields by the large freshwater flow into the lagoon. One of the major reasons for high nitrates in certain pockets during post-monsoon may be due to aquaculture ponds and domestic wastes directly released into the water by the fishing villages around the lagoon. Moreover, in winter the rates of nitrification and denitrification become slow which may have lead to NO_3^- accumulation at some places in the lagoon.

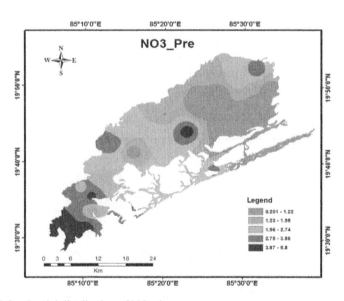

FIGURE 9.8   Spatial distribution of NO_3^- in pre-monsoon season.

9.3.8   PHOSPHATE (PO_4^{3-})

The mean value of PO_4^{3-} was 0.03 mgl⁻¹ (0.0–0.09 mgl⁻¹) in the pre-monsoon season 0.04 mgl⁻¹ (0.01–0.09 mgl⁻¹) in the post-monsoon. Phosphate is one of the major nutrients responsible for biological production in aquatic ecosystems. PO_4^{3-} is also the limiting nutrient in most of the aquatic ecosystems. As compared to NO_3^- the concentration of PO_4^{3-} is quite low in both the seasons (Figure 9.9). Phosphate is mainly derived from agricultural runoff and domestic sewage, part of it is also derived from weathering of P bearing minerals as apatite and fluorapatite. Domestic sewage is a major source of ammonium and phosphate. Phosphate is also added through fertilizers which contain di-ammonium phosphate (Chauhan and Ramanathan, 2008). Low phosphate concentration may be due to its rapid assimilation by the micro and macrophytic vegetation in the lagoon or due to its adsorption on the surface of sediments. Other reasons for PO_4^{3-} removal from the water column is its microbial incorporation. The spatial distribution of PO_4^{3-} shows a higher value in some parts of the northern sector and in small patches in the central sector during pre-monsoon (Figure 9.9). Highest PO_4^{3-} concentrations are ranging from 0.07–0.09 mgl^{-1} observed in the inner portion of the lake during pre-monsoon season might have been primarily influenced by external sources as well as contributions from sediment fluxes. This could be a result of PO_4^{3-} addition from the river runoff in the northern and western boundaries of the lagoon. During post-monsoon moderately higher values ranging from 0.04–0.06 mgl^{-1} prevails in most parts of the lagoon. The values observed in both the seasons are comparable to the values observed in previous studies (Panigrahy et al., 2007; Panigrahy et al., 2009).

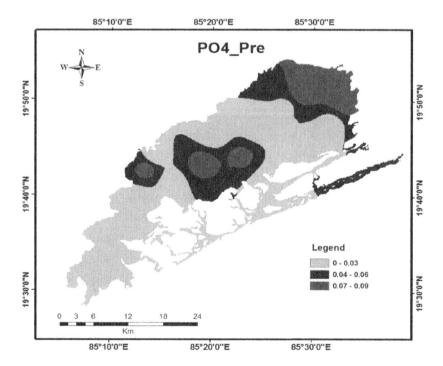

FIGURE 9.9 Spatial distribution of PO_4^{3-} in pre -monsoon season.

9.3.9 DISSOLVED SILICA (H₄SIO₄)

9.3.9   DISSOLVED SILICA (H_4SIO_4)

The mean concentration of dissolved silica (DSi) was 3.33 mgl⁻¹ (0.3–10.3 mgl⁻¹) in pre monsoon seasons. The spatio-temporal variations of DSi in coastal water is influenced by several factors, including the proportional physical mixing of seawater with fresh water, adsorption of reactive silicate into sedimentary particles, chemical interaction with clay minerals (Gouda and Panigrahy, 1992), co-precipitation with humic compounds and iron and biological removal by phytoplankton, especially by diatoms and silico flagellates (Beucher et al., 2004;Sahoo et al., 2014). The DSi was found to be highest in the northern sector due to maximum river discharge in this region (Figure 9.10).

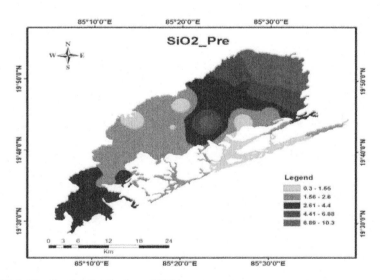

FIGURE 9.10   Spatial distribution of DSi in pre -monsoon season.

9.3.10   CORRELATION ANALYSIS

A bivariate correlation analysis is applied to describe the relationship between the two geochemical parameters. A correlation value above 0.5 between two parameters is considered good whereas a correlation value around 1 (either positive or negative) between two parameters renders them very high relative relationship. The correlation between various parameters has been presented in during the pre-monsoon season has been presented in Table 9.1.

TABLE 9.1 Correlation Analysis for Pre-Monsoon

	pH	EC	Sal	DIC	DOC	DO	NO$_2$	NO$_3^-$	PO$_4^{3-}$	DSi	SO$_4^{2-}$	Cl-	Na+	K+	Ca2+	Mg2+
pH	1															
EC	-0.13	1														
Sal	-0.13	**0.99**	1													
DIC	0.03	-0.01	0.01	1												
DOC	-0.28	-0.35	-0.31	-0.25	1											
DO	-0.13	-0.62	-0.61	-0.06	0.13	1										
NO$_2$	-0.01	-0.37	-0.38	0.18	0.07	0.40	1									
NO$_3^-$	-0.07	-0.44	-0.43	0.18	0.20	0.43	**0.97**	1								
PO$_4^{3-}$	-0.26	-0.45	-0.40	0.05	0.39	0.27	0.18	0.29	1							
DSi	-0.19	**-0.69**	-0.68	-0.24	0.53	0.27	0.27	0.36	**0.62**	1						
SO$_4^{2-}$	-0.17	0.12	0.12	0.12	0.07	0.02	0.21	0.19	-0.05	-0.13	1					
Cl-	-0.12	**0.98**	**0.99**	0.01	-0.30	**-0.60**	-0.36	-0.41	-0.41	**-0.68**	0.11	1				
Na+	-0.16	**0.99**	**0.99**	-0.02	-0.29	**-0.59**	-0.39	-0.43	-0.38	**-0.66**	0.12	**0.98**	1			
K+	-0.05	**0.86**	**0.88**	0.22	-0.41	**-0.51**	-0.34	-0.36	-0.24	**-0.64**	0.05	**0.89**	**0.89**	1		
Ca^{2+}	-0.20	**0.98**	**0.97**	-0.04	-0.32	**-0.58**	-0.44	-0.48	-0.41	**-0.67**	0.13	**0.96**	**0.98**	**0.83**	1	
Mg^{2+}	-0.03	**0.80**	**0.79**	0.06	-0.38	-0.44	-0.23	-0.35	**-0.57**	**-0.64**	0.15	0.79	0.75	0.62	0.77	1

In pre-monsoon season a strong positive correlation between Sal, EC, Cl⁻, Na⁺, K⁺, Ca²⁺ and Mg²⁺ indicates that seawater is the major source of these ions. The correlation value of salinity denotes a moderate negative correlation with DO and DSi. DO decreases with salinity because high salinity waters are low in DO. Negative relation indicates that DSi is mainly brought by the freshwater since tropical rivers have high DSi concentration. A strong positive correlation between NO_2^- and NO_3^- is because NO_2^- being unstable is converted to NO_3^- in aquatic systems. A positive correlation between PO_4^{3-} and DSi indicates their terrigenous source. The probable sources of PO_4^{3-} could be fertilizer residue from the agricultural field. Studies have indicated SiO_2 was reaching water bodies from the agricultural field along with other nutrients. DSi has a negative correlation with Cl⁻, Na⁺, K⁺, Ca²⁺, and Mg²⁺, which shows its freshwater source, as the other ions are predominantly from the seawater. Different sources of PO_4^{3-} (freshwater) and Mg (seawater) could be a reason for the negative correlation between PO_4^{3-} and Mg. Salinity and EC are positively related to Cl⁻, Na⁺, K⁺, Ca²⁺, and Mg²⁺. All these ions are present in major quantity in seawater. This relation is also an indication of seawater dominance in the lake.

9.3.11   PRINCIPAL COMPONENT ANALYSIS

Principal Component Analysis (PCA) is a technique for taking high dimensional data and reducing it to more manageable, lower dimensional form without significant information (Table 9.2). The PCA enables a reduction in data and description of a given multidimensional system by means of a small number of new variables (Loska and Weichula, 2003). Relevant components are those whose eigenvalue is higher than 1. The application of varimax rotation of standardized component loadings enables us to obtain a clear system as a result of the maximization of component loadings variance and elimination of less valid components (Chakrapani and Subramanian, 1993). In the pre-monsoon season, three factors are formed which show 74% of the total variance. PCA 1 is explained by 47.7% of the variance. PC2 and PC3 show 13 % of variance each (Table 9.2).

Factor 1: Factor 1 account for 47.7% of the total variability. Factor 1 shows positive loadings of salinity, EC, Cl⁻, Na⁺, K⁺, Ca²⁺, and Mg²⁺. This factor can be explained by the influence of seawater. EC is governed by the concentration of ions in a solution. Na⁺, Cl⁻, Ca²⁺ and Mg²⁺ are the dominant ions in seawater; hence they show a positive loading with salinity and EC.

TABLE 9.2   Principal Component Analysis for Pre-Monsoon Season

Rotated Component Matrix

	Components		
	1	**2**	**3**
pH	−0.234	−0.231	−0.715
EC	**0.980**	−0.083	−0.100
Sal	**0.982**	−0.079	−0.076
DIC	0.046	0.468	−0.376
DOC	−0.275	−0.059	**0.734**
DO	−0.586	0.314	0.096
NO_2^-	−0.350	**0.851**	0.010
NO_3^-	−0.399	**0.827**	0.126
PO_4^{3-}	−0.448	0.101	**0.631**
DSi	−0.665	−0.030	**0.569**
SO_4^{2-}	0.235	**0.528**	0.192
Cl⁻	**0.978**	−0.064	−0.078
Na⁺	**0.982**	−0.088	−0.033
K⁺	**0.870**	−0.010	−0.157
Ca²⁺	**0.973**	−0.125	−0.043
Mg²⁺	**0.796**	0.019	−0.287

Factor 2: Factor 2 shows a positive loading of NO_2^-, NO_3^- and SO_4^{2-}. A high positive loading between NO_2^- and NO_3^- is an indication of the oxidizing environment which oxidizes unstable nitrite to nitrate. Slightly positive loading of NO_2^- NO_3^- and SO_4^{2-} could be an indication of their anthropogenic source. Like NO_3^-, SO_4^{2-} can also be added into the system from agricultural fields (Singh et al., 2015).

Factor 3: This factor depicts high positive loadings of DOC, PO_4^{3-} and DSi. A positive relation between these components suggests their allochthonous source. DOC is mainly transported into estuarine systems by rivers which carry carbon in the dissolved form to the coastal waters (Gupta et al., 2008). Similarly PO_4^{3-} and DSi are usually brought by the land and river runoff into the lagoon.

9.4   CONCLUSIONS

Distinct spatial variation of water quality parameters was observed in different parts of the lagoon. The lagoon water was alkaline. The salinity was largely governed by freshwater discharge and mixing of the Bay of Bengal water with the lagoon. The water was well oxygenated owing to the photosynthetic activities of various micro and macrophytes. The NO_3^- and PO_4^{3-} in some parts were higher due to runoff from agricultural fields and wastewater discharge. DSi is mainly from land through river discharge. A strong positive correlation between Sal, EC, Cl^-, Na^+, K^+, Ca^{2+} and Mg^{2+} indicates that seawater is the major source of these ions. The correlation value of salinity denotes a moderate negative correlation with DO and DSi. DO decreases with salinity because high salinity waters are low in DO. Negative relation indicates that DSi is mainly brought by the freshwater since tropical rivers have high DSi concentration. The PCA shows that some parameters are governed by seawater whereas others are brought from the land by river discharge. The macro and microphytic vegetation also play a significant role in governing the water quality of the lagoon.

KEYWORDS

- **Chilika lagoon**
- **dissolved nutrients**
- **DOC**
- **salinity**

REFERENCES

Balasubramanian, C. P., Pillai, S. M., & Ravichandran, P., (2004). Zero-water exchange shrimp farming systems (extensive) in the periphery of Chilka lagoon, Orissa, India. *Aquaculture International, 12,* 555–572.

Beltrame, M. O., De Marco, S. G., & Marcovecchio, J. E., (2009). Dissolved and particulate heavy metals distribution in coastal lagoons. A case study from Mar Chiquita Lagoon, Argentina. *Estuar. Coast. Shelf Sci., 85,* 45–56.

Beucher, C., Tréguer, P., Corvaisier, R., Hapette, A. M., & Elskens, M., (2004). Production and dissolution of biosilica, and changing microphytoplankton dominance in the Bay of Brest (France). *Marine Ecology Progress Series, 267,* 57–69.

Bose, R., De, A., Sen, G., & Mukherjee, A. D., (2012). Comparative study of the physico-chemical parameters of the coastal waters in rivers Matla and Saptamukhi: Impacts of coastal water coastal pollution. *Journal of Water Chemistry and Technology, 34*(5), 246.

Burford, M. A., Alongi, D. M., McKinnon, A. D., & Trott, L. A., (2008). Primary production and nutrients in a tropical macrotidal estuary, Darwin Harbor, Australia. *Estuarine, Coastal and Shelf Science*, *79*, 440–448.

Chakrapani, G. J., & Subramanian, V., (1993). Heavy metal distribution and fractionation in sediments of the Mahanadi River basin, India. *Environ. Geol.*, *22*, 80–87.

Chauhan, R., & Ramanathan, A. L., (2008). Evaluation of water quality of Bhitarkanika mangrove ecosystem, Orissa, east coast of India. *Indian Journal of Marine Sciences*, *37*(2), 153–158.

Gouda, R., & Panigrahy, R. C., (1992). Seasonal distribution and behavior of silicate in the Rushikulya estuary, East coast of India. *Indian Journal of Marine Sciences*, *24*, 111–115.

Gupta, G. V. M., Sarma, V. V. S. S., Robin, R. S., Raman, A. V., Kumar, M. J., Rakesh, M., & Subramanian, B. R., (2008). Influence of net ecosystem metabolism in transferring riverine organic carbon to atmospheric $CO2$ in a tropical coastal lagoon (Chilka Lake, India). *Biogeochemistry*, *87*, 265–285.

Hennemann, M. C., & Petrucio, M. M., (2011). *Spatial and Temporal Dynamic of Trophic Relevant Parameters in a Subtropical Coastal Lagoon in Brazil*, 347–361.

Jayakumar, R., Steger, K., Chandra, T. S., & Seshadri, S., (2013). An assessment of temporal variations in physicochemical and microbiological properties of barmouths and lagoons in Chennai (Southeast coast of India). *Marine Pollution Bulletin*, *70*(1/2), 44–53.

Kanuri, V. V., Muduli, P. R., Robin, R. S., Kumar, B. C., Lovaraju, A., Ganguly, D., et al., (2013). Plankton metabolic processes and its significance on dissolved organic carbon pool in a tropical brackish water lagoon. *Continental Shelf Research*, *61/62*, 52–61.

Loska, K., & Wiechuła, D., (2003). Application of principal component analysis for the estimation of source of heavy metal contamination in surface sediments from the Rybnik Reservoir. *Chemosphere*, *51*, 723–733.

Mohanty, P. K., & Panda, U. S. B., (2009). Circulation and mixing processes in Chilka lagoon. *Indian Journal of Marine Sciences*, *38*(2), 205–2014.

Muduli, P. R., Kanuri V. V., Robin, R. S., Kumar, B. C., Patra, S., Raman, A. V., et al., (2013). Distribution of dissolved inorganic carbon and net ecosystem production in a tropical brackish water lagoon, India. *Continental Shelf Research*, *64*, 75–87.

Muduli, P. R., Kanuri, V. V., Robin, R. S., Kumar, C., Patra, S., Raman, A. V., Rao G. N., & Subramanian, B. R., (2012). Spatio-temporal variation of $CO2$ emission from Chilika Lake, a tropical coastal lagoon, on the east coast of India. *Estuarine, Coastal and Shelf Science*, *113*, 305–313.

Newton, A., & Mudge, S. M., (2005). Lagoon-sea exchanges, nutrient dynamics and water quality management of the Ria Formosa (Portugal). *Estuarine, Coastal and Shelf Science*, *62*, 405–414.

Panigrahi, S. N., Wikner, J., Panigrahy, R. C., Satapathy, K. K., & Acharya, B. C., (2009). Variability of nutrients and phytoplankton biomass in a shallow brackish water ecosystem (Chilka Lagoon), India. *Limnology*, *10*, 73–85.

Panigrahi, S., Acharya, B. C., Panigrahy, R. C., Nayak, B. K., Banarjee, K., & Sarkar, S. K., (2007). Anthropogenic impact on water quality of Chilika lagoon RAMSAR site: A statistical approach. *Wetlands Ecol. Manage.*, *15*, 113–126.

Raju, N. J., (2012). Evaluation of hydrogeochemical processes in the Pleistocene aquifers of Middle Ganga Plain, Uttar Pradesh, India. *Environ. Earth Sci.*, *65*, 1291–1308.

Raju, N. J., Ram, P., & Dey, S., (2009). Groundwater quality in the lower Varuna River Basin, Varanasi District, Uttar Pradesh. *Journal Geological Society of India*, *73*, 178–192.

Sahoo, S., Baliarsingh, S. K., Lotliker, A. A., & Sahu, K. C., (2014). Imprint of cyclone Phailin on water quality of Chilika lagoon. *Current Science, 107*(9), 1380–1381.

Singh, S., Raju, N. J., & Nazneen, S., (2015). Environmental risk of heavy metal pollution and contamination sources using multivariate analysis in the soils of Varanasi environs, India. *Environ. Monit. Assess., 187*, 345.

Statham, P. J., (2011). Nutrients in estuaries: An overview and the potential impacts of climate change. *Science of the Total Environment, 434*, 213–227.

PRECISION IRRIGATION AND FERTIGATION FOR THE EFFICIENT WATER AND NUTRIENT MANAGEMENT

P. K. SINGH

Professor, Department of Irrigation and Drainage Engineering, College of Technology, G. B. Pant University of Agriculture & Technology, Pantnagar – 363145, Uttarakhand, India E-mail: singhpk68@gmail.com

ABSTRACT

Significant progress has been made in irrigation and fertilizer application technologies and in the implementation of water and nutrient management practices such as scientific irrigation and fertilizer scheduling under conventional and modern irrigation techniques. However, scientists report that irrigation inefficiency remains the rule rather than the exception. Gains in water and nutrient use efficiency can be achieved when water application is precisely matched to the site-specific (spatially distributed) crop demand, a central principle underlying *precision irrigation* and *fertigation*. This site-specific crop water demand is present in agricultural fields mainly because of variability in soil properties and topography but may also result from variable rainfall or crop variation associated with multiple crops planted in the same field or plants growing at different phonological stages induced by natural or manmade causes. This chapter deals with the various aspect of the precise application of irrigation water and nutrients for the efficient utilization of the two most important agricultural inputs.

10.1   INTRODUCTION

Application of irrigation water and nutrients at the right time, in the right amount, in the right manner at the right place, is the crux of precision irrigation water and nutrient management. Micro irrigation, a technique that provides crops with water through a network of pipelines at a high frequency but with a low volume of water (drips) applied directly to the root zone in a quantity that approaches consumptive use of the plants, can be combined with fertilizer application, to offer fertigation. Fertigation enables the farmer to meet the specific water and nutrient needs of the crops with great precision, thus minimizing losses of both precious water and nutrients. The direct delivery of fertilizers through drip irrigation demands the use of soluble fertilizers and pumping and injection systems for introducing the fertilizers directly into the irrigation system. Fertigation allows an accurate and uniform application of nutrients to the wetted area, where the active roots are concentrated. The nutrients are applied as per the crop need at different growth stages in a split manner. The problem of mobility of non-mobile nutrients is also addressed using fertigation. Planning the irrigation system and nutrient supply to the crops according to their physiological stage of development, and consideration of the soil and climate characteristics, result in high yields and high-quality crops with minimum pollution. In India more than 9.0 Mha of land have been brought under sprinkler and micro-irrigation; i.e., pressurized irrigation (Malhotra, 2017). Most of the crops irrigated under micro-irrigation are horticultural crops. However, field crops such as sugarcane, groundnut, cotton, etc., are also being brought under micro-irrigation. Efforts are going on to develop the economical design of a micro-fertigation system for the efficient water and nutrient management in a crop like paddy and wheat. The fertilizer applications to these micro-irrigated crops are partially through fertigation in open field condition. However, most of the crops under polyhouse condition are under fertigation. In this chapter, an effort has been made to discuss various issues of fertigation for the precision nutrient management for achieving high nutrient use efficiency.

Applications of agricultural inputs at uniform rates across the field without due regard to in-field variations in soil fertility and crop conditions does not yield desirable results in terms of crop yield. The management of in-field variability in soil moisture, fertility, and crop conditions for improving the crop production and minimizing the environmental impact is the crux of precision farming. It's about doing the right thing, in the right place, in the right way, at the right time. It requires the use of new technologies, such as global positioning system (GPS), sensors, satellite or aerial images, and information management tools (GIS) to assess and understand variations. Precision farming may be used to improve a field or a farm management from several perspectives. There are many examples of precision irrigation over the last two decades in the western world particularly in the USA and also in Israel. In India and other developing countries, the term precision irrigation means efficient methods of water application through the sprinkler and drip irrigation. However, few systems of micro irrigation and sprinkler irrigation have been installed in an automated mode based on time, volume, and real-time soil moisture feedback system.

10.1.1   PRECISION IRRIGATION

Better irrigation (precision irrigation) is one of the engineering perspectives of precision farming. We first need to establish a common understanding to clarify or explain the term precision irrigation. The traditional meaning of precision irrigation has been that of given in the literature, is referred to as irrigation scheduling. That is, schedule based on environmental data, whether that data comes from local field sensors or from more global sources such as regional meteorological information at precise locations (within the soil profile) or at precise times. Perhaps a good example of this traditional definition of drip irrigation, which is generally accepted as a very precise irrigation technique because water can be precisely controlled with regard to application rate, timing, and location with respect to the plant. This definition continued to be used today in many countries except in USA and western world where more than 60% irrigated area is under sprinkler irrigation (center pivot system). However, in this paper we define precision irrigation as site-specific irrigation water management, specifically the application of water to a given site (right place) in a given volume (right amount) at right time (when) in a right manner (irrigation method) needed for optimum crop production, profitability, and other management objectives at the specific site. This is in contrast with a simultaneous application of the single amount of water to the entire area of the irrigation system/methods. During nineties in USA precision irrigation concept have been initiated at few locations (Camp et al. 2006) mostly concerning to the hardware development and only a few concerning to the site-specific irrigation (Lu et al., 2004(a); Lu et al., 2004(b); Lu et al., 2005). This method of water management continues to be more or less research issues. Development of hardware is mostly in the area of variable rate applicators (sprinkler nozzles, control valves, pumps, sensors, etc.), and software to operate the system. However, an existing commercial self-propelled system such as center-pivot and lateral-move machines are particularly amenable to cite specific approaches because of their central level of automation and a large area of coverage with a single pipe lateral. This is reflected in some commercial irrigation systems that have recently been modified for precision irrigation. In addition to irrigation, these machines offer an outstanding platform for mounting sensors that can provide real-time monitoring of plant and soil conditions and serve as a transport device for nutrients and other agro-chemical application systems. Adoption of micro (drip/trickle) irrigation and fertigation systems are at an accelerated rate in the developed and developing world for the wide spared horticultural and row crops. This system also offers site-specific management of water and nutrients in a precise manner with the application of controllers and sensors. The development of efficient and cost-effective hardware and software shall be able to accelerate the adoption of a complete precision irrigation system by all category growers.

10.1.1.1   ADVANTAGES AND LIMITATIONS

Applying precision irrigation practices offers significant potential for saving water, nutrient energy, and money. Further, it has the potential to increases crop yield. There is an additional positive environmental impact from precision irrigation in that farm runoff, a major source of water pollution, can be reduced. The major limitation associated with precision irrigation is the limited application of self-propelled center-pivot sprinklers for small land holdings, high initial cost, operation, and maintenance needs a skilled workforce.

10.1.1.2   PRECISION IRRIGATION APPLICATION SYSTEM

In conventional types of irrigation valves, emission devices, sprinkler nozzles, application rates have been altered by manual operations. Similarly, the movement of the lateral line, travel speed were also controlled/ adjusted manually. In newer systems use of controller and software has made these jobs automatically in dynamic mode as per the requirement of the crop and field, i.e., site-specific.

Sprinkler Irrigation: Numerous innovative technologies have been developed to apply the irrigation water in dynamic mode (variable rate) to meet anticipated whole-field management needs in precision irrigation primarily with center-pivot and lateral-move irrigation systems. In general, the operation criteria for these systems include the case of retrofit to the existing commercial irrigation system, good water application uniformity within and between management zone, robust electronics, compatibility with existing irrigation system equipment, bi-directional communication, and flexible expansion for future development and functional requirement. In addition, management of precision water application must include the interactions between individual sprinkler wetted diameters, the start/stop movement of towers, and solenoid valve cycling. These new precision water application technologies generally can be classified as either (1) a multiple of discrete fixed-rate application devices operated in combination to provide a range of application depths, (2) flow interruption to fixed-rate devices to provide a range of application depths that depend upon pulse frequency, or (3) a variable aperture sprinkler with time proportional control. Multiple sprinklers, pulsing sprinklers, and variable-orifice sprinklers have been developed in different part of the world for the precision application of water to the crops.

Micro-Irrigation: In low-pressure irrigation system (drip, microsprinkler, etc.) constant and variable discharge emitters have been developed by manufacturers to provide variable flow rate at the specific site in the field. In micro-irrigation there may be single or multiple emitters at a point, the operation of one or more than one emitters could be possible with the precision irrigation system. The operation time of emitters/lateral is possible through automatic hydraulic valves controlled by the microprocessor/controller.

10.2   SYSTEM CONTROL

Various forms of control systems has been developed for the surface, sprinkler, and micro-irrigation system for the control of the self-propelled sprinklers, emitters on the basis of time, volume, and real-time feedback. The control system generally consists of a microprocessor/controller, communication system (wire/wireless), control valve and sensors. For example, Remote Irrigation Monitoring and Control System (RIMCS) have been developed for continuous move irrigation systems that integrate localized wireless sensor networks for monitoring soil moisture and weather and provide control for individual or groups of nozzles with wireless access to the Internet to enable remote monitoring and control. The *RIMCS* uses a Single Board Computer (SBC) using the Linux operating system to control solenoids connected to individual or groups of nozzles based on prescribed application maps. The main control box houses the SBC connected to a sensor network radio, a

GPS unit, and an Ethernet radio creating a wireless connection to a remote server. A C-software control program resides on the SBC to control the on/off time for each nozzle group using a "time on" application map developed remotely. The SBC also interfaces with the sensor network radio to record measurements from sensors on the irrigation system and in the field that monitor performance and soil and crop conditions. The SBC automatically populates a remote database on the server in real time and provides software applications to monitor and control the irrigation system from the Internet. Another example of the irrigation control system is EIT irrigation control system is a data collection and SCADA based control system which utilizes EIT data telemetry products as well as third-party supplied soil sensors for monitoring and scheduling irrigation activities. The system is designed for flexibility and ease of use. The Human Machine Interface (HMI) provides easy to use functions for setting the selecting irrigation modes and soil moisture set points. The system comprises of three main components. These are the central PC, a sensor for monitoring soil moisture and telemetry for data collection, valve, and pump control.

Sensors: Sensors are the most important component of a precision irrigation control system which provides the desired information for the control of the sprinkler nozzles/emitters, control valve, etc. Field environmental sensors and soil moisture sensors are the most common type of environmental sensor employed for determining a crop's water requirements. However, sensors for ambient temperature and humidity in the crop's field are also common. As stated above, full weather stations may even be included in local sensors. Sensors are strategically located at a number of points within a crop's field in a way that covers variations in soil type and climate. Pressure transducers may also be employed in the field for monitoring the water pressure of irrigation zones. For crops that require continuous flood conditions, such as rice, water level sensors at various points in the field may be used. They may be used as direct real-time feedback for automatic controls (discussed below) and/or data collection and logging. Sensor data collection sensors may be queried manually or automatically by a data collection system. Automatic data collection systems will query at regular intervals (generally every 5–15 minutes or so) and then log the data into a database for subsequent reference. Also, automatic data collection systems generally require a wireless communications network of very low power data collection nodes with solar cells and rechargeable batteries. Any node within the network may have one to several sensors attached. Some nodes may be used only as a communication relay within the wireless network. In addition to the wireless nodes, the network may also include switching hubs, routers, and gateways. Viewing of real-time data as well as data in the database archive may be limited to a local network on the farm or may be accessible from the Internet.

10.2.1 OPTIONAL SYSTEM COMPONENTS

10.2.1.1 LOCATION AND ALIGNMENT

The control system for most site-specific application systems use some form of spatiality indexed data to determine the appropriate application rate for specific sites. The basis for these spatially indexed data is typically a widely accepted geo-reference system, such as latitude and longitude. Consequently, it is necessary to know the precise location of all elements of the application system at all times during operation if accurate site-specific applications are expected. Various approaches have been used, but the greatest challenge is cost. Although it is often desirable to have multiple location sensors along the truss length of a moving irrigation system, the cost would be prohibitive. A general solution has been to use one or two sensors to locate one or both ends of the moving system and to calculate the location intermediate location points. Because moving irrigation systems consist of multiple segments or spans, with each end of the span moving independently but within confined limits, the truss is not always linear. This is not a significant problem for small systems, but misalignment can be significant for a large system. Fortunately, in many cases, the shape of the truss is consistent, predictable, and describable for specific operational conditions. The determination of precise locations for lateral move system is similar to that of center pivot systems concept that is more difficult because both ends move. The laser alignment system is also used for such a system. Although the travel path is constrained by the guidance system, some variation usually exists in repeatability. The issues of tissue misalignment are similar for a lateral move and center-pivot systems. In general, more sensors are required for determining locations in lateral move systems than in center pivot systems. Most lateral move precision irrigation systems have used one or more GPS sensor to determine location.

10.2.1.2 MANAGEMENT DATABASE AND DECISION SUPPORT

Any information to be used in the precision management application must be indexed by its geographical location and stored in an electronic format that can be readily accessed by a computer or computer-based controller. As such, it operates as special purpose GIS. This management database houses the data with which the control system operates the irrigation machines, records the actual application amounts for later use or documentation, and provides the framework on which a decision support system can operate. Data that may be stored could include user-entered soil characteristics, cultural operations, or application maps. It could also include historical geo-referenced data such as yield maps, past application maps, or cultural histories. Spatial arrays of sensors either mounted on the system or in the field could potentially feed information directly into the management database.

10.2.1.3   VARIABLE WATER SUPPLY

Most conventional moving irrigation systems are designed for and operate with a constant water flow rate and pressure to the system in which all sprinklers operate most of the time. With precision irrigation, in which variable flow rates are required for several management zones within the total system, water must be applied to the system at constant pressure but at a variable flow rate. The magnitude of the flow rate variance depends upon the system design and operation characteristics, but in extreme cases, it can vary from full design flow rate to almost zero.

10.3   FERTIGATION FOR PRECISION MANAGEMENT OF NUTRIENTS

The precise management of nutrients in the soil is possible through application of the right amount of fertilizers /nutrients and other chemicals at the right time, at the right place and in the right manner to achieve high yield and the quality of produce along with minimum / no loss to the groundwater caused due to nutrient leaching. Fertigation (application of fertilizer/chemical solution with irrigation) has the potential to ensure that the right combination of water and nutrients is available at the root zone, satisfying the plants total and temporal requirement of these two inputs. Fertilizer application through irrigation can be conducted using a micro (drip/trickle), surface (border, basin, and furrow), pipe, and sprinkler irrigation systems. Micro and subsurface system of irrigation can only be used for fertigation of soil-applied agricultural fertilizers/ chemicals. Surface irrigation methods can, at times, present problems with the uniformity of fertilizer/chemical application and may limit some chemical applications. Sprinkler irrigation system (impact, rain gun, pop-up, center pivots, lateral move, etc.) can be used both for soil and over canopy/ foliar application of nutrients/chemicals.

There are various benefit and risk associated with the application of fertilizers, chemicals, and other nutrients under fertigation/chemigation/ nutrition. Proper management and efficient application of these nutrients/ chemicals offer a significant saving of these chemicals, better scheduling as per crop need, improvement in yield and nutrient use efficiency and less impact on the environment which is an important component of sustainable soil management. The main drawbacks associated with fertigation are the initial set-up costs and the need to monitor the operation carefully to ensure that irrigation and injection systems are working correctly. Water quality can limit the use of fertigation; irrigation waters that are high in salts are not suitable for fertigation. Generally, the concentration of salts in the fertigation solution should not exceed 3000 micro Siemens per centimeter (μs/cm). The most significant risk when utilizing fertigation/chemigation is for water source contamination due to back siphoning, backpressure, over-irrigation and untimely application of N fertilizers, which reduces the efficiency of fertilizer use and compounds N losses to the environment (Ng Kee Kwong and Devile, 1987). In addition, nutrient depletion within plant root zone and soil acidification were reported (Peryea and Burrows, 1999; Mmolawa and Or, 2000; Neilsen et al., 2004) if proper care has not taken during fertigation. Fertigation may favor NO_3-N leaching, which requires a careful calculation of the fertilizer dose to minimize the risk of groundwater contamination.

A well-designed fertigation system can reduce fertilizer application costs considerably and supply nutrients in precise and uniform amounts to the wetted irrigation zone around the tree where the active feeder roots are concentrated. Applying timely doses of small amounts of nutrients to the trees throughout the growing season has significant advantages over conventional fertilizer practices. Fertigation saves fertilizer as it permits applying fertilizer in small quantities at a time matching with the plants nutrient need. Besides, it is considered eco-friendly as it avoids leaching of fertilizers. Liquid fertilizers are best suited for fertigation. In India, inadequate availability and the high cost of liquid fertilizers restrict their uses. Fertigation using granular fertilizers poses several problems namely, their different levels of solubility in water, compatibility among different fertilizers and filtration of undissolved fertilizers and impurities. Different granular fertilizers have different solubility in water. When the solutions of two or more fertilizers are mixed together, one or more of them may tend to precipitate if the fertilizers are not compatible with each other. Therefore, such fertilizers may be unsuitable for simultaneous application through fertigation and would have to be used separately. This article reports on the various issues of fertigation, i.e., advantages, and limitations, selection of water-soluble fertilizers (granular and liquid), fertigation scheduling in various crops and fertigation system for efficient fertigation programme and response of plants to fertigation; and it is economic.

10.3.1   IMPORTANCE OF FERTIGATION

- The fertigation allows applying the nutrients exactly and uniformly only to the wetted root volume, where the active roots are concentrated, which eliminates the over and underfeeding of nutrients.
- Remarkably increase in the application efficiency of the fertilizer, which saves the significant amount of fertilizers.
- Reduction in the production costs and groundwater pollution caused by the fertilizer leaching.
- Fertigation allows adapting the amount and concentration of the applied nutrients in order to meet the actual nutritional requirement of the crop throughout the growing season. Timely application of fertilizers and chemical significantly increases the crop yield.
- Fertigation reduces the operator's exposures to fertilizers and chemicals.
- Mechanical damage to crop by manual/tractors application of fertilizers is reduced by fertigation.
- Many fertilizers/chemical required to be placed at a particular location in the plant root zone which is possible through fertigation.
- Saving of energy and labor.
- Flexibility of the moment of the application (nutrients can be applied to the soil when crop or soil conditions would otherwise prohibit entry into the field with conventional equipment).
- Convenient use of compound and ready-mix nutrient solutions also containing small concentrations of micronutrients which are otherwise very difficult to apply accurately to the soil, and
- The supply of nutrients can be more carefully regulated and monitored. When fertigation is applied through the drip irrigation system, crop foliage can be kept dry thus avoiding leaf burn and delaying the development of plant pathogens.

10.3.2   SELECTION OF FERTILIZERS

Effective fertigation requires an understanding of plant growth behavior including nutrient requirements and rooting patterns, soil chemistry such as solubility and mobility of the nutrients, fertilizers chemistry (mixing compatibility, precipitation, clogging, and corrosion) and water quality factors including pH, salt, and sodium hazards, and toxic ions. The granular and liquid fertilizers used in fertigation are available in various chemical formulations, solubility, and different types of coatings; therefore the selection of fertilizer is an important issue in fertigation programme. The selection of granular fertilizer will depend on the nutrient that is applied, fertilizer solubility and ease of handling.

An essential pre-requisite for the solid fertilizer use in fertigation is its complete dissolution in the irrigation water. Examples of highly soluble fertilizers appropriate for their use in fertigation are: ammonium nitrate, potassium chloride, potassium nitrate, urea, ammonium monophosphate, and potassium monophosphate. The solubility of fertilizers depends on the temperature. The fertilizer solutions stored during the summer form precipitates when the temperatures decrease in the autumn, due to the diminution of the solubility with low temperatures. Therefore, it is recommended to dilute the solutions stored at the end of the summer. Fertilizer solutions of smaller degree specially formulated by the manufacturers are used during the winter.

To ensure the fertilizers selected will not precipitate in irrigation pipe, mix the fertilizer solution with a sample of irrigation water in the same proportions as and when they are mixed in the irrigation system. If the chemical stays in the solution, then the product is safe to use under fertigation. Liquid fertilizers offer many intrinsic advantages over granular fertilizers for fertigation. It is considered as most suitable for of fertigation under micro and sprinkler irrigation. These are available as fertilizers solutions and suspension, both of which may contain single or multi-nutrient materials. To avoid damage to the plant roots high fertilizer concentrations, the fertilizer concentration in irrigation water should not exceed 5%. Although, the susceptibility to root burning from concentrated fertilizers varies with crops, fertilizers, and accompanying irrigation practices. Therefore, it is safer to keep the fertilizer concentration of 1–2% in the irrigation water during fertigation.

10.3.3   IMPORTANT POINTS TO BE CONSIDERED DURING FERTIGATION

10.3.3.1   WATER QUALITY

The interaction of water having pH values (7.2–8.5) with fertilizers can cause diverse problems, such as the formation of precipitates in the fertilization tank and clogging of the drippers and filters. In waters with high calcium content and bicarbonates, use of sulfate fertilizers causes the precipitation of $CaSO_4$ obtruding drippers and filters. The use of urea induces the precipitation of $CaCO_3$ because the urea increases pH. The presence of high concentrations of calcium and magnesium and high pH values lead to the precipitation of calcium and magnesium phosphates. Recycled waters are particularly susceptible to precipitation due to its high bicarbonate and organic matter content. The resultant precipitates are deposited on pipe walls and in orifices of drippers and can completely plug the irrigation system. At the same time, P supply to the roots is impaired. When choosing P fertilizers for fertigation with high calcium and magnesium concentrations, acid P fertilizers (phosphoric acid or monoammonium phosphate) are recommended.

Fertigation under saline conditions: Crops vary widely in their tolerance to plants, reference tables are available defining individual crop sensitivity to total soluble salts and individual toxic ions (Maas and Hoffman, 1977). When brackish waters are used for irrigation, we must bear in mind that fertilizers are salts and Therefore, they contribute to the increase of the EC of the irrigation water. Nonetheless, calculation of the contribution of chloride from KCl to the overall load of chloride from irrigation water shows its relative by low share (Tarchitzky and Magen, 1997). When irrigation water has an EC > 2 dS/m (with high salinization hazard), and crop is sensitive to salinity, we must decrease the amount of accompanying ions added with the N or K. For example, in avocado—a very sensitive crop to chloride—KNO_3 is preferred on KCl to avoid Cl accumulation in the soil solution. This practice diminishes leaf burning caused by Cl excess. Also in greenhouse crops grown in containers with a very restricted root volume, we must choose fertilizers with low salt index. Sodium fertilizers as $NaNO_3$ or NaH_2PO_4 are unsuitable due to the adverse effect of sodium on the hydraulic conductivity and the performance of the plant. A correct irrigation management under saline conditions includes water application over the evaporation needs of the crop, so that there is excess water to pass through and beyond the root zone and to carry away salts with it. This leaching prevents excessive salt accumulation in the root zone and is referred to as leaching requirement (Rhoades and Loveday, 1990).

10.3.3.2 FERTILIZERS COMPATIBILITY

When preparing fertilizer solutions for fertigation, some fertilizers must not be mixed together. For example, the mixture of $(NH_4)_2SO_4$ and KCl in the tank considerably reduce the solubility of the mixture due to the K_2SO_4 formation. Other forbidden mixtures are:

- Calcium nitrate with any phosphates or sulfates.
- Magnesium sulfate with di- or mono- ammonium phosphate.
- Phosphoric acid with iron, zinc, copper, and manganese sulfates.

10.3.4 FERTIGATION SYSTEM

System EU shall not be less than 85 percent where fertilizer or pesticides are applied through the system. Injectors (chemical, fertilizer, or pesticides) and other automatic operating equipment shall be located adjacent to the pump and power unit, placed in accordance with the manufacturer's recommendation and include integrated backflow prevention protection. Fertigation/nutrigation/chemigation shall be accomplished in the minimum length of time needed to deliver the chemicals and flush the pipelines. Application amounts shall be limited to the minimum amount necessary, as recommended by the chemical label. A number of different techniques are used to introduce the fertilizers into the irrigation system. Generally, fertilizers are injected into irrigation systems by three principal methods namely, (1) fertilizer tank (the by-pass system), (2) the venturi pump and (3) the injection pump (piston or destron pump). Non-corrosive material should be used for the fertilizer containers and for the injection equipment.

10.4 CONCLUSIONS

The precision farming and hi-tech agriculture for the improved input use efficiency, more yield, and quality produce in a sustainable manner is incomplete without efficient irrigation and fertilizer application techniques. Application of water at the right time, in the right amount at the right place with right manner is the crux of precision irrigation. Fertigation is the most efficient techniques of fertilizer application in a split manner as per crop need at different stages of crop development. The fertilizers used for fertigation must be 100% water soluble, and it should not precipitate while making the solution. The selection of fertilizers should be in accordance with the pH of the soil. The concentration of fertilizer should not be more than 2% during fertigation. The fertilizer requirement should be determined based on soil analysis and if soil analysis not possible certain correction factors must be applied depending on the soil texture. The fertilizer use efficiency can be maximized by adopting drip-fertigation to the level of 95%. The drip- fertigation system should be appropriately designed to achieve at least 85% of uniformity of water and fertilizer application. The piston pump type of fertigation unit is the most efficient system of fertilizer application. However, fertilizer tank and ventury type of fertigation unit are most common at the farmer's field because of its low cost.

KEYWORDS

- **fertigation**
- **micro irrigation**
- **saline condition**
- **sprinklers**

REFERENCES

Camp, C. R., Sadler, E. J., & Evans, R. G., (2006). Precision water management: Current realities, possibilities and trends In: *Handbook of Precision Agriculture*, Haworth, 153–185.

Lu, Y. C., Camp, C. R., & Sadler, E. J., (2004a). Efficient allocation of irrigation water and nitrogen fertilizer in corn production. *Journal of Sustainable Agriculture, 24*(4), 97–111.

Lu, Y. C., Camp, C. R., & Sadler, E. J., (2004b). Optimal levels of irrigation in corn production in the southeast coastal plains. *Journal of Sustainable Agriculture, 24*(1), 95–106.

Lu, Y. C., Sadler, E. J., & Camp, C. R., (2005). Economic feasibility study of variable irrigation of corn production in southeast coastal plains. *Journal of Sustainable Agriculture, 26*(3), 69–81.

Maas, E. V., & Hoffman, G. J., (1977). Crop salt tolerance - current assessment. *J. Irrig. Drainage Div. ASEC, 103*, 115–134.

Malhotra, S. K., (2017). *Initiative and Option in Transition for Doubling Farmer's Income*. Keynote lecture delivered at the occasion of National Conference on "Technological changes and innovations in agriculture for enhancing farmer's income," held at JAU, Junagadh, Gujrat during May 28–30.

Mmolawa, K., & Or, D., (2000). Root zone salute dynamics under drip irrigation: A review. *Plant Soil, 222,* 163–190.

Neilsen, G. H., Neilsen, D., Herbert, L. C., & Hogue, E. J., (2004). Response of apple to fertigation of N and K under conditions susceptible to the development of K deficiency. *J. Am. Soc. Hortic. Sci., 129,* 26–31.

Ng Kee, K. K. F., & Devile, J., (1987). Residual nitrogen as influenced by the timing and nitrogen forms in silty clay soil under sugarcane in Mauritius. *Fertil. Res., 14,* 219–226.

Peryea, F. J., & Burrows, R. L., (1999). Soil acidification caused by four commercial nitrozen fertilizer solutions and subsequent soil pH rebound. *Commun. Soil Sci. Plant Anl., 30,* 525–533.

Rhoades, J. D., & Loveday, J., (1990). Salinity in irrigated agriculture. In: Stewars, B. A., & Nielsen, D. R., (eds.), *Irrigation of Agricultural Crops* (pp. 1089–1142). ASA-CSAA-SSSA, Madison, WI.

Tarchitzky, J., & Magen, H., (1997). *Status of Potassium in Soils and Crops in Israel, Present K Use Indicating the Need for Further Research and Improved Recommendations.* Presented at the IPI Regional Workshop on Food Security in the WANA Region, May, Bornova, Turkey.

CHAPTER 11

IDENTIFICATION OF URBAN HEAT ISLANDS FROM MULTI-TEMPORAL MODIS LAND SURFACE TEMPERATURE DATA: A CASE STUDY OF THE SOUTHERN PART OF WEST BENGAL, INDIA

PRITI KUMARI[1], NAVAL KISHOR YADAV[2], ABHISEK SANTRA[2], and UTKARSH UPADHAYAY[3]

[1]*Department of Civil Engineering, GGSESTC, Bokaro, Jharkhand, India, E-mail: priti.9407@gmail.com*

[2]*Department of Civil Engineering, Haldia Institute of Technology, Haldia, West Bengal, India*

[3]*Centre for Water Engineering and Management, Central University of Jharkhand, Ranchi, Jharkhand, India*

ABSTRACT

Land surface temperature (LST) is the temperature of the earth's ground surface and is crucial for climate studies in different aspects like hydrology, geology, engineering, phenology, etc. Research on Urban Heat Island (UHI) is mainly dependent upon the LST studies and is highly useful for studying its impact over the surrounding environment, precipitation, and water quality of the area. An attempt has been made here to study the spatiotemporal dynamics of the LST and associated UHIs in the southern part of West Bengal, India. The authors extracted the major UHIs from mean monthly time series MODIS Land Surface Temperature datasets for the period from 2010 to 2015 and identified the pattern of change of LST in the study area.

It has been observed that the temperature values of the four summer months have less variance with the mean summer temperature values. The range of temperature adopted is below 30°C, 30–35°C, 35–40°C, and 40°C onwards. The prevailing aridity shows the high LST in and around Purulia district. The year 2010 and 2014 shows the high spatial distribution of the very high-temperature class probably due to environmental and climatic reasons. The result also reveals a four-yearly cyclic pattern of LST change in the study area and two heat islands in and around Kolkata and Haldia areas. Among the urban agglomerations, only two Kolkata-Howrah and Haldia areas can be visible as UHI at the coarse spatial resolution from the thermal image. The UHI patterns have also been correlated with the classified land use and land cover information of the same area.

11.1   INTRODUCTION

The temperature of an area is greatly affected by its land use/land cover. Over the past several decades, the global process of urbanization has progressed dramatically rapid, thus gave rise to many problems for the urban environment and climate. When a large fraction of natural land cover in an area are replaced by a built surface, it traps incoming solar radiation during the day and reradiates at night, the resulting phenomenon is known as Urban Heat Island (UHI). UHI was considered as the most well-documented example of anthropogenic climate modification within the field of urban climate (Arnfield, 2003). In recent years UHI has become a topic of great interest both among the academicians and the governing bodies. Researchers are interested in understanding the various aspects of this phenomenon including its causes (Huang et al., 2011), impacts (Imhoff et al., 2010) and complexity (Mirzaei and Haghighat, 2010). UHI is increasingly gaining interest as it directly affects both environmental (Ferguson and Woodburry, 2007; Sharma et al., 2012) as well as human health (Lo and Quattrochi, 2003; Tomilson et al., 2011). UHI affects the environment through heat pollution (Papanasta-siou and Kittas, 2012), it increases the number of smog events (Sham et al., 2012), higher energy consumptions (Kolokotroni et al., 2012), while from human health perspectives it causes a larger number of heat-related health problems (Harlan and Ruddell, 2011) and adversely impact human comfort (Steeneveld et al., 2011). Therefore, it is very essential to understand the UHI phenomena and its ill effect on the environment in order to mitigate this effect to some extent for the betterment of the environment and the human race. UHI manifests itself in two basic forms (i) the Surface UHI (SUHI) and

(ii) the Atmospheric UHI (AUHI). SUHI is the phenomenon of the temperature difference between surfaces of urban and surrounding rural areas. The phenomenon revealed high spatial and temporal variability (Stathopoulou and Cartalis, 2009). SUHI is studied using land surface temperature (LST) regained from thermal satellite sensors (Schwarz et al., 2011). AUHI includes the difference in the pattern of air temperature between urban and rural settings. AUHI further falls in one of the two categories viz., Canopy layer or Boundary layer. The atmosphere is extending from the surface to mean building height or tree canopy are influenced by Canopy layer UHI, while the Boundary layer UHI accounts for air beyond canopy layer (Weng, 2003). AUHI is studied using meteorological data (Saaroni et al., 2000). Such data have frequently been explored to study monthly or seasonal variations in AUHI (Jongtanom et al., 2011; Cayan and Douglas, 1984; Fujibe, 2009; Jauregui, 1997; Liu et al., 2007; Gaffin et al., 2008; Hua et al., 2008). Gallo and Owen (1999) for illustration analyzed urban-rural temperature (using observation stations) and Normalized Difference Vegetation Index (NDVI) (using NOAA-AVHRR, National Oceanic and Atmospheric Administration–Advanced Very High-Resolution Radiometer) of 28 cities on a monthly and seasonal basis to study UHI. UHI intensity in Bangkok, Chiang Mai, and Songkhla cities were investigated using urban and rural meteorological station data (Jongtanom et al., 2011). Cayan and Douglas (1984), Fujibe (2009), Jauregui (1997), Liu et al., (2007), Gaffin et al., (2008), and Hua et al., (2008) have also conducted similar researches. Thus it is evident that detailed work is available on seasonal studies of AUHI, but such detailed literature for SUHI is lacking. However, a dearth of such literature exists for SUHI studies. Seasonal analysis of SUHI requires temporal LST (Hu and Brunsell, 2013; Buyantuyev and Wu, 2010). Very few and recent work is available on SUHI seasonal analysis with a huge gap in this research existing for Indian cities. Researchers have proposed various approaches to quantify UHI in terms of intensity or area. Keramitsoglou et al., (2011) have used the difference between LST and reference LST (RLST) to assess UHI intensity. Zhang et al., (2009) calculated LST differences between different impervious surface area categories and water as an estimate for UHI intensity. Zhang and Wang (2008) proposed hot island area (HIA) as UHI intensity estimate that is based on standard deviation segmentation of LST image. In this chapter, an attempt has been made to identify the spatio-temporal pattern of summertime UHI of the southern part of West Bengal. Also, the surface extents of the UHIs have been compared from the land use land cover information of the area.

11.2   MATERIALS AND METHODS

11.2.1   STUDY AREA

The study area is south Bengal and mainly focused on Kolkata, North, and South 24 Parganas, East Midnapur, West Midnapur, Birbhum, Bankura, Purulia, Hooghly, Howrah, and Durgapur districts. In south Bengal, there are several town and villages with a high population (such as Kolkata) and industrial hubs like Haldia, Kharagpur, and Durgapur, etc. The study area (West Bengal) is located about 17 feet above MSL, and it ranks 14th as per land area and 4th in population (India). The highest day temperature of the area ranges from 38°C to 45°C. This high temperature in summer is a combined consequence of high population density and never-ending emission of industries. As urbanization and industrialization will continue, it is very important to identify Heat Islands because the further increase in population and industries in those areas may lead to bad consequences over agriculture, living life and climatic cycle (Figure 11.1).

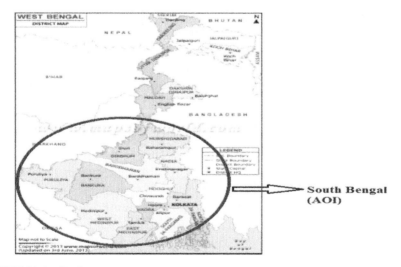

FIGURE 11.1   Study area.

11.2.2   DATABASE AND METHODOLOGY

LST of the southern part of the Bengal is estimated from the MODIS monthly LST images acquired for the period from 2010 to 2015. The data has been freely acquired from NASA Earth Observation (NEO) website. The spatial resolution of the data is 11.132 km (0.1 degrees). The data is a geo-referenced product in geographic latitude/longitude coordinate system and WGS84 datum plane. Since UHI is distinct in the summer season, only four months from March to June have been considered for the identification and understanding of the dynamics of UHI for the study area. The downloaded data were stacked considering four months for each year. After that, data for each year consisting of four layers have been clipped with the exact boundary of the study area using the ERDAS Imagine software package. The exact boundary of the study area (AOI) has been digitized and extracted from the published maps of the Census of India (2011). ArcGIS software package has been used to reclassify the set of images in terms of different LST classes. After that, the classified image of the study area collected from ENVISAT MERIS Globcover was considered to cross-validate the spatial locations of UHIs. The classified data also clipped using the same study area of interest (AOI) layer in Erdas Imagine software. The methodology flow diagram (Figure 11.2) simplifies the broad framework of the research.

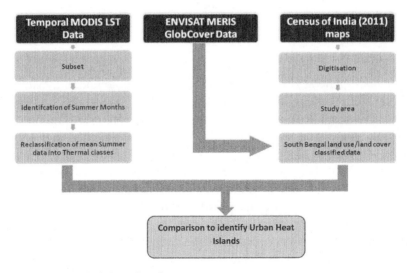

FIGURE 11.2   Methodology flowchart.

11.3   RESULTS AND DISCUSSIONS

As a good indicator of the earth's energy balance at the surface of the earth, LST controls the physics of the land surface processes directly or indirectly both on a global and regional scale. It integrates the surface atmospheric interactions and energy fluxes between the atmosphere and the ground surface. The reclassified area gives the number of pixels contained in each temperature range. It has been observed that the temperature values of the four summer months have less variance with the mean summer temperature values. The mean summer temperature of the study area for 2010 to 2015 was calculated using the modeler tool of Erdas Imagine software package. The range of temperature adopted is below 30°C, 30–35°C, 35–40°C and 40°C onwards. Here it would be more relevant to draw the comparison among the class 4 area values as this class is a high-temperature range (Table 11.1).

TABLE 11.1   Year Wise Land Surface Temperature Values for the Highest Land Surface Temperature Class

Year	Area under class 4 (km2)
2010	16481.36
2011	991.36
2012	1239.20
2013	7806.96
2014	15985.68
2015	247.84

The maps generated (Figure 11.3) show the spatiotemporal dynamics of the LST. The prevailing aridity shows the high LST in and around Puruliya district. The years 2010 and 2014 show the high spatial distribution of the very high-temperature class probably due to environmental and climatic reasons. The results also reveal a four-yearly cyclic pattern of LST change in the study area. However, it is clear from the Figure 11.3; two heat islands have evolved in and around Kolkata and Haldia areas surrounded by cooler temperature. These two are the possible results of high urbanization and industrial development in those areas.

The land use and land cover data of the area depicts mostly cultivated and irrigated lands (Figure 11.4). Some patches of forested lands can also be seen. Mangroves dominate at the lower parts of South 24 Parganas district. Four urban areas can be distinctly located at Kolkata – Howrah, Barddhaman – Asansol – Durgapur, Kharagpur – Medinipur and Haldia. These four areas are of primary concern from UHI point of view. However, among these four urban agglomerations, only two – Kolkata – Howrah, and Haldia areas can be visible as UHIs at the coarse spatial resolution from the thermal images. The prevailing aridity and small scale of the data are probably hindering the visibility of the other two urban regimes and heat islands. However, high spatial resolution thermal images at the larger scale may isolate them as heat islands.

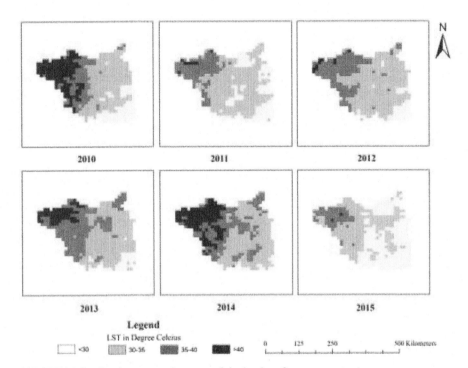

FIGURE 11.3   Spatio-temporal pattern of the land surface temperature.

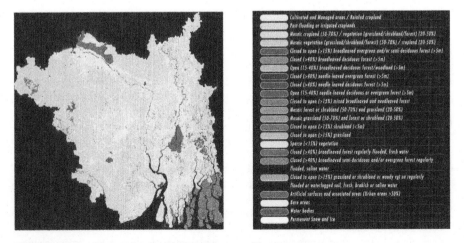

FIGURE 11.4   (See color insert.) Land use and land cover.

11.4 CONCLUSIONS

The present work demonstrates the importance of thermal remote sensing as a valuable source of temperature information for cities and urban agglomerations. However, the scale of the images is the first and foremost criteria in this regard to fill up the gaps of lack of information. The paper identifies the spatiotemporal changes of LST in this area. It also pinpoints the impact of summer months on LST. However, for a large area, the coarser resolution images may be used to identify probable UHIs. After that, the high-resolution satellite images with thermal bands may be applied to see the small UHIs within an urban area. In this way, the temperature characteristics of the cities can be identified in relation to the urban land use and expansion. The relevance of such studies is not immediately obvious and deserves closer inspection.

KEYWORDS

- Arc-GIS
- globcover
- land surface temperature
- MODIS
- Urban Heat Island

REFERENCES

Arnfield, A. J., (2003). Two decades of urban climate research: A review of turbulence, exchanges of energy and water, and the urban heat island. *International Journal of Climatology, 23,* 1–26.

Buyantuyev, A., & Wu, J., (2010). Urban heat islands and landscape heterogeneity: Linking spatiotemporal variations in surface temperatures to land-cover and socioeconomic patterns. *Landscape Ecol., 25,* 17–33.

Cayan, D. R., & Douglas, A. V., (1984). Urban influences on surface temperatures in the southwestern United States during recent decades. *J. Climate Appl. Meteorol., 23,* 1520–15330.

Ferguson, G., & Woodbury, A. D., (2007). Urban heat island in the subsurface. *Geophys. Res. Lett., 34,* 2192–2201.

Fujibe, F., (2009). Urban warming in Japanese cities and its relation to climate change monitoring. In: *The Seventh International Conference on Urban Climate,* Yokohama, Japan.

Gaffin, S. R., Rosenweig, C., Khanbilvardi, R., Parshall, L., & Mahani, S., (2008). Variations in New York city's urban heat island strength over time and space. *Theor. Appl. Climatol.*, *94*, 1–11.

Gallo, K. P., & Owen, T. W., (1999). Satellite-based adjustments for the urban heat island temperature bias. *J. Appl. Meteorol.*, *38*, 806–813.

Hu, L., & Brunsell, N. A., (2013). The impact of temporal aggregation of land surface temperature data for surface urban heat island (SUHI) monitoring. *Remote Sens. Environ.*, *134*, 162–174.

Hua, L. J., Ma, Z. G., & Guo, W. D., (2008). The impact of urbanization on air temperature across China. *Theor. Appl. Climatol.*, *93*, 179–194.

Huang, G., Zhou, W., & Cadenasso, M. L., (2011). Is everyone hot in the city? Spatial pattern of land surface temperatures, land cover and neighborhood socioeconomic characteristics in Baltimore, MD. *J. Environ. Manage.*, *92*, 1753–1759.

Imhoff, M. L., Zhang, P., Wolfe, R. E., & Bounoua, L., (2010). Remote sensing of the urban heat island effect across biomes in the continental USA. *Remote Sens. Environ.*, *114*, 504–513.

Jauregui, E., (1997). Heat island development in Mexico City. *Atmos. Environ.*, *31*, 3821–3831.

Jongtanom, Y., Kositanont, C., & Baulert, S., (2011). Temporal variations of urban heat island intensity in three major cities, Thailand. *Mod. Appl. Sci.*, *5*, 105–110.

Keramitsoglou, I., Kiranoudis, C. T., Ceriola, G., Weng, Q., & Rajasekar, U., (2011). Identification and analysis of urban surface temperature patterns in Greater Athens, Greece, using MODIS imagery. *Remote Sens. Environ.*, *115*, 3080–3090.

Liu, W., Ji, C., Zhong, J., Jiang, X., & Zheng, Z., (2007). Temporal characteristics of the Beijing urban heat island. *Theor. Appl. Climatol.*, *87*, 213–221.

Lo, C. P., & Quattrochi, D. A., (2003). Land-use and land-cover change, urban heat island phenomenon, and health implications: A remote sensing approach. *Photogramm. Eng. Remote Sens.*, *69*, 1053–1063.

Mirzaei, P. A., & Haghighat, F., (2010). Approaches to study urban heat island – abilities and limitations. *Build. Environ.*, *45*, L23743(1–4).

Papanastasiou, D. K., & Kittas, C., (2012). Maximum urban heat island intensity in a medium-sized coastal Mediterranean city. *Theor. Appl. Climatol.*, *107*, 407–416.

Sham, J. F. C., Lo, T. Y. L., & Memon, S. A., (2012). Verification and application of continuous surface temperature monitoring technique for investigation of nocturnal sensible heat release characteristics by building fabrics. *Energy Build.*, *53*, 108–116.

Sharma, R., & Joshi, P. K., (2012). Monitoring urban landscape dynamics over Delhi (India) using remote sensing (1998–2011) inputs. *J. Ind. Soc. Remote Sens.*, 1–10.

Tomilson, C. J., Chapman, L., Thomes, J., & Baker, C. J., (2011). Including the urban heat island in spatial heat health risk assessment strategies: a case study for Birmingham, UK. *Int. J. Health Geographics.*, *10*, 14.

Weng, Q., (2003). Fractal analysis of satellite-detected urban heat island effect. *Photogramm. Eng. Remote Sens.*, *69*, 555–566.

Zhang, Y., Odeh, I. O. A., & Han, C., (2009). Bi-temporal characterization of land surface temperature in relation to impervious surface area, NDVI and NDBI, using a sub-pixel image analysis. *Int. J. Appl. Earth Obs. Geoinf.*, *11*, 256–264.

CHAPTER 12

DERIVATION OF AN OPTIMAL OPERATION POLICY OF A MULTIPURPOSE RESERVOIR

PRABEER KUMAR PARHI

Centre for Water Engineering and Management, Central University of Jharkhand, Jharkhand, India
E-mail: prabeer11@yahoo.co.in

ABSTRACT

Zone-wise operating policy for conservation and flood control zones has been developed for Maithon Reservoir, on Barakar River, a major tributary of Damodar River, which is managed by Damodar Valley Corporation (DVC) in India. The operating policy for conservation and flood control zones has been designed respectively to confirm adequate water supply to different sectors and for real-time flood protection/management. For design, the concept of hedging rule, i.e., rationing of water through Monte-Carlo simulation (MCS) is applied. Further three storage performance indicators, i.e., reliability, resilience, and vulnerability have been tested. It is observed that the three system performance indicators are satisfactory.

12.1 INTRODUCTION

Multipurpose reservoirs are designed and operated to satisfy domestic and industrial water supply, irrigation, hydropower generation, navigation, recreation, environmental flow as well as flood control requirements. Most of the reservoirs are divided into three storage zones (a) dead storage zone, (b) conservation zone, and (c) flood control zone having each zone have its specific objective and operating characteristic. Among the above storage zones, the conservation zone is responsible for water supply to different sectors and organizations whereas flood control zone is used to control and mitigate the flood in monsoon period.

In the available literature, many researchers have discussed on a different policy of reservoir operations like Standard Operating Policy (SOP), Linear Decision Rule (LDR), Pack Rule, Space Rule and different forms of the Hedging Rule in the area of water supply for conservation zone of the multipurpose reservoir. Among the above policies, SOP (Loucks et al., 1981; Stedinger, 1984; Marien et al., 1994) is the simplest policy which aims at releasing the water according to the demand (if possible) and preserve water for future needs if the water is available only after meeting the demands. However, SOP neglects the consideration of potential shortage vulnerability during later periods which is the main limitation of this rule. To overcome this limitation hedging rule (Masse, 1946; Bower et al., 1962; Klemes, 1977; Loucks et al., 1981; Shih and ReVelle, 1994, 1995) came into operation which is based on equal value margin principle. In a recent discussion, Draper and Lund (2004) explained the hedging rule analytically as, at optimality, the marginal benefit of storage (S) must equal the marginal benefits of release (D), which can be expressed as:

$$\frac{\partial C(S)}{\partial S} = \frac{\partial B(D)}{\partial D} \tag{1}$$

where, C(S) = carryover storage function, and B(D) = benefit delivery function. This shows that if B(D) is linear, then SOP is the best optimal policy for reservoir operation and if B(D) is non-linear then hedging rule is the optimal policy. Extending this work of Draper and Lund, (2004), You and Cai (2008) explicitly included uncertain future reservoir inflow in the future marginal value expression and discussed factors that influence hedging rule. Incorporating Monte-Carlo Simulation (MCS) with hedging rule with a different parameter of hedging, Srinivasan and Philipose (1998) investigated reservoir storage performance indicators. In the case of reservoirs with limited capacities and significant perennial water supply demands, the conservation zone usually overlaps with the flood control zone. This overlap yields a normal level that is higher than the initial level, thus prompting the necessary pre-release prior to the actual beginning of a flood. One of the most important aspects of mitigating the damaging impacts of floods is the real-time operation of flood control systems. According to Chow and Wu (2010), flood control includes three flood stages, (a) stage prior to flood arrival, in which water releases are to be managed in such a manner that enough reservoir capacity is available for the upcoming flood; (b) stage preceding peak flow, in which floodwater releases are for disaster mitigation; and (c) stage after peak flow, in which releases are to be regulated such that the storage at the end of the flood are available for future use. Further, the performance of a reservoir operation policy is usually expressed in terms of three performance indicators such as reliability, resiliency, and vulnerability (Vogel and Bolognese, 1995; Kundzewicz and Kindler, 1995; Jain and Bhuniya, 2008; Raje and Majumdar, 2010).

In the above context, the present study attempts to: (a) formulate a zone-wise reservoir operation policy such that the available water is rationally used in such a manner (both in conservation zone and flood control zone) that the water requirement for different purpose is satisfied adequately along with flood mitigation; and (b) evaluate different performance indices such as reliability (volume reliability, time reliability, and annual reliability), resiliency, and vulnerability with respect to two important reservoir functions (water supply and flood control). As a case study, the proposed methodology is applied to the Maithon Reservoir, situated in the States of Jharkhand and West Bengal under Damodar Valley Corporation (DVC) in India.

12.2   MATERIALS AND METHODS

For the purpose of analysis the Maithon Reservoir on Barakar River, a major tributary of Damodar River is considered. The reservoir is situated in the States of Jharkhand and West Bengal and managed by DVC in India. The catchment area of the reservoir is 5309.4 km^2 and is primarily designed to control flood during the monsoon and provide irrigation, domestic, and industrial water supply along with hydropower generation. For reservoir analysis, the inflow and outflow time series has been collected and the combined time-series over the time period 1981 to 2013 has been plotted. Figure 12.1 shows the combined time-series of inflow and outflow for the Maithen Reservoir over the time period 1981 to 2013.

12.2.1   METHODOLOGY

12.2.1.1   MONTE-CARLO SIMULATION-BASED HEDGING MODEL

In the present study, a hedging based simulation model has been developed for the Maithon reservoir which states that the marginal benefit of release is equal to the marginal benefit of storage. For a water supply reservoir, the available storage (AS_t) is calculated which is equal to initial storage (S_t) for time period *t* plus current inflow (I_t) for time period *t* minus evaporation loss (Et) for that time which is represented by Eq. (2).

$$ASt = St + It - Et \qquad (2)$$

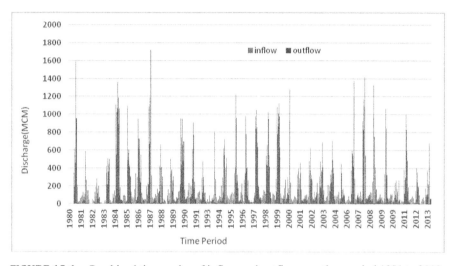

FIGURE 12.1   Combined time-series of inflow and outflow overtime period 1981 to 2013.

The time *t* lies between 0 and K, where K is active reservoir capacity.

This AS_t can be multiplied by hedging factor (H_f) so that some amount of AS_t water can be hedged for future use. The hedging factor can be taken as a discrete random variable (Rv) for the respective month. However, due to the availability of low storage during the months of May to June Rv may be taken as less and relatively more during July to April. In the present study the MCS model has been developed taking hedging factor as random variable divided by 10,000 such that:

$$H_f = Rv/10,000 \qquad (3)$$

If 5000< Rv <10,000, than for July to April 0.5< H_f<1 and if 1000 < Rv < 2000, than for May to June 0.1< H_f<2.

Using MCS, 15 runs are being conducted for different random variables. Figure 12.2 shows the changing trajectory of random variable in MCS. Now available water for release (A_w) is equal to hedging factor multiplied by AS_t.

$$A_w = ASt * H_f \qquad (4)$$

If these random variable change then the value of available water for release will also change, and this can be done through MCS. Now if available water

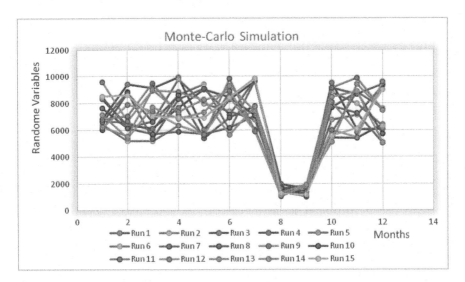

FIGURE 12.2   (See color insert.) Changing the trajectory of a random variable in Monte-Carlo simulation.

for release is less than the total demand (D_t), then water to be released (R_t) is equal to available water for release, and if available water for release is greater than the total demand, then water released is equal to total demand. Mathematically it can be represented as:

If $A_w \leq D_t$, then $R_t = Aw$, thus the hedged water (Hw) = $A_w - ASt *H_f$. And also

If $A_w \geq D_t$, then $R_t = D_t$, thus the hedged water (Hw) = $A_w - R_t$

The above theory implies that the water available at time t is either allocated to the projected demand at a time t or stored in the reservoir for possible future use. The water available after water release (A_{w1}) at a time t is equal to the sum of the water hedged for storage plus available water for release (A_w) minus water released (R_t) in the reservoir at the end of time 't.' This can be expressed as:

$$A_{w1} = A_w - R_t + Hw \tag{5}$$

Now the reservoir storage for next period (S_{t+1}) can be calculated using the hedged water as:

If, $A_{w1} \geq S_{max,}$ the outflow (O_t) = $A_{w1} - S_{max,}$ where S_{max} is the reservoir capacity. And

if, $A_{w1} \leq S_{max}$, the outflow (O_t) = $A_{w1} - S_{max}$. Hence

$$S_{t+1} = A_{w1} - R_t - O_t \tag{6}$$

Eq. (6) predicts AS_t for next time period. This storage will be used to estimate the AS_t of the reservoir for further next time period and so on. Thus the model has been developed through hedging rule for 12 months through MCS by changing the random variable, i.e., hedging factor by 5000 times. The best operating policy for water supply has been chosen among the 5000 iteration results with maximum reliability of Maithon Reservoir.

12.2.1.2   STATISTICAL MODEL FOR FLOOD MITIGATION

The chances of flood occurrence become more when the flood control zone provided is not sufficient to accommodate the expected flood. The flood control zone can be managed optimally by managing the outflow of water to reduce the water level with respect to inflow.

A relationship can be derived between water level (L) and the ratio of outflow to inflow (O_{t+1}/I_t) by multiple linear regression equation from historical data as:

$$(Y_i) = \beta\,(X_i) \tag{7}$$

where i = 1, 2, 3,…., n, Y = water level, $X = (O_{t+1}/I_t)$ and β is constant which is calculated by multiple linear regression equation. The ratio of outflow to inflow at any time t (O_{t+1}/I_t) can be represented by a function 'α' known as flood reduction ratio and is represented by:

$$\alpha = \beta * (O_{t+1}/I_t)$$
$$\text{Or, } O_{t+1} = (\alpha/\beta)*I_t \tag{8}$$

The value of 'α' varies from 0 to 1, such that when the value of 'α' is '0,' $Ot_{+1} = 0$, i.e., all the inflow is retained in the reservoir and when the value of 'α' is '1,' $Ot_{+1} = I_t$ i.e., all the inflow is released as the outflow from reservoir. For flood control, the value of 'α' is calculated by linear interpolation between 0 and 1 at a different level of flood control zone using Eq. (8).

12.2.1.3   PERFORMANCE EVALUATION OF THE MODEL

The performance of the simulation model is evaluated using three statistical performance indices, viz. reliability with respect to water supply and flood control and vulnerability with respect to flood damages and water supply.

Reliability with respect to water supply: The reliability of a system can be described by the frequency or probability that a system is in a satisfactory state. Three reliability indices are generally considered in water resources planning and management.

Time reliability (R_t) is estimated by a number of failure periods (fp) for a particular demand out of the total periods (Tp).

$$R_t = 1 - (fp/Tp) \text{ such that } fp \leq Tp \tag{9}$$

Volume or quantity-based reliability (R_v) is expressed as:

$$R_v = V_s/V_d \tag{10}$$

where V_s is the volume of water supplied and V_d the volume of water demanded during a given period. Annual reliability is an analogous way to express time reliability. But the difference is that the time period is one year. Annual reliability can be estimated by:

$$Ra = 1 - (Fy/Ty) \tag{11}$$

where Fy is the number of failure years when the annual supply is less than the annual demand over a total duration of Ty years.

A vulnerability with respect to flood damages and water supply: The mean vulnerability was simplified by Kjeldsen and Rosbjerg (2004) as:

$$V_{mean} = \frac{1}{M} \sum_{j=1}^{M} Vj \tag{12}$$

where, vj is the excess volume over the maximum live storage (for flood control purpose)/deficit volume than the particular demand (water supply purpose) of the jth failure event and M is the number of failure events.

12.3 RESULTS AND DISCUSSION

12.3.1 MONTE-CARLO SIMULATION-BASED HEDGING

Applying the hedging rule based on MCS to the Maithon reservoir in which 75% of dependable monthly inflow has been taken for model development, 5000 iterations are carried out to choose the best hedging policy among 5000 results. The derived hedging policy based on MCS is shown in Figure 12.3 for Maithon Reservoir. Accordingly, an operation rule curve is designed based on the derived policy through an algorithm that is altered according to the alternative adaptation strategies. The rule curve is designed based on this derived policy through an algorithm that is altered according to the alternative adaptation strategies. This is based on the algorithm as $A = f(Q)$, where

A is a set of monthly reservoir storage levels and $Q = \{q_1, q_2, ..., q_n\}$ is a set of daily inflow sequences. Thus based on this policy, a rule curve is developed for the conservation zone of Maithon Reservoir.

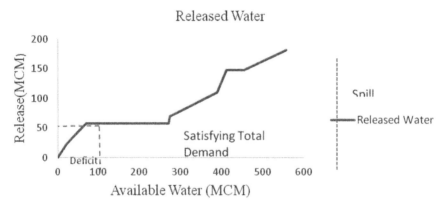

FIGURE 12.3   Hedging policy based on Monte-Carlo simulation for Maithon reservoir.

12.3.2   FLOOD MITIGATION POLICY FOR MAITHON RESERVOIR

Eq. (8) has been derived from a statistical approach is used to derive a flood mitigation policy. In Eq. (8), the value of β and α is calculated according to interpolation at a different level of flood control zone, i.e., between 146.31m and 151m. After putting the ratio of α/β and inflow at a particular height, the value of outflow can be estimated. Table 12.1 shows the different levels of flood control and their corresponding α/β value.

This method of flood mitigation is very simple and easy. As the occurrence of the flood is mainly due to flood peak, this model reduces the flood peak in a manner that the reservoir storage is reduced to optimum storage and optimal water is released to mitigate downstream flooding without allowing water surface level of the reservoir to exceed the acceptable safety level of the flood control zone.

12.3.3   EVALUATION OF PERFORMANCE INDICATORS

The performance of the simulation model has been assessed by estimating various statistical performance indices, i.e., time reliability, volume reliability and annual reliability with respect to water supply, resilience, and vulnerability in connection with water supply and flood damages. From the results, water supply reliability for fulfilling municipal and industrial demands was found to be 100%. The results of other parameters for Monte-Carlo based hedging model are presented in Table 12.2.

TABLE 12.1 Level of Flood Control and Their Corresponding α/β Value

Level (m)	Value of α/β	Level (m)	Value of α/β	Level (m)	Value of α/β	Level (m)	Value of α/β	Level (m)	Value of α/β
146.3	0.000	147.3	0.219	148.3	0.4387	149.3	0.658	150.3	0.877
146.4	0.022	147.4	0.241	148.4	0.4606	149.4	0.680	150.4	0.899
146.5	0.044	147.5	0.255	148.5	0.4826	149.5	0.702	150.5	0.921
146.6	0.066	147.6	0.277	148.6	0.504	149.6	0.724	150.6	0.943
146.7	0.088	147.7	0.298	148.7	0.526	149.7	0.746	150.7	0.965
146.8	0.110	147.8	0.319	148.8	0.548	149.8	0.768	150.8	0.987
146.9	0.132	147.9	0.351	148.9	0.570	149.9	0.790	150.9	1.008993
147	0.154	148	0.373	149	0.592	150	0.812	151	1.030928
147.1	0.175	148.1	0.395	149.1	0.614	150.1	0.834		
147.2	0.197	148.2	0.417	149.2	0.636	150.2	0.855		

TABLE 12.2 Various Performance Indicators for Monte-Carlo Based Hedging Model

Performance Indicators	Reliability (%)		Resilience	Vulnerability (For water supply in MCM)		Vulnerability (For Flood damages in MCM)	
	Volume	Time	Annual	Mean	Max	Mean	Max
MCS based Hedging Model	86.74	83.33	82.35	0.83	126.60	87.80	33.656	33.65

From the above analysis it is found that as more water becomes available to meet the demands, the reliability of the system increases and the system becomes less vulnerable with respect to water supply but more vulnerable with respect to flood damages. As the percentage dependability of inflow increases that is the availability of water decreases the slope of reliability with respect to water supply also decreases but vulnerability with respect to water supply increases and with respect to flood damages decreases. However, the resiliency does not follow the same trend; it measures the relative recovery speed of the system from an unsatisfactory state. Hence as more water becomes available, the unsatisfactory periods are eliminated or shortened. But, as resiliency is a relative measure, if the recovery time is long, relative to the total unsatisfactory period, the resiliency decreases even though the reliability of the system improves.

12.4   CONCLUSIONS

This paper discusses the optimality conditions for reservoir operation with evaluation of performance indicators including optimal release policy for conservation zone and flood mitigation policy for flood control zone. Implications for optimal reservoir operation for conservation zone have been derived through Monte Carlo Simulation-based hedging model whereas flood mitigation policy has derived on the basis of statistical approach and testing with Maithon reservoir under Damoder Valley Corporation in India. The result shows that the Maithen Reservoir is capable of satisfying the water requirement in terms of volume is 86.74%, and the reservoir can satisfy the demands 83.33% times, and its annual reliability is 82.35%. Similarly, the resilience which stands for the ratio of the number of times the system moved from failure to success, to the total number of periods the system was in failure state is 83. The vulnerability, i.e., the maximum period deficit is 87.87 MCM for water supply and 33.65 MCM for the flood.

KEYWORDS

- **hedging rule**
- **Monte-Carlo simulation**
- **performance indicators**
- **reservoir operation policy**
- **statistical method**

REFERENCES

Bower, B. T., Hufschmidt, M. M., & Reedy, W. W., (1962). In: Maass, A., et al., (eds.), *Operating Procedures: Their Role in the Design of Water-Resources Systems by Simulation Analyses, in Design of Water-Resource Systems* (pp. 443–458). Harvard Univ. Press, Cambridge, Mass.
Chou, F. N. F., & Wu, C. W., (2010). Stage-wise optimizing operating rules for flood control in a multi-purpose reservoir. *Journal of Hydrology, 521*, 245–260.

Draper, A. J., & Lund, J. R., (2004). Optimal hedging and carryover storage value. *J. Water Resour. Plan Manag.*, *130*(1), 83–87.

Jain, S. K., & Bhunya, P. K., (2008). Reliability, resilience, and vulnerability of a multipurpose storage reservoir. *Hydrological Sciences Journal*, *53*(2), 434–447.

Klemes, V., (1977). Value of information in reservoir optimization, *Water Resour. Res.*, *13*(5), 850–857, doi: 10.1029/WR013i005p00837.

Kundzewicz, Z. W., & Kindler, J., (1995). Multiple criteria for evaluation of reliability aspects of water resources systems. In: *Modeling and Management of Sustainable Basin-scale Water Resources* (pp. 217–224). IAHS Publ. 231. IAHS Press, Wallingford, UK.

Loucks, D. P., Stedinger, J. R., & Haith, D. A., (1981). *Water Resources Systems Planning and Analysis*, Prentice-Hall, Englewood Cliffs, NJ.

Marien, J. L., Damáio, J. M., & Costa, F. S., (1994). Building flood control rule curves for multipurpose multireservoir systems using controllability conditions, *Water Resour. Res.*, *30*(4), 1135–1144, doi: 10.1029/93WR03100.

Masse, P., (1946). *Les Reserves et la Regulation de l'Avenir Dans la vie Economique* (Vol. I), Avenir Determine (in French), Hermann and Cie, Paris.

Raje, D., & Mujumdar, P. P., (2010). Reservoir performance under uncertainty in hydrologic impacts of climate change. *Adv. Water Resour.*, *33*, 312–326.

Shih, J. S., & ReVelle, C., (1994). Water supply operations during drought: Continuous hedging rule. *J. Water Resour. Plan. Manage*, *120*(5), 613–629.

Shih, J. S., & ReVelle, C., (1995). Water supply operations during drought: A discrete hedging rule, *Eur. J. Oper. Res.*, *82*, 163–175.

Srinivasan, K., & Philipose, M. C., (1998). Effect of hedging on over-year reservoir performance, *Water Resources Management*, *12*(2), 95–120.

Stedinger, J. R., (1984). The performance of LDR models for preliminary design and reservoir operation, *Water Resource Research*, *20*(2), 215–224.

Vogel, R. M., & Bolognese, R. A., (1995). Storage–reliability–resiliency–yield relations for over-year water supply systems. *Water Resour. Res.*, *31*(3), 645–654.

You, J. Y., & Cai, X., (2008). Hedging rule for reservoir operations: A numerical model. *Water Resources Research*, *44*, 1–11.

CHAPTER 13

GLACIERS AND GLACIAL LAKE OUTBURST FLOOD RISK MODELING FOR FLOOD MANAGEMENT

NITY TIRKEY[1], P. K. PARHI[2], and A. K. LOHANI[3]

[1]Research Scholar, Central University of Jharkhand, Ranchi, India,
E-mail: nitytirkey@gmail.com

[2]Assistant Professor, Centre for Water Engineering and Management,
Central University of Jharkhand, Ranchi, India

[3]Scientist G, National Institute of Hydrology, Roorkee, Jharkhand, Ranchi,
India

ABSTRACT

Global temperature rise has been responsible for the depletion of glaciers and consequently creation of lakes on their terminus. Several of these lakes had burst and caused flooding or Glacial Lake Outburst Floods (GLOFs) in the recent past. GLOFs have the potential of releasing millions of cubic meters of water in a very small time period causing catastrophic flooding downstream and damaging whatever comes into their way. Study of GLOFs hazard in Sutlej River basin using geospatial techniques consisting of satellite remote sensing, geographical information system, is proposed in this chapter. The outcomes of the proposed study will be helpful for GLOFs risk management and for developing an overall strategy to address possible risks from future GLOF events in the country.

13.1 INTRODUCTION

Worldwide receding of mountain glaciers is one of the most reliable evidence of the changing global climate. Globally, the impacts of climate change include rising temperatures, shifts in rainfall pattern, melting of glaciers and sea ice, the risk of glacial lake outburst floods (GLOFs), sea level rise and increased intensity and frequency of extreme weather events (Ganguly et al., 2010). The climatic change/variability in recent decades has made considerable impacts on the glacier lifecycle in the Himalayan region. The Himalayas are geologically young and fragile and are vulnerable to even insignificant changes in the climatic system (Lama et al., 2009). Glaciers and glacial lakes play an important role in maintaining ecosystem stability as they act as buffers and regulate runoff water supply to plains during both dry and wet seasons. The glaciers and glacial lakes are generally located in remote and inaccessible areas. The inventories are only possible using time series remote sensing data and geographic information system (GIS) technology. The mountain ecosystems are fragile and highly susceptible to global climate changes. GLOF occurs when a dam containing a glacial lake fails. This is mainly due to the glaciers retreat. As glaciers retreat, glacial lakes are formed behind moraine or ice dams or inside the glaciers. A sudden breach in its walls may lead to a discharge of huge volumes of water and debris. Several of such lakes have been burst in the recent past resulting in a loss of human lives and destruction and damages of infrastructure in the valleys below. Glacier-outburst floods cannot be predicted, and therefore, continuous monitoring and mapping, both spatial and temporal, as opposed to a limited frequency point measurement can reduce the devastating impact of such hazards. Sometimes it is not easy to avoid natural phenomena causing disasters such as GLOFs, but a prior knowledge about their nature and possible extent can develop a capacity of disaster management authorities to respond and recover from emergency and disaster events. Similarly, hazard maps cannot stop a disastrous event from happening, but an effective use of hazard maps can prevent an extreme event from becoming a disaster. Himachal Pradesh is a mountain state in Indian Himalayas covering an area of 55,673 km². Himachal Pradesh has four major river basins namely Satluj, Beas, Chenab, and Ravi. Satluj basin alone covers 45% of the total geographical area of the state (923,645 km²). The basin is very active and experiences regular floods causing widespread damage in the down valleys.

Due to global climate fluctuations, the water resources of the river basin are going to be altered over time. Hence a systematic study of water resources in the basin is pre-requisite for embarking on development plans. Keeping these facts in view the present study on the inventory of glaciers and glacial lakes in Satluj basin was undertaken to see the changes in the glacier lakes.

13.2 MATERIALS AND METHODS

13.2.1 STUDY AREA

The study area of the Satluj river basin lies between 30°22′ to 32°42′ N Longitude and 75°57′ to 78°51′ East Latitude in Himachal Pradesh. The basin constitutes parts of the districts of Lahaul & Spiti, Kinnaur, Shimla, Kullu, Mandi, Bilaspur, Solan, Sirmour, and Una. The basin exists in the topographic maps published by the Survey of India (SOI) vide numbers 53A, 53E, 52H, 52L, 53I, and 53F published in the 1960's–1970's on the scale of 1:50,000. The river Sutlej is one of the main tributaries of Indus and has its origin near Manasarowar and Rakas lakes in Tibetan plateau at an elevation of about 4,500 m (approx.). The entire Satluj basin has been divided into three sub-basins viz. Spiti as sub-basin number 1, Upper Tibet as 3 and Lower Satluj as sub-basin number 2 (Figure 13.1).

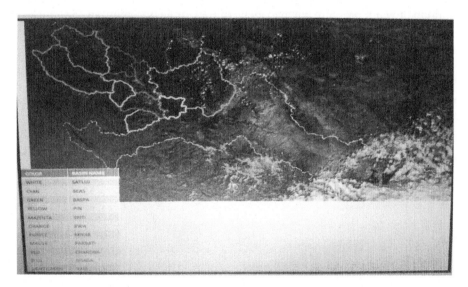

FIGURE 13.1 Different river basin.

13.2.2 METHODOLOGY

A digital database of glaciers and glacial lakes is necessary to identify the potentially dangerous glacial lakes. To identify the individual glaciers and glacial lakes, different image enhancement techniques are useful. The

ERDAS imagine 9.3 and Arc GIS 10.2.2 have been used for the processing of satellite data and GIS analysis.

13.3 RESULTS AND DISCUSSION

In the Himalayas, during the retreating phase, a large number of lakes are being formed either at the snout of the glacier as a result of damming of the moronic material known as moraine-dammed lakes or supraglacial lakes formed in the glacier surface area. Most of these lakes are formed by the accumulation of vast amounts of water from the melting of snow and by blockade of end moraines located in the down valleys close to the glaciers. In addition, the lakes can also be formed due to landslides causing artificial blocks in the waterways. The sudden break of a moraine/block may generate the discharge of large volumes of water and debris from these glacial lakes and water bodies causing flash floods namely GLOF. The sudden bursts of lakes can happen due to erosion, a buildup of water pressure, an avalanche of rock or heavy snow, an earthquake, or if a large enough portion of a glacier breaks off and massively displaces the waters in a glacial lake at its base (Figure 13.2).

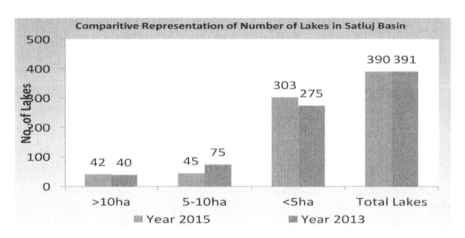

FIGURE 13.2 Number of lakes formed during the year 2013 and 2015.

There is a considerable increase in the number of moraine-dammed lakes (GLOFs) in Satluj basin which reflects that formation of such lakes in the Higher Himalayan region is indicating an increasing trend. The higher number of smaller lakes, i.e., lakes with an area less than 5 hectares indicates that the effect of the climatic variations is more pronounced on the glaciers of the Himalayan region resulting in the formation of small lakes in front of the glacier snouts due to the damming of the moronic material. The lakes with an area more than 10 hectares and those with the area between 5–10 hectares are more vulnerable sites for causing damage in case of the bursting of any one of them. Therefore, a proper monitoring and change analysis of all such lakes in the higher Himalayan region of the State is critical for averting any future eventuality in Himachal Pradesh, so that the precious human lives are saved.

13.4   SUMMARY

The study indicated that eighty percent of Satluj river catchment is snow fed. The glaciers were found to be mostly distributed in the northeastern part of the basin. A total of 38 moraine-dammed lakes were identified in the Satluj basin (Kulkarni et al., 2001), out of which 14 lakes were in the Himachal part, and the remaining 24 lakes were in the Tibetan part which has been increased to 391 which includes the Spiti and the Baspa basins.

KEYWORDS

- **GIS**
- **glacier lake outburst flood**
- **glacier lakes**

REFERENCES

Ganguly, K., & Panda, G. R., (2010). *Adaptation to Climate Change in India: A Study of Union Budgets*. Oxfam India working papers series – OIWPS – I.

Randhawa, S. S., Sood, R. K., & Kulkarni, A. V., (2001). Delineation of moraine-dammed lakes in Himachal Pradesh using high-resolution IRS LISS III satellite data. *Proc. National Symposium on Advances in Remote Sensing Technology With Special Emphasis on High-Resolution Imagery*, SAC Ahmadabad.

CHAPTER 14

DETERMINATION OF DESIGN PARAMETERS FOR THE BORDER IRRIGATION METHOD

GARIMA JHARIYA[1], RAJEEV RANJAN[2], PRATIBHA WARWADE[3], K. L. MISHRA[2], and V. K. JAIN[4]

[1]Institute of Agricultural Sciences, Banaras Hindu University, Varanasi (U.P.), 221005, India, E-mail: garima2304@gmail.com

[2]Collage of Agricultural Engineering, Jawaharlal Nehru Krishi Vishva Vidhyala Jabalpur (M.P.) – 480661, India

[3]Center for Water Engineering and Management, Central University of Jharkhand, Brambe, Ranchi–835205, India

[4]Subject Matter Specialist, KVK, RVSKVV, Ashok Nagar, Madhya Pradesh, India

ABSTRACT

Surface irrigation analysis and design require the knowledge of the variation of the cumulative infiltration water (per unit area) into the soil as a function of the infiltration time. The purpose of this study is to evaluate Design parameters for border irrigation system water infiltration and storage under surface irrigation in a cultivated field. The factors affecting the design parameter are infiltration, slop, roughness time of pounding, etc. The infiltration characteristics are shown by plotting graph between accumulated infiltration and average infiltration rate against elapsed time by the data obtained from concentric cylindrical infiltrometer and the values of b = –2.42, α = 0.3131 and a = 0.2268 which is desirable in sandy loam soil. The slope is found out to be 0.20%, the time of ponding is approximately 27.03 min, hydraulic resistance is 0.0274, the Manning's roughness coefficient is 0.01219. The application efficiency is found out to be 55.99%.

14.1   INTRODUCTION

Border irrigation is an old surface irrigation system used in the western part of the United States to irrigate alfalfa, wheat, other small grains, and sometimes close growing row crops. The concept is to flush a large volume of water over a relatively flat field surface in a short period of time. Borders are raised beds or levees constructed in the direction of the field's slope. The idea is to release water into the area between the borders at the high end of the field. The borders guide the water down the slope as a shallow sheet that spreads out uniformly between the borders.

The crop should be flat planted in the direction of the field slope or possibly at a slight angle to the slope. The spacing between borders is dependent on soil type, field slope, pumping capacity, and field length and field width. A clay soil that cracks is sometimes difficult to irrigate, but with borders, the cracking actually helps as a distribution system between the borders. The border-spacing on sandy and silt loam soils that tend to seal or crust over is more of a challenge than with the cracking clays. The pumping capacity and field dimensions (length and width) are used to determine the number of borders needed and how many can be irrigated in a reasonable time.

The general most of the methods of surface irrigation include four phases advance, storage, depletion, and recession (Walker and Skogerboe, 1987; Alazba, 1999; Amer, 2004) with the objective to maximize a measure of merit (performance criterion) while minimizing some undesirable consequences. This border irrigation has no of advantage over other surface irrigation methods If can be used on a field that is usually flood irrigated. The aim of this work is to study the suitability of border irrigation system for the dusty area farm of J.N.K.V.V. Jabalpur and to determine the different parameters of the border irrigation system of that area.

14.2   MATERIALS AND METHODS

The area proposed for the study is dusty area farm of J.N.K.V.V. Jabalpur at the distance of about 6 km from district headquarter. The area received an average annual rainfall 1354 mm; most of which occurs July to be the south-west monsoon. This indicated that the area has a sub-tropical climate. Maximum temperature is 45°C in the month of May and minimum of 9.3°C in December. Longitude −78°21' E to 80°58'E & Latitude −22°29' N to 24°48'N. The surface texture of majority of the area is Sandy loam soil.

14.2.1 INFILTRATION MEASUREMENT

The infiltration rate (I) is the volume of water infiltrating through a horizontal unit area of soil surface at any instant (infinitely small period of time) (the unit is L T^{-1}, cmh^{-1}). The main aim of preparing infiltration curve through this test is to obtain basic infiltration which is constant infiltration rate for medium. Using the field data as t_1 and t_2 obtained from infiltration measurement experiment the rectifying value of t is found from the relation

$$t_3 = \sqrt{(t_1 * t_2)} \tag{1}$$

where, t is the elapsed time. The corresponding value of Y_3 was determined from the graph plotted between accumulated infiltration rates against elapsed time based on data presented in Table 14.1. The value of constant b is obtained as follows:

$$b = \frac{\left(y_1 y_2 - y_3^2\right)}{\left(y_1 + y_2 - 2 * y^3\right)} \tag{2}$$

The value of b is subtracted from each value of y. The logarithmic form of the equation

$$y = at^a + b \tag{3}$$

$$\text{Log } (y - b) = \log a + \alpha \log t$$

Initial soil moisture content was measured before measuring the infiltration rate. Infiltration (cumulative and/or rate) of the soil was measured using double ring method (Ankeny, 1992; Reynolds et al., 2002) before irrigation for more than location along border furrow.

14.2.2 LAND SLOPE

Irrigation border and furrows have a uniform downfield gradient. The maximum land slope is limited by consideration of soil erosion by the irrigation stream and rainfall. A dumpy level, a level rod (staff) and a measuring chain or tape are required for the survey to determine land slope border strips. When the land slope is uniform, the percentage slope is determined as follows:

$$\text{Percentage slope} = \frac{\text{Difference in elevation between the first and last point}}{\text{Distance between the first and last point}}$$

14.2.3   OPPORTUNITY TIME

The time interval during which infiltration of water into the soil can occur is bounded by the advance and recession functions and is defined as the infiltration opportunity time (Holizapfel et al., 1984; DeTar, 1989; Foroud et al., 1996; Rodriguez, 2003). The infiltration opportunity time of the selected area is calculated by plotting advance and recession curves. The difference between the times, the waterfront reaches a particular point along the border (or furrow) and the time at which the tailwater recesses from the assumed point is the infiltration opportunity time.

14.2.4   HYDRAULIC RESISTANCE

The Darcy-Weisbach resistance coefficient for the non-vegetated border strips was calculated from the modified form of the Blasiu's formula applicable to border strips which are expressed as follows:

$$f = \frac{0.316}{\left(49q|\upsilon\right)^{\frac{1}{4}}} \tag{4}$$

where, f is Darcy-Weisbach resistance coefficient, q is the discharge (liter/sec), and υ is the Kinematics' viscosity of water (cm²/sec).

$$\upsilon = \mu/\rho$$

where, μ is given as dynamic viscosity of water (dyne-sec/cm²), and ρ is the mass density of water (gm/cm³)

$$\mu = \frac{0.0179}{\left(1+0.03368T+0.000221T^2\right)} \tag{5}$$

where, T = temperature of water.

14.2.5   MANNING'S ROUGHNESS CO-EFFICIENT

In non-vegetated borders, when the hydraulic resistance is expressed by the Darcy-Weisbach friction factor 'f,' the equivalent value of Manning's 'n' is calculated using the following relationship:

$$n = \frac{\left(R^{2/3} \times \sqrt{f}\right)}{\left(\sqrt{8} \times \sqrt{g}\right)} \tag{6}$$

where, R is the hydraulic radius in meters, g is the acceleration due to gravity in ms^{-2}, and t is the Darcy-Weisbach friction factor.

14.2.6   APPLICATION EFFICIENCY

The application efficiency is determined by conducting the experiment in the field, by using the method to take moisture sample in borders strips before and after irrigations, and find out the moisture reach in the zone supplied in the field. The water application efficiency is influenced by the factors like amount of water applied in unit time (stream size), infiltration characteristics of the soil, and the rate of advancing front in the furrow.

$$\text{Application efficiency} = \frac{\text{Water stored in the root zone}}{\text{Water applied to the field}}$$

14.2.7   SOILBULK DENSITY

For calculating bulk density core cutter method was used and is given by:

$$\text{Bulk Density} = \text{Mass/Volume (gm cm}^{-3})$$

14.3   RESULTS AND DISCUSSION

The result obtained from the investigation carried on the sites stated as before are categorized into infiltration characteristics of the project site, opportunity time, slope, hydraulic resistance, meanings roughness co-efficient, and application efficiency.

14.3.1   INFILTRATION CHARACTERISTICS

Infiltration test was performed at two different places in the same field at dusty area J.N.K.V.V. farm. The infiltration started with the very high infiltration rate (18 cm h^{-1}), which decreased very rapidly to 9 cm h^{-1} within 5 min. It further went on decreasing till the constant rate of 1.65 cm h^{-1}

was achieved at the end of 70 min, time period of the test. Studies made on infiltration are presented in Figures 14.1 and 14.2. Figure 14.1 indicates the two curves, i.e., average infiltration rate & accumulated infiltration depth as a function of time was found out to be 18.70 and the constants b = –2.42, α = 0.3131 and a = 2.2268. The values observed and calculated accumulated infiltration rate are given in Table 14.1.

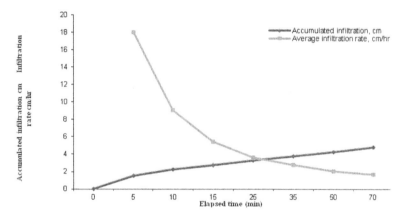

FIGURE 14.1   Plots of average accumulated infiltration and average infiltration rate against elapsed time-based.

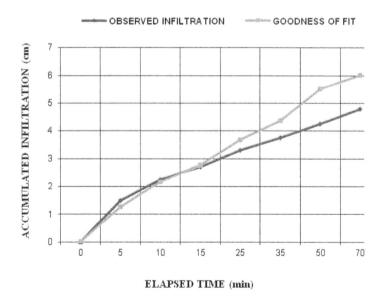

FIGURE 14.2   Compare of observed accumulated infiltration and goodness of fit against elapsed time.

TABLE 14.1   Average Infiltration Test Performed in Two Different Plots

S. No.	Elapsed Time t (Min)	Average Infiltration Rate (cm/hr)	Average Accumulated Infiltration 'Y' (cm)	Goodness of Fit 'Y' (cm)
1	0	-	-	-
2	5	18	1.5	1.266
3	10	9	2.25	2.159
4	15	5.4	2.7	2.770
5	25	3.6	3.3	3.680
6	35	2.7	3.75	4.358
7	50	2.0	4.25	5.506
8	70	1.65	4.8	6.001

14.3.2   HYDRAULIC RESISTANCE

The hydraulic resistance in non-vegetated border strip is expressed as Darcy-Weisbach resistance coefficient "f." The value of "f" is obtained 0.0274 by using Blasius empirical equation.

14.3.3   ROUGHNESS CO-EFFICIENT

The Manning's 'n' Roughness coefficient is found in the dusty area of JNKVV farm is 0.01219 and bulk density obtained in this field is 1.7 gm cc^{-1} which is more clearly explained from the Table 14.2 sowing the data obtained for bulk density.

TABLE 14.2   Average Bulk Density Performed in Two Different Sites for Different Depth

S. No	Depth (cm)	Weight of Wet Soil (gm)	Weight of Dry Soil (gm)	Bulk Density (gm cm^{-3})
1	0–17.5	2663	2324	1.657
2	17.5–35	2680	2392	1.7060
3	35–52.5	2.715	2433	1.735
Average Bulk Density				1.6994 ≈1.70

14.3.4   INFILTRATION OPPORTUNITY TIME

The infiltration opportunity time at any point along a border is the vertical distance between the advance and recession curve at the point. In the

Figure 14.3, the average time of pounding is 27 min 3 sec at the distance of 11 m. The graph represented shows that the relationship between the time of advance and time of recession. When irrigation started in border strip, pump discharge is 3.37 l/sec first water advance time up to 10m is very rapid and it took only 63 sec. After that the time of advance is slow and the total time taken by the water to reach the end of the strip is 16 min. 3 sec. Avg. Recession time is 27 min. 3 sec (Table 14.3).

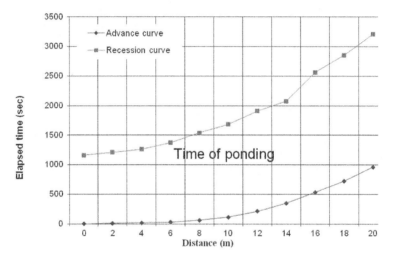

FIGURE 14.3   Advance and recession curves against distance.

TABLE 14.3   Water Advanced Front for Border Size (Average Depth of Water = 7.72 cm)

S. No.	Border Length (M)	Water Advance Time (Min-Sec)	Recession Time (Min-Sec)	Time of Ponding (Min-Sec)	Deviation From Average time of Ponding (Min-Sec)
1	0	0	19–22	19–22	+7–41
2	2	0–8	20–08	20–00	+7–03
3	4	0–16	21–02	20–46	+6–17
4	6	0–29	22–52	22–23	+4–40
5	8	1–03	25–37	24–34	+2–29
6	10	1–52	28–03	26–11	+0–52
7	12	3–32	31–52	28–20	−1–17
8	14	5–48	34–35	29–13	−2–10
9	16	8–54	42–42	33–48	−6–45
10	18	12–05	47–33	35–28	−8–25
11	20	16–03	53–30	37–27	−10–24

14.3.5   APPLICATION EFFICIENCY

The application efficiency determined by conducting an experiment in dusty area field of JNKVV farm is found to be 55.99%.

14.4   CONCLUSIONS

A design procedure for border irrigation system based on the conservation of mass has been developed. The present method presumes that the border has a free overfall outlet and uniform field parameters, slope, roughness, and infiltration. The design criterion is to select the appropriate inflow rate and time of cutoff so that the maximum or possibly desired efficiency is obtained efficiency which is the primary criterion in border irrigation system design and management. The study sites selected for the present study in the dusty area of J.N.K.V.V. farm, which come under Jabalpur (M.P.). The study area composed of mostly by cultivated field, which uses for border irrigation easily adopted in the field. The infiltration characteristics are shown by plotting graph between accumulated infiltration and average infiltration rate against elapsed time by the data obtained from concentric cylindrical infiltrometer are desirable in sandy loam soil. Designed hydraulics of border irrigation system shows it is well adapted for this area.

KEYWORDS

- **application efficiency**
- **border irrigation**
- **infiltration rate**
- **time of pounding**

REFERENCES

Alazba, A. A., (1999). Dimensionless advance curves for infiltration families. *Agricultural Water Management, 41*, 115–131.

Amer, A. M., (2004). *Soil Hydro-Physics* (2nd Part, p. 438, in Arabic). Agricultural irrigation and drainage. El-Dar Al Arabia for Publish. & Distrib. Cairo, Egypt.

Ankeny, M. D., (1992). Methods and theory for unconfined infiltration measurements. In: Topp, G. C., (ed.), *Advances in Measurement of Soil Physical Properties: Bringing Theory Into Practice* (pp. 123–141). SSSA Spec. Publ. 30. SSSA, Madison, WI.

DeTar, W. R., (1989). Infiltration functions from furrow stream advance. *Journal of Irrigation and Drainage Engineering – ASCE, 115*(4), 722–730.

Foroud, N., George, E. S., & Entz, T., (1996). Determination of infiltration rate from border irrigation advance and recession trajectories. *Agricultural Water Management, 30,* 133–142.

Holizapfel, E. A., Marino, M. A., & Morales, J. C., (1984). Comparison and selection of furrow irrigation models. *Agricultural Water Management, 9,* 105–111.

Reynolds, W. D., Elrick, D. E., & Young, E. G., (2002). Single-ring and double-ring or concentric-ring infiltrometers. In: Dane, J. H., & Topp, G. C., (eds.), *Methods of Soil Analysis* (Part 4, pp. 821–826). Physical Methods, SSSA, Madison, WI.

Rodriguez, J. A., (2003). Estimation of advance and infiltration equations in furrow irrigation for untested discharges. *Agricultural Water Management, 60,* 227–239.

Walker, W. R., & Skogerboe, G. V., (1987). *Surface Irrigation: Theory and Practice* (p. 386). Prentice-Hall, Englewood Cliffs, NJ 07632.

CHAPTER 15

ENSO ASSOCIATION WITH RAINFALL

PRATIBHA WARWADE

Assistant Professor, Center for Water Engineering and Management, Central University of Jharkhand, Brambe Ranchi–835205, India, E-mail: pratibhawarwade@gmail.com

ABSTRACT

This study is mainly focused on Dikhow catchment which is a part Brahmaputra river basin, located between 94°28'49"E to 95°09' 52" E longitude and 26°52' 20"N to 26°03' 50" N latitude. The geographical area of Dikhow catchment is about 3100 km^2, which encompasses around 85% of Nagaland, ten percent of Assam and five percent of Arunachal Pradesh. This study is more precise to find out the relationship using between El Niño-Southern Oscillation (ENSO) and rainfall for Dikhow catchment, using statistical techniques. Monthly rainfall data of six stations of study area from 1950–2002 were acquired from the Indian Meteorological Department Pune and monthly SST anomalies of Niño of 3.4 regions (5°N to 5°S and 170°W to 120°W) during the study period (1950–2002) were downloaded from the NOAA, National Weather Service (NWS), and National Center for Environmental Prediction (NCEP), Climatic Prediction Center (CPC) (http://www.cpc.noaa.gov/data/indices), to classify ENSO events. Results showed that the significant correlation was found for monsoon rainfall only while winter and summer rainfall showing insignificant correlation. Sibsagar, Tirap, and Mon stations portrayed the significant correlation between monsoon rainfall with ENSO and insignificant correlation observed between remaining three stations. Further, the probability of exceedance of long-term average monsoon precipitation was also lower during El Niño episodes than La Niña and neutral years for these stations (Sibsagar, Tirap, and Mon).

15.1   INTRODUCTION

The El Niño-Southern Oscillation (ENSO) pattern is driven partly by alternating warmer and cooler temperatures of the sea surface in the eastern and central tropical Pacific Ocean, which themselves are caused by changes in upwelling currents. It changes to rain and temperatures over a large portion of the globe, with severe consequences to human activities like agriculture and fishing, which depend on meteorological conditions and ocean fluxes. In turn, the changes in weather and atmospheric circulation change in the ocean currents (Ward and Richardson, 2011). ENSO is a set of anomalously warm ocean water temperatures that infrequently develops off the western coast of South America and can cause climatic changes across the Pacific Ocean. Many of the researchers (Khandekar and Neralla, 1984; Ropelewaski and Halpert, 1987; Glantz et al., 1991; Diaz and Markgraf, 1992) found that the ENSO associated with climate irregularities all over the globe. Kousky et al. (1984) depicted that in Australia, Indonesia, India, and West Africa drought is strongly associated with ENSO. Investigators have been working on this issue (ENSO's connection with climate anomalies) since the late 1800s. El Niño (La Niña) is one of the environment related phenomena in the equatorial Pacific Ocean characterized by a five consecutive three-month running mean of sea surface temperature (SST) anomalies in the Niño 3.4 region of the threshold of $\pm$ 0.5°C. This standard of measure is known as the Oceanic Niño Index (ONI). As per the Census 2011, 54 percent of the Indian workforce is still occupied in agriculture, and 53 percent of the gross cropped area is about rainfed. ENSO has been known to exert the most important external forcing on Indian summer monsoon rainfall (ISMR) (Kumar et al., 1999; Rasmusson and Carpenter, 1983; Webster and Yang, 1992; Ropelewaski and Halpert, 1987; Chiew et al., 1998; Khole, 2000; Lau and Nath, 2000; Ashok et al., 2001; Krishnamurthy and Kirtman, 2009). The interannual variations of ISMR have motivated studies of the ENSO since the turn of the twentieth century (Walker, 1923; Barnett, 1984). ENSO studies have been used either to mitigate the impacts of adverse conditions or to take advantage of favorable conditions (Selvaraju, 2003).

Historically, scientists have classified the intensity of El Niño based on SST anomalies exceeding a pre-selected threshold in a certain region of the equatorial Pacific. The most commonly used region is the Niño 3.4 region, and the most commonly used threshold is a positive SST departure from normal greater than or equal to +0.5°C. Since this region encompasses the western half of the equatorial cold tongue region, it provides a good measure of important changes in SST and SST gradients that result in changes in the pattern of deep tropical convection and atmospheric circulation. The criterion that is often used to classify El Niño episodes is that five consecutive 3-month running mean SST anomalies exceed the threshold.

Studies have shown that a necessary condition for the development and persistence of deep convection (enhanced cloudiness and precipitation) in the Tropics is that the local SST be 28°C or greater. Once the pattern of deep convection has been altered due to anomalous SSTs, the tropical and subtropical atmospheric circulation adjusts to the new pattern of tropical heating resulted in anomalous patterns of precipitation and temperature that extend well beyond the region of the equatorial Pacific. SST anomaly of +0.5°C in the Niño 3.4 region is sufficient to reach this threshold from late March to mid-June. During the remainder of the year a larger SST anomaly, up to +1.5°C in November-December-January, is required in order to reach the threshold to support persistent deep convection in that region.

Many investigations regarding ENSO Association with rainfall over Indian region are discussed here: Kirtman and Shukla, (2000) found a strong negative correlation between ENSO and ISMR using 100 years of historical record. A negative correlation is strongest during the months of December to March for east Pacific SSTA. Results show that monsoon variability and ENSO variability related to each other, strong (weak) monsoon results in a strengthening (weakening) of the trade winds over the tropical Pacific. Kripalani et al. (2003) examined the decadal variability and inter-annual for monsoon rainfall over India and its teleconnections using observed data for a period of 131 years (1871–2001). The study indicated that the inter-annual variability showed year-to-year variation, and the decadal variability showed distinct alternate epochs of above and below normal rainfall. Further, they studied the links between the ENSO phenomenon, the surface temperature of the Northern Hemisphere and Eurasian snow with Indian monsoon rainfall and stated that the correlations are not only weak but have altered signs in the early 1990s, suggesting that the IMR (Indian Monsoon Rainfall) has linked not only with the Pacific but with the Northern Hemisphere/Eurasian continent also. The fact that temperature/snow relationships with IMR are weak further suggests that global warming may not be responsible for the recent ENSO-Monsoon weakening. Further, they conveyed that warm phase (El Nino) is connected with the weakening of the Indian monsoon, and cold phase (La Niña) is connected with the strengthening of the Indian monsoon. Kumar et al. (2006) exposed that severe droughts in India always go along with El Niño events using 132 data of past rainfall. El Niño events along with SST anomalies in the central equatorial Pacific are very effective for producing drought subsidence in India than the events of warmest SSTs in the eastern equatorial Pacific. Kumar et al. (2007) reported the inverse relationship between ENSO and southwest monsoon rainfall has weakened during the current years and positive relationship between ENSO and North East Monsoon (NEM) rainfall, which has strengthened and developed statistically significant after the mid-1970s. Epochal changes in the regional circulation features are one of the causes of this variation in the relationship. During the recent El Niño years, above normal NEM rainfall is experienced due to stronger easterly wind anomalies and anomalous low-level moisture convergence along with associated changes in the circulation regime throughout the troposphere and across the southern parts of India and Sri Lanka. Ihara et al., (2007) observed the relationship between the ENSO, state of the equatorial Indian Ocean, and ISMR data from 1881 to 1998 of 36 stations in India. SST anomalies and zonal wind anomalies over the equatorial Indian Ocean were used to reflect the situation of the Indian Ocean. Negative correlation with ISMR was observed, while significant with wind index during El Niño years. Krishnamurthy and Kirtman (2009) studied the relationship between the intra-seasonal modes of the South Asian monsoon and the SST in the tropical oceans on a daily timescale. The strong relation of the persistent modes, which mainly determine the seasonal mean monsoon, when the SST leads, provides hope for long-term prediction of the seasonal mean monsoon. The strong relationship between the monsoon and the SST, when the monsoon leads, points toward the strong influence of the monsoon on the variability of ENSO and IOD. Kumar et al. (2013) suggested that the major portion of the drought variability is influenced by the ENSO. Global warming, especially the warming of the equatorial Indian Ocean represents the second coupled mode and is responsible for the observed increase in the intensity of droughts during the recent decades over India. Krishnamurthy and Krishnamurthy (2014) studied decadal-scale oscillations and trend in the Indian monsoon rainfall: Using a long record of high-resolution Indian rainfall data, this study has established the existence of three decadal scale nonlinear oscillations and a nonlinear trend in the IMR. The monsoon rainfall decadal oscillations were shown to be associated with the decadal variability of the North Atlantic and North Pacific oceans. This paper associate's only rainfall with the variability of SST. Parida and Oinam, (2015) conveyed that drought associated with El Niño was not so strong; however, increasing temperature and increased monsoon season rainfall variability have an impact on global climate change. This may cause warming-induced drought leading to an adverse impact on agriculture and food security in the NER (North Eastern Region) of India.

The research on the links between ENSO and rainfall (Achuthavarier et al., 2012), at the regional scale only began in recent years. However, the ENSO does not display the same degree of correlation with different homogeneous regions of India on a longer timescale. Some regions are least impacted by ENSO, while others are moderate to strongly correlated with the phenomenon (Ashok and Saji, 2007). For managing water resources systems such study of regional scale is essential for water authority over the region as well as the country. Hence this study is more precise to present the relationship between ENSO and rainfall using statistical analysis.

15.2 MATERIALS AND METHODS

15.2.1 STUDY AREA

The study area is Dikhow catchment which is a part of the larger Brahmaputra river basin, Dikhow river which originates from the hills of the state Nagaland. It is a south bank tributary of river Brahmaputra contributing 0.7 % runoff. A lower Brahmaputra river basin, a region where the hydrological impact of climate change is expected to be particularly strong, and population pressure is high (Gain and Giupponi, 2015). Brahmaputra river is the biggest trans-Himalayan river basin (Sharma and Flugel, 2015). Figure 15.1 presents the location map and demonstrates with the drainage map of the study area situated between 94°28'49"E to 95°09' 52" E longitude and 26°52' 20"N to 26°03' 50" N latitude. The geographical area of Dikhow catchment is about 3100 km², which encompasses around 85% of Nagaland, 10% of Assam and 5% of Arunachal Pradesh. The river traverses towards the north along the border of Mokokchung and Tuensang districts. The main tributaries of river Dikhow are Yangyu of Tuensang district and Nanung in the Langpangkong range in Mokokchung district. The river flows further northward and leaves the hill near Sibsagar and finally merges with the Brahmaputra River in the plains of Assam.

The study area is remote and largely inaccessible, land cover changes by practicing *shifting* cultivation (slash and burn) predominantly. Stations considered for the present study are Sibsagar, Mokokchung, Mon, Tirap, Tuensang, and Zunheboto portrayed in Table 15.1 with their geographical location and elevation. Altitude of the study area varies from 98m to 1818 m (Sibsagar to Zunheboto). It is highly diversified in terms of elevation hence divided into three physiographic zones namely; Low Elevated Zone (LEZ), Moderate Elevated Zone (MEZ) and High Elevated Zone (HEZ), because it is expected that each zone will respond differently. Stations below than 100 m from m.s.l., fall under LEZ also known as alluvial plain, stations between 101–1000 m elevation from m.s.l., covers the MEZ and stations between 1001 to 2000 m elevations from m.s.l., were classified as HEZ.

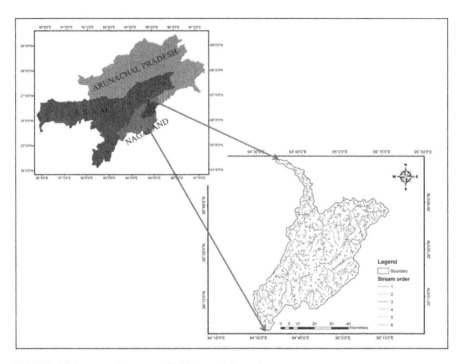

FIGURE 15.1   Location map of Dikhow catchment.

TABLE 15.1   Selected Stations of Study Area

S.N.	State	Station	Longitude (° 'N)	Altitude (m) (° 'E)	Latitude
1	Assam	Sibsagar	26° 5' 42"	94° 37' 42"	98
2	Nagaland	Mokokchung	26° 19' 20"	94° 30' 53"	1323
3	Nagaland	Mon	26° 43' 21"	94° 01' 52"	720
4	Arunachal Pradesh	Tirap	26° 59' 43"	95° 32' 27"	776
5	Nagaland	Tuensang	26° 14' 9"	94° 48' 46"	1570
6	Nagaland	Zunheboto	26° 00' 35"	94° 31'42"	1818

15.2.2 DETAILS OF DATA

The monthly precipitation data of six stations of Dikhow catchment for the years 1950–2002 were obtained from Indian Meteorological Department Pune. The monthly SST anomalies of Niño of 3.4 regions (5°N to 5°S and 170°W to 120°W) during the study period (1950–2002) were downloaded from the NOAA, National Weather Service (NWS), and National Centre for Environmental Prediction (NCEP), Climatic Prediction Centre (CPC) (http://www.cpc.noaa.gov/data/indices), to classify ENSO events.

15.3 METHODOLOGY

The methodology adopted for to find out the relationship between ENSO and rainfall are Pearson's Correlation Analysis, and Cumulative Distribution Frequency (CDF) discussed below.

15.3.1 PEARSON'S CORRELATION ANALYSIS

The correlation coefficient measures the strength of association between two continuous variables. Among several correlation analyses, three methods are common in use, i.e., Kendall's tau (τ), Spearman's rho and Pearson's correlation (r). The first two are based on ranks and measure all monotonic relationships. Also, these are resistant to the effect of outliers. In this study, Pearson's (r) correlation coefficient was calculated. Pearson's correlation (Crichton, 1999) is the most commonly used method to measure the correlation. As (r) measure the linear association between two variables, so it is also called linear correlation coefficient. Pearson's correlation was calculated using the formula:

$$r = \frac{1}{n-1}\sum_{i=1}^{n}(\frac{x_{i-\bar{x}}}{S_x})\left(\frac{y_i - \bar{y}}{S_y}\right) \tag{1}$$

where x and y are precipitation and SST anomalies, respectively. The $\bar{x}$ and $\bar{y}$ are the mean, and S_x and S_y are the standard deviation of x and y, respectively, and n is the total number of years of analysis.

15.3.2   CUMULATIVE DISTRIBUTION FREQUENCY (CDF)

Cumulative frequency distribution (CDF) is a function that gives the probability that a random variable is less than or equal to the independent variable of the function. Steps to calculate CDF are given below:

Step 1: In the first step, precipitation data were first standardized by deducting the mean value from the observed value and dividing by the standard deviation. Note that the standardized time series has a zero mean and a unit standard deviation.

Step 2: In the next step, precipitation data for each El Nino, La Niña and neutral year time series were separated for each station and probability of exceedance for each event was calculated using a Weibull's formula. To calculate the probability of exceedance, the standardized data were arranged in descending order and rank (m) was assigned. For the first entry m = 1: for second m = 2 and so on till the last value as m = n. Then the experience of probability can be calculated by using many methods, but the most common formula was used, i.e., Weibull's formula of exceedance of probability as $(100 \times m)/(n+1)$, where m is the mth value in order of magnitude of the series and n is the number of data series.

Step 3: Now a plot was drawn between the probability of exceedance (in %) on the vertical axis and standardized precipitation data on the horizontal axis and the exceedance probability corresponding to zero precipitation was identified for each El Nino, La Niña and neutral years time series.

Step 4: The same procedure was repeated for each station to calculate the CDF. It was computed for monsoon season.

15.4   RESULTS AND DISCUSSIONS

The Southern Oscillation Index (SOI) and the SST are the two most widely used indicators of ENSO. The monthly time series of SST anomaly are used in this study. Teleconnection between ENSO and rainfall first requires the definition of "El Niño/Lal Niña" as years. Trenberth, 1997 stated that El Niño is an ocean-atmosphere phenomenon where the cooler Eastern Pacific warms up once in every two to seven years. Increase in Eastern Pacific SST is due to the weakening of the easterly trade wind that resulted in the warm water from the western Pacific moving to the east. NOAA, 2008 classified an ENSO event on the basis of SST. An average deviation of $\pm 0.5°C$ from the historical mean of SST and for three consecutive months result shows an ENSO event. A positive anomaly indicates El Niño and a negative anomaly indicates La Niña.

15.4.1 ENSO EPISODES

Since in the 1950s, globally there has been 23 El Niño, 15 La Niña and 16 Neutral years occurred as presented in Table 15.2. Out of which we categorized as strong, moderate, and weak El Niño years on the basis of sea surface temperature anomalies (SSTA) presented in Table 15.3. SSTA of greater than 1.5 categorize as strong, SSTA ranges from 1–1.5 as moderate and SSTA ranges 0.5–1.0 categorized as weak El Niño years. Twelve strong, thirteen moderate, and nine weak El Niño years were obtained and are shown in Table 15.3.

TABLE 15.2 ENSO Episodes

S.N.	El Niño	La Niña	Neutral
1	1951	1950	1952
2	1953	1954	1960
3	1956	1955	1961
4	1957	9561	1962
5	1958	1964	1966
6	1959	1970	1967
7	1963	1971	1979
8	1965	1974	1980
9	1968	1975	1981
10	1969	1985	1984
11	1972	1988	1990
12	1973	1989	1993
13	1976	1998	1994
14	1977	1999	1995
15	1978	2000	1996
16	1982		2001
17	1983		
18	1986		
19	1987		
20	1991		
21	1992		
22	1997		
23	2002		

TABLE 15.3   Classification of El Niño Years.

Categories of El Niño years from 1950–2002			
S.N.	SSTA>1.5	SSTA Ranges 1–1.5	SSTA Ranges 0.5–1
	Strong	**Moderate**	**Weak**
1	1957	1951	1953
2	1958	1952	1954
3	1965	1963	1958
4	1966	1964	1959
5	1972	1968	1969
6	1973	1969	1970
7	1982	1986	1976
8	1983	1987	1977
9	1987	1991	1978
10	1988	1992	
11	1997	1994	
12	1998	1995	
13		2002	

15.4.2   PRECIPITATION RELATIONSHIP WITH ENSO

In the study area, three seasons viz., winter (November–February), summer (March–May), and monsoon (June–October) were defined to analyze the rainfall relationship with El Niño. The running mean of monthly anomalies of SST for each station was used to identify the number of years during El Niño, La Niña and neutral phase for the span of 1950–2002. Number of years in each ENSO phase during the study period in different seasons are presented in Table 15.4, and the number of El Niño episodes was higher during the winter followed by monsoon and summer.

TABLE 15.4   Number of Years in Each El Niño, La Niña, and Normal Phase During 1950–2002

Phase	Winter	Summer	Monsoon
El Niño	18	9	13
La Niña	17	11	15
Neutral	18	33	25

15.4.2.1   VARIATION OF PRECIPITATION DURING ENSO PHASES

The mean annual and seasonal precipitation during El Niño, La Niña and Neutral phase at different stations revealed in Table 15.5. The comparison of results of three ENSO phases shows that the mean of El Niño precipitation was less than the mean of La Niña and Neutral phase precipitation during monsoon season at all the stations. However, in winter season the mean of El Niño precipitation was more than La Niña and neutral phase precipitation. Moreover, the mean of El Niño precipitation was less than the mean of Neutral phase's precipitation in the summer season at all the stations. Results showed that the ENSO influences the seasonal precipitation in the study area which is also revealed by the study of Mooley and Parthasarathy (1983).

TABLE 15.5   Precipitation Variability in Each El Niño, La Niña and Normal Phase During 1950–2002

| Station | Mean Precipitation During ENSO | | | | | | | | |
| | Winter | | | Summer | | | Monsoon | | |
	El Niño	La Niña	Normal	El Niño	La Niña	Normal	El Niño	La Niña	Normal
Sibsagar	114.1	82.8	92.9	516.1	508.5	537.6	1450.4	1560.8	1499.8
Mokokchung	105.8	73.1	85.5	472.6	466.8	494.5	1460.9	1517.6	1479.7
Mon	104.4	75.2	86.1	453.9	444.8	472.5	1426.6	1504.9	1462.3
Tirap	100.4	74.8	83.8	420.6	412.6	438.2	1364.7	1463.0	1414.3
Tuensang	99.1	69.2	81.8	391.8	385.3	412.3	1454.4	1487.8	1464.7
Zunheboto	103.1	70.4	83.7	426.8	421.6	450.7	1487.4	1509.4	1486.8

15.4.2.2   CORRELATION ANALYSIS

The results of Pearson's correlation coefficient between ENSO and precipitation at 5% level of significance are shown in Table 15.6. The bold values are indicating a significant correlation. The results indicate a significant negative correlation between precipitation and ENSO during the monsoon season at Sibsagar, Mon, and Tirap. Remaining stations showed a non-significant positive correlation. During winter and summer, non-significant positive correlation between ENSO and precipitation were observed at all the stations. Negative correlations mean lower precipitation during El Niño and higher during the La Niña and vice-versa during positive correlation. Further analysis was performed only for monsoon season precipitation at all the stations because of significant correlation obtained for monsoon precipitation.

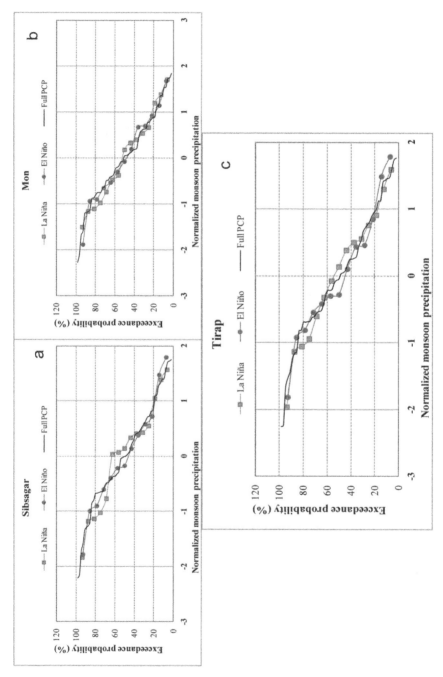

FIGURE 15.2   (See color insert.) Graph of the cumulative distribution of normalized monsoon precipitation during El Niño, La Niña and Normal phase for (a) Sibsagar, (b) Mon and (c) Tirap.

TABLE 15.6   Result of Pearson's Correlation Between ENSO and Precipitation

Station	Winter	Summer	Monsoon
Mokokchung	0.411	0.032	0.195
Sibsagar	0.321	0.023	−0.275
Tuensang	0.406	0.05	0.182
Zunheboto	0.418	0.043	0.158
Tirap	0.306	0.026	−0.270
Mon	0.353	0.036	−0.273

15.4.2.3   CUMULATIVE FREQUENCY DISTRIBUTION (CDF)

The CDF of normalized monsoon precipitation for El Niño, La Niña, and Normal series are shown in Figure 15.2a–c and Figure 15.3a–c, the black line represents the probability of normal series, while the red line with a circle and green line with square shows the probability of El Niño and La Niña series. For the normalized time series, the long-term average monsoon precipitation corresponds to zero. Figure 15.2a shows the probability of exceedance during El Niño years drops up to 45%, while for the La Niña years, this exceedance probability of long-term average rainfall reaches to 62% at Sibsagar. At Mon, the probability of exceedance during El Niño years was 46%, and during La Niña years it was 55% (Figure 15.2b), similarly at Tirap probability of exceedance during El Niño years was less (45%) than the exceedance probability during La Niña years (56%).

On the contrary, Figure 15.3a–c shows the exceedance probability of long-term average precipitation at Mokokchung, Tuensang, and Zunheboto. The probability of exceedance is 50%, 50% and 49% for El Niño, Neutral, and La Niña series for Mokokchung respectively. For Tuensang it was 52%, 49% and 45% for El Niño, Neutral, and La Niña series respectively. And for Zunheboto 54%, 48% and 45% probability of exceedance for El Niño, Neutral, and La Niña series respectively were obtained. It can be concluded that the probability of exceedance of average precipitation drops at Sibsagar, Mon, and Tirap during El Niño. Whereas for Mokokchung, Tuensang, and Zunheboto does not show the same (although the correlation of these stations between ENSO and precipitation were not significant). These results show that the effect of ENSO on the first three stations (Sibsagar, Mon, and Tirap). Although theses station were classified topographically into different zones, on this basis, it can be concluded that LEZ and HEZ shows the ENSO effect as reported in the study of Kripalani and Kulkarni (1997); Krishnamurthy

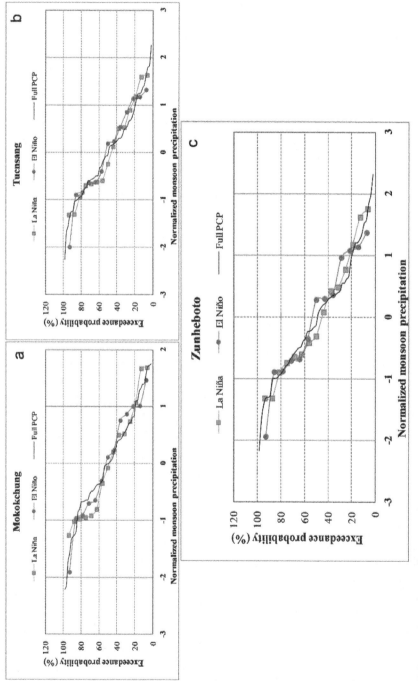

FIGURE 15.3 **(See color insert.)** Graph of the cumulative distribution of normalized monsoon precipitation during El Niño, La Niña and Neutral phase for (a) Mokokchung, (b) Tuensang and (c) Zunheboto.

and Goshwami (2000); Ashok et al. (2001) and Maity and Kumar (2006), wherein they reported that El Niño (La Niña) is associated with lower (higher) normal precipitation. Results are also in conformity with Walker (1924) in which he found that the relation between the Southern Oscillation and seasonal (June–September, JJAS) mean summer monsoon rainfall over India is one of the earliest observed teleconnections. The result shows, the influence of ENSO on monsoon precipitation observed over LEZ and MEZ of study region but not over the HEZ. There can be other factors too, such as mid-latitude circulation feature (Raman and Rao, 1981; Tanaka, 1982), wind speed at a point in the Arabian Sea (10° N 65° E) (Shukla and Mishra, 1977), Darwin pressure anomaly before monsoon season (Shukla and Paolino, 1983) and topography might have some role over the variability of rainfall in the region.

15.5   CONCLUSIONS

Region-specific study of the most important climatic variable, i.e., rainfall is essential to reduce the adverse effects of climate change in developing countries. Correlation analysis between ENSO and precipitation indicated a significant negative correlation in monsoon season for three stations (Sibsagar, Mon, and Tirap) in the catchment. Therefore, further analysis was carried out for monsoon season precipitation only. The CDF of standardized monsoon precipitation during El Niño, La Niña, and full-time series shows that the exceedance probability for average monsoon precipitation is less in El Niño years and more in La Niña years. It can also be concluded in terms of elevation that the LEZ and HEZ show the ENSO effect while there is no effect of ENSO over HEZ.

KEYWORDS

- **cumulative distribution frequency**
- **ENSO**
- **Nino 3.4 region**
- **Pearson's correlation coefficient**
- **SST**

REFERENCES

Achuthavarier, D., Krishnamurthy, V., Kirtman, B. P., & Huang, B., (2012). Role of the Indian Ocean in the ENSO-Indian summer monsoon teleconnection in the NCEP climate forecast system. *Journal of Climate*, *25*(7), 2490–2508.

Ashok, K., & Saji, N. H., (2007). On the impacts of ENSO and Indian Ocean dipole events on sub-regional Indian summer monsoon rainfall. *Natural Hazards*, *42*(2), 273–285.

Ashok, K., Guan, Z., & Yamagata, T., (2001). Impact of the Indian Ocean dipole on the relationship between the Indian monsoon rainfall and ENSO. *Geophysical Research Letters*, *28*(23), 4499–4502.

Barnett, T. P., (1984). Interaction of the monsoon and Pacific trade wind system at interannual timescales. Part III: A partial anatomy of the Southern Oscillation. *Monthly Weather Review*, *112*(12), 2388–2400.

Chiew, F. H., Piechota, T. C., Dracup, J. A., & McMahon, T. A., (1998). El Nino/Southern Oscillation and Australian rainfall, streamflow and drought: Links and potential for forecasting. *Journal of Hydrology*, *204*(1), 138–149.

Diaz, H. F., & Markgraf, V., (1992). *El Niño: Historical and Paleoclimatic Aspects of the Southern Oscillation.* Cambridge University Press.

Glantz, M. H., Katz, R. W., & Nicholls, N., (1991). *Teleconnections Linking Worldwide Climate Anomalies* (p. 535). Cambridge: Cambridge University Press.

Ihara, C., Kushnir, Y., Cane, M. A., & De La Peña, V. H., (2007). Indian summer monsoon rainfall and its link with ENSO and Indian Ocean climate indices. *International Journal of Climatology*, *27*(2), 179–187.

Khandekar, M. L., & Neralla, V. R., (1984). On the relationship between the sea surface temperatures in the equatorial Pacific and the Indian monsoon rainfall. *Geophysical Research Letters*, *11*(11), 1137–1140.

Khole, M., (2000). Anomalous warming over the Indian Ocean during 1997 El-Nino. *Meteorology and Atmospheric Physics*, *75*(1/2), 1–9.

Kirtman, B. P., & Shukla, J., (2000). Influence of the Indian summer monsoon on ENSO. *Quarterly Journal of the Royal Meteorological Society*, *126* (562), 213–239.

Kousky, V. E., Kagano, M. T., & Cavalcanti, I. F., (1984). A review of the Southern Oscillation: Oceanic-atmospheric circulation changes and related rainfall anomalies. *Tellus A.*, *36*(5), 490–504.

Kripalani, R. H., & Kulkarni, A., (1997). Climatic impact of El Nino/La Niña on the Indian monsoon: A new perspective. *Weather*, *52*(2), 39–46.

Kripalani, R. H., Kulkarni, A., Sabade, S. S., & Khandekar, M. L., (2003). Indian monsoon variability in a global warming scenario. *Natural Hazards*, *29*(2), 189–206.

Krishnamurthy, L., & Krishnamurthy, V., (2014). Decadal-scale oscillations and trend in the Indian monsoon rainfall. *Climate Dynamics*, *43*(1/2), 319–331.

Krishnamurthy, V., & Goswami, B. N., (2000). Indian monsoon-ENSO relationship on interdecadal timescale. *Journal of Climate*, *13*(3), 579–595.

Krishnamurthy, V., & Kirtman, B. P., (2009). Relation between Indian monsoon variability and SST. *Journal of Climate*, *22*(17), 4437–4458.

Kumar, K. K., Rajagopalan, B., & Cane, M. A., (1999). On the weakening relationship between the Indian monsoon and ENSO. *Science*, *284*(5423), 2156–2159.

Kumar, K. K., Rajagopalan, B., Hoerling, M., Bates, G., & Cane, M., (2006). Unraveling the mystery of Indian monsoon failure during El Niño. *Science*, *314*(5796), 115–119.

Kumar, K. N., Rajeevan, M., Pai, D. S., Srivastava, A. K., & Preethi, B., (2013). On the observed variability of monsoon droughts over India. *Weather and Climate Extremes, 1,* 42–50.

Kumar, P., Kumar, K. R., Rajeevan, M., & Sahai, A. K., (2007). On the recent strengthening of the relationship between ENSO and northeast monsoon rainfall over South Asia. *Climate Dynamics, 28*(6), 649–660.

Lau, N. C., & Nath, M. J., (2000). Impact of ENSO on the variability of the Asian-Australian monsoons as simulated in GCM experiments. *Journal of Climate, 13*(24), 4287–4309.

Maity, R., & Nagesh, K. D., (2006). Bayesian dynamic modeling for monthly Indian summer monsoon rainfall using El Nino–Southern Oscillation (ENSO) and Equatorial Indian Ocean Oscillation (EQUINOO). *Journal of Geophysical Research: Atmospheres, 111*(D7), 1–12.

Mooley, D. A., & Parthasarathy, B., (1983). Variability of the Indian summer monsoon and tropical circulation features. *Monthly Weather Review, 111*(5), 967–978.

CHAPTER 16

EFFICIENT RESERVOIR OPERATION WITH A MULTI-OBJECTIVE ANALYSIS

PRITI SAGAR[1] and PRABEER KUMAR PARHI[2]

[1]*Research Scholar, Central University of Jharkhand, Centre for Water Engineering & Management, Bramby, 835205, Jharkhand, India, E-mail: sagarpriti68@gmail.com*

[2]*Assistant Professor, Central University of Jharkhand, Centre for Water Engineering & Management, Brambe, 835205, Jharkhand, India*

ABSTRACT

During the past several decades numerous major multiple-purpose reservoir systems have been constructed throughout the nation. About 5,500 large reservoir system exists in India, and generally, their operation is done through the rule curve. As demands on reservoir system is increasing day by day and along with this a continuously changing pattern of various factors like temperature, rainfall, land use land cover, etc. has been observed since the past several decades, evaluation of refinements and modifications in the existing operating rule or development of new policy of the reservoir systems is becoming an increasingly important activity. In the present study, an optimal operation policy has been developed for a multi-purpose reservoir, namely, Maithon Reservoir in order to maximize the hydropower production subject to the condition of satisfying the irrigation water demands (IWDs), domestic and industrial water demand (DIWD) with an eye on sustainable sediment management using a non-linear programming model (NLP). The operation policies have been derived by dividing the entire operation year into three segments, i.e., pre-monsoon (April to May), monsoon (June to October) and post-monsoon (November to March) in order to analyze the different possibilities during real-time operation. In each segment the percentage fulfillment of various purposes like IWD,

DIWD, and corresponding hydropower production are determined. The fulfillment of various purposes has been determined under 50%, 60%, and 80% and dependable inflow conditions. The distinguishing feature of this model is that it maximizes hydropower generation along with sustainable sediment management strategy also.

16.1   INTRODUCTION

Water is the most precious gift of nature, as the existence of no any flora or fauna is possible without water on this earth. It is a basic human need and a prime natural resource with finite dimensions which cannot be created and has no any other substitute. Although there is plenty of water present on the surface of the earth, only 3% of total water resources are fresh, and the rest 97% is contained in the seas and ocean as saline water. Out of the total fresh water, 75% occurs as polar ice and glaciers, and 24% remains in the subsoil. Only 1% of fresh water accounts for water in lakes and rivers.

The average annual rainfall in India is about 1170 mm, which corresponds to an annual precipitation (including snowfall) of 4000 billion m^3. Nearly 75% of this, i.e., 3000 billion m^3 occurs during monsoon season (June-September) in a year. Regional variations are also extreme in the country as the rainfall varies from 100 mm in Western Rajasthan to over 11000 mm in Meghalaya in Northeastern side of India. The extreme variability in the distribution of water (both temporally and spatially) makes the situation more complicated and seeks greater concern among water resource engineers and scientists, encouraging them for in-depth study and development of efficient surface water management systems and techniques.

During the past several decades, one of the most important advances made in the field of water resources engineering has been the development of optimization techniques for planning, design, and management of complex water resource system. Reservoirs are the most important component of a water resources development scheme, which serve to regulate natural stream-flow thereby modifying the temporal and spatial availability of water according to human needs. Optimal operation of reservoirs is crucial in the present context of water scarcity being faced by the country, due to a perceivable overall increase in water demands from various sectors like agricultural water demand, industrial and domestic water demand, hydropower generation demand, water pollution, and environmental concerns, flood mitigation concern, etc.

Due to continuously changing the pattern of land cover and land use in the upstream catchment the reservoir systems are becoming more vulnerable to increased sediment load which reduces the live storage capacity of the reservoir. The multipurpose reservoir systems are subjected to intensive water management efforts due to competing demands from irrigation, domestic, industrial water supply, and hydropower production. With an eye on the above needs of the hour, there is a necessity to rethink the way for the refinement and modification to the existing operation policy accounting for changing hydrologic conditions and also for a sustainable sediment management strategy.

16.2   MATERIALS AND METHODS

16.2.1   STUDY AREA DESCRIPTION

In the present study, an optimal operation reservoir operation model has been developed and applied on the Maithon multi-purpose reservoir system, located at Maithon, 48km from Dhanbad district, in the state of Jharkhand, India. The reservoir is built on the river Barakar, a tributary of Damodar River (catchment area of 6294 km^2) and designed for flood control, hydropower generation, irrigation, navigation, municipal water supply, etc. for the states of Jharkhand and West Bengal State. Figure 16.1 describes the details of Damodar Basin.

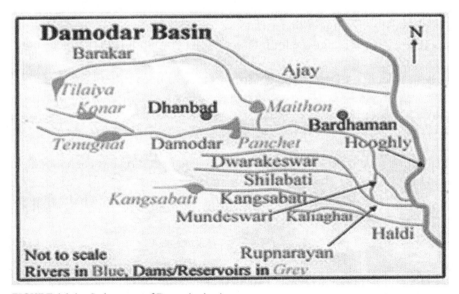

FIGURE 16.1   Index map of Damodar basin.

16.2.2 METHODOLOGY FOR HYDROPOWER OPTIMIZATION USING NLP

The present study attempts to develop an optimal operating policy which satisfies domestic, industrial, irrigation, and environmental flow demands optimally simultaneously producing optimum hydropower, optimum flood space with sustainable sediment management strategy. Optimal policy has been developed considering different dependable inflow levels for dry scenario (50% dependable inflow), normal scenario (60% dependable inflow) and wet scenario (80% dependable inflow), so that different possible operating policies can be generated, which will be helpful to the reservoir authorities in managing the real-time operation of the reservoir.

16.2.2.1 MODEL DESCRIPTION

In past time several optimization techniques have been applied to derive the optimal reservoir operational rules like linear programming (LP), non-linear programming (NLP), goal programming (GP), chance constraint linear programming (CCLP), dynamic programming (DP), and recently, the soft computing techniques. Among several techniques, NLP is widely applied for optimizing hydropower systems (Gagnon et al., 1974; Tejada-Guibert et al., 1990), since it is most accurate, involves no approximation and uses the physically based non-linear function (Barros et al., 2003). The software package LINGO (Language for Interactive General Optimizer) has been used to optimize hydropower generation from the reservoirs under consideration as both objective function and constraints are nonlinear.

16.2.2.2 OBJECTIVE FUNCTION

The various needs fulfilled by the Maithon dam operation has been modeled using the following objective function and constraints.

Objective: Maximize monthly hydropower production. Production of hydropower has been defined as:

$$Maximize \ (HP_t) = \sum_{t=1}^{12} [H_t \times R_t \times \eta \times \Upsilon_w] \tag{1}$$

where,

HP_t = monthly hydropower production in terms of kilowatt hours; (kWh) during month t;

H_t = monthly average head available during month t;
R_t = monthly release to powerhouse during month t;
η = combined turbine and generation efficiency;
Y_w = specific weight of water.

16.2.2.3   CONSTRAINTS

Reservoir Water Mass Balance Constraint: The theory of conservation of mass has to be satisfied as mass can neither be created nor be destroyed.

$$S_{t+1} = S_t + I_t - R_t - Q_t - E_t - S_{pt} \tag{2}$$

where,

S_{t+1} = Final storage in MCM during month t;
S_t = Initial storage in MCM during month t;
I_t = Dependable inflow in MCM during month t;
R_t = Release in MCM during month t;
Q_t = Release through gates in MCM during month t;
E_t = Evaporation Loss in MCM during month t;
S_{pt} = Spill in MCM during month t.

Municipal and Industrial Demand Constraints: Generally, the average water requirements for Municipal and Industrial (M&I) purposes are quite constant throughout the year, as compared to the water requirements for irrigation or hydropower.

$$Q_t \geq IR_t + O_t \tag{3}$$

where,

O_t = Release made during any month t in MCM;
IR_t = Irrigation water demand (IWD) during any month t in MCM;
Q_t = Water requirement of sub-basin other then IWD at any time t in MCM.

Flood Control Constraint: The strategy adopted for controlling flood is that every month some free space must be allocated in the reservoir to handle uncertainties.

$$S_{max} - S_t \geq 50\% \ of \ Flood \ Zone \ Space \tag{4}$$

where,

S_t = Storage at the beginning of month 't';
S_{MAX} = Maximum reservoir capacity.

Channel Capacity Constraint: The water release through the power plant should not exceed the channel carrying capacity. This can be written as:

$$R_t \leq C \tag{5}$$

where,

C   = Maximum carrying capacity of penstock during a month;

R_t = Release made during a month through penstock.

Bounds on Storage Constraint: The reservoir storage in any time should not be more than the maximum capacity of the reservoir and cannot fall below the dead storage level (DSL).

$$S_{MIN} \leq S_t \leq S_{MAX} \tag{6}$$

where,

S_{MAX} = Maximum capacity of the reservoir;

S_t = Storage capacity of the reservoir at the beginning of month t;

S_{MIN} = Minimum head required to produce electricity which is above the dead storage level.

Sediment Management Constraint: In reservoir planning, allowance is made for the storage of sediment. The volume provided for is known as dead storage. In this study, the allowance provided is equal to the minimum drawdown level (MDDL).

$$S_6 = 93\text{MCM} \tag{7}$$

where, S_6 = Storage at the starting of month June.

16.3   RESULTS AND DISCUSSION

In the present study, a monthly time step basis NLP model has been used to optimize the operational plans of Maithon reservoir to maximize the hydro-power production with an eye on sustainable sediment management strategy. The above formulated NLP model is solved using LINGO/Global solver (Lingo User Guide Lindo Systems, Inc., Illinois, 2011). The developed NLP model is optimized for 50%, 60% and 80% dependable inflows, estimated by Weibull's method (Subramanya, 2008). Based on the constraints of the above formulated NLP model, several policies have been generated for each dependable inflow condition by varying the irrigation demands in each policy. These policies will be helpful to the reservoir operators in assessing the full potential of the reservoir system.

16.3.1   UNDER 50% DEPENDABLE INFLOW CONDITION

Under this condition, five policies have been made. Policy 1 says, keeping all the demands fully satisfied in all three segments except irrigation demand in monsoon and pre-monsoon season which satisfies 95% times. Policy 2 says, keeping all the demands fully satisfied in all three segments except irrigation demand in monsoon and pre-monsoon season which satisfies 85% times. Policy 3 says, keeping all the demands fully satisfied in all three segments except irrigation demand in monsoon and pre-monsoon season which satisfies 75% times. Policy 4 says, keeping all the demands fully satisfied in all three segments except irrigation demand in monsoon and pre-monsoon season which satisfies 65% times. Policy 5 says, keeping all the demands fully satisfied in all three segments except irrigation demand in monsoon and pre-monsoon season which satisfies 55% times.

16.3.2   UNDER 60% DEPENDABLE INFLOW CONDITION

Under this condition also five policies have been generated. Policy 1 says, keeping all the demands fully satisfied in all three segments except irrigation demand in monsoon and pre-monsoon season which satisfies 95% and 45% times respectively. Policy 2 says, keeping all the demands fully satisfied in all three segments except irrigation demand in monsoon and pre-monsoon season which satisfies 65%& 35% times, respectively. Policy 3 says, keeping all the demands fully satisfied in all three segments except irrigation demand in monsoon and pre-monsoon season which satisfies 55% and 25% times respectively. Policy 4 says, keeping all the demands fully satisfied in all three segments except irrigation demand in monsoon and pre-monsoon season which satisfies 35% and 15% times respectively. Policy 5 says, keeping all the demands fully satisfied in all three segments except irrigation demand in monsoon and pre-monsoon season which satisfies 25% and 0% times respectively.

16.3.3   UNDER 80% DEPENDABLE INFLOW CONDITION

For 80% inflow condition only one policy was found to be feasible. Keeping all the demands fully satisfied in all three segments except irrigation demand which satisfies 15% and 0% times in monsoon and pre-monsoon season, respectively.

16.3.4   OPTIMAL RESERVOIR OPERATION POLICY

The results shows that at 50% dependable inflow, the policy 1 satisfies all the requirements during all the segments 100% times (except pre-monsoon irrigation demand which is satisfied 90% times) and produces minimum annual hydropower of 94653.45 mWh and trial 5 produces maximum annual hydropower of 98270.39 mWh satisfying all the requirements during all the segments 100% times (except pre-monsoon and monsoon irrigation demand which is satisfied 55% times). In a similar way considering different dependable inflows (60% and 80%) various feasible trials have been derived and corresponding annual hydropower generation have been shown in Table 16.1.

TABLE 16.1   Annual Hydropower Production From Various Policies for Different Inflow Conditions

Policy	Annual hydropower (mWh/yr) for dependable inflow conditions		
	50%	60%	80%
1	94653.45	64503.42	42312.06
2	95410.64	65310.83	-
3	97009.48	66057.95	-
4	97536.36	66805.95	-
5	98270.39	66819.27	-

For 80% inflow condition only one policy was found to be feasible. On comparing all the five policies, it was found that on relaxing the irrigation release slightly the power production can be increased. The derived policy of tradeoffs between irrigation release and hydropower generation will be helpful to the reservoir operators and engineers in assessing the full potential of the reservoir system.

16.4   SUMMARY

A wide range of researches has been carried out in the field of reservoir operation policies. Among those researches, optimizing hydropower production was found to be the main point of focus by many researchers. One of the key feature of this model which distinguishes it from the other one is that it focuses on optimizing hydropower production with the aim of sediment management strategy also. The study also indicates that there is a good scope for further research and development in this field.

Following conclusions are derived based on the above study:

- In the present study, the Maithon reservoir operation policies have been optimized for hydropower production while satisfying all other demands using the NLP model.
- The derived policy is capable of producing maximum hydropower of 98270.39, 66819.27, and 42312.06 mWh/year for 50%, 60%, and 80% dependable inflow conditions, respectively.
- The derived model can be useful in minimizing sediment deposit and thus increasing the reservoir life.

KEYWORDS

- **efficient reservoir operation**
- **operation policy**
- **optimal hydropower generation**
- **sustainable sediment management**

REFERENCES

Arunkumar, R., & Jothiprakash, V., (2012). Optimal reservoir operation for hydropower generation using Non-linear programming model. *Journal of Institution of Engineers, India, 93*(2), 111–120.

Barros, M. T. L., Tsai, F. T. C., Yang, S. L., Lopes, J. E. G., & Yeh, W. W. G., (2003). Optimization of large-scale hydropower system operations. *Journal of Water Resource Planning and Management, ASCE, 129*(3), 178–188.

Gagnon, C. R., Hicks, R. H., Jacoby, S. L. S., & Kowalik, J. S., (1974). A nonlinear programming approach to a very large hydroelectric system optimization. *Math Program, 6*, 28–41.

Husain, A., (2012). An overview of reservoir systems operation techniques. *International Journal of Engineering Research and Development, 4*(10), 30–37.

Lingo User Guide, (2011). (Lindo Systems, Inc., Illinois), p. 834.

Subramanya, K., (2008). *Engineering Hydrology* (3rd edn., p. 452). Tata McGraw-Hill Education Pvt. Ltd.

Tejada-Guibert, J. A., Stedinger, J. R., & Staschus, K., (1990). Optimization of value of CVP's hydropower production. *Journal of Water Resource Planning and Management, ASCE, 116*(1), 52–70.

CHAPTER 17

AN ANALYSIS OF FLOOD CONTROL IN EASTERN SOUTH ASIA

AMARTYA KUMAR BHATTACHARYA

Chairman and Managing Director, MultiSpectra Consultants, 23, Biplabi Ambika Chakraborty Sarani, Kolkata – 700029, West Bengal, India, E-mail: dramartyakumar@gmail.com

ABSTRACT

This chapter studies the causes and features of floods and measures for the control of floods in eastern India and Bangladesh. Structural as well as non-structural measures are being emphasized for flood management in both India and Bangladesh. It has been proved that non-structural measures have a significant effect on flood damage minimization.

17.1 INTRODUCTION

The northeastern region of India comprises seven states, namely Assam, Meghalaya, Mizoram, Manipur, Nagaland, Tripura, and Arunachal Pradesh and the region borders the countries of Bhutan, China, Bangladesh, and Myanmar. The entire region is one of the most hazard-prone regions in the Asian continent, with different areas being prone to multi-hazards like earthquakes, floods, landslides, and cyclonic storms. The rivers Brahmaputra and Barak drain the region. The Brahmaputra river has a catchment area of 5,80,000 km in Tibet, Bhutan, India, and Bangladesh and in terms of discharge is the third largest river in the world, in terms of sediment load it is second after the Hwang-Ho river of China. The river flows for a length of 918 km in India of which 720 km is in through the plains of Assam. In this valley, about 20 major tributaries on its North bank and about 13 on its South Bank join the river Brahmaputra. The precipitation here is mainly due to the South-West monsoon. Heavy rainfall occurs from June to September. Average annual rainfall in the region is very high and ranges from 1750 mm in the plains to about 6400 mm in the hills, this huge volume of water rushes through the narrow bowl-shaped valley of Assam to the Bay of Bengal ravaging the area through floods and land erosion. The recurring floods on an average devastate about 20% of the total area of the plain districts of the state of Assam, and in the high floods years, the devastation has been recorded to be as high as 67%. The region lies at the junction of the Himalayan arc to the north and the Burmese arc to the east and is one of the six most seismically active regions of the world.

The monsoon in the region normally commences around the months of April, May, and is active until the end of October. The pre-monsoon period is often marked by severe cyclonic storms and hailstorms. The annual cyclonic depressions in the Bay of Bengal along the coast of Bangladesh cause severe storms to hit the bordering states of Meghalaya and Tripura. Bangladesh stretches between latitudes 20°34'N and 26°38'N and longitudes 88°01'E and 92°41'E. The country contains the confluence of a distributary of the Ganga (the other distributary, also called Ganga, passes through West Bengal and drains into the sea at Ganga Sagar), Brahmaputra, and Meghna Rivers and their tributaries, which originate in the Himalayas (except the Meghna, which in its upstream portion is called the Barak) and discharge into the Bay of Bengal. The terrain is mainly flat, and with 90% of its landmass, up to 10 meters above the mean sea level, is a primarily low-lying riverine country. It is frequently hit by natural disasters, particularly floods, riverbank erosion, cyclones, and droughts. Each affects the livelihoods of those affected, but with different severity. Displacement due to flood, erosion, and inadequate facilities during and after major floods, as shown in Figure 17.1, can create major hardship and health problems.

The climate of Bangladesh is a tropical monsoon, influenced by the Himalaya Mountains in the north and the Bay of Bengal in the south. High monsoon rains associated with Bangladesh's unique geographical location in the eastern part of the delta of the world's second largest river basin make it extremely vulnerable to recurring floods. Agriculture is the dominant land use in the country covering about 59% of the land, rivers, and other water bodies constitute about 9% (BBS, 2002). Monsoons with varying degrees of associated flooding are anticipated annual events in Bangladesh.

The state of West Bengal lies in the eastern part of India and is flood-prone with floods occurring with a depressing regularity. A number of factors combine to cause floods in southern West Bengal. There is extremely high rainfall in the monsoon season. The seaward slope of southern West Bengal is very low, and the Ganga delta is tidal in nature. There are several low-lying areas where water lies stagnant. There is silting of several outlet canals reducing carrying capacity. In addition, there is human encroachment on some channels hampering renovation of those channels (Figure 17.2).

FIGURE 17.1   Erosion and inadequate facilities during and after major floods in Bangladesh (*Source*: The Daily Prothom Alo, 2004).

FIGURE 17.2   Flood in Assam, India (*Source*: Assam_Disaster_Management.htm).

17.2   TYPES OF FLOODS

The term flood is generally used when the flows in the rivers and channel cannot be contained within natural or artificial riverbanks. By spilling the riverbanks, when water inundates floodplains and adjoining high lands to some extent or when the water level in the river or channels exceeds certain stage, the situation then termed as flood (Hossain, 2004). Important river basins and type of floods are shown in Figures. 17.3–17.8.

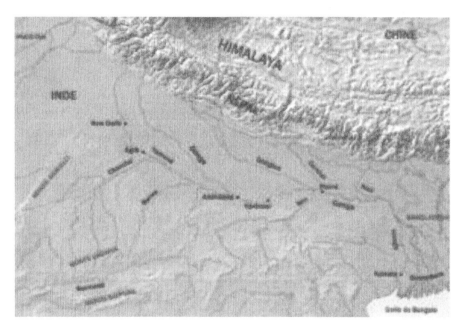

FIGURE 17.3   Ganga Basin in India and Bangladesh.

17.2.1   FLASH FLOOD FROM HILLY AREAS

Flash flood-prone areas of the India and Bangladesh are at the foothills. Intense local and short-lived rainfall often associated with mesoscale convective clusters is the primary cause of flash floods. These are characterized by a sharp rise followed by a relatively rapid recession. Often with high velocities of on-rush flood damages crops, properties, and fish stocks of the wetland. Flash flood can occur within a few hours and are particularly frequent in the months of April and May.

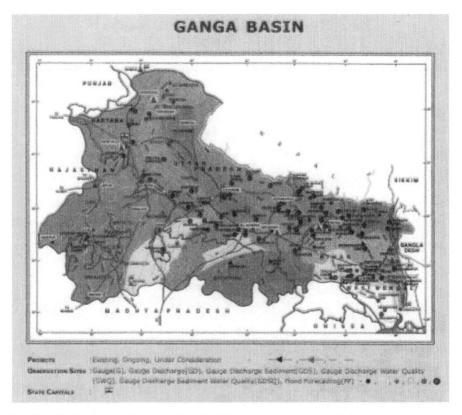

FIGURE 17.4 Ganga river basin in India (*Source*: www.wrmin.nic.in).

17.2.2 MONSOON FLOODS OR NORMAL FLOOD FROM MAJOR RIVERS

River flood is a common phenomenon in India and Bangladesh and is caused by bank overflow. Of the total flow, around 80% occurs in the 5 months of monsoon from June to October (WARPO, 2004). A similar pattern is observed in case of rainfall also. Therefore, to these skewed temporal distributions of river flow and rainfall, India, and Bangladesh suffer from an abundance of water in monsoon, frequently resulting into floods and water scarcity in other parts of the year, developing drought conditions.

In the Brahmaputra, maximum discharge occurs in an early monsoon in June and July whereas in the Ganga maximum discharge occurs in August and September. Synchronization of the peaks of these rivers results in devastating floods in India and Bangladesh. The rivers of Bangladesh drain about 1.72

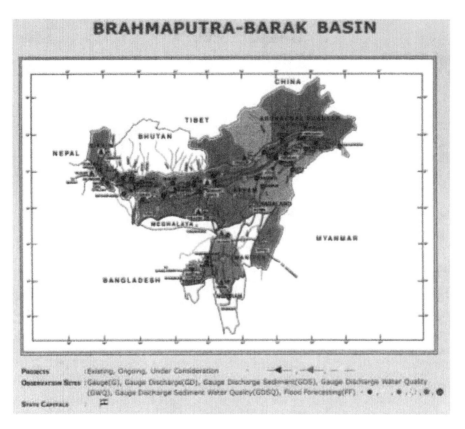

FIGURE 17.5   Brahmaputra and Barak river basins in India (*Source*: www.wrmin.nic.in).

million sq km area of which 93% lies outside its territory in India, Nepal, Bhutan, and China. The annual average runoff of the transboundary rivers of Bangladesh is around 1200 cubic kilometers (WARPO, 2004). A major impediment is the lack of accurate data on a real-time basis on stream flow in the vast upper reach of the Brahmaputra in Tibet in China (the river is called Tsangpo there) and the IRS (Indian Remote Sensing) series of satellites launched from, first the ASLV, and now the PSLV series of rockets from Shriharikota in southern India have to be used to get data. Considering that the floods in the Brahmaputra affect the state of Assam in India as well as Bangladesh, very accurate data on streamflow into India from China on a real-time basis would indubitably be very helpful. However, some streamflow data is being made available to India by China. A cause of concern for the subcontinent as a whole is the Chinese River Linking Plan in which the waters of the Brahmaputra would be diverted northwards via the Yangtze-Kiang River to the Hwang-Ho

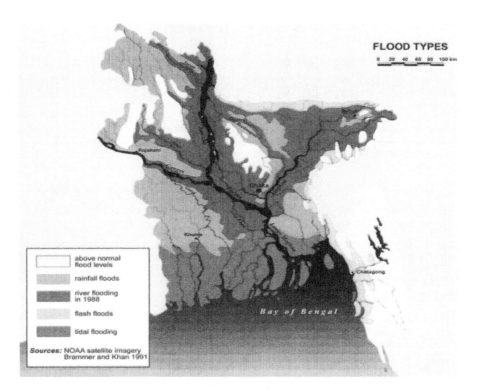

FIGURE 17.6   Types of the flood in Bangladesh (*Source*: FFWC, Dhaka).

(**Note:** Rivers outside the boundaries of Bangladesh are not shown in the figure).

river. This plan, if implemented excluding the water needs of the subcontinent, would result in severe water deficit in the subcontinent.

It may be mentioned that India too has a River Linking Plan (NWP, 2002) to ensure equitable distribution of water and control of floods but much debate is on this issue and Bangladesh, Nepal, and Bhutan would certainly be included in the plan, if it is at all implemented, so that all the countries can get their share of the benefits of this integrated and holistic sub-continental River Linking Scheme. With reference to Figures 17.3 and 17.8, the Indus River Basin and the Ganga River Basin are separated by a low watershed. As per the India-Pakistan Treaty, 1961, the waters of the Indus River Basin are to be shared by the two countries such that India gets the full share of water flowing through the basin via the Sutlej, Beas, and Ravi rivers while Pakistan gets the full share of water flowing through the basin via the Chenab, Jhelum, and Indus Rivers. It is envisaged to construct a canal connecting the Sutlej River in the Indus basin with the Ganga basin so as to divert some water

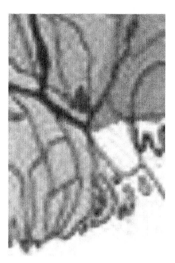

FIGURE 17.7 Confluence of a distributary of the Ganga, Brahmaputra, and Meghna (lower Barak) River Basins in Bangladesh (*Source*: FFWC, Dhaka).

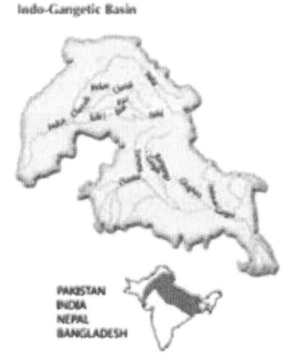

FIGURE 17.8 Indo-Gangetic basin (*Source*: IWMI).

from the Indus Basin to the Ganga Basin without, in any way, impinging on the water rights of Pakistan because the waters of the Sutlej are fully allocated to India. The objective is to augment and increase the discharge in the Ganga to meet the needs of the Ganga basin fully.

17.2.3 FLOODS DUE TO STORM SURGES

This kind of flood mostly occurs along the coastal areas of Bangladesh and West Bengal. Continental shelves in this part of the Bay of Bengal are shallow and extend to about 20–50 kms. Moreover, the coastline in the eastern portion is conical and funnel-like in shape. Because of these two factors, storm surges generated due to any cyclonic storm is comparatively high compared to the same kind of storm in several other parts of the world. In case of the super-cyclones maximum height of the surges were found to be 10–15 m, which causes flooding in the entire coastal belt. The worst kind of such flooding was on 12 Nov 1970 and 29 April 1991, which caused loss of 300,000 and 138,000 human lives respectively (FFWC, 2005). Coastal areas are also subjected to tidal flooding during the months from June to September when the sea is in spate due to the southwest monsoon wind.

17.3 GENERAL PATTERN

The Brahmaputra starts rising in March due to snowmelt on the Himalayas, which causes the first peak in May or early June. It is followed by subsequent peaks up to the end of August caused by the heavy monsoon rains over the catchments. The response to rainfall is relatively quick, resulting in rapid increases in the water level. The Ganga starts rising gradually in May-June to a maximum sometime in August. High water levels are normally sustained until mid-September. The Meghna may not attain its annual peak until August-September. The upper Meghna carries only about 10% of the flow in the Ganga and Brahmaputra. The total volume of runoff in the GBM is determined by the net precipitation over all the catchments. The normal sequence of floods with flash floods in the eastern hill streams during the pre-monsoon period in the months of April and May. High floods occur if the peaks of the Ganga and Brahmaputra coincide; this may happen during August-September (Rahman et al., 2007).

17.3.1   FLOOD DAMAGES IN BANGLADESH

The terrain has experienced seventeen highly damaging floods in the 20th century. Since independence in 1971, Bangladesh has experienced floods of vast magnitudes in 1974, 1984, 1987, 1988, 1998, 2000, and 2004 (FFWC, 2005). The largest recorded flood in depth and duration of flooding in its history occurred in 1998 when about 70% of the country was under water for several months (FFWC, 2005; Nishat et al., 2000). The area affected in percent of the total area during major flood event inundating more than 20% of the country's land area of the country is presented in Figure 17.9. The damages during some severe floods are presented in Table 17.1.

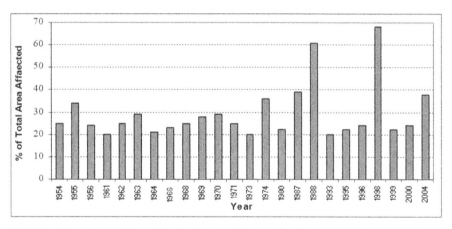

FIGURE 17.9   Area of Bangladesh affected during major flood events (in percent of the total area).

17.4   FLOOD MANAGEMENT IN BANGLADESH

Bangladesh tries to deal with flood and disaster with structural and non-structural measures. Systematic structural measures began by implementing flood control projects in the sixties after the colossal flood of 1963. Non-structural measures have introduced in the seventies. Flooding is a natural phenomenon, which cannot be prevented. Complete flood control is not in the interests of most Bangladeshi farmers. The flood control measures and policies should be directed to mitigation of flood damage, rather than flood prevention. Resources should be allocated to help people adopt a lifestyle that is conformable to their natural environment. Indigenous solutions such as changing the housing structures and

TABLE 17.1   Some Notable Flood and Cyclone Induced Storm Surge Events in Bangladesh

Event	Impact
1974 flood	Inundated 36% of the country (FFWC, 2005), estimated damages US$ 57.9 Million, over 28,700 deaths, (http://www.em-dat.net/disasters/Visualisation/profiles/natural-table-emdat.php?country=Bangladesh dated 2.3.2006)
1987 flood	Inundated over 57,000 sq-km area, estimated damage US$ 1.0 billion and human death 2055 (The World Bank, 2002)
1988 flood	Inundated 61% of the country, persons affected 45 million, 2300 deaths, damage worth about US$ 1.2 billion (The World Bank, 2002)
1998 flood	Inundated 100,250 sq-km (68%) of the country, 1100 deaths, persons affected 31 million, damaged 500,000 homes, 23,500 km roads and 4500 km embankment, destroyed crops of 500,000 ha of land, damage worth about US$ 2.8 billion (The World Bank, 2002)
2004 flood	Inundated 38% of the country, 750 deaths, persons affected 36 million, damaged 58,000 km roads and 3,100 km embankment, crop damage 1.3 million ha, damage worth about US$ 2.2 billion (ADB-World Bank, 2004)

crop patterns can help reduce flood damage. Moreover, good governance, appropriate environmental laws, acts, and ordinances will be necessary to achieve sustainable economic development and to reduce any environmental degradation. In addition, implementation of an improved real-time flood and drought control warning system can reduce the damage caused by floods (http://www.bytesforall.org/8th/control_flood.htm dated 2.3.2006). In recent years, improved forecasting, early warning system, and preparedness measures have helped to reduce the number of lives lost by natural disasters.

17.4.1   FLOOD MANAGEMENT BY STRUCTURAL MEASURES

The structural option provided some benefits especially increase in agricultural production (BWDB, 2005; BBS, 2002) at an earlier period but some adverse effects were observed later on (Nishat et al., 2000). Notably, the construction of a high embankment along both banks of the rivers in some cases resulted in a rise in bed levels due to siltation causing obstruction to drainage. In the coastal areas, although the construction of polders prevented salinity intrusion, but resulted in restriction of the movement of the tidal prism, sedimentation of tidal rivers and obstruction to the gravity drainage. Another important impact on agriculture was found that the farmers in most cases opted for production of cereal crops, especially HYV rice enjoying a flood free situation rather than going for crop diversification. Structural measure caused many adverse effects on the aquatic lives, especially on open water fisheries. National and regional highways and railways, to the extent feasible, have been raised above flood level. Raising feeder and rural roads will be determined in the context of disaster management plans. River maintenance through dredging is also going on in a limited case due to the high cost. Efforts are continued for erosion control on medium and small rivers. Several Flood Control, Drainage, and/or Irrigation (FCD/I) projects have been constructed. FCD/I project are of two types, namely, (i) full flood control facilities; and (ii) partial flood control. Until date, FCD/I projects provide facilities in about 5.38 million ha which is about 59% of the country's net cultivated land (BWDB, 2000). Flood control and drainage structures have also been provided in major cities to make the cities flood free.

17.4.2   FLOOD MANAGEMENT BY NON-STRUCTURAL MEASURES

Introduction of non-structural option, i.e., Flood Forecasting and Warning System I Bangladesh started from the early '70s and contributed to the improvement of the capacity for flood preparedness and mitigation/minimization of flood losses. Other non-structural measures are discussed in the following.

Flood cum Cyclone Shelter: School buildings are so constructed that they can be used as flood-cum cyclone shelter especially in the coastal zone with the highest risk of flood and storm surge. These structures are not intended to change the flood regime, and therefore, considered as no-structural measures of flood management.

Floodproofing: Efforts have been made to provide vulnerable communities with mitigation by raising homesteads, schools, and marketplaces in low-lying areas (rather than flood control) and in the charred land so that peasants can save their livestock and foodstuff.

The concept of *flood zoning* and *flood insurance* are not practiced in the country until date. Flood zoning will facilitate development in a co-coordinated way to avoid expensive investments in vulnerable areas. Proper land development rules need to be developed based on the flood-zoning map.

Other non-structural measures practiced are:

*   Working with communities to improve disaster awareness.
*   Developing disaster management plans.
*   Relief and evacuation.

17.4.3   FLOOD FORECASTING AND WARNING IN BANGLADESH

A flood warning is concerned to reduce sufferings to human life and damages of economy and environment. Flood Forecasting and Warning Service of Bangladesh was established in 1972 as a permanent entity under Bangladesh Water Development Board (BWDB). Initially, co-axial correlation, gauge-to-gauge relationship, and Muskingum-Cunge Routing Model were used for forecasting. From the early nineties, a numerical modeling based approach has been applied for flood forecasting and warning. Using the principal concept of mass transfer based on the continuity and momentum equations, dynamic computation has been used in this method. Very briefly, it comprises of estimating water levels using hydrodynamic simulation model (MIKE 11). Research on Modeling System and capacity building in the forecasting is currently emphasized. During the severe flood in Bangladesh and West Bengal, India, in 1998, loss of lives and damage of FCD/I projects in Bangladesh were minimum mainly because of flood forecasting and early warning (Islam and Dhar, 2000).

17.5   DISASTER MITIGATION

Disaster management (including disaster preparedness) involves prevention and mitigation measures, preparedness plans and related warning systems, emergency response measures, and post-disaster reconstruction and rehabilitation. The main aims for water-related disaster management are to provide the means by which, through a combination of structural and non-structural measures and to the extent feasible and affordable, people are adequately warned of an approaching disaster, and are adequately supported in rebuilding their lives thereafter. The vulnerability to natural disasters combined with the socio-economic vulnerability of the people living in the different states of India poses a great challenge for the government machinery and underscores the need for a comprehensive plan for disaster preparedness and mitigation. The Government of India since the last decade has been actively supporting programs for the reduction of vulnerabilities and risks. UNDP has been a partner of the Government of India in such efforts. Vulnerability reduction and linking with sustainable development efforts have been one of the key approaches of UNDP. Strengthening capacities for disaster risk reduction and sustainable recovery process across the country and bringing together skills and resources for making communities disaster resistant is one of the first steps taken in the long term for achieving a reduction in loss of lives and protecting the development gains.

Quite a few measures may be taken to reduce floods in West Bengal. The network of drainage canals is to be increased, and silted drainage canals are to be dredged to augment channel capacity and allow free flow of excess water through those channels. More dykes are to be built to prevent floodwater from entering low-lying areas, and existing dykes are to be strengthened to prevent their breaching. If possible, human habitation is to be evacuated from flood-prone areas. Pumps of adequate capacity are to be kept on standby to pump out water particularly from low areas. Better meteorological forecasting is necessary so that the water levels in the Damodar Valley Corporation reservoirs can be brought down early enough to accommodate high inflows from upstream in flood periods. Adequate discharge channels are to be provided in the lower Damodar basin; this area is suffering from flood due to inadequate discharge channels. The capacity of the Mayurakshi river also needs to be augmented. Floods in West Bengal can be prevented or reduced by taking adequate structural and non-structural measures.

The Disaster Management Department, Government of West Bengal, India, (MOFM, 2006) has emphasized that during floods, large tracts of land get inundated and, thereby, disconnected from the adjoining areas resulting in disruption of normal day-to-day activity in that area. Though natural calamities like a flood cannot be avoided, its impact in terms of loss of lives and damage to properties can be minimized by undertaking appropriate management practices for preparedness, prevention, and mitigation measures. This constitutes a holistic approach towards management of flood with emphasis not only on the traditional post-disaster response; but also on pre-disaster preventive/mitigation preparedness as well, thereby, laying down a Standard Operating Procedure (SOP) for a Disaster Manager for flood management (www.wbgov.com).

The Government of Bangladesh (GoB) established the Disaster Management Bureau (DMB) in 1993, which has prepared comprehensive Disaster Management Plans. DMB is working under the Ministry of Disaster Management and Relief. Standing orders on Disaster have been prepared in 1997 and upgraded in 1999 by the DMB (Chowdhury, 2003). At the central level, a National Disaster Management Council (NDMC) has formed headed by the Honorable Prime Minister including Ministers from different ministries as a member. Inter-Ministerial Disaster Management Co-ordination Council (IMDMCC) has also been formed which guided by the NDMC. Beside this,

District, Thana (area under the jurisdiction of a Police Station) and Union (lowest level of local government) level committees have also formed with the participation of local community for post-disaster management and mitigation. Task and responsibilities of each committee are stated in the standing order (MoDMR, 1997). By all these steps GoB has strengthened the disaster response capacity through institutional capacity building activities; community disaster response simulation drills; and stockpiling of essential relief items.

Forecasting facilities, preparedness planning, during, and post-disaster relief efforts have reduced the severity of flood disaster impacts. Non-Government Organizations (NGOs) have also responded in an important way. It has been observed that emergency flood fighting during peak flood, evacuation, and relief operation can best be achieved with peoples' participation along with deployment of the army.

17.6 CONCLUSIONS

Structural as well as non-structural measures are being emphasized for flood management in both India and Bangladesh. It has been proved that non-structural measures have a significant effect on flood damage minimization. Flood and disaster cannot fully be controlled, prevented or eliminated, but damages can be reduced significantly by the integration of measures and coordination of agencies. Flood forecasting and early warning are very important. Co-operation is needed at all levels for research and development for improvement of flood mitigation measures.

KEYWORDS

- **Bangladesh**
- **flood control**
- **flood damages**
- **flood mitigation**
- **India**

REFERENCES

Assam Disaster Management. www.AssamDisasterManagement.htm (accessed on XX-XX-XXXX).

Bangladesh Bureau of Statistics (BBS), (2002). *Ministry of Planning*, Bangladesh.

Bangladesh Water Development Board, (BWDB), (2000). Annual Report, Dhaka, Bangladesh.

Chowdhury, J. R., (2003). *Technical Paper Presented in the 47th Annual Convention of the Institution of Engineers Bangladesh (IEB)*, Chittagong, Bangladesh.

FFWC, (2005). *Consolidation and Strengthening of Flood Forecasting and Warning Services* (Vol. II). Final Report, Monitoring, and evaluation, Bangladesh Water Development Board, Dhaka, Bangladesh.

Flood Forecasting and Warning Centre (FFWC), (2005). *Annual Flood Report*, BWDB, Dhaka, Bangladesh.

Hossain, A. A. N. H., (2004). *Flood Management: Issues and Options*. Presented in the International Conference organized by the Institution of Engineers, Bangladesh.

India-Pakistan Water Treaty, (1961). New Delhi, India and Islamabad, Pakistan.

International Water Management Institute (IWMI), (2007). Colombo, Sri Lanka.

Islam, S. R., & Dhar, S. C., (2000). *Bangladesh Floods of 1998: Role of Flood Forecasting & Warning Centre*. BWDB, Dhaka, Bangladesh.

Joint ADB-World Bank, (2004). *Emergency Flood Damage Rehabilitation Project*. Joint ADB-World Bank damage and need assessment, Dhaka, Bangladesh.

Ministry of Disaster Management and Relief (MoDMR), (1997). *Standing Order for Disaster Management*. Dhaka, Bangladesh.

Monograph on Flood Management, (2006). *Department of Disaster Management*. Government of West Bengal, Kolkata, India.

National Water Policy, (2002). India.

Nishat, A., (2000). *The 1998 Flood: Impact on Environment of Dhaka City*. Ministry of Environment & Forest and IUCN Bangladesh, Dhaka, Bangladesh.

Rahman, M. M., (2005). Geo-informatics approach for augmentation of lead time of flood forecasting- Bangladesh Perspective, Proceedings of International Conference on Hydrological Perspectives for Sustainable Development in the Department of Hydrology. Indian Institute of Technology, Roorkee, Uttaranchal, India.

Rahman, M. M., Hossain, M. A., & Bhattacharya, A. K., (2007). Flood management in the floodplain of Bangladesh. *Proceedings, International Conference on Civil Engineering in the New Millennium: Opportunities and Challenges*. Bengal Engineering and Science University, Shibpur, Howrah, India, Paper No. WRE 015.

The World Bank, (2002). Bangladesh Disaster & Public Finance, Paper no 6, Dhaka, Bangladesh.

Water Resources Planning Organization (WARPO), (2004). National Water Management Plan (NWMP), Ministry of Water Resources, Bangladesh, Dhaka, Bangladesh.

www.em-dat.net/disasters/Visualisation/profiles/natural-table-emdat.php?country=Bangladesh dated 2.3.2006 (accessed on XX-XX-XXXX).

www.wbgov.com (accessed on XX-XX-XXXX).

www.wrmin.nic.in (accessed on XX-XX-XXXX).

PART III
Groundwater Management

HYDRO-GEOLOGICAL STATUS OF THE CORE AND BUFFER ZONE OF BEEKAY STEEL INDUSTRIES LIMITED, ADITYAPUR INDUSTRIAL AREA, SARAIKELA, KHARSAWAN, JHARKHAND

UTKARSH UPADHYAY[1], NISHANT KUMAR[2], RANDHIR KUMAR[3], and PRITI KUMARI[4]

[1]Research Scholar (PhD), Center for Water Engineering and Management, Central University of Jharkhand, Ranchi, India, E-mail: utkarsh.wem.cuj@gmail.com

[2]Technical Specialist, Drinking Water and Sanitation Department, Government of Jharkhand, Ranchi, India

[3]Research Scholar (PhD), Center for Water Engineering and Management, Central University of Jharkhand, Ranchi, India

[4]Assistant Professor, Department of Civil Engineering, GGSESTC, Bokaro, Jharkhand, India

ABSTRACT

It has become imperative on the part of water planners to adopt techniques for quantifying the available groundwater resources for sustainable development and management keeping in mind the scarcity of available water resources versus its demand in the near future. The apparent heterogeneities and complexities present in the hard rock aquifers make it a challenge to tackle groundwater problems. The intricacy increases manifold for the management of groundwater when the hard rock aquifers are situated in arid or semi-arid regions. This study was taken up to generate hydro-geological data through field investigations to prepare a groundwater assessment report for the expansion and increase in groundwater extraction of Beekay Steel Plant, Adityapur Industrial Area, Saraikela-Kharswana, so as to study the existing groundwater scenario and impact of the withdrawal of groundwater on groundwater regime in and around the plant area.

The geology and subsurface conditions were studied and interpreted based on the groundwater exploratory data and geological studies made by GSI and AMD. Well inventory with a collection of water samples were done to evaluate the status of groundwater level and water quality in the buffer zone within 10 Km radius from the plant area. The average seasonal fluctuation of the water table works out to be 3.96m. The perusal of concentration of various chemical constituents of water indicates that groundwater is suitable for all the purposes viz. drinking, and domestic purposes. Pumping tests were conducted within the plant area to know the aquifer characteristics, and subsequently, aquifer parameters were calculated. ΔS calculated from drawdown-time curve has been found to be 2.00 m and transmissivity calculated from Jacob's Method has been found to be 12.096 m^2d^{-1}. The value of the radius of influence (R) comes as 169 m. Low permeability and small radius of influence indicate the marginal impact of pumping on local water regime. It is found that the stage of groundwater development in which the core zone fall, does not exceed 20% and marginal changes occur in the buffer zone after the proposed increase in groundwater extraction, which is a positive factor for taking up groundwater abstraction for the expansion.

18.1   INTRODUCTION

M/s Beekay Steel Industries Ltd. (BSIL) a Public Limited Company incorporated under 'Companies Act, 1956' having its registered office at Lansdowne Towers, 2/1A, Sarat Bose Road, 4th Floor, Kolkata, West Bengal, India intends to expand its Rolling Mill unit at its existing premises which is located Large Scale Sector, Adityapur Industrial Area, Gamharia, and District-Seraikela - Kharswan, in Jharkhand. BSIL believes in producing its quality products in an eco-friendly manner. BSIL will be in a very competitive position having control over all the major inputs required to produce TMT Bar. Power and Billets are the major inputs for production of TMT Bar. Looking at the projection of demand for Steel in the present market, it can be realized that there remains a considerable gap between demand and supply of steel by 2015–2016. To meet the demand, BSIL is expanding its production and as a resulting requirement of groundwater will increase from 40 cumd⁻¹ to 160 cumd⁻¹.

Presently there is moderate utilization of water within the plant area for domestic and industrial consumption only. However, it is proposed to meet the primary requirement of industrial as well as domestic water from groundwater sources. The daily water requirement is estimated to increase from the present requirement of 40 cumd⁻¹to 160 cumd⁻¹. A per prevailing groundwater legislations it is mandatory to obtain the permission of the concerned regulating authority prior to any abstraction of water from the sub-soil sources beyond 100 cumd⁻¹. In order to obtain the permission and fulfill the statutory requirements hydro-geological report over an area of 10 km radius surrounding the project site is a prerequisite.

18.1.1   STATUTORY COMPLIANCE

As per prevailing groundwater legislation, it is mandatory to obtain due permission of the concerned Government controlling/regulating authority prior to any abstraction of water from the sub-soil sources. In order to obtain the required permission, submission of a hydro-geological report with evaluation of groundwater potential over an area of 10 km radius surrounding the project site is a prerequisite. The present report has been prepared after field studies and analysis of various hydrological data collected through secondary sources to fulfill the conditions for submitting the application to obtain permission for groundwater abstraction. As per the prevailing groundwater regulations, i.e., guidelines/criteria for evaluation of proposals/requests for groundwater abstraction (with effect from November 15, 2012), it is mandatory to obtain due permission of the concerned Government controlling/ regulating authority prior to any abstraction of water from underground. In order to get the environmental clearance from the concerned authorities' environmental impact assessment studies are being carried out by different consulting agencies.

A field visit was made for collecting hydro-geological data of key wells for studying the present groundwater conditions, estimating the long-term groundwater recharge, the present status of groundwater development. The post-monsoon hydro-geological data from 12 well was collected for November 2017. The present report has been prepared after field studies for core and buffer zones of Beekay Steel Industries Limited and the impact of groundwater abstraction on the water regime and analysis of various hydro-geological data collected through secondary sources to fulfill the condition for submitting, the application to obtain permission for required groundwater abstraction.

18.1.2   OBJECTIVE

The following objectives were taken into account for hydro-geological investigation of the study area:

1. To assess the present hydrological scenario of the study area.
2. To find out aquifer geometry in the area.
3. To evaluate the status of groundwater condition in the area.
4. To evaluate the hydraulic behavior of the aquifer system in the area.
5. To assess the groundwater resources of the area.
6. To assess the impact of present withdrawal on groundwater regime.
7. To find out the hydrochemical character of water resources in the present area.

18.2   METHODOLOGY

To achieve the goal, it becomes essential to evaluate the exact hydro-geological conditions, aquifer parameters, and aquifer geometry, etc. for this purpose data were collected from the reports available in Central and State Government departments. The geology of the area and subsurface conditions has been interpreted based on the exploratory data and geological studies made by G.W.D & C.G.W.B. Intensive well inventory of the area have been undertaken to measure the status of the water table in the study area. The water samples were collected for analysis in order to establish the water quality. To evaluate the aquifer parameter pump test have been conducted in a surrounding area where the rate of depletion in water level at constant pumping rate were observed. The aquifer parameters were calculated using the standard analytical technique to field conditions of the testing site. The groundwater resources and its utilization have been worked out as per the norms prescribed by the groundwater estimation committee, Govt. of India. The impact of groundwater abstraction, on groundwater storage, has been estimated based on field data analysis and interpretation. The hydro-chemical behavior of groundwater has been evaluated based on the analytical results of the water sample collected in the field.

18.2.1   STUDY AREA (BUFFER ZONE)

For hydro-geological and hydrochemical study point of view an area of 314 Km² has been chosen as circular area of 10 km radius keeping place of large industrial area in Adityapur Industrial Area (22° 49' 39.44"N and 86° 04' 30.55"E) of Gamharia block of Saraekela- Kharsawa district in the centre. The area falls under Survey of India Toposheet No.73/F/13, 73J/1 & 73 J/2. The Latitude and Longitude of the proposed site are 22° 49' 39.44"N and 86° 04' 30.55"E, respectively.

It is bounded by coordinates 85°59'02" E to 86°09'56" E longitudes and from 22°44'14" N to 22°55'04" N latitudes. It is bounded by Katia, Manikul, Thakrudiand Salgodih villages in the north and Aunlataur, Nayadih villages and Tamadungri (pahar) in the south. In the east, it is bounded Kagainagar and Shatrinagar urban township of Jamshedpur city and Satnala, while by the western flank of Patubara hills in the west. A major part of the area falls under Gamharia block, but also covers a very little part of Rajnagar and Chandil blocks of Sareikela district and Gulmuri and Potka blocks of East Singhbhum district. The block-wise area coverage is given in Table 18.1. The total area under study is 314 Km², and here it is called a buffer zone.

TABLE 18.1   Block Wise Area Coverage in Buffer Zone

Sl. No.	District	Block	Area	% of Area
1	Sareikela	Gamharia	233.37	74.28
2		Rrajnagar	23.68	7.57
3		Chandil	18.72	5.96
4	East Singhbhum	Gulmuri	25.88	8.24
5		Potka	12.52	3.99

18.2.2   CLIMATE AND METEOROLOGY

The mean annual temperature remains at about 26°C. The temperature ranges from 16°C in winter months to 44°C in the summer months. It experiences tropical and humid climate which shows three distinct seasons viz. hot and dry summer season, humid, and warm rainy season and winter season. The summer is very hot and dry starting from March and continues till mid of June (Figure 18.1). The rainy season lasts from mid-June to end to September October, and November are the post-monsoon months followed by a cold winter which lasts till the middle of March. The district belongs

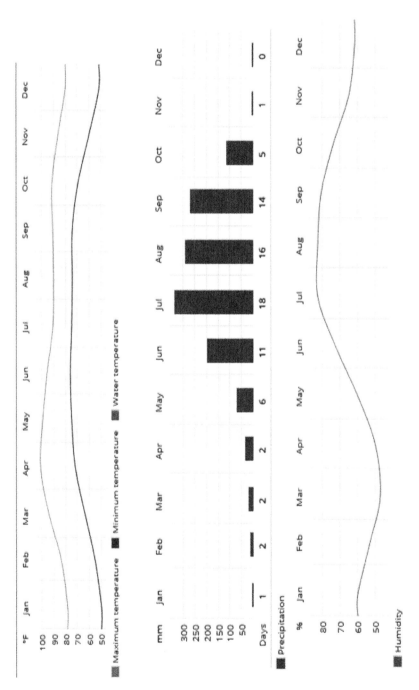

FIGURE 18.1 Monthly temperature, rainfall, and humidity pattern of Gamharia block, Saraikela-Kharsawan.

to one of the 13 districts of Jharkhand, which falls in the Agro climatic sub-zone-IV. The district falls in the rain shadow of the Santhal Pargana plateau. The average annual precipitation is 1307.6 mm and the average number of rainy days is 59. Even this meager precipitation is erratic which coupled with long inter-spell forces the district to suffer from drought (Figure 18.2).

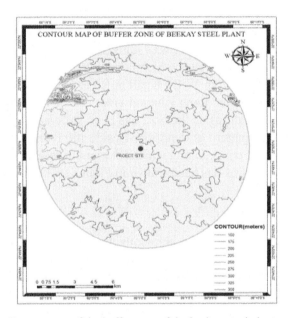

FIGURE 18.2   Contour map of the buffer zone of the beekay steel plant.

18.2.3   GEOMORPHOLOGY AND SOIL TYPES

The predominant physical feature over the major part of the district is the rolling topography dotted with isolated inselbergs except in the Borijore and Sundarpahari blocks. A substantial part of Borijore and Sundarpahari block is under forest cover. The altitude of the land surface increases from west to the east. The major hills are confined to the eastern part of the district comprising the Gandeshwari Pahar (238.41m) and Kesgari Pahar (268.29 m) while in the western part of the district isolated hills are in the form of the inselbergs and other small hillocks. The soil is mostly acidic, reddish yellow; light textured and highly permeable with poor water holding capacity.

18.2.4   DRAINAGE

The present area under study falls under the Subarnarekha river basin. Its main tributaries Sanjai and Kharkai River Subarnarekha originates 15 kms south of Ranchi on the Chhotanagpur plateau draining the states of Jharkhand, Orissa, and West Bengal before entering the Bay of Bengal. The total length of the river is 450 kms, and its important tributaries include the Raru, Kanchi, Kharkai, and Garra rivers. River Subernarekha flows from NW – SE in the northern part of the buffer zone. Its main tributary Kharkai River which flows in the southern part of study area joins Subernarekha river near village Saharbera. Other river and nalas which constituted the drainage in the area are- Sanjai River (flow from W-E), Kharkai River (W-E), Kanki jhor (W-E, and SW-NE) Sanjai River joins Kharkai river. The rivers and nalas show highly meandering features as the area has been suffered prolonged weathering. The drainage pattern is dendritic. Beside the rivers and nalas, there are many small water tanks among which Sitarampur reservoir is prominent. Drainage map of the area has been shown in Figure 18.3.

18.2.5   GEOLOGY

The area is underlain by rocks belonging to Precambrian age. The most common rock types are Granite and granitic -gneisses, Quartzites, Mica Schist, and quartz-mica-schists. At places, sporadic distribution of unclassified rocks such as amphibolites and epidiorites are found in narrow strips. Recent alluvium occurs as thin discontinuous, elongated manner along both sides of the river. A generalized stratigraphic sequence of the area is given Table 18.2.

TABLE 18.2   Geological Succession in the Area

Age	Series	Stage	Lithology
Recent	Alluvium		Sand, silt, and clay
---------------------------- Unconformity -------------------------			
Pre-Cambrian	Iron Ore Series	Chaibasa Stage	Quartzites, Mica Schist, and Quartz-mica-schist
		Singhbhum shear zone	Granite and granitic –gneisses

18.2.6   HYDRO-GEOLOGY

Geological setting, climate, and topography play important roles in occurrence and movement of groundwater. Depending upon the varied geological setup, the hydro-geological features in the area too register wide spatial

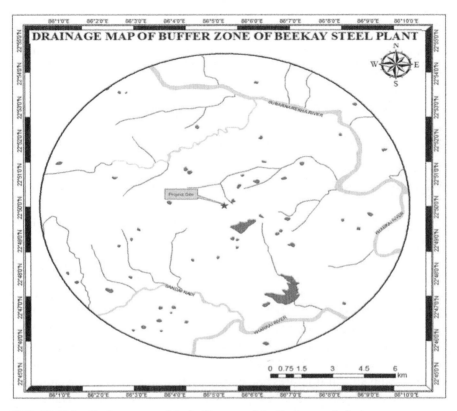

FIGURE 18.3   Drainage map of the buffer zone of the Beekay steel plant.

variations. The geological setting primarily controls the occurrence and movement of groundwater. The composition and structure of the geological formation influence certain inherent properties like porosity, permeability, and hence water holding and water-yielding capacity of aquifers, thereby playing a vital role in the hydro-geological regime.

The occurrence and movement of, by and large, depend on the hydro-geological condition of the subsurface formation. These natural deposits vary greatly in their lithology, thickness of weathering, texture, and structure which in turn influence their hydro-geological characteristics. Depending upon the geological setup of the study area, and water-yielding properties, two major hydro-geological units have been identified in the area. These are:

1. Consolidated formation; and
2. Un-consolidated formations.

As far as the hydro-geological property of underlying rocks is concerned, it is depicted through various groundwater abstractions structures. Most of the open wells are circular in shape. Total 12 nos. of open wells have established an observation well in the area. Hydro-geological properties of each unit are discussed in the following subsections.

18.2.6.1   CONSOLIDATED FORMATIONS

The major part of the area is occupied by consolidated formations comprising granites and granite-gneisses, quartzites, and mica-schists, etc. of Iron Ore series belonging to Precambrian age. These rocks are very hard and compact and lack primary porosity. Groundwater is stored mainly in the secondary porosity resulting from weathering and fracturing of the rocks. The aquifer materials are highly heterogeneous in character showing both vertical and lateral variations. The weathered residuum from the main repository of groundwater, which occurs under water table conditions and circulates through deeper fractures and fissures. Groundwater occurs under an unconfined condition in phreatic aquifers and semi-confined to confined conditions in the deeper fractures zones. The water-yielding capacity of fractured rocks largely depends on the extent of fracturing, openness, and size of fractures and extent of their interconnections into the near-surface weathered zone. These interconnected joints and fractures in the underlying hard rock's facilitate circulation of groundwater and in turn from deeper aquifers.

18.2.6.2   WATER BEARING PROPERTIES OF MAJOR LITHO UNITS

Hydro-geological characteristics of different rock formations are described in the following paragraphs based on the data collected and generated through groundwater survey and investigations.

Granite Gneiss: These are the most predominate formations occupying the northern part of Subernarekha River and central part of the buffer zone. These rocks occur in pediments and pediplains and denudational hills and other smaller hillocks and undulating plains. The rocks are well foliated, jointed, and are weathered easily. Weathering in these rocks is pronounced and fissures and joints, etc., are well developed and can be observed even in exposures. Granite gneisses are also traversed by numerous veins of quartz and pegmatites. The depth of weathering is found in a range of 21.40 to 24.30 meters in general. The weathered zone forms the main repository of groundwater in the hard rocks occupying area of present study and are tapped by dug wells. Groundwater can be developed through dug well, dug-cum-bore well and bore well. The available data generated during a survey reveals that the development of groundwater is being done through all types of groundwater structures, i.e., dug wells and as well as bore wells. Both structures are used for drinking as well as irrigation and industrial purposes.

Quartzites: Quartzite is the second dominating formations in the buffer zone. It is generally associated with Quartz mica-schists of Iron Ore Series of rock. These rocks are quite massive and compact and occupy hill ranges. Weathering is quite prominent. The weathered residuum and the fracture zones constitute the main repository of groundwater. Weathered part of quartzite forms the repository of groundwater and facilitates supply and storage of groundwater for dug well.

Mica schist: These are the least dominating rock types in the area and found generally in association with quartzites and granitic formations and occupy the lower parts of denudational hills and pediplanes in central parts also. Due to its sporadic and poor occurrence, no open wells were found in the area. However, shallow as well as deeper bore wells data indicate underlying formation as mica-schists. It is also confirmed by its outcrops exposed along nalas. These rocks are moderately weathered. Schistosity in these rocks itself controls the movement of groundwater in weathered residuum and at depth formed deeper aquifer. Details of the observation wells are given in the Table 18.3.

18.2.6.3   UNCONSOLIDATED FORMATION

Unconsolidated formation comprises alluvium deposits are brought by Subernarekha River and its tributaries in the area. Alluvium deposits are in a narrow strip and elongated way along the river courses. Due to its limited occurrence that too along river coarse only two monitoring wells were established for measurement. The depth of alluvium is from 15 to 40 m and forms a good potential zone for the shallow and middle aquifer. Hydro-geological features of the buffer zone have been shown in Figure 18.4.

18.2.7   DEPTH TO WATER TABLE

To decipher the status of groundwater detailed inventory of the existing wells falling within the study area has been carried out. The limited well inventory of 12 nos. of open wells has been given in Table 18.3.

TABLE 18.3   Post-Monsoon Water Level Data of Open Wells

S. No.	Location	Elevation (mamsl)	Dia. of Well (ft)	Water Level (mbgl)
1.	Barkatand (22°48'36.530" N, 86 °03'25.190" E)	176.0	10	2.2
2.	Ramjivanpur (22°48'31.67" N, 86 °03'06.890" E)	167.3	10	0.9
3.	Ramjivanpur (22°48'28.240" N, 86 °03'04.220" E)	173.9	3	1
4.	Ramjivanpur (22°48'25.00" N, 86 °03'03.400" E)	162.5	3	2.1
5.	Gamharia (22°49'01.300" N, 86°05'55.760" E)	148.5	6	4
6.	Gamharia (22°48'58.785" N, 86°06'00.25" E)	167.0	8	2.41
7.	Ramchandrapur (22°49'49" N, 86°04'46.550" E)	160.0	6	3.9
8.	Ramchandrapur (22°49'50" N, 86°04'45.400" E)	175.5	6	4.2
9.	Srirampur (22°49'58.200" N, 86°03'47.115" E)	204.35	4	2.85
10.	Hariharpur (22°49'52.750" N, 86°03'36.135" E)	186	5	1.35
11.	Hariharpur (22°49'54.700" N, 86°03'28.950" E)	209.9	10	1.95
12.	Birajpur (22°49'16.750" N, 86°03'57.900" E)	191.5	4	2.08

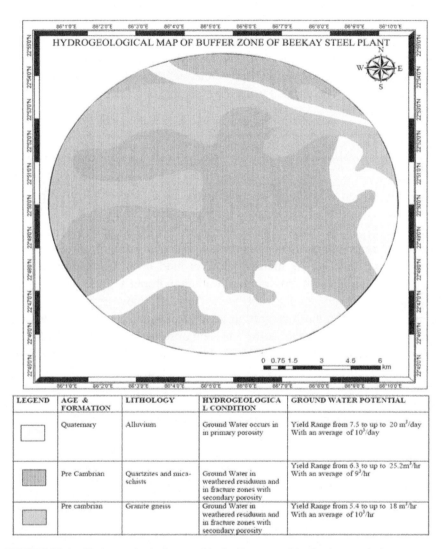

FIGURE 18.4   Hydro-geological map of the buffer zone of the Beekay steel plant.

18.2.7.1   GROUNDWATER FLOW REGIME

The depth to water levels was observed during Nov-2017 reflect post-monsoon water levels. The review of the data indicates moderate variation in depth to water table in the area. Such a variation is mainly due to variation in topography, and partly it is structurally controlled. Using the observed water level, depth to water level map has been prepared. The water table elevation contour map has been drawn based on observed pre-monsoon level and elevation of observation well. The review of the map shows a regional flow towards south-east over the study area that is towards the flow direction of Khakhai River. The surface water flow regime and the groundwater flow regime are similar.

18.2.7.2   FLUCTUATION OF WATER TABLE AND LONG-TERM TREND

The pre-monsoon water level was evaluated by a local enquiry by people having owned well and found to be in the range of 5.5 to 10 meters below ground level, being minimum at Hariharpur village and maximum at Ramchandrapur Village. Post- monsoon water level varies from 0.9 meters below ground level to 4.2 meters below ground level, being minimum at Ramjivanpur village and maximum at Ramchandrapur village. The average water level (post-monsoon period) is 2.41 meters below ground level. The minimum fluctuation has been observed at Ramchandrapur with 3.1 m, and maximum groundwater fluctuation has been observed at Gamharia with 7.59 m (Table 18.4 and Figures 18.5–18.7). As per the local enquiry, significant water table declining trend has not been observed in the area.

18.2.8   EVALUATION OF AQUIFER PARAMETERS

The aquifer character is the vital parameter of the groundwater study especially to evaluate flow regime and quantify subsurface groundwater flow. The Pumping test is the most effective tool to estimate the aquifer parameters. Consequently, short duration pumping test was conducted on bore well existing within the project area, which is presently used for domestic purposes. The pump was kept shut for 12 hours before the start of pumping test. The test details and the observations are discussed subsequently. During the test period, the water levels were observed in the well at regular intervals which are given in Tables 18.5–18.7.

18.2.9   QUALITY OF GROUNDWATER

Quality of water is as important as quantity. A suitable quality of water is whose characteristics make it acceptable to the needs of a particular purpose,

TABLE 18.4 Pre-Monsoon Water Level Data of Open Wells

S. No.	Location	Elevation (mamsl)	Dia. of Well (ft)	Water Level (mbgl)
1.	Barkatand (22°48'36.530" N, 86 °03'25.190" E)	176.0	10	7.3
2.	Ramjivanpur (22°48'31.67" N, 86 °03'06.890" E)	167.3	10	5.5
3.	Ramjivanpur (22°48'28.240" N, 86 °03'04.220" E)	173.9	3	5.5
4.	Ramjivanpur (22°48'25.00" N, 86 °03'03.400" E)	162.5	3	7.5
5.	Gamharia (22°49'01.300" N, 86°05'55.760" E)	148.5	6	9.5
6.	Gamharia (22°48'58.785" N, 86°06'00.25" E)	167.0	8	10.0
7.	Ramchandrapur (22°49'49" N, 86°04'46.550" E)	160.0	6	7.0
8.	Ramchandrapur (22°49'50" N, 86°04'45.400" E)	175.5	6	8.0
9.	Srirampur (22°49'58.200" N, 86°03'47.115" E)	204.35	4	6.0
10.	Hariharpur (22°49'52.750" N, 86°03'36.135" E)	186	5	7.0
11.	Hariharpur (22°49'54.700" N, 86°03'28.950" E)	209.9	10	6.0
12.	Birajpur (22°49'16.750" N, 86°03'57.900" E)	191.5	4	8.5

be it industrial or domestic. Whenever any industry is established towards serving the mankind, it is always a prerequisite to determining the quality of water which will be utilized for the industry as well for domestic use for people engaged in the industry. During the field study, two number of water samples from dug and bore well were collected to know the suitability of water for domestic as well as industrial use. The location and source of water samples collected have given in Table 18.8.

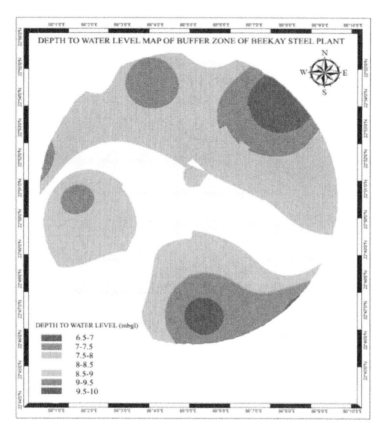

FIGURE 18.5   Depth to water level map of the buffer zone of the Beekay steel plant.

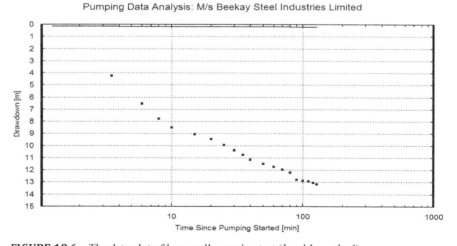

FIGURE 18.6   The data plot of bore well pumping test (Jacob's method).

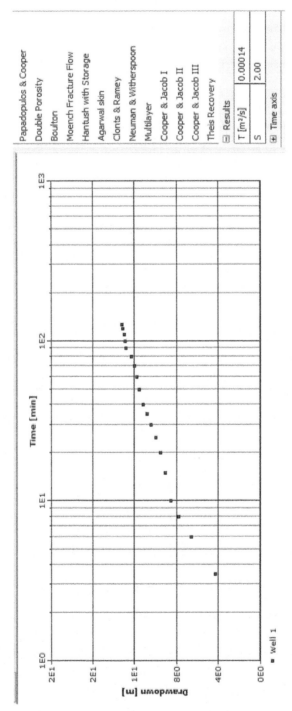

FIGURE 18.7 The data plot of bore well pumping test (Jacob's straight-line method).

TABLE 18.5   Details of Bore Well at Beekay Steel Industries Limited, Adityapur Industrial
Area, Gamharia, Saraikela-Kharsawan, Jharkhand

Site Name	M/S Beekay Steel Industries Private Ltd.	Location	Plant Site (22°49'32.32" N, 86°4'33.1" E)
Well Depth	77 mbgl	Pump lowered at	60 mbgl
Dia of well	6"	pump (HP)	3 HP
MP	0.76 magl	Pumping rate	1.532 LPS
DTWL	8.08 mbgl	Temp	28°C

TABLE 18.6   Pumping Test Data of Bore Well at Beekay Steel Industries Limited, Adityapur
Industrial Area, Gamharia, Saraikela-Kharsawan, Jharkhand

Sl. No.	Time (min)	Hold (m)	Cut (m)	DTWL (mbmp)	DTWL (mbgl)	Drawdown (m)
1.	0	25	16.16	08.84	8.08	0
2.	3.5	25	11.90	13.10	12.34	4.26
3.	6	25	9.61	15.39	14.63	6.55
4.	8	25	8.39	16.61	15.85	7.77
5.	10	25	7.66	17.34	16.58	8.50
6.	15	25	7.11	17.89	17.13	9.05
7.	20	25	6.71	18.29	17.53	9.45
8.	25	25	6.20	18.80	18.04	9.96
9.	30	25	5.77	19.23	18.47	10.39
10.	35	25	5.41	19.59	18.83	10.75
11.	40	25	5.01	19.99	19.23	11.15
12.	50	25	4.67	20.33	19.57	11.49
13.	60	25	4.40	20.60	19.84	11.76
14.	70	25	4.19	20.81	20.05	11.97
15.	80	25	3.94	21.06	20.30	12.22
16.	90	25	3.36	21.64	20.88	12.80
17.	100	25	3.27	21.73	20.97	12.89
18.	110	25	3.24	21.76	21.00	12.92
19.	120	25	3.12	21.88	21.12	13.04
20.	127	25	3.00	22.00	21.24	13.16

TABLE 18.7 Recuperation Data of Bore Well at Beekay Steel Industries Limited, Adityapur Industrial Area, Gamharia, Saraikela-Kharsawan, Jharkhand

Sl. No.	Time Since Pumping Started (t) in min	Time Since Pumping Stopped (t') in min	(t/t')	Hold (m)	Cut (m)	D.T.W.L (mbmp)	D.T.W.L (mbgl)	Residual Draw-down (m)
1.	127	0		25	3.00	22.00	21.24	13.16
2.	130	3	43.33	25	3.90	21.10	20.34	12.26
3.	132	5	26.4	25	6.57	18.43	17.58	9.5
4.	134	7	19.14	25	8.11	16.89	16.10	8.02
5.	136	9	15.11	25	19.91	15.09	14.33	6.25
6.	140	13	10.76	25	10.22	14.78	14.02	5.94
7.	145	18	8.05	25	10.73	14.27	13.51	5.43
8.	150	23	6.52	25	11.32	13.68	12.92	4.84
9.	155	28	5.53	25	11.97	13.03	12.27	4.19
10.	160	33	4.84	25	12.55	12.45	11.69	3.52
11.	170	43	3.95	25	12.77	12.23	11.47	3.39
12.	180	53	3.39	25	13.50	11.50	10.74	2.66
13.	190	63	3.01	25	14.09	10.91	10.15	2.07
14.	200	73	2.73	25	14.54	10.46	9.70	1.62
15.	210	83	2.53	25	14.90	10.10	9.34	1.26
16.	217	90	2.41	25	15.02	9.98	9.22	1.14

$T = (15.8 \times Q) / \Delta S$

ΔS calculated from drawdown-time curve (Theis recovery method) has been found to be

$T = (15.8 \times 1.532)/2.00 = 12.096$

TABLE 18.8 Location of Water Samples Collected in Buffer Zone of M/s Beekay Steel Industries Limited, Gamharia, Saraikela-Kharsawan, Jharkhand

Sl. No.	Type of Well	Location/Village Name	Depth (mbgl)	Temp (°C)	Parameters for chemical analysis: pH, E.C, Ca, Mg, K, F, NO_3, HCO_3, Cl, SO_4, CO_3, As & TDS
1.	Bore Well	22°49'48.900" N, 86°04'46.850" E	77	26	—do—
2.	D1	22°49'32.520" N, 86°04'33" E	31	26	—do—

Quality of water is as important as quantity. pH of groundwater is to slight alkaline in nature. pH varies between 6.47–6.57. Total dissolved solids are within permissible limits prescribed by Bureau of Indian Standards (BIS). Total dissolved slid of groundwater varies between 310.50 to 320.80 mg/l. The value of chloride varies between 31.80 to 32.78 mg/l. The value of sulfate varies from 15.26 to 16.27 mgl⁻¹. The value of calcium varies from 46.60 to 47.42 mgl⁻¹. The value of magnesium varies from 3.97 to 4.96 mgl⁻¹. The total hardness as calcium carbonate varies between 132.70 to 138.72 gl⁻¹. The range of different chemical constituents has been given in given Table 18.9.

Groundwater and surface water of core and buffer zone has been found suitable for domestic as well as industrial uses. The remarks on different samples collected have been given in Table 18.10 and result summarized in table.

A perusal of concentration of various chemical constituents of water indicates that groundwater is suitable for all the purposed viz. drinking, and domestic purposes (Table 18.11).

TABLE 18.9 Comparison of Bis Standards with Chemical Analysis Results of Water Samples Collected From Buffer Zone of M/S Beekay Steel Industries Limited, Gamharia, Saraikela-Kharsawan, Jharkhand

S. No.	Chemical Constituent	Range	Unit	BIS
1.	pH	6.5–8.5		6.47–6.57
2.	TDS	500.0	mgl⁻¹	310.50–320.80
3.	Cl	250.0	mgl⁻¹	31.80–32.78
4.	SO₄	200.0	mgl⁻¹	15.26–16.27
5.	Total Alkalinity as CaCO₃	200.0	mgl⁻¹	106.82–111.18
6.	Total Hardness as CaCO₃	300.0	mgl⁻¹	132.70–138.72
7.	Ca	75.0	mgl⁻¹	46.60–47.42
8.	Mg	30.0	mgl⁻¹	3.97–4.96
9.	Fe	0.3	mgl⁻¹	0.08–0.10
10.	Turbidity	5.0 NTU	NTU	<1.0

TABLE 18.10 Remarks on Chemical Analysis of the Water Samples Collected

S. No.	Well No.	Location in Buffer Zone	Remarks
1.	Bore Well	Beekay Steel (Plant Area)	Water sample after chemical analysis found satisfactory
2.	D1	Sikardih	Water sample after chemical analysis found satisfactory

TABLE 18.11 Results of Chemical Analyses of Water Samples Collected in and Around Plant of M/S Beekay Steel Industries Limited, Adityapur Industrial Area, Gamharia, Saraikela-Kharsawan, Jharkhand

Sample No. Location/Village	D1 Sikardih	Bore Well Beekay Steel (Plant Area)	Testing Method	BIS
pH	320.80	310.50	Digital pH	500.0
TDS	32.78	31.80	Digital TDS Meter	250.0
Cl	16.27	15.26	EDTA Titration Method	200.0
SO_4	111.18	106.82	PhotoMetric Method	200.0
Total Alkalinity as $CaCO_3$	138.72	132.70	EDTA Titration Method	300.0
Total Hardness as $CaCO_3$	47.42	46.60	EDTA Titration Method	75.0
Ca	4.96	3.97	Photometric Method	30.0
Mg	0.10	0.08	Photometric Method	0.3
Fe	<1.0	<1.0	3500 Fe –B	5.0 NTU
Turbidity	6.57	6.47	3500 Mg –B	6.5–8.5
T. Coliform	Not Detected	Not Detected	9221 B	
E. Coliform	Absent	Absent	9221 F	

18.3 GROUNDWATER RESOURCES

Precise quantification of exploitable groundwater resource is essential before any program for its development. It involves quantification and identification of various factors affecting groundwater recharge, discharge, and demarcation of areas suitable for recharge. The principal sources of recharge to groundwater are rainfall, seepage from canals, return flow from applied irrigation seepage from tanks and ponds. The buffer zone of the project area of M/s Beekay Steel Industries Limited, which is the area with 10 km radius from the proposed site, falls in Gamharia, Rajnagar, and Chandil block of Saraikela-Kharsawan district and Potka block of East Singhbhum district. Availability of groundwater resources of these blocks has been estimated based on norms recommended by Groundwater Estimation Committee (G.E.C 1997) by CGWB, which has been presented below in Table 18.12. It is found that the stage of groundwater development in which the core zone of the proposed project fall, does not exceed 20 %, which is a positive factor for taking up groundwater abstraction for the increased demand.

TABLE 18.12   Groundwater Resources in Gamharia Administrative Block Falling in the Buffer Zone of the Proposed Plant

Block	Gamharia (ham)	Rajnagar (ham)	Chandil (ham)	Potka (ham)
Total Annual Groundwater Recharge	2138.52	3521.70	2647.55	4012.70
Net Groundwater Availability	1924.66	3345.62	2382.79	3894.50
Annual Net Draft Through Existing Structures for Irrigation	87.894	90.63	73.53	96.05
Annual Net Draft Through Existing Structures for Domestic and Industrial Supply	282.2	197.1	192.6	287.72
Allocation for Domestic and Industrial Requirement Supply	376.20	262.84	256.82	435.48
Net Groundwater Availability for Future Irrigation	1460.58	2992.15	2052.44	3362.98
Stage of Groundwater Development (%)	19.23	8.60	11.17	9.85

The block in which core zone of the project is situated has been categorized as safe by CGWB, Govt. of India depending upon water table behaviors and stage of groundwater development which has been given in Table 18.13.

TABLE 18.13   Categorization of The Block Falling in the Core Zone of M/s Beekay Steel Industries Limited, Adityapur Industrial Area, Gamharia, Saraikela-Kharsawan, Jharkhand

S. No.	Assessment Unit	Stage of Groundwater Development	Is there any significant decline of pre-monsoon level	Is there any significant decline of post-monsoon level	Categorization of future groundwater development
1.	Gamharia	19.23%	No	No	Safe
2.	Rajnagar	8.60	No	No	Safe
3.	Chandil	11.17	No	No	Safe
4.	Potka	9.85	No	No	Safe

18.3.1   GROUNDWATER RESOURCE OF THE BUFFER ZONE OF THE PROPOSED PROJECT

1. Area of the buffer zone = 314 km^2
2. Area of the water bodies = Nil
3. Hilly area (area not suitable for recharge) = 28.90 km^2
4. Area suitable for recharge under different rock types

  (a) Singhbhum Granite – 44.78 km²
  (b) Mica Schist – 216.93 km².
  (c) Shale, Phyllites – 23.39 km²
5. Groundwater Fluctuation in (m)
  (a) Average Fluctuation in Singhbhum Granite – 3.65 m
  (b) Average Fluctuation in Mica Schist – 3.89 m
  (c) Average Fluctuation in Shale, Phyllites – 5.00 m
6. Specific Yield –
  (a) Shale, Phyllited – 0.03
  (b) Mica Schist – 0.015
  (c) Singhbhum Granite – 0.03
7. Monsoon Recharge under different formation in the Buffer Zone (Water Table Fluctuation Method)
  = Singhbhum Granite + Mica Schist + Shale & Phyllites
  = (44.78 x 3.65 x 0.03) + (216.93 x 3.89 x 0.015) + (23.39 x 5.00 x 0.03)
  = 4.90 + 12.66 + 3.50
  = 21.06 MCM
8. Rainfall Infiltration factor of different rock types in the buffer zone
  (a) Shale, Phyllites – 0.04
  (b) Mica Schist – 0.08
  (c) Singhbhum Granite – 0.11
9. Annual Rainfall = 1307.60 mm
  (a) Monsoon Rainfall = 1131.70 mm
  (b) Non-monsoon Rainfall = 175.90 mm
10. Groundwater Recharge by Rainfall Infiltration Factor (Monsoon Season)
  (a) R_{rf} (Singhbhum Granite) = Infiltration factor of singhbhum granite x Area x Normal rainfall in the monsoon season
    = 0.11 x 44.78 x 1.131
    = 5.571 mcm
  (b) R_{rf} (Mica Schist) = Infiltration factor of mica schist x Area x Normal rainfall in the monsoon season
    = 0.08 x 216.93 x 1.131
    = 19.627 mcm
  (c) R_{rf} (Shale, Phyllites) = Infiltration factor of Shale, Phyllites x Area x Normal rainfall in the monsoon season
    = 0.04 x 23.39 x 1.131
    = 1.05 mcm
  Monsoon Recharge by Rainfall Infiltration Factor = 26.248 mcm

Monsoon Recharge due to water table fluctuation = 21.06 mcm

The rainfall recharge for normal monsoon season rainfall is finally adopted as per criteria are given below:

(a) If PD is greater than or equal to –20%, and less than or equal to +20%; R_{rf} (normal) is taken as the value estimated by the water table fluctuation method.
(b) If PD is less than –20%; R_{rf} (normal) is taken as equal to 0.8 times the value estimated by the rainfall infiltration factor method.
(c) If PD is greater than +20%; R_{rf} (normal) is taken as equal to 1.2 times the value estimated by the rainfall infiltration factor method.

$$P.D. = (21.06–26.248)/26.248 \times 100$$
$$= -19.76\,\%$$

Because PD is greater than or equal to –20%, and less than or equal to +20%; R_{rf} (normal) is taken as the value estimated by the water table fluctuation method

R_{rf} (normal) = 21.06 mcm

11. Groundwater Recharge by Rainfall Infiltration Factor (Non-monsoon Season)
    (a) R_{rf} (Singhbhum Granite) = Infiltration factor of singhbhum granite x Area x Normal rainfall in non-monsoon season
        = 0.11 x 44.78 x 0.175
        = 0.862 mcm
    (b) R_{rf} (Mica Schist) = Infiltration factor of mica schist x Area x Normal rainfall in non-monsoon season
        = 0.08 x 216.93 x 0.175
        = 3.037 mcm
    (c) R_{rf} (Shale, Phyllites) = Infiltration factor of Shale, Phyllites x Area x Normal rainfall in non-monsoon season
        = 0.04 x 23.39 x 0.175
        =0.163 mcm

Total Non-monsoon Recharge by Rainfall Infiltration Factor = 4.062 mcm

Total Annual recharge = Monsoon Recharge + Non-monsoon Recharge
= 21.06 +4.062
= 25.122 mcm

12. Draft

Total Draft = Total draft of Gamharia + Rajnagar + Chandil + Potka
= 13.62 + 0.86 + 0.35 + 0.06
= 14.89 MCM (including industrial/drinking/irrigation)

13. Net Annual Groundwater Availability
= Total replenishable resources – Natural non-monsoon discharge
= 25.122 – 1.48
= 23.642

14. Stage of Groundwater Development
= (Total Draft/Net Annual Groundwater Availability) x 100
= (14.89/23.642) x 100
= 62.98 % (Say 63%)

18.4 IMPACT OF PUMPING ON LOCAL GROUNDWATER SYSTEM

18.4.1 PREVIOUS AND INCREASED WATER DEMAND OF THE PLANT

Total water requirement for the proposed project is estimated to be $160 m^3 d^{-1}$. Present water requirement and the additional requirement will be obtained from groundwater through bore well within the premises (Table 18.14).

Water System
- The proposed project requires water for manufacturing as well as domestic consumption.
- The company will use groundwater through bore well to meet its requirement. The company has implemented the Rain Water Harvesting System for recharge of groundwater through 3 Nos. of recharge pit.
- Water is mainly required for cooling operations, where it is re-used
- As such wastage of water will be negligible, and there is no need of any water treatment arrangement.

TABLE 18.14 Water Requirements (Existing and Additional) of Beekay Steel Plant

S. No.	Particulars	Existing in KLD	Proposed in KLD	Total in KLD
1.	Industrial	33	99	132
2.	Domestic	4	12	16
3.	Others	3	9	12
	TOTAL	40	120	160

18.4.2   EFFECT ON STAGE OF GROUNDWATER DEVELOPMENT

The stage of groundwater development as assessed in the buffer zone of the project is 62.98 %. For meeting the existing as well as the increased water demand, it is proposed to withdraw water of $160 m^3$/day from the bore wells to be drilled within the plant premises.

Daily Water Requirement – 160 $m^3 d^{-1}$

Annual Consumption – $160 \times 365 = 58{,}400$ $m^3 = 0.0584$ MCM

Total Draft = existing total draft + draft from the proposed bore wells

   $= 14.89$ MCM + 0.0584 MCM = 14.9484 MCM

Expected stage of groundwater development

   = Total Draft/Net Groundwater Resources x 100

   = 14.9484/23.642 x 100

   = 63.22%

So the stage of groundwater development after withdrawal of required quantity of groundwater for meeting the increased water demand of project will be increased to only 63.22 % from existing 62.98%.

18.4.3   RADIUS OF INFLUENCE

As per the studies carried out by CGWB in the district and pumping test carried out within the plant premises the storativity has been found to be 3.2 $\times 10^{-4}$ and the transmissivity has been found to be 12.096 m^2/day Due to the low permeability of the aquifer units, the impact of pumping on local water regime will be marginal and the radius of influence will be limited to a small distance. However, to estimate a probable zone of influence, the finding of CGWN studies have been considered for assessing the radius of influence. To evaluate the radius of influence the c-efficient of storage (S) value of 3.2 $\times 10^{-4}$ has been taken from the studies of Central Groundwater Board in the adjoining area of the buffer zone of the plant. The radius of influence has been arrived at using the following equation: $R^2 = (2.25Tt)/S$.

The time (t) has been considered as 8 hours of continuous pumping at a time and transmissivity 12.096 $m^2 d^{-1}$. The value of the radius of influence (R) comes as 169 m. So it is found that there will be no effect of pumping in the local groundwater system at a distance of 169 m from pumping well in and around the plant area.

18.5   CONCLUSIONS AND RECOMMENDATIONS

The proposed project of M/s Beekay Steel Industries Limited falls within the administrative jurisdiction of Gamharia, Rajnagar, and Chandil Block of Saraikela-Kharsawan district and Potka Block of East-Singhbhum district. The stage of groundwater development in Gamharia block in which the core zone falls is 19.23% as assessed by CGWB (2013). There is no significant decline of pre-monsoon water level, as well as post-monsoon water levels in all the blocks and blocks, are categorized as "Safe" for future groundwater development as per CGWB published a report of Central Groundwater Board, Govt. of India on "Dynamic groundwater resources of Jharkhand." This leaves the scope of groundwater development for meeting the water demand of the proposed project. Again it is to mention that the stage of groundwater development after withdrawal of a required quantity of ground-water for meeting the water demand of the proposed project will be increased to only 63.22% from existing 62.98%.

To assess the impact on water levels in time and space, it is recommended to develop a close monitoring network in the zone of influence and quarterly monitoring of the water levels. For observing the impact on the aquifer system, shallow, and deeper piezometers will be constructed for monitoring the unconfined and confined aquifers respectively. The location and design of the piezometers will be finalized in close coordination with CGWB. They will be constructed in a protective place and will be monitored periodi-cally. The water quality is monitored under routine monitoring. Creation of awareness among workers and local people about rainwater harvesting and conservation of water will be helpful. Monitoring the water quality of local nala and in domestic water intake structures will be regular. On analyzing the monitoring data, if any area is found to have any deleterious impact (qualitative or quantitative) due to industrial activity, suitable control and remedial measures will be taken by the project authorities.

KEYWORDS

- BSIL
- CGWB
- TMT

REFERENCES

ATLAS of Jharkhand State, (2008).

Dynamic Groundwater Resource of Jharkhand, Published by CGWB, (2011). Govt. of India and GWID, Govt. of Jharkhand.

Groundwater Information Booklet, Saraikela District, Published by CGWB, (2011). Govt. of India, State Unit Office-Ranchi, Mid-Eastern Region, Patna.

Karanth, K. R., (2016). *Groundwater Assessment Development and Management*, Tata McGraw Hill, New Delhi.

CHAPTER 19

SIMULTANEOUS BIOLOGICAL REMOVAL OF ARSENIC, IRON, AND NITRATE FROM GROUNDWATER BY A TERMINAL ELECTRON ACCEPTING PROCESS

ARVIND KUMAR SHAKYA and PRANAB KUMAR GHOSH

Department of Civil Engineering, IIT Guwahati, India – 781039,
E-mail: a.shakya@iitg.ernet.in

ABSTRACT

Groundwater contamination, natural, and/or anthropogenic is a major concern to human health. There are several reports of arsenic and iron co-contamination of groundwater from many parts of the world including northeastern states of India. Along with arsenic and iron, there are reports on co-occurrence of nitrate in groundwater of Assam, Jharkhand (India) and many other areas of the world. Several groundwaters contain increased concentrations of these pollutants that are observed either isolated or in pairs, or all three together. Although several physicochemical and/or biological processes have been established for the removal of one of the above-mentioned contaminants, but till now very few studies have been performed on the efficient and cost-effective simultaneous removal of two or more contaminants from groundwater.

In the present study, the performance of suspended growth batch bioreactors on simultaneous removal of arsenic, iron, and nitrate by mixed bacterial culture was evaluated by means of terminal electron accepting process in the presence of sulfate. A series of conical flasks were used as batch reactors were inoculated with mixed bacterial culture mainly collected from a wastewater treatment plant and acclimatized in the presence of arsenic, nitrate, and sulfate. The reactors were fed with real contaminated groundwater containing 30–125 μgl^{-1} of arsenic, 1.5–3.0 mgl^{-1} of iron, 50 mgl^{-1} of nitrates, 25 mgl^{-1} of sulfate, along with 105 mgl^{-1} of COD. The reactors were operated for a period of 7 days at 30°C in an incubator shaker. The removal approach consists of reduction of arsenic and nitrate coupled with oxidation of an electron donor (acetic acid). Complete nitrate removal was observed whereas arsenic and iron were below drinking water permissible limits of 10 μgl^{-1} and 0.3 mgl^{-1}, respectively within 3–4 days of operation. The insoluble bio-precipitates of arsenic and/or iron sulfides were the main arsenic and iron removal mechanism in the studied system.

19.1   INTRODUCTION

Arsenic is a common groundwater pollutant affecting more than 150 million people around the world. Long-term consumption of arsenic in drinking water causes adverse effects on human health including cancer (skin, lung, bladder), neurological problems, gastrointestinal disorders, reproductive problems and muscular weakness (Singh et al., 2015). Due to health risks associated with arsenic, the World Health Organization and other state agencies have imposed a maximum contaminant level (MCL) guideline of 10 μgl^{-1}. Under reducing conditions, those prevailing in groundwater, arsenite [As (III)] is the prevalent form of arsenic. Arsenic (III) is 25–60 times more toxic than arsenic (V). Iron is common contaminant often present in groundwater with arsenic, for example, in groundwater of Mekong Delta in Vietnam (Buschmann et al., 2008) and in several districts of Assam (India). Iron is not considered to cause severe health problems in humans; rather its presence in potable water can cause different types of nuisance problems. Recently, due to regular intake of high iron containing groundwater, some adverse health effects like hemochromatosis, liver cirrhosis, and siderosis is observed in people of Assam, India (Chaturvedi et al., 2014). The MCL of iron in drinking water is 0.3 mgl^{-1} (WHO, 2011). Nitrate is another common contaminant of groundwater often coexists with arsenic and iron (Rezaie-Boroon et al., 2014). High nitrate concentrations are reported elsewhere, for example, Australia, USA, China, and India (Kapoor and Viraraghavan, 1997). The major source of nitrate in groundwater are extensive use of nitrogenous fertilizers, discharge from septic tanks, spreading of sewage sludge and seepage from pit latrines. Ingestion of high nitrate levels in drinking water causes "blue baby syndrome" in infants and cancers, reproductive problems, infectious diseases and diabetes (Bhatnagar and Sillanpaa, 2011). Keeping in mind the link between serious health problems and excessive concentration of nitrate in drinking water, WHO recommended nitrate concentration limit of 50 mg NO_3^-/L (WHO, 2011). The combined occurrence of arsenic, iron, and nitrate is reported from many regions of the world including Argentina (Gimenez et al., 2013) India (Chakrabarty and Sarma, 2011), and Myanmar (Bacquart et al., 2015). The co-occurrence of these contaminants in groundwater is either due to natural geogenic sources or anthropogenic origin such as industrial activities in that nearby area, for example, high level of iron, nitrate, and fluoride in Angul-Talcher (Orissa) and Drain Basin Area, Najafgarh (Delhi) (CPCB, 2007). Although much work has been done on the individual removal of these common pollutants with different conventional methods to provide a suitable mitigation approach, little attention was paid on combined removal of them using an effective and sustainable process. Conventional methods for removing nitrate, arsenic, and iron include sorption on to various materials, ion exchange, membrane filtration and electrocoagulation (Bhatnagar and Sillanpaa, 2011; Singh et al., 2015; Khatri et al., 2017). All these methods have specific disadvantages. For example, ion exchange and membrane filtration are generally costly and produces large quantities of spent sludge which requires further treatment and disposal problems (Chung et al., 2007). Similarly, electrocoagulation suffers from the replacement of electrodes at a regular interval, high cost and anode passivation (Kumar and Goel, 2010). The research for sustainable and effective treatment options has focused attention on the biologically mediated process. Terminal electron accepting process based bioreactors is an emerging alternative option to existing methods used to remove multi-contaminants from groundwater (Brown et al.,2008). Chung et al. (2007) demonstrated simultaneous bio-reduction of nitrate, arsenic, and other contaminants in a membrane bioreactor. Membrane bioreactors inoculated with mixed bacterial culture removed nearly 100% nitrate and 80% arsenic from an initial concentration of 21 mg-Nl^{-1} and 7.3 μgl^{-1}, respectively. Zhao et al., (2014) studied the performance of a two-stage membrane biofilm reactor treating groundwater containing nitrate, perchlorate, sulfate, and oxygen. They reported 100% nitrate and 97% perchlorate removal from an initial concentration of 9 mgl^{-1} and 200 μgl^{-1}, respectively. Upadhyaya et al., (2012) demonstrated two stage fixed film bioreactor system for combined removal of arsenic and nitrate. At an influent arsenic concentration of 300 μgl^{-1} and nitrate of 50 mgl^{-1}, the nitrate removal efficiency was 100 %. However, the arsenic removal efficiency was 90% and treated water was not meeting drinking water MCL of 10 μgl^{-1}. In their study, the main arsenic removal mechanism was precipitation of biogenic arsenic and iron sulfides resulting from biosulfidogenesis.

Although few studies have investigated the combined removal of arsenic and nitrate using indigenous mixed microbial culture, the combined removal of arsenic, nitrate, and iron from real contaminated groundwater has not been yet studied. Hence, the present study investigates the efficiency of simultaneous biological removal of arsenic, iron, and nitrate in acetate fed suspended growth reactors in which sulfate serves as an electron acceptor for the biogenic sulfide and the subsequent removal of arsenic and iron. Particular importance was attached to the effects of real groundwater on mixed bacterial culture performance on contaminant removal. The biogenic precipitate was characterized by field emission scanning electron microscopy (FESEM).

19.2 MATERIALS AND METHODS

19.2.1 EXPERIMENTAL SETUP

High-density polyethylene conical flasks with the effective volume of 250 mL were used as the bioreactors. The cap was butyl synthetic rubber, with air tightness, and the solution volume was 100 mL, including feed medium (real groundwater) and inoculation sludge. Real groundwater was added into the reactor with a sterile syringe. Residual air in the reactor was expelled by purging oxygen-free nitrogen gas for 5 minutes. The reactors were covered with black paper and incubated in a shaking incubator at 120 rpm.

19.2.2 REAL GROUNDWATER COLLECTION

Real groundwater was collected from the well of two locations at ($26°16.45'N$ and $90°41.22'E$; $26°16.53'N$ and $90°40.79'E$, depth of well 160 ft.) near New Bongaigaon district, Assam (India), where arsenic and iron concentrations were varied in the range of 30–125 μgl^{-1} and 1.5–3.0 mgl^{-1}, respectively. Groundwater collected after adequate purging (10–15 min.) were stored in pre-acidified (with HCl at pH 2) high-density polyethylene containers for metal analysis. The characterization of other water quality parameters is done with non-acidified groundwater sample. The real groundwater quality parameters are given in Table 19.1. Containers were filled to overflowing and sealed with thread seal tape immediately to further minimize contact of oxygen. All groundwater samples were transported to the Environmental Engineering Laboratory, IIT Guwahati, within 6 h and subsequently kept at 10°C until analysis.

TABLE 19.1 Composition of the Real Groundwater

Components	Unit	Concentration	
		RGW–1	RGW–2
Temperature	°C	27.8°C	27.5
pH	-	6.6	6.62
EC	Mscm^{-1}	862	872
ORP	mv	87.3	84
Fe$_{tot}$	mgl^{-1}	1.68	3.2
SO$_4^{2-}$	mgl^{-1}	5.2	4.4
As$_{tot}$	μgl^{-1}	29.2	124.4
NO$_3^{2-}$	mgl^{-1}	2.6	3.2
PO$_4^{3-}$	mgl^{-1}	1.14	0.86
Cl$^-$	mgl^{-1}	10.6	12.4
HCO$_3^-$	mgl^{-1}	176	170
CO$_3^{2-}$	mgl^{-1}	19.2	17.6
Na	mgl^{-1}	9.8	8.2
Ca	mgl^{-1}	51.8	38.2
Mg	mgl^{-1}	4.82	5.1
D.O.	mgl^{-1}	4.1	4.3

19.2.3 INOCULUM AND STARTING UP

Anoxic sludge was collected from the anoxic pond of IIT Guwahati Sewage Treatment Plant (Guwahati, India). Seed culture for inoculation of reactors was obtained by mixing sludge collected from IIT Guwahati sewage treatment plant and a bench scale sulfate anoxic removal bioreactor. Acclimatized sludge (100 mgl^{-1} as MLVSS) was added as inoculum in the experiments. Based on carbon required for complete removal of all electron acceptors (i.e., residual DO, nitrate, arsenate, and sulfate) and average net yield of 0.4 g biomass/g COD acetate, 105 mgl^{-1} COD as carbon was supplemented in synthetic groundwater with a safety factor of 1.5. All the chemicals used were of analytical grade and purchased from Merck. To investigate the potential for bio-adsorption of arsenic (V) and iron on to biomass and any volatilization loss of contaminants a control group was examined containing only real groundwater (RGW–2) mixed bacterial culture and no carbon source. Two test groups were investigated with three replications, fed with type–1 and type–2 groundwater, to ensure the reliability of results. The reactors were fed solely with real groundwater except for the addition of 50 mgl^{-1} nitrate, 25 mgl^{-1} sulfate and 105 mg^{-1} COD. Experiments were also conducted with simulated groundwater at higher concentrations of arsenic (200 and 300 µgl^{-1}), iron (4 and 5 mg^{-1}) and nitrate (75 and 100 mg^{-1}) to evaluate mixed culture performance. COD in the influent media was increased at higher concentrations of nitrate to meet carbon requirements. All groups were examined at natural pH of real groundwater and 35°C for 7 days. The experiments were conducted in an incubator shaker in self-scarifying mode.

19.2.4 ANALYTICAL METHODS

Liquid samples were centrifuged at 6000 rpm for 10 min using a REMI R–24 centrifuge, and then filtered prior to nitrate, sulfate, COD, iron, and arsenic measurements in the supernatant. In general, standard techniques as given in Standard Methods for the Examination of Water and Wastewater (APHA, 2005) have been followed unless otherwise stated. Samples for pH, nitrate, sulfate, and residual COD were analyzed promptly on the same day. The pH of the medium from the reactor was measured using pH meter (Thermo Scientific). Nitrate was measured using an UV-visible spectrophotometer (Varian, Cary 50 Bio) at 220 and 275 nm. Sulfate analysis was done using digital nephelo-turbidity meter (132, Systronics). Residual COD was measured by using closed reflux titrimetric method (Hach DRB 200). Prior to COD determination, samples were acidified to a pH less than 2 by addition of concentrated H_2SO_4 and then purged with N_2 gas for approximately 5 minutes to remove H_2S. Samples for total arsenic and iron were acidified to a final concentration of 0.02 N HCl to solubilize any precipitates and stored at 4°C after filtering through 0.2 µm filters. Total arsenic samples were analyzed within 48 h by using an atomic absorption spectrometer (Varian, SpectrAA Analyst 800). A continuous flow hydride generation system was used for detection of arsenic concentration with detection limits of 1 µgl^{-1} As (T). Iron was measured using the phenanthroline method. All measurements were performed in duplicate and mean values of the results are presented. Biogenic precipitates were analyzed using field emission scanning electron microscope-energy dispersive X-ray spectroscopy (FESEM-EDX). After centrifugation the pellets were collected and frozen at –20°C for 12 h, and then freeze-dried. The freeze-dried samples were examined using a FESEM (Sigma, Carl Zeiss, Germany). The elemental composition of precipitate was analyzed using EDX (Oxford Instruments, Germany).

19.3   RESULTS AND DISCUSSION

19.3.1   BIOMASS ADSORPTION STUDIES

The adsorption of arsenic and iron on to 100 mgl^{-1} of biomass (MLVSS) from the "control" is shown in Figure 19.1. After 7 days, from an initial 124 µgl^{-1}, only about 11%, removal of arsenic was noticed from control. The total amount of iron adsorbed was found to be 26.8%. Reduction in arsenic concentration in control might be due to loss of arsenic by adsorption on to biomass, adsorption on to ferrous iron and/or volatilization. Teclu et al., (2008) reported about 6.5% and 10% bio-sorption of As (III) and As(V), respectively, from an initial of 1000 and 5000 µgl^{-1}.

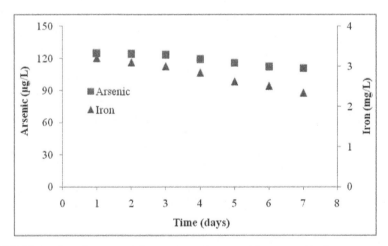

FIGURE 19.1   Arsenic removal due to adsorption on to biomass and/or iron.

Similarly, the loss of iron might be associated with biomass adsorption and/or formation of adsorption complex with arsenic. Low removal of arsenic through bio-sorption could have been due to the fact that the isoelectric point of most microbes is around pH 2. Therefore, bio-sorption of arsenic near neutral pH was very low (Teclu et al., 2008).

19.3.2   PROFILES OF NITRATE, SULPHATE, COD, PH, AND MLVSS

Figure 19.2(a) represents the performance of mixed bacterial culture in batch reactors on nitrate removal. Irrespective of initial iron and arsenic concentration, nitrate in the treated water was always less than the detection limit within 24 h of reactor operation. These results also confirmed that the high nitrate concentration of 75 and 100 mg/L, had not affected the arsenic and iron removal (Figure 19.2a). The sulfate and COD were reduced gradually during an entire period of 7 days operation; reduction rate decreased after about 3–4 days of reaction. Sulfate of only 3–5 mgl⁻¹ and 2–3 mgl⁻¹ was left out in the treated water after 4 days and 7 days of operation respectively (Figure 19.2a). COD removal was rapid on the first day, and about 70±2% COD was removed (Figure 19.2b). This is in compliance with the complete nitrate removal. Most of the COD was removed in the reactors and only 22–24 mgl⁻¹ and 16–18 mgl⁻¹ remained in the treated water after 7 days of operation, respectively. The pH of the treated water always remained between 7.25–8.0. pH of the treated water (Figure 19.2c) increased with the removal of nitrate and sulfate. This could have been due to the formation of alkalinity due to the reduction of nitrate and sulfate. Variation in biomass (MLVSS) concentration is shown in Figure 19.2(d). Biomass concentration started decreasing after an initial increase for the first 3–4 days of operation. The increase in MLVSS suggests that mixed microbial consortia grew well-using nitrate, sulfate, and arsenate as the terminal electron acceptor under an anoxic condition in the presence of acetate as an electron donor.

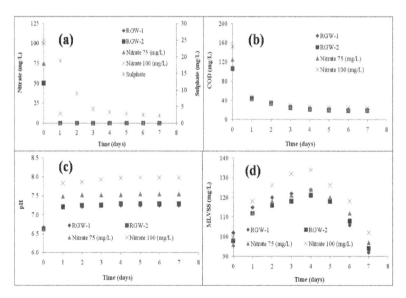

FIGURE 19.2 (a) Change of nitrate and sulfate (b) COD, (c) pH, and (d) MLVSS concentrations in the medium inoculated with mixed culture as a function of incubation time in the presence of variable concentrations of nitrate in the batch reactors.

The decrease in MLVSS after day 5 of reactor run could be associated with unavailability of carbon source in the medium as a result of the reduction of all electron acceptors present in the system.

19.3.3   PROFILES OF ARSENIC AND IRON

Figure 19.3(a) represents the performance of mixed bacterial culture in batch reactors at four different initial arsenic concentrations of 29.2 μgl^{-1} (RGW–1), 124 $\mu g/L$ (RGW–2), 200 μgl^{-1} and 300 μgl^{-1}, respectively. Irrespective of initial concentration, arsenic in the treated water was reduced below the permissible limit (10 μgl^{-1}) in 4–5 days of reaction and finally averaged 4±2 μgl^{-1} with 98.5% removal efficiency. Figure 19.3(b) represents effects of initial iron concentration of 1.68, 3.2, 4.0 and 5.0 mgl^{-1}, respectively. Irrespective of initial iron concentration (up to 5.0 mgl^{-1}), nitrate, and arsenic removal was not affected in batch reactors. Up to an initial 3.2 mgl^{-1}, iron in the treated water was always below the detection limit. However, 0.52 mgl^{-1} (87% removal) and 0.84 mgl^{-1} (83% removal) of iron concentration was observed in the treated water at 4.0 mgl^{-1} and 5.0 mgl^{-1} of initial iron. The possible reason for this higher iron concentration (0.84 mgl^{-1}) in treated water might be due to unavailability of sufficient sulfides for precipitation as iron sulfides. As it can be clearly seen in Figures 19.3a and 3b arsenic and iron removal are associated with sulfate reduction. Furthermore, with respect to the effect of initial arsenic concentration on arsenic removal, arsenic removal was always with in accordance with drinking water guidelines.

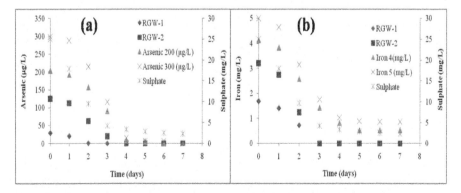

FIGURE 19.3   (a) Change of arsenic, sulfate (b) iron and sulfate concentrations in the medium inoculated with mixed culture as a function of incubation time in the presence of various concentrations of arsenic and iron.

Arsenic removal percentages by the mixed microbial consortia were not affected by high arsenic concentration (up to 300 µgl⁻¹), also indicating the mixed culture viability was not inhibited by arsenite toxicity in the experimental concentration range. The high arsenic and iron concentration had no impact on the nitrate removal. As most of the arsenic and iron was removed during sulfate reduction period which confirmed the precipitation of arsenic and iron with the biogenic sulfides formed as a result of biosulfidogenesis. In summary, mixed culture and sulfate were essential. Once these conditions were met, the arsenic and iron removal was possible.

19.3.4   CHARACTERIZATION OF BIOGENIC PRECIPITATES

The FESEM image of precipitates formed in the inoculated medium is shown in Figure 19.4a. EDS analysis showed that the precipitates were mainly composed of As (39.9%), iron (49.7%) and sulfur (10.3%) (Figure 19.4b). The results indicated that reduced arsenic was probably precipitated with sulfide and/or get adsorbed on to iron sulfide and biomass.

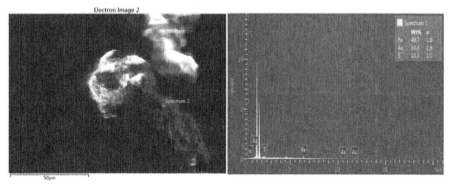

FIGURE 19.4   (See color insert.) FESEM image (a) and EDS analysis of precipitate (b) formed in the batch reactor.

The reduction of electron acceptors (DO, nitrate, sulfate, and arsenate) was coupled to the oxidation of an electron donor (acetate). Thus batch reactors employed in this study relies on terminal electron accepting processes (TEAPs) similar to those occurring in deep subsurface groundwater environment. (Lovley and Chapelle, 1995). The acetic acid was selected as a carbon source as it has been approved for drinking water treatment (National

Sanitation Foundation product and service listings, www.nsf.org) and was previously found to be effective for perchlorate removing bioreactors (Choi et al., 2007). In addition, acetic acid can be effectively utilized by many nitrate and sulfate-reducing bacteria as an electron donor (Calderer et al., 2010, Rabus et al., 2013). Many reports are available on the presence of anaerobic nitrate and sulfate-reducing bacteria in habitats such as activated sludge and sewage waters (Nielsen and Nielsen, 1998, Liu et al., 2008). After nitrate is depleted from the reactors, arsenic, and sulfate was removed from the reactors which were the next most thermodynamically favorable TEAP in the studied system. This segregation of TEAPs also explains the presence of different type of microbes in the bioreactors (Lovley and Chapelle, 1995, Macy et al., 1996). In treated water sample analyses, sulfate reduction corresponded with arsenic and iron removal. Hence, it is possible that arsenic and iron was removed through the precipitation of their respective metal sulfides. The EDS results were consistent with results found by other investigators. They reported arsenic precipitation as orpiment (As_2S_3) (Newman et al., 1997) and realgar (AsS) (O'Day et al., 2004), and confirmed the formation of arsenic sulphides by extensive X-ray absorption near edge structure and thermodynamic modeling (Onstott et al., 2011, Rodriguez-Freire et al., 2014). Some amount of arsenic may also get adsorbed on to biomass as reported by Teclu et al. (2008). In the present study, although the exact arsenic removal processes are not clear, the possible removal mechanism of iron was removal as iron sulfides precipitation and arsenic was removed either as arsenosulfides precipitation and or adsorption on to iron sulfides.

19.4   SUMMARY

In conclusion, the TEAP based batch bioreactors were successfully applied for simultaneous removal of arsenic, iron, and nitrate removal from polluted drinking water supplies to levels below the MCL of 10 μgl^{-1} and 0.3 mgl^{-1}, respectively. Complete nitrate removal was achieved up to 150 mgl^{-1} of nitrate. The presence of high nitrate did not affect arsenic and iron removal in TEAP based process in studied concentration ranges of contaminants. Together with high efficiency, sustainability of the process, environment-friendliness, and the ease of multi-contaminant removal, the present method is having great potential for rural applications in developing countries.

ACKNOWLEDGMENTS

The authors acknowledge the funding by the Ministry of Drinking Water and Sanitation (Government of India), through the project (Ref. No. W.11017/44/2011-WQ) to carry out this work at IIT Guwahati. Authors are also thankful to the central instrument facility of IIT Guwahati for providing facilities for FESEM/EDS analysis of some samples.

KEYWORDS

- **biogenic sulfide**
- **co-contaminant**
- **TEAP**

REFERENCES

APHA, (2005). *Standard Methods for the Examination of Water & Wastewater* (21st edn.). American Public Health Association, Washington DC.

Bacquart, T., Frisbie, S., Mitchell, E., Grigg, L., Cole, C., Small, C., & Sarkar, B., (2015). Multiple inorganic toxic substances contaminating the groundwater of Myingyan Township, Myanmar: Arsenic, manganese, fluoride, iron, and uranium. *Science of the Total Environment, 517*, 232–245.

Bhatnagar, A., & Sillanpaa, M., (2011). A review of emerging adsorbents for nitrate removal from water. *Chemical Engineering Journal, 168*, 493–504.

Brown, J. C., (2008). Biological treatments of drinking water.

Buschmann, J., Berg, M., Stengel, C., Winkel, L., Sampson, M. L., Trang, P. T. K., & Viet, P. H., (2008). Contamination of drinking water resources in the Mekong delta floodplains: Arsenic and other trace metals pose serious health risks to the population. *Environment International, 34*, 756–764.

Calderer, M., Gibert, O., Marti, V., Rovira, M., De Pablo, J., Jordana, S., Duro, L., Guimera, J., & Bruno, J., (2010). Denitrification in presence of acetate and glucose for bioremediation of nitrate-contaminated groundwater. *Environmental Technology, 31*, 799–814.

Chakrabarty, S., & Sarma, H. P., (2011). Fluoride, iron and nitrate-contaminated drinking water in Kamrup district, Assam, India. *Archives of Applied Science Research, 3*, 186–192.

Chaturvedi, R., Banerjee, S., Chattopadhyay, P., Bhattacharjee, C. R., Raul, P., & Borah, K., (2014). High iron accumulation in hair and nail of people living in iron affected areas of Assam, India. *Ecotoxicology and Environmental Safety, 110*, 216–220.

Choi, Y. C., Li, X., Raskin, L., & Morgenroth, E., (2007). Effect of backwashing on perchlorate removal in fixed bed biofilm reactors. *Water Research, 41*, 1949–1959.

Chung, J., Rittmann, B. E., Wright, W. F., & Bowman, R. H., (2007). Simultaneous bio-reduction of nitrate, perchlorate, selenate, chromate, arsenate, and dibromochloropropane using a hydrogen-based membrane biofilm reactor. *Biodegradation, 18*, 199–209.

CPCB, (2007). *Status of Groundwater Quality in India*. Central Pollution Control Board, Ministry of Environment and Forests, Govt. of India.

Gimenez, M. C., Blanes, P. S., Buchhamer, E. E., Osicka, R. M., Morisio, Y., & Farias, S. S., (2013). Assessment of heavy metals concentration in arsenic contaminated groundwater of the Chaco Plain, Argentina. *ISRN Environmental Chemistry.*

Kapoor, A., & Viraraghavan, T., (1997). Nitrate removal from drinking water—Review. *Journal of Environmental Engineering, 123*, 371–380.

Khatri, N., Tyagi, S., & Rawtani, D., (2017). Recent strategies for the removal of iron from water: A review. *Journal of Water Process Engineering, 19*, 291–304.

Kumar, N. S., & Goel, S., (2010). Factors influencing arsenic and nitrate removal from drinking water in a continuous flow electrocoagulation (EC) process. *Journal of Hazardous Materials, 173*, 528–533.

Liu, X., Gao, C., Zhang, A., Jin, P., Wang, L., & Feng, L., (2008). The nos gene cluster from gram-positive bacterium Geobacillus thermodenitrificans NG80–2 and functional characterization of the recombinant NosZ. *FEMS Microbiology Letters, 289*, 46–52.

Lovley, D. R., & Chapelle, F. H., (1995). Deep subsurface microbial processes. *Reviews of Geophysics, 33*, 365–381.

Macy, J. M., Nunan, K., Hagen, K. D., Dixon, D. R., Harbour, P. J., Cahill, M., & Sly, L. I., (1996). Chrysiogenes arsenatis gen. nov., sp. nov., a new arsenate-respiring bacterium isolated from gold mine wastewater. *International Journal of Systematic Bacteriology, 46*, 1153–1157.

Newman, D. K., Kennedy, E. K., Coates, J. D., Ahmann, D., Ellis, D. J., Lovley, D. R., & Morel, F. M., (1997). Dissimilatory arsenate and sulfate reduction in Desulfotomaculum auripigmentum sp. nov. *Archives of Microbiology, 168*, 380–388.

Nielsen, J. L., & Nielsen, P. H., (1998). Microbial nitrate-dependent oxidation of ferrous iron in activated sludge. *Environmental Science & Technology, 32*, 3556–3561.

ODay, P. A., Vlassopoulos, D., Root, R., & Rivera, N., (2004). The influence of sulfur and iron on dissolved arsenic concentrations in the shallow subsurface under changing redox conditions. *Proceedings of the National Academy of Sciences of the United States of America, 101*, 13703–13708.

Onstott, T. C., Chan, E., Polizzotto, M. L., Lanzon, J., & DeFlaun., M. F., (2011). Precipitation of arsenic under sulfate-reducing conditions and subsequent leaching under aerobic conditions. *Applied Geochemistry, 26*, 269–285.

Rabus, R., Hansen, T. A., & Widdel, F., (2013). Dissimilatory sulfate-and sulfur-reducing prokaryotes. *The Prokaryotes* (pp. 309–404). Springer.

Rezaie-Boroon, M. H., Chaney, J., & Bowers, B., (2014). The source of arsenic and nitrate in Borrego valley groundwater aquifer. *Journal of Water Resource and Protection, 6*, 1589.

Rodriguez-Freire, L., Sierra-Alvarez, R., Root, R., Chorover, J., & Field, J. A., (2014). Biomineralization of arsenate to arsenic sulfides is greatly enhanced at mildly acidic conditions. *Water Research, 66*, 242–253.

Singh, R., Singh, S., Parihar, P., Singh, V. P., & Prasad, S. M., (2015). Arsenic contamination, consequences, and remediation techniques: A review. *Ecotoxicology and Environmental Safety, 112*, 247–270.

Teclu, D., Tivchev, G., Laing, M., & Wallis, M., (2008). Bioremoval of arsenic species from contaminated waters by sulfate-reducing bacteria. *Water Research, 42*, 4885–4893.

Upadhyaya, G., Clancy, T. M., Snyder, K. V., Brown, J., Hayes, K. F., & Raskin, L., (2012). Effect of air-assisted backwashing on the performance of an anaerobic fixed-bed bioreactor that simultaneously removes nitrate and arsenic from drinking water sources. *Water Research, 46,* 1309–1317.

WHO Guidelines for Drinking-Water Quality, (2011). *World Health Organisation,* Geneva, *4,* 315–318.

Zhao, H. P., Ontiveros-Valencia, A., & Tang, Y., (2014). Removal of multiple electron acceptors by pilot-scale, two-stage membrane biofilm reactors. *Water Research, 54,* 115–122.

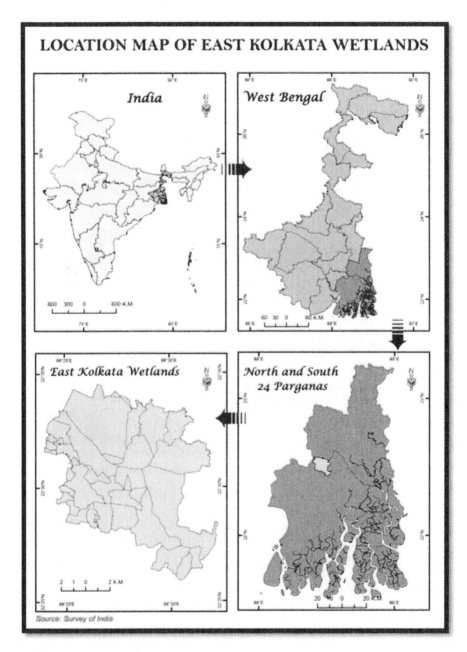

FIGURE 3.1 Location map of EKW, Kolkata.

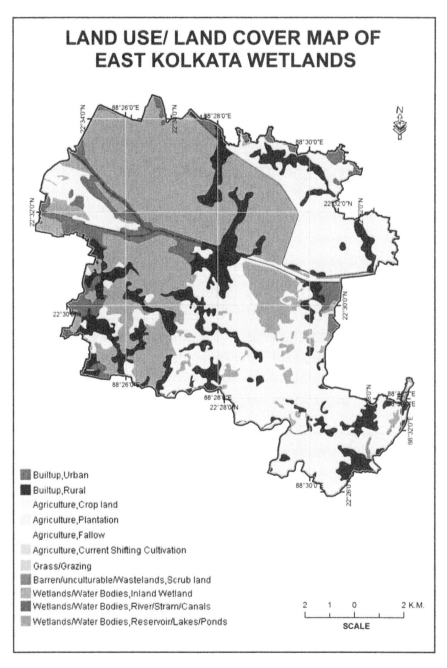

FIGURE 3.2 Current land use and land cover of EKW.

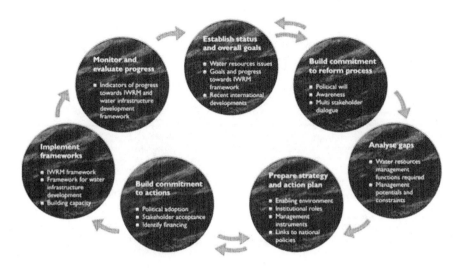

FIGURE 4.3   IWRM planning cycle.

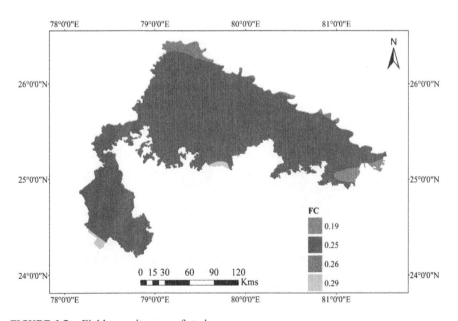

FIGURE 6.2   Field capacity map of study area.

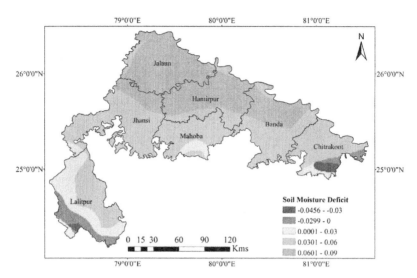

FIGURE 6.4   Soil moisture deficit (SMD) of the study area.

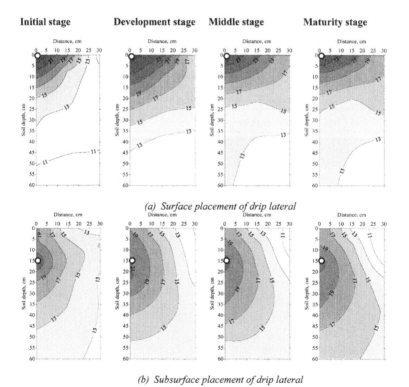

FIGURE 7.1   Observed soil water (volumetric in percent) distribution in the field.

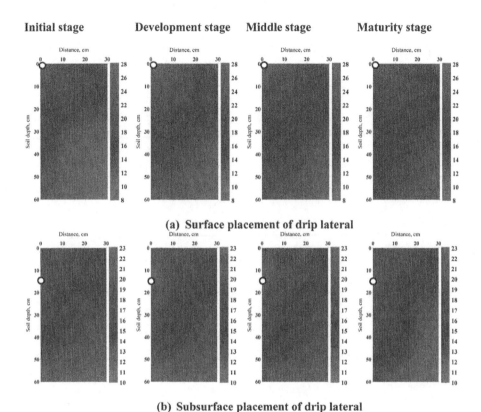

Initial stage   Development stage   Middle stage   Maturity stage

(a) **Surface placement of drip lateral**

(b) **Subsurface placement of drip lateral**

FIGURE 7.2   Simulated soil water (volumetric in percent) distribution with the model.

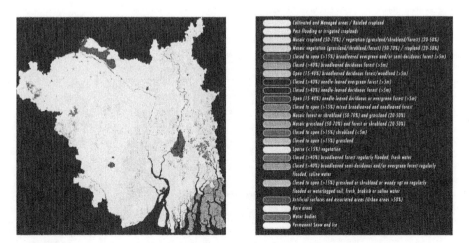

FIGURE 11.4   Land use and land cover.

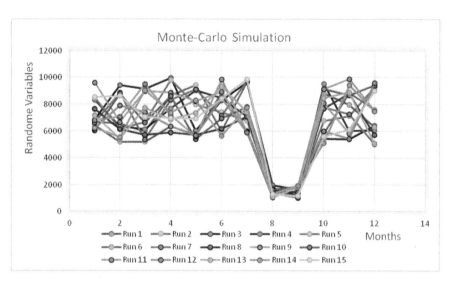

FIGURE 12.2   Changing the trajectory of a random variable in Monte-Carlo simulation.

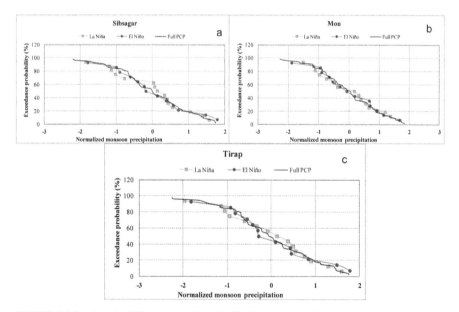

FIGURE 15.2   Graph of the cumulative distribution of normalized monsoon precipitation during El Niño, La Niña and Normal phase for (a) Sibsagar, (b) Mon and (c) Tirap.

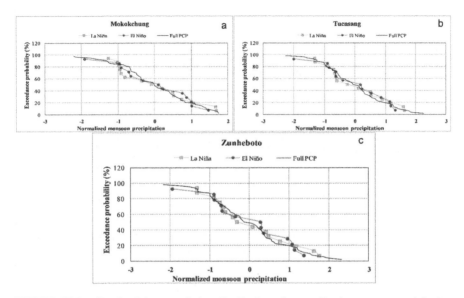

FIGURE 15.3 Graph of the cumulative distribution of normalized monsoon precipitation during El Niño, La Niña and Neutral phase for (a) Mokokchung, (b) Tuensang and (c) Zunheboto.

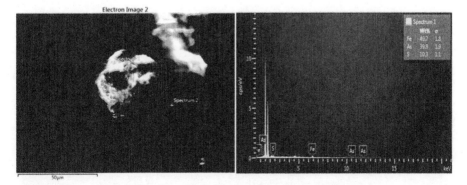

FIGURE 19.4 FESEM image (a) and EDS analysis of precipitate (b) formed in the batch reactor.

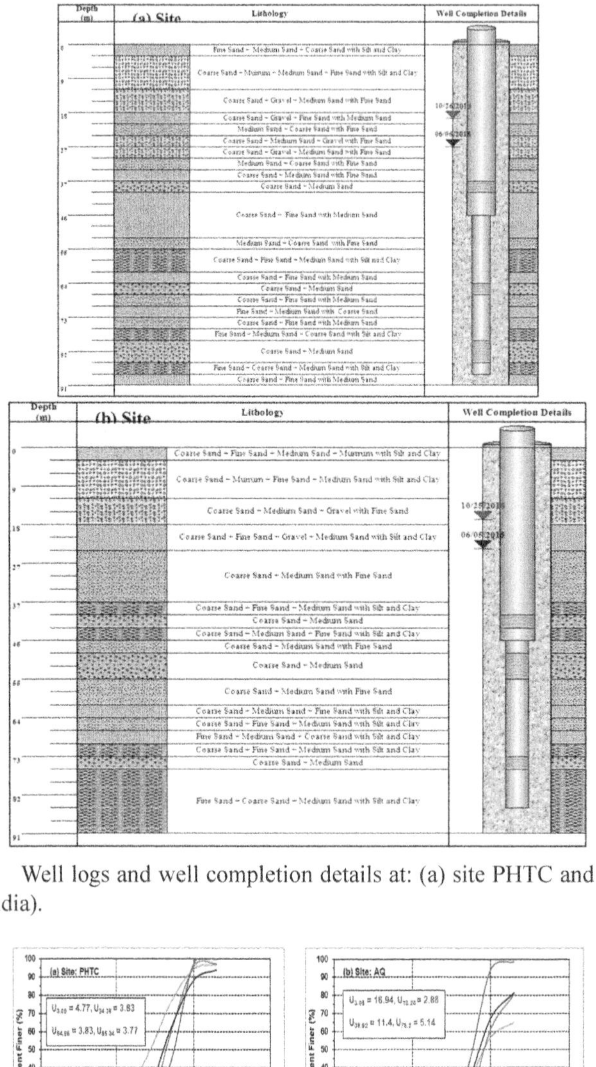

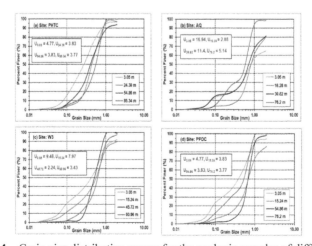

FIGURE 20.3   Well logs and well completion details at: (a) site PHTC and (b) site PFDC (Kharagpur, India).

FIGURE 20.4   Grain-size distribution curves for the geologic samples of different depths.

CHAPTER 20

SCIENTIFIC FRAMEWORK FOR SUBSURFACE CHARACTERIZATION AND EVALUATION OF GRAIN-SIZE ANALYSIS METHODS

SABINAYA BISWAL and MADAN KUMAR JHA

AGFE Department, Indian Institute of Technology Kharagpur, Kharagpur – 721 302, West Bengal, India, E-mail: sabin9937@gmail.com

ABSTRACT

Subsurface formations are highly variable with depth and space, which poses challenges in their characterization at a catchment scale. This study demonstrates a scientifically sound approach for lithology exploration and the relative performance of salient grain-size analysis (GSA) methods in estimating hydraulic conductivity of subsurface formations. Test drillings were carried out at eight locations, and geologic samples were collected at a regular interval of 3.05 m (10 feet). All the geologic samples were subjected to sieve analysis. Thereafter, grain-size distribution curves (GSDC) were prepared for individual geologic samples. The physical characteristics of subsurface formations such as porosity, uniformity coefficient, and effective grain diameter were computed using GSDC and empirical formulae. Hydraulic conductivity of subsurface formations was estimated by using four popular GSA methods, and their relative performance was evaluated. The analysis of GSDC revealed that the geologic samples at deeper depths (≥ 40 m) have S-shape curves compared to the geologic formations present at shallow depths (<40 m). The values of uniformity coefficient range from 2.43 to 16.94 for shallow-depth geologic samples, whereas they vary from 2.24 to 11.40 for deeper depth geologic samples. The lithologic analysis indicated that shallow aquifer layers exist at depths ranging from 9 to 40 m with thickness varying from 3 to

6 m. Besides shallow aquifers, deep aquifer layers of 3 to 34 m thickness exist at 40 to 79 m depths. The average values of hydraulic conductivities (K) based on the four GSA methods range from 2.52 to 208.38 m/day for the deeper aquifer layers and 0.98 to 285.5 m/day for the shallow aquifer layers, thereby suggesting considerable aquifer heterogeneity. Among the four GSA methods, the Hazen and Kozeny methods consistently overestimate the mean K with higher standard deviation at all the sites, while the Slitcher method provides the least mean K values with a less standard deviation. These estimates are useful for a preliminary assessment of subsurface K in the absence of field-based estimates.

20.1   INTRODUCTION

Groundwater is a globally important and valuable renewable natural resource, which supports, human life and health, economic development and ecological diversity. Among various sectors, agriculture accounts for 89% of the total water consumption and domestic consumption account for about 5%. According to the fourth United Nations World Water Development Report, India is the largest consumer of groundwater in the world. Due to excessive human activities and climate change impacts, over-exploitation of groundwater resources has been a major issue in the present condition. In addition, groundwater pumping is often unmonitored and unregulated particularly in the developing country like India resulting a drastic decrease in groundwater levels in several parts of the country. These facts pose a challenging task for the researchers and scientific community to prevent aquifer depletion due to over-exploitation and to supply water in a sustainable way to the society. Thus, it is needed to manage the groundwater resources in a proper way for long-term use.

Hydrogeologic investigation is the prerequisite for groundwater studies and hence plays a central role in the development and management of the vital groundwater resources. The use of empirical grain-size analysis (GSA) models to compute hydraulic conductivity is an indirect and cost-effective method, which has attracted several researchers (Lu et al., 2012; Rosas et al., 2015; Devlin, 2015). The GSA methods estimate subsurface hydraulic conductivity (K) based on the statistical distribution of the soil particles of different sizes. Also, during the sampling procedure, the geologic samples get distorted from its natural structure and don't represent the actual field conditions. Due to these facts, estimates obtained from GSA methods contain some error. Despite the well-accepted limitations of K estimates from grain-size analyses, there are no simpler or more economical measurement-based techniques for obtaining K estimates from aquifer samples. The results obtained from GSA methods are pretty much important for the areas, where other standard methods like pumping test, slug test, etc. are not easily accessible.

In the present study, an attempt has been made to develop a methodology to characterize the hydrogeology of subsurface formations in a more accurate way as compared to many past studies where feel method (without the use of sieve analysis) or indirect method like resistivity survey were used for classifying the subsurface formations. In addition, the performance evaluation of widely used GSA methods has been carried out for estimating the hydraulic conductivity of different subsurface formations.

20.2   MATERIALS AND METHODS

20.2.1   COLLECTION OF GEOLOGIC SAMPLE

Borehole drilling was conducted up to 91 m depth at eight sites namely, PHTC, AQ, RB, TG, W1, W2, W3 and PFDC over the study area. The area selected for this study was Experimental Farm of Indian Institute of Technology Kharagpur, West Bengal, India, and is located 22° 19' 10.97" N latitude and 87° 18' 35.87" E longitude (Figure 20.1). During drilling, the geologic samples were collected at a regular depth interval of 3.05 m (10 feet) up to the drilling depth. Properly designed well screens were placed at appropriate depths to tap the potential aquifer layers and to prevent entering of fine particles to the borehole. After installation of these observation wells, development was carried out with the help of air compressor. The installation and development procedure is shown in Figures 20.2 (a–d).

20.2.2   ANALYSIS OF GEOLOGIC SAMPLES

Sieve analysis of the geologic samples was carried out with the sieve of different sizes varying from 0.05 to 2 mm. Thereafter, Grain-size distribution curves (GSDC) were prepared for each geologic sample and thus, the physical parameters like characteristic grain diameter, porosity, and uniformity coefficients were estimated. Finally, well logs were constructed for the drilled sites by using the state-of-the-art software package "Hydro-GeoAnalyst" developed by Waterloo Hydrogeologic, Inc. (WHI), Canada.

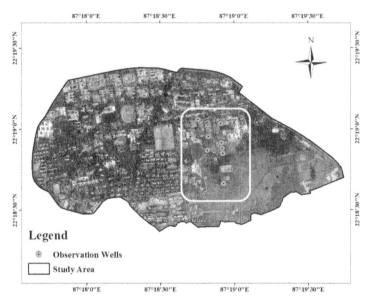

FIGURE 20.1   Location map of the study area.

FIGURE 20.2   (a) Installation of observation well, (b) Development of observation well, (c) Field view of developed observation well with a nest of piezometers and (d) Field view of developed pumping well.

20.2.3   ESTIMATION OF HYDRAULIC CONDUCTIVITY BY GSA METHODS

In this study, four widely used GSA methods having varying applicability criteria were selected for the estimation of hydraulic conductivity of different subsurface formations. According to Vukovic and Soro (1992), porosity (n) value may be derived from the empirical relationship with the coefficient of grain uniformity (U) as follows:

$$n = 0.255 \times (1 + 0.83^{U})$$

(1)

where, U = coefficient of grain uniformity, and it is given by:

$$U = \frac{d_{60}}{d_{10}}$$

(2)

Here, d_{60} and d_{10} in the Eq. (2) represent the grain diameter in (mm) for which, 60% and 10% of the sample respectively, are finer than. A succinct description of the selected GSA methods is given below.

20.2.3.1   HAZEN METHOD

The Hazen method represents a simpler relationship between the hydraulic conductivity and the effective grain diameter and was initially developed for the design of sand filter (i.e., loose and clean sand) for water purification (Hazen, 1892). The equation is given as:

$$K = 6 \times 10^{-4} \times [1 + 10 \times (n - 0.26)] \times d_{10}^{2}$$

(3)

where, K = hydraulic conductivity (cm/s), d_{10} = effective grain diameter (cm), and n = porosity of the soil sample.

20.2.3.2   HAZEN SIMPLIFIED METHOD

Freeze and Cherry (1979) developed an empirical relation due to Hazen (1892) for estimating hydraulic conductivity. The equation is given as:

$$K = A \times d_{10}^{2}$$

(4)

where, K = hydraulic conductivity (cm/s), d_{10} = effective grain diameter (mm), and A = constant (usually taken as 1).

20.2.3.3   SLITCHER METHOD

Slitcher (1898) developed a simple relationship between grain size and hydraulic conductivity which depends on the porosity of the soil sample. The equation is given as:

$$K = \frac{\rho g}{\mu} \times 1 \times 10^{-2} \times n^{3.287} \times d_{10}^{2} \tag{5}$$

where, K = hydraulic conductivity (cm/s), d_{10} = effective grain diameter (cm), and n = porosity of the soil sample.

20.2.3.4   KOZENY METHOD

Kozeny (1953) has specified the simpler non-linear equation to calculate the value of hydraulic conductivity, which is a function of effective grain diameter (d_{10}) and porosity.

$$K = \frac{\rho g}{\mu} \times 8.3 \times 10^{-3} \times \frac{n^{3}}{(1-n)^{2}} \times d_{10}^{2} \tag{6}$$

where, K = hydraulic conductivity (cm/s), and d_{10} = effective grain diameter (cm).

20.3   RESULTS AND DISCUSSION

20.3.1   AVAILABILITY OF AQUIFER LAYERS

Well logs were constructed for all the drilled sites over the study area. As an example, the developed well logs for the sites PHTC and PFDC are shown in Figure 20.3 (a, b). The analysis of the developed well logs indicated the presence of *murrum* (reddish brown colored hard geologic material) along with coarse sand at different depths. The first aquifer layer in the study area exists at depths ranging from 9 to 21 m with varying thickness (3 to 6 m). Following the first aquifer layer, the second aquifer layer resides at a depth ranging from 15 to 40 m, and the thickness varies between 3 to 6 m. Beside these two shallow aquifer layers, deeper aquifer layers also exist at depths varying from 40 to 79 m. The thickness of the deep aquifer layer varies from 3 to 34 m. Overall, it can be seen that the thickness of the deeper aquifer layers is higher than the shallow aquifer layers. It was found that there is an appreciable variation in the lithology over the study area, which suggests significant heterogeneity in subsurface formations.

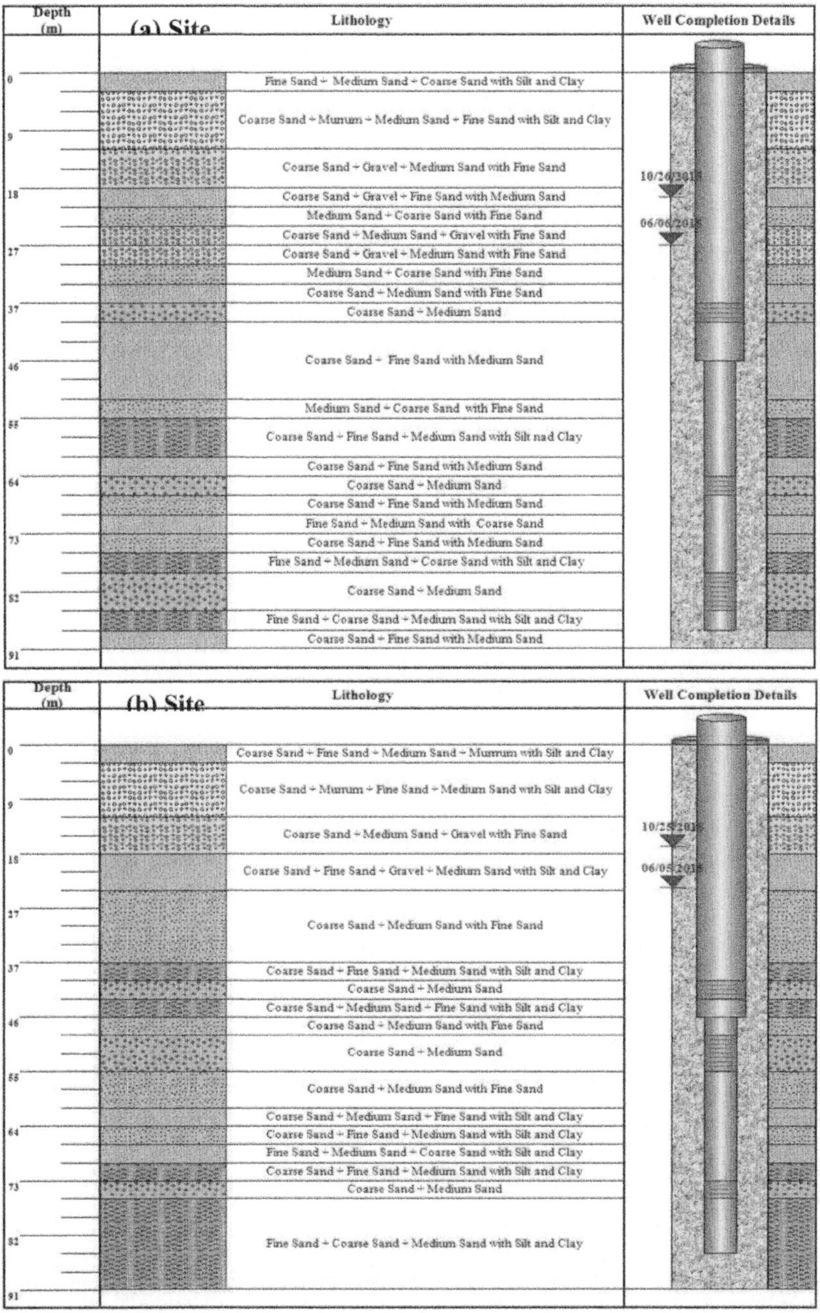

FIGURE 20.3 (See color insert.) Well logs and well completion details at: (a) site PHTC and (b) site PFDC (Kharagpur, India).

20.3.2 CHARACTERISTICS OF SUBSURFACE FORMATIONS

GSDC for the top geologic sample, i.e., 3.05 m depth and at selected depths are shown in Figure 20.4 (a–d) for the four sites as an example. It is apparent from these figures that the uniformity coefficient (U) varies from a minimum value of 2.24 at a depth of 45.72 m (for site W7) to a maximum value of 16.94 at 3.05 m depth (for site W2). It is also evident from these figures that the shallow depth geologic samples (< 40 m) have a low value of U ranging from 2.43 to 16.94 as compared to the U values (2.24 to 11.40) of geologic samples of deeper depth ($\geq$ 40 m). Thus, the deeper depth geologic samples can be considered as uniform or poorly graded compared to the shallow depth geologic samples. Thus, the deep aquifers are supposed to have a high value of porosity as compared to the shallow aquifer layers.

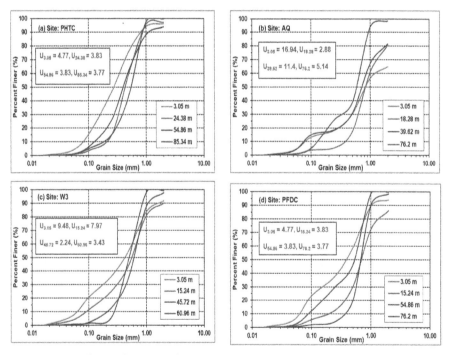

FIGURE 20.4   (See color insert.) Grain-size distribution curves for the geologic samples of different depths.

20.3.3 HYDRAULIC CONDUCTIVITY OF SUBSURFACE FORMATIONS

Hydraulic conductivity (K) values estimated by the four GSA methods vary from a minimum value of 0.98 m/day to a maximum value of. 285.50 m/day for the aquifer layers, whereas it ranges from 0.67 to 165.41 m/day for the non-aquifer layers. As the non-aquifer layers contain a higher percentage of fine particles compared to coarser particles, low values of K is obvious. Further, the shallow aquifer layers have K values varying from 0.98 to 285.50 m/day, while the deeper aquifer layers have K values in the range of 2.52 to 208.38 m/day.

20.3.4   EFFICACY OF GSA METHODS

The performance of the GSA methods is evaluated in terms of statistical indicators and is tabulated in Table 20.1. It is inferred from this table that the Hazen and Kozeny methods have estimated consistently high mean K values (11.91 to 122.88 m/day) with higher standard deviation (25.4 to 144.65 m/day) as compared to the other two methods for all the sites. On the other hand, lowest K values are obtained from the Slitcher method which varies from 0.27 m/day at site W1 to 171.44 m/day at site TG over the study area. Also, the Slitcher method has the lowest standard deviation ranging from 9.12 m/day at site W1 to 43.05 m/day at site TG. Overall, it could be considered that the Slitcher method and Hazen Simplified methods are performing reasonably well for the study area.

20.4   CONCLUSIONS

Based on the findings of this study, the following conclusions can be drawn:

- The shallow aquifer layers are present at depths varying from 9 to 40 m with the thickness 3 to 6 m, and the deep aquifer layers exist at 40 to 79 m depths with a considerable variation in the thickness (3 to 34 m).
- The geologic samples of the deeper depth (> 40 m) were found to be uniform (U = 2.24 to 11.4) as compared to those of the shallow depth (U = 2.43 to 16.94).
- The average values of K for the shallow and deep aquifer layers vary from 0.98 to 285.50 m/day and 2.52 to 208.38 m/day, respectively.
- Out of the four GSA methods used in this study, the Slitcher method and Hazen Simplified methods were found to estimate K values reasonably.

TABLE 20.1 Basic Statistics of the GSA Methods

Methods	Statistics	Sites							
		PHTC	AQ	RB	TG	W1	W2	W3	PFDC
1. Hazen Simplified	Min (m/day)	3.95	4.09	3.42	5.08	1.77	3.22	3.30	3.42
	Max (m/day)	95.94	179.83	101.23	245.12	80.86	96.63	64.18	97.94
	Mean (m/day)	23.11	33.75	17.23	74.49	10.81	31.57	23.28	18.73
	SD (m/day)	22.60	45.48	22.08	65.88	14.82	28.62	23.17	21.28
2. Hazen	Min (m/day)	4.06	3.35	3.03	4.40	1.32	3.22	3.07	3.46
	Max (m/day)	157.15	322.53	170.16	423.44	136.61	169.88	113.40	171.41
	Mean (m/day)	33.04	48.40	22.02	**113.98**	**14.19**	50.18	35.47	27.37
	SD (m/day)	37.30	74.41	34.11	108.34	**25.40**	51.15	41.16	36.87
3. Slitcher	Min (m/day)	0.97	0.71	0.67	0.93	**0.27**	0.76	0.69	0.82
	Max (m/day)	54.48	122.65	60.53	**171.44**	48.84	63.26	42.45	63.54
	Mean (m/day)	10.40	15.71	6.51	39.66	4.24	17.37	11.96	8.82
	SD (m/day)	13.03	26.91	11.75	43.05	9.12	19.01	15.31	13.40
4. Kozeny	Min (m/day)	2.38	1.65	1.58	2.16	0.62	1.81	1.66	2.00
	Max (m/day)	164.96	395.52	186.41	578.87	150.93	201.01	135.38	201.25
	Mean (m/day)	29.95	46.46	18.22	**122.88**	**11.91**	52.98	36.07	25.76
	SD (m/day)	39.60	84.81	35.74	**144.65**	28.21	60.14	48.53	41.87

- It is recommended that the scientific framework should be adopted for the efficient characterization of subsurface formations, which in turn can ensure improved information about the potential of groundwater in an area or a region.

KEYWORDS

- **grain-size analysis methods**
- **grain-size distribution curve**
- **hydraulic conductivity**
- **sieve analysis**
- **subsurface characterization**
- **test drilling**

REFERENCES

Devlin, J. F., (2015). Hydrogeosieve XL: An Excel-based tool to estimate hydraulic from the grain-size analysis. *Journal of Hydrogeology, 23,* 837–844.

Freeze, R. A., & Cherry, J. A., (1979). *Groundwater* (p. 604). Prentice-Hall Inc., Englewood Cliffs, New Jersey.

Hazen, A., (1892). *Some Physical Properties of Sands and Gravels* (pp. 539–556). Massachusetts State Board of Health, 24th Annual Report.

Kozeny, J., (1953). Das wasser im boden. Grundwasserbewegung. *Hydraulik: Ihre Grundlagen und Praktische Anwendung ,* 380–445. The water in the ground: groundwater flow, Springer, Heidelberg, Germany.

Lu, C., Chen, X., Cheng, C., Ou, G., & Shu, L., (2012). Horizontal hydraulic conductivity of shallow streambed sediments and comparison with the grain-size analysis results. *Hydrological Processes, 26*(3), 454–466.

Rosas, J., Jadoon, K. Z., & Missimer, T. M., (2015). New empirical relationship between grain-size distribution and hydraulic conductivity for ephemeral streambed sediments. *Environmental Earth Science, 73*(3), 1303–1315.

Slichter, C. S., (1898). *Theoretical Investigations of the Motion of Groundwaters* (pp. 295–384). 19th Annual Report, US Geological Survey, Reston, VA.

Vukovic, M., & Soro, A., (1992). *Determination of Hydraulic Conductivity of Porous Medium from Grain-Size Composition.* Water Resources Publications, Littleton, Colorado, USA.

CHAPTER 21

STUDY OF CHEMICAL NATURE OF GROUNDWATER IN THE WESTERN PARTS OF JHARKHAND WITH A FOCUS ON FLUORIDE

NEETA KUMARI and GOPAL PATHAK

Department of Civil and Environmental Engineering,
B.I.T. Mesra, Ranchi–835215, India,
E-mail: neetak@bitmesra.ac.in, neeta.sinha2k7@gmail.com

ABSTRACT

The present study aims to find the groundwater chemical nature with respect to the fluoride in the groundwater. The sample data analysis indicated that fluoride content in groundwater increases with depth. The comparison of fluoride concentration of groundwater from the shallow dug wells, and deeper bore wells from the same location indicated that the deeper aquifers have a higher concentration of fluoride than the shallow aquifers. Due to continuous rock-water interaction in aquifers the fluoride concentration is found higher in dug well (DW) and hand pump (HP) than in shallow aquifers. The shallow aquifers have more discharge and recharge rate but is prone to anthropogenic pollution. The pH of the study area belongs to the alkaline in origin. An alkaline pH ranging from 7.4–8.8 resulted in high fluoride concentration (1.7–6.1 mgl^{-1}) in groundwater sources. The high pH of water displaces fluoride ions from mineral surface. In this study minimum 330 μmhos cm^{-1} to maximum 2580 μmhos cm^{-1} is observed. Thus all samples have a high electrical conductivity (EC).

In this sample analysis, total hardness has a minimum value of 105 mgl^{-1} and a maximum of 890 mgl^{-1}. So in hard water, calcium, and magnesium carbonate and bicarbonate are found. Therefore, water will be calcium deficient, and fluoride leeching will be more. The carbonates are completely absent in the study area which suggests the occurrence of bicarbonates due to weathering of silicate minerals. In this study, calcium content has minimum-maximum value range as 36–198 mgl^{-1}. Magnesium content has minimum-maximum value range as 2.4–353 mgl^{-1}. Sodium content has minimum-maximum value range as 14–565 mgl^{-1}. Sodium and calcium are the major cations that influence the fluoride level in groundwater. The average chloride content of sedimentary rocks is about the same as the evaporate rocks 150 ppm and indicating sedimentary depositional environment in the study area. The significance of this study is that it helps in understanding the fluoride mobility in groundwater of the study area.

21.1   INTRODUCTION

Groundwater is a primary natural resource which supports all types of life forms (Singhal et al., 2013). Fluoride mainly presents in groundwater in the form of dissolved ions. Major sources of fluoride in groundwater are fluoride-bearing rocks such as fluorspar, cryolite, fluorite, fluorapatite, and hydroxyapatite (Agarwal et al., 1997). The fluoride content in groundwater is a function of many factors such as the presence of fluoride minerals and their solubility in water, the velocity of flowing water, pH, temperature, and concentration of Ca and HCO_3^- ions in water (Chandra et al., 1981; Largent et al., 1961). The problem of high fluoride concentration in ground water is an international problem. Many countries, like India, Srilanka, and China, East Africa, the rift valley countries, Turkey, and parts of South Africa (Pol et al., 2012) are facing this problem at a large scale. In India, the most affected states are Rajasthan, Gujarat, and Andhra Pradesh (Susheela and Mujumdar, 1992) but the problem is most pronounced in Andhra Pradesh, Bihar, Gujarat, Madhya Pradesh, Punjab, Rajasthan, Tamil Nadu, and Utter Pradesh. (Rao et al., 1992). As per the CGWB survey in 2010, in Jharkhand, districts like Palamu, Ranchi, Gumla, Godda, Giridih, Bokaro are affected with fluoride problem in their groundwater. Studies pertaining to fluoride concentration in groundwater of Jharkhand have been conducted by various researchers (Srikanth and Priyadarshi, 2008; Avishek et al., 2010; MacDonald et al., 2011; Pandey et al., 2012). The fluoride pollution is mainly geogenic (because of the geology of the area) in origin, and less rainfall (arid climate) tends to provide longer residence time to the groundwater with fluoride in aquifers.

In 1984, WHO estimated that more than 260 million people living all over the world consume water with fluoride concentration above 1mg/l (WHO, 1984; Smet et al., 1990). High concentrations of F⁻ in groundwater are found in many places of the world, but notably in Asia and Africa (Appelo and Postma, 2005). Fluorosis is the most severe and widespread in the two largest countries – India and China. As per UNICEF, the disease fluorosis is endemic in at least 25 countries, and around 200 million people from 25 nations have health risks because of high fluoride in groundwater (Ayoob and Gupta, 2006). As per the WHO standards, the prescribed maximum level for fluoride in drinking water is 1.5 mgl⁻¹and the permissible limit is 1.0 mgl⁻¹ (WHO, 1971). Indian standard specifications for drinking water IS 10500 specifies required a desirable limit of fluoride concentration in drinking water as 0.6–1.0 mgl⁻¹and if the limit is below 0.6 mgl⁻¹, water should be rejected, the maximum limit is extended to 1.5 mgl⁻¹. Bureau of Indian Standards (BIS, 1992) has recommended a desirable upper limit of 1.0 mgl⁻¹of fluoride as a desired concentration in drinking water, which can be extended to 1.5 mgl⁻¹of fluoride in case no alternative source of water is available.

21.2 MATERIALS AND METHODS

The water samples were collected from the dug well (DW), hand pump (HP), and shallow aquifers. They were brought to the laboratory and analyzed as per the standard methods. The compiled data is then plotted using Origin software.

21.3 RESULTS AND DISCUSSION

Observed fluoride values associated with the type of well is shown in Figure 21.1. On an average, shallow aquifers have less fluoride level than DW and HP. Although many points have reported above 1.5 mgl⁻¹ but the fluoride range is high in DW and HP in comparison to shallow aquifers.

21.3.1 FLUORIDE AND DEPTH

In Figure 21.2, a variation of fluoride values with depth is shown. The fluoride content is increasing with depth. The comparison of fluoride concentration of groundwater from the shallow DWs and deeper bore wells from the same location indicated that the deeper aquifers have a higher concentration of fluoride than the shallow aquifers (Wodeyar and Sreenivasan, 1996). The variation in the concentration of fluoride in groundwater with respect to the depth of the aquifer may depend upon the lithology, amount, and duration of rainfall, and the level of groundwater exploitation of the area.

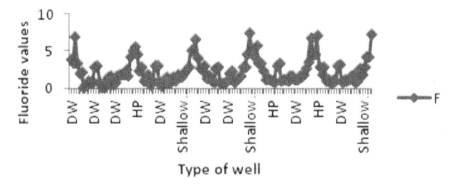

FIGURE 21.1   Observed fluoride values associated with the type of well (X-axis- a type of well and on Y-axis –fluoride level).

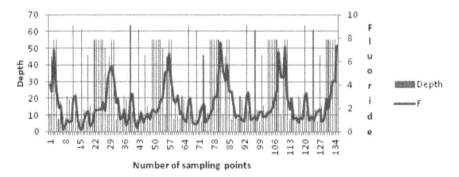

FIGURE 21.2   Variation of fluoride with respect to depth.

At higher depth groundwater contact time with aquifer media increases as velocity decreases. Due to geology and less recharge of groundwater may also cause a high level of fluoride. This is also a reason why high fluoride groundwaters are associated with the arid and semi-arid environment (Kumar et al., 2014). The study area geology has aquifers surrounded with granitic and gneisses rocks which have fluoride-containing minerals like fluorite and fluorapatite (Srikanth et al., 2008). Due to continuous rock-water interaction in aquifers the fluoride concentration is found higher in DW and HP than in shallow aquifers. The shallow aquifers have more discharge and recharge rate but is prone to anthropogenic pollution. Groundwater velocity is also high in comparison to the deeper aquifer (www.igrac.com). The concentration of fluoride largely depends upon the geology of the area. Geologically the area is comprised of hard rock and sedimentary formations. Groundwater in the hard rock area is controlled by the depth and degree of weathering (Kumar et al., 2014).

21.3.2   FLUORIDE AND pH

pH is considered as an important ecological factor and provides an important piece factor and piece of information on many types of geochemical equilibrium or solubility calculation (Shyamala et al., 2008). When compared with the standard values of WHO and IS 10500, 6.5–8.5, the water samples are found to be in the permissible limit at all locations. In groundwater, the fluoride solubility is pH dependent (Chandio et al., 2015). The pH of the study area belongs to the alkaline in origin. Saxena and Ahmed (2001) reported that an alkaline pH ranging from 7.4–8.8 resulted in high fluoride concentration (1.7–6.1 mgl⁻¹) in groundwater sources in India. Most of the samples in the study area have pH in this range only (Figure 21.3). The high pH of water displaces fluoride ions from the mineral surface (Laxen and Harrison, 2005).

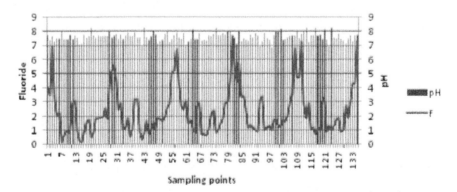

FIGURE 21.3   Variation of fluoride with respect to pH.

21.3.3   FLUORIDE AND HCO3

The carbonates are completely absent in the study area which suggests the occurrence of bicarbonates due to weathering of silicate minerals (Rajmohan and Elango, 2004). It is composed primarily of carbonate $(CO3)^{-2}$ and bicarbonate $(HCO3)^{-}$; alkalinity acts as a stabilizer for pH. Alkalinity, pH, and hardness affect the toxicity of many substances in the water. In Figure 21.4, fluoride values with higher range have bicarbonates 400–600 mgl⁻¹.

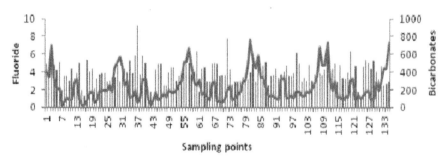

FIGURE 21.4   Variation of fluoride with respect to HCO_3

21.3.4   FLUORIDE VS. CA

Calcium is directly related to hardness and is the chief cation in the water. Low Magnesium and Calcium values indicate the non-carbonaceous aquifers as the source of the groundwater. The ionic strength of the water increases the solubility of fluoride minerals (Kumar et al., 2014). In this study, calcium content has minimum-maximum value range as 36–198 mgl⁻¹. In Figure 21.5, where more calcium is found, fluoride values are in the lower range.

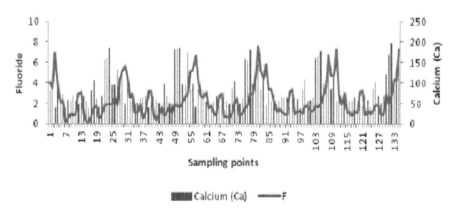

FIGURE 21.5   Variation of fluoride with respect to calcium.

21.3.5   FLUORIDE AND SODIUM

Sodium content has minimum-maximum value range as 14–565 mgl[-1]. Sodium and calcium are the major cations that influence the fluoride level in groundwater (Rao et al., 2003). The range of sodium content is higher than the Ca and Mg in groundwater, which is a favorable point for fluoride leeching (Rao et al., 1993). This is may be due to the precipitation of $CaCO_3$ and $MgCO_3$. Ca is negatively correlated with the fluoride; therefore, calcium gets precipitated out in the form of $CaCO_3$. Sodium readily forms NaF and remains in solution. It favors fluoride enrichment in the unconfined aquifers (Rao et al., 1993). Most of the samples have high sodium content (Figure 21.6).

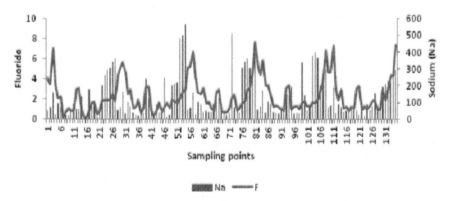

FIGURE 21.6   Variation of fluoride with respect to sodium.

21.3.6   FLUORIDE AND TOTAL HARDNESS

The possible source of Ca^{+2} and Mg^{+2} ions in the groundwater is the geology which suggests the presence of Ca-Mg rich minerals like feldspar, amphiboles (Subramani et al., 2010). The hardness of water mainly depends upon the amount of calcium or magnesium salts or both. Water with hardness up to 75 mgl[-1] is classified as soft, 76–150 mgl[-1] is moderately soft, 151–300 mgl[-1] as hard and more than 300 mgl[-1] as very hard (Saravanakumar and Ranjith Kumar, 2011). The total hardness is due to the presence of calcium and magnesium carbonate and bicarbonate. The decrease in hardness, resulting in higher fluoride concentration contributed to calcium complexion effect (Kumar and Seema, 2016). In this sample analysis, total hardness has a minimum value of 105 mgl⁻¹ and a maximum of 890 mgl⁻¹ (Figure 21.7). So in hard water, calcium, and magnesium carbonate and bicarbonate are found. Therefore, water will be calcium deficient, and fluoride leeching will be more (Rao et al., 1993). WHO has recommended the safe permissible limit for hardness, i.e., 100–500 mgl⁻¹.

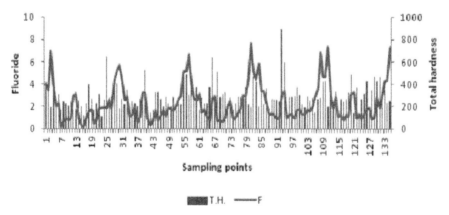

FIGURE 21.7   Variation of fluoride with respect to total hardness.

21.3.7   FLUORIDE VS. ELECTRICAL CONDUCTIVITY (EC)

EC is a numerical expression ability of an aqueous solution to carry electric current (Alagumuthu and Rajan, 2008). It signifies the amount of total dissolved salts (Dahiya and Kaur, 1999) and is a useful tool to evaluate the purity of water (Acharya et al., 2008), indicating the presence of high amount of dissolved inorganic substances in ionized form. Permissible limit for total dissolved solids as per BIS: 10500 is 500 mg/l. USPH recommended a permissible limit for electrical conductivity (EC) is 300 μmhos/cm (Alagumuthu and Rajan, 2008). In this study minimum 330 μmhos/cm to maximum 2580 μmhos/cm is observed (Figure 21.8). Thus all samples have high EC.

21.3.8   FLUORIDE VS. CHLORIDE

Chloride in water sources resulted from agricultural activities, industries, and chloride rich rocks (Dahiya and Kaur, 1999). The chloride concentration serves as an indicator of pollution by sewage. People accustomed to higher chloride in water are subjected to laxative effects (Dahiya and Kaur, 1999; Guruprasad et al., 2005). This chloride may be supplied by the local leaching of sedimentary rocks (Prasad et al., 2014). The average chloride content of sedimentary rocks is about the same as the evaporate rocks 150 ppm and indicating sedimentary depositional environment in the study area (Prasad et al., 2014). The fluoride values in Figure 21.9, varies with total hardness in equal proportionate. Barring one or two points, chloride is within permissible limit. Fluoride and chloride values increase/decrease in a proportionate way.

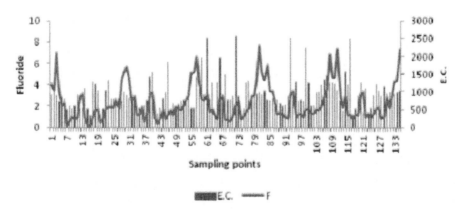

FIGURE 21.8   Variation of fluoride with respect to E.C.

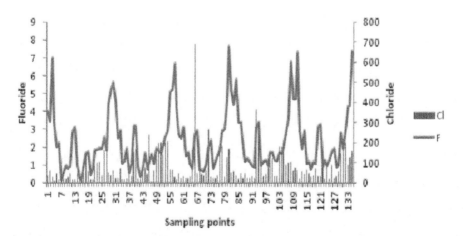

FIGURE 21.9   Variation of fluoride with respect to chloride.

21.3.9   FLUORIDE VS. MAGNESIUM

Magnesium content has minimum-maximum value range as 2.4–353 mgl[-1] (Figure 21.10). Calcium and Magnesium showed a negative correlation with fluoride. This is probably due to the low solubility of fluoride with these ions (Nemade and Srivastava, 1996). Samples with less magnesium have higher values of fluoride and vice versa.

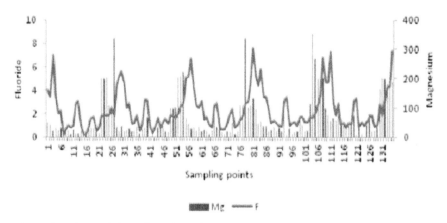

FIGURE 21.10   Variation of fluoride with respect to magnesium.

21.3.10   FLUORIDE VS. POTASSIUM (K)

In Figure 21.11, the higher value of fluoride is associated with lower values of potassium.

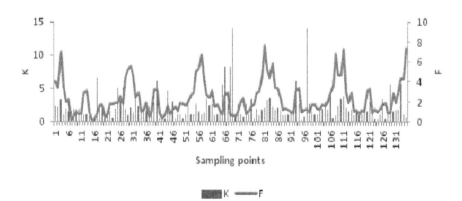

FIGURE 21.11   Variation of fluoride with respect to potassium.

21.4 SUMMARY

With the increasing depth, fluoride content was found to be increasing. The comparison of fluoride concentration of groundwater from the shallow, DWs and deep bore wells from the same location indicated that the deeper aquifers have a higher concentration of fluoride than the shallow aquifers. The pH of the study area belongs to the alkaline in origin. All the samples have high EC. The carbonates are completely absent in the study area which suggests the occurrence of bicarbonates due to weathering of silicate minerals. Soft to hard water is found in the area. The range of sodium content is higher than the Ca and Mg in groundwater, which is a favorable point for fluoride leeching from minerals.

KEYWORDS

- **bicarbonates**
- **carbonates**
- **chloride**
- **depth**
- **electrical conductivity**
- **pH**
- **rock-water interaction**

REFERENCES

Acharya, G. D., & Mathi, M. V., (2010). Fluoride contamination in groundwater sources of Modasa Tehsil of Sabarkantha district, Gujarat, India. *Pollution Research, 29*(1), 43–45.

Agrawal, V., Vaish, A. K., & Vaish, P., (1997). Groundwater quality: Focus on fluoride and fluorosis in Rajasthan., *Current Science, 73*, 743–746.

Alagumuthu, G., & Rajan, M., (2008). Monitoring of fluoride concentration in groundwater of Kadayam block of Tirunelveli district, India: Correlation with physic-chemical parameters. *Rasayan Journal of Chemistry, 1*(4), 920–928.

Appelo, C. A. J., & Postma, D., (2005). *Geochemistry, Groundwater, and Pollution* (2nd edn.). Balkema, Amsterdam, the Netherlands.

Avishek, K., Pathak, G., Nathawat, M. S., Jha, U., & Kumari, N., (2010). Water quality assessment of Majhiaon block of Garhwa District in Jharkhand with special focus on fluoride analysis. *Environmental Monitoring Assessment, 167*, 617–623.

Ayoob, S., & Gupta, A. K., (2006). Fluoride in drinking water: A review on the status and stress effects. *Critical Reviews in Environmental Science and Technology, 36*, 433–487.

BIS (Bureau of Indian Standards), (1992). *Indian Standard Specifications for Drinking Water*, IS: 10500.

Chandio, T. A., Khan, M. N., & Sarwar, A., (2015). Fluoride estimation and its correlation with other physicochemical parameters in drinking water of some areas of Balochistan, Pakistan. *Environment Monitoring Assessment, 187*, 537–539.

Chandra, S. J., Thergaonkar, V. P., & Sharma, R., (1981). Water quality and dental fluorosis. *Indian Journal of Public Health, 25*, 47–51.

Chate, G. T., Yun, S. T., Mayer, B., Kim, K. H., Kim S. Y., Kwon, J. S., Kim, K., & Koh, Y. K., (2007). Fluorine geochemistry in bedrock groundwater of South Korea." *Science of Total Environment, 385*, 272–283.

Dahiya, S., & Kaur, A., (1999). Physicochemical characteristics of underground water in rural areas of Tosham subdivisions, Bhiwani district, Haryana. *Journal of Environmental Pollution, 6*(4), 281.

Guruprasad, B., (2005). Assessment of water quality in canals of Krishna delta area of A. P. *Nature of Environment and Pollution Technology, 4*(4), 521–523.

Kumar, P. J. S., Jegathambal, P., & James, E. J., (2014). *Factors Influencing the High Fluoride Concentration in Groundwater of Vellore District*, South India, 2437–2446.

Largent, E. J., (1961). *Fluorosis: The Health Aspects of Fluoride Compounds.* Ohio State University Press, Columbia, OH. 2nd edition, pp. 140.

Laxen, D. P. H., & Harrison, R. M., (2005). Cleaning methods for polythene containers prior to the determination of trace metals in freshwater samples. *Analytical Chemistry, 53*, 345–352.

MacDonald, L. H., Pathak, G., Singer, B., & Jaffe, P. R., (2011). An integrated approach to address endemic fluorosis in Jharkhand, India. *Journal of Water Resource and Protection, 3*, 457–472.

Nemade, P. N., & Shrivastava, V. S., (1996). Radiological skeletal changes due to chronic fluoride intoxication in Udaipur. *Journal of Environmental Protection, 16*(12), 43–46.

Nezli, I. E., Achour, S., Djidel, M., & Attalah, S., (2009). Presence and origin of fluoride in the complex terminal water of Ouargla Basin (Northern Sahara of Algeria). *American Journal of Applied Sciences, 6*(5), 876–881.

Pandey, A. C., Shekhar, S., & Nathawat, M. S., (2012). Evaluation of fluoride in groundwater sources in Palamu district of Jharkhand. *Journal of Applied Sciences, 12*(9), 882–887.

Pol, P. D., Sangannavr, M. C., & Yadawe, M. S., (2012). Fluoride contamination status of groundwater in Mudhol taluk, Karnataka, India: Correlation of fluoride with other physicochemical parameters. *Rasayan Journal of Chemistry, 5*(2), 186–193.

Prasad, S., Anoop, A., Riedel, N., Sarkar, S., Menzel, P., Basavaiah, N., et al., (2014). Prolonged monsoon droughts and links to the Indo-Pacific warm pool: A Holocene record from Lonar Lake, central India. *Earth and Planetary Science Letters, 391*, 171–182. doi:10.1016/j.epsl.2014.01.043.

Priyadarshi, N., (2008). *Fluoride Toxicity in the Jharkhand State of India.* www. fluoridealert. org.

Rao, N. S., (2003). Groundwater quality: Focus on fluoride concentration in rural parts of Guntur district, Andhra Pradesh, India. *Hydrological Sciences Journal – Des Sciences Hydrologiques, 48*(485), 835–847.

Rao, N. V. R., Rao, N., Rao, K. S. P., & Schuiling, R. D., (1993). Fluorine distribution in waters of Nalgonda District, Andhra Pradesh, India. *Environmental Geology, 21*, 84–89.

Saravanakumar, K., & Ranjith, K. R., (2011). Analysis of water quality parameters of groundwater near Ambattur Industrial Area, Tamil Nadu, India. *Indian Journal of Science and Technology, 4*(5), 560–562.

Saxena, V. K., & Ahmed, S., (2001). Dissolution of fluoride in groundwater: A water-rock interaction study." *Environmental Geology, 40*, 1084–1087.

Shyamala, R., Shanthi, M., & Lalitha, P., (2008). Physicochemical analysis of Borewell water samples of Telungupalayam area in Coimbatore District, Tamil Nadu, India. *Environmental Journal of Chemistry, 5*(4), 924 929.

Singhal, D. C., (2013). Groundwater resource assessment in India- some emerging issues. *Journal of Groundwater Research, 2*(2), 43–48.

Smet, J., (1990). Fluoride in drinking water. In: Frencken, J. E., (ed.), *Proc. Symposium on Endemic Fluorosis in Developing Countries: Causes, Effects and Possible Solutions* (pp. 51–85). Chapter 6, NIPG-TNO, Leiden.

Srikanth, R., Tripathi, R. C., & Kumar, B. R., (2008). Endemic fluorosis in five villages of the Palamau district, Jharkhand, India. *Fluoride, 41*(3), 206–211.

Susheela, A. K., & Majumdar, K., (1992). Fluorosis control programme in India. *Water Environment and Management: 18th WEDC Conference* (pp. 229–233). Kathmandu, Nepal.

Umarani, P., & Ramu, A., (2014). Fluoride contamination status of groundwater in an east coastal area in Tamil Nadu, India. *International Journal of Innovative Research in Science, Engineering and Technology, 3*(3), 10045–10051. ISSN: 2319–8753.

WHO, (1971). *International Standards for Drinking Water*" (3rd edn.). WHO, Geneva.

WHO, (1984). Guidelines for drinking water quality. In: *Health Criteria and Other Supporting Information*" (2nd edn.). World Health Organization, Geneva.

Wodeyar, B. K., & Sreenivasan, G., (1996). Occurrence of fluoride in the ground waters and its impact in Peddavankahalla basin, Bellary District, Karnataka – A preliminary study. *Current Science, 70*, 71–73.

www.igrac.com, https://www.un-igrac.org/sites/default/files/resources/files/IGRAC-SP2007–1_ Fluoride-removal.pdf

CHAPTER 22

GEOHYDROLOGICAL INVESTIGATION USING VERTICAL ELECTRICAL SOUNDING AT CHINAMUSHIDIWADA VILLAGE IN VISAKHAPATNAM, ANDHRA PRADESH, INDIA

KIRAN JALEM

Assistant Professor, Centre for Land Resource Management,
Central University of Jharkhand, Ranchi, India,
E-mail: jalemkiran@gmail.com

ABSTRACT

Geohydrological investigations are performed to assess the groundwater parameters for locating suitable sites for groundwater exploration and resource management at Chinamushidiwada Village in Visakhapatnam, Andhra Pradesh, India. Thirty vertical electrical soundings (VES) using Schlumberger configurations were carried out at selected locations in the vicinity of Chinamushidiwada Village. The interpretation of sounding data has been accomplished using both curves matching as well as computer-assisted automatic iterative resistivity sounding ipi2win (2008) software. On the basis of interpreted sounding results, three to four geoelectrical cross sections have been generated along the profiles. The interpretation of data revealed four layers, generally one top thin layer overlying the other three thick layers. Interpreted results are corroborated with the borehole data. The results depict proper geohydrological conditions for the existence of good aquifers suggesting the continued supply of groundwater in the study area for the extended period.

22.1   INTRODUCTION

Groundwater is very important natural resources for sustainable development of a region. It is the only viable source of water in many areas where the development of surface water is not economically viable. Groundwater in alluvial and sedimentary rocks occurs in pore spaces between grains, while in hard rocks, it is largely due to secondary porosity and permeability resulting from weathering, fracturing, jointing, and faulting activities. The area of investigation is the Chinamushidiwada village of Visakhapatnam urban environment. The recent urban expansion in Visakhapatnam (India) following the increased population growth has resulted in increased developmental activities and shrinking of surface water bodies in Visakhapatnam urban environment.

With increased infrastructure development and irregularity and failure of monsoon, it has been vaguely reported that the groundwater level in Visakhapatnam urban environment is depleting fast. Therefore, with aim of examining the groundwater level and locating the potential aquifers for their management, geohydrological investigations at suitably chosen sites in and around Chinamushidiwada village were carried out with vertical electrical sounding (VES) at thirty sites in the vicinity of Chinamushidiwada village microwatershed of Visakhapatnam, Andhra Pradesh, India urban environment and have interpreted the results for estimating the parameters which may be useful for management of groundwater aquifers in the Chinamushidiwada village microwatershed vicinity to cope with the sustained development of groundwater source in the study area.

22.2   MATERIALS AND METHODS

22.2.1   DESCRIPTION OF THE STUDY AREA

Geographical location of the study area of investigation lies in the Chinamushidiwada village microwatershed of Visakhapatnam district of Andhra Pradesh, India (Figure 22.1). The study area lies between 170 47' 02" N to 170 48' 38" N latitudes and 830 11' 51" E to 830 14' 45" E longitudes, covering an aerial extent of 9.18 km^2. It is a part of Survey of India (SOI) Toposheets – 65 O/1 SE (Figure 22.2) of 1:25,000 scale and is within the administrative boundaries of Greater Visakhapatnam Municipal Corporation (GVMC) of Visakhapatnam, Andhra Pradesh.

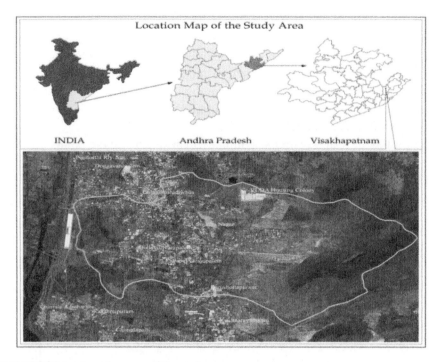

FIGURE 22.1 Location map of the study area.

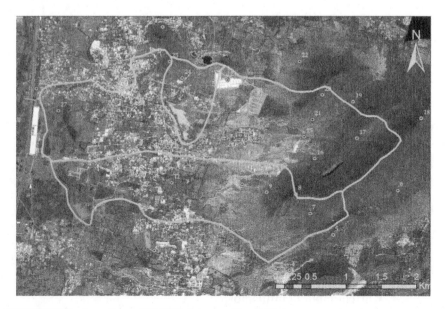

FIGURE 22.2 Locations of 30 VES data sampling stations.

In this study, the VES method has been adopted, and VES surveys have been carried out at thirty locations in the area (Figure 22.2) using Schlumberger electrode configuration. For resistivity sounding using Schlumberger array, the current electrodes are spaced much farther apart as compared to the potential electrodes. Although, a number of electrode arrangements for current electrodes (C1, C2) and potential electrodes (P1, P2) have been suggested for this purpose, we used the symmetrical Schlumberger arrangement for the VES survey. In the symmetrical arrangement the points C1, P1, P2, C2 are taken on a straight line such that points P1 and P2 are symmetrically placed about the center of the spread (Figure 22.3).

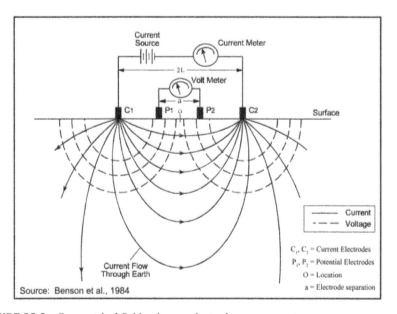

FIGURE 22.3   Symmetrical Schlumberger electrode arrangements.

$$\rho_a = \pi * \left(\left(L^2 - a^2 \right) / 2a \right) * \frac{\Delta V}{I} \tag{1}$$

where,

ρ_a =   Apparent resistivity (Ωm);
L   =   Distance between center point and a current electrode;
ΔV =   Potential deference in millivolts;
a   =   Distance between Potential electrodes;
I   =   Current input in milliamps;
O   =   VES location.

In order to determine the exact thickness and true resistivity of each layer, field curves are interpreted and matched with the set of model curves for two layer, three layer and multilayer cases prepared by Ipi2Win 2008 and broadly 3 to 4 layers are demarcated, which are classified into topsoil, weathered rock, fractured rock, hard fractured rock and hard rock. In total, 30 numbers of VES data has been interpreted to determine the subsurface lithology, and the corresponding elevations for weathered, fractured, hard fractured and hard rock surfaces are determined, which are tabulated in Table 22.1.

TABLE 22.1 The Results of the Fieldwork That Was Carried Out During the Vertical Electrical Sounding Experiment for Exploring Groundwater for Part of Visakhapatnam Urban Area, Andhra Pradesh, India

VES No	Type of Curve	Resistivity Ω-m	Thickness (Meter)	Depth (meter)	Lithological Units
1	H	38	1.19	1.19	Red Loamy Soil
		13.2	12	13.2	Weathered Rock
		220	-	-	Fractured Rock
2	QH	35.4	0.5	0.5	Red Loamy Soil
		18.7	2.46	2.96	Gravel
		8.01	16.9	19.9	Weathered Rock
		2219	-	-	Hard Rock
3	H	37.1	2	2	Red Loamy Soil
		14.7	19.4	21.4	Weathered Rock
		3804	-	-	Hard Rock
4	HA	137	1.34	1.34	Gravel & Boulder
		41.2	2.23	3.57	Weathered Rock
		181	23.5	27.07	Fractured Rock
		269	-	-	Hard Fractured Rock
5	H	1436	0.5	0.5	Gravel & Boulder
		244	12	12.5	Fractured Rock
		486	-	-	Hard Fractured Rock
6	H	247	1.92	1.92	Gravel & Boulder
		123	10.6	12.52	Weathered Rock
		357	-	-	Hard Fractured Rock
7	QH	563	0.524	0.52	Gravel & Builder
		236	5.92	6.44	Gravel
		47.2	8.14	14.58	Weathered Rock

TABLE 22.1   *(Continued)*

VES No	Type of Curve	Resistivity Ω-m	Thickness (Meter)	Depth (meter)	Lithological Units
		364	-	-	Hard Fractured Rock
8	HA	1872	0.8629	0.86	Gravel & Boulder
		75.159	0.9943	1.86	Gravel
		558.21	49.121	50.98	Hard Fractured Rock
		1071	-	-	Hard Rock
9	H	355	1.29	1.29	Gravel & Boulder
		88.4	9.38	10.67	Weathered Rock
		327	-	-	Hard Fractured Rock
10	QH	122	0.672	0.67	Gravel & Boulder
		54.9	1.84	2.51	Gravel
		29.9	14.1	16.61	Weathered Rock
		148	-	-	Fractured Rock
11	HA	28.9	1.34	1.34	Red Loamy Soil
		9.69	2.23	3.57	Clay
		21	22.2	25.77	Weathered Rock
		3070	∞	-	Hard Rock
12	QHK	27.7	0.91	0.91	Red Loamy Soil
		895	7.89	8.6	Weathered Rock
		2926	∞	-	Hard Rock
13	H	48.9	1.01	1.01	Red Loamy Soil
		20.2	18.1	19.11	Weathered Rock
		382	∞	-	Fractured Rock
14	HA	128	0.5	0.5	Gravel & Boulder
		17.9	4.54	5.04	Weathered Rock
		47.9	61.7	66.74	Fractured Rock
		6480	∞	-	Hard Rock
15	H	85.1	2.59	2.59	Gravel & Boulder
		14.7	3.13	5.72	Weathered Rock
		146	∞	-	Fractured Rock
16	HA	191	1.7	1.7	Gravel & Boulder
		28.9	3.7	5.4	Weathered Rock
		89.9	43.8	49.2	Fractured Rock
		4663	∞	-	Hard Rock
17	HKH	816	1.02	1.02	Gravel & Boulder

TABLE 22.1 *(Continued)*

VES No	Type of Curve	Resistivity Ω-m	Thickness (Meter)	Depth (meter)	Lithological Units
		44.4	1.25	2.27	Weathered Rock
		1280	2.3	4.57	Boulder
		119	7.68	12.25	Fractured Rock
		722	∞	-	Hard Fractured Rock
18	HK	1677	1.81	1.81	Gravel & Boulder
		661	11.5	13.3	Fractured Rock
		1356	∞	-	Hard Fractured Rock
19	H	468	1.99	1.99	Gravel & Boulder
		53.7	2.38	4.37	Fractured Rock
		1168	∞	-	Hard Fractured Rock
20	H	2138	0.961	0.96	Gravel & Boulder
		436	7.84	8.8	Fractured Rock
		1442	∞	-	Hard Fractured Rock
21	HA	249	1.34	1.34	Gravel & Boulder
		71.1	2.23	3.57	Fractured Rock
		363	21.9	25.47	Hard Fractured Rock
		629	-	-	Hard Rock
22	HA	203	1.58	1.58	Gravel & Boulder
		47.9	2.29	3.87	Weathered Rock
		216	35.5	39.37	Fractured Rock
		24848	-	-	Hard Rock
23	QH	203	0.744	0.74	Gravel & Boulder
		136	4.41	5.15	Hard Fractured Rock
		28.8	8.81	13.96	Fractured Rock
		318	-	-	Hard Fractured Rock
24	HA	162	1.34	1.34	Gravel & Boulder
		30.3	2.23	3.57	Weathered Rock
		117	22.2	25.77	Fractured Rock
		435	-	-	Hard Fractured Rock
25	H	1046	1.11	1.11	Boulder
		162	6.07	7.18	Fractured Rock
		498	-	-	Hard Fractured Rock
26	HA	12.3	0.938	0.94	Red Loamy Soil
		2.93	1.36	2.3	Clay

TABLE 22.1   *(Continued)*

VES No	Type of Curve	Resistivity Ω-m	Thickness (Meter)	Depth (meter)	Lithological Units
		27.1	63.6	65.9	Weathered Rock
		1164	-	-	Hard Fractured Rock
27	H	3.01	2.04	2.04	Clay
		7.82	15.2	17.24	Weathered Rock
		11377	-	-	Hard Rock
28	H	47.9	1.2	1.2	Gravel
		13.4	4.46	5.66	Weathered Rock
		229	-	-	Fractured Rock
29	QHA	46.3	0.5	0.5	Red Loamy Soil
		23.6	1.53	2.03	Gravel
		4.1	2.2	4.23	Clay
		25.3	16.1	20.33	Weathered Rock
		14887	-	-	Hard Rock
30	H	370	1.26	1.26	Gravel & Weathered
		161	6.38	7.64	Fractured Rock
		608	-	-	Hard Fractured Rock

22.3   RESULTS AND DISCUSSIONS

For better understanding, the results of the investigation are usually presented in the form of geoelectrical cross-sections and isopach map (Figure 22.10) of the aquifer in the study area. Therefore, in accordance with above fact, four geoelectrical cross-sections (Figures 22.5–22.8) are prepared along AA″, BB″, CC″ and DD″ shown in the study area map (Figure 22.4) are interpreted as given in the following subsections.

22.3.1   GEOELECTRICAL CROSS-SECTION AA'

Along the cross section AA,' the area represented in blue in the Figure 22.5 is weathered rock extending from VES location 27 to 11 and this area identified as potential aquifer zone. The green and yellow color in the figure represents Fracture and Hard fracture rock respectively. The red color having high apparent resistivity in the figure is hard rock.

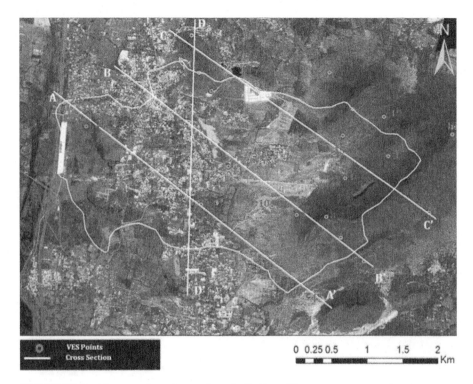

FIGURE 22.4   Traverse along the sections A, B, C, and D.

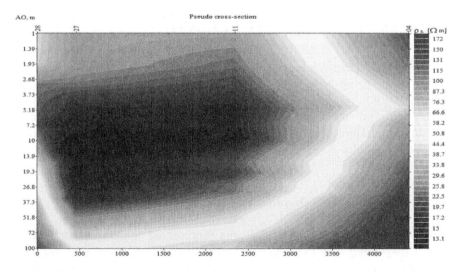

FIGURE 22.5   Geoelectrical cross-section along the line AA.'

22.3.2   GEOELECTRICAL CROSS-SECTION BB'

Along the cross-section BB,' the area represented in blue in the Figure 22.6 is weathered rock extending from VES location 29 to 14 and this area identified as potential aquifer zone. The green and yellow color in the figure represents Fractured and Hard fractured rock, respectively. The red color having high apparent resistivity in the figure is hard rock. The VES location 10 and 9 are in the foothill region of Yerrakonda hill; hence the area has low aquifer potential.

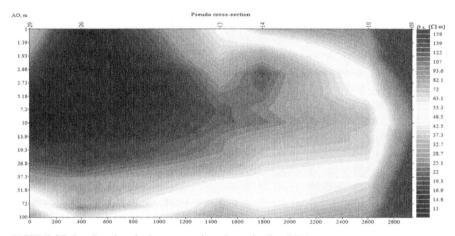

FIGURE 22.6   Geoelectrical cross-section along the line BB.'

22.3.3   GEOELECTRICAL CROSS-SECTION CC'

Along the cross section CC,' the area represented in blue in the Figure 22.7 is weathered rock extending from VES location 24 to 16 and this area identified as potential aquifer zone, but if compared to other cross-sections, the area is not highly potential. The green and yellow color in the figure represents Fractured and Hard fractured rock respectively. The red color having high apparent resistivity in the figure is hard rock.

22.3.4   GEOELECTRICAL CROSS-SECTION DD'

Along the cross section DD,' the area represented in blue in the Figure 22.8 is weathered rock extending from VES location 13 to 03 and this area identified as potential aquifer zone, but if compared to other cross-sections, the area has high potential aquifer zone. The green and yellow color in the figure represents Fractured and Hard fractured rock respectively. The red color having high apparent resistivity in the figure is hard rock.

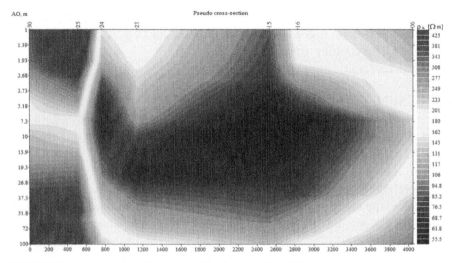

FIGURE 22.7   Geoelectrical cross-section along the line CC.'

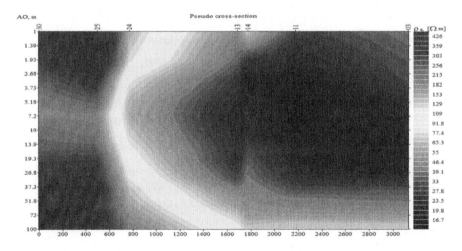

FIGURE 22.8   Geoelectrical cross-section along the line DD.'

From all the vertical cross-sections, it is clear that the thickness of the water-bearing formations, i.e., weathered, and fractured rock zones have increased from the foothill regions to the low lying areas of the study area. Over the hill slopes and hill ridges, hard rock is present immediately below the top soil. Even though there are some fractured rock zones noticed over the hill slopes, these may not contain aquifer system but may guide the rainwater to percolate from the topsoils into the aquifer system down below. However, the hill slopes are useful to retain rainwater for some time and release it into the aquifer system existing down below the foothill region. Therefore, this zone is considered to be suitable for constructing harvesting structures, like contour trenches.

22.4 SUMMARY

22.4.1 BOREHOLE LITHOLOGICAL INFORMATION IN AND AROUND THE STUDY AREA

The borehole lithological information is collected from various locations in and around the study area. Total four borehole investigations were conducted at VUDA Colony, Laxmipuram, Drivers Colony and Sujatha Nagar, respectively (Figure 22.9).

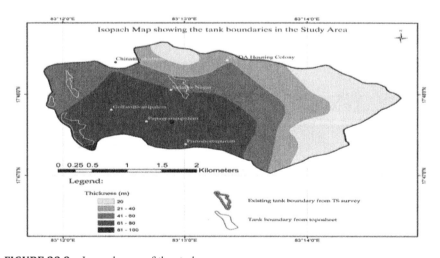

FIGURE 22.9   Isopach map of the study area.

Borehole in VUDA Colony is done for Vishakha urban development authority revealed four layers, first layer extending from ground to 10 meters below revealed Gravel and Boulder, second layer extending till 50 meters reveal weathered rock, third layer extending from 50 to 80 meter below revealed fractured rock and last layed which beyond 80 meter depth is hard rock. Borehole in Laxmipuram is done for Rural Water Supply

Scheme of Panchayati Raj (RWSP) Vishakhapatnam, revealed four layers, first layer extending from ground to 3 meters below revealed soil, second layer extending till 30 meters revealed weathered rock, third layer extending from 30 to 90 meters below revealed fractured rock and last layer beyond 90 meter depth is hard rock. Borehole data collected from local inhabitant Mr N. Nookaraju for household purpose, revealed four layers, first layer extending from ground to 0.5 meter below revealed soil, second layer extending till 15 meters revealed gravel, third layer extending from 15 to 50 meter below revealed fractured rock and last layer beyond 50 meters depth is hard rock. Borehole data from Sujatha Nagar is done for Bharat Heavy Plates & Vessels (BHPV) Vishakhapatnam, revealed four layers, first layer extending from ground to 2 meters below revealed soil, second layer extending till 20 meters reveal weathered rock, third layer extending from 80 to 90 meters below revealed fractured rock and last layed which beyond 80 meter depth is hard rock. The collected borehole data is given below in Table 22.2 with bore log profiles of four locations in Figure 22.10. The borehole data were analyzed with VES data nearest to the location of borewell and identified that the VES data is almost similar to borewell data. The result is strongly proven that VES data are more accurate for identifying potential groundwater aquifers.

TABLE 22.2   Lithological Information of Borehole Stations

Borehole Stations	Location	Depth (m)	Lithological Units
1	VUDA Colony	0 to 10	Gravel& Boulder
	(VUDA)	10 to 50	Weathered Rock
		50 to 80	Fractured Rock
		> 80	Hard Rock
2	Laxmipuram	0 to 3	Soil
	(RWSP Scheme, VSP)	3 to 30	Weathered Rock
		30 to 90	Fractured Rock
		> 90	Hard Rock
3	Drivers Colony	0 to 0.5	Soil
	(N. Nookaraju	0.5 to 15	Gravel
	House Purpose)	15 to 50	Fractured Rock
		> 50	Hard Rock
4	Sujatha Nagar	0 to 2	Soil
	(BHPV-VSP)	2 to 20	Weathered Rock
		20 to 80	Fractured Rock
		> 80	Hard Rock

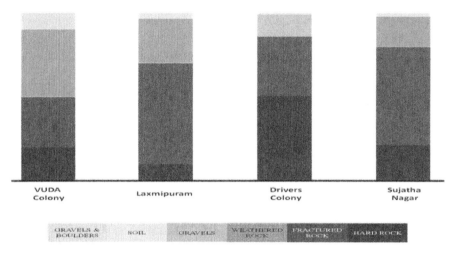

FIGURE 22.10   Borelog profile of four locations.

KEYWORDS

- **borehole**
- **geoelectrical cross section**
- **geohydrological investigations**
- **groundwater**
- **vertical electrical sounding**

REFERENCES

Breusse, J. J., (1963). Modern geophysical methods for subsurface water exploration. *Geophysics, 28,* 663.

Davis, G. H., (1982). Prospect risk analysis applied to groundwater reservoir evaluation, *Groundwater, 20*(6), 657–662.

El-kadi, A. I., Oloufa, A. A., Eltahan, A. A., & Malic, H. U., (1994). Use of a geographic information system in site-specific groundwater modeling. *Ground Water, 32,* 617–625.

Goyal, S., Bharawadaj, R. S., & Jugran, D. K., (1999). *Multicriteria Analysis Using GIS for Groundwater Resource Evaluation in Rawasen and Pilli Watershed,* U.P., http://www.GIS development net Cited 17 Dec 2003.

Jaiswal, R. K., Mukherjee, S., Krishnamurthy, J., & Saxena, R., (2003). Role of remote sensing and GIS techniques for generation of groundwater prospect zones towards rural development-an approach. *International Journal of Remote Sensing, 24,* 993–1008.

Krishnamurthy, J. N., Venkatesa, K., Jayaraman, V., & Manivel, M., (1996). An approach to demarcate potential groundwater zones through remote sensing and geographical information system. *International Journal of Remote Sensing, 17*, 1867–1884.

Masvopo, T. H., David, L., & Hodson, M. *Evaluation of the Ground Water Potential of the Malala Alluvial Aquifer*, Lower mzingwane river, Zimbabwe.

Musa, G., Abdullahi, M. E., Toriman, G., & Mohd, B., (2014). Application of vertical electrical sounding (VES) for groundwater exploration in Tudun Wada Kano state, Nigeria. *International Journal of Engineering Research and Reviews, 2*(4), 51–55.

Nageswara Rao, K., Rao, U. B., & Venkateswara Rao, T., (2008). Estimation of sediment volume through geophysical and GIS analysis – A case study of the red sand deposit along Visakhapatnam coast. *Journal of Indian Geophysical Union, 1*(12), 23–30.

Novaline, J., Saibaba, J., & Prasada, R. P. V. S. P., (1999). *Groundwater Modeling for Sustainable Development Using GIS Techniques*. Preconference volume, Geoinformatics Beyond 2000, Dehradun, India.

Obi Reddy, G. P., Chandra, M. K., Srivastav, S. K., Srinivas, C. V., & Maji, A. K., (2000). Evaluation of groundwater potential zones using remote sensing data: a case study of Gaimukh watershed, Bhandara district, Maharashtra. *Journal of Indian Society of Remote Sensing, 281*, 19–32.

Orellana, E., & Mooney, H. M., (1966). *Master Tables and Curves for Vertical Electrical Soundings Over Layered Structures: Interciencia*, Madrid.

Pratap, K., Ravindran, K. V., & Prabakaran, B., (2000). Groundwater prospect zoning using remote sensing and geographical information system: A case study in Dala-Renukoot Area, Sonbhadra district Uttar Pradesh. *Journal of Indian Society of Remote Sensing, 284*, 249–263.

Rokade, V. M., Kundal, P., & Joshi, A. K., (2007). Groundwater potential modeling through remote sensing and GIS: A case study from Rajura Taluka, Chandrapur district, Maharastra. *Journal of Geological Society of India, 69*, 943–948.

Saraf, A. K., Choudhury, P. R., Roy, B., Sarma, B., Vijay, S., & Choudhury, S., (2004). GIS-based surface hydrological modeling in the identification of groundwater recharge zones. *International Journal of Remote Sensing, 25*(24), 5759–5770.

Saraf, A., & Choudhary, P. R., (1998). Integrated remote sensing and GIS for groundwater exploration and identification of artificial recharge sites. *International Journal of Remote Sensing, 19*(10), 1825–1841.

Schueler, T. (2011). *The Impervious Cover Model: Stream Classification*, Urban Subwatershed Management, and Permitting, Chesapeake Stormwater Network. www.cwp.org/Our_Work/Training/Institutes/icm_and_watershed_mgmt.pdf.

Shahid, S., & Nath, S. K., (1999). GIS Integration of remote sensing and electrical sounding data for hydrogeological exploration. *Journal of Spatial Hydrology, 21*, 1–12.

Shahid, S., Nath, S. K., & Roy, J., (2000). Groundwater potential modeling in soft rock area using GIS. *Journal of Remote Sensing, 21*, 1919–1924.

Shaver, E., Maxted, J. Curtis, G., & Carter, D., (1994). Watershed protection using an integrated approach. In: *Stormwater NPDES Related Monitoring Needs. Proc. of an Eng. Foud. Conf. held in Mount Crested Butte*. Colorado.

Singh, A. K., & Prakash, S. R., (2002). *An Integrated Approach of Remote Sensing, Geophysics and GIS to Evaluation of Groundwater Potentiality of Ojhala Sub Watershed*. Mirzapur District, U. P., India http://www.GISdevelopment.net).

Srinivasa, R. Y., & Jugran, K. D., (2003). Delineation of groundwater potential zones and zones of groundwater quality suitable for domestic purposes using remote sensing and GIS. *Hydro-geology Society Journal, 48*, 821–833.

Subba, R. N., Chakradhar, G. K. J., & Srinivas, V., (2001). Identification of groundwater potential zones using remote sensing techniques in and around Guntur town, Andhra Pradesh, India. *Journal of the Indian Society of Remote Sensing, 29*(12), 69–78.

Teeuw, R. M., (1995). Groundwater exploration using remote sensing and a low-cost geographical information system. *Hydrogeology Journal, 3*, 21–30.

Tuinhof, A., Olsthoorn, T., Heederik, J. P., & De Vries, J., (2002). *Management of Aquifer Recharge and Subsurface Storage: A Promising Option to Cope with Increasing Storage Needs* (p. 4). Netherlands National Committee of theInternational Association of Hydrogeologists.

Venkateswa, R. V., Srinivasa, R., Prakasa, R. B. S., & Koteswara, R. P., (2004). Bedrock investigation by seismic retraction method – A case study. *Journal of Indian Geophysical Union, 8*(3), 223–228.

Venkateswara, R. V., Amminedu, E., Venkateswara, R. T., & Ramprasad, N. D., (2009). *Water Conservation Structures in the Urban Environment, Management of Urban Runoff Through Water Conservation Methods.*

Venkateswaran, S., & Jayapal, P., (2013). Geoelectrical Schlumberger investigation for characterizing the hydrogeological conditions using GIS in Kadavanar sub-basin, Cauvery River, Tamil Nadu, India. *International Journal of Innovative Technology and Exploring Engineering, 3*(2).

CHAPTER 23

DELINEATION OF GROUNDWATER POTENTIAL ZONES IN HARD ROCK TERRAIN USING REMOTE SENSING AND GEOGRAPHICAL INFORMATION SYSTEM (GIS) TECHNIQUES

D. NANDI[1], P. C. SAHU[2], and S. GOSWAMI[3]

[1]*Assistant Professor, Department of RS & GIS: North Orissa University, Baripada, Odisha, India, E-mail: debabrata.gis@gmail.com*

[2]*Reader, Department of Geology: MPC Autonomous Colleges, Baripada, Odisha, India*

[3]*Professor, Department of Earth Science, Sambalpur University, Sambalpur, Odisha, India*

ABSTRACT

Integration of Remote Sensing data and the Geographical Information System (GIS) for targeting of groundwater resources has become an advanced technique in the field of hydrological research, which assists in measuring, monitoring, and conserving groundwater resources. In the present chapter, various groundwater potential zones in Rairangpur block have been delineated using Remote Sensing and GIS techniques. Survey of India (SOI) toposheets and LISS-III satellite imageries are used to preparing various thematic layers viz. Lithology, slope, landuse, lineament, drainage, soil, and geomorphology and were transformed to raster data using the feature to raster converter tool in ArcGIS. The raster map of these factors is allocated a fixed score and weight computed from Multi Influencing Factor (MIF) technique.

Moreover, each weighted thematic layer is statistically computed to get the potential groundwater potential zones. Thus, five different groundwater potential zones were identified, namely 'very good,' 'good,' 'moderate,' 'poor,' and 'very poor.' The villages under good groundwater potential zone and the villages under very good groundwater potential zone have found out in our study area. The above study has clearly demonstrated the capabilities of Remote Sensing and GIS in the demarcation of the different groundwater potential zones in hard rock terrain.

23.1   INTRODUCTION

Groundwater is a purest, dynamic, and replenishing natural resource. It meets the overall demand for water supplies in all climatic regions including developed and developing countries (Todd and Mays, 2005). In India, more than 90% of the rural and nearly 30% of the urban population depend on groundwater for meeting their drinking and domestic requirements (Reddy et al., 1996). Mapping groundwater potential zones are essential for planning the location of new abstraction wells to meet the increasing demand for water. The groundwater occurrence, distribution is dependent upon the geological and hydro-geomorphological features of the area. Remote sensing techniques are used for groundwater exploration, especially for delineating hydrogeomorphological units (Anonymous, 1979, 1986, 1988; Baldev et al., 1991; Krishnamurthy and Srinivas, 1995). GIS has been found to be one of the most powerful techniques, and it is easier to establish the baseline information for groundwater potential zones (de Zurria et al., 1994; Kadam et al., 2016; Abdul et al., 2016). In the past, several researchers have used RS and GIS techniques for the delineation of groundwater potential zones (Nag and Anindita, 2011; Basavaraj and Nijagunappa, 2011; Krishnamurthy and Srinivas 1995; Kamaraju et al., 1995; Krishnamurthy et al., 1996; Sander et al., 1996; Edet et al., 1998; Saraf and Choudhury, 1998; Jaiswal et al., 2003; Rao and Jugran, 2003; Sener et al., 2005; Ravi Shankar and Mohan, 2006; Solomon and Quiel, 2006; Madrucci et al., 2008; Chowdhury et al., 2009; Jha et al., 2010; Sahu et al., 2017)

The study was conducted to find out the potential groundwater zones in Rairangpur block of Mayurbhanj district, Odisha, India. Different types of thematic maps such as geology, geomorphology, soil texture, land use/ land cover, drainage, lineament map were prepared for the study area. The groundwater potential zones were obtained by overlaying all the thematic maps in terms of weighted overlay methods using the spatial analysis tool in Arc Gis10.2.2.

23.2   MATERIALS AND METHODS

23.2.1   STUDY AREA

Rairangpur block of Mayurbhanj District lies between 22° 11′ 30″ to and 22° 26′ 30″ latitude and 86° 06′ 30″ to and 86° 21′ 15″ longitude (Figure 23.1). The block falls in the Survey of India (SOI) topographic sheet no 73J/3, 73J/4, 73J/7, 73J/08. The block is covering an area of 258 km². According to the 2011 Indian census, the total population is 69374. The average rainfall of this area is 445.47 mm. The block is characterized by the presence of Granite and Epidiorite of pre-Cambrians age. The maximum temperature of the block is 45°C and minimum temperature is 30°C. During the summer, the groundwater level in this block lowers beyond the economic lift, which constitutes the main source of drinking water for this region. The study area is severely suffering from water scarcity, and the sustainability of water supply is threatened. The water scarcity has a direct impact on the livelihood, health, and sanitation of the local people.

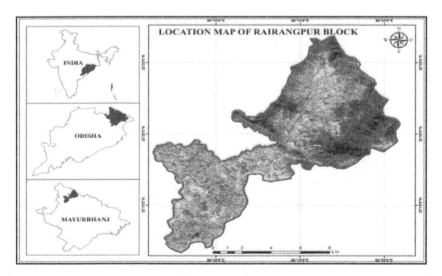

FIGURE 23.1   Location map of the study area with a satellite image.

23.2.2   METHODOLOGY

The administrative base map of Rairangpur block was prepared based on SOI toposheets 73J/3, 73J/4, 73J/7, 73J/08 on a 1:50,000 scale. The drainage system for the study area was digitized in ArcGIS 10.2.2 from SOI toposheets. The slope map was prepared from CARTOSAT DEM data in ArcGIS spatial analyst module. The drainage density and lineament density maps were prepared using the line density analysis tool in ArcGIS. For the preparation of thematic layers such as land-use, lithology, lineament, and soil types satellite images from IRS–1C, LISS-III sensor has been used. All the thematic layers were converted to a raster format. The groundwater potential zones were calculated by overlaying all the thematic maps in terms of weighted overlay methods using the spatial analysis tool in ArcGIS 10.2.2. During weighted overlay analysis, the ranking was given for each individual parameter of each thematic map, and weights were assigned according to the information in groundwater occur acne (Figure 23.2). Cumulative score index (CSI) was used for this classification. CSI was calculated by multiplying the rank and weight age of each thematic layer as expressed in the following equation.

CSI = ∑ (Geology rank x weight + Geomorphology rank x weight + Soil rank x weight, Lineament density rank x weight + Drainage density rank x weight + Slope rank x weight + Land use rank x weight)

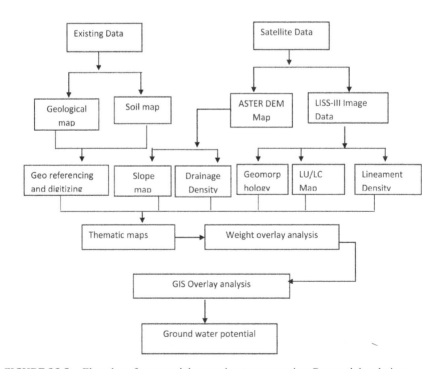

FIGURE 23.2   Flowchart for potential groundwater zone using Geospatial techniques.

23.3 RESULT AND DISCUSSION

23.3.1 *GEOLOGY*

The study area is characterized by the presence of Epidiorite and Granite of Precambrian age. The main rock type granite and epidiorite (Figures 23.3 and 23.4). Percentage of Granite and Epidorite is 83% and 17%, respectively. These rocks lack Primary porosity. Groundwater occurrence is restricted to weathered and fractured zone. Groundwater occurs in unconfined and confined aquifer condition.

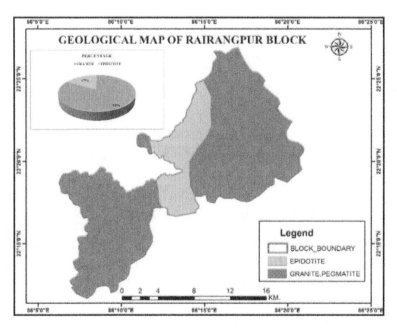

FIGURE 23.3 Geological map of Rairangpur.

23.3.2 *HYDRO GEOMORPHOLOGY*

The hydrogeomorphological study shows that there is a close relationship between the hydrogeomorphic units and groundwater resources (Rao and Devada, 2003). Geomorphological units are extremely helpful for delineating potential groundwater zones and artificial recharge sites (Elango et al., 2003). By taking image interpretation characteristics such as tone, texture, shape, color, and association over the geocoded FCC image, the

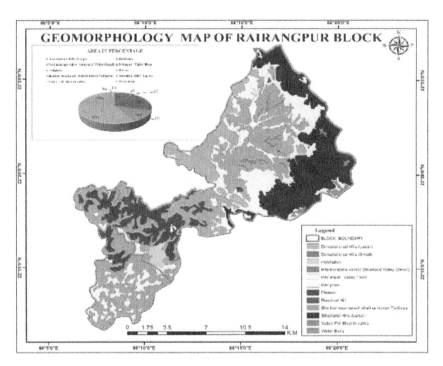

FIGURE 23.4   Geomorphological map of Rairangpur.

geomorphologic units and landforms are interpreted. The geomorphological feature of the Rairangpur block are Denotational Hill (8%), Habitation (1%), Intermontane Valley (1%), Paddy plain (1%), Plateau (10%), Shallow weathered/ shallow buried Pediplain (36%), Structural Hills (Large) (20%), Valley Fill/ filled-in valley (0.74%), Water Body (1%). The Physical Characters of the different landform is present in Table 23.1.

23.3.3   LINEAMENT DENSITY

Lineaments are structurally controlled linear or curvilinear surface expression of zones of weakness or structural displacement in the crust of the earth features, which are identified from the satellite imagery by their relatively linear alignments. Lineament density of an area has a major role for the groundwater potential. High lineament density is good for potential groundwater zones. In hard rock terrain lineaments and fractures act as principal conduits in movement and storage of groundwater. In the Rairangpur block, 33% of the area has very low density, 45% has low density, 18% has moderate density, and 2% has high-density lineament (Figures 23.5 and 23.6).

23.3.4 DRAINAGE DENSITY

Drainage density is indicated as the closeness of spacing of stream channels. It is a measure of the total length of all the orders per unit area. The drainage density is inversely proportional to permeability. The less permeable rock is which conversely tends to be concentrated in surface runoff. Drainage density of the study area is calculated using Arc GIS Spatial analysis tools of line density tool. The study area is divided into five groups, i.e., very high, high, moderate, low, and very low.

TABLE 23.1 Image and Physical Characteristics of Different Landform in the Study

Geomorphic unit	Image elements	Landform Characteristics (Ground observation)	Area (km2)
Structural hills	Dark red tone coarse texture irregular shape	Linear to arcuate hills dissected, khondalite group rocks mostly dendritic drainage, jointed ridges, average height 300 m. strong to very steep slopes	52.4999
Denudational hills	Dull red tone, coarse texture irregular shape	Weathered khondalite, dendritic drainage, moderate to steep slopes, sparse vegetation	19.8984
Residual hills	Dark grey tone, coarse texture shape and size-irregular and rounded	Erosional surfaces, isolated mounds which have undergone the process of denudation, Steep slopes, radial drainage act as runoff zones	0.16309
Pediments	Light red to red tone, moderate to fine texture	Gentle to moderate slopes, devoid of vegetation with various depths of weathering material, shallow sediment cover rocky and gravely surfaces, dendritic to sub-dendritic drainage, mostly vegetated or cultivated lying at foothills	52.1056
Shallow Weathered pediplains	Green-bluish mixed tone moderate to fine texture	These units are characterized by the presence of relatively thicker weathered material. The thickness of the weathered material is (up to 5 m. These hydrogeomorphic units are developed mostly upon Mayurbhanj Granite	93.0976
Intermontane Valley	Green-bluish mixed tone moderate to fine texture	A linear or curvilinear depression valley within the hills, filled with colluvial deposits of IOG sediments	1.9768
Plateau	Dark red tone coarse texture irregular shape	Tableland shaped hill with a flat surface at the top with sloping sides	26.8315

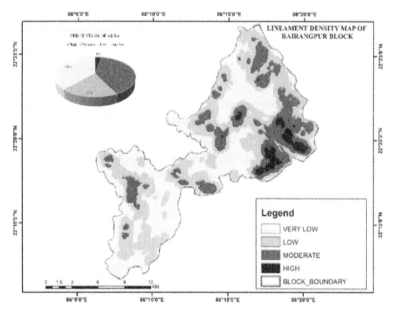

FIGURE 23.5   Lineament density map of the study area.

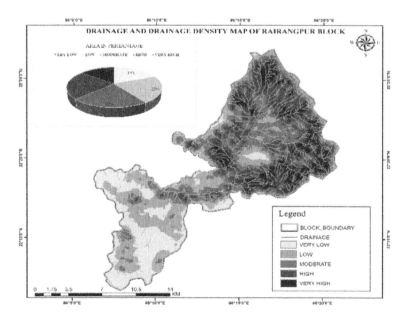

FIGURE 23.6   Drainage map of study area.

23.3.5   SLOPE

The slope has an important role in the identification of potential groundwater zones. Steep slope area facilitates high runoff allowing increased erosion rate with feeble recharge potential whereas, less residence time for rainwater to percolate and hence comparatively less infiltration. The slope map of the study area was prepared based on SRTM data using the 3D analysis tool in Arc Gis10.2.2. Base on a slope the study area is divided into four classes. The area under 0 degree to 4 degrees is very low, 4 degrees to 11 degrees is low, 11 degrees to 21 degree highly moderate, > 21 degrees considered as very poor due to high slope and runoff.

23.3.6   SOIL

The soil is one of the important factors for delineating potential groundwater potential zone. The soil acts as a natural filter and penetration of surface water into an aquifer system and directly related to rates of infiltration, percolation, and permeability (Donahue et al., 1983). The movement and penetration of surface water into the ground is based on the porosity and absorbency of soil. The result of soil classification found that the study area has six types of soils such as, laterite soil, sandy loam, sandy clay, clay, clay loam, and sticky clay (Figure 23.7).

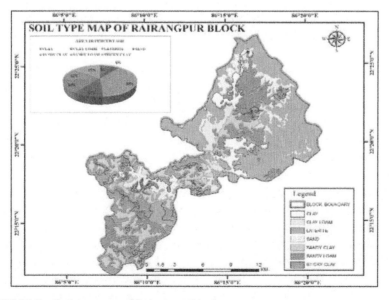

FIGURE 23.7   Soil type map of Rairangpur block.

23.3.7   LAND USE LAND COVER

Land use/land cover map was prepared using LISS-III image Carto Sat Remote Sensing Data. The data was digitally classified in ERDAS 10.0 Software package using a supervised classification technique. The parallelepiped supervised classification technique was applied to extract different types of thematic layers. The study area has five major land use class, i.e., forest (28%), agriculture land (61%) Wasteland (3%), Built-up land (7%) and water bodies (1%). The weights were assigned according to the influence on groundwater occurrence. Water bodies are coming under good categories and lands which are not used for any purpose classified as a wasteland, and built-up land is categorized as poor for groundwater prospects. Agriculture is categorized as moderate categories groundwater occurrence, holding, and recharge (Figure 23.8).

23.3.8   WEIGHT ASSIGNMENT AND GEOSPATIAL MODELING

Suitable weights were assigned to the seven themes according to their hydrogeological importance in groundwater occurrence in the study area. The normalized weights of the individual themes and their different features were obtained through the Saaty's analytical hierarchy process (AHP). The weights assigned to different themes are presented in Table 23.2. After deriving the normal weights of all the thematic maps are converted into raster format and superimposed by weighted overlay method, and all the thematic layers were integrated with one another using spatial analysis of Arc-GIS software to delineate potential groundwater potential zones in the study area. The final integrated layer was divided into five classes, i.e., 'very good,' 'good' 'moderate,' 'poor' and 'very poor' in order to delineate potential groundwater potential zones (Figure 23.9).

23.4   SUMMARY

Delineation of groundwater potential zones in Rairangpur block of Mayurbhanj district using remote sensing and GIS techniques is found efficient to minimize the time, labor, and money and thereby enables quick decision-making for sustainable water resources management. Satellite imageries, topographic maps, and conventional data were used to prepare the thematic layers of lithology, lineament density, drainage density, slope, soil, land-use, and slope. The various thematic layers are

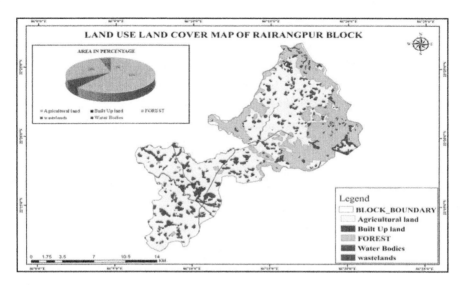

FIGURE 23.8 Land use land cover map of Rairangpur block.

TABLE 23.2 Ranks Assigned to Different Parameters Used for Overlay

Sl No	Parameters	Classes	Feature score	Map Weight
1	Geomorphology	Denudational Hills (Large)	1	25
		Denudational Hills (Small)	1	
		Habitation	1	
		Intermontane valley/ Structural Valley (Small)	8	
		Pediment/ Valley Floor	8	
		Pediplain	2	
		Plateau	7	
		Shallow weathered/ shallow buried Pediplain	2	
		Structural Hills (Large)	6	
		Valley Fill/ filled-in valley	6	
		Water Body	2	
			8	
			10	
2	Slope classes (Degree)	0 to 4	8	15
		4 to 11	6	
		11 to 21	3	
		>21	1	

TABLE 23.2 *(Continued)*

Sl No	Parameters	Classes	Feature score	Map Weight
3	Drainage density (Km/Km^2)	0–93	9	10
		93–186	7	
		186–280	6	
		280–373	4	
		373–467	3	
4	Lineament density (Km/Km^2)	0–0.9	9	15
		0.9–1.9	8	
		1.9–2.9	6	
		2.9–3.9	3	
5	Land use/land cover	Agriculture Land	8	15
		Built-up land	2	
		Forest land	6	
		Waterbody	9	
		Wasteland	3	
6	Geology	Granite	2	15
		Epidote	3	
7	Soil	Clay	1	5
		Clay loam	2	
		Laterite	4	
		Sand	8	
		Sandy clay	3	
		Sandy loam	2	
		Sticky clay	1	

assigned proper weight age through MIF technique and then integrated into the GIS environment to prepare the groundwater potential zone map of the study area. According to the groundwater potential zone map, the block is categorized into four different zones, namely 'very good,' 'good,' 'moderate,' 'poor,' and 'very poor.' The villages potential underground zone are Dangapani, Raunsi, Guhaldangri, Mochianetra, Mahedebdihi, Dhatikidihi, Kusumghaty, Jamuban, Teleijhari, Dandbose, Pokhoria, Sansimila, Palasbani, Badpakhana, Tamalbandh, Niranjan, and the villages under very good groundwater potential zone are Naupada, Udayapur, Tolak, Kukudimundi, Sanmauda, Kalsibhanga, Chhatra-mandal, Badsimila, Kalarda, Katupit, Petepani, and Purunapani. The results of the present study can serve as guidelines for planning future

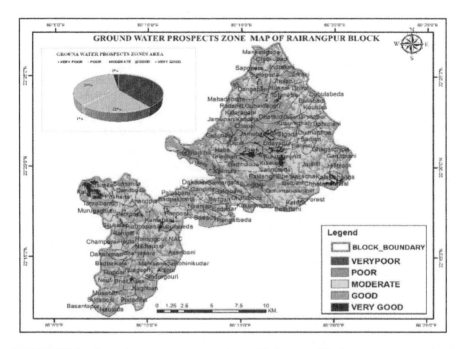

FIGURE 23.9   Groundwater prospect zone map of Rairangpur Block.

artificial recharge projects in the study area in order to ensure sustainable groundwater utilization. This is an empirical method for the exploration of groundwater potential zones using remote sensing and GIS, and it succeeds in proposing potential sites for groundwater zones. This method can be widely applied to a vast area with rugged topography for the exploration of groundwater resources.

KEYWORDS

- **geospatial technology**
- **groundwater**
- **hydrogeomorphology**
- **lineament**

REFERENCES

Anonymous, (1979). Satellite remote sensing survey: southern part of Tamil Nadu (Vol. I and II). Project Report, National Remote Sensing Agency, Department of Space, Hyderabad, India.

Anonymous, (1986). Report on the groundwater potential maps of Karnataka prepared based on visual interpretation of Landsat Thematic Mapper data, NRSA, Department of Space, Government of India, and Hyderabad.

Anonymous, (1988). Preparation of hydrogeomorphological maps of India on 1: 250000 scale using satellite imagery, project report, Department of Space, Bangalore, India.

Bhattacharyaa, B., & Hegde, V. S., (1991). IRS–1A application for groundwater targeting, special issue on remote sensing for National Development. *Journal of Current Science, 61,* 172–179.

Burrough, P. A., (1989). *Principles of Geographical Information Systems for Land Resources Assessment,* Oxford: Oxford University Press.

Chi, K. H., & Lee, B. J., (1994). Extracting potential groundwater area using remotely sensed data and GIS techniques. In: *Proceedings of the Regional Seminar on Integrated Application of Remote Sensing and GIS for Land and Water Resource Management* (pp. 64–69). Bangkok.

Chowdhury, A., Jha, M. K., Chowdary, V. M., & Mal, B. C., (2009). Integrated remote sensing and GIS-based approach for assessing groundwater potential in West Medinipur district, West Bengal, India. *Int. J. Remote Sens., 30*(1), 231–250.

Clavre, B., Maldonado, J. O., & Valenzuela, C. R., (1994). A conceptual approach to evaluating watershed hazards: The Tunari watershed, Cochabamba, Bolivia. *ITC Journal, Special GIS Issue Latin America, No.3,* 283–291.

Elango, L., Kumar, S., & Rajmohan, N., (2003). Hydrochemical studies of groundwater in Chengalpet region, South India. *Indian Journal of Environmental Protection, 23*(6), 624–632.

Hussein, A. A., Govindu, V., & NiGusse, A. G. M., (2016). Evaluation of Groundwater potential using Geospatial Techniques. *Journal of Applied Water Science.*

Hutti, B., & Nijagunappa, R., (2011). Identification of potential groundwater zone using geoinformatics in Ghataprabha basin, North Karnataka, India. *International Journal of Geomatics and Geosciences, 1*(2), 91–109.

Jaiswal, R. K., Mukherjee, S., Krishnamurthy, J., & Saxena, R., (2003). Role of remote sensing and GIS techniques for generation of groundwater prospect zones towards rural development – an approach. *International Journal of Remote Sensing, 24*(5), 993–1008.

Jha, M. K., Chowdary, V. M., & Chowdhury, A., (2010). Groundwater assessment in Salboni Block, West Bengal (India) using remote sensing, geographical information system, and multi-criteria decision analysis techniques. *Hydrogeology Journal, 18*(7), 1713–1728.

Kadam, A. K., Sankhua, R. N., & Umrikar, B. N., (2016). *Assessment of Groundwater Potential Zones Using GIS Technique: A Case Study of Shivganga River Basin* (pp. 70–77). Pune, Maharashtra, India Conference: IGWC–2015, At Chennai.

Kamaraju, M. V. V., Bhattacharya, A., Reddy, G. S., Rao, G. C., Murthy, G. S., & Rao, T. C. M., (1995). Groundwater potential evaluation of West Godavari District, Andhra Pradesh State, India—a GIS approach. *Ground Water, 34*(2), 318–325.

Krishnamurthy, J., & Srinivas, G., (1995). Role of geological and geomorphological factors in groundwater exploration: A study using IRS LISS data. *International Journal of Remote Sensing, 16*(4), 2595–2618.

Krishnamurthy, J., Venkatesa, K. N., Jayaraman, V., & Manuvel, M., (1996). An approach to demarcate potential groundwater zones through remote sensing and a geographical information system. *International Journal of Remote Sensing, 17*(10), 1867–1884.

Krishnamurty, J., & Srinivas, G., (1995). Role of geological and geomorphological factors in groundwater exploration a study through remote sensing technique. *International Journal of Remote Sensing, 16*, 2595–2618.

Madrucci, V., Taioli, F., & De Araújo, C. C., (2008). Groundwater favorability map using GIS multicriteria data analysis on crystalline terrain, São Paulo State, Brazil. *Journal of Hydrology, 357*, 153–173.

Nag, S. K., & Anindita, L., (2011). An integrated approach using remote sensing and GIS techniques for delineating potential groundwater zones in Dwarakeswar watershed, Bankura district, West Bengal. *International Journal of Geomatics and Geosciences, 2*(2), 430–442.

Nandi, D., & Mishra, S. R., (2014). Groundwater quality mapping by using geographic information system (GIS): A case study of Baripada city, Odisha, India. *International Journal of Conservation Science, 5*(1), 79–84.

Rao, Y. S., & Jugran, D. K., (2003). Delineation of groundwater potential zones and zones of groundwater quality suitable for domestic purposes using remote sensing and GIS. *Hydrology Science Journal, 48*(5), 821–833.

Ravi, S. M. N., & Mohan, G., (2006). Assessment of the groundwater potential and quality in Bhatsa and Kalu river basins of Thane district, western Deccan Volcanic Province of India. *Environmental Geology, 49*, 990–998.

Sahu, P. C., (2017). Groundwater resource conservation and augmentation in hard rock terrain: An integrated geological and geospatial approach. *International Journal of Conservation Science, 8*(1), 145–156.

Sander, P., Chesley, M. M., & Minor, T. B., (1996). Groundwater assessment using remote sensing and GIS in a rural groundwater project in Ghana: Lessons learned. *Hydrogeology Journal, 4*(3), 40–49.

Saraf, A. K., & Choudhury, P. R., (1998). Integrated remote sensing and GIS for groundwater exploration and identification of artificial recharge sites. *International Journal of Remote Sensing, 19*(10), 1825–1841.

Sener, E., Davraz, A., & Ozcelik, M., (2005). An integration of GIS and remote sensing in groundwater investigations: A case study in Burdur, Turkey. *Hydrogeology Journal, 13*(5–6), 826–834.

Shaban, A., Khawlie, M., & Abdallah, C., (2006). Use of remote sensing and GIS to determine potential recharge zone: The case of Occidental Lebanon. *Hydrogeology Journal, 14*, 433–443.

Solomon, S., & Quiel, F., (2006). Groundwater study using remote sensing and geographic information systems (GIS) in the central highlands of Eritrea. *Hydrogeology Journal, 14*, 729–741.

Subba, R. N., & John, D. D., (2003). Fluoride incidence in groundwater in an area of peninsular India. *Environmental Geology, 45*, 243–251.

Suja, R. R. S., & Krishnan, N., (2009). Spatial analysis of groundwater potential using remote sensing and GIS in the Kanyakumari and Nambiyar Basins, India. *Journal of the Indian Society of Remote Sensing, 37*, 681–692.

Todd, D. K., & Mays, L. W., (2005). *Groundwater Hydrology* (3rd edn.). New Jersey: John Wiley & Sons.

Zuviria, M. D., & Valenzuela, C. R., (1994). Mapping land suitability for coffee with ILWIS. *ITC Journal, Special GIS Issue Latin America, 3*, 301–307.

CHAPTER 24

AN ANALYSIS OF SALINE WATER INTRUSION INTO COASTAL NIGERIA

AMARTYA KUMAR BHATTACHARYA

Chairman and Managing Director, MultiSpectra Consultants, 23, Biplabi Ambika Chakraborty Sarani, Kolkata – 700029, West Bengal, India, E-mail: dramartyakumar@gmail.com

ABSTRACT

The present chapter gives an overview of saline water intrusion into coastal Nigeria. This chapter places saline water intrusion into coastal Nigeria in the context of geology, hydrogeology, meteorology, and land use of coastal Nigeria. This chapter suggests ways and means to manage the problem of saline water intrusion into coastal Nigeria.

24.1 INTRODUCTION

Groundwater dynamics that is the movement of groundwater through geological formations and it can be linked to different phenomena that occur in deep aquifers such as the upward advance of saline waters of geologic origin; surface waste discharges into shallow aquifers and the invasion of saline water into coastal aquifers. In recent years, considerable interest has been evinced in the study of flow through porous media, because of their natural occurrence and importance in many engineering problems such as movement of water, oil, and natural gas through the ground, saline water intrusion into the coastal aquifers, flow through packed towers in some chemical processes and filtration. Therefore, the physical understanding of flow through porous media is essential to scientists and engineers working in the related areas. The proposed work as the reported herein is on the movement of groundwater as related to saltwater intrusion into coastal aquifers and submarine discharge of freshwater into the sea. The mechanism responsible for this phenomenon involves the reduction or reversal of groundwater gradients, which permits denser saline water to displace freshwater and vice-versa. This situation commonly occurs in a coastal aquifer in hydraulic continuity with the sea when pumping of wells disturbs the natural hydrodynamic balance.

Saline water intrusion in fresh groundwater takes place in the vicinity of coastal regions whenever saline water displaces or mixes with freshwater. This situation usually occurs in coastal regions having hydraulic continuity with the sea when the pumping rate in the wells disturbs the natural hydrodynamic balance. The intrusion of saline water into the freshwater coastal aquifer is likely to cause a problem when such aquifer is tapped for domestic water supply or for irrigation. The coast of Nigeria is located in West Africa, part of the Gulf of Guinea. The coastline of Nigeria stretches for about 850km, from the Republic of Benin in the west to Cameroon in the east. It forms part of the West African groundwater region. River Niger discharges into the Gulf of Guinea via a delta as the major river, thereby, dividing the coastline into two parts. Other rivers discharge through estuaries either into the lagoon of directly into the ocean. The surface strata of the coastline of Nigeria consist of unconsolidated coastal plain sands. Lagoons, estuaries, creeks, and delta dominate the Nigerian coastline. The vegetation of the area is mangrove swamp forest that is being cleared for wood and to give room for other economic activities. The western of the coastline is highly populated with total dependence on groundwater to meet domestic and industrial demands. Thousands of private boreholes have been drilled in the area to extract groundwater for drinking. Continuous pumping or heavy extractions of groundwater have led to increment of chloride content in groundwater. The chloride/bicarbonate ratio is a very good indicator of saline water contamination in groundwater. High chloride content in groundwater may also be caused by the presence of chloride-containing minerals in the area, or there are pollutants being discharged into groundwater. The major cause of saline water pollution into groundwater in this area is as a result of saline water intrusion. Upcoming occurs below wells as a result of continuous pumping of freshwater. The entire coastline of Nigeria is divided into seven administrative units (Lagos, Ogun, Ondo, Delta, Bayelsa, Rivers, and Akwa Ibom states).

Saline water intrusion is one of the most common forms of groundwater contamination in coastal areas (Bear et al., 1999). Saline water intrusion is the movement of salty water into an aquifer or surface water with the consequent mixing or displacement of fresh water. When groundwater is pumped from aquifers that are in hydraulic connection with the sea, and the piezometric surface of the fresh water is lowered, the balance between the saline water/freshwater interface is disturbed, the hydraulic gradients so created induce seawater to encroach the freshwater aquifer (Domenico and Schwartz, 1998; Walton, 1970).

Saline water intrusion has been a major water resource problem in the urban coastal areas of the Niger Delta, and it has therefore been and continues to be a focus of considerable research efforts over time. This is because a scientific understanding of the occurrence of salt and fresh waters in a particular coastal area is essential for the development and management of the water resource. Nigeria has a coastline that is about 1000 km long with the Atlantic Ocean bordering eight States. These are: Rivers, Bayelsa, Akwa Ibom, Cross Rivers State, Delta, Ondo, Ogun, and the Lagos States. Potable water supply to inhabitants in some of the communities in the coastal belt (especially in the saline mangrove swamp) has been a major problem due to saltwater intrusion. Water wells and boreholes drilled have been abandoned in many communities due to high salinity. They, therefore, depend on rain harvesting and purchasing water from water merchants. In Nigeria, there have been few studies aimed at assessing freshwater resources in coastal areas of the country. This forms the main thrust of the proposed research.

The quality of water is the main constraint of groundwater in the coastal aquifers in Nigeria. Like all coastal aquifers worldwide, saline water intrusion into aquifers is the major source of quality impairment in the coastal aquifers in Nigeria. This is followed by pollution from spillages of crude and refined petroleum and lastly by leachates from municipal and other industrial wastes. Most of the confined aquifers in the coastal areas of Nigeria have high iron concentration such that treatment for the removal of iron has to be undertaken. Saline water intrusion into unconfined and confined aquifers occurs in Niger Delta. In Coastal beach ridges or sandy islands within the saline mangrove belt, freshwater lens floating above saline water-bearing sands are found to occur in the unconfined aquifers (Oteri, 1990). The growth of industrial development within the urban coastal areas of the Niger Delta and the attendant population explosion places a very heavy demand on the lean supply of fresh water. According to Frank Briggs (2003) "indiscriminate abstraction of groundwater from the first aquifer has resulted in saltwater intrusion in several coastal wells...." This limits the supply of potable drinking water in the area, which can have detrimental effects on human health, wildlife habitat and increase the cost of water treatment (Domenico and Schwartz, 1998).

24.1.1   GEOGRAPHY AND METEOROLOGY OF COASTAL NIGERIA

In Nigeria, eight states are bounded to the south by the Atlantic Ocean. They include from West to East – Ogun, Lagos, Ondo, Delta, Bayelsa, Rivers, Akwa Ibom, and the Cross River States. The states are contiguous and are in direct contact with the sea. They are located between latitude 7° and 4° 10' N and longitudes 2° 30' and 8 30' East. There are a number of highly populated cities situated within the coastal area of these states which include Port Harcourt, Warri, Yenagoa, Lagos, Uyo, and Calabar. The coastal states of Nigeria fall within the humid tropical zone of the country. Rainfall is copious and lasts for eight to nine months of the year (mid-March to early-November) with a mean annual value often exceeding 3000 mm at the coastal fringes at Akasa and Brass. Seasonal changes of wet and dry are as a result of the interplay of two contrasting air masses – the moisture-laden SW monsoon winds blowing into the country from the Atlantic Ocean and the dry North East Trade winds from the Sahara desert. The two main rivers in the southwestern part of Nigeria are Ogun and Oshun which drain into the ocean.

24.1.2   GEOLOGY OF COASTAL NIGERIA

Geologically, coastal Nigeria is made up of two sedimentary basins-the Dahomey Basin and the Niger Delta Basin separated by the Okitipupa Ridge. The Dahomey Basin covers the southern areas of Lagos, Ogun, and the Ondo States in Nigeria and stretches into the neighboring countries of Benin, Togo, Ghana, and Ivory Coast. The present study is centered on the Niger Delta area of Nigeria. The geological formations of primary interest for the evaluation of the groundwater resources potential of the study area are the sedimentary formations deposited in the Niger Delta during the Eocene-Quaternary periods. The stratigraphic sequence under consideration comprises, from oldest to youngest formations, the Imo Shale Group, the Ameki Formation, the Benin Formation, and the Quaternary deposits as described by Amajor and Agbaire (1989).

The Imo Shale Group is made up of marine clay, shale, and limestone. The formation is estimated to have a thickness of up to 1,000 m, and it outcrops in a belt more than 100 km north of the study area. The formation, due to its impermeable texture, is believed to constitute the base of the groundwater aquifers under consideration in the study area. The Ameki Formation overlies the Imo Shale Group. It comprises numerous alternations of marine shales and fine to coarse, very heterogeneous, coastal – deltaic sands and sandstones. The formation is estimated to have a thickness of up to 1,700m and is out-cropping in an east-west and north-west – south-east oriented belt, approximately 100km north of the study area. The groundwater resources potential of the formation is believed to be good (Tahal, 1998). The present-day Niger Delta is defined geologically by three subsurface sedimentary sequences consisting of Benin, Agbada, and Akata Formations (Short and Stauble, 1967).

24.2   BENIN FORMATION

The Benin Formation overlies the Ameki Formation. The formation is an extensive stratigraphic unit in the Southern Sedimentary Basin, with an average thickness of about 1,900 m (Short and Stauble, 1967). The formation is recognized throughout the delta due to its few shale streaks, the absence of brackish water and a high percentage of sand. The Benin Formation, which consists mainly of coastal plain sands, extends from the west across the whole Niger Delta area and southward beyond the present coastline. The sediments comprise yellow and white sands with pebbly gravels. The clays and sandy clays occur in lenses of 3m to as much as 10m, and they make the groundwater formation a multi-aquifer system. The formation is massive porous fresh water-bearing sandstone with localized thin beds. The formation is, due to its coarse texture and huge outcrop area, believed to constitute a very good groundwater aquifer. The aquifer in this basin has a southwest gradient towards the Delta, and it thickens seawards in the same direction of groundwater movement. It is thus the most prolific aquifer in the region. It is overlain by the Quaternary deposits, which ranges in thickness in between 133 – 500ft (Etu – Efeotor and Akpokodje, 1990). The Quaternary deposit, which comprises recent deltaic sediments made up of sand, silt, and clay beds, overlies the Benin Formation in the swampy delta areas. The formations have a seaward dip resulting in confined aquifers.

The sandstones of the Benin Formation are coarse-grained, locally fine-grained, poorly sorted and sub angular to well rounded. The age of the Benin Formation lies between Oligocene and Recent. This formation constitutes the major aquifer in the Niger Delta area (Udom et al., 1999). Whiteman (1982) notes that a formation such as the Benin Formation, deposited in a continental fluviatile environment has a highly variable lithology, can be recognized because of its high sand percentage (70–100%), few minor shale intercalations and the absence of brackish water and marine faunas.

To date, very little oil has been found in the Benin Formation (mainly minor oil shows).

24.3   AGBADA FORMATION

The Agbada Formation underlies the Benin Formation and forms the second of the three strongly diachronous Niger Delta complex formations. It comprises mainly of alterations of sands, silts, and shales in various thicknesses and proportions, indicative of cyclic sequences of off-lap cycles, better called off-lap rhythms (Weber, 1971). The characteristic features of the sandstones here are poor sorting, calcareous matrix and shell fragment occurrences. The grain size varies from fine to coarse. The approximate total thickness of the Agbada Formation lies between 10,000 ft and 12,000 ft (Kogbe, 1976). The top of the Agbada Formation is sandy whereas the bottom is shaley because it grades into the Akata shales gradually. The shales are denser at the base of the formation due to the compaction process. The sandy parts constitute the main hydrocarbon reservoirs in the delta oil fields. The shales constitute seals to the reservoirs and as such are very important because the formation is rich in hydrocarbon. Paralic clastics of the Agbada Formation represent the true deltaic portions of the delta top-set and fluvio-deltaic environment (Short and Stauble, 1967; Weber and Dakoru, 1975). However, Whiteman (1982) notes that the Agbada Formation, as defined by Short and Stauble (1967), contains beds laid down in a variety of sub-environments grouped together under the heading-paralic environment. The age of the Agbada Formation ranges from Eocene to Recent.

24.4   AKATA FORMATION

This formation is the basal major time-transgressive lithological unit of the Niger Delta Complex. It is composed mainly of marine shales, but contains sands and silty beds which are thought to have been laid down as turbidites, and continental slope channel fills. The shales are typi-cally under-compacted and over-pressured, forming diapric structures. Whiteman (1982) contributes that the Akata Formation is rich in planktonic foraminifera, which indicates deposition on a shallow marine shelf environ-ment. The formation occurs between depths of 0–6000 m below the Agbada Formation and ranges in age from Paleocene to recent. It represents the pro-delta mega facies (Tahal, 1998).

24.5   HYDROGEOLOGY AND AQUIFERS OF THE STUDY AREA

In the study area, groundwater is abstracted from the Benin Formation, mainly from its upper section. Lenses of silty clay of some few meters thickness have been recorded in the borehole penetrating the Benin Formation (Tahal, 1998). To the south of Port Harcourt, belts rich in shales lying at a depth of 10m to 200m have been observed. These lenses create several sub-aquifers in the Benin Formation, the upper sub-unconfined, while the deeper aquifers range from leaky to confined and are isolated from the ground surface. The natural recharge comes mainly from the northern high coastal plain. Generally, the sediments of the Benin Formation dip seawards at a low gradient. The upper section of these sediments is being utilized in the development of groundwater.

The sandy components, in most layers, are about 90% of the lithological sequences (Tahal, 1998). The size distribution of the sedimentary particulates does not vary significantly from place to place in the study area. There is a steady gradation in quantity between coarse sands and clay beds. Locally to some extent, the clay beds separate hydraulically between the sub-aquifers.

Groundwater abstracted from the upper Benin Formation is characterized by low p^H (acidity), high carbon dioxide (CO_2) content, and hence it is corrosive and soft. In addition, the groundwater contains high concentrations of iron and therefore requires treatment. The water tables of all sub-aquifers penetrated by boreholes are relatively shallow and range between 5m to 15m where hydrostatic water levels increase with the depth (Tahal, 1998). According to Etu – Efeotor, and Odigi (1983), three main zones have been differentiated as follows:

i)   A northern zone is consisting shallow aquifer of predominantly continental deposits.
ii)   A transitional zone of marine and continental materials.
iii) A coastal zone of predominantly marine deposits.

Aquifer distributions in the Niger Delta are controlled by the geology. Table 24.1 is the summary of the properties and behavior of hydrostratigraphic units in the Niger Delta Basin. In most parts of the Niger Delta, including the study area, a multi -aquifer system is encountered, and the aquifer lies within the arenaceous Benin Formation. The depth to water table of the Benin Formation ranges between 3–5 m below ground level (Offodile, 1984). Ngah (1990) also identified three main aquifer zones in the Niger Delta Viz:

i)   An upper unconfined aquifer is extending throughout the Benin Formation with its thickness ranging between 15–80 m while the Static Water Level (SWL) varies between 4m and 21 m.

ii) A middle aquifer system, semi -confined and consisting of thick medium to coarse-grained, sometimes pebbly sands with thin clay lenses. Its thickness varies between 30–60m.

iii) A lower aquifer system that extends from 220–300 m and consists of coarse -grained sands and gravels with some interlayer clay. The majority of the groundwater wells abstract water from the first and second aquifers (<100 m deep). The very few industrial and municipal groundwater supply wells tap deeper aquifers.

24.5.1   AQUIFER RECHARGE IN THE STUDY AREA

Precipitation is the main source of recharge to aquifers. The rate of recharge depends to a large extent on the infiltration capacity of the soil, evaporation rate and the overland drainage characteristics. In places where sandy clay forms a part of soil layers, recharge usually occurs mainly through a distant outcrop of the porous formation and partially through the lateritic sand. The northern and the southern movement of air masses and pressure belts characterize the rainfall regime of the coastal plain. The amount of rainfall decreases inland from the coast. The isohyets run parallel to the coast up to a line through Ahoada-Degema-Port Harcourt. From this line inland, the decrease in rainfall is much slower. According to the lithologic profile of some boreholes, in some places, there are few meters of lateritic sandy clay under the soil. Tahal (1998) observed that about 30% to 40% of the yearly average of rainfall (2,280mm) could infiltrate and recharge the Benin Formation. To the north, sandy, and porous outcrops exist, where the replenishment can be 60% to 70% of rainfall. The estimated annual recharge to the aquifer in the study area from the Northern High Coastal Plain is about 100 mcm to 150 mcm (Tahal, 1998).

24.5.2   PROBLEMS AFFECTING THE COASTAL AQUIFER ECOSYSTEM IN NIGERIA

The major constraint to the water resource in the aquifers of the coastal Nigeria is quality. The saline water intrusion is both natural and man-made. For the natural saline water intrusion, the problem is exacerbated by:

TABLE 24.1 Summary of Properties and Behavior of Hydrostratigraphic Units in the Niger Delta Basin

Hydrogeological basin	Hydrostratigraphic units	Lithologic details	Aquifer type and characteristics	Water quality	Economic importance
Niger Delta Basin	Alluvial plains aquifer	Sands, clays, silt	Unconfined	Saline water	Domestic Municipal Industrial uses
	Meander Belt aquifer	Sands, gravel, clays		Corrosion	
	Mangrove swamp aquifer	Sands, clay, swamps	Good, water table is confined	Iron-rich saline water intrusion	
	Abandoned Ridges aquifer, Sombrero	Sands and pebbles, yellow sands, clays	Poor aquifer		
	Benin aquifer	Sand, clays	Prolific		
	Delta Ogwashi – Asaba Ameki aquifer	Clays, shale, lignite, silty sand, clay siltstone	Aquitard	Problematic with iron and saltwater encroachment, Low Ph	
	Imo shales Deep oceanic Agbada – Akpata Shales	Shalestone, claystone lenses of sands	Aquitard Aquitard		

- The lack of willingness to carry out necessary studies or utilize such studies when carried out.
- Lack of appreciation of the need for water resources assessment and management groundwater resources of the coastal aquifers resulting in over-exploitation of the aquifers.
- The activities of man have also led to increased saline water intrusion through.
- Uncontrolled development of both unconfined and confined aquifers especially in Port Harcourt Metropolis, Lagos, Warri, and Bonny. In Lagos which is the commercial capital of Nigeria, the problem is particularly acute as many boreholes which were producing fresh water after drilling becomes salty a few months after especially in Ikoyi, Victoria Island and Apapa.
- Lack of proper sealing of disused boreholes or those abandoned due to saltwater intrusion.
- The groundwater in most of the confined aquifers in coastal Nigeria is corrosive, and casing corrosion is a major source of borehole failures. There is a need to determine the best material for casing and screens and also best completion techniques that will prevent saline water ingress into fresh water yielding boreholes especially in situations where the freshwater aquifers underlie saline water-bearing sands.
- Sea water level rise (as a result of global warming); the building of dams in upstream areas of the rivers leading to low flows and decrease in river sediments to the coast by the rivers have all led to increased coastal erosion.
- Construction of drainage canals, transportation canals.
- Urbanization – In the Lagos area for example. The rate of Urbanization is so high that the recharge area has been converted to concrete zones, while swamps and streams are being reclaimed and turned into cities. This decreases the amount of recharge to the aquifers, thereby increasing saline water intrusion.
- Oil spillage from both upstream and downstream operations of the oil industry is the next major problem affecting both the aquifers and the environment of coastal areas in general.
- Finally, disposal of wastes from industries and municipal areas is a threat to the coastal aquifer ecosystem.

24.6 CONCLUSIONS

The control of saline water intrusion demands knowledge of the hydraulic conditions within aquifer; it also demands knowledge of the source of the saline water. It is, therefore, necessary to identify the extent of the problem and to assess the behavior of the saline water body under various conditions of recharge and discharge, such that efficient water resources management plans can be implemented. The optimum solution to the problem of saline water intrusion is prevention, by which the encroachment of saline water is controlled to an acceptable degree. But in many cases, the problem is a legacy of the past; therefore, management must concentrate on minimizing further intrusion, and/or reducing the extent of the existing saline water, it may be that the aquifer in question is too badly polluted so reclamation may be the only viable option. In extreme cases, if the resource for potable water supply, it may be abandoned, although the water may still be utilized for certain industrial or agricultural applications.

KEYWORDS

- **coastal regions**
- **Nigeria**
- **saline water intrusion**

REFERENCES

Amajor, L. C., & Agbaire, D. W., (1989). Depositional history of the reservoir sandstones, Akpor and Apara Oilfields, Eastern Niger Delta, *Nigeria Journal of Petroleum Geology, 12*(4), 453–464.

Bear, J., Cheng, A. H. D., Sorek, S., Ouazar, D., & Herrer, T. J. B., (1999). *Saltwater Intrusion in Coastal Aquifers–Concepts, Methods, and Practices, in Theory, and Application in Porous Media* (p. 625). Kluwer Academic Publishers. Dordrecht.

Domenico, P. A., & Shwartz, F. W., (1998). *Physical and Chemical Hydrogeology,* John Wiley and Sons Inc., New York.

Etu–Efeotor, J. O., & Akpokodje, G. E., (1990). Aquifer systems of the Niger delta. *Journal of Mining and Geology, 26*(2), 264–266.

Etu–Efeotor, J. O., & Odigi, M. I., (1981). Water supply problems in the Eastern Niger Delta. *Journal of Mining and Geol., 20*(1), 182–192.

Frank-Briggs, I. N., (2003). The geology of some Island Towns in the Eastern Niger Delta, Nigeria. *An Unpublished PhD Thesis Submitted to the Department of Geology*, University of Port Harcourt.

Kampsax–Krüger, (1985). *Final Report on Hydrological Site Studies for Water Well Development, Phase 1*, 118.

Kogbe, C. A., (1976). *The Cretaceous and Paleocene Sediments of Southern Nigeria* (pp. 237–252). Elizabethan Publishing Coy, Lagos.

Ngah, S. A., (1990). *Groundwater Resource Development in the   Niger Delta: Problems and Prospects.* 6th IAEC Congress.

Offodile, M. E., (1982). The Problems of Water Resources Management in Nigeria, *Journal of Mining and Geology, 20th Anniversary edn., 19*(1).

Oteri, A. U., (1990). Delineation of seawater intrusion in a coastal beach ridge of Forcados. *Journal of Mining and Geology, 26*(2), 225–229.

Short, K. C., & Stauble, A. J., (1967). Outline of the geology of the Niger Delta. *AAPG Bull., 51*, 761–779.

Tahal Consultant Engineers Ltd., (1998). Final report on multistate water supply project, feasibility study of rivers state capital (Port Harcourt and Environs and selected Urban Communities), 2–4.

Udom, G. J., Etu-Efeotor, J. O., & Esu, E. O., (1999). Hydrochemical evaluation of groundwater in parts of Port Harcourt and Tai-Eleme Local Government Area, rivers state: *Global Journal of Pure and Applied Sciences, 5*, 546–552.

Walton, W. C., (1970). *Groundwater Resource Evaluation*, McGraw–Hill, 375.

Weber, K. J., & Dakoru, E., (1975). *Petroleum Geology of the Niger Delta* (pp. 109–221). Ninth World Petroleum Congress.

Weber, K. J., (1971). Sedimentological aspects of oil fields in the Niger Delta. *Journal of Environmental Geology, Minbouw, 50*(3), 559–576.

Whiteman, A., (1982). *Nigeria: Its Petroleum Resources and Potentials*. Graham and Trotman, 1, 63–78.

CHAPTER 25

SALINE WATER INTRUSION IN COASTAL AREAS: A CASE STUDY FROM INDIA

AMARTYA KUMAR BHATTACHARYA

Chairman and Managing Director, MultiSpectra Consultants, 23, Biplabi Ambika Chakraborty Sarani, Kolkata – 700029, West Bengal, India, E-mail: dramartyakumar@gmail.com

ABSTRACT

An innovative method of control of saltwater intrusion into the coastal aquifers has been suggested in this paper. A new method consists of withdrawal by Qanat-well structures with reasonable compensation by rainwater harvesting by means of recharge ponds and recharges well. The salient features of the methodology are described by considering a design example adopted in the Contai Polytechnic Institute Campus of the district of Purba Midnapur in the state of West Bengal, India.

25.1 INTRODUCTION

Groundwater is the second largest reserve of fresh water on earth. About 2 billion people, approximately one-third of the world's population, depend on groundwater supplies, withdrawing about 20% of global water (600–700 km) annually—much of it from shallow aquifers (Patra et al., 2006). India cities located near coastlines with a large population can experience saline water intrusion due to over-exploitation of groundwater, causing this significant threat to freshwater resources. Sustainability strategies adopted to retard or halt the rate of saline water intrusion are necessary to protect the resources from further damage. The complexity of hydrogeological setup in the concerned area calls for scientific management techniques to be adopted in the groundwater development. This requires a clear understanding of the hydrogeology of the area concerned, appreciation of the possible consequence of over or under developments and a coordinated approach of the planners, hydrogeologists, irrigation engineers, social scientists in the field (Bhattacharya et al., 2004). Excessive withdrawal of groundwater in coastal zones will lead to depression of water table with associated hazards like putting the well out of use, rendering the abstraction uneconomic with increased lift. A sustained regional groundwater drawdown below sea level runs the risk of saline water intrusion, even for confined coastal aquifers. A careful pumping and rest schedule may help in avoiding many such problems. An uncontrolled groundwater development may lead to a reversal of fresh water gradient thereby resulting in saline water ingress into the coastal aquifers. For example, reports of salinization of wells subjected to continuous heavy pumpage in the coastal district of Purba Midnapur, West Bengal, India though few, are not uncommon.

As suggested by various Scientists and Engineers, there are several established methodologies to control and minimize the problems associated with groundwater extraction followed by saline water intrusion. Some of the popular methods adopted are: the creation of a hydraulic barrier, rainwater harvesting, artificial recharge, canal irrigation, desalination, and reverse osmosis, etc. However, the author has developed an innovative, cost-effective technique to control saltwater intrusion into coastal aquifers; the techniques include withdrawal of coastal fresh water by means of qanat-well structures associated with artificial recharge through rainwater harvesting aided with percolation pond and recharge well. A case study on a selected location of the coastal zone of Purba Midnapur has been carried out. Adequate quantifications of the effectiveness of this new methodology have been incorporated, and relevant conclusions are drawn therefrom.

25.2   MATERIALS AND METHODS

25.2.1   NEW APPROACH FOR CONTROL OF SALINE WATER INTRUSION IN COASTAL AREAS

While a numerical simulation method like the finite difference or finite element scheme might be very illuminating from a research point of view, it is unlikely that field engineers will be able to capture the intricacies of such an approach. Keeping this in view; an innovative analytical method which is appropriate and at the same time easily implemental; is adopted as a new approach for groundwater withdrawal and control of saline water intrusion in the study area.

It is evident that any withdrawal of groundwater from a coastal aquifer results in the advancement of the saltwater-freshwater interface from the shoreline towards to the point of withdrawal (Todd, 1976) unless the withdrawal is compensated by an equivalent artificial recharge. The rainwater harvesting is one of the most popular recharge techniques followed worldwide (UNEP Report, 2009). In the proposed methodology for reduction of saline water intrusion into the coastal aquifer and subsequent safe withdrawal of groundwater, the adoption of qanat-well structure associated with artificial recharge by rainwater harvesting through recharge ponds and recharge wells is hereby studied as one of the useful and cost-effective techniques. The salient features of the methodology are described by considering a design example adopted by the author in the Contai Polytechnic Institute Campus of the district of Purba Medinipur in the state of West Bengal.

25.2.1.1 ADOPTION OF QANAT-WELL STRUCTURE

The inland aquifers are suffering from the maladies of over-exploitation of groundwater by way of unscrupulous pumping; the coastal aquifers encounter the danger of seawater intrusion and saline water upconing. Due to saltwater intrusion, deep tube-well is not recommended because of the upconing problem. However, adoption of the shallow well is also inappropriate because of significantly lower discharge. It is well established (Raghu Babu et al., 2004) that adoption of qanats in such conditions not only yields higher discharge but also reduces the upconing problem significantly. Horizontal wells are more efficient than conventional vertical wells for environmental remediation of groundwater for a number of reasons such as:

- Greater reservoir contact with the well screen increases the productivity of the well.
- Geometry of the groundwater zone is conducive to greater access with a horizontal well than a series of vertical wells.
- Access to groundwater zones with vertical wells are often hindered by obstacles such as buildings, paved surfaces, or other topographical obstructions.

Beljin and Losonsky (1992) provided a generalized solution, based on the work of Joshi (1986), for estimating steady-state discharge for withdrawal of groundwater from a vertical aquifer by means of horizontal water well. The solution provided was given by:

$$Q = \cfrac{2\pi k_h Hs}{\ln\left[\left(\cfrac{\sqrt{1+\sqrt{1+64R^4/L^4}}+\sqrt{-1+\sqrt{1+64R^4/L^4}}}{\sqrt{2}}\right)\cfrac{\left(\cfrac{\beta H^2}{2}+2\beta\delta^2\right)^{\beta H}}{Hr_w}}{L}\right]} \qquad (1)$$

where,

Q = Steady state discharge.
s = Drawdown above the well center.
L = Length of the horizontal well.
r_w = Well radius.
k_h = Hydraulic conductivity of the aquifer along the horizontal direction.
H = Aquifer thickness.

$$\beta = \sqrt{\frac{k_h}{k_v}}$$

k_v = Hydraulic conductivity of the aquifer along the vertical direction.
δ = Off-centered eccentricity of the well-center in the vertical aquifer plane.
R = Radius of the influence of the equivalent vertical well in the same aquifer for the same drawdown, which can be reasonably estimated using the available correlations (for example, Sichardt's formulae).

The above equation has been modified, applying the method of super-imposition, to reasonably estimate the steady-state discharge by means of a 4-legged qanat. The final expression is obtained as:

$$Q_q = \cfrac{2\pi k_h Hs}{\ln\left[\left(\cfrac{\sqrt{1+\sqrt{1+4R^4/L_q^4}}+\sqrt{-1+\sqrt{1+4R^4/L_q^4}}}{\sqrt{2}}\right)\cfrac{\left(\cfrac{\beta H^2}{2}+2\beta\delta^2\right)^{2\beta H}}{Hr_q}}{L_q}\right]} \qquad (2)$$

where, Q_q = Steady state discharge from the qanat.
           L_q = Length of a qanat leg (Figure 25.1)
           r_q = Inner radius of qanat legs

The maximum discharge from the qanat under full flow condition can be obtained by putting $s = d - 2r_q$ (neglecting the wall thickness of the qanat legs) in the Eq. (2). Thus,

$$Q_q(max) = \cfrac{2\pi k_h H(d-2r_q)}{In[(\cfrac{\sqrt{1+\sqrt{1+4R^4/L_{q^4}}}+\sqrt{-1+\sqrt{1+4R^4/L_{q^4}}}}{\sqrt{2}})\cfrac{\cfrac{\beta H^2}{2}+2\beta\delta^2}{Hr_q})^{2\beta H}}{L_q}]} \qquad (3)$$

where, $Q_{q(max)}$ = Maximum possible discharge from the qanat.
   d = Depth of the bottom surface of the qanat legs below the undisturbed water table.

Using Sichardt's formulae $R = 3000(d-2r_q)k_h$ in Eq. (3), where all terms should essentially be in SI units, the Eq. (3) can be written as:

$$Q_q(max) = \cfrac{2\pi k_h H_f(d-2r_q)}{In[\xi\cfrac{\cfrac{BH^2}{2}+2\beta\delta^2}{H_f r_q})_f^{2\beta H}}{L_q}]} \qquad (4)$$

where $\xi = \cfrac{\sqrt{1+\sqrt{1+4[3000(d-r_q)\sqrt{k_h}]^4/L_q}}+\sqrt{-1+\sqrt{1+4[3000(d-r_q)\sqrt{k_q}]^4/L_{q\Leftarrow}}}}{\sqrt{2}}$

Although upconing in case of qanat-well structures is apparently not practical specifically when the saltwater-freshwater interface is situated at a significant depth below the bottom of the structures, the upconing problem may be catastrophic for shallow depth of interface in the area near the sea. Therefore, the present analysis is extended considering upconing as well (Figure 25.2). After the recommendation of Dagan and Bear (1968):

$$Z_\alpha = \cfrac{Q_q}{2\pi\left(\cfrac{\rho_s}{\rho_f}-1\right)K_h(H_f-d)} \qquad (5)$$

where, Z_∞ = the value of Z at infinite time. The value of Q attains the maximum value when $Z_\infty = H_f - d$

$$Q_{q(max)} = 2\pi k_h\left(\cfrac{\rho_s}{\rho_f}-1\right)(H_f-d) \qquad (6)$$

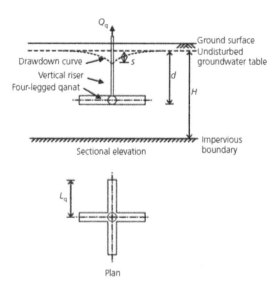

FIGURE 25.1   A typical 4-legged qanat.

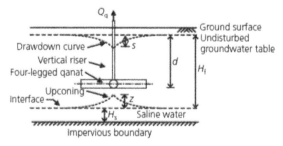

FIGURE 25.2   Withdrawal of groundwater by qanat in salinity affected aquifer.

The optimum values of $Q_{q(max)}$ shall be the least of the two values obtained from the Eq. (4) and (6). It is hereby mentioned that the design values of the depth of qanat-well structures are calculated as unrealistically high, therefore not feasible, the number of such qanat-well structures may be reasonably increased as required.

25.2.2   GROUNDWATER RECHARGE BY RAINWATER HARVESTING

A hybrid method considering ponds and recharge wells is adapted to: (i) combine the best of both, (ii) providing only ponds would eat up a huge amount of unnecessary spaces, and (iii) providing only wells would necessitate pressure injection into the aquifer. Therefore, since due to space constraint in the locality under consideration, the full recharge may not be affected by the pond, recharge wells are needed. The withdrawal of groundwater by qanat should be suitably compensated by a recharge with rainwater harvesting, the salient features of this new approach with adequate quantification is described in the following subsections.

25.2.2.1   RECHARGE AREA

If fresh water in a coastal area is withdrawal regularly, the saltwater-freshwater interface is progressively advanced horizontally as well as vertically unless the withdrawal is subsequently compensated by a suitable artificial groundwater recharge techniques. The method proposed herein includes rainwater harvesting by means of recharge ponds and recharge wells design.

Usually for a particular community, neglecting the area of recharge well,

$$A_t = A_{roof} + A_{road} + A_{pond} + A_1 \tag{7}$$

where, A_t = Total area of the community;
A_{roof} = Total roof cover area for all building in the community;
A_{road} = Total road area in the community;
A_{pond} = Total pond area of the community;
A_1 = Total area of vacant land in the community.

25.2.2.2   FACTOR OF SAFETY FOR RAINFALL RECHARGE

For a particular community in a coastal area, the net volume of freshwater withdrawal in a certain period of time should not exceed the available volume of recharge for that period. With this conception, the corresponding factor of safety for the particular community has been formulated for the volumetric constancy as:

$$F = \frac{Volume\ of\ water\ annually\ available\ for\ recharge}{Volume\ of\ water\ annually\ extracted}$$

$$= \frac{\left[(A_t - A_{roof})\eta + \alpha A_{roof}\right]R}{365WP} \tag{8}$$

where,
   W = Average water consumption of people in the community in liter per capita per day = 140 lit/capita/day;

P = Population of the community;
R = Design annual rainfall in mm;
η = Recharge coefficient;
α = Fraction of rainwater collected at roof which is directed towards the recharge well.

It is hereby mentioned that this technique is most effective when the value of the factor of safety is slightly higher than unity. The excessively high value of the factor of safety may be necessitated adequate drainage facility in the area under consideration to avoid the undesirable circumstance like water logging and flooding. Conversely, when the value of the factor of safety is less than unity, the situation can be compensated either by reducing the withdrawal of groundwater or by increasing the catchments area for rainfall recharge.

25.2.2.3   PERCOLATION POND

The design precipitation chosen depends on the design return period of the precipitation. The longer the design returns period, the greater the precipitation. After recommendation by Sarkar (2007), the water collected from the roof area in the community is partly allowed to percolate through recharge chamber cum recharge well, and the remaining portion is stored for future usages like firefighting and domestic use, etc. Therefore, the total area of recharge pond for the community under consideration may be estimated reasonably considering the net volume of water to be stored in the pond during the monsoon period. Therefore,

$$A_p = \frac{\left[\left(A_t - A_{roof} - A_{road}\right)\eta_1 + A_{road}\right]R_m}{1000\, H_p - (1-\eta_1)R_m} \tag{9}$$

where, A_p = Area of the pond;
   R_m = Design monsoon rainfall in mm;
   η_1 = Runoff coefficient relevant to the area;
   H_p = Depth of pond to be excavated.

25.2.2.4   RECHARGE CHAMBER WITH RECHARGE WELL

As already mentioned earlier Sarkar (2007) designed a rainwater harvesting scheme for TCS building Salt Lake, Kolkata, India. Following his recommendations, the dimensions and number of recharge chamber fitted with 100 mm diameter recharge well for the community under consideration may be reasonably estimated by:

$$\frac{V_w \, N_w}{V_{ws} \, N_{ws}} = \frac{\alpha \, R \, A_{roof}}{\alpha_s \, R_s \, A_{roof \, s}} \tag{10}$$

where V_w = Volume of the recharge chamber in the community.

N_w = Nos. of recharge chamber fitted with 100 dia. recharge well adopted in the community.

The suffix 's' denotes the corresponding parameter for the relevant to Sarkar (2007). From the field investigation, SP reading is observed to move towards negative side indicating the presence of sand aquifer below. As observed, the average depth freshwater-saltwater interface is situated at a depth of about 40m below the ground surface. It is advisable not to go for 'Deep Tube Well' construction; Qanat-Well Structures is preferable in this situation. The input parameters necessary for design in the locality under consideration are summarized below in Table 25.1.

TABLE 25.1 Values of Variables Determined From Hydrogeological Investigation

Input Parameter	Values
Maximum design discharge, $Q_{q(max)}$	Calculated using Eqs. (4) and (6)
Aquifer thickness, H	40 m
Horizontal hydraulic conductivity, K_h	3.512 x 10^{-4} m s^{-1} (Laboratory test)
Vertical hydraulic conductivity, K_v	3.614 x 10^{-4} m s^{-1} (Laboratory test)
Conductivity contrast, $\beta = \sqrt{\dfrac{K_h}{K_v}}$	0.9857
Well eccentricity, δ	0
Population in Contai Polytechnic College Campus, P	300
Total Area of the College Campus, A_t	90,169 m^2
Total Roof Area including Student Hostels, A_r	12,000 m^2
Total road area, A_r	1010 m^2
Total area of pond, A_p	1204 m^2
Area of vacant land, A_1	75,955 m^2
Design annual precipitation, R	Various values are taken
Design monsoon precipitation, R_m	Various values are taken
Per capita water consumption, W	140 liters/capita/day
Recharge coefficient, η	Chosen from available literature
Runoff coefficient, η_1	-do-

25.3 APPLICATION OF NEW APPROACH

From the available groundwater data, the coastal areas in the district of Purba Medinipur mostly consist of unconfined aquifers, with the average freshwater table of 2–3 m below the ground surface (Goswami, 1968; Sarkar, 2005) and the average saline water and fresh water interface varies in the range of 0–100 m. The aquifer is unconfined having an average hydraulic conductivity of k_h =3.512 x 10^{-4} m s^{-1} and k_v =3.614 x 10^{-4} m s^{-1}. From the available data (UNDP Report, 2006; WHO/UNICEF Report, 2010), the value of W (water demand) has been chosen as 140 liters/capita/day. The average population in the campus is 250. Considering a 20% increase, the value of P is taken as 300. On the basis of water demand per day for the college campus with pumping operation per day such as 2, 3, 4, and 5 hours, the qanat-well structures with a different parameter such as length, radius, and depth of qanat is calculated using the Eq. (4). The $Q_{q(max)}$ is calculated as follows:

$$Q_{q(\text{max})} = \frac{W\ P}{t \times 3600 \times 10^3} \tag{11}$$

where, t = hourly pumping rate in the college campus per day.

Using Eq. (4), the values of the depth of the qanat base for the chosen value of L_q and r_q is calculated. This value of 'd' is back-figured in the Eq. (6) to check for upconing. If the discharge calculated from the Eq. (6) exceeds the design discharge as estimated previously, upconing does not take place. Otherwise, the depth 'd' may be calculated by using Eq. (6). It is observed that at the college campus, the upconing does not occur. As observed from Figure 25.3, the parameter d decreases following a curvilinear pattern with the length of leg L_q. The variation is quite sharp in the range of $1m \leq L_q \leq 3m$ and assumes a linear pattern for $L_q > 3$ m. The above curves will be helpful for Design Engineer to adopt suitable values of the qanat parameter r_q, L_q, d, and t considering other design aspects such as, maximum depth of water table, the feasibility of construction, etc.

25.3.1 RECHARGE STRUCTURES

25.3.1.1 RAINFALL RECHARGE

From the available literature (State Forest Report, 2008–2009 and Annual Climate Summary, 2010), the total annual rainfall in the district of Purba Medinipur of the state of West Bengal, India for the decade 2001–2010 varies in the range of 1296–2259 mm. For the design of recharge pond equipped with recharge-wells for the given community, the factor of safety F for recharge may be estimated using the Eq. (8), using the following data:

$A_t = 90,169$ m²
$A_{ROOF} = 12000$ m²

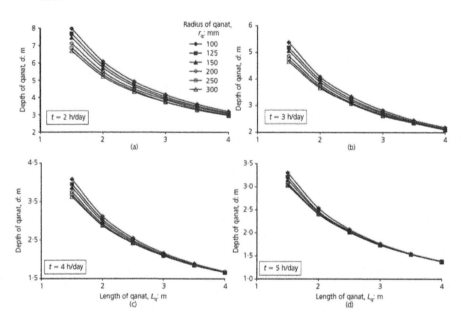

FIGURE 25.3   Variation of the depth of qanat at base versus length of qanat leg for pumping duration (in hours per day), t: (a) 2; (b) 3; (c) 4; and (d) 5.

The value of the recharge coefficient η may be reasonably estimated from the available literature (Wu and Zhang, 1994; Chaturvedi, 1936; Kumar, 2002). Wu and Zhzng (1994) calculated the effective precipitation P_e and the total amount of recharge P_e produced by P_e. By performing a regression analysis on P_e and R data available in China, a functional relationship between them was obtained as:

$$R_e = 0.87(P_e - 27.4) \qquad (12)$$

where R_e = Infiltration Recharge in mm;
        P_e = Effective Rainfall in mm.

Chaturvedi (1936) derived an empirical relationship to arrive at the recharge as a function of annual precipitation in Ganga-Yamuna doab basin as follows:

$$R = 2.0 \, (P - 15)^{0.4} \tag{13}$$

where, R = net recharge due to precipitation during the year (inch);
        P = annual precipitation (inch).

The formula of Chaturvedi (1936) was later modified further by Kumar and Seethapathi (2002), and the modified form of the formula is:

$$R = 1.35 \, (P - 14)^{0.5} \tag{14}$$

As observed from Figure 25.4, the factor of safety increases linearly with an increase in annual precipitation, which is well in agreement with the Eq. (8). Also, the curves relevant to those of Chaturvedi (1936) and Kumar and Seethapathi (2002) almost coincide. Both the magnitudes and the slope of the curve relevant to that of Wu and Zhzng (1994) are significantly high.

25.3.1.2   PERCOLATION POND

The runoff should be assessed accurately for designed the recharge structures and may be assessed by the following formula.

Runoff = Catchments area x Rainfall x Runoff coefficient

Runoff coefficient plays an important role in assessing runoff availability, and it depends upon the catchments characteristics. It is the factor that accounts for the fact that not all rainfall falling on catchments can be collected. Some rainfall will be lost from the catchments by evaporation and retention on the surface itself. The required area of recharge pond decreases in a hyperbolic manner with the depth of pond to be excavated, which is well in agreement with Eq. (9).

Recharge chamber with recharge well: The relevant calculations for recharge chamber with recharge well are described below. As observed from Figure 25.5, the number of recharge chambers N_W decreases fairly exponentially with the volume V_W of the recharge chambers. The rate of decrease is pronounced in the range of V_W, 5, beyond which a stabilizing tendency is noted.

Appropriate Engineering Design: The appropriate site-specific engineering design is suggested by the author following the analysis described above in details.

Qanat-well Structure: The following parameters are suggested:

- t = 3 hours.
- r_q = 125 mm

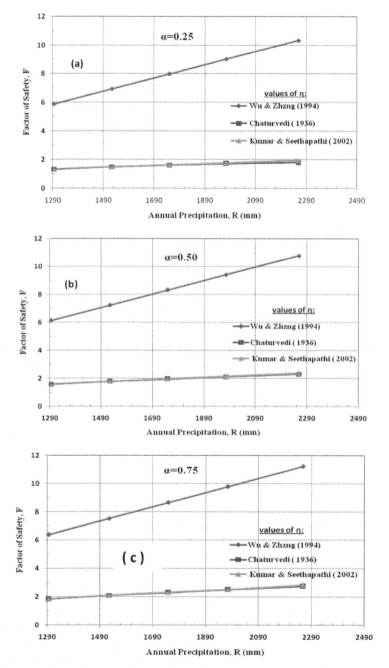

FIGURE 25.4   Variation of the factor of safety F with annual precipitation R for: (a) α = 0.25. (b) α = 0.50, and (c) α = 0.75.

- $L_q = 2.5$ m
- $d = 3.25$ m

(For future safety provision, 2 qanat-well structures are recommended for alternative use). Therefore, design depth of the qanat below G.L. = d + maximum depth of G.W.T. = 3.25 + 3 = 6.25 m.

Factor of Safety: The factor of safety for minimum and maximum rainfall are obtained as

Rainfall	Factor of Safety
Minimum	6.135 (after Wu and Zhzng, 1994)
	1.586 (after Chaturvedi, 1936)
	1.565 (after Kumar and Seethapathi, 2002)
Maximum	10.786 (after Wu and Zhzng, 1994)
	2.321 (after Chaturvedi, 1936)
	2.387 (after Kumar and Seethapathi, 2002)

For the Indian condition, the recommended value as per Kumar and Seethapathi (2002) is mostly suitable. It is also mentioned here that the value of factor of safety under minimum rainfall in the last 10 years should not fall below 1.0., therefore satisfactory in terms of reasonable compensation of withdrawal of groundwater. Also, under maximum rainfall in the last 10 years, the factor of safety exceeds 2.0, which may introduce sufficient pushback of saline water interface. It should also be mentioned here that for the excessively high value of factor of safety, adequate drainage should be facilitated at the site towards nearby stream channel.

Area and depth of Percolation Pond: The design rainfall data has been chosen as $R_M = 367.1$ mm. The depth of the pond is chosen as $H_p = 3$ m. Therefore, the area of pond required may be interpolated as, $A_p = 1202.436$ m^2. Hence, in the Contai Polytechnic College Campus, 4 ponds of area 301 m^2 each are provided, the depth of each pond is 3.0 m. Figure 25.6 illustrates the plots of recharge pond area versus pond depth for monsoon precipitation of (a) 350 mm, (b) 500 mm, and (c) 750 mm.

Recharge Chamber with Recharge Well: Adopting a total roof area in the site as 12000 m^2, $\alpha = 0.5$ and R = 2259 mm (maximum rainfall in last 10 years), keeping the recharge chamber dimensions as: L= 2 m, B= 2 m, H= 1.2 m, the number of recharge chamber with recharge well has been obtained as $N_W = 12$. It is also mentioned that the dimension of the recharge wells adopted here are as per the recommendation of Sarkar (2007) and are as follows:

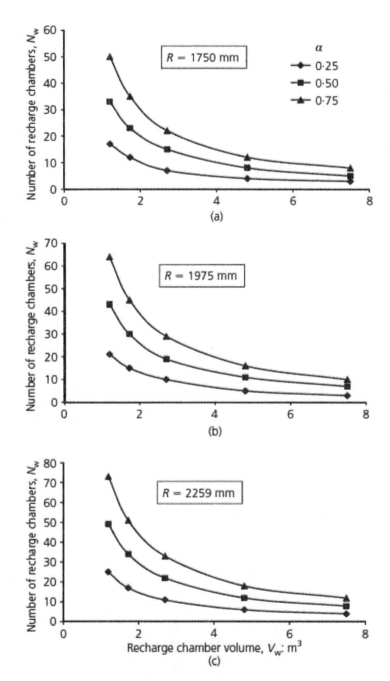

FIGURE 25.5 Variation of recharge chamber volume with number of chambers for (a) R = 1750 mm, (b) R = 1975 mm, (c) R = 2249 mm.

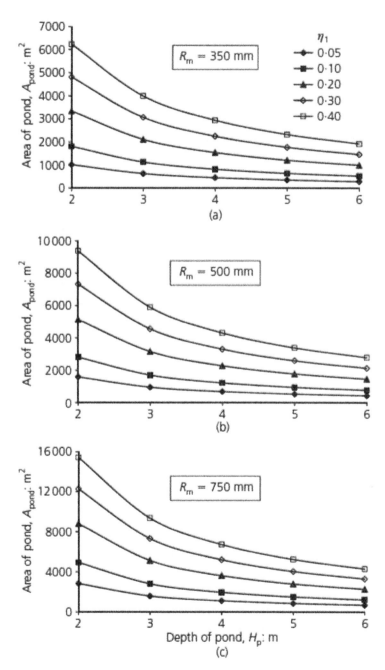

FIGURE 25.6   Plots of recharge pond area versus pond depth for monsoon precipitation of (a) 350 mm, (b) 500 mm, and (c) 750 mm.

Diameter of the well	100 mm
Strainer dia in the aquifer	100 mm
Length of strainer	12 m
Dia of the inlet strainer placed in the recharge chamber	150 mm
Length of 150 mm strainer	800 mm

As per design recommendation, the plan of the site provided with the new methodology has been given in Figure 25.7. The Contai Polytechnic College Campus area is shown with the location of qanat-well structures, percolation ponds and recharge chambers with recharge wells. The cross-section of recharge well adopted and methodology of recharge through recharge wells and recharging chambers are also shown inthe Figure 25.7. It is hereby mentioned that the parameters for recharge, well, etc. used in the design are highly site-specific. Successful application of the entire methodology is possible only when the design parameters are adequately chosen for a specific site.

25.4   CONCLUSIONS

The following significant conclusions are drawn:

An innovative method has been developed by the author for coastal zone groundwater management which involves withdrawal by qanat-well structures associates with equivalent artificial recharge by rainwater harvesting through percolation pond and recharges well. If adequately applied and design the proposed methodology is expected to be quite useful and convenient, for the unconfined condition of the coastal aquifers. As the suggested methodology has been applied to a selected coastal site located at Purba Midnapur district to study the various quantitative aspects of the method, it was observed that depth of qanat d decreases following a curvilinear pattern with the length of leg L_q. The variation is quite sharp in the range of $1m \leq L_q \leq 3m$ and assumes a linear pattern for $L_q > 3$ m.

The factor of safety for rainwater recharge in the selected location increases linearly with an increase in annual precipitation. Also, the curves relevant to those of Chaturvedi (1936), and Kumar and Seethapathi (2002) almost coincide. Both the magnitudes and the slopes of the curves relevant to that of Wu and Zhzng (1994) are significantly high. This technique is most effective when the value of the factor of safety is slightly higher than unity. The required area of recharge pond decreases in a hyperbolic manner with the depth of pond to be excavated.

The number of recharge chambers N_W decreases fairly exponentially with the volume V_W of the recharge chambers. The rate of decrease is pronounced

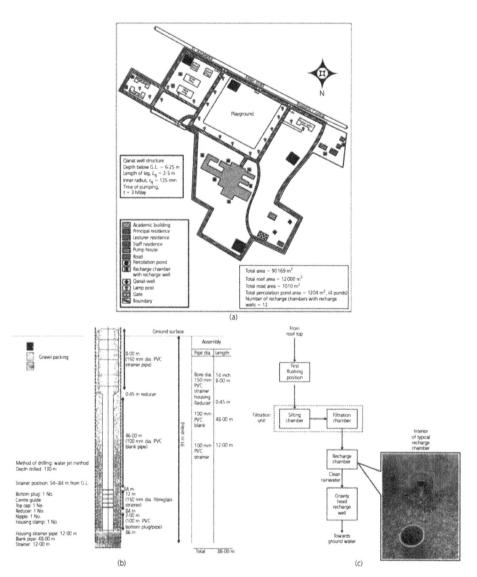

FIGURE 25.7 (a) Contai Polytechnic College Campus area showing the location of qanat-well structures, percolation ponds and recharge chambers with recharge wells; (b) Cross section of recharge well adopted; (c) Methodology of recharge through recharge wells and recharging chambers.

in the range of $V_W < 5$, beyond which a stabilizing tendency is noted. It is hereby mentioned that the model suggested is highly site-specific. Successful application of the entire methodology is possible only when the design parameters are adequately chosen for the particular site under consideration.

KEYWORDS

- coastal environment
- geotechnical investigation
- management and control of saline water intrusion
- mathematical analysis
- quantitative analysis

REFERENCES

Annual Climate Summery (2010). India Meteorological Department, Ministry of Earth Sciences, Government of India.

Bhattacharya, A. K., (2002). Saline water Intrusion into Coastal Aquifers of West Bengal, India, *International Conference on Low Lying Coastal Areas Hydrology and Coastal Zone Management, Bremerhaven* (pp. 197–200). The Federal Republic of Germany.

Bhattacharya, A. K., & Basack, S., (2003). *Analysis and Control of Saline Water Intrusion in Coastal Aquifers with Special Emphasis on East Coast of India.* Final Technical Report, AICTE No. 8022/RID/NPROJ/RPS–82–4 dated 22/3/2004.

Bhattacharya, A. K., & Basack, S., (2003). Chloride and arsenic contamination in groundwater in coastal areas of West Bengal. *Tenth West Bengal State Science and Technology Congress* (2nd edn.), Midnapur, West Bengal, India.

Bhattacharya, A. K., Basack, S., & Maity, P., (2004). A feasible method of groundwater extraction in the United Arab Emirates and Qatar. *Proceedings of, the International Conference on Water Resources – Flood Control, Irrigation, Drinking Water, Waterways, Electric Power and Transmission System* (pp. 170–175). New Delhi.

Bhattacharya, A. K., Basack, S., & Maity, P., (2004). Groundwater extraction in the United Arab Emirates under the constraint of saline water intrusion, *Journal of Environmental Hydrology, 12*(6), 1–5.

Bhattacharya, A. K., Basack, S., & Maity, P., (2010). Groundwater extraction in coastal arid regions-A design methodology using qanats, *Electronic Journal of IETECH, Journal of Civil and Structures, 3*(1), 017–020.

Blair, D. A., Spronz, W. D., & Ryan, K. W., (1999). Brackish groundwater desalination—A community's solution to water supply and aquifer protection, *Journal of the American Water Resources Association, 35*(5), 1201–1212.

Chaturvedi, R. S., (1973). A note on the investigation of groundwater resources in western districts of Uttar Pradesh. *Annual Report, U.P. Irrigation Research Institute,* 86–122.

Dagan, G., & Bear, J., (1968). Solving the problem of local interface upconing in a coastal aquifer by the method of small perturbations, *Journal Hydraulic Research, 6*(1), 15–44.

Drabbbe, J., & Ghyben, B., (1888). *Nota in verband met de voorgenoment putboring nabji Amsterdam, Tijdschrift van het koninklijk institute van ingenieurs*, The Hague, Netherlands.

Ghyben, W. B., (1888). Nota in verband met de voorgenomen putboring nabij Amsterdami, *Tijdschrift Kon. Inst. Ing.,* 8–22.

Herzberg, B., (1901). Wasserversorgung Einiger Nordseebader, *Journal of Gasbeleuchtung and Wasserversorgung,* Munich, *44.*

http://www.rainwaterharvesting.org/.

http://www.tn.gov.in./dtp/rainwater.htm.

Joshi, S. D., (1986). Augmentation of well productivity using slant and horizontal wells, *Paper SPE 15375, Presented at 61st Annual Technical Conference and Exhibition of the Society of Petroleum Engineers.* New Orleans, LA.

Karanth, K. R., (1990). *Groundwater Assessment Development and Management,*" Tata McGraw-Hill Publishing Co. Ltd., New Delhi.

Kumar, C. P., & Seethapathi, P. V., (2002). Assessment of natural groundwater recharge in upper Ganga canal command area, *Journal of Applied Hydrology, Association of Hydrologists of India, XV*(4), 13–20.

Lee, C. H., & Chang, T. S., (1974). On sea water encroachment in coastal aquifers, *Water Resources Research, 10,* 1039–1043.

Mays, L. W., (2001). *Water Resources Engineering*, John Wiley & Sons, N. Y.

Patra, M. N., (2006). *Localized and Generalized Subsidence and Swelling of the Ground Surface Due to Change in Groundwater Piezometric Level With a Special Reference to Calcutta.* PhD thesis, Bengal Engineering and Science University, Shibpur, Howrah, India.

Raghu, B. M., Prasad, B. R., & Srikanth, I., (2004). Subsurface skimming techniques for coastal sandy soils, *Bulletin No. 1/04: 18. A I C R P Saline Water Scheme*, Bapatla.

Raghunath, H. M., (1987). *Ground Water* (2nd edn.), Wiley Eastern Ltd., New Delhi.

Sarkar, A., (2007). Design of rainwater harvesting scheme for water conservation for firefighting and others and artificial recharge to groundwater at TCS main building campus, Salt Lake, Kolkata, *A Scheme Design Report as a Consultancy Submitted to Tata Consultancy Services Limited, Dec.*

Sarkar, S., (2005). *A Field Study on Saline Water Intrusion Into Coastal Aquifer of Purba Medinipur, M.E. Thesis.* Bengal Engineering and Science University, Howrah, India.

State Forest Report 2008–2009 Office of the Principal Chief Conservator of Forests, Kolkata, Directorate of Forests, Government of West Bengal, India.

United Nations Development Programme Report, (2006). Human Development Report 2006 Beyond Scarcity: Power, Poverty and Global Water Crisis, ISBN 0-230-50058-7.

United Nations Environment Programme Report, (2009). *Rainwater Harvesting: A Lifeline for Human Well-Being*, A report prepared for UNEP by Stockholm Environment Institute, ISBN: 978-92-807-3019-7, Job No. DEP/1162/NA.

Wood, E. F., Sivapalan, M., & Beven, K., (1986). Scale effects in infiltration and runoff production, *Proceedings of the Budapest Symposium, IAHS, No., 156,* 375–387.

World Health Organization Report, (2010). *Progress on Sanitation and Drinking Water – 2010 Update, WHO/UNICEF Joint Monitoring Programme for Water Supply and Sanitation.* ISBN 978-92-4-156395-6, NLM classification: WA 670.

Wu, J., & Zhang, R., (1994). Analysis of rainfall infiltration recharge to groundwater. *Proceedings of Fourteenth Annual American Geophysical Union: Hydrology Days* (pp. 420–430). Colorado, USA.

PART IV

Watershed Development and Management

CHAPTER 26

MORPHOMETRIC ANALYSIS AND PRIORITIZATION OF SUB-WATERSHEDS IN THE KOSI RIVER BASIN FOR SOIL AND WATER CONSERVATION

RAJANI K. PRADHAN, SWATI MAURYA, and PRASHANT K. SRIVASTAVA

Institute of Environment and Sustainable Development, Banaras Hindu University, Varanasi–221005, India, E-mail: rkpradhan462@gmail.com

ABSTRACT

Remote sensing and Geographical Information System are efficient techniques for prioritization of sub-watershed through morphometric analysis and positioning water harvesting structures. The morphometric analysis provides a quantitative description of a watershed for evaluating and estimation of drainage characteristics. The present study focused on Kosi River Basin (KRB), situated in Indo-Gangetic plains, Bihar state, India for soil and water conservation. The Digital Elevation Model (DEM) estimated from Shuttle Radar Topographic Mission (SRTM) is utilized for calculating the morphometric parameters in terms of basic, linear, and shape parameters and compound factor. KRB have been divided into 13 sub-watersheds and prioritized accordingly based on compound factor analysis to meet the need of water and soil conservation. The overall results indicate that the sub-watershed nos. 5, 8 and 4 require utmost attention and conservative practices because of their high erodibility characteristics. This study can be very useful in the identification of erosion-prone areas and implementation of conservation practices for water resources.

26.1   INTRODUCTION

The morphometric analysis involves the mathematical analysis of the configuration of the earth surface, form, and dimension of landforms (Ganie et al., 2016). Morphometric analysis of subbasin can be estimated through the measurement of liner, aerial, and relief aspects of the basin and slope contribution (Nag and Chakraborty, 2003). It is well established that the influence of drainage morphometry is very useful in understanding the landform process, soil physical properties and erosional characteristics (Rai et al., 2017). Morphometric analysis of the river basin also provides a quantitative description of the study area, which is an important aspect of the characterization of the basin (Magesh et al., 2013; Strahler, 1964). In addition, the hydrological process of a basin can be correlated with various morphometric parameters like size, shape, drainage density, size, and shape of the contributories (Rama, 2014).

Nowadays, remote sensing and GIS evolved as a promising technique for various watershed management and soil water conservation studies. It is a very effective tool for integration of various spatial data, preparation of various thematic layers and to derive useful outputs for modeling (Patel et al., 2013; Srivastava et al., 2012). Satellite products like digital elevation model (DEM) obtained from Shuttle Radar Topographic Mission (SRTM) can be used for the generation of the elevation model and stream network of the landscape. Number of studies have been conducted in morphometric analysis of river basin using both convention (Horton, 1945; Smith, 1950; Strahler, 1957; Strahler, 1964) and GIS and Remote sensing approach (Harini et al., 2017; Patel et al., 2013; Rai et al., 2017; Rao et al., 2010; Siddaraju et al., 2017). The morphometric analysis will provide a quantitative description of the basin for hydrological investigations like groundwater potential, watershed management, and environmental assessment. Thus, the detailed analysis of morphometric parameters of the river basin is of immense help in understanding the influence of drainage morphometry on landforms and their characteristics.

The main objectives of the present study are to prioritize the KRB for soil and water conservation. The linear and aerial parameters were estimated using various mathematical equations to analyze the morphometric parameters for the better planning and sustainable management of the river basin. It may also be useful in integrated decision making studies like soil erosion assessment, water management, and flood management.

26.2   MATERIALS AND METHODS

26.2.1   STUDY AREA

The river Kosi was also known as the "Sorrow of Bihar" is a tributary of river Ganga. It originates at an altitude of 7000 m from the Himalayas and meets the river Ganga near Kursela, in district Katihar of Bihar (Figure 26.1). Generally the entire catchment of river is divided into two parts, i.e., Upper catchment which is lies in Tibet and Nepal characterized by hilly area (constitute about 80% of total catchment area), whereas the lower catchment, falls in floodplain of north Bihar, India (about 20% of total catchment area). The ridge lines surround the Kosi river basin (KRB), and these ridgelines it from the Mahananda catchment in the east and from Gandak and Tsangpo (Brahmaputra) catchments in west and north respectively, whereas river Ganga bounds it in the in the south. From the past records, it concluded that that the river has laterally shifted westward about 150 km in last 200 years (Bapalu and Sinha, 2005; Gole and Chitale, 1966; Wells and Dorr, 1987) and this shifting has caused extensive damage to local inhabitants, their livelihood, infrastructure, and property. Following Nepal and Himalaya, it finally enters into the alluvial plains of state Bihar and travels nearly 320 kilometers from Chatra before it joins with Ganga. The current research work is carried out in part of the lower catchment of the Kosi which geographically lies between 86° 20' to 87° 10' East longitude and 25° 30' to 26° 30' north latitude and covering an area of 4062 km^2 in Bihar. The high intense rainfall (1200–2000 mm) in most parts (Sinha and Friend, 1994) is the main reason for the most of hydrological processes like soil erosion and thus a high amount of sediment load in the Kosi basin.

26.2.2   METHODOLOGY

The SRTM based digital elevation model (DEM; 90m resolution) is used for the extraction of topographic characteristics (relief) in this study. With the help of Arc GIS 10.1 spatial analysis extension and Arc Hydro tool, further analysis is carried out. The detail methodologies followed is presented in Figure 26.2. The pre-processing of DEM was performed and followed by post-processing as generating the DEM fills and flow direction, and flow accumulation was calculated for the individual pixel of the DEM using the Arc Hydro tool of ArcGIS10.1. Later a total of 13 sub-watersheds were delineated from the study area, and the morphometric parameters were analyzed for each subwatershed. Due to the simplicity and wide application of the Strahler's (1964) method of stream ordering, the same method was applied in this study. Further details of the formula and method used in this study for analysis of various morphometric parameters were listed in Table 26.1. Finally, a compound factor is used to rank the watershed for the prioritization of the various sub-watersheds.

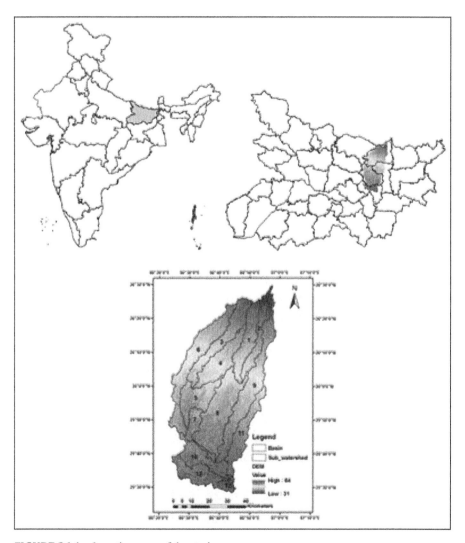

FIGURE 26.1   Location map of the study area.

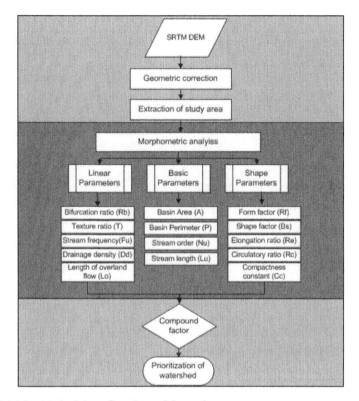

FIGURE 26.2 Methodology flowchart of the study.

26.3 RESULTS AND DISCUSSION

Morphometric analysis is a quantitative process, which constitutes the measurement of configurations of earth surface and dimensions of landforms. The morphometric parameters considered in this study for characterization of the Kosi watershed includes basic parameters, linear parameters and shape parameters. The formulas and equation used in this study for the analysis of various morphometric parameters were listed in Table 26.1.

26.3.1 BASIC PARAMETERS

The main basic parameters considered in this study for morphometric analysis includes drainage area, perimeter, stream order, stream length, and basin length.

TABLE 26.1 Morphometric Parameters Estimated for Kosi River Basin

S. No.	Parameters	Formula	References
1	Stream order (u)	Hierarchical rank	(Strahler, 1964)
2	Number of streams (Nu)	Total number of stream segment of the order u	(Strahler, 1957)
3	Stream length (L_u)	Total length of the stream segment of a particular order	(Horton, 1945)
4	Stream frequency (Fu)	$Fu = \sum Nu/A$	(Horton, 1932)
5	Length of overland flow (Lo)	$Lo = 1/2 * Dd$	(Horton, 1945)
6	Bifurcation ration (R_b)	$(R_b) = N_u/N_u + 1$	(Schumm, 1956)
7	Drainage density (D_d)	$D_d = \sum L_u/A$	(Horton, 1932)
8	Texture ratio (T)	$T = Nu/p$	(Horton, 1945)
9	Elongation ratio (R_e)	$R_e = D/L = 1.128\sqrt{A}/L$	(Schumm, 1956)
10	Circularity ratio (Rc)	$Rc = 4\pi A/P^2$	(Strahler, 1964)
11	Form factor (Rf)	$R_f = A/L^2$	(Horton, 1945)
12	Shape factor (Bs)	$Bs = L^2/A$	(Horton, 1932)
13	Basin length (Lb)	$Lb = 1.312 \times A^{0.5}$	(Ratnam, et al. 2005)
14	Compactness Constant (Cc)	$Cc = 0.282IP/A^{0.5}$	(Horton, 1945)

26.3.1.1   DRAINAGE AREA (A) AND PERIMETER (P)

Drainage area is the most important parameters of any watershed, and it is used to estimate the total volume of runoff and sediment load can be created from the basin. According to the result sub-watershed, no. 8 has the maximum area of 598.21 km^2 while the sub-watershed no. 13 has the minimum area of 51.54 km^2. The basin perimeter cab defined as the length of the line that defines the surface divide of the basin. In the present case, the maximum value of 376 km is found in sub-watershed no. 9, whereas the minimum of 51.08 km in sub-watershed no. 13.

26.3.1.2   STREAM ORDER (N_u)

Hierarchical position of the streams within the drainage basin can be defined through analysis of stream order. In this research work, stream order analysis is performed using Strahler method: where the first order streams does not consist of any tributaries, and the confluence of the two first-order streams formed the second order streams and so on. However, as the stream orders increase the total number of streams of the particular order decreases. The result outlines that the study area is a fifth order drainage basin and having a total number of 502 streams, sprawl over 4021.41 km². Among the 13 subwatershed, the subwatershed no.8 has the maximum number of streams of 79 followed by 70 in sub-watershed 9, while the sub-watershed no.13 has the lowest number of streams of 7. Further details about the number of streams in each order are presented in Table 26.2.

TABLE 26.2   Stream Order of Kosi River Basin

Sub-Watershed	I Order	II Order	III Order	IV Order	V Order	Total no. of streams
1	22	11	10	0	0	43
2	12	9	0	0	0	21
3	20	10	11	0	0	41
4	23	12	9	0	0	44
5	21	13	1	8	0	43
6	24	11	5	0	0	40
7	8	2	0	4	0	14
8	40	18	13	8	0	79
9	37	16	14	3	0	70
10	19	7	0	6	2	34
11	23	15	5	0	0	43
12	16	5	2	0	0	23
13	3	2	0	0	2	7

26.3.1.3   STREAM LENGTH (L_U) AND BASIN LENGTH (L_B)

Stream length has been computed by adding all the stream lengths of a particular order. Generally, streams having relatively smaller length indicate of the high slope, while the length of the large streamlength is of flatter gradient. In the present study subwatershed, no. 8 has the maximum stream length of about 268.60 km while the minimum of 15.52 km in sub-watershed no. 13. Basin length can be defined as the distance measured along the main channel from the watershed outlet to the basin divide. It is an important factor in the estimation of various morphometric analysis, and it is proportional to the drainage area. As per the result, the basin length of sub-watersheds varies from 49.5 km (sub-watershed no. 8) to 12.3 km (sub-watershed no. 13).

26.3.2   LINEAR PARAMETERS

It includes bifurcation ratio, drainage density, stream frequency, length of overland flow and texture ratio.

26.3.2.1   BIFURCATION RATIO (R_B)

It is the ratio between the number streams of a given order to a number of streams of the next higher order. It is a dimensionless property which measures the degree of distribution of stream network in the basin (Mesa, 2006; Soni, 2016) and is highly influenced by the geological characteristics of the drainage basin. The higher value of bifurcation ratio indicates a strong structural control on the drainage pattern, whereas the lower value reflects the basin less affected by structural disturbances. In the present study the maximum Bf ratio of 4.9 found in subwatershed no. 5, whereas the minimum of 1.2 in sub-watershed no. 13, which indicates low structural control on the basin drainage pattern. Table 26.3 shows the list of sub-watersheds and their corresponding bifurcation ratio in KRB.

26.3.2.2   DRAINAGE DENSITY (D_D) AND STREAM FREQUENCY (F_U)

Drainage density (Dd) is the ratio of the total length of the streams of the watershed to the total area of that watershed. It is an important factor for characterizing the degree of drainage development of the basin. High Dd of basin indicates impermeable surface, sparse vegetation, and steep slope while low Dd values reflect the permeable surface, dense vegetation, and flat relief. As per the result, the Dd values range from 0.6 to 0.29 which indicates permeable sub-surface material with low to intermediate drainage and relief. Stream frequency is the ratio of the total number of streams of all orders to the total area of the basin (Horton, 1932). Mainly stream frequency depends on the lithology of the basin, and the higher values of stream frequency indicate higher runoff and the steep ground surface, and a higher chance of flood occurrences. The values of stream frequency show a positive correlation with drainage density. In the present study stream frequency values varies from 0.158 to 0.05 and further details are presented in Table 26.3.

TABLE 26.3 Analyzed Morphometric Parameters

Sub-Water-shed	Area (km²) (A)	Perimeter (km) (P)	Length (km) (Lb)	Bifurcation ratio (Rb)	Texture ratio (T)	Stream frequency (Fu)	Drainage density (Dd)	Length of over land flow (Lo)	Form factor (Rf)	Shape factor (Bs)	Elongation ratio (Re)	Circulatory ratio (Schaap et al.)	Compactness constant (Cc)
1	274.32	177.80	31.832	1.550	0.124	0.157	0.466	0.233	0.271	3.694	0.587	0.109	2.067
2	203.42	182.75	26.859	1.333	0.066	0.103	0.600	0.300	0.282	3.547	0.599	0.077	2.518
3	333.61	228.87	35.574	1.455	0.087	0.123	0.455	0.227	0.264	3.793	0.579	0.080	2.381
4	317.81	180.27	34.607	1.625	0.128	0.138	0.485	0.242	0.265	3.768	0.581	0.123	1.928
5	286.79	169.31	32.646	4.913	0.124	0.150	0.509	0.255	0.269	3.716	0.585	0.126	1.920
6	514.93	318.83	45.521	2.191	0.075	0.078	0.356	0.178	0.249	4.024	0.562	0.064	2.592
7	127.53	144.22	20.602	2.250	0.055	0.110	0.539	0.269	0.300	3.328	0.618	0.077	2.591
8	598.21	253.62	49.566	1.744	0.158	0.132	0.457	0.228	0.243	4.107	0.557	0.117	1.894
9	568.29	376.80	48.143	2.707	0.098	0.123	0.473	0.236	0.245	4.078	0.559	0.050	2.897
10	238.96	166.13	29.432	1.905	0.114	0.142	0.401	0.200	0.276	3.625	0.592	0.109	2.089
11	269.91	207.49	31.540	2.267	0.111	0.159	0.368	0.184	0.271	3.686	0.588	0.079	2.435
12	236.09	168.08	29.231	2.850	0.095	0.097	0.327	0.164	0.276	3.619	0.593	0.105	2.128
13	51.54	51.08	12.315	1.250	0.059	0.136	0.297	0.149	0.340	2.943	0.658	0.248	1.535

26.3.2.3   LENGTH OF OVERLAND FLOW (LO) AND TEXTURE RATIO (T)

Length of overland flow is the length of the water flow over the ground before it combines into the mainstream and it is also expressed as half of reciprocal of the drainage density (Horton, 1945). It is one of the important independent variable, which affecting both hydrologic and physiographic development of the drainage in the basin. The lower the values of length of overland flow, the faster the runoff from the streams (Rama, 2014). In the present study, the values of length of overland flow varies from 0.3 to 0.14, and the mean is 0.22 which indicate ground slope with moderate infiltration and runoff. Texture ratio is defined as the ratio of first-order streams segments to the perimeter of the basin (Horton, 1932). It is an important factor in the morphometric analysis of drainage and is depend on the underlying lithology, infiltration capacity and relief of the basin or terrain. The values of texture ratio vary from 0.158 to 0.05 reflect low erosivity.

26.3.3   SHAPE PARAMETERS

26.3.3.1   SHAPE FACTOR (B_s)

Shape factor is the ratio of the square of basin length to the area of the basin. It reflects the shape irregularity of the drainage basin, and it is inversely related with the form factor. In this study area, shape factor ranges from 2.9 to 4.10, which indicates the shape of the basin is elongated. The form factor can be defined as the ratio between the areas of the basin to the square of the basin length. The form factor varies from 0 to 1, however mostly the values are less than 0.79 (for a perfectly circular basin) (Chopra et al., 2005). The lower the value of form factor the more the elongated of the basin, and have lower peak flow in longer duration, while the higher value indicates the more circular shape of the basin with higher peak flow in short duration. In the present case, the values of form factor vary from 0.24 to 0.33, which indicates the more elongated shape of the basin and thus may take a longer time duration for peak flow.

26.3.3.2   ELONGATION RATIO (R_e)

It is defined as the ratio of the diameter of a circle of the same area as the basin to the maximum basin length (Schumm, 1956). It is a most significant index for the analysis of the basin shape. According to Strahler these values varies from 0.6 to 1.0 for a wide range of climatic and geological conditions (Strahler, 1964). Further the lower the value of this factor means the more elongated the basin shape. In the present case, the values of elongation ratio lie between 0.55 to 0.65 reflect the elongated shape of sub-basins

26.3.3.3   COMPACTNESS COEFFICIENT (C_c) AND CIRCULATORY RATIO (R_c)

Compactness coefficient can be expressed as, basin perimeter divided by the circumference of a circle to the same area of the basin. It is quite the opposite (inversely related) to the elongation ratio a basin and responsible for causing erosion in the basins. The lesser the value of compactness coefficient means the more elongated the shape of the basin and less erosion while, a higher value indicates less elongated and more erosion-prone of the basin. It was also observed that the values of compactness coefficient exhibit a variation from 2.89 to 1.535, shown in Table 26.3. Circularity ratio is the ratio between the basin areas to the area of a circle having the same circumference as the perimeter of the basin. It is equivalent to 1 when the basin is perfectly circular, and it lies between 0.4–0.5 when the basin shape is more elongated in shape and very permeable with homogeneous materials. As per result, the maximum circularity ratio was observed in subwatershed no.13 (0.248) while the minimum in subwatershed no.9 (0.05), which reflects the highly elongated shape of the basins.

26.3.4   COMPOUND FACTOR AND PRIORITIZED RANKS

In this study, a compound factor is used for the prioritization of subwatershed in the KRB. For this, both Shape parameters and linear parameters are taken into consideration. Linear parameters are directly correlated to the erosion (higher the value, the more erodible), whereas as in the case of shape parameters it is vice versa (lower the value, the more erodible). First of all individual ranking is assigned to both linear and shape parameters depending on the parameters values and afterward, finally, a compound factor is obtained by summing up all the parameters ranking divided by the number of parameters. From the compound factor, the first rank is assigned to the lowermost value, and the last rank to the higher most value. In the present case, watershed no.

5 ranked first (4.9) followed by watershed no. 8 and 4 with second and third respectively, whereas the watershed no. 13 have last rank (9.7) and details are demonstrated in Table 26.4 and Figure 26.3 as well.

TABLE 26.4 Calculation of Compound Factor and Prioritized Ranks

Sub-Watershed	Rb	Dd	Fu	T	Lo	Rf	Bs	Re	Rc	Cc	Compound factor	Prioritized rank
1	10	6	2	4	6	7	7	7	9	5	6.5	5
2	12	1	11	11	1	11	3	11	3	10	7.4	9
3	11	8	9	9	8	4	10	4	6	8	7.7	10
4	9	4	5	2	4	5	9	5	11	4	5.8	3
5	1	3	3	3	3	6	8	6	12	3	4.9	1
6	6	11	13	10	11	3	11	3	2	12	8.2	11
7	5	2	10	13	2	12	2	12	4	11	7.3	8
8	8	7	7	1	7	1	13	1	10	2	5.4	2
9	3	5	8	7	5	2	12	2	1	13	6.0	4
10	7	9	4	5	9	9	5	9	8	6	7.1	7
11	4	10	1	6	10	8	6	8	5	9	6.7	6
12	2	12	12	8	12	10	4	10	7	7	8.3	12
13	13	13	6	12	13	13	1	13	13	1	9.7	13

26.4 CONCLUSIONS

The quantitative analysis of morphometric parameters of the river basin is an immense help in evaluations of the hydrological process of the basin (runoff, infiltration capacity, etc.), effective management of the natural resources and prioritization of the sub-watershed. Remote sensing satellite data integrated with GIS evolved as a promising technique for various water management studies at the watershed level. In the present study, a detailed morphometric analysis of the KRB is carried out using the integration of GIS and Remote sensing techniques for the prioritization sub-watershed. There are a total of 13 sub-basin delineated in the catchment. According to the result, the river basin has a well-drained structure with a fourth order basin. The basin characterized with an elongated shape and flattered slope, which indicates a lower chance of flooding. The overall analysis indicates that the sub-watershed nos. 5, 8, and 4 require utmost attention for the conservation measure, while the sub-watershed nos. 13, 12 and 6 have lowered priority. This study can be utilized for implementation effective conservation measures to the vulnerable sites, which will reduce the high runoff and soil erosion, thus for the flood control in the KRB.

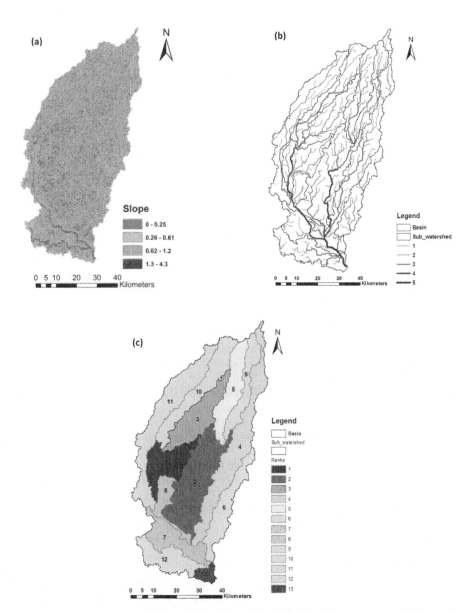

FIGURE 26.3 (a) Slope (b) Drainage order and (c) Prioritization of sub-watersheds from Compound factor.

KEYWORDS

- **GIS**
- **Kosi river basin**
- **morphometric analysis**
- **prioritization**
- **SRTM (DEM)**

REFERENCES

Bapalu, G. V., & Sinha, R., (2005). GIS in flood hazard mapping: A case study of Kosi River Basin, India. *GIS Development Weekly, 1*(13), 1–3.

Chopra, R., Dhiman, R. D., & Sharma, P., (2005). Morphometric analysis of sub-watersheds in Gurdaspur district, Punjab using remote sensing and GIS techniques. *Journal of the Indian Society of Remote Sensing, 33*(4), 531–539.

Ganie, P. A., Posti, R., Kumar, P., & Singh, A., (2016). Morphometric analysis of a Kosi River Basin, Uttarakhand using geographical information system. *Int. J. of Multidisciplinary and Current Research, 4.*

Gole, C. V., & Chitale, S. V., (1966). Inland delta building activity of Kosi river, *Journal of the Hydraulics Division, Proceedings of the American Society of Civil Engineers, 111–126.*

Harini, P., Manikyamba, C., Kumar, S. D., Durgaprasad, M., & Nandan, M., (2017). Geographical information system based morphometric analysis of Krishna River Basin, India. *Journal of Applied Geochemistry, 19*(1), 44.

Horton, R. E., (1932). Drainage-basin characteristics. *Eos, Transactions American Geophysical Union, 13*(1), 350–361.

Horton, R. E., (1945). Erosional development of streams and their drainage basins, hydrophysical approach to quantitative morphology. *Geological Society of America Bulletin, 56*(3), 275–370.

Magesh, N., Jitheshlal, K., Chandrasekar, N., & Jini, K., (2013). Geographical information system-based morphometric analysis of Bharathapuzha river basin, Kerala, India. *Applied Water Science, 3*(2), 467–477.

Mesa, L., (2006). Morphometric analysis of a subtropical Andean basin (Tucuman, Argentina). *Environmental Geology, 50*(8) 1235–1242.

Nag, S., & Chakraborty, S., (2003). Influence of rock types and structures in the development of drainage network in hard rock area. *Journal of the Indian Society of Remote Sensing, 31*(1), 25–35.

Patel, D. P., Gajjar, C. A., & Srivastava, P. K., (2013). Prioritization of malesari mini-watersheds through morphometric analysis: A remote sensing and GIS perspective. *Environmental Earth Sciences, 69*(8), 2643–2656.

Rai, P. K., Mohan, K., Mishra, S., Ahmad, A., & Mishra, V. N., (2017). A GIS-based approach in drainage morphometric analysis of Kanhar River Basin, India. *Applied Water Science, 1–16.*

Rama, V. A., (2014). Drainage basin analysis for characterization of 3rd order watersheds using geographic information system (GIS) and ASTER data. *J. Geomatics, 8*(2), 200–210.

Rao, N. K., Latha, S. P., Kumar, A. P., & Krishna, H. M., (2010). Morphometric analysis of Gostani River basin in Andhra Pradesh state, India using Spatial information technology. *International Journal of Geomatics and Geosciences, 1*(2), 179.

Ratnam, K. N., (2005). Check dam positioning by prioritization of micro-watersheds using SYI model and morphometric analysis—remote sensing and GIS perspective. *Journal of the Indian Society of Remote Sensing, 33*(1), 25.

Schumm, S. A., (1956). Evolution of drainage systems and slopes in badlands at Perth Amboy, New Jersey. *Geological Society of America Bulletin, 67*(5), 597–646.

Siddaraju, K., Nagaraju, D., Bhanuprakash, H., Shivaswamy, H., & Balasubramanian, A., (2017). Morphometric evaluation and sub-basin analysis in Hanur watershed, Kollegal Taluk, Chamarajanagar district, Karnataka, India, using remote sensing and GIS techniques. *International Journal of Advanced Remote Sensing and GIS*, 2178–2191.

Sinha, R., & Friend, P. F., (1994). River systems and their sediment flux, Indo-Gangetic plains, Northern Bihar, India. *Sedimentology, 41*(4), 825–845.

Smith, K. G., (1950). Standards for grading texture of erosional topography. *American Journal of Science, 248*(9), 655–668.

Soni, S., (2016). Assessment of morphometric characteristics of Chakrar watershed in Madhya Pradesh India using the geospatial technique. *Applied Water Science,* 1–14.

Srivastava, P. K., Han, D., Gupta, M., & Mukherjee, S., (2012). Integrated framework for monitoring groundwater pollution using a geographical information system and multivariate analysis. *Hydrological Sciences Journal, 57*(7), 1453–1472.

Strahler, A. N., (1964). Quantitative geomorphology of drainage basin and channel networks. In: Ven Te Chow, editor. *Handbook of Applied Hydrology.* New York (NY): McGraw-Hill; p. 4–39.

Strahler, A. N., (1957). Quantitative analysis of watershed geomorphology. *Eos, Transactions American Geophysical Union, 38*(6), 913–920.

Wells, N. A., & Dorr, J. A., (1987). Shifting of the Kosi river, northern India. *Geology, 15*(3), 204–207.

CHAPTER 27

ANALYSIS OF URBAN DRAINAGE SIMULATIONS OF AN IMMENSELY URBANIZED WATERSHED USING THE PCSWMM MODEL

SATISH KUMAR[1], D. R. KAUSHAL[2], and A. K. GOSAIN[2]

[1]*PhD Scholar, Indian Institute of Technology Delhi, Hauz Khas, New Delhi–110016, India, E-mail: satish.kumar140@gmail.com*

[2]*Professor, Indian Institute of Technology Delhi, Hauz Khas, New Delhi–110016, India*

ABSTRACT

Flooding has caused immense damage to the people as well as to the property. Flooding in urban areas mostly occurs due to increased urbanization, low rate of infiltration and poor infrastructure for stormwater drainage network. Stormwater Management Model (SWMM) is found to be very dynamic hydrology-hydraulic water quality simulation model for modeling of the urban stormwater drainage network. In the present study, PCSWMM model is used for modeling the stormwater drainage network for the southern part of Delhi, the capital city of India. PCSWMM is developed by Computational Hydraulics International (CHI), Canada. PCSWMM uses the same SWMM engine for the modeling work; the only advantage is that it is GIS compatible software which makes this model more efficient. The model required following input information for simulation, i.e., land-use for calculating impervious and previous area, soil type, 15-minute interval precipitation data, temperature, humidity, and three-dimension cross-sectional geometry of the existing drainage network. A field survey was carried out for data collection, and in the process, it was found that most of the storm-water drains are choked, have improper flow gradient or damaged. All the collected field details of the storm-water drains were incorporated in ArcMap 10.1 and then imported in PCSWMM to develop a hydrology-hydraulic model for surface runoff. The simulated results of the model were further calibrated and validated with the available flooding locations data obtained from the Delhi Traffic Police Department. The simulated results were in close agreement with the observed flooding locations. Thus PCSWMM model can be applied to any urban/rural areas for designing stormwater drains or drainage network.

27.1   INTRODUCTION

Urban drainage is described as the process of collecting and transporting wastewater, rain /stormwater or a combination of both. Urban sewer or stormwater is a part of the urban infrastructure. Urban drainage is gaining importance in recent years. Properly designed and operated urban drainage systems with its interactions with other urban water systems are a crucial element of the healthy and safe urban environment. Urban flooding problems range from minor one where water enters the basements of a few houses to major incidents where large parts of cities are inundated for several days. Most modern cities in the industrialized part of the world usually experience small-scale local problems mainly due to insufficient capacity in their drainage systems during heavy rainstorms (Schmitt et al., 2004). Cities in other regions, including those in South/South-East Asia, often have more severe problems because of much heavier local rainfall and lower drainage standards. This situation continues to get worse because many cities in the developing countries are growing rapidly but without the funds to extend and rehabilitate their existing drainage systems. Moreover, New Delhi, capital of India, due to fast urbanization and rapid migration without simultaneous progress in drainage condition is suffering from serious waterlogging issues which result in loss of the property, infrastructure, and severe traffic jams. For solving such big and complex watershed problems, various type of mathematical models is used nowadays by the researchers/designers.

For managing runoff generated in the urban catchments, several models are available like MOUSE (DHI), HEC–1(US army), Hydro works (HRWL), SWMM (Storm Water Management Model) and PCSWMM (CHI). Out of this PCSWMM was identified as the GIS-based powerful model for modeling urban drainage networks because it precisely matches the simulated results with the observed data. PCSWMM used SWMM model for modeling stormwater drainage systems. SWMM is a dynamic rainfall-runoff model for the simulation of quantity and quality complications related to urban catchments runoff (Huber and Dickinson, 1992). EPA had developed SWMM (Storm Water Management Model) in the year 1791 (Metcalf and Eddy Inc., 1971), which is extensively used to simulate all features of urban hydrologic and water cycles. The features include surface runoff, rainfall, and drainage network's flow routing, snowmelt, and pollution concentrations (Huber and Dickinson, 1992). Further, simulation through single or continuous events can be executed for the catchments where combined sewer drains or natural drains or stormwater drain exists.

For managing the urban stormwater system, distributed rainfall-runoff models are used. But the main difficulties in using these models are that developing of these models is a time taking the process and also about the model accuracy. PCSWMM is the most dynamic rainfall-runoff model in which runoff and pollution are generated from the subcatchments goes directly to manholes or to the downstream subcatchments. Thus, spatial discretization of the catchments is required by the rainfall-runoff model for developing runoff model computationally. Yu et al. (2001) reported that for developing rainfall-runoff model three essential elements are considered, i.e., characteristic of catchments, loss due to abstraction and simple equations of flow. In the urban subcatchments complicated processes like infiltration rate of surfaces, flows through overflow, the concentration of runoff due to depression is involved in modeling surface runoff. The factors like a watershed, surface area, flow streams, and outlets play an important role in forecasting the runoff in urban catchments. Traditional methods of discretizing in which the urban catchments were done manually on the basis of topographic maps which is a very slow process as well as not accurate. Geographic Information System (GIS) is a very powerful tool which has the spatial analysis function. GIS can be used for developing the hydrological model. For modeling urban stormwater, the GIS application can be used for storage, handling, evaluating, and demonstrating data in GIS form (Seth et al., 2006). GIS technology can help us in generating input parameters like digital elevation models (DEM), land-use maps, soil imperviousness maps and drainage network maps for developing a rainfall-runoff model (Seth et al., 2006). However, some of the parameters can be taken directly from GIS layers which can be used in SWMM for modeling rainfall-runoff. For developing a rainfall-runoff model, many researchers had used GIS for generating subcatchments (Yu et al., 2001; Du, 2007). But the above concepts cannot be used for the urban areas due to the complexity of the surface. Zaghloul (1981) reported that for obtaining better modeling results from SWMM largely depends on the catchment's discretization. PCSWMM is found to be the efficient tools for modeling urban stormwater system because it is a GIS-based tool. All the GIS layers can be imported into PCSWMM and also subcatchments discretization for an urban area is done very accurately.

In this present study, PCSWMM has been used for the first time for modeling stormwater drains of the urban area (Delhi city, India) and the results obtained from simulation is in good agreement with the observed data.

27.2 MATERIALS AND METHODOLOGY

27.2.1 STUDY AREA

Delhi is located in northern India between the latitudes of 28°24'17" and 28°53'00" North and longitudes of 76°50'24" and 77°20'37" East having an average elevation of 233 m (ranging from 213 to 305 m) above the mean sea level. A study area is a small catchment in Barapullah basin which come under Aravalli hills ranges in the southern part of Delhi is shown in Figure 27.1. Delhi is a monsoon-influenced humid subtropical with high variation between summer and winter temperatures and precipitation. Summers start in early April and peak in May, with average temperatures near 32°C, although occasional heat waves can result in highs close to 45°C on some days and therefore higher apparent temperature. The monsoon starts in late June and lasts until mid-September, with about 797.3 mm of rain. The average temperatures are around 29°C, although they can vary from around 25°C on rainy days to 32°C during dry spells. The monsoons recede in late September, and the post-monsoon season continues until late October, with average temperatures sliding from 29 C to 21°C. The catchment covers an area of 5.79 Sq Kms and has a population of about 1.8 lacs and is a typical commercial/residential area. The Digital Elevation Model (DEM), land-use layer and soil type used in the present study was provided by the Geospatial Delhi Limited (GSDL). The lands of the study area are completely urbanized, and soil type is loam soil. Rainfall data for the last 5 years (2009 to 2013) obtained from the Indian Meteorological Department, Lodhi Road, New Delhi, India was used for the simulation process.

27.2.1.1 STORMWATER DRAINAGE SYSTEM

From stormwater drainage point of view, Delhi can be divided into six drainage basins, ultimately discharging into river Yamuna, namely–Najafgarh Drain, Barapulaah Nallah, Wildlife sanctuary area discharging thro' Haryana, Drainage of Shahdara area and other drains directly out falling into river Yamuna. The NCT of Delhi is prone to flooding from river Yamuna via Najafgarh drain. The low-lying Yamuna floodplains (Khadar) are also prone to recurrent floods.

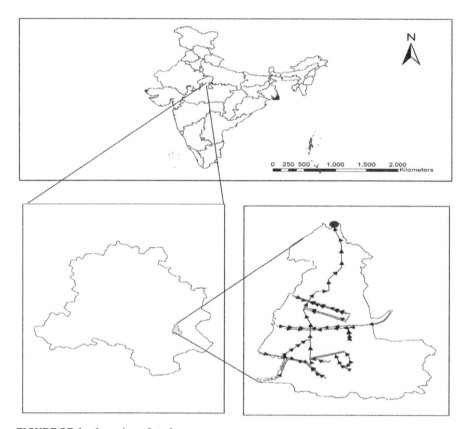

FIGURE 27.1   Location of study area.

Due to fast urbanization in Delhi during last four decades resulting in an increase in paved area and a decrease in the agricultural land which used to act as a percolation (Net Agriculture area shown in 1950–51 was 97,067 ha in 2005–2006 is just 25,000 ha out of total 148,300 ha). Delhi normally remained flooded to the extent of 70,000 ha (50% of its geographical area of 148,300 ha from 1953 to 1984). The Capital of India has suffered floods as back as in 1924, 1947, 1967, 1971, 1975, 1976, 1978, 1988, 1993, 1995, 1998, 2010, etc. The 1978 was the worst ever flood in Delhi when the water level reached at 207.49 m (danger level is 204.83 m) with discharge 2.53 lac cusec at old railway bridge (7.0 lac cusec discharge was released from Tajewala).

Barapullah Nallah is the main stormwater drain which is carrying all the water from Barapullah basin of south Delhi to Yamuna River. The stormwater drainage system in South Delhi is distributed between South Delhi Municipal Corporation, New Delhi Municipal Council, Public Works Department, and Delhi Development Authority. The major challenges caused by the stormwater drainage system in the city is due to the poor infrastructure of the drainage networks.

27.2.2 *PCSWMM*

PCSWMM software was developed by Computational Hydraulics Int. (CHI) in collaboration with the Environmental Protection Agency (EPA). PCSWMM is the graphical decision support system for the USEPA SWMM program. This software offers a wide range of files and time series management, model development and calibration, demonstration of dynamic output obtained from simulation and other related tools to the stormwater designer. More GIS attributes had been added to the new version of PCSWMM. In the latest version, anything related to stormwater modeling can be obtained. PCSWMM is developed by six elements, i.e., rainfall, runoff, temperature, extran, and storage. PCSWMM helps in dividing the urban areas into subcatchments on the basis of DEM and landuse. Infiltration takes place when the water is received by the subcatchments can be explained by Green Ampt, SCS curve number or Horton methods. Surface runoff takes place if water is not taken by the subcatchments and transported to the final outfall points via conduits and manholes. Steady flow, kinematics flow, and dynamic flow routing are the three types of flow routing processes which are used in the PCSWMM are same as in SWMM. Manning's equation explains the relation among cross-section area, discharge, gradients, and hydraulic radius. A full description of PCSWMM can be found in the theoretical documentation by James et al. (2010). The incident rainfall intensity is the input to the control volume on the surface of the plane; the output is a combination of the runoff Q and the infiltration f. Considering a unit breadth of the catchment the continuity and dynamic equations which have to be solved are as shown in equations below.

$$Q = B\frac{C_m}{n}S^{1/2}(y-y_d)^{5/2} \text{ (Dynamic equation)} \tag{1}$$

$$L = (fL+\frac{Q}{B})+L\frac{\Delta_y}{\Delta_t} \text{ (Continuity equation)} \tag{2}$$

where, L = overland flow length, B = catchment breadth, C_m = 1.0 for metric units = 1.49 for Imperial or US customary units, n = Manning roughness coefficient, y_d = surface depression storage depth and f = infiltration.

27.2.3   MODEL SETUP

The surveyed data of the stormwater drains for the study area was converted into conduits layers and manholes layers with the help of ArcGIS 10.2. The drains information like invert level, width, depth, latitude, and longitude, conduits length was available in the attributes table. Then the outfall point layers of the drains were generated in ArcGIS. All these layers were imported into the PCSWMM. In PCSWMM, watershed, and sub-catchments were generated from the DEM on the basis of manholes. The subcatchments was delineated in such a fashion that runoff generated from each subcatchments will contribute to single manholes. Further, the land-use and soil were given to the catchments. And finally, rainfall station information and rainfall data were given to the PCSWMM. After loading all the necessary input data required by the PCSWMM, the model was initially run for the day having peak rainfall.

27.3   RESULTS AND DISCUSSION

The rainfall-runoff model was developed using PCSWMM is shown in Figure 27.2. Precipitation data from the years 2008 to 2012 of 15 minutes interval received from the Indian Meteorological Department, New Delhi was used for the simulation. Between 2008 to 2012 years, the maximum rainfall recorded was 109.2 mm on 20th August, 2010. Since, our objective is to find out whether the existing infrastructure is sufficient for extreme events, so the model was simulated and the results obtained was evaluated for the above days. Time series plot of the rate of flow of the drains, flooding rate and velocity of runoff is shown in Figure 27.3. For validating the model efficiency, the waterlogging locations identified by the model was checked with the past records. The waterlogging locations information is available on the Delhi police website. It was found that the simulated results are in close agreement as shown in Figure 27.4. The waterlogging may be caused due to various reasons like inadequate drainage infrastructure like drains are undersized due to which waterlogging takes place at the junctions or downstream. The other reasons are the rainfall intensity and duration of rainfall. The most important reasons are the improper gradient of drains, i.e., invert levels of the drains is not uniform due to which flow is not able to go forward and hence, backflow takes place causing waterlogging in the upstream.

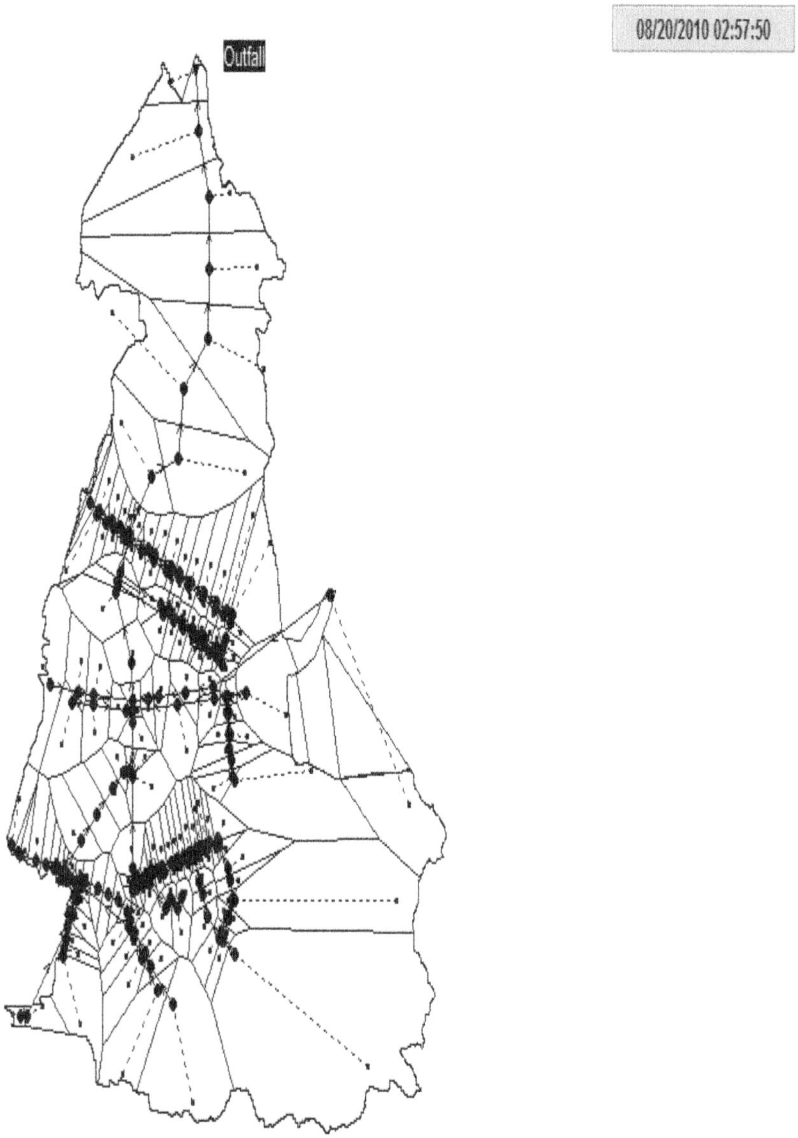

FIGURE 27.2 Rainfall-runoff model.

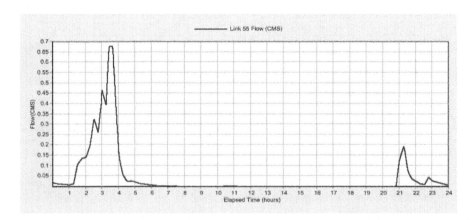

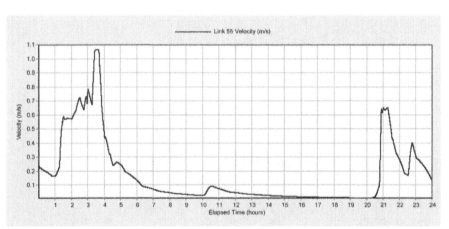

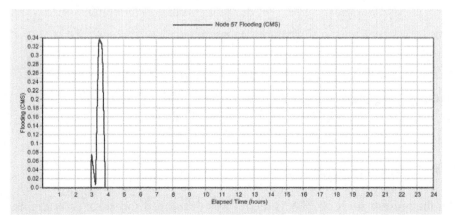

FIGURE 27.3   Showing rate of flow, flooding rate, and velocity.

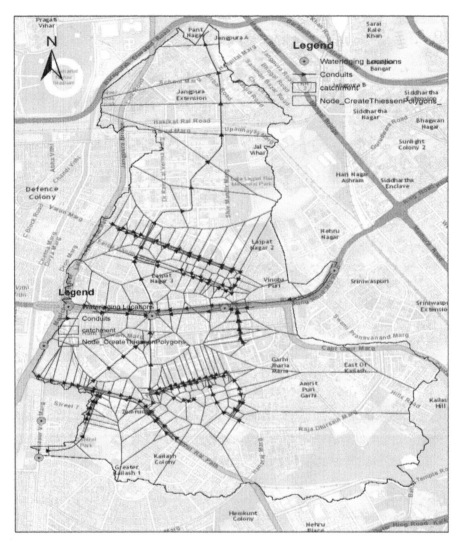

FIGURE 27.4   Simulated results with observed data.

27.4   CONCLUSIONS

In the present study, PCSWM model was developed for stormwater drains of the small watershed of Delhi city. The model is also identifying the exact location of waterlogging locations. The watershed was divided into

192 subcatchments using DEM which makes PCSWMM a unique model. Thus PCSWMM found to be GIS friendly model which is very efficient for modeling any size of the urban drainage system.

ACKNOWLEDGMENT

The authors would like to thanks, Delhi Government, India for providing data to do this research. The authors would also like to thanks, Computational Hydraulics International (CHI), Canada for providing PCSWMM software as an honorarium to Indian Institute of Technology Delhi. Last but not the least, the author's thanks, Indian Institute of Technology Delhi for all the financial support given during the research.

KEYWORDS

- **GIS**
- **PCSWMM**
- **stormwater drainage**
- **surface runoff**
- **urban flooding**

REFERENCES

Du, J. K., Xie, S. P., Xu, Y. P., Xu, C. Y., & Singh, V. P., (2007). Development and testing of a simple physically-based distributed rainfall-runoff model for storm runoff simulation in humid forested basins. *Journal of Hydrology, 306*, 334–346.

Huber, W. C., & Dickinson, R. E., (1992). *Storm Water Management Model User's Manual, Version 4.* Environmental Protection Agency, Georgia.

Huber, W., (1992). Contaminant transport in surface water. In: Maydment, Dr., (ed.), *Handbook of Hydrology.* McGraw-Hill, New York.

James, W., (2005). *Rules for Responsible Modeling* (4th edn.). Computational Hydraulic International Press, Guelph, Ontario, Canada.

James, W., Rossman, L. A., & James, W. R. C., (2010). *User's Guide to SWMM 5*, Computational Hydraulic International Press, Guelph, Ontario, Canada.

Metcalf and Eddy Inc., (1971). *Storm Water Management Model, Final Report.* US Environmental Protection Agency.

Santhi, C., Arnold, J. G., Williams, J. R., Dugas, W. A., Srinivasan, R., & Hauck, L. M., (2001). Validation of the SWAT model on a large river basin with point and nonpoint sources. *Journal of the American Water Resources Association, 37*(5), 1169–1188.

Schmitt, T. G., Thomas, M., & Ettrich, N., (2004). Analysis and modeling of flooding in urban drainage systems. *Journal of Hydrology*, 300–311.

Seth, I., Soonthornnonda, P., & Christensen, E. R., (2006). Use of GIS in urban storm-water modeling. *Journal of Environmental Engineering, 132*(12), 1550–1552.

Tsihrintzis, V., & Hamid, R., (1998). Runoff quality prediction from small urban catchments using SWMM. *Hydrological Processes, 12*(2), 311–329.

Yu, P. S., Yang, T., & Chen, S. J., (2001). Comparison of uncertainty analysis methods for a distributed rainfall-runoff model. *Journal of Hydrology, 244*, 43–59.

Zaghloul, N. A., (1981). SWMM model and level of discretization. *Journal of the Hydraulics Division, 107*(11), 1535–1545.

CHAPTER 28

RAINFALL FORECASTING USING A TRIPLE EXPONENTIAL SMOOTHING STATE SPACE MODEL

SWATI MAURYA and PRASHANT K. SRIVASTAVA

Institute of Environment and Sustainable Development, Banaras Hindu University, Varanasi–221005, India, E-mail: mauryaswati35@gmail.com

ABSTRACT

The prediction of rainfall is too complex due to its high spatial and temporal variability. Study of variation in rainfall is important as it also influences the hydrological cycle of the Earth. While in India, 60–90% rainfall occurs in the monsoon period (June-September), and our agricultural system is dependent on this for rainfall for most of the water requirements. Therefore, this study focuses on the quantitative method for rainfall forecasting such as triple exponential smoothing state space model (tESM) for rainfall prediction in Dhariawad catchment of Mahi River Basin in Rajasthan, India. The tESM is enriched with model level, trend, and seasonal decomposition for efficient forecasting of meteorological parameters. In this study, the long-term data of rainfall during the period 1997–2012 (16 years) were used for optimization and forecasting of the rainfall. In total 12 years (1997–2008) were used for model calibration and 4 years (2009–2012) were kept separately for validation purpose. The overall data analysis indicates that the tESM performed better in terms of root-mean-square error (RMSE = 75.31 mm), index of agreement (d = 0.86) and percent bias (PBIAS = –35.5). The application of tESM forecast methods indicates that it is easy to implement and provide results that are more appropriate in lesser time.

28.1   INTRODUCTION

Rainfall plays a major role in the hydrological system. Rainfall mainly influences the water resource development, agricultural activities, disasters, economy of the region, etc. (Gajbhiye et al., 2016). Thus, the forecast of rainfall is critically important for understanding the variation in rainfall. The variation in rainfall alters stream flow, groundwater resources, distribution of rainfall, soil moisture content, etc. Therefore, regular monitoring of rainfall is necessary for proper management of natural hazards such as flood, drought, landslide, cloud bursting and in agriculture for improving productivity, irrigation scheduling, etc.

For rainfall forecasting, time series analysis is the most common method used in the recent past, and various technique proved to provide more appropriate results for evaluating the variation of rainfall over the region (Tularam et al., 2010). In time series analysis two forms are very common such as trend and seasonality (Kalekar, 2004). The trend defines as a systematic linear or nonlinear component that changes over time and not repeats in the period of time range captured by the data. The seasonality almost similar nature; but it repeats itself in regular intervals over time. These two basic classes of time series components may coexist in the data. The present works focus on the triple Exponential Smoothing Models (tESM) for rainfall forecasting. The main reason for the popularity of these methods that they are simple and easy to put in practice and don't need a fitting parameter (Gelper et al., 2010). It was based on a repetitive computing scheme for update the forecasts for each new upcoming observation.

28.2   STUDY AREA

The Dhariawad catchment geographical lies in 74.48^0 longitudes, 24.09^0 latitudes of Mahi River Basin in Rajasthan, India. The average annual rainfall is 813 mm. The area of this catchment is 1510 sq km. Maximum rainfall occurs in the monsoon period, and climatic conditions are tropical to sub-tropical. Geographical diversity varies from highly dense forests to hilly terrain and plateau. Major crops are wheat, maize, soya bean, gram, garlic, and opium. The soil types are silt loam to clay loam and greyish brown in color. Further small-scale mining activities are in operation extracting mainly Red Ochre, Calcite, Dolomite, Quartz, Feldspar, and Soapstone. Marble, Building-stone, and Limestone are also available in small quantities. The Shuttle Radar Topography Mission (SRTM) is a collaboration of National

Aeronautics and Space Administration (NASA) and the National Geospatial-Intelligence Agency (NGA) for topographical analysis. It provides elevation datasets for the globe at 1 arcsec (approx. 30 m) and at 3-arcsec (approx. 90 m). For evaluating the drainage, patterns freely available SRTM DEM (90 m) datasets are used (http://www.cgiarcsi.org). Geographical location of the study area shown in Figure 28.1.

28.3   MATERIALS AND METHODS

28.3.1   DSSAT

Decision Support System for Agrotechnology Transfer (DSSAT) of precipitation data of 16 years during the period (1997–2012). DSSAT is a NASA POWER (Prediction of Worldwide Energy Resources) database have 1° latitude by 1° longitude grid with global coverage and provides global modeled meteorological dataset in the standard format of International Consortium for Agricultural Systems Applications (ICASA) and American Standard Code for Information Interchange (ASCII) (White et al., 2011). DSSAT database measures daily meteorological parameters based on satellite observations and assimilation models. The main advantage of DSSAT are required only latitude and longitude of particular regions and provide all meteorological data. Precipitation data were downloaded from http://power.larc.nasa.gov/cgibin/cgiwrap/solar/agro.cgi. In the 16 years of monthly datasets, 12 years (1997–2008) were used for model calibration and 4 years (2009–2012) for validation.

28.3.2   TRIPLE EXPONENTIAL SMOOTHING MODEL

This method is used when the data shows level, trend, and seasonality. These set of equations is called the "Holt-Winters" (HW) method after the names of the inventors (Kalekar, 20004; Willmott, 1981). For Additive seasonality measurement, the following equations are used.

In this model, the time series is represented as below.

$$y_t = b_1 + b_{2t} + S_t + \varepsilon_t \tag{1}$$

where; b_1 is the base signal also called the permanent component; b_2 is a linear trend component; S_t is a seasonal additiveseasonal factor; ε_t is the

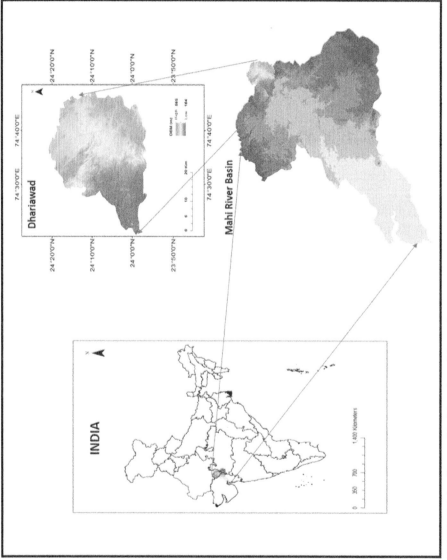

FIGURE 28.1   Geographical location of the study area.

random error component. Let the length of the season be L periods. The seasonal factors are the sum to the length of the season, i.e.

$$\Sigma_{1 \le t \le L} \, S_t = 0 \tag{2}$$

The trend component b_t if deemed unnecessary, may be deleted from the model. After selecting a model, the next step is its specification. The process of specifying a forecasting model first of all select the variables to be included, selected the form of the equation of relationship, and estimated the values of the parameters in that equations.

28.3.2.1   PROCEDURE FOR UPDATING THE ESTIMATES OF MODEL PARAMETERS

28.3.2.1.1   Overall Smoothing

$$R_t = \alpha(y_t - S_{t-L}) + (1 - \alpha) * (R_{t-1} + G_{t-1}) \tag{3}$$

R_t estimate of the deseasonalized level.
G_t estimate of the trend
S_t estimate of seasonal component (seasonal index)

where $0 < \alpha < 1$ is a smoothing constant.

Dividing y_t by S_{t-L}. It is the seasonal factor for T period calculated the one season (L periods) ago, deseasonalizes the data, and the prior value of the permanent component enters into the updating process for R_t.

28.3.2.1.2   Smoothing of the Trend Factor

$$G_t = \beta * (S_t - S_{t-1}) + (1 - \beta) * G_t - 1 \tag{4}$$

where $0 < \beta < 1$ is a second smoothing constant. Trend component estimated by smoothing the difference between two successive estimates of the deseasonalized level. The estimate of the trend component is the simply smoothed difference between two successive estimates of the deseasonalized level.

28.3.2.1.3   Smoothing of the Seasonal Index

$$S_t = \gamma * (y_t - S_t) + (1 - \gamma) * S_{t-L} \tag{5}$$

where $0 < \gamma < 1$ is the third smoothing constant.

The estimate of the seasonal component is a combination of the most recently observed seasonal factor given by the demand y_t divided by the deseasonalized series level estimate R_t and the previous best seasonal factor estimate for this time period. Since seasonal factors represent deviations above and below the average, the average of any L consecutive seasonal factors should always be 1. Thus, after estimating S_t, it is good practice to renormalize the L most recent seasonal factors such that

$$\sum\nolimits_{i=t-q+1}^{t} S_i = q \tag{6}$$

28.3.2.1.4   Value of Forecast

The forecast for the next period given by

$$y_t = R_t - 1 + G_{t-1} + S_{t-L} \tag{7}$$

28.3.3   STATISTICAL EVALUATIONS

The three statistics, root-mean-squared error (RMSE), index of agreement (d) and percent bias (%Bias) are used. RMSE defined as the average of error magnitude. The index of agreement (d) is the ratio between the mean square error and the "potential error" (Willmott, 1981). The potential error is the sum of the squared absolute values of the distances from the predicted values to the mean observed value and distances from the observed values to the mean observed value. Its range varies from 0 to 1. The calculated value 1 express a perfect agreement between the measured and predicted values, while 0 expresses the no agreement at all (Willmott, 1981). Percent bias (%Bias) measures the average tendency of the simulated values to be larger or smaller than their observed value. The optimal value of %Bias is 0.0; with low-magnitude, values express that accurate model simulation. Root-mean-square error (RMSE) can be defined as;

$$RMSE = \sqrt{\frac{1}{n}\sum\nolimits_{i=1}^{n}[y_i - x_i]^2} \tag{8}$$

Degree of Agreement can be defined as;

$$d = 1 - \frac{\sum\nolimits_{i=1}^{n}(x_i - y_i)^2}{\sum\nolimits_{i=1}^{n}(|y_i - x| + |x_{i-x}|)^2} \tag{9}$$

Percent Bias can be defined as;

$$\%Bias = 100 * [\Sigma(y_i - x_i) \Sigma(x_i)] \tag{10}$$

Where n is the number of observations, x is observed values and y is forecasted values.

28.4   RESULTS AND DISCUSSIONS

The trend of rainfall during the period (1997–2012) is shown in Figure 28.2 (a). The pattern of rainfall is variable throughout the years. But the value of rainfall is almost same in 1997–2002 while the value of rainfall suddenly drops down in 2003. After this, the value of rainfall gradually increases as the year progresses from 2003 to 2012. In the seasonal pattern, up, and down steps occur in a regular time interval of whole 17 years. That means repeating pattern obtained over consecutive periods. While scattering plots for validation in between observed and model value shown in Figure 28.2 (b). In this plots, the value indicates a moderate deviation between the observed and model values from the equiline thus resultant in a moderate positive correlation. The rainfall forecasting by tESM is shown in Figure 28.3, with the prediction intervals at 80 and 90%. The values of prediction intervals are necessary because it's defined the uncertainty in the forecasts. Further visual inspection of Figure 28.3 clear that narrow confidence boundary is evident in tESM. That means parameters are concentrated in a smaller area and perform better.

The performances of the tESM model are also evaluated by the three statistical parameters RMSE, d, PBIAS for forecasting the rainfall as shown in Table 28.1. The value of RMSE for tESM is 75.3, which indicate a better agreement in between the observed and forecasted values. Further the value of the degree of agreement (d = 0.8) close to 1, which shows a perfect agreement between the observed and forecasted value. Moreover, the value of percent bias (PBIAS = –33.45) is close to zero; it is indicated that low value had better perform in a model simulation. Thus, the overall analysis indicates that the chosen model provided almost accurate results for forecast rainfall and suggested that rainfall increases gradually in the Dhariawad catchment.

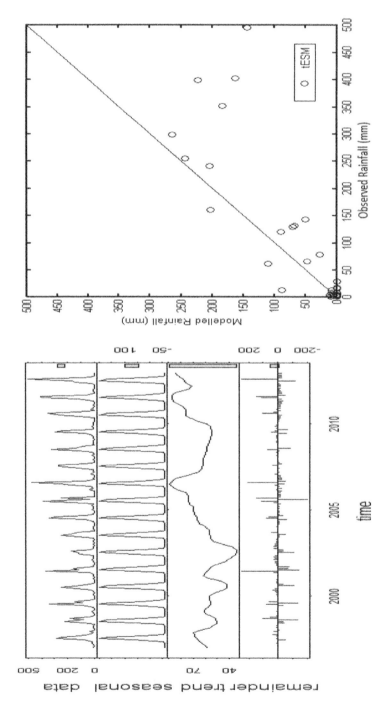

FIGURE 28.2 (a) Rainfall trend of Dhariawad catchment during the period of 1997–2012. (b) Scatter plot of validation datasets.

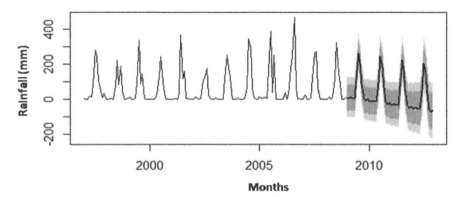

FIGURE 28.3 Forecast results of rainfall (1997–2012) using tESM modeling approach with prediction intervals.

TABLE 28.1 Performance Statistics of the tESM Modeling Approaches for Forecasting Rainfall

tESM					
RMSE(mm)	75.31	PBIAS	−35.5	d	0.86

RMSE, root-mean-square error; *d*, index of the agreement; PBIAS, Percent bias.

28.5 CONCLUSIONS

Rainfall is a major factor in agriculture and in recent years, interest has increased in learning about precipitation variability and predictability for periods of months to years. The effects of climatic change and variability have been analyzed by many researchers and proved that rainfall pattern uncertain in various region. Therefore, our prime concerned to understand the variability of rainfall time to time. In this study, an overall analysis of long-term series (1997–2012) of rainfall by tESM in the Dhariawad catchment indicated that rainfall time series fluctuate randomly and after 2004 increase gradually. The reason behind the rainfall increases are climate change but not assure because of some other factors also influence the rainfall such as land use, land cover, unlimited growth of population and meteorological parameters, etc. The future aspect of this study helped in various field of hydrology such as flood forecasting, water resource management, etc.

KEYWORDS

- **forecasting accuracy**
- **monsoon rainfall**
- **tESM**

REFERENCES

Gajbhiye, S., (2016). Precipitation trend analysis of Sindh River basin, India, from the 102-years record (1901–2002). *Atmospheric Science Letters*, *17*(1), 71–77.

Gelper, S., Fried, R., & Croux, C., (2010). Robust forecasting with exponential and Holt-Winters smoothing. *Journal of Forecasting*, *29*(3), 285–300.

Kalekar, P. S., (2004). Time series forecasting using holt-winters exponential smoothing. *Kanwal Rekhi School of Information Technology. 4329008*, 1–13.

Tularam, G. A., & Ilahee, M., (2010). Time series analysis of rainfall and temperature interactions in coastal catchments. *Journal of Mathematics and Statistics. 6*, 372–380.

White, J. W., (2011). Evaluation of satellite-based, modeled-derived daily solar radiation data for the continental United States. *Agronomy Journal, 103*(4), 1242–1251.

Willmott, C. J., (1981). On the validation of models. *Physical Geography*, *2*(2), 184–194.

CHAPTER 29

IMPROVING IRRIGATION WATER USE EFFICIENCY: A SOLUTION FOR FUTURE WATER NEED

PRABEER KUMAR PARHI

Center for Water Engineering and Management, Central University of Jharkhand, Brambe, Ranchi – 835205, India,
E-mail: prabeer11@yahoo.co.in

ABSTRACT

As productive irrigated land base and available fresh water are declining day by day, there is a need for optimal use of land and water resources so that crop yield is maximized per unit volume of water per unit land area per unit time. There is a greater possibility to raise the productivity of water through the management of soil water than with genetic approaches (biotechnology and breeding). Properly and carefully managed deficit irrigation strategies on agronomic crops would provide the greatest potential for substantially reducing agricultural water use. Scientific irrigation scheduling has the potential to improve the ratio of yield to consumptive use primarily because of the optimized quantity and timing of water applications. High-value and early mature crops may also produce some water savings through various deficit irrigation strategies, Geographically relocating certain crops to their most productive areas and soils, thereby minimizing irrigation amounts and maximizing overall efficiencies would be the most economically efficient use of water resources. Advanced irrigation technologies, precision agriculture tools and state-of-the-art delivery systems, needs to be fully implemented for successful deficit irrigation management as water savings under nonstress conditions (maximum yield) using site-specific technologies have a potential to bring down the water need to the order of 15% to 30%.

29.1   INTRODUCTION

The good quality, non-saline fresh water suitable to support human health and enterprise constitutes only about 1% of the water available worldwide. It is estimated that as a major consumer of the total freshwater, irrigation alone accounts for about four-fifths of the total freshwater. With the rapid increase in human population, more, and more water has to be allocated towards irrigated agriculture to satisfy the increasing food needs. Further, the increasing water needs due to rapid urbanization, environmental consciousness, recreation, tourism, and related concerns are to be taken care of. In this context, it is important to mention that an estimated 60% of the global population is going to suffer from water scarcity by 2025 (Qadir et al., 2007). Recently, the United Nations in its report (UNESCO, 2006) has estimated that increased cropping intensity to meet world food needs will require an increase of 40% in the area of harvest crops by 2030, and that the amount of water allocated to irrigated agriculture must increase correspondingly by 14%. However, it not assured that the needed water will be available by that time.

Further, Global climate change is a source of uncertainty having its potential impacts on temperatures, annual precipitation levels, and regional rainfall distribution patterns and increased water demand. Crop producers could face a decrease in the quantity and temporal availability of water supplies. The timing of precipitation and runoff from mountain snowmelt would differ from historical norms, leading to more frequent and sustained intense weather events such as droughts in some areas and floods in others. Hydrologic uncertainties would be compounded leading to modifications in precipitation and temperature having disproportionately larger effects on crop evapotranspiration (ET). The combination of these factors would force changes in the distribution of existing cropping patterns. Hence, it is free from any doubt that the alarming pressure on water resources ensures that water will be the primary natural resource issue of the 21st century (Seckler and Amarasinghe, 2000). Already there are economic and social pressures on all water users to reduce irrigation water amounts. This obviously calls fora novel approaches to water management and systems will be needed to address the declining land base and water allocations to balance production needs. Hence it is the need of the hour to identify measures through which agricultural users can economically adopt advanced irrigation schemes and implement practices to improve irrigation efficiency and water productivity in sustainable agriculture.

In the above context, the objectives of the present study are to review (1) the ability of irrigated agriculture to meet the growing population food needs, (2) the concepts of water use efficiency, and (3) the ways to improve water use efficiency (WUE).

29.2   BASIC CONCEPTS AND DEFINITIONS OF WUE

The present discussion emphasizes on methods to use less water for crops while maintaining or even increasing total crop productivity by enhancing the efficient use of water through improved management and advanced irrigation technologies. Various strategies to minimize water loss include redesigning total irrigation systems for higher efficiency, successfully treating and reusing degraded waters, reducing evaporation losses, introducing site-specific water applications, implementing managed-deficit irrigations, and employing engineering techniques to minimize leaching.

So far as WUE or the crop water productivity is concerned, it is basically defined as an input/output ratio to measure productivity. In other words, it is also defined as the crop productivity per unit volume of water used. Though many different definitions of WUE have been offered by large number agricultural scientists, WUE is the generally considered as the ratio of the harvested biomass to the water consumed to achieve that yield (Steduto, 1996). According to Viets (1962), WUE is the yield of interest (e.g., grain, biomass) divided by the water used to produce that yield. Howell (2001) pointed out that the denominator (water used) is difficult to measure and suggested that water used to be estimated from effective rainfall plus irrigation plus the change in soil water content. Bos (1980) suggested WUE as the yield benefit from irrigation divided by the irrigation water applied. WUE has also been defined in terms of the dry matter harvest index (ratio of yield biomass to the total cumulative biomass at harvest) as the yield of interest divided by water use (Howell et al., 1990)

Moreover, WUE is primarily considered as biological response ratio rather than an efficiency term, due to which many people are now referring to this concept as crop water production (Howell, 2006; Steduto et al., 2007). Monteith (1984, 1993) criticized the WUE term and pointed out that no theoretical limits exist as a reference, as should be the case for efficiency in an engineering sense. The WUE is alsobe greatly influenced by the timing of water applications as is evident in supplemental irrigations in humid areas. Irrigation at just the right time to avoid water stress can have a very large increase in WUE. Hence as a management practice, this term is very often used as a key indicator to compare values of WUE from different places with similar climatic conditions, geospatial characteristics and crop characteristics.

29.3   OPTIONS TO IMPROVE WATER USE EFFICIENCY (WUE)

29.3.1   IMPROVING MANAGEMENT CAPACITY

Minimization of the negative effects of water deficits on yields and quality through management of soil water (management of appropriate irrigation timing) can raise the productivity of water in irrigated agriculture. This can be achieved through the proper design of water delivery and farm irrigation system, which can help farmers to apply the right amount of water at the right place for all irrigations. While deciding irrigation timing optimal use of available rainwater is important.

Management of irrigation water under severe to moderate soil water deficit conditions during the growing period of crops can increase crop productivity while reducing the amount of water applied. This implies economically optimization of production for each unit of water used can maximize crop productivity while reducing the amount of water applied. Hence it is clear that properly managed deficit irrigation strategies can reduce agricultural and urban water use and conserve water to an appreciable extent. However, it needs excellent control of the timing and amounts of the applied water.

29.3.2   ADOPTING SCIENTIFIC IRRIGATION SCHEDULING

Scientific irrigation scheduling refers to the application of an optimum quantity of water at a most appropriate time so that crop stress and over-application is avoided. The method improves the ratio of yield to consumptive use (water productivity), primarily because of improved timing of water applications. However, it is always essential to combine the effects of scheduling with improved farming practices that typically accompany an on-farm irrigation-scheduling program. The irrigation scheduling can be further improved by converting irrigation system from gravity surface irrigation to pressurized drip or sprinkler systems.

29.3.3   CROP SELECTION

In arid and semiarid areas where there is a scarcity of water resulting in frequent partial season droughts, it is always advisable to choose crops that mature more quickly, such as small grains, cool season oilseeds (e.g., mustards, camelina), or various pulse crops such as peas and lentils. Shifts to deep-rooted, drought-resistant crops such as sunflower and safflower may also occur to maximize use of precipitation stored in the soil. However, longer season crops such as maize (corn) may have reduced yields.

29.3.4   SUPPLEMENTAL IRRIGATION PRACTICE

Supplemental irrigation is a tactical measure to complement reasonably sufficient rainfall and stabilize production despite short-term droughts. However, this practice is important primarily in arid and semiarid areas where it may be possible to apply only one or two irrigations per season. This is a form of managed deficit irrigation where the impact of the timing and applications of limited water supplies relative to only rain-fed agriculture can be very positive (Sojka et al., 1981; Zhang and Oweis, 1999). These techniques imply applying water during critical growth stages so that there is an optimal benefit per unit volume of water per unit quantity of crops of interest produced.

29.3.5   SPATIALLY OPTIMIZING PRODUCTION

Another option for improving the productivity of water is its spatial optimization. Spatially optimal land use includes geographically relocating certain crops to their most productive areas and soils, thereby minimizing irrigation amounts and maximizing overall efficiencies. Relocating specified crops to climatic regions and soil types best suited to maximal output would be the most economically efficient use of resources.

29.3.6   ADVANCED IRRIGATION TECHNOLOGIES

Almost every aspect of irrigation has seen significant innovation, including diversion works, pumping, filtration, conveyance, distribution, application methods, drainage, power sources, scheduling, erosion control, land grading, soil water measurement, and water conservation. Originally, irrigation was accomplished by methods utilizing gravity to distribute and apply water. Sprinkler technology was enhanced by the development of low-cost aluminum and later PVC pipe, and currently, more land is irrigated in the United States by sprinklers than by gravity methods. High-frequency drip irrigation and other micro-irrigation methods have been shown to increase the yield and quality of fruit and vegetable crops through reduced water and nutrient stresses. Tied to an effective soil water monitoring program, good design, and appropriate management practices, micro irrigation can have an application efficiency of 95% or better without drought stress, and is now used on about 5% of the irrigated area in the United States (NASS, 2002).

29.3.7   IMPROVED IRRIGATION SYSTEMS

The major reasons to go for system improvements are to reduce labor by automation, minimizing water costs by conservation (higher irrigation efficiencies) and expanding irrigated area with the same diverted water volume (irrigation capacity). Further, there are several management options for reducing water losses. Making small pits or basins (mini reservoirs), commonly called furrow diking, in sprinkler irrigated fields to hold water can be beneficial. Irrigation at night can reduce evaporation losses. Weeds are a major nonbeneficial use of water, and their control is critical, but chemical control is costly and may have unwanted environmental consequences. The use of mulches for weed control may reduce non-beneficial ET and soil evaporation. Reduced tillage techniques can reduce soil evaporation losses. Drip irrigation technologies can conserve water by greatly reducing soil evaporation and maximizing crop water productivity. These strategies could also incorporate alternative cropping systems including winter crops and deep-rooted cultivars that maximize use of stored soil water and some nutrients.

29.3.8   SITE-SPECIFIC IRRIGATION

Site-specific technologies to maximize crop yield per unit volume of water have a great potential in arid and semi-arid regions. Experiments show that potential water savings under non-stress conditions (maximum yield) using site-specific technologies are probably on the order of about 5% or less, but maybe in the range of 15% to 30% (Sadler et al., 2005). By aligning irrigation water application with variable water requirements in the field, total water diversions may be reduced and, almost certainly, deep percolation and surface runoff can be reduced.

29.3.9   MICRO IRRIGATION

Micro-irrigation is an extremely flexible irrigation method, and it offers the potential for high levels of water savings because of precise, high-level management. Due to its high cost and intensive management requirement currently, its use is restricted to relatively small fields. This method can be applied to almost all cropping situation, climatic zone and over a wide range of terrain. This type of irrigation suits best where soils are of very low or very high infiltration rates and salt-affected soils. However micro irrigation is used on less than 1% of lands worldwide, primarily because of its recent development and high initial capital cost. Micro-irrigation has the potential for use on most agricultural crops, although it is most often used with high-value specialty crops such as vegetables, ornamentals, vines, berries, olives, avocados, nuts, fruit crops, and greenhouse plants because of its relatively high cost and management requirements. The use of micro irrigation is increasing around the world, and it is expected to continue to be a viable irrigation method for agricultural production in the near future. With increasing demands on limited water resources and the need to minimize environmental consequences of irrigation, this technology will undoubtedly play an important role in the future irrigation scenario.

29.3.10   SYSTEM-LEVEL OPTIMIZATION

To maximize WUE, integration of all the above components and optimization of the whole system on a specific zone level is needed. The maximum production per unit volume of water is possible if all terms that do not produce yield were eliminated, leaving all water for productive use. However, in the systems perspective, some uses that do not contribute immediately to yield increases can lead to long-term yield. For example, leaching requirement in arid and semiarid areas never contribute to yield immediately, but irrigation without some leaching can eventually lead to soil salination. Similar is the case of evaporation as it helps meet the energy balance, and loss from one system may be the water supply to another system at watershed or catchment level.

29.4   CONCLUSIONS

Irrigated agriculture contributes nearly 40% of total food and fiber production worldwide. But, the paradox is that the productive irrigated land base and available water is declining day by day. This necessitates the need for optimal use of land and water resources so that crop yield is maximized per unit volume of water per unit land area. Studies show that there is a greater possibility to raise the productivity of water and minimize of the negative effects of water deficits on yields and quality through management of soil water than with genetic approaches (biotechnology and breeding). Properly managed deficit irrigation strategies can reduce agricultural and urban water use and conserve water to an appreciable extent; however, it needs excellent control of the timing and amounts of the applied water.

Scientific irrigation scheduling can improve the ratio of yield to consumptive use primarily because of the optimized quantity and timing of water applications. However, the scientific scheduling and improved farming practices can be combined together to yield best results. Carefully managed deficit irrigation on agronomic crops would provide the greatest potential for substantially reducing agricultural water use. High-value and early mature crops may also produce some water savings through various deficit irrigation strategies. Geographically relocating certain crops to their most productive areas and soils, thereby minimizing irrigation amounts and maximizing overall efficiencies would be the most economically efficient use of water resources.

Advanced irrigation technologies, precision agriculture tools and state-of-the-art delivery systems, needs to be fully implemented for successful deficit irrigation management. Water savings under non-stress conditions (maximum yield) using site-specific technologies have a potential to bring down the water need to the order of 15% to 30%. Hence, in the era of increasing water needs, there is an urgent need to explore the specific knowledge and technologies required to minimize water use while maintaining reasonable production levels to satisfy all the needs for food, fiber, feed, and fuels in addition to environmental, recreation, and municipal requirements.

KEYWORDS

- **advanced irrigation technology**
- **irrigated agriculture**
- **irrigation management**
- **water productivity**
- **water use efficiency**

REFERENCES

Bos, M. G., (1980). Irrigation efficiencies at the crop production level, *ICID Bull.*, *29*(2), 18–25, 60.

Howell, T. A., (2001). Enhancing water use efficiency in irrigated agriculture, *Agron. J.*, *93*(2), 281–289.

Howell, T. A., (2006). *Challenges in Increasing Water Use Efficiency in Irrigated Agriculture, Paper Presented at International Symposium on Water and Land Management for Sustainable Irrigated Agriculture*, Adana, Turkey.

Howell, T. A., Cuenca, R. H., & Solomon, K. H., (1990). Crop yield response, in the management of farm irrigation systems, *Am. Soc. of Agric. Eng.*, 93–122.

Monteith, J. L., (1984). Consistency and convenience in the choice of units for agricultural science, *Exp. Agric*, *20*, 105–117.

Monteith, J. L., (1993). The exchange of water and carbon by crops in a Mediterranean climate, *Irrig. Sci.*, *14*(2), 85–91, doi: 10.1007/ BF00208401.

National Agricultural Statistic Service (NASS), (2002). *Ranch Irrigation Survey* (Vol. 3). Special studies Part 1, U.S. Department of Agriculture, Washington, D.C.

Qadir, M., Sharma, B. R., Bruggeman, A., Choukr-Allah, R., & Karajeh, F., (2007). Non-conventional water resources and opportunities for water augmentation to achieve food security in water-scarce countries, *Agric. Water Manage.*, *87*(1), 2–22.

Sadler, E. J., Evans, R. G., Stone, K. C., & Camp, C. R., (2005). Opportunities for conservation with precision irrigation, *J. Soil Water Conserv.*, *60*(6), 371–379.

Seckler, D., & Amarasinghe, A., (2000). Water supply and demand, 1995 to 2025: Water scarcity and major issues, in Annual Report 1999–2000. *World Water Vision, 3*, 9–17.

Sojka, R. E., Stolzy, L. H., & Fischer, R. A., (1981). Seasonal drought responses of selected wheat cultivars, *Agron.*, *73*, 838–845.

Steduto, P., (1996). In: Pereira, L. S., et al., (eds.), *Water Use Efficiency, in Sustainability of Irrigated Agriculture* (Vol. 312, pp. 193–209). NATO ASI Ser., Ser. E, Kluwer Acad. Dordrecht, Netherlands.

Steduto, P., Hsiao, T. C., & Fereres, E., (2007). On the conservative behavior of biomass water productivity, *Irrig. Sci.*, *25*(3), 189–208.

U.N. educational, scientific, and cultural organization, water, a shared responsibility, Rep. 2, World water assess. The programme, Berghann, New York, 2006.

Viets, F. G. Jr., (1962). Fertilizers and the efficient use of water, *Adv. Agron.*, *14*, 223–264.

Zhang, H., & Oweis, T., (1999). Water-yield relations and optimal irrigation scheduling of wheat in the Mediterranean region, *Agric. Water Manage*, *38*(3), 195–211.

CHAPTER 30

RAINFALL VARIABILITY AND EXTREME RAINFALL EVENTS OVER JHARKHAND STATE

R. S. SHARMA and B. K. MANDAL

Meteorological Centre Ranchi, India Meteorological Department,
B. M. Airport Road, Hinoo, Ranchi–834002, India,
E-mail: radheshyam84@rediffmail.com

ABSTRACT

Rainfall is the primary source of surface and groundwater recharge. The state receives 91% of its annual rainfall due to the southwest monsoon, which is its principal rainy season. The contribution of winter, Pre-Monsoon, and Post-Monsoon season's rainfall amount to about 2%, 3%, and 4%, respectively, of the annual total rainfall. Therefore, the temporal and spatial distribution of rainfall plays a vital role not only in the agricultural community but also in water resources management. This study investigates the rainfall variability over Jharkhand by using past 116 years data from 1901 to 2016. The extreme annual one-day rainfall depths investigated by using normal frequency distribution. The period examined for extreme rainfall events is 1986–2016. The data showed that the annual daily maximum rainfall received at any time ranged between 34.2 mm (minimum) to 341.0 mm (maximum) over Jamshedpur, indicating a very large range of fluctuation during the period of study. The magnitude of one-day annual maximum rainfall corresponds to return period were investigated. The depth of maximum daily rainfall is found highest over Jamshedpur. The highest rainfall observed in a day is: 224.3 mm at Ranchi, 341.0 mm at Jamshedpur and 240.4 mm at Dalton-ganj during the study period. The return period of extremely heavy rainfall (greater than or equal to 204.5 mm) in a day is 6 years, and the probability of exceedance is found 18% at Jamshedpur. Heavy rainfall events with more than 64.5 mm rain in a day was examined by using past 31 years daily rainfall data of Ranchi, Jamshedpur, and Daltonganj stations. To study the presence of a trend in rainfall, a widely used nonparametric Mann-Kendall test is applied. The linear trend line suggests that there is a decreasing trend in seasonal rainfall. The value of Kendall Score is found −2.82, which reveals that the decreasing trend in seasonal rainfall over Jharkhand is significant. An increasing trend is found in annual heavy rainy days over Jamshedpur, and the value of Kendall test statistics Z_s for Jamshedpur is +2.43, which reveals that the increasing trend in an annual number of heavy rainy days is significant at 5% significance level. However, no significant trend is found over Ranchi and Daltonganj.

30.1 INTRODUCTION

Jharkhand is located in the eastern region of India, and it is generously endowed with mineral wealth. It has some of the richest deposits of iron and coal in the world. Forests and woodlands occupy more than 29% of the state. Most of the state lies on the Chota Nagpur Plateau, which is the main source of the Koel, Damodar, Brahmani, Kharkai, and Subarnarekha rivers, whose upper watersheds lie within Jharkhand. Although the territory of Jharkhand holds a rich store of minerals; yet agriculture in Jharkhand is the mainstay for most of the tribal communities. In fact, about 80% of the total population is dependent on agriculture and allied activities for their livelihood. Rainfall is the primary source of surface and groundwater recharge.

The state receives 91% of its annual rainfall due to the southwest monsoon, which is its principal rainy season. The contribution of winter, pre-monsoon, and post-monsoon season's rainfall amount to about 2%, 3%, and 4%, respectively of the annual total rainfall. Therefore, the temporal and spatial distribution of rainfall plays a vital role not only in the agricultural community but also in water resources management.

Hydrological processes are usually regarded as stationary; however, there is growing evidence of trends, which may be related to anthropogenic influences and natural features of the climate system. Serious concerns are drawn on the catastrophic nature of floods, droughts, and storms, caused due to the significant variations in the regional climate including the rainfall pattern taking place on a regional level. The IPCC's (Intergovernmental Panel on Climate Change) fifth climate Assessment Repo (IPCC, 2013) shows that the globally averaged combined land and ocean surface temperature increased by 0.85°C, over the period 1880 to 2012. It is likely that in a warmer climate heavy rainfall will increase and be produced by fewer more intense events. This could lead to longer dry spells and a higher risk of floods.

The studies of seasonal and annual rainfall on global and local scales suggest that the total rainfall is highly variable and extreme rainfall is highly variable over many regions of the world. Several studies have been made in the past to determine the presence of trend or variability in precipitation and temperature all over the globe. Trends in rainfall in the regional scale over India were investigated by Parthasarathy and Dhar (1974), where the trends in rainfall over 31 subdivisions of India were investigated using the sixty years data. Parthasarathy and Dhar (1974) witnessed positive trends over the central India and parts of Northeast and Northwest India (Mehfooz, 2005). Therefore, there is an urgent need to investigate the rainfall pattern of the state for making the long-term strategic plan to minimize or restore the ecosystem.

30.2 DATA AND METHODOLOGY

This study investigates the South West Monsoon rainfall variability and trend over Jharkhand state by using past 116 years (1901–2016) data. The year wise monthly and seasonal (JJAS) rainfall data has been taken from Meteorological Centre Ranchi & National Data Centre, India Meteorological Department. To investigate the trend a widely used nonparametric Mann-Kendall test is applied at 5% significance level. Daily rainfall data at three different stations distributed in Jharkhand with longtime series were used for heavy rainfall analysis. Period examined for heavy rainfall events is 1986–2016. The extreme annual one-day rainfall depths for selected return periods on the basis of frequency analysis were investigated. The days with more than 64.5 mm was considered as heavy rainy days. Yearly variations of a number of heavy rainy days analyzed correspond to each station.

In the extreme rainfall analysis, first of all, one-day maximum rainfall data were extracted for each year from daily data and tabulated for each selected stations. The annual one-day maximum rainfall data were analyzed by using a software package RAINBOW (Raes et al., 2006). In this study, a commonly used probability distribution functions namely: normal distribution is applied for frequency analysis. The main application of frequency distribution in water resource management involves the assignment of an exceedance prob-ability P_e, of the design event. The average Probability of exceedance and return period is estimated by Weibull method (Weibull, 1939).

To bring out major aspects of rainfall variability and trend in time series data, the common nonparametric Mann-Kendall test is applied. The Mann-Kendall test (Kendall, 1975) is the most common one used in studying hydrologic time series trends. There are two advantages of using non-parametric test over parameter test. First, the non-parametric tests do not require the data to be normally distributed. Second, the test has low sensitivity to abrupt breaks due to inhomogeneous time series.

30.3   DATA ANALYSIS

The design and water management for agricultural activities, flood control system, etc. should be average and extreme rainfall pattern. The seasonal rainfall variability is an important parameter for the agricultural community and water resources management. The southwest monsoon season (JJAS) is a principal rainy season for Jharkhand.

30.3.1   MANN KENDALL TEST

It is a statistical test widely used for the analysis of the trend in climatological and in hydrologic time series. Mann-Kendall test had been formulated by Mann (1945) as a non-parametric test for trend detection, and the test statistic distribution had been given by Kendall (1975) for testing non-linear trend and turning point. According to this test, the null hypothesis H_0 assumes that there is no trend (the data is independent and randomly ordered) in precipitation and this is tested against the alternative hypothesis H_1, which assumes that there is a trend (increasing or decreasing).

The computational procedure for the Mann Kendall test considers the time series of n data points and x_i and x_j as two subsets of data where $i = 1, 2, 3, \ldots\ldots n - 1$ and $j = i + 1, i + 2, i + 3 \ldots\ldots n$. Each of the data point x_i is taken as a reference point which is compared with the rest of data points x_j. If a data value from a later time period is higher than a data value from an earlier time period, the statistic S is incremented by 1. On the other hand, if the data value from a later time period is lower than a data value sampled earlier, S is decremented by 1. The net result of all such increments and decrements yields the final value of S.

The Mann-Kendall S Statistic is computed as follows:

$$S = \sum_{i=1}^{n-1} \sum_{j=i+1}^{n} sgn\left(x_j - x_i\right) \tag{1}$$

where,

$$sgn(x_j - x_i) = \begin{cases} +1, \Delta x > 0 \\ 0, \Delta x = 0 \\ -1, \Delta x < 0 \end{cases}$$

For moderate (n about 10) or larger series lengths, the sampling distribution of the test static in equation (1) is approximately Gaussian, and if the null hypothesis (no trend) is true, this Gaussian null distribution will have zero mean. The variance of this distribution is depends on whether all the $x's$ are distinct, or if some are repeated values. If there are no ties, the variance of the sampling distribution of S is

$$Var(S) = \frac{n(n-1)(2n+5)}{18} \tag{2}$$

Otherwise, the variance is

$$Var(S) = \frac{n(n-1)(2n+5) - \sum_{j=1}^{J} t_j (t_j - 1)(2t_j + 5)}{18}$$

where J indicates the number of groups of repeated values, and t_j is the number of repeated values in the jth group.

At certain probability level H_0 is rejected in favor of H_1 if the absolute value of S equals or exceeds a specified value $S_{\alpha/2}$, where $S_{\alpha/2}$ is the smallest S which has the probability less than $\alpha/2$ to appear in case of no trend. A positive (negative) value of S indicates an upward (downward) trend.

The test statistic Z_s is used a measure of the significance of the trend. The test statistic Z_s is given by

$$Z_s = \begin{cases} \dfrac{S-1}{[Var(S)]^{1/2}}, S > 0 \\ \dfrac{S+1}{[Var(S)]^{1/2}}, S < 0 \end{cases} \tag{3}$$

Test statistic Z_s is used to test the null hypothesis, H_0. If $|Z_s|$ is greater than $Z_{\alpha/2}$, where α represents the chosen significance level (e.g., 5% with Z $0.025 = 1.96$) then the null hypothesis is invalid implying that the trend is significant.

30.4   RESULTS

30.4.1   SOUTH-WEST MONSOON RAINFALL VARIABILITY

The mean rainfall observed over the state Jharkhand is 1084.6mm for the period of 116 years from 1901 to 2016 during South-West Monsoon season (JJAS). The seasonal rainfall varied between 1539.0 mm and 578.4 mm in the year 1971 and 2010, respectively. Standard deviation is 165.39 mm for the study period. Seasonal SW-Monsoon rainfall variability and trend over Jharkhand state is represented in Figure 30.1. The linear trend line suggests that there is a decreasing trend in seasonal rainfall. The value of Kendall Score is found –2.82, which reveals that the decreasing trend in seasonal rainfall over Jharkhand is significant at 95% confidence level.

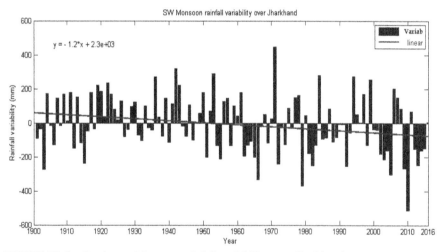

FIGURE 30.1   Southwest Monsoon rainfall variability over Jharkhand (1901–2016).

30.4.2   EXTREME RAINFALL ANALYSIS

Daily rainfall data at three different stations distributed in Jharkhand were analyzed for extreme rainfall analysis for the period of 1986–2016. Annual one-day maximum rainfall data extracted from daily rainfall data of Ranchi, Jamshedpur, and Daltonganj stations. Box plot is shown in Figure 30.2.

Particular rainfall depths that can be expected for a specific period is called the return period, and it is a vital parameter for the management of irrigation and flood control system. Return period for a station is obtained by using frequency analysis of longtime series rainfall data. In the frequency analysis of extreme rainfall events, first of all, one-day maximum rainfall data extracted for each year from daily data. In this study, the normal distribution is applied for frequency analysis. The average probability of exceedance and return period is estimated by Weibull method (Weibull, 1939).

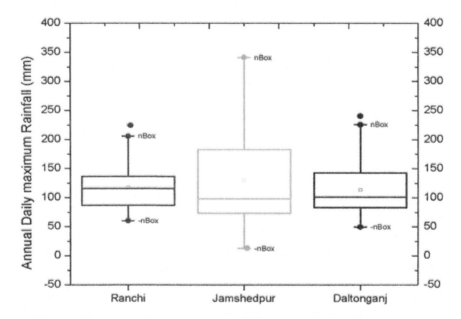

FIGURE 30.2 Box plot of annual daily maximum rainfall (1986–2016).

Weibull estimates the probability of exceedance as:

$$P_e = \frac{r}{n+1}$$

where, r is the rank number and n is the number of observations.

The return period T in years is related to the annual exceedance probability.

$$T = \frac{1}{P_e}$$

Table 30.1 represents the probability of exceedance and return period for all the stations. The return period of extremely heavy rainfall (greater than or equal to 204.5 mm) in a day is 6 years, and the probability of exceedance is about 18% at Jamshedpur.

TABLE 30.1   Probability of Exceedance and Return Period (1986–2016)

Probability of Exceedance (%)	Return Period (Year)	Rainfall (mm)		
		Ranchi	Jamshedpur	Daltonganj
—	25	180.6	267.9	194.7
10	10	163.6	230.7	172.8
20	5	147.6	195.9	152.2
30	3.33	136.1	170.8	137.4
40	2.5	126.3	149.3	124.8
50	2	117.1	129.3	113
60	1.67	108	109.3	101.1
70	1.43	98.2	87.9	88.5
80	1.25	86.7	62.7	73.7
90	1.11	70.7	27.9	53.1

For estimation of the probability of exceedance of annual one-day maximum rainfall, the normal distribution is used in this study. A probability distribution and counts of extreme rainfall events are plotted in Figure 30.3.

To analyses the trend in heavy rainfall events, annual heavy rainy days (more than 64.5 mm) were estimated from daily rainfall data for all the study stations. Highest 13 heavy rainy days in a year were found during 2016 at Jamshedpur. Variability of annual heavy rainy days is represented in Figure 30.4.

The study reveals that there is an increasing trend in annual heavy rainy days over Jamshedpur however; no significant trend is seen over Ranchi and Daltonganj. The linear trend line suggests that there is an increasing trend in annual heavy rainy days over Jamshedpur. The Kendall score for Daltonganj is +2.43 which reveals that the increasing trend in a number of heavy rainy days is significant at 5% significance level. A slight linear decreasing trend in annual heavy rainy days is observed over Ranchi and Daltonganj; however, Man-Kendall suggests that the decreasing trend is not significant. Variations of annual heavy rainy days are represented in Figure 30.5.

30.4.3   NON-PARAMETRIC TESTS

In this study, a non-parametric Mann-Kendall test is applied to investigate the trend in seasonal rainfall and extreme events over Jharkhand. The value of Kendall score for South West Monsoon rainfall over Jharkhand is

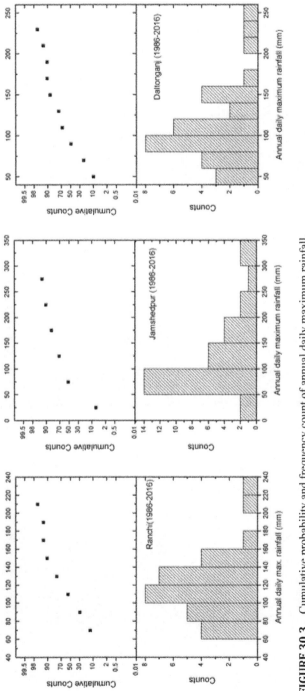

FIGURE 30.3 Cumulative probability and frequency count of annual daily maximum rainfall.

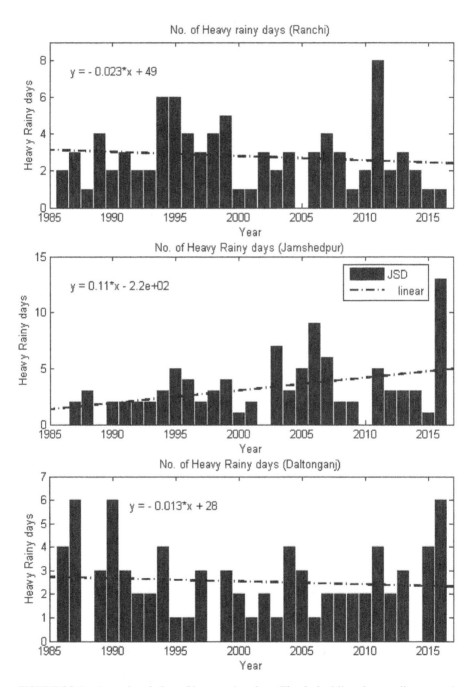

FIGURE 30.4   Annual variation of heavy rainy days. The dashed line shows a linear trend line.

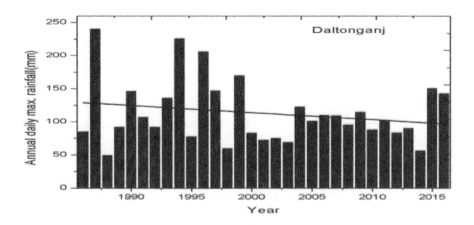

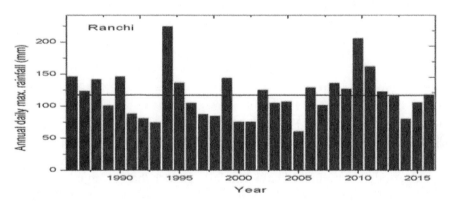

FIGURE 30.5   Variation of annual daily maximum rainfall. The straight line shows a linear trend line.

−1082 and test statistics Z_s is −2.58, which reveals that there is a decreasing trend in seasonal rainfall at significance level 5%. The test is also applied for extreme rainfall events correspond to the stations. The Kendal score of annual daily maximum rainfall is negative over Ranchi and Daltonganj. The negative value of Kendall score represents a decreasing trend; however, the decreasing trend is not significant. The value of Kendall score of annual one-day maximum rainfall over Jamshedpur is 144 and test statistics Z_s is 2.43, which reveals that the increasing trend in extreme rainfall events is significance at level 5%. No other parameters show a significant trend. Various Kendall parameters are shown in Table 30.2.

TABLE 30.2   Mann-Kendall Parameters

Sr. No.	Particulars	Variable	Kendall Score 'S'	Test Statistic Z_s	Hypothesis Test
1.	Jharkhand	SW Monsoon rainfall variability (1901–2016)	−1082	−2.58	H_o accepted
2.	Ranchi	Annual daily max. rainfall (1986–2016)	−13	−0.20402	H_o rejected
		Annual heavy rainy days (1986–2016)	−61	−1.04558	H_o rejected
3.	Jamshedpur	Annual daily max. rainfall (1986–2016)	144	2.432245	H_o accepted
		Annual heavy rainy days (1986–2016)	103	1.77264	H_o rejected
4.	Daltonganj	Annual daily max. rainfall (1986–2016)	−28	−0.45897	H_o rejected
		Annual heavy rainy days (1986–2016)	−14	−0.22628	H_o rejected

30.5   CONCLUSIONS

The depth of maximum daily rainfall is found highest over Jamshedpur. The highest rainfall observed in a day is: 224.3 mm at Ranchi, 341.0 mm at Jamshedpur and 240.4 mm at Daltonganj during the study period. The return period of extremely heavy rainfall (greater than or equal to 204.5 mm) in a day is 6 years, and the probability of exceedance is found 18% at Jamshedpur. The value of Kendall Score is found −2.82, which reveals that the decreasing trend in seasonal rainfall over Jharkhand is significant. An increasing trend is found in annual heavy rainy days over Jamshedpur, and the value of Kendall test statistics Z_s for Jamshedpur is +2.43, which reveals that the increasing trend in a number of heavy rainy days is significant at 5% significance level. However, no significant trend is found over Ranchi and Daltonganj.

ACKNOWLEDGMENT

The authors are grateful to Dr. K. J. Ramesh, Director General of Meteorology, India Meteorological Department for continuously encourage and his support. Authors express their sincere thanks to all staff members and officers of Meteorological Centre Ranchi for their kind and sincere assistance in preparing this paper. The authors are also grateful to Dr. Sanjiv Bandyopadhyay, DDGM, RMC Kolkata for valuable guidance and support.

KEYWORDS

- **extreme events**
- **Mann-Kendall**
- **normal frequency distribution**
- **probability distribution**
- **rainfall variability**
- **return period**

REFERENCES

IPCC (2013). Summary for Policymakers. In: Stocker, T. F., Qin, D., Plattner, G.-K., Tignor, M., Allen, S. K., Boschung, J., Nauels, A., Xia, Y., Bex, V., & Midgley, P. M., (eds.), *Climate Change, the Physical Science Basis. Contribution of Working Group I to the Fifth Assessment Report of the Intergovernmental Panel on Climate Change.* Cambridge University Press, Cambridge, United Kingdom and New York, NY, USA.

Kendall, M. G., (1975). *Rank Correlation Methods* (4th edn.). Charles Griffin, London, U.K.

Mann, H. B., (1945). Non-parametric tests against trend. *Econometrica, 33,* 245–259.

Mehfooz, A., Joardar, D., & Loe, B. R., (2005). Variability of South monsoon over Rajasthan and Kerala, *Mausam., 56*(3), 593–600.

Parthasarathy, B., & Dhar, O. N. (1974). Secular variations of regional rainfall over India. *Quart. J. R. Met. Soc., 100,* 245–257.

Raes, D., Willems, P., & Baguidi, G. F., (2006). RAINBOW – a software package for analyzing data and testing the homogeneity of historical data sets. *Proceedings of the 4th International Workshop on 'Sustainable Management of Marginal Drylands.'* Islamabad, Pakistan. Fourth Project Workshop Islamabad Pakistan (27–31 January 2006), pp. 41–55.

Scnetzler, A. E., Market, P. S., & Zeitler, J. W. (2008). *Analysis of Twenty Five Years of Heavy Rainfall Events in the Texas Hill Country.* MS Thesis, University of Missouri.

Zain, Al-Houri, Abbas Al-Omari, & Osama, S., (2014). "Frequency analysis of Annual one-day Maximum rainfall at Amman Zarqa Basin, Jordan." *Civil and Environmental Research, 6*(3), 44–57.

CHAPTER 31

VARIATION OF CLIMATOLOGICAL PARAMETERS INSIDE AND OUTSIDE NATURALLY VENTILATED POLYHOUSES

RAVISH CHANDRA[1] and P. K. SINGH[2]

[1]*Central Agricultural University, Pusa, Samastipur (Bihar), India,*
E-mail: ravish.cae@gmail.com

[2]*Department of Irrigation and Drainage Engineering, College of*
Technology, Govind Ballabh Pant University of Agriculture and
Technology, Pantnagar, India

ABSTRACT

A field experiment was conducted to study the various climatological parameters taken under the naturally ventilated polyhouse and outside polyhouse during 19th October 2013 to 10th April 2014. Two types of polyhouses, Double Span Naturally Ventilated Polyhouse (DS NVPH) and Walking Tunnel Type Polyhouse were selected for the study. Both the polyhouses were planted with capsicum (*Capsicum annuum L.*). The climatic parameters such as temperature, and solar radiation were measured at 09:00 hrs, and 16:00 hrs. The comparisons were made with the ambient conditions prevailing in outside the open field. Solar intensity was found highest in the open field condition followed by solar radiation inside Walking Tunnel Naturally Ventilated Polyhouse (WT NVPH) and then solar radiation received by DS NVPH. The percentage of solar radiation received by the DS NVPH ranges from 38.40% to 72.13% with an average value of 50.86% and WT NVPH ranges from 55.58% to 87.07% with an average value of 70.58%, of the solar radiation received outside field condition. The maximum and minimum temperatures were higher in WT NVPH followed by DS NVPH and then open field condition. It was observed that the maximum temperature in DS NVPH varied from 16 to 43.6°C with an average value of 33.4°C and minimum temperature varied from 5.5 to 22.5°C with an average value of 14°C. Similarly, the maximum temperature in WT NVPH varied from 17 to 44°C with an average value of 34.8°C and minimum temperature varied from 6 to 23.5°C with an average value of 15.4°C.

31.1 INTRODUCTION

Among these protective cultivation practices, greenhouse/polyhouse cum rain shelter is useful for the off-season cultivation of vegetables. The greenhouse is generally covered by transparent or translucent material such as glass or plastic. The greenhouse covered with a simple plastic sheet is termed as polyhouse. The greenhouse generally reflects back 43% of the net solar radiation incident upon it allowing the transmittance of the "photosynthetically active solar radiation" in the range of 400–700 Nm wavelength. The sunlight admitted to the greenhouse is absorbed by the crops, floor, and other objects. These objects, in turn, emit longwave thermal radiation in the infrared region for which the glazing material has lower transparency. As a result, the solar energy remains trapped in the greenhouse, thus raising its temperature. This phenomenon is called the "Greenhouse Effect." This condition of a natural rise in greenhouse air temperature is utilized in the cold regions to grow crops successfully. However in the summer season due to the above-stated phenomenon ventilation and cooling is required to maintain the temperature inside the structure well below 35C. The ventilation system can be natural or a forced one. Greenhouses provide protection against biotic and abiotic stresses and ensure high quality produce (Peet and Welles, 2005). Low cost naturally ventilated polyhouse offers a great scope for the off-season cultivation of capsicum for round the year supply in the cold climatic areas of India. A greenhouse is a framed structure made of GI Pipe/MS angle/Wood/Bamboo and covered with transparent material or translucent material fixed to the frame with grippers. It has control/monitoring equipment, which is considered necessary for controlling environmental factors such as temperature, light, relative humidity, etc., which is considered necessary for maximizing plant growth and productivity. Thus, the greenhouse is an enclosed area, in which crops are grown under partially or fully controlled conditions. The cladding material is of plastic (polyethylene) film and acts like a selective radiation filter that allows solar radiation to pass through it but traps thermal radiation emitted by the inside objects to create a greenhouse effect. Greenhouse technology since it protects the crop from adverse weather conditions, attack from insects, pests, diseases, and thus helps in increasing yield and quantity. At the same time since the inside environment remains under control, carbon dioxide released by the plants during the night is consumed by the plants itself in the morning. Thus plants get about 8–10 times more food than the open field condition. The regulation of temperature, ventilation, adjustment of the amount of entering sunlight, to provide soil moisture, fertilizer, and even facilitate pollination are important for the crop growth. Environmental control systems in controlled environment plant production facilities (including greenhouses and growth chambers) traditionally focus on maintaining the indoor climate according to pre-defined set points (Fleisher, 2002). It is necessary to know the climatic conditions inside and outside of the greenhouse to control the favorable climate for the proper growth of the particular crop. So the objective was to study the various climatological parameters inside and outside the polyhouse.

31.2   MATERIALS AND METHODS

A field experiment was conducted to study the various climatological parameters taken under the naturally ventilated polyhouse and outside polyhouse during 19th October 2013 to 10th April 2014. The study was conducted at experimental field of the Department of Irrigation and Drainage Engineering, College of Technology, G. B. Pant University of Agriculture and Technology, Pantnagar. Two types of polyhouses, Double Span Naturally Ventilated Polyhouse (DS NVPH) and Walking Tunnel Type Polyhouse were selected for the study. Both the polyhouses were planted with capsicum (*Capsicum annuum L.*). The climatic parameters such as temperature, and solar radiation were measured at 09:00 hrs, and 16:00 hrs. The observations of maximum and minimum temperature were taken from Crop Research Centre, 0.5 km away from the experimental field.

31.3   RESULTS AND DISCUSSION

31.3.1   TEMPERATURE

The daily variation of maximum and minimum temperature in DS NVPH, Walking Tunnel Naturally Ventilated Polyhouse (WT NVPH) and open field condition are shown in Figure 31.1. The analysis revealed that the maximum temperature was higher in WT NVPH followed by DS NVPH and then open field condition. The minimum temperature followed the same trend. It was observed that the maximum temperature in DS NVPH varied from 16 to 43.6C with an average value of 33.4°C and minimum temperature varied from 5.5 to 22.5°C with an average value of 14°C. Similarly, the maximum temperature in WT NVPH varied from 17 to 44°C with an average value of 34.8°C and minimum temperature varied from 6 to 23.5°C with an average value of 15.4°C. The maximum temperature was recorded on 8th April 2014, and the minimum temperature was recorded on 4th January 2014.

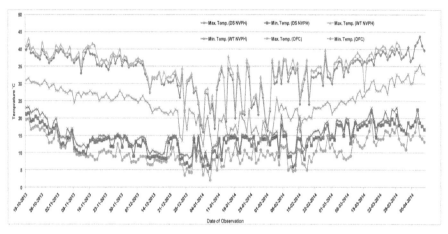

FIGURE 31.1   Daily variation of maximum and minimum temperature in DS NVPH, WT NVPH, and open field condition.

The analysis shows that the lowest maximum temperature inside DS NVPH was recorded during the 4th segment, i.e., from 1st January 2014 to 31st January 2014 with an average maximum temperature value of 26.03°C and the highest maximum temperature was recorded during the last segment, i.e., from 1st April 2014 to 30th April 2014 with an average maximum value of 39.8°C (Table 31.1). When comparison was made for minimum temperature it was found that the lowest minimum temperature inside DS NVPH was recorded during the 3rd segment, i.e., from 1st December 2014 to 31st December 2014 with an average minimum value of 10.64°C and the highest minimum temperature was recorded during the first segment, i.e., from 19th to 31st October 2014 with an average minimum value of 18.8°C. The similar trend was observed for WT NVPH. The lowest maximum temperature inside WT NVPH was recorded during the 4th segment, i.e., from 1st January 2014 to 31st January 2014 with an average maximum temperature value of 27.47°C and the highest maximum temperature was recorded during the last segment,

TABLE 31.1 Maximum and Minimum Temperature in Different Time Segments in DS NVPH, WT NVPH and in Open Field Condition

Time segment		Max. Temp. (°C) DS NVPH	Min. Temp. (°C) DS NVPH	Max. Temp. (°C) WT NVPH	Min. Temp. (°C) WT NVPH	Max. Temp.(°C) Open field	Min. Temp. (°C) Open field
19-31	Lowest	36.00	15.50	37.00	16.80	28.60	12.80
1 October	Highest	41.20	21.50	43.00	23.20	31.60	19.70
2013	Average	**38.27**	**18.78**	**39.47**	**20.78**	**30.16**	**16.38**
1–30 November	Lowest	33.00	9.50	35.00	11.50	24.40	7.80
2013	Highest	40.50	18.00	41.60	20.00	29.00	17.00
	Average	**37.22**	**13.07**	**38.67**	**14.42**	**26.88**	**10.55**
1–31 December	Lowest	22.10	6.00	22.70	8.50	13.50	3.40
2013	Highest	36.80	14.80	38.00	15.00	26.50	12.20
	Average	**31.83**	**10.64**	**32.96**	**11.97**	**22.58**	**7.66**
1–31 January	Lowest	16.00	5.50	17.00	6.00	10.80	2.00
2014	Highest	36.00	14.80	38.00	15.30	22.50	12.60
	Average	**26.03**	**11.89**	**27.47**	**12.73**	**16.96**	**8.09**
1–28 February	Lowest	16.40	6.00	17.40	9.70	12.40	4.60
2014	Highest	38.00	18.00	39.80	21.00	25.90	14.90
	Average	**31.19**	**12.94**	**32.83**	**14.78**	**21.06**	**9.17**
1–31 March	Lowest	30.00	12.80	28.00	14.00	21.20	8.00
2014	Highest	40.10	22.00	42.00	23.00	32.60	19.40
	Average	**36.59**	**17.75**	**37.79**	**19.06**	**27.35**	**13.01**
1–10 April 2104	Lowest	36.50	15.60	37.00	17.60	30.00	10.90
	Highest	43.60	22.50	44.00	23.50	35.50	19.90
	Average	**39.80**	**18.15**	**40.96**	**20.26**	**32.41**	**14.54**

i.e., from 1st April 2014 to 30th April 2014 with an average maximum value of 40.9°C. The similar comparison for minimum temperature revealed that the lowest minimum temperature inside WT NVPH was recorded during 3rd segment, i.e., from 1st December 2014 to 31st December 2014 with an average minimum value of 11.97°C and the highest minimum temperature was recorded during the first segment, i.e., from 19th to 31st October 2014 with an average minimum value of 20.8°C. The maximum temperature decreases from 1st segment to the 4th segment and then increases up to the last segment. The minimum temperature first decreases from 1st segment to 3rd segment and then increases up to the last segment in both DS NVPH and WT NVPH.

31.3.2 *SOLAR INTENSITY*

The daily variation of solar intensity recorded at 9:00 AM and 2:00 PM in DS NVPH, WT NVPH, and open field condition are shown in Figures 31.2 and 31.3. The analysis revealed that the solar intensity at 9:00 AM was higher in WT NVPH followed by DS NVPH.

The DS NVPH receives solar radiation ranging from 33.33% to 67.14% with an average value of 47.56% of total solar radiation received in open field condition. On the other hand, WT NVPH receives solar radiation ranging from 40.15% to 86.10% with an average value of 69.68% of total solar radiation received in open field condition. The maximum solar intensity observed for WT NVPH was 732 Klux on 9th April 2014 and minimum was 2.70 Klux on January 3, 2014. Similarly, the maximum value for DS NVPH was 59.5 Klux on April 11, 2014, and minimum was 2.0 Klux on January 3, 2014.

The similar trend was observed for solar intensity measured at 2:00 PM. The figure revealed that the solar intensity was highest in the open field condition followed by solar radiation inside WT NVPH and then solar radiation received by DS NVPH. The percentage of solar radiation received by the DS NVPH ranges from 38.40% to 72.13% with an average value of 50.86% and WT NVPH ranges from 55.58% to 87.07% with an average value of 70.58%, of the solar radiation received outside field condition. The maximum solar intensity observed for WT NVPH was 784 Klux on March 21, 2014, and the minimum was 6.8 Klux on January 30, 2014, when measurements were made at 2:00 PM (Figure 31.3). Similarly, the maximum value for DS NVPH was 66.5 Klux on March 21, 2014, and the minimum was 4.8 Klux on January 31, 2014. When comparison was made between solar radiation received at 9:00 AM and 2:00 PM inside DS NVPH, it was found that the minimum

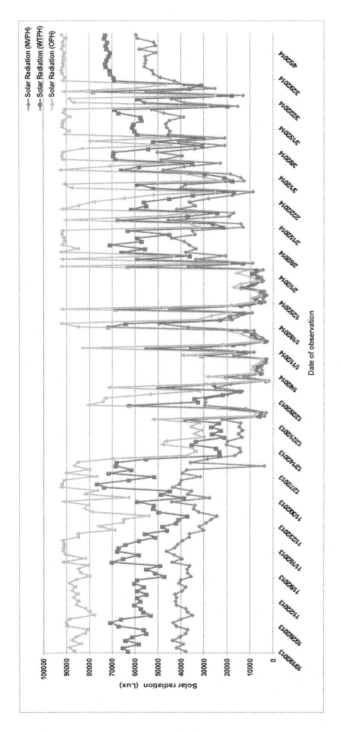

FIGURE 31.2   Daily variations in solar intensity at 9:00 AM in Double Span NVPH, Walking Tunnel NVPH and open field condition.

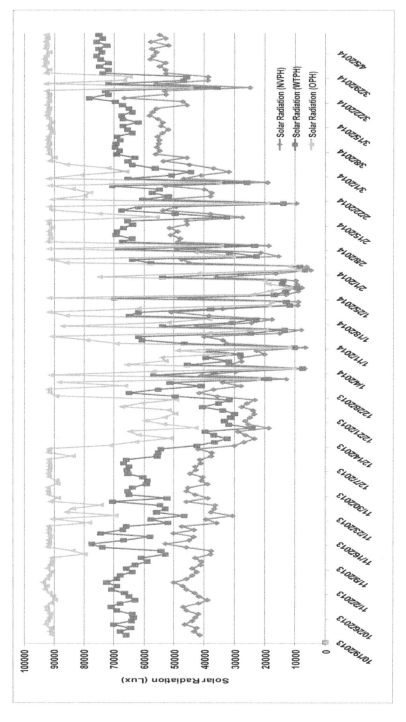

FIGURE 31.3 Daily variation in solar intensity at 2:00 PM in Double Span NVPH. Walking Tunnel NVPH and open field condition.

TABLE 31.2 Solar Radiation Received in Different Time Segments in DS NVPH, WT NVPH and in Open Field Condition

Time segment			Solar Radiation (Lux) inside DS NVPH (9:00 AM)	Solar Radiation (Lux) inside WT NVPH (9:00 AM)	Solar Radiation (Lux) outside (9:00 AM)	Solar Radiation (Lux) inside DS NVPH (2:00 PM)	Solar Radiation (Lux) inside WT NVPH (2:00 PM)	Solar Radiation (Lux) outside (2:00 PM)
19–31		Minimum	35000.00	53000.00	78000.00	39000.00	63000.00	90000.00
October		Maximum	43600.00	70500.00	90500.00	47000.00	71000.00	93100.00
2013		Average	39515.38	60915.38	85115.38	43223.08	66376.92	91700.00
	1–30	Minimum	24450.00	37200.00	54300.00	30800.00	46800.00	69600.00
November 2013		Maximum	46500.00	70000.00	91900.00	52900.00	77500.00	94000.00
		Average	37688.33	55026.67	82943.33	42870.00	63070.00	89043.33
1–31		Minimum	2900.00	5000.00	7000.00	13000.00	19000.00	22600.00
December 2013		Maximum	43750.00	78000.00	92400.00	44800.00	66900.00	92300.00
		Average	24248.39	39174.19	56767.74	32180.65	46387.10	72641.94
1–31		Minimum	2000.00	2500.00	3000.00	4800.00	6800.00	8800.00
January		Maximum	49700.00	71500.00	92400.00	51000.00	70000.00	92000.00
2014		Average	12814.52	19045.16	26196.77	23422.58	33190.32	48083.87
1–28		Minimum	4550.00	6500.00	9800.00	6200.00	8500.00	12000.00
February		Maximum	45500.00	70800.00	92600.00	53800.00	70600.00	92400.00
2014		Average	26482.14	42228.57	62957.14	38682.14	51850.00	73800.00
1–31		Minimum	13000.00	17500.00	25400.00	25000.00	36000.00	45000.00
March		Maximum	55000.00	73500.00	92800.00	66500.00	78400.00	93000.00
2014		Average	36707.69	51334.62	73703.85	51569.23	65373.08	88234.62
1–10		Minimum	50900.00	70500.00	90600.00	52000.00	72000.00	92000.00
April		Maximum	59500.00	73200.00	92400.00	60800.00	76000.00	93800.00
2104		Average	54775.00	71883.33	91725.00	56241.67	74541.67	93041.67

values increased by 15.25%, the maximum value increased by 7.42% and the average value increased by 6.95%. The higher amount of radiation received in WT NVPH might be due to lower control height compared to DS NVPH. Besides this, the inner surface area of WT NVPH is lesser compared to DS NVPH, and hence probably fewer energy losses were there in heating the system. The other researchers reported on similar results for different kind of playhouses (Dhandare et al. 2006, Dhandare et al. 2008). Table 31.2 shows that the minimum average radiation inside DS NVPH was received during the 4th segment, i.e., from 1st January 2014 to 31st January 2014 with a value of 12.8 Klux and the maximum average radiation was received during the last segment, i.e., from 1st April 2014 to 30th April 2014 with a value of 54.8 Klux, when measurement was made at 9:00 AM (Table 31.2). The analysis revealed that average solar radiation was higher for WT NVPH compared to DS NVPH. The minimum average radiation inside WT NVPH was received during the 4th segment, i.e., from 1st January 2014 to 31st January 2014 with a value of 19.0 Klux and the maximum average radiation was received during the last segment, i.e., from 1st April 2014 to 30th April 2014 with a value of 71.8 Klux, when measurement was made at 9:00 AM. A similar trend was found when measurements were made at 2:00 PM. The average solar radiation based on time segment shows that the average solar radiation first decreases from the 1st segment up to 4th segment and then increases gradually in both DS NVPH and WT NVPH. The study suggests that proper ventilation is required in the 1st segment, 2nd segment, 6th and last segment of the study. The study also suggests that there is a need for increasing the photoperiod and strength of radiation in the month of January.

31.4 CONCLUSION

The maximum and minimum temperatures were observed higher in WT NVPH followed by DS NVPH and then open field condition. The solar radiation was observed high in the open field followed by WT NVPH than DS NVPH, also observed high at 16:00 hrs than 09:00 hrs. The percentage of solar radiation received by the DS NVPH ranges from 38.40% to 72.13% with an average value of 50.86% and WT NVPH ranges from 55.58% to 87.07% with an average value of 70.58%, of the solar radiation received outside field condition.

KEYWORDS

- **double span naturally ventilated poly-house**
- **walking tunnel naturally ventilated poly-house**

REFERENCES

Dhandare, K. M., (2006). *Response of Capsicum (Capsicum annuum L.) to Cyclic Irrigation and Fertigation Under Environmental Controlled and Naturally Ventilated Polyhouse* (p. 152). MTech thesis, Department of Irrigation and Drainage Engineering, G. B. Pant University of Agriculture and Technology, Pantnagar, India.

Dhandare, K. M., Singh, K. K., Singh, P. K., Singh, M. P., & Bayissa, G., (2008). Variation of climatological parameters under environmental controlled and naturally ventilated polyhouses. *Pantnagar. Journal of Research, 6*(1), 142–147.

Fleisher, D. H., (2002). Preliminary analysis of plant response to environmental disturbances in controlled environments. *ASAE Paper No: 024076 St. Joseph, Mich.*

Peet, M. M., & Welles, G. W. H., (2005). In: Heuvelink, E., (ed.), *Greenhouse Tomato Production (In) Tomatoes* (pp. 257–304). CABI Publishing, Wallingford, U.K.

CHAPTER 32

DESIGN OF A RAIN WATER HARVESTING STRUCTURE AT ALLAHABAD MUSEUM, ALLAHABAD CITY, U.P., INDIA

VIVEK TIWARI[1], SARIKA SUMAN[2], and H. K. PANDEY[3]

[1]*Banaras Hindu University, Varanasi, India,*
E-mail: viveektiwary@gmail.com

[2]*Banaras Hindu University, Varanasi, India*

[3]*Department of Civil Engineering, MNNIT Allahabad, India*

ABSTRACT

A study has been conducted to design the Rain Water Harvesting system at Allahabad Museum. The Allahabad Museum is centrally located in Allahabad city in the picturesque of Chandrashekhar Azad Park and is a monument of national importance since the pre-historic era. The study was aimed to carry out the assessment of The Allahabad Museum's potential towards Roof Top Rain Water Harvesting (RTRWH) and to design an effective groundwater recharge structure for its main building which can serve as the model site for the scientific study and inspiration to the visitors, apart from the groundwater recharge. It has a sprawling green campus of 21850 m^2 and lies under the zone in which underground water level is declining at the rate of 0.49 m/year since last few years. It receives 950 mm of annual rainfall every year. It is estimated that about 3568.34 m^3 volume of water can be contributed towards groundwater recharging through RTRWH structure at the main building of The Allahabad Museum. It was observed that the Allahabad Museum has enough potential and resources available which can significantly contribute towards the groundwater recharge through cost-effective techniques.

32.1   INTRODUCTION

Groundwater, which is the source for more than 85% of India's rural domestic water requirements, 50% of its urban water requirements and more than 50% of its irrigation requirements is depleting fast in many areas due to its large-scale withdrawal for various sectors (CGWB, 2006). In India even though an average rainfall is 1100 mm there is a scarcity of water that results in water crisis (Hajare et al., 2003). There have been continued efforts in India for development of groundwater resources to meet the increasing demands of water supply. The artificial recharge has now been accepted worldwide as a cost-effective method to augment groundwater resources. Groundwater sources are depleting in quantity due to unregulated extraction of water and reduced replenishment of the groundwater. Quality of existing sources is also deteriorating (Biswas, 2007). Artificial recharge efforts are basically aimed at augmentation of the natural movement of surface water into groundwater reservoir through suitable civil construction techniques (ASCE, 2001). Artificial recharge is achieved by storing water on the land surface where it infiltrates into the soil and moves downward to under-lying groundwater (Bouwer, 1997; 1999). Such techniques interrelate and integrate the source water to groundwater reservoir which depends on the hydro-geological situation of the area. Domestic Rainwater harvesting or RTRWH is the technique through which rain water is captured from roof catchments and stored in tanks/reservoirs/groundwater aquifers. It consists of conservation of rooftop rainwater in urban areas and utilized to augment groundwater by artificial recharge. It requires connecting the outlet pipe from rooftop to divert collected water to existing well/tube well/ bore well or a specially designed well. Our study was concerned to carry out an assessment of Allahabad Museum's potential towards Rainwater harvesting and to design an effective groundwater recharge structure for its main building which can serve a model site for scientific study and motivation.

32.2   STUDY AREA

The Allahabad Museum is centrally located in Allahabad city in the picturesque of Chandrashekhar Azad Park and is a monument of national importance (Figure 32.1). It has a sprawling green campus of 5.4 acres and lies under the zone in which underground water level is declining at the rate of 0.49 m/year. Total average precipitation in Allahabad is 950mm, which is equivalent to 0.950 lm^{-2}.

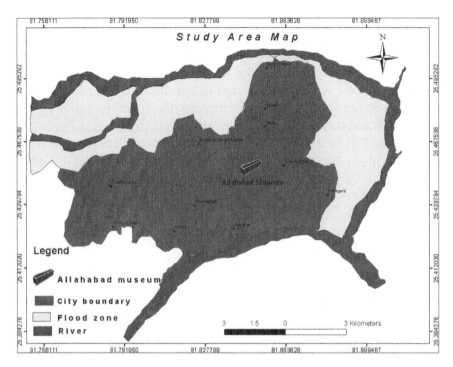

FIGURE 32.1   Location of Allahabad Museum.

32.3   HYDROGEOLOGICAL SETUP

Exploratory drilling data of CGWB and state tube well department show that there are three distinct granular zones, i.e., shallow aquifer ranging from 20 to 50 mbgl, middle aquifer ranging from 70 to 120 mbgl and deeper aquifer lies below 150 mbgl down to 300 mbgl. The extension of individual zones varies over the Allahabad district. A comparative analysis of various hydrological parameters (pre-monsoon groundwater level, post-monsoon groundwater level, and water table fluctuation) has been done for better understanding of changes in groundwater (Singh et al., 2014).

32.3.1   GROUNDWATER CONDITION

The area is characterized by quaternary to recent sediments which act as aquifer (sand of various grades) with intercalations of clay. The groundwater occurs in the porous formation and has different groundwater level and fluc-tuation in different parts of the city. The groundwater behavior in and around the study area is explained below.

32.3.1.1   GROUNDWATER ELEVATION MAP (2017)

Figures 32.2 and 32.3 present pre-monsoon and post-monsoon groundwater levels for the Allahabad city. The pre-monsoon groundwater elevation map shows that the Allahabad Museum falls in the range of 68 m to 72 m water level elevation and in the post-monsoon, the water level shows slight improvement and comes in a range of 64 m to 72 m. The seasonal fluctuation map (Figure 32.4) for the year 2016–2017 shows that the Allahabad Museum lies in the region where annual groundwater fluctuation is 3 mbgl to 4 mbgl.

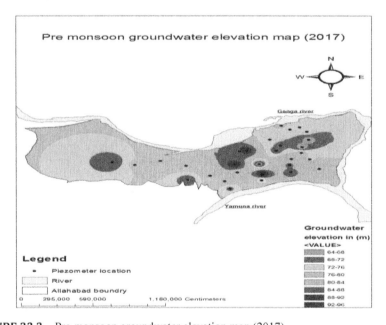

FIGURE 32.2   Pre-monsoon groundwater elevation map (2017).

32.4   MATERIALS AND METHOD

The study aimed at carrying out an assessment of Allahabad Museum's potential towards Rainwater harvesting and to design an effective ground-water recharge structure for its main building which can serve a model site for scientific study and motivation. The Allahabad museum doesn't have a supply of water from Water Board of Allahabad City and is completely dependent on groundwater for its needs. Salient features of the museum campus (Table 32.1) are used to plan the layout of recharge structure network (Figure 32.5).

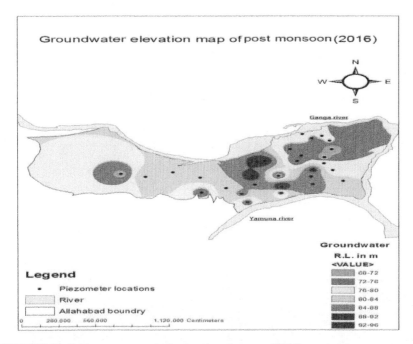

FIGURE 32.3   Post monsoon groundwater elevation map (2016).

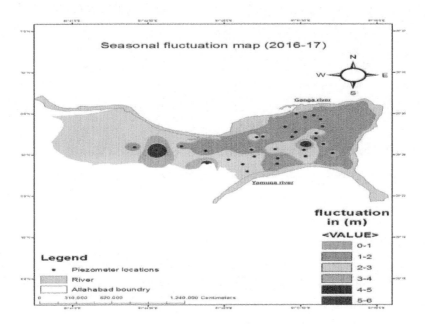

FIGURE 32.4   Seasonal fluctuation groundwater map (2016–17).

TABLE 32.1 Salient Features of the Museum Campus

Natural slope:	towards the west side
Rate of decline of groundwater level:	0.49m year^{-1}
Average depth to groundwater level:	16.0 mbgl
Intake capacity of recharge well:	12 m^{3}h^{-1}
Type of aquifer:	Unconfined
Total Roof Area (Main Building area):	4910 m^2 – 10% of the total roof area (Open area) = 4419 m^2
Average annual Rainfall Intensity:	30mm h^{-1}
Runoff Coefficient:	0.85
Water Available for Recharge:	4419 x 0.03 x 0.85 = 112.68 ≈ 113m^3
Expected recharge:	0.950 x 4419 x 0.85 = 3,568.34 m^{3}year^{-1}

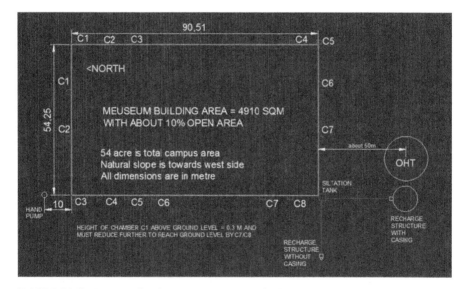

FIGURE 32.5 Layout of recharge structure network plan.

32.4.1 DIAMETER OF PIPE REQUIRED TO CARRY RAIN WATER COLLECTED

We know that Discharge, $Q = \pi r^2 * v$

where, r = radius, v = velocity of flow of water

Here, Q = 113 m^{3}h^{-1}, r =?, v =1 m^{2}s^{-1}

Therefore, $r = \dfrac{\sqrt{Q}}{\sqrt{\pi v}} = \dfrac{\sqrt{113}}{\sqrt{\pi * 1 * 3600}}$ or, $r = 0.09$ m

Or, Diameter, $d = 2 \times r = 2 \times 0.09 = 0.19m \approx 8"$

32.4.2 DESIGN OF RECHARGE STRUCTURE HAVING RECHARGE WELL WITH CASING

Location: Adjacent to Over Head Tank at about 50m from Main Building in its South direction Shape of structure: Concentric Cylinder (Figure 32.6).

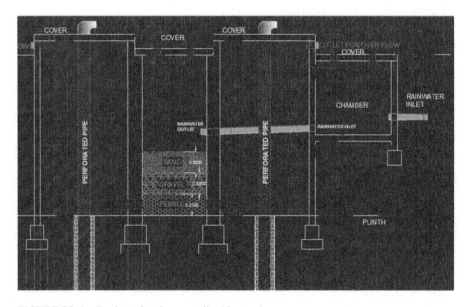

FIGURE 32.6   Design of recharge well with a casing.

32.4.2.1   DESIGN OF FILTER

Total discharge, $Q = 113$ m^{3}h^{-1}

Intake capacity of recharge well with casing = 145 lpm = 8.7m^{3}h$^{-1} \approx 9$ m^{3}h^{-1}

Intake capacity of 4 number of recharge well with casing = $9 \times 4 = 36$ m^{3}h^{-1}

Intake capacity of recharge well without casing = 80 lpm = 4.8 m^{3}h^{-1}

Intake capacity of 10 number of recharge well without casing = $4.8 \times 10 = 48$ m^{3}h^{-1}

Available water as discharge = $113 - (36 + 48) = 29$ m³h⁻¹

Hydraulic conductivity in medium, K = 110 lpm = 6.60 m³h⁻¹

Height of aquifer, L = 4m; Depth of filter, h = 2.5 m

Depth of filter media = to be filled up to 40% the height of filter = $2.5 \times 0.4 = 1$m

Sand, Gravel, and Pebble are supposed to be in equal length, i.e., 0.33 m each

Provide Nylon Mesh after layer of sand Area, $A = \dfrac{QL}{kh} = (29 \times 4)/ (6.6 \times 2.5)$ = 7.03 m²

Radius of Filter, r = 0.525 m

32.4.2.2   DESIGN OF RECHARGE WELL

Number of recharge wells = 4

Intake Capacity of recharge well with casing = 145 lpm = 8.7 m³h⁻¹ ≈ 9 m³h⁻¹

Intake Capacity of 4 number of recharge well with casing = $9 \times 4 = 36$ m³h⁻¹

Diameter of Recharge well = 8" = 0.2m

Reaming of bore hole = up to 13" in diameter = 0.325 m in diameter

Depth of Recharge well = 40 m bgl

Depth of pilot hole= 45 m bgl

32.4.2.3   DESIGN OF RECHARGE TANK

Capacity of tank = $113 - (36 + 48) = 29$ m³h⁻¹

Radius of Recharge tank, R = 2.3m

32.4.2.4   COVER

Provide a RCC slab as the cover of 0.1 m in thickness to cover the whole structure.

Here, Radius, R = 2.3 m.

32.4.2.5   CHAMBER

To check the velocity of incoming flow. Its dimension = 1.25 m x 1.25 m x 1 m.

32.4.3 DESIGN OF RECHARGE STRUCTURE HAVING RECHARGE WELL WITHOUT CASING

Location: Southwest of Main building
Shape of structure: Cylindrical (Figure 32.7)

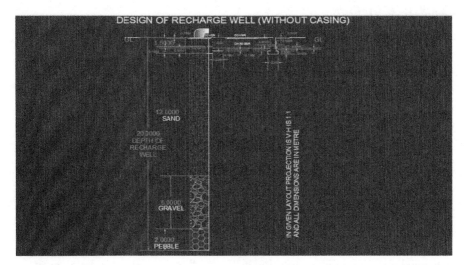

FIGURE 32.7 Design of recharge well without casing.

32.4.3.1 DESIGN OF FILTER AND STRUCTURE

Intake Capacity of recharge well without casing = 80 lpm = 4.8 m³h⁻¹

Intake Capacity of 10 number of recharge well without casing = 4.8 x 10 = 48 m³ h⁻¹

Depth of filter, h = 19 m

Total depth of well = 20 m

Diameter of bore well = 12" = 0.3 m

Reaming of bore hole= up to 16" in diameter = 0.4 m in diameter

Provide a lining of 0.1m by brickwork around the circumference of well for 1.5 m in depth

Depth of Sand layer = 12 m

Depth of Gravel layer = 5 m

Depth of Pebble Layer = 2 m

32.4.3.2   COVER

Provide a RCC slab as the cover of 0.1m in thickness to cover the whole structure. Here, Radius, R = 0.25m.

32.4.3.3   CHAMBER

To check the velocity of incoming flow, its dimension = 1.25 m x 1.25 m x 1 m.

32.4.4   OTHER COMPONENTS

Siltation tank: Dimension 2.5m x 2.5m x 1 m (underground).
    Leaf screens: Provide screen at all the outlets of rooftop and over holes provided in covers of structures to check impurities.

32.5   RESULT

The study at the Allahabad Museum and its nearby area revealed that there is a need for immediate attention towards the groundwater level as it is declining at an appreciable rate. The premises of Allahabad Museum can be considered to be best in potential to yield significant of rainfall runoff to recharge groundwater table. It is estimated that from the main building about = 3,568.34 m^{3}year^{-1} water will be recharged every year. The Allahabad Museum has an enormous and sprawling campus of about 5.4 acres or 21853.02 m^2 (carpet area). If effectively planned then about 17,646.31 m^3 quantity of water can be harvested every year.

32.6   CONCLUSION

There is a need for RTRWH at The Allahabad Museum to ameliorate the groundwater stress condition in the study area. It is rich in potential to yield towards groundwater recharge significantly. Based on preliminary survey and detailed study over the Allahabad Museum, 11 structures under two categories were proposed, i.e., recharge structure having recharge well with casing and recharge structure having well without casing. There is 4 well with casing and 10 well without casing is proposed. The institution of National importance and being completely dependent on groundwater supply to meet the water demands, must install, promote, and advocate for groundwater recharge practices and techniques.

KEYWORDS

- **groundwater**
- **recharge structure**
- **RTRWH**

REFERENCES

Biswas, R., (2007). Water management in Delhi: Issues, challenges, and options, *Jour. Indian Water Works Assoc*, *39*, 89–96.

Bouwer, H., (1997). Role of groundwater recharge and water reuse in integrated water management, *Arabian Jour. Sci. Engg., 22*, 123–131.

Bouwer, H., (1999). Chapter 24: Artificial recharge of groundwater: systems, design, and management. In: Mays, L. W., (ed.), *Hydraulic Design Handbook* (Vol. 24, pp. 1–24, 44). McGraw Hill, New York.

Central Ground Water Board, (2007). Manual on Artificial Recharge of Ground Water.

Hajare, H. V., Wadhai, P. J., & Khode, B. V., (2003). *Rainwater Conservation, All India Seminar on Challenges in Environmental Protection and Possible Solutions*, 46–50.

Singh, S., Samaddar, A. B., Srivastava, R. K., & Pandey, H. K., (2014). Groundwater recharge in Urban areas – the experience of rainwater harvesting. *Journal of the Geological Society of India, 83*(3), 295–302.

Standard Guidelines for Artificial Recharge of Ground Water, (2001). American Society of Civil Engineers, EWRI/ASCE 34–01.

CHAPTER 33

DOWNSCALING OF PRECIPITATION USING LARGE SCALE VARIABLES

PRATIBHA WARWADE[1], SURENDRA CHANDNIHA[2], GARIMA JHARIYA[3], and DEVENDRA WARWADE[4]

[1,3]*Assistant Professor, Centre for Water Engineering and Management, Central University of Jharkhand, Brambe, Ranchi-835205, India, E-mail: pratibhawarwade@gmail.com / Institute of Agricultural Sciences, Banaras Hindu University, Varanasi (U.P.) 221005, India, E-mail: garima2304@gmail.com*

[2]*Research Associate, National Institute of Technology, Roorkee, Uttrakhand, India*

[4]*Assistant Professor, Department of Physics, Government College Sehore, Barktulla University, Bhopal, M.P., India*

ABSTRACT

Present study carried to downscale the monthly precipitation using large-scale variables for Dikhow Catchment, North Eastern region of India. Statistical downscaling is performed in this study by employing multiple linear regression (MLR) approach followed by Bias correction. A daily reanalysis dataset of NCEP (scale of 2.5° latitude × 2.5° longitude) re-gridded on a scale of 2.5° latitude × 3.75° longitude, simulated data of HadCM3 on a scale 2.5° latitude × 3.75° and monthly precipitation data from IMD Pune were used for the analysis. Six potential predictors; wind at 500 hPa, wind direction at 500 hPa, geopotential height at 500 hPa, relative humidity at 500 hPa, near-surface relative humidity, and air temperature at 2m are selected. Nash-Sutcliffe Efficiency (NSE) value varies from 0.88 to 0.89 for calibration and 0.88 to 0.92 for validation period. MLR model has performed well over the region. Future precipitation is increasing at all stations for both A2 and B2 scenarios, the magnitude of increase in precipitation, is higher for A2 scenario than that of B2 scenario (monsoon precipitation 1857.5 mm and 1715.7 mm, summer precipitation 606.3 mm and 559.7 mm whereas winter precipitation 139.8 mm and 144.05 mm).

33.1   INTRODUCTION

Climate change affects the Earth's environments which leads to effect the people's livelihoods and wellbeing. Apart from climate change, economic development, current demographic trends, and land use changes have a direct impact on the growing demand for freshwater resources (Gain et al., 2012). Sustainable development will secure the availability of resources for further generations (Pechstädt et al., 2011). Rainfall is pertinent parameters that are affected by climate change which in turn influences the hydrological cycle, and the rainfall-runoff process has significant importance in Hydrology (Sahu et al., 2007). The constantly increasing demand and probable changes in the way water resources are distributed in future will be a challenge for water resources managers around the world. Local-scale information about the change in precipitation obtained from large-scale GCMs outputs is used to provide as an input for hydrological modeling, which operates on a daily, weekly, and monthly scales (Rosenberg et al., 2003). Direct use of GCM outputs is not acceptable for estimation of hydrological response to climate change due to its coarser spatial resolution.

However, the GCMs are extensively accepted for future climate change prediction (Dibike and Coulibaly, 2005). Downscaling is the methodology used to fill the gap between large-scale GCM outputs and finer-scale (local-scale) requirements (Chen et al., 2010). Basically, statistical downscaling and dynamical downscaling are two downscaling techniques applied to derive the local-scale information from GCM outputs. Dynamical downscaling has a high complexity, high computational cost and require detailed awareness of physical processes of the hydrological cycle, based on nesting RCM into an existing GCM (Anandhi et al., 2008), identification of bias from GCM to RCM is another problem (Giorgi et al., 2001).

Statistical downscaling methods are most widely used to derive the climate information at a regional scale in which statistical relationships are developed between the large scale (predictors) variables and local scale (pedcitands) parameters (Chen et al., 2010). Statistical downscaling has a low computational cost and can apply without knowing information about the physical processes of the hydrological cycle and geography of the study. Statistical downscaling requires long-term observed data for the model calibration and validation (Heyen et al., 1996) and it assumes that the relationship between coarse-scale and finer-scale climate are constant to generate point/station data of a specific the region using GCMs output variables (Anandhi et al., 2014). Several statistical approaches are available to the description of these relationships are multiple linear regression (MLR), canonical correlation analysis (CCA), and support vector machines (SVMs), etc. (Anandhi et al., 2008, 2009; Chen et al., 2010; Chu et al., 2010). The present study focused on downscale rainfall over a part of the Brahmaputra river basin, India, using the statistical downscaling technique for estimation of average monthly rainfall at six selected stations.

33.2   MATERIALS AND METHODS

33.2.1   STUDY AREA

The study area is Dikhow catchment which is a part of the larger Brahmaputra river basin, Dikhow river which originates from the hills of the state Nagaland. It is a south bank tributary of river Brahmaputra contributing 0.7% runoff. A lower Brahmaputra river basin, a region where the hydrological impact of climate change is expected to be particularly strong, and population pressure is high (Gain & Giupponi, 2015). Brahmaputra river is the biggest trans-Himalayan river basin (Sharma & Flügel, 2015). Study area situated between 94° 28'49"E to 95° 09' 52" E longitude and 26° 52' 20"N to 26° 03' 50" N latitude. The geographical area of Dikhow catchment is about 3100 km².

33.2.2   DATA USED

Statistical downscaling assumes the validity of relationships between predictors (GCM outputs) and predictands (observed meteorological variables) under future climate change (Yan et al., 2011). For the present study, NCEP reanalysis datasets used for the predictor selection to downscale precipitation at selected stations. And observed data of monthly precipitation were procured from IMD website.

33.2.2.1   LARGE SCALE ATMOSPHERIC VARIABLES

A daily reanalysis dataset of NCEP (scale of 2.5° latitude × 2.5° longitude) re-gridded on a scale of 2.5° latitude × 3.75° longitude and simulated data of HadCM3 on a scale 2.5° latitude × 3.75° longitude were downloaded from the website http://www.cics.uvic.ca/scenarios/sdsm/select.cgi and described in Table 33.1. The data were downloaded for two different emission scenarios A2 and B2; both are regionally focused, but the priority of first to economic issues (A2 scenario) and other to environmental issues (B2 scenario). HadCM3 GCM is a complex model of land surface processes; however, it is unique, most popular and does not require flux adjustments to produce a realistic scenario (Toews and Allen, 2009). The A2 and B2 scenarios are applied in this study. The description of both scenarios: A2 describes cultural identities separate for the different regions, making the world more heterogeneous and international cooperation is less likely. "Family values," local traditions and high population growth (0.83% per year) are emphasized with less focus on economic growth (1.65%/year) and material wealth. And B2 describes a heterogeneous society that emphasizes local solutions to economic, social, and environmental sustainability rather than global solutions. Human welfare, equality, and environmental protection have high priority, specified in the Special Report on Emissions Scenarios (SRES).

TABLE 33.1   List of 26 Predictor Variables

Variable	Description	Variable	Description
P5_f	Geostrophic air flow velocity at 500hPa	P8_f	Geostrophic air flow velocity at 850 hPa
P5_u	Horizontal wind at 500 hPa	P8_u	Horizontal wind at 850 hPa
P5_v	Zonal wind at 500hPa	P8_v	Zonal wind at 850 hPa
P5_z	Vorticity at 500 hPa	P8_z	Vorticity at 850 hPa
P5zh	Divergence at 500 hPa	P8zh	Divergence at 850 hPa
P500	Geopotential height at 500 hPa	P850	Geopotential height at 850 hPa
R500	Relative Humidity at 500 hPa	R850	Relative Humidity at 850 hPa
Pf	Surface geostrophic airflow	mslp	Mean sea level pressure
Pu	Surface horizontal wind	rhum	Near-surface relative humidity
P_v	Surface zonal wind	shum	Near-surface specific humidity
P_z	Surface vorticity	Temp2	2 m air temperature
P_zh	Surface divergence	P5th	500 hpa wind direction
P8th	850 hpa wind direction	Pth	Surface wind direction

The daily data of A2 and B2 scenarios for nine grid points, latitude varies from 25° N to 30° N and longitude varies from 89° 59' 6" E to 97° 29' 6" E revealed in Figure 33.1 were downloaded. As suggested by Tripathi et al. (2006) and Anandhi et al. (2008), they used 6 × 6and 3 × 3 grid respectively in their study, for downscaling (of precipitation and temperature) purpose. Though, there is no rule for the size of atmospheric domain selection; large domain size may allow to identify better relationship (correlations) between predictors and predictand it may also have a drawback to increase computational cost. Predictor variables are listed below. The predictor variables are available at a grid resolution of 2.5° latitude × 3.75° longitude.

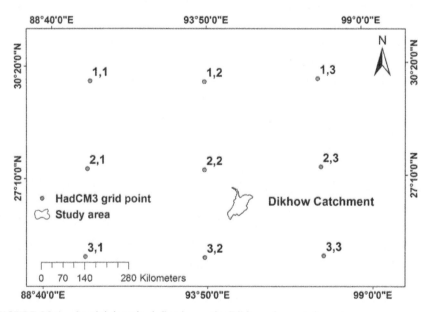

FIGURE 33.1   Spatial domain defined over the Dikhow river catchment.

33.2.3   STATISTICAL DOWNSCALING

Statistical downscaling is performed in this study by employing MLR approach. The procedure of downscaling of precipitation and temperature using MLR are explained in the following subsections.

33.2.3.1   POTENTIAL AND PROBABLE PREDICTORS FOR DOWNSCALING

Selection of predictor variables along with spatial domain plays a key role in statistical downscaling performance; predictor selection should be on first priority (Wilby and Wigly, 1997; Fowler et al., 2007). Detailed knowledge about the physics of predictors and its relation with predictand may help for the selection procedure. In the present study, twenty-six different atmospheric variables were (listed in Table 33.1), extracted from a daily reanalysis dataset of the NCEP for the years 1961–2001 based on the literature on similar studies. Predictor selection depends on the selected region and physical characteristics of the high-resolution atmospheric circulation. Wetterhall et al., (2005) suggested that predictors selection is strongly based on relationship existence (between the predictor and the predictand).

33.2.3.2   BRIEF DESCRIPTION OF MULTIPLE LINEAR REGRESSION METHOD

MLR technique is most popular in downscaling of large-scale GCM outputs (e.g., Huth, 1999; Murphy, 1999; Schoof and Pryor, 2001; Hay and Clark, 2003) of climate change impact studies. MLR is a statistical technique, which is used to develop a linear relationship between a predictand (dependent variable) and predictors (one or more independent variables), it is a least square-based method, and assumes a linear relationship between variables. Therefore, the MLR model can be expressed as a linear function (Eq. 1).

$$y = \beta_0 + \beta_1 x_1 + \beta_2 x_2 + \beta_3 x_3 \ldots \ldots \ldots + \beta_n x_n \tag{1}$$

where $\{i = 1 \ldots \ldots \ldots n\}$, y is the predictand, β_0 is the intercept, x_i is the ith predictor variable, and β_i is the ith predictor variable coefficient. MLR attempts to find a plane of best fit, and the fit can be assessed by the correlation coefficient (R) expresses the degree to which two or more predictors are related to the predictand.

Bias Correction: The bias correction techniques were applied to remove any bias in future predictands as per need. For precipitation a simple linear correction was used (Eq. 2):

$$P_{meancorr} = P_{gcm} \times \left(\frac{\overline{P_{obs}}}{\overline{P_{gcm}}} \right) \tag{2}$$

where, $P_{meancorr}$ = bias-corrected precipitation monthly, P_{gcm} = uncorrected precipitation monthly, $\overline{P_{obs}}$ = mean of monthly observed precipitation, $\overline{P_{gcm}}$ = mean of uncorrected precipitation monthly.

33.2.3.3 MODEL PERFORMANCE

Monthly reanalysis data of National Center for Environmental Prediction (NCEP) and the predictands (Tmin, Tmax, and precipitation) data from the years 1961 to 1990 were used in the calibration (training) and validation (testing) of the downscaling models. The predictands and NCEP data (treated as an output of a typical GCM) were divided into two data sets (Cannon and Whitfield, 2002): one for the training of the model and another for testing, 70% of data were chosen for training, and remaining 30% for testing of the model. Nash-Sutcliffe Efficiency (NSE), Coefficient of Correlation (R), Root Mean Square Error (RMSE) and graphical representation were used as performance indicators of the model during training and testing period for each station of the study area.

33.3 RESULTS AND DISCUSSIONS

33.3.1 SELECTED PREDICTOR VARIABLES

Pearson correlation coefficients between each probable predictor at each grid point in the spatial domain and predictands (Precipitation, Tmax, and Tmin) were calculated. The probable variables that showed the best, statistically significant (95% confidence level, $p = 0.05$) correlations with the predictands, consistently over the above mentioned period were selected as potential variables. Table 33.2 shows the potential predictors with their grid locations (according to Figure 33.1) used as the final inputs to the MLR based downscaling model for respective predictand variables. Potential predictors used in MLR model consist of horizontal wind at 500 hPa, wind direction at 500 hPa, geopotential height at 500 hPa (Devak and Dhanya, 2014; Anandhi et al., 2008), relative humidity at 500 hPa, near-surface relative humidity, and air temperature at 2 m.

33.3.2 CALIBRATION AND VALIDATION OF MODEL

The selected potential variables were introduced to the MLR model, which showed the best correlation coefficients. The model performances in the calibration and validation were monitored with the original NSE formula, in addition, the correlation coefficient (R) and RMSE were also used. The model that displayed the best NSE in both calibration and validation was selected as the best model. The model performances were monitored using the raw values of observed and predicted values for each predictand variable. The performance of the downscaling method was also evaluated by comparing data distributions (Cheng et al., 2008) to produce visual comparisons of model predicted values with observed values. The same process was repeated for each of the predictand variable (precipitation, Tmin, and Tmax) for calibration and validation of the MLR model and discussed below separately.

TABLE 33.2 Summary of Selected Large-Scale Predictor Variables Corresponding to Each of the Predictands

Precipitation			
S.N.	Potential predictor selected	Grid location	Correlation
1	Horizontal wind at 500 hPa	(2,1), (2,2), (2,3),(3,2), (3,3)	0.601–0.712
2	Air temperature at 2 m	(2,2), (2,3), (3,1),(3,2), (3,3)	0.791 – 0.812
3	Geopotential height at 500 hPa	(2,1), (2,2), (2,3),(3,2), (3,3)	0.710–0.737
4	Relative Humidity at 500 hPa	(2,2), (2,3), (3,1),(3,2), (3,3)	0.0732 –.794
5	Horizontal wind at 850 hPa	(2,2), (2,3), (3,1),(3,2), (3,3)	0.681–0.704
6	Wind direction at 500 hPa	(2,1), (2,2), (2,3),(3,2), (3,3)	0.412–0.582

33.3.2.1   PRECIPITATION

Rain occurrence and amount of precipitation are stochastic processes; therefore, the downscaling of precipitation is always a difficult problem (Doyle et al., 1997). Table 33.3 shows that for MLR model NSE of 0.88 to 0.89 was obtained for calibration and 0.88 to 0.92 for the validation period. The corresponding RMSE for calibration and validation varies from 14.31 to 18.89 and 14.41 to 19.21 respectively. Correlation coefficient (R) 0.88 to 0.89 was obtained for calibration and 0.87 to 0.92 for validation respectively.

Figure 33.2-f shows the graphical comparison of observed and model predicted monthly precipitation for the calibration (1961–1990) and validation (1991–2001) periods for each station, and graphs were plotted between precipitation in mm at y-axis and time in months at the x-axis. The Red dotted line shows the observed precipitation values, whereas green and blue line shows simulated precipitation for calibration and validation period. The results show under-prediction of the highest precipitation values during the calibration phase (except 1961, 1967, 1997, and 1979) and some of the years of validation phase (except 1994, 1997, 1999 and 2001) at almost all the stations (Sibsagar, Mokokchung, Mon, Tirap, Tuensang, and Zunheboto). This can be the explanation that the MLR model is unable to capture the extreme values of precipitation but follows good behavior for non-monsoon months.

TABLE 33.3 The Performance of Model for Precipitation During Calibration and Validation Period

Precipitation Calibration (1961–1998)			
Station	**NSE**	**R**	**RMSE**
Sibsagar	0.88	0.88	14.31
Mokokchung	0.89	0.89	15.83
Mon	0.88	0.89	18.84
Tirap	0.88	0.89	15.8
Tuensang	0.89	0.89	17.02
Zunheboto	0.89	0.89	18.89
Precipitation Validation (1991–2001)			
Station	NS	R	RMSE
Sibsagar	0.88	0.87	15.69
Mokokchung	0.91	0.9	15.89
Mon	0.9	0.9	14.66
Tirap	0.91	0.91	14.73
Tuensang	0.92	0.92	19.21
Zunheboto	0.91	0.91	14.41

33.3.3 FUTURE PROJECTION OF PRECIPITATION AND TEMPERATURE TIME SERIES FOR A2 AND B2 SCENARIOS

Figure 33.3 presents the projected precipitation, for the period of 2010–2099 at each station of the study area for both A2 and B2 scenarios. The middle line of the box shows the median values, whereas upper and lower edges show the 75%ile and 25%ile of the dataset, respectively. The difference between 75%ile and 25%ile is known as Inter Quartile (IQR). The red dotted line with circle represents the mean value. Figures 4.6 to 4.8 demonstrate the projected output for A2 and B2 scenarios, A2 having a higher magnitude than the B2 scenario for precipitation. Precipitation was projected uniformly over all the stations of the catchment.

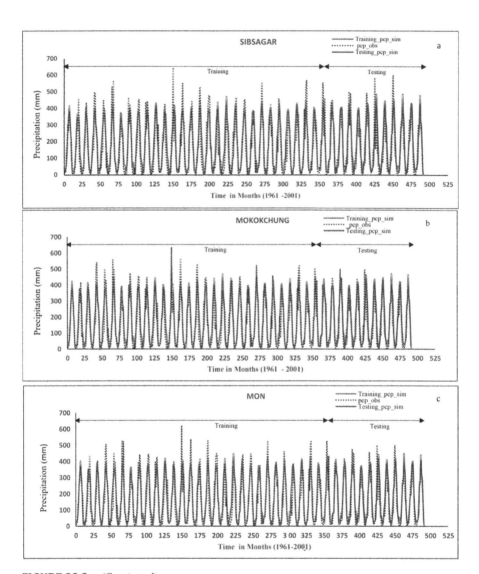

FIGURE 33.2   *(Continued)*

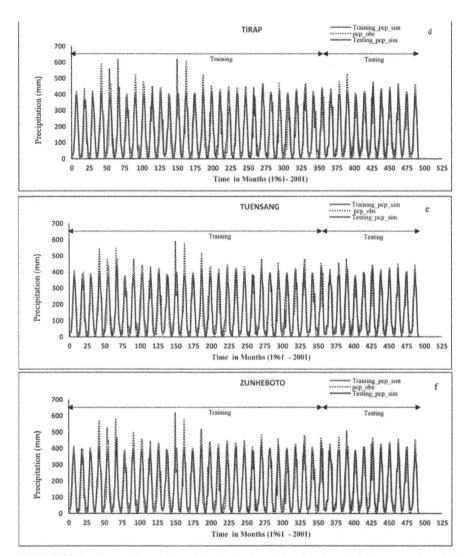

FIGURE 33.2 Comparison of monthly observed and simulated precipitation at stations (a) Sibsagar, (b) Mokokchung, (c) Mon, (d) Tirap, (e) Tuensang, and (f) Zunheboto for training (1961–1990) and validation period (1991–2001).

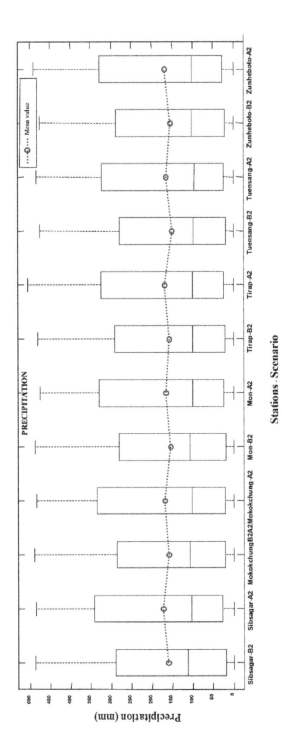

FIGURE 33.3   Projected scenarios A2 and B2 for monthly precipitation from 2010–2099 at each station of Dikhow catchment.

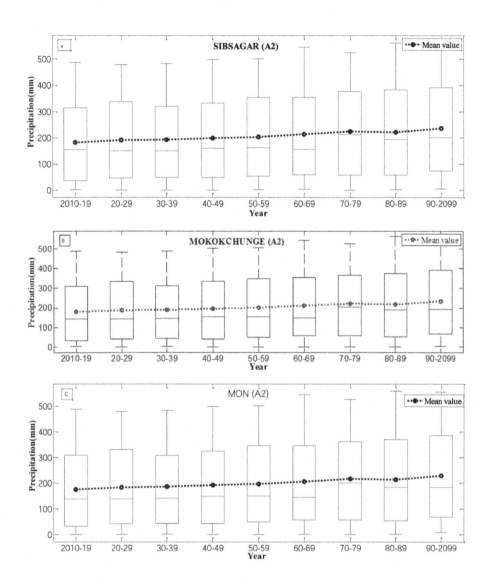

FIGURE 33.4 *(Continued)*

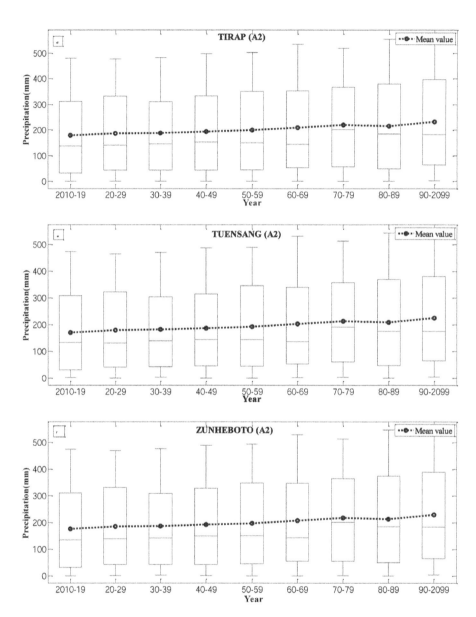

FIGURE 33.4 Box plot depicts the decadal changes in downscaled precipitation from 2010–2099 at (a) Sibsagar, (b) Mokokchung, (c) Mon, (d) Tirap, (e) Tunesang and (f) Zunheboto of Dikhow catchment for emission scenario A2. The red horizontal line in the boxes denotes the median. The red dotted line with circle represents the mean value.

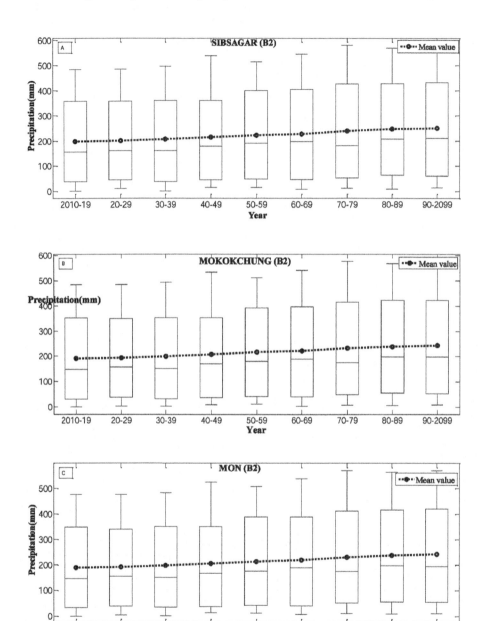

FIGURE 33.5 *(Continued)*

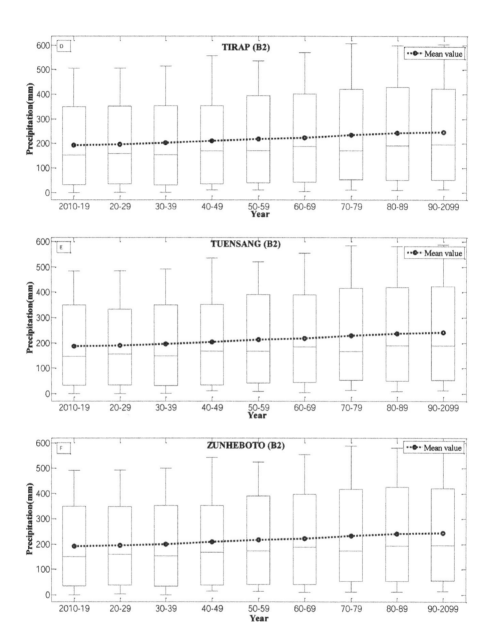

FIGURE 33.5 Box plot depicts the decadal changes in downscaled precipitation from 2010–2099 at (A) Sibsagar, (B) Mokokchung, (C) Mon, (D) Tirap, (E) Tunesang and (F) Zunheboto of Dikhow catchment for emission scenario B2. The red horizontal line in the boxes denotes the median. The red dotted line with circle represents the mean value.

33.3.4   DECADAL CHANGES OF CLIMATE PARAMETER

The box plots of 10-years' time slice are used to determine patterns in predictand. Figures 33.4a–f and 33.5a–f demonstrates the rise in future precipitation at all stations for both A2 and B2 scenarios. Annual future projection of precipitation using MLR model has figured out 2420.3 mm for both A2 and B2 Scenarios over the catchment. Seasonal future projection of precipitation was also figured out; monsoon precipitation 1857.5 mm and 1715.7 mm, summer precipitation 606.3 mm and 559.7 mm whereas winter precipitation 139.8 mm and 144.05 mm were projected at the catchment for A2 and B2 scenarios respectively. Seasonal precipitation was projected more for A2 scenario than the B2 scenario.

Statistical downscaling (Goubanova et al., 2011; Lee et al., 2011) techniques are an effective tool to fill the gap between large-scale climate change and local-scale hydrological response (Chu et al., 2010). Downscaling of local/station-scale monthly meteorological parameters is important to better extend climate change impacts analysis in a variety of environmental assessments (Cheng et al., 2005). MLR (Yan et al., 2011; Sachindra et al., 2013) method was presented to derive station-scale monthly future climate scenarios in terms of various meteorological variables, including precipitation. Performance of the downscaling method was evaluated by (1) analyzing model R, RSME, and NSE, (2) validating downscaling method using cross-validation, (3) comparing data distributions of downscaled GCM historical runs with observations over the same period of 1961–2001. The results showed that MLR-based downscaling methods used in the study performed well in deriving station-scale monthly climate scenarios. However, these methods implicitly assume that historical relationships between GCM-scale synoptic field and station scale responses of the past years would remain constant in the future (Cheng et al., 2008; Ghosh and Mujumdar, 2006). Downscaling shows an increasing trend in precipitation, in general, it was noted that the results show an agreement with the findings of other research works like Duhan et al. (2015); Saraf and Regulwar (2016) over India, and other region. Water resource management (Mahanta, 2006) and planning increasingly need to incorporate the effects of global climate change on regional climate variability in order to assess future water supplies accurately. Therefore, future climate projections, particularly of rainfall is utmost interest to water resource management and water-users (Fu et al., 2011).

33.4   CONCLUSIONS

The following conclusions are drawn from the present study:

1. Six potential predictors wind at 500 hPa, wind direction at 500 hPa, geopotential height at 500 hPa, relative humidity at 500 hPa, near-surface relative humidity, and air temperature at 2 m are selected from 26 large-scale variables.
2. The results of calibration and validation show that the predictands can be downscaled potentially using MLR models, the model has performed well over the region.
3. The future projection shows that precipitation may increase in future at all stations for both A2 and B2 scenarios.
4. The magnitude of increase in precipitation is higher for A2 scenario than that of B2 scenario.

KEYWORDS

- **downscaling techniques**
- **GCM**
- **HadCM3**
- **IPCC 2007**

REFERENCES

Anandhi, A., Srinivas, V. V., Kumar, D. N., & Nanjundiah, R. S., (2009). Role of predictors in downscaling surface temperature to river basin in India for IPCC SRES scenarios using support vector machine. *International Journal of Climatology, 29*(4), 583–603.

Anandhi, A., Srinivas, V. V., Kumar, D. N., Nanjundiah, R. S., & Gowda, P. H., (2014). Climate change scenarios of surface solar radiation in data sparse regions: a case study in Malaprabha River Basin, India. *Climate Research, 59*(3), 259.

Anandhi, A., Srinivas, V. V., Nanjundiah, R. S., & Nagesh, K. D., (2008). Downscaling precipitation to river basin in India for IPCC SRES scenarios using support vector machine. *International Journal of Climatology, 28*(3), 401–420.

Chen, S. T., Yu, P. S., & Tang, Y. H., (2010). Statistical downscaling of daily precipitation using support vector machines and multivariate analysis. *Journal of Hydrology, 385*(1), 13–22.

Cheng, C. S., Li, G., Li, Q., & Auld, H., (2008). Statistical downscaling of hourly and daily climate scenarios for various meteorological variables in south-central Canada. *Theoretical and Applied Climatology*, *91*(1–4), 129–147.

Cheng, Y. S., Zhou, Y., Irvin, C. M., Pierce, R. H., Naar, J., Backer, L. C., & Baden, D. G., (2005). Characterization of marine aerosol for assessment of human exposure to brevetoxins. *Environmental Health Perspectives*, 638–643.

Chu, J. T., Xia, J., Xu, C. Y., & Singh, V. P., (2010). Statistical downscaling of daily mean temperature, pan evaporation and precipitation for climate change scenarios in Haihe River, China. *Theoretical and Applied Climatology*, *99*(1–2), 149–161.

Devak, M., & Dhanya, C. T., (2014). Downscaling of precipitation in Mahanadi basin, India. *International Journal Civil Engineering Research*, *5*, 111–120.

Dibike, Y. B., & Coulibaly, P., (2005). Temporal neural networks for downscaling climate variability and extremes. In: *Neural Networks, 2005. IJCNN'05, Proceedings, 2005 IEEE International Joint Conference* (Vol. 3, pp. 1636–1641).

Doyle, J. D., & Foley, D. M., (1999). *Veneziani* (2ndpart, p. 438), CA: Edwards.

Duhan, D., & Pandey, A., (2015). Statistical downscaling of temperature using three techniques in the Tons River basin in Central India. *Theoretical and Applied Climatology*, *121*(3/4), 605–622.

Fowler, H. J., Blenkinsop, S., & Tebaldi, C., (2007). Linking climate change modeling to impacts studies: Recent advances in downscaling techniques for hydrological modeling. *International Journal of Climatology*, *27*(12), 1547–1578.

Fu, G., Charles, S. P., Chiew, F. H. S., & Teng, J., (2011). Statistical downscaling of daily rainfall for hydrological impact assessment. In: *MODSIM2011, 19th International Congress on Modeling and Simulation* (pp. 12–16). Perth, Australia.

Gain, A. K., & Giupponi, C. A., (2015). Dynamic assessment of water scarcity risk in the lower Brahmaputra river basin: An integrated approach. *Ecological Indicators*, *48*, 120–131.

Gain, A. K., Giupponi, C., & Renaud, F. G., (2012). Climate change adaptation and vulnerability assessment of water resources systems in developing countries: A generalized framework and a feasibility study in Bangladesh. *Water*, *4*(2), 345–366.

Ghosh, S., & Mujumdar, P. P., (2006). Future rainfall scenario over Orissa with GCM projections by statistical downscaling. *Current Science*, *90*(3), 396–404.

Giorgi, F., Whetton, P. H., Jones, R. G., Christensen, J. H., Mearns, L. O., Hewitson, B., & Jack, C., (2001). Emerging patterns of simulated regional climatic changes for the 21st century due to anthropogenic forcings. *Geophysical Research Letters*, *28*(17), 3317–3320.

Goubanova, K., Echevin, V., Dewitte, B., Codron, F., Takahashi, K., Terray, P., & Vrac, M., (2011). Statistical downscaling of sea-surface wind over the Peru–Chile upwelling region: Diagnosing the impact of climate change from the IPSL-CM4 model. *Climate Dynamics*, *36*(7/8), 1365–1378.

Hay, L. E., & Clark, M. P., (2003). Use of statistically and dynamically downscaled atmospheric model output for hydrologic simulations in three mountainous basins in the western United States. *Journal of Hydrology*, *282*(1), 56–75.

Heyen, H., Zorita, E., & Von Storch, H., (1996). Statistical downscaling of monthly mean North Atlantic air-pressure to sea level anomalies in the Baltic Sea. *Tellus. A*, *48*(2), 312–323.

Huth, R., (1999). Statistical downscaling in central Europe: Evaluation of methods and potential predictors. *Climate Research*, *13*(2), 91–101.

Lee, T. C., Chan, K. Y., & Ginn, W. L., (2011). Projection of extreme temperatures in Hong Kong in the 21st century. *Acta Meteorologica Sinica*, *25*, 1–20.

Mahanta, C., (2006). Water resources of the Northeast: State of the knowledge base. *Background Paper*, 2.

Murphy, J., (1999). An evaluation of statistical and dynamical techniques for downscaling local climate. *Journal of Climate, 12*(8), 2256–2284.

Pechstädt, J. O. R. G., Bartosch, A. N. I. T. A., Zander, F. R. A. N., Schmied, H. M., & Flugel, W. A., (2011). *Development of a River Basin Information System for a Sustainable Development in the Upper Brahmaputra River Basin.*

Rosenberg, N. J., Brown, R. A., Izaurralde, R. C., & Thomson, A. M., (2003). Integrated assessment of Hadley Centre (HadCM2) climate change projections on agricultural productivity and irrigation water supply in the conterminous United States: I. Climate change scenarios and impacts on irrigation water supply simulated with the HUMUS model. *Agricultural and Forest Meteorology, 117*(1), 73–96.

Sachindra, D. A., Huang, F., Barton, A., & Perera, B. J. C., (2013). Least square support vector and multi-linear regression for statistically downscaling general circulation model outputs to catchment streamflows. *International Journal of Climatology, 33*(5), 1087–1106.

Sahu, R. K., Mishra, S. K., Eldho, T. I., & Jain, M. K., (2007). An advanced soil moisture accounting procedure for SCS curve number method. *Hydrological Processes, 21*(21), 2872–2881.

Saraf, V. R., & Regulwar, D. G., (2016). Assessment of climate change for precipitation and temperature using statistical downscaling methods in upper Godavari River Basin, India. *Journal of Water Resource and Protection, 8*(01), 31.

Schoof, J. T., & Pryor, S. C., (2001). Downscaling temperature and precipitation: A comparison of regression-based methods and artificial neural networks. *International Journal of Climatology, 21*(7), 773–790.

Toews, M. W., & Allen, D. M., (2009). Evaluating different GCMs for predicting spatial recharge in an irrigated arid region. *Journal of Hydrology, 374*(3), 265–281.

Wetterhall, F., Halldin, S., & Xu, C. Y., (2005). Statistical precipitation downscaling in central Sweden with the analog method. *Journal of Hydrology, 306*(1), 174–190.

Wilby, R. L., & Wigley, T. M. L., (1997). Downscaling general circulation model output: A review of methods and limitations. *Progress in Physical Geography, 21*(4), 530–548.

Yan, G., Jian-Ping, L., & Yun, L., (2011). Statistically downscaled summer rainfall over the middle-lower reaches of the Yangtze River. *Atmospheric and Oceanic Science Letters, 4*(4), 191–198.

CHAPTER 34

GROUNDWATER PROSPECT MAPPING IN UPPER SOUTH KOEL RIVER BASIN JHARKHAND (INDIA) BASED ON GIS AND REMOTE SENSING TECHNOLOGIES

STUTI[1], ARVIND CHANDRA PANDEY[2], and SAURABH KUMAR GUPTA[1]

[1]*Research Scholar, Centre for Land Resource Management, School of Natural Resource Management, Central University of Jharkhand, Brambe, Jharkhand, India, E-mail: stuti@cuj.ac.in*

[2]*Professor, School of Natural Resource Management, Centre for Land Resource Management, Central University of Jharkhand, Brambe, Jharkhand, India*

ABSTRACT

Groundwater is dynamic natural resources which depend on various parameters mainly hydrogeomorphology of an area. In hard rock terrain availability of groundwater is limited. In such terrains groundwater is essentially confined to fractured and weathered zones. Because of undulating topography and hard rock terrain in Upper South Koel river basin the availability of ground water is limited. Multispectral satellite image from LISS-III sensor having resolution 23.5 m is used for the creation of various thematic layers like land-use landcover, lithology, lineaments, and hydrogeomorphology. Cartosat 1 DEM data is used for the creation of a slope map of the area. Field survey data and GPS point records were used to create groundwater yield map and depth to water level map. Overlaying various weighted thematic maps prepared in Arc GIS platform and validated with field surveying data help to demarcate potential groundwaterpotential zones in the area. Very good groundwater prospect zones are mainly located along the valley fills, covering an area of 1.92 sq. km. whereas very poor groundwater prospect zones lies along dissected pediments, structural hills, relict hills, covering an area of 15 sq. km. Groundwater management and planning must be done in the areas having very poor groundwater prospects to enhance groundwater potential and to combat future drought. Geoinformatics techniques facilitate effective evaluation of groundwater potential for effective watershed development planning.

34.1   INTRODUCTION

Groundwater is the main source of drinking water in India which helps to sustain life on earth. In order to maintain the sustainability of groundwater, there is a need to understand the groundwater prospects, hydrogeomorphology as well as the geological setting of that area. Water scarcity is the major trouble in the modern-day situation, to overcome this proper water management via underground or surface water management need to be done to reduce water shortage and fight drought. It is now felt that to deal with and overcome those issues, the drinking water supply schemes need to be evolved considering the hydrogeological records.

Jharkhand is hard rock terrain regions and the area under study, i.e., part of upper South Koel river basin not having good groundwater prospect zones and faces severe water scarcity. Surface water assets in the area inadequate to satisfy the nearby need, consequently to meet the current need, exploration, and exploitation of groundwater resources require thorough expertise of geology, hydrology, and geomorphology of the location. Hence an assessment of this useful resource is extremely huge for the sustainable control of the groundwater system in this area. A number of works considered within the area of groundwater potential zones. Application of remote sensing (RS) and geographical information systems (GIS) for the exploration of potential groundwater zones is achieved by some of the researchers around the world. Teeuw (1995) relied handiest on the lineaments for groundwater exploration, even as others merged various factors aside from the lineaments like drainage density, geomorphology, geology, slope, land use, rainfall intensity and soil texture (Sander et al., 1996; Sener et al., 2005; Ganapuram et al., 2009). The satellite statistics gives quick and beneficial baseline records about various factors controlling directly or indirectly the occurrence and movement of groundwater which includes geomorphology, soil, land slope, land use/land cover, drainage patterns and lineaments (Waters et al., 1990; Meijerink, 1996; Jha et al., 2007). RS research offers an opportunity for better observation and extra systematic analysis of various hydrogeomorphic units/ landforms/ lineaments capabilities following the synoptic, multispectral, and repetitive coverage of the terrain (Horton, 1945; Kumar and Srivastava, 1991; Sharma and Jugran, 1992). Remotely sensed information are generally price effective as compared to the traditional techniques of hydrological surveys and specifically are of first-rate significance in remote areas (Machiwal et al., 2011). GIS gives an exquisite framework for efficiently managing big and complicated spatial data for natural asset control, accordingly, it has proved to be a useful tool for groundwater studies (Krishnamurthy et al., 1996; Meijerink, 1996; Nour, 1996; Sander et al., 1996). In the beyond, several researchers have used RS and GIS techniques for the delineation of groundwater potential zones (Chi and Lee, 1994; Kamaraju et al., 1995) with a successful result.

In last two decades, many researchers have found that multi-criteria decision-making (MCDM) presents a powerful tool for water management by using structure, shape, auditability, transparency to the decision (Flug et al., 2000; Joubert et al., 2003). Recently, Hajkowicz and Higgins (2008) counseled that at the same time as the choice of a MCDM approach is important for water resources management, extra emphasis is required at the preliminary structuring of a selective problem which entails selecting criteria and decision alternatives. Integration of RS with GIS for making numerous thematic layers, together with lithology, drainage density, lineament density, rainfall, slope, soil, and land use with assigned weightage in a spatial domain will aid the identification of ability groundwater zones. The present study aims at the identification of groundwater potential zones of the study area as part of upper South Koel river basin, Jharkhand, India using RS, GIS, and weighted overlay analysis for groundwater resources exploration. The predominant goal of this study is to construct groundwater potential map of the hard rock terrain of the study area. This study targets to spatial analyze the connection among groundwater potential and terrain and hydrological parameters which manipulate groundwater accumulation.

34.2 MATERIALS AND METHODS

34.2.1 STUDY AREA

The watershed is a part of South Koel basin with an area of 772 km^2 bounded by latitude 23°17'16"N & 23°32'16"N and longitude 84°14'15"E & 85°46'51"E (Figure 34.1). It lies in SOI toposheet no 73 A/14, 73 A/15, 73 E/2, 73 E/3 covering Lohardaga and Ranchi districts of Jharkhand (Figure 34.2). South Koel is the main river in the study area with Kandani and Saphi River as its major tributaries. The drainage pattern is mainly dendritic. The climate of the area is subtropical. The annual rainfall in the region is

1400 mm, on an average of which 82.1% is received during the periods June to September and the rest 17.9% in remaining months. Temperature is lowest during December and January with a mean minimum of 9°C and highest during April and May with a mean maximum of 37.2°C.

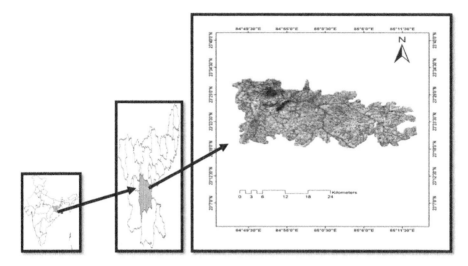

FIGURE 34.1   Map showing a satellite image of study area of LISS III data.

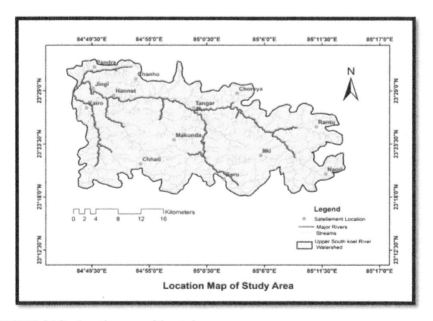

FIGURE 34.2   Location map of the study area.

34.2.2   METHODOLOGY

The details of the steps followed in the study are illustrated in Figure 34.3. Multispectral Satellite image of LISS-III sensor has been used for preparing thematic maps of land-use landcover, lithology, lineament, and geomorphology of the study area with the help of image interpretation keys, i.e., color, shape, size, texture, pattern, shadow, and association. Nine major class of LULC has been identified as Intense agriculture (143.65 km^2), Sparse agriculture (148.72 km^2), Open forest (39.59 km^2), Degraded forest (46.54 km^2), Barren land (91 km^2), Barren land rocky (80.37 km^2), Water bodies/reservoir (9.24 km^2), Built up areas (81.68 km^2) as shown (Figure 34.4) which is further validated with google earth data and survey of India toposheet of 1:50,000. Drainage order map has been prepared with the help of toposheet of 1.50,000 (SOI) which shows 6th order watershed. Lithology map has been prepared which shows mainly five classes as Alluvium, Granite Gneiss, Hornblende Schist and Amphibolite, Schist, metabasic dykes and laterite as shown (Figure 34.5). Lineament map has been prepared which basically shows major lineaments and dissected lineaments as shown in Figure 34.7. Hydrogeomorphology of study area shows nine major types of landforms, i.e., Denudational hills, Structural hills, Relict hills, Inselberg, Pediment inselberg complex, Plateau weathered shallow, Plateau weathered moderate, Pediment shallow buried, Valley fills, lineaments, Waterbodies as shown in Figure 34.6. Cartosat 1 DEM data has been used for slope map. Study area is mainly having a high slope between 0–45 degrees and elevation 766 m as highest and 677 m as lowest. Soil map is taken from National Bureau of soil science mainly showing two types of soil for the area of 772 km^2 i.e., fine loamy, coarse loamy soil which is further divided into seven classes i.e., Fine loamy (Aeric Haplaquents 309.52 km^2), coarse loamy (Haplaquents 230.02 km^2) Fine loamy (Typic Ustochrepts 91.01 km^2) Fine (Ustrochrepts 11.92 km^2), Fine loamy (Aeric 11.09 km^2), Fine loamy (Haplaustalfs 45.89 km^2), Fine (Vertic Ustochrepts 53.62 km^2) as shown in Figure 34.7. A weighted overlay analysis of various thematic maps of various layers has been done in the study area for suggesting the site to identify the areas facing severe water scarcity or less groundwater prospect. The spatial variability of groundwater in upper south koel river basin was obtained by assigning weightage to the various themes such as land-use landcover, geomorphology, slope, soil map and drainage which was overlaid in the platform of Arc GIS software. On the basis of groundwater prospect, each thematic maps were assigned weights from 1 to 10 based on an understanding of groundwater potential with 1 being considered least significant in regard to water potential zone and highest number, i.e., 10 being assigned for the most potential zone in consideration of groundwater prospect. Maximum weight to reservoir i.e., 10 which is having more probability of water contained and minimum weight to barren land rocky as 2 (Krishnamurthy et al., 1996; Malczewski, 1999; Balachandar, 2010). The given weights for each map were normalized so that the difference in the number of classes in all maps can be brought to the same scale with the value of weights in the range of 0 to 1. For normalization, weights of each class were recalculated by dividing the class weight by cumulative weight of all the classes. The total weight was calculated by multiplying the normalized weight with the theme weight, and on the basis of the total weight, all the themes were rasterized and overlaid in raster calculator of ArcGIS for estimation of groundwater prospect zones. The groundwater prospects map obtained were classified into five zones (viz., very poor, poor, moderate, good, and very good). The assignment of theme weight, unit weight, normalized weight, and total weight for the various thematic layers has been shown in Table 34.1. For validating this result ground truthing and field validation has been done. Field GPS points of different wells and their depth in different geomorphological units and data from Rajiv Gandhi Ground Water Atlas has been taken to prepare a map showing depth of water table and water yield capacity in different places of the Upper south koel river basin as shown in Figures 34.10 and 34.11 which help to validate our work.

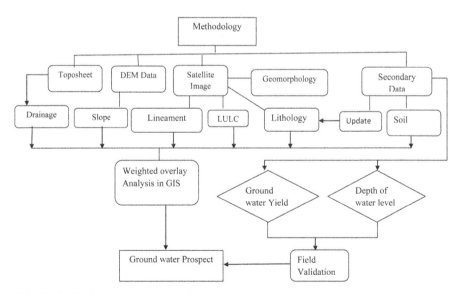

FIGURE 34.3   Flowchart of methodology.

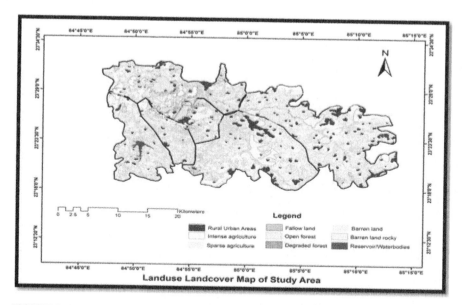

FIGURE 34.4 Map showing land use landcover of study area using LISS III data.

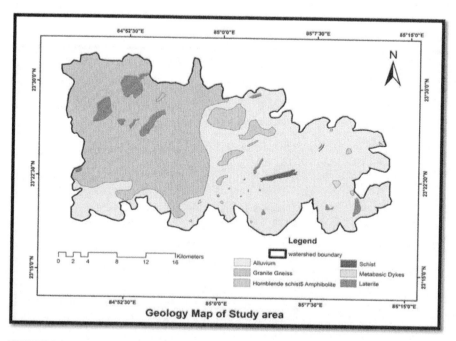

FIGURE 34.5 Map showing geology of study area using LISS III data.

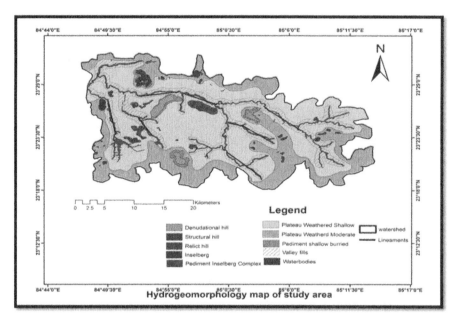

FIGURE 34.6 Map showing hydrogeomorphology of study area using LISS III satellite data.

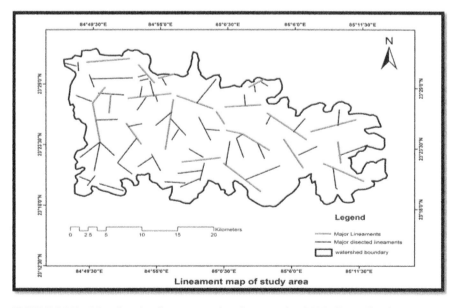

FIGURE 34.7 Map showing lineaments of study area using LISS III satellite data.

TABLE 34.1 Theme and Theme Weight, Unit Weight, Normalized Weight and Total Weight of Different Land Classes

Theme and theme weight	S. No	Class	Unit Weight	Normalized Weight	Total Weight
Landuse Land cover(5)	1	Rural-urban areas	5	0.10	0.5
	2	Intense agriculture	6	0.12	0.6
	3	Sparse agriculture	6	0.12	0.6
	4	Fallow land	3	0.06	0.3
	5	Open forest	7	0.14	0.7
	6	Degraded forest	7	0.14	0.7
	7	Barren land	3	0.06	0.3
	8	Barren land rocky	2	0.04	0.2
	9	Reservoir/water bodies	8	0.17	0.85
Hydrogeomorphology (9)	1	Denudation hills	2	0.03	0.27
	2	Structural hills	3	0.05	0.45
	3	Relict hills	3	0.05	0.45
	4	Inselberg	4	0.06	0.54
	5	Pediment inselberg complex	5	0.08	0.72
	6	Plateau weathered shallow	6	0.10	0.9
	7	Plateau weathered moderate	6	0.10	0.9
	8	Plateau shallow buried	7	0.11	0.99
	9	Valley fills	7	0.11	0.99
	10	Lineaments	8	0.13	1.17
	11	Waterbodies	8	0.13	1.17
Slope (7)	1	0–5	8	0.14	0.98
	2	5–10	7	0.12	0.84
	3	10–15	6	0.10	0.7
	4	15–20	5	0.09	0.63
	5	20–30	5	0.09	0.63
	6	30–40	3	0.05	0.35
	7	40–50	2	0.03	0.21
Stream (7)	1	1st order	4	0.03	0.21
	2	2nd order	4	0.06	0.42
	3	3rd order	5	0.12	0.84
	4	4th order	6	0.19	1.33

TABLE 34.1 *(Continued)*

Theme and theme weight	S. No	Class	Unit Weight	Normalized Weight	Total Weight
	5	5th order	7	0.25	1.75
	6	6th order	8	0.32	2.24
Lithology (6)	1	Alluvium	4	0.26	1.56
	2	Granite Gneiss	1	0.06	0.36
	3	Hornblende schist	2	0.13	0.78
	4	Amphibolite schist	2	0.13	0.78
	5	Metabasic dykes	3	0.2	1.2
	6	Laterite	3	0.2	1.2
Soil (7)	1	Coarser loamy	3	0.75	5.25
	2	Fine loamy	1	0.25	1.75

34.3   RESULT AND DISCUSSION

Different classes of Landuse landcover were digitized with the help of Arc-GIS 10.3 software based on visual interpretation by using image interpretation keys. Each land-use landcover classes having different percolation capacity of water, i.e., because of hard surface and cemented construction total 81.68 km^2 area is mainly covered with built-up areas shows nil infiltration capacity. Area of 143.65 km^2 is mainly covered with intensive agriculture where infiltration capacity is good as compared to built-up areas hence which is one of the prospects to enhance or increase groundwater recharge condition. Different land-use classes and their infiltration capacity based on soil texture of different classes as Intense agriculture (good), sparse agriculture (good), open forest (good), degraded forest (fair), reservoir/water bodies (good), lake (fair), fallow land (poor), barren land rocky (poor), barren land (poor), built up (poor). As per earlier study runoff is high in these plateau terrain areas because of that most of agricultural land is converted into fallow land which is having less infiltration capacity of water. Infiltration capacity mainly depends on soil and geology of areas. Most of the areas are mainly covered by fine loamy soil having less infiltration rate and water transmission rate.

34.3.1   LITHOLOGY

Groundwater prospect of any area is not only controlled by climatic conditions, but also the lithology and geologic structure have great control as they influence the nature of flow, erosion, and sediment transportation. The basic rocks present in the area is mainly oldest rock as unclassified metamorphic represented by Mica Schist, Hornblende schist, and amphibolites which form the basement rocks in the study area. The overlying Chotanagpur Gneissic Complex comprising Granite gneiss forms the most widespread outcrop in the study area. Laterite, metabasic dykes, and recent alluvial deposits are other rock types found in the area. Lineaments are found mainly along the valley fills and having an excellent source of groundwater. There are mainly two types of lineaments as major lineaments and major dissected lineament. Rainfall in these areas is high still facing water scarcity may be because of undulating topography or presence of basic and meta basic dykes, and quartz reefs in the areas acted as barriers for flows of water (Singh et al., 1997).

Alluvium comprising of sand silt, clay, and gravel has been given higher weightage as compared to sedimentary rocks like sandstone, shale. Among the metamorphic rocks, granitoid gneiss because of its high weathering due to a high fractured condition in the area has been given higher weightage as compared to fractured granite gneiss.

34.3.2   GEOMORPHOLOGY

Landforms observed in the study areas are Structural hill, Denudation hill, Relict hill, Inselberg, Pediment Inselberg complex, Valley fills, Plateau dissected shallow, and Plateau dissected moderate. The study area is having varied hydrogeological characteristics due to which groundwater potential differs from one region to another. Lineaments acting as fracture zones generally act as conduits for movement of groundwater in hard rocks. Along these zones, the yield is significantly higher, and wells are likely to be sustainable for a longer duration. There are dykes, quartz which generally act as barriers for groundwater movement. Drainage ordering map has been prepared which shows upper South Koel watershed having 6th order streams. Highest order is mainly along valley fills, and small order streams are mainly dominated along the structural hills and denudational hills areas as shown in Figure 34.8.

The hydrogeomorphology in the hard rock terrain is highly influenced by the lithology and structure of the underlying formations and is one of the most important features in evaluating the groundwater potential and prospect (Kumar et al., 2008). Material associated with river/ waterbodies and active and active floodplain has higher water retention capability and therefore constitutes best landforms for high groundwater potential.

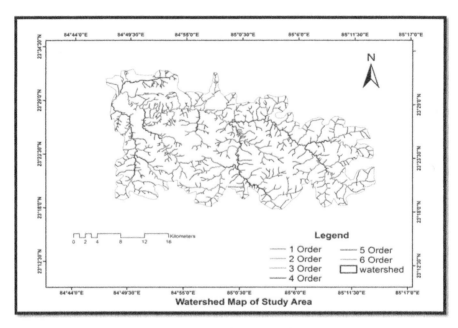

FIGURE 34.8   Map showing streams order of Upper South Koel river basin using toposheet (1:50,000).

34.3.3   SOIL

The soil for the study area is represented by four main soil categories namely fine, fine loamy, loamy, and coarse loamy (Figure 34.9). Rank of soil has been assigned on the basis of their infiltration rate; sandy soil has high infiltration rate. Hence coarse loamy soil dominated by high sand proportion has been given higher priority, whereas the fine soil exhibiting least infiltration rate due to higher clay proportion were assigned a low priority (Figure 34.10).

34.3.4   SLOPE

Slope and relief map is prepared using DEM data shows slope between 0–25 degrees in which 1.92 sq. km. of area covered with slope between (0–5) degrees show very good prospect of groundwater and along 14–92 km² of an area having slope >25 is mainly affected by water scarcity, and groundwater prospect is not too good along these areas.

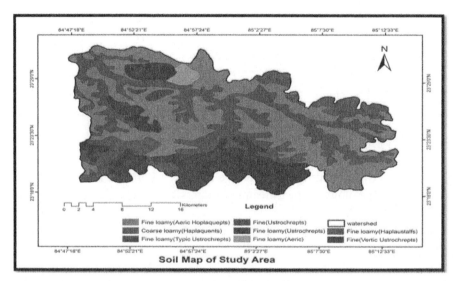

FIGURE 34.9   Soil map of study area using NBSS data source.

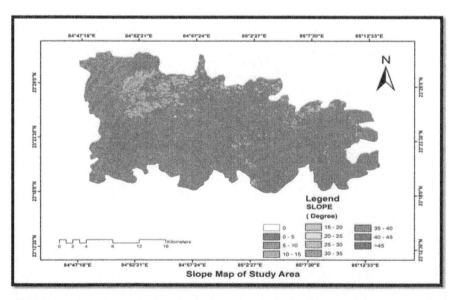

FIGURE 34.10   Slope map of the study area using Cartosat 1 DEM data.

Groundwater prospect with the help of Slope Map		
S.NO	Slope (Degree)	Area (sq km)
1	0–5	1.92
2	5–10	516.29
3	10–15	197.52
4	15–20	32.61
5	20–25	9.19
6	>25	14.92

34.3.5   WEIGHTED OVERLAY

On the basis of integrating all the above map, i.e., Landuse and cover map, Hydrogeomorphology map, Lineament map, slope, streams, lithology, and soil map on the Arc-GIS platform, the final map of Groundwater prospect is prepared which show groundwater status of the study area as shown in Figure 34.11.

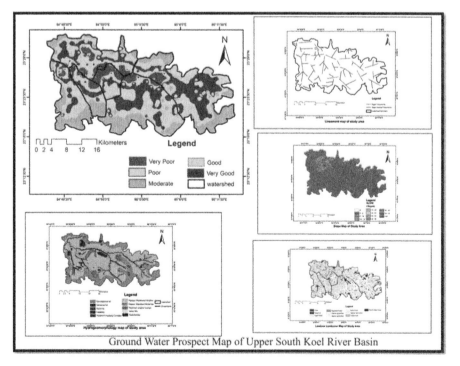

Ground Water Prospect Map of Upper South Koel River Basin

FIGURE 34.11   Groundwater prospect map of the study area using weighted overlay techniques.

34.3.6 FIELD VALIDATION

Various field data has been collected during field checks to validate our work. Number of wells in differentlanduse, i.e., no built-up areas along hills hence no wells and hand pumps were found in landforms as Structural hills, denudation hills, relict hills, and inselbergs. Depth of water level map is mainly prepared with the help of ground surveying data, i.e., the number of wells and their depth lying in the study area and the number of wells observed in different landforms of the study area. Depth of water level mainly in summer and pre-monsoon season is >14 m to 11 m. Maximum 25–35 bore wells found in plateau dissected moderate having depth of 3–6 m shown in Table 34.2. Different wells have different yield range. Yield range in this studyis varies between > 10 to >80 Lpm which shows >10 Lpm in relict hills and maximum water can be yield from valley fills, i.e., >80 Lpm. Depth rangeis mainly varies from 30–80m of the water table. Water quality is portable and having different recharge conditions shown in Table 34.3. Depth of water table and groundwater yield map has been prepared based on these data which shows the depth of water table is from < 3 m to > 14 m and having high water yield range shown in Figures 34.12 and 34.13. Water harvesting structures or recharge conditions must be developed in these areas to enhance the groundwater prospect so that maximum water yield from these tube wells. Map prepared with the help of field data shows an approx similar result with the map prepared by using Arc GIS software in

TABLE 34.2 Showing Depth of Water Level (Summer/Pre-Monsoon/Post-Monsoon) of Different Wells and Dug Wells Locations of Study Area

		Field data		
S. No	Landforms	No of Wells	Types of Wells	Depth of water level (Summer/Pre-monsoon)
1	Structural hill	No wells	-	>14 meters
2	Denudational hill	No wells	-	11–14 meters
3	Relict hill	No wells	-	11–14 meters
4	Inselberg	No wells	-	11–14 meters
5	Pediment Inselberg complex	6–7	Borewells	9–11 meters
6	Plateau weathered shallow	20–25	Borewells	6–9 meters
7	Plateau weathered moderate	25–35	Dug wells and Borewells	3–6meters
8	Valley fills	No wells	-	<3 meters

GIS platform by using weighted overlay analysis of various thematic maps which used to validate our work.

34.3.7   GROUND WATER PROSPECT

The final map generated by using overlay analysis in Arc-GIS software mainly shows five zones of groundwater prospect, i.e., very good to very poor.

TABLE 34.3   Yield Range of Wells Depth of Different Wells and Dug Wells of Different Locations of Study Area

Yield Ranges of wells				
S. No	Landforms	Yield range of wells(Lpm)	Depth range of wells(meter)	Water Quality	Recharge condition
1	Structural hill	-	-	-	-
2	Denudational hill	-	-	-	-
3	Relict hill	>10	-	-	-
4	Inselberg	10–25	30–80	Portable	Limited
5	PedimentInselberg complex	25–40	30–80	Portable	Limited
6	Plateau weathered shallow	40–65	30–80	Portable	Low priority
7	Plateau weathered Moderate	65–80	30–80	Portable	Moderate
8	Valley fills	>80	30–80	Portable	Good

1) **Very Poor**: These areas are covered with open forest and characterized by hills with a high slope greater than 25 degrees resulting in high runoff and covering an area of 15 sq. km. Soil comprised of gravel- loamy soil and not suitable for groundwater development. Depth of water level is >14 meters not having any recharge condition. Dominated with lower order streams. Therefore, groundwater prospect are nil to very poor mainly along the major cracks and weathered zones.

2) **Poor**: Mainly comprised of steep slope (20–25) degrees and covered with fallow land with very high runoff zone. Covering an area of 10 sq. km. groundwater prospects are very poor to poor in these Inselbergs. Depth of water level is 11–14 meters not having any wells and recharging condition.

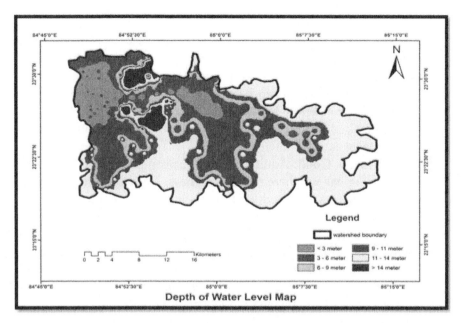

FIGURE 34.12   Map showing the depth of water level of study area using field data.

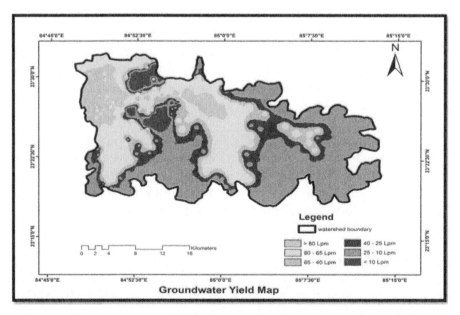

FIGURE 34.13   Map showing a groundwater yield of study area using field data.

3) **Poor – Moderate**: These comprises of isolated hills, knobs, and ridges having slope between 15–20 degrees that rise abruptly from a gently sloping or virtually level surrounding plain. The unit covering an area of 33 sq. km. represents barren land with shallow well-drained gravelly sandy soil on sloping landscape with severe erosion. Groundwater prospect are poor to moderate. Limited yield may be expected from fractured zone in these pediments a weathered and fractured rock comprises the aquifer material having depth of water level as 11–14 meters.

4) **Moderate – Good**: Aquifer material constituted by fractured rock with marginal weathered rock thickness. Because of gentle slope (10–15 degrees) covered with mainly intense agriculture with less erosion in the weathered zones, groundwater prospects are good in these zones covering an area of 197 sq. km. In hard rock's they form very good recharge and storage zones based upon the thickness of weathering/accumulated material, its composition and recharge conditions. 20–25 number of dug wells and bore wells having depth of water level as 6–9 meter of yield 40–60 Lpm.

5) **Good**: Covered by open forest and sparse agricultural land having slopes lying between 5–10 degree with aquifer material largely constituted by weathered rock, clay, and fractured rock. As weathering is comparatively more than in plateau weathered shallow, slightly higher yield may be expected from the deeper fractures under thick weathered zones. Groundwater prospects are moderate to good covering an area of 516 sq km. Depth of water level is 3–6 meters having 25–35 dug wells and bore wells of 65–80 Lpm yield range.

6) **Very Good**: Aquifer material comprises mainly of loose sediments, weathered, and fractured rock fragments mainly along higher order streams, i.e., of 6th order. Covering an area of 1.92 sq. km. valley fills are located along lineaments and in pediment areas having slope between 0–25 degrees. Due to better recharge conditions in the valley from surrounding uplands groundwater prospect are very good. Depth of water level is <3-meter having good source of recharge.

34.4   CONCLUSIONS

The final integrated map as generated by applying weighted overlay analysis in Arc GIS platform which shows five prospect grades in terms of water resources potentiality viz.; very good, good, moderate-poor, poor to moderate, poor, and very poor. Good water prospect shows very good zones cover only along valley fills (cover 45% of the area), very good zones cover only 12% of the study area. They are located along lineaments and in pediment areas. Moderate prospect zone covers 15% of the area. Features like buried pediplains, peneplains, and denundational hills come under this category, whereas poor regions cover up to 22%. These features are mainly confined to undulating upland and buried pediments with the intermountain valley. Lastly, very poor groundwater prospect zones are approximately 10%. These features include dissected pediments, inselberg complex and residual hill. Geologically it is observed that the groundwater is mainly confined to secondary porosity, i.e., fractured zone, fault, joint, and weathered column. It is observed from field survey and also from various wells located in the region the hard granite gneisses, and Meta basic dykes sometimes act as barriers for the groundwater flow in the region. The geoinformatics based groundwater potential mapping help to enhance the zones required specific watershed management to increase groundwater potential and combat future drought. This will serve as an insurance against future droughts in the study area.

ACKNOWLEDGMENTS

Authors are thankful to Bhuvan for online availability of LISS-III Satellite data and Cartosat 1 DEM data. The first author acknowledges the receipt of financial assistance under the Fellowship (DST INSPIRE) from Ministry of Science and Technology and Department of Science and Technology.

KEYWORDS

- **Arc GIS**
- **field validation**
- **groundwater prospect**
- **overlay analysis**
- **Upper South Koel river**

REFERENCES

Amadi, A. N., Nwawulu, C. D., Unuevho, I., Okoye, N. O., Okunlola, I. A., Egharevba, N, A., Ako, T. A., & Alkali, Y. B., (2011). Evaluation of the groundwater potential in Pompo Village, Gidan Kwano, Minna using vertical electrical resistivity sounding. *British J. Appl. Sci. Technol.*, *1*, 53–66.

Chi, K. H., & Lee, B. J., (1994). Extracting potential groundwater area using remotely sensed data and GIS techniques. In: *Proceedings of the Regional Seminar on Integrated Application of Remote Sensing and GIS for Land and Water Resource Management* (pp. 64–69). Bangkok.

Flug, M., Seitz, H. L. H., & Scott, J. F., (2000). Multicriteria decision analysis applied to Glen Canyon Dam. *J. Water Resour. Plan Manage ASCE.*, *126*, 270–276.

Ganapuram, S., Kumar, G. T. V., Krishna, I. V. M., Kahya, E., & Demirel, M. C., (2009). Mapping of groundwater potential zones in the Musi basin using remote sensing data and GIS. *Adv. Eng. Softw.*, *40*, 506–518.

Hajkowicz, S., & Higgins, A., (2008). A comparison of multiple criteria analysis techniques for water resource management. *Eur. J. Oper. Res.*, *184*, 255–265.

Horton, R. E., (1945). Erosional development of streams and their drainage basins, hydrophysical approach to quantitative morphology. *Geol. Soc. Am. Bull.*, *56*, 275–370.

Jaiswal, R., Mukherjee, S., Krishnamurthy, J., & Saxena, R., (2003). Role of remote sensing and GIS techniques for generation of groundwater prospect zones towards rural development – an approach. *Int. J. Remote Sens.*, *24*, 993–1008.

Jha, M. K., Chowdhury, A., Chowdary, V. M., & Peiffer, S., (2007). Groundwater management and development by integrated remote sensing and geographic information systems: Prospects and constraints. *Water Resour. Manage.*, *21*, 427–467.

Joubert, A., Stewart, T. J., & Eberhard, R., (2003). Evaluation of water supply augmentation and water demand management options for the City of Cape Town. *J. Multi-Criteria Decis. Anal.*, *12*, 17–25.

Kamaraju, M. V. V., Bhattacharya, A., Reddy, G. S., Rao, G. C., Murthy, G. S., & Rao, T. C. M., (1995). Groundwater potential evaluation of West Godavari District, Andhra Pradesh State, India – a GIS approach. *Ground Water*, *34*, 318–325.

Krishnamurthy, J., Srinivas, G., Jayaram, V., & Chandrasekhar, M. G., (1996). Influence of rock types and structures in the development of drainage networks in typical hard rock terrain. *ITC Journal*, *3/4*, 252–259.

Kumar, A., & Pandey, A. C., (2016). Geoinformatics based groundwater potential assessment in hard rock terrain of Ranchi urban environment, Jharkhand State (India) using MCDM–AHP techniques. *Groundwater for Sustainable Development*, *2/3*, 27–41.

Kumar, A., & Srivastava, S. K., (1991). Geomorphological units, their geohydrological characteristic and vertical electrical sounding response near Munger, Bihar. *J. Indian Soc. Remote Sens.*, *19*, 205–215.

Kumar, M. G., Agarwal, A. K., & Bali, R., (2008). Delineation of potential sites for water harvesting structures using remote sensing and GIS. *J. Indian Soc. Remote Sens.*, *36*, 323–334.

Machiwal, D., Jha, M. K., & Mal, B. C., (2011). Assessment of groundwater potential in a semi-arid region of India using remote sensing, GIS and MCDM techniques. *Water Resour. Manage.*, *25*, 1359–1386.

Machiwal, D., Mishra, A., Jha, M. K., Sharma, A., & Sisodia, S. S., (2012). Modeling short-term spatial and temporal variability of groundwater level using geostatistics and GIS. *Arab J. Geosci.*, *21*(1), 117–136.

Meijerink, A. M. J., (1996). Remote sensing applications to hydrology: Groundwater. *Hydrol. Sci. J.*, *41*, 549–561.

Mondal, M. S., Pandey, A. C., & Garg, R. D., (2007). Groundwater prospects evaluation based on hydrogeomorphological mapping using high-resolution satellite images: A case study in Uttarakhand. *J. Indian Soc. Remote Sens.*, *36*, 69–76.

Nour, S., (1996). Groundwater potential for irrigation in the East Oweinat area, Western Desert. *Egypt Environ. Geol.*, *27*, 143–154.

Pandey, A. C., & Stuti, (2017). Geospatial technique for runoff estimation based on SCS-CN method in upper South Koel River Basin of Jharkhand (India). *Int. J. Hydro.*, *1*(7), 00037. doi: 10.15406/ijh.2017.01.00037].

Parveen, R., Kumar, U., & Singh, V. K., (2012). Geomorphometric characterization of upper South Koel Basin, Jharkhand: A remote sensing & GIS approach, *Journal of Water Resource and Protection*, *4*, 1042–1050.

Rajiv Gandhi National Drinking Water Mission Atlas, (2005). *Rajiv Gandhi National drinking water mission Atlas for Jharkhand State* (Vol. I & II). Hyderabad: National Remote Sensing Centre, Department of Space, Govt. of India.

Sharma, D., & Jugran, D. K., (1992). Hydromorphological studies around Pinjaur-Kala Amb area, Ambala district (Haryana), and Sirmour district (Himachal Pradesh). *J. Indian Soc. Remote Sens.*, *29*, 281–286.

Shekhar, S., & Pandey, A. C., (2014). Delineation of groundwater potential zone in hard rock terrain of India using remote sensing, geographical information. *Geocarto. International*, doi: 10.1080/10106049.2014.894584.

Shekhar, S., Pandey, A. C., & Nathawat, M. S., (2012). Evaluation of fluoride contamination in groundwater sources in hard rock terrain in Garhwa district, Jharkhand, *India. Int. Journal of Environmental Sciences*, *3*(3). doi:10.6088/ijes.2012030133010.

Srivastava, V. K., Giri, D. N., & Bharadwaj, P., (2012). Study and mapping of groundwater prospect using remote sensing, GIS and geoelectrical resistivity techniques – a case study of Dhanbad district, Jharkhand, India, *J. Ind. Geophys. Union*, *16*(2), 55–63.

Teeuw, R. M., (1995). Groundwater exploration using remote sensing and a low-cost geographical information system. *Hydrol. Sci. J.*, *3*, 21–30.

Tirkey, A. S., Ghosh, M., & Pandey, A. C., (2016). Soil erosion assessment for developing suitable sites for artificial recharge of groundwater in a drought-prone region of Jharkhand state using geospatial technique. *Arab J. Geosci.*, *9*, 362. doi: 10.1007/s12517-016-2391-0.

Waters, P., Greenbaum, D., Smart, P. L., & Osmaston, H., (1990). Applications of remote sensing to groundwater hydrology. *Remote Sens. Rev.*, *4*, 223–264.

CHAPTER 35

GEO-PROCESSING BASED HYDROLOGICAL SENSITIVITY ANALYSIS AND ITS IMPACT ON FOREST AND TERRAIN ATTRIBUTES IN THE HAZARIBAGH WILDLIFE SANCTUARY, JHARKHAND, INDIA

SAURABH KUMAR GUPTA[1], A. C. PANDEY[2], and STUTI[1]

[1]*Research Scholar, Centre for Land Resource Management, School of Natural Resource Management, Central University of Jharkhand, Brambe, Jharkhand, India*

[2]*Professor, School of Natural Resource Management, Centre for Land Resource Management, Central University of Jharkhand, Brambe, Jharkhand, India, E-mail: arvindchandrap@yahoo.com*

ABSTRACT

The health of a watershed is vital to ensure the sustainability of water resources. Understanding and recognizing the spatial changes in the sensitive hydrological zones (HSZs) in a watershed is of prime importance to maintain water sustainability in the forest regions. In the present research satellite data-based spatial technique known as topographic wetness index (TWI) was used to identify HSZs in the forest landscape. This study was conducted in the Hazaribagh Wildlife Sanctuary, Jharkhand, India. Further, the relationship of HSZ with a slope length (LS) factor, flow accumulation, and forest cover were observed. The result shows that the watershed that contains high wetted area is the healthiest watershed and has good forest productivity. The LS factor and flow accumulation negatively related to HSZs. The present study emphasizes the protection of stream corridors through land use management. Also, the forest cover in the areas of degraded and forest blank (FB) needs to be given the highest priority for decreasing the hydrological sensitivity by improving soil moisture retention capacity and water demand of the landscape.

35.1   INTRODUCTION

The variable source area VSA (Hewlett and Hibbert, 1967) is the hydrology idea to survey commitments of various parts of a watershed to run-off generation (Wu, 2016). The sensitive hydrological zones (HSZs) are characterized as parts of VSAs which are more susceptible to create runoff contrasted with different parts of the watershed (Walter et al., 2000). HSZs assume a basic part in watershed hydrology. The idea of HSZs relates the watershed scale issues to different regions in the watershed that possibly contribute water saturation and soil moisture retention capacity. The extensive research has been utilizing HSZs based way to deal with comprehend watershed hydrology and prioritize the watershed based on water demand of landscape. The spatial variation of HSZs would additionally encourage organizing precise runoff generation areas, and their land uses. Forest degradation by human association changes the land uses in a part of a watershed, which prompts water asset degradation. However, its impacts on water assets shift crosswise over various parts of the watershed. For instance, urbanization, and agriculture that happened near the streams has a more prominent effect than far from the streams (Wu, 2016). The program named Healthy Watersheds Initiative (HWI) presented by the United States Environmental Protection Agency (USEPA, 2009), the dynamic attributes of water their interconnections with the land cover such as in forest region and secures all vital hydrologic, geomorphic, and different procedures all in the interconnected framework. Ensuring healthy watersheds gives various advantages, including adequate water to influence vegetation cover and better human wellbeing. Additionally, it lessens the vulnerability of water assets to future land use and environmental change impacts (USEPA, 2011).

To delineate HSZs topographic indices was used based on Digital Elevation Models (DEM) such as Topographic Wetness Indices (TWI). The TWI values have some predictive power concerning soil and vegetation characteristics, but scatter among the relationships is high, and TWIs usually explain less than 40 % of the variation in soil and vegetation properties (Thompson and Moore, 1996; Florinsky et al., 2002; Sorenson et al., 2006; Seibert et al., 2007). The 20 and 30 m DEMs produced spatial distributions of TWI values that most closely matched with the distribution of mapped soil types, the relative wetness determined from depth to groundwater maps, and the spatial distribution of forest stands managed partly on the basis of wetted and drainage networks (Wu, 2016). In the present study, we utilized the Topographic Wetness Index (TWI) for calculating HSZs areas. The HSZz was then related to LS factor, flow accumulation, and forest cover.

35.2 MATERIALS AND METHODS

35.2.1 STUDY AREA

This study was performed in the Hazaribagh Wildlife Sanctuary, Jharkhand, India (Figure 35.1). It lies between 24°45'22" N to 24°08'20"N latitude and 85° 30'13" E to 85°21'58"E longitude and also includes bird Sanctuary and other biodiversity parks. The climate of the region is tropical having hot summers and chilly winters. In summer, the greatest temperature rises to up to 41 degrees and low of 19 degrees. In winter, most extreme high and low temperature decreased is 19 degrees and 7 degrees respectively. The sanctuary has Sambhar, Deer, Bison, and various mammalian fauna. The Cheetah, Kakar, Nilgai, Sambar, and Wild Boar are among the most effective and frequently spotted creatures, especially close to the waterholes at the time of the sunset.

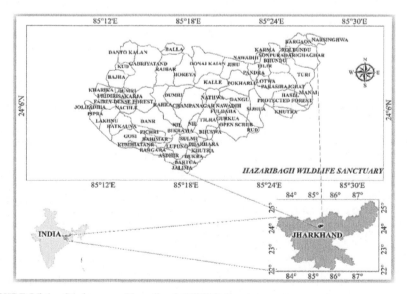

FIGURE 35.1   Study area – Hazaribagh Wildlife Sanctuary.

35.2.2 METHODOLOGY

The flow chart of all processing steps and analysis was given in Figure 35.2 which indicates the methodology adopted in the present study. It includes the preprocessing SRTM DEM 1 arc second resolution (https://earthex-plorer.usgs.gov/) for HSZ analysis and estimation of terrain attributes. The Sentinel–2A satellite data was used for finding the forest cover area. Each part of the methodology is illustrated below.

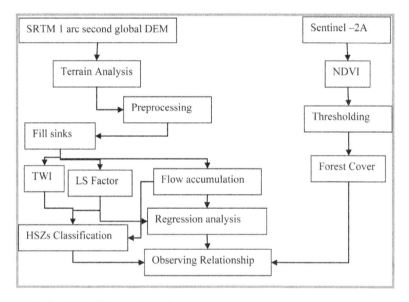

FIGURE 35.2   Methodology adopted.

35.2.3   TOPOGRAPHIC WETNESS INDEX (TWI)

The application of TWIs to explain the spatial distribution of moisture and soil and vegetation characteristics has encountered problems of several types. The Quantum GIS (QGIS) and System for Automated Geoscientific Analyses GIS (SAGA GIS) software were used for the calculation of TWI and its processing. Both GIS software are freely available and user-friendly. TWI used in QGIS was suggested by Beven and Kirk by (1979) as

$$TWI = \ln (A_s/\tan\beta) \qquad (1)$$

where, As is the (upslope) flow accumulation area (or drainage area) per unit contour length (Wilson and Gallant, 2000) and β is the angle of the slope.

TWI has found a wide range of application in hydrology (Moore et al., 1991; Quinn et al., 1995; Sorenson et al., 2006).

35.2.4 HYDROLOGICAL SENSITIVE ZONE CLASSIFICATION

The TWI indicates the spatial distribution of surface saturation and soil moisture. The HSZ was classified based on TWI index values, the positive value of TWI represents the high wetted area, and low HSZs and its negative parts represent HSZ. The histogram of TWI shows that TWI is the high number of negatives values and low in positive values. Therefore, all the negative values (−18.55 to 0) were classified as HSZs classes, representing very low to high HSZ and the positive part (7.9 to 0) was separated as high wetted area as given in Figure 35.3. The ARC GIS was used for the classification of HSZs based on the natural breaks method (Jenks) of classification. The statistics of the classified HSZ were calculated by pixel count and DEM resolution.

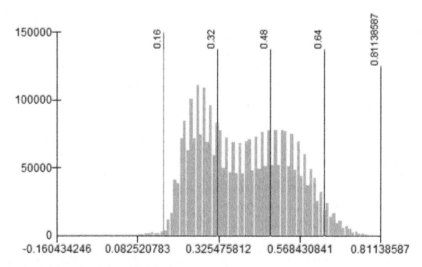

FIGURE 35.3   NDVI threshold for forest classification.

35.2.5 TERRAIN ATTRIBUTES

35.2.5.1 LS FACTOR

Length-Slope (LS) factor (Moore, 1992) implemented in the revised universal soil loss equation (RUSLE) was used in this study. It gives a value for the water erosion potential (A^s) to a slope of 22.13 m length and a slope angle of 5 degrees. It can be represented as the following equation:

$$LS = (m+1)(\frac{As}{22.13})^m\left((\frac{sin\beta}{0.0896})^n\right) \tag{2}$$

where $m = 0.4$ and $n = 1.3$ for a slope length (LS) <100 m and a slope angle <14°. LS and slope gradient (sin β) are most often considered topographic factors are influencing soil loss. The SAGA GIS was used for the calculation of the LS Factor.

35.2.5.2   FLOW ACCUMULATION

The flow accumulation operation plays out a collective count of the number of pixels that normally drain into outlets. The process can be used to discover drainage pattern. The flow accumulation is calculated by taking flow direction as input parameters which helps as to demarcate a total number of pixels that will drain into an outlet. The flow direction operation decides the characteristics of drainage direction for all pixels in a digital elevation model (DEM). The output derived as flow accumulation map contains hydrological flow values that indicate the number of input pixels which contribute to a water body and the outlets of the largest water body have the largest value.

35.2.5.3   NDVI BASED FOREST CLASSIFICATION

NDVI: Vegetation cover is possibly the most crucial element in the process of soil erosion study and management. It is the most dynamic factor in a watershed which can be readily altered to control the loss of water and soil. It is a numerical indicator, which uses the visible and near-infrared bands of the remote sensing data for identifying green vegetation in the study site. The NDVI algorithm subtracts the red reflectance values from the near-infrared and divides it by the sum of near-infrared and red bands. Theoretically, NDVI values are represented as a ratio ranging in value from −1 to 1, but in practice, extreme negative values represent water, values around zero represent bare soil and values over 0.5 represent dense green vegetation.

$$NDVI = \frac{NIR - RED}{NIR + RED} \tag{3}$$

Forest Classification: The NDVI values were assessed (0.81 to –0.16) in the forest to gain output of the forest cover distribution in the forest. The NDVI values of 0.64–0.8 named dense forest (DF), very DF having a small area (greater than 0.8) which was not included because of its small size, 0.48–0.64 as moderately dense forest (MDF), 0.32–0.48 as open forest (OF), 0.16 to 0.32 as degraded forest (DeF) and less than these comprises forest blank (FB). The classification threshold of NDVI with its histogram is given in Figure 35.3.

35.3 RESULTS AND DISCUSSION

35.3.1 TOPOGRAPHIC WETNESS INDEX AND HYDROLOGICAL SENSITIVE ZONE

35.3.1.1 TWI

The TWI index used for the measurement of wetness and shows the area of maximum water accumulation potential. The high TWI area represents the highest amount of precipitation, low water demand and the presence of soil type that hold large amounts of water, means less infiltration rate. The figure that represents TWI index is given in Figure 35.4. The figure indicates that the high TWI mainly found in the path of drainage basins.

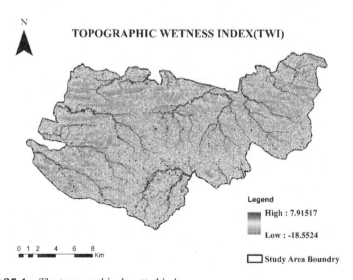

FIGURE 35.4 The topographical wetted index.

35.3.1.2   *HSZs CLASSIFICATION AND HIGH WETTED AREA*

The previous research on the TWI shows that it is directly proportional to hydrological sensitive zone determination. The HSZ was classified in four zones (high, medium, low, and very low HSZ) given in Figure 35.5. It was found that medium HSZ was in greater magnitude compared to other zones of area equal to 110.39 km^2 (40%). The very low and low HSZs was low in the magnitude of 16.3 km^2 (6%) and 38.8 km^2 (14.29%) respectively. The wetted area that was calculated is very low of about 7 km^2 (2.65%). It means the study area has lacked of clay soil and a large quantity of sandy soil. The High HSZ was 98.89km^2 in the area and has contributed 36.39% of the study area which has larger in magnitude. The statistical part of HSZs classification and the high wetted zone is given (Figure 35.6). The analysis shows that the area encounters the problem of low precipitation and run-off and a large part of the area has a water deficiency (high water demand) for the good forest productivity. These problems have a direct effect on forest vegetation health and their growths.

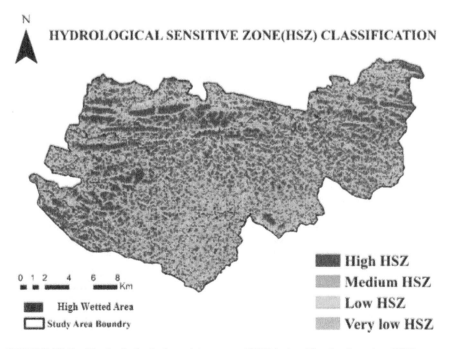

FIGURE 35.5   The hydrological sensitive zones (HSZs) classification based on TWI.

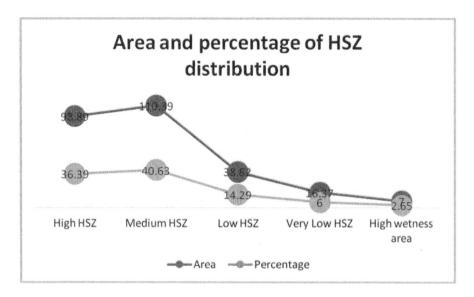

FIGURE 35.6 The statistics of HSZs and high wetted area.

35.3.1.3 THE TERRAIN ATTRIBUTES AND HSZS

LS-Factor: LS factor is used for identifying the net erosion area (Figure 35.7). The high value of LS factor represents the high soil erosion-prone area, and low LS factor shows low soil erosion prone area. The LS-factor is found in the drainage area. The relationship between LS factor and HSZs was observed and concluded that high wetted area has high LS factor and high HSZs mean low LS factor. In Figure 35.8, the scatter plot is given which shows the relationship of LS factor with HSZs and high wetted zone. The negative linear relationship was found between HSZ and LS factor. It shows that here that the HSZ is the area having the low soil erosion, which was found scattered around the drainage boundary.

Flow Accumulation: A sample usage of the flow accumulation tool with an input weighted raster might be to determine how much rain has fallen within a given watershed. In such a case, the weight raster may be a continuous raster representing average rainfall during a given storm. The output from the tool would then represent the amount of rain that would flow through each cell, assuming that all rain became runoff and there was no interception, evapotranspiration, or loss to groundwater. This could also be viewed as the amount of rain that fell on the surface, upslope from each cell. The results of flow accumulation can be used to create a stream network

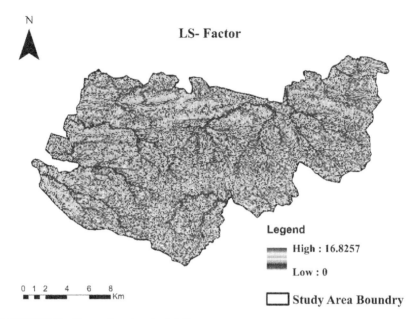

FIGURE 35.7 Figure showing the LS factor.

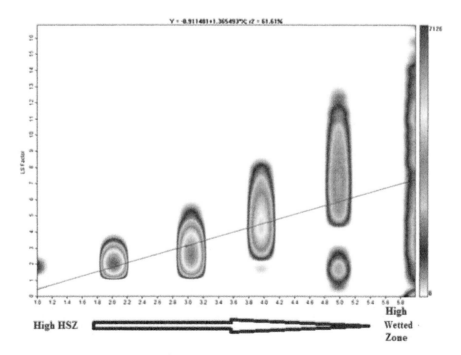

FIGURE 35.8 The relationship between HSZs and high wetted zone area with LS factor.

by applying a threshold value to select cells with a high accumulated flow. The flow accumulation map is given (Figure 35.11). The effort was applied to get information about the relationship between HSZs and Flow accumulation. The observation shows that more the flow accumulation more the TWI index. The result indicated that area having a low flow accumulation has high HSZ and the HSZ decreases with more the flow accumulation. The high wetted area has a high flow accumulation. In Figure 35.9, flow accumulation relationship with the decreasing level of HSZ and increasing level of the wetted area was quite low ($r^2 = 19.48\%$) (Figures 35.10 and 35.11).

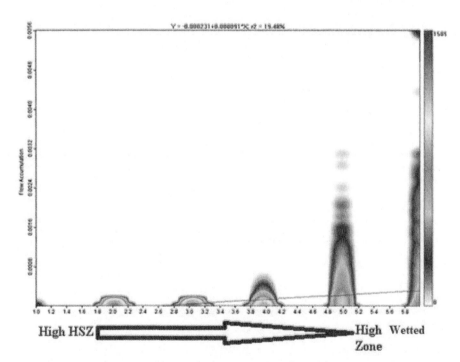

FIGURE 35.9   The relationship between the LS factor and HIGH HSZ and High wetted zone.

Forest Cover and HSZS Distribution: The landscape position and soil moisture exert first-order control on soil properties (e.g., texture, organic matter, and chemistry) and vegetation characteristics, these characteristics correlate with TWI values (e.g., Florinsky et al., 2002, 2004; Sariyildiz et al., 2005; Seibert et al., 2007). The forest cover in the study area is given (Figure 35.12). Forest cover type is a factor that influences watershed hydrology. Forest cover types help in regulation of surface runoff. The Land use changes

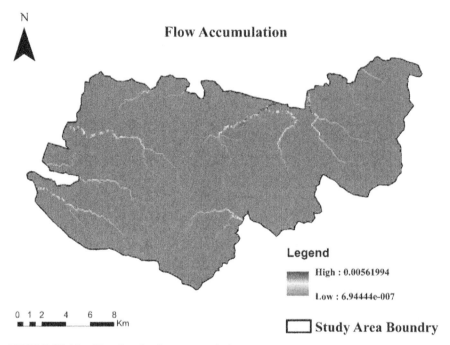

FIGURE 35.10 Showing the flow accumulation.

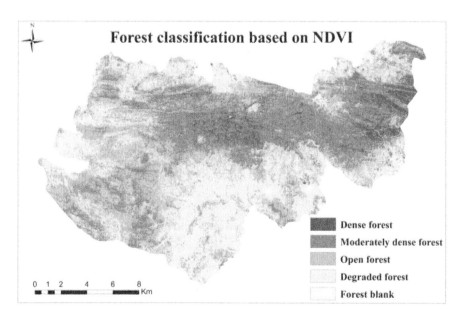

FIGURE 35.11 Forest cover classification in the study area based on NDVI values.

in forest affect water conservation practices in the past several decades. The degraded forest and FB lies on medium HSZ. The OF and some part of MDF lies on high HSZ showing that the forest in the region can tolerate water deficiency condition. They do not depend on the low water presence and need less water for their growth. The DF found scattered around the high wetted zone or drainage basin. The analysis shows that different forest cover lies on different HSZ. The relationship between decreasing wetted and increasing HSZ does not affect forest productivity in some part of the forest cover like MDF and OF. The attention is needed for rebuilding sound watershed and improving soil water retention capacity in degraded and FB part of the forest cover.

Validation: In the Figure 35.12, the false color composite of the Sentinel 2A imagery is given, which indicates the true condition of vegetation and can be taken for validating the obtained results of the classified forest cover. At the FCC, the DF area appears in dark red color. Sometimes, it is misunderstood by the shadow, as the forest canopy cover creates shadows depends upon the sun geometry. The darkness in red color decreases in MDF followed by open and degraded cover area. The degraded forest and FB appears pink and gray-white, respectively. Thus, the above validation of the obtained result is satisfactory.

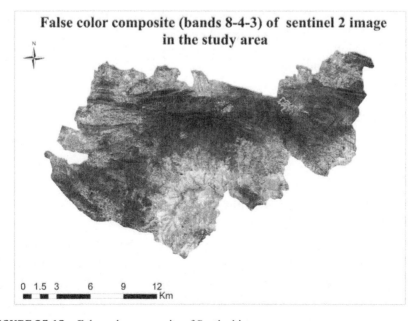

FIGURE 35.12 False color composite of Sentinel imagery.

35.4 CONCLUSIONS

Maintaining healthy watershed is a difficult task due to increased anthropogenic activities in the forest landscape. Different innovative programs such as HWI help to overcome the overall water deficiency. With limited resources, the understanding of the spatial distribution of HSZs in the landscape and their land uses helps prioritize conservation measures and support these programs. In this study, the TWI technique was used to delineate HSZs in the study area. A TWI layer was developed by using SRTM DEM which represents topography. The TWI threshold was used to create different categories of HSZs scenarios for detailed analysis in this study. The whole process was performed using SAGA GIS in R- platform. The relationship between the LS factor and Flow accumulation with HSZ found negative; it is clear that the HZS is the area having the highest water deficiency. Such detailed representation of HSZs would increase the efficacy of land use control measures in protecting water deficiency in the soil and improving the soil moisture retention capacity. The attention is needed for rebuilding sound watershed and improving soil water retention capacity in degraded and FB part of the forest cover. The analysis of land use patterns in HSZs provides insights on how and where to target the management efforts to overcome water deficiency. The significant intrusion of high-intensity land uses into HSZs would signify the deterioration of landscape conditions that would lead to high water demand for vegetation growth. Land planning measures should be taken to protect HSZs from the intrusion of high-intensity land uses.

KEYWORDS

- forest
- forest cover
- hydrological sensitive zones
- TWI
- watershed

REFERENCES

Beven, K. J., & Kirkby, M. J., (1979). A physically-based, variable contributing area model of basin 25 hydrology, *Hydrological Sciences Bulletin, 24*, 43–69.

Blyth, E. M., Finch, J., Robinson, M., & Rosier, P., (2004). Can soil moisture be mapped onto the terrain? *Hydrol. Earth Syst. Sci., 8*, 923–930, doi: 10.5194/hess-8-923-2004.

Dunne, T., & Black, R. D., (1970). Partial area contributions to storm runoff in a small New England watershed. *Water Resour. Res., 6*, 1296–1311.

Dunne, T., Moore, T. R., & Taylor, C. H., (1975). Recognition and prediction of runoff-producing zones 10 in humid regions, *Hydrological Sciences Bulletin, 20*, 305–327.

Florinsky, I. V., Eilers, R. G., Manning, G. R., & Fuller, L. G., (2002). Prediction of soil properties by digital terrain modeling. *Environ. Model. Softw., 17*, 295–311.

Florinsky, I. V., McMahon, S., & Burton, D. L., (2004). Topographic control of soil microbial activity: A case study of denitrifiers. *Geoderma, 119*, 33–53.

Güntner, A., Seibert, J., & Uhlenbrook, S., (2004). Modeling spatial patterns of saturated areas: An evaluation of different terrain indices, *Water Resour. Res., 40*, W05114, doi:10.1029/2003WR002864.

Hewlett, J. D., & Hibbert, A. R., (1967). Factors affecting the response of small watersheds to precipitation in humid regions, In: Sopper, W. E., & Lull, H. W., (eds.), *Forest Hydrology* (Vol. 20, pp. 275–290). Pergamon Press, Oxford.

Jenson, S. K., & Domingue, J. O., (1988). Extracting topographic structure from digital elevation data for geographic information system analysis. *Photogrammetric Engineering and Remote Sensing, 54*(11), 1593–1600.

Moore, I. D., Grayson, R. B., & Ladson, A. R., (1991). Digital terrain modeling: A review of hydrological, geomorphological and biological applications. *Hydrol. Process., 5*, 3–30.

Moore, I. D., & Wilson, J. P., (1992). Length-slope factors for the revised universal soil loss equation: Simplified method of estimation. *Journal of Soil & Water Conservation, 47*(5), 423–428.

Quinn, P., Beven, K., & Lamb, R., (1995). The $\ln(\alpha s/\tan\beta)$: How to calculate it and how to use it within the TOPMODEL framework, *Hydrol. Process., 9*, 161–182.

Sariyildiz, T., Anderson, J. M., & Kucuk, M., (2005). Effects of tree species and topography on soil chemistry, litter quality, and decomposition in Northeast Turkey. *Soil Biol. Biochem., 37*(5), 1695–1706.

Seibert, J., Stendahl, J., & Sorenson, R., (2007). Topographical influences on soil properties in boreal forests. *Geoderma, 14*, 139–148.

Sørensen, R., Zinko, U., & Seibert, J., (2006). On the calculation of the topographic wetted index: Evaluation of different methods based on field observations, *Hydrol. Earth Syst. Sci., 10*(10), 101–112. doi:10.5194/hess–10–101–2006.

Thompson, J. C., & Moore, R. D., (1996). Relations between topography and water table depth in a shallow forest soil. *Hydrol. Process., 10*, 1513–1525.

U. S. Department of the Interior | U. S. Geological Survey, URL: http://water.usgs.gov/edu/watershed.html.

USEPA, (2010). *Identifying and Protecting Healthy Watersheds: A Technical Guide. Washington, DC*. Available at: (http://acwi.gov/monitoring/conference/2010/G3/ G3_Gabanski & Godfrey.pdf). Accessed on 14.03.16 USEPA (2011). Healthy watersheds initiative: National framework and action plan. Washington, DC. Available at: (https://www.epa.gov/sites/production/files/2015–10/documents/hwi_action_plan.pdf). Accessed on 14.03.16.

Walter, M. T., Steenhuis, T. S., Mehta, V. K., Thongs, D., Zion, M., & Schneiderman, E., (2002). Refined conceptualization of topmodel for shallow subsurface flows. *Hydrological Process, 16,* 2041–2046.

Walter, M. T., Walter, M. F., Brooks, E. S., Steenhuis, T. S., Boll, J., & Weiler, K., (2000). Hydrologically sensitive areas: Variable source area hydrologically implications for water quality risk assessment. *Journal of Soil and Water Conservation, 3,* 277–284.

Wilson, J. P., & Gallant, J. C., (2000). Digital terrain analysis. *Terrain Analysis: Principles and Applications,* 1–27.

INDEX

HANDBOOK OF PTSD

Handbook of PTSD

Science and Practice

edited by
Matthew J. Friedman
Terence M. Keane
Patricia A. Resick

THE GUILFORD PRESS
New York   London

Paperback edition 2010

Printed in the United States of America

This book is printed on acid-free paper.

Last digit is print number: 9 8 7 6 5 4 3

The authors have checked with sources believed to be reliable in their efforts to provide
information that is complete and generally in accord with the standards of practice that
are accepted at the time of publication. However, in view of the possibility of human error
or changes in medical sciences, neither the authors, nor the editor and publisher, nor any
other party who has been involved in the preparation or publication of this work warrants
that the information contained herein is in every respect accurate or complete, and they
are not responsible for any errors or omissions or the results obtained from the use of such
information. Readers are encouraged to confirm the information contained in this book
with other sources.

Library of Congress Cataloging-in-Publication Data

Handbook of PTSD : science and practice / edited by Matthew J. Friedman,
Terence M. Keane, Patricia A. Resick.
    p. ; cm.
  Includes bibliographical references and index.
  ISBN 978-1-59385-473-7 (hardcover : alk. paper)
  ISBN 978-1-60918-174-1 (paperback : alk. paper)
  1. Post-traumatic stress disorder—Handbooks, manuals, etc. I. Friedman, Matthew J.
II. Keane, Terence Martin. III. Resick, Patricia A.
  [DNLM: 1. Stress Disorders, Post-Traumatic. WM 170 H2357 2007]
  RC552.P67H353 2007
  616.85'21—dc22
                                                          2006036543

About the Editors

Matthew J. Friedman, MD, PhD, is Executive Director of the National Center for PTSD, White River Junction VA Medical Center, and Professor of Psychiatry and Pharmacology, Dartmouth Medical School. He is a recipient of the International Society for Traumatic Stress Studies (ISTSS) Lifetime Achievement Award, among many other honors.

Terence M. Keane, PhD, is Director of the National Center for PTSD, Behavioral Sciences Division, VA Boston Healthcare System, where he is also Associate Chief of Staff for Research and Development, and Professor and Vice Chairman of Psychiatry at Boston University School of Medicine. He is a recipient of the ISTSS Lifetime Achievement Award, among many other awards.

Patricia A. Resick, PhD, is Director of the National Center for PTSD, Women's Health Sciences Division, VA Boston Healthcare System, and Professor of Psychiatry and Psychology at Boston University. She is a recipient of the ISTSS Robert S. Laufer Memorial Award for Outstanding Scientific Achievement. She is past president of both the Association for Cognitive and Behavioral Therapies and the International Society for Traumatic Stress Studies and has won awards for her research.

Contributors

Deane E. Aikins, PhD, National Center for PTSD, VA Connecticut Healthcare System, West Haven, Connecticut, and Department of Psychiatry, Yale University, New Haven, Connecticut

Jill Barron, MD, Yale Child Study Center and Department of Psychiatry, Yale University, New Haven, Connecticut

Chris R. Brewin, PhD, Department of Psychology, University College London, London, United Kingdom

Deborah J. Brief, PhD, VA Boston Healthcare System, Boston, Massachusetts

Shawn P. Cahill, PhD, Department of Psychiatry and Center for the Treatment and Study of Anxiety, University of Pennsylvania, Philadelphia, Pennsylvania

Joan M. Cook, PhD, New York State Psychiatric Institute and Department of Psychiatry, Columbia University, New York, New York

Jonathan R. T. Davidson, MD, Department of Psychiatry and Behavioral Sciences, Duke University Medical Center, Durham, North Carolina

Lori L. Davis, MD, VA Medical Center, Tuscaloosa, Alabama

Joop T. V. M. de Jong, MD, Transcultural Psychosocial Organization, Amsterdam, The Netherlands

Anne P. DePrince, PhD, Department of Psychology, University of Denver, Denver, Colorado

B. Heidi Ellis, PhD, Department of Psychiatry, Boston University School of Medicine and Boston Medical Center, Boston, Massachusetts

John A. Fairbank, PhD, UCLA–Duke University National Center for Child Traumatic Stress and Department of Psychiatry, Duke University Medical Center, Durham, North Carolina

Edna B. Foa, PhD, Department of Psychiatry and Center for the Treatment and Study of Anxiety, University of Pennsylvania, Philadelphia, Pennsylvania

Jennifer J. Freyd, PhD, Department of Psychology, University of Oregon, Eugene, Oregon

Matthew J. Friedman, MD, PhD, National Center for PTSD, White River Junction VA Medical Center, White River Junction, Vermont, and Department of Psychiatry, Dartmouth Medical School, Hanover, New Hampshire

Joel Gelernter, MD, National Center for PTSD, VA Connecticut Healthcare System, West Haven, Connecticut, and Department of Psychiatry, Yale University, New Haven, Connecticut

Laura Gibson, PhD, Department of Psychology, University of Vermont, Burlington, Vermont

Bonnie L. Green, PhD, Department of Psychiatry, Georgetown University Medical Center, Washington, DC

Cassidy Gutner, BA, National Center for PTSD, VA Boston Healthcare System, Boston, Massachusetts

William W. Harris, PhD, Children's Research and Education Institute, Belmont, Massachusetts

Shannan Henry, BS, National Center for PTSD, VA Connecticut Healthcare System, West Haven, Connecticut, and Department of Psychiatry, Yale University, New Haven, Connecticut

Stacey Kaltman, PhD, Department of Psychiatry, Georgetown University, Washington, DC

Terence M. Keane, PhD, National Center for PTSD, VA Boston Healthcare System, and Department of Psychiatry, Boston University School of Medicine, Boston, Massachusetts

Rachel Kimerling, PhD, National Center for PTSD, VA Palo Alto Health Care System, Menlo Park, California, and Department of Psychiatry, University of California, San Francisco, California

Daniel W. King, PhD, National Center for PTSD, VA Boston Healthcare System, and Department of Psychiatry, Boston University, Boston, Massachusetts

Lynda A. King, PhD, National Center for PTSD, VA Boston Healthcare System, and Department of Psychiatry, Boston University, Boston, Massachusetts

John H. Krystal, MD, National Center for PTSD, VA Connecticut Healthcare System, West Haven, Connecticut, and Department of Psychiatry, Yale University, New Haven, Connecticut

Heidi A. J. La Bash, BS, National Center for PTSD, VA Boston Healthcare System, Boston, Massachusetts

Christopher M. Layne, PhD, National Center for Child Traumatic Stress, University of California, Los Angeles, California

Brett T. Litz, PhD, National Center for PTSD, VA Boston Healthcare System, and Department of Psychiatry, Boston University School of Medicine, Boston, Massachusetts

Helen Z. MacDonald, MA, Department of Psychology, Boston University, Boston, Massachusetts

Shira Maguen, PhD, National Center for PTSD, VA Boston Healthcare System, Boston, Massachusetts

Mark W. Miller, PhD, National Center for PTSD, VA Boston Healthcare System, and Departments of Psychiatry and Psychology, Boston University, Boston, Massachusetts

Candice M. Monson, PhD, National Center for PTSD, VA Boston Healthcare System, Boston, Massachusetts

Charles A. Morgan III, MD, National Center for PTSD, VA Connecticut Healthcare System,

West Haven, Connecticut, and Department of Psychiatry, Yale University School of Medicine, New Haven, Connecticut

Alexander Neumeister, MD, National Center for PTSD, VA Connecticut Healthcare System, West Haven, Connecticut, and Department of Psychiatry, Yale University School of Medicine, New Haven, Connecticut

George Niederehe, PhD, Neuro Science Center, National Institute of Mental Health, Bethesda, Maryland

Fran H. Norris, PhD, National Center for PTSD, White River Junction VA Medical Center, White River Junction, Vermont, and Department of Psychiatry, Dartmouth Medical School, Hanover, New Hampshire

Janet E. Osterman, MD, Department of Psychiatry, Boston University, Boston, Massachusetts

Paige Ouimette, PhD, Syracuse VA Medical Center, Syracuse, New York

Roger K. Pitman, MD, Department of Psychiatry, Massachusetts General Hospital, Harvard Medical School, Boston, Massachusetts

Elizabeth M. Pratt, PhD, National Center for PTSD, VA Boston Healthcare System, and Department of Psychology, Boston University, Boston, Massachusetts

Frank W. Putnam, MD, Mayerson Center for Safe and Healthy Children, Children's Hospital Medical Center, Cincinnati, Ohio

Ann Rasmusson, MD, National Center for PTSD, VA Connecticut Healthcare System, West Haven, Connecticut, and Department of Psychiatry, Yale University School of Medicine, New Haven, Connecticut

Patricia A. Resick, PhD, National Center for PTSD, VA Boston Healthcare System, Boston, Massachusetts

Barbara Olasov Rothbaum, PhD, Department of Psychiatry, Emory University School of Medicine, Atlanta, Georgia

Josef I. Ruzek, PhD, National Center for PTSD, VA Palo Alto Health Care System, Menlo Park, California

Glenn N. Saxe, MD, Department of Psychiatry, Children's Hospital, Boston, Massachusetts

Paula P. Schnurr, PhD, National Center for PTSD, White River Junction VA Medical Center, White River Junction, Vermont, and Department of Psychiatry, Dartmouth Medical School, Hanover, New Hampshire

Ronnen Segman, MD, Department of Psychiatry, Hadassah University Hospital, Jerusalem, Israel

Arieh Y. Shalev, MD, Department of Psychiatry, Hadassah University Hospital, Jerusalem, Israel

Laurie B. Slone, PhD, National Center for PTSD, White River Junction VA Medical Center, White River Junction, Vermont, and Department of Psychiatry, Dartmouth Medical School, Hanover, New Hampshire

Steven M. Southwick, MD, National Center for PTSD, VA Connecticut Healthcare System, West Haven, Connecticut, and Department of Psychiatry and Yale Child Study Center, Yale University School of Medicine, New Haven, Connecticut

Landy F. Sparr, MD, Department of Psychiatry, Oregon Health and Science University, Portland, Oregon

Bessel A. van der Kolk, MD, The Trauma Center at Justice Resource Institute, Brookline, Massachusetts

Dawne S. Vogt, PhD, National Center for PTSD, VA Boston Healthcare System, Boston, Massachusetts

Jared S. Warren, PhD, Department of Psychology, Brigham Young University, Provo, Utah

Patricia J. Watson, PhD, National Center for PTSD, White River Junction VA Medical Center, White River Junction, Vermont, and Department of Psychiatry, Dartmouth Medical School, Hanover, New Hampshire

Julie C. Weitlauf, PhD, National Center for PTSD, VA Palo Alto Health Care System, Menlo Park, California

Stacy Shaw Welch, PhD, Anxiety and Stress Reduction Center of Seattle, Seattle, Washington

Preface

We had several goals when we designed this book. First, we wanted to offer to graduate students, interns, fellows, scientists, and practitioners a comprehensive handbook that would provide a sophisticated introduction to the trauma field. Second, we thought that the time had come to erect a milestone, marking all the progress that has been achieved since 1980, when posttraumatic stress disorder (PTSD) first appeared as an official diagnosis in the American Psychiatric Association's third edition of its *Diagnostic and Statistical Manual of Mental Disorders* (DSM-III). Last, we wanted to highlight the work of some of the major contributors to the field of psychological trauma.

This handbook was developed to serve as the main textbook for an advanced-level curriculum on PTSD and trauma. In our opinion, there is no authoritative source currently available to serve such a purpose. The challenge in constructing this volume was to accurately reflect how the field has expanded across many discrete academic and clinical disciplines. As just one example, we found it necessary to include a chapter on forensic issues because PTSD-related litigation has become so prominent. As readers will discover, there are not only many different scientific and clinical topics to consider, but many different levels of analyses to understand. At the microscopic level, we had to consider gene–environment interactions, synaptic plasticity, neurocircuitry, and neurobiological mechanisms. At the macroscopic level, we needed to present information on epidemiology, cross-cultural issues, and public mental health. In between are chapters encompassing psychological models, memory, dissociation, gender, human development, and a wide variety of clinical approaches to diagnosis and treatment.

Our second goal was to provide a benchmark with sufficient breadth and depth so that scientists and practitioners could gauge both how far we have come since 1980 and how much further we need to go in the future. For example, a textbook published only 12 years ago had no chapters on gene–environment interactions, brain imaging, memory, dissociation, gender, human development, early intervention, cultural issues, forensic issues, resilience, or public mental health (Friedman, Charney, & Deutch, 1995).

The present volume opens with a look at the past that includes two historical chapters and an overview that briefly documents major challenges and controversies that have demanded attention since PTSD was recognized as an official psychiatric diagnosis. The final section, "Uncharted Territory," focuses on emerging treatments, resilience, public mental health, and an agenda for future advances in science and practice. Between them, the two sections that constitute the heart of this volume address "Scientific Foundations and Theoretical Perspectives," on the one hand, and "Clinical Practice: Evidence-Based State of the Art," on the other. Each chapter within these two sections (except for one on psychological theories of PTSD) has the same structure. Each begins with a section titled "Methodological Considerations" that presents the scientific techniques needed to acquire knowledge pertinent to that specific topic. In other words, readers will receive thoughtful descriptions of the very different techniques needed to acquire scientific knowledge in fields as diverse as genetics, brain imaging, cognitive psychology, psychosocial treatments, psychopharmacology, epidemiology, and other areas. The second section in each chapter, "Current State of the Art," provides a comprehensive and rigorous review of the published literature in that particular field. Third, each author discusses the "Generalizability of Current Findings" so that readers can evaluate the relevance of the empirical literature to scientific or clinical questions that concern them most. Each chapter concludes by identifying important directions for future science and practice.

Finally, the contributing authors we have had the privilege of working with on this project represent some of the leading PTSD scholars and clinicians in the world today. We encourage students entering the field to model our contributors' dedication and talent.

Many of our authors are colleagues within the National Center for PTSD, U.S. Department of Veterans Affairs. Most of the others are experts affiliated with the top clinical and research centers in the field, and most of them have worked closely with Center investigators on key initiatives. The National Center for PTSD is our institutional home. Established in 1989 by a mandate from the U.S. Congress, it has become the global leader in research and education concerning the psychological impact of exposure to traumatic events and PTSD. Since it was established, Center research, education, and consultation has influenced the national and international scientific, clinical, and public policy agenda, especially since the September 11, 2001, terrorist attacks on the World Trade Center and the Pentagon, Hurricane Katrina, and the current wars in Iraq and Afghanistan.

We hope that this book serves as a useful textbook for graduate-level and continuing education curricula. We hope that it helps PTSD investigators conceptualize and design studies that will have a significant impact on the field. And we hope that it will enable practitioners to select and utilize the best evidence-based approaches for their clientele.

REFERENCE

Friedman, M. J., Charney, D, S., & Deutch, A. Y. (Eds.). (1995). *Neurobiological and clinical consequences of stress: From normal adaptation to post-traumatic stress disorder.* Philadelphia: Lippincott-Raven.

Contents

Part III. Clinical Practice: Evidence-Based State of the Art

Part IV. Uncharted Territory

Part I

HISTORICAL OVERVIEW

Chapter 1

PTSD

TWENTY-FIVE YEARS
OF PROGRESS AND CHALLENGES

**Matthew J. Friedman, Patricia A. Resick,
and Terence M. Keane**

Although symptoms reminiscent of posttraumatic stress disorder (PTSD) are illustrated in the world's great literature and have been noted following war or catastrophe over the centuries (Kilpatrick et al., 1998; Saigh, 1992; Shay, 1994; see van der Kolk, Chapter 2, and Monson, Friedman, & La Bash, Chapter 3, this volume), the actual term "posttraumatic stress disorder" did not appear in our nosology until 1980. In the late 1800s, in the attempt to categorize psychological disorders, Kraepelin (1896, translated by Jablensky, 1985, p. 737) used the term "fright neurosis" (*schreckneurose*) to capture anxiety symptoms following accidents and injuries. After World War II and during the Korean Conflict, the American Psychiatric Association produced the first *Diagnostic and Statistical Manual of Mental Disorders* (DSM-I; American Psychiatric Association, 1952), which included "gross stress reaction." This first DSM did not list the detailed criteria that we see today but did propose the diagnosis for people who were previously relatively normal, but who had symptoms resulting from their experiences with extreme stressors such as civilian catastrophe or combat. Strangely, at the height of the Vietnam War, the DSM-II (American Psychiatric Association, 1968) was published, and this category was eliminated. Some psychiatrists of that era assumed political motivations in the sudden disappearance of this diagnostic category (Bloom, 2000). According to Bloom, John Talbott, future president of the American Psychiatric Association, called for the return of the diagnostic category by the next year, 1969, because of his observations as a psychiatrist who had served in Vietnam, that there was no way to capture the symptoms he was observing with the current diagnostic system.

During the 1970s, a number of social movements in the United States and around the world converged to bring attention to reactions following interpersonal violence, as well as combat. The women's movement focused attention on sexual and physical assault of women from speak-outs and consciousness raising groups by the National Organization for Women. Laws were changed to reflect the understanding that abuse incidents within the family were crimes and of societal concern, not merely family matters. Mandatory reporting of child abuse was enacted in all U.S. states. Rape shield laws, marital rape laws, and the legal recognition that rape could happen to boys and men, as well as girls and women, also changed attitudes and services provided. Landmark studies by Kemp and his colleagues (Gray, Cutler, Dean, & Kempe, 1977; Schmitt & Kempe, 1975) Burgess and Holmstrom (1973, 1974) and Walker (1979) resulted in descriptions of the child abuse syndrome, the rape trauma syndrome, and the battered woman syndrome, respectively, and spawned a generation of research on those topics. The descriptions of responses to these forms of interpersonal traumas were much like those being described by the millions of Vietnam veterans who had returned from the war. As a result, when the revision of the DSM was considered, reactions to all traumatic events were pooled into one category.

In 1980, DSM-III (American Psychiatric Association, 1980) included PTSD for the first time as an official diagnosis. PTSD, now classified as an anxiety disorder, had four criteria: (1) the existence of a recognizable stressor that would evoke distress in nearly anyone; (2) at least one of three types of reexperiencing symptoms (recurrent and intrusive recollections, recurrent dreams, or suddenly acting as if the traumatic event were recurring); (3) at least one indicator of numbing of responsiveness or reduced involvement in the world (diminished interest in activities, feeling of detachment and disinterest, or constricted affect); and (4) at least two of an array of other symptoms, including hyperarousal or startle, sleep disturbance, survivor guilt, memory impairment or trouble concentrating, avoidance of activities reminiscent of the trauma, or intensification of symptoms when exposed to reminiscent events. Two subtypes were distinguished: acute, within the first 6 months, and chronic or delayed, with duration or onset occurring beyond 6 months. Interestingly, this earlier version of the DSM had separated numbing from effortful avoidance, a finding that has been established repeatedly, with factor analyses of DSM-IV symptoms (American Psychiatric Association, 1994; King, Leskin, King, & Weathers, 1998). Following the introduction of the diagnosis, there was a wave of prevalence studies to determine who develops the disorder and under what conditions, along with the development of valid and reliable assessment instruments for these criteria. Publications on treatment outcome studies began to appear by the mid- to late 1980s.

The introduction of PTSD into the DSM was not without controversy, which continues to this day. On the one hand, clinicians, who had been seeking an appropriate nosological category for psychiatrically incapacitated Holocaust survivors, rape victims, combat veterans, and other traumatized individuals, were delighted. They finally had a DSM-III diagnosis that validated the unique clinical phenomenology of their clientele. Recognition of the deleterious impact of traumatic stress provided a conceptual tool that transformed mental health practice and launched decades of research. For the first time, interest in the effects of trauma did not disappear with the end of a war.

On the other hand, critics of the diagnosis claimed and still claim that (1) people have always had reactions to events, and there is no need to pathologize it; (2) it is not a legitimate syndrome but a construct created by feminist and veteran special interest groups; (3) it serves a litigious rather than a clinical purpose, because the explicit causal relationship between traumatic exposure and PTSD symptoms has opened the door to a multitude of frivolous lawsuits and disability claims in which the financial stakes are enormous; (4) verbal reports of both traumatic exposure and PTSD symptoms are unreliable; (5) traumatic memories are not valid; (6) the diagnosis is a European American culture-bound syndrome that has no applicability to posttraumatic reactions within traditional cultures; and (7) it needlessly pathologizes the normal distress experienced by victims of abusive violence.

The next revision, DSM-III-R (American Psychiatric Association, 1987) produced the criteria, which, for the most part, exist today. Five criteria were established: (A) the stressor criterion; (B) reexperiencing symptoms (at least one), (C) avoidance symptoms (at least three), (D) arousal symptoms (at least two), and (E) duration criterion of 1 month. The acute designation was dropped from this iteration. The stressor criterion continued to define eligible stressors to be events "outside the range of usual human experience (i.e., outside the range of such common experiences as simple bereavement, chronic illness, business losses, and marital conflict)" and usually experienced with intense fear, terror, and helplessness (p. 247).

Avoidance symptoms included both efforts to avoid thoughts and reminders, and numbing. However, it also included a sense of foreshortened future and amnesia for parts of the event. The arousal criterion included both direct (startle, hypervigilance and/or physiological reactivity upon stimulus exposure) or indirect (irritability/anger, sleep problems and/or difficulty concentrating) indicators of physiological arousal. Once these reconfigured symptoms and clusters were established, another wave of research began to examine the individual symptoms, the clusters, and the configuration of the symptoms themselves. The committee assigned to conduct field trials for DSM-IV was asked to focus on a few specific questions (Kilpatrick et al., 1998). One was whether criterion A, the stressor criterion, should be changed or dropped entirely. After the first wave of prevalence studies, it had become evident that "outside the range of normal experience" was inaccurate, because most people experience at least one qualifying traumatic event in their lives, and some events, although infrequent in one person's life, are all too common across the population. Researchers asked whether people who experienced other stressful events, such as divorce, the loss of a job, or the natural death of a loved one, would also develop PTSD. They found that it made little difference whether the definition in the rates of PTSD was strict or nonrestrictive; few people developed PTSD unless they had experienced an extremely stressful life event. They did find more support for including a subjective distress component in criterion A because of consistent findings that the level of panic, physiological arousal, and dissociation present at the time of the event are predictors of later PTSD. Other questions posed in the field trial concerned placement of various symptoms and the threshold for criterion C, the avoidance criterion (Kilpatrick et al., 1998).

DSM-IV was published in 1994 and slightly revised in 2000 (American Psychiatric Association, 1994, 2000). Several changes in PTSD diagnosis were formalized, along with the introduction of a new disorder, acute stress disorder. Despite some strong interest by the PTSD subcommittee to move the disorder out of the anxiety disorders group, the diagnosis remained where it was. Criterion A now had two parts: (1) The person experienced, witnessed, or was confronted with an event or events that involved actual or threatened death or serious injury, or a threat to the physical integrity of self or others; and (2) the person's response involved intense fear, helplessness, or horror. An item that had been listed under the arousal category (physiological reactivity on exposure to trauma cues) was moved to the reexperiencing criterion. The only other significant change was that the symptoms must cause significant distress or impairment in some realm of functioning (criterion F).

The bigger development was the introduction of acute stress disorder (ASD), which emerged at the recommendation of the DSM-IV Dissociative Disorders Subcommittee, with the observation that people who had dissociative symptoms during or immediately after the traumatic event were most likely to develop PTSD. ASD was also introduced to bridge the diagnostic gap between the occurrence of traumatic event and 1 month later, when PTSD could be introduced. Criteria for ASD include the same stressor criterion as PTSD, and the presence of reexperiencing , avoidance, and arousal symptoms, although not in the 1, 3, 2 configuration required by PTSD. ASD differs, however, in that the person must experience at least three types of dissociative responses (amnesia, depersonalization, derealization, etc.). Like PTSD before it, ASD has proven to be controversial.

Criticisms of PTSD have not abated with the passage of time (Brewin, 2003; Rosen, 2004). Some have probably been exacerbated by concerns about the escalating number of PTSD disability claims recently filed by Canadian and American veterans. The cross-cultural argument currently rages within the context of Indonesian and Sri Lankan survivors of the 2005 Tsunami. These arguments also appear currently within the popular culture, due to increased attention from the mass media following the September 11, 2001, terrorist attacks, the South Asian Tsunami, Hurricane Katrina, and the wars in Iraq and Afghanistan. As a result, scientific debates about PTSD, previously restricted to professionals, have found their way into daily newspapers, popular magazines, radio talk shows, and televised documentaries.

We believe that these criticisms demand a thoughtful and balanced response, because they reflect concerns about PTSD that are shared both by the professional community and the public. Before we address these criticisms, however, it is necessary to review briefly the wealth of scientific information that has accrued since 1980, because of the new conceptual context provided by PTSD. Such research has not only transformed our understanding of how environmental events can alter psychological processes, brain function, and individual behavior but it has also generated new approaches to clinical treatment. Indeed, the translation of science into practice during the past 25 years is the major impact of the PTSD diagnosis.

SCIENTIFIC FINDINGS AND CLINICAL IMPLICATIONS

Epidemiology

When PTSD was first operationalized in DSM-III, exposure to traumatic stress was defined as "a catastrophic event beyond the range of normal human experience." Epidemiological surveys conducted since 1980 have shown otherwise (see Chapter 5, this volume, by Norris & Slone). More than half of all U.S. adults (50% female and 60% male) are exposed to traumatic stress during the course of their lifetimes (Kessler, Sonnega, Bromet, Hughes, & Nelson, 1995). In nations at war or subject to internal conflict, such as Algeria, Cambodia, Palestine, or the former Yugoslavia, traumatic exposure is much higher, ranging from 70 to 90% (de Jong et al., 2001). Surveys of American military veterans have shown, as might be expected, high rates of exposure to war-zone stress, although prevalence estimates have varied in magnitude depending on the specific nature of each war and the war-specific demands of each deployment (Hoge et al., 2004; Kang, Natalson, Mahan, Lee, & Murphy, 2003; Schlenger et al., 1992).

One of the most robust findings in epidemiological research on PTSD is a dose–response relationship between the severity of exposure to trauma and the onset of PTSD. Therefore, in the United States, where lifetime trauma exposure is 50–60%, PTSD prevalence is 7.8%, whereas in Algeria, where trauma exposure is 92%, PTSD prevalence is 37.4% (de Jong et al., 2001; Kessler et al., 1995). This dose–response association has held up whether the traumatic experience has been sexual assault, war-zone exposure, a natural disaster, or a terrorist attack (Galea et al., 2002; Kessler et al., 1995; Norris, Friedman, & Watson, 2002; Norris, Friedman, Watson, Byrne, et al., 2002; Schlenger et al., 1992). Within this context, however, in the United States, the toxicity of interpersonal violence, such as that in rape, is much higher than that in accidents; whereas wherein 45.9% female rape victims are likely to develop PTSD, only 8.8% female accident survivors develop the disorder (Kessler et al., 1995; Resnick, Kilpatrick, Dansky, Saunders, & Best, 1993). In developing nations, however, natural disasters are much more likely to produce PTSD because of the magnitude of resource loss associated with such exposure (Norris, Friedman, & Watson, 2002; Norris, Friedman, Watson, Byrne, et al., 2002; see Norris & Slone, Chapter 5, this volume).

It is also important to recognize that PTSD is not the only clinically significant consequence of traumatic exposure. Other psychiatric consequences include depression, other anxiety disorders, and alcohol or drug abuse/dependency (Galea et al., 2002; Shalev et al., 1998). Finally, accumulating evidence indicates that when traumatized individuals develop PTSD, they are at greater risk to develop medical illnesses (Schnurr & Green, 2004; see Schnurr, Green, & Kaltman, Chapter 20, this volume).

The clinical implications of these data are clear. Given that exposure to traumatic experiences occurs in at least half of the adult American population (and much more frequently within nations in conflict), mental health and medical clinicians should always take a trauma history as part of their routine intake. If there is a positive history of such exposure, the next step is to assess for the presence or absence of PTSD (see Keane, Brief, Pratt, & Miller, Chapter 15, this volume).

Risk Factors

Most people exposed to traumatic stress do not develop persistent PTSD. Even among female victims of rape, the most toxic traumatic experience, 54.1% will not exhibit full PTSD after 3 months, and 91.2% female accident survivors never develop PTSD (Riggs, Rothbaum, & Foa, 1995; Rothbaum, Foa, Riggs, Murdock, & Walsh, 1992). This means that most people have sufficient resilience to protect them from developing the disorder. Research on risk factors generally divides them into pretraumatic, peritraumatic, and posttraumatic factors (see Vogt, King, & King, Chapter 6, this volume). Pretraumatic factors include age, gender, previous trauma history, personal or family psychiatric history, educational level, and the like. Although a great deal of research has identified such factors, all have relatively low power to predict the likelihood of PTSD onset following traumatic exposure (Brewin, Andrews, & Valentine, 2000).

In addition to limited predictive power, it is not clear why certain pretraumatic risk factors are associated with PTSD prevalence. For example, female rather than male gender predicts greater likelihood of developing PTSD following exposure to trauma. It is possible that this is just due to the greater likelihood of women to have experienced the events most likely to be associated with PTSD, such as child sexual abuse, rape, or intimate partner violence (Kessler et al., 1995). However, such apparent gender differences may actually represent more complex phenomena, such as gender differences in how trauma is conceptualized, potential gender-related differences in the PTSD construct itself, or how comorbid disorders contribute to this difference. Finally, there is evidence that whereas female gender predicts greater risk of PTSD, it may also predict more favorable responsivity to PTSD treatment (see Kimerling, Ouimette, & Weitlauf, Chapter 12, this volume).

With the recent characterization of the human genome, it will not be long before pretraumatic factor research includes genotype assessment (see Segman, Shalev, & Gelernter, Chapter 11, this volume). Indeed, in two studies on depression that have shown a clear gene–environment interaction, people exposed to three or more adverse events and also have two copies of the short form of the serotonin transporter gene are much more likely than those with two copies of the long form to report depressive symptoms or suicidal behavior (Caspi et al., 2003; Kaufman et al., 2004). Given the nature of this gene–environment interaction, it would not be surprising if comparable results were found with PTSD.

Peritraumatic risk factors concern the nature of the traumatic experience itself, as well as one's reaction to it. The dose–response relationship between trauma exposure and PTSD onset, mentioned previously, applies here, so that the severity of traumatic exposure predicts the likelihood of PTSD symptoms. Other peritraumatic risk factors include exposure to atrocities, peritraumatic dissociation, panic attacks, and other emotions (Bernat, Ronfeldt, Calhoun, & Arias, 1998; Davis, Taylor, & Lurigio, 1996; Epstein, Saunders, & Kilpatrick, 1997; Galea et al., 2002; Ozer, Best, Lipsey, & Weiss, 2003).

The major *post*traumatic factor is whether the traumatized person received social support, followed by other posttraumatic stressors (Brewin et al., 2000). Indeed, receipt of social support, which appears to be the most important risk factor of all, can protect trauma-exposed individuals from developing PTSD. Social support appears to be such a powerful factor that in one of the genetic depression studies mentioned earlier, social support significantly reduced the prevalence of depression among children with the greatest genetic vulnerability to adverse life events (Kaufman et al., 2004).

Schnurr, Lunney, and Sengupta (2004) have distinguished between risk factors for the onset of PTSD and those that predict maintenance of PTSD. Risk factors for persistence of PTSD emphasize current rather than past factors. They include current emotional sustenance, ongoing social support, and recent adverse life events. The clinical significance of these findings is noteworthy. Assessment of risk factors, especially the strength and availability of social support, should be a routine part of any PTSD diagnostic interview. Furthermore, mobilization of social support, whenever possible, should be a part of any treatment plan. This applies whether the client has either chronic PTSD or an acute posttraumatic reaction, and whether the clinician is providing treatment within a traditional clinical setting or an early intervention following a mass casualty within a public mental health context (see Watson, Gibson, & Ruzek, Chapter 25, this volume).

Psychological Theory and Practice

PTSD invites explication in terms of classic experimental psychological theory to a far greater degree than any other psychiatric syndrome. It is one of the more interesting and unique disorders as well, inasmuch as researchers, theorists, and clinicians have the rare opportunity to be present at the genesis of a disorder that began at a precise moment in time. Hence, there is a rich conceptual context within which to understand the disorder (see Monson et al., Chapter 3, and Cahill & Foa, Chapter 4, this volume).

Both conditioning and cognitive models have been proposed. Pavlovian fear conditioning, either as a unitary model (Kolb, 1989) or within the context of Mowrer's two-factor theory (Keane & Barlow, 2002; Keane, Zimering, & Caddell, 1985) has influenced research and treatment. Such models have inspired animal, psychophysiological, and brain imaging research, in addition to psychological investigations with clinical cohorts. Emotional processing theory (Foa & Kozak, 1986) has also been very influential. This theory proposes that pathological fear structures (Lang, 1977) activated by trauma exposure produce cognitive, behavioral, and physiological anxiety. Finally, cognitive models derived from classical cognitive theory (Beck, Rush, Shaw, & Emery, 1979) postulate that it is the interpretation of the traumatic event, rather than the event itself, that precipitates clinical symptoms.

A number of cognitive-behavioral therapies (CBTs) have been derived from the aforementioned theories and tested with patients with PTSD. What all CBT approaches have in common is that they elegantly translate theory into practice. As reviewed by Resick, Monson, and Gutner (Chapter 17, this volume), the most successful treatments for PTSD are CBT approaches, most notably prolonged exposure, cognitive therapy, cognitive processing therapy, and stress inoculation therapy. Indeed, all clinical practice guidelines for PTSD identify CBT as the treatment of choice (American Psychiatric Association, 2004; Foa, Keane, & Friedman, 2000; National Collaborating Centre for Mental Health, 2005; Veterans Administration/Department of Defense [VA/DoD], Clinical Practice Guideline Working Group, 2004).

It is noteworthy that CBT has also been shown to be effective in treating acutely traumatized patients with ASD within weeks of exposure to a traumatic event (see Litz & Maguen, Chapter 16, this volume). This approach utilizes briefer versions of the prolonged exposure and cognitive restructuring protocols that have been so effective for chronic PTSD. Also, CBT protocols have been modified so that they can be delivered through the Internet or with the aid of virtual reality (see Welch & Rothbaum, Chapter 23, this volume).

Although such progress is gratifying, it is important to recognize that there is still much work ahead. Almost all randomized clinical trials for PTSD have only tested components of CBT or single medications. Such studies have shown that approximately half of all CBT patients achieve full remission of symptoms, leaving another half that experience partial or no improvement after a course of CBT. Clearly there is room for new treatments, a better understanding of how to combine medications, combined medication and psychosocial treatment, and tests of whether these therapies work in real-world settings. Also, questions about optimal strategies for specific phasing of treatments may benefit those who typically drop out of therapy early or do not benefit from a standard course of treatment. Indeed, future research will need to investigate systematically which treatment (or combination of treatments) is most effective for which patients with PTSD under what conditions. Finally, it is imperative that we focus now on dissemination of evidence-based practices for the treatment of PTSD in clinical settings.

There has also been recent progress in developing clinical approaches for PTSD among children and adolescents (see Fairbank, Putnam, & Harris, Chapter 13, and Saxe, MacDonald, & Ellis, Chapter 18, this volume), thanks in part to establishment of the National Child Traumatic Stress Network in the United States. Progress with regard to older adults has lagged further behind (see Cook & Niederehe, Chapter 14, this volume). In short, there is a real need for better understanding of the consequences of traumatic exposure and for developmentally sensitive treatment approaches for people at either end of the lifespan.

Biological Theory and Practice

Thanks to advances in technology, biological research has progressed beyond animal models and neurohormonal assays to brain imaging and genetic research. It is noteworthy that a book on the neurobiology of PTSD, published in 1995, had neither a chapter on brain imaging nor one on genetics, as in this volume (Friedman, Charney, & Deutch, 1995). The neurocircuitry that processes threatening stimuli centers on the amygdala, with major reciprocal connections to the hypothalamus, hippocampus, locus coeruleus, raphe nuclei, mesolimbic, mesocortical, and downstream autonomic systems. Major restraint on the amygdala is ordinarily exercised by the medial prefrontal cortex. In PTSD, amygdala activation is excessive, whereas prefrontal cortical restraint is diminished (Charney, 2004; Davis & Whalen, 2001; see Neumeister, Henry, & Krystal, Chapter 9, and Southwick et al., Chapter 10, this volume).

Many different neurohormones, neurotransmitters, and neuropeptides play important roles in this stress-induced fear circuit. Thus, there are many potential opportunities to translate such basic knowledge into pharmacological practice. At present, two medications, both selective serotonin reuptake inhibitors (SSRIs) have received U.S. Food and Drug Administration (FDA) approval as indicated treatments for PTSD. There is growing research with other medications affecting different mechanisms, but few randomized clinical trials have been carried out so far. Given our growing knowledge in this area, and the fact that only 30% patients receiving SSRIs achieve full remission, there is reason to expect that newer agents will prove more effective in the future (Friedman, 2002; see Friedman & Davidson, Chapter 19, this volume).

Another significant translation of science into practice concerns the association between PTSD and physical illness (see Schnurr et al., Chapter 20, this volume). Given the dysregulation of major neurohormonal and immunological systems among individuals with PTSD, it is perhaps not surprising that patients with PTSD are at greater risk for medical illness (Schnurr & Green, 2004) and for increased mortality due to cancer and cardiovascular illness (Boscarino, 2006). Again, as a mark of recent progress, such relationships were merely hypothesized in 1995 (Friedman & Schnurr, 1995). Now there is a compelling and rapidly growing database to verify these hypotheses.

Resilience, Prevention, and Public Health

Two epidemiological findings have had a profound effect on our understanding about the risk of exposure to trauma, and about the consequences of such exposure. First, as noted earlier (see "Epidemiology"), exposure to catastrophic stress is not unusual in the course of a lifetime. Second, most exposed individuals are resilient; they do not develop PTSD or some other disorder in the aftermath of traumatic events. Recent world events have thrust such scientific findings into the context of public policy and public health, including terrorist attacks in New York, Madrid, Moscow, London, and elsewhere; the tsunami of 2005; Hurricane Katrina; and many other man-made and natural disasters. The scientific question is: Why are some individuals resilient, while others develop PTSD following such catastrophic stressful experiences? The clinical question is: What can be done to fortify resilience among individuals who might otherwise be vulnerable to PTSD following traumatic exposure? And the public mental health question is: Following mass casualties or large-scale disasters, what can be done to prevent psychiatric morbidity in vulnerable populations?

From a historical perspective, these three questions are remarkable. Only because of recent scientific progress can such questions even be conceptualized. The new interest in resilience is emblematic of both maturity in the field and technological advances. Resilience is a multidimensional construct that includes genetic, neurohormonal, cognitive, personality, and social factors (see Layne, Warren, Watson, & Shalev, Chapter 24, this volume). From the clinical and public health perspective, the major question is: Can we teach vulnerable individuals to become more resilient? Our emergent understanding of the multidimensional mechanisms underlying resilience has given the term "stress inoculation" a new meaning in the 21st century. This in turn has raised public policy and public mental health questions about the feasibility of preventing posttraumatic distress and PTSD in the population at large.

In the United States, the September 11, 2001, terrorist attacks instigated a national initiative to understand the longitudinal course of psychological distress and psychiatric symptoms following exposure to mass casualties. In this regard, civilian disaster mental health found much in common with military mental health. In both domains, it is recognized that most posttraumatic distress is a normal, transient reaction from which complete recovery can be expected. A significant minority of both civilian and military traumatized individuals, however, do not recover but go on to develop clinical problems that demand professional attention. Thus, there are two trajectories following traumatic stress: normal transient distress or chronic clinical morbidity. The second trajectory requires treatment by traditional mental health professionals; indeed, evidence-based early interventions have also been developed for acutely traumatized individuals (see Litz & Maguen, Chapter 16, this volume). On the other hand, the first trajectory, affecting most of the population, demands a public mental health approach that fortifies resilience (see Ritchie, Watson, & Friedman, 2006; Watson et al., Chapter 25, this volume).

It is very exciting to consider the conceptual and clinical advances that have been made in this area during the last few years. Future research should produce a wide spectrum of scientific advances that will enhance our understanding of resilience (at the genetic, molecular, social, etc., levels), thereby providing needed tools to foster prevention and facilitate recovery at both individual and societal levels.

CRITICISMS OF THE PTSD CONSTRUCT

PTSD Is Not a Legitimate Diagnosis

We agree that men, women, and children have been exposed to traumatic events since prehistoric times. Indeed, a literary record of the adverse impact of such exposure is recorded by Homer, Shakespeare, Dickens, Remarque, up to and including contemporary authors. A recent article using American Civil War archival data indicates that high rates of traumatic exposure were associated with high rates of physical and psychological morbidities (Pizarro, Silver, & Prause, 2006). Attempts to record and understand such events and their consequences within a scientific or medical context are much more recent, dating back to the mid-19th century. These latter observations have generated a number of somatic (e.g., soldier's heart, effort syndrome, shell shock, neurocirculatory asthenia) and psychological (nostalgia, combat fatigue, traumatic neurosis) conceptual models (see van der Kolk, Chapter 2, and Monson et al., Chapter 3, this volume). Reviewing some of the rich clinical (and literary) reports provided prior to 1980, it is clear that many authors were describing what would now be labeled PTSD. So what has been gained by this new conceptual and diagnostic construct?

It is evident that the explication and official adoption of PTSD as a DSM-III diagnosis ushered in a significant paradigm shift in mental health theory and practice. First, it highlighted the etiological importance of traumatic exposure as the precipitant of stress-induced alterations in cognition, emotion, brain function, and behavior. Dissemination of this model has provided a coherent context within which practitioners have been able to understand the pathway from traumatic exposure to clinical abnormalities. Second, the PTSD model has stimulated basic research (both human and animal), in which it has been possible to investigate the causal impact of extreme stress on molecular, hormonal, behavioral, and social expression. Recently, investigators have begun to explore gene–environment interactions within this paradigm. Third, as noted earlier, the traumatic stress model has invited the elaboration of therapeutic strategies that have successfully ameliorated PTSD symptoms. Finally, PTSD was a unifying principle at a time when investigators were describing symptoms across a range of traumatic events, such as child abuse syndrome, battered women's syndrome, rape trauma syndrome, and Vietnam veterans syndrome. The important inductive leap of the DSM-III PTSD diagnosis was recognition that the reactions to these different types of events had more commonalities than differences. Subsequent research has shown that the same therapies can be used successfully across different types of traumas. All of these extraordinary advances could not have occurred before posttraumatic distress and dysfunction were reconceptualized as PTSD.

Some objections to the PTSD diagnosis are historical. It is certainly possible that it would not have been included in DSM-III without strong support from veteran and feminist advocacy groups. Unlike depression, schizophrenia, and other anxiety disorders, PTSD emerged from converging social movements rather than academic, clinical, or scientific initiatives. As a result, PTSD received an ambivalent, if not hostile, reception in many prominent psychiatric quarters when it was first introduced in 1980.

The response to this negative reception was an outpouring of research to test rigorously the legitimacy of PTSD as a diagnosis. This entire volume documents that early research and its more recent elaborations. The bottom line is that people who meet PTSD diagnostic criteria exhibit significant differences from nonaffected individuals, as well as from individuals with depression, other anxiety disorders, or other psychiatric disorders. Such research spans the spectrum from brain imaging to cognitive processing to clinical phenomenology to interpersonal dynamics. Factor analysis of the PTSD symptom clusters has generally validated the DSM-III–DSM-IV construct, although there are questions about whether a four-factor solution that splits avoidant from numbing symptoms is a better construct than the current three-factor model (Friedman & Karam, in press).There can no longer be any doubt about the legitimacy of PTSD as a diagnosis.

PTSD Needlessly Pathologizes Normal Reactions to Abusive Violence

This criticism asserts that normal reactions to the abnormal conditions of political repression and torture (or interpersonal violence, e.g., domestic violence) should be understood as appropriate coping responses to extremely stressful events. The argument further states that a psychiatric label such as PTSD removes such reactions from their appropriate sociopolitical–historical context and thrusts them into the inappropriate domain of individual psychopathology. We reject this argument because it fails to acknowledge that some people cope successfully with such events and manifest normal distress, whereas others exhibit clinically significant symptoms. This is another area in which both public health and individual psychopathology models are applicable to different segments of a population exposed to the same traumatic stressor (see "Resilience, Prevention, and Public Health").

As we have learned during the post–September 11, 2001, era of posttraumatic public mental health, most people exposed to severe stress have sufficient resilience to achieve full recovery. A significant minority, however, develop acute and/or chronic psychiatric disorders, among which PTSD is most prominent. The purpose of any medical diagnosis is to inform treatment decisions, not to "pathologize." Therefore, we reiterate that it is beneficial to detect PTSD among people exposed to traumatic stress to provide a treatment that may ameliorate their suffering.

PTSD Is a Culture-Bound European American Syndrome

The PTSD construct has been criticized from a cross-cultural perspective as an idiosyncratic European American construct that fails to characterize the psychological impact of traumatic exposure in traditional societies (Summerfield, 2004). We acknowledge that there may be culture-specific idioms of distress around the world that may do a better job describing the expression of posttraumatic distress in one ethnocultural context or another (Green et al., 2004; Marsella, Friedman, Gerrity, & Scurfield, 1996). On the other hand, PTSD has been documented throughout the world (Green et al., 2004). De Jong and colleagues (2001) documented the high prevalence of PTSD in non-Western nations subjected to war or internal conflict, such as Algeria, Cambodia, Palestine, and the former Yugoslavia. An important recent report has a unique bearing on this issue, because it compared people from widely different cultures who were exposed to a similar traumatic event. North and colleagues (2005) compared Kenyan survivors of the bombing of the American embassy in Nairobi with American survivors of the bombing of the Federal Building in Oklahoma City. Both events were remarkably similar with respect to death, injury, destruction, and other consequences. Similar, too, was PTSD prevalence among Africans and Americans exposed to these different traumatic events.

We agree with Osterman and de Jong (Chapter 21, this volume) that the time has come for the fields of mental health and anthropology to end the debate about the validity of the PTSD diagnosis. What is needed is a "culturally competent model of traumatic stress" that addresses how culture may differentially influence explanatory models of traumatic stress, how it is implicated in the appraisal of risk–protective factors, and how such understanding might contribute to diagnosis and treatment.

PTSD Primarily Serves a Litigious Rather Than a Clinical Purpose

One of the reasons PTSD has played so prominently in disability and legal claims is that it has been assumed that the traumatic event is causally related to PTSD symptom expression and, hence, functional impairment. Although traumatic exposure is a necessary condition for the development of PTSD, it is not a sufficient condition. For example, the event most likely to result in PTSD is rape, yet only a minority of rape victims are diagnosable with PTSD after a few months. Other risk factors play a role in symptom onset and duration, as described earlier in the section on risk factors (and in Vogt et al., Chapter 6, this volume). Despite the etiological complexity of PTSD onset, the stressor criterion is fundamental in personal injury litigation, and in compensation and pension disability claims. This is because traumatic exposure establishes liability or responsibility for psychiatric sequelae in a context that puts PTSD in a category by itself with respect to other psychiatric diagnoses.

As noted by Sparr and Pitman (see Chapter 22, this volume) the geometric increase in PTSD claims in civil litigation is due to society's growing recognition that traumatic exposure can have significant and long-lasting consequences. There is also concern that the redefinition of the stressor criterion in DSM-IV has opened the door to frivolous litigations in which PTSD-related damages or disabilities are dubious at best. Another important factor driving much of this criticism is the sheer magnitude of money awarded for successful personal injury suits or compensation and pension disability claims.

There is a significant difference, however, between challenging the utility of PTSD as a clinical diagnosis and questioning the quality of forensic or disability evaluations performed by mental health professionals. We believe that minimal standards for such evaluations must be developed and enforced, so that people who have a legitimate claim for compensation because of their PTSD are not penalized because of misuse or abuse of this diagnosis in civil litigation or in the disability claims process.

Traumatic Memories Are Not Valid

An important scientific question concerns the validity of traumatic memories. A review of the literature on memory and dissociation (see Brewin, Chapter 7, and DePrince & Freyd, Chapter 8, respectively, this volume) indicates that trauma-related alterations in physiological arousal and information processing may affect how such input is encoded as a memory. Furthermore, the retrieval of such information may be affected by both current emotional state and the presence of PTSD. Such appropriate concerns notwithstanding, when external verification has been possible, it appears that most traumatic memories are appropriate representations of the stressful event in question. A particularly newsworthy manifestation of questions about the accuracy of trauma-related memories was sensationalized in the popular media as "the false memory syndrome." The issue concerned formerly inaccessible memories of childhood sexual abuse that later were "recovered." Some individuals who recovered such memories went on to sue the alleged perpetrator, thereby transforming a complex, controversial, and relatively obscure scientific and clinical question into a very public debate argued in the courtroom and mass media. It is now being documented that accurate traumatic memories may be lost and later recovered, although it is also clear that some recovered memories are not accurate. The veracity of any specific, recovered memory must be judged on a case-by-case basis (see Roth & Friedman, 1998; see Brewin, Chapter 7, this volume).

Verbal Reports Are Unreliable

A major theme throughout modern psychiatry has been the search for pathophysiological indicators that do not rely on verbal report. This is a challenge not only to PTSD assessment but also to assessment of all DSM-IV diagnoses. We recognize the importance of this concern in some circles but see no reason why it should be cited as a specific problem for PTSD, and not for any other psychiatric diagnosis.

Several laboratory findings hold promise as potential non-self-report assessment protocols for refining diagnostic precision (see Neumeister et al. and Southwick et al., Chapters 9 and 10, respectively, this volume). These include psychophysiological assessment with script-driven imagery or the startle response, or utilization of pharmacological probes, such as yohimbine or dexamethasone. At the moment, however, none have sufficient sensitivity or specificity for routine utilization in clinical practice.

In the meantime, we should not overlook the remarkable progress we have made in diagnostic assessment through development of structured clinical interviews and self-report instruments with excellent psychometric properties. In addition to improving diagnostic precision, such instruments have been utilized as dimensional measures to quantitate symptom severity and to monitor the effectiveness of therapeutic interventions (Wilson & Keane, 2004; see Keane et al., Chapter 15, this volume).

A remarkable recent study by Dohrenwend and colleagues (2006) indicates the high reliability of retrospective self-report data among a representative sample of 260 Vietnam Theater veterans who participated in the National Vietnam Veterans Readjustment Study (NVVRS). They compared verbal reports of combat exposure recorded by NVVRS investigators with a military–historical measure comprising military personnel files, military archival sources, and historical accounts. Results showed a strong positive relationship between the documented military–historical measure of exposure and the dichotomous verbal report–based assessment of high versus low to moderate war-zone stress previously constructed by NVVRS investigators. In short, this meticulous study indicates that verbal reports are usually quite reliable.

Summary

PTSD has been at the center of a number of controversies. Close examination of these contentious issues indicates that the arguments are generally not about PTSD per se, but about the appropriateness of invoking PTSD within a controversial or adversarial context. Because the issue of causality or etiology is so clearly specified in PTSD, as in few other diagnoses, it is likely that it will continue to be applied or misapplied in a number of clinical, forensic, and disability scenarios. An important goal is to respect the scientific evidence to ensure appropriate applications in the future. It is also useful to recognize that, as in the recovered memory controversy, such contentious issues have spawned important basic and clinical research that has resulted in better mental health assessment and treatment.

The purpose of this volume is to document how far we have come during the past 25 years, so that we can generate forward momentum in the right directions. Translating the science concerning traumatic stress into better clinical practice is the underlying process. The goal is to understand the disorder, to optimize assessment and treatment for people with PTSD and other posttraumatic problems, and to identify processes that facilitate recovery from exposure to traumatic events.

REFERENCES

American Psychiatric Association. (1952). *Diagnostic and statistical manual: Mental disorders.* Washington, DC: Author.

American Psychiatric Association. (1968). *Diagnostic and statistical manual of mental disorders* (2nd ed.). Washington, DC: Author.

American Psychiatric Association. (1980). *Diagnostic and statistical manual of mental disorders* (3rd ed.). Washington, DC: Author.

American Psychiatric Association. (1987). *Diagnostic and statistical manual of mental disorders* (3rd ed., rev.). Washington, DC: Author.

American Psychiatric Association. (1994). *Diagnostic and statistical manual of mental disorders* (4th ed.). Washington, DC: Author.

American Psychiatric Association. (2000). *Diagnostic and statistical manual of mental disorders* (4th ed., rev.). Washington, DC: Author.

American Psychiatric Association. (2004). Practice guidelines for the treatment of acute stress and posttraumatic stress disorder. *American Journal of Psychiatry, 161,* 1–31.

Beck, A. T., Rush, A. J, Shaw, B. F., & Emery, G. (1979). *Cognitive therapy of depression*. New York: Guilford Press.

Bernat, J. A., Ronfeldt, H. M., Calhoun, K. S., & Arias, I. (1998). Prevalence of traumatic events and peritraumatic predictors of posttraumatic stress symptoms in a nonclinical sample of college students. *Journal of Traumatic Stress, 11*, 645–664.

Bloom, S. L. (2000). Our hearts and our hopes are turned to peace: Origins of the International Society for Traumatic Stress Studies. In A. Y. Shalev, R. Yehuda, & A. C. McFarlane (Eds.), *International handbook of human responses to trauma* (pp. 27–50). New York: Kluwer Academic/Plenum Press.

Boscarino, J. A. (2006). Posttraumatic stress disorder and mortality among U.S. Army veterans 30 years after military service. *Annals of Epidemiology, 16*, 248–256.

Brewin, C. R. (2003). *Posttraumatic stress disorder: Malady or myth?* New Haven, CT: Yale University Press.

Brewin, C. R., Andrews, B., & Valentine, J. D. (2000). Meta-analysis of risk factors for posttraumatic stress disorder in trauma-exposed adults. *Journal of Consulting and Clinical Psychology, 68*, 748–766.

Burgess, A. W., & Holmstrom, L. L. (1973). The rape victim in the emergency ward. *American Journal of Nursing, 73*, 1740–1745.

Burgess, A. W., & Holmstrom, L. L. (1974). Rape trauma syndrome. *American Journal of Psychiatry, 131*, 981–986.

Caspi, A., Sugden, K., Moffitt, T. E., Taylor, A., Craig, I. W., Harrington, H., et al. (2003). Influence of life stress on depression: Moderation by a polymorphism in the 5HTT gene. *Science, 301*, 386–389.

Charney, D. S. (2004). Psychobiological mechanisms of resilience and vulnerability: Implications for the successful adaptation to extreme stress. *American Journal of Psychiatry, 161*, 195–216.

Davis, M., & Whalen, P. J. (2001). The amygdala: Vigilance and emotion. *Molecular Psychiatry, 1*, 13–34.

Davis, R. C., Taylor, B., & Lurigio, A. J. (1996). Adjusting to criminal victimization: The correlates of postcrime distress. *Violence and Victims, 11*, 21–38.

De Jong, J. T., Komproe, I. H., Van Ommeren, M., El Masri, M., Araya, M., Khaled, N., et al. (2001). Lifetime events and posttraumatic stress disorder in 4 postconflict settings. *Journal of the American Medical Association, 286*, 555–562.

Dohrenwend, B. P., Turner, J. B., Turse, N. A., Adams, B. G., Koenen, K. C., & Marshall, R. (2006). The psychologic risks of vietnam for U.S. veterans: A revisit with new data and methods. *Science, 313*, 979–982.

Epstein, J. N., Saunders, B. E., & Kilpatrick, D. G. (1997). Predicting PTSD in women with a history of childhood rape. *Journal of Traumatic Stress, 10*, 573–588.

Foa, E. B., Keane, T. M., & Friedman, M. J. (Eds.). (2000). *Effective treatments for PTSD: Practice guidelines from the International Society of Traumatic Stress Studies*. New York: Guilford Press.

Foa, E. B., & Kozak, M. J. (1986). Emotional processing of fear: Exposure to corrective information. *Psychological Bulletin, 99*, 20–35.

Friedman, M. J. (2002). Future pharmacotherapy for PTSD: Prevention and treatment. *Psychiatric Clinics of North America, 25*, 427–441.

Friedman, M. J., Charney, D. S., & Deutch, A. Y. (Eds.). (1995). *Neurobiological and clinical consequences of stress: From normal adaptation to post-traumatic stress disorder*. Philadelphia: Lippincott–Raven.

Friedman, M. J., & Karam, E. G. (in press). PTSD: Looking toward DSM-V and ICD-11. In G. Andrews, D. Charney, P. Sirovatka, & D. Regier (Eds.), *Stress-induced fear circuitry disorders: Refining the research agenda for DSM-V*. Washington, DC: American Psychiatric Association.

Friedman, M. J., & Schnurr, P. P. (1995). The relationship between trauma and physical health. In M. J. Friedman, D. S. Charney, & A. Y. Deutch (Eds.), *Neurobiological and clinical consequences of stress: From normal adaptation to post-traumatic stress disorder* (pp. 507–526). Philadelphia: Lippincott–Raven.

Galea, S., Ahern, J., Resnick, H. S., Kilpatrick, D. G., Bucuvalas, M. J., Gold, J., et al. (2002). Psychological sequelae of the September 11 terrorist attacks in New York City. *New England Journal of Medicine, 346*, 982–987.

Gray, J. D., Cutler, C. A., Dean, J. G., & Kempe, C. H. (1977). Prediction and prevention of child abuse and neglect. *Child Abuse and Neglect, 1*, 45–58.

Green, B. L., Friedman, M. J., de Jong, J. T. V. M., Solomon, S. D., Keane, T. M., Fairbank, J. A., et al. (Eds.). (2003). *Trauma interventions in war and peace: Prevention, practice, and policy*. Amsterdam: Kluwer Academic/Plenum Press.

Hoge, C. W., Castro, C. A., Messer, S. C., McGurk, D., Cotting, D. I., & Koffman, R. L. (2004). Combat duty in Iraq and Afghanistan, mental health problems, and barriers to care. *New England Journal of Medicine, 351*, 13–22.

Jablensky, A. (1985). Approaches to the definition and classification of anxiety and related disorders in European psychiatry. In A. H. Tuma & J. D. Maser (Eds.), *Anxiety and the anxiety disorders* (pp. 735–758). Hillsdale, NJ: Erlbaum.

Kang, H. K., Natelson, B. H., Mahan, C. M., Lee, K. Y., & Murphy, F. M. (2003). Post-traumatic stress disorder and chronic fatigue syndrome-like illness among Gulf War veterans: A population-based survey of 30,000 veterans. *American Journal of Epidemiology, 157*, 141–148.

Kaufman, J., Yang, B-Z., Douglas-Palumberi, H., Houshyar, S., Lipschitz, D., Krystal, J. H., et al. (2004). Social supports and serotonin transporter gene moderate depression in maltreated children. *Proceedings of the National Academy of Sciences USA, 101*, 17316–17321.

Keane, T. M., & Barlow, D. H. (2002). Posttraumatic stress disorder. In D. H. Barlow (Ed.), *Anxiety and its disorders: The nature and treatment of anxiety and panic* (2nd ed., pp. 418–453). New York: Guilford Press.

Keane, T. M., Zimering, R. T., & Caddell, J. M. (1985). A behavioral formulation of posttraumatic stress disorder in Vietnam veterans. *Behavior Therapist, 8*, 9–12.

Kessler, R. C., Sonnega, A., Bromet, E., Hughes, M., & Nelson, C. B. (1995). Posttraumatic stress disorder in the National Comorbidity Survey. *Archives of General Psychiatry, 52*, 1048–1060.

Kilpatrick, D. G., Resnick, H. S., Freedy, J. R., Pelcovitz, D., Resick, P. A., Roth, S., et al. (1998). Posttraumatic stress disorder field trial: Evaluation of the PTSD construct—criteria A through E. In T. A. Widiger (Ed.), *DSM-IV sourcebook* (pp. 803–838). Washington, DC: American Psychiatric Association.

King, D. W., Leskin, G. A., King, L. A., & Weathers, F. W. (1998). Confirmatory factor analysis of the Clinician-Administered PTSD Scale: Evidence for the dimensionality of posttraumatic stress disorder. *Psychological Assessment, 10*, 90–96.

Kolb, L. C. (1989). Heterogeneity of PTSD. *American Journal of Psychiatry, 146*, 811–812.

Lang, P. J. (1977). Imagery in therapy: An information processing analysis of fear. *Behavior Therapy, 8*, 862–886.

Marsella, A. J., Friedman, M. J., Gerrity, E. T., & Scurfield, R. M. (Eds.). (1996). *Ethnocultural aspects of post-traumatic stress disorder: Issues, research and clinical applications.* Washington, DC: American Psychological Association.

National Collaborating Centre for Mental Health. (2005). *Post-traumatic stress disorder: The management of PTSD in adults and children in primary and secondary care.* London: Gaskell and the British Psychological Society.

Norris, F. H., Friedman, M. J., & Watson, P. J. (2002). 60,000 disaster victims speak: Part II. Summary and implications of the disaster mental health research. *Psychiatry, 65*, 240–260.

Norris, F. H., Friedman, M. J., Watson, P. J., Byrne, C. M., Diaz, E., & Kaniasty, K. Z. (2002). 60,000 disaster victims speak: Part I. An empirical review of the empirical literature, 1981–2001. *Psychiatry, 65*, 207–239.

North, C. S., Pfefferbaum, B., Narayanan, P., Thielman, S. B., McCoy, G., Dumont, C. E., et al. (2005). Comparison of post-disaster psychiatric disorders after terrorist bombings in Nairobi and Oklahoma City. *British Journal of Psychiatry, 186*, 487–493.

Ozer, E. J., Best, S. R., Lipsey, T. L., & Weiss, D. S. (2003). Predictors of posttraumatic stress disorder and symptoms in adults: A meta-analysis. *Psychological Bulletin, 129*, 52–73.

Pizarro, J., Silver, R. C., & Prause, J. (2006). Physical and mental health costs of traumatic war experiences among Civil War veterans. *Archives of General Psychiatry, 63*, 193–200.

Resnick, H. S., Kilpatrick, D. G., Dansky, B. S., Saunders, B. E., & Best, C. L. (1993). Prevalence of civilian trauma and posttraumatic stress disorder in a representative national sample of women. *Journal of Consulting and Clinical Psychology, 61*, 984–991.

Riggs, D. S., Rothbaum, B. O., & Foa, E. B. (1995). A prospective examination of symptoms of posttraumatic stress disorder in victims of nonsexual assault. *Journal of Interpersonal Violence, 10*, 201–214.

Ritchie, E. C., Watson, P. J., & Friedman, M. J. (Eds.). (2006). *Interventions following mass violence and disasters: Strategies for mental health practice.* New York: Guilford Press.

Rosen, G. M. (2004). *Posttraumatic stress disorder: Issues and controversies.* Chichester, UK: Wiley.

Roth, S., & Friedman, M. J. (1998). *Childhood trauma remembered: A report on the current scientific knowledge base and its applications.* Northbrook, IL: International Society for Traumatic Stress Studies.

Rothbaum, B. O., Foa, E. B., Riggs, D. S., Murdock, T. B., & Walsh, W. (1992). A prospective examination of post-traumatic stress disorder in rape victims. *Journal of Traumatic Stress, 5,* 455–475.

Saigh, P. A. (1992). History, current nosology, and epidemiology. In *Posttraumatic stress disorder: A behavioral approach to assessment and treatment* (pp. 1–27). Boston: Allyn & Bacon.

Schlenger, W. E., Kulka, R. A., Fairbank, J. A., Hough, R. L., Jordan, B. K., Marmar, C. R., et al. (1992). The prevalence of post-traumatic stress disorder in the Vietnam generation: A multimethod, multisource assessment of psychiatric disorder. *Journal of Traumatic Stress, 5,* 333–363.

Schmitt, B. D., & Kempe, C. H. (1975). Prevention of child abuse and neglect. *Current Problems in Pediatrics, 5,* 35–45.

Schnurr, P. P., & Green, B. L. (Eds.). (2004). *Trauma and health: Physical health consequences of exposure to extreme stress.* Washington, DC: American Psychological Association.

Schnurr, P. P., Lunney, C. A., & Sengupta, A. (2004). Risk factors for the development versus maintenance of posttraumatic stress disorder. *Journal of Traumatic Stress, 17,* 85–95.

Shalev, A.Y., Freedman, S. A., Peri, T., Brandes, D., Sahar, T., Orr, S. P., et al. (1998). Prospective study of posttraumatic stress disorder and depression following trauma. *American Journal of Psychiatry, 155,* 630–637.

Shay, J. (1994). (Ed.). *Achilles in Vietnam: Combat trauma and the undoing of character.* New York: Atheneum.

Summerfield, D. A. (2004). Cross-cultural perspectives on the medicalization of human suffering. In G. M. Rosen (Ed.), *Posttraumatic stress disorder: Issues and controversies* (pp. 233–245). Chichester, UK: Wiley.

VA–DoD Clinical Practice Guideline Working Group, Veterans Health Administration, Department of Veterans Affairs and Health Affairs, Department of Defense. (2004). *Management of post-traumatic stress* (Publication No. 10Q-CPG/PTSD-04 2003). Washington, DC: Office of Quality and Performance. (*www.oqp.med.va.gov/cpg/ptsd/ptsd_base.htm*)

Walker, L. E. (1979). *The battered woman.* New York: Harper & Row.

Wilson, J. P., & Keane, T. M. (Eds.). (2004). *Assessing psychological trauma and PTSD* (2nd ed.). New York: Guilford Press.

Chapter 2

The History of Trauma in Psychiatry

Bessel A. van der Kolk

The subject of neurotic disturbances consequent upon war has . . . been submitted to a good deal of capriciousness. . . . The public does not sustain its interest, which was very great after WWI, and neither does psychiatry. . . . It is a deplorable fact that each investigator who undertakes to study these conditions considers it his sacred obligation to start from scratch and work at the problem as if no one had done anything with it before.

—KARDINDER AND SPIEGEL (1947, p. 25)

People have always been aware that exposure to overwhelming terror can lead to troubling memories, arousal, and avoidance: This has been a central theme in literature from the time of Homer (Alford, 1992; Shay, 1994) to the present (Caruth, 1995). However, psychiatry, as a profession, has had a troubled relationship with the idea that reality can profoundly and permanently alter people's psychology and biology. Psychiatry has periodically suffered from marked amnesias, in which well-established knowledge was abruptly forgotten, and the psychological impact of overwhelming experiences ascribed to constitutional or intrapsychic factors alone. Mirroring the intrusions, confusion, and disbelief of victims whose lives are suddenly shattered by traumatic experiences, the psychiatric profession periodically has been fascinated by trauma, followed by stubborn disbelief about the relevance of our patients' stories.

From the earliest encounters between psychiatry and posttraumatic syndromes, there have been vehement arguments about etiology: whether it is organic or psychological; whether it is the event itself or its subjective interpretation that causes the pathology; and whether trauma itself or preexisting vulnerabilities are responsible for psychological disintegration. Do these patients malinger, and do they suffer from moral weaknesses? To what degree is their failure to take charge of their lives voluntary? Do they make up their memories? Are they accurate; can they be repressed and retrieved at a later time? Is dissociation always present in response to trauma? Is it a dissociative disorder or an anxiety disorder? Do multiple personalities spontaneously arise as a consequence of trauma, or are they iatrogenically induced? All these questions have been raised repeatedly since the 1880s and thought to be settled, only to be raised again repeatedly at later times. None of them have been definitively settled in the beginning of the 21st century.

TRAUMATIC STRESS: EMOTIONAL OR ORGANIC?

The conflict between organic and psychological origins, and between malingering and genuine breakdown, was at the center of the earliest scientific discussions about the effects of trauma: in whiplash injuries and "railroad spines." The English surgeon John Eric Erichsen (1866, 1886) ascribed the psychological problems of severely injured patients to organic causes and warned against confusing these symptoms with those of hysteria, a condition that he, and most of his contemporaries, claimed only occurred in women. Not unlike today, physicians in those days struggled with trying to understand body–mind relationships: Physical signs of anxiety then, as now, were easily misdiagnosed as symptoms of organic illness. Erichsen's fellow surgeon Page (1883) disagreed with him, and proposed that the symptoms of railroad spine had psychological origins. He claimed that "many errors in diagnosis have been made because fright has not been considered of itself sufficient."

The German neurologist Herman Oppenheim, the first to use the term "traumatic neurosis" (1889), was an organicist who proposed that functional problems were produced by subtle molecular changes in the central nervous system. The frequent occurrence of cardiovascular symptoms in traumatized persons, particularly in combat soldiers, started a long tradition of associating posttraumatic problems with "cardiac neuroses." This began with names such as "irritable heart" and "soldier's heart" (Da Costa, 1871; Myers, 1870), and progressed to "disorderly action of the heart" or "neurocirculatory asthenia" during World War I (Merskey, 1991).

The issue of psychological versus organic origins of traumatic neuroses was particularly relevant for combat soldiers. When issues of cowardice and shirking are raised, ascribing posttraumatic symptoms to organic problems offers an honorable solution: The soldier preserves his self-respect, the doctor stays within his professional role and does not have to get involved in disciplinary actions, and military authorities do not have to explain psychological breakdowns in previously brave soldiers. Organic causes help them to avoid addressing troublesome issues such as cowardice, low unit morale, poor leadership, or the meaning of the war effort itself.

But if it were an illness, how could it be defined? Charles Samuel Myers (1915), a British military psychiatrist, was the first to use the term "shell-shock" in the medical literature. However, because "shell-shock" could be found in soldiers who had never been directly exposed to warfare, it gradually became clear that often the cause was purely emotional. Meyers, like so many after him, emphasized the close resemblance of the war neuroses to hysteria (Myers, 1940). He proposed that emotional disturbance alone was enough of an explanation and rejected a relationship between battle neuroses and an organic, "molecular commotion in the brain." However, this did not end the controversy: In his memoirs of his service in World War I, Churchill's doctor, Lord Moran

(1945), confessed that doctors found it very difficult to distinguish between shell-shock and cowardice. During World War I, more than 2,200 British soldiers were condemned to death for cowardice and desertion, though only about 200 were actually executed.

Psychological explanations for traumatic neuroses were easier to pursue in civilian settings. American neurologist James J. Putnam (1881) developed a theory based on Hughlings Jackson's notion of illness, in which psychic traumatization was considered a functional regression toward earlier, simple, reflexive, automated modes of functioning (MacLeod, 1993; Putnam, 1898). These notions were similar to those held by Pierre Janet at the Salpêtrière, who in his doctoral thesis, "L'Automatisme Psychologique" (1889), had documented the relationships between trauma and various automatic, involuntary behaviors. When Harvard Medical School inaugurated its new buildings on Longwood Avenue in 1906, Putnam was instrumental in inviting Janet to Boston to lecture on these common interests under the rubric *The Major Symptoms of Hysteria* (1907).

TRAUMA, SUGGESTIBILITY, AND SIMULATION

Since the beginning of psychiatry's attempt to become a scientific discipline, people have noted an association between psychological trauma and hysterical symptoms: As early as 1859, the French psychiatrist Briquet (1859) elucidated an association between childhood histories of trauma and symptoms of "hysteria," such as somatization, intense emotional reactions, dissociation, and fugue states. Of the 501 patients with hysteria he described, Briquet reported specific traumatic origins as the cause of illness in 381 patients (Crocq & De Verbizier, 1989). Sexual abuse of children was well documented during the second half of the 19th century in France by researchers such as Tardieu (1878), a professor of forensic medicine.

Almost as soon as the issue of sexual trauma in children was recognized, the thorny issue of false memories was raised by people like Alfred Fournier, who described "pseudologica phantastica" in children who were thought to falsely accuse their parents of incest. Similar problems arose when the first systematic explorations of the relationships between trauma and psychiatric illness were conducted at the Salpêtrière Hospital in Paris. The great neurologist Jean-Martin Charcot (1887) described how traumatically induced "choc nerveux" could put patients into a mental state similar to that induced by hypnosis. This so-called "hypnoid" state was a necessary condition of what Charcot called "hystero-traumatic auto-suggestion." Thus, Charcot became the first to describe the problems of both suggestibility in these patients and the fact that hysterical attacks are dissociative, the result of having endured unbearable experiences.

Charcot had a deep interest in the issues of trauma, memory, and dissociation, and his pupils split into two groups: those who continued to study adaptation to trauma, and those who studied suggestibility and false memories. Pierre Janet became a great researcher of the nature of dissociation and of traumatic memories (1887,1889, 1894), whereas Gilles de la Tourette and Babinski focused their research on hysterical suggestibility. When Babinski took over as head of the Salpêtrière in 1905, Charcot's notions about the traumatic origins of hysteria were rejected as worthless (Ellenberger, 1970). After that, simulation and suggestibility were considered the hallmarks of the now emphatically neurological disease entity "hysteria" (Babinski, 1901, 1909). This development set the stage during World War I for the overriding interest among Babinski and his French students, as well as among many German psychiatrists, in the treatment of "simulation" rather than the alleviation of the horror of traumatic memories (Babinski

& Froment, 1918; Nonne, 1915). For many French and German neurologists and psychiatrists the treatment of war syndromes became a battle against simulation.

With the focus on simulation, the notion of "will" became the dominant issue; war neurosis (war hysteria) for many psychiatrists was essentially a disease of the will, a *Willenskrankheit* (Fischer-Homberger, 1975). Hence, largely for political reasons, the medical diagnosis of posttraumatic stress disorder (PTSD) in Germany during World War I and subsequent decades was recast as a failure of individual soldiers' willpower (*Willenversagung, Willenshemmung, Willensperrung, Wille zur Krankheit*). As a result, treatment comprised "causal will therapies": Patients' "desire for health" had to be stimulated and bolstered by physiological exercises. Because the treatment was so painful, many patients preferred frontline duty and were considered "cured."

In the period following World War I, the leading German psychiatrist Bonhoeffer (1926) and his colleagues founded a school of thought that regarded traumatic neuroses as social illnesses that could only be cured by social remedies. However, these social cures did not consist of an amelioration of social conditions. Because Bonhoeffer found that practically all of his 142 patients with traumatic neurosis had been predisposed to develop the disorder, dealing with these patients' inherent weakness, rather than preventing and ameliorating their misery, became the critical issue. He and his colleagues believed that the real cause of traumatic neurosis was the availability of compensation (*Das Gesetz ist die Ursache der Unfallsneurosen* [the law is the cause of traumatic neuroses]). In other words, the disorder was caused by secondary gain: Traumatic neurosis was not an illness, but an artifact of the insurance system, a *Rentenneurose*, a compensation neurosis. The Reichversicherungs Ordnung (RVO) of 1926 (National Health Insurance Act) cemented this position in Germany. Traumatic neurosis was not to be compensated. The philosophy was that traumatic neurosis was incurable as long as patients were awarded pensions or other compensations. To this day German compensation practices continue to be more restrictive than those of most other countries (Venzlaff, 1975).

VULNERABILITY AND PREDISPOSITION

The issue of predisposition could be studied with more detachment in civilian settings, where national pride was not at stake, and compensation was a less potentially massive economic issue. The Swiss psychiatrist Edouard Stierlin (1909, 1911), can be considered the first researcher in disaster psychiatry, with his study of nonclinical populations from the Messina earthquake in 1907, and a mining disaster. Nonclinical populations allowed him to address issues of vulnerability and resiliency. He came to the same conclusions as his predecessors at the Salpêtrière that violent emotions are the most important etiological factors of "fright neuroses" or posttraumatic neuroses. He expressed concern that physicians had little awareness that emotions can cause serious long-term psychoneurotic problems, whereas laymen tended to equate posttraumatic psychological problems with simulation.

Stierlin found that a substantial proportion of victims developed long-lasting posttraumatic stress symptoms; for example, after the Messina earthquake killed 70,000 of the town's inhabitants, 25% of the survivors suffered from sleep disturbances, including nightmares. Stierlin made the important observation that traumatic neurosis was the only psychogenic symptom complex for which no psychopathological predisposition was required. He also suggested that the term "neurosis" was not a good descriptor and took issue with Kraepelin, who in his famous textbook of psychiatry (1899) claimed that the type of traumatic neurosis in which fear played the dominant etiological role was rare and "atypical."

THE PSYCHOLOGICAL PROCESSING OF TRAUMA: REALITY IMPRINT VERSUS INTRAPSYCHIC ELABORATION

In retrospect, the issue of the traumatic origins of hysteria is probably the most important legacy of psychiatry of the last decades of the 19th century. At the Hôpital du Salpêtrière in Paris, Charcot (1887) had first proposed that the symptoms of patients with hysteria their origins in histories of trauma. Pierre Janet, who ran the psychological laboratory at that hospital, noted that patients with hysteria were unable to attend to their internal processes as guides for adaptive action. In line with the prevailing thinking of his days (e.g., Bergson, 1896), Janet considered personal consciousness to be the central issue in psychological health. He believed that awareness of one's personal past, combined with accurate perceptions of current surroundings, determined the capacity to respond appropriately to stress. Janet coined the word *subconscious* to describe the collection of memories forming the mental schemes that guide a person's interaction with the environment (Janet, 1904; van der Kolk & van der Hart, 1989). Appropriate categorization and integration of memories of past experience allow people to develop meaning schemes that prepare them to cope with subsequent challenges.

Janet proposed that when people experience "vehement emotions," the mind may not be able to match what is going on with existing cognitive schemes. As a result, memories of the experience cannot be integrated into personal awareness. Instead, they are split off (dissociated) from conscious awareness and from voluntary control. Thus, the first comprehensive formulation of the effects of trauma on the mind was based on the notion that failure to integrate traumatic memories due to extreme emotional arousal results in the symptoms of what we call today PTSD. Janet stated: "They are unable to make the recital which we call narrative memory, and yet they remain confronted by (the) difficult situation" (1919/1925, p. 661). This results in "a phobia of memory" (p. 661) that prevents the integration ("synthesis") of traumatic events and splits these traumatic memories off from ordinary consciousness (Janet, 1909, p. 145). The memory traces of the trauma linger as unconscious fixed ideas that cannot be "liquidated" as long as they have not been translated into a personal narrative and, instead, continue to intrude as terrifying perceptions, obsessional preoccupations, and somatic reexperiences such as anxiety reactions (Janet, 1889, 1930).

Janet observed that traumatized patients reacted to reminders of the trauma with responses that had been relevant to the original threat, but currently had no adaptive value. Upon exposure to reminders, the somatosensory representations of the trauma predominated (Janet, 1889). He proposed that when patients fail to integrate the traumatic experience into the totality of their personal awareness, they become "attached" (Freud would later use the term "fixated") to the trauma: "Unable to integrate traumatic memories, they seem to have lost their capacity to assimilate new experiences as well. It is . . . as if their personality has definitely stopped at a certain point, and cannot enlarge any more by the addition or assimilation of new elements" (Janet, 1911, p. 532). "All [traumatized] patients seem to have had the evolution of their lives checked; they are attached to an insurmountable obstacle" (1919/1925, p. 660). Janet proposed that the efforts to keep the fragmented traumatic memories out of conscious awareness eroded the psychological energy of these patients. This in turn interfered with their capacity to engage in focused and creative actions, and to learn from experience. Unless the dissociated elements of the trauma were integrated into personal consciousness, the patient was likely to experience a slow decline in personal and occupational functioning (van der Kolk & van der Hart, 1989).

Until psychoanalysis, the doctrine of intrapsychic conflict and repressed infantile sexuality, crowded out competing schools of thought, Janet's clinical observations were widely accepted as the correct formulations of the effects of trauma on the mind. William James, Jean Piaget, Henry Murray, Carl Jung, Charles Myers, William MacDougal, and students of dissociation such as Ernest Hilgard all acknowledged the importance of Janet's work on their understanding of mental processes. In accepting the centrality of dissociation as the central pathogenic process that gives rise to post-traumatic stress, they all disagreed with the psychoanalytic notion of catharsis and abreaction as the treatment of trauma, emphasizing instead the role of synthesis and integration (van der Hart & Brown, 1992). Despite Janet's large body of work and his profound influence on both his contemporaries and the next generation of psychiatrists, his legacy was slowly forgotten. Janet's extensive work on trauma, memory, and the treatment of dissociative states was not integrated with the body of knowledge of PTSD, until the role of dissociation in the origins of PTSD was rediscovered in the 1980s (van der Hart & Friedman, 1989; van der Kolk & van der Hart, 1989; Putnam, 1989).

Freud and Trauma

When Sigmund Freud visited Charcot in late 1885, he adopted many of the ideas then current in that hospital, which he expressed and acknowledged in his early papers on hysteria (Breuer & Freud, 1893–1895/1962; Freud, 1896a; MacMillan, 1980, 1991). In much that he wrote between 1892 and 1896, Freud followed the notions of the Salpêtrière: that the "subconscious" contains affectively charged events encoded in an altered state of consciousness. In "Physical Mechanism of Hysterical Phenomena," Breuer and Freud (1893/1955) wrote on the nature of hysterical attacks:

> We must point out that we consider it essential for the explanation of hysterical phenomena to assume the presence of a dissociation—a splitting of the content of consciousness. . . . The regular and essential content of a (recurrent) hysterical attack is the recurrence of a psychical state which the patient has experienced earlier. (p. 30)

When Breuer and Freud expanded this work in 1895 in "Studies on Hysteria," they acknowledged their debt to Janet and stated that "hysterics suffer mainly from reminiscences. . . . The traumatic experience is constantly forcing itself upon the patient and this is proof of the strength of that experience: the patient is, as one might say, fixated on his trauma" (Breuer & Freud, 1893/1955). Like Janet (1909, 1919/1925), Breuer thought that something becomes traumatic because it is dissociated and remains outside conscious awareness. He called this state "hypnoid hysteria." As late as 1896, Freud proposed in "Heredity and the Aetiology of Neuroses" (1896b) that "a precocious experience of sexual relations . . . resulting from sexual abuse committed by another person . . . is the specific cause of hysteria . . . not merely (as Charcot had claimed), an agent provocateur."

In "The Aetiology of Hysteria," Freud (1896a) began to make his original contributions by claiming that repressed instinctual wishes form the foundation of the neuroses. Freud later claimed: "I have never in my own experience met with a genuine hypnoid hysteria" (Masson, 1984). Although he again became interested in the traumatic neuroses during World War I, the relationship between actual childhood trauma and the development of psychopathology was henceforth ignored. In Freud's view, it was not the actual memories of childhood trauma that were split off from consciousness, but rather the unacceptable sexual and aggressive wishes of the child that threatened the ego and motivated defenses against the conscious awareness of these wishes. In "An Autobiographical Study" (1925), Freud wrote:

> I believed these stories (of childhood sexual trauma) and consequently supposed that I had discovered the roots of the subsequent neurosis in these experiences of sexual seduction in childhood. If the reader feels inclined to shake his head at my credulity, I cannot altogether blame him. . . . I was at last obliged to recognize that these scenes of seduction had never taken place, and that they were only fantasies which my patients had made up. (p. 34)

Freud henceforth argued that the memory disturbances and reenactments seen in hysteria were not the result of a failure to integrate new data into existing meaning schemes, but the active repression of conflict-laden sexual and aggressive ideas and impulses centering around the oedipal crisis at about age 5 (Freud, 1905a; van der Kolk & van der Hart, 1991). Psychiatry, as a discipline, came to follow Freud in his explorations of how the normal human psyche functioned; real-life trauma was ignored in favor of fantasy.

However, just as his early patients with hysteria seemed incapable of getting rid of their traumatic memories, Freud kept coming back to the issue of "fixation on the trauma." World War I temporarily confronted the world, including Freud, with the inescapable reality of the effects of trauma on people's spirits. During this time, he revived Janet's notion that "vehement emotions" were at the root of traumatic neuroses: The overwhelming intensity of the stressor, the absence of abreactive verbal or motoric channels, and the unpreparedness of the individual caused a failure of the stimulus barrier (*Reitschutz*). The organism was unable to deal with the excitement flooding the mental apparatus, resulting in mental paralyses and intense affect storms (Freud, 1920). In 1920 Freud testified in the case against Wagner-Jauregg, the leading Viennese psychiatrist and subsequent Nobel laureate, who was accused of torturing patients with war neuroses by applying brutal electrical treatments in his clinic. Eissler's (1986) book on Freud's statements and Wagner-Jauregg's explanations during the inquest is most informative about the concepts of war neurosis at that time. Freud stated before the Commission that every neurosis (1) had a purpose and (2) constituted a flight into illness by subconscious intentions; he also believed (3) that at the end of the war the war neuroses would disappear. He was wrong on all three counts.

Like Janet, Freud kept being fascinated with the issue of the compulsion to repeat the trauma. In the "Introductory Lectures on Psychoanalysis" (1917), he stated:

> The traumatic neuroses give a clear indication that a fixation to the traumatic incident lies at their root. These patients regularly repeat the traumatic neuroses in their dreams, where . . . we find that the attack conforms to a complete transplanting of the patient into the traumatic situation. It is as if the patient has not finished with the traumatic situation.

Freud proposed that the compulsion to repeat was a function of repression itself: "We therefore concluded that the keeping away from consciousness is the main characteristic of hysterical repression" (1920, p. 18). Because the memory is repressed the patient "is obliged to repeat the repressed material as a contemporary experience, instead of . . . remembering it as something belonging to the past" (p. 18).

In "Beyond the Pleasure Principle" (1920), Freud was close to reintegrating his earliest observations with his later understanding of intrapsychic reality: "The symptomatic picture presented by traumatic neurosis approaches that of hysteria . . . but surpasses it as a rule in its strongly marked signs of subjective ailment as well as the evidence it gives of a far more comprehensive general enfeeblement and disturbances of mental capacities" (p. 12). He was struck by the fact that patients with traumatic neuroses often experienced a lack of conscious preoccupation with the memories of their accident. He postulated that "perhaps they are more concerned with NOT thinking of it," but he did not connect this observation with the notion of "la belle indifference" in people with hysteria.

The acceptance of psychoanalytic theory went hand in hand with a total lack of research on the effects of real traumatic events on children's lives. From 1895 until very recently, no studies were conducted on the effects of childhood sexual trauma. Although psychoanalysts, including Freud, tended to acknowledge sexual trauma as tragic and harmful (Freud, 1905b, 1917), the subject seems to have been too awful to consider seriously in civilized company. One notable exception, Sandor Ferenczi, presented a paper entitled "Confusion of Tongues between the Adult and the Child: The Language of Tenderness and of Passion" (1955), to the Psychoanalytic Congress in 1932. In this presentation he talked about the helplessness of the child when confronted with an adult who uses the child's vulnerability to gain sexual gratification. Ferenczi talked with more eloquence than any psychiatrist before him about the helplessness and terror experienced by children who were victims of interpersonal violence, and he introduced the critical concept that the predominant defense available to children so traumatized is "identification with aggressor." The response of the psychoanalytic community seems to have been one of embarrassment, and the paper was not published in English until 1949, 17 years after Ferenczi's death (Masson, 1984).

THE BEGINNINGS OF INTEGRATION: ABRAM KARDINER

Although several psychiatrists after World War I tried to apply what they had learned to intervention in civilian settings, their influence on psychiatry was minor and did not result in institutional change (Merskey, 1991). One notable exception was Abram Kardiner, who began his career treating traumatized U.S. World War I veterans. Starting in 1923, after finishing his analysis with Freud, he first unsuccessfully tried to create a theory of war neuroses based on the concepts of early psychoanalytic theory. From 1939 on, as World War II was breaking out, Kardiner reassessed the meaning of his entire body of careful clinical observations, which he published in *The Traumatic Neuroses of War* (1941). Like the previous great pioneers of psychological trauma, Kardiner was a master of detailed descriptions of the complex and unusual symptoms of his patients, and a chronicler of the plethora of previous diagnoses these patients had received before a connection was made between the trauma and current symptoms, including hysteria, malingering, or epileptiform disorders. More than anyone else, Kardiner defined PTSD for the remainder of the 20th century.

Kardiner noted that people with "traumatic neuroses" develop an enduring vigilance for and sensitivity to environmental threat, and stated that "the nucleus of the neurosis is a physioneurosis. This is present on the battlefield and during the entire process of organization; it outlives every intermediary accommodative device, and persists in the chronic forms. The traumatic syndrome is ever present and unchanged" (p. 95). He described extreme physiological arousal in these patients: They suffered from sensitivity to temperature, pain, and sudden tactile stimuli: "These patients cannot stand being slapped on the back abruptly; they cannot tolerate a misstep or a stumble. From a physiologic point of view there exists a lowering of the threshold of stimulation; and, from a psychological point of view a state of readiness for fright reactions" (p. 95).

Aside from the physiological alterations, Kardiner (1941) noted that the "pathological traumatic syndrome" comprises an altered conception of the self in relation to the world, based on being fixated on the trauma and having an atypical dream life, with chronic irritability, startle reactions, and explosive aggressive reactions. He believed that this was the result of the fact that "the ego dedicates itself to the specific job of ensuring the security of the organism, and of trying to protect itself against recollection of the trauma." Patients became "stuck" in the trauma, and frequently had what he called "the Sisyphus dream": "Whatever activity he engages upon is greeted with a certain stereotyped futility." This sense of futility often overtook patients; they became withdrawn and detached, even when they had functioned well prior to combat.

Kardiner (1941) acknowledged that it often is difficult to differentiate between hysterical and organic origins of symptoms, and he noted the variety of ways to store traumatic memories. Describing many patients who developed unusual physical symptoms, he recognized that this may sometimes be an adaptive mode of "remembering," because medical complaints are both socially acceptable and financially compensated. However, he cautioned that these symptoms cannot be explained by secondary gain alone (van der Kolk, Herron, & Hostetler, 1994).

Kardiner (1941) also appreciated that the memories of traumas can be generalized and triggered by a large array of experiences. For example, any loud noise over time came to be experienced as a threat, regardless of its origin. He thought that this generalization of the traumatic experience accounted for the fixation on the trauma. Central in Kardiner's thinking, as in that of Janet and Freud, is that fact that "the subject acts as if the original traumatic situation were still in existence and engages in protective devices which failed on the original occasion. This means in effect that his conception of the outer world and his conception of himself have been permanently altered" (p. 82). Sometimes the fixation takes the form of dissociative fugue states: For example, in a reaction triggered by a sensory stimulus, a patient might lash out, employing language suggestive of his trying to defend himself during a military assault. Many patients, while riding the subway, had flashbacks to being back in the trenches, especially upon entering a tunnel. In other cases, people had panic attacks in response to stimuli reminiscent of the trauma, while failing to make a conscious connection between their emotional states and their prior traumatic experience.

Kardiner (1941) was aware of not only the healing power of the process of psychotherapy but also the dangers and difficulties of talking about traumas. One of the issues that all therapists of "traumatic neuroses" continue to grapple with is how and whether to help patients bring unconscious traumatic material into awareness. In one case study, Kardiner urged the patient to discuss his combat traumas, believing this to be the cause of his frequent and severe headaches and dissociative spells. However the patient was not amenable to this:

He showed unusual strength and emphatically refused to enter into any discussion about (the trauma), although he denied that any discussion would be painful for him. In other words, the original trauma and all its secondary ramifications seemed to be entirely encapsulated and to have no apparent connection with the patient's other psychic spheres. The prognosis, therefore, appeared to be practically hopeless, since no bridge remained between the patient's conscious life and the activity of the trauma in unconsciousness. (van der Kolk et al., 1994)

WORLD WAR II AND ITS AFTERMATH

Although Kardiner's work was available for practical application when World War II broke out, most of the lessons of forward psychiatry from World War I had been forgotten and needed to be rediscovered. Thus, initially, the same inadequate treatment procedures were practiced during World War I, including evacuation from the front, at great cost to the individual soldier, and of manpower loss for the military (Ahrenfeldt, 1958; Social Science Research Council, 1949). However, the essential elements of forward psychiatry were soon practiced, even at the front: the principles of proximity, immediacy, and expectancy (PIE). For the first time, there was research on protective factors such as training, group cohesion, leadership, motivation, and morale (Belenky, 1987; Grinker & Spiegel, 1945).

In the United States, many of the best minds of that generation tried to apply Kardiner's lessons in the combat theater. Lawrence Kubie, Roy Grinker, Herbert Spiegel, John Spiegel, Walter Menninger, and Lawrence Kolb are just a few of the pioneers in American psychiatry who were actively involved in the treatment of combat neuroses, both in the field and at home. They confirmed Kardiner's observations about the profound conditioned biological responses that persisted in traumatized patients. In response, they pioneered the use of somatic therapies. In the process of trying to find an effective cure, they rediscovered that patients "remember" the somatosensory aspects of the traumatic experience in an altered state of consciousness. Following this observation, they reintroduced, for the first time in four decades, hypnosis and narcosynthesis to help patients "remember" and to abreact the trauma. They also confirmed Janet's observation that abreaction without transformation and substitution did not help: Grinker and Spiegel (1945) noted the lasting imprint that traumatic memories left on the psyche "is not like the writing on a slate that can be erased, leaving the slate like it was before. Combat leaves a lasting impression on men's minds, changing them as radically as any crucial experience through which they live." The U.S. Army pioneered the use of group stress debriefing (Shalev & Ursano, 1990).

As a result of their war experiences, psychiatrists such as Walter Menninger in the United States, and Wilfred Bion and his British colleagues at the Tavistock Clinic, discovered the use of group psychotherapy and the therapeutic community (Main, 1989). Clearly, the group had become a focus for psychiatric interest. War, as well as disasters, made the mental health profession aware that under extreme conditions, the group, rather than the individual, is the basic unit of study and treatment. Given the vast experience gained during the war, the dedication of the practitioners and the solid collection of data on combat neuroses, it is astounding how the memory of war trauma was again completely forgotten for the subsequent quarter-century. An interesting example is that Roy Grinker, coauthor of one of the two most important books to come out of World War II (Grinker & Spiegel, 1945), went on the become a pioneer in the study of borderline personality disorder (Grinker, Werble, & Drye, 1968) without ever seeming to make a connection between these two areas of interest.

STUDIES OF CONCENTRATION CAMP SURVIVORS

After World War II, an independent line of investigation emerged with the study of the long-term effects of trauma in survivors of the Holocaust and other war-related trauma. Studies by Eitinger and Strøm (Eitinger, 1964; Eitinger & Strøm, 1973) showed that concentration camp survivors were a representative sample in terms of prewar health of their national populations. Their increased mortality, general somatic morbidity, and psychiatric morbidity was thoroughly documented (Bastiaans, 1970; Hocking, 1970; Venzlaff, 1966). These investigators coined the term "concentration camp syndrome," which included not only the symptoms currently listed under PTSD but also enduring personality changes. Perhaps the most consistent finding from these studies, however, was the devastating effect of extreme and long-lasting stress on subsequent health. This was also shown in the so-called "war sailor syndrome" from the Allied Convoy Service (Askevold, 1976–1977), and in survivors of Japanese concentration camps (Archibald & Tuddenham, 1965). The studies of people who had undergone concentration camp experiences once again showed that extreme trauma has severe biological, psychological, social, and existential consequences, including a diminished capacity to cope with both psychological and biological stressors later in life.

Henry Krystal (1968, 1978, 1988), a psychoanalyst who studied the long-term outcome of the massive traumatization in concentration camp victims, suggested that the core experience of being traumatized consists of "giving up" and accepting death and destruction as inevitable. Like Janet and Kardiner before him, but cast in the language of psychoanalysis, Krystal noted that the trauma response evolves from a state of hyperalert anxiety to a progressive blocking of emotions and behavioral inhibition. He noted that trauma leads to a "dedifferentiation of affects": Whereas, developmentally, children learn to interpret bodily states as emotions that are indicators of personal significance, and that come to serve as guides for subsequent action, the chronic hyperarousal of traumatized people leads to a loss of ability to grasp the personal meaning of bodily feelings. Traumatized patients come to experience emotional reactions merely as somatic states, without being able to interpret the meaning of what they are feeling. Unable to "know" what they feel, they become prone to undifferentiated affect storms and psychosomatic reactions that are devoid of personal meaning and cannot lead to adaptive responses. According to Krystal, this development of "alexythymia" is central to the psychosomatic symptoms typical of chronically traumatized individuals.

TRAUMATIC STRESS SINCE THE 1970S

Over the past 30 years, much of the impetus for the development of an integrated understanding of the effects of trauma on social, psychological, and biological functions has continued to come from the participation of individuals who themselves were exposed to trauma, such as Vietnam War veterans (e.g., Figley, 1978), and from people who were working with two hitherto entirely neglected traumatized populations: women and children. Between 1895 and 1974, the study of trauma centered almost exclusively on its effects on males. In "Rape Trauma Syndrome" (1974), Ann Burgess and Linda Holstrom at Boston City Hospital described the syndrome, noting that the terrifying flashbacks and nightmares seen in these women resembled the traumatic neuroses of war. Around the same time, the Kempes (1978) started their work on battered children, and Lenore Walker (1979), Elaine Carmen [Hilberman] (1978), Murray Strauss (1977), and Richard Gelles (Gelles & Strauss, 1979) published the first systematic research on trauma and family violence. Although in 1980 the leading American textbook of psychiatry still claimed that incest occurred in less than 1 in 1 million women, and that its impact was not particularly damaging (Kaplan, Freedman, & Saddock, 1980), people like Judith Herman (1981) began to document the widespread sexual abuse of children and the devastation that it caused. Sarah Haley, one of the people most directly involved in the acceptance of PTSD as a diagnostic category in DSM-III, was both the daughter of a WW II veteran with severe "combat neurosis" and an incest victim herself. Haley (1974) wrote the first comprehensive paper on the problems in tolerating reports of atrocities in the therapeutic setting.

During the Vietnam War, in 1970, New York psychiatrists Chaim Shatan and Robert J. Lifton started "rap groups" with recently returned veterans belonging to "Vietnam Veterans against the War," in which they talked about their war experiences. These "rap sessions" rapidly spread around the country and formed the nucleus for an informal network of professionals concerned about the lack of recognition of the effects of the war on these men's psychological health. They started to read Kardiner, the literature on Holocaust survivors, and the existing work on burn and accident victims (Andreasen, 1980). Based on this, they made a list of the 27 most common symptoms of "traumatic neuroses" reported in the literature. These they compared with over 700 clinical records of Vietnam veterans, from which they distilled what they thought were the most critical elements.

Not by accident given that Kardiner had served as the beacon for this enterprise, the final classification system was very close to the one Kardiner described in 1941. As the DSM-III process unfolded, numerous committee meetings and presentations at the American Psychiatric Association culminated, in 1980, with the inclusion of PTSD in DSM-III. All the different syndromes—the "rape trauma syndrome," the "battered woman syndrome," the "Vietnam veterans syndrome," and the "abused child syndrome"—were subsumed under this new diagnosis. However, all of these different syndromes originally had been described with considerable variations from the eventual definition of PTSD.

In the United States, four pivotal works that appeared in the mid-1970s made critical linkages between the trauma of war and the traumas of civilian life. One was Mardi Horowitz's *Stress Response Syndromes* (1978), which built a model for effective psychotherapy of acute life-threatening experiences. Building on Erich Lindemann's observations after the Coconut Grove Fire (1944), Horowitz defined the biphasic responses to trauma: the alternating phases of intrusion and numbing (which we now know do not alternate but coexist), and presented a systematic dynamic psychotherapy for acute trauma. Lenore Terr introduced a developmental focus on the effects of trauma on psychological functioning when she started to publish her research on the Children of Chowchilla (1979, 1983). Henry Krystal's article "Trauma and Affects" (1978) spelled out the effect of trauma on the capacity to verbalize inner experience, and the resulting somatization and impairment of symbolic functioning. Charles Figley (1978), a Vietnam War combat veteran, edited the first significant book on Vietnam War trauma. Most of the findings in these publications appeared too late for inclusion into the DSM-III defini-tion of PTSD, but the revised DSM-III-R (1987) incorporated some of this work.

The DSM-III PTSD diagnosis was not a result of careful factor-analytic studies of the symptom picture of people with "traumatic neuroses," but a compilation of symptoms arrived at on the basis of literature searches, scrutiny of clinical records of veterans, and a thoughtful political process. Only later was the relevance of PTSD as a diagnostic classifi-cation subjected to closer scrutiny, and both its advantages and limitations were more fully explored. Scientific field trials were not conducted until the PTSD diagnosis was reconsid-ered for DSM-IV (American Psychiatric Association, 1994), and the results of those stud-ies, which presented a much more complex impact of trauma across various victim popu-lations and developmental periods than does the PTSD definition (van der Kolk, Roth, Pelcovitz, Sunday, & Spinazzola, 2005), were not incorporated into DSM-IV, except for the study that focused on the A (stressor) criterion (Kilpatrick et al., 1998).

As part of the DSM process, another group of researchers and clinical psychiatrists created a diagnostic system for dissociative disorders, without any known communica-tion with the PTSD work group. Initially, there simply seems to have been no awareness about the relation between dissociation and trauma, and an entirely separate classifica-tion for the dissociative disorders was set up (Nemiah, 1980). Once these two commit-tees understood that they were essentially entrusted with creating diagnostic systems of overlapping phenomena, there were several attempts to merge both the committees and the diagnostic categories. However, their unanimous recommendation to combine and to create a broader diagnostic system was tabled by both the DSM-III-R and the DSM-IV Committees.

Further Developments since the Acceptance of the Diagnosis of PTSD

Over the 25 years since the acceptance of the PTSD diagnosis in the formal psychiatric nomenclature, there has been a veritable explosion of studies on the effects of trauma on individuals and communities. The recognition of PTSD fostered the emergence of a large group of basic and clinical researchers who could devote their professional lives to the study and treatment of psychological trauma. Presently, there exists at least one journal devoted exclusively to the study of psychological trauma: the *Journal of Traumat-ic Stress*. Another, the *Journal of Trauma and Dissociation*, is devoted to specialized issues regarding those topics; a number of other peer-reviewed journals, including *Child Abuse and Neglect* and *Developmental Psychopathology*, focus exclusively on traumatized children. Particularly in the area of childhood trauma, important efforts have been made to inte-grate the research on failures in early parental attachment patterns with the impact of specific traumatic events.

Starting in 1985, a variety of professional organizations focused on the study of the effects of trauma on children and adults were founded in the United States, Europe, Australia, Israel, Japan, and Argentina. In the United States, the National Institutes of Mental Health founded a Violence and Traumatic Stress branch, the U.S. Veterans Administration founded the National Center for PTSD, and the U.S. Department of Health and Human Services created the National Child Traumatic Stress Network.

Since 1980, significant advances have been made in understanding the epidemiol-ogy of PTSD, and the issues of vulnerability, course, and phenomenology. Many of these advances are highlighted in this book, but three of the most significant advances concern (1) understanding of the impact of trauma on a variety of developmental competencies through the life cycle (Putnam, 1995; Pynoos, Steinberg, Ornitz, & Goenjian, 1998; van der Kolk et al., 2005a, 2005b), (2) elucidation of the several of the underlying neurobiological processes of trauma (Friedman, Charney, & Deutch, 1995; Yehuda & McFarlane, 1997), and (3) systematic exploration of treatment outcome in various trauma populations (Foa, Keane, & Friedman, 2000). These lines of research have intersected, representing a true biopsychosocial approach to the study of trauma.

CONCLUSIONS

Perhaps the most important lesson from the history of psychological trauma is the intimate connection between cultural, social, historical, and political conditions, and the ways that people approach traumatic stress. History demonstrates psychiatry's imbeddedness in social forces, possibly more so than any other branch of medicine. These cultural forces include the status of women and children, issues of compensation, availability of funding for particular scientific endeavors, forensic issues, and other economic and political processes.

Until the current generation brought the study of trauma into full bloom, historically, there have always been brilliant students of trauma who described a rich and complex tapestry of adaptation to trauma in men, women, and children that transcends the relatively narrow definition of the PTSD diagnosis. Although biological and treatment outcome research has made vast strides over the past quarter-century, much remains to be learned about those issues, as well as about other complex posttraumatic phenomena that have been repeatedly observed over the past 130 years but received relatively little attention at this point: automatic behaviors, dissociative states, problems with intimacy, focus, and attention; helplessness and a persistent sense of victimization; as well as debilitating, ill-defined, and shifting somatic problems.

It is likely that a quarter-century from now we will look back on our current knowledge of PTSD with a combination of bemusement about how little we once understood, and a sense of awe about the brilliance of the clinical and scientific observations that for more than a century have helped our knowledge to gradually become more and more refined.

ACKNOWLEDGMENTS

I wish to acknowledge the significant contributions by Onno van der Hart, Lars Weisæth, and Alexander McFarlane to earlier versions of this chapter.

REFERENCES

Ahrenfeldt, R. H. (1958). *Psychiatry in the British army in the Second World War.* New York: Columbia University Press.

Alford, C. F. (1992). *The psychoanalytic theory of Greek tragedy.* New Haven, CT: Yale University Press,.

American Psychiatric Association. (1980). *Diagnostic and statistical manual of mental disorders* (3rd ed.). Washington, DC: Author.

American Psychiatric Association. (1987). *Diagnostic and statistical manual of mental disorders* (3rd edition, rev.). Washington, DC: Author.

American Psychiatric Association. (1994). *Diagnostic and statistical manual of mental disorders* (4th ed.). Washington, DC: Author.

Andreasen, N. C. (1980), Post-traumatic stress disorder. In H. I. Kaplan, A. M. Freedman, & B. J. Saddock (Eds.), *Comprehensive textbook of psychiatry* (pp. 1517–1525). Baltimore: Williams & Wilkins.

Archibald, H., & Tuddenham, R. (1965). Persistent stress reactions after combat. *Archives of General Psychiatry, 12,* 475–481.

Askevold, F. (1976–1977). War sailor syndrome. *Psychotherapy and Psychosomatics, 27,* 133–138.

Babinski, J. (1901). Définition de l'hystérie. *Revue Neurologique, 9,* 1074–1080.

Babinski, J. (1909). Démembrement de l'hystérie traditionelle. Pithiatisme. *La Semaine Médicale, 59*(1), 3–8.

Babinski, J., & Froment, J. (1918). *Hystérie-pithiatisme et troubles nerveux d'ordre reflexe en neurologie de guerre.* Paris: Masson & Cie.

Bastiaans, J. (1970). Over de specificiteit en de behandeling van het KZ-syndroom [On the specifics and the treatment of the concentration camp syndrome]. *Nederlands Militair Geneeskunde Tijdschrift, 23,* 364–371.

Belenky, G. (Ed.). (1987). *Contemporary studies in combat psychiatry.* New York: Greenwood Press.

Bergson, H. (1896). *Matière et mémoire.* Paris: Alcan.

Bonhoeffer, D. (1926). Beurteilung, Begutachtung und Rechtsprechung bei den sogenannten Unfallsneurosen. *Deutsche Medicinische Wochenschrift, 52,* 179–182.

Breuer, J., & Fred, S. (1955). On the psychical mechanisms of hysterical phenomena: Preliminary communication. In J. Strachey (Ed. & Trans.), *The standard edition of the complete psychological works of Sigmund Freud* (Vol. 2, pp. 1–181). London: Hogarth Press. (Original work published 1893)

Breuer, J., & Freud, S. (1962). Studies on hysteria. In J. Strachey (Ed. & Trans.), *The standard edition of the complete psychological works of Sigmund Freud* (Vol. 2). London: Hogarth Press. (Original work published 1893–1895)

Briquet, P. (1859). *Traité clinique et thèrapeutique de l'hysterie* [Clinical and therapeutic aspects of hysteria]. Paris: Ballière.

Burgess, A. W., & Holstrom, L. (1974). Rape trauma syndrome. *American Journal of Psychiatry, 131,* 981–986.

Carmen [Hilberman], E., & Munson, M. (1978). Sixty battered women. *Victimology, 2,* 460–471.

Caruth, C. (Ed.). (1995). *Trauma and memory.* Baltimore: Johns Hopkins University Press.

Charcot, J. M. (1887). *Leçons sur les maladies du système nerveux faites à la Salpêtrière* [Lessons on the illnesses of the nervous system held at the Salpêtrière] (Tome III). Paris: Progrès Médical en A. Delahaye & E. Lecrosnie.

Crocq, L., & De Verbizier, J. (1989). Le traumatisme psychologique dans l'oeuvre de Pierre Janet. *Annales Médico-Psychologiques, 147*(9), 983–987.

Da Costa, J. M. (1871). On irritable heart: A clinical study of a form of functional cardiac disorder and its consequences. *American Journal of the Medical Sciences, 61,* 17–52.

Eissler, K. R. (1986). *Freud as an expert witness: The discussion of war neuroses between Freud and Wagner-Jauregg.* Madison, CT: International Universities Press.

Eitinger, L. (1964). *Concentration camp survivors in Norway and Israel.* Oslo: Universitetsforlaget.

Eitinger, L., & Strøm, A. (1973). *Mortality and morbidity after excessive stress: A follow-up investigation of Norwegian concentration camp survivors.* New York: Humanities Press.

Ellenberger, H. F. (1970). *The discovery of the unconscious: The history evolution of dynamic psychiatry.* New York: Basic Books.

Erichsen, J. E. (1866). *On railway and other injuries of the nervous system.* London: Walton & Maberly.

Erichsen, J. E. (1886). *On concussion of the spine, nervous shock and other obscure injuries to the nervous system in their clinical and medico-legal aspects.* New York: William Wood.

Ferenczi, S. (1955). Confusion of tongues between adults and the child: The language of tenderness and of passion. In *Final contributions to the problems and methods of psychoanalysis.* New York: Basic Books.

Figley, C. (1978). *Stress disorders among Vietnam veterans: Theory, research and treatment implications.* New York: Brunner/Mazel.

Fischer-Homberger, E. (1975). *Die Traumatische Neurose, von somatischen zum sozialen Leiden.* Bern: Verlag Hans Huber.

Foa, E. B., Keane, T. M., & Friedman, M. J. (2000). *Effective treatments for PTSD: Practice guidelines from the International Society of Traumatic Stress Studies.* New York: Guilford Press.

Freud, S. (1896a). The aetiology of hysteria. In J. Strachey (Ed. & Trans.), *The standard edition of the complete psychological works of Sigmund Freud* (Vol. 3, pp. 189–221). London: Hogarth Press.

Freud, S. (1896b). Heredity and the aetiology of the neuroses. In J. Strachey (Ed. & Trans.), *The standard edition of the complete psychological works of Sigmund Freud* (Vol. 3, pp. 142–156). London: Hogarth Press.

Freud, S. (1905a). The interpretation of dreams. In J. Strachey (Ed. & Trans.), *The standard edition of the complete psychological works of Sigmund Freud* (Vols. 4–5). London: Hogarth Press.

Freud, S. (1905b). Three essays on the theory of sexuality. In J. Strachey (Ed. & Trans.), *The standard edition of the complete psychological works of Sigmund Freud* (Vol. 7, pp. 125–243). London: Hogarth Press.

Freud, S. (1917). Introductory lectures on psychoanalysis. In J. Strachey (Ed. & Trans.), *The standard edition of the complete psychological works of Sigmund Freud* (Vol. 16, p. 369). London: Hogarth Press.

Freud, S. (1920). Beyond the pleasure principle. In J. Strachey (Ed. & Trans.), *The standard edition of the complete psychological works of Sigmund Freud* (Vol. 18, pp. 7–64). London: Hogarth Press.

Freud, S. (1925). An autobiographical study. In J. Strachey (Ed. & Trans.), *The standard edition of the complete psychological works of Sigmund Freud* (Vol. 20). London: Hogarth Press.

Freud, S. (1926). Inhibitions, symptoms and anxiety. In J. Strachey (Ed. & Trans.), *The standard edition of the complete psychological works of Sigmund Freud* (Vol. 20, pp. 77–174). London: Hogarth Press.

Friedman, M. J., Charney, D. S., & Deutsch, A. Y. (Eds.). (1995). *Neurobiological and clinical consequences of stress: From normal adaptation to post-traumatic stress disorder*. Philadelphia: Lippincott-Raven.

Gelles, R. J., & Straus, M. A. (1979). Determinants of violence in the family: Toward a theoretical integration. In W. R. Burr, R. Hill, & F. I. Nye (Eds.), *Comtemporary theories about the family*. New York: Free Press.

Grinker, R. R., & Spiegel, J. P. (1945). *Men under stress*. Philadelphia: Blakiston.

Grinker, R. R., Werble, B., & Drye, R. C. (1968). *The borderline syndrome: A behavioral study of ego functions*. New York: Basic Books.

Haley, S. (1974). When the patient reports atrocities. *Archives of General Psychiatry, 30*, 191–196.

Herman, J. L. (1981). *Father–daughter incest*. Cambridge, MA: Harvard University Press.

Hocking, F. (1970). Psychiatric aspects of extreme environmental stress. *Diseases of the Nervous System, 31*, 1278–1282.

Horowitz, M. J. (1978). *Stress response syndromes*. New York: Aronson.

Janet, P. (1887). L'anesthésie systématisée et al dissociation des phénomènes psychologiques. *Revue Philosophique, 23*(1), 449–472.

Janet, P. (1889). *L'Automatisme psychologique*. Paris. Alcan.

Janet, P. (1894). Histoire d'une idee fixe. *Revue Philosophique, 37*, 21–163.

Janet, P. (1904). L'amnesie et la dissociation des souvenirs par l'emotion. *Journal de Psychlogie, 1*, 417–453.

Janet, P. (1909). *Les nervoses*. Paris: Flammarion.

Janet, P. (1925). *Psychological healing* (Vols. 1–2). New York: MacMillan. (Original work published 1919)

Janet, P. (1930). Autobiography. In C. A. Murchinson (Ed.), *A history of psychology in autobiography* (Vol. 1). Worcester, MA: Clark University.

Janet, P. (2007). *The major symptoms of hysteria* (2nd ed.). London: Macmillan. (Original work published 1907)

Kaplan, H. I., Freedman, A. M., & Saddock, B. J. (Eds.). (1980). *Comprehensive textbook of psychiatry*. Baltimore: Williams & Wilkins.

Kardiner, A. (1941). *The traumatic neuroses of war*. New York: Hoeber.

Kempe, R. S., & Kempe, C. H. (1978). *Child abuse*. Cambridge, MA: Harvard University Press.

Kilpatrick, D. G., Freedy, J. R., Resnick, H. S., Pelcovitz, D., Resick, P. A., Roth, S., et al. (1998). Posttraumatic stress disorder field trial: Evaluation of the PTSD construct—criteria A through E. In *DSM-IV sourcebook* (4th ed., pp. 803–844). Washington, DC: American Psychiatric Press.

Kraepelin, É. (1899). *Psychiatrie, 6*. Leipzig: Auflage.

Krystal, H. (Ed.). (1968). *Massive psychic trauma*. New York: International Universities Press.

Krystal, H. (1978). Trauma and affects. *Psychoanalytic Study of the Child, 33*, 81–116.

Krystal, H. (1988). *Integration and self-healing: Affect, trauma, and alexithymia.* Hillsdale, NJ: Analytic Press.

Lindemann, E. (1944). Symptomatology and management of acute grief. *American Journal of Psychiatry, 101,* 141–148.

MacLeod, A. D. (1993). Putnam, Jackson and post-traumatic stress disorder. *Journal of Nervous and Mental Disease, 181*(11), 709–710.

MacMillan, M. (1980). *Freud evaluated: The completed arc.* Amsterdam: North-Holland.

MacMillan, M. (1991). Freud and Janet on organic and hysterical paralyses: A mystery solved? *International Review of Psychoanalysis, 17,* 189–203.

Main, T. (1989). *"The Ailment" and other psychoanalytic essays.* London: Free Association Press.

Masson, J. (1984). *The assault on truth.* New York: Farrar, Strauss & Giroux.

Merskey, H. (1991). Shell-shock. In G. E. Berrios & H. Freeman (Eds.), *150 years of British psychiatry 1841–1991* (pp. 245–267). London: Gaskell, The Royal College of Psychiatrists.

Moran, C. (1945). *Anatomy of courage.* London: Constable.

Myers, A. B. R. (1870). *On the aetiology and prevalence of disease of the heart among soldiers.* London: Churchill.

Myers, C. S. (1915). A contribution to the study of shell shock. *Lancet,* pp. 316–320.

Myers, C. S. (1940). *Shell shock in France 1914–18.* Cambridge, UK: Cambridge University Press.

Nemiah, J. C. (1980). Psychogenic amnesia, psychogenic fugue, and multiple personality. In A. M. Freedman, H. I. Kaplan, & B. J. Saddock (Eds.), *Comprehensive textbook of psychiatry* (Vol. 2., pp. 942–957). Baltimore: Williams & Wilkins.

Nonne, M. (1915). Zur therapeutischen Verwendung der Hypnose bei Fällen von Kriegshysterie. *Medizinische Klinik, 11*(51), 1391–1396.

Oppenheim, H. (1889). *Die traumatische Neurosen.* Berlin: Hirschwald.

Page, H. (1883). *Injuries of the spine and spinal cord without apparent mechanical lesion and nervous shock in their surgical and medio-legal aspects.* London: Churchill.

Putnam, F. W. (1989). Pierre Janet and modern views on dissociation. *Journal of Traumatic Stress, 2*(4), 413–430.

Putnam, F. W. (1997). *Dissociation in children and adolescents: A developmental perspective.* New York: Guilford Press.

Putnam, J. J. (1881). Recent investigations into patients of so-called concussion of the spine. *Boston Medical and Surgical Journal, 109,* 217.

Putnam, J. J. (1898). On the etiology and pathogenesis of the posttraumatic psychoses and neuroses. *Journal of Nervous and Mental Disease, 25,* 769–799.

Pynoos, R. S., Steinberg, A. M., Ornitz, E. M., & Goenjian, A. (1997). Issues in the developmental neurobiology of traumatic stress. *Annals of the New York Academy of Sciences, 21,* 176–193.

Shalev, A., & Ursano, R. J. (1990). Group debriefing following exposure to traumatic stress. In J. E. Lundeberg, U. Otto, & B. Rybeck (Eds.), *War medical services.* Stockholm: FOA.

Shay, J. (1994). *Achilles in Vietnam: Combat trauma and the undoing of character.* New York: Atheneum.

Social Science Research Council. (1949). *Studies in social psychology in WW II: Vol. 2. The American soldier: Combat and its aftermath.* Princeton, NJ: Princeton University Press.

Stierlin, E. (1909). *Über psychoneuropathische Folgezustände bei den Überlebenden der Katastrophe von Courrières am 10. Marz 1906* [On the psychoneuropathic consequences among the survivors of the Courrières catastrophe of 10 March 1906]. Doctoral dissertation, University of Zürich, Zürich, Switzerland.

Stierlin, E. (1911). Nervöse und psychische Störungen nach Katastrophen [Nervous and psychic disturbances after catastrophes]. *Deutsches Medizinische Wochenschrift, 37,* 2028–2035.

Strauss, M. A. (1977). Sociological perspective on the prevention ad treatment of wife-beating. In M. Roy (Ed.), *Battered women: A psychological study of domestic violence.* New York: Van Nostrand Reinhold.

Tardieu, A. (1878). *Etude medicolegale sur les attentats aux moeurs* [A medico-legal study of assaults on decency]. Paris: Balliere.

Terr, L. (1979). Children of Chowchilla: A study of psychic trauma. *Psychoanalytic Study of the Child, 34,* 552–623.

Terr, L. C. (1983). Chowchilla revisited: The effects of psychic trauma four years after a school-bus kidnapping. *American Journal of Psychiatry, 140,* 1543–1550.

van der Hart, O., & Brown, P. (1992). Abreaction re-evaluated. *Dissociation*, *5*(4), 127–140.

van der Hart, O., & Friedman, B. (1989). A reader's guide to Pierre Janet on dissociation: A neglected intellectual heritage. *Dissociation*, *2*(1), 3–16.

van der Kolk, B. A., Herron, N., & Hostetler, A. (1994). The history of trauma in psychiatry. *Psychiatric Clinics of North America*, *17*, 583–600.

van der Kolk, B. A., Roth, S., Pelcovitz, D., Sunday, S., & Spinazzola, J. (2005). Disorders of extreme stress: The empirical foundation of a complex adaptation to trauma. *Journal of Traumatic Stress*, *18*(5), 389–399.

van der Kolk, B. A., & van der Hart, O. (1989). Pierre Janet and the breakdown of adaptation in psychological trauma. *American Journal of Psychiatry*, *146*, 1530–1540.

van der Kolk, B. A., & van der Hart, O. (1991). The intrusive past: the flexibility of memory and the engraving of trauma. *American Imago: Psychoanalysis Culture*, *48*, 425–454.

Venzlaff, U. (1966). Das akute und das chronische Belastungssyndrom. *Medizinsche Welt*, *17*, 369–376.

Venzlaff, U. (1975). Aktuelle Probleme der forensischen Psychiatrie. In K. P. Kisker, J. E. Meyer, C. Müller, & E. Strømgren (Eds.), *Psychiatrie der Gegenwart* (pp. 920–932). Berlin: Springer-Verlag.

Walker, L. (1979). *The battered women*. New York: Harper & Row.

Yehuda, R., & McFarlane, A. C. (1997). *Psychobiology of posttraumatic stress disorder*. New York: New York Academy of Sciences.

Chapter 3

A Psychological History
of PTSD

**Candice M. Monson, Matthew J. Friedman,
and Heidi A. J. La Bash**

Relative to other scientific disciplines, psychology has a relatively short history, and the application of psychological theory and research to psychological trauma and posttraumatic stress disorder (PTSD) has an even shorter history. Although the history is brief, it is replete with a number of seminal offerings to the scientific understanding and treatment of PTSD. This chapter provides a historical journey into the origins of psychological, as contrasted with psychiatric, theories of PTSD. We briefly revisit the contributions of psychoanalysts in the late 1800s and early 1900s (see van der Kolk, Chapter 2, this volume), who provided the first psychological explanations of posttraumatic sequelae, and reframe some of their theoretical approaches within a more contemporary psychological framework. We describe the rise of the first wave of behaviorism and its contributions to understanding anxiety reactions, the second wave of cognitive behaviorism, which involved the cognitive revolution and its application to PTSD, and the most recent third wave of cognitive behaviorism. Based on this review, we conclude with several theories and findings that might be further explored to enhance our conceptual and clinical approaches to PTSD.

In the late 1800s, psychoanalytically oriented psychiatrists dedicated to studying and treating trauma victims from a psychological perspective gained increasing recognition. Their theories and related treatment approaches ran in opposition to the *zeitgeist* of organic explanations of these reactions at the time. Elements of current evidence-based theories explaining the development and maintenance of PTSD can be traced to writings by these individuals. Based on his work with Josef Breuer, Sigmund Freud hypothesized that hysterical reactions were a product of early traumatic experiences (Freud & Breuer, 1895). His theory of trauma, referred to as the "seduction theory," held that childhood sexual abuse experiences resulted in the use of the most primitive defense mechanisms (e.g., dissociation, denial, repression). Freud's pioneering therapeutic approach for these individuals involved retelling the traumatic event in order to promote emotional catharsis and abreaction (i.e., release of repressed emotions). This approach can be seen as a precursor to current cognitive-behavioral therapy (CBT), especially prolonged exposure techniques (see below).

Although Freud stood strong against the winds of Victorian culture pertaining to organic versus psychological explanations of psychopathology, he unfortunately later wavered against these winds regarding the "nonexistence" of childhood sexual abuse and attributed hysteria to repressed psychosexual drives in early developmental processes rather than to the toxic impact of external stressors.

Freud's contemporary, Pierre Janet, was also instrumental in bringing a psychological approach to posttraumatic reactions. Although Janet's writings are often cited as they relate to current conceptions of dissociation, his works include some precursor elements of CBT. For example, Janet (1925) theorized that people develop meaning schemes based on past experiences that prepare them to cope with subsequent challenges. He argued that when people experience "vehement emotions," their minds are not capable of integrating their frightening experiences with existing cognitive maps. When these memories cannot be integrated into personal awareness, something akin to dissociation occurs. He also introduced the notion of patients experiencing a "phobia of memory," which foreshadowed the anxiety-based conceptualization of PTSD. Janet postulated that memory traces linger as long as they are not translated into a personal narrative. In his conception of trauma, synthesis and integration of the traumatic information were the goals of treatment, as they are today.

Among classic psychoanalysts, the work of Alfred Adler also stands out as prescient with respect to contemporary cognitive theory and therapy (see Hyer, 1994). What Adler called "methods of operation," modal personality styles, and self-perpetuated behavioral patterns closely approximate the core schemas invoked by modern cognitive therapists. Adlerian psychotherapy, conducted within the safety of the therapeutic relationship, included "facilitative confrontation," which greatly resembles the Socratic questioning and cognitive restructuring practiced by modern cognitive therapists (see below).

PRECURSOR PSYCHOPHYSIOLOGICAL UNDERSTANDING

Between the late 19th and early 20th century, military combat was the major arena within which advances were made in the field of traumatic stress studies. Competing organic and psychological theories were proposed to explain such observations. Organic theories, first proposed during the American Civil War, focused mostly on cardiovascular function, with descriptions of syndromes such as soldier's heart, Da Costa's syndrome, and neurocirculatory asthenia. During World War I, hypothesized disruption of neurocircuitry in the brain was labeled "shell-shock." At the same time, psychological theories were proposed to explain the same phenomena. These also date back to the late 19th century and include constructs such as nostalgia, war neurosis, combat fatigue, and combat exhaustion (Hyams, Wignall, & Roswell, 1996).

The first person to formally integrate organic and psychological explanations of traumatic reactions was the psychoanalytically trained psychiatrist Abram Kardiner (1941), who should be distinguished as the founder of biopsychosocial approaches to trauma. Based on his brilliant insight regarding the critical importance of the startle response as a manifestation of combat stress and war neurosis among traumatized World War I veterans, he insisted that there were both psychological and physiological components of traumatic reactions. Indeed, he argued that "traumatic neurosis" was actually a "physioneurosis."

Kardiner (1941) was also among the first to identify behavioral and cognitive disturbances that occur as a result of traumatization. In describing the traumatized World War I veterans he treated, Kardiner documented these patients' tendency to be triggered by sensory stimuli and to act out as if they were still involved in military combat. He characterized them as having flashbacks and panic attacks evoked by situations reminiscent of their traumatic experiences. Kardiner also suggested that these patients often failed to make conscious connections between their emotional states and their prior traumatic experience. All of these observations have threads to the classical fear conditioning conceptualization of traumatic stimuli and responses. Kardiner also noted in his writings about a war veteran, "[H]is conception of the outer world and his conception of himself have been permanently altered" (p. 82), and underscored changes in the survivor's perception of threat as a result of trauma. These writings foreshadow trauma-related changes in cognition later theorized, studied, and directly targeted in treatment.

THE FIRST WAVE OF BEHAVIORISM

Behavioral theories of psychopathology sprang up in the early 1900s as a reaction against the dominant psychoanalytic approaches at that time. This movement, dubbed the first wave of behavior therapy, was fueled by the conflicting data produced by various laboratories. The behaviorists were committed to establishing an empirically based, scientific approach focused on observable behavior. Influenced by advances in animal models and clinical applications of learning theory, proponents of behaviorism put little or no credence in constructs such as motivation and intrapsychic factors (O'Donnell, 1985).

Classical Conditioning

Ivan Petrovitch Pavlov stumbled upon one of the founding principles of behaviorism while conducting a study on canine digestive systems, for which he received a Nobel Prize. Pavlov realized that his dogs would salivate when they heard creaking stairs as their caretakers were bringing them food. Pavlov set up formal experiments to test his serendipitous observation, and the theory of classical conditioning was born. In short, he showed that a neutral or "meaningless" stimulus could evoke a response in an organism, if the response was conditioned by pairing the previously neutral stimulus with one that naturally evokes a response in the organism. This research, and that of later researchers who expanded the bounds of this theory and associated neurobiology, made important contributions to understanding anxious responding. Classical conditioning theory, and especially the principles of stimulus generalization, extinction, conditioned inhibition, and spontaneous recovery, has important implications for understanding PTSD.

There have since been efforts to explicate the underlying neurobiology associated with classical conditioning in PTSD. Lawrence Kolb was among the first to apply the work of Kardiner and Pavlov explicitly to PTSD within a neurobiological context. In his seminal paper (Kolb, 1987), he postulated that PTSD could be understood as a manifestation of Pavlovian fear conditioning. Kolb also integrated late twentieth century neuroscience into his theory by postulating that the amygdala was the major brain nucleus that mediated the hyper-reactive responses to traumatic stimuli associated with PTSD. In an important paper, van der Kolk, Greenberg, and Boyd (1985) also attempted to merge experimental psychology (e.g., inescapable shock) and neuropharmacology (e.g., altered noradrenergic and opioid mechanisms) as a conceptual formulation for PTSD.

A book edited by Friedman, Charney, and Deutch (1995) represented the first comprehensive and integrated effort to explicate PTSD within the context of the three separate but complementary scientific traditions: learning theory, stress research, and neuroscience. More recent neurobiological elaborations of both fear conditioning as the origin of PTSD and resistance to extinction of such conditioned responses as the basis for maintenance of PTSD have been proposed by Charney (2004).

Operant Conditioning

Clark Hull and Edward Thorndike each contributed elements culminating in a second theory seminal in behaviorism that competed with classical conditioning at the time—operant conditioning. Hull's (1943) theory of motivation, or drive-reduction theory, was based on the premise that the goal of human behavior is to reduce biological drives, and when a stimulus–response relationship is followed by a reduced biological drive, behavior is likely to be repeated. In a similar vein, Thorndike (1927) pioneered the law of effect, which held that responses resulting in satisfaction are strengthened, and those followed by discomfort or annoyance are weakened. Both Hull and Thorndike had trial-and-error approaches that highlighted the role of the organism in the ability to adapt responses to different stimuli to avoid adverse effects.

Burrhus Frederick (B. F.) Skinner is ultimately credited with the theory of operant conditioning, which was based on the notion of an organism "operating" on the environment. He theorized that an "operant," or behavior, would increase or decrease based on the consequences of the operant. He went beyond the basic stimulus–response conceptualization of operant conditioning to identify schedules of reinforcement, the processes of stimulus discrimination and extinction, and the notion of shaping. Skinner, as a radical behaviorist, believed all behavior was malleable under the right environmental conditions. Operant conditioning theory has been especially relevant to understanding the processes that maintain PTSD and other anxiety disorders. Specifically, avoidant behavior has been identified as a negative reinforcer, because of its anxiety-reducing effect. Meanwhile, avoidance maintains the originally learned anxiety associations, because it prohibits exposure and extinction or habituation of conditioned responses.

By virtue of his translational research applying behavioral principles to anxiety and phobic responses, as well as his famous "Behavioral Manifesto" delivered at Columbia University, John Broadus Watson (1913) is one of the most prominent figures in behaviorism. Watson and his colleague Rosalie Rayner's case of Little Albert is a staple example in abnormal psychology texts of the learning processes involved in anxious responding. This 11-month-old infant was presented with a white rat, and eagerly reached for it. As his hand touched the rat, the experimenters made a loud sound with a metal bar and hammer above his head. The infant was described as having "jumped violently and fell forward, burying his head in the mattress. A second pairing brought the same response in addition to a whimper" (Watson & Rayner, 1920, p. 4). Seven days after this initial pairing, Little Albert was found to have generalized his fear to a wide range of objects, including Watson's white hair, a Santa mask, a dog, and wooden blocks.

Little Albert's behavior demonstrates both the classical and operant conditioning processes postulated to account for disturbances in anxiety. The mere sight of the rat, absent the sound, elicited the response (i.e., classical conditioning), and the infant scampered away to avoid the rat as soon as possible (i.e., operant conditioning through behavioral avoidance).

Orval Hobart Mowrer is famous for articulating this complementary relationship between classical and operant conditioning learning processes in his two-factor explanation of conditioned fears. While debates raged about the "true" method of learning, Mowrer asserted that neither classical nor operant theories of learning alone could satisfy all the questions that lay in the research. While seeing validity in both approaches, Mowrer also saw serious limitations in both for explaining behavior. He capitalized on the strengths of each theory with his two-factor theory of anxious responding.

Two-factor theory was translated to the understanding of posttraumatic sequelae by Kilpatrick, Veronen, and Resick (1979), who first proposed classical conditioning, stimulus generalization, and subsequent avoidance to explain the fear responses they observed in a longitudinal prospective study of rape victims compared with nonvictims. This study was conducted prior to the introduction of the PTSD diagnosis in the third edition of the *Diagnostic and Statistical Manual of Mental Disorders* (American Psychological Association, 1980). Others have also discussed the application of two-factor theory to the development and maintenance of PTSD (e.g., Foa & Kozak, 1986, 1991; Keane, Zimering, & Caddell, 1985). The first part of the two-factor theory is that classical conditioning explains the origins of the posttraumatic response, in that the stimuli originally paired with traumatic events elicit the emotional, physiological, cognitive, and behavioral responses at subsequent presentations of the stimuli. Examples of specific stimuli occurring independent of a traumatic situation that may elicit classically conditioned responses include specific sights, such as movie scenes of rape, sounds of Huey helicopters, smells of aftershave worn by perpetrators, or sexual touch. Internal experiences, such as intrusive thoughts and images of the traumatic event, may also elicit the learned response. The process of generalization accounts for how traumatic reactions can spread to situations that are not ostensibly related to the traumatic event. For example, any form of transportation might elicit trauma-related reactions for motor vehicle accident survivors.

The second part of the two-factor theory is that operant conditioning maintains the maladaptive posttraumatic response. Negative reinforcement of fear through behavioral avoidance is the primary process that is postulated to sustain, and even promote, the maladaptive fear response. Typical behavioral avoidance manifested by traumatized individuals includes avoidance of stimuli associated with the traumatized event, not disclosing or discussing the traumatic event with others, social isolation, and dissociation.

The development of systematic desensitization by Joseph Wolpe was a natural extension of behavioral conceptualizations of anxiety reactions. Thus, he might be acclaimed as the first behavior therapist. Wolpe began his career practicing psychoanalysis as a captain in the South African Army Medical Corps in World War I, but became discouraged and disillusioned by the lack of successful clinical outcomes. As a result, he sought to develop alternative treatments based on the theoretical advances of that time. Watson and Rayner's (1920) work involving the induction of fear were instrumental to his developing a treatment that modified fear reactions.

Wolpe (1954) asserted that anxiety disorders could be successfully treated through "reciprocal inhibition," or exposure to the source of fears with anything "physiologically antagonistic." Through counterconditioning, a new response, in opposition to the fear response, is conditioned to the original stimuli. In Wolpe's systematic desensitization, a patient is taught deep muscle relaxation, which is incompatible with anxiety. Based on the patient's hierarchy of anxiety-provoking stimuli, the patient is repeatedly, progressively exposed to imaginal exposures while being instructed in how to use the relaxation techniques, until the stimuli are no longer anxiety provoking. Systematic desensitization, a landmark in treating human anxiety and fear, is evident in various treatments since developed that vary with regard to the dimensions of exposure type (i.e., imaginal vs. in vivo), exposure length (i.e., short vs. long), and arousal level during exposure (low vs. high; Foa & Rothbaum, 1997).

After a large number of analogue and clinical studies in the 1970s and 1980s, it was determined that neither progressing through a hierarchy nor simultaneous relaxation were necessary for counterconditioning to occur. Exposure therapies evolved into in vivo and/or imaginal exposure interventions in which clients expose themselves to the "top of the hierarchy," the worst event or feared stimulus, for an extended period of time until anxiety decreases within and across sessions. Fairbank, Keane, and Kaloupek were the first to examine exposure treatment for PTSD with combat veterans following the Vietnam War (Fairbank & Keane, 1982; Keane & Kaloupek, 1982). In the field, the most studied exposure protocol has been prolonged exposure (PE; Foa et al., 1999, 2005; Foa, Rothbaum, Riggs, & Murdock, 1991), which combines imaginal exposure of the worst traumatic event and in vivo exposures to feared, but safe conditioned stimuli.

While systematic desensitization and exposure treatments were being examined, Stress inoculation training (SIT; Meichenbaum & Novaco, 1985) was also being adapted and systematically tested with trauma victims (Kilpatrick, Veronen, & Resick, 1982). SIT was unique in that it offered a behavioral alternative that focused on symptom management versus trauma processing. SIT comprises teaching patients a variety of techniques for managing anxiety, including controlled breathing, deep muscle relaxation, thought stopping, preparation for stressors, covert modeling, and role play. The rationale of these interventions is that management of the generalized fear to many situations and experiences leads to decreased avoidance, which ultimately results in diminished posttraumatic symptomatology. A few early trials substantiated SIT's efficacy in improving PTSD symptoms, at least in PTSD secondary to sexual assault (Foa & Rothbaum, 1997; Kilpatrick et al., 1982).

THE SECOND WAVE OF COGNITIVE BEHAVIORISM: THE COGNITIVE REVOLUTION

Cognition as a topic of scholarly inquiry can be traced back to the earliest writings of the Greek thinkers. Socrates is credited with developing a particular method to facilitate introspection and the acquisition of knowledge. This method, "Socratic questioning," is a mainstay technique in cognitive therapies for various conditions. Plato was an early proponent of the notion of "nativism," which holds that individuals have inborn or innate knowledge structures and cognitive abilities present before birth. This was in opposition to the Empiricists' notion of the mind being a tabula rasa determined by external factors. Aristotle later emphasized the degree to which a person's knowledge and thinking abilities came from that person's own individual experiences (Taylor, Hare, & Barnes, 1998).

In the late 19th century, when psychology began to be considered a science and not just a branch of philosophy, the first experimental psychology laboratory established by Wilhelm Wundt in Germany in 1879 was dedicated to studying the content and processes of the conscious mind (i.e., structuralism). Wundt pioneered the objective measurement of mental events in relation to observable and measurable stimuli and reactions, and was instrumental in developing a psychophysiological branch of psychology. Famous psychologists associated with the school of functionalism, William James and G. Stanley Hall, were subsequently concerned with the purposes of cognition and how it functioned to adapt the organism to the environment. The importance of cognition for coping, adaptation, and survival of the species was a logical extension of Charles Darwin's theory of evolution (1859) that was prominent at the time. Thus, exploration of cognitive constructs that arose in the mid-20th century was not a new development, but a rediscovery and expansion of the crucial role of cognition in behavior and psychopathology.

Cognitive explanations burgeoned in the 1960s, in part as an antithesis to radical behaviorism but also because of developments in computing science and methodological advances that made the study of cognitive constructs more observable and measurable (Neisser, 1967). Ardent, radical behaviorists had disavowed the notion of cognition. They insisted that psychological science should only consider phenomena that were publicly observable. Nothing mental, subjective, or private could be included. The so-called "mind" was considered a "black box," absent any activity beyond learned associations to environmental stimuli and consequences (O'Donohue & Krasner, 1999).

Noam Chomsky's review of Skinner's (1957) book *Verbal Behavior* is considered one of the famous turning points for cognitive psychology. Chomsky, a linguist at the Massachusetts Institute of Technology, argued that language could not be explained through stimulus–response processes, as Skinner maintained, because such processes could not account for some of the common facts about language. He argued that psychology is the "science of mind" and pointed out that behavioral science was analogous to calling physics a "science of meter readings." He considered human behavior to be evidence for the laws of the operation of the mind, and contended one should not use behavior itself as evidence for laws of behavior (Chomsky, 1968).

Thinking "inside" the black box, cognitive scientists began exploring activities at work within the mind. Several research contributions that evolved prior to the inception of the PTSD diagnosis have since been drawn upon extensively in the PTSD literature. One of the most frequently discussed cognitive constructs is that of the "schema." Bartlett (1958), an English psychologist, is credited with first proposing the concept of schema. Bartlett's schema theory offered an innovative concept at the time: that memory involves active reconstruction, and is not the simple laying down of percepts and verbal information in the mind as experienced in the environment. He argued that individual schemata provide a mental framework for understanding and remembering information based on life experience, culture, and knowledge.

Jean Piaget (1962), a Swiss psychologist, was also at work studying the cognitive development of children, including his own three children, in the first half of the 1900s. In addition to his delineation of the various stages of cognitive development, he had a primary interest in the processes by which knowledge grows and changes with age and experience. This interest resulted in a theory of schema and the processes involved in handling new information. In Piaget's theory, when external information is perceived to be congruent with the content of existing schemata, the information is readily incorporated into the meaning structure (i.e., assimilation). If, on the other hand, information is discrepant with existing schema content, there may be schema accommodation in which the existing schema is modified to take into account the new information.

Later cognitive psychologists expanded on the processes surrounding schemas, and incorporated research innovations from computing science about associative networks and parallel processes. Collins and Loftus's (1975) spreading activation model of human memory and Kintsch and Keenan's (1973) propositional network of memory both accounted for processes governing cognitive content.

Meanwhile, social psychologists focused on the content of cognition, which has since been applied to the study of traumatic reactions. Attribution theory, associated with Fritz Heider (1958), Edward Jones and Keith Davis (1965), and Harold Kelley (1967), has been extensively drawn upon in social cognitive theories of PTSD. In essence, attribution theory is about causal explanations, or how individuals answer questions beginning with "why." The dimensions upon which individuals make attributions are theorized to be the perceived locus of the actor's control (internal vs. external), stability of the event, and its global versus specific nature.

Translational research that ensued drew on this basic cognitive science and made applications to psychopathology in general, and the encoding, retrieval, and processing of traumatic events more specifically. Lang (1977) is credited with extending behavioral theories of anxiety disorders by integrating cognitive structures that account for the storage, retrieval, and modification of stimuli and response information in his bio-informational theory of fear. Lang, Melamed, and Hart's (1970) investigation of the psychophysiological correlates of systematic desensitization established the relationship between the physiology of evoked fear-related imagery and behavioral change. Lang integrated this work with the research and theoretical writings of the cognitive psychologist Pylyshyn (1973), who, like Barlett (1958), argued that humans do not simply store unprocessed, raw visual information. Rather, these images are a result of constructive mental processes. Lang, translating this to fear-related images, held that "fear is a network of specific propositional units between stimuli and response, which have designating and action functions. These propositional units are considered to be fluid in nature, allowing for additions or subtractions to the protean cognitive structure as it unfolds over time" (p. 867).

Lang also attended to the meaning elements of information attached to these cognitive structures. For example, the fear structure in the case of PTSD related to rape might include an image of the trauma scene, various physiological and behavioral fear responses (i.e., increased heart rate, running, sweating), as well as meanings of threat associated with the event (e.g., "Parking lots are dangerous"). The fear structure is activated by incoming information that is congruent with the information contained in the structure, and activation of the structure leads to the triggering of other, associated representations. The implication of his theory for behavioral treatments was that activation of the fear structure is necessary to make modifications in this constructed image, and related fear and psychophysiolocial reactions.

Foa and Kozak (1986) specifically applied and expanded Lang's theory to traumatic events with their emotional processing theory of PTSD, which Cahill and Foa articulate in more depth in Chapter 4, this volume. A specific strength of the evolution of this theory was the additional attention paid to normal versus pathological fear structures, as well as the common and discriminating elements of the fear structures for various anxiety disorders. They also argued that modifications in the fear structure involved weakening ever-present erroneous associations, while acquiring new, accurate, and less fearful associations. This notion is consistent with work on extinction and spontaneous recovery of fear responses (Bouton, 2000; Rescorla & Mahwah, 2001), which suggests that extinction does not eliminate or replace previous associations, but rather results in new learning that competes with the old information. A number of experimental psychopathology studies using modifications of the Stroop task provide evidence that these cognitive processes have a high degree of automaticity (Foa, Feske, Murdock, Kozak, & McCarthy, 1991; McNally, 1998; Thrasher, Dalgleish, & Yule, 1994).

The thematic schema content discussed in cognitive theories of trauma can be traced to classic writings about anxiety by Beck, Emery, and Greenberg (1985). They proposed that anxiety disorders are a product of a general overestimation of danger and threat, in tandem with an underestimation of one's ability to cope. Janoff-Bulman (1989), a social psychologist building upon attribution theory, also contributed to the repertoire of cognitive content. She applied the "just world theory," originally articulated by Lerner (1965), to trauma victims. Just world theory holds that people generally believe in "just" and deserved cause–effect relationships between events in the world. Janoff-Bulman (1989) argued that positive, but illusory, just world beliefs about the benevolence of the world, meaningfulness of experiences, and intrinsic self-worth are shattered with traumatization and consequently cause PTSD.

The thematic content pertinent to PTSD has also been drawn from more psychodynamic, interpersonal, and feminist theories of trauma. For example, Herman's (1992) landmark book, *Trauma and Recovery*, contributes important cognitive themes across trauma populations, especially those related to power and control issues that arise during and subsequent to victimization. McCann and Pearlman (1990) also outlined cognitive content salient to trauma survivors in their theory of trauma, which integrates psychodynamic and social cognitive constructs. These themes include disruptions in one's sense of safety, trust, power, esteem, and intimacy. An important additional contribution that McCann, Sackheim, and Abrahamson (1988) made to the cognitive understanding of PTSD is the possibility that traumatic events do not always necessarily shatter previously held positive beliefs, but can reinforce preexisting negative beliefs about self, others, and the world.

Resick and Schnicke (1993) specifically applied Piaget's concepts of assimilation and accommodation to conceptualize how traumatic content is integrated (or not) into existing belief structures. In the case of traumatic material, individuals can alter their perceptions of the circumstances surrounding the traumatic event to maintain their existing belief systems. Self-blame, hindsight bias, and just world thinking are examples of assimilation that contribute to PTSD symptomatology. As noted earlier, in some cases, clients hold preexisting negative beliefs about themselves and/or others, and the traumatic event is easily assimilated into, and reinforces, their already problematic schemas. In schema accommodation, external information is perceived to be discrepant with existing schema content, and the schema changes to integrate such new information. Appropriate processing of traumatic material is believed to occur when accommodation of traumatic information is achieved in a balanced manner (e.g., "The world is mostly safe"). Resick and Schnicke also propose that individuals can overaccommodate their schemas to account for external information, and in the case of trauma, schemas may be radically altered in an effort to reconcile traumatic experiences that are incongruent with existing schemas (e.g., "No one can be trusted"). Overaccommodation is believed to contribute to the development and maintenance of PTSD.

It is worth noting that cognitive theories of PTSD, with their relative focus on thought processes or content, are complementary to one another. Social cognitive theories identify and explicate the content of the myriad distorted schemas associated with posttraumatic pathology, whereas emotional/information processing theories help to account for the dysfunctional processes handling the problematic content. Schema content selectively guides attention to information, influences interpretation of that information, and influences retrieval of schema-congruent information.

Using similar language, Brewin, Dalgleish, and Joseph (1996) discussed this integration of more and less effortful cognitive dimensions in their "dual representation theory." They posited that there are both conscious and nonconscious emotional aspects to traumatic memories. The conscious aspects have to do with meaning making of traumatic events, whereas nonconscious emotional reactions are considered to be conditioned during the events and relatively automatically activated. As Brewin and colleagues noted, this theory suggests that a different prescription of behavioral and cognitive interventions may be needed in some cases to address the different aspects. It also explains why behavioral interventions induce cognitive changes, and vice versa (e.g., Foa, Molnar, & Cashman, 1995).

One implication of cognitive conceptualizations for theory and treatment of PTSD is that their focus has been on not only anxiety but also a range of emotions, including sadness, grief, anger, guilt, and shame. They do not presuppose that PTSD is only a disorder of anxiety, but rather a disorder that results in myriad emotions associated with disturbances in the ability to make meaning of one's experience and disrupted beliefs about oneself, others, and the world.

Cognitive interventions for PTSD have been a natural outgrowth of these cognitive theories of posttraumatic reactions. Dysfunctional and/or irrational interpretations that lead to maladaptive emotional responses are targeted for cognitive restructuring. Like exposure therapies, the depth and trauma-focused nature of cognitive interventions have varied. Cognitive therapies aim to correct core beliefs and assumptions that have developed or been disrupted by traumatic events (Ehlers & Clark, 2003; Hamblen, Gibson, Mueser, & Norris, 2006; Resick & Schnicke, 1993). Cognitive interventions may also include restructuring of more "shallow" thoughts related to day-to-day events to facilitate symptom management (Mueser, Rosenberg, Jankowski, Hamblen, & Descamps, 2004).

Also within this phase of history, Francine Shapiro (1989) developed eye movement desensitization and reprocessing (EMDR). She documents in her book published in 1995 that she discovered the procedure based on a chance observation while she was walking through a park one day. She described having some disturbing thoughts and observed that her eyes were spontaneously moving very rapidly back and forth in an upward diagonal. She later began making the eye movements deliberately, while concentrating on a variety of disturbing thoughts and memories, and found that these thoughts also disappeared and "lost their charge" (Shapiro, 1995, p. 2). She subsequently had patients replicate these eye movements by following her fingers while she moved her hand back and forth. This gave rise to the procedure she named eye movement desensitization (EMD), and subsequently renamed eye movement desensitization and reprocessing after her reported realization that the optimal procedure included cognitive restructuring of memories and personal attributions related to traumatic events.

Empirical investigations have consistently shown that EMDR is effective in treating PTSD. All current practice guidelines for PTSD recommend EMDR as an evidence-based treatment for PTSD. And some, but not all, head-to-head randomized trials have shown that EMDR is as effective as CBT (e.g., Rothbaum, Astin, & Marsteller, 2005).

Despite positive findings, EMDR remains controversial on a number of fronts. There has been much debate about the active ingredients of EMDR, the theory upon which it is based, and the science that supports it. Shapiro (1995) contends that the eye movements (or alternative repetitive motor movements) used in EMDR trigger a physiological mechanism that activates "accelerated information processing." This activation is postulated to be a product of dual attention to present stimuli and past trauma, a differential effect of "neuronal bursts" caused by the various repetitive movements or deconditioning caused by a relaxation response. However, no neurobiological or psychophysiological studies have corroborated the hypothesized mechanisms of action proposed to be responsible for the treatment's successful effects.

Dismantling studies raise questions about the mechanisms of action responsible for improvements. Studies that have compared EMDR with and without eye movements have found no differences in treatment outcome between the conditions, providing little support for the notion that eye movements are critical to the effects of EMDR (e.g., Pitman et al., 1996a, 1996b; Renfrey & Spates, 1994). Given these results, and the nature of the interventions, some argue that EMDR is a variant of CBT (Rothbaum et al., 2005; Shapiro, 1995).

On the other hand, it has been asserted that EMDR may have unique aspects that account for its therapeutic efficacy and appeal among clinicians and patients. For example, the client-directed nature of the therapy has been noted, because clients choose the traumatic material on which they wish to focus, and process this material in their own manner, and at their own pace (Hyer & Brandsma, 1997).

THE THIRD WAVE OF COGNITIVE BEHAVIORISM

In the last decade, the cognitive-behavioral tradition has witnessed a movement to focus on not only behavior and cognitive change but also the acceptance of one's circumstances, internal experiences, behavioral patterns, and the characteristics and behaviors of others. It has been referred to as the third wave of behaviorism (Hayes, Follette, & Linehan, 2004). This form of cognitive behaviorism has been influenced by Zen Buddhist teachings and mindfulness approaches that embrace acceptance of self, internal experiences, environment, and others. Dialectical behavior therapy (DBT; Linehan, 1993), mindfulness-based cognitive therapy (MBCT; Segal, Williams, & Teasdale, 2002), and acceptance and commitment therapy (ACT; Hayes, Strosahl, & Wilson, 1999) are therapies emblematic of this approach.

In this brand of CBT, negative emotions, experiences, and circumstances are not considered problematic. Rather, the behavior in which an individual engages to avoid these negative experiences is, paradoxically, considered the cause of psychopathology. This tendency to avoid private experiences, such as feelings, memories, behavioral predispositions, and thoughts that are construed to be negative has been named "experiential avoidance" (Hayes, Wilson, Gifford, Follette, & Strosahl, 1996). The notion of experiential avoidance may be well suited to understanding and treating posttraumatic reactions, because trauma survivors often try desperately to get away from major distressing stimuli within themselves, the memories, sensations, and feelings associated with the traumatic event(s) (Batten, Orsillo, & Walser, 2005). At present treatment out-come data supporting the utility of these approaches for PTSD are limited (Batten & Hayes, 2005; Walser, Westrup, Rogers, Gregg, & Lowe, 2003).

FUTURE DIRECTIONS

Though much progress has been made in developing both theory and an empirical basis for understanding PTSD, the work is far from complete. In writing this chapter, we noted a couple of landmark discoveries within the field of psychology that can be capitalized upon to advance PTSD research and practice. One example is the minimal to absent attention paid to social learning theory. Albert Bandura's (1962) original the-ory of social learning has been applied to a range of psychopathology, including basic research regarding other anxiety disorders. Despite research suggesting that it may have important applicability to PTSD, it has yet to be invoked to enhance our theoretical understanding of this disorder. For example, in research with rhesus monkeys, Mineka and Cook (1986) demonstrated that exposing monkeys to a relaxed monkey model in the presence of a stimulus that reliably produces a classically conditioned fear response (e.g., a snake) resulted in reduced acquisition of fear in the observing monkeys. Social learning theory may help explain why some, but not all, individuals develop PTSD after exposure to traumatic events. Modeled reactions by significant others to stressful events both prior to and in the wake of trauma would seem to have an effect on survivors' own reactions to their traumatic experience. It would seem that these notions might be par-ticularly applicable to child victims of trauma, given that the response of caretakers is a strong predictor of a child's posttraumatic reaction. The implications for treatment are that conjoint interventions should be designed for both caretakers and children, as exemplified by the work of Deblinger, Lippmann, and Steer (1996). This type of model-ing procedure might also be applied in prevention efforts in which a relaxed response could be modeled prior to trauma exposure with high-risk individuals (e.g., military personnel) (Feldner, Monson, Friedman, & Bouton, 2007).

In this chapter we have described some of the research efforts that have integrated the biological and psychological correlates of PTSD. The future of mental health, in general, is about continued interface of these areas; PTSD is no exception. In this vein, most of the basic neurobiological research to date has focused on the classical condi-tioning processes involved in PTSD. We encourage more investigation of the operant conditioning processes that likely serve to maintain PTSD. Moreover, elucidating bio-logical and psychological markers associated with resilience against the development of PTSD will be an important avenue of future research. Determining biological markers associated with psychotherapy response will also facilitate future treatment develop-ment, and possibly treatment-matching efforts.

It is noteworthy that few treatment outcome studies have investigated SIT as a symptom-focused treatment for PTSD since its early application. SIT was chosen as a comparison condition by Foa and Kozak (1991) in their early study of PE, because SIT was considered a standard-of-care psychotherapy for rape victims at the time. They found that SIT performed as well as PE. This was even after removing an important component of the SIT protocol: practice of anxiety management skills in the environ-ment along a hierarchy of stressors. Because all other current, evidence-based CBTs for trauma are trauma-focused, it would be advantageous to further investigate SIT and other here-and-now focused interventions for individuals who are not sufficiently stable or willing to engage in trauma-focused interventions. A promising area for future research would be to identify active ingredients in effective therapies that do not process traumatic material. Related to such initiatives would be investigations to determine whether there are individuals with PTSD who achieve greater benefit from trauma-focused approaches versus SIT or some other present-centered therapy and vice versa. Non-trauma-focused interventions may also lend themselves well to broad dissemination efforts.

In the future, it will be important to examine staged approaches to trauma treatment, such as that examined by Cloitre, Koenen, Cohen, and Han (2002), as well as the mindfulness-based treatments mentioned earlier. The field will also be advanced by examination of the possibility of improved efficacy and tolerance of treatment by simultaneously applying different types of treatment interventions (e.g., additive effect of psychopharmacological regimens to psychosocial treatments, combination of family or social support interventions to trauma-focused interventions). Studies that further elucidate the essential elements of existing evidence-based treatments will also help make treatment more efficient and potentially effective.

Psychology has made a number of extraordinary contributions to the scientific understanding and treatment of PTSD. Its distinctive offerings include its methodologically rigorous documentation of posttraumatic reactions, development of experimental designs that aid investigation of the mechanisms responsible for the development and maintenance of PTSD, and theoretically driven and systematic treatments that have been empirically tested. We hope that exploration of these historical contributions will serve as a catalyst for future innovations in the conceptualization, study, and treatment of PTSD.

REFERENCES

American Psychiatric Association. (1980). *Diagnostic and statistical manual of mental disorders* (3rd ed.). Washington, DC: Author.

Bandura, A. (1962). *Social learning through imitation*. Lincoln: University of Nebraska Press.

Bartlett, F. (1958). *Thinking: An experimental and social study*. New York: Basic Books.

Batten, S. V., & Hayes, S. C. (2005). Acceptance and commitment therapy in the treatment of comorbid substance abuse and posttraumatic stress disorder: A case study. *Clinical Case Studies, 4*, 246–262.

Batten, S. V., Orsillo, S. M., & Walser, R. D. (2005). Acceptance and mindfulness-based approaches to the treatment of posttraumatic stress disorder. In S. M. Orsillo & L. Roemer (Eds.), *Acceptance and mindfulness-based approaches to anxiety: Conceptualization and treatment* (pp. 241–269). New York: Springer.

Beck, A. T., Emery, G., & Greenberg, L. S. (1985). *Anxiety disorders and phobias: A cognitive perspective*. Philadelphia: Basic Books.

Bouton, M. E. (2000). A learning theory perspective on lapse, relapse, and the maintenance of behavior change. *Health Psychology, 19*, 57–63.

Brewin, C. R., Dalgleish, T., & Joseph, S. (1996). A dual representation theory of posttraumatic stress disorder. *Psychological Review, 103*, 670–686.

Charney, D. S. (2004). Psychobiological mechanisms of resilience and vulnerability: Implications for successful adaptation to extreme stress. *American Journal of Psychiatry, 161*, 195–216.

Chomsky, N. (1968). *Language and mind*. New York: Harcourt, Brace, and World.

Cloitre, M., Koenen, K. C., Cohen, L. R., & Han, H. (2002). Skills training in affective and interpersonal regulation followed by exposure: A phase-based treatment for PTSD related to childhood abuse. *Journal of Consulting and Clinical Psychology, 70*, 1067–1074.

Collins, A. M., & Loftus, E. F. (1975). A spreading-activation theory of semantic processing. *Psychological Review, 82*, 407–428.

Darwin, C. (1859). *On the origin of species by means of natural selection*. London: Murray.

Deblinger, E., Lippmann, J., & Steer, R. A. (1996). Sexually abused children suffering posttraumatic stress symptoms: Initial treatment outcome findings. *Child Maltreatment, 1*, 310–321.

Ehlers, A., & Clark, D. (2003). Early psychological interventions for adult survivors of trauma: A review. *Biological Psychiatry, 1*, 817–826.

Fairbank, J. A., & Keane, T. M. (1982). Flooding for combat-related stress disorders: Assessment of anxiety reduction across traumatic memories. *Behavior Therapy, 13*, 499–510.

Feldner, M. T., Monson, C. M., Friedman, M. J., & Bouton, M. E. (2007). A critical analysis of approaches to targeted PTSD prevention: Current status and theoretically-derived future directions. *Behavior Modification, 31*, 80–116.

Foa, E. B., Dancu, C. V., Hembree, E. A., Jaycox, L. H., Meadows, E. A., & Street, G. P. (1999). A comparison of exposure therapy, stress inoculation training, and their combination for reducing posttraumatic stress disorder in female assault victims. *Journal of Consulting and Clinical Psychology, 67*, 194–200.

Foa, E. B., Feske, U., Murdock, T. B., Kozak, M. J., & McCarthy, P. R. (1991). Processing of threat-related information in rape victims. *Journal of Abnormal Psychology, 100*, 156–162.

Foa, E. B., Hembree, E. A., Cahill, S. E., Rauch, S. A. M., Riggs, D. S., Feeny, N. C., et al. (2005). Randomized trial of prolonged exposure for posttraumatic stress disorder with and without cognitive restructuring: Outcome at academic and community clinics. *Journal of Consulting and Clinical Psychology, 73*, 953–964.

Foa, E. B., & Kozak, M. J. (1986). Emotional processing of fear: Exposure to corrective information. *Psychological Bulletin, 99*, 20–35.

Foa, E. B., & Kozak, M. J. (1991). Emotional processing: Theory, research, and clinical implications for anxiety disorders. In J. D. Safran & L. S. Greenberg (Eds.), *Emotion, psychotherapy, and change* (pp. 21–49). New York: Guilford Press.

Foa, E. B., Molnar, C., & Cashman, L. (1995). Change in rape narratives during exposure therapy for posttraumatic stress disorder. *Journal of Traumatic Stress, 8*, 675–690.

Foa, E. B., & Rothbaum, B. O. (1997). *Treating the trauma of rape: Cognitive-behavioral therapy for PTSD*. New York: Guilford Press.

Foa, E. B., Rothbaum, B., Riggs, D., & Murdock, T. (1991). Treatment of posttraumatic stress disorder in rape victims: A comparison between cognitive-behavioral procedures and counseling. *Journal of Consulting and Clinical Psychology, 59*, 715–723.

Friedman, M. J., Charney, D. S., & Deutch, A. Y. (1995). *Neurobiological and clinical consequences of stress: From normal adaptation to post-traumatic stress disorder*. Philadelphia: Lippincott-Raven.

Freud, S., & Breuer, J. (1895). *Studies on hysteria*. Vienna: Franz Deuticke.

Hamblen, J. L., Gibson, L. E., Mueser, K. T., & Norris, F. H. (2006). Cognitive behavioral therapy for prolonged disaster distress. *Journal of Clinical Psychology, 62*(8), 1043–1052.

Hayes, S. C., Follette, V. M., & Linehan, M. M. (2004). *Mindfulness and acceptance: Expanding the cognitive-behavioral tradition*. New York: Guilford Press.

Hayes, S. C., Strosahl, K. D., & Wilson, K. G. (1999). *Acceptance and commitment therapy: An experiential approach to behavior change*. New York: Guilford Press.

Hayes, S. C., Wilson, K. G., Gifford, E. V., Follette, V. M., & Strosahl, K. (1996). Experimental avoidance and behavioral disorders: A functional dimensional approach to diagnosis and treatment. *Journal of Consulting and Clinical Psychology, 64*, 1152–1168.

Heider, F. (1958). *The psychology of interpersonal relations*. New York: Wiley.

Herman, J. L. (1992). *Trauma and recovery*. New York: Basic Books.

Hull, C. (1943). *Principles of behavior*. New York: Appleton–Century–Crofts.

Hyams, K. C., Wignall, F. S., & Roswell, R. (1996). War syndromes and their evaluation: From the U.S. Civil War to the Persian Gulf War. *Annals of Internal Medicine, 125*, 398–405.

Hyer, L. (1994). *Trauma victim: Theoretical issues and practical suggestions*. Muncie, IN: Accelerated Development.

Hyer, L., & Brandsma, J. M. (1997). EMDR minus eye movements equals good psychotherapy. *Journal of Traumatic Stress, 10*, 515–522.

Janet, P. (1925). *Psychological healing* (Vols. 1–2). New York: Macmillan.

Janoff-Bulman, R. (1989). Assumptive worlds and the stress of traumatic events: Applications of the schema construct. *Social Cognition, 7*, 113–136.

Jones, E. E., & Davis, K. E. (1965). From acts to dispositions: The attribution process in person perception. In L. Berkowitz (Ed.), *Advances in experimental social psychology* (Vol. 2, pp. 219–266). Orlando, FL: Academic Press.

Kardiner, A. (1941). *The traumatic neuroses of war*. New York: Hoeber.

Keane, T. M., & Kaloupek, D. G. (1982). Imaginal flooding in the treatment of a posttraumatic stress disorder. *Journal of Consulting and Clinical Psychology, 50,* 138–140.

Keane, T. M., Zimering, R. T., & Caddell, J. M. (1985). A behavioral formulation of posttraumatic stress disorder in Vietnam veterans. *Behavior Therapist, 8,* 9–12.

Kelley, H. H. (1967). Attribution in social psychology. *Nebraska Symposium on Motivation, 15,* 192–238.

Kilpatrick, D. G., Veronen, L. J., & Resick, P. A. (1979). The aftermath of rape: Recent empirical findings. *American Journal of Orthopsychiatry, 49,* 658–669.

Kilpatrick, D. G., Veronen, L. J., & Resick, P. A. (1982). Psychological sequelae to rape: Assessment and treatment strategies. In D. M. Dolays & R. L. Meredith (Eds.), *Behavioral medicine: Assessment and treatment strategies* (pp. 473–497. New York: Plenum Press.

Kintsch, W., & Keenan, J. (1973). Reading rate and retention as a function of the number of propositions in the base structure of sentences. *Cognitive Psychology, 5,* 257–274.

Kolb, L. C. (1987). A neuropsychological hypothesis explaining posttraumatic stress disorders. *American Journal of Psychiatry, 144,* 989–995.

Lang, P. J. (1977). Imagery in therapy: An information processing analysis of fear. *Behavior Therapy, 8,* 862–886.

Lang, P. J., Melamed, B. G., & Hart, J. (1970). A psychophysiological analysis of fear modification using an automated desensitization procedure. *Journal of Abnormal Psychology, 76,* 220–234.

Lerner, M. J. (1965). Evaluation of performance as a function of performer's reward and attractiveness. *Journal of Personality and Social Psychology, 1,* 355–360.

Linehan, M. M. (1993). *Cognitive-behavioral treatment of borderline personality disorder*. New York: Guilford Press.

McCann, I. L., & Pearlman, L. A. (1990). *Psychological trauma and the adult survivor: Theory, therapy, and transformation*. New York: Brunner/Mazel.

McCann, I. L., Sakheim, D. K., & Abrahamson, D. J. (1988). Trauma and victimization: A model of psychological adaptation. *Counseling Psychologist, 16,* 531–594.

McNally, R. J. (1998). Information-processing abnormalities in anxiety disorders: Implications for cognitive neuroscience. *Cognition and Emotion, 12,* 479–495.

Meichenbaum, D., & Novaco, R. (1985). Stress inoculation: A preventative approach. *Issues in Mental Health Nursing, 7,* 419–435.

Mineka, S., & Cook, M. (1986). Immunization against the observational conditioning of snake fear in rhesus monkeys. *Journal of Abnormal Psychology, 95,* 307–318.

Mueser, K. T., Rosenberg, S. D., Jankowski, M. K., Hamblen, J. L., & Descamps, M. (2004). Cognitive behavioral treatment program for posttraumatic stress disorder in persons with severe mental illness. *American Journal of Psychiatric Rehabilitation, 7,* 107–146.

Neisser, U. (1967). *Cognitive psychology*. New York: Appleton–Century–Crofts.

O'Donnell, J. (1985). *The origins of behaviorism: American psychology, 1870–1920*. New York: New York University Press.

O'Donohue, W., & Krasner, L. (1999). *Theories of behavior therapy: Exploring behavior change*. Washington, DC: American Psychological Association.

Piaget, J. (1962). The stages of the intellectual development of the child. *Bulletin of the Menninger Clinic, 26,* 120–128.

Pitman, R. K., Orr, S. P., Altman, B., Longpre, R. E., Poiré, R. E., Macklin, M. L., et al. (1996a). Emotional processing during eye movement desensitization and reprocessing (EMDR) therapy of Vietnam veterans with post-traumatic stress disorder. *Comprehensive Psychiatry, 37,* 419–429.

Pitman, R. K., Orr, S. P., Altman, B., Longpre, R. E., Poiré, R. E., Macklin, M. L., et al. (1996b). Emotional processing and outcome of imaginal flooding therapy in Vietnam veterans with chronic posttraumatic stress disorder. *Comprehensive Psychiatry, 37,* 409–418.

Pylyshyn, Z. W. (1973). What the mind's eye tells the mind's brain: A critique of mental imagery. *Psychological Bulletin, 80,* 1–24.

Renfrey, G., & Spates, C. R. (1994). Eye movement desensitization: A partial dismantling study. *Journal of Behavior Therapy and Experimental Psychiatry, 25,* 231–239.

Rescorla, R. A., & Mahwah, R. R. (2001). Experimental extinction. In R. R. Mowrer & S. B. Klein (Eds.), *Handbook of contemporary learning theories* (pp. 119–154). Mahwah, NJ: Erlbaum.

Resick, P. A., & Schnicke, M. K. (1993). *Cognitive processing therapy for rape victims: A treatment manual.* Newbury Park, CA: Sage.

Rothbaum, B. O., Astin, M. C., & Marsteller, F. (2005). Prolonged exposure versus eye movement desensitization and reprocessing (EMDR) for PTSD rape victims. *Journal of Traumatic Stress, 18,* 607–616.

Segal, Z. V., Williams, J. M. G., & Teasdale, J. D. (2002). *Mindfulness-based cognitive therapy for depression: A new approach to preventing relapse.* New York: Guilford Press.

Shapiro, F. (1989). Eye movement desensitization: A new treatment for post-traumatic stress disorder. *Journal of Behavior Therapy and Experimental Psychiatry, 20,* 211–217.

Shapiro, F. (1995). *Eye movement desensitization and reprocessing (EMDR): Basic principles, protocols, and procedures.* New York: Guilford Press.

Skinner, B. F. (1957). *Verbal behavior.* East Norwalk, CT: Appleton–Century–Crofts.

Taylor, C. W., Hare, R. M., & Barnes, J. (1998). *Greek philosophers–Socrates, Plato, and Aristotle.* New York: Oxford University Press.

Thorndike, E. L. (1927). The law of effect. *American Journal of Psychology, 39,* 212–222.

Thrasher, S. M., Dalgleish, T., & Yule, W. (1994). Information processing in post-traumatic stress disorder. *Behaviour Research and Therapy, 32,* 247–254.

van der Kolk, B., Greenberg, M., & Boyd, H. (1985). Inescapable shock, neurotransmitters, and addiction to trauma: Toward a psychobiology of post traumatic stress. *Biological Psychiatry, 20,* 314–325.

Walser, R. D., Westrup, D., Rogers, D., Gregg, J., & Lowe, D. (2003, November). *Acceptance and commitment therapy for PTSD.* Paper presented at the International Society of Traumatic Stress Studies, Chicago.

Watson, J. B. (1913). Psychology as the behaviorist views it. *Psychological Review, 20,* 158–177.

Watson, J. B., & Rayner, R. (1920). Conditioned emotional reactions. *Journal of Experimental Psychology, 3,* 1–14.

Wolpe, J. (1954). Reciprocal inhibition as the main basis of psychotherapeutic effects. *Archives of Neurology and Psychiatry, 72,* 205–226.

Part II

SCIENTIFIC FOUNDATIONS AND THEORETICAL PERSPECTIVES

Chapter 4

Psychological Theories of PTSD

Shawn P. Cahill and Edna B. Foa

Exposure to traumatic events is relatively common, with estimates from epidemiological studies in the United States ranging between 37 and 92% of respondents, depending on the sample (see Breslau, 1998), reporting one or more events that would meet the *Diagnostic and Statistical Manual of Mental Disorders*, fourth edition (DMS-IV; American Psychiatric Association, 1994) "objective" criteria (A1) for trauma. Posttraumatic stress disorder (PTSD), a common anxiety disorder occurring in approximately 8% of the population (Kessler, Sonnega, Bromet, Hughes, & Nelson, 1995), is frequently a chronic disorder that may persist for years following the traumatic event and is highly comorbid with other psychiatric conditions, such as other anxiety disorders, depression, and substance use or dependence (Kessler et al., 1995). Moreover, PTSD is associated with poor functioning and low quality of life (Kessler, 2000; Malik et al., 1999). As such, it is of great importance to understand this disorder to better prevent and treat its occurrence. In this chapter, we review several prominent theories of PTSD.

WHAT NEEDS TO BE EXPLAINED?

For any theory to provide an adequate psychological account of PTSD, at the very least it must address three well established areas of research findings. First, it must address the *phenomenology* of PTSD, including the specific symptoms of PTSD and important associated features, such as trauma-related cognitions about the dangerous nature of the world and incompetence of the self (Foa, Ehlers, Clark, & Tolin, 1999; Janoff-

Bulman, 1992; McCann & Pearlman, 1990). Second, an adequate theory of PTSD must account for the *natural course of posttrauma reactions*. Specifically, it must account for the fact that PTSD symptoms are very common in the immediate aftermath of a traumatic event, but most trauma survivors experience a rapid reduction in those symptoms within the first 3 months following the trauma (Riggs, Rothbaum, & Foa, 1995; Rothbaum, Foa, Riggs, Murdock, & Walsh, 1992) and do not develop chronic PTSD. However, a significant minority of trauma survivors do not fully recover from their symptoms. Why do most people recover from exposure to a traumatic event, but some people do not? What mechanisms are involved in natural recovery, and what mechanisms interfere with it? An adequate theory of PTSD must explain both natural recovery and failure of recovery. Third, several forms of cognitive-behavioral therapy (CBT) have been demonstrated to be highly efficacious in reducing PTSD symptom severity, with concomitant improvements on depression and general anxiety (Cahill & Foa, 2004). An adequate theory of PTSD must be able to account for the efficacy of CBT.

THEORIES OF PTSD

Conditioning Theories

Several PTSD researchers (e.g., Keane, Zimering, & Caddell, 1985; Kilpatrick, Veronen, & Best, 1985) have proposed that Mowrer's (1960) two-factor learning theory of fear and anxiety (and its application to phobias by Dollard and Miller in 1950) can explain the clinical symptoms of PTSD. According to this theory, in the first stage, fear is acquired via classical conditioning and avoidance, via instrumental conditioning.

To explain the symptoms of PTSD in Vietnam War veterans, Keane and colleagues (1985) suggested that a person exposed to a life-threatening experience may become conditioned to a wide variety of stimuli that were present during the trauma (e.g., sounds, time of day, odors) through the process of classical conditioning; as a result, these previously neutral stimuli come to elicit intense anxiety. Keane and colleagues further suggested that anxiety is evoked not just by stimuli that were present during the trauma. Rather, via the processes of higher order conditioning and stimulus generalization, a wide range of situations acquire fear-inducing capacities. The authors proposed that characteristic PTSD responses—such as reexperiencing the traumatic event via thoughts, recollections, and nightmares—are part of the normal recovery process following a traumatic experience. However, when a high degree of generalization and higher order conditioning occur, these symptoms become chronic.

Keane and colleagues (1985) discussed an apparent contradiction between their conceptualization of PTSD and the empirical finding that repeated exposure to feared stimuli extinguishes the anxiety associated with them. To explain this seeming contradiction, they proposed that spontaneous exposures to combat-related stimuli via reexperiencing are incomplete (i.e., do not include all the conditioned stimuli) and are of short duration; therefore, they do not extinguish. To understand why survivors often do not recall all the cues involved in the original trauma, Keane and colleagues put forth several explanations. First, the traumatic event is so aversive that individuals with PTSD attempt to avoid engaging with the memory of the event for any prolonged period. Second, society discourages expression of emotion by men. Therefore, war veterans have limited opportunity to expose themselves to the upsetting event through discussing it with others in the military or at home. Consistent with this explanation, Resick (1986) found that female victims of robbery tended to express more emotions soon after the event than did male victims; furthermore, the emotional reactions of the female victims subsided within 1–3 months, whereas the male victims' symptoms persisted. Third, many traumatized individuals report long periods of time for which they are unable to account. This, Keane and colleagues suggest, happens because of the discrepancy between the mood state that occurred during the trauma and the mood state present at the time of recall. In support of this proposition, they cite Bower's (1981) summary of research demonstrating that mood discrepancy hinders recall.

On the basis of their conceptualization, Keane and colleagues (1985) hypothesized that providing more cues during imaginal exposure to the traumatic event may improve memory and thereby enhance confrontation with the trauma related stimuli. Foa, Steketee, and Rothbaum (1989) suggested that the increased arousal during the imagining of the traumatic event may be a mediator for improved recall. Increasing the arousal level, they proposed, provides a better match between the affect at the time of the trauma and that at the time of recall; this increased match facilitates the recall of details of the event. Indeed, fear activation during confrontation with the feared situation or memory has been repeatedly found to correlate with degree of benefit from treatment by exposure (Borkovec & Sides, 1979; Foa, Riggs, Massie, & Yarczower, 1995; Kozak, Foa, & Steketee, 1988; Lang, Melamed, & Hart, 1970; Pitman et al., 1996).

Keane and colleagues (1985) also invoked conditioning theory to explain the anger and irritability often demonstrated by combat veterans with PTSD. These behaviors, they suggested, were acquired during military training. In civilian life, such aggressive behaviors are maintained by both positive reinforcement (i.e., attaining one's goals) and negative reinforcement (i.e., the reduction of anxiety when anger is expressed). Foa and colleagues (1989) commented that this explanation does not seem to account for aggressive responses that follow other types of traumas (e.g., rape, accidents), nor does it address the continuation of angry reactions in the face of their aversive consequences. Keane and colleagues suggested that the diminished interest in social and leisure activities is a result of a contrast effect: Vietnam veterans returning from war are less interested in civilian activities, because they are not as stimulating as wartime events. Again, this explanation cannot account for numbness and withdrawal evidenced in people with PTSD after exposure to other traumas (e.g., rape, accident).

Becker, Skinner, Abel, Axelrod, and Chichon (1984) utilized Mowrer's theory (1960) to explain the development and persistence of sexual problems in rape victims, viewing the assault situation as an unconditioned stimulus that evokes fear and anxiety. They suggested that sexual activities associated with the rape become conditioned stimuli for anxiety. Via generalization and higher order conditioning, other sexual activities also come to elicit fear. To avoid discomfort, the victim may inhibit sexual feelings or abstain from sex. This theory appears to be a parsimonious explanation for the sexual symptoms of rape victims.

Kilpatrick and his colleagues (1985) have also used Mowrer's two-factor theory (1960) to explain rape victims' reactions. Rape victims, they proposed, perceive the rape situation as life-threatening and consequently react with terror and extremely high autonomic arousal. Stimuli associated with the rape acquire the capacity to elicit fear via classical conditioning. Some cues, such as sexual activity and men, are common to all rape victims, whereas other cues are idiosyncratic, depending on the specific rape situation. They also invoke the concepts of stimulus generalization and second-order conditioning to account for the wide range of circumstances that evoke anxiety in the rape victim. Thoughts and words associated with the rape experience, Kilpatrick and colleagues suggest, also acquire the capacity to provoke anxiety; thus, when the rape victim describes her rape experience, she becomes anxious. The therapeutic context, therefore, often provokes considerable discomfort and may consequently be avoided in the same way that other rape-related stimuli are avoided.

More recently, Keane and Barlow (in Barlow, 2002) described an etiological model of PTSD that retains the basic features of the Keane and colleagues (1985) conditioning model but situates PTSD within Barlow's (1988) broader theory of pathological anxiety. Central to this theory is the idea of two generalized vulnerabilities to psychopathology. First is a biological vulnerability that is largely a genetic trait to experience intense, negative affective states such as panic and depression. Related to the biological vulnerability is the distinction between true and false alarms, which reflect triggering of the "fight or flight" (also freezing) reaction. True alarms occur when the stimuli triggering the reaction are actually threatening; a false alarm occurs when the reaction is triggered by nonthreatening stimuli. The second vulnerability is an acquired psychological vulnerability of a diminished sense of control and the related construct of anxious apprehension, a future-oriented mood state characterized by hypervigilance and cognitive biases toward threat, both external and internal (i.e., interoceptive stimuli associated with strong emotions). Anxious apprehension intensifies negative affect, promotes avoidance of both external triggers and internal states of negative affect, and engages worry.

Applying these concepts to the development of PTSD, Keane and Barlow (in Barlow, 2002) maintained that the experience of a traumatic event by individuals with the prerequisite generalized biological and psychological vulnerabilities results in the triggering of a true alarm at the time of the trauma (similar to the concepts of unconditioned stimuli and responses in classical conditioning). Subsequently, these individuals learn alarms in response to internal and external cues associated with trauma, and their initial response to it. These learned alarms trigger the process of anxious apprehension, particularly focused on the reexperienced emotions, which then leads to avoidance of both the triggers of the learned alarms and strong emotions that result in emotional numbing.

Critical Comments

The use of Mowrer's theory (1960) to explain PTSD is compelling because it is uncomplicated and parsimonious. It accounts for the acquisition of fear in response to trauma-related cues that were previously neutral. It also explains why individuals with PTSD avoid nondangerous situations, and why such avoidance persists despite its disruption of daily functioning. The forgoing expositions of conditioning theory address the extensive generalization of fear cues but do not provide an explanation for the greater generalization in PTSD relative to phobics. As noted earlier, individuals with PTSD fear and avoid a wider range of cues than do people with phobias, even agoraphobia. Perhaps this greater generalization is due to the greater severity of the trauma that precedes the onset of PTSD. Animal experiments indicate that longer duration and greater intensity of the unconditioned stimulus (UCS) lead to stronger avoidance and escape responses (Baum, 1970; Kamin, 1969; Overmeier, 1966), are more difficult to extinguish (Baum, 1970), and promote greater generalization. Another explanation for the extensive generalization in PTSD may lie in the greater complexity of the conditioned stimulus (CS) during traumas that lead to PTSD than during traumas that lead to simple phobias.

Not addressed by these authors are startle responses, which, again, are characteristic of PTSD but not of phobia. If this startle response is a result of the high tonic arousal found in PTSD (Blanchard, Kolb, Gerardi, Ryan, & Pallmeyer, 1986), then learning theory at first glance, does not appear to account for this symptom, because it is not commonly found among either people with agoraphobia (who also exhibit high tonic arousal) (Ehlers et al., 1986) or phobia when highly aroused in the presence of their feared stimulus. But, if these reactions are viewed as a component of the unconditioned response (UCR) that occurred in the original trauma, then learning theory can account for them by predicting the reoccurrence of this conditioned response (CR) in the presence of conditioned stimuli (CSs). A normal startle response occurs when an individual is confronted with a sudden, unexpected, external stimulus (such as a car backfiring). It follows that such responses will be more likely to occur when the original trauma includes unpredictable, surprising elements. For example, a Vietnam War veteran will more probably show startle responses than an airplane crash survivor. In war, some degree of hypervigilance to sudden noises or movements is adaptive and is therefore more likely to occur, whereas vigilance by an airplane passenger is irrelevant to his or her safety. Similarly, the victim of a "blitz" rape, in which the attack was sudden and unexpected, would be more likely to exhibit startle reactions to sudden stimuli than the victim of a "con" rape, in which the assailant used persuasion to coerce the victim. Startle, then, appears to be a conditioned response whose occurrence or absence can be predicted from the nature of the trauma.

More problematic for traditional learning theories are the symptoms characterized as reexperiencing of the traumatic event. Keane and colleagues (1985) attempted to explain flashbacks, nightmares, and intrusive images of the event via the high degree of generalization that renders avoidance of traumatic memories impossible. Although this may plausibly explain intrusive thoughts and perhaps flashbacks, it does not adequately account for nightmares. Notably, factors that contribute to the content, occurrence, and function of nightmares in the general population are largely unknown.

Schema Theories

A very different approach to understanding posttrauma reactions has been advanced by scholars employing theories from personality and social psychology (e.g., Epstein, 1991; Horowitz, 1976, 1986; Janoff-Bulman, 1992; McCann & Pearlman, 1990). To explain the psychological effects of traumatic experiences these theorists invoke the concept of schemas, that is, core assumptions and beliefs that guide the perception and interpretation of incoming information. Common to these theories are the suppositions that (1) traumatic events are usually discrepant with existing assumptions; and (2) processing a traumatic experience requires modification of existing assumptions. Based on Piaget's (1971) model of cognitive development, it is thought that such modification is accomplished through two mechanisms: assimilation and accommodation.

Horowitz provided an integration of psychoanalytic and information-processing concepts to account for posttrauma psychopathology, suggesting that people have a basic need to match trauma-related information with their "inner models based on old information" (1986, p. 92). The process of recovery entails the repetitive "revision of both [sources of information] until they agree" (p. 92), which Horowitz referred to as the "completion tendency," and which explains the reexperiencing (intrusive) symptoms observed in individuals with PTSD. Horowitz further noted that if the trauma information matches existing inner models, then that information will serve to strengthen these models. On the other hand, if the trauma information is incongruent, then each time the information is processed, "alterations of inner working models and plans for adaptive actions are accomplished" (p. 96). Competing with the completion tendency is the tendency to avoid the distress caused by the reexperiencing symptoms, which accounts for the various avoidance symptoms of PTSD. However, avoidance strategies prevent resolution of the discrepancy between existing inner models and the new information provided by the trauma. An implication of Horowitz's model is that avoidance maintains persistent incongruity between the traumatic experience and internal mental structures, which is central to maintenance of posttrauma psychopathology.

Horowitz's theory and the importance it assigned to reexperiencing (intrusions) and avoidance symptoms is reflected in the two factors of the Impact of Event Scale, the symptom measure he and his colleagues developed to measure posttrauma reactions (Horowitz, Wilner, & Alvarez, 1979). The recognition that reexperiencing and avoidance symptoms are central to trauma-related psychopathology had a profound influence on the definition of PTSD, which was introduced in DSM-III and DSM-III-R as an anxiety disorder.

Several other theorists hypothesized which schemas were particularly relevant to posttrauma reactions. For example, in his cognitive–experiential self-theory, Epstein (1985) proposed that "the essence of a person's personality is the implicit theory of self and world that the person constructs" (p. 283). Building on this foundation, Epstein (1991) suggested that four core beliefs change after a traumatic experience: the belief that the world is benign, that the world is meaningful, the self is worthy, and people are trustworthy. Influenced by Epstein's concepts, Janoff-Bulman (1992) suggested that the basic assumptions held by people in general are "the world is benevolent, the world is meaningful, and self is worthy" (p. 6). She further proposed that these assumptions are incompatible with a traumatic experience, which "shatters" these fundamental assumptions. Therefore, after a traumatic event, the victim must struggle either to assimilate the traumatic experience into the old set of assumptions or, more often, to change the assumptions, such that they can accommodate the traumatic experience. An example of assimilation is when a rape victim blames her own behavior as the cause of the assault to maintain her assumption that the world is meaningful and benevolent. Accommodation occurs when a rape victim changes her pretrauma assumptions and adopts the belief that the world is ruthless rather than benevolent.

Based on an extensive review of the literature on adaptation to trauma, McCann and Pearlman (1990) proposed seven fundamental psychological needs: frame of reference, safety, dependency/trust of self and others, power, esteem, intimacy, and independence. Frame of reference is viewed as a superordinate need, which is similar to Epstein's (1991) and Janoff-Bulman's (1992) notion of "meaningful world." McCann and Pearlman further suggested that individuals develop schemas that include beliefs, assumptions, and expectations in each of these fundamental need areas. Similar to the ideas developed by Horowitz, Epstein, and Janoff-Bulman, McCann and Pearlman emphasized that trauma causes disruptions in any or all of these need areas. Accordingly, several of the therapeutic implications that McCann and Pearlman discuss appear to focus on helping trauma survivors accommodate their schemas to the new information. Although less emphasized and less well developed, McCann and Pearlman suggest that traumatic events can sometimes also cause troublesome emotions, thoughts, or images when they strengthen existing negative schemas, such as in the case of a person who experiences repeated traumas. Influenced by McCann and Pearlman's ideas,

Resick and Schnicke (1993) introduced the idea that posttrauma psychopathology may result from not only a failure to accommodate trauma-relevant information but also overaccomodation of trauma-relevant information. Accordingly, they developed cognitive processing therapy, a treatment designed to correct the disruptions in these schemas.

Critical Comments

Schema theories have made several contributions to the understanding of posttrauma reactions, including the ideas that traumatic experiences change people's views about themselves, others, and the world in general; that the distressing and intrusive images that commonly follow a traumatic event reflect a discrepancy between pretrauma schemas and information provided by the traumatic event; and that recovery from trauma requires resolution of the discrepancy. In addition, using concepts from psychoanalytic, schema, and information-processing theories, Horowitz (1976, 1986) provides an interesting account of how people recover or fail to recover from trauma.

In general, however, the schema theories we have reviewed have two general weaknesses. First, these theories have generally not concerned themselves with clinical issues surrounding factors involved in the development of specific postrauma psychopathology following a traumatic experience. Instead, they have focused on the impact of trauma on beliefs more generally. However, as noted earlier, not all trauma survivors develop posttrauma psychopathology (PTSD), and traumas vary in their likelihood of producing PTSD (Kessler et al., 1995). Schema theories do not account for these observations. Second, the primary mechanisms by which traumatic events produce posttrauma reactions is "shattering" of positive assumptions about the nature of the world, others, and the self. This assumption might be true for the few people who had not experienced major stressors prior to the index trauma. However, as noted earlier, epidemiological studies indicate that many people experience repeated traumas (e.g., Kessler et al., 1995). How would a new trauma violate a preexisting schema in individuals with histories of multiple traumas? According to these schema theories, such individuals should experience a match between their inner models of the world and the new trauma; thus, their inner models need not undergo alterations, and they should consequently show fast recovery. Research findings, however, do not support this prediction. The experience of multiple traumas increases, rather than decreases, the probability of chronic PTSD (Kessler et al., 1995). Moreover, Bryant and Guthrie (2005) found that negative self-schemas prior to the trauma increased, rather than decreased, severity of PTSD symptoms 20 months after exposure to a trauma.

Janoff-Bulman (1992) has been aware of the problems inherent in her supposition that the negative effects of a trauma stem from the violation of pretrauma positive assumptions about the nature of the world. She therefore suggested that the holding of positive core assumptions will be a risk factor for greater initial psychological disruption but may be associated with enhanced long-term psychological recovery. This argument does not explain why individuals with histories of traumas are more likely to develop chronic PTSD, nor does it correspond to findings that initial severity of PTSD is strongly associated with later severity of the disorder (Rothbaum et al., 1992). The findings that prior trauma and prior negative cognitions are risk factors for the development of PTSD are less problematic for McCann and Pearlman (1990) and Resick and Schnicke (1993), who implicitly or explicitly permit the strengthening of existing negative beliefs to cause posttrauma psychopathology. Moreover, McCann and Pearlman even note there are times when traumatic events can promote personal growth rather than psychopathology, and that preexisting positive beliefs can serve as a protective factor rather than a risk factor for posttrauma psychopathology. The difficulty here is that it is not clear when, for example, the effects of trauma will shatter preexisting positive assumptions and when they will serve as protective factor. What McCann and Pearlman appear to have gained in explanatory value by moving away from a strong "shattered assumptions" hypothesis, they would appear to have lost in precision and testability.

Emotional Processing Theory

Basic Premises

Emotional processing theory was initially proposed by Foa and Kozak (1985, 1986) to explain the anxiety disorders and the process and outcome of exposure therapy for these disorders. The theory rests on two basic premises. The first premise is that anxiety disorders reflect the presence of pathological fear structures in memory. A fear structure includes interrelated representations of feared stimuli, fear responses, and the meanings associated with them. A fear structure is activated when information in the environment matches some of the information represented in the structure, resulting in spreading activation to associated elements, thereby producing cognitive, behavioral, and physiological anxiety reactions. When a fear structure accurately represents dangerous situations in the world, it serves as a blueprint for effective action, such as hurrying across a busy street when an approaching car does not appear to be slowing down. However, a fear structure becomes maladaptive or pathological when (1) associations among stimulus elements do not accurately represent the world, (2) physiological and escape/avoidance responses are evoked by harmless stimuli, (3) excessive and easily triggered response elements interfere with adaptive behavior, and (4) harmless stimulus and response elements are erroneously associated with threat meaning. Foa and Kozak (1985) hypothesized that different anxiety disorders reflect characteristically different fear structures.

The second basic premise of emotional processing theory is that successful treatment modifies the pathological elements of the fear structure such that information that once evoked anxiety symptoms no longer does so. Two conditions are necessary for modification of the fear structure: (1) The fear structure must be activated, and (2) new information that is incompatible with the erroneous information embedded in the structure must be available and incorporated into the fear structure. Intentional exposure to safe but feared stimuli during exposure therapy meets these two conditions. Specifically, exposure to feared stimuli results in the activation of the fear structure and provides corrective information about the probability and the cost of feared consequences. In addition, erroneous beliefs the person may have about the nature of anxiety are disconfirmed, such as the belief that anxiety will continue unabated unless one escapes the situation, or that the anxiety will cause one to "lose control" or "go crazy." This new information is encoded during the exposure therapy session, altering the fear structure and mediating between-session habituation upon subsequent exposure to the same or similar stimuli, thereby resulting in symptom reduction.

This original theoretical exposition of emotional processing has undergone subsequent refinement and elaboration, resulting in a comprehensive theory of PTSD that accounts for the acquisition of PTSD, natural recovery, and the efficacy of cognitive behavioral therapy in the treatment and prevention of chronic PTSD (Foa & Cahill, 2001; Foa, Huppert, & Cahill, 2006; Foa & Jaycox, 1999; Foa & Riggs, 1993; Foa et al., 1989).

The Fear Structure Underlying PTSD and the Symptoms of PTSD

Emotional processing theory posits that the fear structure underlying PTSD is characterized by a particularly large number of harmless stimulus representations that are erroneously associated with the meaning of danger, as well as representations of physiological arousal and of behavioral reactions that result in the symptoms of PTSD. Because a large number of stimuli can activate the fear structure, individuals with PTSD perceive the world as entirely dangerous. In addition, representations of how the person behaved during the trauma and his or her subsequent symptoms erroneously become associated with the meaning of self-incompetence. These two broad sets of negative cognitions ("The world is entirely dangerous" and "I am completely incompetent") further promote the severity of PTSD symptoms, which in turn reinforce the erroneous cognitions (for more details, see Foa & Rothbaum, 1998). Additional development of the emotional processing theory of PTSD has focused on the nature of the trauma memory, how prior information about the self and the world influences the interpretation of the traumatic experience, and the subsequent symptoms, which then influence the kinds of posttrauma experiences the person will have and how they are interpreted (Foa & Jaycox, 1999; Foa & Riggs, 1993).

Trauma survivors' narratives of their trauma have been characterized as fragmented and disorganized (e.g., Kilpatrick, Resnick, & Freedy, 1992). Foa and Riggs (1993) proposed that the disorganization of trauma memories is the result of several mechanisms known to interfere with the processing of information encoded under intense distress. Consistent with the hypotheses that PTSD would be associated with a disorganized memory for the trauma, Amir, Stafford, Freshman, and Foa (1998) found that the level of articulation of the trauma memory shortly after an assault was negatively correlated with PTSD symptom severity 12 weeks after the trauma. In a complementary finding, Foa, Molnar, and Cashman (1995) reported that treatment of PTSD with prolonged exposure (PE), a treatment that has been found highly efficacious in the treatment of PTSD (Foa & Meadows, 1997), was associated with increased organization, and that reduced fragmentation was associated with reduced anxiety, whereas increased organization was associated with reduced depression.

Preexisting knowledge about the safety (or dangerousness) of the world and competence (or incompetence) of the self also influences how the traumatic memory is encoded and how posttrauma experiences and symptoms are perceived. Two pathways are of relevance to the development of PTSD. In the first pathway, preexisting knowledge about safety is violated by the trauma, such as when an assault occurs in the safety of one's own home. This pathway to PTSD is essentially the same as the "shattered assumptions" hypothesis advanced by schema theorists discussed earlier, which assumes the presence of positive schemas about the self and others. The second pathway to PTSD involves the trauma strengthening existing negative perceptions about the self and the world (for more detailed discussion, see Foa & Rothbaum, 1998). The results of the previously mentioned study by Bryant and Guthrie (2005)—that negative perceptions of self predict later severity of PTSD symptoms—are consistent with this pathway.

Natural Recovery versus Development of Chronic PTSD

Whereas high levels of PTSD symptoms are common immediately following a traumatic event, most individuals show a decline in their symptoms, a pattern Foa and Cahill (2001) termed "natural recovery." However, as noted earlier, a significant minority of trauma victims fail to recover and continue to suffer from PTSD symptoms for years (Kessler et al., 1995). Foa and Cahill (2001) proposed that natural recovery results from emotional processing that occurs in the course of daily life by repeated activation of the trauma memory through engagement with trauma-related thoughts and feelings, sharing them with others, and being confronted with trauma reminders. In the absence of additional traumas, these natural exposures contain information that disconfirms the common posttrauma associations within the fear structure, such as "The world is entirely dangerous" and "I am incompetent." In addition, repeatedly talking about the event with supportive others can help the survivor organize the memory in a meaningful way. Why then do some trauma victims go on to develop chronic PTSD?

Within the framework of emotional processing theory, the development of chronic PTSD is conceptualized as a failure to process the traumatic memory adequately. Therefore, evidence for activation of the trauma memory in the posttrauma period, along with talking about the trauma with supportive others, should be associated with recovery, whereas failure to engage emotionally with the memory, such as through avoidance coping, emotional numbing, and dissociation, should be associated with the development of chronic PTSD. Several sources of evidence are consistent with the emotional processing theory of natural recovery. Creamer, Burgess, and Pattison (1992) found that early reexperiencing symptoms are associated with better subsequent outcome. Similarly, Gilboa-Schectman and Foa (2001) found that individuals whose peak in PTSD symptoms occurs within approximately 2 weeks of the trauma have less severe PTSD several months later compared to those whose peak occurs 3 weeks after the trauma or later. Lepore, Silver, Wortman, and Wayment (1996) found that mothers who lost their babies to sudden infant death syndrome (SIDS) and were able to share their loss with a supportive social network had more effectively resolved their grief 18 months later than those who were not able to do so.

Therapy That Promotes Emotional Processing

As in natural recovery, PE treatment of PTSD is assumed to work through activation of the fear structure as patients intentionally confront thoughts and reminders of the trauma via imaginal and *in vivo* exposure, respectively, and incorporate new information. Foa and Jaycox (1999) suggested several mechanisms and sources of information that explain the efficacy of PE in ameliorating PTSD symptoms. Avoidance of trauma memories and related reminders is maintained through the process of negative reinforcement, that is, through the reduction of anxiety in the short run. In the long run, however, avoidance impedes emotional processing. By confronting trauma memories and reminders, PE blocks negative reinforcement of cognitive and behavioral avoidance, thereby reducing one of the factors that maintains PTSD.

Another mechanism involved in the promotion of emotional processing is extinction of anxiety, which disconfirms erroneous beliefs that in the absence of avoidance or escape anxiety will not diminish. Patients also learn that they can tolerate their symptoms, and that having them does not result in "going crazy" or "losing control," thereby altering their perceptions of themselves as lacking personal mastery and courage rather than further supporting evidence of incompetence. Imaginal and *in vivo* exposure also help patients to differentiate between the traumatic event and other, similar but nontraumatic events, thereby allowing them to see the trauma as specific event occurring in space and time rather than evidence that the world is entirely dangerous and that the self is completely incompetent. Importantly, patients with PTSD often report that thinking about the traumatic event makes them feel as if it is "happening right now." Repeated imaginal exposure to the trauma memory promotes discrimination between the past and present by helping patients realize that although remembering the trauma can be emotionally upsetting, the trauma is not happening again; therefore, thinking about the event is not dangerous.

Repeated imaginal exposure also gives patients the opportunity to evaluate accurately aspects of the event that are contrary to their beliefs about danger and self-incompetence that may otherwise be overshadowed by the more salient, threat-related elements of the memory. For example, individuals who feel guilty about not having done more to resist an assailant may come to the realization that the assault likely would have been more severe had they resisted.

Critical Comments

Emotional processing theory has had enduring appeal to many researchers in the area of anxiety disorders. Its basic associative network structure has much in common with conditioning theories of PTSD, but by incorporating associations between stimuli and responses, associations among stimuli, and the meanings associated with both stimuli and responses, the theory is better able to model the phenomenology of PTSD. The developments offered by Foa and Riggs (1993) provide a way for the theory to incorporate the impact of the victim's experiences prior to the target trauma that may serve as risk or resilience factors, and leads to a coherent explanation for why both presence and absence of prior traumas may amplify the effects of trauma. Two clear strengths of the theory are its explicit attention to differentiating among various anxiety disorders in terms of underlying fear structures and an explicit account of the processes and mechanisms responsible for the efficacy of exposure therapy in the treatment of PTSD and other anxiety disorders. Indeed, emotional processing theory was originally introduced to account for the efficacy of exposure therapy and to generate hypotheses about the conditions under which the efficacy of exposure therapy is enhanced or impeded. Nevertheless, it is important to remember that the efficacy of a treatment by itself does not constitute proof of the validity of theory. An additional strength is that the theory enlists the same mechanisms to explain both natural recovery and that through exposure therapy, and specifies the mechanisms through which some individuals recover and others do not. Finally, the theory provides a unified account of why both individuals with extreme pretrauma beliefs about safety of the world (and, by extension, competence of the self) and people who have prior negative perceptions of the world and self are at greater risk for developing PTSD compared to individuals without such perceptions.

One limitation is that whereas emotional processing theory focuses on the consequences of numbing and dissociation for recovery, it does not carefully specify how these symptoms are explained within the theory. At least three mechanisms are readily consistent with emotional processing theory. One possibility is that several of the numbing symptoms of PTSD are the consequence of avoidance and the resulting reduction in exposure to emotion-eliciting experiences, as has been proposed in behavioral theories of depression (e.g., Lewinsohn, Hoberman, Teri, & Hautzinger, 1985). A second possibility is the general avoidance of strong emotional states because the physiological responses involved in even strong positive emotional states (e.g., elevated heart rate) becomes associated with the meaning of danger, as envisioned in the Barlow and Keane (in Barlow, 2002) model discussed previously. A third mechanism, discussed by Foa, Zinbarg, and Rothbaum (1992), emerges from animal research, demonstrating that analgesia is induced by uncontrollable/unpredictable shock. Research has shown that exposure to both UCSs and CSs results in temporary reduction in sensitivity to pain that, at least in some instances, has been shown to be mediated through activation of the endorphin system.

Another potential limitation of emotional processing theory is its focus on fear as the primary emotion and danger as the primary meaning associated with PTSD. This focus is consistent with the DMS-IV definition of a trauma as involving real or perceived danger to self or others and a subjective response of intense fear, horror, or helplessness. Moreover, one of the symptom clusters of PTSD is arousal symptoms, which include mostly symptoms of anxiety. Nevertheless, as noted by Dalgleish and Power (2004), other emotions, such as bereavement and disgust, may also be associated with PTSD-like emotional disorders that have in common reexperiencing distressing events and avoidance of memories and reminders of these events. We agree with Dalgleish and Power that a theory that can accommodate PTSD can be extended to account for emotional states that result from extremely distressing events in which the primary emotion is sadness or disgust. In emotional processing theory, an emotional state such as bereavement would be represented by an emotional structure in which the meaning of the stimuli or response representations is extreme sadness and loss rather than fear and danger. Such structures still lead to the perception of the world as extremely negative and, to the extent that the person struggles with chronic symptoms, the self as incompetent.

Cognitive Theory

Classical Cognitive Theory

Cognitive therapy (Beck, 1972; Beck, Rush, Shaw, & Emery, 1979; Ellis, 1977) was first developed for the treatment of depression and later extended to the treatment of anxiety disorders (e.g., Beck, Emery, & Greenberg, 1985). This therapy is based on the assumption that rather than events themselves, one's interpretation of events is responsible for the evocation of emotional reactions. Accordingly, an event can be interpreted in different ways and consequently evoke different emotions. An example is the woman who hears a noise at the window in the middle of the night. If she thinks that the noise is made by a burglar, she will immediately become anxious. However, if she thinks it is just the wind, she will perhaps be slightly annoyed to be awakened but not anxious. Cognitive theory further assumes that each emotion is associated with a particular class of thoughts. In anxiety, the characteristic thoughts revolve around the perception of danger. The thoughts that produce anger involve the perception that other people have behaved in a wrong or unfair way. The thoughts that produce guilt involve the perception that one has behaved in a wrong or unfair way. The thoughts that produce sadness involve the perception of fundamental loss.

In everyday life, people experience a wide range of events that evoke negative emotions. However, sometimes the emotional responses are more intense and/or more prolonged than would be expected, interrupting the individual's daily functioning. These exaggerated emotional responses are thought to originate from distorted or dysfunctional interpretations. The aim of cognitive therapy is to make people aware of their dysfunctional thoughts, to challenge them, and to replace them with more functional ones.

The originators of cognitive theory for anxiety disorders gave little attention to PTSD. They did suggest, however, that people with traumatic neuroses do not discriminate between safe and unsafe signals, and that the concept of danger dominates their thinking. They also suggested that traumatic fear can be maintained through a sense of incompetence to handle stressful events (Beck et al., 1985).

Ehlers and Clark's Cognitive Model of PTSD

The central tenet of Ehlers and Clark's (2000) model of PTSD follows the classical cognitive theory described earlier. Accordingly, PTSD, like other anxiety disorders, is a result of appraisals related to impending threat. However, PTSD is a disorder associated with an event that happened in the past; therefore, the threat related to the event also belongs to the past. To explain how this past event generates a sense of current threat, Ehlers and Clark proposed that individuals with chronic PTSD process the traumatic event and/or its consequences in a way that gives rise to a sense of current threat. Two key processes lead to the current sense of threat: the individual's appraisals of the traumatic event and/or its consequence, and the nature of the traumatic memory and how it is integrated with other episodic memories of the individual.

With regard to appraisals, Ehlers and Clark's (2000) model follows Beck's (1976) model in emphasizing the causal role of negative (threat-relevant) cognitions in emotional disorders. In the case of PTSD, Ehlers and Clark concur with emotional processing theory that relevant appraisals can be either about external threat (viewing the world as a dangerous place) or internal threat (viewing oneself as incapable). However, emotional processing theory takes the position that cognitions underlying PTSD may or may not be available to introspection and self-report. In contrast, by using the concept of appraisals, it seems that Ehlers and Clark assign a prominent place to the thoughts and beliefs in the realm of one's awareness that can therefore be reported and directly challenged through verbal discourse. Like emotional processing theory, Ehlers and Clark suggest that the negative cognitions can be about what happened during or after the traumatic event.

The second core process in Ehlers and Clark's (2000) PTSD model is the unique nature of the traumatic memory in individuals with PTSD. The starting point here is the premise (see also Foa & Riggs, 1993) that individuals with PTSD relate a fragmented and poorly elaborated trauma narrative and when they recall the memory, feel as if the trauma is happening now rather than in the past (see also Foa & Jaycox, 1999). Ehlers and Clark suggest that the fragmented nature of the trauma memory, the perception of the memory as if it is happening in the present rather than in the past, as well as the lack of incorporation of the traumatic memory with other autobiographical memories, all explain how an event that happened in the past causes a sense of present threat.

Ehlers and Clark (2000) include concepts from associative network theories in their model and suggest that an additional aspect that distinguishes a traumatic memory is the presence of particularly strong S-S and S-R associations that render the memory easily retrievable by a large numbers of triggers (see also Foa et al., 1989). They further suggest that retrieval from associative memory is cue driven and unintentional, so that the individual may not be aware of the relationship between his or her emotional reactions and the trauma memory. Failure to identify the trigger of the reexperiencing symptoms may prevent the individual from learning that triggers are not themselves dangerous.

Critical Comments

As noted by Ehlers and Clark (2000), their model of persistent PTSD is greatly influenced by previous theories, including those of Beck and colleagues (1985) and Foa and colleagues (1989). As such, because much of the model reiterates concepts introduced and developed in other theories, it shares their strengths. What is innovative in this model is the view that there is a reciprocal relationship between the nature of the trauma memory and the appraisals of the trauma and its sequelae. In this regard, Ehlers and Clark propose that "when individuals with persistent PTSD recall the traumatic event, their recall is biased by their appraisals and they selectively retrieve information that is consistent with these appraisals" (pp. 326–327). Such selective retrieval, which prevents the individual from remembering aspects of the trauma that are inconsistent or contradict the appraisals, thereby prevents changes in the appraisals. Conversely, the nature of the trauma memory can influence appraisals. For example, a disorganized, fragmented memory can produce negative appraisals of oneself ("Something is wrong with me if I am unable to remember details of the memory").

Ehlers and Clark's (2000) model of PTSD invokes two factors: negative appraisals about the trauma and the nature of the traumatic memory. The reciprocal relationship between these two factors produces a vicious cycle that fosters the maintenance of PTSD symptoms by promoting a sense of current threat by preserving negative appraisals and other aspects of the trauma memory. Unlike emotional processing theory, which gives cognitive and behavioral avoidance a central role in impeding natural recovery, thereby maintaining PTSD symptoms, for Ehlers and Clark the presence of avoidance takes a backseat. Avoidance strategies are viewed as the result rather than the cause of a persistent sense of current threat. Accordingly, the treatment developed by Ehlers and Clark emphasizes cognitive procedures, whereas the treatment emerging from emotional processing theory focuses on confrontation with the traumatic memory and trauma-related reminders. What the model does not explain is that the addition of cognitive therapy to exposure does not enhance treatment efficacy (Foa et al., 2005; Marks, Lovell, Noshirvani, Livanou, & Thrasher, 1998; Paunovic & Ost, 2001), but the addition of exposure therapy does enhance the efficacy of cognitive therapy (Foa & Cahill, 2006).

Theories Invoking Multiple Representation Structures

Several researchers studying basic human cognition have postulated the existence of multiple representational systems in memory, for example, the familiar distinctions between short- and long-term memory, declarative and nondeclarative memory, and implicit and explicit memory. The starting point for theorists of psychopathology who take this perspective is the assumption of at least two (but possibly more) separable memory systems, thereby potentially providing greater explanatory power for the content area of interest (e.g., PTSD). The burden for these theorists is to specify clearly characteristics of the different representational systems and how they interact with one another. Moreover, it is important that these theories be able to explain a greater range of phenomena related to PTSD than can be explained by existing theories; otherwise, the principle of parsimony would guide us to prefer simpler theories. In addition, to justify their complexity, these theories need to generate new, testable predictions.

Dual-Representation Theory

Brewin, Dalgliesh, and Joseph (1996; Brewin & Holmes, 2003) have elaborated on an earlier model of cognitive change processes in psychotherapy (Brewin, 1989) to provide an account of PTSD related to concepts and findings in contemporary cognitive neuroscience (Brewin, 2001). Similar to many general cognitive models, Brewin assumes two separate representational systems in memory that operate in parallel, both during the trauma and subsequently: the verbally accessible memory (VAM) and situationally accessible memory (SAM) systems. VAM memories "contain information that the individual has attended to before, during, and after the traumatic event, and that received sufficient conscious processing to be transferred to a long-term memory store in a form that can later be deliberately retrieved" (Brewin & Holmes, 2003, p. 356) and easily communicated verbally to others. Thus, the VAM system is responsible for the ability of trauma victims to provide a narrative account of what happened. Because VAM memories include a record of what happened during the trauma, they can include information about the emotions they experienced at the time of the trauma (called primary emotions; typically, fear and helplessness). In addition, because VAM memories can be intentionally retrieved and deliberated upon, secondary emotions (e.g., anger, shame, guilt) can be generated by subsequent appraisals of that information. Moreover, information about the trauma can be integrated with other VAM memories.

By contrast, SAM memories contain "information that has been obtained from far more extensive, lower level perceptual processing of the traumatic scene, such as the sights and sounds [also the person's bodily responses to the trauma] that were too briefly apprehended to receive much conscious attention" (Brewin & Holmes, 2003, p. 357). SAM memories cannot be deliberately recalled; rather, they are triggered involuntarily by external and internal reminders of the trauma. Thus, the SAM system is responsible for symptoms such as flashbacks and cued physiological arousal. Because SAM memories are not verbally coded and cannot be deliberately retrieved, they are more difficult to communicate to others and to integrate with other memories. Unlike Foa and Kozak (1986), who define "emotional processing" as changes in the fear structure that promote recovery, Brewin and colleagues (1996) view emotional processing as "a largely conscious process in which representations of past and future events, and awareness of associated bodily states, repeatedly enter into and are actively manipulated within working memory" (p. 677), regardless of whether symptoms improve. Indeed, Brewin and colleagues posit three potential outcomes of emotional processing: completion/integration (i.e., "recovery), chronic emotional processing, and premature inhibition of processing. Recovery involves two goals: (1) altering negative secondary emotions resulting from unhelpful appraisals of the trauma and the person's reactions during and afterward, and (2) preventing the automatic activation of SAM memories of the trauma. Theoretically, this would seem to be accomplished most effectively through a combination of cognitive restructuring and exposure therapy. By its focus on identifying and evaluating unhelpful beliefs that can be accessed verbally, cognitive restructuring would facilitate the process of elaborating and integrating trauma information stored in the VAM system. Cognitive restructuring would also help the patient generate new, more adaptive appraisals that would compete with earlier ones and serve to inhibit the SAM system. Exposure to information stored in the SAM system can further facilitate the process of restructuring information in the VAM system by providing access to information encoded in the SAM system but not consciously processed at the time of the trauma and, therefore, not included in the VAM system. In addition, exposure therapy, in which the SAM system is activated in the presence of new information (e.g., in the absence of intense physiological arousal or combined with effective coping strategies), results in the formation of new SAMs that then compete with the old SAMs. However, as noted previously, recovery is not inevitable, and failure to recover can take two different forms. In chronic emotional processing, the trauma-related SAM and VAM memories are repeatedly retrieved, but with little or no change occurring. As a result of this state of continuous processing of trauma-related information, the survivor continues to experience intense, trauma-related symptoms and distress, and may develop dysfunctional ways of coping with symptoms (e.g., substance abuse). This outcome may occur because of a discrepancy between the trauma and previous beliefs, competing demands for cognitive resources, inadequate cognitive development (e.g., the survivor is too young to understand the meaning of the event), lack of social support, or the inability to prevent intrusions of SAMs into consciousness (e.g., because of ongoing trauma or threat).

In premature inhibition of processing, trauma survivors have developed avoidance strategies that generally serve to limit activation of the trauma-related SAMs and VAMs. Here again, there is little change in these underlying trauma memories. However, because they are not regularly accessed, the individual may have relatively few symptoms of PTSD but may instead have memory impairments, maintenance of an attentional bias to permit early detection of threat (to engage efficiently in avoidance behaviors), pervasive avoidance, and dissociation. Moreover, because the trauma memories have not been altered, nor new memories developed, such individuals are at high risk for relapse upon exposure to situations or emotional states similar to those that were present during the trauma. The conditions under which premature inhibition of processing is likely to occur are the same for the development of the chronic process, with the addition of dissociation during the trauma and a previously existing tendency to avoid processing negative information (e.g., repressive coping).

The SPAARS Model

Dalgleish (1999; see also Power & Dalgleish, 1997, 1999) formulated a model of normal emotional experience that was subsequently extended to provide an account of PTSD (Dalgleish, 2004). The model postulates four levels or formats of mental representation: the schematic, propositional, analogue, and associative representational systems, which provide the basis for the acronym SPAARS. Information at the propositional level is verbally accessible, similar to VAMs, whereas information at the analogue level is stored as images across all sensory systems (including internal sensations), similar to SAMs. Associative representations are described as being similar to the fear structures hypothesized in emotional processing theory, in that they denote that information encoded at one level (or in one format) can be associated with other information in the same or different levels (or formats). Although processing at all levels of representation mutually influence one another, the analogical and propositional systems are limited to relatively basic manipulations of information, whereas the schematic level involves a higher level of mental representation. Specifically, information from the analogical and propositional levels can enter into the schematic level, either directly or through associative links, where it can be integrated and evaluated in light of the person's goals. In addition, processing at the schematic level influences what information is activated or inhibited at the lower levels of representation and serves to filter new information that is consistent with the person's dominant schemas.

In the SPAARS model, there are two routes to the generation of emotion. One is automatic elicitation of emotion through associative representations, similar to the emotional processing theory. The other is the result of an appraisal process at the schematic level. Such appraisals take into consideration expectations of what will happen in light of the person's goals. For example, an appraisal of threat occurs when a person anticipates the disruption or failure to complete a valued goal, generating the emotion of fear. As applied to PTSD, traumatic events represent threats to the goal of personal survival. Similarly, the SPAARS model provides two routes to reexperiencing symptoms following trauma. Reexperiencing may be triggered automatically via associative representations or reflect schema-level processing as the person attempts to assimilate the new information. It is assumed that trauma information is encoded in parallel at the schematic, propositional, and analogical levels at the time of the trauma and its aftermath. Furthermore, associations are formed to connect trauma information within each level and across levels. It is assumed that all this new information is initially unassimilated into the existing memory system, and that recovery involves such assimilation. To account for the finding that not all individuals recover, Dalgleish invokes the construct of pretrauma "personality type" (in terms of pretrauma schemas about the world, the self, and others), hypothesizing five personality types that differ in their outcome following exposure to trauma. The concepts here are quite similar to those developed by Foa and Jaycox (1999). The normative case involves individuals with *balanced* pretrauma schematic representations, in which the world is relatively safe, others are relatively benign, and the self is relatively invulnerable. Individuals with such balanced schematic representations are likely to recover from a trauma without any need for specific intervention, because they are able to assimilate the new information into their existing schemas without requiring excessive accommodation. By contrast, other individuals may hold *overvalued* or inflexible positive schemas. Some of these individuals may have led fairly protected lives, and the overvalued positive schemas result from direct experience and are therefore viewed as *valid* overvalued positive schemas. In this case, the experience of trauma is likely to cause severe and chronic PTSD, in a manner consistent with the shattered assumptions hypothesis. Other individuals, however, may have had prior experience with trauma but hold an *illusory* overvalued positive schema that is maintained by the systematic inhibition of information about their negative experiences. This case is similar to Brewin's premature inhibition of processing concept, which hypothesizes that the inhibitory processes in place prior to the most recent trauma continue to be utilized to minimize acute distress. Instead, they are likely to experience symptoms such as numbing, amnesia, and dissociation. Moreover, these individuals are vulnerable to delayed-onset PTSD should the inhibitory processes fail.

Both of the remaining cases involve preexisting negative schemas about the world, but differ in their schemas regarding the self. Some individuals may hold negative pretrauma schemas about both the world and the self. For such individuals, the experience of trauma is consistent with their existing schemas; therefore, they do not have the appraisal-based reexperiencing symptoms that follow attempts to assimilate the schema-discrepant information. However, they would still experience associatively based reexperiencing symptoms. Alternatively, individuals with negative pretrauma schemas about the world could nonetheless have developed positive pretrauma schemas about the self, such as in the case of emergency responders (police, firemen, paramedics). Although these individuals repeatedly face instances of trauma that support having negative schemas about the world, to be able to perform their job, they are able to maintain positive schemas about themselves, thereby protecting themselves from the development of severe PTSD symptoms. However, if the positive self-schemas are rigid, then such individuals are at risk for developing severe PTSD if faced with a significant discrepancy between their self-schemas and their performance (e.g., suddenly freezing in a critical situation, leading to severe consequences for the self or others).

Critical Comments

The potential benefits of PTSD theories invoking multiple representational structures include greater explanatory power, guidance in the acquisition of new knowledge, and enhancement of treatments. Moreover, the constructs of these models are based on research in cognitive neuroscience, in which knowledge about basic memory processes has the potential to advance our knowledge about psychopathology and its treatment. However, such outcomes are not guaranteed by increased theoretical complexity, and there is a price to pay for such complexity. For example, one of the key insights of schema theory has been the notion that the psychological effects of trauma are the results of extreme violation of preexisting schemas about the world and the self. However, the shattered assumptions hypothesis has difficulty accounting for empirical evidence indicating that prior trauma is a risk factor for developing PTSD on exposure to a subsequent trauma. By contrast, emotional processing theory adequately accounts for how subsequent traumas can strengthen and elaborate existing fear structures, thereby increasing the likelihood of developing PTSD. Thus, it might be assumed that by combining emotional processing theory and schema theory, one automatically increases the range of phenomena that can be explained by the new multirepresentational theory. In fact, emotional processing theory can account for both pathways to PTSD, and the marriage between schema theories and emotional processing theory does not add to the explanatory power of the latter. It is not yet apparent that either dual representation theory or SPAARS can better account for the psychopathology and treatment of PTSD than what can be explained by simpler theories.

Research designed to provide specific empirical tests of the dual representation theory and SPAARS is currently in its infancy, therefore it is too soon to tell the extent to which these theories will yield significant, testable hypotheses , and whether such hypotheses will be supported. Findings from Holmes, Brewin, and Hennessy (2004) with college students and Hellawell and Brewin (2002, 2004) with individuals with PTSD suggest that dual representation theory has the potential to make novel, testable hypotheses that are not readily derivable from other theories. Notably, multirepresentationl theories of PTSD have not yielded significant advances in the efficacy of psychotherapy. In particular, one would expect interventions that separately target verbal representations (e.g., cognitive restructuring) and sensory representations (e.g., imaginal exposure) or situationally accessible memories (e.g., *in vivo* exposure) to result in better outcome. Yet, as note earlier, the addition of cognitive restructuring to a treatment that combined imaginal and *in vivo* exposure failed to improve the latter treatment (Foa et al., 2005; Marks et al., 1998; Paunovic & Ost, 2001).

DISCUSSION

Earlier we proposed that a satisfactory theory of PTSD should account for the psychopathology of the disorder, why some people recover and others do not, and how treatment reduces PTSD symptoms. All the theories reviewed here account for the conscious reexperiencing of the trauma in the form of intrusive, unwanted thoughts or recollections of the traumatic event that are a hallmark of PTSD. Across the different theories, three different mechanisms have been proposed to explain intrusive thoughts. One mechanism, present in all theories except the schema theories, is based on the formation of associations between stimuli present at the time of the trauma and fear responses. Variations on this associative mechanism include associations between trauma reminders and the trauma memory, or the meaning of danger. A second mechanism, emphasized in the schema theories, is that reexperiencing is a result of the discrepancy between current knowledge and prior schemas about safety and competence that persist until the discrepancy has been resolved through assimilation and/or accommodation. The third mechanism, emphasized in the Ehlers and Clark (2000) model and present in the two multirepresentational theories discussed, is through cognitive appraisals of the traumatic event and its sequelae. Emotional processing theory, on the other hand, does not invoke a separate appraisal construct; instead, it incorporates meaning of the event and its sequelae directly into the fear structure.

Each of these theoretical strategies appears able to account adequately for the basic reexperiencing symptoms. A possible exception to the preceding generalization is flashbacks—depending on whether one views flashbacks as being quantitatively or qualitatively different from other reexperiencing symptoms. It would appear that all but the multirepresentational theories assume that flashbacks are simply more intense versions of reexperiencing, and that both types of reexperiencing are explained by the same mechanism. By contrast, both dual representation theory and SPAARS differentiate between these two kinds of reexperiencing and explicitly assume that different memory systems are operating in each. Thus, one important area that would help clarify whether multirepresentational theories provide a better account of PTSD than single-representation theories is how flashbacks and other reexperiencing symptoms differ from another.

Active avoidance of trauma-related thoughts and reminders as a method of coping with the distress associated with reexperiencing is implicitly or explicitly incorporated into all models. Conditioning theory, emotional processing theory, and Ehlers and Clark's (2000) cognitive theory also explicitly acknowledge a second function of avoidance behavior: preventing change in the underlying memory structure that is responsible for the maintenance of PTSD symptoms. This idea is also discussed within multirepresentational theories in their explanation of who recovers from trauma and who does not (e.g., premature inhibition of processing in dual representation theory). Similarly, most theories seem to view numbing/dissociative symptoms of PTSD as a kind of avoidance that reduces acute distress but contributes to the persistence of PTSD. However, few theories provide an explicit account of numbing phenomena. Barlow and Keane's (in Barlow, 2002) model explicitly suggests that the numbing symptoms reflect the avoidance of emotion. This account of the numbing symptoms would appear readily compatible with all the other theories of PTSD discussed in this chapter. However, it is not clear whether such an account addresses all of the phenomena covered under the umbrella of numbing. In an alternative view of numbing advanced by

Litz (1992), numbing is a result of acute reexperiencing symptoms causing a temporary depletion of cognitive and emotional resources. On the basis of an animal model of PTSD, Foa and colleagues (1992) suggested that numbing reflects stress-induced analgesia. At present, none of the current theories provide a satisfactory account for all of the numbing symptoms.

Several theories provided an explanation for the mechanisms underlying recovery and specify who will recover and who will not. Emotional processing theory emphasizes the role of avoidance in preventing exposure to corrective information, whereas Ehlers and Clark's (2000) model places a greater emphasis on the role of cognitive appraisals and the nature of the trauma memory in promoting current symptoms. However, both theories differentiate between trauma victims who have recovered and those who develop persistent PTSD. Multirepresentational theories propose more complex typologies of recovery. Whether these more complex typologies add to the understanding of the mechanism involved in natural recovery awaits further research.

All the theories described in this chapter, with the exception of schema theories, provide an explanation for the mechanisms underlying cognitive-behavioral therapy for PTSD. Conditioning theory emphasizes the role of exposure in modifying pathological associations, and emotional processing theory emphasizes the role of exposure in modifying both maladaptive associations and cognitions. As noted by Dalgleish (2004), one primary aim of the Ehlers and Clark (2000) model is "to provide a theoretical context for the development of a new cognitive behavioral treatment package for PTSD" (p. 241). Although the theory succeeds in achieving this aim, it is inconsistent with studies showing that the addition of cognitive restructuring is superfluous when exposure therapy is provided. The same point may be applied to dual-representation theory and the SPAARS model. Thus, it remains incumbent upon the more complex theories to yield more effective treatment interventions.

REFERENCES

American Psychiatric Association. (1994). *Diagnostic and statistical manual of mental disorders* (4th ed.). Washington, DC: Author.

Amir, N., Stafford, J., Freshman, M. S., & Foa, E. B. (1998). Relationship between trauma narratives and trauma pathology. *Journal of Traumatic Stress, 11,* 385–392.

Barlow, D. H. (1988). *Anxiety and its disorders: The nature and treatment of anxiety and panic.* New York: Guilford Press.

Barlow, D. H. (2002). *Anxiety and its disorders: The nature and treatment of anxiety and panic* (2nd ed.). New York: Guilford Press.

Baum, M. (1970). Extinction of avoidance responding through response prevention (flooding). *Psychological Bulletin, 74,* 276–284.

Beck, A. T. (1972). *Depression: Causes and treatment.* Philadelphia: University of Pennsylvania Press.

Beck, A. T. (1976). *Cognitive therapy and the emotional disorders.* New York: International Universities Press.

Beck, A. T., Emery, G., & Greenberg, R. L. (1985). *Anxiety disorders and phobias.* New York: Basic Books.

Beck, A. T., Rush, A. J., Shaw, B. F., & Emery, G. (1979). *Cognitive therapy of depression.* New York: Guilford Press.

Becker, J. V., Skinner, L. J., Abel, G. G., Axelrod, R., & Chichon, J. (1984). Sexual problems of sexual assault survivors. *Women and Health, 9,* 5–20.

Blanchard, E. B., Kolb, L. C., Gerardi, R. J., Ryan, P., & Pallmeyer, T. P. (1986). Cardiac response to relevant stimuli as an adjunctive tool for diagnosing post-traumatic stress disorder in Vietnam veterans. *Behavior Therapy, 17,* 592–606.

Borkovec, T. D., & Sides, J. K. (1979). The contribution of relaxation and expectancy to fear reduction via graded, imaginal exposure to feared stimuli. *Behaviour Research and Therapy, 17*, 529–540.

Bower, G. H. (1981). Mood and memory. *American Psychologist, 36*, 129–148.

Breslau, N. (1998). Epidemiology of trauma and posttraumatic stress disorder. In R. Yehuda (Ed.), *Psychological trauma* (pp. 1–29). Washington, DC: American Psychiatric Press.

Brewin, C. R. (1989). Cognitive change processes in psychotherapy. *Psychological Review, 96*, 379–394.

Brewin, C. R. (2001). A cognitive neuroscience account of posttraumatic stress disorder and its treatment. *Behaviour Research and Therapy, 39*, 373–393.

Brewin, C. R., Dalgleish, T., & Joseph, S. (1996). A dual representation theory of posttraumatic stress disorder. *Psychological Review, 103*, 670–686.

Brewin, C. R., & Holmes, E. A. (2003). Psychological theories of posttraumatic stress disorder. *Clinical Psychology Review, 23*, 339–376.

Bryant, R. A., & Guthrie, R. M. (2005). Maladaptive appraisals as a risk factor for posttraumatic stress: A study of trainee firefighters. *Psychological Science, 16*, 749–752.

Cahill, S. P., & Foa, E. B. (2004). A glass half empty or half full?: Where we are and directions for future research in the treatment of PTSD. In S. Taylor (Ed.), *Advances in the treatment of posttraumatic stress disorder: Cognitive-behavioral perspectives* (pp. 267–313). New York: Springer.

Creamer, M., Burgess, P., & Pattison, P. (1992). Reaction to trauma: A cognitive processing model. *Journal of Abnormal Psychology, 101*, 425–459.

Dalgleish, T. (1999). Cognitive theories of post-traumatic stress disorder. In W. Yule (Ed.), *Post-traumatic stress disorders: Concepts and therapy* (pp. 193–220). New York: Wiley.

Dalgleish, T. (2004). Cognitive approaches to posttraumatic stress disorder: The evolution of multi-representational theorizing. *Psychological Bulletin, 130*, 228–260.

Dalgleish, T., & Power, M. J. (2004). Emotion specific and emotion-non-specific components of posttraumatic stress disorder (PTSD): Implications for a taxonomy of related psychopathology. *Behaviour Research and Therapy, 42*, 1069–1088.

Dollard, J., & Miller, N. E. (1950). *Personality and psychotherapy: An analysis in terms of learning, thinking, and culture.* New York: McGraw-Hill.

Ehlers, A., & Clark, D. M. (2000). A cognitive model of posttraumatic stress disorder. *Behaviour Research and Therapy, 38*, 319–345.

Ehlers, A., Margraf, J., Roth, W. T., Taylor, C. B., Maddock, R. J., Sheikh, J., et al. (1986). Lactate infusions and panic attacks: Do patients and controls respond differently? *Psychiatry Research, 17*, 295–308.

Ellis, A. (1977). The basic clinical theory of rational–emotive therapy. In A. Ellis & R. Grieger (Eds.), *Handbook of rational-emotive therapy* (pp. 3–34). New York: Springer.

Epstein, S. (1985). The implications of cognitive-experiential self-theory for research in social psychology and personality. *Journal of the Theory of Social Behavior, 15*, 283–310.

Epstein, S. (1991). The self-concept, the traumatic neurosis, and the structure of personality. In D. Ozer, J. M. Healy, Jr., & A. J. Stewart (Eds.), *Perspectives on personality* (Vol. 3, Part A, pp. 63–98). London: Jessica Kingsley.

Foa, E. B., & Cahill, S. P. (2001). Psychological therapies: Emotional processing. In N. J. Smelser & P. B. Bates (Eds.), *International encyclopedia of the social and behavioral sciences* (pp. 12363–12369). Oxford, UK: Elsevier.

Foa, E. B., & Cahill, S. P. (2006). Psychosocial treatments for PTSD: An overview. In Y. Neria, R. Gross, R. Marshall, & E. Susser (Eds.), *9/11: Public health in the wake of terrorist attacks* (pp. 457–474). Cambridge, UK: Cambridge University Press.

Foa, E. B., Ehlers, A., Clark, D., & Tolin, D. F. (1999). Posttraumatic Cognitions Inventory (PTCI): Development and comparison with other measures. *Psychological Assessment, 11*, 303–314.

Foa, E. B., Hembree, E. A., Cahill, S. P., Rauch, S. A., Riggs, D. S., Feeny, N. C., et al. (2005). Randomized trial of prolonged exposure for PTSD with and without cognitive restructuring: Outcome at academic and community clinics. *Journal of Consulting and Clinical Psychology, 73*, 953–964.

Foa, E. B., Huppert, J. D., & Cahill, S. P. (2006). Emotional processing theory: An update. In B. O. Rothbaum (Ed.), *Pathological anxiety: Emotional processing in etiology and treatment* (pp. 3–24). New York: Guilford Press.

Foa, E. B., & Jaycox, L. H. (1999). Cognitive-Behavioral theory and treatment of posttraumatic stress disorder. In D. Spiegel (Ed.), *Efficacy and cost-effectiveness of psychotherapy* (pp. 23–61). Washington, DC: American Psychiatric Press.

Foa, E. B., & Kozak, M. J. (1985). Treatment of anxiety disorders: Implications for psychopathology. In A. H. Tuma & J. D. Maser (Eds.), *Anxiety and the anxiety disorders* (pp. 421–452). Hillsdale, NJ: Erlbaum.

Foa, E. B., & Kozak, M. J. (1986). Emotional processing of fear: Exposure to corrective information. *Psychological Bulletin, 99,* 20–35.

Foa, E. B., & Meadows, E. A. (1997). Psychosocial treatments for post-traumatic stress disorder: A critical review. In J. Spence, J. M. Darley, & D. J. Foss (Eds.), *Annual review of psychology* (Vol. 48, pp. 449–480). Palo Alto, CA: Annual Reviews.

Foa, E. B., Molnar, C., & Cashman, L. (1995). Change in rape narratives during exposure therapy for posttraumatic stress disorder. *Journal of Traumatic Stress, 8,* 675–690.

Foa, E. B., & Riggs, D. S. (1993). Post-traumatic stress disorder in rape victims. In J. Oldham, M. B. Riba, & A Tasman (Eds.), *American Psychiatric Press review of psychiatry* (Vol. 12, pp. 285–309). Washington, DC: American Psychiatric Press.

Foa, E. B., Riggs, D. S., Massie, E. D., & Yarczower, M. (1995). The impact of fear activation and anger on the efficacy of exposure treatment for posttraumatic stress disorder. *Behavior Therapy, 26,* 487–499.

Foa, E. B., & Rothbaum, B. O. (1998). *Treating the trauma of rape: Cognitive-behavioral therapy for PTSD.* New York: Guilford Press.

Foa, E. B., Steketee, G. S., & Rothbaum, B. O. (1989). Behavioral/cognitive conceptualizations of posttraumatic stress disorder. *Behavior Therapy, 20,* 155–176.

Foa, E. B., Zinbarg, R., & Rothbaum, B. O. (1992). Uncontrollability and unpredictability in posttraumatic stress disorder: An animal model. *Psychological Bulletin, 112,* 218–238.

Gilboa-Schechtman, E., & Foa, E. B. (2001). Patterns of recovery after trauma: Individual differences and trauma characteristics. *Journal of Abnormal Psychology, 110,* 392–400.

Hellawell, S. J., & Brewin, C. R. (2002). A comparison of flashbacks and ordinary autobiographical memories of trauma: Cognitive resources and behavioural observations. *Behaviour Research and Therapy, 40,* 1143–1156.

Hellawell, S. J., & Brewin, C. R. (2004). A comparison of flashbacks and ordinary autobiographical memories of trauma: Content and language. *Behaviour Research and Therapy, 42,* 1–12.

Holmes, E. A., Brewin, C. R., & Hennessy, R. G. (2004). Trauma films, information processing, and intrusive memory development. *Journal of Experimental Psychology, 133,* 3–22.

Horowitz, M. J. (1976). *Stress response syndromes.* New York: Aronson.

Horowitz, M. J. (1986). *Stress response syndromes* (2nd ed.). Northvale, NJ: Aronson.

Horowitz, M. J., Wilner, N., & Alvarez, W. (1979). Impact of Event Scale: A measure of subjective distress. *Psychosomatic Medicine, 41,* 209–218.

Janoff-Bulman, R. (1992). *Shattered assumptions: Towards a new psychology of trauma.* New York: Free Press.

Kamin, L. J. (1969). Predictability, surprise, attention, and conditioning. In B. A. Campbell & R. M. Church (Eds.), *Punishment and aversive behavior* (pp. 279–296). New York: Appleton–Century–Crofts.

Keane, T. M., Zimering, R. T., & Caddell, J. M. (1985). A behavioral formulation of posttraumatic stress disorder. *Behavior Therapist, 8,* 9–12.

Kessler, R. C. (2000). Posttraumatic stress disorder: The burden to the individual and to society. *Journal of Clinical Psychiatry, 61*(Suppl. 5), 4–14.

Kessler, R. C., Sonnega, A., Bromet, E., Hughes, M., & Nelson, C. B. (1995). Posttraumatic stress disorder in the National Comorbidity Survey. *Archives of General Psychiatry, 52,* 1048–1060.

Kilpatrick, D. G., Resnick, H. S., & Freedy, J. R. (1992, May). *Post-traumatic stress disorder field trial report: A comprehensive review of the initial results.* Paper presented at the annual meeting of the American Psychiatric Association, Washington, DC.

Kilpatrick, D. G., Veronen, L. J., & Best, C. L. (1985). Factors predicting psychological distress among rape victims. In C. R. Figley (Ed.), *Trauma and its wake* (pp. 113–141). New York: Brunner/Mazel.

Kozak, M. J., Foa, E. B., & Steketee, G. S. (1988). Process and outcome of exposure treatment with obsessive–compulsives: Psychophysiological indicators of emotional processing. *Behavior Therapy, 19,* 157–169.

Lang, P. J., Melamed, B. G., & Hart, J. (1970). A psychophysiological analysis of fear modification using an automated desensitization procedure. *Journal of Abnormal Psychology, 76*, 220–234.

Lepore, S. J., Silver, R. C., Wortman, C. B., & Wayment, H. A. (1996). Social constraints, intrusive thoughts, and depressive symptoms among bereaved mothers. *Journal of Personality and Social Psychology, 70*, 271–282.

Lewinsohn, P. M., Hoberman, H. M., Teri, L., & Hautzinger, M. (1985). An integrative theory of depression. In S. Reiss & R. R. Bootzin (Eds.), *Theoretical issues in behavior therapy* (pp. 331–359). Orlando, FL: Academic Press.

Litz, B. T. (1992). Emotion numbing in combat-related post-traumatic stress disorder: A critical review. *Clinical Psychology Review, 12*, 417–432.

Malik, M. L., Connor, K. M., Sutherland, S. M., Smith, R. D., Davison, R. M., & Davidson, J. R. T. (1999). Quality of life and posttraumatic stress disorder: A pilot study assessing changes in SF-36 scores before and after treatment in a placebo-controlled trial of fluoxetine. *Journal of Traumatic Stress, 12*, 387–393.

Marks, I., Lovell, K., Noshirvani, H., Livanou, M., & Thrasher, S. (1998). Treatment of posttraumatic stress disorder by exposure and/or cognitive restructuring. *Archives of General Psychiatry, 55*, 317–325.

McCann, I. L., & Pearlman, L. A. (1990). *Psychological trauma and the adult survivor: Theory, therapy, and transformation.* New York: Brunner/Mazel.

Mowrer, O. H. (1960). *Learning theory and the symbolic processes.* New York: Wiley.

Overmier, J. B. (1966). Differential transfer of control of avoidance responses as a function of UCS duration. *Psychonomic Science, 5*, 25–26.

Paunovic, N., & Ost, L. G. (2001). Cognitive-behavior therapy vs. exposure therapy in the treatment of PTSD in refugees. *Behaviour Research and Therapy, 39*, 1183–1197.

Piaget, J. (1971). *Psychology and epistemology: Towards a theory of knowledge.* New York: Viking.

Pitman, R. K., Orr, S. P., Altman, B., Longpre, R. E., Poiré, R. E., Macklin, M. L., et al. (1996). Emotional processing and outcome of imaginal flooding therapy in Vietnam veterans with chronic posttraumatic stress disorder. *Comprehensive Psychiatry, 37*, 409–418.

Power, M. J., & Dalgleish, T. (1997). *Cognition and emotion: From order to disorder.* Hove, UK: Psychology Press.

Power, M. J., & Dalgleish, T. (1999). Two routes to emotion: Some implications of multi-level theories of emotion for therapeutic practice. *Behavioural and Cognitive Psychotherapy, 27*, 129–141.

Resick, P. A. (1986). *Reactions of female and male victims of rape or robbery* (Grant No. MH37296, Final Report). Washington, DC: National Institutes of Mental Health.

Resick, P. A., & Schnicke, M. K. (1993). *Cognitive processing therapy for rape victims: A treatment manual.* Newbury Park, CA: Sage.

Riggs, D. S., Rothbaum, B. O., & Foa, E. B. (1995). A prospective examination of symptoms of posttraumatic stress disorder in victims of nonsexual assault. *Journal of Interpersonal Violence, 10*, 201–214.

Rothbaum, B. O., Foa, E. B., Riggs, D. S., Murdock, T., & Walsh, W. (1992). A prospective examination of post-traumatic stress disorder in rape victims. *Journal of Traumatic Stress, 5*, 455–475.

Chapter 5

The Epidemiology of Trauma and PTSD

Fran H. Norris and Laurie B. Slone

Epidemiology is the science concerned with estimating and describing the prevalence and distribution of health and illness in the population. Research on the epidemiology of posttraumatic stress disorder (PTSD) has focused on three interrelated concepts: prevalence of exposure to potentially traumatic events, total prevalence of PTSD in the population, and conditional risk, which is the prevalence of PTSD given exposure. Studies estimate these prevalence rates for a defined period, typically a lifetime, but sometimes for the past year or some other interval of time.

In the last two decades, there has been an explosion of interest in the epidemiology of trauma. As a result, knowledge of the distribution and impact of traumatic events in the population has progressed dramatically. Our primary purpose in this chapter is to review findings regarding the prevalence of trauma exposure and PTSD. Methodological progress and evolving definitions have significantly influenced epidemiological findings; thus, we begin this chapter by reviewing these methodological and definitional changes. We next describe what is currently known about the epidemiology of PTSD based on research results, including what is known about trauma exposure and a review of findings for PTSD, including lifetime, chronic, current, and subsyndromal forms of the disorder. Although epidemiological studies have also been important for identifying risk factors for exposure and PTSD, these findings receive minimal attention in this chapter, because they are addressed by Vogt, King, and King, Chapter 6, this volume. However, sex differences are so pervasive in this area that they must be considered to provide an accurate overview of the epidemiology of trauma and PTSD. In fact, many studies have presented key results separately for men and women. Age and ethnicity are also key demographic variables that receive attention in epidemiological research.

In summarizing findings for trauma exposure and PTSD, we begin with studies of adult men and women in "Western" or developed countries; these studies comprise the bulk of the literature and have shaped mainstream thought. We then examine age trends in the results, then review findings for American minorities and populations in developing or non-Western countries. This organization helps to clarify the generalizability of current findings and to identify populations that have not been studied. We conclude this chapter by speculating on the challenges for future epidemiological research.

METHODOLOGICAL CONSIDERATIONS

Sampling and Representativeness

By definition, epidemiology is concerned with populations, typically (although not always) defined according to geographic boundaries, be it a specific city, state, or nation. Populations may be further defined by sex (e.g., the National Women's Study), ethnicity (e.g., the Mexican American Prevalence Study in Fresno County, California), or other characteristic (e.g., the National Vietnam Veterans Readjustment Survey [NVVRS]). Epidemiological studies are conducted in community rather than clinical samples. Because these studies aim for precision in their estimates, sample sizes are usually quite large relative to other types of trauma research. However, size alone does not make an estimate correct, unless the sample was selected to be representative of the population. Epidemiologists must always be able to identify the group to whom their work generalizes, and to show that their methods yielded a sample that is highly representative of that group.

Unless the population is small enough for all members to participate, epidemiologists usually begin by drawing a random sample of the population of interest. Probability sampling means that each member of the population has a known, nonzero chance of selection. In its purest form (simple random sampling), each member's chance of selection is equal to and independent of any other member's chance, but it is common for these two assumptions of equality and independence to be violated. As long as the probability of selection is known, data can be weighted to correct for unequal or nonindependent probabilities.

Response rates are highly important in epidemiological research, making methods of participant recruitment crucial. When response rates are high, investigators may be quite confident that a probability sample approximates the population. Low response rates undermine the randomness of selection; thus, epidemiologists must attend to these rates carefully and document the extent to which their sample matches the population on key variables available in census data. The technical issues in sampling, sampling error, and sampling bias can become quite complex, making collaboration with experts in survey research important.

For all epidemiological research, including that on trauma and PTSD, a critical challenge is to draw samples that represent the diversity of the population. A number of investigators have argued that health data should be disaggregated by use of subethnic groups (e.g., African Caribbean within the African Americans in the United States) because of considerable differences within groups (e.g., Srinivasan & Guillermo, 2000). For example, whereas Asian Americans as a group may appear similar to European

Americans on a number of health-related and socioeconomic indicators, such statistics disguise higher rates of health problems and poverty among Asian American subgroups, such as the Vietnamese.

Assessment

In addition to sampling, assessment is a critical methodological issue for epidemiology. Because of their large samples, epidemiological studies typically employ structured interviews designed for use by lay interviewers. Ideally, researchers establish through their pilot work that these instruments provide results similar to those provided by clinician-administered instruments. For example, results by Breslau, Kessler, and Peterson (1998) showed that there is good agreement between the PTSD module of the Composite International Diagnostic Interview (CIDI 2.1; World Health Organization, 1993) and clinicians' evaluations.

Even when reliability is high, however, assessment challenges remain. Most often, epidemiologists rely on cross-sectional designs and retrospective reporting of trauma experiences. People are asked to report on events that may have happened long ago, to recount their reactions to them at the time, and to estimate how long these responses lasted. There are inherent limitations in this approach. Recall of objective events may be inaccurate or even biased by the consequences of those events (distressing events are remembered better than others).

For population-based studies, it is essential to use instruments that are cross-culturally valid. Responses to screener items in diagnostic batteries may vary as a function of ethnicity/race, gender, education, and socioeconomic status of the respondent (Alegria & McGuire, 2003), highlighting the importance of more in-depth studies to see how these differences might be understood. Because DSM diagnoses originated in Western psychiatry, they are inherently biased, although the newest revisions address cultural considerations. A strict focus on traditional diagnoses may cause the clinician to miss "culture-bound syndromes" and somaticized distress (Kirmayer, 1996; Norris et al., 2001; Paniagua, 2000). Zheng and colleagues (1997) provided an excellent example of this in their research on *neurasthenia*, a condition that originated in China and is characterized by fatigue or weakness accompanied by an array of physical and psychological complaints, such as diffuse pains, gastrointestinal problems, memory loss, irritability, and sleep problems. In the Zheng and colleagues study, over half of Chinese Americans meeting criteria for neurasthenia did not meet criteria for any DSM-III-R diagnoses. Another example is *ataques de nervios*. In a Puerto Rican disaster study, 14% of the sample reported experiencing these acute episodes of emotional upset and loss of control, although the rate of disaster-specific PTSD according to DSM-III criteria was quite low (Guarnaccia, Canino, Rubio-Stipec, & Bravo, 1993). Specific examples aside, myriad broad issues in cross-cultural assessment, including willingness to disclose and scale equivalence, are beyond the scope of this chapter (Keane, Kaloupek, & Weathers, 1996; Manson, 1997).

Epidemiological findings are profoundly dependent upon assessment strategies. Despite the short history of research on PTSD, we are clearly in our third generation of diagnostic measures, perhaps even our fourth. First-generation measures, perhaps unavoidably, were flawed. The PTSD module included in the original Diagnostic Interview Schedule (DIS) conformed to the third edition of the *Diagnostic and Statistical Manual for Mental Disorders* (DSM-III; American Psychiatric Association, 1980). The DIS first assessed symptoms of posttraumatic stress, then probed for the cause. The Epidemiologic Catchment Area Survey, one of the first epidemiological studies to assess PTSD in the United States, used this module. The study yielded low estimates of PTSD and little data about trauma itself (Davidson, Hughes, Blazer, & George, 1991; Helzer, Robins, & McEvoy, 1987).

The second generation of measures, primarily a revision of the DIS for DSM-III-R (American Psychiatric Association, 1987), began with a single-item screen that provided examples of unusually stressful events that sometimes happen to people. Respondents were asked whether these or similar events had ever happened to them and, if so, they were asked about criterion symptoms that followed the worst and up to three of the events. Studies that used these measures (e.g., Breslau, Davis, Andreski, & Peterson, 1991) appear to have yielded quite reliable estimates of the prevalence of PTSD. At the same time, they appear to have underestimated the prevalence of potentially traumatic events and to have overestimated the conditional risk for PTSD associated with particular events.

Third-generation measures replaced single-item screens with more detailed event inventories. Studies using these measures, most notably the National Comorbidity Survey (NCS; Kessler, Sonnega, Bromet, Hughes, & Nelson, 1995), yielded estimates of the overall prevalence of PTSD that were quite similar to those provided by second-generation measures. However, they yielded higher rates of trauma exposure and, accordingly, lower rates of conditional risk (PTSD rates given exposure). Because symptom questions were anchored to the worst event, estimates of conditional risk, although lower, remained biased.

The Detroit Area Survey of Trauma (Breslau, Kessler, Chilcoat, et al., 1998) inaugurated a fourth generation of measures, applying DSM-IV criteria. This study employed an expanded version of the PTSD Module from the CIDI that corrects for the reporting of multiple traumas associated with the same occasion and provides estimates of PTSD for both the worst event and a randomly selected event, thereby providing unbiased estimates of conditional risk. This module is very long and complex but, as computerized interviewing becomes more common, it may become the standard for epidemiological assessment in the field.

Definitions and Criteria

Interpretation of the database that is emerging over time is further complicated by changes in DSM diagnostic criteria for PTSD. In DSM-III, a *trauma* was defined as a "recognizable stressor that would evoke significant symptoms of distress in almost anyone" (American Psychiatric Association, 1980, p. 238). In 1987, The DSM-III-R definition of trauma was revised to mean an event that is "outside the range of usual human experience and that would be markedly distressing to almost anyone" (American Psychiatric Association, 1987, p. 250). These two definitions were intended to capture catastrophic events that happen with low frequency and to exclude more common events, such as simple bereavement, chronic illness, business loss, and marital conflict. DSM-IV (American Psychiatric Association, 1994, pp. 427–428) defines a *traumatic event* as one in which "(1) the person experienced, witnessed, or was confronted with an event or events that involved actual or threatened death or serious injury, or a threat to the physical integrity of self or others (criterion A1), and (2) the person's response involved intense fear, helplessness, or horror" (criterion A2). Thus, the DSM-IV definition was expanded to include events that would not have been considered in earlier versions because of their frequency, such as sudden and unexpected death of a loved one and life-threatening illness. On the other hand, criterion A2 was added to require that the event be experienced with helplessness, terror, or horror. As Breslau (2002, p. 924) observed, "The DSM-IV revision—the broader range of qualifying traumatic events and the added criterion of a specific emotional response—deemphasizes the objective features of the stressors and highlights the clinical principle that people may perceive and respond differently to outwardly similar events."

Another crucial change in DSM-IV relative to previous definitions of PTSD was that symptoms had to cause significant distress or functional impairment (criterion F). Therefore, although criterion A1 expanded the number of traumatic events included, criteria A2 and F are more stringent. Prevalences estimated on the basis of DSM-IV criteria are lower than those using DSM-III-R criteria (Breslau, 2002).

For estimating exposure to trauma and the conditional risk for PTSD associated with particular events, this chapter relies most heavily on studies that have used third- or fourth-generation measures (i.e., event inventories rather than single-item screens). For estimating the overall prevalence of PTSD, results from second-generation measures are included as well. Although we do not include studies that used first-generation measures, we acknowledge their pioneering contributions to the field by identifying the complex issues surrounding the measurement of trauma and PTSD.

We turn now to discussion of the current status of the literature on the epidemiology of PTSD, beginning with the prevalence of traumatic events. Trauma takes many forms. Categories of traumatic events are not standard across studies, but it is common for investigators to distinguish between violence (including physical and sexual assault and sometimes combat) and other types of trauma (including accidental injuries, natural disasters, and witnessing traumatic events).

CURRENT STATE OF THE ART

Epidemiology of Trauma Exposure

Prevalence and Types of Traumatic Events (Criterion A1)

STUDIES OF ADULT MEN AND WOMEN IN WESTERN/DEVELOPED SOCIETIES

The NCS (Kessler et al., 1995) provided our only estimates of trauma exposure that are based on a nationwide probability sample of adult residents of the United States. Over 2,800 men and 3,000 women, ages 15–54, were interviewed in their homes and asked about 12 specific types of trauma, such as life-threatening accident, sexual assault, sexual molestation, witnessing, fire/disaster, combat, or physical assault. Previous studies had prompted a new understanding of trauma as a frequent rather than rare occurrence (e.g., 69% in Norris, 1992; 69% in Resnick, Kilpatrick, Dansky, Saunders, & Best, 1993), but this study made the point unequivocally: 61% of men and 51% of women (a significant difference) reported at least one DSM-III-R traumatic event during their lives. Among persons exposed to any trauma, multiple traumatization was more common than not. The most prevalent events were witnessing someone being injured or killed (36% men, 15% women), being involved in a fire or natural disaster (19% men, 15% women), and being involved in a life-threatening accident (25% men, 14% women). More women than men reported rape, sexual molestation, sexual assault, and child abuse, but more men than women reported fire/disaster, life-threatening accident, physical assault, combat, being threatened with a weapon, and being held captive.

Creamer, Burgess, and McFarlane (2001) reported similar findings from 10,000 adults who participated in the Australian National Survey of Mental Health and Well-

Being. Using a list of events similar to that used by Kessler and colleagues (1995), Creamer and colleagues estimated that 65% of Australian men and 50% of Australian women had experienced at least one qualifying event over their lives. Again, multiple events were more common than not among adults who had experienced at least one event. These investigators also found witnessing someone being badly injured or killed (38% men, 16% women), life-threatening accidents (28% men, 14% women), and disasters (20% men, 13% women) to be the most prevalent events. Men were more likely than women to experience these three events, physical assault, and combat; women were more likely than men to experience rape and sexual molestation.

Using an expanded DSM-IV inventory of qualifying events, Breslau, Kessler, Chilcoat, and colleagues (1998) found an even higher lifetime prevalence of exposure (90%) in the Detroit Area Survey. In this study, approximately 2,200 adults ages 18–45 were randomly selected and interviewed by telephone. Persons who experienced at least one qualifying event averaged five events over their lifetimes. The most prevalent event was sudden, unexpected death of a love one; 60% of the sample had experienced this event over the course of their lives. More women than men reported rape and sexual assault, but more men than women reported being threatened with a weapon, being shot or stabbed, or being badly beaten up. More men than women experienced other forms of injury or shock, such as accidents or fires, as well. Similarly, Stein, Walker, Hazen, and Forde's (1997) telephone survey of 1,000 randomly selected Canadian adults from Winnipeg yielded prevalence rates for lifetime exposure of 74% of women and 81% of men, with more men (55%) than women (46%) experiencing multiple events.

Research on military populations focuses primarily, although not exclusively, on combat trauma. The 2001 National Survey of Veterans (NSV; U.S. Department of Veterans Affairs, 2003) provided data for over 20,000 veterans living in the United States or Puerto Rico. Across wars and eras, 39% of veterans (41% men, 12% women) reported exposure to combat, and 36% reported exposure to the dead, dying, or wounded. Rates of exposure to combat were 54% for World War II and 19% for Korean Conflict veterans (Spiro, Schnurr, & Aldwin, 1994). War-zone exposure was also reported in the NVVRS (Kulka et al., 1990). Because female military personnel were mostly nurses, war-zone exposure included different criteria for men and women. Of theater veterans, 34% of men reported high as opposed to moderate/low war-zone stress as defined for men, and 39% of women veterans reported high levels of war-zone stress as defined for women.

AGE TRENDS IN EXPOSURE PREVALENCE

On the basis of adults' retrospective reports, Breslau, Kessler, Chilcoat, and colleagues (1998) documented that people's risk for trauma exposure in the Detroit area peaked between the ages of 16 and 20. However, age trends varied across four classes of traumatic events. An age-related decline in risk after the age of 20 was strongest for the category of violence. Exposure to sudden, unexpected death was highest in the 41–45 age group, the oldest group included in the Detroit Area Study. These age trends illustrate the importance of gaining more information about exposure in children and youth. Boney-McCoy and Finkelhor (1995) conducted an important longitudinal study of interpersonal violence in a U.S. national, random telephone sample of 1,042 boys and 958 girls ages 10–16. Completed incidents were reported by 26% of the girls and 44% of the boys. When attempted victimizations are added, these rates increase to 33% and 47%, respectively. Consistent with the adult literature, more boys than girls had experienced physical assaults, and more girls than boys had experienced sexual assaults. Roughly 15 months later, these same young people were reinterviewed (Boney-McCoy & Finkelhor, 1996). Over the interim, 20% of the girls and 22% of the boys had experienced one of the studied events. Similarly, Kilpatrick and colleagues (2003) assessed prevalence of victimization in a random telephone sample of 4,023 U.S. adolescents ages 12–17. Boys had significantly higher lifetime prevalence rates of physical assault and witnessing violence (among boys, 21% and 44%, respectively; among girls, 13% and 35%, respectively), girls had significantly higher rates of sexual assault (13% among girls; 3% among boys), and boys and girls had equal rates of exposure to physically abusive punishment (8.5% for boys; 10.2% for girls). Singer, Anglin, Song, and Lunghofer (1995) likewise found boys to be more exposed to violence than girls in their sample of 3,735 youth ages 14–19 from six U.S. public schools in three cities. Exceptions were victimization at home, sexual assault, and abuse.

Very few studies have examined the prevalence of traumatic events other than violence (e.g., serious accidents, sudden death of loved ones) among youth. Costello, Erkanli, Fairbank, and Angold (2002) examined exposure to potentially traumatic events from middle childhood through adolescence in a general population study of 1,420 youth in western North Carolina. Approximately 25% of this sample had experienced a potentially traumatic event by the age of 16. The most common events were death of a loved one, witnessing a traumatic event, and learning about a traumatic event (each approximately 5%). Perkonigg and Wittchen (1999) interviewed over 3,000 persons ages 14–24 in metropolitan Munich, Germany. They used the CIDI and DSM-IV criteria. Before criterion A2 was taken into account, the prevalence of exposure was 25% for males and 18% for females. Elkit (2002) studied a nationally representative sample of 390 eighth-grade students (ages 13–15) in Denmark using a list of 20 potentially traumatizing events (e.g., threat of physical assault) and distressing events (e.g., having an absent parent). Data were collected by self-report questionnaire. At least one traumatic or distressing event was experienced by 87% of girls and 78% of boys; thus, on the basis of this broader array of events, this study did not replicate the common finding of increased risk for exposure among male respondents. The most common events were death of a family member (52%), threat of physical assault (41%), and accident (24%).

Data are just as meager at the other end of the age continuum. Older people have often been excluded from general population studies of adult mental health. The upper age limit for the NCS was 55, and for the Detroit Area Survey, 45. Their exclusion leaves gaps in our knowledge, because older people constitute a significant, and growing, portion of the population. In the Norris study (1992) of adults in the southeastern United States, the prevalence of past-year exposure was 27% in adults ages 18–39, 21% in adults ages 40–59, and 14% in adults age 60 and older.

TRAUMA EXPOSURE IN ETHNIC MINORITIES AND DEVELOPING OR NON-WESTERN COUNTRIES

The research on the role of ethnicity in trauma has produced inconsistent results. Breslau and colleagues (1991) found no difference in the prevalence of trauma exposure between black (42%) and white participants (39%) in their study of 1,000 young adults from a health maintenance organization (HMO) in Detroit. Likewise, in the Great Smoky Mountains Study in North Carolina, nonwhite youth did not differ from white youth in their risk for exposure (Costello et al., 2002). In the Norris (1992) southeastern U.S. study, white participants had a higher overall lifetime prevalence of exposure (77%) than did black participants (61%). Black participants were neither more nor less likely than white participants to report robbery, sexual assault, fire, motor vehicle crash, or combat in their lifetimes, but white participants were more likely than black participants to report physical assault, disaster exposure, or traumatic bereavement. In contrast, in the Detroit Area Survey, the lifetime prevalence of trauma exposure, especially assaultive violence, was higher among nonwhite participants than among white participants (Breslau, Kessler, Chilcoat, et al., 1998).

Few data exist about the prevalence of trauma exposure in general population samples of Latinos or Hispanics. Norris and colleagues (2003b) estimated the prevalence of exposure to trauma in Mexico by using the CIDI for DSM-IV and a probability sample of 2,509 adults from four cities representing different regions of the country. Lifetime rates of exposure (76% overall, 83% of men, 71% of women) were in the range of previous reports from North America. For the sample as a whole, the most prevalent events were traumatic bereavement (loss of a loved one due to homicide, suicide, or accident), witnessing someone injured or killed, life-threatening accident, and physical assault. A striking 45% of men had experienced at least one form of violence, compared to 27% of women.

Information about the prevalence of trauma exposure in the general population of Asian Americans or Asians is also sparse, especially relative to their proportion of the world's population. The Chinese American Psychiatric Epidemiology Study (CAPES) used the CIDI for DSM-III-R to assess trauma exposure in a sample of 1,747 participants, of whom 95% were immigrants. In this study, 32% of the women and 42% of the men experienced one or more traumatic events. (These findings were made available to Norris, Foster, and Weisshaar [2002] by David Takeuchi, Principal Investigator, National Institutes of Mental Health Grant No. 47460, and his associate, Lisa Tracey.)

Regardless of ethnicity, inner-city residents may be disproportionately exposed to some forms of trauma. National and local statistics reported by Osofsky (1997) suggest that American children and adolescents are exposed to considerable stress in the form of community violence, and that this exposure is related to where they live. Studies examining regional differences in violence exposure have found that the percentage of students witnessing a shooting ranges from 5% in suburban communities to over 50% in large cities. In a sample of 2,248 urban students in sixth through 10th grades, Schwab-Stone and colleagues (1995) found that 46% of boys and 38% of girls had been exposed to some form of community violence.

De Jong and colleagues (2001) studied exposure to trauma in four postconflict, low-income countries (Cambodia, $n = 610$; Algeria, $n = 653$; Ethiopia, $n = 1200$; and Gaza, $n = 585$). Adults were randomly selected from specific communities and interviewed with an adapted version of the Life Events and Social History Questionnaire, as well as the CIDI for DSM-IV. Potentially traumatic events were grouped into five domains: torture, youth domestic stress, death or separation within the family before age 12, and conflict-related events before and after age 12. Trauma exposure varied across countries. The prevalence of torture ranged from 8% (Algeria) to 26% (Ethiopia); youth domestic stress, from 29% (Ethiopia) to 55% (Algeria); death or separation within the family before age 12, from 5% (Gaza) to 18% (Cambodia); conflict events before age 12, from 3% (Cambodia) to 72% (Algeria); and conflict events after age 12, from 59% (Gaza) to 92% (Algeria). These shockingly high rates of severe trauma exposure underscore the importance of conducting epidemiological research in poor and war-torn countries.

Several studies have used epidemiological methods to describe exposure to political violence in countries around the world. Mollica, Poole, and Tor (1998) used multistage area probability sampling from several campsites inhabited by Cambodian refugees. In their sample of nearly 1,000 persons, the authors found that men and women had been exposed to means of 14 and 12 traumatic events, respectively. Ninety-nine percent of their sample reported exposure to at least one war-related trauma, and over 90% had been subject to lack of food or water. In a study of 209 young Cambodian refugees ages 13 to 25 in Portland, Oregon, and Salt Lake City, Utah (Sack et al., 1994), almost 100% had been exposed to atrocities, such as witnessing executions and being separated from family. Similarly high exposure rates have been found among children living in other war-zone areas. Kuterovac, Dyregrov, and Stuvland (1994) found an exposure rate of over 90% in a purposive sample of 134 youth, ages 10–15, living in Croatia. Working in Zenica in central Bosnia, Goldstein, Wampler, and Wise (1995) identified all Bosnian children ages 6–12 living in centers for displaced families with three or more children. These 364 youth had a 100% exposure rate to war violence.

Criterion A2

As noted earlier, DMS-IV introduced a two-part definition of trauma. Criterion A1 described the range of qualifying events, and criterion A2 required that the person's response to that event involve intense fear, horror, or helplessness. A few recent studies have specifically addressed the impact of this subjective criterion on the epidemiology of trauma and PTSD. Using data from the Detroit Area Survey of Trauma and an expanded list of 19 qualifying events, Breslau and Kessler (2001) found that 77% of persons who reported an event met the A2 criterion for that event. Across trauma types, percentages ranged from 34% meeting criterion A2 for military combat, to 93% and 94% for rape and a child's life-threatening illness, respectively. Use of the criterion did not change the overall prevalence of PTSD in their data very much. However, traumatic experiences that did not involve these subjective experiences rarely produced PTSD.

The conditional probability of meeting the A2 criterion did not differ between white (77%) and nonwhite participants (75%) in the Detroit study. For most events for which such comparisons could be made, women were more likely to meet criterion A2 than were men. Accordingly, overall, women were more likely than men to meet the A2 criterion (82 vs. 73%), a finding that represents a reversal of the findings summarized for criterion A1. Perkonigg and Wittchen (1999) reported similar results for their sample of adolescents and young adults in Germany. Of respondents meeting criterion A1, 74% of males and 87% of females met criterion A2. Norris and colleagues (2003b) also found that the relative exposure risks for Mexican men and women reversed when criterion A2 was considered. Of those reporting an event, 73% of men and 80% of women experienced terror, horror, or helplessness.

Summary of Findings about Trauma Exposure

The most general conclusion to be drawn from these data is that exposure to potentially traumatic events (criterion A1) is exceedingly common. By the onset of adulthood, at least 25% of the population will have experienced such an event, and by the age of 45, most of the population will have experienced such an event. A significant subset of the population will experience multiple events. Not all of these events are perceived as traumatic, but on the basis of available data, it appears that a majority of adults will experience an event that involves intense fear, horror, or helplessness (criterion A2) at least once during their lives.

Research also shows quite consistently that men (especially young men) experience potentially traumatic events more frequently than do women. However, of persons that meet criterion A1, women are more likely than men to meet criterion A2. Therefore, overall, men and women (and girls and boys) differ little or not at all in their prevalence of the subjective experience of trauma.

Results for ethnicity have been inconsistent, and much more research on subpopulations is needed to determine whether differential exposure is a reality across cultures and for American minority groups, including immigrant and refugee populations. Clearly, prevalence rates of trauma exposure may be expected to increase dramatically in the context of war, political violence, or community violence.

Epidemiology of PTSD

Lifetime Risk and Conditional Risk for PTSD

STUDIES OF ADULT MEN AND WOMEN IN WESTERN/DEVELOPED SOCIETIES

Estimates of the rate of lifetime PTSD in the U.S. population have been quite consistent since the advent of DSM-III-R. The Detroit HMO study yielded a 9% prevalence (11% women, 6% men) of lifetime DSM-III-R PTSD (Breslau et al., 1991), the National Women's Study (N = ~4,000) yielded a 12% prevalence for lifetime DSM-III-R PTSD (Resnick et al., 1993), the NCS yielded an 8% prevalence (10% women, 5% men) of lifetime DSM-III-R PTSD (Kessler et al., 1995), and the revised NCS-R yielded a 7% prevalence for lifetime DSM-IV PTSD (Kessler, Berglund, Demler, Jin, & Walters, 2005).

Conditional risk is the probability of having PTSD given exposure to a qualifying stressor. In the NCS, 20% of exposed women and 8% of exposed men developed PTSD. On the basis of DSM-IV criteria, the Detroit Area Survey (Breslau et al., 1998a) found the conditional probability of lifetime PTSD to be 13% in women and 6% in men, when estimated on the basis of a randomly selected event, compared to 18% in women and 10% in men, when estimated on the basis of the respondent's worst event. These results confirm suspicions that estimates of conditional risk made on the basis of "most upsetting" events are biased.

Events vary considerably in the probability of precipitating PTSD. Resnick and colleagues (1993) showed that women's rate of PTSD is much higher among crime victims (26%) than among survivors of other types of trauma (9%). In the NCS (Kessler et al., 1995), the event with the highest conditional risk among both men (65%) and women (46%) was rape. Other events associated with a high probability of lifetime PTSD included combat, childhood abuse/neglect, sexual molestation, and physical assault. Accidents, natural disasters, and witnessing were associated with a lower probability of lifetime PTSD. Sexual violence accounted for almost half of cases of PTSD among women, and combat accounted for 29% of cases of PTSD among men. The category of assaultive violence (which includes combat, sexual violence, and physical violence) accounted for almost 40% of PTSD cases in the Detroit Area Survey (Breslau, Kessler, Chilcoat, et al., 1998). Sudden unexpected death accounted for almost 30% of cases, indicating that this event is much more important in the epidemiology of trauma than was previously thought.

The impact of combat on PTSD has been especially well researched. The U.S. Congress mandated the NVVRS in 1983 to estimate the prevalence and effects of PTSD in the Vietnam War veteran population (Kulka et al., 1990). The sample comprised over 2,300 Vietnam War veterans (both those that served in the theater of Vietnam directly and other era veterans) as well as over 600 civilian counterparts. The NVVRS used a composite of measures to diagnose PTSD (Kulka et al., 1990; Schlenger et al., 1992). Lifetime rates among theater veterans were 31% for men and 27% for women. Prevalence of PTSD was higher for those in the Army as opposed to other branches of the military, and diagnoses were more likely for those who served more than 12 months. In addition, entering service between ages 17–19 also increased the likelihood of developing PTSD. In 2006, a reanalysis of the NVVRS data using military records to reduce any recall bias for combat exposure found lifetime PTSD prevalence of 18.7% (Dohrenwend, Turner, Turse, Adams, & Marshall, 2006). Vietnam veterans have also been studied in Australia (Australia Commonwealth Department of Veterans' Affairs, 1998; O'Toole et al., 1996). The prevalence of PTSD in Australian veterans was 19%, as estimated on the basis of the Structured Clinical Interview for DSM-III-R (SCID; Spitzer, Williams, Gibbon, & First, 1990).

AGE TRENDS IN PTSD PREVALENCE

The conditional risk for PTSD, given exposure, may decline modestly as age increases (Kessler et al., 1995; Norris, 1992), and childhood trauma may be especially likely to lead to PTSD. Using the Detroit HMO sample of persons ages 21–30, Breslau, Davis, Andreski, Peterson, and Schulz (1997) estimated the cumulative incidence of PTSD separately for childhood and adulthood trauma. For childhood events (occurring at or before age 15), lifetime conditional prevalences of PTSD were approximately 35% for women and 10% for men. For adulthood events, conditional prevalences were approximately 25% for women and 15% for men. The greater impact of childhood events was not explained by differences in the types of traumas experienced.

Kilpatrick and colleagues (2003) assessed PTSD in a sample of over 4,000 adolescents using a modified version of the DIS for DSM-III-R designed for the National Women's Survey. Lifetime prevalences of PTSD were 10% among girls and 6% among boys. Youth who had experienced multiple sexual assaults were at highest conditional risk for PTSD, with lifetime rates of 34% for girls and 41% for boys. Youth who had experienced multiple physical assaults or abusive punishments were also at high risk, with PTSD prevalences of 40% for girls and 20% for boys. Giaconia and colleagues (1995) estimated the prevalence of PTSD within a general population sample of 194 boys and 190 girls age 18, who had been studied periodically since age 5. They used the DSM-III-R version of the DIS. The lifetime prevalence of PTSD was 11% for girls and 2% for boys. This sex difference is particularly striking, as were sex differences in the conditional probability of PTSD (24% of girls vs. 5% of boys).

In the Munich study (Perkonigg & Wittchen, 1999) lifetime prevalence rates of DSM-IV PTSD were 2.2% for young women (1.1% for those ages 14–17; 2.8% of those ages 18–24) and 0.4% for young men (0.2% for those ages 14–17; 0.5% for those ages 18–24). These rates are lower than the NCS (Kessler et al., 1995) DSM-III-R rates for the cohort ages 15–24 (10% of young women; 3% of young men). Higher rates emerged in the Denmark youth study (Elkit, 2002). There the estimated lifetime prevalence of PTSD was 9%. Again, girls (12%) were more likely than were boys (6%) to meet PTSD criteria.

PTSD PREVALENCE IN ETHNIC MINORITIES
AND DEVELOPING OR NON-WESTERN COUNTRIES

Estimating the relative vulnerability of culturally diverse groups to PTSD is challenging. The NCS did not detect ethnic differences in the prevalence of PTSD (Kessler et al., 1995); nor did Norris (1992) in a survey of black and white residents of four mid-size southeastern cities. The Detroit Area Survey of Trauma (Breslau, Kessler, Chilcoat, et al., 1998) showed blacks to be at increased risk for PTSD relative to whites, but this effect dropped out when central city residence was controlled. CAPES found extraordinarily low rates of PTSD—1.1% of men and 2.2% of women (reported by Norris et al. [2002], with the permission of CAPES investigators).

Within veteran samples, ethnic differences in rates of lifetime PTSD have been reported. A summary of PTSD rates for several minority groups was provided in the Matsunaga Vietnam Veterans Project (National Center for American Indian and Alaska Native Mental Health Research, 1997). Lifetime PTSD rates were 45% for Southwest Plains American Indians, 57% for Northern Plains American Indians, 38% for Native Hawaiians, 9% for Americans of Japanese ancestry, 34% for Hispanics, 35% for blacks, and 20% for whites. Nevertheless, it appears that ethnic differences are largely explained by degree of direct combat exposure (Beals et al., 2002; Friedman, Schnurr, Sengupta, Holmes, & Ashcraft, 2004).

As noted earlier, few data exist on the prevalence of PTSD in general population samples drawn from non-Western or developing countries. Norris and colleagues (2003b) reported a lifetime prevalence rate of 11% for the Mexican adults included in their four-city epidemiological study. This DSM-IV rate was higher than the DSM-IV rate obtained in the Detroit Area Survey (Breslau, Kessler, Chilcoat, et al., 1998) and the DSM-III-R rate obtained in the NCS (Kessler et al., 1995). When criterion F was excluded, the Mexican rate (13%) was 70% higher than the NCS U.S. rate (8%). Consistent with other North American studies, the prevalence of PTSD was approximately twice as high among Mexican women (15%) as among Mexican men (7%). The prevalence of PTSD also varied across cities, being far higher in Oaxaca, the poorest city (17%), than in the other three cities surveyed (9–10%).

De Jong and colleagues (2001) found exceptionally high population rates of lifetime PTSD in their study of four postconflict settings. DSM-IV rates were 16% in Ethiopia, 18% in Gaza, 28% in Cambodia, and 37% in Algeria. These findings are invaluable for showing the relevance of PTSD to understanding the public health of poor, war-torn countries. Sex differences were not consistent in this study: Women had more PTSD than men in Cambodia and Algeria, whereas men had more PTSD than women in Gaza. Conflict-related events after the age of 12 years were most consistently related to PTSD across samples. In Mollica, Poole, and Tor's (1998) sample of nearly 1,000 Cambodian refugees living in camps along the Thai–Cambodian border, rates of PTSD varied: 17% among refugees reporting four or fewer trauma events, increasing to 80% among refugees reporting 25 or more traumatic events.

Chronic and Current PTSD

For research purposes, "chronic PTSD" is usually defined as an episode of PTSD that lasts 1 year or longer. For many people, PTSD is transient, but it fails to remit in more than one-third of individuals who develop it, even after many years (Kessler et al., 1995). Among adults in the NCS who developed PTSD but were not treated, the average duration of the episode was over 5 years; the survival curve declined sharply in the first 12 months but continued to decline gradually for 5 years thereafter (Kessler et al., 1995). In the absence of treatment, the prognosis for recovery is generally considered to be quite poor among persons who continue to meet PTSD criteria 1 or 2 years postevent. Kessler and colleagues (p. 1059), however, challenged this opinion, noting that "even after two years, the average person with PTSD who has not been in treatment still has a 50% chance of eventual remission." The Detroit Area Survey (Breslau, Kessler, Chilcoat, et al., 1998) showed that the median time from onset to remission was 4 years for women, compared to only 1 year for men. Using data from the first wave of the earlier Detroit HMO study of young adults, Breslau and Davis (1992) also found that women were overrepresented among lifetime PTSD cases of more than 1-year duration. Given exposure, 22% of women developed chronic PTSD compared to 6% of men. In the Mexico study (Norris et al., 2003b), 62% of all lifetime cases, or 7% of the total sample (10% women, 4% men), met study criteria for chronic PTSD (all DSM-IV criteria and the symptoms lasting 1 year or longer). Residents of the poorest city, Oaxaca, were twice as likely to suffer from chronic PTSD (12%), as were residents of the other three Mexican cities (5–6%), suggesting that characteristics of the setting may influence the chronicity of PTSD.

Of course, the prevalence of current or recent PTSD is much smaller than the prevalence of lifetime and chronic PTSD. Resnick and colleagues (1993) reported that 5% of the National Women's Study sample had had PTSD within the past 6 months. Kessler, Chiu, Demler, and Walters (2005) estimated 12-month prevalence of DSM-IV PTSD in the United States at 3.5%. Creamer and colleagues (2001) estimated that the 12-month prevalence of DSM-IV PTSD was 1.3% in Australia. Stein and colleagues (1997) estimated the prevalence of current DSM-IV PTSD to be 2.7% for women and 1.2% for men in Winnipeg, Manitoba, Canada. Perkonigg, Kessler, Storz, and Wittchen (2000) found similar, current DSM-IV rates (2.2% for females, 1% for males) in their sample of German adolescents and young adults. Costello and colleagues (1996) found that less than 1% of the children and youth in the Great Smoky Mountains Study had had PTSD within the past 3 months. On the basis of their national survey of adolescents, Kilpatrick and colleagues (2003) reported past 6-month DSM-III-R rates of 6% for girls and 4% for boys.

Members of veteran populations often show higher prevalence of current PTSD than do civilian populations, especially if they were exposed to combat. Although Schnurr and colleagues (Schnurr, Spiro, Aldwin, & Stukel, 1998; Spiro et al., 1994) reported current rates of PTSD for World War II and Korean Conflict veterans at around 1% (3.5% for those highly exposed to combat), the NVVRS (Kulka et al., 1990) determined that 15% of male theater veterans and 9% of female theater veterans had current PTSD. Nontheater veterans had substantially lower prevalences of current PTSD: 2.5% for men and 1.1% for women. NVVRS current rates also varied across ethnic groups: 28% for Hispanics, 21% for blacks, and 14% for whites and others. The 2006 reanalysis reported current prevalence (11–12 years postwar) of PTSD at 9.1% (Dohrenwend et al., 2006). In a representative sample of 15,000 veterans of the Gulf War (Kang, Natelson, Mahan, Lee, & Murphy, 2003), 10% of those deployed as military personnel had current PTSD, compared to 4% of those who were in the service but not deployed at that time. Among those who saw combat in the Gulf, 23% met criteria for current PTSD.

Another exception to the general rule of low rates of current or recent PTSD arises in studies of disasters and other contexts in which the population has shared exposure to an overwhelming and recent stressor. Several disaster studies have used methods consistent with an epidemiological approach (probability sampling, diagnostic measures), and they tend to find abnormally high population rates of current or recent PTSD. A few examples serve to make this point. Shore, Tatum, and Vollmer (1986), who studied 1,025 adults 38–42 months after the volcano erupted at Mt. St. Helens in 1980, presented combined results from the depression, generalized anxiety, and PTSD modules of the DIS for DSM-III, and used date of onset to distinguish between post- and pre-disaster disorders. Among women, 21% of the high-exposure group developed new disorders compared to 6% of the low-exposure group and 2% of the control group. These rates were approximately twice those of men: 11% in the high-exposure group, 3% in the low-exposure group, and 1% of the control group. Hanson, Kilpatrick, Freedy, and Saunders (1995) surveyed 1,200 randomly selected adults in Los Angeles County after the 1992 civil disturbance that began when police officers were acquitted of beating Rodney King. Current prevalence of PTSD for the entire sample was 4%. Galea and colleagues (2002) surveyed 1,000 residents of lower Manhattan 1 month after the terrorist attacks on the World Trade Center. The prevalence of current (past month) PTSD was 7.5%. This prevalence declined to 2% by 4 months postdisaster (Galea et al., 2003). Norris, Murphy, Baker, and Perilla (2004) interviewed 561 adults representative of Teziutlán, Puebla, and Villahermosa, Tabasco 6 months after the devastating 1999 floods and mudslides in Mexico. DSM-IV PTSD for the 6 months following the floods was assessed with the CIDI. At Wave 1, PTSD was highly prevalent (24% combined), especially in Teziutlán (46%), which had experienced mass casualties and displacement. Rates of current PTSD decreased sharply over the next 18 months but remained higher than population norms at 2 years postevent.

Subsyndromal PTSD

Epidemiological studies seldom present data on proportions of the population meeting specific symptom criteria but rather focus on proportions meeting all diagnostic criteria. Nevertheless, a few studies have shown that much larger proportions of the population develop substantial levels of posttraumatic stress that are below diagnostic criteria. For example, Norris (1992) found a 13% prevalence of PTSD among assault victims. Of these assault victims, 68% met criterion B and 53% met criterion D, but only 16% met criterion C (which requires three symptoms). In this group, 73% showed at least one criterion C symptom, and 31% showed two. Across the variety of events studied by Norris, conditional rates of PTSD would have doubled—in some cases, tripled—if two rather than three criterion C symptoms were required. Thus, depending on how it is defined, at least as many persons in the population have severe subsyndromal PTSD as do those who have full PTSD, and possibly many more. Stein and colleagues (1997) addressed the issue of subsyndromal PTSD and its relation to functionality. They defined "full PTSD" as meeting all DSM-IV criteria, and "partial PTSD" as experiencing at least one symptom in each of the three criterion categories of intrusion (B), avoidance/numbing (C), and arousal (D). The authors demonstrated that persons with partial PTSD still show significant decreases in work and school functionality compared to those who have no PTSD. Those with full PTSD, as expected, show even greater interference with functioning than the other groups.

Summary of the Epidemiology of PTSD

It is clear than only a fraction of North Americans (10–20%) who are exposed to trauma develop the full syndrome of PTSD. Thus, despite the high likelihood of trauma exposure (50–90%), the prevalence rates of lifetime PTSD have been in the range of 8–12%. At any given point in time, 1–3% of the civilian population and higher proportions of military populations will have currently active cases. Much larger proportions develop symptoms below criterion level. There are at least one to two times as many persons in the current population with severe subsyndromal PTSD as those with full PTSD. Moreover, it should also be kept in mind that rates that seem fairly low can produce overwhelmingly large numbers when applied to populations. A 2% prevalence of current PTSD in a metropolitan area of 1,000,000 adults yields 20,000 *active* cases presumably in need of treatment.

These data also allow several tentative conclusions about relative risk to be drawn. Violence is the type of trauma most likely to lead to PTSD. The data quite consistently show women and young adults to be at greater conditional risk for PTSD than men and older adults. Much more work is needed to understand the epidemiology of PTSD in childhood. The data do not allow clear conclusions about the relative risks of majority and minority groups in North America or across cultures, but they do suggest that youth and adults from inner-city, violence-infested, or impoverished areas are at greater conditional risk for PTSD than others. The few data that have emerged from developing or non-Western countries suggest that PTSD may be much more prevalent in poor or war-torn countries than elsewhere in the world.

CHALLENGES FOR THE FUTURE

Epidemiology is a relatively mature field, with well-defined purposes and methods. Specific to trauma and PTSD, there are two primary challenges. The first concerns definitions, criteria, and assessment of trauma; the second concerns inclusion of omitted populations.

Assessment Challenges

Assessment issues emerge for both trauma and PTSD. Content validity should receive much more attention than it has in the development of checklists that assess potentially traumatic events. Any list of life events, traumatic or otherwise, is a sample representing a larger population of life events (e.g., Dohrenwend, Krasnoff, Askenasy, & Dohrenwend, 1978). Decisions made in constructing the list ultimately determine the kinds of inferences and generalizations that can be made. Life-event scale developers seldom have described *explicitly* the population of events that the items on their scales purportedly represent. Consensus has not emerged with regard to just where to draw the line between traumatic events and other undesirable events. This is not to say that other events are not important in the lives of individuals, but simply that they are beyond the domain of concern for these measures. This is a critical issue for content validity, which, like construct validity, is often established more on conceptual than on empirical grounds (Wilson, 1994). This review also points out the need for more prospective studies involving PTSD and trauma. Time course analyses are needed, and retrospective reporting biases need to be more thoroughly addressed.

Lack of Diversity

A disappointment in this literature is the lack of attention to diversity in validation samples, and one of the key challenges for epidemiology is cross-cultural assessment of trauma and PTSD. Mollica and colleagues (1995) noted that it is important to *adapt* rather than merely to translate measures for each trauma population and culture. According to Mollica and colleagues, the "core" PTSD section should be kept equivalent across languages, but the remaining symptom questions should vary, so that they are specific and relevant to the culture of respondents. These items should be identified by ethnographic studies, clinical experience, key informants, and healers in the setting of interest.

For this reason and others, the greatest flaw in the epidemiological PTSD literature is the paucity of solid research on non-English-speaking American minorities and persons who live outside of the developed world. The NCS Hispanic, Asian, and Native American samples were small in size, heterogeneous in terms of national origin, and limited to English-speaking persons. Supplementary surveys provided good data for specific subpopulations (i.e., Chinese Americans) but can be generalized past them only with the utmost caution. The results quite obviously do not apply to the various smaller populations of Asian, African, Latino, and European refugees who live in the United States precisely because of violence and trauma in their home countries (e.g., Cervantes, Salgado de Snyder, & Padilla, 1989; Kinzie et al., 1990).

We need improved methods that distinguish between the impact of culture and the influence of minority status or poverty. Data exist that suggest urbanicity may increase risk for trauma exposure and PTSD. One group in particular, the homeless, may be particularly understudied in epidemiological research due to the restriction of probability sampling by household. According to the existing research (Bassuk, Buckner, Perloff, & Bassuk, 1998; Buhrich, Teesson, & Hodder, 2000; Leda, Rosenheck, & Gallup, 1992; North & Smith, 1992; North, Smith, & Spitznagel, 1994; Rosenheck, Frisman, Fontana, & Leda, 1997; Rosenheck, Leda, & Gallup, 1992; Smith, North, & Spitznagel, 1992, 1993), this group is undoubtedly one that experiences high rates of trauma and PTSD, yet most of this research has examined special subgroups of the homeless (e.g., substance abusers, severely mentally ill persons) or has not used epidemiological sampling techniques that allow wider generalizations. The importance of further research on trauma in youth has been delineated, especially due to the severity of impact of traumatic events during this developing life stage. Populations that have experienced political trauma and terror are also understudied.

Public Health Implications

Epidemiological studies typically have a descriptive rather than an explanatory purpose. This area of research has been instrumental in documenting the significance of trauma and PTSD from a public health perspective. The field has shown that proposed solutions to this public health problem must match a stress/trauma process that occurs over time, involving four stages: (1) objective stressors or events, (2) subjective interpretations, (3) acute distress, and (4) chronic disorder. Since the seminal writings of Caplan (1964), population-based solutions have distinguished between primary, secondary, and tertiary prevention. Consequently, it has become traditional to distinguish between interventions that take place before the crisis (primary prevention), during the crisis (secondary prevention), or after the crisis (tertiary prevention). Different approaches may be called for depending on whether the goal is to prevent objective stressors, experienced trauma, acute posttraumatic stress, or chronic disorders. A variety of approaches—ranging from individual psychotherapy to political action—is necessary to tackle the problem at different points (Dohrenwend, 1978; Norris & Thompson, 1995).

Perhaps no single objective would do as much to reduce the prevalence of PTSD in the population as curtailing violence. Whether political or interpersonal, sexual or nonsexual, violence is the single leading cause of PTSD in both men and women. Humans also play a role in causing, and can therefore also play a role in preventing, unintentional trauma, such as that experienced in the context of disasters or accidents.

Even natural disasters are often the result of individual and societal practices that are difficult, but not impossible, to alter.

The second stage in the sequence is the transition from objective stressor to the subjective experience of trauma. Certain events almost uniformly elicit terror, horror, or helplessness. In many ways, these are natural responses to danger that, overall, have been quite adaptive in our species and others. Breslau and Kessler (2001) expressed concern that the A2 criterion confounds objective exposure with risk factors for PTSD (i.e., with characteristics, such as gender, that influence the likelihood of developing PTSD given exposure) and proposed that it might be conceptualized better as a separate criterion describing an acute response rather than as an objective feature of the stressor. Suggesting that people should feel less frightened or horrified in the face of threats to life makes little sense, for that same rush of adrenalin could, in fact, save their lives. The only fruitful possibility here is to empower persons in advance of exposure in the hope of reducing the profound powerlessness that may be elicited by traumatic events (Harvey, 1990).

The next stage is the transition from traumatic stress to acute posttraumatic distress. It has been said often, but bears repeating, that some distress is a normal reaction to abnormal events. Transient stress reactions are not, in themselves, pathological, and most people can and do "get over" stressful events (Dohrenwend, 1978; Norris, Murphy, Baker, & Perilla, 2003a). It is just as accurate to say that 90% of men and 80% of women do *not* develop criterion-level psychiatric problems following trauma exposure, as it is to say that 10% of men and 20% of women do. This observation bears witness to the resilience of most men and most women, but we need to continue to search for ways to bolster and facilitate access to naturally occurring resources.

Of course, our greatest concern is for those individuals who develop chronic, enduring PTSD. It is here that the gender gap grows particularly large, with women being three to four times as likely as men to develop chronic conditions that in all likelihood require medical or psychotherapeutic intervention. Treatments must be sensitive to issues of culture and gender. Advances in treatment are extremely important, but it also must be remembered that remediation constitutes a population-level solution (tertiary prevention) only if conducted on a very large scale (Caplan, 1964).

ACKNOWLEDGMENTS

Preparation of this chapter was funded in part by Grant Nos. K02 MH63909 and R01 MH51278 from the National Institute of Mental Health, Fran H. Norris, Principal Investigator.

REFERENCES

Alegria, M., & McGuire, T. (2003). Rethinking a universal framework in the psychiatric symptom–disorder relationship [Special Issue: Race, Ethnicity and Mental Health]. *Journal of Health and Social Behavior, 44*(3), 257–274.

American Psychiatric Association. (1980). *Diagnostic and statistical manual of mental disorders* (3rd ed.). Washington DC: Author.

American Psychiatric Association. (1987). *Diagnostic and statistical manual of mental disorders* (3d ed., rev.). Washington, DC: Author.

American Psychiatric Association. (1994). *Diagnostic and statistical manual of mental disorders* (4th ed.). Washington, DC: Author.

Australia Commonwealth Department of Veterans' Affairs. (1998). *Morbidity of Vietnam veterans: A study of the health of Australia's Vietnam veteran community: Vol. 1. Male Vietnam veterans: Survey and community comparison outcomes.* Canberra, Australia: Department of Veterans' Affairs.

Bassuk, E. L., Buckner, J. C., Perloff, J. N., & Bassuk, S. S. (1998). Prevalence of mental health and substance use disorders among homeless and low-income housed mothers. *American Journal of Psychiatry, 155*(11), 1561–1564.

Beals, J., Manson, S. M., Shore, J. H., Friedman, M. J., Ashcraft, M., Fairbank, J. A., et al. (2002). The prevalence of posttraumatic stress disorder among American Indian Vietnam veterans: Disparities and context. *Journal of Traumatic Stress, 15*(2), 89–97.

Boney-McCoy, S., & Finkelhor, D. (1995). Psychosocial sequelae of violent victimization in a national youth sample. *Journal of Consulting and Clinical Psychology, 63*(5), 726–736.

Boney-McCoy, S., & Finkelhor, D. (1996). Is youth victimization related to trauma symptoms and depression after controlling for prior symptoms and family relationships?: A longitudinal, prospective study. *Journal of Consulting and Clinical Psychology, 64*(6), 1406–1416.

Breslau, N. (2002). Epidemiologic studies of trauma, posttraumatic stress disorder, and other psychiatric disorders. *Canadian Journal of Psychiatry, 47*(10), 923–929.

Breslau, N., & Davis, G. C. (1992). Posttraumatic stress disorder in an urban population of young adults: Risk factors for chronicity. *American Journal of Psychiatry, 149*(5), 671–675.

Breslau, N., Davis, G. C., Andreski, P., & Peterson, E. L. (1991). Traumatic events and posttraumatic stress disorder in an urban population of young adults. *Archives of General Psychiatry, 48*(3), 216–222.

Breslau, N., Davis, G. C., Andreski, P., Peterson, E. L., & Schultz, L. R. (1997). Sex differences in posttraumatic stress disorder. *Archives of General Psychiatry, 54*(11), 1044–1048.

Breslau, N., & Kessler, R. C. (2001). The stressor criterion in DSM-IV posttraumatic stress disorder: An empirical investigation. *Biological Psychiatry, 50*(9), 699–704.

Breslau, N., Kessler, R. C., Chilcoat, H. D., Schultz, L. R., Davis, G. C., & Andreski, P. (1998). Trauma and posttraumatic stress disorder in the community: The 1996 Detroit Area Survey of Trauma. *Archives of General Psychiatry, 55*, 626–631.

Breslau, N., Kessler, R. C., & Peterson, E. (1998). Posttraumatic stress disorder assessment with a structured interview: Reliability and concordance with standardized clinical interview. *International Journal of Methods in Psychiatric Research, 7*, 121–127.

Buhrich, N., Teesson, M., & Hodder, T. (2000). Lifetime prevalence of trauma among homeless people in Sydney. *Australian and New Zealand Journal of Psychiatry, 34*(6), 963–966.

Caplan, G. (1964). *Principles of preventive psychiatry.* Oxford, UK: Basic Books.

Cervantes, R. C., Salgado de Snyder, V. N., & Padilla, A. M. (1989). Posttraumatic stress in immigrants from Central America and Mexico. *Hospital and Community Psychiatry, 40*(6), 615–619.

Costello, E. J., Angold, A., Burns, B. J., Stangl, D. K., Tweed, D. L., Erkanli, A., et al. (1996). The Great Smoky Mountains Study of youth: Goals, design, methods, and the prevalence of DSM-III-R disorders. *Archives of General Psychiatry, 53*(12), 1129–1136.

Costello, E. J., Erkanli, A., Fairbank, J. A., & Angold, A. (2002). The prevalence of potentially traumatic events in childhood and adolescence. *Journal of Traumatic Stress, 15*(2), 99–112.

Creamer, M., Burgess, P. M., & McFarlane, A. C. (2001). Post-traumatic stress disorder: Findings from the Australian National Survey of Mental Health and Well-Being. *Psychological Medicine, 31*(7), 1237–1247.

Davidson, J. R. T., Hughes, D. C., Blazer, D. G., & George, L. K. (1991). Post-traumatic stress disorder in the community: An epidemiological study. *Psychological Medicine, 21*(3), 713–721.

de Jong, J. T., Komproe, I. H., Van Ommeren, M., El Masri, M., Araya, M., Khaled, N., et al. (2001). Lifetime events and posttraumatic stress disorder in 4 postconflict settings. *Journal of the American Medical Association, 286*(5), 555–562.

Dohrenwend, B. S. (1978). Social stress and community psychology. *American Journal of Community Psychology, 6*(1), 1–14.

Dohrenwend, B. S., Krasnoff, L., Askenasy, A. R., & Dohrenwend, B. P. (1978). Exemplification of a method for scaling life events: The PERI Life Events Scale. *Journal of Health and Social Behavior, 19*(2), 205–229.

Dohrenwend, B. P., Turner, J. B., Turse, N., Adams, B. G., & Marshall, R. (2006). The psychological risks of Vietnam for U.S. veterans: A revisit with new data and methods. *Science, 313*, 979–982.

Elkit, A. (2002). Victimization and PTSD in a Danish national youth probability sample. *Journal of the American Academy of Child and Adolescent Psychiatry, 41*(2), 174–181.

Friedman, M. J., Schnurr, P. P., Sengupta, A., Holmes, T., & Ashcraft, M. (2004). The Hawaii Vietnam Veterans Project: Is minority status a risk factor for posttraumatic stress disorder? *Journal of Nervous and Mental Disease, 192*(1), 42–50.

Galea, S., Ahern, J., Resnick, H. S., Kilpatrick, D. G., Bucuvalas, M. J., Gold, J., et al. (2002). Psychological sequelae of the September 11 terrorist attacks in New York City. *New England Journal of Medicine, 346*(13), 982–987.

Galea, S., Vlahov, D., Resnick, H. S., Ahern, J., Susser, E. S., Gold, J., et al. (2003). Trends of probable post-traumatic stress disorder in New York City after the September 11 terrorist attacks. *American Journal of Epidemiology, 158*(6), 514–524.

Giaconia, R. M., Reinherz, H. Z., Silverman, A. B., Pakiz, B., Frost, A. K., & Cohen, E. (1995). Traumas and posttraumatic stress disorder in a community population of older adolescents. *Journal of the American Academy of Child and Adolescent Psychiatry, 34*(10), 1369–1380.

Goldstein, R. D., Wampler, N. S., & Wise, P. H. (1995). War experiences and distress symptoms of Bosnian children. *Pediatrics, 100*(5), 873–878.

Guarnaccia, P. J., Canino, G., Rubio-Stipec, M., & Bravo, M. (1993). The prevalence of *ataques de nervios* in the Puerto Rico Disaster Study: The role of culture in psychiatric epidemiology. *Journal of Nervous and Mental Disease, 181*(3), 157–165.

Hanson, R. F., Kilpatrick, D. G., Freedy, J. R., & Saunders, B. E. (1995). Los Angeles County after the 1992 civil disturbances: Degree of exposure and impact on mental health. *Journal of Consulting and Clinical Psychology, 63*(6), 987–996.

Harvey, M. (1990, November). *An ecological view of psychological trauma and recovery from trauma.* Paper presented at the International Society of Traumatic Stress Studies, New Orleans, LA.

Helzer, J. E., Robins, L. N., & McEvoy, L. (1987). Post-traumatic stress disorder in the general population: Findings of the epidemiologic catchment area survey. *New England Journal of Medicine, 317*(26), 1630–1634.

Kang, H. K., Natelson, B. H., Mahan, C. M., Lee, K. Y., & Murphy, F. M. (2003). Post-traumatic stress disorder and chronic fatigue syndrome-like illness among Gulf War veterans: A population-based survey of 30,000 veterans. *American Journal of Epidemiology, 157*(2), 141–148.

Keane, T. M., Kaloupek, D. G., & Weathers, F. W. (1996). Ethnocultural considerations in the assessment of PTSD. In A. J. Marsella, M. J. Friedman, E. T. Gerrity, & R. M. Scurfield (Eds.), *Ethnocultural aspects of posttraumatic stress disorder: Issues, research, and clinical applications* (pp. 183–205). Washington, DC: American Psychological Association.

Kessler, R. C., Berglund, P., Demler, O., Jin, R., & Walters, E. E. (2005). Lifetime prevalence and age-of-onset distributions of DSM-IV disorders in the National Comorbidity Survey Replication. *Archives of General Psychiatry, 62*, 593–602.

Kessler, R. C., Chiu, W. T., Demler, O., & Walters, E. E. (2005). Prevalence, severity, and comorbidity of 12-month DSM-IV disorders in the National Comorbidity Survey Replication. *Archives of General Psychiatry, 62*, 617–627.

Kessler, R. C., Sonnega, A., Bromet, E., Hughes, M., & Nelson, C. B. (1995). Posttraumatic stress disorder in the National Comorbidity Survey. *Archives of General Psychiatry, 52*(12), 1048–1060.

Kilpatrick, D. G., Ruggiero, K. J., Acierno, R., Saunders, B. E., Resnick, H. S., & Best, C. L. (2003). Violence and risk of PTSD, major depression, substance abuse/dependence, and comorbidity: Results from the National Survey of Adolescents. *Journal of Consulting and Clinical Psychology, 71*(4), 692–700.

Kinzie, J. D., Boehnlein, J. K., Leung, P. K., Moore, L. J., Riley, C. M., & Smith, D. (1990). The prevalence of posttraumatic stress disorder and its clinical significance among Southeast Asian refugees. *American Journal of Psychiatry, 147*(7), 913–917.

Kirmayer, L. J. (1996). Confusion of the senses: Implications of ethnocultural variations in somatoform and dissociative disorders for PTSD. In A. J. Marsella, M. J. Friedman, E. T. Gerrity, & R. M. Scurfield (Eds.), *Ethnocultural aspects of posttraumatic stress disorder: Issues, research, and clinical applications* (Vol. 22, pp. 131–163). Washington, DC: American Psychological Association.

Kulka, R. A., Schlenger, W. E., Fairbank, J. A., Hough, R. L., Jordan, B. K., Marmar, C. R., et al. (1990). *Trauma and the Vietnam War generation: Report of findings from the National Vietnam Veterans Readjustment Study* (Vol. 29). New York: Brunner/Mazel.

Kuterovac, G., Dyregrov, A., & Stuvland, R. (1994). Children in war: A silent majority under stress. *British Journal of Medical Psychology, 67*(4), 363–375.

Leda, C., Rosenheck, R. A., & Gallup, P. (1992). Mental illness among homeless female veterans. *Hospital and Community Psychiatry, 43*(10), 1026–1028.

Manson, S. M. (1997). Cross-cultural and multiethnic assessment of trauma. In J. P. Wilson & T. M. Keane (Eds.), *Assessing psychological trauma and PTSD* (pp. 239–266). New York: Guilford Press.

Mollica, R., Caspi-Yavin, Y., Lavelle, J., Tor, S., Yang, T., Chan, S., et al. (1995). *Manual for the Harvard Trauma Questionnaire.* Brighton, MA: Indochinese Psychiatry Clinic.

Mollica, R. F., Poole, C., & Tor, S. (1998). Symptoms, functioning, and health problems in a massively traumatized population: The legacy of the Cambodian tragedy. In B. P. Dohrenwend (Ed.), *Adversity, stress, and psychopathology* (pp. 34–51). New York: Oxford University Press.

National Center for American Indian and Alaska Native Mental Health Research. (1997). *Matsunaga Vietnam Veterans Project: Final report.* White River Junction, VT: National Center for PTSD.

Norris, F. H. (1992). Epidemiology of trauma: Frequency and impact of different potentially traumatic events on different demographic groups. *Journal of Consulting and Clinical Psychology, 60*(3), 409–418.

Norris, F. H., Foster, J. D., & Weisshaar, D. L. (2002). The epidemiology of sex differences in PTSD across developmental, societal, and research contexts. In R. Kimerling, P. Ouimette, & J. Wolfe (Eds.), *Gender and PTSD* (pp. 3–42). New York: Guilford Press.

Norris, F. H., Murphy, A. D., Baker, C. K., & Perilla, J. L. (2003a). Severity, timing, and duration of reactions to trauma in the population: An example from Mexico. *Biological Psychiatry, 53*(9), 767–778.

Norris, F. H., Murphy, A. D., Baker, C. K., & Perilla, J. L. (2004). Postdisaster PTSD over four waves of a panel study of Mexico's 1999 flood. *Journal of Traumatic Stress, 17,* 283–292.

Norris, F. H., Murphy, A. D., Baker, C. K., Perilla, J. L., Gutiérrez Rodriguez, F., & Gutiérrez Rodriguez, J. D. J. (2003b). Epidemiology of trauma and posttraumatic stress disorder in Mexico. *Journal of Abnormal Psychology, 112*(4), 646–656.

Norris, F. H., & Thompson, M. P. (1995). Applying community psychology to the prevention of trauma and traumatic life events. In J. R. Freedy & S. E. Hobfoll (Eds.), *Traumatic stress: from theory to practice* (pp. 49–71). New York: Plenum Press.

Norris, F. H., Weisshaar, D. L., Conrad, M. L., Diaz, E. M., Murphy, A. D., & Ibanez, G. E. (2001). A qualitative analysis of posttraumatic stress among Mexican victims of disaster. *Journal of Traumatic Stress, 14*(4), 741–756.

North, C. S., & Smith, E. M. (1992). Posttraumatic stress disorder among homeless men and women. *Hospital and Community Psychiatry, 43*(10), 1010–1016.

North, C. S., Smith, E. M., & Spitznagel, E. L. (1994). Violence and the homeless: An epidemiologic study of victimization and aggression. *Journal of Traumatic Stress, 7*(1), 95–110.

Osofsky, J. D. (1997). *Children in a violent society.* New York: Guilford Press.

O'Toole, B. I., Marshall, R. P., Grayson, D. A., Schureck, R. J., Dobson, M., Ffrench, M., et al. (1996). The Australian Vietnam Veterans Health Study: III. Psychological health of Australian Vietnam veterans and its relationship to combat. *International Journal of Epidemiology, 25*(2), 331–340.

Paniagua, F. A. (2000). Culture-bound syndromes, cultural variations, and psychopathology. In I. Cuellar & F. A. Paniagua (Eds.), *Handbook of multicultural mental health* (pp. 139–169). San Diego, CA: Academic Press.

Perkonigg, A., Kessler, R. C., Storz, S., & Wittchen, H.-U. (2000). Traumatic events and post-traumatic stress disorder in the community: Prevalence, risk factors and comorbidity. *Acta Psychiatrica Scandinavica, 101*(1), 46–59.

Perkonigg, A., & Wittchen, H. U. (1999). Prevalence and comorbidity of traumatic events and posttraumatic stress disorder in adolescents and young adults. In A. Maercker, Z. Solomon, & M. Schützwohl (Eds.), *Post-traumatic stress disorder: A lifespan developmental perspective* (pp. 113–133). Seattle: Hogrefe & Huber.

Resnick, H. S., Kilpatrick, D. G., Dansky, B. S., Saunders, B. E., & Best, C. L. (1993). Prevalence of civilian trauma and posttraumatic stress disorder in a representative national sample of women. *Journal of Consulting and Clinical Psychology, 61*(6), 984–991.

Rosenheck, R. A., Frisman, L., Fontana, A., & Leda, C. (1997). Combat exposure and PTSD among homeless veterans of three wars. In C. S. Fullerton & R. J. Ursano (Eds.), *Posttraumatic stress disor-*

der: Acute and long-term responses to trauma and disaster (pp. 191–207). Washington, DC: American Psychiatric Press.

Rosenheck, R. A., Leda, C., & Gallup, P. (1992). Combat stress, psychosocial adjustment, and service use among homeless Vietnam veterans. *Hospital and Community Psychiatry, 43*(2), 145–149.

Sack, W. H., McSharry, S., Clarke, G. N., Kinney, R., Seeley, J. R., & Lewinsohn, P. (1994). The Khmer Adolescent Project: I. Epidemiologic findings in two generations of Cambodian refugees. *Journal of Nervous and Mental Disease, 182*(7), 387–395.

Schlenger, W. E., Kulka, R. A., Fairbank, J. A., Hough, R. L., Jordan, B. K., Marmar, C. R., et al. (1992). The prevalence of post-traumatic stress disorder in the Vietnam generation: A multimethod, multisource assessment of psychiatric disorder. *Journal of Traumatic Stress, 5*(3), 333–363.

Schnurr, P. P., Spiro, A., Aldwin, C. M., & Stukel, T. A. (1998). Physical symptom trajectories following trauma exposure: Longitudinal findings from the Normative Aging Study. *Journal of Nervous and Mental Disease, 186*(9), 522–528.

Schwab-Stone, M., Ayers, T., Kasprow, W., Voyce, C., Barone, C., Shriver, T., et al. (1995). No safe haven: A study of violence exposure in an urban community. *Journal of the American Academy of Child and Adolescent Psychiatry, 34*(10), 1343–1352.

Shore, J. H., Tatum, E. L., & Vollmer, W. M. (1986). Evaluation of mental effects of disaster, Mount St. Helens eruption. *American Journal of Public Health, 76*(3), 76–83.

Singer, M. I., Anglin, T. M., Song, L. Y., & Lunghofer, L. (1995). Adolescents' exposure to violence and associated symptoms of psychological trauma. *Journal of the American Medical Association, 273*(5), 477–482.

Smith, E. M., North, C. S., & Spitznagel, E. L. (1992). A systematic study of mental illness, substance abuse, and treatment in 600 homeless men. *Annals of Clinical Psychiatry, 4*(2), 111–120.

Smith, E. M., North, C. S., & Spitznagel, E. L. (1993). Alcohol, drugs, and psychiatric comorbidity among homeless women: An epidemiologic study. *Journal of Clinical Psychiatry, 54*(3), 82–87.

Spiro, A., Schnurr, P. P., & Aldwin, C. M. (1994). Combat-related posttraumatic stress disorder symptoms in older men. *Psychology and Aging, 9*(1), 17–26.

Spitzer, R. L., Williams, J. B. W., Gibbon, M., & First, M. B. (1990). *User's guide for the Structured Clinical Interview for DSM-III-R: SCID.* Washington, DC: American Psychiatric Association.

Srinivasan, S., & Guillermo, T. (2000). Toward improved health: Disaggregating Asian American and Native Hawaiian/Pacific Islander data. *American Journal of Public Health, 90*(11), 1731–1734.

Stein, M. B., Walker, J. R., Hazen, A. L., & Forde, D. R. (1997). Full and partial posttraumatic stress disorder: Findings from a community survey. *American Journal of Psychiatry, 154*(8), 1114–1119.

U.S. Department of Veterans Affairs. (2003). *2001 National Survey of Veterans, Final Report.* Washington, DC: Author.

Wilson, J. (1994). The historical evolution of PTSD diagnostic criteria: From Freud to DSM-IV. *Journal of Traumatic Stress, 7,* 681–689.

World Health Organization. (1993). *Composite International Diagnostic Interview.* Geneva: Author.

Zheng, Y. P., Lin, K. M., Takeuchi, D., Kurasaki, K. S., Wang, Y., & Cheung, F. (1997). An epidemiological study of neurasthenia in Chinese-Americans in Los Angeles. *Comprehensive Psychiatry, 38*(5), 249–259.

Chapter 6

Risk Pathways for PTSD

MAKING SENSE OF THE LITERATURE

Dawne S. Vogt, Daniel W. King, and Lynda A. King

Much of the early research on posttraumatic stress disorder (PTSD) was based on the assumption that PTSD is a natural consequence of trauma exposure. However, a growing body of research indicates that many individuals exposed to traumatic events do not develop PTSD, or they recover quickly from stress symptomatology experienced in the immediate aftermath of trauma exposure (Brewin, Andrews, & Valentine, 2000). This has fostered the recognition that some people may be more vulnerable to the effects of trauma and resulted in a large literature aimed at elucidating risk factors for PTSD.

After several decades of research, it has become evident that there is no "magic bullet" to explain who will and will not develop PTSD. Instead, researchers have amassed evidence for a number of risk factors for PTSD following trauma exposure (Brewin et al., 2000; Ozer, Best, Lipsey, & Weiss, 2003). Nevertheless, the literature is rife with conflicting results, with researchers finding evidence for some risk factors in some studies and failing to support the same risk factors in other studies (Bremner, Southwick, & Charney, 1995; Brewin et al., 2000; Creamer & O'Donnell, 2002). Relatedly, there is growing evidence that associations between risk factors and PTSD may vary depending on the particular population under study or as a function of other study attributes (Brewin et al., 2000; Kazdin, Kraemer, Kessler, Kupfer, & Offord, 1997). These findings suggest that PTSD is unlikely to have a single cause but may instead have multiple causal pathways.

Our goal in this chapter is to facilitate clearer thinking about how risk factors may work together to influence PTSD following trauma exposure. This we accomplish by applying a risk factor framework proposed by Kraemer and her colleagues (Kraemer et al., 1997; Kraemer, Stice, Kazdin, Offord, & Kupfer, 2001) to the literature on risk factors for PTSD. The goal of Kraemer and colleagues' framework is to provide a common language for thinking and communicating about risk factors that draws attention to potential pathways through which risk factors may influence outcomes and is responsive to the limitations researchers often face in their ability to draw causal inferences from study findings. Throughout this discussion, we highlight the advantages of longitudinal, and especially experimental, designs in advancing knowledge about possible causal mechanisms.

We begin by providing a brief overview of the PTSD risk literature to emphasize risk factors that have received the greatest amount of theoretical and empirical attention. We then present Kraemer and colleagues' (1997) risk factor classification system and apply this system to risk factors for PTSD. After that, we describe five different ways that risk factors may work together, as proposed by Kraemer and colleagues (2001), and offer examples for each proposed pathway from the PTSD risk literature.

CURRENT STATE OF THE ART: LITERATURE ON PSYCHOSOCIAL RISK FACTORS FOR PTSD

Risk factors for PTSD may be classified into psychosocial, genetic, and biological categories. The focus of this chapter is psychosocial risk factors. We base our brief overview on our earlier summary of this literature (King, Vogt, & King, 2004), as well as two meta-analyses of risk factors for PTSD (Brewin et al., 2000; Ozer et al., 2003).

Psychosocial risk factors may be categorized into features of the traumatic event, preexisting attributes or experiences of the trauma victim, and posttrauma circumstances. The majority of studies have quite understandably focused on identifying features of the traumatic event that make PTSD more or less likely. A number of investigators have found evidence for a dose–response relationship between the severity of a traumatic event and PTSD (e.g., Fairbank, Keane, & Malloy, 1983; Foy, Carroll, & Donahoe, 1987; March, 1993; Rodriguez, van de Kemp, & Foy, 1998), and this is consistent with the results of Brewin and colleagues' (2000) meta-analysis, which documented a modest association between event severity and PTSD. It is also consistent with Ozer and colleagues' (2003) finding that the degree of "life threat" experienced during a traumatic event demonstrated a modest association with PTSD.

Several researchers have elaborated on the severity domain, finding evidence that traumatic events that involve injury (e.g., Acierno, Resnick, Kilpatrick, Saunders, & Best, 1999; Green, 1990, 1993; Green, Grace, & Gleser, 1985; March, 1993); that are more malicious and grotesque events (e.g., Gallers, Foy, Donahoe, & Goldfarb, 1988; Green et al., 1985; Kessler, Sonnega, Bromet, Hughes, & Nelson, 1995; Laufer, Gallops, & Frey-Wouters, 1984), in which one is actively involved rather than merely a witness (e.g., Breslau & Davis, 1987; Laufer et al., 1984; Lund, Foy, Sipprelle, & Strachan, 1984); that involve subjective distress (e.g., King, King, Gudanowski, & Vreven, 1995; Solomon, Mikulincer, & Hobfoll, 1987) and especially dissociation at the time of the trauma (e.g., Bremner & Brett, 1997; Ozer et al., 2003; Shalev, Peri, Canetti, & Schreiber, 1996); and that are accompanied by lower magnitude stressors (e.g., King et al., 1995) are most likely to lead to a dysfunctional response. Ozer and colleagues (2003) found especially strong evidence for the role of peritraumatic dissociation in determining who develops PTSD in the aftermath of trauma exposure. Among all of the risk factor categories examined in their meta-analysis, this factor demonstrated the strongest association with

PTSD, with an average weighted effect size of $r = .35$. It is interesting to note that this effect size, while the most powerful predictor of PTSD, is still not especially large, underscoring the importance of looking beyond aspects of the traumatic event to other factors that may increase risk for PTSD. As a growing body of research indicates, trauma exposure is a necessary, but not sufficient, condition for PTSD (Creamer & O'Donnell, 2002; King et al., 2004; Resick, 2001).

To this end, researchers have also examined the role that preexisting attributes or experiences play in PTSD. With regard to preexisting attributes, there is evidence that female gender, younger age at the time of trauma exposure, lower socioeconomic status (SES), lower education, lower intelligence, and minority racial/ethnic status may serve as risk factors for PTSD in trauma-exposed individuals (Brewin et al., 2000; King et al., 2004). Yet, these influences generally appear to be quite modest, with effect sizes ranging from $r = .06$ to $r = .18$ in Brewin and colleagues' (2000) meta-analysis.

Another preexisting attribute that may increase vulnerability to PTSD is one's psychiatric history; in both Brewin and colleagues' (2000) and Ozer and colleagues' (2003) meta-analyses this factor demonstrated a modest association with PTSD. A number of preexisting experiences also appear to put individuals at risk for developing PTSD. As several researchers have noted, exposure to prior trauma can increase the likelihood that one will exhibit posttraumatic symptomatology in response to a later trauma (Andrykowski & Cordova, 1998; King, King, Foy, & Gudanowski, 1996; King, King, Foy, Keane, & Fairbank, 1999; Koopman, Classen, & Speigel, 1994; Moran & Britton, 1994; Peretz, Baider, Ever-Hadani, & De-Nour, 1994; van der Kolk & Greenberg, 1987), although there is also a small literature suggesting that prior trauma may inoculate individuals against later trauma (e.g., Bolin, 1985; Burgess & Holmstrom, 1979; Cohen, 1953; Dougall, Herberman, Inslicht, Baum, & Delahanty, 2000; Norris & Murrell, 1988; Quarantelli, 1985; Warheit, 1985). In particular, researchers have attended to the effects of childhood sexual and physical abuse (e.g., Andrews, Brewin, Rose, & Kirk, 2000; Bremner, Southwick, Johnson, Yehuda, & Charney, 1993), as well as other adverse childhood experiences, such as family instability and poor family functioning (Fontana & Rosenheck, 1994; King et al., 1996, 1999). Relatedly, researchers have demonstrated that family psychiatric history may serve as a risk factor for PTSD among trauma-exposed individuals (e.g., Breslau, Davis, Andreski, & Peterson, 1991; Bromet, Sonnega, & Kessler, 1998; Emery, Emery, Sharma, Quiana, & Jassani, 1991). Findings from both Brewin and colleagues' and Ozer and colleagues' meta-analyses of the literature revealed slightly higher, albeit quite modest, associations for prior trauma exposure compared with the preexisting attributes described previously (e.g., gender, age at time of trauma). Effect sizes corresponding to these results averaged between $r = .12$ and $r = .21$.

With respect to posttrauma characteristics, researchers have generally focused on two categories that may increase the likelihood of an adverse response to trauma: lack of social support and exposure to additional life stressors. It is well-established that a lack of social support is a risk factor for PTSD among individuals exposed to trauma (e.g., Egendorf, Kadushin, Laufer, Rothbart, & Sloan, 1981; Keane, Scott, Chavoya, Lamparski, & Fairbank, 1985; King et al., 1999; Solomon & Mikulincer, 1990; Solomon, Mikulincer, & Avitzur, 1988; Solomon, Mikulincer, & Flum, 1989), and social support demonstrated the strongest association with PTSD in Brewin and colleagues' (2000) meta-analysis, with an effect size of $r = .40$ (with this variable scored as the presence of social support). Ozer and colleagues (2003) also found a moderate, albeit slightly weaker, association between social support and PTSD ($r = .28$). Findings regarding the impact of exposure to additional life stressors suggest that this is also a risk factor for PTSD (Brewin et al., 2000; King, King, Fairbank, Keane, & Adams, 1998). Although not included in Ozer and colleagues' meta-analysis, additional life stressors demonstrated the second highest effect size (r = .32) in Brewin and colleagues' meta-analysis.

Importantly, a growing body of literature suggests that the risk factors for the acquisition of PTSD may differ from the risk factors for the maintenance of PTSD. For example, in one study that examined risk factors for PTSD among Vietnam veterans, Schnurr, Lunney, and Sengupta (2004) found that the development of PTSD was related to factors occurring before, during, and after a traumatic event, whereas maintenance of PTSD was related to factors occurring only during and after the event. Another study of Vietnam veterans revealed that high combat exposure, perceived negative homecoming reception, and higher depression and anger in the postdeployment period predicted both the development and course of PTSD. Discomfort in disclosing Vietnam experiences uniquely predicted the development of PTSD, whereas minority status and less community involvement uniquely predicted the course of PTSD (Koenen, Stellman, Stellman, & Sommer, 2003). A study of cognitive risk factors for PTSD revealed that appraisals of the assault, appraisals of the sequelae of the assault, dysfunctional strategies, and global beliefs were associated with both the onset and maintenance of PTSD, whereas only detachment during assault, failure to perceive positive responses from others, and mental undoing were uniquely related to the onset of PTSD (Dunmore, Clark, & Ehlers, 1999). These results suggest the need for additional attention in future studies to differential risk factors for the development and maintenance of PTSD.

METHODOLOGICAL CONSIDERATIONS: APPLICATION OF THE KRAEMER ET AL. RISK FACTOR FRAMEWORK TO THE PTSD RISK LITERATURE

As this review of the literature demonstrates, there are a number of well-documented risk factors for PTSD following trauma exposure. However, as Kraemer and her colleagues (1997, 2001) have noted, the accumulation of a laundry list of risk factors does little either to increase our understanding of etiological processes underlying associations with outcomes or to inform decision making about how interventions can be optimally timed, constructed, and delivered to prevent or treat mental disorders. Thus, to advance the literature, it is necessary to shift research attention from the question of "What are the risk factors for PTSD?" to the question of "What are the pathways through which risk factors are associated with PTSD?" This transition requires a clear understanding of both the different types of risk factors that may be implicated in the development of PTSD and the different pathways through which they may have their impact.

Below, we briefly review Kraemer and colleagues' (1997) framework for categorizing risk factors and provide examples of each type of risk factor from the previously summarized literature. We then present five ways that variables may work together, as proposed by Kraemer and her colleagues (2001), describe the conditions that must be met to make a strong case for each scenario, and provide examples of each possible scenario from the PTSD risk literature. It is our hope that these examples suggest avenues for future investigations in the field.

Risk Factor Terminology

Kraemer and her colleagues (1997) have argued that an adequate understanding of the role of any given factor in increasing or decreasing risk for an outcome is necessary for valid causal inference, scientific communication, and appropriate clinical and policy applications. According to these researchers, risk factors can be grouped into several categories that reflect the current state of knowledge regarding their causal role in the outcome. According to the terminology used by Kraemer and colleagues and in the context of PTSD risk research, a factor that demonstrates a positive association with PTSD (i.e., covariation between the putative causal agent and the outcome) in a sample of individuals who have experienced a traumatic event may or may not qualify as a risk factor. If temporal precedence cannot be affirmed—that is, if researchers cannot demonstrate that the correlate precedes the outcome—then the correlate should be termed either a "concomitant" or a "consequence." To the extent that temporal precedence can be demonstrated, as might be the case for a study employing a longitudinal or experimental design, the factor can be appropriately labeled a "risk factor." Once risk factor status is established, several additional distinctions are useful (Kraemer et al., 1997). Risk factors that do not vary within individuals over time, and that cannot be altered to affect a change in an outcome, are termed "fixed markers." By definition, fixed markers are considered antecedents to outcomes such as PTSD. On the other hand, "variable risk factors" are factors that either change within individuals naturally over time or can be manipulated in some way. To the extent that a variable risk factor is manipulable, and when manipulated results in a change in an outcome, Kraemer and colleagues assert that then, and only then, can one claim "causal risk factor" status.

It is important to note that even when a risk factor meets criteria for causal risk factor status, there may be uncertainty regarding the actual causal mechanism underlying the observed relationship. Kraemer and colleagues (1997) give the example of unsafe sex practices as a risk factor for AIDS. Even if one is able to demonstrate that unsafe sex is a causal risk factor for AIDS, the actual mechanism accounting for the relationship is likely to be the transmission of the human immunodeficiency virus (HIV). In addition, causality is a probabilistic concept and as such, one can never be sure of the true causal mechanism underlying associations (see King & King, 1991). Even with random assignment, there is still the chance that the treatment group may a priori differ systematically on a factor related to the outcome.

Although the assumption typically underlying the presentation of proposed risk factors for PTSD appears to be that these variables represent causal risk factors (i.e., that they effect some change in the risk for PTSD), the majority of risk factors put forth in the literature are unlikely to meet Kraemer and colleagues' (1997) requirements to be classified as causal risk factors. In fact, it is likely that a number of commonly asserted risk factors do not even meet criteria for risk factor status, and should more appropriately be classified as concomitants or consequences of PTSD. Any proposed risk factor that has only been evaluated in cross-sectional studies, and in which one therefore cannot establish temporal precedence (or that does not precede PTSD by definition; e.g., female gender or other such fixed markers) would most appropriately be classified as concomitants or consequences of PTSD. For example, we are unaware of any study that has examined associations between poor childhood family functioning and adult-onset PTSD using a longitudinal design that involves following individuals from childhood to adulthood. Thus, the possibility remains that reports of childhood family functioning are a consequence of PTSD rather than a risk factor for PTSD. It is well known that current psychiatric status can impact reporting of earlier events or circumstances (Brewin et al., 2000; King et al., 1996), and this would certainly be a concern for studies requiring adult respondents to provide ratings of childhood experiences.

Similarly, to the extent that studies examining lack of social support as a risk factor for PTSD are based on cross-sectional rather than longitudinal or experimental designs, the possibility that PTSD results in a reduction of available social support rather than vice versa (i.e., individuals with less social support are more at risk for PTSD) cannot be dismissed. In fact, a recent study by two of the authors of this chapter provides some support for this possibility. Specifically, King, Taft, King, Hammond, and Stone (2006) examined longitudinal associations between social support and PTSD in a sample of Gulf War I veterans. The goal of this study was to determine whether social support is a risk factor for PTSD, and can thus be demonstrated to precede PTSD, or a consequence of PTSD, such that the development of PTSD leads to a depletion in social support over time. Their findings revealed a moderately strong negative relationship between PTSD at Time 1 and social support at Time 2, and no association between social support at Time 1 and PTSD at Time 2, suggesting that, at least in this case, social support was more appropriately classified as a consequence than as a risk factor for PTSD.

Other factors from the literature may be best considered concomitants rather than risk factors for PTSD. These factors may coincide with PTSD but neither precede nor result from PTSD. An example of a variable that is often comorbid with PTSD is depression. A great deal of debate has centered on the relationship between depression and PTSD (Erickson, Wolfe, King, King, & Sharkansky, 2001; Kessler, 1997). Yet the majority of studies thus far have been cross-sectional, limiting conclusions about the temporal precedence of depression and PTSD. Findings from a prospective analysis of PTSD and depression symptomatology among Gulf War I veterans found a reciprocal relationship, suggesting that depression may both precede and follow from PTSD (Erickson et al., 2001). Yet the causal link is unclear: Is it that PTSD causes depression, and depression in turn causes PTSD, or are they both simply co-occurring consequences of a third variable, trauma exposure? Perhaps a more clear-cut example of a possible concomitant of PTSD would be posttraumatic growth. Several studies have suggested that traumatic experiences may result in both negative consequences, such as PTSD, and positive consequences, such as posttraumatic growth; thus, these variables may coincide (Lev-Wiesel & Amir, 2003; Salter & Stallard, 2004). At the same time, there is little reason to expect that posttraumatic growth would lead to PTSD or vice versa.

A number of the risk factors reviewed in our overview of the PTSD risk literature represent factors that do not vary within individuals over time, or as Kraemer and colleagues (1997) term them, fixed markers. Specifically, demographics such as female gender, age at trauma exposure, and race/ethnicity are good examples of fixed markers for PTSD. To the extent that intelligence and SES are stable over time, they may also be classified as fixed markers for PTSD. These factors are useful for identifying who may be most likely to develop the condition as a consequence of trauma exposure, but they cannot be manipulated. On the other hand, factors that would be classified as variable risk factors are those that are associated with PTSD, that can change within an individual over time, but that have not been (and may never be) demonstrated to effect a change in PTSD.

Whether a factor is considered a fixed marker or a variable risk factor depends largely on the population under study (H. Kraemer, personal communication, May 9, 2005). If the population under study includes victims of trauma exposure, as is often the case in PTSD risk research, key aspects of trauma exposure (e.g., trauma severity, peritraumatic dissociation), as well as a number of pretrauma characteristics (e.g., prior life trauma, family psychiatric history), can be considered fixed markers. On the other hand, if the study is prospective and the population under study is military personnel who may or may not be deployed to war and possibly exposed to trauma, variables such as severity of trauma exposure may instead be classified as variable risk factors. In this case, the severity of trauma exposure may change over time (e.g., trauma exposure might change from "none" to "some" for an individual deployed to a region of potentially hazardous duty) or be manipulated (e.g., unit commanders could decide to reposition military troops to a location where trauma exposure is more or less likely).

Taking this example a step further illustrates an important distinction between variable risk factors and causal risk factors. Recall that to be a candidate for causal risk factor status a variable must be demonstrated to effect a change in the outcome. Yet a number of the variables of most interest to trauma researchers may not be easily manipulable. For example, although it is theoretically possible to manipulate trauma severity to study its effect on PTSD, there are serious ethical and practical constraints associated with doing so. Imagine the ethical issues that would arise from randomly assigning individuals to trauma exposure, never mind the difficulty of getting Institutional Review Board approval for such a study! To the extent that such studies are not done, we are necessarily limited in the conclusions we can draw about trauma severity as a causal risk factor for PTSD. Another potential variable risk factor for PTSD that may not be easily manipulable is peritraumatic dissociation. A number of studies have found that peritraumatic dissociation predicts later symptoms of PTSD (e.g., Birmes et al., 2003; Ehlers, Mayou, & Bryant, 1998; Shalev et al., 1996). For a sample of individuals assessed prior to assault, dissociation may be considered a variable risk factor. However, unless, or until, researchers are able to manipulate peritraumatic dissociation experimentally, this variable will not qualify for causal risk factor status.

A more promising candidate for causal risk factor status, based on our ability to manipulate it experimentally, may be social support after trauma exposure. To the extent that social support can be manipulated such that individuals are randomly assigned either to receive or not receive an infusion of social support after experiencing a traumatic event, and results indicate that individuals who receive social support show a decrease in PTSD relative to those who do not, one could conclude that social support is a causal risk factor for PTSD. Although few interventions aimed at enhancing social support for trauma victims have undergone rigorous empirical evaluation, the literature on the effectiveness of cognitive-behavioral therapies could be relevant to the extent that therapy is viewed as a special form of social support. The results of treatment–outcome studies suggest that experimentally manipulating exposure to one or more forms of cognitive-behavioral therapy results in a reduction in PTSD (Creamer & O'Donnell, 2002; Resick, Nishith, Weaver, Astin, & Feuer, 2002). Thus, one might conclude that the lack of this form of "social support" is a causal risk factor for PTSD. Similarly, to the extent that efforts aimed at reducing trauma victims' exposure to additional stressors that often follow exposure to traumatic events were to result in a decrease in PTSD, one could conclude that additional life stressors are a causal risk factor for PTSD. Unfortunately, few rigorous experimental designs have yet to be applied to evaluate the impact of early intervention efforts of this nature.

We conclude our discussion of risk factor terminology by emphasizing to both researchers and research consumers that the majority of identified risk factors for PTSD cannot be demonstrated to be causal risk factors. Thus, conclusions regarding their causal role in PTSD are premature. Moreover, as we discussed previously, even when causal risk factor status is confirmed, uncertainty regarding the precise causal mechanisms underlying associations may remain. In addition, a number of so-called "risk factors" for PTSD may not be risk factors at all, but may instead be better classified as concomitants or consequences of PTSD. Finally, some of the most studied risk factors in the PTSD literature (e.g., gender, age at time of exposure, and for many populations, even severity of trauma exposure) are best considered fixed markers that identify individuals who may be at risk for PTSD but cannot, according to Kraemer and colleagues' (1997) framework, be labeled as causal risk factors.

These observations have clear implications for practice, because interventions that focus on reducing risk by effecting a change in risk factors for which evidence for causality is weaker (e.g., variable risk factors) may be less useful than efforts aimed at manipulating factors for which evidence for causality is stronger (e.g., causal risk factors). Similarly, efforts to manipulate variables that are better considered concomitants or consequences of PTSD are quite simply a waste of time, whereas fixed markers may be best used for the purpose of identifying those who are most vulnerable to PTSD and most likely to need PTSD support services in the aftermath of trauma exposure. Additional discussion of the implications of Kraemer and colleagues' (1997) taxonomy of risk factors for the PTSD risk literature can be found in King and colleagues (2004).

Risk Factor Mechanisms

Whereas the previous section focused on the different types of factors that may be implicated in PTSD, this section focuses on different pathways through which risk factors may be related to PTSD. As discussed previously, attention to potential pathways is critical given the growing recognition that no single risk factor is implicated in PTSD. Instead, there are likely to be multiple pathways of influence.

Below we summarize five different ways that risk factors may work together to influence an outcome and the conditions that must be met for each, as proposed by Kraemer and colleagues (2001). Please note that the criteria put forth by Kraemer and her colleagues represent a strong case for each scenario; that is, even when certain conditions are not met, it is still possible that the variables may work together as proposed. However, one's confidence in the proposed pathway must necessarily be diminished. The benefit of the Kraemer and colleagues framework is that, to the extent that these conditions are met, confidence in the proposed causal mechanism is increased.

For each scenario, we provide examples from the previously reviewed literature on risk factors for PTSD, suggesting ways that these factors may work together to influence PTSD. In addition, we integrate Kraemer and colleagues' (1997) risk terminology throughout this discussion, suggesting risk factor categories that may be particularly likely to be implicated in each possible scenario. We hope that this discussion of the different ways that risk factors may work together in their associations with PTSD will foster greater sophistication in the research designs implemented in studies of PTSD and more precise interpretation of existing PTSD research findings.

One Variable May Be a Proxy Risk Factor for Another Variable

Some factors that may initially appear to be causal risk factors for PTSD may actually turn out to be proxy risk factors. A "proxy risk factor" is a variable that is correlated with another risk factor but not causally implicated in the outcome (Kraemer et al., 2001). According to Kraemer and colleagues (2001), and using their symbols and terminology, one can operationally confirm that B is a proxy risk factor for variable A when (1) A and B are correlated; (2) either A precedes B or there is no temporal precedence of either variable; and (3) A demonstrates a stronger relationship with the outcome in the presence of B. Pretty much any variable that is correlated with a strong risk factor may appear to be a risk factor itself (Kraemer et al., 2001). For example, to the extent that individuals of minority racial/ethnic status have less access to resources, it may appear that minority racial/ethnic status is a risk factor for PTSD when the "real" causal risk factor is lack of access to resources.

Similarly, a composite variable may appear to be an important risk factor, when only one small component of the variable is actually associated with the outcome (Kraemer et al., 2001). For example, there is some research suggesting that low IQ is a risk factor for PTSD (Macklin et al., 1998). Closer examination could reveal, however, that only one component of IQ (e.g., analytical ability) is related to PTSD, whereas other components of IQ (e.g., verbal ability, math ability) are unrelated to PTSD. Similarly, there is ample evidence that social support is negatively related to PTSD (e.g., Egendorf et al., 1981; Keane et al., 1985; King et al., 1999; Solomon & Mikulincer, 1990; Solomon et al., 1988, 1989). It may turn out, however, that certain facets of social support (e.g., structural social support) are less relevant to PTSD than are other aspects of social support (e.g., emotional social support), and some research supports this conclusion (e.g., King et al., 1998; Solomon et al., 1987). As demonstrated in these examples, proxy risk factors may be either fixed markers (e.g., minority racial/ethnic status, math intelligence) or variable risk factors (e.g., structural social support), but they cannot, by definition, be causal risk factors.

Variables Are Overlapping Risk Factors

To the extent that two or more risk factors address a single, overlapping construct and are similarly related to the outcome, they can be considered overlapping risk factors (Kraemer et al., 2001). According to Kraemer and her colleagues, and again using their symbols and terminology, one can confirm overlapping risk factor status when (1) A and B are correlated; (2) neither A nor B has temporal precedence; and (3) A and B are codominant (i.e., the strongest association with the outcome is achieved by using A and B simultaneously). To borrow from the structural equation modeling framework, overlapping risk factors are akin to manifest effect indicators of a single underlying latent variable (Bollen, 1989; Hoyle, 1995; Loehlin, 1998). For example, to the extent that a lack of warmth in one's childhood family and exposure to fighting among family members in childhood are part of the larger construct of dysfunctional childhood family environment and are similarly related to PTSD, both could be considered overlapping risk factors.

Similarly, factors such as education and job status, which are often invoked as potential risk factors for PTSD, may be better considered indicators of a broader "access to resources" factor that more accurately represents the likely causal mechanism underlying observed associations. Researchers would be well advised to combine overlapping risk factors when they can. Combining indicators in meaningful ways can provide more reliable assessments of focal constructs and enhance power to detect effects.

As this discussion should illustrate, only factors that have meaning beyond their particular measurement can serve as overlapping risk factors. Fixed markers that have no meaning beyond their measurement (i.e., that are not considered indicators of an underlying latent variable), such as gender and race/ethnicity, are less likely to qualify as overlapping risk factors (and more likely to serve as proxy risk factors). Causal risk factors cannot be overlapping risk factors either, given that, by definition, overlapping risk factors are manifestations of some underlying latent variable that itself may serve as a causal factor. To take the example of "access to resources," having a higher status job in and of itself may not enhance one's resistance to PTSD, but the resources that are available to an individual with a more prestigious job (e.g., financial resources) may be causally implicated in PTSD.

Variables May Be Independent Risk Factors

To the extent that two risk factors are unrelated but both demonstrate associations with the outcome, they can be considered independent risk factors (Kraemer et al., 2001). According to Kraemer and her colleagues (2001), two variables, A and B, may be considered independent risk factors if (1) they are uncorrelated; (2) there is no temporal precedence of A or B; and (3) they are codominant (i.e., the strongest association with the outcome is achieved by using A and B together). For example, two independent risk factors for PTSD might be emotional social support and subsequent trauma exposure. In some instances, perhaps the occurrence of a natural disaster, there may be little reason to expect that the amount of emotional social support available following trauma exposure would be related to subsequent trauma exposure. To the extent that these factors are independent and multivariate analyses indicate that both are risk factors for PTSD, one could conclude they are independent risk factors for PTSD. Because of the requirement for no temporal precedence among them, both independent risk factors must be fixed markers or both must be variable or causal risk factors. One variable cannot be a fixed marker when the other is a variable or causal risk factor.

A "weaker" case for independent risk factor status based on loosening the requirement for no temporal precedence would allow for additional possible scenarios. For example, two independent risk factors for PTSD might be gender and how much social support was available following the traumatic event. There may be little reason to expect gender to be associated with the quantity of social support available in the aftermath of a trauma exposure. Thus, evidence that both of these variables are related to PTSD might suggest that they are independent risk factors. However, as noted by Kraemer (personal correspondence, April 11, 2005), when there is temporal precedence for two variables, it is possible that one variable may moderate the effect of the other. For example, the effect of social support in reducing PTSD could be stronger for women than for men, and in fact, some evidence suggests this may be the case (e.g., King et al., 1999). The case of moderation is discussed in more detail below.

One Variable Mediates Another Variable

According to Baron and Kenny (1986), mediation explains how or why a variable is related to an outcome. The recent reformulation of mediation proposed by Kraemer and her colleagues (2001) builds on Baron and Kenny's (1986) work in this area to suggest a "strong" case for mediation. According to Kraemer and colleagues, and using their symbols and terminology, one can conclude that a variable (B) is a mediator of another variable (A) if (1) there is a correlation between A and B; (2) A demonstrates temporal precedence relative to B; and (3) when A and B are considered simultaneously, there is either domination of A by B (the association between A and the outcome disappears in the presence of B) or codomination by A and B.

A key difference between the Kraemer and colleagues (2001) reformulation and that initially proposed by Baron and Kenny (1986) is that Kraemer and her colleagues require temporal precedence of the predictor relative to the mediator. With respect to the kinds of variables that may be implicated, predictors can be either fixed markers, or variable or causal risk factors. On the other hand, mediators must be variable or causal risk factors; that is, mediators must be variables that are free to change within an individual over time either naturally or via manipulation. Fixed markers cannot be mediators, because they do not change or cannot be manipulated to effect a change in PTSD.

A number of studies have explored potential mediators of risk factors on PTSD. However, the majority of these studies have relied on cross-sectional designs that cannot address issues of temporal precedence. Although an increasing number of PTSD risk studies have applied longitudinal strategies, few have assessed predictors and mediators at different time points, leaving uncertainty with regard to the temporal precedence of risk factors relative to proposed mediators. As we have asserted elsewhere (King et al., 2004), a number of demonstrated risk factors for PTSD may carry risk through their impact on the resources available in the aftermath of trauma exposure, as well as their impact on one's ability to cope with the event and its aftermath. For example, one might hypothesize that more severe trauma exposure challenges one's ability to make sense of and to recover from the event, which in turn increases the risk of PTSD. Future studies that explore such proposed mechanisms using longitudinal designs that allow researchers to begin to disentangle issues of temporal precedence among predictors and mediators are needed.

One Variable Moderates Another Variable

A moderator specifies on whom or under what conditions another variable will operate to produce an outcome (Baron & Kenny, 1986). In other words, moderation implies that the relationship between a predictor variable and an outcome varies across different levels of the moderator. Both previous formulations of moderation (e.g., Baron & Kenny, 1986) and the more recent reformulation proposed by Kraemer and her colleagues (2001), suggest that for moderation to obtain: (1) There must be a statistical interaction between a moderator variable A and a predictor variable B. To meet the requirement for their "strong" case for moderation, Kraemer and colleagues further assert that (2) there is no correlation between A and B, and (3) the moderator A precedes the predictor variable B.

Brewin and colleagues' (2000) meta-analysis provides evidence for a number of possible moderators of risk factor–PTSD associations. Briefly, results indicated that associations between a number of risk factors and PTSD differed depending on whether the sample is a military or civilian sample (effects were generally stronger for military compared to civilian samples), as well as the gender of the participants (although female gender was a stronger risk factor for PTSD, associations between other risk factors and PTSD were generally stronger for men compared with women). Of course, in a number of these situations there may be associations between risk factors and proposed moderators; thus, evidence would not meet the "strong" case for moderation. For example, it is well-known that women are more likely than men to experience sexual assault (Wolfe & Kimerling, 1997); thus, evidence for an interaction between gender and sexual assault trauma exposure (present, absent) cannot be confirmed to represent moderation according to the Kraemer and colleagues (2001) reformulation. In addition, the finding that gender interacts with combat exposure to predict PTSD might be considered questionable evidence for moderation given that men are more likely than women to experience combat (Vogt, Pless, King, & King, 2005). When risk factors and proposed moderators are related, it is possible that effects may be more appropriately labeled as "mediation" (H. Kraemer, personal correspondence, March 17, 2005).

As these examples illustrate, moderators are often fixed markers that identify different subgroups (e.g., women vs. men). However, moderators may also be variable risk factors. For example, to the extent that coping style was demonstrated to moderate the impact of the severity of a natural disaster on PTSD, evidence for coping style as a moderator would obtain. At the same time, moderators are not causal risk factors; that is, changing a moderator should not impact PTSD directly, but manipulating a moderator could potentially reduce the impact of the focal risk factor on PTSD. Taking the preceding example, efforts aimed at enhancing coping strategies could reduce the impact of exposure to natural disasters on PTSD.

As this discussion of moderation should highlight, no single set of risk mechanisms operate across trauma categories and trauma populations. Importantly, the presence of moderation can obscure meaningful phenomena and lead to null results. For example, to the extent that the impact of a particular risk factor on PTSD is different for women and men, and this effect operates in opposite directions (i.e., the effect is positive for women but negative for men), a researcher who does not consider gender in the analysis of the association between that risk factor and PTSD is likely to obtain null results. Extensive evidence for moderation, including, perhaps most notably, Brewin and colleagues' (2000) finding of significant heterogeneity among effect sizes for different subgroups, suggests that researchers might best expend their energy in carefully delineating potential moderators of associations between risk factors and PTSD in their research programs.

SUMMARY AND CHALLENGES FOR THE FUTURE

The application of the Kraemer and colleagues (1997, 2001) framework to the PTSD literature reveals a number of promising avenues for further investigation. Future studies that apply longitudinal designs to the study of potential risk factors, such as childhood family functioning, depression, and social support, will be especially beneficial. The majority of the literature on these potential risk factors has been based on cross-sectional designs and, as a consequence, questions remain regarding the extent to which they are causal risk factors. As discussed previously, many of the factors that have received the greatest amount of research attention are likely to turn out to be proxy risk factors. For example, there is ample evidence that gender is a risk factor for PTSD. However, gender is likely to be a proxy for other factors that put women at higher risk for PTSD. A more fruitful avenue of inquiry would involve identifying potential causal risk factors that underlie such associations. One such candidate is exposure to childhood stress and trauma. A better understanding of classes of risk factors can lead to advances in our understanding of who develops and who does not develop PTSD in the aftermath of trauma exposure.

At the same time, it is important to note that even when causal risk factors for PTSD are identified, underlying causal mechanisms may still not be well understood. This point is raised in the work of Kraemer and her colleagues (1997, 2001), as well as that of Rutter (2000a, 2000b), who emphasizes the importance of attending to the risk mechanisms that underlie demonstrated associations between risk factors and outcomes. Thus, even were one to demonstrate that social support, for example, is a causal risk factor for PTSD, the question of the risk mechanism through which social support is related to PTSD remains. It could be that individuals with more social support are those who have access to better resources that are protective against PTSD. Similarly, the risk mechanism underlying evidence that additional life stressors are a causal risk factor for PTSD (should it obtain) could be that individuals who experience fewer additional life stressors as they attempt to make sense of a traumatic event are better able to cope than those who must divide their coping energy between dealing with the aftermath of a traumatic event and dealing with additional life stressors. Once evidence for causal risk factors for PTSD is in place, the next step will be to explore in more depth the possible causal mechanisms that may underlie these associations.

Researchers are also urged to move beyond research designs that involve regressing PTSD on a number of possible risk factor candidates to designs that allow for tests of theoretically driven models of mediation and moderation. As discussed previously, a number of risk factors for PTSD, such as additional life stressors and lack of social support, may carry risk through their impact on the resources available in the aftermath of trauma exposure, as well as one's ability to cope with the event and its aftermath. Other risk factors may carry risk by enhancing the likelihood that others factors will be associated with PTSD. For example, there is some evidence that coping strategies may moderate the impact of combat exposure on PTSD (Sharkansky, King, King, Wolfe, Erickson, & Stokes, 2000), with individuals who employ more adaptive coping strategies less likely to develop PTSD as a consequence of combat exposure. Careful attention to delineating possible pathways through which risk factors are associated with PTSD can enhance our knowledge and strengthen assertions about causation.

Underscored by this discussion is the relative inability of studies employing cross-sectional designs to generate new knowledge about how risk factors are related to PTSD. Additional attention needs to be focused on applying experimental and longitudinal designs in PTSD research. For example, the majority of studies of mediators and moderators of risk factors for PTSD have not involved assessments of risk factors and potential moderators or mediators at different time points, and this restricts the conclusions researchers can draw regarding proposed associations. Of course, trauma researchers will always be limited in the variables they can ethically and practically manipulate in their work. Thus, the next best option to experimental manipulations may be longitudinal studies that involve repeated assessments of key variables over time. Such improvements in research design, although expensive and often logistically challenging to employ, are critical to facilitate advances in knowledge regarding risk factors for PTSD. Finally, although not the focus of this chapter, additional attention to how PTSD is operationalized is also needed. As we discussed previously, risk factors for the initial onset of PTSD may differ from risk factors for chronic PTSD. Moreover, risk factors for exposure to traumatic events are likely to differ from risk factors for PTSD (e.g., King et al., 1996).

In closing, we reiterate the need for PTSD risk research studies to move beyond the question "What are the risk factors for PTSD?" to the question "What are the pathways through which risk factors are associated with PTSD?" by using precise terminology and research designs that can begin to shed light on possible causal mechanisms.

REFERENCES

Acierno, R., Resnick, H., Kilpatrick, D. G., Saunders, B., & Best, C. L. (1999). Risk factors for rape, physical assault, and posttraumatic stress disorder in women: Examination of differential multivariate relationships. *Journal of Anxiety Disorders, 13*(6), 541–563.

Andrews, B., Brewin, C. R., Rose, S., & Kirk, M. (2000). Predicting PTSD symptoms in victims of violent crime: The role of shame, anger, and childhood abuse. *Journal of Abnormal Psychology, 109*(1), 69–73.

Andrykowski, M. A., & Cordova, M. J. (1998). Factors associated with PTSD symptoms following treatment for breast cancer: Test of the Anderson model. *Journal of Traumatic Stress, 11*, 189–203.

Baron, R. M., & Kenny, D. A. (1986). The moderator–mediator variable distinction in social psychological research: Conceptual, strategic, and statistical considerations. *Journal of Personality and Social Psychology, 51*, 1173–1182.

Birmes, P., Brunet, A., Carreras, D., Ducasse, J. L., Charlet, J. P., Lauque, D., et al. (2003). The predictive power of peritraumatic dissociation and acute stress symptoms for posttraumatic stress symptoms: A three-month prospective study. *American Journal of Psychiatry, 160*(7), 1337–1339.

Bolin, R. (1985). Disaster characteristics and psychosocial impacts. In B. Sowder (Ed.), *Disasters and mental health: Selected contemporary perspectives* (pp. 3–28). Rockville, MD: National Institute of Mental Health.

Bollen, K. A. (1989). *Structural equations with latent variables.* New York: Wiley.

Bremner J. D., & Brett, E. (1997). Trauma-related dissociative states and long-term psychopathology in posttraumatic stress disorder. *Journal of Traumatic Stress, 10*(1), 37–49.

Bremner, J. D., Southwick, S. M., & Charney, D. S. (1995). Etiological factors in the development of posttraumatic stress disorder. In C. M. Mazure (Ed.), *Does stress cause psychiatric illness?* (pp. 149–185). Washington, DC: American Psychiatric Press.

Bremner, J. D., Southwick, S. M., Johnson, D. R., Yehuda, R., & Charney, D. S. (1993). Childhood physical abuse and combat-related posttraumatic stress disorder in Vietnam veterans. *American Journal of Psychiatry, 150*, 235–239.

Breslau, N., & Davis, G. C. (1987). Posttraumatic stress disorder: The stressor criterion. *Journal of Nervous and Mental Disease, 175*, 255–264.

Breslau, N., Davis, G. C., Andreski, P., & Peterson, E. (1991). Traumatic events and posttraumatic stress disorder in an urban population of young adults. *Archives of General Psychiatry, 48*, 216–222.

Brewin, C. R., Andrews, B., & Valentine, J. D. (2000). Meta-analysis of risk factors for posttraumatic stress disorder in trauma-exposed adults. *Journal of Consulting and Clinical Psychology, 68*(5), 748–766.

Bromet, E., Sonnega, A., & Kessler, R. C. (1998). Risk factors for DSM-III-R posttraumatic stress disorder: Findings from the National Comorbidity Survey. *American Journal of Epidemiology, 147*, 353–361.

Burgess, A. W., & Holmstrom, L. L. (1979). *Rape: Crisis and recovery.* Bowine, MD: Brady.

Cohen, E. A. (1953). *Human behavior in the concentration camp.* New York: Grosset & Dunlap.

Creamer, M., & O'Donnell, M. (2002). Post-traumatic stress disorder. *Current Opinion in Psychiatry, 15*(2), 163–168.

Dougall, A. L., Herberman, H. B., Inslicht, S. S., Baum, A., & Delahanty, D. L. (2000). Similarity of prior trauma exposure as a determinant of chronic stress responding to an airline disaster. *Journal of Consulting and Clinical Psychology, 68*(2), 290–295.

Dunmore, E., Clark, D. M., & Ehlers, A. (1999). Cognitive factors involved in the onset and maintenance of posttraumatic stress disorder (PTSD) after physical or sexual assault. *Behaviour Research and Therapy, 37*, 809–829.

Egendorf, A., Kadushin, C., Laufer, R. S., Rothbart, G., & Sloan, L. (1981). *Legacies of Vietnam: Comparative adjustment of veterans and their peers.* New York: Center for Policy Research.

Elhers, A., Mayou, R., & Bryant, B. (1998). Psychological predictors of chronic post-traumatic stress disorder after motor vehicle accidents. *Journal of Abnormal Psychology, 107,* 508–519.

Emery, V. O., Emery, P. E., Sharma, D. K., Quiana, N. A., & Jassani, A. K. (1991). Predisposing variables in PTSD patients. *Journal of Traumatic Stress, 4,* 325–343.

Erickson, D. J., Wolfe, J., King, D. W., King, L. A., & Sharkansky, E. J. (2001). Posttraumatic stress disorder and depression symptomatology in a sample of Gulf war veterans: A prospective analysis. *Journal of Consulting and Clinical Psychology, 69*(1), 41–49.

Fairbank, J. A., Keane, T. M., & Malloy, P. F. (1983). Some preliminary data on the psychological characteristics of Vietnam veterans with posttraumatic stress disorder. *Journal of Consulting and Clinical Psychology, 51,* 912–919.

Fontana, A., & Rosenheck, R. (1994). Posttraumatic stress disorder among Vietnam theater veterans: A causal model of etiology in a community sample. *Journal of Nervous and Mental Disease, 182*(12), 677–684.

Foy, D. W., Carroll, E. M., & Donahoe, D. P., Jr. (1987). Etiological factors in the development of PTSD in clinical samples of Vietnam combat veterans. *Journal of Clinical Psychology, 43*(1), 17–27.

Gallers, J., Foy, D. W., Donahoe, C. P., & Goldfarb, J. (1988). Post-traumatic stress disorder in Vietnam combat veterans: Effects of traumatic violence exposure and military adjustment. *Journal of Traumatic Stress, 1,* 181–192.

Green, B. L. (1990). Defining trauma: Terminology and generic stressor dimensions. *Journal of Applied Social Psychology, 20,* 1632–1642.

Green, B. L. (1993). Identifying survivors at risk. In J. P. Wilson & B. Raphael (Eds.), *International handbook of traumatic stress syndromes* (pp. 135–144). New York: Plenum Press.

Green, B. L., Grace, M. C., & Gleser, G. C. (1985). Identifying survivors at risk: Long-term impairment following the Beverly Hills supper club fire. *Journal of Consulting and Clinical Psychology, 53*(5), 672–678.

Hoyle, R. H. (1995). *Structural equation modeling: Concepts, issues and applications.* Newbury Park, CA: Sage.

Kazdin, A. E., Kraemer, H. C., Kessler, R. C., Kupfer, D. J., & Offord, D. R. (1997). Contributions of risk-factor research to developmental psychopathology. *Clinical Psychology Review, 17*(4), 375–406.

Keane, T. M., Scott, W. O., Chavoya, G. A., Lamparski, D. M., & Fairbank, J. A. (1985). Social support in Vietnam veterans with posttraumatic stress disorder: A comparative analysis. *Journal of Consulting and Clinical Psychology, 53,* 95–102.

Kessler, R. C. (1997). The effects of stressful life events on depression. *Annual Review of Psychology, 48,* 191–214.

Kessler, R. C., Sonnega, A., Bromet, E., Hughes, M., & Nelson, C. B. (1995). Posttraumatic tress disorder in the National Comorbidity Survey. *Archives of General Psychiatry, 52,* 1048–1060.

King, D. W., & King, L. A. (1991). Validity issues in research on Vietnam veteran adjustment. *Psychological Bulletin, 109,* 107–124.

King, D. W., King, L. A., Foy, D. W., & Gudanowski, D. M. (1996). Prewar factors in combat-related posttraumatic stress disorder: Structural equation modeling with a national sample of female and male Vietnam veterans. *Journal of Consulting and Clinical Psychology, 64,* 520–531.

King, D. W., King, L. A., Foy, D. W., Keane, T. M., & Fairbank, J. A. (1999). Posttraumatic stress disorder in a national sample of female and male Vietnam veterans: Risk factors, war-zone stressors, and resilience–recovery variables. *Journal of Abnormal Psychology, 108*(1), 164–170.

King, D. W., King, L. A., Gudanowski, D. M., & Vreven, D. L. (1995). Alternative representations of war zone stressors: Relationships to posttraumatic stress disorder in male and female Vietnam veterans. *Journal of Abnormal Psychology, 104*(1), 184–196.

King, D. W., Taft, C. T., King, L. A., Hammond, C., & Stone, E. R. (2006). Directionality of the association between social support and posttraumatic stress disorder: A longitudinal investigation. *Journal of Applied Social Psychology, 36,* 2980–2992.

King, D. W., Vogt, D. S., & King, L. A. (2004). Risk and resilience factors in the etiology of chronic posttraumatic stress disorder. In B. T. Litz (Ed.), *Early interventions for trauma and traumatic loss in children and adults: Evidence-based directions* (pp. 34–64). New York: Guilford Press.

King, L. A., King, D. W., Fairbank, J. A., Keane, T. M., & Adams, G. A. (1998). Resilience–recovery factors in post-traumatic stress disorder among female and male Vietnam veterans: Hardiness, postwar social support, and additional stressful life events. *Journal of Personality and Social Psychology, 74*(2), 420–434.

Koenen, K. C., Stellman, J. M., Stellman, S. D., & Sommer, J. F. (2003). Risk factors for course of posttraumatic stress disorder among Vietnam veterans: A 14-year follow-up of American Legionnaires. *Journal of Consulting and Clinical Psychology, 71*(6), 980–986.

Koopman, C., Classen, C., & Spiegel, D. (1994). Predictors of posttraumatic stress symptoms among survivors of the Oakland/Berkeley California firestorm. *American Journal of Psychiatry, 151*, 888–894.

Kraemer, H. C., Kazdin, A. E., Offord, D. R., Kessler, R. C., Jensen, P. S., & Kupfer, D. J. (1997). Coming to terms with the terms of risk. *Archives of General Psychiatry, 54*, 337–343.

Kraemer, H. C., Stice, E., Kazdin, A., Offord, D., & Kupfer, D. (2001). How do risk factors work together?: Mediators, moderators, and independent, overlapping, and proxy risk factors. *American Journal of Psychiatry, 158*, 848–856.

Laufer, R. S., Gallops, M. S., & Frey-Wouters, E. (1984). War stress and trauma: The Vietnam veteran experience. *Journal of Health and Social Behavior, 25*, 65–85.

Lev-Wiesel, R., & Amir, M. (2003). Posttraumatic growth among Holocaust child survivors. *Journal of Loss and Trauma, 8*(4), 229–237.

Loehlin, J. C. (1998). *Latent variable models: An introduction to factor, path, and structural analysis.* Mahwah, NJ: Erlbaum.

Lund, M., Foy, D., Sipprelle, C., & Strachan, A. (1984). The Combat Exposure Scale: A systematic assessment of trauma in the Vietnam War. *Journal of Clinical Psychology, 40*, 1323–1328.

Macklin, M. L., Metzger, L. J., Litz, B. T., McNally, R. J., Lasko, N. B., & Orr, S. P. (1998). Lower precombat intelligence is a risk factor for posttraumatic stress disorder. *Journal of Consulting and Clinical Psychology, 66*(2), 323–326.

March, J. S. (1993). What constitutes a stressor?: The "Criterion A" issue. In J. R. T. Davidson & E. B. Foa (Eds.), *Posttraumatic stress disorder: DSM-IV and beyond* (pp. 37–54). Washington, DC: American Psychiatric Press.

Moran, C., & Britton, N. R. (1994). Emergency work experience and reactions to traumatic incidents. *Journal of Traumatic Stress, 7*, 575–585.

Norris, F. H., & Murrell, S. A. (1988). Prior experience as a moderator of disaster impact on anxiety symptoms in older adults. *American Journal of Community Psychology, 16*, 665–683.

Ozer, E., Best, S., Lipsey T., & Weiss, D. (2003). Predictors of posttraumatic stress disorder and symptoms in adults: A meta-analysis. *Psychological Bulletin, 129*(1), 52–73.

Peretz, T., Baider, L., Ever-Hadani, P., & De-Nour, A. K. (1994). Psychological distress in female cancer patients with Holocaust experience. *General Hospital Psychiatry, 16*, 413–418.

Quarantelli, E. L. (1985). What is disaster?: The need for clarification in definition and conceptualization in research. In B. J. Sowder (Ed.), *Disasters and mental health: Selected contemporary perspectives* (pp. 41–73). Rockville, MD: National Institute of Mental Health.

Resick, P. A. (2001). *Stress and trauma.* Philadelphia: Taylor & Francis.

Resick, P. A., Nishith, P., Weaver, T. L., Astin, M. C., & Feuer, C. A. (2002). A comparison of cognitive processing therapy, prolonged exposure and a waiting condition for the treatment of posttraumatic stress disorder in female rape victims. *Journal of Consulting and Clinical Psychology, 70*, 867–879.

Rodriguez, N., van de Kemp, H., & Foy, D. W. (1998). Posttraumatic stress disorder in survivors of childhood sexual and physical abuse: A critical review of the empirical research. *Journal of Child Sexual Abuse, 7*(2), 17–45.

Rutter, M. (2000a). Psychosocial influences: Critiques, findings, and research needs. *Development and Psychopathology, 12*, 375–405.

Rutter, M. (2000b, July–August). Resilience in the face of adversity. *World Congress on Medicine and Health.* Retrieved November 8, 2002, from *www.mh-hannover.de/aktuelles/projectte/mmm/english-version/fs_programme/speech/rutter_v.html*

Salter, E., & Stallard, P. (2004). Posttraumatic growth in child survivors of a road traffic accident. *Journal of Traumatic Stress, 17*(4), 335–340.

Schnurr, P. P., Lunney, C. A., & Sengupta, A. (2004). Risk factors for the development versus maintenance of posttraumatic stress disorder. *Journal of Traumatic Stress, 17*(2), 85–95.

Shalev, A. Y., Peri, T., Canetti, L., & Schreiber, S. (1996). Predictors of PTSD in injured trauma survivors: A prospective study. *American Journal of Psychiatry, 155*, 219–225.

Sharkansky, E. J., King, D. W., King, L. A., Wolfe, J., Erickson, D. J., & Stokes, L. R. (2000). Coping with
Gulf War combat stress: Mediating and moderating effects. *Journal of Abnormal Psychology, 109*(2),
188–197.

Solomon, Z., & Mikulincer, M. (1990). Life events and combat-related posttraumatic stress disorder:
The intervening role of locus of control and social support. *Military Psychology, 2*, 241–256.

Solomon, Z., Mikulincer, M., & Avitzur, E. (1988). Coping, locus of control, social support, and combat-
related posttraumatic stress disorder: A prospective study. *Journal of Personality and Social Psychology, 55*, 279–285.

Solomon, Z., Mikulincer, M., & Flum, H. (1989). The implications of life events and social integration
in the course of combat-related post-traumatic stress disorder. *Social Psychiatry and Psychiatric Epidemiology, 24*, 41–48.

Solomon, Z., Mikulincer, M., & Hobfoll, S. E. (1987). Objective versus subjective measurement of stress
and social support: Combat-related reactions. *Journal of Consulting and Clinical Psychology, 55*(4),
577–583.

van der Kolk, B. A., & Greenberg, M. S. (1987). The psychobiology of the trauma response:
Hyperarousal, constriction, and addiction to traumatic reexposure. In van der Kolk (Ed.), *Psychological trauma* (pp. 63–87). Washington, DC: American Psychiatric Press.

Vogt, D. S., Pless, A. P., King, L. A., & King, D. W. (2005). Deployment stressors, gender, and mental
health outcomes among Gulf War I veterans. *Journal of Traumatic Stress, 18(2), 115–127.*

Warheit, G. T. (1985). A prepositional paradigm for estimating the impact of disasters on mental
health. In B. J. Sowder (Ed.), *Disasters and mental health: Selected contemporary perspectives* (pp. 196–
214). Rockville, MD: National Institute of Mental Health.

Wolfe, J., & Kimerling, R. (1997). Gender issues in the assessment of posttraumatic stress disorder. In J.
P. Wilson & T. M. Keane (Eds.), *Assessing psychological trauma and PTSD* (pp. 192–238). New York:
Guilford Press.

Chapter 7

Remembering and Forgetting

Chris R. Brewin

Posttraumatic stress disorder (PTSD) is often described as a disorder of memory, mainly because of the intrusive trauma recollections that are such a prominent feature of the disorder. Historically, however, exposure to trauma has frequently been linked at the same time to the opposite problem, an impoverished memory for the distressing event or events. In the most extreme case, it is suggested that extremely stressful experiences, particularly in childhood, may be completely forgotten for a period of years. Consistent with these clinical observations, DSM-IV describes PTSD as being characterized by high frequency, distressing, involuntary memories that individuals are unable to forget and make great efforts to prevent coming to mind. Among these are the traumatic "flashback," memories characterized as being triggered spontaneously by exposure to trauma cues, as being fragmented, as containing prominent perceptual features, and as involving an intense reliving of the event in the present. At the same time, DSM-IV describes an inability to recall important aspects of the trauma as one of the symptoms of PTSD.

The contradictory nature of claims that victims are unable to forget their traumas but at the same time show amnesic gaps in their memory has not gone unnoticed. Apart from a natural skepticism concerning this claim, there are several other enduring sources of controversy in the field. One of these concerns is whether memory for trauma is "special" (i.e., has unique characteristics or involves unique processes not seen in memory for everyday events). Another controversy concerns whether people are ever able to forget extreme or repeated trauma, and what mechanisms might account for such observations.

This chapter reviews methodological issues and empirical research on remembering and forgetting relevant to these claims. Where appropriate, samples of people diagnosed with PTSD are distinguished from nonclinical samples selected for exposure to trauma. Comparisons of trauma memories in these groups are valuable to determine what is attributable to simple exposure and what to the specific presence of disorder. The last few years have also seen publication of a number of experimental studies using trauma-related stimuli such as words or films. These have provided useful insights and are reviewed separately from memory for actual traumatic events.

METHODOLOGICAL CONSIDERATIONS

The topic of traumatic memory is one in which inappropriate conclusions abound. Most prominence has probably been given to the mistaken assumptions of some trauma clinicians, who have at times been too ready to take their patients' accounts at face value and to assume that phenomena such as flashbacks or recovered memories necessarily correspond to some kind of objective reality (McNally, 2003). But nonclinicians attempting to do research in this area have also fallen prey to mistaken assumptions. For example, it is sometimes assumed that any kind of negative experience or experimental stimulus can yield data relevant to memory for trauma, ignoring the profound biological changes that accompany extremely stressful events and impact on key brain structures supporting memory processes (Vasterling & Brewin, 2005). All research in the area has had a tendency to ignore the evidence that trauma and PTSD may be associated with a widespread impact on many different aspects of attention and memory, and to place findings on memory for traumatic events in this context.

Another problem is that the vast majority of studies of trauma and memory have focused on traditional measures of recall and recognition, and have ignored the tendency for emotional memories to come to mind involuntarily. This is important, because involuntary memories are a particular problem in clinical disorders such as PTSD. Ever since the last century trauma theorists have distinguished between "traumatic memories" (intrusions) and "narrative memories" (memories available for conscious recall and recounting to others) (e.g., van der Kolk & van der Hart, 1991). Consistent with this, several authors (Brewin & Saunders, 2001; Halligan, Clark, & Ehlers, 2002; Holmes, Brewin, & Hennessy, 2004) studied voluntary and involuntary memories produced by exposure to a stressful film, and found that traditional measures of recall and recognition are unrelated to the number of involuntary memories reported over the following week. Accordingly, recent reviews (e.g., Brewin, 2003; Ehlers, Hackmann, & Michael, 2004) have distinguished between voluntary and involuntary remembering of trauma. Further progress in understanding the relations between emotion and memory is likely to depend on studies describing more carefully the nature of what is recalled, and measuring voluntary and involuntary memories separately. Beyond this, studies need to consider that there may be different types of intrusive memory, some corresponding to ordinary autobiographical memories, and others to the intense reliving experiences known as "flashbacks."

These distinctions are relevant to studies that have encouraged patients to write trauma narratives, or deliberately recall their trauma, and have drawn conclusions from these data about the nature of the underlying memory. As Hopper and van der Kolk (2001) have noted, the method of memory retrieval is critical in these studies. Although respondents may be asked to recall their trauma intentionally, it is likely that remembering also incorporates information from involuntary memories triggered during this process (Hellawell & Brewin, 2002, 2004). In addition, any narrative that has been produced may reflect characteristics of the output process (e.g., problems in controlling motor movements and translating the contents of memory into words) rather than characteristics of the underlying memory representations. Van der Kolk and Fisler (1995) attempted to isolate the narrative qualities of memory by asking people whether they could "tell others the story" of their trauma. Even this method does not unambiguously discriminate among narrative memories, however, because individuals may possess a personal narrative but be unable or unwilling to communicate it to others. This is likely to be a particular issue when the trauma occurred in childhood. Similar problems apply to most studies that ask people to rate or describe their trauma memories, without requiring them to distinguish between voluntary and involuntary memories.

Another key question involves study design. On the whole, clinicians working in the trauma field have not considered the broader question of whether other extreme emotions might produce similar memory phenomena. They have focused on describing the detailed phenomenology of trauma memories, and have considered whether there are different types of memory. But if trauma memories are thought to be special, they must be compared with memories for other events. Autobiographical memory researchers have mostly contrasted trauma memories with memories of happy or extremely positive events, but have not always distinguished between different kinds of trauma memory. This strategy also leaves open the possibility that both extremely positive and negative memories may differ from neutral memories. Both detailed phenomenological work and controlled contrasts with other memory types are necessary if sound conclusions about traumatic memory are to be drawn.

Measuring people's capacity to remember, however, is simple in comparison to measuring their capacity to forget. Data from surveys of patients reporting their forgetting of traumatic events are open to the obvious objection that respondents might be mistaken in thinking these were genuine memories. Beyond this, can we rely on people's account that they have completely forgotten some traumatic event? In his series of individual cases of forgotten traumatic memories, Schooler (2001) noticed that a few respondents who thought they had forgotten the events *had* apparently talked about them in the relatively recent past. This was clearly described by friends and family, although the person who recovered the memory found this hard to believe. Schooler speculated that if the experience of recovering the memory is shocking, people assume that they must previously have completely forgotten it. He suggests that recovered memories may be telling us about not only the operation of the memory system but also how well we monitor the system ("metamemory"); that is, people who are asked if they have forgotten something have to try to retrieve any occasion on which they might have remembered it in the past. If these occasions do not come to mind, they might falsely infer that the event had been forgotten for a long period of time.

Another methodological issue is also concerned with how people answer questions about the operation of their own memories. One group of patients reports complete memory gaps lasting for several years, a phenomenon that, again, has been linked to the existence of early trauma. But how do people judge the adequacy of their childhood memories in the first place? It has been argued (Belli, Winkielman, Read, Schwarz, & Lynn, 1998; Read & Lindsay, 2000) that people rely partly on the ease or difficulty with which they can bring instances to mind. In the experiment by Belli and colleagues (1998), participants were asked to report either four, eight, or 12 events from when they were 5–7 and 8–10 years old, after which they had to evaluate the adequacy of their childhood memory. Those who were instructed to retrieve more events paradoxically rated their childhood memory as worse than the groups who had to retrieve fewer events, presumably because they attributed the difficulty of the task to deficiencies in their memory. In a later study, Winkielman, Schwarz, and Belli (1998) showed that the effect of retrieving 12 events was abolished by telling participants that the task was difficult, consistent with the suggestion that without such explicit instructions, participants used the difficulty of the task to make inferences about the quality of their memory.

On the basis of these reports, both Belli and colleagues (1998) and Winkielman and colleagues (1998) suggested that psychotherapy patients' reports of incomplete childhood memory might be a mistaken consequence of difficulty in trying to recall large numbers of events, rather than a reflection of genuine problems with memory. They warned that such processes might lead clients to infer, wrongly, that they were amnesic for parts of their childhood and might therefore have forgotten or repressed traumatic experiences. These arguments imply that judgments about the adequacy of memory for childhood may bear no relation to actual memory performance, but this proposition had never been tested directly.

Brewin and Stokou (2002) investigated whether ordinary individuals who judge themselves to have poor memory for their childhood do in fact score lower on a standardized test of autobiographical memory. They found that group members who thought they had poor memory for childhood did in fact score lower than a control group on tests of memory for both the facts and events of their own life. Using the same test of autobiographical memory, Hunter and Andrews (2002) found that women with recovered memories of child abuse found it harder to recall facts about childhood, such as home addresses and names of teachers, friends, and neighbors than did women who had never been abused. Similar results have been obtained with traumatized adolescents by Meesters, Merckelbach, Muris, and Wessel (2000). These studies suggest that some people have genuine deficits in autobiographical memory for their childhood, and that memory judgments do have some basis in reality.

CURRENT STATE OF THE ART

Naturalistic Studies of Trauma Memory in Nonclinical Groups

Although a number of studies have focused on "most traumatic" or "most stressful" experiences, they typically include some events that would not qualify as traumas according to DSM-IV and are not reviewed here. Relatively few studies have focused explicitly on memory for highly traumatic events in samples not seeking treatment, and these have returned inconsistent results. In one of the studies, Porter and Birt (2001) found that memories involving sexual violence were rated as more vivid and contained more sensory components than memories involving other forms of violence. In contrast, Koss, Figueredo, Bell, Tharan, and Tromp (1996) found that rape memories, compared to other unpleasant memories, were rated as less clear and vivid, less likely to occur in a meaningful order, less well-remembered, and less thought about and discussed. Similarly, Byrne, Hyman, and Scott (2001) reported that the trauma memories of their participants, relative to positive memories, had less visual and olfactory detail.

Morgan and colleagues (2004) studied servicemen exposed to an extremely stressful survival course that included confinement and interrogation. At the conclusion of the course, men exposed to the most extreme stress, compared to those exposed to lower levels of stress, were significantly poorer at visually identifying their interrogator, whether in a live or photo lineup. Another of the few systematic studies in this area assessed memory in a sample of recent rape victims (Mechanic, Resick, & Griffin, 1998). Two weeks postrape, approximately two-thirds of the women had a clear memory of the event, whereas one-third had difficulty remembering at least a few aspects of it. About 10% of the women said that they were unable to recall many or most aspects of the event. Ten weeks later, 82% reported a clear memory, and none of the original 10% with problematic recall were still having problems remembering the event. Mechanic and colleagues noted that in remembering the rape, there appeared to be a specific problem that improved over time.

Experimental Studies of Trauma Memory and Forgetting in Nonclinical Groups

There is evidence that postevent processing can affect the development of intrusive memories in individuals exposed to a trauma film. Wells and Papageorgiou (1995) had participants engage in verbal worrying after watching a trauma film, and found that they experienced significantly more intrusions than controls during the next 3 days. This suggestion that some kinds of mental activity are either conducive or detrimental to the successful processing of traumatic material has been supported by more recent studies that have focused on the conditions under which films involving traumatic events are encoded.

Halligan and colleagues (2002, Experiment 1) instructed participants either to become absorbed in the images and sounds of the film (data-driven processing) or to focus on the overall story (conceptual processing). The data-driven processing group had poorer explicit recall of the film but did not differ in their experience of intrusive, film-related memories. In contrast, a group selected for its general tendency to engage in data-driven processing showed more disorganized explicit memory for the film and more subsequent intrusions than a group selected for its high conceptual processing (Halligan et al., 2002, Experiment 2).

In another approach Brewin and Saunders (2001) and Holmes and colleagues (2004) had college student volunteers watch a traumatic film involving scenes of real-life injury and death, while carrying out a secondary, concurrent task. Both studies showed that a concurrent visuospatial tapping task had the effect of reducing later intrusions relative to a control no-task condition. The effect could not be due simply to distraction, because a concurrent verbal task had the opposite effect of increasing subsequent intrusions relative to a control condition (Holmes et al., 2004, Experiment 3).

Holmes and colleagues (2004, Experiments 1 and 2) found that the more participants reported dissociative experiences (derealization and depersonalization) while they watched the film, the more likely they were to have intrusive memories of the film over the next week. Also, the lower participants' heart rates while they watched the film, the more likely they were to report later intrusions. Moreover, the specific scenes that intruded for any individual were associated with a lower heart rate during the film.

Dissociative experiences have been linked to the freezing response to threat (Nijenhuis, Vanderlinden, & Spinhoven, 1998), and to lowered heart rate (Griffin, Resick, & Mechanic, 1997). These findings, consistent with retrospective clinical studies, have repeatedly shown that high levels of peritraumatic dissociation are an important predictor of later PTSD symptoms (Ozer, Best, Lipsey, & Weiss, 2003). However, attempts so far to manipulate dissociation or data-driven processing experimentally have been ineffective, suggesting that this represents a relatively stable aspect of individual cognitive style. Responses at encoding do seem to be involved in determining whether intrusive memories develop, but it appears easier to alter encoding conditions indirectly, through the use of concurrent tasks.

Interestingly, individuals with high levels of dissociative symptoms also appear to be particularly good at intentionally not attending to or forgetting negative stimuli, a pattern that is particularly evident when they are under a cognitive load (DePrince & Freyd, 2001, 2004). Individuals with a repressive coping style, who deny negative emotions, such as anxiety, and deny antisocial characteristics, are also superior at selectively forgetting negative material (Myers & Brewin, 1995; Myers, Brewin, & Power, 1998). These results underscore the functional distinction between intrusions and intentional memory processes. They are also significant because they contradict the commonsense view that individuals with more negative experiences in life must be better able to recall negative stimuli. Both these groups are more likely to have experienced trauma or adverse parenting, yet they are better able to forget the negative. These results are consistent with clinical views about the importance of defensive mental processes that affect attention and memory.

Naturalistic Studies of Trauma Memory in PTSD

The trauma memories of patients with PTSD have been distinguished from the memories of people without PTSD in terms of them containing prominent perceptual features being highly emotional, and involving intense reliving of the event in the present (Berntsen, Willert, & Rubin, 2003; Bremner, Krystal, Southwick, & Charney, 1995; Ehlers et al., 2002; Ehlers & Steil, 1995; van der Kolk & Fisler, 1995; van der Kolk, Hopper, & Osterman, 2001). Compared to individuals without PTSD, their memories are more likely to have an observer perspective, in which the events are viewed from outside their body, rather than a field perspective, in which events are seen through their own eyes (Berntsen et al., 2003; Reynolds & Brewin, 1999). Patients with PTSD who report field memories recall more emotion and physical sensations, whereas those who report observer memories recall more spatial information, self-observations, and peripheral details (McIsaac & Eich, 2004).

Interestingly, relations between memory characteristics may be different in individuals who do and do not have PTSD, supporting the suggestion that the disorder is accompanied by qualitative differences in memory. For example, severity of trauma was strongly related to self-reported memory fragmentation in people with PTSD, but these variables were unrelated in people without PTSD (Berntsen et al., 2003). As noted earlier, recent studies have attempted to investigate the phenomenology of PTSD in more detail, for example, by distinguishing between involuntary memories and memories characterized by reliving, measuring them separately, or comparing them with deliberately retrievable trauma and nontrauma memories. They have also investigated the timing of intrusive memories and whether they may act as warning signals (Ehlers et al., 2002).

Involuntary Memories

Reynolds and Brewin (1998) compared matched groups of nonpatients and patients with either PTSD or depression, and asked them about current images or thoughts related to a stressful event that were most frequently on their minds. Flashbacks, either on their own or in combination with other images and thoughts, were reported as most frequent by 43% of the patients with PTSD, 9% of the patients with depression, and none of the nonpatients. Reynolds and Brewin noted that flashback content sometimes involved not only a literal record but also an imaginative extension of what had been experienced, so that patients had intrusive images of scenes that had not actually happened ("worst outcome scenarios"; Merckelbach, Muris, Horselenberg, & Rassin, 1998). Respondents were clearly able to distinguish between intrusive memories, typically involving visual images, and more general evaluative thoughts about the trauma.

Hackmann, Ehlers, Speckens, and Clark (2004) specifically assessed involuntary intrusive memories in their sample of patients with PTSD. Patients typically described between one and four highly repetitive memories, mainly comprising sensory experiences. Only 17% of these were about the worst moment of the trauma; the majority comprised prior moments as the trauma unfolded. Like Reynolds and Brewin (1998), they also noted a small proportion of images that did not correspond to actual events. Over a small number of reliving sessions the frequency of the intrusions diminished, as did their vividness, the associated distress, and the sense of how much the events appeared to be happening all over again.

Another study compared involuntary trauma and nontrauma memories, using diary methods, in a small sample of 12 individuals with PTSD (Berntsen, 2001, Study 2). Even when the traumatic event had occurred more than 5 years earlier, intrusive trauma memories were more vivid and more likely to be accompanied by physical reactions than were nontrauma memories. Trauma memories were also more likely to have the qualities of flashbacks, although Berntsen reported that some experiences similar to flashbacks occurred in relation to highly positive events. Similarly, trauma memories have been found to be more intrusive than other unpleasant memories in patients with PTSD, but this difference is much attenuated in trauma survivors without PTSD (Halligan, Michael, Ehlers, & Clark, 2003).

Hellawell and Brewin (2004) described the difference between flashbacks, involving a marked sense of reliving in the present, and ordinary memories to people with PTSD, then had them write a detailed narrative of their traumatic event. At the completion of the narrative, participants retrospectively identified periods of writing during which they experienced both types of memory. All the participants reported recognizing and being able to distinguish between the two types of memory as they wrote about their trauma, but there was great individual variation in how many reliving periods they identified, how long these lasted, and where in the narrative they occurred. Consistent with prediction, during parts of the narrative involving reliving, they used more words that described seeing, hearing, smelling, tasting, and bodily sensations, as well as more verbs and references to motion, than they did during ordinary memory sections. Again in line with predictions, fear, helplessness, horror, and thoughts of death were more prominent during the reliving sections, and secondary emotions, such as sadness, were more prominent during the ordinary memory sections.

Several authors have observed that some aspects of traumatic events seem to become fixed in the mind, unaltered by the passage of time, and are continually reexperienced in the form of images or "video clips" (Hackmann et al., 2004; Herman, 1992; Ehlers & Steil, 1995; van der Kolk & Fisler, 1995). Although this represents PTSD patients' own views of their memories, there is as yet little objective evidence for the stability of involuntary intrusions. Nor is it clear that the properties of intrusive memories are unique to PTSD. Involuntary, emotion-laden memories also occur, albeit less often, in response to extremely positive events (Berntsen, 2001, Study 4; Pillemer, 1998), and intrusive memories in depression are also vivid, characterized by reliving, and accompanied by bodily sensations (Reynolds & Brewin, 1999). Further research is needed to find out what, if anything, is unique about memories associated with trauma.

Intentional Recall

In contrast to the claims made concerning intrusions, some systematic studies have found that intentionally recalled trauma memories do show variability and errors in recall across time (Schwarz, Kowalski, & McNally, 1993; Southwick, Morgan, Nicolaou, & Charney, 1997). Consistent with the idea that intentional recall of trauma is different from involuntary recall, DSM-IV describes PTSD as being characterized by amnesia for the details of the event. Patients typically remember that the traumatic event happened but describe blanks or periods during which their memory for the details of the event is vague and unclear. In addition to the endorsement of this symptom on diagnostic measures, trauma narratives intentionally recalled by individuals with clinical disorders have been described as being disorganized and containing gaps (Foa, Molnar, & Cashman, 1995; Harvey & Bryant, 1999). Halligan and colleagues (2003) confirmed that patients with PTSD rated trauma memories as much more disorganized than other unpleasant memories, and noted that this was not true of trauma survivors without PTSD. This greater level of memory disorganization in patients with a clinical disorder is present as early as the first week posttrauma and predicts the subsequent course of their disorder, even when initial symptoms are controlled (Jones, Harvey, & Brewin, 2007).

Higher levels of fragmentation in trauma narratives have often been found to be related to self-reported dissociation either during or after the traumatic event (Engelhard, van den Hout, Kindt, Arntz, & Schouten, 2003; Halligan et al., 2003; Harvey & Bryant, 1999; Murray, Ehlers, & Mayou, 2002; but see Kindt & van den Hout, 2003). During psychotherapy it is common for patients to say that details are returning to them, and that they now recall numerous aspects of the event that had been forgotten. However, the evidence that fragmentation and disorganization of trauma memories decrease as patients recover from PTSD is inconsistent (Foa et al., 1995; Halligan et al., 2003; Jones et al., 2007; van Minnen, Wessel, Dijkstra, & Roelofs, 2002).

Studies of Recovered Memory in Traumatized Samples

The claim that involuntary and intentional trauma recall are different finds its ultimate expression in the theory that some individuals can deliberately choose to forget traumatic events and prevent them from coming to mind for long periods. This has been proposed mainly on the basis of clinical work with individuals abused in childhood. How robust is the claim that memories of sexual abuse can be forgotten and later recovered? Many longitudinal and retrospective studies have now apparently found that a substantial proportion of people reporting child sexual abuse (somewhere between 20 and 60%) say they have had periods in their lives (often lasting for several years) when they had less memory of the abuse or could not remember that it had taken place (for reviews, see Freyd, 1996; Mollon, 1998). A great deal of skepticism has been expressed about these findings (e.g., McNally, 2003). Broadly, there are two objections. The first is that investigators have not been putting the right questions to respondents, and have misinterpreted what they are really saying. The second is that the respondents mistakenly believe they have recovered "memories," but these are actually false memories suggested or implanted by their therapists.

The first objection is that even if people agree that there was a time when they did not remember the abuse, this does not rule out the possibility that they *would* have remembered, if only they had been asked. In other words, they had not forgotten the abuse, they had just not thought about it. But even when more probing questions are asked—such as "Was there ever a period when you would not have remembered this event, even if you were asked about it directly?"—similar results are obtained (Joslyn, Carlin, & Loftus, 1997). Also, memory recovery is often accompanied by extreme shock or surprise (Brewin, 2003). This, again, is inconsistent with the idea that individuals remained aware of the abuse but simply did not think about it.

The false memory argument (e.g., Loftus, 1993) contains some key propositions, including (1) the content of recovered memories is usually stereotypical, conforming to therapists' preconceptions about child sexual abuse as a ubiquitous cause of psychological disorder, or highly implausible (e.g., Satanic rituals with human sacrifices); (2) the age at which the events are supposed to have occurred may precede the development of explicit event memory; (3) there is typically no independent corroboration of the events; (4) recall generally occurs within therapy; and (5) the idea that trauma can be forgotten is contrary to established knowledge about how memory works. These claims have been systematically evaluated elsewhere (Brewin, 2003), and the findings are now briefly reviewed.

Content of Recovered Memories

Individuals reporting recovered memories do from time to time claim to have experienced implausible events, but these do not account for more than a small proportion of cases (Andrews et al., 1995, 1999; Gudjonsson, 1977). The largest estimate is provided by the False Memory Syndrome Foundation, which reported that 11% of accused parents calling them mentioned Satanic or ritual abuse spontaneously, and 18% did so in response to a closed question (see Morton et al., 1995). It should be noted that there is no independent corroboration of callers' claims, however. Moreover, several studies have reported the existence of recovered memories of events that have nothing to do with sexual abuse. These most frequently involved other child maltreatment, traumatic medical procedures, and witnessing violence or death (Andrews et al., 1999; Elliott, 1997; Feldman-Summers & Pope, 1994; Melchert, 1996).

Age of Memory

Psychological research indicates that there is little evidence for the later retrieval of narrative memories from the first 2 years of life, and that memories from the third year of life are rare. What memories there are from this period are likely to involve significant personal or family events. Consistent with this, in the available surveys, no more than 6% of apparent recovered memories of abuse are claimed to involve events falling completely within the first 5 years of life (Andrews, 1997; Andrews et al., 1999; Morton et al., 1995).

Corroboration

Schooler and his colleagues described a very interesting series of individual cases in which high-quality corroboration of forgotten trauma was available, and other explanations for apparent memory recovery were systematically considered (Schooler, 1994,

2001; Schooler, Bendiksen, & Ambadar, 1997). Some of the traumas were protracted in time, contradicting the argument that although single events can be forgotten, people are unable to forget repeated or protracted events. The Recovered Memory Project (see Cheit, 1998; *www.brown.edu/departments/taubman_center/recovmem/archive.html*) is an Internet-based resource detailing over 100 corroborated recovered memory accounts drawn from legal, academic, and other sources. Other case studies and group studies presenting corroboration for recovered memories include Corwin and Olafson (1997), Williams (1995), Feldman-Summers and Pope (1994), Andrews and colleagues (1999), and Chu, Frey, Ganzel, and Matthews (1999).

When and How Are Memories Recalled?

Surveys indicate that between one-half and one-third of recovered memories are recalled prior to any therapy or in a nontherapeutic context (Andrews et al., 1995; Feldman-Summers & Pope, 1994). In Elliott and Briere's (1995) community survey, being in therapy was unrelated to whether sexual abuse had been forgotten and later remembered, or had been continuously recalled.

Is Forgetting of Trauma Contrary to Our Knowledge of How Memory Works?

Just as many studies indicate that it is possible to implant some types of false memory in some individuals (McNally, 2003), numerous studies also show that people can deliberately forget a wide variety of material when they choose to do so, and that some individuals are particularly adept at forgetting negative material (see Brewin, 2003; Freyd, 1996; for reviews, see Gleaves, Smith, Butler, & Spiegel, 2004).

Experimental Studies of Remembering and Forgetting in Traumatized Patients

As part of their study comparing flashbacks and ordinary memories during trauma narratives, Hellawell and Brewin (2002) investigated whether the former were predominantly image-based, using visuospatial resources, and the latter predominantly verbal. They reasoned that if flashbacks are visuospatial, then they should interfere with performance on other tasks that also make visuospatial demands but not interfere with unrelated tasks. So, while writing their narratives, participants were stopped on two occasions, once in a reliving phase and again in an ordinary memory phase, and made to carry out two tasks. One task, trail making, involved visuospatial abilities and the other, counting backwards in threes, involved more verbal abilities. The results showed that trail-making performance was much worse when participants had been halted during a reliving phase of their narrative than during an ordinary memory phase, whereas counting backwards in threes was adversely affected to an equal extent in both phases. This supports the idea that there is a qualitative difference between flashbacks and ordinary memories.

It has also been found that a trauma history affects people's ability to generate specific autobiographical memories in response to cue words, such as *successful* or *lonely*. In this situation patients with PTSD, like patients with depression, tend to produce an excess of overgeneral memories relating to repeated experiences in their lives (McNally, Lasko, Macklin, & Pitman, 1995; McNally, Prassas, Shin, & Weathers, 1994). Even in samples without PTSD, people with a history of trauma are particularly likely to produce overgeneral memories (de Decker, Hermans, Raes, & Eelen, 2003; Hermans et al., 2004; Kuyken & Brewin, 1995; Wessel, Merckelbach, & Dekkers, 2002). Samples with less severe trauma (Wessel, Meeren, Peeters, Arntz, & Merckelbach, 2001) or borderline personality disorder (Arntz, Meeren, & Wessel, 2002; Kremers, Spinhoven, & Van der Does, 2004) may be exceptions, however. Williams, Stiles, and Shapiro (1999) have suggested that the association between trauma history and more overgeneral memory may reflect the existence of defensive processes that reduce the probability of specific painful memories being retrieved.

Clancy and McNally have tested one aspect of the false memory position by comparing the suggestibility of women who think they have been abused but have no memory of it (the repressed memory group), women who have recovered memories of abuse (the recovered memory group), women who have always known they were abused (the continuous memory group), and women who were never abused (the control group). In one study women were asked to rate their confidence that certain nontraumatic events from childhood had happened to them, such as finding a $10 bill in a parking lot, then on a later occasion were asked to imagine vividly a subset of these events and rerate their confidence that they had actually occurred. Although this procedure led to a slight increase in the belief that the imagined events had happened, this effect was larger for the controls than for the recovered memory group (Clancy, McNally, & Schachter, 1999).

Another study (Clancy, Schacter, McNally, & Pitman, 2000) involved showing participants a list of related words (e.g., *candy, bitter, sour, sugar*) and testing whether they would later falsely remember having seen another word (e.g., *sweet*) that was highly associated with all of them but never actually shown. In this experiment, the recovered memory group was more prone than the other groups to agree they had seen the nonpresented word, but the repressed memory group did not differ from the controls. However, as Freyd and Gleaves (1996) have pointed out, it is hazardous to draw inferences from this kind of word experiment, with its compelling associative cues to the situation of people with recovered memories of actual incidents of abuse, especially because other techniques have failed to show that they are any more suggestible than nonabused women (Leavitt, 1997).

McNally, Metzger, Lasko, Clancy, and Pitman (1998; McNally et al., 2001) have also to investigated the proposition that traumatized individuals, including those reporting recovered memories of trauma, are able to forget trauma-related material. They used a directed forgetting task in which participants are shown a list of words, and instructions to forget or remember are given after each one (item method). Ordinary volunteers find it difficult to remember words they have been told to forget, and this is thought to be due to the way the words are encoded in the first place. In two studies using this method, McNally found no evidence that people with PTSD related to child sexual abuse, or with repressed or recovered memories of abuse, were any more able to forget trauma-related words than people who had not been abused.

In some ways these results are not surprising, because one would not expect to find forgetting in individuals with active PTSD or in individuals who willingly come forward to take part in research. The idea that trauma can be forgotten has mainly been applied to young people faced with intolerable family stresses, who are isolated and have nobody in whom they can safely confide, or who have been betrayed by attachment figures (Freyd, 1996). Forgetting provides a way of coping with their situation, and it is the breakdown in this strategy, when distressing thoughts and images can no longer be prevented from reaching consciousness, that is thought to result in PTSD. However, when the trauma is very recent, individuals meeting criteria for acute stress disorder, largely based on a marked dissociative reaction, have been shown to forget trauma-related material more easily than controls (Moulds & Bryant, 2002, 2005). These results emphasize the need for more longitudinal studies that investigate changes in cognitive processing as posttraumatic disorders unfold.

SUMMARY AND GENERALIZABILITY OF CURRENT FINDINGS

Empirical findings strongly support clinical observations of traumatized individuals. Among nonclinical samples exposed to extreme stress, memory for the traumatic event is sometimes enhanced but at other times, at least in the short term, impaired. Similar contradictory findings among patients with PTSD can largely be resolved by distinguishing between intrusive memories and narrative memories that are under greater conscious control. Evidence is mounting that these two forms of memory are functionally distinct, both in nonclinical and clinical samples.

There is considerable agreement that intrusive memories in PTSD are repetitive, vivid, perceptually based, emotion-laden, and involve a reliving of events in the present. Descriptive and experimental studies support the view that intrusions commonly comprise an image-based rather than verbal form of memory. A considerable amount of naturalistic and experimental evidence suggests that intrusions are strengthened by dissociation at the point of encoding, even though dissociation is accompanied by a reduction in arousal. Intrusive images are not necessarily veridical, and the impression that they remain unchanged for long periods of time has not been tested. These investigations of the nature of trauma memories have now involved a wide variety of different kinds of trauma victims, producing similar findings in each case.

There are several indications, however, that although these characteristics of intrusive memories are most commonly found following trauma, they are not unique to trauma. Extreme positive emotions may also generate vivid intrusive memories accompanied by bodily reactions, although these are unlikely to occur so frequently. Moreover, it is not apparent that anyone has reported the more extreme forms of dissociative reactions in connection with positive events. Emergent evidence indicates that traumatic memory is most distinct when examined against a background of PTSD. It is possible that memory processing in nonclinical samples is distinct for a brief period of time after the trauma but then normalizes. This can only be confirmed by longitudinal studies. It seems fairest at present to conclude that traumatic memory is *unusual* rather than *special*, and becomes increasingly unusual with greater severity of PTSD.

There is also good evidence that intentionally retrieved memories in traumatized samples are often deficient. This rests on four sets of findings: that there are at least short-term memory deficits following extreme stress, that autobiographical memories are often overgeneral, that trauma narratives are likely to be fragmented and to contain gaps, and that trauma can be forgotten. As far as the latter is concerned, it is clear that empirical research and detailed case studies have produced convincing examples of genuine memories recovered after a long delay, as well as examples of false memories, and it is now widely accepted that both phenomena may occur (e.g., Gleaves et al., 2004; Lindsay & Read, 1995). Again, the most unusual phenomena tend to occur in the context of severe PTSD. One important exception is that patients with PTSD find it very difficult to forget trauma words. Nonclinical and clinical research indicates that a more specific group of individuals, those with high levels of dissociative symptoms or a repressive coping style, has a particular facility for forgetting negative or trauma-related material.

Some of the available evidence is limited and cannot be generalized with confidence to all groups of traumatized individuals. Until recently, for example, most of the studies of severe trauma in nonclinical samples have involved rape victims deliberately recalling their ordeal. Studies of servicemen under interrogation have provided a very welcome addition. Similarly, many of the experimental studies of forgetting and suggestibility have involved survivors of child sexual abuse. In contrast, investigations of the nature of trauma memories in clinical samples have now involved a wide variety of different kinds of trauma victims, producing similar findings in each case. Of particular interest are findings in which parallel results have been obtained in clinical samples and among volunteers in the laboratory. There is support of this kind for the idea that memory for trauma can be both exceptionally vivid and vague or disorganized, depending on how the study is conducted and what type of memory is targeted. There is also support for flashbacks or reliving experiences depending on some kind of visuospatial rather than verbal processing, and for dissociation at the time of the trauma increasing the risk of later intrusions.

CHALLENGES FOR THE FUTURE

Studies of trauma memory need to be much more rigorous and inventive if they are to reflect the complexity of the processes that have already been demonstrated to occur. They are also likely to benefit from drawing more explicitly on concepts and methods developed by cognitive and experimental psychologists. For example, the diary methods used by autobiographical memory researchers have already proved to be valuable, coupled with comparisons of trauma memories with positive and neutral memories. It is essential to distinguish clearly among the different types of memory (intentional, involuntary episodic memories, flashbacks) that may contribute to a given set of observations. Follow-up studies that measure the change in memories over time (Mechanic et al., 1998) or before and after treatment (Hopper & van der Kolk, 2001) will be particularly informative.

Studies also need to be guided by theory to a much greater extent than has previously been the case. For example, the dual representation theory of PTSD (Brewin, 2003; Brewin, Dalgleish, & Joseph, 1996) explicitly distinguishes between verbal and image-based forms of memory, detailing their characteristics and how they interact. In this theory the extent of encoding into the two memory systems is critical for both the development of intrusions and the quality of intentional recall. Similarly, in their cognitive model, Ehlers and Clark (2000) distinguish between episodic memories and associative processes, and have also suggested that encoding processes (specifically, data-driven vs. conceptual processing) are critical to the development of intrusions. Their model draws on Conway and Pleydell-Pearce's (2000) model of autobiographical memory, which is arranged hierarchically and distinguishes between more general autobiographical knowledge and event-specific knowledge comprising largely sensory information. All these theories are generating novel findings and currently provide a stimulus to ask new questions and use new methods in the quest to understand traumatic memory.

It is likely that neuroimaging methods will also assist in this understanding. Existing functional magnetic resonance imaging (fMRI) studies of patients with PTSD have often used a provocation technique, "script-driven imagery," in which patients are presented with brief scripts they have previously generated about their own trauma and asked to imagine the events as vividly as possible while they listen to the script being played. Positron emission tomography (PET) and fMRI studies have suggested that the retrieval of trauma memories in patients with PTSD, compared to controls is characterized by increased activity of limbic and paralimbic areas, including the amygdala (Hull, 2002; Shin, Rauch, & Pitman, 2005). Other replicated findings include deactivation of medial prefrontal areas and Broca's area (Hull, 2002; Shin et al., 2005). Two PET studies have found reduced hippocampal activity in patients with PTSD when they processed emotional rather than neutral material (Bremner et al., 1999, 2003). As noted by Hull (2002), the findings are not consistent across all studies. One problem is that different patients may react in a variety of ways to the requirement to retrieve trauma memories deliberately in a scanning context: Some experience flashbacks (e.g., Osuch et al., 2001; Shin et al., 1999), whereas others have strong dissociative reactions (e.g., depersonalization) associated with a different pattern of neuronal responses (Lanius et al., 2002).

It is important to distinguish between the effect of processing information that is specific to the trauma and that of processing any information that has been encoded in an emotional context. Bremner and colleagues (2003) and Lanius and colleagues (2003) have shown that in PTSD, neural processing differs for not only traumatic memories but also sad and anxious memories, and emotional words likely to be high in personal relevance. Other work by Maratos and Rugg (2001) and Maratos, Dolan, Morris, Henson, and Rugg (2001) shows that neural responses in normal participants are affected by the emotional context of a word, even when this context is retrieved incidentally. What is not known is whether there is a more fundamental disturbance in PTSD, such that an emotional context is processed differently, even when the stimuli are neutral and uncontaminated by personal meanings.

To date, studies have not been designed to distinguish between the retrieval of ordinary autobiographical memories of trauma and involuntary flashbacks, even though they should have a different neural basis. Conscious autobiographical memories are widely regarded as dependent on hippocampal processing. Brewin (2001, 2003) suggested that high levels of neurotransmitters and neurohormones released during traumatic experiences interfere with the normal operation of the hippocampus, resulting in the trauma being poorly represented in the autobiographical memory system. In contrast, lower-level representations of sensory (primarily visuospatial or image-based) information about the trauma remain intact, because they depend on processing that has little hippocampal involvement, relying instead on pathways that link other cortical and subcortical areas directly to the amygdala. These lower-level representations are informationally encapsulated, encode temporal information poorly, and, when triggered, are experienced as flashbacks.

To summarize, a wide range of different approaches are now being brought to bear on the study of traumatic memory. Phenomenological inquiries, experimental methods, and neuroimaging all have an important part to play and are already yielding significant new insights. What is clear, however, is that the questions are subtle and complex. Effective research will be based on the understanding gained from all these methods, and no single approach on its own is likely to be sufficient.

REFERENCES

Andrews, B. (1997). Can a survey of British False Memory Society members reliably inform the recovered memory debate? *Applied Cognitive Psychology, 11,* 19–23.

Andrews, B., Brewin, C. R., Ochera, J., Morton, J., Bekerian, D. A., Davies, G. M., et al. (1999). Characteristics, context and consequences of memory recovery among adults in therapy. *British Journal of Psychiatry, 175,* 141–146.

Andrews, B., Morton, J., Bekerian, D. A., Brewin, C. R., Davies, G. M., & Mollon, P. (1995). The recovery of memories in clinical practice: Experiences and beliefs of British Psychological Society practitioners. *The Psychologist, 8,* 209–214.

Arntz, A., Meeren, M., & Wessel, I. (2002). No evidence for overgeneral memories in borderline personality disorder. *Behaviour Research and Therapy, 40,* 1063–1068.

Belli, R. F., Winkielman, P., Read, J. D., Schwarz, N., & Lynn, S. J. (1998). Recalling more childhood events leads to judgments of poorer memory: Implications for the recovered false memory debate. *Psychonomic Bulletin and Review, 5,* 318–323.

Berntsen, D. (2001). Involuntary memories of emotional events: Do memories of traumas and extremely happy events differ? *Applied Cognitive Psychology, 15,* S135–S158.

Berntsen, D., Willert, M., & Rubin, D. C. (2003). Splintered memories or vivid landmarks?: Qualities and organization of traumatic memories with and without PTSD. *Applied Cognitive Psychology, 17,* 675–693.

Bremner, J. D., Krystal, J. H., Southwick, S. M., & Charney, D. S. (1995). Functional neuroanatomical correlates of the effects of stress on memory. *Journal of Traumatic Stress, 8,* 527–553.

Bremner, J. D., Narayan, M., Staib, L. H., Southwick, S. M., McGlashan, T., & Charney, D. S. (1999). Neural correlates of memories of childhood sexual abuse in women with and without posttraumatic stress disorder. *American Journal of Psychiatry, 156,* 1787–1795.

Bremner, J. D., Vythilingam, M., Vermetten, E., Southwick, S. M., McGlashan, T., Nazeer, A., et al. (2003). MRI and PET study of deficits in hippocampal structure and function in women with childhood sexual abuse and posttraumatic stress disorder. *American Journal of Psychiatry, 160,* 924–932.

Brewin, C. R. (2001). A cognitive neuroscience account of posttraumatic stress disorder and its treatment. *Behaviour Research and Therapy, 39,* 373–393.

Brewin, C. R. (2003). *Posttraumatic stress disorder: Malady or myth?* New Haven, CT: Yale University Press.

Brewin, C. R., Dalgleish, T., & Joseph, S. (1996). A dual representation theory of post traumatic stress disorder. *Psychological Review, 103,* 670–686.

Brewin, C. R., & Saunders, J. (2001). The effect of dissociation at encoding on intrusive memories for a stressful film. *British Journal of Medical Psychology, 74,* 467–472.

Brewin, C. R., & Stokou, L. (2002). Validating reports of poor childhood memory. *Applied Cognitive Psychology, 16,* 509–514.

Byrne, C. A., Hyman, I. E., & Scott, K. L. (2001). Comparisons of memories for traumatic events and other experiences. *Applied Cognitive Psychology, 15,* S119–S133.

Cheit, R. E. (1998). Consider this, skeptics of recovered memory. *Ethics and Behavior, 8,* 141–160.

Chu, J. A., Frey, L. M., Ganzel, B. L., & Matthews, J. A. (1999). Memories of childhood abuse: Dissociation, amnesia, and corroboration. *American Journal of Psychiatry, 156,* 749–755.

Clancy, S. A., McNally, R. J., & Schacter, D. L. (1999). Effects of guided imagery on memory distortion in women reporting recovered memories of childhood sexual abuse. *Journal of Traumatic Stress, 12,* 559–569.

Clancy, S. A., Schacter, D. L., McNally, R. J., & Pitman, R. K. (2000). False recognition in women reporting recovered memories of sexual abuse. *Psychological Science, 11,* 26–31.

Conway, M. A., & Pleydell-Pearce, C. W. (2000). The construction of autobiographical memories in the self-memory system. *Psychological Review, 107,* 261–288.

Corwin, D. L., & Olafson, E. (1997). Videotaped discovery of a reportedly unrecallable memory of child sexual abuse: Comparison with a childhood interview videotaped 11 years before. *Child Maltreatment, 2,* 91–112.

de Decker, A., Hermans, D., Raes, F., & Eelen, P. (2003). Autobiographical memory specificity and trauma in inpatient adolescents. *Journal of Clinical Child and Adolescent Psychology, 32,* 22–31.

DePrince, A. P., & Freyd, J. J. (2001). Memory and dissociative tendencies: The roles of attentional context and word meaning. *Journal of Trauma and Dissociation, 2*, 67–82.

DePrince, A. P., & Freyd, J. J. (2004). Forgetting trauma stimuli. *Psychological Science, 15*, 488–492.

Ehlers, A., & Clark, D. M. (2000). A cognitive model of posttraumatic stress disorder. *Behaviour Research and Therapy, 38*, 319–345.

Ehlers, A., Hackmann, A., & Michael, T. (2004). Intrusive re-experiencing in post-traumatic stress disorder: Phenomenology, theory, and therapy. *Memory, 12*, 403–415.

Ehlers, A., Hackmann, A., Steil, R., Clohessy, S., Wenninger, K., & Winter, H. (2002). The nature of intrusive memories after trauma: The warning signal hypothesis. *Behaviour Research and Therapy, 40*, 995–1002.

Ehlers, A., & Steil, R. (1995). Maintenance of intrusive memories in posttraumatic stress disorder: A cognitive approach. *Behavioural and Cognitive Psychotherapy, 23*, 217–249.

Elliott, D. M. (1997). Traumatic events: Prevalence and delayed recall in the general population. *Journal of Consulting and Clinical Psychology, 65*, 811–820.

Elliott, D. M., & Briere, J. (1995). Posttraumatic stress associated with delayed recall of sexual abuse: A general population study. *Journal of Traumatic Stress, 8*, 629–647.

Engelhard, I. M., van den Hout, M. A., Kindt, M., Arntz, A., & Schouten, E. (2003). Peri-traumatic dissociation and posttraumatic stress after pregnancy loss: A prospective study. *Behaviour Research and Therapy, 41*, 67–78.

Feldman-Summers, S., & Pope, K. S. (1994). The experience of forgetting childhood abuse: A national survey of psychologists. *Journal of Consulting and Clinical Psychology, 62*, 636–639.

Foa, E. B., Molnar, C., & Cashman, L. (1995). Change in rape narratives during exposure to therapy for posttraumatic stress disorder. *Journal of Traumatic Stress, 8*, 675–690.

Freyd, J. J. (1996). *Betrayal trauma: The logic of forgetting childhood abuse.* Cambridge, MA: Harvard University Press.

Freyd, J. J., & Gleaves, D. H. (1996). "Remembering" words not presented in lists: Relevance to the current recovered/false memory controversy. *Journal of Experimental Psychology: Learning, Memory, and Cognition, 22*, 811–813.

Gleaves, D. H., Smith, S. M., Butler, L. D., & Spiegel, D. (2004). False and recovered memories in the laboratory and clinic: A review of experimental and clinical evidence. *Clinical Psychology: Science and Practice, 11*, 3–28.

Griffin, M. G., Resick, P. A., & Mechanic, M. B. (1997). Objective assessment of peritraumatic dissociation: Psychophysiological indicators. *American Journal of Psychiatry, 154*, 1081–1088.

Gudjonsson, G. H. (1997). Accusations by adults of childhood sexual abuse: A survey of the members of the British False Memory Society (BFMS). *Applied Cognitive Psychology, 11*, 3–18.

Hackmann, A., Ehlers, A., Speckens, A., & Clark, D. M. (2004). Characteristics and content of intrusive memories in PTSD and their changes with treatment. *Journal of Traumatic Stress, 17*, 231–240.

Halligan, S. L., Clark, D. M., & Ehlers, A. (2002). Cognitive processing, memory, and the development of PTSD symptoms: Two experimental analogue studies. *Journal of Behavior Therapy and Experimental Psychiatry, 33*, 73–89.

Halligan, S. L., Michael, T., Ehlers, A., & Clark, D. M. (2003). Posttraumatic stress disorder following assault: The role of cognitive processing, trauma memory, and appraisals. *Journal of Consulting and Clinical Psychology, 71*, 419–431.

Harvey, A. G., & Bryant, R. A. (1999). A qualitative investigation of the organization of traumatic memories. *British Journal of Clinical Psychology, 38*, 401–405.

Hellawell, S. J., & Brewin, C. R. (2002). A comparison of flashbacks and ordinary autobiographical memories of trauma: Cognitive resources and behavioural observations. *Behaviour Research and Therapy, 40*, 1139–1152.

Hellawell, S. J., & Brewin, C. R. (2004). A comparison of flashbacks and ordinary autobiographical memories of trauma: Content and language. *Behaviour Research and Therapy, 42*, 1–12.

Herman, J. L. (1992). *Trauma and recovery.* London: Pandora Books.

Hermans, D., Van den Broeck, K., Belis, G., Raes, F., Pieters, G., & Eelen, P. (2004). Trauma and autobiographical memory specificity in depressed inpatients. *Behaviour Research and Therapy, 42*, 775–789.

Holmes, E. A., Brewin, C. R., & Hennessy, R. G. (2004). Trauma films, information processing, and intrusive memory development. *Journal of Experimental Psychology: General, 133*, 3–22.

Hopper, J. W., & van der Kolk, B. A. (2001). Retrieving, assessing, and classifying traumatic memories: A preliminary report on three case studies of a new, standardized method. *Journal of Aggression, Maltreatment and Trauma, 4*, 33–71.

Hull, A. M. (2002). Neuroimaging findings in post-traumatic stress disorder—systematic review. *British Journal of Psychiatry, 181*, 102–110.

Hunter, E. C. M., & Andrews, B. (2002). Memory for autobiographical facts and events: A comparison of women reporting child sexual abuse and nonabused controls. *Applied Cognitive Psychology, 16*, 575–588.

Jones, C., Harvey, A. G., & Brewin, C. R. (2007). The organisation and content of trauma memories in survivors of road traffic accidents. *Behaviour Research and Therapy, 45*, 151–162.

Joslyn, S., Carlin, L., & Loftus, E. F. (1997). Remembering and forgetting childhood sexual abuse. *Memory, 5*, 703–724.

Kindt, M., & van den Hout, M. (2003). Dissociation and memory fragmentation: Experimental effects on meta-memory but not on actual memory performance. *Behaviour Research and Therapy, 41*, 167–178.

Koss, M. P., Figueredo, A. J., Bell, I., Tharan, M., & Tromp, S. (1996). Traumatic memory characteristics: A cross-validated mediational model of response to rape among employed women. *Journal of Abnormal Psychology, 105*, 421–432.

Kremers, I. P., Spinhoven, P., & Van der Does, A. J. W. (2004). Autobiographical memory in depressed and non-depressed patients with borderline personality disorder. *British Journal of Clinical Psychology, 43*, 17–29.

Kuyken, W., & Brewin, C. R. (1995). Autobiographical memory functioning in depression and reports of early abuse. *Journal of Abnormal Psychology, 104*, 585–591.

Lanius, R. A., Williamson, P. C., Boksman, K., Densmore, M., Gupta, M., Neufeld, R. W. J., et al. (2002). Brain activation during script-driven imagery induced dissociative responses in PTSD: A functional magnetic resonance imaging investigation. *Biological Psychiatry, 52*, 305–311.

Lanius, R. A., Williamson, P. C., Hopper, J., Densmore, M., Boksman, K., Gupta, M. A., et al. (2003). Recall of emotional states in posttraumatic stress disorder: An fMRI investigation. *Biological Psychiatry, 53*, 204–210.

Leavitt, F. (1997). False attribution of suggestibility to explain recovered memory of childhood sexual abuse following extended amnesia. *Child Abuse and Neglect, 21*, 265–272.

Lindsay, D. S., & Read, J. D. (1995). "Memory work" and recovered memories of childhood sexual abuse: Scientific evidence and public, professional, and personal issues. *Psychology, Public Policy, and Law, 1*, 846–908.

Loftus, E. F. (1993). The reality of repressed memories. *American Psychologist, 48*, 518–537.

Maratos, E. J., & Rugg, M. D. (2001). Electrophysiological correlates of the retrieval of emotional and non-emotional context. *Journal of Cognitive Neuroscience, 13*, 877–891.

Maratos, E. J., Dolan, R. J., Morris, J. S., Henson, R. N. A., & Rugg, M. D. (2001). Neural activity associated with episodic memory for emotional context. *Neuropsychologia, 39*, 910–920.

McIsaac, H. K., & Eich, E. (2004). Vantage point in traumatic memory. *Psychological Science, 15*, 248–253.

McNally, R. J. (2003). *Remembering trauma*. Cambridge, MA: Harvard University Press.

McNally, R. J., Clancy, S. A., & Schacter, D. L. (2001). Directed forgetting of trauma cues in adults reporting repressed or recovered memories of childhood sexual abuse. *Journal of Abnormal Psychology, 110*, 151–156.

McNally, R. J., Lasko, N. B., Macklin, M. L., & Pitman, R. K. (1995). Autobiographical memory disturbance in combat-related posttraumatic stress disorder. *Behaviour Research and Therapy, 33*, 619–630.

McNally, R. J., Metzger, L. J., Lasko, N. B., Clancy, S. A., & Pitman, R. K. (1998). Directed forgetting of trauma cues in adult survivors of childhood sexual abuse with and without posttraumatic stress disorder. *Journal of Abnormal Psychology, 107*, 596–601.

McNally, R. J., Prassas, A., Shin, L. M., & Weathers, F. W. (1994). Emotional priming of autobiographical memory in posttraumatic stress disorder. *Cognition and Emotion, 8*, 351–367.

Mechanic, M. B., Resick, P. A., & Griffin, M. G. (1998). A comparison of normal forgetting, psychopathology, and information-processing models of reported amnesia for recent sexual trauma. *Journal of Consulting and Clinical Psychology, 66*, 948–957.

Meesters, C., Merckelbach, H., Muris, P., & Wessel, I. (2000). Autobiographical memory and trauma in adolescents. *Journal of Behavior Therapy and Experimental Psychiatry, 31*, 29–39.

Melchert, T. P. (1996). Childhood memory and a history of different forms of abuse. *Professional Psychology: Research and Practice, 27*, 438–446.

Merckelbach, H., Muris, P., Horselenberg, R., & Rassin, E. (1998). Traumatic intrusions as "worse case scenarios." *Behaviour Research and Therapy, 36*, 1075–1079.

Mollon, P. (1998). *Remembering trauma: A psychotherapist's guide to memory and illusion*. Chichester, UK: Wiley.

Morgan, C. A., III, Hazlett, G., Doran, A., Garrett, S., Hoyt, G., Thomas, P., et al. (2004). Accuracy of eyewitness memory for persons encountered during exposure to highly intense stress. *International Journal of Law and Psychiatry, 27*, 265–279.

Morton, J., Andrews, B., Bekerian, D., Brewin, C. R., Davies, G. M., & Mollon, P. (1995). *Recovered memories*. Leicester, UK: British Psychological Society.

Moulds, M. L., & Bryant, R. A. (2002). Directed forgetting in acute stress disorder. *Journal of Abnormal Psychology, 111*, 175–179.

Moulds, M. L., & Bryant, R. A. (2005). An investigation of retrieval inhibition in acute stress disorder. *Journal of Traumatic Stress, 18*, 233–236.

Murray, J., Ehlers, A., & Mayou, R. (2002). Dissociation and posttraumatic stress disorder: Two prospective studies of motor vehicle accident survivors. *British Journal of Psychiatry, 180*, 363–368.

Myers, L. B., & Brewin, C. R. (1995). Repressive coping and the recall of emotional material. *Cognition and Emotion, 9*, 637–642.

Myers, L. B., Brewin, C. R., & Power, M. J. (1998). Repressive coping and the directed forgetting of emotional material. *Journal of Abnormal Psychology, 107*, 141–148.

Nijenhuis, E. R. S., Vanderlinden, J., & Spinhoven, P. (1998). Animal defensive reactions as a model for trauma-induced dissociative reactions. *Journal of Traumatic Stress, 11*, 243–260.

Osuch, E. A., Benson, B., Geraci, M., Podell, D., Herscovitch, P., McCann, U. D., et al. (2001). Regional cerebral blood flow correlated with flashback intensity in patients with posttraumatic stress disorder. *Biological Psychiatry, 50*, 246–253.

Ozer, E. J., Best, S. R., Lipsey, T. L., & Weiss, D. S. (2003). Predictors of posttraumatic stress disorder and symptoms in adults: A meta-analysis. *Psychological Bulletin, 129*, 52–73.

Pillemer, D. B. (1998). *Momentous events, vivid memories*. Cambridge, MA: Harvard University Press.

Porter, S., & Birt, A. R. (2001). Is traumatic memory special?: A comparison of traumatic memory characteristics with memory for other emotional life experiences. *Applied Cognitive Psychology, 15*, S101–S117.

Read, J. D., & Lindsay, D. S. (2000). "Amnesia" for summer camps and high school graduation: Memory work increases reports of prior periods of remembering less. *Journal of Traumatic Stress, 13*, 129–147.

Reynolds, M., & Brewin, C. R. (1998). Intrusive cognitions, coping strategies and emotional responses in depression, post-traumatic stress disorder, and a non-clinical population. *Behaviour Research and Therapy, 36*, 135–147.

Reynolds, M., & Brewin, C. R. (1999). Intrusive memories in depression and posttraumatic stress disorder. *Behaviour Research and Therapy, 37*, 201–215.

Schwarz, E. D., Kowalski, J. M., & McNally, R. J. (1993). Malignant memories: Posttraumatic changes in memory in adults after a school shooting. *Journal of Traumatic Stress, 6*, 545–553.

Schooler, J. W. (1994). Seeking the core: The issues and evidence surrounding recovered accounts of sexual trauma. *Consciousness and Cognition, 3*, 452–469.

Schooler, J. W. (2001). Discovering memories of abuse in the light of meta-awareness. *Journal of Aggression, Maltreatment and Trauma, 4*, 105–136.

Schooler, J. W., Bendiksen, M., & Ambadar, Z. (1997). Taking the middle line: Can we accommodate both fabricated and recovered memories of sexual abuse? In M. A. Conway (Ed.), *Recovered memories and false memories* (pp. 251–292). Oxford, UK: Oxford University Press.

Shin, L. M., McNally, R. J., Kosslyn, S. M., Thompson, W. L., Rauch, S. L., Alpert, N. M., et al. (1999). Regional cerebral blood flow during script-driven imagery in childhood sexual abuse-related PTSD: A PET investigation. *American Journal of Psychiatry, 156*, 575–584.

Shin, L. M., Rauch, S. L., & Pitman, R. K. (2005). Structural and functional anatomy of PTSD: Findings from neuroimaging research. In J. J. Vasterling & C. R. Brewin (Eds.), *The neuropsychology of PTSD: Biological, clinical, and cognitive perspectives* (pp. 59–82). New York: Guilford Press.

Southwick, S. M., Morgan, A. C., Nicolaou, A. L., & Charney, D. S. (1997). Consistency of memory for combat-related traumatic events in veterans of Operation Desert Storm. *American Journal of Psychiatry, 154,* 173–177.

van der Kolk, B. A., & Fisler, R. (1995). Dissociation and the fragmentary nature of traumatic memories: Overview and exploratory study. *Journal of Traumatic Stress, 8,* 505–525.

van der Kolk, B. A., Hopper, J. W., & Osterman, J. E. (2001). Exploring the nature of traumatic memory: Combining clinical knowledge with laboratory methods. *Journal of Aggression, Maltreatment and Trauma, 4,* 9–31.

van der Kolk, B. A., & van der Hart, O. (1991). The intrusive past: The flexibility of memory and the engraving of trauma. *American Imago, 48,* 425–454.

Van Minnen, A., Wessel, I., Dijkstra, T., & Roelofs, K. (2002). Changes in PTSD patients' narratives during prolonged exposure therapy: A replication and extension. *Journal of Traumatic Stress, 15,* 255–258.

Vasterling, J. J., & Brewin, C. R. (Eds.). (2005). *The neuropsychology of PTSD: Biological, clinical, and cognitive perspectives.* New York: Guilford Press.

Wells, A., & Papageorgiou, C. (1995). Worry and the incubation of intrusive images following stress. *Behaviour Research and Therapy, 33,* 579–583.

Wessel, I., Meeren, M., Peeters, F., Arntz, A., & Merckelbach, H. (2001). Correlates of autobiographical memory specificity: The role of depression, anxiety and childhood trauma. *Behaviour Research and Therapy, 39,* 409–421.

Wessel, I., Merckelbach, H., & Dekkers, T. (2002). Autobiographical memory specificity, intrusive memory, and general memory skills in Dutch-Indonesian survivors of the World War II era. *Journal of Traumatic Stress, 15,* 227–234.

Williams, J. M. G., Stiles, W. B., & Shapiro, D. A. (1999). Cognitive mechanisms in the avoidance of painful and dangerous thoughts: Elaborating the assimilation model. *Cognitive Therapy and Research, 23,* 285–306.

Williams, L. M. (1995). Recovered memories of abuse in women with documented child sexual victimization histories. *Journal of Traumatic Stress, 8,* 649–673.

Winkielman, P., Schwarz, N., & Belli, R. F. (1998). The role of ease of retrieval and attribution in memory judgments: Judging your memory as worse despite recalling more events. *Psychological Science, 9,* 124–126.

Chapter 8

Trauma-Induced Dissociation

Anne P. DePrince and Jennifer J. Freyd

A man who had seen his greatest friend killed beside him developed the follow-ing symptoms. At first he struck several of his comrades, but later he assume a semi-stuporose condition, in which he would stare curiously at such objects as shining buttons and play with them as a child. He became depressed, tearful, vacant, speechless and heedless of what was said to him. . . . He took not notice of a pin-prick until it had been repeated several times, whereupon he gazed at the spot without attempting to withdraw from the pricking. . . . Two days later, he suddenly sat up and exclaimed: "Where am I." Then he got out of bed and sat by the fire, speaking quite intelligently to the orderly, but with no memory of his military life. After a few minutes he relapsed into his former state. The next day he became very restless, and on being quieted and assured that he was in hospital, he gradually came to himself, but had completely lost all mem-ory of what had occurred since he left the trenches. He had to be evacuated in this condition to England, where, it was considered, he made a complete recov-ery. But after his return to duty in England, he began to complain of shakiness, bad dreams, attacks of headache and dizziness, which, when severe, caused "fainting attacks." Finally after a sudden shock he was readmitted to hospital, suffering from complete "functional paraplegia."

—MYERS (1940, pp. 46–48)

Hysteria, soldier's heart, and shell-shock are among the many terms that signify psychia-try's history of grappling with human responses to trauma. The roots of traumatic stress studies began as early as the 19th century, when psychiatrist Pierre Janet drew a connection between traumatic experiences and "hysteria" in adult women (van der Kolk, Weisæth, & van der Hart, 1996). Janet was the first to articulate the basic princi-ples of dissociative phenomena based on observations of alterations in consciousness in patients with hysteria (Putnam, 1989). Beyond articulating principles of dissociation, Janet was among the first investigators to elucidate the adaptive nature of dissociation for dealing with acute and/or chronic trauma (Putnam, 1989). The foundation for traumatic stress studies established by Janet and his colleagues at the turn of the century was lost to a period of neglect of dissociation and trauma, with limited interest resurfacing after World Wars I and II (see Herman, 1992; Hilgard, 1986; van der Kolk et al., 1996). For example, Myers (1940) described dissociative reactions to combat exposure, as in the quotation opening this chapter, in which a soldier was "vacant" and forgot his combat experience. A sustained interest in dissociation on the part of clinicians and researchers working with trauma began in the 1980s and continues strongly into the present.

METHODOLOGICAL CONSIDERATIONS

Defining Dissociation

As clinical and research interest in dissociation has increased over the last two decades, the need to clearly define the term has arisen. During this period, definitions of dissociation have varied along many dimensions, including the degree of specificity of what we mean by the term "dissociation." Among the issues that need to be considered in defining the phenomenon are continuum–taxon views, state–trait distinctions, and outcome–mechanism discussions. Each of these issues is considered in an effort to define the term "dissociation."

Definitions of Dissociation

Although definitions of "dissociation" have varied, they have generally centered on the assumption that dissociation involves a lack of integration of aspects of information processing that would typically be connected. Beyond an agreement that dissociation involves a lack of integration, theorists vary in estimates of the scope and type of disintegration necessary to characterize experiences meaningfully as trauma-induced dissociation. In a recent commentary, van der Hart, Nihenhuis, Steele, and Brown (2004) argued that many definitions of dissociation are over- and/or underinclusive, and that this definitional issue impedes study of the phenomenon. They argued that dissociation is a "lack of integration among psychobiological systems that constitute personality" (p. 906). Similarly, Putnam (1997) argued that pathological dissociation is "characterized by profound developmental differences in the integration of behavior and in the acquisition of developmental competencies and metacognitive functions" (p. 15).

Continuum–Taxon

Janet's early conceptualization of dissociation suggested that a subset of individuals experience dissociative states that nondissociative individuals do not experience (see Putnam, 1997). In spite of Janet's view that dissociation involves a distinct category of experience, the prevailing view placed dissociation on a continuum; that is, theorists assumed that everybody dissociates to some degree. Common forms of dissociation were thought to include highway hypnosis or absorption in a movie/book. When the most widely used measure of adult dissociation, the Dissociative Experiences Scale (DES; see the section "Observing Dissociation" for more information on this scale), was developed, the prevailing assumption was that dissociation exists on a continuum. Factor analysis of the DES reveals an absorption–imaginative factor (Ross, Ellason, & Anderson, 1995; Sanders & Green, 1994) that seemingly encompasses normative experiences that are more normally distributed in the population than pathological dissociation.

In recent years, taxometric analyses have been used to justify treating dissociation as a taxon instead of a dimensional variable (e.g., Waller, Putnam, & Carlson, 1996). In this view, dissociation exists as a taxon, in which individuals display behaviors that are or are not consistent with pathological dissociation. This shift is important, because it affects not only theories about the development and maintenance of dissociation but also measurement. For example, existing measures include non-pathological experiences that may not be informative or related to pathological degrees of dissociation. The taxon view influences theory building by assuming that those individuals who pathologically dissociate differ in their basic cognitive organization (Putnam, 1997).

The issue of whether dissociative phenomena fall on a continuum or a taxon necessarily invokes issues of consciousness. Arguably, many experiences (e.g., absorption, daydreaming, trance states) can cause alterations in consciousness; however, the quality of such an experience may be better described as something other than dissociation. For example, van der Hart and colleagues (2004) argued that experiences such as daydreaming or trance can involve alterations in the level of consciousness (the degree to which the individual has awareness of consciousness) and the field of consciousness (the stimuli available to consciousness), and that it is structural dividedness that separates nondissociative experiences (e.g., absorption) from dissociation. "Structural dividedness" involves alterations between an apparently normal part of the personality and an emotional part (van der Hart et al., 2004).

For our purposes of this chapter, we treat trauma-induced dissociation as pathological dissociation; that is, we do not deal with understanding alterations in consciousness that are more typically distributed in the population (e.g., absorption) or that are not trauma-induced (e.g., neurologically based alterations in consciousness).

State–Trait

Inherent in continuum–taxon issues are also temporal issues. From the continuum view of dissociation, it is easy to imagine relatively transient periods of dissociation during a traumatic event. Reports of dissociation at the time of the event—called "peritraumatic dissociation"—have been made across a variety of traumas. Early reports of dissociation were noted by Myers during World War I, who described soldier's dissociative responses as varying "from a slight, momentary, almost imperceptible dizziness or 'clouding' to profound and lasting unconsciousness" (as cited in Brewin, 2003, p. 53). Peritraumatic dissociation has been found to be predictive of later posttraumatic stress disorder (PTSD; e.g., Gershuny, Cloitre, & Otto, 2003; Tichenor, Marmar, Weiss, Metzler, & Ronfeldt, 1996; Weiss, Marmar, Metzler, & Ronfeldt, 1995), leading theorists to question how adaptive dissociation is at the time of the event. In turn, recent work suggests that peritraumatic dissociation may be a common response that is not necessarily associated with later psychopathology (e.g., Bryant & Harvey, 2000). Panasetis and Bryant (2003) argued that persistent rather than peritraumatic dissociation may actually predict later psychopathology, such as PTSD. In a sample of participants who entered the hospital following motor vehicle accidents or nonsexual assaults, Panasetis and Bryant found that "persistent" dissociation was associated more strongly with acute stress disorder (ASD) severity and intrusion symptoms than with peritraumatic dissociation. The authors defined "persistent dissociation" as dissociation at the time of the assessment rather than at the time of the event. In other work, Gershuny and colleagues (2003) found that the relationship between peritraumatic dissociation and later PTSD was mediated by fears of death and loss of control during the event, which are central cognitive components of panic, raising the possibility that peritraumatic dissociation may be related to panic and not necessarily to pathological dissociation.

Outcome–Mechanism

Dissociation is referred to as both a psychological outcome of trauma and a mechanism of trauma-related problems (e.g., of memory problems) in the literature. For example, dissociative processes have been used to explain trauma-related memory impairment. It becomes difficult to distinguish whether dissociation is a static phenomenon that describes the status of integration of parts of a person's personality, or a process by which information is disintegrated. Van der Hart and colleagues (2004) shed light on this issue, arguing that experiences such as depersonalization and derealization may be alterations in consciousness but are not necessarily dissociative symptoms. The authors argue that to qualify as dissociative symptoms, the experience must involve structural dissociation; for example, the experience must involve dissociation between observing and experiencing ego (van der Hart et al., 2004).

Development of Dissociation: Motivation

The discrete behavioral states (DBS) model of dissociation argues that pathological dissociation is the result of developmental processes whereby children do not learn to integrate across behavioral states (Putnam, 1997). Putnam (1997) links the development of dissociation to early childhood abuse and notes three primary defensive functions of dissociation: automatization of behavior, compartmentalization of information and affect, and alteration of identity and estrangement from self.

Maldonado, Butler, and Spiegel (1998) stated that dissociative symptoms "should be understood as failures in integration, defects in control systems, rather than the creation of multiple identities . . ." that result in distress and dysfunction (p. 463). This statement captures a common viewpoint: that dissociation is a deficit with negative consequences. An alternative viewpoint is that dissociation is a creative adaptation to external insult and may even be seen as a positive set of skills. For example, dissociative automatization of behavior may allow a child to endure painful abuse without full awareness of what is happening and/or his or her own actions (Putnam, 1997). These two perspectives in their extremes may have profoundly differing implications for those who experience high levels of dissociation that necessitate treatment.

One issue implicit in this distinction between dissociation as a deficit and dissociation as an adaptation is the origin or motivation for developing dissociation. Theorists have long argued that dissociation may serve a protective or defensive function at the time of the trauma, or later, to keep trauma-related information out of awareness. Some authors have observed, though, that dissociation at the time of an event predicts later distress, including PTSD (Ozer, Best, Lipsey, & Weiss, 2003), raising the question of how effectively dissociation protects the individual. The key to evaluating the adaptive–maladaptive nature of dissociation lies in thinking about the function of dissociation given the individual's context. Betrayal trauma theory (Freyd, 1996), discussed below in more detail, argues that dissociation enables victims who are dependent on an abusive caregiver to maintain necessary attachments. Under conditions in which survival depends on structural dissociation—that is, lack of awareness of the trauma-related information by the part of the personality that must manage tasks necessary to survival, such as attachment with caregivers—dissociation may very well serve an adaptive function. In the long run, dissociation may play different roles in later distress, perhaps mediating or moderating the relationship between some traumas (e.g., abuse) and later psychological symptoms. There may be contexts, too, in which so-called "pathological" dissociation puts individuals at a distinct disadvantage. For example, the dissociation of emotion information from the personality acting in day-to-day situations may result in individuals missing danger cues or otherwise increasing risk of problems, such as revictimization or HIV risk (DePrince, Freyd, & Malle, 2007; Zurbriggen & Freyd, 2004).

Seeing dissociation as a creative adaptation may have benefits for the dissociative client seeking treatment. Rather than pathologizing the trauma survivor, this viewpoint more likely empowers the client because of the implicit respect it offers. However, there is some danger in ignoring real suffering if dissociation is seen as a "normal" response. Some might conclude that because it is a normal response, there is no need for intervention. However, this may be mistaken. By analogy, if one were to slice off a part of the body, bleeding would be a normal response, yet intervention might be very much needed.

One more interesting aspect of this distinction between dissociation as a deficit or an adaptation is how one models individual differences in the tendency to respond to trauma with dissociation. If individuals do differ, perhaps due to heredity (Becker-Blease et al., 2004), in their tendency to dissociate, then this can be viewed in terms of a diathesis–stress model; that is, the underlying tendency may be a vulnerability that is provoked by trauma. An alternative would be to see the underlying tendency as a resilience factor that is awakened by trauma. In this view, dissociation protects the individual from greater harm. Additional research is needed to provide more evidence on these issues.

A dialectical view may help resolve issues of how adaptive or maladaptive dissociation is viewed. Specifically, dissociation may be both a creative adaptation to an environmental insult that threatens survival (e.g., child abuse by a caregiver) and a deficit that causes problems in other domains of life (e.g., difficulty in school). We have the classic problem of looking at "survivor data" when we examine adults who are high in dissociation and evaluate whether dissociation has been adaptive or maladaptive; that is, we are not able to see what these individuals would be like had they not dissociated. Perhaps consequences for some individuals would have been far worse had they not engaged in dissociation, so although dissociation is linked to negative consequences, we have no way of evaluating whether those consequences are better or worse than if the individuals had not chronically dissociated. Furthermore, a dialectical view of dissociation as both adaptive and maladaptive invokes the importance of examining context. In some contexts, dissociation may be the most helpful thing the person could do (e.g., under some conditions of child abuse); in others, it may increase potential harm (e.g., revictimization risk). By viewing dissociation dialectically, practitioners and researchers are likely to examine both the adaptation in the response (and seek to teach alternative skills given the person's current context) and the negative consequences that cause problems for the individual.

Observing Dissociation

Measuring Dissociation

Measuring dissociation requires thought about both the definition of dissociation (e.g., pathological vs. normative) and conditions under which it occurs. We have argued that trauma-induced dissociation should include pathological dissociation (as opposed to alterations in consciousness that are more normally distributed in the population). Several reliable and validated self-report measures of dissociative experiences in children, adolescents and adults are available (see Table 8.1 for a listing of several widely used measures).

The vast majority of the literature has focused on negative symptoms of dissociation, such as amnesia, loss of skills, and loss of awareness (van der Hart et al., 2004). In recent years, theorists have argued that dissociation also includes positive symptoms, such as flashbacks and intrusions (e.g., van der Hart et al., 2004). Only recently have dissociative symptoms related to movement, sensation, and perception been noted. Using the Somatoform Dissociation Questionnaire (SDQ), researchers were able to discriminate between individuals diagnosed with dissociative disorders and those diagnosed with other psychiatric disorders (Nijenhuis, Spinhoven, van Dyck, van der Hart, & Vanderlinden, 1998).

TABLE 8.1. Self-Report Measures of Dissociative Experiences

Measure name	Relevant references	Respondent	Comments
Adolescent Dissociative Experiences Scale (A-DES)	Armstrong, Putnam, Carlson, Libero, & Smith (1997); Putnam (1997)	Adolescent	Emphasis on dissociation of mental functions (vs. movement, sensation, and perception).
Child Dissociative Checklist (CDS)	Putnam, Helmers, & Trickett (1993); Putnam (1997)	Parent	Emphasis on dissociation of mental functions.
Dissociative Experiences Scale (DES)	Bernstein & Putnam (1986); Putnam (1997)	Adult	Emphasis on dissociation of mental functions.
Peritraumatic Dissociative Experiences Questionnaire (PDEQ)	Marmar, Weiss, & Metzler (1997); Marshall, Orlando, Jaycox, Foy, & Belzberg (2002)	Adult	Assesses retrospective reports of dissociative experiences at the time of the event.
Somatoform Dissociation Questionnaire (SDQ)	Nijenhuis, Spinhoven, van Dyck, van der Hart, & Vanderlinden (1998)	Adult	Emphasis on somatoform dissociation symptoms. Five- and 20-item measures available.
Multidimensional Inventory of Dissociation (MID)	Dell (2006)	Adult	Assesses 14 facets of dissociation and includes validity items.

Observing Dissociation Posttrauma

Numerous researchers have documented a correlation between dissociative symptoms and self-reported trauma (e.g., Francia-Martinez, de Torres, Alvarado, Martinez-Taboas, & Sayers, 2003; Irwin, 1999; Putnam, 1997). Generally this correlation is interpreted as an indication that trauma is a causal factor in the development of dissociative symptoms. Some researchers, however, have questioned the assumption of causality. Merckelbach, Horselenberg, and Schmidt (2002), for instance, argue that structural equation modeling analyses applied to self-report data from a sample of undergraduate students show equally good fit for both a model assuming that dissociation causes self-reports of trauma and one that assumes trauma causes dissociation. It is important, given this inherent difficulty in interpreting correlational data, to look at samples in which trauma history is documented independent of self-reports. Studies using samples of documented trauma survivors have revealed that, indeed, dissociation is present as a function of trauma experience. For example, Putnam and Trickett (1997) compared 77 sexually abused girls to 72 control girls in a longitudinal study of the biological and psychological effects of sexual abuse. The sexually abused girls were referred by child protective service agencies. The control girls were matched on age, race, socioeconomic status, and family constellation. Putnam and Trickett found that, compared with the controls, the sexually abused girls had significantly elevated dissociation scores at three different testing times during the study. Similarly, using a longitudinal design with children at risk due to poverty, Ogawa, Sroufe, Weinfeld, Carlson, and Egeland (1997) reported that age of onset, chronicity, and severity of trauma predicted level of dissociation, as measured at four time points across 19 years.

Several studies have examined dissociation in populations in which trauma is more easily documented or verified than in cases of child abuse or assault (e.g., Bremner, Southwick, & Brett, 1992; Carlson & Rosser-Hogan, 1991; Koopman, Classen, & Spiegel, 1996; Marmar, Weiss, & Metzler, 1997; Yehuda et al., 1996). In Carlson and Rosser-Hogan's (1991) study, for example, 50 Cambodian refugees who had settled in the United States participated in a study involving the administration of a series of questionnaires. DES scores in the sample were strikingly high (mean = 37.1); notably, only two of the 50 participants scored under 10 on the scale, which is considered to be within the range of normal adults.

Even if the converging evidence provided by these documented trauma studies makes us confident that trauma can be a causal factor in the development of dissociation, how much societal and cultural expectations play a role in this relationship is still an open question. For instance, a trauma survivor may learn from others or from the culture at large to evidence dissociative symptoms as a socially accepted response to trauma. Might the correlation between dissociation and trauma be at least partially a result of suggestion by therapist or media exposure? If so, we should see lower correlations between trauma and dissociation in societal contexts in which individuals would be less likely to be exposed to suggestive influences regarding this relationship. Dalenburg and Palesh (2004) evaluated the relationship between dissociative symptoms and trauma in a Russian population that was relatively unexposed to these suggestive sources: 301 Russian university students, who completed measures of dissociative symptoms, and history of violent trauma and child abuse. The relationship between trauma and dissociation was discovered in this sample and, if anything, rates of dissociation were higher than in comparable American samples, suggesting that suggestive influences do not explain the correlation.

Observing Dissociation in Other Psychiatric Contexts

Dissociative symptoms have been observed in conjunction with a range of diagnostic categories, including ASD (e.g., Bryant & Harvey, 2000), PTSD (e.g., Brewin, 2003), complex PTSD (Herman, 1992), eating disorders (see Putnam, 1997), and the dissociative disorders (Putnam, 1997). For our purposes in this chapter, we focus on the co-occurrence of PTSD and trauma-induced dissociation.

The co-occurrence of dissociation and PTSD has received attention in terms of both describing the phenomenon of co-occurrence and what that co-occurrence may mean conceptually for understanding posttraumatic responses. Several studies have observed relations between PTSD and dissociation; for example, people who meet criteria for PTSD score higher on the DES than those who do not (e.g., Carlier, Lamberts, Fouwels, & Gersons, 1996; Maldonado & Spiegel, 1998; Putnam, 1997; Yehuda et al., 1996).

Some researchers have suggested that, conceptually, dissociation may play a central role in the onset and/or maintenance of PTSD. For example, van der Kolk and Fisler (1995) suggested that dissociation is at the core of the development of PTSD. In addition, Braun (1988) and van der Hart (2000) suggested that intrusive symptoms may in fact be dissociative phenomena. Van der Hart likened intrusive PTSD symptoms to positive dissociation symptoms (e.g., presence of intrusive memories), whereas avoidance symptoms reflect negative dissociation symptoms (e.g., feeling detached from others). Indeed, the experience of a flashback fits many definitions of dissociation, in which normally integrated aspects of consciousness are not integrated (one's mental experience may not be integrated with conscious awareness of current surroundings, passage of time, etc.).

With interest in dissociation, some speculation about PTSD as a dissociative disorder can be found in the literature. Support for such a move is drawn from the observation that both PTSD and dissociative disorders are reactions to extreme stress and therefore have similar etiologies (Brett, 1993). Furthermore, both PTSD and dissociative disorders include alterations in memory among their criteria. In spite of this support, researchers have argued that PTSD includes anxiety that is more consistent with other anxiety disorders than with the dissociative disorders; in addition, some people with PTSD do not experience amnesia or dissociative episodes (Brett, 1993). Taken together, these observations raise the question of whether there may be subtypes of PTSD that vary in their involvement of dissociative processes.

DISSOCIATION AND INFORMATION PROCESSING

As research has progressed to the point that factors associated with dissociation have been observed repeatedly (e.g., history of child abuse), the literature has moved in exciting directions that focus on identifying both the motivation for and mechanisms underlying the relationship between trauma, dissociation, and associated outcomes (e.g., autobiographical memory impairment). Such basic research that identifies and tests mechanisms—emotional, cognitive, and social—that may underlie dissociation and related outcomes is necessary to advancing treatment approaches. To the extent that the mechanisms underlying dissociative problems are better understood, interventions can be fine-tuned to target particular mechanisms. We now review various information-processing approaches to dissociation.

Dissociation, Forgetting, and Betrayal Trauma Theory

One aspect of dissociation is amnesia. Betrayal trauma theory predicts that forgetting abuse will be greater when the relationship between perpetrator and victim involves closeness, trust, and/or caregiving. In these cases the potential for a conflict between need to stay in the relationship and awareness of betrayal is greatest; thus, this is where we should see the greatest amount of forgetting or memory impairment. Freyd (1996), in reanalyses of a number of relevant data sets, reported that incestuous abuse is more likely to be forgotten than nonincestuous abuse. These data sets included the prospective sample assessed by Williams (1994, 1995), and retrospective samples assessed by Cameron (1993) and Feldman-Summers and Pope (1994). Using new data collected from a sample of undergraduate students, Freyd, DePrince, and Zurbriggen (2001) found that physical and sexual abuse perpetrated by a caregiver is related to higher levels of self-reported memory impairment for the events compared to noncaregiver abuse. Research by Schultz, Passmore, and Yoder (2003) and a doctoral dissertation by Stoler (2001) revealed similar results. For instance, in their abstract, Schultz and colleagues (2003, p. 67) state: "Participants reporting memory disturbances also reported significantly higher numbers of perpetrators, chemical abuse in their families, and closer relationships with the perpetrator(s) than participants reporting no memory disturbances." Sheiman (1999) reported that, in a sample of 174 students, participants who reported memory loss for child sexual abuse were more likely to have experienced abuse by people well known to them, compared to those who did not have memory loss. Similarly Stoler noted in her dissertation abstract: "Quantitative comparisons revealed that women with delayed memories were younger at the time of their abuse and more closely related to their abusers" (p. 5582). Interestingly, Edwards, Fivush, Anda, Felitti, and Nordenberg (2001) reported that general autobiographical memory loss measured in a large epidemiological study was strongly associated with a history of childhood abuse, and that one of the specific factors associated with this increased memory loss was sexual abuse by a relative.

Dissociation has long been implicated in trauma-related memory disruption. Betrayal trauma theory predicts that dissociating information from awareness is mediated by the threat that the information poses to the individual's system of attachment (Freyd, 1996). Consistent with this, Chu and Dill (1990) reported that childhood abuse by family members (both physical and sexual) is significantly related to increased DES scores in psychiatric inpatients, whereas abuse by nonfamily members is not. Similarly, Plattner and colleagues (2003) reported that they found significant correlations between symptoms of pathological dissociation and intrafamilial (but not extrafamilial) trauma in a sample of delinquent juveniles. DePrince (2005) found that the presence of betrayal trauma before the age of 18 is associated with pathological dissociation and with revictimization after age 18. She also found that individuals who report being revictimized in young adulthood following an interpersonal assault in childhood, compared to individuals who have not been revictimized, perform worse on reasoning problems that involve interpersonal relationships and safety information.

DePrince (2001) found that self-reported betrayal predicted dissociation (across multiple self-report measures) above and beyond self-reported fear in a sample of trauma survivors, the vast majority of whom reported childhood physical, sexual, and/or emotional abuse. Freyd, Klest, and Allard (2005) found that a history of betrayal trauma was strongly associated with physical and mental health symptoms, including dissociative symptoms, in a sample of ill individuals. Goldsmith, Freyd, and DePrince (2004) reported similar results in a sample of college students.

Some researchers have presumably failed to find a statistically significant relationship between betrayal trauma and memory impairment. For instance, Goodman and colleagues (2003) reported that that "relationship betrayal" was not a statistically significant predictor for forgetting in their unusual sample of adults who had been involved in child abuse prosecution cases during childhood. It is not clear whether the relationship truly does not exist in this sample (which is possible given how unusual a sample it is) or whether there was simply insufficient statistical power to detect the relationship (see Zurbriggen & Becker-Blease, 2003). Future research will need to clarify these issues. At this point we know that betrayal effects on memorability of abuse have been found in at least seven data sets (see the previous paragraph).

Dissociation and Cognitive Mechanisms

Phenomenologically, dissociation involves alterations in attention and memory. Thus, it seems that basic cognitive processes involved in attention and memory most likely play an important role in dissociating explicit awareness of betrayal traumas. Across several studies, empirical support for the relationship between dissociation and knowledge isolation in laboratory tasks has been found. Using the classic Stroop task, Freyd, Martorello, Alvarado, Hayes, and Christman (1998) found that participants who scored high on the DES showed greater Stroop interference than individuals with low DES scores, suggesting that they had more difficulty with the selective attention task than low dissociators. Freyd and colleagues' results suggest a basic relationship between selective attention and dissociative tendencies. A follow-up study tested high- and low-scoring DES groups using a Stroop paradigm, with both selective and divided attention conditions; participants saw stimuli that included color terms (e.g., *red* in red ink), baseline strings of *x*'s, neutral words, and trauma-related words, such as *incest* and *rape* (DePrince & Freyd, 1999). A significant DES × attention task interaction revealed that high-scoring DES participants' reaction time was worse (slower) in the selective attention task than in the divided attention task compared to low-scoring dissociators' performance (replication and extension of Freyd et al.). A significant interaction of dissociation by word category revealed that high-scoring DES participants recalled more neutral and fewer trauma-related words than did low-scoring DES participants, who showed the opposite pattern. Consistent with betrayal trauma theory, the free recall finding supports the argument that dissociation may help to keep threatening information from awareness. DeRuiter, Phaf, Elzinga, and van Dyck (2004) extended observations of attention–dissociation relationships to examine working memory in an undergraduate sample; working memory has been observed to be closely related to attention. As predicted, they found the verbal span of the high-scoring dissociative group was larger than the medium- or low-scoring dissociative groups.

In two follow-up studies that used a directed forgetting paradigm (a laboratory task in which participants are presented with items and told after each item, or a list of items, whether to remember or forget the material), high-scoring DES participants recalled fewer charged and more neutral words relative to low-scoring DES participants, who showed the opposite pattern for items they were instructed to remember when divided attention was required (item method: DePrince & Freyd, 2001; list method: DePrince & Freyd, 2004). The high-scoring participants report significantly more trauma history (Freyd & DePrince, 2001) and betrayal trauma (DePrince & Freyd, 2004). Two additional studies have replicated this pattern in undergraduate samples, revealing an average effect size for the interaction across studies of $d = 0.67$ (DePrince et al., 2007). Similar findings have been found in children by researchers using pictures instead of words as stimuli. Children who had trauma histories and who were highly dissociative recognized fewer charged pictures relative to nontraumatized children under divided attention conditions; no group differences were found under selective attention conditions (Becker-Blease, Freyd, & Pears, 2004).

Other research using the standard (selective attention) directed forgetting paradigm converges on these findings. Moulds and Bryant (2002) compared participants diagnosed with ASD and nontraumatized controls on a directed forgetting task. ASD includes dissociation among the diagnostic criteria. All participants with ASD had been exposed to some form of physical threat. The ASD group showed poorer recall of to-be-forgotten trauma-related words than the nontraumatized group. Elzinga, de Beurs, Sergeant, van Dyck, and Phaf (2000) examined directed forgetting performance for neutral and sex words among undergraduate volunteers and patients with dissociative disorder. Under the standard selective attention instructions, directed forgetting of sex words decreased with higher levels of dissociation. Furthermore, dissociative patients and highly dissociative students remembered more overall compared to the low-scoring dissociative group. Elzinga and colleagues argued that the highly dissociative participants may demonstrate special learning abilities. In particular, drawing on activation/elaboration theory, the authors argued that highly dissociative participants may be skilled at elaboration and construction of conscious experiences. Furthermore, elaboration can be used to detect discrepancies; in the case of the directed forgetting paradigm, it may actually be discrepant to try to forget threatening information, such as sex words. Thus, the dissociative participants showed better recall of sex words relative to the low dissociative group. Elzinga and colleagues argued that highly dissociative individuals may use their capacity to construct separate conscious experiences to keep threatening or painful memories from current awareness. Thus, the same skill that results in increased recall in the directed forgetting paradigm may also underlie memory impairment.

As exemplified by this research, dissociation is one theoretically viable route to memory impairment, though many routes exist. For example, memories may be impaired due to incomplete or fragmented encoding; such routes would be consistent with the concept of dissociative amnesia. Alternatively, forgetting can occur due to retrieval blockage. This sort of forgetting (Anderson et al., 2004) may not involve dissociative processes, as currently conceptualized. In future research, it will be important to examine dissociative-related and -unrelated routes to memory disruption for trauma.

In addition to the research reviewed here, several other studies have focused on memory in individuals diagnosed with dissociative identity disorder (DID) and other dissociative disorders. This work has included examinations of working memory (e.g., Dorahy, Irwin, & Middleton, 2003; Dorahy, Middleton, & Irwin, 2004), as well as interidentity memory in DID (e.g., Elzinga, Phaf, Ardon, & Dyck, 2003; Huntjens, Postma, Peters, Woertman, & van der Hart, 2003). Taken together, the advancement of the use of cognitive methods to examine dissociation, memory, and attention points to exciting discoveries that we hope will add to the growing literature on intervention for trauma-induced dissociation.

CURRENT STATE OF THE ART

Research to date has examined dissociative responses across a broad range of traumas (e.g., Bremner et al., 1992; Bryant & Harvey, 2000; Carlson & Rosser-Hogan, 1991; Freyd, 1996), developmental stages (e.g., Putnam, 1997), and cultures (e.g., Carlson & Rosser-Hogan, 1991; Dorahy & Paterson, 2005). Although culture-specific dissociative reactions exist, the core components of pathological dissociation appear similar across cultures (see Putnam, 1997). The generalizability of findings at any given time is tied to the field's ongoing struggle to better define the construct of dissociation. Findings based on a continuum view of dissociation, for example, may or may not fully generalize to our knowledge of pathological dissociation. We are hopeful that as we more precisely define dissociative symptoms, we reduce the risk of pathologizing experiences that include alterations in consciousness that do not involve structural dissociation. For example, trance experiences, or certain religious experiences, in other cultures are not viewed as pathological dissociation with our advances in defining dissociative symptoms. Furthermore, normally distributed attributes, such as absorption, are at less risk of being defined as pathological. For example, imaginary play has, at times, been suspected of correlating with problematic outcomes in children. Play involves absorption. Recently, however, Taylor, Carlson, Maring, Gerow, and Charley (2004) found that imaginary friends are very common in children (65% of children up to 7 years of age had an imaginary companion at one point in their lives) and the lack of impersonation of imaginary characters was associated with poorer emotion understanding.

CHALLENGES FOR THE FUTURE

Types of Dissociation

We applaud recent work that has involved stepping back from the past two decades of observation to reevaluate the definition of dissociation. Continued work is needed to fine-tune what experiences we include in the category of trauma-induced dissociation. With conceptual clarity about the operationalization of dissociation comes the promise of increased capacity to identify dissociative developmental pathways and mechanisms.

The Evolving Definition of Dissociation

Several challenges remain in the quest to specify further the operationalization of dissociation, including continued work to examine differences in pathological versus nonpathological views of dissociation. At the same time that we work to exclude normative phenomena (e.g., absorption), we must work to ensure that we do not exclude relevant phenomena. For example, much of the contemporary literature on dissociation has focused on dissociation of mental functions (e.g., memory and attention). Work by Nijenjuis and colleagues (1998) points to the importance of including dissociation of perceptual, movement, and sensory information. In addition, constructs that may be dissociative in nature, such as alexithymia, have not yet been included routinely in analyses. Alexithymia is the inability to label emotions, a phenomenon that may be consistent with the lack of integration observed in dissociation.

   As researchers and clinicians improve on the scope of definitions of dissociation, we will be in a better position to evaluate the relationship between dissociation and other psychiatric phenomena. With more precise definitions and measurement of dissociation, researchers can begin to untangle the complicated picture of comorbidity between dissociation and other forms of trauma-related distress. For example, PTSD and dissociation have long been observed as frequently co-occurring phenomena. There are several reasons that this overlap might be observed; for example, the comorbidity could be due to symptom overlap and/or common underlying mechanisms. Recent work in the PTSD literature suggests that the PTSD cluster of avoidance symptoms may actually comprise two distinct symptom clusters: avoidance and numbing (for a review, see Asmundson, Stapleton, & Taylor, 2004). If this is the case, the extent to which dissociation and numbing overlap must be evaluated.

REFERENCES

Anderson, M. C., Ochsner, K. N., Kuhl, B., Cooper, J., Robertson, E., Gabrieli, S. W., et al. (2004). Neural systems underlying the suppression of unwanted memories. *Science, 303,* 232–235.

Armstrong, J. G., Putnam, F. W., Carlson, E. B., Libero, D. Z., & Smith, S. R. (1997). Development and validation of a measure of adolescent dissociation: The adolescent dissociative experiences scale. *Journal of Nervous and Mental Disease, 185,* 491–497.

Asmundson, G. J. G., Stapleton, J. A., & Taylor, S. (2004). Are avoidance and numbing distinct PTSD symptom clusters? *Journal of Traumatic Stress, 17,* 467–477.

Becker-Blease, K. A., Deater-Deckard, K., Eiley, T., Freyd, J. J., Stevenson, J., & Plomin, R. (2004). A genetic analysis of individual differences in dissociative behaviors in childhood and adolescence. *Journal of Child Psychology and Psychiatry, 45,* 522–532.

Becker-Blease, K. A., Freyd, J. J., & Pears, K. C. (2004). Preschoolers' memory for threatening information depends on trauma history and attentional context: Implications for the development of dissociation. *Journal of Trauma and Dissociation, 5,* 113–131.

Bernstein, E., & Putnam, F. W. (1986). Development, reliability and validity of a dissociation scale. *Journal of Nervous and Mental Disease, 174,* 727–735.

Braun, B. G. (1988). The BASK model of dissociation. *Dissociation, 1,* 4–21.

Bremner, J. D., Southwick, S., & Brett, E. (1992). Dissociation and posttraumatic stress disorder in Vietnam combat veterans. *American Journal of Psychiatry, 149,* 328–332.

Brett, E. A. (1993). Classifications of posttraumatic stress disorder in DSM-IV: Anxiety disorder, dissociative disorder, or stress disorder? In J. R. T. Davidson & E. B. Foa (Eds.), *Posttraumatic stress disorder: DSM-IV and beyond* (pp. 191–206). Washington, DC: Author.

Brewin, C. (2003). *Post-traumatic stress disorder: Malady or myth?* New Haven, CT: Yale University Press.

Bryant, R. A., & Harvey, A. G. (2000). *Acute stress disorder: A handbook of theory, assessment, and treatment.* Washington, DC: American Psychological Association.

Carlier, I. V. E., Lamberts, R. D., Fouwels, A. J., & Gersons, B. P. R. (1996). PTSD in relation to dissociation in traumatized police officers. *American Journal of Psychiatry, 153,* 1325–1328.

Carlson, E. B., & Rosser-Hogan, R. (1991). Trauma experiences, posttraumatic stress, dissociation and depression in Cambodian refugees. *American Journal of Psychiatry, 148,* 1548–1551.

Cameron, C. (1993, April). *Recovering memories of childhood sexual abuse: A longitudinal report.* Paper presented at the Western Psychological Association Convention, Phoenix, AZ.

Chu, J. A., & Dill, D. L. (1990). Dissociative symptoms in relation to childhood physical and sexual abuse. *American Journal of Psychiatry, 147,* 887–892.

Dalenberg, C. J., & Palesh, O. G. (2004). Relationship between child abuse history, trauma, and dissociation in Russian college students. *Child Abuse and Neglect, 28,* 461–474.

Dell, P. F. (2006). The Multidimensionality Inventory of Dissociation (MID): A comprehensive measure of pathological dissociation. *Journal of Trauma and Dissociation, 7*(2), 77–106.

DePrince, A. P. (2001). *Trauma and posttraumatic responses: An examination of fear and betrayal.* Unpublished doctoral dissertation, University of Oregon.

DePrince, A. P. (2005). Social cognition and revictimization risk. *Journal of Trauma and Dissociation, 6,* 125–141.

DePrince, A. P., & Freyd, J. J. (1999). Dissociative tendencies, attention, and memory. *Psychological Science, 10,* 449–452.

DePrince, A. P., & Freyd, J. J. (2001). Memory and dissociative tendencies: The roles of attentional context and word meaning in a directed forgetting task. *Journal of Trauma and Dissociation, 2*(2), 67–82.

DePrince, A. P., & Freyd, J. J. (2004). Forgetting trauma stimuli. *Psychological Science, 15*, 488–492.

DePrince, A. P., Freyd, J. J., & Malle, B. (2007). A replication by another name: A response to Devilly et al. (2007). *Psychological Science, 18*, 218–219.

de Ruiter, M. B., Phaf, R. H., Elzinga, B. M., & van Dyck, R. (2004). Dissociative style and individual differences in verbal working memory span. *Consciousness and Cognition, 13*, 821–828.

Dorahy, M. J., Irwin, H. J., & Middleton, W. (2003). Assessing markers of working memory function in dissociative identity disorder using neutral stimuli: A comparison with clinical and general population samples. *Australian and New Zealand Journal of Psychiatry, 38*, 47–55.

Dorahy, M. J., Middleton, W., & Irwin, H. J. (2004). Investigating cognitive inhibition in dissociative identity disorder compared to depression, posttraumatic stress disorder and psychosis. *Journal of Trauma and Dissociation, 5*, 93–110.

Dorahy, M. J., & Paterson, M. D. (2005). Trauma and dissociation in Northern Ireland. *Journal of Trauma Practice, 4*, 221–243.

Edwards, V. J., Fivush, R., Anda, R. F., Felitti, V. J., & Nordenberg, D. F. (2001). Autobiographical memory disturbances in childhood abuse survivors. *Journal of Aggression, Maltreatment, and Trauma, 4*, 247–264.

Elzinga, B. M., deBeurs, E., Sergeant, J. A., van Dyck, R., & Phaf, R. H. (2000). Dissociative style and directed forgetting. *Cognitive Therapy and Research, 24*, 279–295.

Elzinga, B. M., Phaf, R. H., Ardon, A. M., & van Dyck, R. (2003). Directed forgetting between, but not within, dissociative personality states. *Journal of Abnormal Psychology, 112*, 237–243.

Feldman-Summers, S., & Pope, K. S. (1994). The experience of 'forgetting' childhood abuse: A national survey of psychologists. *Journal of Consulting and Clinical Psychology, 62*, 636–639.

Francia-Martinez, M., de Torres, I. R., Alvarado, C. S., Martinez-Taboas, A., & Sayers, S. (2003). Dissociation, depression and trauma in psychiatric inpatients in Puerto Rico. *Journal of Trauma and Dissociation, 4*, 47–61.

Freyd, J. J. (1996). Blind to betrayal: New perspectives on memory for trauma. *Harvard Mental Health Letter, 15*, 4–6.

Freyd, J. J., & DePrince, A. P. (2001). Perspectives on memory for trauma and cognitive processes associated with dissociative tendencies. *Journal of Aggression, Maltreatment, and Trauma, 4*(2), 137–163.

Freyd, J. J., DePrince, A. P., & Zurbriggen, E. L. (2001). Self-reported memory for abuse depends upon victim–perpetrator relationship. *Journal of Trauma and Dissociation, 2*, 5–17.

Freyd, J. J., Klest, B., & Allard, C. B. (2005). Betrayal trauma: Relationship to physical health, psychological distress, and a written disclosure intervention. *Journal of Trauma and Dissociation, 6*(3), 83–104.

Freyd, J. J., Martorello, S. R., Alvarado, J. S., Hayes, A. E., & Christman, J. C. (1998). Cognitive environments and dissociative tendencies: Performance on the Standard Stroop task for high versus low dissociators. *Applied Cognitive Psychology, 12*, S91–S103.

Gershuny, B. S., Cloitre, M., & Otto, M. W. (2003). Peritraumatic dissociation and PTSD severity: Do event-related fears about death and control mediate their relation? *Behaviour Research and Therapy, 41*, 157–166.

Goldberg, L. R., & Freyd, J. J. (2006). Self-reports of potentially traumatic experiences in an adult community sample: Gender differences and test-retest stabilities of the items in a Brief Betrayal-Trauma Survey. *Journal of Trauma and Dissociation, 7*(3), 39–63.

Goldsmith, R. E., Freyd, J. J., & DePrince, A. P. (2004, February). *Health correlates of exposure to betrayal trauma.* Poster presented at the annual meeting of the American Association for the Advancement of Science, Seattle, WA.

Goodman, G. S., Ghetti, S., Quas, J. A., Edelstein, R. S., Alexander, K. W., Redlich, A. D., et al. (2003). A Prospective study of memory for child sexual abuse: New findings relevant to the repressed-memory debate. *Psychological Science, 14*, 113–118.

Herman, J. L. (1992). *Trauma and recovery.* New York: Basic Books.

Hilgard, E. R. (1986). *Divided consciousness: Multiple controls in human thought and action.* New York: Wiley.

Huntjens, R. J. C., Postma, A., Peters, M. L., Woertman, L., & van der Hart, O. (2003). Interidentity amnesia for neutral, episodic information in dissociative identity disorder. *Journal of Abnormal Psychology, 112,* 290–297.

Irwin, H. J. (1999). Pathological and nonpathological dissociation: The relevance of childhood trauma. *Journal of Psychology, 133,* 157–164.

Koopman, C., Classen, C., & Spiegel, D. (1996). Predictors of posttraumatic stress symptoms among survivors of the Oakland/Berkeley, Calif., firestorm. *American Journal of Psychiatry, 151,* 888–894.

Maldonado, J. R., Butler, L. D., & Spiegel, D. (2002). Treatments for dissociative disorders. In P. E. Nathan & J. M. Gordon (Eds.), *A guide to treatments that work* (2nd ed., pp. 463–496). New York: Oxford University Press.

Maldonado, J. R., & Spiegel, D. (1998). Trauma, dissociation, and hypnotizability. In J. D. Bremner & C. R. Marmar (Eds.), *Trauma, memory and dissociation* (pp. 57–106). Washington, DC: American Psychiatric Press.

Marmar, C. R., Weiss, D. S., & Metzler, T. J. (1997). The Peritraumatic Dissociative Experiences Questionnaire. In J. P. Wilson & T. M. Keane (Eds.), *Assessing psychological trauma and PTSD* (pp. 412–428). New York: Guilford Press.

Marshall, G. N., Orlando, M., Jaycox, L. H., Foy, D. W., & Belzberg, H. (2002). Development and validation of a modified version of the Peritraumatic Dissociative Experiences Questionnaire. *Psychological Assessment, 14*(2), 123–134.

Merckelbach, H., Horselenberg, R., & Schmidt, H. (2002). Modeling the connection between self-reported trauma and dissociation in a student sample. *Personality and Individual Differences, 32*(4), 695–705.

Moulds, M. L., & Bryant, R. A. (2002). Directed forgetting in acute stress disorder. *Journal of Abnormal Psychology, 111,* 175–179.

Myers, C. S. (1940). *Shell shock in France 1914–18.* Cambridge, UK: Cambridge University Press.

Nijenhuis, E. R. S., Spinhoven, P., van Dyck, R., van der Hart, O., & Vanderlinden, J. (1998). Psychometric characteristics of the Somatoform Dissociation Questionnaire: A replication study. *Psychotherapy and Psychosomatics, 67,* 17–23.

Ogawa, J. R., Sroufe, L. A., Weinfield, N. S., Carlson, E. A., & Egeland, B. (1997). Development and the fragmented self: Longitudinal study of dissociative symptomatology in a nonclinical sample. *Development and Psychopathology, 9,* 855–879.

Ozer, E. J., Best, S. R., Lipsey, T. L., & Weiss, D. S. (2003). Predictors of posttraumatic stress disorder and symptoms in adults: A meta-analysis. *Psychological Bulletin, 129,* 52–73.

Panasetis, P., & Bryant, R. A. (2003). Peritraumatic versus persistent dissociation in acute stress disorder. *Journal of Traumatic Stress, 16,* 563–566.

Plattner, B., Silvermann, M. A., Redlich, A. D., Carrion, V. G., Feucht, M., Friedrich, M. H., et al. (2003). Pathways to dissociation: Intrafamilial versus extrafamilial trauma in juvenile delinquents. *Journal of Nervous and Mental Disease, 191,* 781–788.

Putnam, F. W. (1989). Pierre Janet and modern views of dissociation. *Journal of Traumatic Stress, 2,* 413–428.

Putnam, F. W. (1997). *Dissociation in children and adolescents: A developmental perspective.* New York: Guilford Press.

Putnam, F. W., Helmers, K., & Trickett, P. K. (1993). Development, reliability, and validity of a child dissociation scale. *Child Abuse and Neglect, 17,* 731–741.

Putnam, F. W., & Trickett, P. K. (1997). The psychobiological effects of sexual abuse: A longitudinal study. *Annals of the New York Academy of Sciences, 821,* 150–159.

Ross, C. A., Ellason, J. W., & Anderson, G. (1995). A factor analysis of the Dissociative Experiences Scale (DES) in dissociative identity disorder. *Dissociation: Progress in the Dissociative Disorders, 8,* 229–235.

Sanders, B., & Green, J. A. (1994). The factor structure of the Dissociative Experiences Scale in college students. *Dissociation: Progress in the Dissociative Disorders, 7,* 23–27.

Schultz, T. M., Passmore, J., & Yoder, C. Y. (2003). Emotional closeness with perpetrators and amnesia for child sexual abuse. *Journal of Child Sexual Abuse, 12,* 67–88.

Sheiman, J. A. (1999). Sexual abuse history with and without self-report of memory loss: Differences in psychopathology, personality, and dissociation. In L. M. Williams & V. L. Banyard (Eds.), *Trauma and memory* (pp. 139–148). Thousand Oaks, CA: Sage.

Stoler, L. R. (2001). Recovered and continuous memories of childhood sexual abuse: A quantitative and qualitative analysis. *Dissertation Abstracts International, Section B: The Sciences and Engineering, 61*(10-B), 5582.

Taylor, M., Carlson, S. M., Maring, B. L., Gerow, L., & Charley, C. M. (2004). The characteristics and correlates of fantasy in school-age children: Imaginary companions, impersonation, and social understanding. *Developmental Psychology, 40*, 1173–1187.

Tichenor, V., Marmar, C. R., Weiss, D. S., Metzler, T. J., & Ronfeldt, H. M. (1996). The relationship of peritraumatic dissociation and posttraumatic stress: Findings in female Vietnam theater veterans. *Journal of Consulting and Clinical Psychology, 64*, 1054–1059.

van der Hart, O. (2000, November). *Dissociation: Toward a resolution of 150 years of confusion*. Plenary presented at the annual meeting of the International Society for the Study of Dissociation, San Antonio, TX.

van der Hart, O., Nijenhuis, E., Steele, K., & Brown, D. (2004). Trauma-related dissociation: Conceptual clarity lost and found. *Australian and New Zealand Journal of Psychiatry, 38*, 906–914.

van der Kolk, B. A., & Fisler, R. (1995). Dissociation and the fragmentary nature of traumatic memories: Overview and exploratory study. *Journal of Traumatic Stress, 8*, 505–525.

van der Kolk, B. A., Weisaeth, L., & van der Hart, O. (1996). History of trauma in psychiatry. In B. A. van der Kolk, A. C. McFarlane, & L. Weisaeth (Eds.), *Traumatic stress: The effects of overwhelming experience on mindy, body, and society* (pp. 47–76). New York: Guilford Press.

Waller, N. G., Putnam, F. W., & Carlson, E. B. (1996). Types of dissociation and dissociative types: A taxometric analysis of dissociative experiences. *Psychological Methods, 1*, 300–321.

Weiss, D. S., Marmar, C. R., Metzler, T. J., & Ronfeldt, H. M. (1995). Predicting symptomatic distress in emergency services personnel. *Journal of Consulting and Clinical Psychology, 63*, 361–368.

Williams, L. M. (1994). Recall of childhood trauma: A prospective study of women's memories of child sexual abuse. *Journal of Consulting and Clinical Psychology, 62*, 1167–1176.

Williams, L. M. (1995). Recovered memories of abuse in women with documented child sexual victimization histories. *Journal of Traumatic Stress, 8*, 649–674.

Yehuda, R., Elkin, A., Binder-Brynes, K., Kahana, B., Southwick, S. M., Schmeidler, J., et al. (1996). Dissociation in aging Holocaust survivors. *American Journal of Psychiatry, 153*, 935–940.

Zurbriggen, E. L., & Becker-Blease, K. (2003). Predicting memory for childhood sexual abuse: "Nonsignificant" findings with the potential for significant harm. *Journal of Child Sexual Abuse, 12*, 113–121.

Zurbriggen, E. L., & Freyd, J. J. (2004). The link between childhood sexual abuse and risky sexual behavior: The role of dissociative tendencies, information-processing effects, and consensual sex decision mechanisms. In L. J. Koenig, L. S. Doll, A. O'Leary, & W. Pequegnat (Eds.), *From child sexual abuse to adult sexual risk: Trauma, revictimization, and intervention* (pp. 135–158). Washington, DC: American Psychological Association.

Chapter 9

Neurocircuitry and Neuroplasticity in PTSD

Alexander Neumeister, Shannan Henry, and John H. Krystal

The past decade has seen a rapid progression in our knowledge of the neurobiological basis of fear and anxiety. Specific neurochemical and neuropeptide systems have been demonstrated to play important roles in the behaviors associated with fear and anxiety-producing stimuli. Long-term dysregulation of these systems appears to contribute to the development of anxiety disorders, including posttraumatic stress disorder (PTSD). These neurochemical and neuropeptide systems have been shown to have effects on distinct cortical and subcortical brain areas that are relevant to the mediation of the symptoms associated with PTSD. Moreover, advances in molecular genetics portend the identification of the genes that underlie the neurobiological disturbances that increase vulnerability to anxiety disorders, including PTSD. This chapter reviews clinical research pertinent to the neurobiological basis of PTSD, with an emphasis on neurocircuitry and neuroplasticity. The implications of these novel discoveries on our understanding of the pathophysiology of PTSD are discussed.

METHODOLOGICAL CONSIDERATIONS

Functional Neurocircuitry of Fear and Anxiety

Evidence from a large body of preclinical studies provides a basis for proposing a neurocircuit of anxiety and fear. The brain structures constituting a neural circuit of anxiety or fear should have several features:

151

1. There must be sufficient afferent sensory input to permit assessment of the fear- or anxiety-producing nature of external or internal stimuli.
2. The neuronal interactions among the brain structures must be capable of incorporating an individual's prior experience (memory) into the cognitive appraisal of stimuli. These interactions are important in the attachment of affective significance to specific stimuli and the mobilization of adaptive behavioral responses.
3. The efferent projections from the brain structures should be able to mediate an individual's neuroendocrine, autonomic, and skeletal motor responses to threat both to facilitate survival and to account for the pathological reactions that result in anxiety-related signs and symptoms.

The major afferent arms of the neural circuitry of anxiety include the exteroceptive sensory systems of the brain (auditory, visual, somatosensory), comprising serially organized relay channels that convey directly or through multisynaptic pathways information relevant to the experience of fear or anxiety. The sensory information contained in a fear- or anxiety-inducing stimulus is transmitted from peripheral receptor cells to the dorsal thalamus (LeDoux et al., 1987). An exception is the olfactory system, which does not relay information through the thalamus, and whose principal targets in the brain are the amygdala and entorhinal cortex (Turner, Gupta, & Mishkin, 1978). Visceral afferent pathways alter the function of the locus coeruleus and the amygdala, either through direct connections or via the nucleus paragigantocellularis (PGI) and the nucleus tractus solitarius (Elam, Svensson, & Thoren, 1986; Saper, 1982; Whitlock & Nauta, 1956).

The thalamus relays sensory information to the primary sensory receptive areas of the cortex. In turn, these primary sensory regions project to adjacent unimodal and polymodal cortical association areas (Jones, 1983; Jones & Powell, 1970; Mesulam, Van Hoesen, Pandya, & Geschwind, 1977). The cortical association areas of visual, auditory, and somatosensory systems send projections to other brain structures, including the amygdala, entorhinal cortex, orbitofrontal cortex, and cingulate gyrus (Turner, Mishkin, & Knapp, 1980; Van Hoesen, Pandya, & Butters, 1972; Vogt & Miller, 1983). The hippocampus receives convergent, integrated inputs from all sensory systems by way of projections from entorhinal cortex (Swanson, 1983).

Thus, much of the sensory information of fear- and anxiety-inducing stimuli are first processed in the sensory cortex prior to transfer to subcortical structures, which are more involved in affective, behavioral, and somatic responses. It is noteworthy that the amygdala also receives sensory information directly from the thalamus. The medial geniculate nuclei of the thalamus (acoustic thalamus) send projections to the amygdala and hypothalamus. The thalamic areas associated with the visual system also innervate the amygdala. These data support a pivotal role for the amygdala in the transmission and interpretation of fear- and anxiety-inducing sensory information, because it receives afferents from thalamic and cortical exteroceptive systems, as well as subcortical visceral afferent pathways (Amaral, Price, Pitanken, & Carmichael, 1992). The neuronal interactions between the amygdala and cortical regions, such as the orbitofrontal cortex, enable the individual to initiate adaptive behaviors to threat based on the nature of the threat and prior experience. Current research indicates that the production of emotions may be more dependent on a neural network comprising cortical and limbic regions, than on the activity of a single region (Anand et al., 2005).

The efferent pathways of the anxiety–fear circuit mediate autonomic, neuroendocrine, and skeletal–motor responses. The structures involved in these responses include the amygdala, locus coeruleus, hypothalamus, periaqueductal gray (PAG), and striatum.

Many of the autonomic changes produced by anxiety- and fear-inducing stimuli are produced by the sympathetic and parasympathetic neural systems. Stimulation of the lateral hypothalamus results in sympathetic system activation—increases in blood pressure and heart rate, sweating, piloerection, and pupil dilatation. Activation of the paraventricular nucleus of the hypothalamus promotes the release of a variety of hormones and peptides. The hypothalamus integrates information it receives from a variety of brain structures into a coordinated pattern of sympathetic responses. The sympathetic activation and hormonal release associated with anxiety and fear are probably mediated in part by stimulation of the hypothalamus via projections from the amygdala and locus coeruleus (LeDoux, Iwata, Cicchetti, & Reis, 1988; Sawchenko & Swanson, 1982, 1983). In addition, the PGI also plays an important role in regulating sympathetic function and may account for the parallel activation of the peripheral sympathetic system and the locus coeruleus.

The vagus and splanchnic nerves are major projections of the parasympathetic nervous system. Afferents to the vagus include the lateral hypothalamus, paraventricular nucleus, locus coeruleus, and the amygdala. There are afferent connections to the splanchnic nerves from the locus coeruleus (Clark & Proudfit, 1991). This innervation of the parasympathetic nervous system may relate to visceral symptoms associated with anxiety, such as gastrointestinal and genitourinary disturbances.

The regulatory control of skeletal muscle by the brain in response to emotions is complex. Both subtle movements involving a few muscle groups (facial muscles) and fully integrated responses requiring the entire musculoskeletal system for fight or flight may be required. Adaptive mobilization of the skeletal motor system to respond to threat probably involves pathways between the cortical association areas and motor cortex, cortical association areas and the striatum, and the amygdala and striatum (see Figure 9.1).

The amygdala also has strong projections to most areas of the striatum, including the nucleus accumbens, olfactory tubercle, and parts of the caudate and putamen. The portion of the striatum that is innervated by the amygdala also receives efferents from the orbitofrontal cortex and the ventral tegmental area. The amygdalocortical and amygdalostriatal projections are topographically organized. Individual areas of the amygdala, and in some cases individual amygdaloid neurons, can integrate information from the cortial–striatal–pallidal systems. The dense innervation of the striatum and prefrontal cortex by the amygdala indicates that the amygdala can powerfully regulate both of these systems (McDonald, 1991a, 1991b). These interactions between the amygdala and the extrapyramidal motor system may be very important for generating motor responses to threatening stimuli, especially those related to prior adverse experiences. Other functional neurocircuits that are anatomically related include the hippocampus and the prefrontal cortex. Neuroimaging abnormalities of these areas, which are closely connected to the amygdala, have been described in the literature in individuals with stress-related disorders and are also discussed in this chapter.

Functional Neurocircuitry of Reexperiencing

Integration of conditioned stimuli (CSs) and unconditioned stimuli (UCSs), and the resultant fear conditioning and response are mediated by the amygdala and its projections (LeDoux, 2000). Cue CSs are transmitted by external and visceral sensory path-

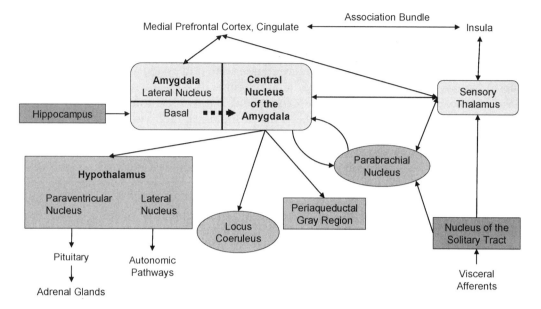

FIGURE 9.1. The amygdala, with its connections to the hippocampus and cortical regions, and the anterior and posterior cingulated cortices and the orbitofrontal cortex are responsible for determining the emotional "content" of information and processing emotionally valenced stimuli. Given its central role in the circuit, considerable research has attempted to define the role of the amygdala in the regulation of fear and anxiety, and its role in the pathogenesis of PTSD.

ways to the thalamus. Afferents then reach the basolateral nucleus of the amygdala (BLA) via two parallel neural circuits: a rapid subcortical path ("short loop") directly from the dorsal ("sensory") thalamus, and a slower regulatory, cortical pathway ("long loop") encompassing the primary somatosensory cortices, insula, and anterior cingulate/prefrontal cortex (LeDoux, 2000). Context CSs are projected to the lateral amygdala from the hippocampus (Phillips & LeDoux, 1992) and perhaps also from the bed nucleus of the stria terminalis (BNST; Davis, Walker, & Lee, 1997). Human subjects with lesions of the amygdala and adjacent regions show impaired fear conditioning. Pathways conveying the UCSs have not been studied as much but are believed to reach the (basolateral/central nucleus [CE]) amygdala from the thalamus, parabrachial area, spinal cord, and somatosensory cortical regions (LeDoux, 2000).

Less is currently known about neuroanatomy and mechanisms of extinction. Opinions differ regarding the role of the medial prefrontal cortex (LeDoux, 2000) or amygdala (Davis et al., 1997) in this process. Activity of the right prefrontal cortex (PFC) was shown to be inversely correlated with amygdala activity, suggesting an inhibitory modulatory influence.

Human functional magnetic resonance imaging (fMRI) studies in healthy subjects show fear-conditioning-related activation in the amygdala, anterior cingulate, and precentral regions (Buchel, Dolan, Armony, & Friston, 1999; LaBar, Gatenby, Gore, LeDoux, & Phelps, 1998). Positron emission tomographic (PET) studies of aversive conditioning do not depict amygdala involvement in fear conditioning, but describe activation of diverse cortical regions, such as anterior cingulate precentral and premotor regions, and orbitofrontal, prefrontal, and temporal cortices.

The amygdala is also implicated in memory processing of particularly emotionally arousing events (McGaugh, 2000). Lesions of the amygdala block the enhancing effects of emotional arousal on memory consolidation, as well as the memory modulatory effects of systemic administration of catecholamines and cortisol (Roozendaal, 2000). These effects are mediated by the BLA and BNST.

Functional imaging studies in PTSD showed right amygdala activation when patients and controls were exposed to traumatic imagery and pictures, whereas the left amygdala was activated in response to sounds. Such studies also showed decreased perfusion in the anterior cingulate in patients with PTSD after traumatic memory provocation. fMRI studies similarly showed reduced perfusion in the anterior cingulate in patients with PTSD compared to controls after emotional provocation (see Pitman, Shin, & Rauch, 2001, for review).

Neuroimaging Techniques to Study PTSD

The increasing availability and advancement of neuroimaging technology provide a solid backbone for ongoing studies aimed at deciphering neurological differences between subjects diagnosed with psychiatric disorders, such as PTSD, and healthy controls (defined as individuals without a history of major psychiatric illness). Existing neuroimaging techniques, including PET, single photon emission computed tomography (SPECT), and fMRI, offer the ability to assess regional cerebral blood flow and glucose metabolism in PTSD subjects *in vivo*. Previous studies have included scanning subjects in a resting state, during pharmacological challenges, engaged in cognitive tasks, or while experiencing functional stimuli (e.g., viewing faces depicting various emotional states). Functional imaging allows pairing the visualization of brain activity with paradigms and tasks designed to elicit activation in specific brain areas, thus parsing out hypothesized differences between subject groups. PET and SPECT additionally provide a modality for examining neurotransmitter systems. By studying specific neuronal pathways throughout the brain that distribute neurotransmitters such as serotonin (5-HT) and dopamine (DA), we are able to gain an understanding of how alterations in brain function may contribute to the manifestation of anomalous behavioral traits (i.e., PTSD symptomology).

Imaging studies have already begun helping researchers to determine brain pathways and neuronal circuits that may be associated with behavior, such as the involvement of the amygdala in emotional processing. Techniques differ in resolution, radioligand availability, and analysis methods, providing an array of imaging options suitable for examining a wide range of questions regarding neuronal structure and function. PET and SPECT allow study of neurotransmitters but are inadequate for distinguishing differences in some smaller brain regions, such as the raphe and amygdala (due to poor temporal resolution). In contrast, fMRI allows for detection of signals in smaller areas, thus determining distinctions in cerebral blood flow in small regions. fMRI studies in larger samples may provide definitive answers regarding cerebral blood flow differences between subjects with PTSD and controls. MRI studies may also focus on brain volume comparisons, revealing possible structural abnormalities (i.e., atrophy) in brain regions associated with PTSD.

Ultimately, neuronal circuits are not independently operating mechanisms, each translating into one specific behavioral phenomenon; rather, they work in conjunction with multiple pathways and neurotransmitter systems. Thus, utilizing various neuroimaging techniques is useful in identifying regions of altered activity and may contribute to understanding the underlying pathophysiology of PTSD.

CURRENT STATE OF THE ART

Neuroimaging Abnormalities in PTSD

Hippocampal Volumetric Studies

Clinical investigation of hippocampal volume in PTSD was largely stimulated by numerous preclinical studies reporting hippocampal neuronal loss and dendritic atrophy following exposure to hydrocortisone or psychosocial stress in rats (Sapolsky, 2000; Watanabe, Gould, & McEwen, 1992). The first report by Bremner and colleagues (1995) of a small but significant decrease (8%) in the body of the right hippocampus in patients with combat-related PTSD was confirmed by three studies from two independent groups (Gurvits et al., 1996; Stein, Koverola, Hanna, Torchia, & McClarty, 1997). The decrease in hippocampal volume was also seen in patients with PTSD related to childhood sexual and/or physical abuse (see Figure 9.2). More recently, Bremner and colleagues (2003) reported a significant 18% reduction in left hippocampal volume in patients with early childhood abuse and current major depressive disorder (MDD) compared to patients with current MDD without abuse after controlling for age, race, education, whole-brain volume, alcohol use and PTSD. Most recently, Bonne and colleagues report hippocampal volume loss in noncombat patients with PTSD compared to healthy controls (Omer Bonne, personal communication, November 4, 2004). Notably, the volume reduction was confined to the posterior hippocampus, which may be relevant for the pathophysiology of PTSD given that the posterior hippocampus has been associated with processing, storage, and retrieval of spatiotemporal information.

Other studies failed to find a reduction in hippocampal volume in Holocaust victims (Golier et al., 2000), combat veterans (Schuff et al., 1997), or women with PTSD secondary to domestic violence. Longitudinal MRI studies in adult subjects with PTSD immediately and 6 months after a motor vehicle accident (Bonne et al., 2001) and a 2-year study in children with sexual and/or physical abuse–related PTSD (De Bellis, Hall, Boring, Frustaci, & Moritz, 2001) did not find reductions in hippocampal volume. The largest volumetric study to date performed in 44 abused children with PTSD and 61 healthy children found a significant decrease in intracranial and corpus callosal volume but not hippocampal volume (DeBellis et al., 1999). A post hoc analysis in patients with alcohol dependence showed that the volume of the hippocampus in women with alcohol dependence and PTSD was similar to that of women with alcohol dependence without PTSD (Agartz, Momenan, Rawlings, Kerich, & Hommer, 1999).

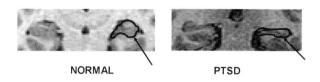

NORMAL   PTSD

FIGURE 9.2. Hippocampal volume reduction in PTSD. MRI scan of the hippocampus in a normal control and a patient wiht PTSD secondary to childhood abuse. The hippocampus (outlined) is visibly smaller in PTSD. Overall, there was a 12% reduction in volume in PTSD.

Several factors may explain the contradictory findings in hippocampal volume studies in PTSD. Trauma variables include differences in the kinds of trauma (sexual abuse/rape, physical abuse, witnessing violence, motor vehicle accident, combat, victim of mugging, etc.), duration of trauma (repeated episodes over a period of years vs. a single episode), severity of trauma, and the timing of trauma in development (prepubertal vs. postpubertal). Differences in the prevalence of comorbid disorders such as major depression and alcohol and substance use might also explain the variance in hippocampal volume. Possibly, exposure to antidepressants could enhance dendritic branching (Duman, Mahlberg, Nakagawa, & D'Sa, 2000; Duman, Mahlberg, & Thome, 1999) and contribute to differences in hippocampal volume in humans. However, a recent hippocampal volume study in people with major depression suggests that maintenance treatment with antidepressants may be necessary to prevent hippocampal volume loss (Neumeister, Wood, et al., 2005). Ongoing twin studies in PTSD have just begun to elucidate the contribution of genetic differences to hippocampal volumetric changes in patients with PTSD. Future studies in patients with PTSD should take into account the factors that might explain heterogeneity in biological markers and study more homogenous groups of traumatized patients, comparing them to control subjects without current PTSD who are exposed to similar kinds of trauma and healthy subjects. An excellent example of this type of study is an article by Gilbertson and colleagues (2002) who studied homozygotic twins discordant for trauma exposure. They found that PTSD severity in patients exposed to trauma was negatively correlated with the hippocampal volume of both the patients and their trauma-unexposed identical co-twins. Furthermore, twin pairs with severe PTSD—both the trauma-exposed and unexposed members—had significantly smaller hippocampi than pairs without PTSD. These data suggest that smaller hippocampal volume may in fact be a vulnerability marker for PTSD rather than a consequence of exposure to trauma or PTSD.

Putative pathophysiological mechanisms for smaller hippocampal volume include increased levels of cortisol at the time of the trauma and increased levels of excitatory amino acids such as glutamate. Evaluating whether patients with PTSD have increased central glucocorticoid sensitivity may help to address whether this phenomenon is related to smaller hippocampal volume. If patients with PTSD have increased central glucocorticoid sensitivity and/or frequent episodes of stress-induced hypercortisolemia, cumulative hippocampal neuronal loss may ensue over time.

Abnormalities in the Amygdala and Prefrontal Cortex

Several functional imaging studies have confirmed a connection between abnormalities in the amygdala and PFC in patients with PTSD (Figure 9.1). Although preclinical and clinical literature have supported the observation that adverse experiences early in development lead to increased hypothalamic–pituitary–adrenal (HPA) responsiveness to stress in adulthood, it is not clear whether patients with PTSD have greater central sensitivity to "stress" levels of hydrocortisone compared to trauma controls and healthy subjects. Because glucocorticoid receptors are abundantly distributed in these regions, it is important to determine whether glucose utilization as a measure of brain function in the amygdala and PFC regions is inhibited, in addition to the hippocampal region in patients with PTSD after hydrocortisone administration. Therefore, it will be relevant to evaluate central glucocorticoid sensitivity by measuring changes in glucose metabolism, memory, and PTSD symptoms following hydrocortisone administration. Besides enhancing our understanding of the effects of the role of possibly increased HPA responsiveness in at least subgroups of patients with PTSD, these studies may ultimately lead to the identification of novel treatments for PTSD via reversing HPA axis hyper-responsiveness.

Although the role of the amygdala in fear conditioning has been demonstrated repeatedly in preclinical studies, few human studies have demonstrated activation of the amygdala during acquisition and extinction phases of fear conditioning (LaBar et al., 1998). Studies of amygdala activation in PTSD are mixed, and are confounded by different behavioral paradigms of symptom provocation used in the various studies. Right amygdala activation was reported in combat-related PTSD when patients and controls were exposed to traumatic imagery and combat pictures (Rauch et al., 1996; Shin et al., 1997), whereas the left amygdala was activated in response to combat sounds (Liberzon et al., 1999). However, other studies that used similar paradigms were unable to replicate the finding of greater amygdala activation in patients with PTSD (Bremner, Staib, et al., 1999).

An fMRI study by Rauch and colleagues (2000) extended the masked faces paradigm to patients with PTSD. This study was an important step in our understanding of how the amygdala respond to emotionally valenced stimuli. Specifically, activity within the amygdala is increased in response to pictures of fearful versus neutral versus happy faces. Use of the masked-face paradigm highlights the automaticity of the amygdala response, because activation of other brain regions outside the amygdala is minimized. In particular, nonmasked faces paradigms lead to amygdala and medial frontal activation, which is important in understanding the circuit involved in processing emotional information but does not tell us about amygdala function itself. Using a modified approach by exposing individuals to pictures of fearful versus neutral or happy faces presented below the level of awareness, Whalen and colleagues (1998) demonstrated activation of the amygdala. In contrast, overt presentation of emotional faces resulted in significant medial frontal activation (Morris et al., 1998; Whalen et al., 1998). Rauch and colleagues (2000) found exaggerated amygdala responses to mask-fearful versus mask-happy faces in patients with PTSD compared to combat-exposed veterans without PTSD, suggesting that these patients exhibit exaggerated amygdala activation to general, threat-related stimuli presented at the subliminal level. Studies that use similar symptom provocation designs, but with fMRI techniques in larger samples of patients and controls, may provide definitive answers with respect to the role of the amygdala in PTSD symptoms.

Abnormalities in the functioning of subregions of the medial PFC of patients with PTSD have been shown in PET and SPECT studies using personalized scripts, combat slides, or sounds. The PFC is reciprocally connected to the amygdala, and inhibits acquisition of the fear response and promotes extinction of behavioral response to fear-conditioned stimuli that are no longer reinforced (Morgan & LeDoux, 1995; Quirk, Russo, Barron, & Lebron, 2000). Various subregions of the medial PFC mediate different responses. Lesions of the ventral medial PFC or the orbital cortex prolong the extinction phase (Morgan, Romanski, & LeDoux, 1993), whereas lesions of the dorsomedial PFC (anterior cingulate) facilitated the fear response during the acquisition and extinction phases of fear conditioning, resulting in a generalized increase in fear response (Morgan & LeDoux, 1995). Suppression of the neuronal firing in the subgenual prefrontal cortical equivalent in the rat (prelimbic cortex) is inversely correlated with increase in the amygdala neuronal activity (Garcia, Vouimba, Baudry, & Thompson, 1999). Based on the lesions studies, it can be hypothesized that a dysfunction in medial PFC function can result in a disinhibition of amygdala activity. This in turn promotes acquisition of a fear response to traumatic stimuli; such a dysfunction also results in a failure in extinction of the fear response, even after traumatic stimulation has ceased.

Receptor Imaging Studies

Compared to studies of glucose metabolism and cerebral blood flow, relatively few studies have addressed abnormalities of receptors and transporters that play key roles in the regulations of neurochemical functions in PTSD. Pertinent for the pathophysiology of PTSD are benzodiazepine (BZD) receptor and 5-HT systems.

A recent neuroimaging study assessing BZD receptor binding using SPECT and the radiotracer [^{123}I]iomazenil showed reduced BZD receptor binding potential in the medial PFC in patients with combat-related PTSD relative to controls. Such studies argue for the role of BZD receptor dysfunction in the pathogenesis of PTSD (Bremner et al., 2000).

Altered BZD receptor number or function may be related to alterations in receptor expression of serotonergic receptors. Several recent studies suggest close interactions between serotonergic and gamma-aminobutyric acid (GABA-ergic) systems. Mice lacking the 5-HT$_{1A}$ receptor display marked anxiety (Heisler et al., 1998; Parks, Robinson, Sibille, Shenk, & Toth, 1998; Ramboz et al., 1998), and animals exposed to stress exhibit down-regulation of 5-HT$_{1A}$ receptors (McKittrick et al., 1995). Subordinate rats in a dominance hierarchy show severe anxiety accompanied by reduced 5-HT$_{1A}$ receptor levels (McKittrick, Blanchard, Blanchard, McEwen, & Sakai, 1995). It has been shown that 5-HT$_{1A}$ receptor knock-out mice show (1) a reduction in the alpha$_1$ and alpha$_2$ subunits of the GABA$_A$ receptor function, (2) reduced binding of both BZD and non-BZD GABA$_A$ receptor–ligand, and (3) BZD-resistant anxiety (Sibille, Pavlides, Benke, & Toth, 2000). This suggests a pathological pathway originating from a 5-HT$_{1A}$ receptor deficit leading toward dysfunctions within GABA-ergic systems, resulting in increased levels of anxiety.

These preclinical studies are supported by a recent study in patients with panic disorder, who were enrolled in a randomized, double-blind study comparing the effects of a selective serotonin reuptake inhibitor (SSRI) and a BZD (clonazepam) versus SSRI and placebo (Goddard et al., 2001). This strategy was designed to take advantage of the rapid anxiolytic effects of the BZD prior to the delayed onset of SSRI treatment, thereby accelerating and possibly enhancing overall treatment effects. The study proved to be successful, showing that subjects treated with both the SSRI and the BZD showed an earlier onset of anxiolytic action at Week 1 than those treated with the SSRI alone.

Therefore, a logical next step in the evaluation of brain systems possibly involved in the pathophysiology of PTSD was a study to determine 5-HT$_{1A}$ receptor expression in patients with PTSD versus controls. We acquired PET images of 5-HT$_{1A}$ receptor binding using PET imaging on 12 unmedicated subjects with PTSD and 11 healthy controls without a history of trauma, using [^{18}F]fluorocarbonyl-WAY-100635 (FCWAY), a highly selective 5-HT$_{1A}$ receptor radioligand (Bonne et al., 2005). Unexpectedly, we found no difference in 5-HT$_{1A}$ receptor expression between the groups (Figure 9.3). This result suggests no direct role for the 5-HT$_{1A}$ receptor in the pathophysiology of PTSD; however, it does not exclude its relevance in mediating the effects of SSRIs in the treatment of PTSD by involving other transmitter systems and neurotrophic systems (e.g., brain-derived neurotrophic factor [BDNF] and 5-HT).

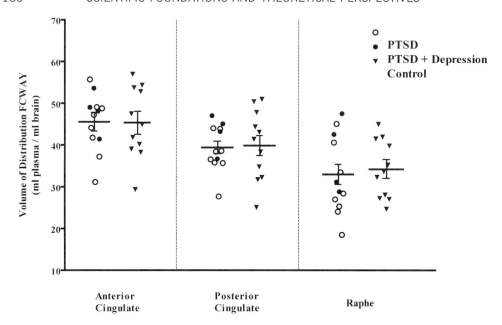

FIGURE 9.3. PET study of 5-HT$_{1A}$ receptors with the radioligand FCWAY. No difference in receptor expression was found in PTSD or PTSD and depression versus normal controls.

GENERALIZIBILITY OF CURRENT FINDINGS

Functional imaging studies in PTSD are confounded by the number of various paradigms used for recreating symptoms, small sample sizes, heterogeneous clinical populations across and within studies, and the presence of comorbid alcohol and or substance dependence and major depression. However, a consistent finding among the various PET studies is the failure of activation of the left anterior cingulate cortex in response to personalized scripts of the abuse (Bremner, Narayan, et al., 1999; Shin et al., 1999) or exposure to combat slides and sounds (Bremner, Staib, et al., 1999). In contrast, cerebral blood flow increased in the right pregenual anterior cingulate cortex in response to combat sounds and script-driven imagery (Liberzon et al., 1999; Rauch et al., 1996), suggesting laterality of medial PFC regulation of emotional behavior in patients with PTSD. In keeping with the hypothesis that a dysfunctional medial PFC, particularly the anterior cingulate cortex, inadequately inhibits the amygdala, increased activation of the amygdala has been demonstrated in response to trauma-related stimuli (Liberzon et al., 1999; Rauch et al., 1996; Shin et al., 1997). Although cerebral blood flow in other limbic and paralimbic cortical structures such as the posterior orbital cortex, anterior temporal lobe, and the anterior insular increased in both patients with PTSD and those with trauma without symptoms of PTSD, these findings did not distinguish between these two groups.

These findings support the possibility that neural processes mediating extinction to trauma-related stimuli may be impaired in patients with PTSD. Psychophysiological data have confirmed that, compared to subjects without PTSD, patients with PTSD acquire conditioned responses more readily and take a longer time to extinguish such responses, despite the absence of the primary stimulus (Orr et al., 2000; Peri, Ben-

Shakhar, Orr, & Shalev, 2000). However, to date, the underlying biology of these mechanisms remains to be elucidated and is the topic of ongoing studies.

Studies about morphometric abnormalities in PTSD are consistent with growing recognition of the important role of neurotrophic mechanisms in stress-related disorders. Whether morphometric alterations in individuals with PTSD relative to healthy controls reflect a consequence of trauma or a vulnerability factor to PTSD remains a controversial topic in the literature. As noted earlier, there is evidence for either mechanism. In this context, there is increased interest in the role of neurotrophic factors, such as BDNF, which is a member of the neurotrophin family of peptides and has been shown to support neuronal growth, differentiation, and survival in developing and adult neurons. Its effects have been most intensively studied in the hippocampus, which is intriguing because morphological alterations in the hippocampus have been reported in subjects with PTSD relative to controls. Decreased expression of BDNF is hypothesized to play a role in the atrophy of hippocampal neurons in experimental animals in response to stress. Reexposure to trauma-related cues may thus disrupt the formation of associations, such as those required by extinction, undermining the efficacy of exposure therapy in some individuals with PTSD. There is evidence that stress-induced changes in cerebral metabolites, hippocampal volume, and cell proliferation can be prevented by antidepressants, for example, tianeptine (Czeh et al., 2001).

Neurotoxic damage to the hippocampus and suppression of ongoing neurogenesis may also be a direct consequence of glucocorticoid administration (reviewed in McEwen, 2000), augmented by the presence of glutamate and glutamate analogues such as N-methyl-D-aspartate (NMDA). Animal studies have shown immediate glutamate efflux in PFC and hippocampus after induction of acute stress (Bagley & Moghaddam, 1997). These mechanisms may lead to hippocampal neuronal damage and cell loss, which can be determined with the use of structural MRI. Consequently, medications that block NMDA receptors or enhance BDNF function to enhance neurogenesis may well be therapeutics of the next generation that act beyond the currently used medications, which mostly interact with serotonergic and noradrenergic systems.

CHALLENGES FOR THE FUTURE

As described earlier, researchers have successfully determined the neurocircuitry underlying mechanisms, such as fear and anxiety or extinction, that are believed to be abnormal in PTSD. Others have been successful in determining neurochemical correlates of symptoms of PTSD. However, we have not yet sufficiently addressed the potential role of receptors and transporters that have been shown to regulate neurochemical mechanisms important for PTSD, for example, the serotonin transporter or presynaptic alpha-2 receptors that regulate synaptic concentrations of serotonin and noradrenaline, respectively. More recent research has focused on the determination of genes and gene variants that determine transmitter synthesis and release in healthy people and in individuals with stress-related disorders. These studies likely will lead to an enhanced understanding of an individual's vulnerability to environmental stressors and, in combination with modern imaging techniques including fMRI and molecular imaging using PET, will provide novel insight into processes underlying stress-related disorders such as PTSD. Whereas fMRI will provide insight into neural connectivity of circuits involved in PTSD, the strength of PET imaging will be to show the neurochemical processes that underlie altered emotion processing or hyperarousal in patients with PTSD. The ultimate goal of these research efforts is to provide novel, improved treatments for people with trauma-related symptoms.

REFERENCES

Agartz, I., Momenan, R., Rawlings, R. R., Kerich, M. J., & Hommer, D. W. (1999). Hippocampal volume in patients with alcohol dependence. *Archives of General Psychiatry, 56*(4), 356–363.

Amaral, D. G., Price, J. L., Pitanken, A., & Carmichael, S. T. (1992). Anatomical organization of the primate amygdala complex. In J. P. Aggleton (Ed.), *The amygdala: Neurobiological aspects of emotion, memory and mental dysfunction* (pp. 1–66). New York: Wiley-Liss.

Anand, A., Li, Y., Wang, Y., Wu, J., Gao, S., Bukhari, L., et al. (2005). Antidepressant effect on connectivity of the mood-regulating circuit: An fMRI study. *Neuropsychopharmacology, 30*(7), 1334–1344.

Bagley, J., & Moghaddam, B. (1997). Temporal dynamics of glutamate efflux in the prefrontal cortex and in the hippocampus following repeated stress: Effects of pretreatment with saline or diazepam. *Neuroscience, 77*(1), 65–73.

Bonne, O., Bain, E., Neumeister, A., Nugent, A. C., Vythilingam, M., Carson, R. E., et al. (2005). No change in serotonin type 1A receptor binding in patients with posttraumatic stress disorder. *American Journal of Psychiatry, 162*(2), 383–385.

Bonne, O., Brandes, D., Gilboa, A., Gomori, J. M., Shenton, M. E., Pitman, R. K., et al. (2001). Longitudinal MRI study of hippocampal volume in trauma survivors with PTSD. *American Journal of Psychiatry, 158*(8), 1248–1251.

Bremner, J. D., Innis, R. B., Southwick, S. M., Staib, L., Zoghbi, S., & Charney, D. S. (2000). Decreased benzodiazepine receptor binding in prefrontal cortex in combat-related posttraumatic stress disorder. *American Journal of Psychiatry, 157,* 1120–1126.

Bremner, J. D., Narayan, M., Staib, L. H., Southwick, S. M., McGlashan, T., & Charney, D. S. (1999). Neural correlates of memories of childhood sexual abuse in women with and without posttraumatic stress disorder. *American Journal of Psychiatry, 156*(11), 1787–1795.

Bremner, J. D., Randall, P., Scott, T. M., Bronen, R. A., Seibyl, J. P., Southwick, S. M., et al. (1995). MRI-based measurement of hippocampal volume in patients with combat-related posttraumatic stress disorder. *American Journal of Psychiatry, 152*(7), 973–981.

Bremner, J. D., Staib, L. H., Kaloupek, D., Southwick, S. M., Soufer, R., & Charney, D. S. (1999). Neural correlates of exposure to traumatic pictures and sound in Vietnam combat veterans with and without posttraumatic stress disorder: A positron emission tomography study. *Biological Psychiatry, 45*(7), 806–816.

Bremner, J. D., Vythilingam, M., Vermetten, E., Southwick, S. M., McGlashan, T., Nazeer, A., et al. (2003). MRI and PET Study of deficits in hippocampal structure and function in women with childhood sexual abuse and posttraumatic stress disorder. *American Journal of Psychiatry, 160*(5), 924–932.

Buchel, C., Dolan, R. J., Armony, J. L., & Friston, K. J. (1999). Amygdala–hippocampal involvement in human aversive trace conditioning revealed through event-related functional magnetic resonance imaging. *Journal of Neuroscience, 19*(24), 10869–10876.

Clark, F. M., & Proudfit, H. K. (1991). The projection of locus coeruleus neurons to the spinal cord in the rat determined by anterograde tracing combined with immunocytochemistry. *Brain Research, 538,* 231–245.

Czeh, B., Michaelis, T., Watanabe, T., Frahm, J., de Biurrun, G., van Kampen, M., et al. (2001). Stress-induced changes in cerebral metabolites, hippocampal volume, and cell proliferation are prevented by antidepressant treatment with tianeptine. *Proceedings of the National Academy of Sciences USA, 98*(22), 12796–12801.

Davis, M., Walker, D. L., & Lee, Y. (1997). Amygdala and bed nucleus of the stria terminalis: Differential roles in fear and anxiety measured with the acoustic startle reflex. *Philosophical Transactions of the Royal Society of London B, 352,* 1675–1687.

DeBellis, M. D., Hall, J., Boring, A. M., Frustaci, K., & Moritz, G. (2001). A pilot longitudinal study of hippocampal volumes in pediatric maltreatment-related posttraumatic stress disorder. *Biological Psychiatry, 50*(4), 305–309.

De Bellis, M. D., Keshavan, M. S., Clark, D. B., Casey, B. J., Giedd, J. N., Boring, A. M., et al. (1999). A. E. Bennett Research Award: Developmental traumatology: Part II. Brain development. *Biological Psychiatry, 45*(10), 1271–1284.

Duman, R. S., Malberg, J., Nakagawa, S., & D'Sa, C. (2000). Neuronal plasticity and survival in mood disorders. *Biological Psychiatry, 48*(8), 732–739.

Duman, R. S., Malberg, J., & Thome, J. (1999). Neural plasticity to stress and antidepressant treatment. *Biological Psychiatry, 46*(9), 1181–1191.

Elam, M., Svensson, T. H. E., & Thoren, P. (1986). Locus coeruleus neurons in sympathetic nerves: Activation by visceral afferents. *Brain Research, 375*, 117–125.

Garcia, R., Vouimba, R. M., Baudry, M., & Thompson, R. F. (1999). The amygdala modulates prefrontal cortex activity relative to conditioned fear. *Nature, 402*, 294–296.

Gelpin, E., Bonne, O., Peri, T., Brandes, D., & Shalev, A. Y. (1996). Treatment of recent trauma survivors with benzodiazepines: A prospective study. *Journal of Clinical Psychiatry, 57*, 390–394.

Gilbertson, M. W., Shenton, M. E., Ciszewski, A., Kasai, K., Lasko, N. B., Orr, S. P., et al. (2002). Smaller hippocampal volume predicts pathologic vulnerability to psychological trauma. *Nature Neuroscience, 5*, 1242–1247.

Goddard, A. W., Brougette, T., Almai, A., Jetty, P., Woods, S. W., & Charney, D. S. (2001). Early co-administration of clonazepam with sertraline for panic disorder. *Archives of General Psychiatry, 58*(7), 681–686.

Golier, J., Yehuda, R., Grossman, R., De Santi, S., Convit, A., & de Leon, M. (2000). *Hippocampal volume and memory performance in Holocaust survivors with and without PTSD.* Paper presented at the annual meeting of the American College of Neuropsychopharmacology, San Juan, PR.

Gurvits, T. V., Shenton, M. E., Hokama, H., Ohta, H., Lasko, N. B., Gilbertson, M. W., et al. (1996). Magnetic resonance imaging study of hippocampal volume in chronic, combat-related posttraumatic stress disorder. *Biological Psychiatry, 40*(11), 1091–1099.

Heisler, L. K., Chu, H. M., Brennan, T. J., Danao, J. A., Bajwa, P., Parsons, L. H., et al. (1998). Elevated anxiety and antidepressant-like responses in serotonin 5-HT1A receptor mutant mice. *Proceedings of the National Academy of Sciences USA, 95*, 15049–15054.

Jones, E. G. (1983). The thalamus. In P. Emson (Ed.), *Chemical neuroanatomy* (pp. 257–293). New York: Raven Press.

Jones, E. G., & Powell, T. P. S. S. (1970). An experimental study of converging sensory pathways within the cerebral cortex of the monkey. *Brain, 93*, 793–820.

LaBar, K. S., Gatenby, J. C., Gore, J. C., LeDoux, J. E., & Phelps, E. A. (1998). Human amygdala activation during conditioned fear acquisition and extinction: A mixed-trial fMRI study. *Neuron, 20*(5), 937–945.

LeDoux, J. E. (1987). Nervous system: V. Emotion. In F. Blum (Ed.), *Handbook of physiology* (pp. 419–459). Washington, DC: American Physiological Society.

LeDoux, J. E. (2000). Emotion circuits in the brain. *Annual Review of Neuroscience, 23*, 155–184.

LeDoux, J. E., Iwata, J., Cicchetti, P., & Reis, D. J. (1988). Different projections of the central amygdaloid nucleus mediate autonomic and behavioral correlates of conditioned fear. *Journal of Neuroscience, 8*, 2517–2529.

Liberzon, I., Taylor, S. F., Amdur, R., Jung, T. D., Chamberlain, K. R., Minoshima, S., et al. (1999). Brain activation in PTSD in response to trauma-related stimuli. *Biological Psychiatry, 45*(7), 817–826.

McDonald, A. J. (1991a). Organization of amygdaloid projections to prefrontal cortex and associated striatum in the rat. *Neuroscience, 44*, 1–14.

McDonald, A. J. (1991b). Topographical organization of amygdaloid projections to the caudatoputamen, nucleus accumbens, and related striatal-like areas of the rat brain. *Neuroscience, 44*, 15–33.

McEwen, B. S. (2000). The neurobiology of stress: From serendipity to clinical relevance. *Brain Research, 886*(1–2), 172–189.

McGaugh, J. L. (2000). Memory—a century of consolidation. *Science, 287*, 248–251.

McKittrick, C. R., Blanchard, D. C., Blanchard, R. J., McEwen, B. S., & Sakai, R. R. (1995). Serotonin receptor binding in a colony model of chronic social stress. *Biological Psychiatry, 37*, 383–393.

Mesulam, M. M., Van Hoesen, G., Pandya, D. N., & Geschwind, N. (1977). Limbic and sensory connections of the IPL in the rhesus monkey. *Brain Research, 136,* 393–414.

Morgan, M. A., & LeDoux, J. E. (1995). Differential contribution of dorsal and ventral medial prefrontal cortex to the acquisition and extinction of conditioned fear in rats. *Behavioral Neuroscience, 109*(4), 681–688.

Morgan, M. A., Romanski, L. M., & LeDoux, J. E. (1993). Extinction of emotional learning: Contribution of medial prefrontal cortex. *Neuroscience Letters, 163*(1), 109–113.

Morris, J. S., Friston, K. J., Buchel, C., Frith, C. D., Young, A. W., Calder, A. J., et al. (1998). A neuromodulatory role for the human amygdala in processing emotional facial expressions. *Brain, 121*(1), 47–57.

Neumeister, A., Charney, D. S., Belfer, I., Geraci, M., Holmes, C., Alim, T., et al. (2005). Sympatho-neural and adrenomedullary functional effects of a_{2C}-adrenoreceptor gene polymorphism in healthy humans. *Pharmacogenetics and Genomics, 15,* 43–149.

Neumeister, A., Wood, S., Bonne, O., Nugent, A., Luckenbaugh, D., Young, T., et al. (2005). Reduced hippocampal volume in unmedicated, remitted patients with major depression versus controls. *Biological Psychiatry, 57,* 935–937.

Orr, S. P., Metzger, L. J., Lasko, N. B., Macklin, M. L., Peri, T., & Pitman, R. K. (2000). De novo conditioning in trauma-exposed individuals with and without posttraumatic stress disorder. *Journal of Abnormal Psychology, 109*(2), 290–298.

Parks, C. L., Robinson, P. S., Sibille, E., Shenk, T., & Toth, M. (1998). Increased anxiety of mice lacking the serotonin1A receptor. *Proceedings of the National Academy of Sciences USA, 95,* 10734–10739.

Peri, T., Ben-Shakhar, G., Orr, S. P., & Shalev, A. Y. (2000). Psychophysiologic assessment of aversive conditioning in posttraumatic stress disorder. *Biological Psychiatry, 47*(6), 512–519.

Phillips, R. G., & LeDoux, J. E. (1992). Differential contribution of amygdala and hippocampus to cued and contextual fear conditioning. *Behavioral Neuroscience, 106*(2), 274–285.

Pitman, R. K., Shin, L. M., & Rauch, S. L. (2001). Investigating the pathogenesis of posttraumatic stress disorder with neuroimaging. *Journal of Clinical Psychiatry, 62*(Suppl. 17), 47–54.

Quirk, G. J., Russo, G. K., Barron, J. L., & Lebron, K. (2000). The role of ventromedial prefrontal cortex in the recovery of extinguished fear. *Journal of Neuroscience, 20*(16), 6225–6231.

Ramboz, S., Oosting, R., Amara, D. A., Kung, H., Blier, P., Mendelsohn, M., et al. (1998). Serotonin receptor 1A knockout: An animal model of anxiety-related disorder. *Proceedings of the National Academy of Sciences USA, 95*(24), 14476–14781.

Rauch, S. L., van der Kolk, B. A., Fisler, R. E., Alpert, N. M., Orr, S. P., Savage, C. R., et al. (1996). A symptom provocation study of posttraumatic stress disorder using positron emission tomography and script-driven imagery. *Archives of General Psychiatry, 53*(5), 380–387.

Rauch, S. L., Whalen, P. J., Shin, L. M., McInerney, S. C., Macklin, M. L., Lasko, N. B., et al. (2000). Exaggerated amygdala response to masked facial stimuli in posttraumatic stress disorder: A functional MRI study. *Biological Psychiatry, 47*(9), 769–776.

Roozendaal, B. (2000). Glucocorticoids and the regulation of memory consolidation. *Psychoneuroendocrinology, 25*(3), 213–238.

Saper, C. B. (1982). Convergence of autonomic and limbic connections in the insular cortex of the rat. *Journal of Comparative Neurology, 210,* 163–173.

Sapolsky, R. M. (2000). Glucocorticoids and hippocampal atrophy in neuropsychiatric disorders. *Archives of General Psychiatry, 57*(10), 263–263.

Sawchenko, P. E., & Swanson, L. W. (1982). Central noradrenergic pathways for the integration of hypothalamic neuroendocrine and autonomic responses. *Science, 214,* 685–687.

Sawchenko, P. E., & Swanson, L. W. (1983). The organization of forebrain afferents to the paraventricular and supraoptic nucleus of the rat. *Journal of Comparative Neurology, 218,* 121–144.

Schuff, N., Marmar, C. R., Weiss, D. S., Neylan, T. C., Schoenfeld, F., Fein, G., et al. (1997). Reduced hippocampal volume and n-acetyl aspartate in posttraumatic stress disorder. *Annals of the New York Academy of Sciences, 821,* 516–520.

Shin, L. M., Kosslyn, S. M., McNally, R. J., Alpert, N. M., Thompson, W. L., Rauch, S. L., et al. (1997). Visual imagery and perception in posttraumatic stress disorder: A positron emission tomographic investigation. *Archives of General Psychiatry, 54*(3), 233–241.

Shin, L. M., McNally, R. J., Kosslyn, S. M., Thompson, W. L., Rauch, S. L., Alpert, N. M., et al. (1999).

Regional cerebral blood flow during script-driven imagery in childhood sexual abuse-related PTSD: A PET investigation. *American Journal of Psychiatry, 156*(4), 575–584.

Sibille, E., Pavlides, C., Benke, D., & Toth, M. (2000). Genetic inactivation of the serotonin$_{1A}$ receptor in mice results in downregulation of major GABA$_A$ receptor alpha subunits, reduction of GABA$_A$ receptor binding, and benzodiazepine-resistant anxiety. *Journal of Neuroscience, 20*(8), 2758–2765.

Stein, M. B., Koverola, C., Hanna, C., Torchia, M. G., & McClarty, B. (1997). Hippocampal volume in women victimized by childhood sexual abuse. *Psychological Medicine, 27*(4), 951–959.

Swanson, L. W. (1983). The hippocampus and the concept of the limbic system. In W. Seifert (Ed.), *Neurobiology of the hippocampus* (pp. 3–19). London: Academic Press.

Turner, B., Gupta, K. C., & Mishkin, M. (1978). The locus and cytoarchitecture of the projection areas of the olfactory bulb in *Macaca mulatta*. *Journal of Comparative Neurology, 19*, 381–396.

Turner, B. H., Mishkin, M., & Knapp, M. (1980). Organization of the amygdalopetal projections from modality-specific cortical association areas in the monkey. *Journal of Comparative Neurology, 191*, 515–543.

Van Hoesen, G. W., Pandya, D. N., & Butters, N. (1972). Cortical afferents to the entorhinal cortex of the rhesus monkey. *Science, 175*, 1471–1473.

Vogt, B. A., & Miller, M. W. (1983). Cortical connections between rat cingulate cortex and visual, motor, and postsubicular cortices. *Journal of Comparative Neurology, 216*, 192–210.

Watanabe, Y., Gould, E., & McEwen, B. S. (1992). Stress induces atrophy of apical dendrites of hippocampal CA3 pyramidal neurons. *Brain Research, 588*(2), 341–345.

Whalen, P. J., Rauch, S. L., Etcoff, N. L., McInerney, S. C., Lee, M. B., & Jenike, M. A. (1998). Masked presentations of emotional facial expressions modulate amygdala activity without explicit knowledge. *Journal of Neuroscience, 18*(1), 411–418.

Whitlock, D. G., & Nauta, W. J. H. (1956). Subcortical projections from the temporal neocortex in *Macaca mulatta*. *Journal of Comparative Neurology, 106*, 183–212.

Chapter 10

Neurobiological Alterations Associated with PTSD

Steven M. Southwick, Lori L. Davis, Deane E. Aikins, Ann Rasmusson, Jill Barron, and Charles A. Morgan III

Fear and life-threatening situations mobilize numerous brain regions, hormones, and neurotransmitter systems. In the short run, this complex, fear-induced activation allows the organism to recognize and respond to potential threats. However, in some cases, traumatic stress can lead to long-term negative psychological and physiological sequelae. Negative sequelae are most common when traumatic stressors are recurrent, and when they are perceived as uncontrollable and overwhelming.

In this chapter we focus on a number of neurobiological factors believed to be critical in the stress response and in the pathophysiology of posttraumatic stress disorder (PTSD). These include glutamate, gamma-aminobutyric acid (GABA), norepinephrine (NE), neuropeptide Y (NPY), serotonin (5-hydroxytryptamine, or 5-HT), corticotropin-releasing factor (CRF), and psychophysiological reactivity. As such, our review does not encompass the enormous complexity of neurobiological responses to danger and the full array of known alterations in PTSD. For example, alterations have also been reported in systems involving dopamine, opiates, thyroid hormone, and the immunological system (Boscarino, 2004; Friedman, Wang, Jalowiec, McHugo, & McDonagh-Coyle, 2005; Mason et al., 1995; Pitman, van der Kolk, Orr, & Greenberg, 1990). Although our review primarily describes studies in traumatized humans, it is important to emphasize that much of this research grows out of preclinical investigations.

METHODOLOGICAL CONSIDERATIONS

Over the past decade, exceptional preclinical advances have been made in understanding the physiology, biochemistry, neural circuitry, pharmacology, and behavioral manifestations of fear, stress and trauma. These advances have great relevance for understanding trauma-related psychopathology in humans. Current preclinical models of fear conditioning, stress sensitization, stress-related hippocampal dysfunction, altered corpus collosum development, amygdala hyperresponsivity, stress-related prefrontal cortical impairment, and overconsolidation of memory for traumatic events closely approximate the likely underlying pathophysiology of PTSD and have now made it possible to test rational hypotheses regarding the pathophysiology and treatment of stress-related disorders in humans.

The neurobiological investigation of trauma-related psychopathology is fraught with methodological concerns and limitations. Perhaps most important is the enormous complexity of the nervous system and its response to fear and trauma. Detecting and responding to fear involve the elaborate coordination of numerous brain regions and neurotransmitter systems. Most studies in humans with trauma-related disorders have investigated one brain region or neurotransmitter system at a time. However, hormones, neurotransmitters, and neuropeptides are known to interact with one another in a complex fashion, so that alteration in one system affects functioning in other systems. For example, the locus coeruleus is regulated by a variety of neurotransmitters and neuropeptides, with inhibitory effects from NE, epinephrine, endogenous opiates, GABA, and serotonin, and stimulatory effects from CRF and glutamate. Alterations in any of these systems can affect noradrenergic activity.

Even more complex is the relationship between neurotransmitters and behavior. For example, hyperarousal and hypervigilance are commonly experienced by trauma survivors with PTSD. Arousal is influenced by multiple neurotransmitters (e.g., NE, dopamine, acetylcholine, serotonin) that are simultaneously active in varying degrees and in various brain regions. Chronic alterations in arousal systems are likely to be highly complex and involve long-term changes in neural function. Studies that investigate multiple brain regions and neurotransmitter systems in the same subjects tend to be labor intensive and expensive. For example, investigating prefrontal cortical, amygdala, and hippocampal function in the same traumatized individual requires separate brain scans, each with a specialized task designed to activate the specific brain region of interest.

Additionally, the central nervous system (CNS) is difficult to access in humans. As a result, many studies in PTSD have relied on peripheral markers, such as 24-hour urine excretion, peripheral blood element (platelet and lymphocyte) receptor binding, and plasma neurotransmitter, hormone, and metabolite levels. Unlike preclinical studies in which brain tissue is readily available (through sacrificing the animal), the study of specific brain regions in humans must be conducted indirectly through neuropsychological, electrophysiological, and neuroimaging techniques. Although postmortem studies of brain tissue in humans are possible, they are difficult to arrange, and the interpretation of data is challenging.

Neurobiological investigations are also dependent on available technology. Although great technical advances have been made in molecular biology, brain imaging, electrophysiology, and neuropharmacology, the ability to probe specific CNS neurotransmitter systems and to investigate complex interactions in the nervous system currently remains limited. As ligand development, electrophysiological, and brain imaging techniques advance, our understanding of fear- and trauma-induced alterations in nervous system functioning will expand.

Numerous other methodological issues must be considered when investigating the neurobiological underpinnings of trauma-related psychopathology. These include accurate diagnosis of psychopathology and accompanying comorbid disorders, sensitive measurement of past and recent traumas, and meticulous accounting or controlling for factors that potentially affect the biological systems being studied (e.g., medication, tobacco use). Even when these methodological considerations have been addressed, one must be cautious if attempting to attribute symptoms or behaviors to specific neurobiological alterations. Symptoms and behaviors are highly complex phenomena that typically involve numerous interactions between brain regions and neurotransmitter systems.

Even more tenuous is the attempt to relate an isolated neurobiological system to a psychiatric diagnosis such as PTSD. PTSD includes a host of symptoms and behaviors, each of which may have its own characteristic neurobiological profile. In fact, some investigators have suggested that neurobiological studies should focus on specific endotypes or internal phenotypes (i.e., not obvious to the unaided eye) that are relatively specific to the disease being studied, state independent, heritable, and associated with relatives (particularly ill relatives), and that bear some plausible clinical and biological conceptual relationship to the disease (Hasler, Drevets, Manji, & Charney, 2004; Southwick, Vythilingam, & Charney, 2005). Proposed psychopathological endotypes for major depression include depressed mood (mood bias), anhedonia (impaired reward function), impaired learning, and impaired stress sensitivity. Proposed biological endotypes for major depression include rapid eye movement (REM) sleep abnormalities, reduced 5-HT$_{1A}$ receptor binding potential, and left anterior cingulate cortex volume reduction. Endotypes for PTSD might include intrusive memories, hyperarousal, numbing, and reduction in hippocampal volume.

CURRENT STATE OF THE ART

Cardiophysiology

Cardiophysiological measures (e.g., heart rate reactivity to stimuli) have provided some ability to differentiate between PTSD and control samples. The most consistent effect has been demonstrated for accelerated heart rate in individuals with PTSD during presentations of trauma-related sounds, videos, and imagery. In the largest study to date, accelerated heart rate during exposure to combat-related materials has been demonstrated in veterans with combat-related PTSD compared to veterans with past PTSD or nonanxious veterans (Keane et al., 1998). This paradigm has been used in multiple samples of Vietnam Era veterans: combat-exposed with PTSD versus combat-exposed without current PTSD (Blanchard, Kolb, Gerardi, Ryan, & Pallmeyer, 1986; Blanchard, Kolb, Taylor, & Wittrock, 1989; Gerardi, Blanchard, & Kolb, 1989), medication-free combat-exposed with PTSD versus medication-free combat-exposed without PTSD (Pitman, Orr, Forgue, de Jong, & Claiborn, 1987), medication-free combat-exposed with PTSD versus non-combat-exposed non-PTSD veterans with other anxiety disorders (Pitman, Orr, et al., 1990), treatment-seeking combat-exposed with PTSD versus psychiatric controls and healthy combat-exposed without PTSD (Malloy, Fairbank, & Keane,

1983; McFall, Murburg, Ko, & Veith, 1990), and combat-exposed with PTSD versus combat-exposed without PTSD, veterans with other psychiatric disorders, nondeployed veterans, and nonveterans with anxiety disorders (Pallmeyer, Blanchard, & Kolb, 1986). Elevated heart rate in response to affective tasks also has been demonstrated with Vietnam Era nurses with PTSD relative to trauma-equivalent nurses without PTSD (Carson et al., 2000) and samples of medication-free World War II and Korean veterans with either current full or partial combat-related PTSD and a control group of equivalently trauma-exposed veterans with other diagnosed anxiety disorders (Orr, Pitman, Lasko, & Herz, 1993). Thus, it would appear that trauma stimuli, in either standardized or idiosyncratically prepared formats, are capable of generating elevated heart rate in combat-related chronic PTSD samples. Indeed, the only null effect was found in a limited sample of Persian Gulf War veterans with and without PTSD (Davis, Adams, Uddo, Vasterling, & Sutker, 1996). However, the PTSD severity scores were lower than those typically found in other PTSD samples.

Trauma-relevant stimuli have included sights and sounds of combat, as well as scripts of personally experienced traumas. Pooling subjects from all published studies, approximately two-thirds of subjects with PTSD have demonstrated exaggerated reactivity to trauma-associated cues. The percent appears even higher in subjects with severe PTSD (Orr, 1997; Orr, Lasko, Metzger, Berry, Ahern, & Pitman, 1997). On the other hand, most studies have found that subjects with PTSD do not experience exaggerated physiological reactivity in response to generic, non-trauma-related stimuli (e.g., Orr, 1997).

Whereas potentiated cardiophysiological reactivity to trauma-relevant stimuli is the most consistently documented effect in PTSD samples, similar trauma-reactive effects with other physiological systems have been found as well (e.g., skin conductance; McNally et al., 1987). Preliminary evidence exists for hyperresponsive Pavlovian conditioning with skin conductance measures (Orr et al., 2000) but not with startle eyeblink (see Grillon & Morgan, 1999). Additionally, elevated conditioned contextual responses have been found when measuring startle eyeblink (e.g., Grillon, Morgan, Davis, & Southwick, 1998).

Glutamate and GABA

Glutamate is an amino acid and the brain's primary excitatory neurotransmitter. It is rapidly released in response to arousing and dangerous situations, and mediates nearly all fast, excitatory, point-to-point synaptic transmission in the brain. GABA, the brain's primary inhibitory neurotransmitter, counters excitatory glutamatergic synaptic transmission. During resting, nonstressful states, GABA exerts tonic inhibition on glutamate transmission in numerous brain regions, such as the thalamus and amygdala, allowing the brain to filter out a continuous flow of irrelevant and extraneous sensory information. However, when excitation is increased in response to stress or danger, elevated levels of glutamate have the capacity to overcome tonic inhibition by GABA, thereby triggering a cascade of protective responses (Krystal, Bennett, Bremner, Southwick, & Charney, 1995).

Although stress-induced elevations of glutamate facilitate the cortical and subcortical communication necessary for effective responses to danger, failure to modulate heightened glutamatergic activation can lead to extreme changes in intracellular calcium, toxicity, and even cell death (Armanini, Hutchins, Stein, & Sapolsky, 1990; Stein-

Behrens, Mattson, Chang, Yeh, & Sapolsky, 1994). To protect the brain from its own unchecked glutamatergic excitation, additional GABA, as well as neuroactive steroids that facilitate GABA function (Barbaccia et al., 1997), are released during stress. Thus, enhanced GABA receptor activation provides tonic CNS inhibition during nonstressful states and enhanced CNS inhibition during stressful states.

Both GABA and glutamate possess two classes of receptors: (1) ionotropic receptors (GABA$_A$ and the glutamatergic N-methyl-D-aspartate [NMDA] and non-NMDA receptors) that enhance membrane ion conductance, and (2) metabotropic receptors (GABA$_B$ and metabotropic glutamate receptors [mGluR]) that increase intracellular second messengers activity. In addition to its site for binding GABA, the GABA$_A$ receptor complex has binding sites for alcohol, barbiturates, and benzodiazepines. GABA increases the permeability of chloride ions through the GABA$_A$ chloride ion channel, which decreases neuronal excitability by hyperpolarizing the neuronal membrane.

Preclinical research points to GABA as a key neurotransmitter in stress-induced behavioral deficits that mirror depression and PTSD (i.e., the learned helplessness animal model) (Kram, Kramer, Steciuk, Ronan, & Petty, 2000). These data suggest that decreasing GABA transmission renders naive, nonstressed rats helpless. Conversely, increasing GABA in selective brain regions ameliorates many of the harmful effects of stress.

In humans with trauma-related disorders, very little research has focused specifically on GABA. In a study of accident victims, Vaiva and colleagues (2004) reported significantly lower plasma GABA levels in subjects who developed PTSD compared to subjects who did not. The authors suggested that low plasma GABA may have increased vulnerability for development of PTSD, whereas normal or high levels may have served a protective role. In a second study, Bremner and colleagues (2000) found that benzodiazepine receptor density and/or affinity was reduced in the medial prefrontal cortex (PFC) among patients with PTSD compared to controls. This reduction may have been secondary to stress-related alterations in GABA-ergic transmission. On the other hand, Fujita and colleagues (2004) did not find a difference in prefrontal cortical benzodiazepine receptor density in Desert Storm veterans with PTSD compared to healthy controls.

Although decreased GABA activity in a number of brain regions (e.g., medial PFC, amygdala) has been suggested as a potential neurobiological factor associated with PTSD, excessive glutamatergic activity may also play a role in the pathophysiology of PTSD. Recent investigations in healthy subjects have provided evidence that the NMDA glutamate receptor plays a central role in symptoms of dissociation, symptoms that are commonly seen in individuals with PTSD (Krystal et al., 1995). In a series of studies by Krystal and colleagues (1994, 1998, 1999), administration of the NMDA antagonist ketamine, which increases glutamate release, produced significant dose-dependent increases in dissociative symptoms.

Low doses of ketamine caused alterations in the form and content of thought (e.g., paranoia, loosening of associations, tangentiality, and ideas of reference), whereas high doses caused dissociative symptoms commonly reported by trauma victims. Pretreatment with a GABA agonist, benzodiazepine, prior to infusion with ketamine resulted in a significant reduction in some, but not all, dissociative symptoms (Krystal et al., 1998). Similarly, pretreatment with lamotrigine, an anticonvulsant that attenuates glutamate release via inhibition of sodium, calcium, and potassium channels, significantly decreased dissociative and cognitive effects of ketamine (Anand et al., 2000).

Monoamines

Serotonin

GENERAL CHARACTERISTICS

Serotonin (5-hydroxytryptamine, or 5-HT) is a monoamine that is synthesized from tryptophan. Neurons that synthesize and release serotonin are primarily located in the raphe nuclei of the brainstem (Nestler, Hyman, & Malenka, 2001), from which projections extend to limbic structures and all areas of cerebral cortex (e.g., PFC, amygdala, locus coeruleus (LC), hippocampus, nucleus accumbens, and hypothalamus) (Nestler et al., 2001). To date, at least 14 serotonin receptors have been identified, and serotonin is known to have both excitatory and inhibitory actions in the CNS. Serotonin appears to play multiple regulatory roles in the CNS, including regulation of sleep, aggression, cardiovascular and respiratory activity, motor output, anxiety, mood, neuroendocrine activity, and analgesia (Agaganian, 1995; Davis, Astrachan, & Kass, 1980).

In preclinical studies, chronic psychosocial stress has been shown to decrease 5-HT_{1A} receptor density in limbic brain structures. However, adrenalectomy can prevent these stress-induced reductions, suggesting that postsynaptic 5-HT_{1A} gene expression is under tonic inhibition by adrenal steroids. These data point to a possible sequence in which stress-induced increases in corticotropin-releasing hormone (CRH) and cortisol down-regulate 5-HT_{1A} receptors, with an accompanying lowered threshold for anxiogenic stressful life events (Lopez, Chalmers, Little, & Watson, 1998). Alternatively, low 5-HT_{1A} receptor density may have a genetic origin or represent the combined result of inheritance and psychosocial stress (Charney, 2004).

CLINICAL PTSD STUDIES

Direct evidence of 5-HT dysregulation in PTSD comes from clinical studies of traumatized subjects at rest and in response to neuroendocrine challenge paradigms (see Table 10.1). Resting decreased platelet serotonin uptake, as measured by paroxetine binding (Arora, Fichtner, O'Connor, & Crayton, 1993; Bremner, Southwick & Charney, 1999), blunted prolactin response to the serotonin-releasing and uptake inhibitor *d*-fenfluramine (Davis, 1999) and exaggerated panic/anxiety and heart rate reactions to the serotonergic probe meta-chlorophenylpiperazine (MCPP), have all been reported in subjects with PTSD compared with non-PTSD controls (Southwick et al., 1997). The MCPP study provided preliminary evidence for possible neurobiological subgroups of patients with PTSD, one showing increased yohimbine-induced reactivity of the noradrenergic, system, and the other, increased MCPP-induced reactivity of the serotonergic system. Furthermore, Arora, Fitchner, O'Connor, and Crayton (1994) found that patients with PTSD who responded best to paroxetine were those with the highest pretreatment platelet affinity for paroxetine.

Indirect evidence supporting a role for serotonin in the pathophysiology of PTSD comes from studies in subjects with aggression, impulsivity, suicidality, and depression. These symptoms are commonly seen in individuals with PTSD. Aggressive behavior in healthy males has been shown to increase as a result of tryptophan depletion (Moeller et al., 1996), and cerebrospinal fluid (CSF) serotonin metabolite (5-HIAA) has been reported as low in suicide victims who have killed themselves through violent means, in impulsive men, and in aggressive psychiatric patients (Davidson, Putnam, & Larson,

TABLE 10.1. Key Neurotransmitter/Neuroendocrine Findings in PTSD

Serotonin	• Decreased platelet serotonin uptake • Blunted prolactin response to *d*-fenfluramine • Exaggerated reactivity to MCPP • Clinical response to SSRIs
Norepinephrine	• Increased 24-hour urine NE excretion • Increased 24-hour plasma NE levels • Decreased platelet alpha-2 adrenergic receptor number • Exaggerated NE and epinephrine responses to traumatic reminders • Exaggerated MHPG response to yohimbine • Blunted prolactin response to clonidine
HPA axis	• Increased CSF CRF • Abnormal 24-hour urine excretion of cortisol • Abnormal 24-hour plasma levels of cortisol • Increased lymphocyte glucocorticoid number • Exaggerated suppression of cortisol to dexamethasone • Increased cortisol response to CRH and ACTH • Increased ACTH response to metyrapone

Note. This table does not include key findings in PTSD related to dopamine, thyroid hormone, or opiate systems.

2000). Additionally, Handelsman and colleagues (1996) found that prolactin (PRL) response to the partial serotonin agonist MCPP was inversely associated with measures of hostility, irritability, and depression in abstinent alcoholics—further demonstrating the possible relationship a serotonin and hostility.

Pharmacological treatment studies provide perhaps the strongest clinical evidence for serotonin's role in PTSD. In large multicenter, randomized, double-blind treatment trials, selective serotonin reuptake inhibitors (SSRIs) have been shown to improve significantly all three PTSD symptom clusters (reexperiencing, avoidance, arousal) compared to placebo. To date, only two medications have been approved by the U.S. Food and Drug Administration (FDA) for the treatment of PTSD. Both are SSRIs (sertraline and paroxetine). Additionally monoamine oxidase inhibitors, which inhibit degradation of serotonin, and tricyclic antidepressants, which inhibit presynaptic serotonin reuptake, have been effective in treating trauma victims with PTSD (Friedman, Davidson, Mellman, & Southwick, 2000; Friedman & Davidson, Chapter 19, this volume).

STRESS AND PREFRONTAL CORTEX, AMYGDALA, LOCUS COERULEUS, AND HIPPOCAMPUS

The relationship between trauma-related psychopathology and alterations in 5-HT activity are likely to be mediated by both cortical and subcortical regions of the CNS. 5-HT plays an important role in regulating brain regions that have been implicated in the pathophysiology of PTSD, including the PFC, amygdala, LC, and hippocampus. The effect of 5-HT on prefrontal cortical function is undoubtedly complex. Accumulating evidence, particularly from a series of tryptophan depletion studies (which cause a transient reduction in 5-HT stores), suggests that the orbitofrontal cortex may be especially sensitive to the effects of serotonin (Park et al., 1994). The orbitofrontal cortex is known for its role in filtering, processing, and evaluating social and emotional information. Impaired function of the orbitofrontal cortex has been associated with a number of symptoms commonly seen in patients with PTSD, including impulsivity, aggression, misinterpretation of emotional stimuli, and deficits in processing of affective memory. Furthermore, investigations in subjects with PTSD have revealed impaired functioning on neuropsychological tasks mediated by the orbitofrontal cortex (Koenen et al., 2001) and exaggerated decreases in orbitofrontal cortex blood flow during retrieval of emotionally balanced word pairs (Bremner et al., 2003). It has been suggested that deficits in orbitofrontal cortical functioning among subjects with PTSD may, in part, be related to alterations in 5-HT activity (Southwick, Rasmusson, Barron, & Arnsten, 2005).

In the amygdala, when levels of 5-HT are reduced, the threshold of amygdala firing decreases (i.e., increased activation of the amygdala) through effects on inhibitory GABAergic interneurons that modulate excitatory glutamatergic input (Morgan, Krystal, & Southwick, 2003). This ability of 5-HT to modulate glutamatergic activity is dependent on the presence of corticosterone (Stutzmann & LeDoux, 1999; Stutzmann, McEwen, & LeDoux, 1998). On the other hand, increased 5-HT has been found to stimulate GABAergic interneurons, which inhibit glutamatergic activity and increase the threshold of amygdala firing. The result is a decrease in vigilance and fear-related behaviors. Efficacy of SSRIs in patients with PTSD may be related in part to an increased threshold of amygdala firing (Morgan et al., 2003).

Lesion, electrophysiological, and biochemical studies (Aston-Jones et al., 1991) have all demonstrated an inhibitory role of 5-HT on LC activity. For example, increases in tyrosine hydroxylase and firing rate of NE neurons in the LC have been reported following lesions to the raphe nuclei or pretreatment with 5-HT synthesis inhibitors (which effectively release inhibitory control of the LC by 5-HT). In a recent preclinical study, rats with lesions of 5-HT neurons had approximately 50% greater firing activity of NE neurons than intact, nonlesioned animals (Blier, 2001). Blier (2001) also found that administration of the SSRI citalopram for 14 and 21 days led to progressive decrease in the firing activity of NE neurons.

Preclinical and clinical studies have clearly demonstrated a relationship between chronic stress, 5-HT and hippocampal function. In animals, inescapable stress has been associated with hippocampal damage and inhibition of neurogenesis. Fluoxetine administration, on the other hand, has been shown to block this stress-induced decrease in hippocampal cell proliferation. Furthermore, pretreatment with an SSRI has been shown to prevent the development of many fear-induced behaviors in animals. This effect is probably mediated through activation of postsynaptic 5-HT_{1A} receptors (reviewed in Bonne, Grillon, Vythilingam, Neumeister, & Charney, 2004). In human studies, a number of research groups have reported reduced hippocampal volume and deficits in hippocampal-based declarative verbal memory among traumatized individuals with PTSD. Additionally, in a study of women with PTSD, Vermetten, Vythilingam, Southwick, Charney, and Bremner (2003) found a significant increase in hippocampal volume and verbal memory after long-term treatment with the SSRI paroxetine.

SEROTONIN AND GENE POLYMORPHISMS

There is also genetic evidence for a relationship between serotonin, aggression and stress-induced psychopathology (i.e., depression). Individual differences in aggressive behavior have been associated with a polymorphism in the gene that codes for tryptophan hydroxylase (Manuck et al., 1999; Nielsen et al., 1994). An increase in the risk for depression in response to life stressors is also associated with having one or two copies of the short allele of the 5-HT transporter promoter polymorphism (Caspi et al., 2003). Increased amygdala neuronal activity in response to fear-inducing stimuli has recently been reported in healthy subjects with the 5-HT transporter polymorphism that is associated with reduced 5-HT expression and increased fear and anxiety (Bertolino et al., 2005).

Norepinephrine

GENERAL PTSD CLINICAL STUDIES

A large number of physiological, neuroendocrine, receptor binding, pharmacological challenge, brain imaging, and pharmacological treatment studies have provided compelling evidence for exaggerated noradrenergic activity in traumatized humans with PTSD (Friedman & Southwick, 1995; Southwick, Bremner, et al., 1999) (see Table 10.1). This exaggerated reactivity has been observed in response to a variety of stressors, particularly those associated with personally experienced traumas. In contrast, noradrenergic hyperreactivity has not consistently been found under baseline or resting conditions. It has been suggested that a number of PTSD hyperarousal and reexperiencing symptoms are associated with this altered reactivity of noradrenergic neurons (Southwick, Bremner, et al., 1999).

BASELINE NOREPINEPHRINE

Most studies measuring baseline or resting indices of catecholamine activity have found no significant differences between subjects with PTSD and control groups (see Table 10.1). This includes psychophysiology studies, which compare indices of resting heart rate, blood pressure, and galvanic skin conductance, as well as studies measuring plasma NE and epinephrine (Southwick, Bremner, et al., 1999; Southwick, Paige, et al., 1999). In separate studies measuring NE response to combat-related laboratory stimuli, McFall and colleagues (1990) and Blanchard, Kolb, Prins, Gates, and McCoy (1991) found no differences in baseline resting plasma NE levels among subjects with PTSD compared to controls. Similarly, as part of a yohimbine challenge study, Southwick and colleagues (1993) reported no differences in baseline plasma levels of the NE metabolite 3-methoxy-4-hydroxyphenylglycol (MHPG) in combat veterans with PTSD compared to healthy controls.

TWENTY-FOUR-HOUR PLASMA AND URINE CATECHOLAMINES AND PLATELET ADRENERGIC RECEPTORS

Unlike baseline psychophysiology and neuroendocrine data, studies of 24-hour plasma NE levels, 24-hour urine hormone excretion, and platelet adrenergic receptor number have found significant differences between subjects with PTSD and controls (Southwick, Bremner, et al., 1999) (see Table 10.1). In a study conducted under resting or nonstimulated conditions, Yehuda and colleagues (1998) sampled plasma MHPG and NE every hour for a period of 24 hours in subjects with PTSD compared to healthy controls. Combat veterans with PTSD but without comorbid depression had higher mean NE levels at nearly every time point compared to combat veterans with PTSD and comorbid depression, subjects with major depressive disorder (MDD) alone, and healthy controls. No differences in MHPG were found between groups.

Most 24-hour urine excretion studies have also found higher levels of NE in trauma survivors with PTSD compared to controls. This is true for individuals with PTSD secondary to combat and civilian violence, adult women with a history of child abuse, male automobile accident survivors, and traumatized children (reviewed in Southwick, Bremner, et al., 1999). It has been suggested that 24-hour catecholamine levels reflect the summation of both phasic physiological changes in response to meaningful stimuli (e.g., traumatic reminders) and tonic resting levels of catecholamines, whereas single plasma samples taken at rest reflect only tonic activity (Murburg, 1994; Southwick, Yehuda, & Morgan, 1995). Nondiminished central noradrenergic activity, as reflected by exaggerated urinary NE and MHPG excretion, has also been found at night and correlated with sleep disturbances among veterans with chronic, combat-related PTSD (Mellman, Kumar, Kulick-Bell, Kumar, & Nolan, 1995).

The neurochemical message of NE and epinephrine is translated in part by alpha-2 adrenergic receptors. In studies of combat veterans (Perry, 1994) and traumatized children, Perry, Southwick, Yehuda, and Giller (1990) found fewer total alpha-2 adrenergic receptor binding sites per platelet in subjects with PTSD compared to controls. Reduced alpha-2 adrenergic receptor sites have also been reported in congestive heart failure, a condition marked by excessive and chronic catecholamine activity. It has been hypothesized that reduced numbers, or down-regulation, of alpha-2 adrenergic receptors serves as an adaptive response to chronic elevation of circulating catecholamines. Additionally, Maes and colleagues (1999) found that accident survivors with PTSD and comorbid MDD had higher plasma availability of the precursor tyrosine and lower platelet alpha-2 adrenergic receptor affinity than healthy controls.

CATECHOLAMINE CHALLENGE PARADIGMS

Challenge paradigms generally involve exposure to provocative auditory or visual stimuli, or exogenously administered biological substances, such as lactate, clonidine, or yohimbine (see Table 10.1). A number of challenge paradigms have been used to assess reactivity of catecholamine systems in individuals with PTSD.

As noted earlier, a review of the psychophysiological literature suggests that trauma survivors with PTSD experience greater physiological reactivity (particularly heart rate) in response to trauma-relevant stimuli than do trauma survivors without PTSD or nontraumatized, healthy controls. This exaggeration in cardiovascular physiology has been associated with increased catecholamine activity. McFall and colleagues (1990) reported parallel increases in subjective distress, heart rate, blood pressure, and plasma epinephrine in combat veterans with PTSD during and after viewing a combat film, but not in response to the film of an automobile accident. Using auditory, war-related stimuli, Blanchard and colleagues (1991) reported similar parallel increases in heart rate and plasma NE among combat veterans with PTSD, but not among controls. Finally, Liberzon, Abelson, Flagel, Raz, and Young (1999) found exaggerated heart rate, plasma epinephrine, and plasma NE responses to combat sounds in veterans with PTSD compared to combat veterans without PTSD and healthy controls. These findings suggested that heightened physiological reactivity is related to elevated levels of circulating catecholamines.

In a study designed to assess dynamic functioning and regulation of $alpha_2$ adrenergic receptors, Perry and colleagues (1990) incubated intact platelets with high levels of epinephrine, and found a greater and more rapid loss in receptor number among subjects with PTSD compared to controls, suggesting that subjects with PTSD are particularly sensitive to stimulation and adaptive "down-regulation" of alpha$_2$ adrenergic receptors by the agonist epinephrine. Mixed results have been reported among *in vitro* challenge studies assessing the effects of epinephrine on forskolin-stimulated adenylate cyclase activity and the lymphocyte beta-adrenergic receptor mediated cyclic adenosine 3'5'-monophosphate (Southwick, Bremner, et al., 1999).

A series of pharmacological challenge studies that used yohimbine as a probe of the noradrenergic system has shed additional light on catecholamine systems in PTSD. Yohimbine is an alpha$_2$ adrenergic receptor antagonist that blocks postsynaptic alpha$_2$ adrenergic receptors and increases presynaptic release of NE by occupying the alpha$_2$ autoreceptor. Yohimbine has relatively minimal effects when administered to healthy subjects; however, it can cause marked increases in subjective anxiety/panic, heart rate, and biochemical indices of noradrenergic activity (plasma MHPG, a metabolite of NE) when administered to individuals with panic disorder. Similarly, in three studies involving combat veterans with PTSD, yohimbine caused significant increases in subjective anxiety/panic, heart rate, and plasma MHPG (Bremner et al., 1999; Southwick et al., 1993, 1997). Yohimbine provoked panic attacks in over 50% of subjects with PTSD who had a history of panic attacks. However, unlike subjects with panic disorder, subjects with PTSD also experienced significant increases in yohimbine-induced PTSD symptoms, such as intrusive memories and hypervigilance. In fact, nearly 80% of combat veterans with PTSD experienced vivid intrusive memories of war-related traumas, and 40% experienced war-related flashbacks when yohimbine was administered. Consistent with these findings, Morris, Hopwood, Maguire, Norman, and Schweitzer (2004) reported a significantly blunted plasma growth hormone response to clonidine among nondepressed Vietnam veterans with PTSD compared to veteran controls. A blunted growth hormone response to clonidine suggests postsynaptic alpha$_2$ adrenergic receptor subsensitivity, potentially secondary to increased noradrenergic activity.

NE AND GENE POLYMORPHISMS

Genetic factors clearly play an important role in sympathetic nervous system (SNS) reactivity to stress. Recent evidence suggests that alpha$_2$ adrenoreceptor gene polymorphisms play a role in baseline catecholamine levels, intensity of stress-induced SNS activation, and rate of catecholamine return to baseline after stress. For example, exaggerated baseline total body noradrenergic spillover (a measure of noradrenergic turnover) exaggerated yohimbine-induced increases in total body noradrenergic spillover and anxiety, and a slower than normal return of total body noradrenergic spillover to baseline after yohimbine infusion has been found in healthy subjects who are homozygous carriers for the alpha$_{2c}$Del322-325-AR polymorphism (Neumeister et al., 2005). Such individuals, like those mentioned earlier with polymorphisms of the 5-HT transporter, may be more vulnerable to stress-related psychiatric disorders such as PTSD and depression.

Neuropeptide Y

NPY is a 36-amino-acid neurotransmitter that is found in the peripheral SNS and in multiple stress-responsive brain regions, including the LC, amygdala, hippocampus, PAG, and PFC (Heilig & Widerlov, 1995). In most noradrenergic neurons, NPY is co-localized with NE. However, NPY is co-released with NE only during intense activation/stress and not during brief or mild stress. At high levels of stimulation, NPY is released with and potentiates the effects of NE at postsynaptic noradrenergic receptors. A metabolite of NPY then serves to inhibit further release of NE.

In a variety of preclinical studies, NPY has been shown to inhibit the firing rate of the LC and the release of CRF and NE, and to be an anxiolytic (Heilig & Widerlov, 1995). These inhibitory actions are mediated through the presynaptic Y_2 receptor. Among healthy humans exposed to high levels of acute, uncontrollable stress, NPY also appears to have an antistress effect. Morgan and colleagues (2000, 2001) reported both a significant positive association between NPY release and superior performance, and a significant negative relationship between NPY and symptoms of dissociation among a group of elite soldiers engaged in high-intensity survival training. In these soldiers, robust increases in NPY may advantageously modulate the effects of simultaneously observed robust increases in NE.

Unlike the effects of acute, intense stress, chronic stress has been shown to decrease plasma NPY and to increase the noradrenergic response to a novel stressor. Significantly lower elevations of plasma NPY, but significantly greater elevations of plasma NE, have been reported in a study of rats exposed to 12 consecutive days of restraint stress, followed by one episode of footshock compared to naive rats exposed to footshock alone. Similarly, low baseline levels of NPY and blunted NPY responses to yohimbine in the context of high noradrenergic responses to yohimbine have been observed in combat veterans with PTSD (Rasmusson et al., 2000). In this group of veterans, it appeared that stress-induced increases of plasma NPY were insufficient to hold rising levels of NE in check. It is likely that rapid increases in NE contributed to exaggerated increases in heart rate, blood pressure, respiratory rate, anxiety, panic, vigilance, and even intrusive, combat-related memories (Southwick, Bremner, et al., 1999).

The Hypothalamic–Pituitary–Adrenal Axis

General Characteristics

Under conditions of acute and chronic stress, the paraventricular nucleus of the hypothalamus secretes CRF, which then stimulates the anterior pituitary gland to synthesize and release adrenocorticotropic hormone (ACTH), which in turn stimulates the synthesis and release of adrenocortical glucocorticoids, as well as other adrenally derived neuroactive steroids. Whereas the SNS prepares the organism to react to stressful stimuli, the hypothalamic–pituitary–adrenal (HPA) axis appears to facilitate defensive responses, as well as to help restore homeostasis (Munck, Guyre, & Holbrook, 1984; Southwick, Vythilingam, & Charney, 2005; Yehuda, 2002). Cortisol, in particular, helps to mobilize and replenish energy stores, inhibits growth and reproductive systems, contains the immune response and contains sympathetic noradrenergic responses, and affects behavior through actions on multiple neurotransmitter systems and brain regions.

Corticotropin-Releasing Factor

CRF is one of the most important mediators of the stress response, and CRF-containing neurons are located throughout the brain. Centrally administered CRF produces a number of symptoms and behaviors commonly seen in depression and anxiety, such increased heart rate, increased blood pressure, decreased appetite, decreased sexual activity, increased arousal, and a reduction in reward expectations (Dunn & Berridge,

1987). CRH_1 and CRH_2 receptors both play an important role in the stress response. Agonist binding of CRH_1 receptors may be responsible for anxiety-like responses, whereas activation of CRH_2 receptors may produce anxiolytic-like responses. Regulation of these two CRH receptor types in critical brain regions appears to be associated with psychological and physiological responses to stress (Dautzenberg, Kilpatrick, Hauger, & Moreau, 2001).

HPA Axis Studies in PTSD

In humans, a large number of studies have reported abnormalities in HPA axis functioning among trauma survivors with PTSD (see Table 10.1). Abnormalities have included alterations in 24-hour urinary free-cortisol excretion, 24-hour plasma cortisol levels, lymphocyte glucocorticoid receptor number, cortisol response to dexamethasone, ACTH response to CRF, and B-endorphin and ACTH responses to metyrapone (Yehuda, 2002; Rasmusson Vythilingam, & Morgan, 2003). Additionally, elevated resting CSF levels of CRF have been reported in two studies of combat veterans with chronic PTSD (Baker et al., 1999; Bremner et al., 1997).

There is, however, a striking lack of uniformity of findings related to the HPA axis in PTSD. Indeed, a growing body of clinical and basic research suggests that factors such as gender, diagnostic comorbidity, age at the time of trauma exposure, and ethnicity or genetic factors influence the pattern of HPA axis adaptation to traumatic or chronic severe stress and contribute to the variable outcomes of HPA axis studies in PTSD (Rasmusson et al., 2004). In addition, differences among studies in medication use, smoking, alcohol use, and activity among subjects may play a role in producing variable outcomes.

The variability in findings regarding HPA axis function in PTSD is perhaps best exemplified by the results of 24-hour urinary free-cortisol output studies. Low 24-hour urinary cortisol output has been found in populations of male combat veterans and in male and postmenopausal female Holocaust survivors. Low cortisol output may best be explained by enhanced glucocorticoid negative feedback and reduced adrenal capacity even in the context of apparent increases in pituitary or adrenal reactivity in these populations. Reduced adrenal capacity may represent a preexisting risk factor, whereas enhanced negative feedback inhibition and increased HPA reactivity may develop over time in response to trauma and chronic PTSD (Yehuda, 2002). High 24-hour urinary free-cortisol output has been found most consistently in populations of premenopausal women or children with PTSD—particularly when current or past comorbid major depression is present (Heim et al., 2000; Lipschitz et al., 2003; Rasmusson et al., 2003, 2004; Young & Breslau, 2004). Increased 24-hour urinary cortisol output may result from increased pituitary ACTH responses to CRF and secondary increases in adrenal capacity (Rasmusson et al., 2004). These alterations in turn may result from an imbalance in the release of adrenally derived neuroactive steroids other than cortisol and a reduction in delayed negative feedback to the HPA axis (see below).

Other Adrenally Derived Neuroactive Steroids

Dehydroepiandrosterone (DHEA) is another adrenal steroid that is released during stress (Baulieu & Robel, 1998). In response to fluctuating levels of ACTH, DHEA is released synchronously and episodically with cortisol and crosses the blood–brain barrier. In the brain, DHEA can exert antiglucocorticoid effects, antagonize $GABA_A$ receptors, and positively modulate glutamate effects at excitatory NMDA receptors. A sulfated metabolite of DHEA (DHEAS) provides a large storage pool for DHEA in the periphery. In the brain, DHEA also may be sulfated; DHEAS in turn acts even more potently than DHEA at $GABA_A$ and NMDA receptors.

In the brain, DHEA may confer neuroprotection through its antiglucocorticoid activity. Support for DHEA as a possible stress protective factor includes a negative correlation between DHEA reactivity (in response maximal stimulation of the adrenal gland by ACTH administration) and severity of PTSD symptoms in premenopausal women (Rasmusson et al., 2003), as well as a negative relationship between the DHEAS:cortisol ratio and dissociation, and a positive correlation between the DHEAS:cortisol ratio and performance among elite male special forces undergoing intensive survival training (Morgan et al., 2004). Furthermore, low plasma DHEA or DHEAS levels, or a low ratio between DHEA(S) and cortisol have been associated with depressed mood and feelings of reduced vigor and well-being. Of note, DHEA(S) may modulate PFC functioning through antagonism of $GABA_A$ receptors and facilitation of NMDA receptor functioning; in addition, it may facilitate both fear conditioning and extinction through positive modulation of NMDA receptors in the amygdala (Walker & Davis, 2002).

Allotetrahydrodeoxycorticosterone and allopregnanolone are two potent, adrenally derived, neuroactive steroids that positively modulate GABA effects at $GABA_A$ receptors and enhance chloride ion flux into neurons. In response to stress, the adrenal gland releases allopregnanolone, which is believed to exert delayed negative feedback on the HPA axis (e.g., Barbaccia et al., 1997), as well as produce anxiolytic, sedative, and analgesic effects. In a study of premenopausal women with PTSD, who were in the follicular phase of the menstrual cycle, CSF allopregnanolone levels in the women with PTSD were less than 40% of those in the nontraumatized, healthy subjects (Rasmusson et al., 2005); a reduced CSF allopregnanolone:DHEA ratio in these subjects correlated strongly, significantly, and negatively with reexperiencing and depressive symptoms.

Interestingly, the variable patterns of dysregulation of the HPA axis in PTSD (reviewed in Rasmusson et al., 2003; also see Young & Breslau, 2004) may be related to low CNS allopregnanolone levels. Glucocorticoids up-regulate transcription of the gene for the enzyme that synthesizes allopregnanolone (Hou, Lin, & Penning, 1998). Thus, individuals with limited cortisol production during stress may fail to up-regulate allopregnanolone normally and thereby have excessive activation of monoamine systems and CRF release during stress; facilitation of fear conditioning and resistance to extinction of conditioned fear upon reexposure to trauma reminders may result. On the other hand, individuals with a primary problem with allopregnanolone synthesis (e.g., due to a genetic predisposition) might be expected to have heightened HPA axis and monoaminergic responses to traumatic stress.

HPA Axis and the PFC, Amygdala, and Hippocampus

During situations of threat, glucocorticoids are released from the adrenal gland and cross the blood–brain barrier to exert effects on the amygdala, hippocampus, and PFC. Glucocorticoid levels are controlled, in part, by hippocampus-mediated inhibition (i.e., negative feedback) of the HPA axis. Although high levels of glucocorticoids enhance functioning of the amygdala (e.g., fear conditioning and consolidation of emotional memory), they can also impair functioning of the PFC (interfering with working memory and inhibition of the amygdala) and the hippocampus (interfering with inhibition of the HPA axis) (Southwick, Vythilingham, & Charney, 2005). Enhanced glucocorticoid-mediated stimulation of the amygdala, in combination with decreased prefrontal cortical inhibition of the amygdala, can leave the organism in a physiological state dominated by poorly inhibited limbic activity. This may be especially true when glucocorticoid release is relatively unchecked due to compromised inhibition by the hippocampus or an imbalance in the release of neuroactive steroids that provide positive versus negative feedback inhibition of the HPA axis (e.g., a decrease in the ratio of allopregnanolone to DHEA). Enhanced glucocortioid receptor sensitivity in the context of relatively low cortisol output might be expected to have similar functional results.

HPA Axis and Gene Polymorphisms

To date, numerous HPA-axis-related genes with polymorphisms known to enhance either ACTH or cortisol responses to stress have been identified. These include polymorphisms of the catechol-*O*-methyltransferase (COMT) gene (Hernandez-Avila, Wand, Luo, Gelernter, & Kranzler, 2003; Oswald, McCaul, Choi, Yang, & Wand, 2004), the angiotensin I-converting enzyme (ACE-I) gene (Baghai et al., 2002), the glucocorticoid receptor gene (Wust et al., 2004), the ACTH gene (Slawik et al., 2004) and the CRF or CRF receptor gene (Challis et al., 2004; Gonzalez-Gay et al., 2003; Smoller et al., 2003). In addition, more than 65 different functional mutations of the 21-hydroxylase gene are known to affect cortisol and DHEA production at baseline and in response to extreme activation of the adrenal gland. Unfortunately, no studies have yet attempted to link these genes to HPA axis functioning in patients with PTSD. Future work in this area may help to identify factors that influence the risk for PTSD development or response to treatment.

CONCLUSIONS

In this chapter we have focused on several neurotransmitter/neuroendocrine systems known to be involved in the pathophysiology of PTSD. In a subgroup of individuals with PTSD, these neurotransmitter/neuroendocrine systems appear to hyperrespond, with exaggerated reactivity to stress. These systems are characterized by complex interactions with one another and with multiple brain regions, including the amygdala, LC, dorsal raphe nucleus, hippocampus, and PFC.

In recent years, a number of neurobiological models have been proposed to explain these findings in PTSD. Commonly cited models include behavioral sensitization, fear conditioning, failure of extinction, learned helplessness, enhanced encoding and consolidation of emotional memory, abnormal activation of the amygdala and anterior cingulate, altered prefrontal cortical inhibition of limbic activity, and stress-induced structural alterations within a number of brain regions, such as the hippocampus.

Stress sensitization provides a good example of the complex neurobiological interactions involved in any attempt to understand PTSD, or some aspect of PTSD. "Stress sensitization" refers to a stressor-induced increase in behavioral, physiological, and biochemical responding to subsequent stressors of the same or lesser magnitude (Post, Weiss, & Smith, 1995; Sorg & Kalivas, 1995). Biological systems that have become sensitized gradually increase their responsiveness to a given stressor. It has been suggested that stress sensitization may be adaptive and allow the organism to rapidly and robustly respond to future stressors. On the other hand, stress-sensitized neurobiological systems may become maladaptive if organisms begin to overreact to even minor stressors. The consequence might be an inability to modulate arousal, resulting in hypervigilance, exaggerated startle, and a tendency to respond biologically as if a danger exists, even when no real danger is present (Southwick et al., 1995).

A substantial body of evidence suggests that a number of neurotransmitter/ neuroendocrine systems are hyperreactive or sensitized in many individuals with PTSD. In one of the most thoroughly studied systems, the noradrenergic system, clinical studies have repeatedly found higher levels of stress-induced NE in a subgroup of individuals with PTSD compared to controls. Multiple factors may contribute to this increased noradrenergic activity. First, enhanced noradrenergic activity may be related to increased synthesis and subsequent release of NE. This possibility is supported by preclinical findings, where increases in dopamine beta-hydroxylase activity, tyrosine hydroxylase, and synaptic levels of NE (Irwin, Ahluwalia, & Anisman, 1986; Karmarcy, Delaney, & Dunn, 1984) are found in animals exposed to repeated shock. Second, elevated noradrenergic activity may be related to increased release of CRF, which has been reported in PTSD (Bremner et al., 1997). Preclinical studies have clearly shown that CRF and NE participate in a mutually reinforcing feedback loop during stressful situations. NE turnover is increased in several forebrain regions by intracerebral infusion of CRF (Dunn & Berridge, 1987); CRF concentrations in the LC are markedly increased by stress that increases firing rate of LC/NE neurons (Chappel et al., 1990; Valentino & Foote, 1988); and CRF infusion into the LC is anxiogenic and significantly increases MHPG in the amygdala and hypothalamus (Butler, Weiss, Stout, & Nemeroff, 1990). Third, decreases in NPY, as reported in chronic PTSD, may lead to exaggerated noradrenergic activity. Lower NPY may result in a reduced capacity to restrain noradrenergic system reactivity, with a resultant exaggerated release in NE. Since NPY also inhibits release of CRF, lower NPY would contribute to increased CRF release, with a subsequent increase in LC/NE activity. Thus, increased synthesis of NE, hypersecretion of CRF, and decreased NPY may each be factors that contribute to the exaggerated release of NE seen in a subgroup of individuals with PTSD (Southwick, Bremner, et al., 1999).

When an organism with a hyperresponsive noradrenergic/LC system is stressed, the amygdala, hippocampus, and PFC become flooded with NE. In the amygdala, elevated levels of NE enhance fear conditioning and consolidation of memory (McGaugh, 2002). Of note, the relationship between NE and memory is time- and dose-dependent. In the hippocampus, LC activation is known to play an important role in learning and memory. Activation of LC neurons by glutamate affects beta-adrenergic-mediated synaptic plasticity in the dentate gyrus of the hippocampus (Walling & Harley, 2004). Finally, NE has differential effects on functioning of the PFC depending on the levels of NE. Moderate levels of NE in the PFC preferentially engage postsynaptic alpha$_2$ receptors, which enhance executive functioning and inhibit posterior sensory cortical and limbic activity. However, high levels of NE (as reported in a subgroup of individuals with PTSD) preferentially engage postsynaptic alpha$_1$ receptors, which in essence take the PFC "offline" (Arnsten, 2000).

Of course the PFC, amygdala, and hippocampus are also affected by other neurotransmitter/hormones that are altered in PTSD. For example, in the PFC, high levels of stress-induced glucocorticoids tend to augment synaptic catecholamine levels, potentially adding to engagement of postsynaptic alpha₁ receptors and even greater impairment of PFC functioning (Grundemann, Schechinger, Rappold, & Schomig, 1998; Park et al., 1994). Similarly, reduced 5-HT appears to compromise functioning of the orbitofrontal cortex, with effects on emotional processing of affective memories and interpretation of social and emotional information.

Whereas altered levels of NE, cortisol, and 5-HT would decrease effective functioning of the PFC and its inhibition of the amygdala, these same altered levels might augment amygdala activity. The result might be an "unleashed amygdala" where decreased inhibition by the PFC, in combination with amygdala excitation secondary to elevated levels of catecholamines and cortisol, and reduced levels of 5-HT, would leave the amygdala in an activated state (Southwick, Rasmusson, et al., 2005). It is also possible that an overactivated amygdala would stimulate additional release of NE (LC), which might further compromise PFC functioning and contribute to diminished capacity for rational influence on behavior and thought, exaggerate the startle response, increase fear conditioning, enhance consolidation of emotional memory, and increase vigilance, insomnia, impulsivity, intrusive memories, and other fear-related behaviors.

Clearly, altered neurotransmitter/neuroendocrine activity and functioning would also affect other brain regions, such as the hippocampus and anterior cingulate cortex. For example, it is well known that stress-induced elevations in glucocorticoids may cause damage to the hippocampus under conditions of high metabolic demand. Glucocorticoid toxicity appears to be mediated in part through excessive glutamate accumulation and toxicity (McEwen, 1998). SSRIs, on the other hand, have been shown to enhance neurogenesis in the hippocampus (Djavadian, 2004; Duman, 2004).

In summary, we have briefly reviewed clinical data related to several neurotransmitter/neuroendocrine systems known to be involved in the pathophysiology of PTSD and that may contribute to some of the symptoms and neurocognitive deficits reported in this patient population. These neurotransmitters appear to exert their stress-related effects through actions in multiple brain regions, including the PFC, amygdala, hippocampus, dorsal raphe nucleus, and the LC.

As noted in "Methodological Considerations," attempting to relate neurobiological alterations to symptoms or diagnoses is both challenging and risky. As such, the preceding discussion is largely speculative in nature. Furthermore, in our discussion, we have focused on a limited number of neurobiological systems and their interactions with one another.

GENERALIZABILITY OF CURRENT FINDINGS

The conclusions discussed in this chapter should be viewed with caution for a number of reasons. First, most neurobiological studies in PTSD have involved a small number of subjects. It will be important to replicate findings with larger sample sizes. Second, many neurobiological studies in PTSD have involved traumatized combat veterans. It is not known whether the findings in this patient population generalize to civilian trauma populations or necessarily generalize to women (Rasmusson & Friedman, 2002). Third, not all neurobiological investigations in PTSD have required that subjects be medication-free at the time of testing. This is particularly problematic for studies focused on neurobiological systems that are affected by the medications that subjects are taking. Fourth, most neurobiological studies in PTSD have included subjects with a variety of comorbid psychiatric and medical conditions. Again, it is not known how these conditions affect study results. Fifth, adherence to inclusion and exclusion criteria generally makes interpretation of results far easier. However, adherence to exclusion criteria (e.g., history of substance abuse) also limits the generalizability of findings, because many traumatized subjects are typically excluded from neurobiological studies. Finally, most investigations to date have studied subjects with chronic PTSD. Far less is known about neurobiological alterations in individuals with acute stress reactions, and it is not known how findings from populations with chronic PTSD relate to individuals with acute stress reactions.

CHALLENGES FOR THE FUTURE

Future studies in PTSD will benefit from concurrent use of multiple scientific technologies and methodologies. For example, combining DNA testing, brain imaging, neuropsychological testing, and pharmacological treatment in the same subject should dramatically add to our understanding of PTSD. It is likely that stress-related risk and resilience factors have multiple determinants, such as genetic makeup, developmental history, personality, coping style, and history of exposure to traumatic events.

To date, most pharmacological studies in PTSD have tested medications that are known to treat other psychiatric conditions and symptoms such as depression, insomnia, and anger effectively. However, as researchers better understand underlying pathophysiology of PTSD, it will become increasingly possible to develop rational pharmacological interventions that target abnormalities specifically observed in PTSD. It is likely that such targeted approaches will lead to more robust treatment responses.

It is also true that most currently available trauma-related treatments are aimed at reducing symptoms once they have developed. A critical challenge for the future involves early detection and treatment, so that acute stress disorders do not progress into chronic PTSD. Even more exciting and challenging will be the development of neurobiological approaches to enhance resilience and, we hope, prevent the development of trauma-related symptoms. For example, it is possible that SSRIs, tricyclic antidepressants, adrenergic blockers, NPY, CRF antagonists, and DHEA might each play a role in fostering stress resilience.

REFERENCES

Agaganian, G. K. (1995). Electrophysiology of serotonin receptor subtypes and signal transduction pathways. In F. E. Bloom & D. J. Kupfer (Eds.), *Psychopharmacology: The fourth generation of progress.* New York: Raven Press.

Anand, A., Charney, D. S., Oren, D. A., Berman, R. M., Hu, X. S., Cappiello, A., et al. (2000). Attenuation of the neuropsychiatric effects of ketamine with lamotrigine: Support for hyperglutamatergic effects of N-methyl-D-aspartate receptor antagonists. *Archives of General Psychiatry, 57*(3), 270–276.

Armanini, M. P., Hutchins, D., Stein, B. A., & Sapolsky, R. M. (1990). Glucocorticoid endangerment of hippocampal neurons is NMDA-receptor dependent. *Brain Research, 532*(1–2), 7–12.

Arnsten, A. F. (2000). Through the looking glass: Differential noradenergic modulation of prefrontal cortical function. *Neural Plasticity, 7*(1–2), 133–146.

Arora, R. C., Fichtner, C. G., O'Connor, F., & Crayton, J. W. (1993). Paroxetine binding in the blood platelets of post-traumatic stress disorder patients. *Life Sciences, 53*(11), 919–928.

Aston-Jones, G., Shipley, M. T., Chouvet, G., Ennis, M., van Bockstaele, E., Pieribone, V., et al. (1991). Afferent regulation of locus coeruleus neurons: Anatomy, physiology and pharmacology. *Progress in Brain Research, 88,* 47–75.

Baghai, T. C., Schule, C., Zwanzger, P., Minov, C., Zill, P., Ella, R., et al. (2002). Hypothalamic–pituitary–adrenocortical axis dysregulation in patients with major depression is influenced by the insertion/deletion polymorphism in the angiotensin I–converting enzyme gene. *Neuroscience Letters, 328*(3), 299–303.

Baker, D. G., West, S. A., Nicholson, W. E., Ekhator, N. N., Kasckow, J. W., Hill, K. K., et al. (1999). Serial CSF coricotrophin-releasing hormone levels and adrenocortical activity in combat veterans with posttraumatic stress disorder. *American Journal of Psychiatry, 156*, 585–588.

Barbaccia, M. L., Roscetti, G., Trabucchi, M., Purdy, R. H., Mostallino, M. C., Concas A., et al. (1997). The effects of inhibitors of GABAergic transmission and stress on brain and plasma allopregnanolone concentrations. *British Journal of Pharmacology, 120*(8), 1582–1588.

Baulieu, E., & Robel, P. (1998). Dehydroepiandrosterone (DHEA) and dehydroepiandrosteronesulfate (DHEAS) as neuroactiveneurosteroids. *Proceedings of the National Academy of Sciences USA, 95*, 4089–4091.

Bertolino, A., Aciero, G., Rubino, V., Latorre, V., DeCandia, M., Mazzola, V., et al. (2005). Variation of human amygdala response during threatening stimuli as a function of 5'HTTLPR genotype and personality style. *Biological Psychiatry, 57*(12), 1517–1525.

Blanchard, E. B., Kolb, L. C., Gerardi, R. J., Ryan, P., & Pallmeyer, T. P. (1986). Cardiac response to relevant stimuli as an adjunctive tool for diagnosing post-traumatic stress disorder in Vietnam veterans. *Behavior Therapy, 17*, 592–606.

Blanchard, E. B., Kolb, L. C., Prins, A., Gates, S., & McCoy, G. C. (1991). Changes in plasma norepinephrine to combat-related stimuli among Vietnam veterans with posttraumatic stress disorder. *Journal of Nervous and Mental Disease, 179*(6), 371–373.

Blanchard, E. B., Kolb, L. C., Taylor, A. E., & Wittrock, D. A. (1989). Cardiac response to relevant stimuli as an adjunct in diagnosing post-traumatic stress disorder: Replication and extension. *Behavior Therapy, 20*, 535–543.

Blier, P. (2001). Crosstalk between the norepinephrine and serotonin systems and its role in the antidepressant response. *Journal of Psychiatry and Neuroscience, 26*, S3–10.

Bonne, O., Grillon, C., Vythilingam, M., Neumeister, A., & Charney, D. S. (2004). Adaptive and maladaptive psychobiological responses to severe psychological stress: Implications for the discovery of novel pharmacotherapy. *Neuroscience and Biobehavioral Reviews, 28*(1), 65–94.

Boscarino, J. A. (2004). Posttraumatic stress disorder and physical illness: Results from clinical and epidemiologic studies. *Annals of the New York Academy of Sciences, 1032*, 141–153.

Bremner, J., Southwick, S., & Charney, D. (1999). The neurobiology of posttraumatic stress disorder: An integration of animal and human research. In P. A. Saigh & J. D. Bremner (Eds.), *Posttraumatic stress disorder: A comprehensive text* (pp. 103–143). Boston: Allyn & Bacon.

Bremner, J. D., Innis, R. B., Southwick, S. M., Staib, L., Zoghbi, S., & Charney, D. S. (2000). Decreased benzodiazepine receptor binding in prefrontal cortex in combat-related posttraumatic stress disorder. *American Journal of Psychiatry, 157*(7), 1120–1126.

Bremner, J. D., Licinio, J., Darnell, A., Krystal, J. H., Owens, M. J., Southwick, S. M., et al. (1997). Elevated CSF corticotropin-releasing factor concentrations in posttraumatic stress disorder. *American Journal of Psychiatry, 154*(5), 624–629.

Bremner, J. D., Vythilingam, M., Vermetten, E., Southwick, S. M., McGlashan, T., Staib, L. H., et al. (2003). Neural correlates of declarative memory for emotionally valenced words in women with posttraumatic stress disorder related to early childhood sexual abuse. *Biological Psychiatry, 53*(10), 879–889.

Butler, P. D., Weiss, J. M., Stout, J. C., & Nemeroff, C. B. (1990). Corticotropin-releasing factor produces fear-enhancing and behavioral activating effects following infusion into the locus coeruleus. *Journal of Neuroscience, 10*, 176–183.

Carson, M. A., Paulus, L. A., Lasko, N. B., Metzger, L. J., Wolfe, J., Orr, S. P., et al. (2000). Psychophysiologic assessment of posttraumatic stress disorder in Vietnam nurse veterans who witnessed injury or death. *Journal of Consulting and Clinical Psychology, 68*(5), 890–897.

Caspi, A., Sugden, K., Moffitt, T. E., Taylor, A., Craig, I. W., Harrington, H., et al. (2003). Influence of life stress on depression: Moderation by a polymorphism in the 5-HTT gene. *Science, 301*, 386–389.

Challis, B. G., Luan, J., Keogh, J., Wareham, N. J., Farooqi, I. S., & O'Rahilly, S. (2004). Genetic variation in the corticotrophin-releasing factor receptors: Identification of single-nucleotide polymorphisms and association studies with obesity in UK Caucasians. *International Journal of Obesity and Related Metabolic Disorders, 28*(3), 442–446.

Chappell, P. B., Smith, M. A., Kilts, C. D., Bissette, G., Ritchie, J., & Anderson, C. (1990). Alterations in corticotropin-releasing factor-like immunoreactivity in discrete rat brain regions after acute and chronic stress. *Journal of Neuroscience, 6,* 2908–2914.

Charney, D. S. (2004). Psychological mechanisms of resilience and vulnerability: Implications for successful adaptation to extreme stress. *American Journal of Psychiatry, 161,* 195–216.

Dautzenberg, F. M., Kilpatrick, G. J., Hauger, R. L., & Moreau, J. L. (2001). Molecularbiology of the CRH receptors—in the mood. *Peptides, 22,* 753–760.

Davidson, R. J., Putnam, K. M., & Larson, C. L. (2000). Dysfunction in the neural circuitry of emotion regulation—a possible prelude to violence. *Science, 289,* 591–594.

Davis, J. M., Adams, H. E., Uddo, M., Vasterling, J. J., & Sutker, P. B. (1996). Physiological arousal and attention in veterans with posttraumatic stress disorder. *Journal of Psychopathology and Behavioral Assessment, 18,* 1–20.

Davis, M., Astrachan, D. L. & Kass, E. (1980). Excitatory and inhibitory effects of serotonin on sensorimotor reactivity measured with acoustic startle. *Science, 209,* 521–523.

Davis, M. (1999). Functional neuroanatomy of anxiety and fear: A focus on the amygdala. In D. S. Charney, E. J. Nestler, & B. S. Bunney (Eds.), *Neurobiology of mental illness* (pp. 463–474). New York: Oxford University Press.

Djavadian, R. L. (2004). Serotonin and neurogenesis in the hippocampal dentate gyrus of adult mammals. *Acta Neurobiologiae Experimentalis, 64*(2), 189–200.

Duman, R. S. (2004). Depression: A case of neuronal life and death? *Biological Psychiatry, 56*(3), 140–145.

Dunn, A. L., & Berridge, C. W. (1987). Corticotropin-releasing factor administration elicits a stresslike activation of cerebral catecholaminergic systems. *Pharmacology, Biochemistry, and Behavior, 27,* 685–691.

Fichtner, C. G., Arora, R. C., O'Connor, F. L., & Crayton, J. W. (1994). Platelet paroxetine binding and fluoxetine pharmacotherapy in posttraumatic stress disorder: Preliminary observations on a possible predictor of clinical treatment response. *Life Sciences, 54*(3), 39–44.

Friedman, M. J., & Southwick, S. M. (1995). Towards pharmacotherapy for post-traumatic stress disorder. In M. J. Friedman, D. S. Charney, & A. Y. Deutch (Eds.), *Neurobiological and clinical consequences of stress* (pp. 465–482). Philadelphia: Lippincott–Raven.

Friedman, M. J., Davidson, J. R. T., Mellman, T. A., & Southwick, S. M. (2000). Guidelines for treatment of PTSD: Pharmacotherapy. *Journal of Traumatic Stress, 13*(4), 563–568.

Friedman, M. J., Wang, S. Jalowiec, J. E., McHugo, G. J., & McDonagh-Coyle, A. (2005). Thyroid hormone alterations among women with posttraumatic stress disorder due to childhood sexual abuse. *Biological Psychiatry, 57*(10), 1186–1192.

Fujita, M., Southwick, S. M., Denucci, C. C., Zoghbi, S. S., Dillon, M. S., Baldwin, R. M., et al. (2004). Central type benzodiazepine receptors in Gulf War veterans with posttraumatic stress disorder. *Biological Psychiatry, 56*(2), 95–100.

Gerardi, R. J., Blanchard, E. B., & Kolb, L. C. (1989). Ability of Vietnam veterans to dissimulate a psychophysiological assessment for post-traumatic stress disorder. *Behavior Therapy, 20,* 229–243.

Gonzalez-Gay, M. A., Hajeer, A. H., Garcia-Porrua, C., Dababneh, A., Amoli, M. M., Botana, M. A., et al. (2003). Corticotropin-releasing hormone promoter polymorphisms in patients with rheumatoid arthritis from northwest Spain. *Journal of Rheumatology, 30*(5), 913–917.

Grillon, C., & Morgan, C. A., III. (1999). Fear-potentiated startle conditioning to explicit and contextual cues in Gulf War veterans with posttraumatic stress disorder. *Journal of Abnormal Psychology, 108,* 134–142.

Grillon, C., Morgan, C. A., III, Davis, M., & Southwick, S. M. (1998). Effects of experimental context and explicit threat cues on acoustic startle in Vietnam veterans with posttraumatic stress disorder. *Biological Psychiatry, 44,* 1027–1036.

Grundemann, D., Schechinger, B., Rappold, G. A., & Schomig, E. (1998). Molecular identification of the corticosterone-sensitive extraneuronal catecholamine transporter. *Nature Neuroscience, 1*(5), 349–351.

Handelsman, L, Holloway, K., Kahn, R. S., Sturiano, C., Rinaldi, P. J., Bernstein, D. P., et al. (1996). Hostility is associated with a low prolactin response to metachlorophenylpiperazine in abstinent alcoholics. *Alcoholism, Clinical and Experimental Research, 20*(5), 824–829.

Hasler, G., Drevets, W. C., Manji, H. K., & Charney, D. S. (2004). Discovering endophenotypes for major depression. *Neuropsychopharmacology, 29*(10), 1765–1781.

Helig, M., & Widerov, E. (1995). Neurobiology and clinical aspects of neuropeptide Y. *Critical Reviews in Neurobiology, 9*, 115–136.

Heim, C., Newport, D. J., Heit, S., Graham, Y. P., Wilcox, M., Bonsall, R., et al. (2000). Pituitary-adrenal and autonomic responses to stress in women after sexual and physical abuse in childhood. *Journal of the American Medical Association, 284*, 592–597.

Hernandez-Avila, C. A., Wand, G., Luo, X., Gelernter, J., & Kranzler, H. R. (2003). Association between the cortisol response to opioid blockade and the Asn40Asp polymorphism at the mu-opioid receptor locus (OPRM1). *American Journal of Medical Genetics, 118B*(1), 60–65.

Hou, Y. T., Lin, H. K., & Penning, T. M. (1998). Dexamethasone regulation of the rat 3-hydroxysteroid/ dihydrodiol dihydrogenase (3-HSD/DD) gene. *Molecular Pharmacology, 53*, 459–466.

Irwin, J., Ahluwalia, P., & Anisman, H. (1986). Sensitization of norepinephrine activity following acute and chronic footshock. *Brain Research, 379*(1), 98–103.

Karmarcy, N. R., Delaney, R. L., & Dunn, A. L. (1984). Footshock treatment activates catecholamine synthesis in slices of mouse brain regions. *Brain Research, 290*, 311–319.

Keane, T. M., Kolb, L. C., Kaloupek, D. G., Orr, S. P., Blanchard, E. B., Thomas, R. G., et al. (1998). Utility of psychophysiological measurement in the diagnosis of posttraumatic stress disorder: Results from a Department of Veterans Affairs Cooperative Study. *Journal of Consulting and Clinical Psychology, 66*(6), 914–923.

Koenen, K. C., Driver, K. L., Oscar-Berman, M., Wolfe, J., Folsom, S., Huang, M. T., et al. (2001). Measures of prefrontal system dysfunction in posttraumatic stress disorder. *Brain and Cognition, 45*(1), 64–78.

Kram, M. J., Kramer, G. L., Steciuk, M., Ronan, P. J., & Petty, F. (2000). Effects of learned helplessness on brain GABA receptors. *Neuroscience Research, 38*(2), 193–198.

Krystal, J. H., Bennett, A. L., Bremner, J. D., Southwick, S. M., & Charney, D. S. (1995). *Toward a cognitive neuroscience of dissociation and altered memory functions in post traumatic stress disorder.* Philadelphia: Lippincott–Raven.

Krystal, J. H., D'Souza, D. C., Karper, L. P., Bennett, A., Abi-Dargham, A., Abi-Saab, D., et al. (1999). Interactive effects of subanesthetic ketamine and haloperidol in healthy humans. *Psychopharmacology, 145*(2), 193–204.

Krystal, J. H., Karper, L. P., Bennett, A., D'Souza, D. C., Abi-Dargham, A., Morrissey, K., et al. (1998). Interactive effects of subanesthetic ketamine and subhypnotic lorazepam in humans. *Psychopharmacology, 135*(3), 213–229.

Krystal, J. H., Karper, L. P., Seibyl, J. P., Freeman, G. K., Delaney, R., Bremner, J. D., et al. (1994). Subanesthetic effects of the noncompetitive NMDA antagonist, ketamine, in humans. Psychotomimetic, perceptual, cognitive, and neuroendocrine responses. *Archives of General Psychiatry, 51*(3), 199–214.

Liberzon, I., Abelson, J. L., Flagel, S. B., Raz, J., & Young, E. A. (1999). Neuroendocrine and psychophysiologic responses in PTSD: A symptom provocation study. *Neuropsychopharmacology, 21*(1), 40–50.

Lopez, J. F., Chalmers, D. T., Little, K. Y., & Watson, S. J. (1998). A. E. Bennett Research Award. Regulation of serotonin1A, glucocorticoid, and mineralocorticoid receptor in rat and human hippocampus: Implications for the neurobiology of depression. *Biological Psychiatry, 43*(8), 547–573.

Maes, M., Lin, A. H., Verkerk, R., Delmeire, L., Van Gastel, A., Van der Planken, M., et al. (1999). Sertonergic and noradrenergic markers of post-traumatic stress disorder with and without major depression. *Neuropsychopharmacology, 20*(2), 188–197.

Malloy, P. F., Fairbank, J. A., & Keane, T. M. (1983). Validation of a multimethod assessment of post-traumatic stress disorders in Vietnam veterans. *Journal of Consulting and Clinical Psychiatry, 51*(4), 488–494.

Manuck, S. B., Flory, J. D., Ferrell, R. E., Dent, K. M., Mann, J. J., & Muldoon, M. F. (1999). Aggression and anger-related traits associated with a polymorphism of the tryptophan hydroxylase gene. *Biological Psychiatry, 45*(5), 603–614.

Mason, J. W., Wang, S., Yehuda, R., Bremner, J. D., Riney, S. J., Lubin, H., et al. (1995). Some approaches to the study of the clinical implications of thyroid alterations in post-traumatic stress disorder. In M. J. Friedman, D. S. Charney, & A. Y. Deutch (Eds.) *Neurobiological and clinical consequences of stress: From normal adaptation to PTSD* (pp. 367–379). Philadelphia: Lippincott–Raven.

McEwen, B. S. (1998). Protective and damaging effects of stress mediators. *New England Journal of Medicine, 338*(3), 171–179.

McFall, M. E., Murburg, M. M., Ko, G. N., & Veith, R. C. (1990). Autonomic responses to stress in Vietnam combat veterans with posttraumatic stress disorder. *Biological Psychiatry, 27*(10), 1165–1175.

McGaugh, J. L. (2002). Memory consolidation and the amygdala: A systems perspective. *Trends in Neuroscience, 25*(9), 456.

McNally, R. J., Luedke, D. L., Besyner, J. K., Peterson, R. A., Bohn, K., & Lips, O. J. (1987). Sensitivity to stress-relevant stimuli in posttraumatic stress disorder. *Journal of Anxiety Disorder, 1*(2), 105–116.

Mellman, T. A., Kumar, A., Kulick-Bell, R., Kumar, M., & Nolan, B. (1995). Nocturnal/daytime urine noradrenergic measures and sleep in combat-related PTSD. *Biological Psychiatry, 38*(3), 174–179.

Moeller, F. G., Dougherty, D. M., Swann, A. C., Collins, D., Davis, C. M., & Cherek, D. R. (1996). Tryptophan depletion and aggressive responding in healthy males. *Psychopharmacology, 126*(2), 97–103.

Morgan, C. A., III, Krystal, J. H., & Southwick, S. M. (2003). Toward early pharmacological posttraumatic stress intervention. *Biological Psychiatry, 53*(9), 834–843.

Morgan, C. A., III, Wang, S., Rasmusson, A., Hazlett, G., Anderson, G., & Charney, D. S. (2001). Relationship among plasma cortisol, catecholamines, neuropeptide Y, and human performance during exposure to uncontrollable stress. *Psychomatic Medicine, 63*(3), 412–422.

Morgan, C. A., III, Wang, S., Southwick, S. M., Rasmusson, A., Hazlett, G., Hauger, R. L., et al. (2000). Plasma neuropeptide-Y concentrations in humans exposed to military survival training. *Biological Psychiatry, 47*(10), 902–909.

Morris, M., Hopwood, M., Maguire, K., Norman, T., & Schweitzer, I. (2004). Blunted growth hormone response to clonidine in post-traumatic stress disorder. *Psychoneuroendocrinology, 29*, 269–278.

Munck, A., Guyre, P. M., & Holbrook, N. J. (1984). Physiological functions of glucocorticoids in stress and their relation to pharmacological actions. *Endocrine Reviews, 93*, 9779–9783.

Murburg, M. M. (1994). *Catecholamine function in posttraumatic stress disorder: Emerging concepts.* Washington, DC: American Psychiatric Publishing.

Nestler, E. J., Hyman, S. E., & Malenka, R. C. (2001). *Molecular neuropharmacology: A foundation for clinical neuroscience.* New York: McGraw-Hill.

Neumeister, A., Charney, D. S., Belfer, I., Geraci, M., Holmes, C., Sherabi, Y., et al. (2005). Sympathoneural and adrenomedullary functional effects of A2c-adrenoreceptor gene polymorphism in healthy humans. *Pharmacogenetics Genomics, 15*, 143–149.

Nielsen, D. A., Goldman, D., Virkkunen, M., Tokola, R., Rawlings, R., & Linnoila, M. (1994). Suicidality and 5-hydroxyindoleacetic acid concentration associated with a tryptophan hydroxylase polymorphism. *Archives of General Psychiatry, 51*(1), 34–38.

Orr, S. P. (1997). Psychophysiologic reactivity to trauma-related imagery in PTSD: Diagnostic and theoretical implication of recent findings. In R. Yehuda & A. C. McFarlane (Eds.), *Psychobiology of posttraumatic stress disorder* (pp. 114–124). New York: New York Academy of Sciences.

Orr, S. P., Lasko, N. B., Metzger, L. J., Berry, N. J., Ahern, C. E., & Pitman, R. K. (1997). Psychophysiologic assessment of PTSD in adult females sexually abused during childhood. In R. Yehuda & A. C. McFarlane (Eds.), *Psychobiology of posttraumatic stress disorder* (pp. 491–493). New York: New York Academy of Sciences.

Orr, S., Metzger, L. J., Lasko, N. B., Macklin, M. L., Peri, T., & Pitman, R. K. (2000). De Novo conditioning in trauma-exposed individuals with and without posttraumatic stress disorder. *Journal of Abnormal Psychology, 109*, 290–298.

Orr, S. P., Pitman, R. K., Lasko, N. B., & Herz, L. R. (1993). Psychophysiological assessment of posttraumatic stress disorder imagery in World War II and Korean combat veterans. *Journal of Abnormal Psychology, 102*(1), 152–159.

Oswald, L. M., McCaul, M., Choi, L., Yang, X., & Wand, G. S. (2004). Catechol-*O*-methyltransferase polymorphism alters hypothalamic–pituitary–adrenal axis responses to naloxone: A preliminary report. *Biological Psychiatry, 55*(1), 102–105.

Pallmeyer, T. P., Blanchard, E. B., & Kolb, L. C. (1986). The psychophysiology of combat-induced post-traumatic stress disorder in Vietnam veterans. *Behavioral Research Therapy, 24*(6), 645–652.

Park, S. B., Coull, J. T., McShane, R. H., Young, A. H., Sahakian, B. J., Robbins, T. W., et al. (1994). Tryptophan depletion in normal volunteers produces selective impairments in learning and memory. *Neuropharmacology, 33,* 575–588.

Perry, B. D. (1994). Neurobiological sequelae of childhood trauma: PTSD in children. In M. Murburg (Ed.), *Catecholamine function in post-traumatic stress disorders: Emerging concepts, progress in psychiatry* (pp. 253–276). Washington, DC: American Psychiatric Press.

Perry, B. D., Southwick, S. M., Yehuda, R., & Giller, E. L. (1990). *Adrenergic receptor regulation in post-traumatic stress disorder.* Washington, DC: American Psychiatric Publishing.

Pitman, R. K., Orr, S. P., Forgue, D. F., Altman, B., de Jong, J. B., & Herz, L. R. (1990). Psychophysiologic responses to combat imagery of Vietnam veterans with posttraumatic stress disorder versus other anxiety disorders. *Journal of Abnormal Psychology, 99*(1), 49–54.

Pitman, R. K., Orr, S. P., Forgue, D. F., de Jong, J. B., & Claiborn, J. M. (1987). Psychophysiologic assessment of posttraumatic stress disorder imagery in Vietnam combat veterans. *Archives of General Psychiatry, 44*(11), 970–975.

Pitman, R. K., van der Kolk, B. A., Orr, S. P., & Greenberg, M. S. (1990). Naloxone reversible analgesic response to combat-related stimuli in posttraumatic stress disorder. *Archives of General Psychiatry, 47,* 541–544.

Post, R. M., Weiss, S. R. B., & Smith M. A. (1995). Implications for the evolving neural substrates of post-traumatic stress disorder. In M. J. Friedman, D. S. Charney, & A. Y. Deutch (Eds.), *Neurobiological and clinical consequences of stress: From normal adaptation to PTSD* (pp. 203–224). Philadelphia: Lippincott–Raven.

Rasmusson, A. M., & Friedman, M. J. (2002). Gender issues in the neurobiology of PTSD. In R. Kimerling, P. C. Ouimette, & J. Wolfe (Eds.), *Gender and PTSD* (pp. 43–75). New York: Guilford Press.

Rasmusson, A. M., Hauger, R. L., Morgan, C. A., III, Bremner, J. D., Charney, D. S., & Southwick, S. M. (2000). Low baseline and yohimbine-stimulated plasma neuropeptide Y (NPY) in combat-related posttraumatic stress disorder. *Biological Psychiatry, 47,* 526–539.

Rasmusson, A., Pinna, G., Weisman, D., Gottschalk, C., Charney, D., Krystal, J., et al. (2005, May). *Decreases in CSF allopregnanolone levels in women with PTSD correlate negatively with reexperiencing symptoms.* Paper presented at the annual meeting of the Society of Biological Psychiatry, Atlanta, GA.

Rasmusson, A. M., Vasek, J., Lipschitz, D., Mustone, M. E., Vojvoda, D., Shi, Q., et al. (2004). An increased capacity for adrenal DHEA release is associated negatively with avoidance symptoms and negative mood in women with PTSD. *Neuropsychopharmacology, 29,* 1546–1557.

Rasmusson, A. M., Vythilingam, M., & Morgan, C. A., III. (2003). The neuroendocrinology of PTSD: New directions. *CNS, 8,* 651–667.

Slawik, M., Reisch, N., Zwermann, O., Maser-Gluth, C., Stahl, M., Klink, A., et al. (2004). Characterization of an adrenocorticotropin (ACTH) receptor promoter polymorphism leading to decreased adrenal responsiveness to ACTH. *Journal of Clinical Endocrinology and Metabolism, 89*(7), 3131–3137.

Smoller, J. W., Rosenbaum, J. F., Biederman, J., Kennedy, J., Dai, D., Racette, S. R., et al. (2003). Association of a genetic marker at the corticotropin-releasing hormone locus with behavioral inhibition. *Biological Psychiatry, 54*(12), 1376–1381.

Sorg, B. A., & Kalivas, P. W. (1995). Stress and neuronal sensitization. In M. J. Friedman, D. S. Charney, & A. Y. Deutch (Eds.), *Neurobiological and clinical consequences of stress: From normal adaptation to PTSD* (pp. 83–102). Philadelphia: Lippincott–Raven.

Southwick, S. M., Bremner, J. D., Rasmusson, A., Morgan, C. A., III, Arnsten, A., & Charney, D. S. (1999). Role of norepinephrine in the pathophysiology and treatment of posttraumatic stress disorder. *Biological Psychiatry, 46*(9), 1192–1204.

Southwick, S. M., Krystal, J. H., Morgan, C. A., Johnson, D., Nagy, L. M., Nicolaou, A., et al. (1993). Abnormal noradrenergic function in posttraumatic stress disorder. *Archives of General Psychiatry, 50*(4), 266–274.

Southwick, S. M., Morgan, C. A., Bremner, J. D., Grillon, C. G., Krystal, J. H., & Nagy, L. M. (1997). *Neuroendocrine alteration in posttraumatic stress disorder.* New York: New York Academy of Sciences.

Southwick, S. M., Paige, S., Morgan, C. A., III, Bremner, J. D., Krystal, J. H., & Charney, D. S. (1999). Neurotransmitter alterations in PTSD: Catecholamines and serotonin. *Seminars in Clinical Neuropsychiatry, 4*(4), 242–248.

Southwick, S. M., Rasmusson, A., Barron, J., & Arnsten, A. (2005). Neurobiological and neurocognitive alterations in PTSD: A focus on norepinephrine, serotonin, and the hypothalamic–pituitary–adrenal axis. In J. J. Vasterling & C. R. Brewin (Eds.), *Neuropsychology of PTSD: Biological, cognitive, and clinical perspectives* (pp. 27–58). New York: Guilford Press.

Southwick, S. M., Vythilingam, M., & Charney, D. S. (2005). The psychobiology of depression and resilience to stress. *Annual Review of Clinical Psychology, 1,* 255–291.

Southwick, S. M., Yehuda, R., & Morgan, C. A. (1995). *Clinical studies of neurotransmitter alterations in post-traumatic stress disorder.* Philadelphia: Lippincott–Raven.

Stein-Behrens, B., Mattson, M. P., Chang, I., Yeh, M., & Sapolsky, R. (1994). Stress exacerbates neuron loss and cytoskeletal pathology in the hippocampus. *Journal of Neuroscience, 14*(9), 5373–5380.

Stutzmann, G. E., & LeDoux, J. E. (1999). GABAergic antagonists block the inhibitory effects of serotonin in the lateral amygdala: A mechanism for modulation of sensory inputs related to fear conditioning. *Journal of Neuroscience, 19*(11), 1–4.

Stutzmann, G. E., McEwen, B. S., & LeDoux, J. E. (1998). Serotonin modulation of sensory inputs to the lateral amygdala: Dependency on corticosterone. *Journal of Neuroscience, 18*(22), 9529–9538.

Vaiva, G., Thomas, P., Ducrocq, F., Fontaine, M., Boss, V., Devos, P., et al. (2004). Low posttrauma GABA plasma levels as a predictive factor in the development of acute posttraumatic stress disorder. *Biological Psychiatry, 55*(3), 250–254.

Valentino, R. J., & Foote, S. L. (1988). Corticotropin-releasing hormone increases tonic but not sensory-evoked activity of noradrenergic locus coeruleus neurons in unanesthetized rats. *Journal of Neuroscience, 8,* 1016–1025.

Vermetten, E., Vythilingam, M., Southwick, S. M., Charney, D. S., & Bremner, J. D. (2003). Long-term treatment with paroxetine increases verbal declarative memory and hippocampal volume in posttraumatic stress disorder. *Biological Psychiatry, 54*(7), 693–702.

Walker, D. L., & Davis, M. (2002). The role of amygdala glutamate receptors in fear learning, fear-potentiated startle, and extinction. *Pharmacology, Biochemistry, and Behavior, 71,* 379–392.

Walling, S. G., & Harley, C. W. (2004). Locus ceruleus activation initiates delayed synaptic potentiation of perforant path input to the dentate gyrus in awake rats: A novel beta-adrenergic- and protein synthesis-dependent mammalian plasticity mechanism. *Journal of Neuroscience, 24*(3), 598–604.

Wust, S., Van Rossum, E. F., Federenko, I. S., Koper, J. W., Kumsta, R., & Hellhammer, D. H. (2004). Common polymorphisms in the glucocorticoid receptor gene are associated with adrenocortical responses to psychosocial stress. *Journal of Clinical Endocrinology and Metabolism, 89*(2), 565–573.

Yehuda, R. (2002). Current status of cortisol findings in post-traumatic stress disorder. *Psychiatric Clinics of North America, 25,* 341–368.

Yehuda, R., Siever, L. J., Teicher, M. H., Levengood, R. A., Gerber, D. K., Schmeidler, J., et al. (1998). Plasma norepinephrine and 3-methoxy-4-hydroxyphenylglycol concentrations and severity of depression in combat posttraumatic stress disorder and major depressive disorder. *Biological Psychiatry, 44*(1), 56–63.

Young, E. A., & Breslau, N. (2004). Cortisol and catecholamines in posttraumatic stress disorder: An epidemiologic community study. *Archives of General Psychiatry, 61,* 394–401.

Chapter 11

Gene–Environment Interactions

TWIN STUDIES AND GENE RESEARCH IN THE CONTEXT OF PTSD

Ronnen Segman, Arieh Y. Shalev, and Joel Gelernter

Posttraumatic stress disorder (PTSD) is a genetically influenced, complex trait that shows in particularly stark relief some of the issues encountered in studying gene–environment (G × E) interaction. In this chapter, we review some of the basic techniques used to demonstrate a genetic contribution to a trait, discuss specific findings from studies of PTSD, review the few published molecular population genetics studies of PTSD, and assess prospects for future study.

METHODOLOGICAL CONSIDERATIONS

Establishing a Genetic Basis (Heritability)

Establishing heritability provides a basis for the search for susceptibility genes for any disorder. Twin studies provide an optimal design for determining heritability; familial clustering of cases does not distinguish between heritability and shared environmental effects. Comparing affection rates among relatives as a function of their genetic distance may be used to approximate the extent of genetic influence on the phenotype. Tracing familial segregation patterns in successive generations can be informative regarding mode of inheritance (i.e., Mendelian vs. complex). In applying these conventional strategies to the study of PTSD, it is very problematic to identify high-density families, that is, families with multiple affected members, due to factors such as low genetic penetrance. Additionally, exposure to an extreme event is required to unravel an underlying genetic vulnerability to PTSD. Although lifetime exposure to significant traumatization is not uncommon in the population, adequate assessment of the actual severity of exposure is inherently difficult to obtain retrospectively. Particularly for healthy subjects, it is inherently difficult to determine in retrospect whether remaining nonsymptomatic results from lack of sufficient exposure to traumatic events. Family studies can therefore only be informative if probands are carefully selected and assessed for trauma exposure.

Monozygotic (MZ) twins are genetically identical, whereas dizygotic (DZ) twins share half of their genes (on average). MZ twins, therefore, show increased trait concordance relative to DZ twins for genetically determined traits. The extent of the difference in MZ–DZ concordance rates can be taken to measure the heritability of a trait. (Higher MZ concordance rates may also theoretically represent some shared environmental influence, because MZ twins may be treated more similarly than DZ twins). To be informative for PTSD heritability, both twins must share a degree of exposure to a traumatic event. Moreover, focusing on trauma survivors creates inherent selection bias, because it is difficult to differentiate between a heritable component of being prone to trauma exposure (e.g., Eaves & Erkanli, 2003) and a heritable vulnerability to develop PTSD upon exposure. Previously published twin studies illustrate the complexity of the interaction of genetic and environmental factors. For example, Kessler and colleagues (1992) reported that genetic and environmental factors are important in evoking social support, whereas it might commonly have been assumed that support constitutes one component of environmental effects. Adoption studies are an important complementary strategy for separating genetic and environmental effects, but they are exceedingly difficult to implement for PTSD.

Locating Causative Genes

Strategies for locating genes that mediate heritable vulnerability to PTSD include linkage and association methods (the latter including single-nucleotide polymorphism [SNP] whole-genome scanning).

A *linkage*-based genome-wide scan systematically searches for polymorphic marker alleles coinherited with the phenotype, in families with the phenotype of interest. It allows the discovery of unknown genes, detected through their chromosomal proximity to informative markers. Multiple affected family relatives (who have been exposed to trauma) are required for implementing linkage strategies. Parametric linkage analysis makes use of data from unaffected subjects also, but unaffected family members cannot be ascertained without having shared a similar documented trauma exposure with the affected probands; otherwise, their phenotype can only be considered "undefined" from a genetic perspective. This constitutes an obstacle for ascertainment of large pedigree samples informative for genetic linkage.

In contrast, the *association*-based approach explores the involvement of known genes implicated either through a prior hypothesis or, if there have been prior linkage studies, through their position within an informative linkage region. Changes in allele and genotype frequencies of a genomic variant located in a candidate risk gene among affected subjects can be explored employing a case–control sample, or family-based transmission disequilibrium test (TDT) analysis.

Case–Control Association

Case–control association studies compare the relative abundance of a candidate genomic variant among a group of subjects with PTSD and that of a control or comparison group. Excluding latent genetic vulnerability among unaffected controls requires documentation of trauma survivors, with asymptomatic survivors sharing similar exposure serving as controls (see below), although this is not the only possible valid design.

Population stratification or admixture can introduce spurious case–control gene association results (Devlin & Roeder, 1999; Pritchard, Stephens, Rosenberg, & Donnelly, 2000). Unequal distribution of component population subgroups (i.e., admixture) among case and control samples may further lead to false positive errors, if there are differences in the frequency both of the allele of interest and of the phenotype between the component populations. Strategies employing Bayesian methods were devised to control for population structure effects. Such strategies either use multiple markers and data from founding populations to approximate individual admixture (Pritchard et al., 2000) or search for sets of loci that are distributed differently from other markers, to derive a deviate distribution that is then used for correcting the significance of findings (Devlin, Roeder, & Wasserman, 2001).

Transmission Disequilibrium Test

The TDT test (Spielman, McGinnis, & Ewens, 1993) investigates transmission of alleles from parents (who do not need to be phenotypically assessed themselves) to affected offspring. It avoids population structure false-positive errors, as well as false-negative diagnoses among unaffected controls, because it does not require healthy controls. Despite these advantages, there are considerable disadvantages: The requirement for sampling both biological parents of each affected subject is costly and difficult to implement. In contrast, case–control samples are simpler to collect, and ascertaining large samples with higher power to detect small-effect genes is more cost-effective.

Hypotheses-Driven Search for PTSD Risk Genes

Locating susceptibility loci through association relies on a knowledge base implicating underlying molecular pathways. A candidate gene thus implicated can be evaluated regarding whether polymorphic variants in that gene show altered allele frequency among affected subjects compared to unaffected subjects in a case–control sample, or biased transmission from parents to the affected offspring in a TDT sample. The direct association approach explores polymorphic variation in promoter and coding regions that may have functional significance for expression or sequence. The alternative approach examines marker polymorphisms or haplotypes of several marker polymorphisms spanning the same gene, assuming linkage disequilibrium with an unknown functional site (i.e., that the markers or haplotypes studied directly reflect allelic status at functional variants that are not studied directly).

So far, the actual set of genes mediating the heritable component contributing to risk for PTSD has not been defined. A number of biological alterations have been implicated with PTSD and may serve to guide a specific hypotheses-driven candidate gene search, discussed below.

Whole-Genome Association Studies

SNPs are the most abundant form of genetic variation. Because they are diallelic, they are generally less informative than the multiallelic short, tandem repeat markers traditionally employed for linkage analysis. However, use of multiple SNPs compensates for the decreased information content per marker, and their abundance makes them the most informative markers overall (cumulatively), allowing the possibility of a high-resolution linkage disequilibrium (LD) scan of the whole genome. The density of SNP markers required to map a certain chromosomal region depends on the extent of LD in that region. This depends on the extent of recombination between adjacent markers, which in turn depends on the physical distance between markers, the number of generations elapsed (since creation of the more recent variant), and the regional recombination rate in a particular chromosomal region. In the presence of high LD, a small set of markers may provide most of the genetic information for a region, whereas recombination hot spots may result in flanking markers being uninformative for each other. With the growing availability of denser SNP maps and improving knowledge of the haplotype map of the genome, high-resolution SNP mapping for the whole genome should soon become a cost-effective option (Carlson, Eberle, Kruglyak, & Nickerson, 2004). This approach can be implemented in a large case–control design of unrelated subjects and should eventually allow the detection of previously unknown genes contributing to PTSD risk. Such a study recently used the whole-genome associate approach (querying a set of 100,000 SNPs) to identify a gene increasing risk for age-related macular degeneration (Klein et al., 2005).

Phenotype Measures for Genetic Studies

Classifying Unaffected Subjects

Sufficient trauma exposure is a precondition for defining vulnerability for PTSD; excluding latent vulnerability among unaffected controls would ideally involve prospective documentation following trauma exposure. However, retrospective interview assessment is usually more practical, and it is also valid to use an unscreened random population control (cf. Gelernter et al., 1999). Assuming small-effect gene contributions to PTSD (consistent with PTSD being a genetically complex trait) (True et al., 1993), allelic variants that contribute to risk for PTSD, given triggering by trauma exposure, should show higher frequency among patients with PTSD compared to a control population. However, the relatively high incidence of PTSD among survivors of traumatic events (Kessler, Sonnega, Bromet, Hughes, & Nelson, 1995), suggests that a significant proportion of healthy controls not selected for trauma exposure may carry latent PTSD genetic susceptibility. Furthermore, PTSD symptoms may wane with time in a significant proportion of patients. These factors may serve to underestimate the effect of causative genes for PTSD in a population-based case–control sample (especially with use of a random population control). These considerations suggest that it is ideal to control for trauma severity exposure, as well as to assess PTSD at several points (e.g., at 1 and 3 months) after trauma, and not just long after the traumatic event.

Comorbid Disorders Following Trauma

Documentation of additional relevant psychiatric outcome measures is important, because PTSD frequently co-occurs with other disorders, including substance dependence, affective disorders, and additional anxiety disorders. A complex causal interplay may underlie this observed comorbidity. Trauma exposure or additional shared and nonshared environmental factors may facilitate the coexpression of these comorbid disorders. Heritable factors, both additive and specific to each disorder, may also operate. Finally, the expression of one disorder may facilitate the expression of another. The tendency for specific environmental exposures, such as trauma, also to be partly heritable, as discussed below, adds yet an additional layer of complexity. Study of MZ and DZ twin pairs has demonstrated the extent of specific and additive heritability for PTSD and alcohol and drug dependence (Xian et al., 2000), generalized anxiety disorder (GAD), and panic disorder (PD) (Chantarujikapong et al., 2001). A discordant MZ co-twin control paradigm has demonstrated that combat trauma exposure increases risk for major depressive disorder (MDD), GAD, and PD, and that shared familial vulnerability mediated by heritable factors plays a role in comorbidity of PTSD and affective disorders (MDD and dysthymia) (Koenen et al., 2003b). Using the same paradigm, Koenen and colleagues (2003a) further demonstrated the impact of both combat trauma exposure and combat-related PTSD on the expression of comorbid disorders, beyond the effect of shared familial vulnerability. Specifically, they found that combat exposure was associated with alcohol and cannabis dependence risk, whereas combat-related PTSD mediated the association between combat exposure, MDD, and nicotine dependence (Koenen et al., 2003a). Additional environmental influences are also of great importance in shaping the likelihood of certain phenotypic outcomes. For example, striking differences in rates of comorbid substance use among patients with PTSD from different population backgrounds may reflect in part environmentally dependent differences in the availability of substances, or the social desirability of substance use, by severely traumatized patients with PTSD (see below).

Endophenotypes

PTSD comprises a symptom triad of intrusive reexperiencing, avoidance and numbing, and hyperarousal. Clinical practice employs a narrow definition of affection, based on overall severity and number of reported symptoms (Andreasen, 1997; Radant, Tsuang, Peskind, McFall, & Raskind, 2001). Discrete phenotypic dimensions (e.g., intrusion, avoidance, or hyperarousal) among trauma survivors may have discrete heritability (True et al., 1993) and, therefore, higher genetic validity for locating specific small-gene effects (Neiderhiser, Plomin, & McClearn, 1992). Several biomarkers—potential endophenotypes—that have been described among patients with PTSD may serve to search for correlations with specific causative genes. These include, among others, measures of hippocampal volume (Gilbertson et al., 2002), abnormal autonomic response (Orr et al., 2003), and hypothalamic–pituitary–adrenal (HPA) axis reactivity (Yehuda, 2002). Unfortunately, each of these measures has only limited diagnostic specificity.

PTSD as a Complex Phenotype

PTSD likely represents an end result of interplay between multiple small-effect risk loci with appropriate exposure. No small-gene contribution is by itself necessary or sufficient for expressing PTSD. The general issues that arise when investigating genotype–phenotype correlations with complex traits bear relevance to PTSD. "Genetic heterogeneity" refers to the occurrence of different genes or gene combinations that contribute to the phenotype. Among any large sample of unrelated patients with PTSD we may assume that some patients may possess different sets of risk alleles, complicating the search for common-gene allelic effects. From a clinical perspective, etiological heterogeneity for PTSD reflects specific vulnerability to one (or many) of the known risk factors for PTSD (e.g., degree of physiological or unconditioned response during the traumatic event, propensity to extinguish fear-driven learning, capacity to effectively engage in soothing human interaction at the aftermath of traumatic events) (Brewin, Andrews, & Valentine, 2000). Phenotypic heterogeneity is attributable to many causes. It relates in part to the same gene or gene combinations resulting in a different phenotype. This may be because of a modifying effect of other genes, environment, or a gene–environment (G × E) interaction. The low penetrance of PTSD provides a case in point. In any large sample of healthy controls not selected for trauma exposure, a significant proportion will carry any PTSD-associated genotype without manifesting the phenotype, for lack of exposure; that is, outside the context of a specific set of environmental conditions, an underlying vulnerability to PTSD may not be expressed.

PTSD as an Extreme Model for G × E Interaction

The completion of the human genomic reference sequence brings the ultimate quandary of understanding the impact of gene allelic variation on phenotypes to center stage. Parallel efforts are currently devoted to comprehensive documentation of allelic variation in different population groups. For variants that may increase the risk of disease, however, relevant environmental exposure is often required to trigger such latent genetic susceptibility. Deciphering G × E interactions is therefore a final major road block for understanding the underpinning functional relevance of most gene variation.

Determining the small and variable incremental effects of the concert of polygenes that build up complex disorders requires unraveling a parallel concert of similarly small and variable environmental events. These events serve to modify (extenuate or muffle) the overt expression of small-effect genetic variation. Such events may covertly spread over long periods or appear only briefly at crucial time points during ontogenesis and later life. Just as gene allelic effects depend on other gene allelic combinations, triggering events can be enhanced or overridden by other events. Understanding this intricate interplay of parallel or interacting streams of alleles and events is mandatory for understanding how common complex phenotypes are expressed and modified—or averted. A simplified strategy employs models in which diseases are caused by a single gene mutation or by exposure to a single environmental agent. Although this is rarely the case for behavioral phenotypes, such models are important in helping to illustrate straightforward genotype–phenotype correlations. Rare Mendelian disorders occur when highly penetrant single-gene variations shape overt phenotypes. They represent simple models that aid in understanding a single gene's potential to affect a phenotype. PTSD offers a different look at a relatively simplified model, in which an extreme environmental exposure can trigger phenotypic expression of latent genotypic susceptibility. As detailed below, some of the genetic endowment predisposing to PTSD has the potential also to predispose to other psychiatric phenotypes, depending on additional undefined environmental factors, and given additional specific genetic vulnerabilities. The case in defining genotype–phenotype correlations for both affected and unaffected subjects is more readily definable for PTSD than for the other potential phenotypes given knowledge of exposure to an extreme traumatic event.

CURRENT STATE OF THE ART

Twin Data for Determining the Relative Contributions of Genes and Environment

The Vietnam Veterans Twin Registry

Evidence for familial clustering is suggestive of a genetic component for PTSD (Connor & Davidson, 1997), but conclusive support comes from twin data. In order to ascertain large numbers of twins sharing similar trauma exposure, most twin studies on the subject queried a registry of veterans. The Vietnam Era Twin (VET) Registry comprises 7,375 male–male twin pairs, both having been in active service during the Vietnam period (1965–1975). Blood group typing and perceived physical resemblance (reflecting very basic genetic information) were employed as proxies to define zygosity. In 1991–1992, approximately 5,000 twin pairs of the VET Registry were interviewed by telephone using the Diagnostic Interview Schedule, Version III–Revised (DIS-III-R). These data were then used to generate DSM-III-R–defined psychiatric symptoms and diagnoses (cf. Goldberg, Curran, Vitek, Henderson, & Boyko, 2002). This large-scale database was used for a number of informative studies, discussed in detail below.

Genetic Modeling Approach to Twin Data

The genetic modeling approach applied by these studies has been extensively reviewed (Neale & Cardon, 1992). Four categories may impact on individual variation in the phenotype measured, including additive genetic effects, shared family environment, nonadditive genetic factors, and unique environmental effects. Different alternative models may be applied to the observed data. Members of an MZ twin pair are expected to show a full correlation for *additive genetic effects*, whereas members of a DZ twin pair should show such correlation to half this extent on average. *Nonadditive genetic effects* are expected to correlate fully between members of an MZ twin pair but only 25% between members of a DZ twin pair. To the contrary, MZ and DZ twins raised together should be similarly affected by their common environment; thus, *shared environmental effects* should contribute equally to phenotypic measures among MZ and DZ twin pairs. Finally, *unique environmental effects* result from experiences that are differentially shared and may contribute to differences within MZ and DZ twin pairs. The genetic modeling approach relies on the expectation that environmental effects that may affect the phenotype measured are equally shared between MZ and DZ twin pairs (equal environment assumption, see Xian, Scherrer, et al., 2000). However, MZ twins are more alike and could therefore tend to elicit and be exposed to more similar environments. If such a differential exposure has a significant impact on the risk for expressing the phenotype measured, it could lead to spuriously overstating gene effect and understating environmental impact (Xian, Scherrer, et al., 2000). Xian, Scherrer, and colleagues (2000) therefore investigated the validity of the equal-environments assumption in the VET sample by using a specified family environmental factor defined by measures of each twin's perceived zygosity (Kendler, Neale, Kessler, Heath, & Eaves, 1993), across several psychiatric phenotypes measured. Their results confirm that perceived zygosity does not have a major impact on twin similarity for common psychiatric disorders, including PTSD; alcohol, drug, and nicotine dependence; and MDD.

Finally, it should be noted that combat exposure itself has been found to have a substantial heritability estimate of 47% in the VET Registry sample (Lyons et al., 1993).

Combat exposure must therefore be adjusted for before assessing heritability of combat-related PTSD. (This can also be appreciated as an example of a genetic influence on environmental exposure. As such, a genetic factor that influenced propensity to become exposed to extreme stress could also legitimately be a PTSD risk factor.)

Genetic and Environmental Contributions to Combat-Related PTSD

The original VET Registry study of 4,042 male–male veteran twin pairs (2,224 MZ and 1,818 DZ pairs) (Goldberg, True, Eisen, & Henderson, 1990; True et al., 1993) demonstrated that genetic factors account for approximately 30% of the variance in risk for PTSD symptoms, even after differences in trauma exposure between twins are taken into account. As noted earlier, genetic influences on the extent of trauma exposure were also found (Lyons et al., 1993), so that PTSD risk includes partially overlapping risks for (1) trauma exposure, and (2) developing PTSD after such exposure. Combat predicted "reexperiencing" cluster and avoided activities symptoms. The authors concluded that heritable factors may affect individual reactivity to environmental cues and to trauma. To the contrary, premorbid environmental factors shared by siblings were not found to contribute substantially to susceptibility to develop PTSD symptoms (True et al., 1993).

Genetic and Environmental Contribution to Comorbid Anxiety Disorders

Anxiety disorders, including PTSD, often co-occur. Chantarujikapong and colleagues (2001) investigated the overlap of genetic and environmental contributions among symptoms of GAD, PD, and PTSD. Subjects were 3,327 MZ and DZ VET twin pairs. They reported that susceptibility for GAD symptoms was due to a 38% additive heritable genetic contribution (i.e., input) common to PD symptoms and PTSD. Susceptibility for PD symptoms was due to a 21% additive heritable input common to GAD symptoms and PTSD (and a 20% additive heritable input specific to PD symptoms). Additive heritable input common to symptoms of GAD and PD were estimated to account for 21% of the heritable variance in PTSD. Overall, additive heritable input specific to PTSD accounted for 13.6% of the heritable variance in PTSD. Chantarujikapong and colleagues concluded that each of these co-occurring anxiety disorders poses both distinct genetic influences and significant common heritable and unique environmental influences. Similar, shared vulnerability is frequently seen for psychiatric disorders and is expected in the context of a complex genetic trait. The observed comorbidity between combat-related PTSD and other psychopathology could derive from a confounding effect of a shared familial vulnerability to these disorders. Koenen and colleagues (2003a) employed a discordant MZ twin design to investigate this possibility. Within MZ pairs, dissimilar psychiatric outcome would have to originate primarily from the nonshared environmental factors (other factors could include, for example, X-chromosome inactivation mosaicism, in females). Discordant MZ twin paradigms can therefore be used to control for heritable or shared environmental factors as a cause for co-occurrence of combat exposure, combat-related PTSD, and other psychiatric outcomes. Koenen and colleagues employed the MZ co-twin control design to examine the unique contributions of combat exposure and combat-related PTSD to risk for several substance dependence disorders and MDD. They found that combat exposure and combat-related PTSD have nonshared environmentally mediated effects on risk for comorbid psychopathology. Combat exposure was noted to increase risk for substance dependence, whereas combat-related PTSD increased risk for MDD and nicotine dependence. These findings suggest both a sequential effect on risk of one phenotype on acquiring another and common etiological factors are both at play in accounting for observed comorbidity.

Previous data have implicated preexisting individual and familial psychopathology as contributing to risk of both trauma exposure and developing PTSD once exposed (Brewin et al., 2000). Familial psychopathology may contribute to the individual's risk to develop a psychiatric disorder, which may in turn contribute to risk of developing PTSD through either shared genetic, or other familial, factors. Alternatively, such a family history could act to increase the risk for PTSD by putting an individual at a greater risk of more severe trauma exposure. Koenen and colleagues (2002) further attempted to tease apart the differential influences of premorbid individual and familial psychopathology on risk for combat-related trauma exposure and combat-related PTSD. They examined the association of individual and familial risk factors with exposure to trauma and to PTSD among 6,744 VET male twins. Both heritable and environmental factors were noted to contribute to risk for the various components of trauma exposure: military service in Southeast Asia, premorbid conduct disorder, premorbid substance dependence, and a family history of affective disorders. In contrast, premorbid affective disorder in the individual *reduced* risk for subsequent trauma exposure. The risk of developing PTSD after trauma exposure was increased by an earlier age at first trauma, exposure to multiple traumas, paternal depression, limited formal education, service in Southeast Asia, and premorbid conduct disorder, PD, GAD, or MDD. Results of the Koenen and colleagues (2002) study suggest that the association of familial psychopathology and PTSD may be accounted for in part by an augmented risk of trauma exposure, and in part by premorbid psychopathology. However, a complex interaction with heritable factors is also at play. Koenen and colleagues (2003a) further demonstrated the impact of both combat trauma exposure and combat-related PTSD on the expression of comorbid disorders, beyond the effect of shared familial vulnerability. Specifically, they found that combat exposure is associated with alcohol- and cannabis-dependence risk, whereas combat-related PTSD mediates the association between combat exposure, and MDD and nicotine dependence (Koenen et al., 2003b).

Genetic and Environmental Contribution to Comorbid Alcohol and Drug Dependence

Xian, Chantarujikapong, and colleagues (2000) investigated how heritable and environmental factors overlap among PTSD, alcohol dependence, and drug dependence, again studying VET twin pairs. They used genetic model fitting to approximate magnitude of heritable and environmental influences on the lifetime co-occurrence of PTSD, alcohol dependence, and drug dependence. Again, considerable overlap was noted, and the results suggested that PTSD, alcohol dependence, and drug dependence each have both distinct and common heritable and unique environmental inputs (Xian, Chantarujikapong, et al., 2000).

The evidence for a shared heritable vulnerability for PTSD and alcohol dependence reported by Xian, Chantarujikapong, and colleagues (2000) does not indicate whether this shared vulnerability is independent from that for trauma exposure. McLeod and colleagues (2001) used biometrical modeling to address this issue. They demonstrated that the same additive heritable influences affecting the degree of combat exposure also affect severity of alcohol use and of some specific PTSD-related symptoms. These findings support shared vulnerability as opposed to alcohol use resulting from PTSD. McLeod and colleagues therefore suggest that genetically influenced characteristics affect the probability of exposure to a high level of combat, higher PTSD symptoms, and alcohol use severity. This does not exclude the importance of environmental factors in the development of PTSD symptoms or alcohol use.

Whereas rates of comorbid depression are similar among patients with PTSD from different populations, rates of comorbid alcohol and substance use show striking differences. Israeli patients with PTSD show a persistent overlap between PTSD and depression during the year that follows trauma; however, comorbid substance abuse is rare in Israel (Shalev et al., 1998), compared with much higher prevalence of substance abuse in U.S. patients with PTSD. Consistent with twin data suggesting that environmental factors impact on the expression of this phenotype, these differences may reflect in part differential environmental (e.g., cultural) modulation of a similar genetic predisposition (i.e., reflecting the much lower prevalence and availability of alcohol and other psychoactive substances in Israel), but such striking differences may in part reflect ethnic variability in genetic factors predisposing to alcohol and substance use disorders. These differences could also reflect systematic differences in the trauma-inducing stress.

Twin Studies of Civilian, Trauma-Related PTSD

Stein, Jang, Taylor, Vernon, and Livesley (2002) investigated 406 Canadian volunteer twin pairs (222 MZ and 184 DZ). They found that additive genetic, common environmental, and unique environmental effects best explained the variance in exposure to assaultive trauma, but exposure to nonassaultive trauma was better explained by common and unique environmental influences. Symptoms of PTSD were moderately heritable. Genetic vulnerability to exposure to assaultive trauma and to susceptibility to develop PTSD symptoms overlapped significantly. Stein and colleagues concluded that genetic factors influence the risk of exposure to some trauma. This study was important in extending VET results to civilian-related PTSD and to women. An earlier, small study of anxiety disorders (fewer than 50 twin pairs) found PTSD only in co-twins of probands with anxiety disorders (Skre, Onstad, Torgesen, Lygren, & Kringlen, 1993).

Use of Co-Twin Control to Investigate Endophenotypes' Temporal Causality

Twin studies constitute a powerful paradigm for investigating causality of PTSD endophenotypes. Reduced hippocampal volume (Gilbertson et al., 2002) and an abnormal startle response (Orr et al., 2004) are two replicated biological trait markers among PTSD patients. Previous studies, however, could not determine whether such abnormalities preceded and perhaps predisposed to PTSD, or were a consequence of trauma exposure and PTSD. Thus, a study of MZ twins discordant for PTSD addressed this question.

Reduced Hippocampal Volume

Gilbertson and colleagues (2003) investigated 40 pairs of MZ twins discordant for Vietnam Era combat exposure. Of those exposed to combat, 42% had chronic PTSD, demonstrating smaller hippocampal volumes compared to psychiatrically healthy combat veteran controls sharing similar combat exposure. Within subjects with PTSD, symptom severity was inversely associated with hippocampal volumes. Siblings of the combat veterans with PTSD also had small hippocampal volumes, despite not sharing combat exposure, and their hippocampal volumes were equally predictive of the severity of their combat siblings' PTSD. These findings suggest that reduced hippocampal volumes may represent a premorbid trait contributing to PTSD risk given trauma exposure (Sapolsky, 2002). Incomplete fusion of the septum pellucidum has also been reported among patients with PTSD (May, Chen, Gilbertson, Shenton, & Pitman, 2004). Using the discordant twin design, May and colleagues (2004) reported that an abnormally large cavum septum pellucidum may constitute a familial vulnerability factor for PTSD (e.g., representing heritable or shared environmental effects).

Startle Response

Increased response to sudden, loud tones (startle) is a well-replicated psychophysiological marker for PTSD. It may represent a premorbid risk factor or an acquired PTSD biomarker. To address this issue, Orr and colleagues (2004) studied pairs of Vietnam combat veterans and their non-combat-exposed MZ twins. Combat veterans were either diagnosed as having current chronic PTSD or as never having PTSD. Heart rate responses to startle were greater among Vietnam combat veterans with PTSD than among non-combat-exposed co-twins. The responses of the non-combat-exposed co-twins were more similar to those of the combat veterans without PTSD. The results suggest that increased heart rate responses to startle represent an acquired sign of PTSD, not a premorbid familial or heritable risk factor; therefore, they do not represent an endophenotype.

Forays Into Measuring G × E Interaction

Several publications provide striking reports of observation of G × E effects in human subjects, although not for PTSD. Caspi and colleagues (2002) reported that genotype at the MAO-A (monoamine oxidase A) locus moderates development of antisocial problems in response to maltreatment in children. In this study based on a birth cohort sample of 1,037 children, maltreated, male children with one set of MAO-A genotypes were at greater risk for developing antisocial behaviors than children with another set of genotypes. A further study in the same cohort demonstrated that genotype at the serotonin transporter protein (SLC6A4) locus modulates risk for depression in the presence of stressful life events (Caspi et al., 2003). This latter finding was replicated and extended to implicate also an interaction of genotype at SLC6A4 and social support affecting risk for depression (Kaufman et al., 2004).

These studies show that it is in fact feasible to identify G × E interaction in human samples and provide models for conceptualizing similar interactions in the context of PTSD. Some papers have presented updated statistical models for evaluating such interactions (e.g., Eaves & Erkanli, 2003).

Locating Susceptibilities: Association Studies in PTSD

Dopamine D₂ Receptor Gene

Increased plasma and urinary dopamine (DA) have been reported among patients with PTSD, and mouse data suggest genotype effects on stress-induced alterations of central dopaminergic neurotransmission (Segman et al., 2002). (Of course, evidence of alteration in dopaminergic function is observed in many psychiatric disorders.) Two studies examined the association of what has been called the DRD2 "A" polymorphic system (DRD2*A; *Taq*-I A RFLP; 3′ to the D_2 dopamine receptor [DRD2] gene) in PTSD. (It has been shown that this variant actually maps to a different gene, ANKK1—not DRD2 [Neville, Johnstone, & Walton, 2004]). It can still be considered to reflect DRD2 gene effect to some extent because of linkage disequilibrium, however.) The first study compared 37 drug- and alcohol-abusing subjects with PTSD and 19 substance abusers without PTSD, reporting higher frequency of the A1 allele among the former (Comings, Muhleman, & Gysin, 1996). A very small sample size and presence of substance abuse in both subjects with PTSD and controls limit interpretation of this finding. Gelernter and colleagues (1999) could not find association of DRD2*A, DRD2*B, or DRD2*D marker loci in a larger sample consisting of 52 patients with PTSD and 87 controls (this study considered not only the ANKK1 marker ["DRD2*A"] but also two markers that map within the DRD2 locus). In a third study, Young and colleagues (2002) reported increased DRD2*A1 allele frequency among 91 patients with PTSD compared with 51 controls; however, the association in this sample was attributed to a subgroup of the PTSD group with high alcohol consumption. Again, the small sample size greatly limits interpretation of the results, especially because the sample was subdivided into subgroups according to a subjective cutoff regarding the amount of daily alcohol consumption. Moreover, the concept of "carriers" for nonfunctional variants (like DRD2*A) is, in general, not physiologically meaningful. Lawford and colleagues (2003) examined association of DRD2*A genotype and response to a selective serotonin reuptake inhibitor (SSRI), paroxetine, in PTSD treatment. Sixty-three white veterans with PTSD were treated with paroxetine for 8 weeks. Before paroxetine treatment, DRD2*A1 carriers reported both greater initial PTSD severity and greater subsequent responsivity to paroxetine. The small sample and multiple outcome measures (i.e., multiple statistical tests) limit the generalizability of findings. The demonstration of a common DRD2 variant affecting transcription (Duan et al., 2003) could provide a more valid avenue to investigating possible effects of DRD2 variation on PTSD risk.

Dopamine Transporter Gene

Segman and colleagues (2002) examined association of the dopamine transporter SLC6A3 3′ variable number tandem repeat (VNTR) polymorphism with PTSD. They applied a discordant design comparing a sample of 102 patients with chronic PTSD and 104 prospectively followed trauma survivor (TS) controls who were exposed but did not develop PTSD. A significant excess of nine-repeat allele carriers was observed among patients with PTSD (43 vs. 30.5% in TS controls; p = .012). An excess of nine-repeat homozygous genotype carriers was also observed in PTSD (20.43% in PTSD vs. 9.47% in TS controls). Compared to the other genotypes, homozygosity for the nine-repeat allele was associated with an increased risk for chronic PTSD (odds ratio [OR] = 2.45, 95% confidence interval [CI] = .98–6.52).

A normal Jewish population sample, not selected for trauma exposure, has shown 15% and 35% excess for nine-repeat homozygotes and allele carrier frequencies, respectively (Frisch et al., 1999). An intermediate frequency between that observed for PTSD and that in TS subjects (i.e., between 20.4 and 9.5% nine-repeat homozygotes, respectively, for PTSD and TS, and between 43 and 30.5% nine-repeat allele carrier frequency for the same groups; Segman et al., 2002). The VNTR is located in the 3′ noncoding region of the dopamine transporter gene and may be a marker for another functional site. Haplotype analysis of the dopamine transporter gene may further clarify these findings.

Neuropeptide Y Gene

Compared with healthy subjects, individuals with combat-related PTSD reported more anxiety, and had low baseline levels of neuropeptide Y (NPY), and a *blunted* NPY response to the alpha$_2$ antagonist yohimbine (Rasmusson et al., 2000). A study by Lappalainen and colleagues (2002) examined a functional NPY polymorphism (Leu7Pro) for association with alcohol dependence in European Americans (EA), using a design comparing the Leu7Pro allele frequencies in alcohol-dependent subjects, and subjects with other psychiatric diagnoses, and controls. They studied population stratification potential and diagnostic specificity by genotyping individuals from additional populations and psychiatric diagnostic classes (Alzheimer's disease, schizophrenia, posttraumatic stress disorder, and MDD). The frequency of the Pro7 allele was significantly higher in the alcohol-dependent subjects (Sample I, 5.5%; Sample II, 5.0%) compared to the screened EA controls (2.0%; Sample I vs. controls, $p = .006$; Sample II vs. controls, $p = .03$). There was no significant evidence that the association of the Pro7 allele to alcohol dependence was due to association with a comorbid psychiatric disorder; however, the PTSD sample showed Pro7 allele frequency of 3.9%.

GABA$_A$ β$_3$ Subunit Gene

GABA (gamma-aminobutyric acid)ergic neurotransmission has been implicated in the pathogenesis of anxiety disorders and may be involved in PTSD pathogenesis. Feusner and colleagues (2001) examined within-group correlations of dinucleotide repeat polymorphisms of the GABA$_A$ receptor β$_3$ subunit gene with General Health Questionnaire–28 (GHQ) scores, in a sample of 86 patients with PTSD. They arbitrarily divided the patients into G1 and non-G1 allele carriers. They reported that G1–non-G1 heterozygotes had a significantly higher score when compared to either the G1–G1 genotype ($α = 0.01$) or the non-G1–non-G1 genotype ($α = 0.05$), whereas no significant difference was found between the G1–G1 and non-G1–non-G1 genotypes. This observation is difficult to explain, and the authors suggest it may stem from a "heterosis"—a situation in which the heterozygote, rather than one of the homozygotes, represents the extreme phenotype. There is, however, no a priori biological basis for the previous observation, especially with a nonfunctional polymorphism.

Generalizability of Current Findings

The progress made through studies employing the VET Registry pertains to that specific population and that trauma type. Generalizability of the findings to civilian trauma, to women, and to other age groups is limited by the lack of equivalent, large-scale studies on non-combat-related PTSD. (In fact, in general, heritability is population-specific.) However, a recent small-scale civilian twin study demonstrates encouraging similarities in preliminary findings pertaining to non-combat-related PTSD and to both sexes. All PTSD gene association findings reported are currently very preliminary and require further replication. Given the low OR expected for any small-effect gene contribution to risk for PTSD, and the twin data supporting both specific and additive genetic contribution to PTSD and comorbid disorders, using large samples and controlling for comorbid disorders are mandatory for further pursuit of specific gene contribution to PTSD. Further methodological issues include use of healthy trauma survivors as controls and structured association approaches (or related statistical approaches) to correct for admixture effects in case–control samples (e.g., Kaufman et al., 2004). An additional limitation underlying these hypotheses-driven gene association studies is the lack of conclusive biological markers for PTSD that could inform the selection of molecular candidates.

CHALLENGES FOR THE FUTURE

The VET Registry has provided a rich data set for delineating the magnitude of heritable and environmental influences to combat exposure, PTSD, and comorbid disorders. Despite the limitations described earlier, the scale of the database allowed clear conclusions to be drawn regarding differential heritable and environmental effects on combat-related PTSD. Equivalent large-scale studies pertaining to noncombat trauma, different age groups, and both sexes are warranted. Natural disasters—such as the December 2004 tsunami event in Asia—may provide contexts for such studies.

The use of large-scale twin data resulted in significant progress in our understanding of the intricate effects of gene–environment interplay on trauma exposure, PTSD, and comorbid disorders. It was further applied effectively to trace temporal causation of PTSD-related putative endophenotypes. The progress in the delineation of a heritable component sets the stage for further work demonstrating how this genetic liability is transmitted and discovering the specific genes involved. Future studies for locating involved genes should use large samples, with better delineation of the phenotype (for both affected subjects and controls) to facilitate detection of genotype–phenotype correlations. Among patients with PTSD, controlling for comorbid disorders is important, and we foresee that use of endophenotypes is eventually likely to be important as well. For controls, documented trauma exposure is ideal to reduce the possibility of a confounding dormant genotypic vulnerability, but it may be impractical generally. The rapid progress in implementing techniques for whole-genome association studies may allow homing in on the actual genes involved in the near future.

ACKNOWLEDGMENTS

This work was supported in part by the U.S. Department of Veterans Affairs (National Center for PTSD Research, Mental Illness Research, Education and Clinical Center); by National Institute on Drug Abuse Grant Nos. R01 DA12849, R01 DA12690, and K24 DA15105; and by a Horowitz Foundation grant from the Hadassah–Medical Center Hebrew University–Hadassit Research and Development Division.

REFERENCES

Andreasen, N. C. (1997b). Linking mind and brain in the study of mental illnesses: A project for a scientific psychopathology. *Science, 275,* 1586–1593.

Brewin, C. R., Andrews, B., & Valentine, J. D. (2000). Meta-analysis of risk factors for posttraumatic stress disorder in trauma-exposed adults. *Journal of Consulting and Clinical Psychology, 68,* 748–766.

Carlson, C. S., Eberle, M. A., Kruglyak, L., & Nickerson, D. A. (2004). Mapping complex disease loci in whole-genome association studies. *Nature, 429*, 446–452.

Caspi, A., McClay, J., Moffitt, T. E., Mill, J., Martin, J., Craig, I. W., et al. (2002). Role of genotype in the cycle of violence in maltreated children. *Science, 297*, 851–854.

Caspi, A., Sugden, K., Moffitt, T. E., Taylor, A., Craig, I. W., Harrington, H., et al. (2003). Influence of life stress on depression: Moderation by a polymorphism in the 5-HTT gene. *Science, 301*, 386–389.

Chantarujikapong, S. I., Scherrer, J. F., Xian, H., Eisen, S. A., Lyons, M. J., Goldberg, J., et al. (2001). A twin study of generalized anxiety disorder symptoms, panic disorder symptoms and post-traumatic stress disorder in men. *Psychiatry Research, 103*(2–3), 133–145.

Comings, D. E., Muhleman, D., & Gysin, R. (1996). Dopamine D2 receptor (DRD2) gene and susceptibility to posttraumatic stress disorder: A study and replication. *Biological Psychiatry, 40*(5), 368–372.

Connor, K. M., & Davidson, J. R. (1997). Familial risk factors in posttraumatic stress disorder. *Annals of the New York Academy of Sciences, 821*, 35–51.

Devlin, B., & Roeder, K. (1999). Genomic control for association studies. *Biometrics, 55*, 997–1004

Devlin, B., Roeder, K., & Wasserman, L. (2001). Genomic control, a new approach to genetic-based association studies. *Theoretical Population Biology, 60*(3), 155–166.

Duan, J., Wainwright, M. S., Comeron, J. M., Saitou, N., Sanders, A. R., Gelernter, J., et al. (2003). Synonymous mutations in the human dopamine receptor D2 (DRD2) affect mRNA stability and synthesis of the receptor. *Human Molecular Genetics, 12*, 205–216.

Eaves, L., & Erkanli, A. (2003). Markov Chain Monte Carlo approaches to analysis of genetic and environmental components of human developmental change and G × E interaction. *Behavior Genetics, 33*(3), 279–299.

Feusner, J., Ritchie, T., Lawford, B., Young, R. M., Kann, B., & Noble, E. P. (2001). GABA(A) receptor beta 3 subunit gene and psychiatric morbidity in a post-traumatic stress disorder population. *Psychiatry Research, 104*(2), 109–117.

Frisch, A., Postilnick, D., Rockah, R., Michaelovsky, E., Postilnick, S., Birman, E., et al. (1999). Association of unipolar major depressive disorder with genes of the serotonergic and dopaminergic pathways. *Molecular Psychiatry, 4*(4), 389–392.

Gelernter, J., Southwick, S., Goodson, S., Morgan, A., Nagy, L., & Charney, D. S. (1999). No association between D2 dopamine receptor (DRD2) "A" system alleles, or DRD2 haplotypes, and posttraumatic stress disorder. *Biological Psychiatry, 45*(5), 620–625.

Gilbertson, M. W., Shenton, M. E., Ciszewski, A., Kasai, K., Lasko, N. B., Orr, S. P., et al. (2003). Smaller hippocampal volume predicts pathologic vulnerability to psychological trauma. *Nature Neuroscience, 5*(11), 1242–1247.

Goldberg, J., Curran, B., Vitek, M. E., Henderson, W. G., & Boyko, E. J. (2002). The Vietnam Era Twin Registry. *Twin Research, 5*(5), 476–481.

Goldberg, J., True, W. R., Eisen, S. A., & Henderson, W. G. (1990). A twin study of the effects of the Vietnam War on posttraumatic stress disorder. *Journal of the American Medical Association, 263*, 1227–1232.

Kaufman, J., Yang, B.-Z., Douglas-Palumberi, H., Houshyar, S., Lipschitz, D., Krystal, J. H., et al. (2004). Social supports and serotonin transporter gene moderate depression in maltreated children. *Proceedings of the National Academy of Sciences USA, 101*, 17316–17321.

Kendler, K. S., Neale, M. C., Kessler, R. C., Heath, A. C., & Eaves, L. J. (1993). A test of the equal-environment assumption in twin studies of psychiatric illness. *Behavior Genetics, 23*, 21–27.

Kessler, R. C., Kendler, K. S., Heath, A., Neale, M. C., & Eaves, L. J. (1992). Social support, depressed mood, and adjustment to stress: A genetic epidemiologic investigation. *Journal of Personality and Social Psychology, 62*(2), 257–272.

Kessler, R. C., Sonnega, A., Bromet, E., Hughes, M., & Nelson, C. B. (1995). Posttraumatic stress disorder in the national comorbidity survey. *Archives of General Psychiatry, 52*, 1048–1060.

Klein, R. J., Zeiss, C., Chew, E. Y., Tsai, J.-Y., Sackler, R. S., Haynes, C., et al. (2005). Complement Factor H polymorphism in age-related macular degeneration. *Science, 15*, 385–389.

Koenen, K. C., Harley, R., Lyons, M. J., Wolfe, J., Simpson, J. C., Goldberg, J., et al. (2002). A twin registry study of familial and individual risk factors for trauma exposure and posttraumatic stress disorder. *Journal of Nervous and Mental Disease, 190*(4), 209–218.

Koenen, K. C., Lyons, M. J., Goldberg, J., Simpson, J., Williams, W. M., Toomey, R., et al. (2003a). Co-twin control study of relationships among combat exposure, combat-related PTSD, and other mental disorders. *Journal of Traumatic Stress, 16*(5), 433–438.

Koenen, K. C., Lyons, M. J., Goldberg, J., Simpson, J., Williams, W. M., Toomey, R., et al. (2003b). A high risk twin study of combat-related PTSD comorbidity. *Twin Research, 6*(3), 218–226.

Lappalainen, J., Kranzler, H. R., Malison, R., Price, L. H., Van Dyck, C., Rosenheck, R. A., et al. (2002). A functional neuropeptide Y Leu7Pro polymorphism associated with alcohol dependence in a large population sample from the United States. *Archives of General Psychiatry, 59*(9), 825–831.

Lawford, B. R., Young, R., Noble, E. P., Kann, B., Arnold, L., Rowell, J., et al. (2003). D2 dopamine receptor gene polymorphism: Paroxetine and social functioning in posttraumatic stress disorder. *European Neuropsychopharmacology, 13*(5), 313–320.

Lyons, M. J., Goldberg, J., Eisen, S. A., True, W., Tsuang, M. T., Meyer, J. M., et al. (1993). Do genes influence exposure to trauma?: A twin study of combat. *American Journal of Medical Genetics, 48,* 22–27.

May, F. S., Chen, Q. C., Gilbertson, M. W., Shenton, M. E., & Pitman, R. K. (2004). Cavum septum pellucidum in monozygotic twins discordant for combat exposure: Relationship to posttraumatic stress disorder. *Biological Psychiatry, 55*(6), 656–658.

McLeod, D. S., Koenen, K. C., Meyer, J. M., Lyons, M. J., Eisen, S., True, W., et al. (2001). Genetic and environmental influences on the relationship among combat exposure, posttraumatic stress disorder symptoms, and alcohol use. *Journal of Traumatic Stress, 14*(2), 259–275.

Neale, M. C., & Cardon, L. R. (1992). *Methodology for genetic studies of twins and families.* Dordrecht, the Netherlands: Kluwer Academic.

Neiderhiser, J. M., Plomin, R., & McClearn, G. E. (1992). The use of CXB recombinant inbred mice to detect quantitative trait loci in behavior. *Physiology and Behavior, 52*(3), 429–439.

Neville, M. J., Johnstone, E. C., & Walton, R. T. (2004). Identification and characterization of ANKK1: A novel kinase gene closely linked to DRD2 on chromosome band 11q23.1. *Human Mutation, 23,* 540–545.

Orr, S. P., Metzger, L. J., Lasko, N. B., Macklin, M. L., Hu, F. B., Shalev, A. Y., et al. (2003). Physiologic responses to sudden, loud tones in monozygotic twins discordant for combat exposure: Association with posttraumatic stress disorder. *Archives of General Psychiatry, 60,* 283–288.

Pritchard, J. K., Stephens, M., Rosenberg, N. A., & Donnelly, P. (2000). Association mapping in structured populations. *American Journal of Human Genetics, 67,* 170–181.

Radant, A., Tsuang, D., Peskind, E. R., McFall, M., & Raskind, W. (2001). Biological markers and diagnostic accuracy in the genetics of posttraumatic stress disorder. *Psychiatry Research, 102*(3), 203–215.

Rasmusson, A. M., Hauger, R. L., Morgan, C. A., Bremner, J. D., Charney, D. S., & Southwick, S. M. (2000). Low baseline and yohimbine-stimulated plasma neuropeptide Y (NPY) levels in combat-related PTSD. *Biological Psychiatry, 47*(6), 526–539.

Sapolsky, R. M. (2002). Chickens, eggs and hippocampal atrophy. *Nature Neuroscience, 5*(11), 1111–1113.

Shalev, A. Y., Sahar, T., Freedman, S., Peri, T., Glick, N., Brandes, D., et al. (1998). A prospective study of heart rate response following trauma and the subsequent development of posttraumatic stress disorder. *Archives of General Psychiatry, 55*(6), 553–559.

Segman, R. H., Kooper-Kazaz, R., Macciardi, F., Gulcer, T., Chalfon, Y., Dubroborski, T., et al. (2002). Association between the dopamine transporter gene and posttraumatic stress disorder. *Molecular Psychiatry, 7*(6), 903–907.

Skre, I., Onstad, S., Torgesen, S., Lygren, S., & Kringlen, E. (1993). A twin study of DSM-III-R anxiety disorders. *Acta Psychiatrica Scandinavica, 88,* 85–92.

Spielman, R. S., McGinnis, R. E., & Ewens, W. J. (1993). Transmission test for linkage disequilibrium: The insulin gene region and insulin-dependent diabetes mellitus (IDDM). *American Journal of Human Genetics, 52,* 506–516.

Stein, M. B., Jang, K. L., Taylor, S., Vernon, P. A., & Livesley, W. J. (2002). Genetic and environmental influences on trauma exposure and posttraumatic stress disorder symptoms: A twin study. *American Journal of Psychiatry, 159*(10), 1675–1681.

True, W. R., Rice, J., Eisen, S. A., Heath, A. C., Goldberg, J., Lyons, M. J., et al. (1993). A twin study of genetic and environmental contributions to liability for posttraumatic stress symptoms. *Archives of General Psychiatry, 50,* 257–264.

Yehuda, R. (2002). Post-traumatic stress disorder. *New England Journal of Medicine, 346*(2), 108–114.

Young, R. M., Lawford, B. R., Noble, E. P., Kann, B., Wilkie, A., Ritchie, T., et al. (2002). Harmful drinking in military veterans with post-traumatic stress disorder: Association with the D2 dopamine receptor A1 allele. *Alcohol and Alcoholism, 37*(5), 451–456.

Xian, H., Chantarujikapong, S. I., Scherrer, J. F., Eisen, S. A., Lyons, M. J., Goldberg, J., et al. (2000). Genetic and environmental influences on posttraumatic stress disorder, alcohol and drug dependence in twin pairs. *Drug and Alcohol Dependence, 61*(1), 95–102.

Xian, H., Scherrer, J. F., Eisen, S. A., True, W. R., Heath, A. C., Goldberg, J., et al. (2000). Self-reported zygosity and the equal-environments assumption for psychiatric disorders in the Vietnam Era Twin Registry. *Behavioral Genetics, 30*(4), 303–310.

Chapter 12

Gender Issues in PTSD

Rachel Kimerling, Paige Ouimette, and Julie C. Weitlauf

Effective research and treatment of posttraumatic stress disorder (PTSD) requires attention to gender issues regarding trauma exposure, traumatic stress reactions, and treatment of PTSD. Although gender differences in the prevalence of PTSD are well-documented, we propose a general conceptual and methodological framework that can guide efforts to explain these gender differences and address more comprehensively gender issues related to traumatic stress. We use this conceptual framework to review literature regarding exposure to trauma, the prevalence of PTSD, assessment and diagnosis, comorbidity, and treatment of PTSD. The literature review in this chapter is organized to suggest possible explanations for the gender issues observed, with an emphasis on clinical and research applications. We also point to specific directions for future research in an effort to elucidate better the role of gender in the development and treatment of PTSD. Such an analysis of the literature maximizes the utility of empirical PTSD research as it pertains to our understanding of both male and female trauma populations.

METHODOLOGICAL CONSIDERATIONS

Initial investigations of gender issues address the question: How do men and women differ? Clear examples of this line of inquiry are hypotheses of sex-based comparisons in the prevalence of trauma exposure and PTSD that find elevated rates of PTSD in women compared to men (Breslau, Davis, Andreski, & Peterson, 1997; Kessler,

Sonnega, Bromet, Hughes, & Nelson, 1995). Essentially, this research is a direct comparison between men and women that develops a catalog of similarities and differences. These direct comparisons are a necessary, but not sufficient, means to understand gender issues. Such descriptive data alert us to important areas of sex differences. However, explanations for difference derived from these designs can only result in generalized conclusions concerning men and women. Furthermore, a focus solely on these statistical group differences all too often diverts attention away from any overlap of the distributions, which represents the similarities between men and women. Such analyses also lack the ability to factor out shared sources of variance, such as the effects of environmental context. We assert that the conclusions derived from direct comparisons are too fundamental to account adequately or consistently for multidimensional constructs such as gender or posttrauma responses.

This "direct comparison" line of inquiry is often augmented by examining secondary hypotheses regarding control variables or covariates intended to rule out alternative hypotheses or confounders for observed sex differences. For example, gender differences in rates of PTSD remain, even after controlling for differential rates of trauma exposure (Breslau, Davis, Andreski, & Peterson, 1997; Kessler et al., 1995), though exposure to interpersonal violence appears to account for at least some of the discrepancy (Breslau, Chilcoat, Kessler, Peterson, & Lucia, 1999). This "controlled comparison" line of inquiry can elucidate factors that covary with gender and influence trauma outcomes in a given population. Thus, the controlled comparison approach can describe sex differences with greater precision than the direct comparison approach, yet these designs still provide limited explanatory power.

If the framework for understanding sex differences is limited to the direct comparisons and controlled comparisons approaches, then explanations rely on static characteristics of men and women, such as biological bases, social roles, or cognitive style. If we operate on the assumption that all human behavior, including stress processes, occurs within a social context (Mischel, 2004; Moos, 2003), then it follows that gender differences between men and women are most meaningful when interpreted in context (Yoder & Kahn, 2003). In this chapter, we use the word *sex* to refer to the biological fact of being male or female, whereas the word *gender* refers to the social context and psychological experience of a male or female individual in a given society and culture. Gender issues (compared to sex differences) are then best conceptualized as an interaction between sex-based biology and the individual's social context. This definition of "gender differences" accounts for intragender diversity and differences between genders by assuming that these differences are context-dependent. Such an analysis of the literature allows us to integrate seemingly conflicting data on gender differences by examining the conditions under which men and women appear to differ, and acknowledging that the factors that appear to explain these differences may not be the same in different contexts and populations.

This conceptualization of gender then invites another framework for addressing gender issues in PTSD: the identification of moderator variables (King, Orcutt, & King, 2002; Yoder & Kahn, 2003). Research questions can be broadly categorized as follows: Under what conditions are gender differences observed, and under what conditions do they disappear? For example, in a study of U.S. and Mexican survivors of natural disaster (Norris, Perilla, Ibanez, & Murphy, 2001), gender differences in PTSD were moderated by culture, with gender differences more pronounced in the Mexican sample, and more attenuated in the U.S. sample. These findings suggest that cultural contexts that emphasize more traditional gender roles influence gender disparities in rates of PTSD.

This hypothesis is consistent with a conceptualization of gender as an interaction between sex and social context.

This "gender-interactional" model for inquiry into gender issues in PTSD is extremely useful in identifying social-contextual factors that may moderate the extent to which sex differences are observed in PTSD. More consistent implementation of this approach might also identify factors that influence treatment approaches and outcomes for men and women in "real-world" settings. Social roles, such as those defined within a given culture, family structure, or military service are also important potential moderator variables. Life events, especially pre- and posttrauma events, can influence both an individual's perception of his or her social context and influence behavior within a given social context. These variables can be an especially informative direction of research that uses a gender-interactional model.

A final step in developing research hypotheses and designs to address gender issues in PTSD transcends the analysis of gender differences. These designs examine the extent to which relationships among trauma variables differ as a function of gender. Because these designs conceptualize gender as an elemental basis for difference, they have the greatest power to organize data regarding gender differences into gender-informed models of traumatic stress. An extension of the gender-interactional approach can be used to better understand complex relationships between gender and traumatic stress reactions. In some instances gender may moderate the relationship between two or more trauma-related variables. For example, research suggests that children and adolescents with abuse-related PTSD fail to show expected age-related brain growth in the corpus callosum, and that the severity of these neurological effects is moderated by gender; trauma-related brain differences are significantly more pronounced among males than among females (DeBellis & Keshavan, 2003).

Operationalizing the gender-interactional model is accomplished at both conceptual and methodological levels of research. Regarding the former, the inclusion of specific gender-based hypotheses helps us systematically to uncover relationships between gender and PTSD. Regarding the latter, methodological changes, regression, or path models must be constructed and analyzed separately for men and women to disaggregate by gender relationships among multiple variables. Such models can be informative even when no sex differences are observed for the primary outcome variable. For example, analysis of national survey data indicated that males and females with PTSD are equally likely to have a comorbid medical disorder. Differences emerged when separate models were constructed by gender. For women, medical morbidity was multiply determined as a function of poverty, depression, and PTSD, whereas for men, depression and poverty had no effect on morbidity, which was related to PTSD only (Kimerling, 2004).

The broadest extension of these "gender-informed" research questions is to assess whether the constructs we use to account for traumatic stress reactions are identical for men and women. Is PTSD really the same construct in men, as compared to women, given the many gender differences that research using the preceding models has documented? This approach would rely on multiple-group confirmatory factor analysis to determine whether the same latent factors represent PTSD and other trauma-related constructs across gender (King et al., 2002). For example, an analysis of male and female responses to the Childhood Trauma Questionnaire (CTQ; Bernstein et al., 1994) suggests different constructs by gender for childhood trauma. Most notably, physical abuse did not emerge as a factor separate from emotional abuse for women, for whom childhood trauma was best represented by factors for emotional abuse, emotional neglect, physical neglect, and sexual abuse. For men, factors were emotional abuse, emotional neglect, physical abuse, physical neglect, and sexual abuse (Wright et al., 2001). Few studies have explored these broader conceptual models for gender issues, but such research could substantially inform our interpretation of data on gender differences in the rates of PTSD, and our clinical treatment of male and female trauma survivors.

CURRENT STATE OF THE ART

Gender Issues in the Prevalence of Trauma and PTSD

Gender disparities in the prevalence of trauma exposure and PTSD yield an apparent paradox: Although men are more likely to experience a traumatic life event, women are more likely to develop PTSD. To provide a foundation for further consideration of gender issues and PTSD, we review major studies that have established prevalence estimates of PTSD among men and women, and highlight some important issues for future research on gender differences in rates of trauma and PTSD.

Gender and the Prevalence of Trauma

Epidemiological studies have documented rates of trauma exposure between 51 and 92% in U.S. adult populations (Breslau, Davis, Andreski, & Peterson, 1991; Breslau et al., 1998; Helzer, Robins, & McEvoy, 1987; Kessler et al., 1995; Norris, 1992). The National Comorbidity Survey (NCS; Kessler et al., 1995) is important, because it was the first nationally representative, face-to-face, general population survey of DSM-III-R disorders, including PTSD. Lifetime prevalence of exposure to trauma was 61% for men and 51% for women. Significant gender differences were found on specific types of trauma: More women reported sexual molestation, sexual assault, and child physical abuse, whereas more men reported fire/disaster, life-threatening accident, physical assault, combat, being threatened with a weapon, and being held captive.

In a survey of Detroit area residents, Breslau and colleagues (1998, 1999) interviewed 2,181 adults by telephone using an inventory of 19 traumatic life events and PTSD based on DSM-IV criteria. A total of 87% of females and 92% of males reported lifetime exposure to trauma. Women reported less exposure in the categories of assaultive violence (32%; e.g., sexual assault, physical assault) and other injury or shocking events (52%; e.g., car accident, disaster) than men reported in the same categories (43% and 68%). No gender differences emerged on learning about traumas to others or sudden unexpected death of a loved one. For specific events within these categories, women reported more rape and sexual assault, whereas men reported more being shot/stabbed, mugged/threatened with a weapon, or being badly beaten up. These data, based on reports of objective trauma exposure, suggest that whereas men are more likely to be involved in violence or accidents of a physical nature, women bear the burden of sexual violence.

In a review of epidemiological research, Norris, Foster, and Weissharr (2002) noted that studies from countries outside the United States, including Canada, Germany, Israel, New Zealand, China, and Mexico, also document elevated rates of trauma exposure among men compared to women. Moreover, as with U.S. studies, the types of trauma men and women experience are not equivalent. Across studies, a major finding is that women are more likely than men to report exposure to rape or sexual assault. As noted by Norris and colleagues, a major caveat to these findings is that most trauma inventories do not assess PTSD DSM criterion A2, in which an individual must experience fear, horror, or helplessness at the time for the event to qualify as traumatic. Thus, there is probably some proportion of false-positive reporting of PTSD criterion A trauma (i.e., persons experiencing objective events, or criterion A1, that did not incur fear, horror, or helplessness, or criterion A2) in many of these studies.

This raises the question: Are there gender differences in trauma exposure that meet both DSM criteria A1 and A2? In the Detroit Area Survey (Breslau & Kessler, 2001), women were more likely to endorse criterion A2 following exposure. Interestingly, the gender difference in lifetime exposure to trauma reversed when criterion A2 was required (73.3% of males vs. 82% of women). In a study in Mexico, when Norris, Foster, and Weissharr (2002) assessed criterion A2, they found that men and women did not differ on reporting of trauma. In a German study of young adults and adolescents, gender differences in rates of trauma exposure were attenuated (but not reversed) when criterion A2 was used (Perkonigg, Kessler, Storz, & Wittchen, 2000).

The gender differences in the proportion of events that meet the criterion for subjective distress (A2) suggest that women's responses to potentially traumatic life events differ from men's responses, thus placing them at greater risk for PTSD. In line with this finding, in a recent meta-analysis of predictors of PTSD, the strongest predictors of PTSD were emotions and dissociation during or in the immediate aftermath of the trauma (Ozer, Best, Lipsey, & Weiss, 2003). Thus, an individual's response in the face of trauma, or immediately following it, appears to play an important role in predicting future PTSD symptoms.

Few studies report on gender differences in peritraumatic emotions and dissociation. Studies of crime victims and motor vehicle accident survivors provide relevant data. Among crime victims, Brewin, Andrews, and Rose (2000) found that women reported more intense fear and horror at the time of the trauma than did men. Although Breslau and Kessler (2001) reported that more women endorse peritraumatic emotions, or criterion A2 (i.e., fear, horror, or helplessness), than men, trauma rarely led to PTSD when only criterion A1 was endorsed. In examining dissociative responses at the time of the trauma, Bryant and Harvey (2003) found that peritraumatic dissociation was a more accurate predictor of PTSD among female motor vehicle accident victims compared to males (see also Fullerton et al., 2001). More research is needed to better understand gender differences in immediate responses to trauma such as fear, horror, helplessness, guilt, and dissociation, and what implications these reactions have for the development of PTSD.

Gender and the Prevalence of PTSD

General population studies consistently find that women are approximately twice as likely as males to meet criteria for PTSD at some point in their lives. Major studies that have used DSM-III-R criteria have documented lifetime (current) prevalence rates of 10.4–11.3% (3%) in women, and 5–6% (1%) in men (Breslau et al., 1991; Kessler et al., 1995). The NCS (Kessler et al., 1995) documented that among individuals exposed to trauma, 20.4% of women and 8.2% of men developed PTSD, suggesting that the gender difference associated with the conditional risk for PTSD is even stronger. Whereas men were more likely to experience at least one trauma, women were more likely to experience a trauma with a high probability of developing PTSD. There were no gender differences in rates of PTSD following rape (the event with the highest conditional risk for men, 65%, and women, 45.9%). Women had a greater conditional risk for PTSD following molestation, physical attack, being threatened with a weapon, and childhood physical abuse, whereas no gender differences emerged on rates of PTSD following natural disaster, accidents, and witnessing the injury and/or death of another.

In the Detroit Area Survey, which was based on DSM-IV criteria, the conditional risk for lifetime PTSD was 13% in women and 6.2% in men (Breslau et al., 1998). This finding was specific to vulnerability following exposure to assaultive violence; no gender differences in PTSD emerged for the categories of injury/other shocking event, learning about traumas to others, and sudden unexpected death of a loved one.

In a review of studies from both the United States and other cultures, Norris, Foster, and Weissharr (2002) noted that women's differential risk for PTSD stays close to 2:1, regardless of the absolute rate of PTSD reported. In addition, among studies focusing on specific disasters (e.g., Hurricane Hugo), women usually had higher rates of PTSD than men. In community and political violence research compared to studies from other research contexts, a higher proportion of studies yielded no gender differences in PTSD. Norris and colleagues hypothesized that as the context of trauma becomes especially "dire," male PTSD rates may catch up to female rates. This idea—that when the severity of context or characteristics of the trauma is equal, gender differences may disappear—may also explain Kessler and colleagues' (1995) finding of no gender differences in PTSD following rape.

In summary, general population and disaster studies suggest that women are at greater risk than men for PTSD. Women's greater risk for sexual assault or assaultive violence may partially account for these gender differences. However, event type does not fully account for the gender difference. For example, gender differences in rates of PTSD in disaster survivors (see Norris, Friedman, et al., 2002, for a review), and motor vehicle accident survivors and military personnel (Brewin, Andrews, & Valentine, 2000) are inconsistent (Bryant & Harvey, 2003; Freedman, Brandes, Peri, & Shalev, 1999) suggesting a need for further research on the contexts of these situations for men and women. In the Detroit Area Survey, the gender difference in PTSD remained in analyses that controlled for type of event. One notable exception to the gender difference in PTSD may be found in violent community and political situations, in which men's risk for PTSD may be similar to that of women. This latter finding may suggest that when the characteristics of trauma are similar for men and women, gender differences in PTSD may disappear.

Kimerling, Prins, Westrup, and Lee (2004) noted that broad categorization of traumatic events, which can equate events that have different situational characteristics, may obscure important information. For example, a single, physical fight with a stranger and prolonged physical abuse by an intimate partner would be classified under the category of physical assault on most trauma inventories. However, the characteristics of these "physical assaults" (e.g., single vs. multiple occurrences, stranger vs. intimate partner) may differentially predict risk for PTSD.

A recent meta-analysis of risk factors for PTSD revealed that trauma severity is a significant predictor of PTSD (Brewin, Andrews, & Valentine, 2000). Along a similar vein, one perspective on gender and PTSD posits that women's greater risk can be explained by their experience of more severe traumas. In other words, an event from the same category (e.g., physical assault) may have different correlates for men and women, with women's trauma characteristics being more severe. Thus, this approach seeks to understand what characteristics of trauma that raise risk for PTSD are confounded by female gender. Examples of important characteristics to consider include severity of injury, relationship to assailants, and chronicity.

Studies in the area of interpersonal violence generally show that, based on trauma characteristics, women are exposed to more severe traumas than are men. In studies that include male and female sexual assault survivors, females were more likely to be assaulted by an intimate partner (Elliott, Mok, & Briere, 2004) and report penetration, experience injury, and be restrained by the assailant (Kimerling, Rellini, Kelly, Judson, & Learman, 2002). However, Kimerling and colleagues (2002) found that males were more likely to report burns during the assault, and other studies have found that men more often have multiple assailants (e.g., Frazier, 1993).

In the National Violence Against Women Survey (NVAWS; (Tjaden & Thoennes, 2000) a national sample of 16,000 men and women were interviewed via telephone regarding victimization experiences. Results indicated that although men reported more physical assaults, women were more likely to be harmed by an intimate male partner. In addition, the difference between women's and men's rates of partner physical assault became greater as the seriousness of the assault increased (e.g., choked, beat up, threatened with gun). Women were also more likely to be injured during a rape or physical assault.

Pimlott-Kubiak and Cortina (2003) presented a reanalysis of the NVAWS, examining the nature and extent of victimization across genders and comparing gender differences in psychological outcomes within specific victimization groups (thus equating women and men on victimization history). Women reported more severe sexual violence and more multiple events with sexual violence; men reported more childhood physical abuse and multiple events without sexual violence. Pimlott-Kubiak and Cortina found that gender was linked to certain types of situations related to symptoms of psychological distress. The victimization profiles did not relate to psychological outcomes differentially by gender. Thus, the authors argued that gender was a proxy for a history of exposure to severe aggression. Although these results are intriguing, they do not report on PTSD as an outcome, so it is unknown whether these findings hold for PTSD. Although additional research is needed on gender differences in the characteristics and contexts of various traumas, these data indicate that a single type of event may not necessarily be the same experience for men and for women. In addition, research in this area could address whether gender differences in the characteristics of a trauma also explain gender differences in immediate responses to the same event.

The Role of Social Context

Several aspects of the social context of exposure have relevance for explaining the gender difference observed in the development of PTSD. Social roles and contexts may impact risk for trauma exposure, as well as posttrauma responses. In military samples, a stronger association is observed between deployment social support and postdeployment mental health for women than for men (Vogt, Pless, King, & King, 2005), suggesting the important role of social context for women. Sexual assault and harassment in the military occur significantly more frequently during wartime and combat (Wolfe et al., 1998), disproportionately so for women in nontraditional female military occupations. Stressful or adverse environments also appear to increase the likelihood of trauma exposure to interpersonal violence. Rates of childhood abuse escalate in the wake of major natural disasters (Curtis, Miller, & Berry, 2000) and in high-crime environments (Finkelhor, Ormrod, Turner, & Hamby, 2005). Adverse contexts potentiate the impact of direct exposure to traumatic events, and this process may be more salient to the types of trauma exposure experienced by women.

Women are also more often victims of interpersonal violent crimes associated with negative or stigmatizing social responses, such as sexual assault. The greater negative social response experienced by female crime victims may partly explain women's elevated rates of PTSD symptoms (Andrews, Brewin, & Rose, 2003). Recent research suggests there are also gender differences in social responses to sexual trauma. In a study of child sexual abuse, Ullman and Filipas (2005) found that whereas men and women who experienced childhood sexual abuse did not differ in the likelihood of receiving negative reactions upon disclosure, women were more likely to have disclosed the abuse, and the intensity of negative social responses reported by women was nearly twice that of men. These negative responses were associated with greater PTSD symptom severity. Relatively little study has addressed the gender differences, social context, and social responses to trauma, especially sexual trauma, in relation to symptom development and recovery. More research in this area is certainly warranted.

Social roles, such as that of wife, mother, or caretaker, may moderate the impact of exposure and posttrauma responses. In a review of outcomes following natural disaster, Norris, Friedman, and colleagues (2002) found that women are not only consistently at greater risk for PTSD following disaster but also certain social roles function as gender-specific risks for PTSD. Women in the roles of mothers and wives were particularly at risk for poorer outcomes. Marital status was a risk factor for poorer outcome among women but not men. Furthermore, the severity of the husband's PTSD symptoms has a greater impact on the symptoms of his female partner than do the severity of a wife's symptoms on her male partner, suggesting possible gender differences in social support processes following trauma. A similar pattern is observed in other trauma-exposed samples, in which the beneficial impact of marital social support on trauma-related distress is more pronounced when women's partners more accurately appraise her stressor exposure (Ritter, Hobfoll, Lavin, Cameron, & Hulsizer, 2000), whereas women's accurate perceptions of their male partners' PTSD symptoms do not appear to be related to male satisfaction with spousal support (Taft, King, King, Leskin, & Riggs, 1999). Norris, Friedman, and colleagues (2002) also identified mothers as especially at risk for substantial psychological distress following disaster. Because mothers occupy a social role that emphasizes close relationships with others, they may be at risk when the trauma, such as a natural disaster, is experienced with close others.

Cultures that emphasize the gendered nature of these roles to a greater or lesser extent, then, provide an environment in which gender differences following trauma are either exaggerated or attenuated. Baker and colleagues (2005) have also proposed that posttrauma cognitions, such as helplessness and emotional distress, are more consonant for women's gender roles than men's, an effect that is more pronounced in cultures that emphasize traditional gender roles. There seems to be support for these hypotheses: In a study of U.S. and Mexican survivors of natural disaster (Norris et al., 2001), gender differences in PTSD were moderated by culture. Women's rates of PTSD exceeded those of men to a greater extent in the Mexican sample than in the U.S. sample.

Assessment and Diagnosis

Gender-informed assessment and diagnosis are essential to understanding PTSD. In both research and clinical applications, assessment measures must be sensitive to factors that distinguish the characteristics of trauma exposure and associated stress reactions that are most salient for men and for women, while still retaining sufficient criterion-rated validity and generalizability to ensure adequate adherence to the PTSD construct and effective communication with other professionals. If assessment instruments lack sensitivity to gender issues, then we can draw spurious conclusions from the data. If assessment methods are too specialized to a specific trauma population or gender role, then we cannot generalize or compare these data with other PTSD populations.

Measurement of Trauma Exposure

Gender differences in the prevalence and type of trauma exposure are a central hypothesis for women's increased rates of PTSD (Breslau et al., 1999; Norris, Foster, & Weissharr, 2002). Specific, gender-linked forms of exposure, such as child sexual abuse, early abuse, and perpetration by a caregiver or intimate, have long been proposed to explain differences in types of trauma-related symptom presentation (e.g., Battle et al., 2004; DePrince & Freyd, 2002; Herman, 1992). Thus, special care should be taken that the assessment of trauma exposure is sufficiently detailed, so that gender and trauma type are not confounded. Trauma exposure measures are less likely to be influenced by gender-related factors when (1) trauma exposure is queried in behaviorally specific language; (2) qualitative aspects of traumatic events are measured (e.g., severity, chronicity, and age at onset); and (3) a variety of events or experiences relevant to both men and women are examined and distinguished from one another.

The role of assessment language in the accuracy of data first became apparent when researchers observed that women with sexual experiences meeting the legal definition of rape did not endorse questionnaire items such as "Have you ever been raped?" (Kilpatrick et al., 1989; Koss, 1985), though they did endorse more behaviorally specific queries without the label of rape. Behaviorally specific language ensures accurate prevalence rates of trauma and comprehensive clinical assessments with men and women. Qualitative aspects of the trauma, such as age, severity, and chronicity, are especially important in assessing the degree of trauma exposure among men and women.

Although men appear to have more trauma exposure than women when numbers of discrete events are derived from event checklists, the specific characteristics that define the parameters of these events (childhood onset, chronicity, perpetration by an intimate or family member) may explain at least part of observed gender differences in PTSD prevalence. Women are more likely to experience interpersonal violence perpetrated by an intimate and to experience chronic, compared to single-incident, physical assault (Tjaden & Thoennes, 2000). A given traumatic event can differ by gender in age at onset, likelihood of physical harm, or a number of other factors (Kimerling et al., 2002). Thus, trauma assessment must not only capture appropriate qualitative characteristics but also make germane distinctions between events such as physical assault and intimate partner violence. Only with sufficiently detailed measurement can clinical or research-related assessments conceptualize posttrauma reactions as a function of exposure, rather than gender, when appropriate.

Assessing PTSD

The presentation of trauma-related symptoms can overlap with depressive or anxiety disorders, borderline personality disorder, and other Axis II disorders. Studies indicate that women are given an average of four Axis I diagnoses when symptoms of PTSD are evaluated without accounting for trauma histories and diagnoses (Cloitre, Koenen,

Gratz, & Jakupcak, 2002). A comprehensive and gender-sensitive assessment of both trauma characteristics and PTSD symptoms likely yield the most informative and descriptive diagnoses in both research and clinical settings (Kimerling et al., 2004).

Assessment of PTSD symptoms should allow for the experience of multiple traumatic events, because men tend to experience a greater number of events than women, and anchoring PTSD symptoms to a single event may actually restrict estimates of the prevalence of PTSD among men. When assessing PTSD symptoms, measures that require respondents to link symptoms to "before" and "after" the event can be difficult for individuals exposed to childhood sexual or physical abuse, the majority of whom are women (Sachs-Ericsson, Blazer, Plant, & Arnow, 2005). More research is needed that utilizes a gender-informed research approach to test the equivalence of the PTSD construct for men and for women. Confidence in the utility of PTSD measures is generated when psychometric properties of the measure are available from samples of both men and women, and there is no evidence of differential prediction.

Associated Features and Symptoms

Traumatic stressors that occur during key developmental phases, or are sufficiently severe, prolonged, and socially salient, are also associated with alterations in affect regulation, dissociation, and marked difficulties with interpersonal relationships. Accounting for these associated symptoms is a complex diagnostic issue. These symptoms represent core features of both borderline personality disorder and complex PTSD (also referred to as DESNOS, disorders of extreme stress not otherwise specified). The use of the latter diagnostic category has been proposed as a means to indicate a trauma-related, chronic mental health condition that includes symptoms that differ from, or extend beyond, the criteria for PTSD (van der Kolk et al., 1996). The construct of DESNOS is still somewhat experimental, and more research is needed to ascertain the relative merits of the proposed diagnosis.

Little evidence, however, suggests that the process by which early trauma affects development is gender-linked. For both men and women, prolonged and severe early trauma appears to impact the process of learning to experience, identify, and talk about emotions; to observe how they function; and to develop strategies for modulating and utilizing emotions effectively (van der Kolk et al., 1996). This appears to be especially true for interpersonal violence, and when the perpetrator is a family member, or other intimate, who would model or coach the child in emotional regulation strategies. Research that investigates gender issues in these patients should address men's and women's development, experience, and regulation of emotion, because this may be the most chronically impaired domain (Zanarini, Frankenburg, Hennen, & Silk, 2003). For example, child maltreatment is strongly associated with adult diagnoses of personality disorders in both men and women, and there is a strong association between child sexual abuse and later diagnoses of borderline personality disorder (Johnson, Cohen, Brown, Smailes, & Bernstein, 1999; Spataro, Mullen, Burgess, Wells, & Moss, 2004).

Freyd and colleagues (DePrince & Freyd, 2002; Freyd, 1996) has have proposed that PTSD and "complex" features stem from different aspects of trauma exposure: that PTSD symptoms are associated with threat to life, and that affective instability, dissociation, and interpersonal difficulties are associated with the social betrayal inherent in early abuse by a caregiver or family member. Although these symptoms appear to be more prevalent among female patients, Freyd and colleagues posit that women's greater risk for these "betrayal" types of chronic interpersonal violence perpetrated by intimates, such as child sexual abuse or intimate partner violence, is the basis for the prevalence difference in these symptoms.

Comorbidity Issues

PTSD is often comorbid with other psychiatric conditions. In the NCS, 59% of men and 43.6% of women with PTSD had three or more additional diagnoses (Kessler et al., 1995). Comorbidity is important to consider, because it is related to a more chronic course of PTSD (Breslau & Davis, 1992) and more functional impairment among persons with PTSD (Shalev et al., 1998). However, information on how gender affects patterns of comorbidity among individuals with PTSD is very limited. Although research documents that PTSD is often comorbid with other psychiatric conditions, limited available work has examined whether patterns of comorbidity vary by gender. The most research evidence is available to support tentatively gender-related differences in the association between PTSD and mania, substance use disorders, and panic disorder (Orsillo, Raha, & Hammond, 2002). However, replication of these findings is needed. The methodological issues in conducting this research are substantial: The sample size needed to obtain sufficient statistical power to compare men's and women's rates of diagnoses comorbid to PTSD can be found in only the largest epidemiological investigations. Gender comparisons of the rates of comorbid diagnoses across studies are difficult given the variation in methods for assessing PTSD and the differences in trauma types each study might examine.

Mood Disorders

Major depression and dysthymia are among the disorders most frequently comorbid to PTSD. In a review of comorbidity issues in PTSD, Deering, Glover, Ready, Eddleman, and Alarcon (1996) suggested that mood disorders might develop as part of the complicated grieving response to losses that may occur as a by-product of the traumatic event.

Rates of major depression and dysthymia are similar among community samples of women and men with PTSD. Among women, rates of current major depression range from 17 to 23% (lifetime rates, 42–49%), whereas among men, rates are 10–55% (lifetime rates, 26–70%). Lifetime rates of dysthymia among men and women with PTSD are 21–29% and 23–33%, respectively (Kessler et al., 1995; Kulka et al., 1990). In Vietnam Era veterans, however, women have more current major depression (men, 26%; women, 42%), lifetime major depression (men, 26%; women, 42%) and dysthymia (men, 21%; women, 33%) (Kulka et al., 1990) than men.

The similarity of the rates of comorbid depression among men and women with PTSD is surprising given gender differences in depression in the general population. Major life stressors, including trauma, play an etiological role in depression and partly explain women's greater risk for the disorder (Nolen-Hoeksema & Girgus, 1994). The high rates of comorbid depression in men and women may indicate that PTSD, or a traumatic event that leads to PTSD, may create vulnerability toward depression in men that suppresses the protective effect of male gender.

Interestingly, two epidemiological studies have suggested a stronger association between PTSD and mania among men compared to women. In the NCS, men with PTSD were classified as having mania at twice the rate of women (12% vs. 6%; Kessler et al., 1995). In the National Vietnam Veterans Readjustment Study (NVVRS) (Kulka et al., 1988), male veterans with PTSD were more likely than women to have current (4 to

3%) and lifetime mania (6 to 3%). In the general population, men and women do not appear to differ on mania (Kessler et al., 1995). Although there is not a clear explanation for this gender difference, Orsillo and colleagues (2002) suggest this is an artifact of epidemiological misdiagnosis: Among men, PTSD-related hypervigiliance, hyperarousal, or anger and irritability might be mistakenly attributed to mania. Additional research might indicate the extent to which gender differences in PTSD and comorbid mania are observed clinically.

Substance Use Disorders

Substance use disorders are common among individuals with PTSD. Current population estimates for lifetime comorbidity indicate that approximately 30–50% of men and 25–30% of women with lifetime PTSD have a co-occurring substance use disorder (SUD) (Kessler et al., 1995). These numbers actually indicate a stronger risk for comorbid substance use disorders among women with PTSD, compared to men, because women's rates of substance use disorders are far less in the absence of PTSD (Stewart, Ouimette, & Brown, 2002).

Preliminary work suggests that the etiology of PTSD–SUD comorbidity for men and women may differ. For example, women are more likely than men to develop SUDs subsequent to trauma exposure and PTSD (Stewart et al., 2002), with approximately 65–84% of women meeting criteria for PTSD before they develop SUDs. This pattern suggests that women may "self-medicate" or cope with trauma-related symptoms by using substances. In men, the temporal pattern is more consistent with an increased risk for trauma exposure during the course of substance use and abuse, which then leads to PTSD.

Other data suggest gender differences in the functional relationships between PTSD and substance use. Both men and women with PTSD use substances more frequently than SUD-diagnosed individuals without PTSD, and tend do so in negatively reinforcing situations, in response to cues such as negative emotions, interpersonal conflict, or physical discomfort (Sharkansky, Brief, Peirce, Meehan, & Mannix, 1999). Male substance abusers are more likely than women with PTSD–SUD comorbidity to use in situations involving positive emotions, alone or with others. It has been hypothesized (Stewart et al., 2002) that positive emotions occur more rarely among women with PTSD–SUD comorbidity, due to more intensive emotional numbing. As a result, these cues may trigger substance use less often among women.

Anxiety Disorders

Although women bear the burden of most of the anxiety disorders, preliminary evidence suggests that gender differences in anxiety disorders may be attenuated among individuals with PTSD. The exception may be PTSD and comorbid panic, which may be more common among women. In a general community sample (Helzer et al., 1987) that provided direct statistical comparisons of women and men, women with PTSD were more often diagnosed with comorbid panic disorder than men. However, the dearth of research on PTSD and comorbid anxiety disorders limits strong conclusions.

Research has suggested that the rates of panic disorder are 13% among women with PTSD and 5% among men with PTSD (Kulka et al., 1988), with lifetime rates ranging from 7 to 21% among women and from 7 to 18% among men (Breslau, Davis, Peterson, & Schultz, 1997; Kessler et al., 1995; Kulka et al., 1988; Orsillo et al., 1996). In a general community sample (Helzer et al., 1987) that provided direct statistical comparisons of women and men with PTSD, women were more often diagnosed with comorbid panic disorder than men. Panic disorder has been hypothesized to develop with PTSD through a classical conditioning process (Falsetti, Resnick, & Davis, 2005). Unconditioned fear responses to the trauma pair with internal, trauma-related cues, leading the individual to develop fear and avoidance of the physical sensations associated with the fear response. Whether this process differs for men and women, and is possibly related to women's greater peritraumatic emotional response, would be an important area of further inquiry.

Treatment of PTSD

Effective treatment for PTSD must address the unique aspects of men's and women's social context, stressor exposure, and presentation and course of the disorder. Several in-depth reviews of the treatment literature reveal a complex picture of the relationships among gender, PTSD, and its treatment (Cason, Grubaugh, & Resick, 2002; Foa, Keane, & Friedman, 2000). Moreover, taking a broader perspective on the intersection of gender, PTSD, and its effective treatment, one might hypothesize that there are gender issues related to access to services, utilization, and the institutions and means by which men and women receive PTSD-related care. Understanding gender issues in the treatment literature is another step toward more comprehensive conceptualization and treatment for this disorder, including the successful translation of research findings into effective practice.

Psychosocial Treatments

Analysis of gender issues in the effectiveness of treatments for PTSD is complicated by gender differences in types of trauma exposure. For example, clinical trials for the effective treatment of rape and child sexual abuse have been conducted on exclusively female populations (Cloitre, Koenen, Cohen, & Han, 2002; Foa, Dancu, Hembree, Jaycox, & Street, 1999; Resick, Nishith, Weaver, Astin, & Feuer, 2002). Clinical trials of combat-related PTSD have utilized samples of men almost exclusively (Rothbaum, Hodges, Ready, Graap, & Alarcon, 2001; Schnurr et al., 2003). The field is only beginning to investigate men's experience with validated treatments for assault-related PTSD, and we have not yet begun to study the treatment of combat exposure among women, though troops returning from recent conflicts will yield ample opportunity.

Cason and colleagues (2002), who conducted a meta-analysis of gender-specific effect sizes of treatment response across a number of studies comparing commonly utilized treatments for PTSD, found substantial evidence for a trend towards a superior treatment response in women compared to men. For example, in a study by Tarrier, Sommerfield, Pilgrim, and Humphreys (1999) comparing efficacy of imaginal exposure protocol to cognitive therapy, treatment effect sizes were larger for women than for men. Among literature of single-sex studies, average effect size for treatment response is significantly higher in women (1.39) than in men (0.40), maintained through follow-up, where average effect size for women is (0.79) and for men is (0.65). When a wider range of outcomes was included in the analyses (i.e., symptoms of depression), the differences between men and women's responses at the follow-up interval were even more pronounced: men (0.48) and women (0.73) (Cason et al., 2002).

Conclusions about the relative effectiveness of behavioral PTSD treatment for men and for women are complicated by the difficulty of discriminating relationships among gender, trauma type, and treatment outcomes. For example, whereas it may be true that women are generally more responsive than men to treatment, factors such as baseline symptom severity and trauma type (both of which are linked with gender) cannot yet be teased apart from the more general gender-based differences. A more general meta-analysis of psychotherapies for PTSD (Bradley, Greene, Russ, Dutra, & Westen, 2005) has addressed these issues. In the analysis, treatment effect sizes significantly differed as a function of trauma type, in which the lowest effect sizes were observed in studies of combat trauma, and those associated with sexual or physical assault were considerably higher. The authors noted that the studies of combat trauma included patients with more chronic PTSD or with greater severity at baseline, which may have accounted for the lower effect sizes. These findings indicated that the apparent relationship between gender and treatment effectiveness may in fact be mediated or moderated by factors such as trauma type, chronicity, or other gender-related differences.

Several efficacy trials have included samples of both men and women, providing the foundation for gender-informed investigations. For example, Blanchard and colleagues (2003) compared cognitive-behavioral therapy (CBT) to supportive psychotherapy in a study of 98 survivors of motor vehicle accidents. The authors compared 26 men and 72 women in the study by examining both main effects for treatment outcome and effects for gender × group × time, but did not find statistically significant results. Limitations in sample size may have occluded the detection of gender differences in response to treatment. Often, when gender is not a primary hypothesis, it can be difficult to obtain the statistical power required for the two- and three-way interactions necessary to detect these effects.

Other studies that were not powered for a three-way gender × treatment × time interaction have still accounted for gender in the design of their clinical trials. Studies comparing traditional cognitive therapies with imaginal exposure or repeated assessment paradigms on crime or accident victims have included relatively equal numbers of men and women, and used stratified random sampling to account for gender (Ehlers et al., 2003; Tarrier et al., 1999). This technique results in a minimum of potential gender bias. A small, randomized clinical trial comparing the impact of eye movement desensitization and reprocessing (EMDR) to exposure plus cognitive restructuring paradigms (Lee, Gavriel, Drummond, Richards, & Greenwald, 2002; Power et al., 2002), and another comparing EMDR to stress innoculation training with prolonged exposure (Lee et al., 2002) have shown the potential utility of EMDR as a comparable treatment for trauma. Although these studies also stratified their samples by gender, thereby removing gender as a potential confound, and used samples with mixed trauma types, only a very large study would be able to detect difference among trauma types or between genders.

Given the current literature, we draw a general conclusion that many cognitive-behavioral, exposure-based, and well-executed EMDR protocols are all efficacious treatments for PTSD (Bradley et al., 2005; Foa et al., 2000). However, in the many instances in which these validated or promising treatments have been applied to specific types of trauma (i.e., sexual assault) but validated only in a single gender, we can only estimate their potential utility with the opposite gender. Researchers need to conduct parallel, randomized trials that replicate the existing findings for the opposite gender. Although there are inherent difficulties in obtaining sufficiently large samples to conduct randomized trials with, for example, male rape victims or female combat veterans, even smaller evaluation studies could lend confidence to generalizing treatment methods developed with opposite-gender samples.

Translational Science and PTSD Treatment

As empirically supported treatments for PTSD are implemented across treatment settings and patient populations, research on treatment effectiveness and dissemination becomes increasingly important. Data-driven guidelines that describe the circumstances under which a given treatment works, or that identify subgroups of patients for whom treatment may be more or less effective, translate research into practice. Gender-informed approaches would address gender as a moderator of treatment outcome, examine settings in which gender is relevant to treatment outcomes, and identify approaches or techniques that may particularly benefit men and women. Relatedly, many sociodemographic and extradiagnostic factors (i.e., compensation, seeking status, location of treatment) that may be linked with both gender and treatment outcome could contribute to a gender-informed understanding of the treatment literature when elucidated in treatment studies.

Generalization of a treatment's applicability across genders often involves gender-sensitive adaptations of the therapeutic techniques, as well as restructuring around different types of traumatic material. For example, recent research has found that cognitive processing therapy (CPT), a treatment originally developed with female assault-related trauma (Resick et al., 2002), is also effective in the treatment of males veterans with chronic, combat-related PTSD (Monson et al., 2006). A randomized clinical trial that used a wait-list control found similar effect sizes for this treatment compared to those in civilian samples. However, the authors noted a number of salient gender issues in the implementation of the treatment (Monson, Price, & Ranslow, 2005). Males were more likely to hold negative beliefs about emotional expression that could interfere with the cognitive and emotional processing that forms the core of this treatment, unless these beliefs were specifically targeted by the therapist. The authors also noted gender-related advantages of the treatment: The Socratic techniques of the treatment were especially helpful in addressing the cognitive dissonance surrounding symptom improvement and compensation seeking. The treatment was also especially helpful in addressing perpetration of violence, an important issue in the resolution of combat-related PTSD in male patients. Such research highlights the importance of elucidating these translational issues that help "fit" empirically supported treatments to gender, trauma type, and treatment setting.

The relationship of gender and trauma type to the "typical" treatment settings is an important feature of understanding gender issues in PTSD treatment. Women are more likely to seek treatment following a traumatic event (Gavrilovic, Schutzwohl, Fazel, & Priebe, 2005), but the typical locations where men and women seek treatment may have different staffing patterns, characteristics, and types of clinicians, all of which could impact treatment effectiveness. For example, among veterans with PTSD, women are more likely than men to obtain mental health care from non-VA sources (Suffoletta-Maierle, Grubaugh, Magruder, Monnier, & Freuh, 2003). These settings may differ from VA settings in the types of PTSD treatments employed, their familiarity with military trauma, or quality of care. Conversely, in the civilian sector, women may be more likely than men to obtain mental health treatment in primary care (Sherbourne, Weiss, Duan, Bird, & Wells, 2004), which may increase women's access to mental health treatments. More research is certainly needed to understand treatment setting as a contextual factor in gender issues in PTSD treatment.

Health services research would help to identify gender-linked barriers to care and potential gender-specific preferences for treatments and outcomes. For example, women are more likely than men to use specialty mental health services (Freiman & Zuvekas, 2000). The factors associated with access to care differ by gender (Albizu-Garcia, Alegria, Freeman, & Vera, 2001), and women may be especially vulnerable to social context factors linked to gender-related disparities in access to care, such as low educational attainment or poverty (Sherbourne, Dwight-Johnson, & Klap, 2001). However, we do not know whether men or women are more likely to be treated for PTSD in specialty mental health compared to primary care settings, or any gender-related factors that influence the likelihood of a patient being treated in a given setting.

In summary, gender is emerging as an important but complex construct in the development, etiology, expression, and treatment of PTSD. Though the extant literature suggests emerging differences in the treatment of PTSD among men and women, available data are substantially limited with respect to definitive assessment of the impact of gender on core aspects of treatment response and outcome. The development of testable, gender-informed conceptual models and hypotheses regarding the impact of gender on treatment outcomes in PTSD is an important next step, as are clinical trials that formally assess the potential role of gender as a moderator of treatment efficacy. An additional step would be to replicate validated, single-gender efficacy trials (i.e., rape trauma treatments) for both genders. Thus, rather than remaining focused on "big picture" questions (i.e., how to describe gender differences in PTSD), we would be addressing molecular questions, such as when, where, and how gender impacts PTSD and its treatment.

CHALLENGES FOR THE FUTURE

• *There are plausible explanations for the apparent paradox between the greater prevalence of trauma exposure among men and the greater prevalence of PTSD among women.* The extant data speak to the importance of how trauma exposure is conceptualized. Whereas results are generally consistent across samples, studies, and cultures, the studies reviewed here emphasize the importance of qualitative aspects of traumatic events. Most notably, men and women report similar rates of events that yield reactions that meet DSM-IV criteria of fear, horror, or helplessness. Although these data may indicate a greater vulnerability for women to traumatic stress, they might also indicate a need for more sensitive measurement of exposure to inform our current conceptualizations of the impact of traumatic events and risk of PTSD. The broad categories used in most event checklists may equate experiences that are not necessarily similar, or that result in false-positive cases of exposure. For example, single incidents of physical assaults or minor physical altercations can be equated with chronic forms of physical abuse, such as intimate partner violence. Gender-informed models of how trauma characteristics impact risk for PTSD are the natural extension of this line of inquiry. Additional research on how to parse the severity of trauma exposure can also help to elucidate important aspects of gender differences in trauma and PTSD. Finally, research is also needed that explores gender differences in the experience of similar events, such as men's greater likelihood of experiencing sexual assault by a same-sex perpetrator, or the greater severity of violence that women experience in violent intimate relationships.

• *Gender differences in trauma exposure and PTSD have implications for the assessment process with men and women.* Greater detail concerning these qualitative aspects of traumatic evens should be included routinely in both clinical and research assessments. Although most epidemiological studies assess PTSD with respect to a single participant-selected or randomly selected event, an emphasis on PTSD as it pertains to multiple traumatic experiences in needed, because the majority of men and women report more than one event. Such research could particularly benefit our understanding of PTSD among men, and possible underestimation of this prevalence, because men consistently report more traumatic events. Gender-informed models of traumatic stress reactions are an important direction of research, especially with respect to childhood or chronic trauma. Few studies have addressed the potential differences in the PTSD construct among men compared to women, or among individuals exposed to trauma in childhood compared to later in life. Using epidemiological samples for such research would allow adequate sample size to construct separate models by gender. However, clinical studies would also be informative, especially with respect to avoiding potential "epidemiological misdiagnosis," in which chronic or elaborated forms of PTSD may appear as overdiagnoses of comorbid Axis I and II diagnoses.

- *Additional research on gender differences in comorbid diagnoses could inform models of the vulnerability to mental health conditions and the course of PTSD among men and women.* In the general population, women's rates of depression are higher than those observed among men, but among individuals with PTSD, men's rates of depression are more similar to those of women. Thus, it appears that rather than increasing women's risk, PTSD attenuates a protective effect associated with male gender for both depression and a chronic course of PTSD. Much of what we know about comorbid diagnoses and PTSD is based on lifetime rates of these disorders, due to the methodological difficulties in obtaining sample sizes sufficient to examine subgroups of patients with PTSD and comparing these subgroups on the basis of patient characteristics such as gender. However, we can learn much from data that compare base rates of disorders in a given population to those observed among individuals with PTSD. Clinical studies that investigate the functional relationship between PTSD and other disorders, such as those conducted with substance abuse populations, might have important treatment implications for other, frequently observed comorbid disorders, such as panic and depression.

- *Several efficacious treatments for PTSD exist, though little is known about whether the effectiveness of these treatments is moderated by patient gender, implementation factors that could enhance generalization of treatments across gender, or how these treatments impact "real-world" treatment for PTSD.* Despite 10 years of federal regulation requiring the inclusion of women and minorities in clinical research, only recently have these regulations been amended to address the conduct and reporting of specific analyses to determine effects of gender. Few studies of PTSD treatment are designed to examine main effects of gender, or gender as a moderator of treatment outcome. Because treatments are often developed around specific trauma types that vary by gender, many empirically supported treatments have been conducted on samples of women only or men only. Extant reviews and meta-analyses suggest that treatments targeting women may be somewhat more effective than those targeting men; however, research is only beginning to replicate treatment trials to generalize across gender. A greater emphasis on effectiveness research could identify targets for dissemination of these treatments. Similarly, health services research that addresses access to and preferences for treatment of PTSD in "real-world" settings would indicate how these efficacious treatments are utilized by men and women.

- *Investigations of gender issues in PTSD can improve assessment and treatment of both men and women exposed to traumatic stress.* Investigations of gender issues yield a more comprehensive understanding of trauma severity, the range of traumatic stress reactions, and the influence of social context on the response to trauma. For example, research on sexual harassment and sexual assault as a component of war-zone trauma emerged from research on women veterans and PTSD. Recognizing these factors as important aspects of both men's and women's military trauma might result in better treatment for both men and women with PTSD.

Research on gender issues in trauma exposure has illuminated the need for research on the role of multiple traumas with men, including the dearth of research on traumas that occur less frequently among men, such as sexual assault and child trauma. The function of social roles and social context in explaining women's rates of PTSD emphasizes the important potential for community-based interventions focusing on social resources for natural disasters and mass trauma. As research on gender continues to develop beyond sex-based comparisons and incorporates the qualitative aspects of trauma, context in which trauma occurs, and the social roles and experiences that influence the risk for and outcome of traumatic stressors, these results enhance our capacity to address PTSD as it occurs in individual patients and in communities.

ACKNOWLEDGMENT

This work was supported in part by the Office of Academic Affiliations, Department of Veterans Affairs Special Mental Illness Research, Education and Clinical Center Fellowship Program in Advanced Psychiatry and Psychology.

REFERENCES

Albizu-Garcia, C. E., Alegria, M., Freeman, D., & Vera, M. (2001). Gender and health services use for a mental health problem. *Social Science and Medicine, 53,* 865–878.

Andrews, B., Brewin, C. R., & Rose, S. (2003). Gender, social support, and PTSD in victims of violent crime. *Journal of Traumatic Stress, 16,* 421–427.

Baker, C. K., Norris, F. H., Diaz, D. M., Perilla, J. L., Murphy, A. D., & Hill, E. G. (2005). Violence and PTSD in Mexico: Gender and regional differences. *Journal of Social Psychiatry and Psychiatric Epidemiology, 40,* 519–512.

Battle, C. L., Shea, M. T., Johnson, D. M., Yen, S., Zlotnick, C., Zanarini, M. C., et al. (2004). Childhood maltreatment associated with adult personality disorders: Findings from the collaborative longitudinal personality disorders study. *Journal of Personality Disorders, 18,* 193–211.

Bernstein, D. P., Fink, L. A., Handelsman, L., Foote, J., Lovejoy, M., Wenzel, K., et al. (1994). Initial reliability and validity of a new retrospective measure of child abuse and neglect. *American Journal of Psychiatry, 151,* 1132–1136.

Blanchard, E. B., Hickling, E. J., Devineni, T., Veazey, C. H., Galovski, T. E., Mundy, E., et al. (2003). A controlled evaluation of cognitive behavioral therapy for posttraumatic stress in motor vehicle accident survivors. *Behaviour Research and Therapy, 41,* 79–96.

Bradley, R., Greene, J., Russ, E., Dutra, L., & Westen, D. (2005). A multidimensional meta-analysis of psychotherapy for PTSD. *American Journal of Psychiatry, 162*(2), 214–227.

Breslau, N. (2001). The epidemiology of posttraumatic stress disorder: What is the extent of the problem? *Journal of Clinical Psychiatry, 62*(Suppl. 17), 16–22.

Breslau, N., Chilcoat, H. D., Kessler, R. C., Peterson, E. L., & Lucia, V. C. (1999). Vulnerability to assaultive violence: further specification of the sex difference in post-traumatic stress disorder. *Psychological Medicine, 29,* 813–821.

Breslau, N., & Davis, G. C. (1992). Posttraumatic stress disorder in an urban population of young adults: Risk factors for chronicity. *American Journal of Psychiatry, 149,* 671–675.

Breslau, N., Davis, G. C., Andreski, P., & Peterson, E. (1991). Traumatic events and posttraumatic stress disorder in an urban population of young adults. *Archives of General Psychiatry, 48,* 216–222.

Breslau, N., Davis, G. C., Andreski, P., & Peterson, E. L. (1997). Sex differences in posttraumatic stress disorder. *Archives of General Psychiatry, 54,* 1044–1048.

Breslau, N., Davis, G. C., Peterson, E. L., & Schultz, L. (1997). Psychiatric sequelae of posttraumatic stress disorder in women. *Archives of General Psychiatry, 54,* 81–87.

Breslau, N., & Kessler, R. C. (2001). The stressor criterion in DSM-IV posttraumatic stress disorder: An empirical investigation. *Biological Psychiatry, 50,* 699–704.

Breslau, N., Kessler, R. C., Chilcoat, H. D., Schultz, L. R., Davis, G. C., & Andreski, P. (1998). Trauma and posttraumatic stress disorder in the community: The 1996 Detroit Area Survey of Trauma. *Archives of General Psychiatry, 55,* 626–632.

Brewin, C. R., Andrews, B., & Rose, S. (2000). Fear, helplessness, and horror in posttraumatic stress disorder: Investigating DSM-IV criterion A2 in victims of violent crime. *Journal of Traumatic Stress, 13,* 499–509.

Brewin, C. R., Andrews, B., & Valentine, J. D. (2000). Meta-analysis of risk factors for posttraumatic stress disorder in trauma-exposed adults. *Journal of Consulting and Clinical Psychology, 68,* 748–766.

Bryant, R. A., & Harvey, A. G. (2003). Gender differences in the relationship between acute stress disorder and posttraumatic stress disorder following motor vehicle accidents. *Australian and New Zealand Journal of Psychiatry, 37,* 226–229.

Cason, D., Grubaugh, A., & Resick, P. (2002). Gender and PTSD treatment: Efficacy and effectiveness. In R. Kimerling, P. Ouimette, & J. Wolfe (Eds.), *Gender and PTSD* (pp. 305–334). New York: Guilford Press.

Cloitre, M., Koenen, K. C., Cohen, L. R., & Han, H. (2002). Skills training in affective and interpersonal regulation followed by exposure: A phase-based treatment for PTSD related to childhood abuse. *Journal of Consulting and Clinical Psychology, 70,* 1067–1074.

Cloitre, M., Koenen, K. C., Gratz, K. L., & Jakupcak, M. (2002). Differential diagnosis of PTSD in women. In R. Kimerling, P. Ouimette, & J. Wolfe (Eds.), *Gender and PTSD* (pp. 117–149). New York: Guilford Press.

Curtis, T., Miller, B. C., & Berry, E. H. (2000). Changes in reports and incidence of child abuse following natural disasters. *Child Abuse and Neglect, 24,* 1151–1162.

DeBellis, M. D., & Keshavan, M. S. (2003). Sex differences in brain maturation in maltreatment-related pediatric posttraumatic stress disorder. *Neuroscience and Biobehavioral Reviews, 27,* 103–117.

Deering, C. G., Glover, S. G., Ready, D., Eddleman, H. C., & Alarcon, R. D. (1996). Unique patterns of comorbidity in posttraumatic stress disorder from different sources of trauma. *Comparative Psychiatry, 37*(5), 336–346.

DePrince, A. P., & Freyd, J. J. (2002). The intersection of gender and betrayal in trauma. In R. Kimerling, P. Ouimette, & J. Wolfe (Eds.), *Gender and PTSD* (pp. 98–113). New York: Guilford Press.

Ehlers, A., Clark, D. M., Hackmann, A., McManus, F., Fennell, M., Herbert, C., et al. (2003). A randomized controlled trial of cognitive therapy, a self-help booklet, and repeated assessments as early interventions for posttraumatic stress disorder. *Archives of General Psychiatry, 60,* 1024–1032.

Elliott, D. M., Mok, D. S., & Briere, J. (2004). Adult sexual assault: Prevalence, symptomatology, and sex differences in the general population. *Journal of Traumatic Stress, 17*(3), 203–211.

Falsetti, S. A., Resnick, H. S., & Davis, J. (2005). Multiple channel exposure therapy: Combining cognitive-behavioral therapies for the treatment of posttraumatic stress disorder with panic attacks. *Behavior Modification, 29,* 70–94.

Finkelhor, D., Ormond, R., Turner, H., & Hamby, S. L. (2005). The victimization of children and youth: A comprehensive national survey. *Child Maltreatment, 10,* 5–25.

Foa, E. B., Dancu, C. V., Hembree, E. A., Jaycox, L. H., & Street, G. P. (1999). A comparison of exposure therapy, stress inoculation training, and their combination for reducing posttraumatic stress disorder in female assault victims. *Journal of Consulting and Clinical Psychology, 67,* 194–200.

Foa, E. B., Keane, T. M., & Friedman, M. J. (Eds.). (2000). *Effective treatments of PTSD: Practice guidelines for the International Society for Traumatic Stress Studies.* New York: Guilford Press.

Frazier, P. A. (1993). A comparative study of male and female rape victims seen at a hospital-based rape crisis program. *Journal of Interpersonal Violence, 8*(1), 64–76.

Freedman, S. A., Brandes, D., Peri, T., & Shalev, A. (1999). Predictors of chronic post-traumatic stress disorder: A prospective study. *British Journal of Psychiatry, 174,* 353–359.

Freiman, M. P., & Zuvekas, S. H. (2000). Determinants of ambulatory treatment mode for mental illness. *Health Economics, 9*, 423–434.

Freyd, J. J. (1996). *Betrayal trauma: The logic of forgetting childhood abuse.* Cambridge, MA: Harvard University Press.

Fullerton, C. S., Ursano, R. J., Epstein, R. S., Crowley, B., Vance, K., Kao, T. C., et al. (2001). Gender differences in posttraumatic stress disorder after motor vehicle accidents. *American Journal of Psychiatry, 158*, 1486–1491.

Gavrilovic, J. J., Schutzwohl, M., Fazel, M., & Priebe, S. (2005). Who seeks treatment after a traumatic event and who does not?: A review of findings on mental health service utilization. *Journal of Traumatic Stress, 18*, 595–605.

Helzer, J. E., Robins, L. N., & McEvoy, L. (1987). Post-traumatic stress disorder in the general population: Findings of the epidemiologic catchment area survey. *New England Journal of Medicine, 317*, 1630–1634.

Herman, J. L. (1992). *Trauma and recovery.* New York: Basic Books.

Johnson, J. G., Cohen, P., Brown, J., Smailes, E. M., & Bernstein, D. P. (1999). Childhood maltreatment increases risk for personality disorders during early adulthood. *Archives of General Psychiatry, 56*, 600–606.

Kessler, R. C., Sonnega, A., Bromet, E., Hughes, M., & Nelson, C. B. (1995). Posttraumatic stress disorder in the National Comorbidity Survey. *Archives of General Psychiatry, 52*, 1048–1060.

Kilpatrick, D. G., Saunders, B. E., Amick-McMullan, A., Best, C. L., Veronen, L. J., & Resnick, H. S. (1989). Victim and crime factors associated with the development of crime-related post-traumatic stress disorder. *Behavior Therapy, 20*, 199–214.

Kimerling, R. (2004). An investigation of gender differences in non-psychiatric morbidity associated with posttraumatic stress disorder. *Journal of the American Medical Womens Association, 59*, 43–47.

Kimerling, R., Prins, A., Westrup, D., & Lee, T. (2004). Gender issues in the assessment of PTSD. In J. P. Wilson & T. M. Keane (Eds.), *Assessing psychological trauma and PTSD* (2nd ed., pp. 565–600). New York: Guilford Press.

Kimerling, R., Rellini, A., Kelly, V., Judson, P. L., & Learman, L. A. (2002). Gender differences in victim and crime characteristics of sexual assaults. *Journal of Interpersonal Violence, 17*, 526–532.

King, L. A., Orcutt, H. K., & King, D. W. (2002). Gender differences in stress, trauma, and PTSD research: Application of two quantitative methods. In R. Kimerling, P. Ouimette, & J. Wolfe (Eds.), *Gender and PTSD* (pp. 403–433). New York: Guilford Press.

Koss, M. P. (1985). The hidden rape victim: Personality, attitudinal, and situational characteristics. *Psychology of Women Quarterly, 9*, 193–212.

Kulka, R. A., Schlenger, W. E., Fairbank, J. A., Hough, R. L., Jordan, B. K., Marmar, C. R., et al. (1988). The prevalence of other psychiatric disorders and nonspecific distress. In *Contractual report of findings from the National Vietnam Veterans Readjustment Study: Vol. I. Executive summary, description of findings, and technical appendices* (Vol. 6, pp. 1–47): Research Triangle Park, NC: Research Triangle Institute.

Kulka, R. A., Schlenger, W. E., Fairbank, J. A., Hough, R. L., Jordan, B. K., Marmar, C. R., et al. (1990). *Trauma and the Vietnam War generation: Report of findings from the National Vietnam Veterans Readjustment Study.* New York: Brunner/Mazel.

Lee, C. W., Gavriel, H., Drummond, P. D., Richards, J., & Greenwald, R. (2002). Treatment of PTSD: Stress inoculation training with prolonged exposure compared to EMDR. *Journal of Clinical Psychology, 58*, 1071–1089.

Mischel, W. (2004). Towards an integrative science of the person. *Annual Review of Psychology, 55*, 1–22.

Monson, C. M., Price, J. L., & Ranslow, E. (2005, October). Lessons learned in researching an evidence-based PTSD treatment in the VA: The case of cognitive processing therapy. *Federal Practitioner*, pp. 75–83.

Monson, C. M., Schnurr, P. P., Resick, P. A., Friedman, M. J., Young-Xu, Y., & Stevens, S. P. (2006). Cognitive processing therapy for veterans with military-related posttraumatic stress disorder. *Journal of Consulting and Clinical Psychology, 74*, 898–907.

Moos, R. H. (2003). Social contexts: Transcending their power and their fragility. *American Journal of Community Psychology, 31*, 1–13.

Nolen-Hoeksema, S., & Girgus, J. S. (1994). The emergence of gender differences in depression during adolescence. *Psychological Bulletin, 115*, 424–443.

Norris, F. H. (1992). Epidemiology of trauma: Frequency and impact of different potentially traumatic events on different demographic groups. *Journal of Consulting and Clinical Psychology, 60,* 409–418.

Norris, F. H., Foster, J. D., & Weissharr, D. L. (2002). The epidemiology of sex differences in PTSD across developmental, societal, and research contexts. In R. Kimerling, P. Ouimette, & J. Wolfe (Eds.), *Gender and PTSD* (pp. 3–42). New York: Guilford Press.

Norris, F. H., Friedman, M. J., Watson, P. J., Byrne, C. M., Diaz, E., & Kaniasty, K. (2002). 60,000 disaster victims speak: Part I. An empirical review of the empirical literature, 1981–2001. *Psychiatry, 65,* 207–239.

Norris, F. H., Perilla, J. L., Ibanez, G. E., & Murphy, A. E. (2001). Sex differences in symptoms of posttraumatic stress disorder: Does culture play a role? *Journal of Traumatic Stress, 14,* 7–28.

Orsillo, S. M., Raja, S., & Hammond, C. (2002). Gender issues in PTSD with comorbid mental health disorders. In R. Kimerling, P. Ouimette, & J. Wolfe (Eds.), *Gender and PTSD* (pp. 207–231). New York: Guilford Press.

Orsillo, S. M., Weathers, F. W., Litz, B. T., Steinberg, H. R., Huska, J. A., & Keane, T. M. (1996). Current and lifetime psychiatric disorders among veterans with war zone–related posttraumatic stress disorder. *Journal of Nervous and Mental Disease, 184,* 307–313.

Ozer, E. J., Best, S. R., Lipsey, T. L., & Weiss, D. S. (2003). Predictors of posttraumatic stress disorder and symptoms in adults: A meta-analysis. *Psychological Bulletin, 129,* 52–73.

Perkonigg, A., Kessler, R. C., Storz, S., & Wittchen, H. U. (2000). Traumatic events and post-traumatic stress disorder in the community: Prevalence, risk factors and comorbidity. *Acta Psychiatrica Scandinavia, 101,* 46–59.

Pimlott-Kubiak, S., & Cortina, L. M. (2003). Gender, victimization, and outcomes: Reconceptualizing risk. *Journal of Consulting and Clinical Psychology, 71,* 528–539.

Power, K., McGoldrick, T., Brown, K. W., Buchanan, R., Sharp, D., Swanson, V., et al. (2002). A controlled comparison of eye movement desensitization and reprocessing versus exposure plus cognitive restructuring versus waiting list in the treatment of post-traumatic stress disorder. *Clinical Psychology and Psychotherapy, 9,* 299–318.

Resick, P. A., Nishith, P., Weaver, T. L., Astin, M. C., & Feuer, C. A. (2002). A comparison of cognitive-processing therapy with prolonged exposure and a waiting condition for the treatment of chronic posttraumatic stress disorder in female rape victims. *Journal of Consulting and Clinical Psychology, 70,* 867–879.

Ritter, C., Hobfoll, S. E., Lavin, J., Cameron, R. P., & Hulsizer, M. R. (2000). Stress, psychosocial resources, and depressive symptomatology during pregnancy in low-income, inner-city women. *Health Psychology, 19,* 576–585.

Rothbaum, B. O., Hodges, L. F., Ready, D., Graap, K., & Alarcon, R. D. (2001). Virtual reality exposure therapy for Vietnam veterans with posttraumatic stress disorder. *Journal of Clinical Psychiatry, 62,* 617–622.

Sachs-Ericsson, N., Blazer, D., Plant, E. A., & Arnow, B. (2005). Childhood sexual and physical abuse and the 1-year prevalence of medical problems in the National Comorbidity Survey. *Health Psychology, 24,* 32–40.

Schnurr, P. P., Friedman, M. J., Foy, D. W., Shea, M. T., Hsieh, F. Y., Lavori, P. W., et al. (2003). Randomized trial of trauma-focused group therapy for posttraumatic stress disorder: Results from a Department of Veterans Affairs cooperative study. *Archives of General Psychiatry, 60,* 481–489.

Shalev, A. Y., Freedman, S., Peri, T., Brandes, D., Sahar, T., Orr, S. P., et al. (1998). Prospective study of posttraumatic stress disorder and depression following trauma. *American Journal of Psychiatry, 155,* 630–637.

Sharkansky, E. J., Brief, D. J., Peirce, J. M., Meehan, J. C., & Mannix, L. M. (1999). Substance abuse patients with posttraumatic stress disorder (PTSD): Identifying specific triggers of substance use and their associations with PTSD symptoms. *Psychology of Addictive Behaviors, 13,* 89–97.

Sherbourne, C. D., Dwight-Johnson, M., & Klap, R. (2001). Psychological distress, unmet need, and barriers to mental health care for women. *Women's Health Issues, 11,* 231–243.

Sherbourne, C. D., Weiss, R., Duan, N., Bird, C. E., & Wells, K. B. (2004). Do the effects of quality improvement for depression care differ for men and women?: Results of a group-level randomized controlled trial. *Medical Care, 42,* 1186–1193.

Spataro, J., Mullen, P. E., Burgess, P. M., Wells, D. L., & Moss, S. A. (2004). Impact of child sexual abuse on mental health: Prospective study in males and females. *British Journal of Psychiatry, 184,* 416–421.

Stewart, S. H., Ouimette, P., & Brown, P. J. (2002). Gender and the comorbidity of PTSD with substance use disorders. In R. Kimerling, P. Ouimette, & J. Wolfe (Eds.), *Gender and PTSD* (pp. 232–270). New York: Guilford Press.

Suffoletta-Maierle, S., Grubaugh, A. L., Magruder, K., Monnier, J., & Freuh, B. C. (2003). Trauma-related mental health needs and service utilization among female veterans. *Journal of Psychiatric Practice, 9,* 367–375.

Taft, C. T., King, L. A., King, D. W., Leskin, G. A., & Riggs, D. S. (1999). Partners' ratings of combat veterans' PTSD symptomatology. *Journal of Traumatic Stress, 12,* 327–334.

Tarrier, N., Sommerfield, C., Pilgrim, H., & Humphreys, L. (1999). Cognitive therapy or imaginal exposure in the treatment of post-traumatic stress disorder: Twelve-month follow-up. *British Journal of Psychiatry, 175,* 571–575.

Tjaden, P. G., & Thoennes, N. (2000). *Full report of the prevalence, incidence, and consequences of violence against women: Findings from the National Violence Against Women Survey.* Washington, DC: U.S. Department of Justice, Office of Justice Programs, National Institute of Justice.

Ullman, S. E., & Filipas, H. H. (2005). Gender differences in social reactions to abuse disclosures, post-abuse coping, and PTSD of child sexual abuse survivors. *Child Abuse and Neglect, 29,* 767–782.

van der Kolk, B. A., Pelcovitz, D., Roth, S., Mandel, F. S., McFarlane, A., & Herman, J. L. (1996). Dissociation, somatization, and affect dysregulation: The complexity of adaptation to trauma. *American Journal of Psychiatry, 153*(Suppl. 7), 83–93.

Vogt, D. S., Pless, A. P., King, L. A., & King, D. W. (2005). Deployment stressors, gender, and mental health outcomes among Gulf War I veterans. *Journal of Traumatic Stress, 18,* 115–127.

Wolfe, J., Sharkansky, E. J., Read, J. P., Dawson, R., Martin, J. A., & Ouimette, P. C. (1998). Sexual harassment and assault as predictors of PTSD symptomatology among U.S. female Persian Gulf War military personnel. *Journal of Interpersonal Violence, 13,* 40–57.

Wright, K. D., Asmundson, G. J. G., McCreary, D. R., Scher, C., Hami, S., & Stein, M. B. (2001). Factorial validity of the Childhood Trauma Questionnaire in men and women. *Depression and Anxiety, 13,* 179–183.

Yoder, J. D., & Kahn, A. S. (2003). Making gender comparisons more meaningful: A call for more attention to social context. *Psychology of Women Quarterly, 27,* 281–290.

Zanarini, M. C., Frankenburg, F. R., Hennen, J., & Silk, K. R. (2003). The longitudinal course of borderline psychopathology: 6-year prospective follow-up of the phenomenology of borderline personality disorder. *American Journal of Psychiatry, 160,* 274–283.

Chapter 13

The Prevalence and Impact of Child Traumatic Stress

John A. Fairbank, Frank W. Putnam, and William W. Harris

METHODOLOGICAL CONSIDERATIONS

The field of child traumatic stress epidemiology faces daunting problems that present numerous challenges. Precisely estimating the prevalence of children's exposure to traumatic events remains an extremely challenging task. Reporting requirements of abuse and neglect carry appropriately stringent confidentiality concerns. Too often, a survivor of maltreatment is also a victim/witness to other forms of violence and trauma, and if counted, may be counted more than once. State agency computers often are incapable of communicating with one another and also have strict privacy prohibitions. Collecting separate data on trauma exposure from schools, public assistance, and public mental health is extremely difficult for agency and privacy reasons, and all too often impossible to merge because of technical, legal, and administrative requirements. Child maltreatment data are sometimes purposively not collected because of mandatory reporting requirements.

Survivors, clinicians, service providers, and researchers are acutely aware of the adverse impact of childhood trauma. Yet we have been unable to formulate an encompassing narrative that includes the epidemiological and clinical science in a sufficiently compelling way to communicate the scope of the problem to policymakers and the voting public (Harris, Lieberman, & Marans, in press). Without this overarching narrative, we contend, the public resources needed to address the prevalence and the opportunities presented will not become available. Environmental realities, such as long-term, pervasive poverty and teenage pregnancy, have long been implicated as high-risk factors in childhood trauma (e.g., Sameroff, 1998), with effects that often bring forth interconnected and frequently volatile issues of race, class, gender, culture, and inequality that are difficult to address and to incorporate in public discussions. In this chapter, we hope to provide a beginning framework for this narrative, with some specific recommendations that we hope others will critique, revise, expand, and, finally, communicate and disseminate widely to many audiences for comments and suggestions.

Types of Trauma

Children's exposure to trauma is a major public health problem with devastating and costly effects on individual children, families, and communities throughout the world (Harris, Putnam, & Fairbank, 2006; Osofsky, 1999). Children may be exposed to a range of traumatic events, including child maltreatment, domestic violence, community and school violence, traumatic loss, medical trauma, war-zone and refugee trauma, natural disasters, and terrorism.

Child Maltreatment

"Child maltreatment" is a generic term referring to physical abuse, neglect, emotional abuse, and sexual abuse (Wolf & Nayak, 2003). Physical abuse includes causing physical pain or injury; neglect describes failing to give a child the care needed according to his or her age and development (National Child Traumatic Stress Network, n.d.). Child sexual abuse includes a wide range of sexual behaviors between a child and an older person, often involving bodily contact. However, behaviors may be sexually abusive even if they do not involve contact, such as genital exposure, verbal sexual harassment, and exploitation for pornography (National Child Traumatic Stress Network, n.d.). The World Health Organization (WHO; 1999) describes "emotional abuse" as acts that have a high probability of impairing a child's health or physical, mental, spiritual, moral, or social development, such as restriction of movement, patterns of belittling, denigrating, scapegoating, threatening, scaring, discriminating, ridiculing, or other nonphysical forms of hostile or rejecting treatment.

The Child Maltreatment Report from the National Child Abuse and Neglect Data Systems (NCANDS) estimates that 906,000 children were confirmed victims of child abuse and neglect in 2003 (U.S. Department of Health and Human Services, 2005). The victimization rate was 12.4 per 1,000 children, with many children suffering multiple forms of abuse and neglect. Actual prevalence is likely higher: General population surveys typically yield rates two- to threefold higher than official child abuse reports (Edleson, 1999).

Domestic Violence

Domestic violence is often called "intimate partner violence," "domestic abuse," or "battering" and includes actual or threatened physical or sexual violence, or emotional abuse between adults in a child's home environment (National Child Traumatic Stress Network, n.d.). Although the terms "witnessing" domestic violence and "exposure" to domestic violence are often used interchangeably, Moroz (2005) has noted recent efforts to draw a distinction between the two terms, using "witnessing" to describe presence in the room when the violence occurred, and "exposure" to refer to a much broader array of affective, cognitive, and behavioral experiences a child may have in relation to domestic violence.

Neglect and physical and sexual abuse are often found in combination with emotional abuse and exposure to domestic violence (Harris et al., 2006; Spinazzola et al., 2005). Indeed, it has long been recognized that children who are the victims of one form of abuse are more likely to experience multiple forms of abuse and exposure to domestic violence (Edwards, Holden, Felitti, & Anda, 2003; Mullen, Martin, Anderson, Romans, & Herbison, 1996). The total number of children nationwide experiencing maltreatment and exposure to domestic violence is estimated to exceed 3 million cases per year (Carter, Weithorn, & Behrman, 1999; Fantuzzo & Mohr, 1999; Osofsky, 1999).

Complex Trauma

The term "complex trauma" describes the problem of children's exposure to multiple or prolonged traumatic events and the impact of this exposure on their development (Spinazzola et al., 2005). Typically, complex trauma exposure involves the simultaneous or sequential occurrence of child maltreatment, including psychological maltreatment, neglect, physical and sexual abuse, and domestic violence that is chronic, and that begins in early childhood and occurs within the primary caregiving system (National Child Traumatic Stress Network, n.d.).

Traumatic Loss and Grief

Childhood traumatic grief occurs after the death of a loved one, when the child perceives the experience as traumatic (Brown, Pearlman, & Goodman, 2004; Cohen, Goodman, Brown, & Mannarino, 2004). Indeed, the cause of death can be due to traumatic events, such as an act of violence, an accident, disaster or war, or it can be due to natural causes. A defining characteristic of childhood traumatic grief is that trauma symptoms interfere with the child's ability to process the death in ways that are developmentally appropriate (Lieberman, Compton, Van Horn, & Gosh Ippen, 2003; National Child Traumatic Stress Network, n.d.).

Community and School Violence

Community and school violence include violence emanating from personal conflicts between people who are not family members, and may include acts of assault, such as rapes, shootings, stabbings, and beatings (National Child Traumatic Stress Network, n.d.). Children may experience trauma as victims, perpetrators, or witnesses of violence. Adolescents appear to be at much greater risk than any other age group for experiencing community violence (Breslau, Wilcox, Storr, Lucia, & Anthony, 2004). Because many neglected and/or abused children live in unstable, crime-ridden neighborhoods, they are also exposed to violence in their schools and communities (Harris et al., 2006).

Medical Trauma

Medical trauma includes trauma associated with an injury or accident, chronic or life-threatening illness, or painful or invasive medical procedures (National Child Traumatic Stress Network, n.d.). Examples include the event of being told that one has a serious, life-threatening illness, such as cancer or human immunodeficiency virus (HIV) infection, and the experience of major medical procedures, such as undergoing organ transplantation, dialysis, or chemotherapy. In 2002, nearly 10 million children in the United

States under the age of 15 were seen in hospital emergency rooms for injuries (Saxe, Vanderbilt, & Zuckerman, 2003).

Refugee and War-Zone Violence

Refugee and war-zone trauma include exposure to human-perpetrated acts of violence, such as war, political violence, or torture (Shaw, 2003). Refugee trauma can be the result of living in a region affected by bombing, shelling, shooting, land mines, sniper fire, atrocities, or looting, as well as forced displacement from communities by fleeing war, civil strife, and persecution (Barenbaum, Ruchkin, & Schwab-Stone, 2003). Some young refugees have served as soldiers, guerrillas, or other combatants in their home countries, and their traumatic experiences may closely resemble those of combat veterans (National Child Traumatic Stress Network, n.d.). Many refugee children have experienced traumatic loss, and children who survive land mine explosions often confront loss of bodily function, disfiguration, chronic pain, posttraumatic stress disorder (PTSD), and stigmatization (Lamb, Levy, & Reich, 2004). In addition, refugee children have often been exposed to interconnected ecological calamities such as hunger, extreme poverty, environmental degradation, and HIV/AIDS and other public health epidemics that combine negatively with war-related traumatic experiences.

Natural Disasters

A disaster is any natural catastrophe (e.g., tornadoes, hurricanes, and earthquakes) or any fire, flood, or explosion that extensively damages properties and lives, exerting a disruptive impact on the vital daily routines of communities, families, and individuals (Laor, Wolmer, Friedman, Spiriman, & Knobler, 2005; National Child Traumatic Stress Network, n.d.); Norris, Perilla, Riad, Kaniasty, & Lavizzo, 1999). In addition to loss of property and life, major disasters may affect entire communities and families in terms of evacuation and permanent relocation. Disasters can result from a man-made event (e.g., a nuclear reactor explosion), but damage that is caused intentionally is classified as an act of terrorism.

Terrorism

Terrorism is defined in a variety of formal, legal ways, but the essential element is intent to inflict psychological damage on an adversary by creating an atmosphere of danger and threat (National Child Traumatic Stress Network, n.d.; Pynoos, Schreiber, Steinberg, & Pfefferbaum, 2005). Pine, Costello, and Masten (2005) define "terrorism" as a form of undeclared war, fought using the civilian population in addition to, or instead of, the military as a target. Terrorism includes attacks by both individuals acting in isolation (e.g., sniper attacks) and groups or people acting for groups (cf. Pine, Costello, & Masten, 2005).

CURRENT STATE OF THE ART

General Population Studies of Children's Exposure to Trauma

Children's, adolescents', and young adults' exposure to a range of traumatic experiences has been examined in a number of large epidemiological studies in countries throughout the world. In the United States, the National Survey of Adolescents, sponsored by the National Institute of Justice, estimated that 5 million adolescents, ages 12–17, had experienced a serious physical assault, 1.8 million had experienced a sexual assault, and 8.8 million had witnessed interpersonal violence during their lifetimes (Kilpatrick et al., 2000). A recent survey examined a broad spectrum of violence, crime, and victimization experiences in a nationally representative sample of children and youth ages 2–17 years (Finkelhor, Ormrod, Turner, & Hamby, 2005). The results revealed widespread exposure to violence: More than half of the children sampled had experienced a physical assault during the study year; more than 1 in 8 had experienced a form of child maltreatment; more than 1 in 12, a sexual victimization; and more than 1 in 3 had witnessed violence. A child or youth victimized once had a 69% chance of revictimization during a single year.

A representative longitudinal study of children in the primarily rural western counties of North Carolina in the United States found that by age 16, more than 25% of children were exposed to one or more incidents of acute and chronic trauma, such as child maltreatment or domestic violence, traffic injury, major medical trauma, traumatic loss of a significant other, or sexual assault (Costello, Erkanli, Fairbank, & Angold, 2002). Of the 25% of children reporting a traumatic event, the majority (72%) experienced only one during their lives, whereas the remaining 28% reported experiencing two or more traumatic events. Thus, 7% of children in the general population of this largely low-income, rural region of the United States had experienced two or more traumatic events by age 16.

Community studies of American children, teens, and young adults in urban areas report even higher rates of exposure. Among public school students in New York City in grades 4–12, over 60% experienced at least one major traumatic event prior to the World Trade Center terrorist attacks on September 11, 2001 (Hoven et al., 2002). These events included seeing someone killed or seriously injured (39%), and seeing the violent/accidental death of a close friend (29%) or family member (27%). Nearly 25% of New York City school children reported exposure to two or more traumatic events before the September 11, 2001, attacks (Hoven et al., 2005). The likelihood of exposure to multiple traumatic events (e.g., two or more) was roughly three times higher for children and adolescents in urban New York City than for youth in rural western North Carolina.

Several studies have examined the prevalence of traumatic experiences in samples of youth whose ages extend into young adulthood. In the National Comorbidity Survey (NCS), 60.7% of American males and 51.2% of females ages 14–24 reported exposure to one or more traumatic events (Kessler, Sonnega, Bromet, Hughes, & Nelson, 1995). A study of a representative sample of urban youth in a large city on the U.S. eastern seaboard found that by age 22–23 years, the lifetime occurrence of exposure to any trauma was 82.5%, with males (87.2%) more likely to be exposed than females (78.4%; Breslau et al., 2004). The most commonly reported traumatic event was learning about the sudden, unexpected death of a significant other (51.9%). A significantly higher percentage of males (62.2%) than females (33.7%) reported lifetime exposure to interpersonal violence. Examination of the age-specific rates of exposure to trauma revealed that the rate of exposure to violence began to rise after age 15 years and peaked at age 16–17 years with males' rates more than twice those of females: 15–16% compared to 5–6% (Breslau et al., 2004). By age 20–21 years, the rate of exposure to violence returned to what it had been at early adolescence, before the precipitous rise in midadolescence.

Similarly, a retrospective survey of young Japanese women's exposure to different types of traumatic events through four phases of life (preschool, primary school, junior high to high school, and college to present) indicated that 12% reported exposure to at least one traumatic event during their preschool years, with the rate of exposure to at least one traumatic event increasing during primary school (21.2%), peaking during high school (27.5%), and subsequently decreasing in college and young adulthood (23.8%) (Mizuta et al., 2005). Consistent with the findings from the Breslau and colleagues (2004) study, the results of Mizuta and colleagues' (2005) research suggest that mid- to late adolescence may be a time of peak vulnerability for exposure to traumatic events.

A prospective longitudinal study of 14- to 24-year-old adolescents and young adults primarily residing in suburban Munich, Germany, reported that 21.4% of the sample reported at least one lifetime event that met DSM-IV criterion A1 for exposure: 26% of males and 17.7% of females (Perkonigg, Kessler, Storz, & Wittchen, 2000). When the investigators applied the more stringent DSM-IV criterion A2 (i.e., "when the event occurred, did you feel or react with intense fear, hopelessness, horror, or irritability?") to the definition of trauma, 17% of the sample qualified: 18.6% males and 15.5% females. In this sample, the most prevalent events were physical attacks (7.5%), serious accidents (5.4%), witnessing traumatic events happen to others (3.6%), and sexual abuse as a child (2.0%). Cumulative age-of-onset curves revealed a dramatic increase in exposure at about age 11 years, with sexual abuse and rape accounting for much of the increase in females up to the age of 15 years. Between ages 15 and 21 years, physical attacks and witnessing traumatic events were most prevalent, with the increase in males largely accounted for by physical attacks and serious accidents (Perkonigg et al., 2000).

Cuffe and colleagues (1998) surveyed the prevalence of exposure to trauma in a sample of students enrolled in the 12th grade of high school in a suburban community in South Carolina. Findings indicated that a higher percentage of females (16.6% European American, 25.2% African American) than males (11.6% European American, 15.8% African American) reported lifetime exposure to at least one traumatic event. Lifetime exposure to rape and child sexual abuse were reported by 12.9% of African American and 9.4% of European American females, compared to 1.5% of European American males and no African American males. Overall, African American 12th graders reported exposure to more traumatic events than their European American counterparts. A study of 160 Head Start preschool-age children in Michigan found that 65.2% had been exposed to at least one incidence of violence in the community, and 46.7% had been exposed to at least one incident of mild or severe violence in their family, including child maltreatment and interpersonal violence (Graham-Bermann & Seng, 2005).

At-Risk Children

A plurality of community residents experiences exposure to one or more traumatic events in their lifetimes, highlighting the hard but simple reality that the prevalence of exposure to traumatic events has risen to the level of a public health problem (Breslau, 2002; Harris et al., 2006; McFarlane, 2004; Osofsky, 1999). Although the numbers of children exposed to trauma in communities worldwide is high, trauma exposure is unevenly distributed within populations, with certain groups of children, such as children suffering chronic poverty and teenage pregnancy, experiencing dramatically higher rates of trauma exposure than the general population (Sameroff, 1998). Harris and colleagues (2006) identified groups of children at high risk for exposure to trauma, including children known to have been abused or neglected; children in out-of-home placement; children exposed to domestic violence or those who witnessed the violent death of a parent, caregiver, sibling, or friend; children in the juvenile justice system;

children who were victims of catastrophic accidents or mass casualty events, including those associated with school violence, terrorism, or natural disasters; children from countries that have had or are having major armed conflicts or civil disturbances; and children who require residential treatment or hospitalization for certain mental health or behavioral problems, such as substance abuse or suicide attempts.

Regarding adolescents involved in the juvenile justice system, a large study of youth ages 10–18 held pretrial in an urban detention center found that 92.5% of detainees had experienced one or more lifetime traumatic events (mean, 14.6 incidents; median, 6 incidents), with 84% reporting more than one traumatic experience and a majority exposed to six or more such events (Abram et al., 2004). More males (93.2%) than females (84.0%) reported at least one traumatic event, and for both males and females, significantly more youth age 14 years or older reported trauma than did youth ages 10–13 years.

In a recent example of elevated risk in children from countries in armed conflicts, Khamis (2005) studied 1,000 Palestinian school children ages 12–16 years from the West Bank and East Jerusalem, and reported that a majority (54.7%) had experienced at least one lifetime traumatic event. Many children reported experiencing trauma associated with armed conflict in their communities, including personal physical injury (22.9%), traumatic death of a family member (17.6%), being in a motor vehicle accident (30.9%), and nearly drowning (3%), with less than 1% (0.7%) reporting sexual assault (Khamis, 2005).

Service providers, teachers, and caregivers are also valuable sources of information about children's exposure to traumatic events. In a recent survey of clinicians participating in the National Child Traumatic Stress Network (NCTSN), for example, 77.6% of the children and adolescents they had assessed and treated for traumatic stress reactions had experienced prolonged exposure to multiple traumatic events (Spinazzola et al., 2005). Clinicians reported that interpersonal victimization in the home was the most prevalent type of trauma that their treatment-seeking and referred patients experienced.

Factors Increasing Risk for Exposure to Trauma

Sociodemographic correlates of exposure to trauma vary considerably by population subgroup and type of traumatic events studied. In a large sample of primarily rural children, Costello and colleagues (2002) found no sex differences in the mean number of traumatic events reported or in the likelihood of one or more such events during the child's lifetime. However, females were significantly more likely than males to report rape, sexual abuse, or coercion, whereas males more often reported causing death or severe harm to someone else (Costello et al., 2002). Among urban teens and young adults, Breslau and colleagues (2004) found that males', but not females', cumulative occurrence of exposure to assaultive violence varied significantly by socioeconomic status and race/ethnicity, which was not found to be true for other categories of traumatic events. Specifically, subsidized school lunch status and African heritage of male respondents were predictive of exposure to interpersonal violence (Breslau et al., 2004). In Germany, the risk of experiencing traumatic events was found to be significantly associated with being female and older, having low socioeconomic status, and living in an urban environment (Perkonigg et al., 2000). Among Palestinian children in the West Bank and East Jerusalem, the prevalence of exposure to traumatic events was higher among males, refugees, and children who had been working (Khamis, 2005).

Costello and colleagues (2002) reported a strong, graded relationship between the number of family vulnerability factors reported by parent or child and risk of exposure to childhood trauma. Vulnerability factors were parental psychopathology, family relationship problems, and family/community environment. Children with no vulnerability factors had less than a 12% chance of having experienced a traumatic event during their lifetime, whereas the risk to the most vulnerable children rose to almost 60%. The events whose likelihood increased most in vulnerable children were sexual abuse and events occurring to people whom the child knew. Costello and colleagues also found a marked difference between males and females in the factors that increased their risk for exposure to traumatic events. Parental history of mental illness was associated with a significant increase in males' risk. For females, risk factors were more broadly spread across different types of vulnerability.

Historical Trends in the United States: Is Child Trauma Increasing or Decreasing?

Few data speak to the historical trends of the various forms of childhood trauma. Early cases, such as the famous lawsuit brought by the American Society for the Prevention of Cruelty to Animals on behalf of "Little Mary Ellen" in 1874, marked a slowly growing public awareness in the late 19th century of extreme examples of child maltreatment. Child maltreatment, as a medical condition, was "discovered" by Henry Kempe and colleagues in 1962, with the publication of a paper on the "battered child syndrome" (Kempe, Silverman, Steele, Droegmueller, & Silver, 1962). Federal child protection legislation (the Child Abuse Prevention and Treatment Act, or CAPTA) was delayed another 12 years, until 1974 (see Nelson [1984] for a history of early childhood legislation in the United States). Sexual abuse first became widely recognized in the late 1970s, whereas other forms of child trauma, such as exposure to domestic and community violence, although extraordinarily prevalent, are only now beginning to be appreciated. Thus, with the partial exception of child maltreatment, there are almost no historical trend data on most forms of child trauma.

Although incomplete and methodologically limited, some data on child maltreatment provide the best examples of child trauma to examine for historical trends. Melton (2005) notes that when the present child protection system was first conceived and implemented, it was intended to annually address "a few hundred children subjected to the violent behavior of some seriously disturbed parents" (Melton, 2005, p. 9). Indeed, he attributes the dangerous dysfunction in our current system to this initial failure to appreciate and design a child protection system commensurate with the enormous magnitude of the child maltreatment problem. However, only as the various forms of child maltreatment were identified, and their "syndromic" profiles widely disseminated, could we begin to gain awareness of the widespread nature of child maltreatment. This fostered, in turn, a growing recognition of child trauma in general, a process that continues.

Two data sets offer some insight into national historical trends for child maltreatment. The first is a series of studies known as the National Incidence Studies (NIS), mandated by the United States Congress to establish the incidence of various forms of child maltreatment. To date, three NIS studies have been conducted and analyzed (results reported in 1981 [NIS-1], 1988 [NIS-2], and 1996 [NIS-3]) by the research organization Westat of Rockville, Maryland. These three studies represent the "gold standard" for incidence of child maltreatment and provide the only standardized, general population–based, data-collection methodology that systematically tracks changes in maltreatment rates over time. A fourth study (NIS-4) collected data from 122 representative counties across the United States during the period September 2005 to May 2006, although incidence data were collected in any given county for only 3 months. The results of the NIS-4 study are not expected to be released publicly until 2008. The NIS design assumes that the maltreated children investigated by child protective services (CPS) represent only the "tip of the iceberg." Although NIS estimates include children investigated at CPS agencies, they also include maltreated children identified by professionals in a wide range of agencies in representative communities. The NIS studies use a "sentinel" methodology, in which official observers in the field report all cases of suspected child abuse that they encounter during a fixed sampling frame.

The most recent study (NIS-3) findings are based on a nationally representative sample of over 5,600 professionals serving as "sentinels" in 842 agencies, located in 42 demographically selected counties. The study used two sets of standardized definitions of abuse and neglect: The Harm Standard, which identified children as maltreated only when they had already *experienced harm* from abuse or neglect, and the Endangerment Standard, which identified both children who experienced abuse or neglect that put them *at risk for maltreatment* and children who had already been harmed. Reliable retested methods of the multiple NIS studies revealed a rise in child maltreatment from 1986 (NIS-2) to 1993 (NIS-3). Under the Harm Standard definition, the total number of abused and neglected children was two-thirds higher in the NIS-3 published report than that for NIS-2. This means that a child's risk of experiencing harm-causing abuse or neglect in 1993 was 1.5 times higher than the child's risk in 1986. There have been substantial and significant increases in the incidence of child abuse and neglect since the NIS-2, conducted in 1986. Under the Endangerment Standard, the number of abused and neglected children nearly doubled from 1986 to 1993. Findings from the NIS-3 published report revealed that physical abuse nearly doubled, sexual abuse more than doubled, and emotional abuse, physical neglect, and emotional neglect were all more than 2.5 times their NIS-2 levels.

The second source of historical trend data is the data set maintained by the federally sponsored National Child Abuse and Neglect Data System (NCANDS), which collects and analyzes annual data on child abuse and neglect. The most recent data from this system typically lag 2–3 years behind the present. Unfortunately, not all states submit their data annually, so the national totals are based on incomplete data from the states. This incompleteness is taken into account for some of the statistical analyses. The most recent statistics, published in *Child Maltreatment 2003* (U.S. Department of Health and Human Services, 2005), are based on case-level data from 44 states, including the District of Columbia, and aggregate data from the remaining states.

Over the past decade, 1992 to 2003, there was an overall drop in the total number of officially reported cases. The peak year was 1994 (according to the total number of victims reported in *Child Maltreatment 2003*)[1] with 1,031,000 "substantiated" cases ("substantiated" is defined as a type of investigation disposition that concludes the allegation of maltreatment or risk of maltreatment was supported or founded by state law or state policy; highest level of finding by a state agency) for all forms of child maltreatment. Case types were as follows: 530,873 neglect (52.9%), 255,907 physical abuse

---

[1] *Child Maltreatment 1994* and *Child Maltreatment 2003* data report different numbers for 1994 substantiated cases. We used the total number from *Child Maltreatment 2003*; however, we used percentage breakdowns of number of victims in each type from *Child Maltreatment 1994*.

(25.5%), 138,554 sexual abuse (13.8%),[2] and 263,323 miscellaneous (26.1%).[3] In 2003, the last year for which data are available, 906,000 substantiated cases comprised 479,567 neglect cases (60.9%), 148,877 physical abuse cases (18.9%), 78,188 sexual abuse cases (9.9%), and 191,333 miscellaneous cases (24.3%). These data indicate a 12.1% decrease in the total number of substantiated cases, which breaks down into decreases of 9.7% for neglect, 41.8% for physical abuse, 43.6% for sexual abuse, and 27.3% for miscellaneous cases.

The two sources of historical data therefore suggest opposite trends. The NIS studies indicate that child maltreatment is increasing at an alarming rate, whereas NCANDS data suggest that maltreatment, particularly sexual and physical abuse, is showing marked declines. However, the two data sets do not cover the same time periods. The NIS data cover the period 1979 to 1993, whereas NCANDS data cover the period 1990 to 2003.

Reasons for the decrease in officially substantiated cases of child maltreatment found in the NCANDS data set are not well understood. A careful analysis of this apparent decline has been conducted by David Finkelhor and Lisa Jones (2004). They focused on the decline in sexual abuse cases in particular, examining NCANDS data from 1992 to 2000. To control for the varying number of states submitting data each year, sexual abuse data were extrapolated for all 50 states and the District of Columbia based on U.S. Census data. From an estimated peak of approximately 149,800 cases in 1992, sexual abuse declined an average of 2–11% a year, to approximately 89,355 cases in 2000 (Finkelhor & Jones, 2004). This represented a 40% decline over an 8-year interval. On a state-by-state basis, the trend was not universal, with some states showing little or no change. Nineteen states, however, had declines of 50% or greater, accounting for much of the overall effect.

In an effort to explain this apparent decline Finkelhor and colleagues surveyed administrators in CPS systems and arrived at a set of general hypotheses (Jones, Finkelhor, & Kopiec, 2001). The first is that CPS systems have raised their threshold for what constitutes a substantiated case. It was also suggested that CPS systems increasingly were restricting their jurisdiction and specifically excluding sexual abuse cases in which the alleged perpetrator was not a family member. Another possible explanation that was considered is changes in the ways that CPS systems counted their cases. Over the study period, a number of states moved from three-tiered to two-tiered (substantiated/non-substantiated) classification systems. CPS administrators also felt that public and professional backlash against what was viewed by some as an excessive tendency to diagnosis sexual abuse on soft data, and the increased liability of professionals involved in such cases, were contributing to a growing reluctance to report and investigate suspected cases. Finally, there was the possibility that most of the previously undisclosed cases had already been found, diminishing the number of new disclosures but not necessarily decreasing new cases.

Each of these possible explanations is considered in detail by Finkelhor and colleagues and rejected in whole or in part (Finkelhor & Jones, 2004). They conclude that the decline in sexual abuse cases is real, and that it is correlated with declines in out-

---

[2] Numbers of victims of neglect, physical abuse, and sexual abuse came from *Child Maltreatment 1994*, pages 2–4.

[3] "Miscellaneous cases" include medical neglect, emotional maltreatment, other maltreatment, and unknown maltreatment based on Data Table from *Child Maltreatment 1994*, Section IV: Victim Data, Victims by Maltreatment Type, pp. 4–7.

comes commonly associated with child sexual abuse. In particular, they note that crime and violent crime declined over the same period, as did teenage pregnancy, and reports of runaway children and teen suicide, all of which are significantly associated with histories of child sexual abuse. They offer a number of possible explanations for the decline in sexual abuse, including the possibility that improvements in child abuse prevention programs, and greater public and professional awareness, may be eliminating the most preventable cases, especially those involving biological fathers in intact families (Finkelhor & Jones, 2004). Another possibility is that a large percentage of child sexual offenders have already been caught and no longer have access to victims, because they are either in jail or are legally barred.

Although these and other possibilities may account for some of the decline in some regions of the United States, Finkelhor and Jones (2004) opine that, as yet, there is no solid and convincing explanation for the apparent overall decline in sexual abuse cases from 1992 to 2000. Indeed, they acknowledge that many CPS administrators do not believe that it is real and are "resistant to the possibility that the numbers represent a true decline, preferring almost any other explanation as an alternative" (p. 10). In conclusion, they stress the need to understand the causes of this decline—if it is real—to inform policy and program development that will accelerate the decline and extend it to other forms of child maltreatment and trauma.

Consequences of Trauma

Exposure to trauma can have a dramatic impact on the development of children, including very young children, because children react to traumatic experiences in ways that reflect the developmental tasks they are confronting (Graham-Bermann & Seng, 2005; Lieberman & Van Horn, 2004; Salmon & Bryant, 2002; Van Horn & Lieberman, 2004). In a study of 6-year-olds, prior exposure to violence and trauma was associated with substantial decrements in IQ and reading achievement (Delaney-Black et al., 2002). Numerous studies have shown similar results, with child abuse related to delayed language and cognitive development, low IQ, and poor school performance (Veltman & Browne, 2001).

Research has also focused on the consequences of exposure to trauma on the normal developmental path of very young children (Lieberman, 2005). Such violence can have serious consequences when it adversely affects the child's sense of personal safety, predictability, and protection, even when it does not objectively threaten survival (Groves, Zuckerman, Marans, & Cohen, 1993). Children struggling with intense fears and concerns about their primary caregiver are often unable to achieve other, more normal developmental milestones, and may fall behind in their emotional, social, and cognitive growth, and have poorer physical health (Osofsky, 1999). For example, in a study of 160 children enrolled in Project Head Start, children exposed to violence and maltreatment who developed traumatic stress symptoms were more prone to poor physical health than were children without a history of exposure to traumatic stress and symptoms of posttraumatic stress disorder (PTSD) (Graham-Bermann & Seng, 2005). Longitudinal research shows that adverse and traumatic childhood experiences impair mental and physical health into adulthood (Edwards et al., 2003; Felitti, Anda, Nordenberg, & Williamson, 1998).

Studies have identified childhood trauma and adversity as a major risk factor for many serious adult mental and physical health problems (Edwards et al., 2003; Felitti et al., 1998). The emerging epidemiological literature suggests that traumatic life events increase the risk of a range of psychopathological outcomes, including PTSD, substance abuse, depression, and poor health outcomes (Goenjian et al., 1995; Kilpatrick et al., 2003; Putnam, 2003; Sameroff, 1998). The Adverse Childhood Experiences (ACEs) Study found a strong, graded relationship between number of ACEs and increased risk for alcoholism, drug abuse, suicide attempts, smoking, poor general health, poor mental health, severe obesity, sexual promiscuity, and sexually transmitted diseases among adult study participants (Dube et al., 2001; Edwards et al., 2003; Felitti et al., 1998). Trauma, through its effects on health-risk behaviors such as smoking and physical inactivity, contributes to multiple health problems, including heart disease, cancer, and liver disease (Felitti et al., 1998).

Decreased capacity for emotional regulation is one of the adverse effects of significant early exposure to severe interpersonal violence and other forms of trauma (Allen & Tarnowski, 1989; Cheasty, Claire, & Collins, 2002; Levitan et al., 1998; Schwartz & Proctor, 2000). Empirical studies in the developmental literature have indicated that child abuse disturbs the acquisition of appropriate emotion regulation and interpersonal skills (Cloitre, Miranda, Stovall-McClough, & Han, 2005; Manly, Cicchetti, & Barnett, 1994). For example, in a study of 165 treatment-seeking adult women with histories of childhood sexual abuse and/or physical abuse found that PTSD symptoms, affect regulation and interpersonal problems predicted significant functional impairment (Cloitre et al., 2005).

Research findings indicate that traumatic experiences such as sexual abuse affect the development of the brain and impair major hormonal systems (Teicher, Andersen, Pocari, et al., 2003). Affected areas of the brain appear to be those associated with the regulation of emotion and with control of impulses and reasoning, problem solving, and judgment (DeBellis et al., 1999; DeBellis, Keshavan, Frustaci, et al., 2002; DeBellis, Keshavan, Shiflett, et al., 2002). Major hormonal systems, such as the hypothalamic–pituitary–adrenal (HPA) axis, which plays a crucial biological role in buffering the physical effects of stress, are significantly dysregulated in survivors of childhood trauma (DeBellis, Baum, Birmaher, & Ryan, 1997; DeBellis, Baum, et al., 1999; DeBellis et al., 1994; DeBellis, Lefter, Trickett, & Putnam, 1994; DeBellis & Putnam, 1994). In addition, the sympathetic nervous system may become hyperactive, leading to increased arousal and hypervigilance in trauma survivors (DeBellis et al., 1997).

Abused and neglected children have also been found to exhibit significantly poorer performance in school than nonabused children (Veltman & Browne, 2001). Studies have identified significant effects on IQ scores, language ability, and school performance (Shonk & Cicchetti, 2001). Children with maltreatment-related PTSD have significant impairments on attention tasks, abstract reasoning, and executive functioning compared with matched healthy children (Beers & DeBellis, 2002). A population-based sample of over 1,000 twin pairs found that exposure to domestic violence accounted for approximately 4% of the variation in child IQ and was associated with an average decrease of 8 points (Koenen, Moffitt, Caspi, Taylor, & Purcell, 2003). A study of state and local administrative databases for 7,940 children who had received Aid to Families with Dependent Children found that child maltreatment system involvement generally predates and is predictive of entry into special education even after controlling for other factors (Jonson-Reid, Drake, Kim, Porterfield, & Han, 2004).

Because the impact of violence and trauma on the developmental growth of very young children is often overlooked, many clinicians do not recognize the far-reaching impact of these experiences, particularly for children in foster care and child welfare programs, who may have multiple experiences of community and family violence (Stein et al., 2001). As a result of this oversight, many children are misdiagnosed and receive inappropriate treatment or no treatment at all (Burns et al., 2004). According to Lieberman (2005), it is particularly important in the treatment of very young children to include a focus on trauma as part of a developmentally appropriate approach to treatment and to include parents in the treatment protocol. In many instances, the parent also has a trauma history, and it is important to incorporate an understanding of the parent's experiences as well. In Lieberman's work with toddlers and preschoolers who witnessed domestic violence, she found that on average their mothers had experienced 13 traumatic events in their lifetimes, with a range from 8 to 23 events; 40% of the children had been physically abused, in addition to witnessing domestic violence, and many others had been sexually abused, placed in foster home, exposed to neighborhood and community violence, or exposed to other traumas before they came to treatment between ages 3 and 5 years. Similarly, a 2001 study of Head Start participants that examined both maternal and child exposure to violence found that maternal distress symptoms are even more important than community violence exposure in contributing to heightened child behavior problems (Aisenberg, 2001).

Posttraumatic Stress Disorder

From several decades of research, we now know that children and adolescents can develop PTSD. The diagnosis of PTSD is used when reexperiencing, avoidance/numbing, and arousal symptoms are serious, continue, and interfere with the daily functioning of children and adolescents. The prevalence of PTSD has been studied in children, teens, and young adults in the general population, in samples of children exposed to specific types of trauma (e.g., hurricanes, school violence, motor vehicle accidents), and in samples of children in clinical and service settings (e.g., juvenile justice system, foster care, substance abuse treatment).

The prevalence of PTSD in the general population of adolescents and young adults in Munich, Germany, is reported at 1% for males and 2.2% for females (Perkonigg et al., 2000). The conditional probability of a lifetime PTSD diagnosis (percentage of trauma-exposed persons who met PTSD diagnostic criteria) among respondents whose reported traumatic events met DSM-IV A1 and A2 criteria was 7.8%. A national survey of American adolescents ages 12–17 reported PTSD prevalence of 3.7% for males and 6.3% for females (Kilpatrick et al., 2003). Breslau, Davis, Andreski, and Peterson (1991) found that 10.4% of a large Midwestern sample of women ages 16–24 and 6% of their male counterparts had a lifetime history of PTSD. Breslau and colleagues' (2004) more recent study of young adults from a large eastern U.S. city found that 7.9% of females and 6.3% of males met lifetime DSM-IV criteria for PTSD. The overall conditional probability of PTSD for any trauma was 8.8%. The highest conditional probability was 15.1% for young adults exposed to interpersonal violence at some point in their lives. Abram and colleagues (2004) found that 11.2% of youth in an urban juvenile detention center met criteria for PTSD in the past year. There were no significant differences in PTSD diagnosis by sex or race/ethnicity for males and females. More than half of the participants with PTSD reported that witnessing violence was the precipitant event.

Higher PTSD prevalence rates have been reported in studies of children and adolescents who have been exposed to specific events that potentially affected entire communities, such as violent terrorist events, hurricanes, earthquakes, fires, industrial explosions, and armed conflicts. For example, a study of 80 boys and 79 girls examined 1 month after a sniper attack at their school revealed that 60.4% met criteria for PTSD

(Pynoos et al., 1987). The prevalence of PTSD in New York City schoolchildren 6 months after the attack on the World Trade Center was 10.6% (Hoven et al., 2005). A study of 12- to 14-year-old students exposed 6 weeks earlier to a severe earthquake (7.3 on the Richter scale) in Taiwan revealed that 21.7% met criteria for PTSD (Hsu, Chong, Yang, & Yang, 2002). McFarlane (1987) reported that parent ratings of over 800 Australian children exposed to a major brush fire revealed PTSD prevalence estimates of 52.8% at 8 months and 57.2% at 26 months. Teacher ratings were, respectively, 29.5% and 26.3% for these periods. Studies of Lebanese (Saigh, 1988) and Palestinian (Khamis, 2005) children exposed to war revealed that roughly one-third met diagnostic criteria for PTSD. For example, structured diagnostic interviews with 92 Lebanese 13-year-olds exposed to armed conflict revealed that 29.3% met criteria for PTSD (Saigh, 1988).

Risk and Protective Factors for Traumatic Stress Reactions in Children

Risk factors for traumatic stress reactions in children can be genetic, individual, or environmental, and they interact with protective factors in complex ways (Harris et al., 2006). Risk factors often co-occur, particularly in highly traumatizing environments; they are more predictive in combination than in isolation (Sameroff, 1998). Research on risk and resiliency in children includes studies of broad and cumulative risks, of stressful life events, and of acute trauma and chronic adversity (Felitti et al., 1998; Kendler et al., 2000).

Risk factors include exposure to life-threatening situations; experience of loss, separation, and displacement; personal injury during the event; and severe psychological response in parents and other caregivers (cf. Laor et al., 2002); as well as ecological factors such as poverty and other family vulnerabilities (Sameroff, 1998). Maternal symptoms of traumatic stress, depression, and anxiety have been associated with higher levels of PTSD symptoms in children, even after controlling for level of exposure to the traumatic event (Laor, Wolman, & Cohen, 2001). Elevated life threat, family vulnerabilities, and child characteristics (e.g., age, gender, temperament) are associated with PTSD among children exposed to trauma (Fairbank, Klaric, O'Dekirk, Fairbank, & Costello, 2006). Vernberg, Silverman, LaGreca, and Prinstein (1996) reported that the degree of loss and disruption from natural disasters was strongly associated with children's traumatic stress symptoms. Youth ages 7–15 years with traumatic stress symptoms related to exposure to war in Sarajevo were found to have suffered more losses involving family members and deprivation of basic needs than had youth without PTSD symptoms (Husain et al., 1998).

Researchers have also studied resilient children—those children who seem to adapt and thrive despite exposure to traumatic situations—in an attempt to identify factors that potentially protect trauma-exposed children from adverse psychological and functional outcomes (Hughes, Graham-Bermann, & Gruber, 2001). Protective factors are individual or environmental characteristics that predict or are correlated with positive outcomes for children (Masten & Coatsworth, 1995). Researchers have identified a number of protective factors associated with children's resiliency and increased resistance to stress, including intelligence, the capacity for emotional regulation, the presence of social supports provided by caring and competent adults, holding a positive belief about self, the child's belief in the safety and fairness of their situation, and a motivation to act effectively on one's environment (Harris et al., 2006; Lieberman, Padron, Van Horn, & Harris, 2005). Studies that have looked specifically at resilient functioning among children exposed to community violence have identified three key social/environmental factors as particularly important: parent support, school support, and peer support (Lynch, 2003). Resilience, however, is not a fixed characteristic, but changes across time and circumstances. Protective factors may be more accessible to children as they mature (Margolin & Gordis, 2004).

It's possible that significant levels of adverse and traumatic experiences can impair the acquisition of protective factors or can overwhelm the coping capacities of even those children with multiple protective factors (Harris et al., 2006). The ACE Study has examined the cumulative effects of multiple and chronic adverse childhood experiences, such as physical, sexual, and psychological abuse, on physical and mental health (Edwards et al., 2003). Using a metric that counts the types of ACEs occurring before age 18, the research team reported finding a strong dose–response relationship between the number of ACEs experienced and the number of adversely affected adult study participants (Felitti et al., 1998). Compared to subjects reporting no ACEs, individuals with one or more ACE are at much greater risk for a range of serious health problems. For example, the presence of four or more ACEs increases risk two- to 12-fold for alcoholism, drug abuse, smoking, poor general health, having more than 50 sexual partners, and sexually transmitted diseases (Felitti et al., 1998). In addition, there is a strong, graded relationship between the number of ACEs and risk for attempted suicide throughout childhood and adulthood (Dube et al., 2001). Similar results have been obtained in twin studies that control for genetic factors (Dinwiddie et al., 2000; Kendler et al., 2000; True et al., 1993).

With an understanding of resilience and protective factors, as well as stressors and risk factors, public policies can be shaped to help support those characteristics of the child, the family, or the community that help a child overcome adversities of life. Such policies can be focused on risk, with a priority to prevent or eliminate risk factors (e.g., poverty, family violence) and on directing resources toward prevention of trauma and violence, reducing their impact, or supporting positive adaptations; or on processes that emphasize efforts to strengthen children's relationships or self-mastery experiences (Harris et al., 2006; Hughes et al., 2001).

For example, Lieberman and colleagues (2005) have proposed that clinical research should begin to examine the efficacy of therapeutic interventions for child maltreatment that include narratives emphasizing the integration of positive childhood memories with the painful, core-trauma-focused childhood narrative. Lieberman and colleagues raise the empirical question of whether such an approach may more effectively promote a positive treatment outcome than interventions using trauma-focused narratives alone.

Implications for Systems That Care for Traumatized Children and Families

Numerous agencies provide services to children and families experiencing acute and chronic exposure to trauma, including first responders, CPS, law enforcement, family/dependency courts, and domestic violence service providers; child welfare and foster family associations; residential treatment centers; early childhood care providers, schools, and health care providers (including primary care physicians, pediatricians, and mental health and public health services). The way these organizations work together is critically important, because they have the potential to promote child safety and reduce the harmful impact of trauma on children. However, unfortunately, their actions at times may exacerbate the adverse impact of the traumatic experience for children and their families. Although some attention has been given to understanding and improving the interaction of systems that become involved with a child immediately following exposure to trauma (e.g., the CPS and law enforcement systems), less attention has been paid to some of the agencies that become involved later in the process. A few system-specific studies have examined the role of trauma, maltreatment, and violence, for example, in the child welfare system (Smyke, Wajda-Johnson, & Zeanah, 2004), the juvenile justice system (Van Horn & Hitchens, 2004) and adolescent substance abuse treatment (Dennis & Stevens, 2003). Overall, however, there are few studies on the integration of trauma-related information and expertise into the responses of child-serving agencies and systems.

A recent survey by the National Child Traumatic Stress Network (NCTSN) examined the specific issue of how children receive care following exposure to trauma, and how agencies involved in the care of such children communicate and consult with each other (Taylor & Siegfried, 2005). Specifically, in assessing the ways that 53 agencies in 11 communities gathered and shared trauma-related information, the study found major gaps in the way agencies addressed children's histories of violence and trauma. Regardless of the type of service system, agencies rarely received in-depth information about a child's trauma history or exposure to violence history upon the child's first referral to them by another agency or system. Most agencies did not include standardized trauma assessments as part of initial intake procedures or trauma information in staff training. Many organizations working with traumatized children focused only on addressing traumatic reactions, such as anger and irritability, or symptoms, such as avoidance and hyperarousal. They rarely addressed the traumatic experiences that precipitate problematic behavior and the trauma reminders that can trigger posttraumatic reactions. The ways in which systems share information about a child's trauma history and treatment can have a direct impact on the quality of care given to the child and on the child's and family's well-being.

CHALLENGES FOR THE FUTURE

The impact of trauma on children can be pervasive, impairing school performance and subsequent earnings capacity, diminishing cognitive abilities, and leading to chronic substance use, psychological disorders, and physical health problems (Fairbank, Ebert, & Zarkin, 1999; Kaysen, Scher, Mastnak, & Resick, 2005; Perkonigg et al., 2005; Schroeder & Polusny, 2004). Rapid and early identification of children who have been traumatized could lead to more effective interventions and services that diminish the negative impact for children, families, and communities (Balaban et al., 2005; Nader, 2004). The following recommendations build on the extant epidemiological information to address some of the unique challenges in addressing gaps in public and professional awareness of the scope of children's trauma exposure and service needs.

1. *Promote public education.* Increase awareness of the prevalence and serious impact of child traumatic stress. Messages about the role of exposure to trauma should be included as part of mental health and public health education campaigns. Major media campaigns to reduce the stigma associated with mental health treatment should be expanded to address the stigma that continues to surround certain types of trauma, particularly interpersonal traumas, including domestic violence and sexual abuse.

2. *Develop real-world research and monitoring infrastructure.* Improve and develop better child trauma surveillance and epidemiological systems to monitor incidence and

prevalence, and to measure the effects of policy changes and prevention programs (Fairbank, Jordan, & Schlenger, 1996). There is a need to count children across, as well as within, systems.

- Use epidemiological data when developing new intervention systems and revising old ones, such as CPS, to better estimate the scope of the problem, to achieve an appropriate scale for services, and to monetize the cost.
- Integrate data on relevant outcomes (teen pregnancy, attempted suicide, etc.) with epidemiological data to better detect beneficial effects of preventions and treatment programs.

3. *Create capacity.* Engage multiple child-serving systems to make them more trauma-informed. Currently professionals in these systems (including public mental health, education, child welfare, juvenile justice, and health care agencies) have different levels of not only awareness of the importance of child traumatic stress but also training and ability to address the needs of children and families who have experienced trauma. The field of child trauma should work with leaders from these systems to better understand their specific issues. Customized approaches will be needed to address the challenges within different systems, many of which are in crisis, and to identify meaningful incentives for professionals in these systems to prioritize trauma.

- Foster collaboration across disciplines, geographic boundaries, service settings, and trauma populations, as well as across child-serving systems and between professionals in academic and community settings. With few exceptions, providers and systems still operate largely in isolation, and there is a great need to develop interdisciplinary approaches and models for interagency collaboration related to child traumatic stress.
- Promote identification, encourage referral, and avoid retraumatization within different child-serving systems. This includes the identification of trauma exposure as part of intake procedures in all child services, including primary care, schools (including special education and preschool settings), child welfare and foster care, juvenile courts and correctional settings, mental health and addiction services, domestic violence and homeless shelters, and child emergency services following disasters and catastrophic events.
- Finally, a promising example of this approach is the recently launched *Psychological First Aid: Field Operations Guide–Second Edition* for mental health providers, developed collaboratively by the NCTSN and the National Center for PTSD (National Child Traumatic Stress Network and National Center for PTSD, 2006). Psychological first aid is an evidence-informed, modular approach for assisting children, adolescents, adults, and families in the hours and days immediately following disasters and other catastrophic events. Designed for delivery by mental health responders in diverse settings, psychological first aid comprises eight core components that focus on reducing initial posttraumatic distress and fostering short- and longer-term adaptation.

For many children, exposure to new traumatic events is a part of a pattern of chronic exposure to trauma (Costello et al., 2002; Finkelhor et al., 2005). Histories of multiple traumatic experiences provide assessment and treatment challenges for children attempting to cope with trauma, and for those in the service fields attempting to help them recover. Organizations involved in assisting the child should collaborate to put together a full trauma profile, including the disruptions that may have occurred in the child's development. Without understanding these complex histories, providers may miss opportunities to address the causes of behavioral and emotional problems, and the factors that maintain them. A comprehensive assessment helps caregivers and others appreciate the seriousness of the child's experience and understand the context of the full range of violent and traumatic experiences. Examples of integrated psychometrically, developmentally, and clinically sound screening and assessment approaches and instruments for child traumatic stress are described in recent reviews by Balaban and colleagues (2005), Elhai, Gray, Kashdan, and Franklin (2005), Harris and colleagues (in press), and Nader (2004).

REFERENCES

Abram, K. M., Teplin, L. A., Charles, D. R., Longworth, S. L., McClelland, G. M., & Dulcan, M. K. (2004). Posttraumatic stress disorder and trauma in youth in juvenile detention. *Archives of General Psychiatry, 61*, 403–410.

Aisenberg, E. F. (2001). The effects of exposure to community violence upon Latina mothers and pre-school children. *Hispanic Journal of Behavioral Sciences, 23*, 378–398.

Allen, D. M., & Tarnowski, K. J. (1989). Depressive characteristics of physically abused children. *Journal of Abnormal Child Psychology, 17*, 1–11.

Balaban, V. F., Steinberg, A. M., Brymer, M. J., Layne, C. M., Jones, R. T., & Fairbank, J. A. (2005). Screening and assessment for children's psychosocial needs following war and terrorism. In M. J. Friedman & A. Mikus-Kos (Eds.), *Promoting the psychosocial well being of children following war and terrorism* (pp. 121–161). Amsterdam: IOS Press.

Barenbaum, J., Ruchkin, V., & Schwab-Stone, M. (2003). The psychosocial aspects of children exposed to war: Practice and policy initiatives. *Journal of Child Psychology and Psychiatry, 44*, 1–22.

Beers, S. R., & DeBellis, M. D. (2002). Outcomes of child abuse. *Neurosurgery Clinics of North America, 13*, 235–241.

Breslau, N. (2002). Epidemiologic studies of trauma, posttraumatic stress disorder, and other psychiatric disorders. *Canadian Journal of Psychiatry, 547*, 923–929.

Breslau, N., Davis, G. C., Andreski, P., & Peterson, E. (1991). Traumatic events and posttraumatic stress disorder in an urban population of young adults. *Archives of General Psychiatry, 48*, 216–222.

Breslau, N., Wilcox, H. C., Storr, C. L., Lucia, V. C., & Anthony, J. C. (2004). Trauma exposure and posttraumatic stress disorder: A study of youths in urban America. *Journal of Urban Health: Bulletin of the New York Academy of Medicine, 81*, 530–544.

Brown, E. J., Pearlman, M. Y., & Goodman, R. F. (2004). Facing fears and sadness: Cognitive-behavioral therapy for childhood traumatic grief. *Harvard Review of Psychiatry, 12*, 187–198.

Burns, B. J., Phillips, S. D., Wagner, H. R., Barth, R. P., Kolko, D. J., Campbell, Y., et al. (2004). Mental health need and access to mental health services by youths involved with child welfare: A national survey. *Journal of the American Academy of Child and Adolescent Psychiatry, 43*, 960–970.

Carter, L., Weithorn, L., & Behrman, R. (1999). Domestic violence and children: Analysis and recommendations. *The Future of Children, 9*, 4–20.

Cheasty, M., Clare, A. W., & Collins, C. (2002). Child sexual abuse: A predictor of persistent depression in adult rape and sexual assault victims. *Journal of Mental Health, 11*, 79–84.

Cloitre, M., Miranda, R., Stovall-McClough, C., & Han, H. (2005). Emotion regulation and interpersonal problems as predictors of functional impairment in survivors of childhood abuse. *Behavior Therapy, 36*, 119–124.

Cohen, J. A., Goodman, R. F., Brown, E. J., & Mannarino, A. P. (2004). Treatment of childhood traumatic grief: Contributing to a newly emerging condition in the wake of community trauma. *Harvard Review of Psychiatry, 12*, 213–216.

Costello, E. J., Erkanli, A., Fairbank, J. A., & Angold, A. (2002). The prevalence of potentially traumatic events in childhood and adolescence. *Journal of Traumatic Stress, 15*, 99–112.

Cuffe, S. P., Addy, C. L., Garrison, C. Z., Waller, J. L., Jackson, K. L., McKeown, R. E., et al. (1998).

Prevalence of PTSD in a community sample of older adolescents. *Journal of the American Academy of Child and Adolescent Psychiatry, 37,* 147–154.

De Bellis, M. D., Baum, A. S., Birmaher, B., Keshavan, M. S., Eccard, C. H., Boring, A. M., et al. (1999). Developmental traumatology: Part I. Biological stress systems. *Biological Psychiatry, 45*(10), 1259–1270.

De Bellis, M. D., Baum, A. S., Birmaher, B., & Ryan, N. D. (1997). Urinary catecholamine excretion in childhood overanxious and posttraumatic stress disorders. *Annals of the New York Academy of Sciences, 821,* 451–455.

De Bellis, M., Chrousos, G., Dorn, L., Burke, L., Helmers, K., Kling, M. A., et al. (1994). Hypothalamic–pituitary–adrenal axis dysregulation in sexually abused girls. *Journal of Clinical Endocrinology and Metabolism, 78,* 249–255.

De Bellis, M. D., Keshavan, M. S., Clark, D. B., Casey, B. J., Giedd, J. N., Boring, A. M., et al. (1999). Developmental traumatology: Part II. Brain development. *Biological Psychiatry, 45*(10), 1271–1284.

De Bellis, M. D., Keshavan, M. S., Frustaci, K., Shifflett, H., Iyengar, S., Beers, S. R., et al. (2002). Superior temporal gyrus volumes in maltreated children and adolescents with PTSD. *Biological Psychiatry, 51*(7), 544–552.

De Bellis, M. D., Keshavan, M. S., Shiflett, H., Iyengar, S., Beers, S. R., Hall, J., et al. (2002). Brain structures in pediatric maltreatment-related posttraumatic stress disorder: A sociodemographically matched study. *Biological Psychiatry, 52*(7), 1066–1078.

De Bellis, M., Lefter, L., Trickett, P., & Putnam, F. (1994). Urinary catecholamine excretion in sexually abused girls. *Journal of the American Academy of Child and Adolescent Psychiatry, 33,* 320–327.

De Bellis, M., & Putnam, F. (1994). The psychobiology of childhood maltreatment. *Child and Adolescent Psychiatric Clinics of North America, 3,* 1–16.

Delaney-Black, V., Covington, C., Ondersma, S. J., Nordstrom-Klee, B. A., Templin, T. M., Ager, J., et al. (2002). Violence exposure, trauma, and IQ and/or reading deficits among urban children. *Archives of Pediatrics and Adolescent Medicine, 156*(3), 280–285.

Dennis, M. L., & Stevens, S. J. (Eds.). (2003). Maltreatment issues and outcomes of adolescents enrolled in substance abuse treatment [Special issue]. *Child Maltreatment, 8*(1), 3–71.

Dinwiddie, S., Heath, A. C., Dunne, M. P., Bucholz, K. K., Madden, P. A., Slutske, W. S., et al. (2000). Early sexual abuse and lifetime psychopathology: A co-twin-control study. *Psychological Medicine, 30,* 41–52.

Dube, S. R., Anda, R. F., Felitti, V. J., Chapman, D. P., Williamson, D. F., & Giles, W. H. (2001). Childhood abuse, household dysfunction, and the risk of attempted suicide throughout the lifespan: Findings from the Adverse Childhood Experiences Study. *Journal of the American Medical Association, 286,* 3089–3096.

Edleson, J. (1999). The overlap between child maltreatment and woman battering. *Violence Against Women, 5*(2), 134–154.

Edwards, V. J., Holden, G. W., Felitti, V. J., & Anda, R. F. (2003). Relationship between multiple forms of childhood maltreatment and adult mental health in community respondents: Results from the adverse childhood experiences study. *American Journal of Psychiatry, 160,* 1453–1460.

Elhai, J. D., Gray, M. J., Kashdan, T. B., & Franklin, C. L. (2005). Which instruments are most commonly used to assess traumatic event exposure posttraumatic effects?: A survey of traumatic stress professionals. *Journal of Traumatic Stress, 18,* 541–545.

Fairbank, J. A., Ebert, L., & Zarkin, G. A. (1999). Socioeconomic consequences of traumatic stress. In P. A. Saigh & J. D. Bremner (Eds.), *Posttraumatic stress disorder: A comprehensive text* (pp. 180–198). Needham Heights, MA: Allyn & Bacon.

Fairbank, J. A., Jordan, B. K., & Schlenger, W. E. (1996). Designing and implementing epidemiologic studies. In E. B. Carlson (Ed.), *Trauma research methodology* (pp. 105–125). Lutherville, MD: Sidran Press.

Fairbank, J. A., Klaric, J. S., O'Dekirk, J. M., Fairbank, D. W., & Costello, E. J. (2006). Environmental vulnerabilities and posttraumatic stress disorder (PTSD) among children with different personality styles. In J. Strelau & T. Klonowicz (Eds.), *People under stress* (pp. 35–48). New York: Nova Science.

Fantuzzo, J., & Mohr, W. (1999). Prevalence and effects of child exposure to domestic violence. *The Future of Children, 9,* 21–32.

Felitti, V., Anda, R., Nordenberg, D., & Williamson, D. F. (1998). Relationship of childhood abuse and household dysfunction to many of the leading causes of death in adults. *American Journal of Preventive Medicine, 14*, 245–258.

Finkelhor, D., & Jones, L. M. (2004). Explanations for the decline in child sexual abuse cases. In *Juvenile Justice Bulletin–NCJ199298* (pp. 1–12). Washington, DC: U.S. Government Printing Office.

Finkelhor, D., Ormrod, R., Turner, H., & Hamby, S. (2005). The victimization of children and youth: A comprehensive, national survey. *Child Maltreatment, 10*(1), 5–25.

Goenjian, A. K., Pynoos, R. S., Steinberg, A. M., Najarian, L. M., Asarnow, J. R., Karayan, I., et al. (1995). Psychiatric comorbidity in children after the 1988 earthquake in Armenia. *Journal of the American Academy of Child and Adolescent Psychiatry, 34*, 1174–1184.

Graham-Bermann, S. A., & Seng, J. S. (2005). Violence exposure and traumatic stress symptoms as additional predictors of health problems in high-risk children. *Journal of Pediatrics, 146*, 349–354.

Groves, B. M., Zuckerman, B., Marans, S., & Cohen, D. (1993). Silent victims: Children who witness violence. *Journal of the American Medical Association, 269*, 262–264.

Harris, W. W., Lieberman, A. F., & Marans, S. (in press). In the best interests of society. *Journal of Child Psychiatry and Psychology*.

Harris, W. W., Putnam, F. W., & Fairbank, J. A. (2006). Mobilizing trauma resources for children. In A. F. Lieberman & R. DeMartino (Eds.), *Shaping the future of children's health* (pp. 311–339). Calverton, NY: Johnson & Johnson Pediatric Institute.

Hoven, C. W., Duarte, C. S., Lucas, C. P., Mandell, D. J., Cohen, M., Rosen, C., et al. (2002). *Effects of the World Trade Center attack on NYC public school students: Initial report to the New York City Board of Education*. New York: Columbia University Mailman School of Public Health, New York State Psychiatric Institute and Applied Research and Consulting.

Hoven, C. W., Duarte, C. S., Lucas, C. P., Wu, P., Mandell, D. J., Goodwin, R. D., et al. (2005). Psychopathology among New York City public school children 6 months after September 11. *Archives of General Psychiatry, 62*, 545–552.

Hsu, C.-C., Chong, M.-Y., Yang, P., & Yang, C.-F. (2002). Posttraumatic stress disorder among adolescent earthquake victims in Taiwan. *Journal of the American Academy of Child and Adolescent Psychiatry, 41*(7), 875–881.

Hughes, H. M., Graham-Bermann, S. A., & Gruber, G. (2001). Resilience in children exposed to domestic violence. In J. L. Edleson & S. A. Graham-Bermann (Eds.), *Domestic violence in the lives of children: The future of research, intervention, and social policy* (pp. 67–90). Washington, DC: American Psychological Association.

Husain, S. A., Nair, J., Holcomb, W., Reid, J. C., Vargas, V., & Nair, S. S. (1998). Stress reactions of children and adolescents in war and siege conditions. *American Journal of Psychiatry, 155*, 1718–1719.

Jones, L. M., Finkelhor, D., & Kopiec, K. (2001). Why is sexual abuse declining?: A survey of state child protection administrators. *Child Abuse and Neglect, 25*, 1139–1158.

Jonson-Reid, M., Drake, B., Kim, J., Porterfield, S., & Han, L. (2004). A prospective analysis of the relationship between reported child maltreatment and special education eligibility among poor children. *Child Maltreatment, 9*(4), 382–394.

Kaysen, D., Scher, C. D., Mastnak, J., & Resick, P. (2005). Cognitive mediation of childhood maltreatment and adult depression in recent crime victims. *Behavior Therapy, 36*, 235–244.

Kempe, C. H., Silverman, F. N., Steele, B. F., Droegmueller, W., & Silver, H. K. (1962). The battered child syndrome. *Journal of the American Medical Association, 181*, 17–24.

Kendler, K., Bulik, C., Silberg, J., Hettema, J., Myers, J., & Prescott, C. (2000). Childhood sexual abuse and adult psychiatric and substance abuse disorders in women. *Archives of General Psychiatry, 57*, 953–959.

Kessler, R. C., Sonnega, A., Bromet, E., Hughes, M., & Nelson, C. B. (1995). Posttraumatic stress disorder in the National Comorbidity Survey. *Archives of General Psychiatry, 52*, 1048–1060.

Khamis, V. (2005). Post-traumatic stress disorder among school age Palestinian children. *Child Abuse and Neglect, 29*, 81–95.

Kilpatrick, D. G., Acierno, R., Saunders, B. E., Resick, H. S., Best, C. L., & Schnurr, P. P. (2000). Risk factors for adolescent substance abuse and dependence: Data from a national sample. *Journal of Consulting and Clinical Psychology, 68*, 19–30.

Kilpatrick, D. G., Ruggiero, K. J., Acierno, R., Saunders, B. E., Resnick, H. S., & Best, C. L. (2003). Violence and risk of PTSD, major depression, substance abuse/dependence, and comorbidity: Results from the National Survey of Adolescents. *Journal of Consulting and Clinical Psychology, 71*, 692–700.

Koenen, K. C., Moffitt, T. E., Caspi, A., Taylor, A., & Purcell, S. (2003). Domestic violence is associated with environmental suppression of IQ in young children. *Development and Psychopathology, 15,* 297–311.

Lamb, J. M., Levy, M., & Reich, M. R. (2004). *Wounds of war.* Cambridge, MA: Harvard Center for Population and Development Studies, Harvard University.

Laor, N., Wolmer, L., & Cohen, D. J. (2001). Mothers' functioning and children's symptoms 5 years after a SCUD missile attack. *American Journal of Psychiatry, 158,* 1020–1026.

Laor, N., Wolmer, L., Friedman, Z., Spiriman, S., & Knobler, H. Y. (2004). Disaster intervention: An integrative systemic perspective for health and social service professionals. In M. J. Friedman & A. Mikus-Kos (Eds.), *Promoting the psychosocial well-being of children following war and terrorism* (pp. 33–43). Amsterdam: IOS Press.

Laor, N., Wolmer, L., Kora, M., Yucel, D., Spirman, S., & Yazgan, Y. (2002). Posttraumatic, dissociative and grief symptoms in Turkish children exposed to the 1999 earthquakes. *Journal of Nervous and Mental Disease, 190,* 824–832.

Levitan, R. D., Parikh, S. V., Lesage, A. D., Hegadoren, K. M., Adams, M., Kennedy, S. H., et al. (1998). Major depression in individuals with a history of childhood physical or sexual abuse: Relationship to neurovegetative features, mania, and gender. *American Journal of Psychiatry, 155,* 1746–1752.

Lieberman, A. F. (2005, January 25). *What do best practices have in common?* Plenary presentation, Chadwick Center 19th Annual San Diego International Conference on Child and Family Maltreatment, San Diego, CA.

Lieberman, A. F., Compton, N. C., Van Horn, P., & Gosh Ippen, C. (2003). *Losing a parent to death in the early years: Guidelines for the treatment of traumatic bereavement in infancy and early childhood.* Washington, DC: Zero to Three Press.

Lieberman, A. F., Padron, E., Van Horn, P., & Harris, W. W. (2005). Angels in the nursery: The intergenerational transmission of benevolent parental influences. *Infant Mental Health Journal, 26,* 504–520.

Lieberman, A. F., & Van Horn, P. (2004). Assessment and treatment of young children exposed to traumatic events. In J. D. Osofsky (Ed.), *Young children and trauma: Intervention and treatment* (pp. 111–138). New York: Guilford Press.

Lynch, M. (2003). Consequences of children's exposure to community violence. *Clinical Child and Family Psychology Review, 6*(4), 265–274.

Manly, J. T., Cicchetti, D., & Barnett, D. (1994). The impact of subtype frequency, chronicity, and severity of child maltreatment on social competence and behavior problems. *Development and Psychopathology, 6,* 121–143.

Margolin, G., & Gordis, E. B. (2004). Children's exposure to violence in the family and community. *Current Directions in Psychological Science, 13*(4), 152–155.

Masten, A., & Coatsworth, J. (1995). Competence, resilience, and psychopathology. In D. Cicchetti & D. Cohen (Eds.), *Developmental psychopathology* (Vol. 2, pp. 715–752). New York: Wiley.

McFarlane, A. (2004). The contribution of epidemiology to the study of traumatic stress. *Social Psychiatry and Psychiatric Epidemiology, 39,* 874–882.

McFarlane, A. C. (1987). Posttraumatic phenomena in a longitudinal study of children following a natural disaster. *Journal of the American Academy of Child and Adolescent Psychiatry, 26,* 764–769.

Melton, G. B. (2005). Mandated reporting: A policy without reason. *Child Abuse and Neglect, 29,* 9–18.

Mizuta, I., Ikuno, T., Shimai, S., Hirotsune, H., Ogasawara, M., Ogawa, A., et al. (2005). The prevalence of traumatic events in young Japanese women. *Journal of Traumatic Stress, 18,* 33–37.

Moroz, K. J. (2005). Understanding the current mental health needs of children experiencing domestic violence in Vermont: Recommendations for enhancing and improving responses. *Vermont's partnership between domestic violence programs and child protective services* (Publication No. 7). Retrieved on April 5, 2006, from *www.vawnet.org/domesticviolence/publicpolicy/children/vtnetworkdv-cspspub7.pdf*

Mullen, P. E., Martin, J. L., Anderson, J. C., Romans, S. E., & Herbison, G. P. (1996). The long-term impact of the physical, emotional, and sexual abuse of children: A community study. *Child Abuse and Neglect, 20,* 7–21.

Nader, K. O. (2004). Assessing traumatic experiences in children and adolescents: Self-reports of DSM PTSD criteria B–D symptoms. In J. P. Wilson & T. M. Keane (Eds.), *Assessing psychological trauma and PTSD* (2nd ed., pp. 513–537). New York: Guilford Press.

National Child Traumatic Stress Network. (n.d.). *Types of traumatic stress*. Retrieved on April 5, 2006, from *www.nctsnet.org/nccts/nav.do?pid=typ_main*

National Child Traumatic Stress Network and National Center for PTSD. (2006, July). *Psychological first aid: Field operations guide, 2nd edition*. Retrieved on September 24, 2006, from *www.nctsn.org* and *www.ncptsd.va.gov*.

Nelson, B. H. (1984). *Making an issue of child abuse: Political agenda setting for social problems*. Chicago: University of Chicago Press.

Norris, F. H., Perilla, J. L., Riad, J. K., Kaniasty, K. Z., & Lavizzo, E. A. (1999). Stability and change in stress, resources, and psychological distress following natural disaster: Findings from Hurricane Andrew. *Anxiety, Stress, and Coping, 12*(4), 363–396.

Osofsky, J. D. (1999). The impact of violence on children. *The Future of Children, 9*, 33–49.

Perkonigg, A., Kessler, R. C., Storz, S., & Wittchen, H.-U. (2000). Traumatic events and post-traumatic stress disorder in the community: Prevalence, risk factors and comorbidity. *Acta Psychiatrica Scandinavica, 101*, 46–59.

Perkonigg, A., Pfister, H., Stein, M. B., Hofler, M., Lieb, R., Maercker, A., et al. (2005). Longitudinal course of posttraumatic stress disorder and posttraumatic stress disorder symptoms in a community sample of adolescents and young adults. *American Journal of Psychiatry, 162*, 1320–1327.

Pine, D. S., Costello, J., & Masten, A. (2005). Trauma, proximity, and developmental psychopathology: The effects of war and terrorism on children. *Neuropsychopharmacology, 30*, 1781–1792.

Pine, D. S., Mogg, K., Bradley, B. P., Montgomery, L., Monk, C. S., McClure, E., et al. (2005). Attention bias to threat in maltreated children: Implications for vulnerability to stress-related psychopathology. *American Journal of Psychiatry, 162*, 291–296.

Putnam, F. (2003). Ten-year research update review: Child sexual abuse. *Journal of the American Academy of Child and Adolescent Psychiatry, 42*(3), 269–278.

Pynoos, R., Frederick, C., Nader, K., Arroyo, W., Steinberg, A., Eth, S., et al. (1987). Life threat and posttraumatic stress in school-age children. *Archives of General Psychiatry, 44*, 1057–1063.

Pynoos, R. S., Schreiber, M. D., Steinberg, A. M., & Pfefferbaum, B. J. (2005). Impact of terrorism on children. In B. J. Sadock & V. A. Sadock (Eds.), *Comprehensive handbook of psychiatry* (8th ed., pp. 3551–3564). Philadelphia: Lippincott/Williams & Wilkins.

Saigh, P. (1988). The validity of the DSM-III posttraumatic stress disorder as applied to adolescents. *Professional School Psychology, 3*, 283–290.

Salmon, K., & Bryant, R. A. (2002). Posttraumatic stress disorder in children: The influence of developmental factors. *Clinical Psychology Review, 22*, 163–188.

Sameroff, A. J. (1998). Environmental risk factors in infancy. *Pediatrics, 102*(Suppl. 5), 1287–1292.

Saxe, G., Vanderbilt, D., & Zuckerman, B. (2003). Traumatic stress in injured and ill children. *PTSD Research Quarterly, 14*(2), 1–7.

Schroeder, J. M., & Polusny, M. A. (2004). Risk factors for adolescent alcohol use following a natural disaster. *Prehospital and Disaster Medicine, 19*, 122–127.

Schwartz, D. R., & Proctor, L. J. (2000). Community violence exposure and children's social adjustment in the school peer group: The mediating roles of emotion regulation and social cognition. *Journal of Consulting and Clinical Psychology, 68*, 670–683.

Shaw, J. (2003). Children exposed to war/terrorism. *Clinical Child and Family Psychology Review, 6*, 237–246.

Shonk, S. M., & Cicchetti, D. (2001). Maltreatment, competency deficits, and risk for academic and behavioral maladjustment. *Developmental Psychology, 37*, 3–14.

Smyke, A. T., Wajda-Johnston, V., & Zeanah, C. H., Jr. (2004). Working with traumatized infants and toddlers in the child welfare system. In J. Osofsky (Ed.), *Young children and trauma: Intervention and treatment* (pp. 260–284). New York: Guilford Press.

Spinazzola, J., Ford, J. D., Zucker, M., van der Kolk, B. A., Silva, S. G., Smith, S. F., et al. (2005). Survey evaluates complex trauma exposure, outcome, and intervention among children and adolescents. *Psychiatric Annals, 35*, 433–439.

Stein, B. D., Zima, B. T., Elliott, M. N., Burnam, M. A., Shahinfar, A., Fox, N. A., et al. (2001). Violence exposure among school-age children in foster care: Relationship to distress symptoms. *Journal of the American Academy of Child and Adolescent Psychiatry, 40*, 588–594.

Taylor, N., & Siegfried, C. B. (2005). *Helping children in the child welfare system heal from trauma: A systems integration approach* (Report by the National Child Traumatic Stress Network). Retrieved on August 25, 2005, from *www.nctsnet.org*

Teicher, M. H., Andersen, S. L., Polcari, A., Anderson, C. M., Navalta, C. P., & Kim, D. M. (2003). The neurobiological consequences of early stress and childhood maltreatment. *Neuroscience and Biobehavioral Reviews, 27,* 33–44.

True, W. R., Rice, J., Eisen, S. A., Heath, A. C., Goldberg, J., Lyons, M. J., et al. (1993). A twin study of genetic and environmental contributions to liability for posttraumatic stress symptoms. *Archives of General Psychiatry, 50,* 257–264.

U.S. Department of Health and Human Services, Administration on Children Youth and Families. (2005). *Child Maltreatment 2003.* Washington, DC: U.S. Government Printing Office. Retrieved on September 24, 2006, from *www.acf.hhs.gov/programs/cb/publications/cm03/cm2003.pdf*

U.S. Department of Health and Human Services, National Center on Child Abuse and Neglect. (1996). *Child Maltreatment 1994: Reports from the states to the National Center on Child Abuse and Neglect.* Washington, DC: U.S. Government Printing Office.

Van Horn, P., & Hitchens, D. J. (2004). Partnerships for young children in court: How judges shape collaborations serving traumatized children. In J. D. Osofsky (Ed.), *Young children and trauma: Intervention and treatment* (pp. 242–259). New York: Guilford Press.

Van Horn, P., & Lieberman, A. F. (2004). Early intervention with infants, toddlers, and preschoolers. In B. T. Litz (Ed.), *Early intervention for trauma and traumatic loss* (pp. 112–130). New York: Guilford Press.

Veltman, M., & Browne, K. (2001). Three decades of child maltreatment research: Implications for the school years. *Trauma, Violence and Abuse, 2,* 215–239.

Vernberg, E. M., Silverman, W. K., La Greca, A. M., & Prinstein, M. J. (1996). Prediction of posttraumatic stress symptoms in children after hurricane Andrew. *Journal of Abnormal Psychology, 105,* 237–248.

Wolfe, D. A., & Nayak, M. B. (2003). Child abuse in peacetime. In B. L. Green, M. J. Friedman, J. T. V. M. de Jong, S. D. Solomon, T. M. Keane, J. A. Fairbank, et al. (Eds.), *Trauma interventions in war and peace: Prevention, practice and policy* (pp. 75–104). New York: Kluwer Academic/Plenum Press.

World Health Organization. (1999). *Report of the Consultation on Child Abuse Prevention.* Geneva: Author.

Chapter 14

Trauma in Older Adults

Joan M. Cook and George Niederehe

OVERVIEW AND THE AGING POPULATION

Traumatic exposure and its concomitants, including posttraumatic stress disorder (PTSD), have received much less clinical attention and scientific study in older adults than in persons under the age of 65. Recently, however, the importance of understanding trauma-related distress in older adults has begun to increase. Our goals in this chapter are to summarize briefly the extant literature in this area and to aid the scientifically informed practitioner in development and implementation of effective psychosocial treatments for survivors of trauma in the aging population. In this regard, we review demographic projections for industrialized countries; discuss late life developmental tasks and normal aging concerns; report on the epidemiology, course and phenomenology of PTSD; note current psychosocial treatments; discuss methodological challenges; and highlight potential opportunities for future investigation. Although numerous deleterious effects and disorders are related to traumatic exposure, what is known about the long-term psychological consequences of traumatic exposure in older adults is limited. Of the many trauma-related effects and disorders, PTSD has been the most extensively researched and for this reason, we focus this chapter primarily on that disorder.

Demographic projection estimates indicate that the number and proportion of older adults is increasing in industrialized countries (United Nations, 2003). For example, the number of Americans age 65 and older is projected to reach 70.3 million, or 20% of the U.S. population by 2030 (U.S. Bureau of the Census, 2004). Additionally, the heterogeneity of elders with respect to ethnicity and cultural background, socioeconomic and educational standing, sexual orientation, disability, and urban–rural residence is already vast, and the older adult population will likely continue to diversify. For instance, by the year 2050, nonwhite minorities will represent one-third of all older adults in the U.S. (Gerontological Society of America Task Force on Minority Issues in Gerontology, 1994). The changing demographic context will likely translate into an increased need for and range of services for older individuals.

DEVELOPMENTAL TASKS AND NORMAL AGING

Although negative stereotypes regarding older adults and the aging process may be diminishing, some continue to exist and may affect mental health practice and policy. In particular, misconceptions about aging may affect older adults' willingness to enter and participate in mental health services. Included in the misconceptions are the following stereotypes: Older adults are a homogeneous group; they are usually alone, lonely, sick, frail, and dependent on others; they typically live in segregated housing or nursing homes; they are often cognitively impaired, depressed, obstinate, and rigid; and they do not cope well with age-related physical and intellectual changes (American Psychological Association Working Group on the Older Adult, 1998). On the contrary, older adults are generally a heterogeneous group and maintain close relations with family, reside independently, and adapt favorably to the challenges of aging (American Psychological Association Working Group on the Older Adult, 1998). Typically, older adults' personalities stay consistent throughout the lifespan (Costa, Yang, & McCrae, 1998), with proportionally few individuals suffering from major mental health disorders (U.S. Department of Health and Human Services, 2001).

Most aging-related changes in cognitive functioning are mild and do not considerably impede daily functioning (American Psychological Association Working Group on the Older Adult, 1998). These may include slowing in reaction times and speed of information processing (Salthouse, 1996; Sliwinski & Buschke, 1999), as well as a reduction in visuospatial and motor control capacities. Despite the resilience and independence of older adults, most have at least one chronic medical illness or health condition (e.g., arthritis, hypertension, heart disease), are currently taking numerous medications, and have some degree of functional impairment/disability.

Psychosocial developmental tasks include accommodating to physical changes (including greater susceptibility to physical illness and alterations in physical appearance and body composition), functional limitations (e.g., decreased mobility and diminished sensory capacities), and numerous losses (e.g., income shrinkage and financial limitations; loss of family members, friends, and social status; changes in housing and work; and possible widowhood). Erik Erikson (1959, 1982), in his well-known eight-stage model of psychosocial development, postulated that old age is a time when individuals' struggle between ego integrity and despair; that is, if older adults look back on their lives with happiness and contentment, feeling that life has had meaning and that they have made a positive contribution, then they achieve integrity. On the other hand, if older adults struggle to find purpose and focus on their failures, then they suffer from despair. Butler (1963) expanded on this model and proposed that during old age, one prepares for death by reviewing and integrating a lifetime of aspirations, achievements, and failures.

Despite the changes and psychosocial tasks associated with aging, the majority of older adults age successfully. One theory related to "successful aging," termed "selective optimization with compensation" (Baltes & Baltes, 1990), postulates that older adults set priorities and select the goals they see as central, refine their methods to achieve goals, and use alternative approaches to compensate for aging-related losses. This process of adaptation and resilience is discussed in further detail later in this chapter.

With the exception of dementia, older adults have a lower prevalence of psychiatric disorders than do younger adults (U.S. Department of Health and Human Services, 2001), including PTSD. Additionally, they are less likely than younger adults to seek treatment in specialized mental health facilities. It is more common for older adults to be assessed by their primary care professional than by a mental health specialist for difficulties pertaining to their mental health (Goldstrom et al., 1987; Phillips & Murrell, 1994). Although accessibility and reimbursement patterns may partially account for the lack of specialized mental health care within the geriatric community, cohort effects and other issues may also be contributing factors. For example, the current older adult cohort's relative underuse of mental health services and probable lack of familiarity with available services often reflects the strong ethic of self-reliance that was emphasized during their formative years, particularly for those who grew up during the economic depression of the 1930s (Elder, 1999). Likewise, this cohort's reluctance to seek mental health services or to accept psychological explanations for problems can often be traced to early socialization experiences in communities where attitudes toward acknowledging psychiatric difficulties and contacting mental health professionals were overwhelmingly negative. Misinterpretation and manifestation of psychological difficulties as somatic complaints, and reluctance to admit to psychological difficulties due to the associated stigma, are common for this generation.

DEVELOPMENTAL TASKS AND AGING FOR TRAUMA SURVIVORS

Most older adults accommodate and adapt to the changes and losses associated with normal aging. However, for some trauma survivors, in particular those who experienced severe and prolonged trauma, such adaptation may be difficult. With the increasing frequency of loss events in old age, and diminishing ability to control such events, the accompanying physical, psychological, cognitive, and social changes related to aging could potentially be exceptionally stressful for those who have been previously traumatized.

In older trauma survivors, many of whom may be predisposed to depression, anxiety, or other mental disorders, the types of skills that are necessary to navigate predictable aging challenges effectively may be absent or already compromised (Gagnon & Hersen, 2000; Weintraub & Ruskin, 1999). Thus, they may be less likely to complete age-specific tasks. For the current cohort of older adults, trauma that occurred before middle adulthood preceded the 1980 introduction of PTSD into the diagnostic nomenclature, possibly affecting older adults' disclosure and acknowledgment of trauma and its effects. An understanding of the potential effects of traumatic experiences and a terminology with which to articulate these sequelae were not available to many persons in this generation, likely leaving many to suffer in silence or receive inadequate/inappropriate care. Additionally, with the rising awareness of the PTSD category, it is more likely that professionals will be more attentive to/recognize traumatic symptoms in older adults. Older adult or provider misattribution of trauma-related symptoms may have deleterious implications for treatment and recovery in this population, including the design of inappropriate or inadequate treatment plans, administration of ineffective or unnecessary psychotherapy or pharmacotherapy, and costly or inefficient medical intervention (Allers, Benjack, & Allers, 1992).

When considering the age-related differences relevant to assessing and treating older adults, it is important to separate the effects of maturation from the effects of cohort (Knight & Satre, 1999). Maturational effects include similarities that are developmentally common or specific to older adulthood, such as adjusting to chronic illness/disability or loss of friends and family due to death. Cohort effects are attributable to specific birth-year-defined groupings rather than to generic aging. For example, cohorts born earlier have lower educational levels and less exposure to psychological concepts (Knight & Satre, 1999). Mental health professionals need to be aware of both maturational and cohort influences on older adults' expression and handling of psychological problems.

Moreover, older adulthood encompasses at least a 30 year range in age, such that the differences between the relatively younger and relatively older adults within that range can be quite substantial. Thus, in light of these differences, mental health providers should consider conceptualizing older adults as young-old (ages 65–74), middle-old (ages 75–84) and old-old (age 85 and older) rather than simply lumping together as "elderly" all those 65 and older (Neugarten, 1974). The life experiences of these subgroups may differ markedly. For instance, those who are currently in the old-old category lived through the Great Depression as teenagers or young adults; whereas the young-old are unlikely to have any mature memories of that period or may not even have directly experienced it. As well, these age-based groupings are likely to be differentiated in terms of both health status (with the old-old having a greater burden of medical illness and physical frailty) and life functioning, and clinical presentation of psychopathology and approach to treatment.

There has, however, been so little examination of the variations in PTSD among older adults that it remains unclear whether or in what ways PTSD features may differ across these various age groupings. One can anticipate certain group generalities based on the age range (e.g., PTSD in the old-old is more likely to be complicated by comorbid medical conditions and functional limitations), but it is not possible to predict with any precision how the PTSD features or processes are apt to be colored by an individual's specific age. It is probably more useful to know whether the individual has a prior history of PTSD symptoms (and for how long) than it is to know his or her age.

EPIDEMIOLOGY, PHENOMENOLOGY, AND COURSE OF TRAUMA-RELATED DISTRESS IN OLD AGE

Unfortunately, most studies examining the impact of traumatic exposure either have not recruited sufficient numbers of older adults to examine age effects or have failed to include older adults at all. Traditionally, trauma in older adults has been relatively neglected within both trauma research and geriatric communities, leading some to label the effects of trauma as "hidden variables" in the lives of older adults (Nichols & Czirr, 1986; Spiro, Schnurr, & Aldwin, 1994). What is known about trauma in older adults is mainly derived from three groups of survivors: who experienced trauma (1) earlier in life during military combat–captivity in World War II and the Korean Conflict, (2) during the Holocaust, and (3) later in life (i.e., natural and man-made disasters). Two fairly comprehensive reviews of PTSD cover these three older adult trauma groups (Averill & Beck, 2000; Falk, Hersen, & Van Hasselt, 1994), so only highlights are presented here.

Within this age group, the psychological sequelae of earlier physical and sexual abuse or rape have received scant empirical inquiry (e.g., Acierno et al., 2001, 2002; Bechtle-Higgins & Follette, 2002). As well, recent traumatization in older adults remains relatively unexamined, despite some empirical investigation on elder abuse (e.g., Comijs, Penninx, Knipscheer, & van Tilburg, 1999; Pillemer & Finkelhor, 1988) and criminal victimization (e.g., Gray & Acierno, 2002). Taken collectively, the limited studies range from specific sample characteristics according to the trauma and settings assessed (e.g., inpatient, outpatient, non-treatment-seeking; men, women) to the methods used (e.g., self-report or clinician-administered measures); thus, integration of findings across studies is difficult.

Epidemiology

No epidemiological studies utilizing a representative sample have examined the incidence or prevalence of traumatic exposure and PTSD in older adults. The best estimates in older adults come from Norris (1992), who assessed the frequency and impact of 10 traumatic events in a sample of 1,000 adults evenly divided among younger, middle-aged, and older adults. Both traumatic exposure and rates of PTSD were highest among younger adults.

Additional community estimates come from research on older veterans. The Normative Aging Study (NAS), which began in the 1960s, is a large, longitudinal cohort study of community-residing male veterans. In this study sample, traumatic exposure to combat (World War II and the Korean Conflict) was high, with those exposed to moderate or heavy combat in World War II having 13.3 times greater risk of PTSD symptoms decades later (Spiro et al., 1994). Prevalence of PTSD in this sample, however, was relatively low. Possible explanatory factors for these prevalence rates may include the high educational and relatively high socioeconomic levels of many of the men who participated in the NAS, both of which are potentially protective factors against emotional distress. It is also possible that those with PTSD may have tended to die earlier in life, such that current data represent differential survival of the psychologically and physically most robust, thus underestimating true lifetime rates of the incidence of disorder (Kasprow & Rosenheck, 2000).

In a review of the trauma literature, Falk and colleagues (1994) found that many studies utilizing clinical samples report that older adult survivors of combat, natural and man-made disasters, and the Holocaust meet diagnostic criteria for PTSD decades after their trauma. Naturally, estimates derived from clinical samples, patients in psychiatric or medical settings, are higher. For example, in those hospitalized for medical illness, Blake and colleagues (1990) found that the prevalence of current PTSD in World War II and Korean Conflict veterans who never sought psychiatric treatment was 8% and 7%, respectively. Among those who had previously sought psychiatric treatment, 37% of World War II and 80% of the Korean Conflict veterans had current PTSD. In an older adult, nonveteran, primary care sample, Davis, Moye, and Karel (2002) found that 14% screened positive for PTSD symptoms. Thus, treatment-seeking medical and psychiatric patients likely have higher rates of PTSD.

Another clinically relevant point is that whereas older adults may not meet full diagnostic criteria, some may have subthreshold PTSD symptoms that warrant clinical and research attention. For example, although most participants in a large community sample of older survivors of World War II bombardments, persecution, resistance and combat did not meet diagnostic criteria for PTSD, many had subthreshold or other negative, long-term aftereffects (Bramsen & van der Ploeg, 1999). This study highlights the need for the examination of subthreshold PTSD as well as other trauma-related symptoms, such as depression, in older adults.

Phenomenology

Clinical presentation of trauma-related distress in older adults appears to be less intense than in younger populations (e.g., Acierno et al., 2002; Davidson, Kudler, Sunders, & Smith, 1990; Fontana & Rosenheck, 1994). For example, in a sample of veterans with PTSD, those who had served in Vietnam reported more severe PTSD, depression, hostility, guilt, derealization, suicidal tendencies, and impairment at work than those who served in World War II (Davidson et al., 1990). Cross-sectional examination of treatment-seeking World War II, Korean Conflict, and Vietnam War veterans revealed that psychiatric symptoms tend to be more prominent as the severity of traumatic exposure increases, and that symptoms are typically less severe for and differ between older and younger veterans (Fontana & Rosenheck, 1994).

Among victims of assault, no significant differences were reported by older compared to younger women in terms of physical and sexual assault characteristics (e.g., prior acquaintance with the perpetrator; perpetrator or survivor were under the influence of a substance; reporting assault to authorities [Acierno et al., 2001]). Physical and sexual assault prevalence, as well as posttraumatic depression and other psychopathology, were lower for older compared to younger women (Acierno et al., 2002). Whereas sexual assault predicted all forms of PTSD symptomatology, as well as depression, in younger adult women, it predicted only PTSD avoidance symptoms in older adults. Similarly, physical assault predicted only PTSD reexperiencing symptoms in older women, but all forms of PTSD symptoms and depression in younger women.

Numerous explanations for these findings include the fact that older adults may tend to misinterpret psychological difficulties as somatic complaints or be reluctant to admit psychological difficulties due to a generational stigma, or that the degree of veterans' postwar popularity and support may buffer subsequent stress reactions. Additionally, considerations may need to be made for recency or elapsed time effects as possible explanations. For example, early life traumas are necessarily more temporally distant in older than in younger adults, confounding age and temporal distance of events, with all that may imply about possible coping with or waning of the emotional sequelae during the intervening time.

Additionally, in response to recent and temporally remote trauma, older adults may experience different symptoms or exhibit differences in coexisting disorders (Goenjian et al., 1994; Yehuda et al., 1996). In a comparison of older and younger adult earthquake survivors, although overall PTSD severity was comparable, older adults exhibited relatively higher arousal and lower intrusive symptoms. Additionally, some trauma-related symptoms, such as dissociation, may be less persistent over time (Yehuda et al., 1996). The relationships between PTSD and type, recency, or severity of trauma; age at time of trauma; presence of current stressful events; and sociocultural factors require further investigation.

Older adults not only reported fewer symptoms of PTSD, major depressive disorder, and generalized anxiety disorder than younger adults following a hurricane but also their psychological reactions were more closely connected to economic consequences of disasters (Acierno, Ruggerio, Kilpatrick, Resnick, & Galea, 2006). Specifically, postinsurance dollar losses and number of days displaced from their homes predicted distress in older adults, but were unrelated in younger individuals. As Acierno and his colleagues point out, many older individuals with fixed incomes may not have the ability to increase their earnings to address unexpected postdisaster expenses, leading to a sense of helplessness and hopelessness.

Course

With a few exceptions (Clipp & Elder, 1996; Norris, Phifer, & Kaniasty, 1994; Port, Engdahl, & Frazier, 2001; Spiro et al., 1994), most studies do not follow survivors longitudinally in old age or for an extended period of time. The limited information on the course of PTSD symptoms primarily comes from former prisoners of war (POWs). For these survivors of prolonged and extreme stress, retrospective recall of PTSD symptoms indicates that the course is variable: Some survivors are continuously troubled; others having waxing and waning symptoms across the lifespan, and still others remain symptom-free (Zeiss & Dickman, 1989). Other investigations with this population indicate immediate and intense onset shortly after the trauma, followed by a gradual decline for several decades, and resurgence later in life (Port et al., 2001).

Factors that mediate the relationship between trauma and PTSD in late life or influence the ebb and flow of symptoms over the life course are not yet known. Possible factors may include occurrence of other stressful or traumatic life events, cognitive appraisal of trauma, and locus-of-control or coping strategies. Clinical lore suggests that occurrence or reactivation of traumatic stress symptoms may in part be due to aging-related life events (e.g., illness; decrements in functional status; bereavement; and changes in occupational, social, and familial roles). Additionally, aging is often tied to loss of control or increased vulnerability in late life. These changes and losses can elicit traumatic thoughts of death, physical injury, and lack of control.

Empirical investigation finds partial support for this hypothesis. Port, Engdahl, Frazier, and Eberly (2002) examined relationships between socioenvironmental factors occurring in later life and PTSD symptoms in community-dwelling former World War II and Korean Conflict POWs. Negative health changes, social support, and death acceptance were significantly related to current PTSD symptomatology, but negative life events were not. This is consistent with an investigation by Schnurr, Lunney, and Sengupta (2004), who found that failure to recover from PTSD is related to factors that occur during and soon after the traumatic event (e.g., lower social support) and exposure to stressful life events in the past year (e.g., illness or financial difficulties).

In other studies, older adults' perceptions of the effects of traumatic exposure in their lives (i.e., desirable vs. undesirable) mediated the effect of traumatic stress (i.e., combat) on PTSD symptoms in later life (Aldwin, Levenson, & Spiro, 1994). Older male veterans reporting predominately desirable effects of military service (e.g., increased mastery, self-esteem, and coping skills) reported fewer PTSD symptoms. Higher instrumental coping and lower emotional coping were found to be significant predictors of psychological well-being in Holocaust survivors (Harel, Kahana, & Kahana, 1988). A lower sense of mastery, negative perception of self-efficacy, and passive response style were associated with higher psychological distress in older adult abuse victims (Comijs et al., 1999). These studies have potential clinical implications in that strategies designed to increase locus of control, perceived self-efficacy, and positive reappraisal may alleviate distress and increase coping ability.

Coping strategies can become compromised in older trauma survivors by challenges associated with aging. For example, upon retirement, an older veteran who has used overwork as a coping mechanism to deal with his trauma may find himself experiencing more symptoms. Other examples might include an older adult with declining eyesight, who previously used reading to distract herself from traumatic memories, or an older adult who, now arthritic, used to engage in arts and crafts to cope with PTSD symptoms. Institutionalization, such as placement in a nursing home, particularly when the older adult had little or no control over the decision or circumstances, may also initiate or reawaken trauma-related symptoms. In addition, cognitively impaired older adults with losses in recent short-term memory may find that long-term memory for traumatic events that occurred in earlier times comes to the forefront.

An interesting debate in the literature regards whether traumatic exposure has an inoculating effect in older adults, thus promoting resilience (stress evaporation model), or whether it predisposes one to negative reactions in subsequent events (residual stress model). This may depend in part on the type and severity of trauma to which the individual was initially exposed. For example, older adult survivors of less severe trauma, such as natural disasters, appear to exhibit both direct and cross-tolerance for subsequent stressors (Knight, Gatz, Heller, & Bengston, 2000; Norris & Murrell, 1988) due to a possible inoculating effect. These survivors have a less distressed response to future stressors, whether similar to or different from the initial trauma. However, some evidence from both Holocaust survivors and combat veterans supports the "vulnerability" perspective (Danieli, 1997). For example, older adult survivors of severe trauma appear to have a heightened vulnerability to subsequent external and internal stressors (Yehuda et al., 1995), such as war (Solomon & Prager, 1992), perceived discrimination (Eaton, Sigal, & Weinfeld, 1982), or medical disease (Peretz, Baider, Ever-Hadani, & De-Nour, 1994).

It is likely that old age, in and of itself, is not a risk factor for the development or course of maladaptive traumatic stress reactions (Solomon & Ginzburg, 1999). In an examination of PTSD symptoms following disasters in the United States, Mexico, and Poland, Norris and colleagues (2002) found that there was no single, consistent effect of age; rather, there were significant interactions of age with the social, economic, cultural, and historical contexts of the different countries. Additionally, older individuals' experiences prior to trauma, including premorbid personality, and personal and family psychiatric history, are likely risk factors of posttraumatic distress (Weintraub & Ruskin, 1999). Risk factors for development of traumatic stress reactions in male military veterans have included adolescent psychological difficulties, age at time of entry into service, degree of combat exposure, societal and familial responses and support postconflict, and POW status (Clipp & Elder, 1996; Schnurr et al., 2004). Though comparatively less is known about individual differences in biology, these factors may also affect an individual's vulnerability to traumatic stress reactions (Yehuda, 1999).

Other Potential Effects of Trauma

The relationships among trauma, PTSD, and physical health have been examined in older military veterans (Schnurr & Spiro, 1999; Schnurr, Spiro, & Paris, 2000). Both combat exposure and PTSD were related to poorer self-reported physical health (Schnurr & Spiro, 1999). Whereas combat exposure had only an indirect effect on health status via the pathway of PTSD, PTSD had a direct effect on health. The association between physician-diagnosed medical disorders and combat-related PTSD symptoms confirms this link (Schnurr et al., 2000; Schnurr & Green, 2004). PTSD symptoms were associated with increased onset of arterial, lower gastrointestinal, dermatological, and musculoskeletal disorders.

The physical health effects of nonmilitary traumas in older men are under-investigated, as are the physical effects of trauma in older women in general—with some exception. In a large cross-sectional study, Stein and Barrett-Connor (2000) examined the association between self-reported sexual assault history and objective parameters of physical health (coronary heart disease, hypertension, diabetes, osteoporosis, obesity, asthma, migraine, thyroid disease, arthritis, and cancer). Sexual assault history was associated with an increased risk of arthritis and breast cancer in older women, and thyroid disease in older men. More recently, utilizing data from a large nationwide survey of older adults, Krause, Shaw, and Cairney (2004) found that exposure to trauma over the life course is associated with poorer physical health. Furthermore, traumatic events experienced between the ages of 18 and 30, and 31 and 64, appeared to exert the greatest effects on health, with traumatic events arising within the family being the most consequential. Future investigation should examine the mediating role of PTSD in the relationships between trauma and physical health.

Although the most widely researched mental disorder in older adults is depression, rarely has investigation in this area explored associations with prior trauma (Cook, Areán, Schnurr, & Sheikh, 2001; Tyler & Hoyt, 2000). Symptoms of depression are particularly important in terms of their linkage to suicide. Suicide is a major public health problem, with rates increasing with age (Pearson & Brown, 2000). Given trauma's association with suicidality (Adams & Lehnert, 1997), the potential overlap of influences makes this area worthy of further investigation. For example, in a random sample of over 350 charts of women age 55 and older from a medical hospital and two state psychiatric facilities, Osgood and Manetta (2000–2001) found that women with identified suicidal ideation had experienced significantly more prior victimization than women without suicidal ideation. In a retrospective cross-sectional study of consecutive first admissions of Jewish patients to a day hospital, severity of depression and exposure to the Holocaust were independently associated with suicidal ideation (Clarke et al., 2004). These findings are of particular interest, because the relationships remained significant after researchers controlled known risk factors for suicide, such as poor social support, history of suicide attempts, and other negative life events.

Although general rates of distress and PTSD-related morbidity in older adults appear lower than those of younger populations, other aging-related issues, such as difficulty accessing and receiving mental health services due to mobility and sensory impairments, deserve consideration (Zeiss, Cook, & Cantor, 2003). Additionally, evidence has shown the importance of social support in the recovery process in younger populations (Brewin, Andrews, & Valentine, 2000), and it likely has a similar effect in older adults. However, older adults often experience a reduction in their social connections with the loss of a spouse, relatives, and friends, or as a result of cognitive impairment or frailty. As a result of the decline in social support, older adults have more difficulty learning about mental health options, traveling to and from health care visits, and accessing the proper coverage for psychotropic medications.

Populations Rarely Studied

As previously noted, the most widely studied groups of older trauma survivors have been those who experienced combat–captivity or Holocaust-related trauma earlier in life, and those who experienced trauma later in life, as examined primarily in survivors of natural and man-made disasters. There has been limited investigation on interpersonal violence and criminal victimization as types of trauma in later life. Additionally, there has been relatively little examination of trauma and its effects in older adult women; ethnic and racial minorities; or lesbian, gay, and bisexual people. The lack of information in these areas is particularly limiting given that the heterogeneity of aging populations in industrialized countries is expected to increase.

Existing research indicates that treatment-seeking older adult crime victims are often multiply traumatized and experience moderate to severe levels of psychopathology, namely, PTSD, depression, and panic (Gray & Acierno, 2002). D'Augelli and Grossman (2001) examined the lifetime verbal and physical victimization of lesbian, gay, and bisexual older adults based on their sexual orientation. More than 25% had been threatened with violence, and 16% had been punched, kicked or beaten. Those who had been physically attacked reported lower self-esteem, more loneliness, and poorer mental health, including more suicide attempts, than others.

In the abuse literature on older adults (as one form of crime to which some older adults are particularly susceptible), PTSD remains largely unstudied. With few exceptions, empirical evidence on abuse of older adults is derived mainly from surveys of professionals or from highly selected samples, such as those assessed because of reports to adult protective services (Comijs, Pot, Smit, & Jonker, 1998; Pillemer & Finkelhor, 1988). In a large-scale, random sample survey of older adult abuse and neglect, roughly equal numbers of men and women were victims, although women suffered more serious abuse (Pillemer & Finkelhor, 1988). Those who were mistreated (chronic verbal aggression, physical aggression, or financial abuse) had significantly higher levels of psychological distress than those who had not been mistreated (Comijs et al., 1999).

Relationship with Cognitive Functioning and Impairment

The neurobiology of trauma and PTSD in older adults has received limited empirical attention. Investigation of younger adults reveals neurochemical, neurological, and neuropsychological impairments that accompany PTSD (Bremner, Southwick, & Charney, 1999; Neumeister, Henry, & Krystal, Chapter 9, this volume; Southwick et al., Chapter 10, this volume). Prolonged stress or exposure to glucocorticoids can have an adverse effect on cortical function, which may contribute to memory impairment (Sapolsky, 2000). However, because aging individuals with cognitive impairment have typically been excluded from PTSD studies, little is known about the relationship between PTSD, or a history of extreme trauma, and the cognitive impairments typical in later life.

Some evidence demonstrates that individuals exposed to prolonged and extreme trauma, such as former POWs or survivors of Nazi concentration camps, have neurological concomitants decades after traumatic exposure (Golier et al., 2002; Sutker, Galina, West, & Allain, 1990; Sutker, Vasterling, Brailey, & Allain, 1995). For example, severity of POW captivity stress, as reflected by trauma-induced weight loss, was predictive of long-term compromise in cognitive performance. Former POWs from the Korean Conflict and World War II reporting captivity weight losses of greater than 35% performed more poorly on a host of memory tasks than did POWs who reported lower weight loss percentages and non-POW combat veterans (Sutker et al., 1990). More specifically, captivity weight loss was associated with impaired learning and memory performance, whereas PTSD was associated with attention, mental tracking, and executive system deficits (Sutker et al., 1995).

Several investigators have suggested that severe and prolonged trauma or a history of PTSD may place aging individuals at increased risk for cognitive decline and onset of dementia (see, Cook, Ruzek, & Cassidy, 2003). One plausible explanation is that trauma causes both PTSD and vulnerability to subsequent cognitive impairment. Other explanations include the viewpoint that PTSD may moderate or mediate the effects of earlier trauma on cognitive impairment. For example, there was a significant inverse association between age and recall in Holocaust survivors with current PTSD, but not in Holocaust survivors without current PTSD or in healthy Jewish adults not exposed to the Holocaust (Golier et al., 2002). Explicit memory impairments in those with PTSD were not accounted for by depression or education, strengthening the evidence that memory decline may be accelerated in PTSD.

From another perspective, there is the possibility that cognitive impairment may disinhibit symptoms of PTSD that might have been less apparent or more controlled for years. Along such lines, Floyd, Rice, and Black (2002) outlined a cognitive aging explanation for the recurrence of PTSD. They proposed that age-related decreases in attention make the intrusion of trauma-related memories more likely, and that an increase in intrusive memories, combined with age-related decrements in working memory, explicit memory, and prospective memory, are apt to raise the level of subjective distress associated with the memories, resulting in a recurrence of PTSD.

As information on the relationship among traumatic exposure, PTSD, and cognitive impairment in older adults continues to accumulate, there is some clinical basis for alerting mental health professionals to issues in the recognition and management of PTSD in cognitively impaired patients. Scientifically informed mental health providers can play various roles in addressing the needs in this area: from educating and training other health professionals (e.g., primary care physicians, frontline long-term care staff) with respect to a possible association between past trauma, PTSD, and cognitive impairment to teaching trauma and PTSD assessment skills, to providing tips regarding clinical management of PTSD and related behavioral problems (Cook, Cassidy, & Ruzek, 2001).

In a thoughtful discussion of the relationship between PTSD and cognitive dysfunction, Danckwerts and Leathem (2003) cautioned against overgeneralizing from specific populations (e.g., veterans) to the general population, explaining that the existing literature has inconsistencies and anomalies due to a variety of methodological shortcomings, including sample restrictions, comorbidity of PTSD with other mental health conditions, and blurring of cognitive and emotional difficulties. Much of the research in this area has been conducted on small samples comprised primarily of veterans, who typically are older male survivors of combat. Most research involves participants who meet criteria for other psychiatric disorders, as well as PTSD, thus obfuscating what is attributable to PTSD per se. For example, whereas several brain imaging studies have shown abnormalities in the hippocampus, a part of the brain that has a major role in regulating stress response, this has been a finding in several psychiatric disorders, including major depressive disorder and borderline personality disorder, rather than being specific to PTSD (Sala et al., 2004).

The examination of cognitive difficulties in these studies generally fails to distinguish changes caused by emotional distress from those caused by actual brain impairment (Danckwerts & Leathem, 2003). Most of the cognitive information is based on self-report and has not been verified against current observations in everyday settings. Thus, as Danckwerts and Leathem (2003) point out, the relationships between PTSD and cognitive variables are vague and will remain elusive until more sophisticated research is accomplished. Improvements that are needed include better specification of the type, subjective severity, and the time elapsed since the original trauma; exclusion or separate analysis of patients with PTSD and preexisting and comorbid conditions; use of cognitive assessments with demonstrated ecological validity; and more comprehensive neuropsychological testing in association with neuroimaging (Danckwerts & Leathem, 2003).

METHODOLOGICAL CONSIDERATIONS

As noted throughout this chapter, there are numerous methodological challenges in conducting mental health research on older traumatized individuals. These include, but are not limited to, cognitive, sensory, and functional impairments that may affect the experience, expression, or reporting of trauma-related symptoms; the typical presence of one or more chronic medical illnesses; decreased mobility; and unfamiliarity or reluctance with respect to engaging in mental health issues.

There may also be operational difficulties in employing commonly accepted measures with older adults, and researchers may encounter a shortage of adequately validated measures for use with this age group, or for older adults with particular handicaps. Some general guidelines have been proposed for adapting standard measures validated in younger populations for use with older, medically frail, or cognitively impaired older adults (Hunt & Lindley, 1990; Lichtenberg, 1999). One readily accomplished example of adapting self-report measures is to enlarge the font, and increase the spacing and shading of the printed letters to allow for easier reading. Other suggestions for using and adapting specific PTSD measures, such as the PTSD Checklist (PCL; Weathers, Litz, Herman, Huska, & Keane, 1993) and the Mississippi Scale for Combat-Related PTSD (M-PTSD; Keane, Caddell, & Taylor, 1988) in older adults have been made by Cook and colleagues (2005). These ideas include altering the response format to simplify the cognitive demand placed on older adult respondents. For example, the standard M-PTSD may have limited use with older adults who have limited cognitive abilities and may find its response format confusing given that the wording of its response options repeatedly shifts directions. Likewise, the PCL's 5-point scale can be difficult for cognitively impaired older adults to understand. Cook and colleagues suggested that a 3-point scale (e.g., *Not at all/Moderate/Severe* or *Not at all/Once per week or less/Several times per week*) might be more easily administered. When major cognitive impairments, extreme sensory deficits, and/or aphasias are present, any self-reports may need to be limited in length, if used at all. In such circumstances, additional sources of information on PTSD, such as observational measures, existing medical records, and collateral reports, should be utilized.

Methodological challenges similar to those described for assessment situations also must be faced in the psychological treatment of older adult populations.

CURRENT STATE OF THE ART

Treatment

The provision of psychotherapy services to the older adult population had a precarious beginning. Sigmund Freud (1904/1959) alleged that the application of psychological treatment to adults over the age of 50 was ineffective. In support of these claims, he asserted that older adults had limitations in ego or cognitive functioning; that analysis would encompass longer lifetimes, thus going on *ad infinitum*; and that the limited longevity of older patients made it less worthwhile to give them the benefits of psychoanalysis as compared to younger people, who would benefit for a longer time from it. Freud's views dominated clinical thinking for many years and as a result impeded the advancement of psychological treatment for older adults.

In 1959, Rechtschaffen provided a landmark summary of anecdotal and case report data on psychotherapy with older adults. From that time forward, geropsychological treatment research has steadily continued to grow. For a recent review, see Cook, Gallagher-Thompson, and Hepple (2005).

Whereas the treatment of PTSD in younger populations has received a great deal of attention and investigation, available information about the application and provision of psychotherapy to traumatized older adults exists predominantly in the form of uncontrolled case studies and anecdotal reports, mainly with veterans (Boehnlein & Sparr, 1993; Lipton & Schaffer, 1986; Molinari & Williams, 1995; Snell & Padin-Rivera, 1997). Even if taken together, these scattered, scarce sources of information do not provide a strong empirical base, or even an indication of best practices, with respect to providing psychological intervention to traumatized older adults.

Effective PTSD treatment for younger adults often focuses on or includes repeated exposure to images or memories associated with the traumatic events (Foa, Keane, & Friedman, 2000). Because the physical health of older adults is often compromised and direct trauma processing can produce strong physiological effects, such as changes in heart rate and respiration that may exacerbate existing health conditions, there have been questions about the benefit of using exposure or treatments involving disclosure of trauma material in older adults (Coleman, 1999; Hankin, 1997; Hyer & Woods, 1998; Kruse & Schmitt, 1999). Though there is no need to disregard clinical judgment, it is important to note that these warnings are not based on empirical evidence, and that a lack of studies to support the efficacy of these approaches does not prove that these treatments ought not to be used (Shalev, Friedman, Foa, & Keane, 2000). There is one documented case study on the successful use of exposure therapy with a 57-year-old woman with current PTSD related to childhood sexual abuse (Russo, Hersen, & Van Hasselt, 2001). Imaginal exposure was one component of a 60-session psychotherapy package over a 24-month period. This case study at least demonstrates that imaginal exposure can be successfully applied to treat PTSD in a middle-aged individual when used in a well-controlled fashion. Until systematic evaluations of the use of exposure with older adults are conducted (and examine whether there are adverse effects or low adherence rates to exposure in this population), practitioners are advised not to exclude use of exposure techniques automatically, but to proceed with caution (see Cook, Schnurr, & Foa, 2004).

The published cautions about the use of exposure therapy are not an issue unique to older adults. In fact, there appears to be caution in the trauma stress field more broadly about the use of this technique, as evidenced by reports that frontline clinicians in "real-world" settings rarely use this treatment (Becker, Zayfert, & Anderson, 2004; Fontana, Rosenheck, Spencer, & Gray, 2002; Rosen et al., 2004). In addressing the perceived and actual barriers that interfere with adoption of exposure techniques, Cook and colleagues (2004) made specific suggestions for bringing this effective modality into more routine practice.

A case series of older adults with PTSD successfully treated with life review therapy was reported by Maercker (2002). This may be another good therapy for clients who

need to focus on/confront past trauma issues. Life review therapy, a technique originally developed for use in geriatric populations, involves the reworking of previously
experienced conflicts to gain a better understanding and acceptance of one's past. It is
based on the work of Butler (1963) and Erikson (1959, 1982).

There is one manualized psychoeducational treatment program for older combat
veterans, developed at the Cleveland VA Medical Center (Clower, Snell, Liebling, &
Padin-Rivera, 1996, 1998). This program involves therapy education, PTSD education,
life review, stress management, building of social support, anger management, working
through of grief and loss, and forgiveness. A description of its delivery can be found in
Snell and Padin-Rivera (1997).

One intervention that has been effective for depressed older adults is interpersonal
psychotherapy (IPT; Klerman, Weissman, Rounsaville, & Chevron, 1984; for a review,
see Miller et al., 2001). Regardless of etiology of psychopathology, the underlying premise of IPT is to understand and renegotiate current relationships that play a key role in
reducing psychiatric symptoms, restoring function, and possibly preventing future disturbance. IPT focuses on four general relationship problem areas: role transition, role
dispute, abnormal grief, and interpersonal deficit. Given the relationship between
social support and the development and maintenance of PTSD (Brewin et al., 2000), the
potential application of this therapy to older trauma survivors appears to warrant attention.

Special Needs and Concerns

Although some therapeutic interventions are similar to those used in treating PTSD
with younger adults (e.g., education about symptoms, enhancement of social support,
and teaching coping skills to manage symptoms more effectively), special, unique considerations in treating older adults are important. Many mental health clinicians have
insufficient training with respect to aging and older adults. For example, the majority of practicing psychologists lack formal training in geropsychology and perceive
themselves as needing additional education (Qualls, Segal, Norman, Niederehe, &
Gallagher-Thompson, 2002). Although the psychotherapist's generic training in his or
her mental health profession (e.g., psychology, psychiatry, or social work) provides the
broad skills necessary in the assessment and treatment of all adults, often these are not
sufficient, and additional particular knowledge and skills are essential in the provision
of psychotherapeutic interventions to older adults.

As mentioned previously, although similar in some ways to younger age groups,
older adults have unique developmental issues and cohort perspectives. The *Guidelines
for Psychological Practice with Older Adults* (American Psychological Association, 2004)
advise clinicians to gain knowledge about theory and research in aging, including
social-psychological dynamics of the aging process, and biological and health-related
aspects of aging, as well as an understanding of common clinical issues in this age
group, such as cognitive changes and problems in daily living (e.g., ability to function
independently). Those interested in working with older individuals should familiarize
themselves with the myths and actualities of aging, maturational and cohort differences, and potential modifications of assessment and treatment techniques. It is also
advisable to learn about chronic illness and its psychosocial impact, management of
chronic pain, factors influencing adherence to medical treatment, rehabilitation methods, and assessment of behavioral signs of negative medication effects (Knight & Satre,
1999).

The choice of an optimal form of psychotherapy for an older adult patient can be influenced by a number of factors, including the primary and secondary mental disorders present, severity and timing of the problem (e.g., acute vs. chronic), cognitive functioning, likelihood of effectiveness for the particular patient (e.g., existing skills), patient preference and motivation, and ethnic and cultural considerations (Cook, Gallagher-Thompson, & Hepple, 2005). Other important factors to consider include history of previous mental health treatment and response. For example, in treating recurrent psychiatric disorders in late life, Knight and Satre (1999) recommend focusing on symptom management and rehabilitative maximization of function rather than "cure." Because it is questionable whether severe or chronic PTSD in older adulthood can ever be completely eradicated, to treat it successfully, it may be useful to consider and target psychosocial functioning goals that the intervention is intended to accomplish.

The goals of mental health treatment should be highlighted for older adults to reinforce the purpose and facilitate the direction of treatment. As a general matter, psychotherapy with older adults often requires a collaborative approach, organized around a few clearly outlined goals, that employs a more active or task-focused approach than is customary with younger patients (Gallagher-Thompson & Thompson, 1996). Specifically, rather than giving suggestions or expecting the older patient to infer answers, Knight and Satre (1999) suggest that because there is a normal age decline in fluid intelligence, providers may need to lead older adults to conclusions.

Provision of mental health treatments to older adults often occurs at a slower pace due to their possible sensory problems and slowed learning rates (Gallagher-Thompson & Thompson, 1996). Repetition is very important in the learning process; thus, information should be presented in both verbal and visual modalities (i.e., on chalkboards and handouts) to assist older patients in encoding and retaining information. Older clients are frequently encouraged to take notes to help aid memory retention and increase efficacy of psychotherapy (Knight & Satre, 1999). It is also advisable that educational information or assignments be in bold print, or that sessions be tape-recorded for review. To facilitate therapy with older individuals with sensory problems, particularly hearing and vision impairments, additional adaptations should be made available, such as providing pocket talkers to assist in hearing or taking steps to eliminate glare from materials for sight-impaired individuals.

Providing mental health treatment to older individuals often requires flexibility in scheduling, location, and collaboration. Older adults often have reduced mobility or a reluctance to travel in bad weather conditions, a greater chance of hospitalization, or responsibilities to care for infirm relatives, all of which may necessitate more frequent changes in appointments (American Psychological Association Working Group on the Older Adult, 1998). Thus, sometimes brief hospital visits, telephone sessions, or correspondence via letters may need to be arranged to maintain contact and continuity of care. Transportation services and facilitated access to buildings may also be needed. Furthermore, because older adults often have concurrent physical and social problems, consultation and coordination with other health service providers are often critical (American Psychological Association Working Group on the Older Adult, 1998). When an older adult has become dependent on a formal or informal care provider for assistance, it may be crucial to engage the care provider in the treatment process. An example of this is the innovation of treating depression in older patients with dementia via training caregivers in behavioral interventions, such as changing others' responses to a patient's maladaptive behavior or modifying the patient's living environment to decrease contact with objects that might be related to his or her agitation (Teri, Logsdon, Uomoto, & McCurry, 1997).

Older adults may hold negative stereotypes about mental health issues and services that may result in reluctance to accept or engage in therapy or limit self-disclosure and acknowledgement of symptoms. Some of these myths include the following: Only "crazy" people seek mental health treatment; psychological problems indicate moral weakness; therapy constitutes an invasion of privacy; adults do not need to ask for help; and therapy has no relevance (Glantz, 1989). Thus, some geriatric mental health providers advise that mental health treatment for the current cohort of older adults begin with a short, preparatory introduction to treatment (Cook, Gallagher-Thompson, & Hepple, 2005; Gallagher-Thompson & Thompson, 1995). In this introductory "role induction" period, incorrect assumptions can be corrected, and roles and expectations should be clarified.

Although cognitive impairments are not inevitable in older adults, they are more prevalent in this age group than in any other. Persons with moderate to severe memory loss or decreased capacity for judgment and problem solving are generally not deemed suitable for traditional psychotherapy. However, the symptoms and behaviors of persons with dementia can be affected by social, psychological, and environmental contexts; thus, patients with memory impairments are able to derive some benefit from psychological interventions (for a review, see Kasl-Godley & Gatz, 2000). The U.S. Department of Veterans Affairs (1997) published practice guidelines for the assessment of competency and capacity in older adults. Additionally, Lichtenberg and colleagues (1998) articulated standards for psychological services in long-term care facilities.

Like younger adults, older adults who typically present to a mental health provider may not be aware that their current difficulty is related to past traumatic experiences. Accordingly, older adults who are experiencing negative effects of unresolved trauma may present with somatic complaints or other clinical needs; thus, without evaluation, the trauma connections may be missed entirely. Even when mental health needs (e.g., depression, anxiety) are identified, patients and providers may not recognize or focus on potential links to trauma. It is also possible that if older adult trauma survivors do present to a mental health provider, they may request inappropriate or unusual treatments (e.g., "truth serum") to rid themselves of trauma-related distress (Hankin, 1997). Thus, appropriate and effective mental health service provision may depend strongly on increasing providers' ability to recognize trauma-related issues.

Providing preliminary education about trauma and its effects, and assuring older adults that their reactions are understandable in the light of extreme stress, may enhance their engagement in and adherence to treatment. Additionally, providing information that PTSD is treatable may kindle hope for recovery. Clinical endeavors should include teaching methods of managing PTSD and other traumatic stress symptoms that will continue to benefit older adults after formal therapy ends.

The geropsychology literature has discussed innovative ways of improving the availability of mental health services for older adults, such as telehealth or telephonic use of therapies, mental health treatment in managed primary care (Gallo & Lebowitz, 1999), home- and community-based interventions (Rabins et al., 2000), and the inclusion of psychotherapists on integrated, interdisciplinary care teams (Zeiss & Gallagher-Thompson, 2003).

Three particularly innovative mental health programs and their delivery to older adult trauma survivors have been described. Lew (1991) discussed the need for improved access to mental health services for older Cambodian refugees living in California. She pointed out that in addition to high rates of exposure to traumatic experiences, many older refugees experience myriad social, economic, and cultural stressors (e.g., limited education, difficulty speaking English, lack of transportation) that can be viewed as barriers to mental health services. Her proposed interventions include community outreach, hiring intermediaries to act as linguistic translators and cultural interpreters between older adult refugees and their providers, as well as designing and implementing health education classes in accessible and familiar locations (e.g., Buddhist temples). These suggestions have direct applicability to older adult trauma survivors of genocidal atrocities such as Rwanda and Kosovo.

Vinton (1992) noted that only 132 of the 6,026 women in 25 shelters for battered women throughout Florida were age 60 or over. Of those shelters, only two offered special programming for older women. Vinton argued that older women who are victims of spousal or physical abuse by other family members are more often viewed as "abused elders" than as battered women, which commonly leads to adopting a more paternalistic and medicalized approach to intervention. Suggestions for programmatic and policy changes to improve the care of older battered women include dispensing medication, making shelters handicapped accessible, and improving linkages between aging agencies and battered women's shelters.

Acierno, Rheingold, Resnick, and Stara-Riemer (2004) tested the efficacy of a brief video-based intervention in older victims of recent crime. The video included psychoeducation and normalization about common reactions to crime, behavioral coping strategies to manage distress (i.e., exposure-based and behavioral activation interventions), and ways to increase awareness of safety planning strategies. In a randomized, controlled trial, 116 older crime victims received either standard advocate-based services plus the video intervention or standard services alone. Despite the increase in knowledge, those who participated in the video intervention did not significantly reduce their anxiety and depression more than the victims assigned to standard practice of care. This is clearly a first strong step in advancing the geriatric trauma prevention and intervention literatures.

Although this chapter is not specifically designed to address pharmacotherapy, we would be remiss not to discuss this modality briefly. To date, no pharmacotherapy study of PTSD has focused solely on older individuals or separately examined aging as a factor in safety or effectiveness (Weintraub & Ruskin, 1999). The rate at which the body metabolizes medications slows with age, such that older individuals (particularly the old-old and those with a greater burden of comorbid medical conditions) are more susceptible to building up toxic drug blood levels and may experience intolerance or adverse reactions at lower dosages than would be typical in younger adults. Although older adults with PTSD are typically given the same medications as prescribed for younger adults, a general rule of thumb in delivering safe and effective pharmacotherapy to older adults is to start at a lower dose and titrate the dosage more slowly and cautiously. Because older adults are likely to take more medications than younger adults, another important consideration is to pay greater attention to potential drug–drug interactions, given that this risk increases exponentially with the number of medications prescribed. In the general adult population, two selective serotonin reuptake inhibitors, paroxetine and sertraline, have been shown to be efficacious in the treatment of PTSD in randomized controlled trials and are recommended as first-line treatments (for a review, see Friedman, 2003). These medications have generally been found to be safe for and well tolerated by older adults. Although research with other classes of medications is limited, Friedman (2003) provides information to guide effective choices for PTSD pharmacotherapies.

CHALLENGES FOR THE FUTURE

The demographic imperative associated with an aging population should be a call to action for trauma clinicians. Despite limited professional and public awareness or interest in this issue, traumatic exposure can have substantial and pervasive negative effects for older adults, including deleterious changes in physical and mental health, impairment of functional status, and increased utilization of health care services (Schnurr & Spiro, 1999; Spiro et al., 1994). Importantly, PTSD is one of the main concomitants of trauma exposure in older, as well as younger, adults and is strongly associated with degree of traumatic exposure (Fontana & Rosenheck, 1994).

Due to the drastic increase in the number of older people with a psychiatric disorder that is anticipated by 2030, Jeste and colleagues (1999) recommended the formulation of a 15- to 25-year plan for mental health research on older adults. Aging baby boomers are expected to show greater rates of depression, anxiety, and substance abuse than prior generations of older adults. Moreover, given their influence over the expression and experience of psychological distress in general, both age cohort effects and developmental aging issues will likely affect the clinical presentations and syndromes of trauma-related distress in future generations of older adult trauma survivors. Traumatized older individuals in the future may admit to more psychological symptomatology because of either fewer stigmas for members of their generation or their ability to identify psychological problems and seek treatments.

Unfortunately, knowledge in the geriatric trauma field is far from complete. More information is needed regarding the prevalence, symptom expression, and course of trauma-related symptoms, particularly age-specific psychosocial and behavioral responses, mediators and moderators of negative and positive consequences, and assessment techniques. Many questions regarding treatment and delivery remain unexamined or only partially answered.

As long as there continue to be major gaps in fundamental knowledge about the understanding, assessment, and diagnosis of older adult trauma survivors, it is difficult to engage in treatment development or to plan effective means of organizing, implementing, and delivering needed services to these individuals. Questions to be addressed include the following: Who among older adult trauma survivors continue to experience symptoms of distress and require intervention? Where are they located? How do they present clinically?

It is likely that trauma-related distress in older adults frequently goes undetected and untreated, because of either older adult or provider misattribution of trauma-related symptoms or perhaps misperception. One of the first tasks for scientifically informed practitioners is to assess their older patients for traumatic exposure and its effects. Such screening is particularly important for high-risk groups such as older individuals likely to have experienced trauma, whether in the remote past (e.g., veterans, Holocaust survivors, refugees or immigrants) or more recently (e.g., those identified in rape crisis centers or older adult abuse contexts), and probably also those who present with anxiety disorders.

Identifying the needs of older adult trauma survivors, including treatment preferences, is an important next step in the evolution of care. As more complete answers to the preceding questions are revealed, clinicians will need to engage in theory-guided intervention research, including feasibility and efficacy trials to determine optimal methods of intervention, the safety and durability of treatment effects, as well as factors that affect engagement, adherence, and outcome. Additionally, to make these treatments broadly accessible, the research must evaluate their acceptability and tolerability by older adults, and their transportability and deliverability across a variety of settings.

Future directions also include the development of appropriate programs of services (e.g., outreach, education, community networking). The geropsychology and trauma fields need to collaborate to develop/adapt, test, and refine mental health services for older trauma survivors. In addition to addressing their needs, it is important that these programs support survivors' strengths, as well as identify and eliminate barriers (e.g., financial, social) that impede their access to effective treatments.

In summary, the impact and effects of trauma can be long-lasting, and indeed PTSD does occur in older adults. For older adults with trauma in early life, the psychiatric symptom course is variable: Some individuals are continuously troubled; others experience waxing and waning of symptoms over time; and still others remain mostly symptom-free. Trauma-related distress in older adults, though perhaps less intense in some circumstances, resembles PTSD in younger adults. Assessment needs to be comprehensive and, in special circumstances such as cognitive impairment, requires special adaptation of the assessment approach, such as use of behavioral observation and collateral reports. This chapter further illuminates the need for the geropsychology and traumatic stress fields to integrate both the conceptualization of and treatment approaches to affected individuals. Despite growth in the separate fields, the geriatric trauma field remains in need of considerable exploration and greater development.

AUTHORS' NOTE

This chapter was written by George Niederehe in his private capacity. No official support or endorsement by the National Institute of Mental Health, National Institutes of Health, or U.S. Department of Health and Human Services is intended or should be inferred.

REFERENCES

Acierno, R., Brady, K. L., Gray, M., Kilpatrick, D. G., Resnick, H. S., & Best, C. L. (2002). Psychopathology following interpersonal violence: A comparison of risk factors in older and younger adults. *Journal of Clinical Geropsychology, 8,* 13–23.

Acierno, R., Gray, M. J., Best, C. L., Resnick, H. S., Kilpatrick, D. G., Saunders, B. E., et al. (2001). Rape and physical violence: Comparison of assault characteristics in older and younger adults in the National Women's Study. *Journal of Traumatic Stress, 14,* 685–695.

Acierno, R., Rheingold, A. A., Resnick, H. S., & Stara-Riemer, W. (2004). Preliminary evaluation of a video-based intervention for older adult victims of violence. *Journal of Traumatic Stress, 17,* 535–541.

Acierno, R., Ruggiero, K. J., Kilpatrick, D. G., Resnick, H. S., & Galea, S. (2006). Risk and protective factors for psychopathology among older versus younger adults following the 2004 Florida hurricanes. *American Journal of Geriatric Psychiatry, 14,* 1051–1059.

Adams, D. M., & Lehnert, K. L. (1997). Prolonged trauma and subsequent suicidal behavior: Child abuse and combat trauma reviewed. *Journal of Traumatic Stress, 10,* 619–634.

Aldwin, C. M., Levenson, M. R., & Spiro, A., III. (1994). Vulnerability and resilience to combat exposure: Can stress have life-long effects? *Psychology and Aging, 9,* 34–44.

Allers, C. T., Benjack, K. J., & Allers, N. T. (1992). Unresolved childhood sexual abuse: Are older adults affected? *Journal of Counseling and Development, 71,* 14–17.

American Psychological Association. (2004). Guidelines for psychological practice with older adults. *American Psychologist, 59,* 236–260.

American Psychological Association Working Group on the Older Adult. (1998). What practitioners should know about working with older adults. *Professional Psychology: Research and Practice, 29,* 413–427.

Averill, P. M., & Beck, J. G. (2000). Posttraumatic stress disorder in older adults: A conceptual review. *Journal of Anxiety Disorders, 14,* 133–156.

Baltes, P. B., & Baltes, M. M. (1990). Psychological perspectives on successful aging: The model of selective optimization with compensation. In P. B. Baltes & M. M. Baltes (Eds.), *Successful aging: Perspectives from the behavioral sciences* (pp. 1–34). Cambridge, UK: Cambridge University Press.

Bechtle-Higgins, A., & Follette, V. M. (2002). Frequency and impact of interpersonal trauma in older women. *Journal of Clinical Geropsychology, 8,* 215–226.

Becker, C. B., Zayfert, C., & Anderson, E. (2004). A survey of psychologists' attitudes towards and utilization of exposure therapy for PTSD. *Behaviour Research and Therapy, 42,* 277–292.

Blake, D. B., Keane, T. M., Wine, P. R., Mora, C., Taylor, K. L., & Lyons, J. A. (1990). Prevalence of PTSD symptoms in combat veterans seeking medical treatment. *Journal of Traumatic Stress, 3,* 15–27.

Boehnlein, J. K., & Sparr, L. F. (1993). Group therapy with World War II ex-POWs: Long term posttraumatic adjustment in a geriatric population. *American Journal of Psychotherapy, 47,* 273–282.

Bramsen, I., & van der Ploeg, H. M. (1999). Fifty-years later: The long-term psychological adjustment of ageing World War II survivors. *Acta Psychiatrica Scandinavica, 100,* 350–358.

Bremner, J. D., Southwick, S. M., & Charney, D. (1999). The neurobiology of posttraumatic stress disorder: An integration of animal and human research. In P. A. Saigh & J. D. Bremner (Eds.), *Posttraumatic stress disorder: A comprehensive text* (pp. 103–143). Boston: Allyn & Bacon.

Brewin, C. R., Andrews, B., & Valentine, J. D. (2000). Meta-analysis of risk factors for posttraumatic stress disorder in trauma-exposed adults. *Journal of Consulting and Clinical Psychology, 68,* 748–766.

Butler, R. N. (1963). The life review: An interpretation of reminiscence in the aged. *Psychiatry, 26,* 65–76.

Clarke, D. E., Colantonio, A., Heslegrave, R., Rhodes, A., Links, P., & Conn, D. (2004). Holocaust experience and suicidal ideation in high-risk older adults. *American Journal of Geriatric Psychiatry, 12,* 65–74.

Clipp, E. C., & Elder, G., Jr. (1996). The aging veteran of World War II: Psychiatric and life course insights. In P. E. Ruskin & J. A. Talbott (Eds.), *Aging and posttraumatic stress disorder* (pp. 19–51). Washington, DC: American Psychiatric Press.

Clower, M. W., Snell, F. I., Liebling, D. S., & Padin-Rivera, E. (1996). *Senior Veterans Program: A treatment program for elderly veterans with war-related post-traumatic stress disorder: Therapist notes.* Cleveland, OH: Department of Veterans Affairs.

Clower, M. W., Snell, F. I., Liebling, D. S., & Padin-Rivera, E. (1998). *Senior veterans PTSD workbook: A personal journey.* Cleveland, OH: Department of Veterans Affairs.

Coleman, P. G. (1999). Creating a life story: The task of reconciliation. *Gerontologist, 39,* 133–139.

Comijs, H. C., Penninx, B. W., Knipscheer, K. P., & van Tilburg, W. (1999). Psychological distress in victims of elder maltreatment: The effects of social support and coping. *Journals of Gerontology: Series B, Psychological Sciences and Social Sciences, 54,* P240–P245.

Comijs, H. C., Pot, A. M., Smit, H. H., & Jonker, C. (1998). Elder abuse in the community: Prevalence and consequences. *Journal of the American Geriatrics Society, 46,* 885–888.

Cook, J. M., Areán, P. A., Schnurr, P. P., & Sheikh, J. (2001). Symptom differences of older depressed primary care patients with and without history of trauma. *International Journal of Psychiatry in Medicine, 31,* 415–428.

Cook, J. M., Cassidy, E. L., & Ruzek, J. I. (2001). Aging combat veterans in long-term care. *National Center for PTSD Clinical Quarterly, 10,* 25–29.

Cook, J. M., Elhai, J., Cassidy, E. L., Ruzek, J. I., Ram, G. D., & Sheikh, J. I. (2005). Assessment of trauma exposure and posttraumatic stress disorder in older, long-term care veterans: Preliminary data on psychometrics and PTSD prevalence. *Military Medicine, 170,* 862–866.

Cook, J. M., Gallagher-Thompson, D., & Hepple, J. (2005). Psychotherapy across the life cycle: Old age. In G. Gabbard, J. Beck, & J. Holmes (Eds.), *Concise Oxford textbook of psychotherapy* (pp. 381–390). Oxford, UK: Oxford University Press.

Cook, J. M., Ruzek, J. I., & Cassidy, E. L. (2003). Post-traumatic stress disorder and cognitive impairment in older adults: Awareness and recognition of a possible association. *Psychiatric Services, 54,* 1223–1225.

Cook, J. M., Schnurr, P. P., & Foa, E. B. (2004). Bridging the gap between posttraumatic stress disorder research and clinical practice: The example of exposure therapy. *Psychotherapy: Theory, Research, Practice, Training, 41,* 374–387.

Costa, P. T., Jr., Yang, J., & McCrae, R. R. (1998). Aging and personality traits: Generalizations and clinical implications. In I. H. Nordhus, G. R. VandenBos, S. Berg, & P. Fromholt (Eds.), *Clinical geropsychology* (pp. 33–48). Washington, DC: American Psychological Association.

Danckwerts, A., & Leathem, J. (2003). Questioning the link between PTSD and cognitive dysfunction. *Neuropsychology Review, 13,* 221–235.

Danieli, Y. (1997). As survivors age: An overview. *Journal of Geriatric Psychiatry, 30,* 9–26.

D'Augelli, A. R., & Grossman, A. H. (2001). Disclosure of sexual orientation, victimization, and mental health among lesbian, gay, and bisexual older adults. *Journal of Interpersonal Violence, 16,* 1008–1027.

Davidson, J. R. T., Kudler, H. S., Sunders, W. B., & Smith, R. D. (1990). Symptom and comorbidity patterns in World War II and Vietnam veterans with posttraumatic stress disorder. *Comprehensive Psychiatry, 31,* 162–170.

Davis, M. J., Moye, J., & Karel, M. J. (2002). Mental health screening of older adults in primary care. *Journal of Mental Health and Aging, 8,* 139–149.

Eaton, W., Sigal, J., & Weinfeld, M. (1982). Impairment in Holocaust survivors after 33 years: Data from an unbiased community sample. *American Journal of Psychiatry, 139,* 773–777.

Elder, G. H., Jr. (1999). *Children of the Great Depression: Social change in life experience.* Boulder, CO: Westview Press.

Erikson, E. (1982). *The life cycle completed.* New York: Norton.

Erikson, E. H. (1959). *Identity and the life cycle.* New York: Norton.

Falk, B., Hersen, M., & Van Hasselt, V. (1994). Assessment of post-traumatic stress disorder in older adults: A critical review. *Clinical Psychology Review, 14,* 383–415.

Floyd, M., Rice, J., & Black, S. R. (2002). Recurrence of posttraumatic stress disorder in late life: A cognitive aging perspective. *Journal of Clinical Geropsychology, 8,* 303–311.

Foa, E. B., Keane, T. M., & Friedman, M. J. (2000). *Effective treatments for PTSD: Practice guidelines from the International Society for Traumatic Stress Studies.* New York: Guilford Press.

Fontana, A., & Rosenheck, R. (1994). Traumatic war stressors and psychiatric symptoms among World War II, Korean, and Vietnam War veterans. *Psychology and Aging, 9,* 27–33.

Fontana, A., Rosenheck, R., Spencer, H., & Gray, S. (2002). *The long journey home X: Treatment of posttraumatic stress disorder in the department of Veterans Affairs: Fiscal year 2001 service delivery and performance.* West Haven, CT: Northeast Program Evaluation Center, Department of Veterans Affairs.

Freud, S. (1959). On psychotherapy. In J. Riviere (Trans.), *Collected papers: Vol. 1. Early papers on the history of the psycho-analytic movement* (pp. 249–263). New York: Basic Books. (Original work published 1904)

Friedman, M. J. (2003). Pharmacologic management of posttraumatic stress disorder. *Primary Psychiatry, 10,* 66–68, 71–73.

Gagnon, M., & Hersen, M. (2000). Unresolved childhood sexual abuse and older adults: Late-life vulnerabilities. *Journal of Clinical Geropsychology, 6,* 187–198.

Gallagher-Thompson, D., & Thompson, L. W. (1995). Psychotherapy with older adults in theory and practice. In B. Bongar & L. E. Beutler (Eds.), *Comprehensive textbook of psychotherapy: Theory and practice* (pp. 359–379). New York: Oxford University Press.

Gallagher-Thompson, D., & Thompson, L. W. (1996). Applying cognitive-behavioral therapy to the psychological problems of later life. In S. H. Zarit & B. G. Knight (Eds.), *A guide to psychotherapy and aging: Effective clinical interventions in a life-stage context* (pp. 61–82). Washington, DC: American Psychological Association.

Gallo, J. J., & Lebowitz, B. D. (1999). The epidemiology of common late-life mental disorders in the community: Themes for the new century. *Psychiatric Services, 50,* 1158–1166.

Gerontological Society of America Task Force on Minority Issues in Gerontology. (1994). *Minority elders: Five goals toward building a public policy base.* Washington, DC: Gerontological Society of America.

Glantz, M. D. (1989). Cognitive therapy with the elderly. In A. Freeman, K. Simon, L. Beutler, & H. Arkowitz (Eds.), *A comprehensive handbook of cognitive therapy* (pp. 467–489). New York: Plenum Press.

Goenjian, A. K., Najarian, L. M., Pynoos, R. S., Steinberg, A. M., Manoukian, G., Tavosian, A., et al. (1994). Posttraumatic stress disorder in elderly and younger adults after the 1988 earthquake in Armenia. *American Journal of Psychiatry, 151,* 895–901.

Goldstrom, I. D., Burns, B. J., Kessler, L. G., Feuerberg, M. A., Larson, D. B., Miller, N. E., et al. (1987). Mental health services use by elderly adults in a primary care setting. *Journals of Gerontology: Series B, Psychological Sciences and Social Sciences, 42,* 147–153.

Golier, J. A., Yehuda, R., Lupien, S. J., Harvey, P., Grossman, R., & Elkin, A. (2002). Memory performance in Holocaust survivors with posttraumatic stress disorder. *American Journal of Psychiatry, 159,* 1682–1688.

Gray, M. J., & Acierno, R. (2002). Symptom presentations of older adult crime victims: Description of a clinical sample. *Journal of Anxiety Disorders, 16,* 299–309.

Hankin, C. S. (1997). Treatment of older adults with posttraumatic stress disorder. In A. Maercker (Ed.), *Treatment of PTSD* (pp. 357–384). New York: Springer.

Harel, Z., Kahana, B., & Kahana, E. (1988). Psychological well-being among Holocaust survivors and immigrants in Israel. *Journal of Traumatic Stress, 1,* 413–429.

Hunt, T., & Lindley, C. J. (1990). Testing older adults: A reference guide for geropsychological assessments. Austin, TX: Pro-Ed.

Hyer, L. A., & Woods, M. G. (1998). Phenomenology and treatment of trauma in later life. In V. M. Follette, J. I. Ruzek, & F. R. Abueg (Eds.), *Cognitive-behavioral therapies for trauma* (pp. 383–414). New York: Guilford Press.

Jeste, D. V., Alexopoulos, G. S., Bartels, S. J., Cummings, J. L., Gallo, J. J., Gottlieb, G. L., et al. (1999). Consensus statement on the upcoming crisis in geriatric mental health: Research agenda for the next 2 decades. *Archives of General Psychiatry, 56,* 848–853.

Kasl-Godley, J., & Gatz, M. (2000). Psychosocial intervention for individuals with dementia: An integration of theory, therapy, and a clinical understanding of dementia. *Clinical Psychology Review, 20,* 755–782.

Kasprow, W. J., & Rosenheck, R. (2000). Mortality among homeless and nonhomeless mentally ill veterans. *Journal of Nervous and Mental Disease, 188,* 141–147.

Keane, T. M., Caddell, J. M., & Taylor, K. L. (1988). Mississippi Scale for Combat-Related Posttraumatic Stress Disorder: Three studies in reliability and validity. *Journal of Consulting and Clinical Psychology, 56,* 85–90.

Klerman, G. L., Weissman, M. M., Rounsaville, B. J., & Chevron, E. (1984). *Interpersonal psychotherapy of depression.* New York: Basic Books.

Knight, B. G., Gatz, M., Heller, K., & Bengston, V. L. (2000). Age and emotional response to the Northridge earthquake: A longitudinal analysis. *Psychology and Aging, 15,* 627–634.

Knight, B. G., & Satre, D. D. (1999). Cognitive behavioral psychotherapy with older adults. *Clinical Psychology: Science and Practice, 6,* 188–203.

Krause, N., Shaw, B. A., & Cairney, J. (2004). A descriptive epidemiology of lifetime trauma and the physical health status of older adults. *Psychology and Aging, 19,* 637–648.

Kruse, A., & Schmitt, E. (1999). Reminiscence of traumatic experiences in (former) Jewish emigrants and extermination camp survivors. In A. Maercker, M. Schützwohl, & Z. Solomon (Eds.), *Posttraumatic stress disorder: A lifespan developmental perspective* (pp. 155–176). Seattle, WA: Hogrefe & Huber.

Lew, L. (1991). Elderly Cambodians in Long Beach: Creating cultural access to health care. *Journal of Cross-Cultural Gerontology, 6,* 199–203.

Lichtenberg, P. A. (1999). *Handbook of assessment in clinical gerontology.* New York: Wiley.

Lichtenberg, P. A., Smith, M., Frazer, D., Molinari, V., Rosowsky, E., Crose, R., et al. (1998). Standards for psychological services in long-term care facilities. *Gerontologist, 38,* 122–127.

Lipton, M. I., & Schaffer, W. R. (1986). Post-traumatic stress disorder in the older veteran. *Military Medicine, 151,* 522–524.

Maercker, A. (2002). Life-review technique in the treatment of PTSD in elderly patients: Rationale and three single case studies. *Journal of Clinical Geropsychology, 8,* 239–249.

Miller, M. D., Cornes, C., Frank, E., Ehrenpreis, L., Silberman, R., Schlernitzauer, M. A., et al. (2001). Interpersonal psychotherapy for late-life depression: Past, present, and future. *Journal of Psychotherapy Practice and Research, 10*, 231–238.

Molinari, V., & Williams, W. (1995). An analysis of aging World War II POWs with PTSD: Implications for practice and research. *Journal of Geriatric Psychiatry, 28*, 99–114.

Neugarten, B. L. (1974). Age groups in American society and the rise of the young–old. *Annals of the American Academy of Political and Social Science, 415*, 187–198.

Nichols, B. L., & Czirr, R. (1986). Post-traumatic stress disorder: Hidden syndrome in elders. *Clinical Gerontologist, 5*, 417–433.

Norris, F. H. (1992). Epidemiology of trauma: Frequency and impact of different potentially traumatic events on different demographic groups. *Journal of Consulting and Clinical Psychology, 60*, 409–418.

Norris, F. H., Kaniasty, K. Z., Conrad, M. L., Inman, G. L., & Murphy, A. D. (2002). Placing age differences in cultural context: A comparison of the effects of age on PTSD after disasters in the United States, Mexico, and Poland. *Journal of Clinical Geropsychology, 8*, 153–173.

Norris, F., & Murrell, S. (1988). Prior experience as a moderator of disaster impact on anxiety symptoms in older adults. *American Journal of Community Psychology, 16*, 665–683.

Norris, F. H., Phifer, J. F., & Kaniasty, K. Z. (1994). Individual and community reactions to the Kentucky floods: Findings from a longitudinal study of older adults. In R. J. Ursano, B. G. McCaughey, & C. A. Fullerton (Eds.), *Individual and community responses to trauma and disaster: The structure of human chaos* (pp. 378–400). Cambridge, UK: Cambridge University Press.

Osgood, N. J., & Manetta, A. M. (2000–2001). Abuse and suicidal issues in older women. *Omega, 42*, 71–81.

Pearson, J. L., & Brown, G. K. (2000). Suicide prevention in late life: Directions for science and practice. *Clinical Psychology Review, 20*, 685–705.

Peretz, T., Baider, L., Ever-Hadani, P., & De-Nour, A. K. (1994). Psychological distress in female cancer patients with Holocaust experience. *General Hospital Psychiatry, 16*, 413–418.

Phillips, M. A., & Murrell, S. A. (1994). Impact of psychological and physical health, stressful events, and social support on subsequent mental health help seeking among older adults. *Journal of Consulting and Clinical Psychology, 62*, 270–275.

Pillemer, K., & Finkelhor, D. (1988). The prevalence of elder abuse: A random sample survey. *Gerontologist, 28*, 51–57.

Port, C. L., Engdahl, B., & Frazier, P. (2001). A longitudinal and retrospective study of PTSD among older POWs. *American Journal of Psychiatry, 158*, 1474–1479.

Port, C. L., Engdahl, B. E., Frazier, P. A., & Eberly, R. E. (2002). Factors related to the long-term course of PTSD in older ex-prisoners of war. *Journal of Clinical Geropsychology, 8*, 203–214.

Qualls, S. H., Segal, D. L., Norman, S., Niederehe, G., & Gallagher-Thompson, D. (2002). Psychologists in practice with older adults: Current patterns, sources of training, and need for continuing education. *Professional Psychology: Research and Practice, 33*, 435–442.

Rabins, P. V., Black, B. S., Roca, R., German, P., McGuire, M., Robbins, B., et al. (2000). Effectiveness of a nurse-based outreach program for identifying and treating psychiatric illness in the elderly. *Journal of the American Medical Association, 283*, 2802–2809.

Rechtschaffen, A. (1959). Psychotherapy with geriatric patients: A review of the literature. *Journals of Gerontology, 14*, 73–84.

Rosen, C. S., Chow, H. C., Finney, J. F., Greenbaum, M. A., Moos, R. H., Sheikh, J. I., et al. (2004). Practice guidelines and VA practice patterns for treating posttraumatic stress disorder. *Journal of Traumatic Stress, 17*, 213–222.

Russo, S. A., Hersen, M., & Van Hasselt, V. B. (2001). Treatment of reactivated post-traumatic stress disorder: Imaginal exposure in an older adult with multiple traumas. *Behavior Modification, 25*, 94–115.

Sala, M., Perez, J., Soloff, P., Ucelli di Nemi, S., Caverzasi, E., Soares, J. C., et al. (2004). Stress and hippocampal abnormalities in psychiatric disorders. *European Neuropsychopharmacology, 14*, 393–405.

Salthouse, T. A. (1996). The processing speed theory of adult age differences in cognition. *Psychological Review, 103*, 403–428.

Sapolsky, R. M. (2000). Glucocorticoids and hippocampal atrophy in neuropsychiatric disorders. *Archives of General Psychiatry, 57*, 925–935.

Schnurr, P. P., & Green, B. L. (2004). *Trauma and health: Physical health consequences of exposure to extreme stress*. Washington, DC: American Psychological Association.

Schnurr, P. P., Lunney, C. A., & Sengupta, A. (2004). Risk factors for the development versus maintenance of posttraumatic stress disorder. *Journal of Traumatic Stress, 17*, 85–95.

Schnurr, P. P., & Spiro, A., III. (1999). Combat exposure, posttraumatic stress disorder symptoms, and health behaviors as predictors of self-reported physical health in older veterans. *Journal of Nervous and Mental Disease, 187*, 353–359.

Schnurr, P. P., Spiro, A., III, & Paris, A. H. (2000). Physician-diagnosed medical disorders in relation to PTSD symptoms in older male military veterans. *Health Psychology, 19*, 91–97.

Shalev, A. Y., Friedman, M. J., Foa, E. B., & Keane, T. M. (2000). Integration and summary. In E. B. Foa, T. M. Keane, & M. J. Friedman (Eds.), *Effective treatments for PTSD: Practice guidelines from the International Society for Traumatic Stress Studies* (pp. 359–379). New York: Guilford Press.

Sliwinski, M., & Buschke, H. (1999). Cross-sectional and longitudinal relationships among age, cognition, and processing speed. *Psychology and Aging, 14*, 18–33.

Snell, F. I., & Padin-Rivera, E. (1997). Group treatment for older veterans with post-traumatic stress disorder. *Journal of Psychosocial Nursing, 35*, 10–16.

Solomon, Z., & Ginzburg, K. (1999). Aging in the shadow of war. In A. Maercker, Z. Solomon, & M. Schützwohl (Eds.), *Post-traumatic stress disorder: A lifespan developmental perspective* (pp. 137–153). Seattle, WA: Hogrefe & Huber.

Solomon, Z., & Prager, E. (1992). Elderly Israeli Holocaust survivors during the Persian Gulf War: A study of psychological distress. *American Journal of Psychiatry, 140*, 1177–1179.

Spiro, A., Schnurr, P. P., & Aldwin, C. M. (1994). Combat-related posttraumatic stress disorder symptoms in older men. *Psychology and Aging, 9*, 17–26.

Stein, M. B., & Barrett-Connor, E. (2000). Sexual assault and physical health: Findings from a population-based study of older adults. *Psychosomatic Medicine, 62*, 838–843.

Sutker, P. B., Galina, H., West, J. A., & Allain, A. N. (1990). Trauma-induced weight loss and cognitive deficits among former prisoners of war. *Journal of Consulting and Clinical Psychology, 58*, 323–328.

Sutker, P. B., Vasterling, J. J., Brailey, K., & Allain, A. N., Jr. (1995). Memory, attention, and executive deficits in POW survivors: Contributing biological and psychological factors. *Neuropsychology, 9*, 118–125.

Teri, L., Logsdon, R. G., Uomoto, J., & McCurry, S. M. (1997). Behavioral treatment of depression in dementia patients: A controlled clinical trial. *Journals of Gerontology: Series B, Psychological Sciences and Social Sciences, 52*, P159–P166.

Tyler, K. A., & Hoyt, D. R. (2000). The effects of an acute stressor on depressive symptoms among older adults: The moderating effects of social support and age. *Research on Aging, 22*, 143–164.

United Nations. (2003). *Population division of the Department of Economic and Social Affairs of the United Nations Secretariat, World Population Prospects: The 2002 Revision and World Urbanization Prospects*. Retrieved January 1, 2004, from *esa.un.org/unpp*

U.S. Bureau of the Census. (2004). National population projections. Retrieved *www.census.gov/population/www/projections/natsum-T3.html*

U.S. Department of Health and Human Services, Administration on Aging. (2001). *Older adults and mental health: Issues and opportunities*. Washington, DC: Author.

U.S. Department of Veterans Affairs. (1997). *Assessment of competency and capacity of the older adult: A practice guideline for psychologists* (National Center for Cost Containment, NTIS No. PB-97-147904). Milwaukee, WI: Author.

Vinton, L. (1992). Battered women's shelters and older women: The Florida experience. *Journal of Family Violence, 7*, 63–72.

Weathers, F. W., Litz, B. T., Herman, D. S., Huska, J. A., & Keane, T. M. (1993). *The PTSD Checklist: Reliability, validity, and diagnostic utility*. Paper presented at the ninth annual meeting of the International Society for Traumatic Stress Studies, San Antonio, TX.

Weintraub, D., & Ruskin, P. E. (1999). Posttraumatic stress disorder in the elderly: A review. *Harvard Review of Psychiatry, 7*, 144–152.

Yehuda, R. (1999). Biological factors associated with susceptibility to posttraumatic stress disorder. *Canadian Journal of Psychiatry, 44*, 34–39.

Yehuda, R., Elkin, A., Binder-Brynes, K., Kahana, B., Southwick, S. M., Schmeidler, J., et al. (1996). Dissociation in aging Holocaust survivors. *American Journal of Psychiatry, 153*, 935–940.

Yehuda, R., Kahana, B., Schmeidler, J., Southwick, S., Wilson, S., & Giller, E. (1995). Impact of cumulative lifetime trauma and recent stress on current posttraumatic stress disorders symptoms in Holocaust survivors. *American Journal of Psychiatry, 152*, 1815–1818.

Zeiss, A. M., Cook, J. M., & Cantor, D.W. (2003). *Fact sheet: Fostering resilience in response to terrorism: For psychologists working with older adults.* Report to American Psychological Association Task Force on Resilience in Response to Terrorism. Washington, DC: American Psychological Association

Zeiss, A. M., & Gallagher-Thompson, D. (2003). Providing interdisciplinary geriatric team care: What does it really take? *Clinical Psychology: Science and Practice, 10*, 115–119.

Zeiss, R. A., & Dickman, H. R. (1989). PTSD 40 years later: Incidence and person-situation correlates in former POWs. *Journal of Clinical Psychology, 45*, 80–87.

Part III

CLINICAL PRACTICE

EVIDENCE-BASED STATE OF THE ART

Chapter 15

Assessment of PTSD and Its Comorbidities in Adults

Terence M. Keane, Deborah J. Brief, Elizabeth M. Pratt, and Mark W. Miller

The initial inclusion of posttraumatic stress disorder (PTSD) in the third edition of the *Diagnostic and Statistical Manual of Mental Disorders* (DSM-III; American Psychiatric Association, 1980) was accompanied by a discussion of traumatic events as relatively rare experiences and by an understanding of PTSD as a relatively rare psychological condition in the general population. Today, high prevalence rates of traumatic events and PTSD are viewed as worldwide phenomena crossing national, geographic, cultural, and ethnic boundaries. A series of outstanding epidemiological studies examined the prevalence of PTSD, providing information on rates of exposure to traumatic life events, relative rates of comorbidity of PTSD with other psychiatric conditions, and distribution of PTSD among various subgroups of the population (e.g., adults, adolescents, children; males and females; minorities; as well as particular groups at risk, such as war veterans). These initial studies examined and informed us about factors that affect the onset and course of PTSD (see Breslau, Davis, Andreski, & Peterson, 1991; Kessler, Sonnega, Bromet, Hughes, & Nelson, 1995; Kilpatrick, Edmunds, & Seymour, 1992; Kulka et al., 1990; Norris, 1992).

Most epidemiological studies of trauma exposure and PTSD described the condition within the United States, yet the vast majority of the wars, violence, and natural disasters in the 20th century actually occurred in the developing world. With increasing recognition of the health and economic costs associated with psychological morbidity internationally (Murray & Lopes, 1996), there is a growing appreciation for the need for regional estimates of PTSD and related psychiatric disorders (see de Jong et al., 2001).

Importantly, trauma exposure does not always lead to the development of PTSD. Fortunately, most survivors of single, discrete traumatic events do not develop any form of psychopathology. The most common trajectory is for recovery over time (Bonnano, 2004). For a distinct minority of persons exposed to trauma, PTSD, depression, anxiety disorders, or substance abuse may emerge (Kessler et al., 1995). For some of these people, the development of additional comorbid conditions may ensue (Breslau, Chilcoat, Kessler, Peterson, & Lucia, 1999). Assessment of PTSD and its comorbidities, then, becomes a challenge for clinicians and researchers alike. Perhaps as important as the evaluation of comorbidity in clinical and research settings is how we conceptualize this comorbidity. Is it the case that these conditions are independent of one another? Are they a function of overlapping symptom criteria? Can we predict who will develop which comorbid conditions? Perhaps even more importantly, can we treat these symptoms as a single posttraumatic entity, or do these conditions require independent interventions? As recognition of the high degree of comorbidity in PTSD increases, these questions will require the development of new conceptual models and the collection of empirical evidence to further our understanding of the impact of traumatic events and the treatments we provide.

Our purpose in this chapter is to examine and discuss the methods for assessing PTSD in a wide variety of settings. As interest in PTSD and its comorbidities grows internationally (e.g., Keane, Marshall, & Taft, 2006), so also grows the need for sensitive and specific structured diagnostic interviews, psychological tests, questionnaires, and psychophysiological approaches. As progress is made in the trauma field, the consistent use of standardized psychological measures will permit cross-study comparisons, meaningful meta-analyses, the specification of conclusions regarding public policy based on sound empirical methods, and the more rapid use of evidence-based clinical protocols for treatment.

The terrorist attacks on New York City and Washington, D.C., of September 11, 2001, accompanied by the natural disasters in Louisiana, Mississippi, and Texas in August 2005, emphasize the public health importance of arriving at best practices for the management of mass disaster and violence in contemporary society. Understanding the optimal methods to assess the presence of PTSD, related psychiatric conditions, treatment outcomes, and the monitoring of progress in real time are a few of the important topics we address. Because PTSD is often accompanied by considerable comorbidity in clinical and field trials, we also offer a novel approach for conceptualizing and understanding this comorbidity.

METHODOLOGICAL CONSIDERATIONS

Mental health clinicians recognize that a substantial portion of their patients have experienced traumatic events and may require treatment for PTSD. In clinical settings PTSD is assessed for many different reasons, and the goals of any particular assessment determine the methods and measures selected by the professional. For example, a typical objective for clinicians is a diagnostic evaluation that includes a differential diagnosis, a functional assessment, and other related information that helps in treatment planning. On other occasions practitioners may be involved in forensic work, in which diagnostic accuracy is of paramount importance.

Researchers involved in epidemiological or prevalence studies are interested in the rate (i.e., incidence of lifetime and current disorder) of occurrence of PTSD, the risk factors associated with it, and the occurrence of comorbid psychiatric and substance abuse problems. In addition, researchers involved in clinical studies may be interested in which assessment methods offer the highest levels of diagnostic accuracy (i.e., sensitivity and specificity) when they examine the biological and psychological processes associated with the disorder. Case–control study design is typically used in clinical research and is usually accompanied by rigorous diagnostic measures administered by experienced and trained clinicians. Importantly, different clinical and research situations may require different solutions depending on the particular assessment goals of the professional. For this reason, we present a general overview of the means by which a clinician can assess PTSD, and evaluate the quality of measures by providing information on their psychometric properties.

Evidence-Based Assessment of PTSD

Although the application of evidence-based treatments in mental health has received considerable attention across the world, these same standards have only recently been applied consistently to the assessment of psychological disorders (Hunsley & Mash, 2005). An entire special section of the journal *Psychological Assessment* was dedicated to promoting the use of evidence-based assessment across psychological disorders. Included in the guidelines was that contributors provide detailed coverage of indices of reliability, validity, and utility for any measure reviewed. As well, they were instructed to emphasize the importance of factors such as gender, culture, age, and ethnicity when deriving conclusions about the effectiveness of a measure.

Rarely in the clinical setting is only one measure administered. More typically, selected measures comprise an assessment battery. Questions can then arise about the reliability, validity, and utility of individual measures when administered in a more comprehensive battery. Perhaps most importantly, and very relevant to the approach we recommend for the assessment of PTSD, clinician judgment is always a part of the process. Measures are generally evaluated in the context of multiple indicators; the application of clinical judgment, then, becomes an additional variable to examine in the process of assessing the evidence base for a disorder. To date, examination of the process of applying clinical judgment in the context of the administration of multiple measures has not been attempted in the field of PTSD.

In the *Psychological Assessment* special section, Antony and Rowa (2005) reviewed the issues for establishing an evidence base for assessment of the anxiety disorders. Included in their consideration was the use of measures to establish a diagnosis; to measure the presence, absence, or severity of a disorder; to measure features that a structured interview does not assess well; to facilitate treatment planning; to measure treatment outcome; to measure a phenomenon of interest (e.g., heart rate, cognitions); to establish inclusion–exclusion criteria in research; and to predict future behavior.

In addition, Antony and Rowa (2005) set useful standards to review assessment instruments for the anxiety disorders. Important to their perspective was the extent to which an instrument predicts a person's actual *in vivo* behavior, its usefulness across population characteristics, the level of training required for the clinician to use the tool properly, the effectiveness of the tool across administrators and settings, the ease with which it can be disseminated, the acceptability of the instrument to patients, and the cost-effectiveness of any instrument considered. This heuristic framework for examining the usefulness of instruments sets a goal or a standard for future work in the field of psychological assessment. Today, work in PTSD assessment clearly suggests that there are many instruments with considerable empirical support for use across clinics, settings, and purposes, yet more work needs to be done in the field. Comorbidity is common across all psychiatric conditions, and it is especially common in clinic settings. Hunsley and Mash's (2005) challenge to the field is to improve and enhance assessment in much the same way that clinical trials and meta-analyses have enhanced our understanding of psychological and pharmacological treatments that work. It is, indeed, an exciting time to work in the field of PTSD assessment; yet considerable work remains to be done.

Selection of Assessment Measures

Since 1980 there has been excellent progress in developing high-quality measures to assess trauma symptoms and PTSD in adults (Keane & Barlow, 2002; Keane, Weathers, & Foa, 2000; Weathers, Keane, & Davidson, 2001). The process of assessing PTSD may include a range of different approaches, such as a clinician-administered structured diagnostic interview for PTSD, a structured diagnostic interview to assess its related comorbidity, self-report psychological tests and questionnaires, and psychophysiological measures. The clinician may also want to review medical records and check with multiple informants regarding the patient's behavior and experiences when the accuracy of self-report might be questionable. We have referred to this as a multimethod approach to the assessment of PTSD (Keane, Fairbank, Caddell, Zimering, & Bender, 1985). When faced with different assessment contexts, the clinician or researcher can evaluate the quality of the measures used in similar contexts in the past, or if this information is unavailable, use as a guideline the psychometric properties of each instrument. Standards for evaluating these psychological measures are briefly described in the following section.

Psychometric Principles

The quality of a psychological assessment is determined by estimates of its reliability and validity. *Reliability* is the consistency of test scores. Reliability data can be reported as consistency of tests over time (test–retest reliability), over different interviewers (interrater reliability), or over the many items comprising a test (internal consistency). For continuous measures, reliability is reported as a simple coefficient that can vary between 0.0 and 1.0 (for internal consistency this is referred to as Cronbach's alpha). For dichotomous measures such as diagnostic interviews (indicating the presence or absence of a disorder), reliability can be reported as kappa (Cohen, 1960), which also ranges between 0.0 and 1.0 and is interpreted as the percent agreement above chance.

   Validity refers to the extent to which evidence supports the various inferences, interpretations, conclusions, or decisions made on the basis of a test. *Content validity* refers to evidence that test items adequately reflect the construct being assessed. *Criterion-related validity* refers to evidence that the test can predict some variable or criterion of interest (e.g., performance). The criterion may be measured either at the same time the test is administered (concurrent) or at some point after the test is administered (predictive). Finally, *construct validity* refers to evidence that the test measures the construct of interest (e.g., PTSD). This is often demonstrated by examining whether the test correlates with other, known measures of the same construct, but not with measures of other unrelated constructs.

Diagnostic measures are usually evaluated on the basis of their diagnostic utility, a type of criterion-related validity pertaining to a test's capacity to predict diagnostic status (Kraemer, 1992). There are three steps in determining the diagnostic utility of a given instrument. First, a "gold standard" is selected. In psychological research, this is ordinarily a diagnosis based on a clinical interview, but it may also be a composite based on several sources of information. Second, both the gold standard and the newly developed test are administered to the experimental group of participants. Finally, several cutoff scores are examined to determine their diagnostic utility, that is, their ability to predict the diagnosis provided by the gold standard. Optimal cutoff scores for the test are those that predict the greatest number of cases and noncases from the original sample (Keane et al., 2000; Weathers, Keane, King, & King, 1996).

All measures of a psychological disorder are imperfect (Gerardi, Keane, & Penk, 1989). Two measures of the error contained within a test are false positives and false negatives. A false positive occurs when a patient's score is above the cutoff but is not a true case. A false negative occurs when a patient's score falls below the given cutoff but is in fact a true case. Diagnostic utility is often described in terms of a test's *sensitivity* and *specificity*. Sensitivity is the measure of a test's true-positive rate, or the probability that those with the disorder will score above a given cutoff score. Specificity is the true negative rate of a test, or the probability that those without the disorder will score below the cutoff for the test.

CURRENT STATE OF THE ART

Structured Diagnostic Interviews

Clinician-administered structured diagnostic interviews are valuable tools for assessing PTSD (Keane, Kaloupek, & Weathers, 1996). Whereas use of structured diagnostic interviews is standard practice in clinical research settings, use of these interviews in the clinical setting is less common, with perhaps the single exception of clinical forensic practice (Keane, 1995; Keane, Buckley, & Miller, 2003). In general, this may be due to time and cost burdens, as well as a need for specialized training to master the administration of many of these interviews. Nonetheless, increased use of structured diagnostic interviews in clinical settings may well improve diagnostic accuracy and aid in treatment planning (Litz & Weathers, 1994).

Several available structured interviews were developed for the assessment of PTSD either as modules of comprehensive diagnostic assessment tools or as independent measures. These are described below.

Structured Clinical Interview for DSM-IV

The Structured Clinical Interview for DSM-IV (SCID-IV; First, Spitzer, Williams, & Gibbon, 2000) is designed to assess a broad range of psychiatric conditions on Axes I and II. It is divided into separate modules corresponding to DSM-IV (American Psychiatric Association, 1994) diagnostic criteria, with each module providing the interviewer with specific prompts and follow-up inquiries. Symptom presence is rated on a 3-point confidence scale based on the interviewer's evaluation of the individual's responses. To assess PTSD, respondents are asked to frame symptoms in terms of their "worst trauma experience." The SCID is intended for use only by clinicians and other highly trained interviewers.

Whereas the administration of the full SCID-IV can be time consuming, the modular structure allows clinicians to limit their assessment to conditions that are frequently comorbid with PTSD. Within the context of a trauma clinic, inclusion of modules for the anxiety disorders, affective disorders, and substance use disorders is recommended. Administration of the psychotic screen also helps to rule out conditions that may very well require a different set of interventions.

The SCID-PTSD module is considered psychometrically sound. For example, Keane and colleagues (1998) examined the interrater reliability of the SCID by asking a second interviewer to listen to audiotapes of an initial interview. They found a kappa of .68 and agreement across lifetime, current, and never having had PTSD of 78%. Similarly, in a sample of patients reinterviewed within a week by a different clinician, they found a kappa of .66 and diagnostic agreement of 78%. McFall, Smith, Roszell, Tarver, and Malas (1990) also reported evidence of convergent validity, finding significant correlations between the number of SCID-PTSD items endorsed and other measures of PTSD (e.g., Mississippi Scale for Combat-Related PTSD [.65; Keane et al., 1988] and PTSD Scale of the Minnesota Multiphasic Personality Inventory [MMPI-PTSD (PK); .46; Keane, Malloy, & Fairbank, 1984]). The SCID-PTSD module also yielded substantial sensitivity (.81) and specificity (.98), and a robust kappa (.82) when compared to a composite PTSD diagnosis (Kulka et al., 1988), again indicating good diagnostic utility.

The SCID possesses some important limitations that have been described previously. First, the scoring algorithm of the SCID permits only a dichotomous rating of PTSD (e.g., presence or absence). Most clinicians agree that psychological symptoms occur in a dimensional rather than dichotomous fashion (Keane, et al., 2000; Ruscio, Ruscio, & Keane, 2002). A second limitation is that the SCID does not assess for the frequency or severity of symptoms. Third, assessing symptoms in response to the "worst event" experienced may mean a loss of important information regarding the effects of other traumatic events (Cusack, Falsetti, & de Arellano, 2002). Finally, the trauma screen of the SCID may miss significant traumatic events (Falsetti et al., 1996).

Clinician Administered PTSD Scale

Developed by the National Center for PTSD (Blake et al., 1990), the Clinician Administered PTSD Scale (CAPS) is currently the most widely used structured interview for diagnosing and measuring the severity of PTSD (Weathers et al., 2001). The CAPS assesses all DSM-IV diagnostic criteria for PTSD, including criterion A (exposure), criteria B–D (core symptom clusters), criterion E (chronology), and criterion F (functional impairment), as well as associated symptoms of guilt and dissociation. An important CAPS feature is that it contains separate ratings for the frequency and intensity of each symptom, which can be summed to create a severity score for each symptom. This permits flexibility in scoring and analyses. The CAPS also promotes uniform administration and scoring through carefully phrased prompt questions and explicit rating anchors with clear behavioral referents. Once trained, interviewers are able to ask their own follow-up questions and use their clinical judgment in arriving at optimal ratings.

Similar to the SCID, flexibility is built into the administration of the CAPS. Interviewers can administer only the 17 core symptoms, all DSM-IV criteria, and/or add the associated symptoms. If administered completely, the CAPS takes approximately 1 hour, but if only the 17 core symptoms are assessed, the time for administration is cut in half.

Weathers and colleagues (2001) extensively reviewed the psychometric studies on the CAPS. Weathers, Ruscio, and Keane (1999) examined the reliability and validity data of the CAPS across five samples of male Vietnam veterans collected at the National Center for PTSD. Robust estimates were found for interrater reliability over a 2- to 3-day interval for each of the three symptom clusters (.86–.87 for frequency, .86–.92 for intensity, and .88–.91 for severity) and all 17 symptoms (.91 for total frequency, .91 for total intensity, and .92 for total severity). Test–retest reliability for a CAPS-based PTSD diagnosis was also high (kappa = .89 in the first sample and 1.00 in the second sample). Thus, the data indicate that trained and calibrated raters can achieve a high degree of consistency in using the CAPS to rate PTSD symptom severity and diagnose PTSD. Weathers and colleagues also found high internal consistency across all 17 items in a research sample (alphas of .93 for frequency and .94 for intensity and severity) and a clinical sample (alphas of .85 for frequency, .86 for intensity, and .87 for severity), which supports the use of the CAPS in both research and clinical settings.

Strong evidence for validity of the CAPS was again provided by Weathers and colleagues (1999), who found that the CAPS total severity score correlated highly with other measures of PTSD (Mississippi Scale = .91, MMPI-PTSD (PK) Scale = .77, the number of PTSD symptoms endorsed on the SCID = .89, and the PTSD Checklist = .94; Weathers, Litz, Herman, Huska, & Keane, 1993). As expected, correlations with measures of antisocial personality disorder were low (.14–.33). Weathers and colleagues also found strong evidence for the diagnostic utility of the CAPS using three CAPS scoring rules for predicting a SCID-based PTSD diagnosis. The rule having the closest correspondence to the SCID yielded a sensitivity of .91, specificity of .84, and efficiency of .88, with a kappa of .75, indicating good diagnostic utility.

The CAPS has now been used successfully in a wide variety of trauma-exposed populations (e.g., combat veterans; survivors of rape, crime, motor vehicle accidents, incest, the Holocaust, torture, and cancer), has served as the primary diagnostic or outcome measure in hundreds of empirical studies on PTSD, and has been translated into at least 12 languages (Weathers et al., 2001). Thus, the existing data strongly support its continued use in both clinical and research settings.

PTSD Symptom Scale Interview

Developed by Foa, Riggs, Dancu, and Rothbaum (1993), the PTSD Symptom Scale Interview (PSS-I) is a structured interview designed to assess symptoms of PTSD in individuals with a known trauma history. Using a Likert scale, interviewers rate the severity of 17 symptoms corresponding to the DSM-III-R criteria for PTSD. One limitation of the PSS-I is that it measures symptoms over the past 2 weeks rather than 1 month, which the DSM criteria specify as necessary for a diagnosis of PTSD (Cusack et al., 2002). According to the authors, the PSS-I can be administered by lay interviewers trained to recognize the clinical picture presented by traumatized individuals.

The PSS-I, originally tested in a sample of women with a history of rape and nonsexual assault (Foa et al., 1993), was found to have strong psychometric properties. Foa et al. reported high internal consistency (Cronbach alphas = .85 for full scale, .65–.71 for subscales), test–retest reliability over a 1-month period (.80), and interrater agreement for a PTSD diagnosis (kappa = .91, 95% agreement). With respect to validity, the PSS-I was correlated with other measures of traumatic stress (e.g., .69, Impact of Event

Scale Intrusion score [Horowitz, Wilner, & Alvarez, 1979] and .67, Rape Aftermath Symptom Test (RAST) total score [Kilpatrick, 1988]) and demonstrated good diagnostic utility when compared to a SCID-PTSD diagnosis (sensitivity = .88, specificity = .96). The PSS-I appears to possess many strong features that warrant its consideration for clinical and research use, especially with sexual assault survivors.

Structured Interview for PTSD

Originally developed by Davidson, Smith, and Kudler (1989), the Structured Interview for PTSD (SIP) is designed to diagnose PTSD and measure symptom severity. It includes 17 items focused on DSM-IV criteria for PTSD, as well as two items focused on survivor and behavior guilt. Each item is rated by the interviewer on a Likert scale. There are initial probe questions and follow-up questions to promote a more thorough understanding of the respondent's symptom experiences. It can be administered by either clinicians or appropriately trained paraprofessionals. The SIP takes 10–30 minutes to administer, depending on the degree of symptomatology present.

In a sample of combat veterans, Davidson and colleagues (1989) reported high interrater reliability (.97–.99) on total SIP scores and perfect agreement on the presence or absence of PTSD across raters. High alpha coefficients were also reported (.94 for the veteran sample [Davidson et al., 1989] and .80 for patients with PTSD enrolled in a clinical trial [Davidson, Malik, & Travers, 1997]. In the veteran sample, test–retest reliability for the total SIP score was .71 over a 2-week period. With respect to validity, the SIP correlated significantly with other measures of PTSD (.49–.67; Davidson et al., 1989). They also compared the SIP scores of current and remitted PTSD cases and reported good sensitivity (.96) and specificity (.80) against the SCID. At a cutoff score of 25, the SIP correctly classified 94% of cases relative to a structured clinical interview (Davidson, Malik, & Travers, 1997). Overall, the SIP appears to be a sound diagnostic instrument.

Anxiety Disorders Interview Schedule—Revised

Originally developed by Di Nardo, O'Brien, Barlow, Waddell, and Blanchard (1983), the Anxiety Disorders Interview Schedule–Revised (ADIS-R) was designed to provide diagnoses among the DSM-III anxiety and related diagnostic conditions. The interview was revised to accommodate DSM-IV criteria (DiNardo, Brown, & Barlow, 1994). The ADIS-IV includes an assessment of affective disorders, substance use disorders, and selected somatoform disorders, a diagnostic time line, and a dimensional assessment of the key associated features of the disorders. The provision of both dimensional and categorical assessment allows the clinician to describe subthreshold manifestations of each disorder and offers more possibilities for analyses. The ADIS has been translated into numerous languages and has been used in over 150 clinical and research settings around the world. It is recommended only for trained, experienced interviewers.

Psychometric studies on the ADIS-PTSD module indicate mixed results. Originally tested in a small sample of Vietnam War combat veterans, the ADIS-PTSD module yielded strong sensitivity (1.0) and specificity (.91), and 93% agreement with interview-determined diagnoses (Blanchard, Gerardi, Kolb, & Barlow, 1986). Next, DiNardo, Moras, Barlow, Rapee, and Brown (1993) tested the reliability of the ADIS-R in a community sample recruited from an anxiety disorders clinic and found only fair agreement between two independent raters when PTSD was the principal diagnosis or a secondary diagnosis (kappa = .55). In a test of the ADIS-IV, the interrater reliability across two interviews given 10 days apart was also fair for current diagnoses (kappa = .59; Brown, DiNardo, Lehman, & Campbell, 2001) but slightly improved for lifetime diagnoses (kappa = .61). Additional reliability and validity data on the ADIS-IV are needed to ensure its continued use in clinical and research settings.

Self-Report Questionnaires

To obtain information bearing on the presence or absence of PTSD and its severity, psychologists developed several self-report measures that for the most part are continuous indicators of PTSD and reflect symptom severity, but specific cutoff scores in several of these measures can provide a diagnosis of PTSD. These measures are generally more time- and cost-efficient than structured interviews and can be especially valuable when used as screens for PTSD or in conjunction with structured interviews. The data also support the use of self-report questionnaires alone in clinical and research settings, when administering a structured interview is not feasible or practical. Many of the measures can be used interchangeably, because the findings appear to be robust for the minor variations in methods and approaches involved. In selecting a particular instrument, the clinician is encouraged to examine the data for that instrument for use in the population with which it is to be employed. In so doing, the clinician is apt to maximize the accuracy and efficiency of the test employed (Keane & Barlow, 2002).

Impact of Event Scale—Revised

Developed by Horowitz, Wilner, and Alvarez (1979), the Impact of Event Scale (IES) was the first and remains one of the most widely used self-report measures to assess psychological responses to a traumatic stressor. The initial 15-item questionnaire, which focused only on intrusion and avoidance symptoms, was derived from a model of traumatic stress developed by Horowitz (1976). Since the publication of DSM-IV, a revised 22-item version of the scale (IES-R; Weiss & Marmar, 1997) includes symptoms of hyperarousal. Thus, the IES-R more closely parallels DSM-IV criteria for PTSD. To complete the measure, respondents rate on a Likert scale "how distressed or bothered" they were during the past week by each symptom since a traumatic event. The IES has been translated into several languages and used with many different trauma populations; it takes approximately 10 minutes to complete.

Data on the psychometric properties of the revised IES-R are preliminary in nature. In two studies that incorporated four samples of emergency workers and earthquake survivors, Weiss and Marmar (1997) reported good internal consistency for each of the subscales (alphas = .87–.92 for Intrusion, .84–.86 for Avoidance, and .79–.90 for Hyperarousal). Test–retest reliability data from two samples yielded a range of reliability coefficients for the subscales (Intrusion = .57–.94, Avoidance = .51–.89, Hyperarousal = .59–.92). Weiss and Marmar suggested that the shorter interval between assessments and the greater recency of the traumatic event in one of the samples contributed to higher coefficients of stability for that sample.

Convergent and discriminant validity data are not yet available for the IES-R. There were many questions raised about the validity of the original scale when used to measure PTSD, because, having been developed years before the diagnostic criteria were published, it did not assess all DSM criteria for PTSD (see Joseph, 2000; Weathers et al., 1996). Although it now more closely parallels DSM-IV, its items measuring numbing are considered limited by some investigators (Foa, Cashman, Jaycox, & Perry, 1997). In a review of psychometric studies on the IES, Sundin and Horowitz (2002) report a range of correlations between the IES subscales and other self-report measures (e.g., .31–.46 on Symptom Checklist–90 [SCL-90] PTSD items; Arata, Saunders, & Kilpatrick, 1991) and diagnostic interviews (e.g., .32–.49 for SCID [McFall, Smith, Roszell, et al., 1990] and .75–.79 for CAPS [Neal et al., 1994]). Neal and colleagues (1994) reported high sensitivity (.89) and specificity (.88) for the original scale when compared to a CAPS diagnosis. Additional studies with the revised instrument are still clearly needed before its use as a diagnostic tool can be fully supported.

Mississippi Scale for Combat-Related PTSD

Developed by Keane, Caddell, and Taylor (1988), the 35-item Mississippi Scale (MPTSD) is widely used to assess combat-related PTSD symptoms. The scale items were selected from an initial pool of 200 items generated by experts to closely match DSM-III criteria and associated features of the disorder. Respondents are asked to rate, on a Likert scale, the severity of symptoms over the time period "since the event." The Mississippi Scale yields both a continuous score of symptom severity and diagnostic information. It is available in several languages and takes 10–15 minutes to administer.

The Mississippi Scale has excellent psychometric properties. In Vietnam Era veterans seeking treatment, Keane and colleagues (1988) reported high internal consistency (alpha = .94) and test–retest reliability (.97) over a 1-week time interval. In a subsequent validation study, the authors found substantial sensitivity (.93) and specificity (.89), with a cutoff of 107, and an overall hit rate of 90% when the scale was used to differentiate between a PTSD group and two non-PTSD comparison groups.

McFall, Smith, Mackay, and Tarver (1990) replicated these findings, and further demonstrated that patients with PTSD with and without substance use disorders did not differ on the Mississippi Scale. Given the high comorbidity between PTSD and substance use disorders, the authors felt it was important to demonstrate that the test assesses PTSD symptoms rather than effects associated with alcohol and drug use. McFall and colleagues also obtained information on convergent validity, and found significant correlations between the MPTSD and other measures of PTSD, including the total number of SCID-PTSD symptoms (.57), total IES score (.46), and degree of traumatic combat exposure (.40; Vietnam Era Stress Inventory [Wilson & Krauss, 1984]). These findings suggest that the Mississippi Scale is a valuable self-report tool in settings where assessment of combat-related PTSD is needed.

More recently, Keane, Charney, and Orazem (in press) examined the psychometric properties of the Mississippi Scale in over 1,200 Vietnam War veterans participating in a multisite study of the psychophysiology of PTSD (Keane et al., 1998). All subjects received a comprehensive psychological assessment, including the Mississippi Scale. Results indicated that the Mississippi Scale possesses excellent internal consistency (coefficient alpha = .96); item–total score correlations ranging from .33–.77, with a mean of .65; and a correlation of .83 with the PTSD scale (PK) of the MMPI-2. Using the SCID PTSD module as the diagnostic gold standard, the Mississippi Scale possessed a sensitivity of .84 and a specificity of .83, with an area under the curve estimate of .91. A cutoff of 106 or above was the optimal cutoff threshold for a diagnosis of PTSD, suggesting strong support for the use of this test in assessment of combat-related PTSD.

Keane PTSD Scale of the MMPI-2

Originally derived from the MMPI Form R (Keane et al., 1984), the Keane PTSD Scale now consists of 46 items empirically drawn from the MMPI-2 (Lyons & Keane, 1992). The items are answered in a true–false format. The scale is typically administered as part of the full MMPI-2, but it can be useful as a stand-alone scale. The embedded and stand-alone versions are highly correlated (.90; Herman, Weathers, Litz, & Keane, 1996). The PTSD Scale yields a total score that reflects the presence or absence of PTSD.

Psychometric data on the embedded and stand-alone versions of the PTSD Scale are excellent. Herman and colleagues (1996) reported evidence from a veteran sample of strong internal consistency of the embedded and stand-alone versions of the MMPI-2 PTSD Scale (alphas ranging from .95 to .96), and high test–retest reliability coefficients for the stand-alone version over 2–3 days (.95). With regard to validity, the embedded and stand-alone versions of the MMPI-2 PTSD Scale were correlated with other self-report measures of PTSD, including the Mississippi Scale (.81–.85), IES (.65–.71), and PTSD Checklist (PCL, .77–.83), and a diagnostic interview (CAPS; .77–.80). The embedded and stand-alone versions differed slightly in their optimally efficient cutoff score (26 vs. 24, respectively), but both demonstrated good sensitivity (.72 for embedded, .82 for stand-alone), specificity (.82 for embedded, .76 for stand-alone), and efficiency (.76 for embedded, .80 for stand-alone) compared to a CAPS diagnosis. As reported by Keane and colleagues (2006), the correlation of the PTSD Scale with the Mississippi Scale in a large sample of Vietnam War veterans was .83, demonstrating strong construct validity.

More research is needed to determine the generalizability of the findings with other trauma populations (Foa et al., 1997; Watson et al., 1986). Although only a few studies have used the PTSD Scale in nonveteran populations, the data appear promising (Koretzky & Peck, 1990; Neal et al., 1994). The PTSD Scale may be particularly useful in the areas of forensic psychology and disability assessment, in which the MMPI-2 is frequently employed.

Posttraumatic Diagnostic Scale

Developed by Foa and colleagues (1997), the Posttraumatic Diagnostic Scale (PDS) is a 49-item scale designed to measure DSM-IV PTSD criteria and symptom severity. The PDS is a revised version of an earlier self-report scale based on DSM-III-R (American Psychiatric Association, 1987), referred to as the PTSD Symptom Scale–Self-Report Version (PSS-SR; Foa et al., 1993). The PDS reviews trauma exposure and identifies the most distressing trauma. It also assesses criterion A2 (physical threat or helplessness), criteria B–D (intensity and frequency of all 17 symptoms), and functional impairment (criterion F). This scale has been used with several populations, including combat veterans, accident victims, and sexual and nonsexual assault survivors. The PDS can be administered in 10–15 minutes.

The psychometric properties of the PDS were evaluated among 264 volunteers recruited from several PTSD treatment centers, as well as from non-treatment-seeking populations at high risk for trauma (Foa et al., 1997). Investigators reported high internal consistency for the PTSD total score (alpha = .92) and subscales (alphas = .78–.84), and satisfactory test–retest reliability coefficients for the total PDS score and for symptom cluster scores (.77–.85). With regard to validity, the PDS total score correlated highly with other scales that measure traumatic responses (IES; Intrusion = .80 and Avoidance = .66; RAST = .81). In addition, the measure yielded substantial sensitivity (.89), specificity (.75), and high levels of diagnostic agreement with a SCID diagnosis (kappa = .65, 82% agreement). Based on these data, the authors have recommended the PDS as an effective and efficient screening tool for PTSD.

In a more recent examination of the psychometric properties of the PDS, Griffin, Uhlmansiek, Resick, and Mechanic (2004) compared the PDS to scores obtained in a CAPS clinical interview. They found strong intercorrelations between the two measures. Among 138 female survivors of domestic violence these investigators observed a high base rate of PTSD, with the PDS exhibiting excellent sensitivity (.94) and acceptable specificity (.53) when compared to the structured diagnostic interview. These findings suggest that the PDS might serve as an excellent first-tier screening device for identifying cases of PTSD in large samples or studies.

PTSD Checklist

Also developed at the National Center for PTSD (Weathers et al., 1993), the PTSD Checklist (PCL) is a 17-item self-report measure of PTSD symptoms. Different scoring procedures may be used to yield either a continuous measure of symptom severity or a dichotomous indicator of diagnostic status. Furthermore, dichotomous scoring methods include either an overall cutoff score or a cluster score approach. The original scale was based on DSM-III-R criteria for PTSD, but the PCL was updated to assess the 17 diagnostic criteria outlined in DSM-IV. Respondents are asked to rate, on a Likert scale, "how much each problem has bothered them" during the past month. The time frame can be adjusted as needed to suit the goals of the assessment. There are civilian (PCL-C) and military (PCL-M) versions of the measure. With PCL-C, reexperiencing and avoidance symptoms apply to any lifetime stressful event, whereas for the PCL-M, reexperiencing and avoidance symptoms apply to stressful events that are military-related only. The PCL has been used extensively in both research and clinical settings and takes 5–10 minutes to administer. If needed, a 17-item Life Events Checklist, developed as a companion to the CAPS and aimed at identifying exposure to potentially traumatic experiences, can be used with the PCL.

The PCL was originally validated in a sample of Vietnam and Persian Gulf War veterans and found to have strong psychometric properties (Weathers et al., 1993). Keen, Kutter, Niles, and Krinsley (2004) examined the psychometric properties of the updated PCL in veterans with both combat and noncombat traumas, and found evidence for high internal consistency (alpha = .96 for all 17 symptoms). Test–retest reliability was not examined, but the Weathers and colleagues (1993) original study suggested this was robust (.96) over a 2- to 3-day interval; other investigators also documented adequate test–retest reliability of this measure over a 2-week time frame (Ruggiero, Del Ben, Scotti, & Rabalais, 2003).

Several studies now offer evidence for the reliability and validity of the PCL in nonveteran samples, although the optimal cutoff score varies across samples when researchers attempt to obtain the highest level of diagnostic accuracy. The reasons for these discrepancies are unclear (e.g., gender, recency of trauma, severity of trauma, and treatment-seeking status; Manne, DuHamel, Gallelli, Sorgen, & Redd, 1998) and warrant further investigation.

Distressing Event Questionnaire

Developed by Kubany, Leisen, Kaplan, and Kelly (2000), the Distressing Event Questionnaire (DEQ) provides dichotomous and continuous information. It does not assess criterion A-1 (occurrence of the traumatic event), but it has three items that assess criterion A-2 (presence of intense fear, helplessness, and horror at the time of the event) and 17 items that assess the DSM-IV diagnostic symptoms of PTSD (criteria B–D). Respondents are asked to indicate on a Likert scale "the degree to which they experienced each of the symptoms" within the last month. Additional items focus on chronology (criterion E), distress and functional impairment (criterion F), and associated features of guilt, anger, and unresolved grief. The DEQ takes 5–7 minutes to complete.

Kubany and colleagues (2000) conducted a series of studies to evaluate the psychometric properties of the DEQ and the results are excellent. Samples included male Vietnam combat veterans and women with mixed trauma histories (including incest, rape, partner abuse, prostitution, and sexual abuse). In the initial study, they found high internal consistency (alpha = .93 for total score and .88–.91 across symptom clusters). In a second study, they reported high test–retest reliability (.83–.94) over an average of 10 days, using a variety of scoring methods. The third and largest study provided evidence for construct validity. The DEQ total scale score was highly correlated with the CAPS (.82–.90) and Modified PTSD Scale (.86–.94; Falsetti, Resnick, Resick, & Kilpatrick, 1993). Furthermore, when they used the CAPS as a criterion measure, an optimal cutoff score of 26 yielded a sensitivity of .87, specificity of .85, and diagnostic efficiency of .86 in the veteran sample. For women, a cutoff score of 18 yielded a sensitivity of .98, specificity of .58, and overall efficiency of .90. A particular strength of this scale is its ability to classify PTSD correctly in a high percentage of men and women, despite differences in trauma exposure and ethnicities.

Psychophysiological Approaches

Diagnostic Assessment of PTSD

PTSD diagnoses are generally based on whether an individual reports having the requisite severity and number of symptoms on interview and/or questionnaire measures. Yet conferring a valid diagnosis is greatly influenced by the accuracy of the patient's report and the clinician's ability to determine whether self-reported symptoms exceed diagnostic thresholds. Psychophysiological assessment can offer important additional information to the diagnostic process. PTSD is uniquely suited to this type of assessment, because the DSM-IV definition of the disorder specifies that individuals with the disorder may show "physiological reactivity on exposure to internal or external cues that symbolize or resemble an aspect of the traumatic event." The typical psychophysiological assessment for PTSD diagnostic purposes involves the clinician recording physiological responses during presentation of standardized audiovisual stimuli reminiscent of the trauma or script-driven imagery of the event using a paradigm advanced originally by Peter Lang (1979) and his colleagues. With either approach, responses during resting baseline and/or neutral intervals are compared to those obtained during processing of trauma-related stimuli. Theoretically, the physiological response to trauma-related cues reflects activation of the memory network in which a traumatic event is encoded. Once such a memory is cued, emotions associated with it are activated along with their accompanying physiological responses.

Physiological measures relevant to the assessment of PTSD include indices of autonomic activity (i.e., heart rate, blood pressure, and electrodermal responses) and the overt expression of negative affect, which may be collected via recordings of facial muscle activity. Heart rate is obtained through electrocardiogram (ECG) recordings of the interval between beats and may be recorded continuously through baseline, stimulus presentation, and recovery periods. Blood pressure, on the other hand, is typically recorded intermittently (i.e., once during each phase of the procedure), because it is dependent on the temporary inflation of a cuff around the arm. Electrodermal (i.e., sweat gland) activity is recorded continuously by maintaining a very small constant voltage between two electrodes attached to the palm of the hand, and variations in electrical conductance that results from changes in sweat gland activity are recorded. When other factors are held constant (e.g., ambient room temperature) skin conductance responses provide a near-direct measure of general sympathetic activation. Finally, facial muscle activity is recorded using electromyography (EMG) techniques. The corrugator, or "frown muscle" over the eyebrow, is the most common recording site, because this muscle is recruited in the production of many negative facial expressions.

Studies that have compared psychophysiological reactivity of individuals with and without PTSD to trauma-related stimuli have consistently shown that groups of individuals with the disorder exhibit greater mean levels of reactivity than do trauma-exposed controls. These effects have been observed in populations ranging from combat veterans of the Vietnam War, Korean Conflict, and World War II, to survivors of childhood sexual abuse (for a comprehensive summary of this literature, see Orr, Metzger, Miller, & Kaloupek, 2004). The largest and most methodologically rigorous study of this type, conducted by the Department of Veterans Affairs (DVA; Keane et al., 1998), tested the ability of psychophysiological responding to predict SCID-based PTSD diagnosis in a sample of over 1,300 male Vietnam War veterans. Results revealed that an equation derived to predict PTSD status on the basis of four physiological variables correctly classified approximately two-thirds of veterans with a current PTSD diagnosis. Other, smaller studies have typically reported sensitivities and specificities for PTSD classification in the range of 60–90% and 80–100%, respectively (see Orr, 1997). These estimates clearly suggest that heightened physiological responsivity during exposure to trauma cues indicates a strong likelihood that an individual would meet criteria for PTSD. They also suggest that only a small proportion of individuals who have histories of trauma but never met criteria for the disorder show such reactivity, but the correspondence between psychophysiological responses to trauma-related stimuli and a PTSD diagnosis is far from perfect. There was one important methodological difference between the multisite trial funded by the DVA and the remainder of the trials. No patients taking any type of medication could be enrolled. This restriction led to the exclusion of many of the most severe cases of PTSD in the clinics, because neither the patient nor the provider was interested in eliminating medications thought to be effective.

Numerous factors may contribute to disparities between physiological reactivity and the self-reported diagnosis of PTSD, including participant compliance with protocol demands; the appropriateness of trauma cue stimuli; biological influences such as age, sex, race, and fitness level; the presence of pharmacological agents (i.e., benzodiazepines, beta-adrenergic blockers); cognitive avoidance processes, including dissociation (Griffin, Resick, & Mechanic, 1997); and even personality traits that influence the emotional response to aversive stimuli (e.g., antisocial characteristics; Miller, Kaloupek, & Keane, 1999). Given the array of factors that influence the psychophysiological response to trauma-related stimuli in individuals with PTSD, we are not optimistic about the prospect of improving the performance of psychophysiological tests for the clinical diagnostic assessment of PTSD beyond the level achieved by Keane and colleagues (1998) without modifying the diagnostic criteria themselves.

Assessment of PTSD Treatment Outcome

Psychophysiological methods may also be used in the assessment of treatment process and outcome. From a cognitive-behavioral standpoint, the successful treatment of PTSD may involve extinguishing pathological responses to clinically relevant fear stimuli (Foa & Kozak, 1985). Thus, psychophysiological measures can be used to assess the extent to which these responses are (1) activated within a session and (2) extinguished over repeated sessions. Several preliminary treatment studies of this type have produced promising results. First, Shalev, Orr, and Pitman (1992) used systematic desensitization in the treatment of three individuals with PTSD and found that physiological responding to trauma-related imagery diminished from pre- to posttreatment, with reductions in PTSD symptomatology. Second, in a single-case study, Keane and Kaloupek (1982) reported reductions in heart rate and skin conductance during trauma-related imagery both within and between sessions. Third, Boudewyns and Hyer (1990) treated 51 individuals with combat-related PTSD using either exposure-based therapy or conventional counseling. Although there were no group differences on physiological measures in terms of treatment, results did reveal that individuals who showed reductions in physiological arousal posttreatment exhibited greater posttreatment improvement at a 3-month follow-up.

There is also evidence that psychophysiological responses measured pretreatment may be useful predictors of treatment response. For example, Levin, Cook, and Lang (1982) found that anxious patients who responded favorably to CBT showed larger heart rate and skin conductance responses to clinically relevant affective imagery during pretreatment assessment compared to patients who subsequently dropped out of treatment. Similarly, Lang, Melamed, and Hart (1970) found that snake phobics who produced larger heart rate responses while visualizing their phobic scenes during a desensitization procedure show greater fear reduction after treatment compared to phobics with little or no heart rate response. In PTSD, Pitman and colleagues (1996) found that greater improvement in a behavioral measure of intrusive recollections of the traumatic events was associated with larger heart rate increases during the initial exposure session.

In summary, research on the use of psychophysiology for clinical assessment of PTSD suggests that it has value as both a diagnostic tool and assessment of treatment process and outcome. The primary benefit afforded by the use of psychophysiology is the objectivity of the data, which in the context of an otherwise exclusively self-reported assessment is, in principle, quite valuable. Another advantage is that it preserves the possibility that some individuals who have difficulty reporting on their subjective state (e.g., young children, individuals with brain damage) still may provide evidence of reexperiencing. The drawbacks of this approach include the technical skills and equipment required and the imperfect association between physiological reactivity and the subjective experience of psychological distress. Finally, psychophysiological methodologies also hold promise for evaluating the cognitive, affective, and biological mechanisms underlying posttraumatic psychopathology.

GENERALIZABILITY OF FINDINGS

The generalizability of methods to assess PTSD is a function of several features of the assessment setting. Culture, language, race, age, and gender are factors that might influence the use and the interpretation of psychological instruments, whether structured diagnostic interviews or psychological tests. Attention to these variables is fundamental to discerning the presence or absence of PTSD and its co-occurring disorders. Although much more scientific work is needed to determine how well these instruments generalize, the following sections address some of the key variables that might influence generalizability.

Cultural Issues

Clinicians highlight the importance of considering the different populations on which an assessment instrument for PTSD was validated when selecting a measure. The need to develop culturally sensitive instruments has been of great interest for many years as a result of documentation of ethnoculturally specific responses to traumatic events. For example, several researchers have provided evidence of differences in the severity of PTSD symptoms following a traumatic event between ethnic minorities and Caucasians (e.g., Frueh, Brady, & de Arellano, 1998; Green, Grace, Lindy, & Leonard, 1990; Kulka et al., 1990). The need for culturally sensitive instruments is further emphasized by the growing awareness among scholars that prevalence of PTSD in developing countries is higher than that in industrialized nations (De Girolamo & McFarlane, 1996).

To date, the psychological assessment of PTSD has developed primarily within the context of Western, developed, and industrialized countries. Thus, PTSD assessment may be limited by a lack of culturally sensitivity measures and by the tremendous diversity among the cultural groups of interest (Marsella, Friedman, Gerrity, & Scurfield, 1996). However, there is progress in developing culturally sensitive measures.

A good example of a measure that possesses culturally relevant features is the Harvard Trauma Questionnaire (HTQ; Mollica et al., 1992), which is widely used in refugee and displaced populations. The HTQ assesses a range of potentially traumatic events and trauma-related symptoms. The assessment of trauma includes questions about many events to which refugees from war-torn countries may have been exposed, including torture, brainwashing, and deprivation of food or water. Originally developed in English, the HTQ has been translated and validated in Vietnamese, Laotian, and Khmer versions. In addition, the HTQ possesses linguistic equivalence across the many cultures and languages with which it has been used thus far. Mollica and colleagues (1992) have reported good reliability (test–retest = .89; interrater = .93, coefficient alpha = .96) for the HTQ (Cusack, Falsetti, & de Arellano, 2002). Future research will need to document the reliability and validity of new instruments on a wider range of populations and develop instruments that have the culturally sensitive characteristics embodied in the HTQ.

In addition, the CAPS has been studied among culturally different groups with excellent success. As one example, Charney and Keane (in press) recently examined the psychometric properties of the CAPS after it was adapted for use among Bosnian refugees. We applied contemporary methods for translation, back-translation, and then qualitative approaches to reconcile any differences in meaning that might have arisen as a function of this process. We found that the CAPS–Bosnian translation had comparable psychometric properties to prior analyses of the instrument, indicating that when properly adapted, the CAPS can be used successfully to measure PTSD symptomatology in culturally diverse populations, and that PTSD secondary to war in civilians appears to be comparable in nature to other forms of PTSD.

PTSD and Its Comorbidities

Characteristics of PTSD Comorbidity

As we mentioned early in this chapter, studies that have examined comorbidity in individuals diagnosed with PTSD indicate high levels of other conditions. For example, Brown, Campbell, Lehman, Grisham, and Mancill (2001) assessed the comorbidity of current and lifetime DSM-IV anxiety and unipolar mood disorders in 1,126 community outpatients, and found that of all the disorders assessed, PTSD showed the most severe and diverse pattern of comorbidity: Of individuals with a current diagnosis of PTSD, 92% met criteria for another, current Axis I disorder, most frequently major depressive disorder (77%), generalized anxiety disorder (38%), and alcohol abuse/dependence (31%). High rates of Axis I comorbidity were also observed in studies of veterans, including one in which 82% of those with a current diagnosis of PTSD met criteria for another Axis I disorder (Orsillo et al., 1996). In the National Vietnam Veterans Readjustment Study (NVVRS), Kulka and colleagues (1990) found that 50% of veterans with PTSD had an additional Axis I diagnosis. Kessler and colleagues (1995) also found high rates of comorbidity in PTSD in the National Comorbidity Survey (NCS) in the United States. Thus, comorbidity in PTSD is a problem of significant concern.

Fewer studies have assessed the comorbidity of PTSD with personality disorders. In most studies that have, the focus has been on the comorbidity between PTSD and either borderline or antisocial personality disorder, probably because they are the most common comorbid disorders and are characterized by socially problematic behaviors leading to treatment (Bollinger, Riggs, Blake, & Ruzek, 2000). Two studies have assessed the full range of comorbid Axis II disorders in samples of at least 100 individuals with PTSD. In the first, Bollinger and colleagues (2000) examined 107 veteran inpatients with PTSD and found that 79% of the sample met criteria for an Axis II disorder. Most common were avoidant (47%), paranoid (46%), obsessive–compulsive (28%) and antisocial (15%) personality disorders.

In the second study, Dunn and colleagues (2004) reported that nearly half of their male veteran outpatient sample with PTSD also met criteria for at least one personality disorder, and 17% met criteria for two or more personality disorder diagnoses. Although the two studies differ significantly in terms of the base rates of Axis II comorbidity, presumably due to demographic differences between the samples surveyed (i.e., inpatient vs. outpatient), their findings together underscore the heterogeneous mix of Axis II disorders that co-occur with PTSD.

PTSD comorbidity presents problems on multiple levels for clinicians and research-ers. In the clinical arena, individuals with comorbid Axis I or II disorders present with more severe PTSD symptoms (e.g., Back, Sonne, Killeen, Danksy, & Brady, 2003; Brady & Clary, 2003; Zayfert, Becker, Unger, & Shearer, 2002) and are poorer responders to treatment (e.g., Cloitre & Koenen, 2001; Zlotnick et al., 1999). Nosologically, this comorbidity challenges the neo-Kraepelinian idea that PTSD is a discrete syndrome that can be readily distinguished from other DSM-IV disorders. For PTSD researchers, comorbidity often represents a nuisance variable requiring statistical or other method-ological efforts to control (Keane & Kaloupek, 2002).

Conceptualizing Comorbidity in PTSD

One well-established model of psychiatric comorbidity that appears promising in its application to our understanding of patterns of PTSD comorbidity (see Miller, Greif, & Smith, 2003; Miller, Kaloupek, Dillon, & Keane, 2004) proposes that patterns of behavioral disturbance and psychiatric symptoms cohere along latent dimensions of *externalization* and *internalization*. This model, rooted in over 30 years of research in the area of childhood behavior disorders (cf., Achenbach & Edelbrock, 1978, 1984), has recently come to the fore in the adult psychopathology literature as the result of a series of influential factor-analytic studies of the latent structure of adult mental disorders (Cox, Clara, & Enns, 2002; Kendler, Prescott, Myers, & Neale, 2003; Krueger, Caspi, Moffitt, & Silva, 1998; Krueger, McGue, & Iacono, 2001). Krueger and colleagues (1998) reported that patterns of comorbidity tend to cohere along these dimensions, with the alcohol and substance-related disorders and antisocial per-sonality disorder loading on the externalizing dimension, and the unipolar mood and anxiety disorders falling on the internalizing dimension. This structure has demon-strated invariance across genders and multiple samples drawn at random from a larger sample (Krueger, 1999). Furthermore, twin and adoption studies implicate genetic factors in the etiology of externalizing and internalizing (Deater-Deckard & Plomin, 1999; Krueger et al., 2002) and the magnitude of these genetic effects increases with the severity of the behavior problems on a given dimension (Gjone, Stevenson, Sundet, & Eilerstein, 1996). Krueger and colleagues (1998, 2001, 2002) also reported that tendencies toward externalizing or internalizing are stable over time. Thus, one important strength is that this model affords consistency in the con-ceptualization of psychopathology across the lifespan, and it ties the evidence in PTSD to that of psychopathology research more broadly.

The externalizing–internalizing model is conceptually consistent with other mod-els of comorbidity, which state that the overlap among broad classes of disorders is due largely to the fact that they emerge from a common diathesis (e.g., Barlow, 2002; Clark, Watson, & Mineka, 1994). This conceptualization is supported by evidence suggesting that whereas much overlap exists in terms of the predisposing factors with-in a given spectrum of psychopathology, the manifestations of these diatheses vary considerably as a function of exposure to various environmental factors (e.g., trauma exposure, other life stressors, or developmental experiences). In other words, the dif-ferent manifestations of these shared vulnerability dimensions are represented by the various DSM-IV diagnoses. This concept is in accord with a leading theoretical expla-nation for the high rate of co-occurrence of disorders within a spectrum of psycho-pathology (i.e., disorders A and B co-occur because they are both influenced by another underlying or causal factor C; cf. Frances, Pincus, Widiger, Davis, & First, 1990).

Implications for PTSD

The forgoing conceptual model suggests that patterns of psychiatric symptoms tend to cohere along latent dimensions of psychopathology termed externalization and internalization, and that these dimensions appear to have an etiological basis in the personality of the individual. Recent work suggests that this model may be very relevant to our understanding of patterns of PTSD comorbidity (Miller et al., 2003, 2004; Miller & Resick, 2007). Through a series of three cluster-analytic studies of personality inventories completed by individuals with PTSD, Miller and colleagues (2004) have shown evidence of internalizing and externalizing subtypes of PTSD in both male and female samples. Summarizing findings across these three studies, Miller and colleagues found that one subtype of individuals with PTSD, termed "externalizers," was characterized by tendencies to express posttraumatic distress outwardly, through antagonistic interactions with others and conflict with societal norms and values. They endorsed elevated levels of anger and aggression, substance-related disorders, and cluster B personality disorder features, and produced personality inventory profiles defined by high disconstraint coupled with negative emotionality.

Furthermore, externalizers described themselves as prone to act impulsively, with little regard for the consequences of their actions, to become upset easily, and as chronically stressed. They described themselves as tending toward exhibitionistic, manipulative, and unconventional behavior. Also, they reported being emotionally labile, overactive, impulsive, fearless, at times aggressive and intimidating, and feeling chronically betrayed and mistreated by others. In both studies of veterans in which data on premilitary characteristics were available, individuals in this subgroup reported elevated rates of premilitary delinquency, suggesting that these characteristics may reflect the influence of externalizing personality traits that were present prior to the trauma.

In contrast, "internalizers" were characterized by tendencies to direct their posttraumatic distress inwardly, through shame, self-defeating/deprecating and anxious processes, avoidance, depression, and withdrawal. Across these three studies, individuals in this subtype were characterized by high rates of comorbid major depression and panic disorder, schizoid and avoidance personality disorder features, and personality profiles defined by high negative emotionality combined with low positive emotionality. They further described themselves as unenthusiastic, uninspired, easily fatigued, lacking interests, but, like externalizers, prone to experiencing frequent and intense negative emotions. They reported having few friends, being aloof and distant from others, and preferring to spend time alone. In contrast to the externalizers, they reported themselves as self-effacing and humble, and as not feeling particularly special, admirable, or talented. They endorsed a restricted range of emotions in interpersonal settings, and feelings of social inhibition, inadequacy, and hypersensitivity to negative evaluation. Their elevated scores on measures of trauma-related shame reflected a key psychological process of the internalizer: the tendency to incorporate into one's identity shameful aspects of the traumatic experience.

These findings suggest that the model of externalizing and internalizing psychopathology originally developed to account for covariation among broad classes of mental disorders (cf. Krueger et al., 1998, 2001) is relevant to the understanding of the heterogeneity of psychopathology and comorbidity within PTSD. Moreover, the close correspondence between the PTSD subtypes identified in this work, subtypes reported by an independent group of Australian investigators (Forbes et al., 2003), and three major personality "types" identified by developmental psychologists (i.e., resilient, overcontrolled, and undercontrolled; Asendorpf & van Aken, 1999; Hart, Hofmann, Edelstein, & Keller, 1997; Robins, John, Caspi, Moffitt, & Stouthamer-Loeber, 1996) lends support to the validity of this conceptual model of PTSD comorbidity. Moreover, these findings underscore the considerable variability in comorbidity within PTSD populations and suggest that the internalizing–externalizing model may be a useful heuristic for advancing our understanding of why individuals with PTSD may exhibit considerable behavioral differences in the clinic setting.

CHALLENGES FOR THE FUTURE

Assessment of PTSD and its comorbidities is a topic of growing interest and concern in the mental health field (Wilson & Keane, 1997, 2004). Since the inclusion of PTSD in the DSM-III, there has been considerable progress in understanding and evaluating the psychological consequences of exposure to traumatic events. Conceptual models of PTSD assessment have evolved (Keane, Wolfe, & Taylor, 1987; Sutker, Uddo-Crane, & Allain, 1991), psychological tests have been developed (Foa et al., 1997; Norris & Riad, 1997), diagnostic interviews have been validated (Davidson et al., 1989; Foa et al., 1993; Weathers et al., 2001), and subscales of existing tests have been created to assess PTSD (i.e., MMPI-2, Keane et al., 1984). We can rightly conclude that the assessment instruments available to evaluate PTSD are comparable to or better than those available for any DSM disorder. Moreover, multiple instruments that are now available cover the range of needs of the clinician. The psychometric data examining the reliability and validity of many of these instruments are nothing short of excellent.

Clearly, the assessment of PTSD in clinical settings focuses on more than the presence, absence, and severity of PTSD and its comorbidities. A comprehensive assessment strategy that purports to gather information about an individual's family history, life context, symptoms, beliefs, strengths, weaknesses, support system, and coping abilities (Newman, Kaloupek, & Keane, 1996) would assist in the development of an effective treatment plan for the patient. Our primary purpose in this review has been to examine the quality of a range of different instruments used to diagnose and assess PTSD, and to offer a conceptual model of the high prevalence of comorbid conditions in PTSD. This comorbidity, recognized very early in the history of PTSD (Keane et al., 1985) continues to represent a conceptual challenge to the field. The model we present here places the issue directly in the mainstream of considerations in psychopathology research more broadly.

Finally, the comprehensive assessment of a patient for PTSD certainly needs to include indices of social, interpersonal, and occupational functioning. To achieve an optimal examination, the clinical and interpersonal skills of the clinician require a sensitivity to the intensity of the traumatic events, the difficulties many people have in disclosing aspects of traumatic experiences, the recency of the exposure, and the debilitating effects of these symptoms on the individual and his or her family.

Clearly, this review is not intended to be comprehensive in its examination of the psychometric properties of all instruments available for the assessment of PTSD. Our intent has been to provide a heuristic structure that clinicians might employ when selecting a particular instrument for their clinical purposes. By carefully examining the psychometric properties of an instrument, the clinician can make an informed decision about the appropriateness of a particular instrument for the task at hand. Instruments that provide a full utility analysis (i.e., sensitivity, specificity, hit rate, etc.) do much to assist clinicians in making their final judgments. Furthermore, instruments that are developed and evaluated on multiple trauma populations of both genders, and with different racial, cultural, and age groups, are highly desirable; future work should address the needs of clinicians and researchers working in diverse settings with diverse clients. There is an excitement about work in the field of PTSD at this time in our history. The quality of our measurement will ultimately determine the quality of all new knowledge developed. If the strength of the evidence base contained in this review is an indication, we will experience an explosion of new knowledge in the next decade.

REFERENCES

Achenbach, T. M. & Edelbrock, C. S. (1978). The classification of child psychopathology: a review and analysis of empirical efforts. *Psychological Bulletin, 85*(6), 1275–1301.

Achenbach, T. M. & Edelbrock, C. S. (1984). Psychopathology of childhood. *Annual Review of Psychology, 35*, 227–256.

American Psychiatric Association. (1980). *Diagnostic and statistical manual of mental disorders* (3rd ed.). Washington, DC: Author.

American Psychiatric Association. (1987). *Diagnostic and statistical manual of mental disorders* (3rd ed., rev.). Washington, DC: Author.

American Psychiatric Association. (1994). *Diagnostic and statistical manual of mental disorders* (4th ed.). Washington, DC: Author.

Antony, M. M., & Rowa, K. (2005) Evidence-based assessment of anxiety disorders in adults. *Psychological Assessment, 17*, 256–266.

Arata, C. M., Saunders, B. E., & Kilpatrick, D. G. (1991). Concurrent validity of a crime-related post-traumatic stress disorder scale for women within the Symptom Checklist-90–Revised. *Violence and Victims, 6*, 191–199.

Asendorpf, J. B., & van Aken, M. A. (1999). Resilient, overcontrolled, and undercontrolled personality prototypes in childhood: Replicability, predictive power, and the trait-type issue. Journal of Personality and Social Psychology, 77, 815–832.

Back, S. E., Sonne, S. C., Killeen, T., Dansky, B. S., & Brady, K. T. (2003). Comparative profiles of women with PTSD and comorbid cocaine or alcohol dependence. *American Journal of Drug and Alcohol Abuse, 29*, 169–189.

Barlow, D. H. (2002). *Anxiety and its disorders: The nature and treatment of anxiety and panic* (2nd ed.). New York: Guilford Press.

Blake, D. D., Weathers, F. W., Nagy, L. M., Kaloupek, D. G., Charney, D. S., & Keane, T. M. (1990). *The Clinician-Administered PTSD Scale-IV*. Boston: National Center for PTSD, Behavioral Sciences Division.

Blanchard, E. B., Gerardi, R. J., Kolb, L. C., & Barlow, D. H. (1986). The utility of the Anxiety Disorders Interview Schedule (ADIS) in the diagnosis of post-traumatic stress disorder (PTSD) in Vietnam veterans. *Behaviour Research and Therapy, 24*, 577–580.

Bollinger, A., Riggs, D., Blake, D., & Ruzek, J. (2000). Prevalence of personality disorders among combat veterans with posttraumatic stress disorder. *Journal of Traumatic Stress, 13*, 255–270.

Bonanno, G. A. (2004). Loss, trauma, and human resilience: Have we underestimated the human capacity to thrive after extremely aversive events? *American Psychologist, 59*, 20–28.

Boudewyns, P. A., & Hyer, L. (1990). Physiological responses to combat memories and preliminary treatment outcome in Vietnam veteran PTSD patients treated with direct therapeutic exposure. *Behavior Therapy, 21*, 63–87.

Brady, K. T., & Clary, C. M. (2003). Affective and anxiety comorbidity in post-traumatic stress disorder treatment trials of sertraline. *Comprehensive Psychiatry, 44*, 360–369.

Brown, T. A., Campbell, L. A., Lehman, C. L., Grisham, J. R., & Mancill, R. B. (2001). Current and life-time comorbidity of the DSM-IV anxiety and mood disorders in a large clinical sample. *Journal of Abnormal Psychology, 110*, 585–599.

Breslau, N., Chilcoat, H. D., Kessler, R. C., Peterson, E. L., & Lucia, V. C. (1999). Vulnerability to assaultive violence: Further specification of the sex difference in post-traumatic stress disorder. *Psychological Medicine 29*, 813–821.

Breslau, N., Davis, G. C., Andreski, P., & Peterson, E. (1991). Traumatic events and posttraumatic stress disorder in an urban population of young adults. *Archives of General Psychiatry, 48*, 216–222.

Brown, T. A., DiNardo, P. A., Lehman, C. L., & Campbell, L. A. (2001). Reliability of DSM-IV anxiety and mood disorders: Implications for the classification of emotional disorders. *Journal of Abnormal Psychology, 110*, 49–58.

Charney, M. E., & Keane, T. M. (in press). Psychometric analysis of the Clinician Administered PTSD Scale (CAPS)—Bosnian Translation. *Cultural and Ethnic Minority Psychology*.

Clark, L. A., Watson, D., & Mineka, S. (1994) Temperament, personality, and the mood and anxiety disorders. *Journal of Abnormal Psychology, 103*(1), 103–116.

Cloitre, M., & Koenen, K. C. (2001). The impact of borderline personality disorder on process group outcome among women with posttraumatic stress disorder related to childhood abuse. *International Journal of Group Psychotherapy, 51*, 379–398.

Cohen, J. (1960). A coefficient of agreement for nominal scales. *Educational and Psychological Measurement, 20*, 37–46.

Cox, B. J., Clara, I. P., & Enns, M. W. (2002). Posttraumatic stress disorder and the structure of common mental disorders. *Depression and Anxiety, 15*, 168–171.

Cusack, K., Falsetti, S., & de Arellano, M. (2002). Gender considerations in the psychometric assessment of PTSD. In R. Kimerling, P. Ouimette, & J. Wolfe (Eds.), *Gender and PTSD* (pp. 150–176). New York: Guilford Press.

Davidson, J. R. T., Malik, M. A., & Travers, J. (1997). Structured Interview for PTSD (SIP): Psychometric validation for DSM-IV criteria. *Depression and Anxiety, 5*, 127–129.

Davidson, J. R. T., Smith, R., & Kudler, H. (1989). Validity and reliability of the DSM-III criteria for posttraumatic stress disorder: Experience with a structured interview. *Journal of Nervous and Mental Disease, 177*, 336–341.

Deater-Deckard, K., & Plomin, R. (1999). An adoption study of the etiology of teacher and parent reports of externalizing behavior problems in middle childhood. *Child Development, 70*, 144–154.

De Girolamo, G., & McFarlane, A. C. (1996). Epidemiology of posttraumatic stress disorder among victims of intentional violence: A review of the literature. In F. L. Mak & C. C. Nadelson (Eds.), *International review of psychiatry* (Vol. 2, pp. 93–119). Washington, DC: American Psychiatric Press.

de Jong, J. T. V. M., Komproe, I. H., Van Ommeren, M., El Masri, M., Khaled, N., van de Put, W., et al. (2001). Lifetime events and posttraumatic stress disorder in four postconflict settings. *Journal of the American Medical Association, 286*, 555–562.

DiNardo, P. A., Brown, T. A., & Barlow, D. H. (1994). *Anxiety Disorders Interview Schedule for DSM-IV: Lifetime version (ADIS-IV-L)*. San Antonio, TX: Psychological Corporation.

DiNardo, P. A., Moras, K., Barlow, D. H., Rapee, R. M., & Brown, T. A. (1993). Reliability of DSM-III-R anxiety disorder categories: Using the Anxiety Disorders Interview Schedule—Revised (ADIS-R). *Archives of General Psychiatry, 50*, 251–256.

DiNardo, P. A., O'Brien, G. T., Barlow, D. H., Waddell, M. T., & Blanchard, E. B. (1983). Reliability of

DSM-III anxiety disorder categories using a new structured interview. *Archives of General Psychiatry, 40,* 1070–1074.

Dunn, N. J., Yanasak, E., Schilaci, J., Simotas, S., Rehm, L., Souchek, J., et al. (2004). Personality disorders in veterans with posttraumatic stress disorder and depression. *Journal of Traumatic Stress, 17,* 75–82.

Falsetti, S. A., Johnson, M. R., Ware, M. R., Emmanuel, N. J., Mintzer, O., Book, S., et al. (1996, March). *Beyond PTSD: Prevalence of trauma in an anxiety disorders sample.* Paper presented at the 16th annual conference of the Anxiety Disorders Association of America, Orlando, FL.

Falsetti, S. A., Resnick, H. S., Resick, P. A., & Kilpatrick, D. G. (1993). The Modified PTSD Symptom Scale: A brief self-report measure of posttraumatic stress disorder. *Behavior Therapist, 16,* 161–162.

First, M., Spitzer, R., Williams, J., & Gibbon, M. (2000). Structured Clinical Interview for DSM-IV Axis I Disorders (SCID-I). In American Psychiatric Association (Ed.), *Handbook of psychiatric measures* (pp. 49–53). Washington, DC: American Psychiatric Association.

Foa, E. B., Cashman, L., Jaycox, L., & Perry, K. (1997). The validation of a self-report measure of posttraumatic stress disorder: The Posttraumatic Diagnostic Scale. *Psychological Assessment, 9,* 445–451.

Foa, E. B., & Kozak, M. J. (1985). Treatment of anxiety disorders: Implications for psychopathology. In A. H. Tuma & J. D. Maser (Eds.), *Anxiety and the anxiety disorders.* Hillsdale, NJ: Erlbaum.

Foa, E. B., Riggs, D. S., Dancu, C. V., & Rothbaum, B. O. (1993). Reliability and validity of a brief instrument for assessing post-traumatic stress disorder. *Journal of Traumatic Stress, 6,* 459–474.

Forbes, D., Creamer, M., Allen, N., Elliott, P., McHugh, T., Debenham, P., et al. (2003). MMPI-2 subgroups of veterans with combat-related PTSD. *Journal of Nervous and Mental Disease, 191,* 531–537.

Frances, A., Pincus, H. A., Widiger, T. A., Davis, W. W., & First, M. B. (1990). DSM-IV: Work in progress. *American Journal of Psychiatry, 147*(11), 1439–1448.

Frueh, B. C., Brady, K. L., & Arellano, M. A. (1998). Racial differences in combat-related PTSD: Empirical findings and conceptual issues. *Clinical Psychology Review, 18,* 287–305.

Gerardi, R., Keane, T. M., & Penk, W. E. (1989). Utility: Sensitivity and specificity in developing diagnostic tests of combat-related post-traumatic stress disorder (PTSD). *Journal of Clinical Psychology, 45,* 691–703.

Gjone, H., Stevenson, J., Sundet, J. M., & Eilersten, D. E. (1996). Changes in heritability across increasing levels of behavior problems in young twins. *Behavior Genetics, 26,* 419–426.

Green, B. L., Grace, M. C., Lindy, J. D., & Leonard, A. C. (1990). Race differences in response to combat stress. *Journal of Traumatic Stress, 3,* 379–393.

Griffin, M. G., Resick, P. A., & Mechanic, M. B. (1997). Objective assessment of peritraumatic dissociation: Psychophysiological indicators. *American Journal of Psychiatry, 154,* 1081–1088.

Griffin, M. G., Uhlmansiek, M., Resick, P. A., & Mechanic, M. B. (2004). Comparison of the PTSD Diagnostic Scale vs. the Clinician Administered PTSD Scale in domestic violence survivors. *Journal of Traumatic Stress, 17,* 497–503.

Hart, D., Hofmann, V., Edelstein, W., & Keller, M. (1997). The relation of childhood personality types to adolescent behavior and development: A longitudinal study of Icelandic children. *Developmental Psychology, 32,* 195–205.

Herman, D. S., Weathers, F. W., Litz, B. T., & Keane, T. M. (1996). Psychometric properties of the embedded and stand-alone versions of the MMPI-2 Keane PTSD Scale. *Assessment, 3,* 437–442.

Horowitz, M. J. (1976). *Stress response syndromes.* Northvale, NJ: Aronson.

Horowitz, M. J., Wilner, N., & Alvarez, W. (1979). Impact of Event Scale: A measure of subjective stress. *Psychosomatic Medicine, 41,* 209–218.

Hunsley, J., & Mash, E. J. (2005). Introduction to the special section on developing guidelines for the evidence-based assessment (EBA) of adult disorders. *Psychological Assessment, 17,* 251–255.

Joseph, S. (2000). Psychometric evaluation of Horowitz's Impact of Event Scale: A review. *Journal of Traumatic Stress, 13,* 101–113.

Keane, T. M. (1995). Guidelines for the forensic psychological assessment of posttraumatic stress disorder claimants. In R. I. Simon (Ed.), *Posttraumatic stress disorder in litigation: Guidelines for forensic assessment* (pp. 99–115). Washington, DC: American Psychiatric Press.

Keane, T. M., & Barlow, D. H. (2002). Posttraumatic stress disorder. In D. H. Barlow (Ed.) *Anxiety and its disorders: The nature and treatment of anxiety and panic* (2nd ed., pp. 418–453). New York: Guilford Press.

Keane, T. M., Buckley, T., & Miller, M. (2003). Guidelines for the forensic psychological assessment of posttraumatic stress disorder claimants. In R. I. Simon (Ed.), *Posttraumatic stress disorder in litigation: Guidelines for forensic assessment* (2nd ed., pp. 119–140). Washington, DC: American Psychiatric Association Press.

Keane, T. M., Caddell, J. M., & Taylor, K. L. (1988). Mississippi Scale for Combat-Related Posttraumatic Stress Disorder: Three studies in reliability and validity. *Journal of Consulting and Clinical Psychology, 56,* 85–90.

Keane, T. M., Charney, M. E., & Orazem, R. (in press). The interrelationship of evidence based measures of posttraumatic stress disorder: Evidence from a VA cooperative study. *Journal of Rehabilitation Research and Development.*

Keane, T. M., Fairbank, J. A., Caddell, J. M., Zimering, R. T., & Bender, M. E. (1985). A behavioral approach to assessing and treating post-traumatic stress disorder in Vietnam veterans. In C. R. Figley (Ed.), *Trauma and its wake* (pp. 257–294). New York: Brunner/Mazel.

Keane, T. M., & Kaloupek, D. G. (1982). Imaginal flooding in the treatment of posttraumatic stress disorder. *Journal of Consulting and Clinical Psychology, 50,* 138–140.

Keane, T. M., & Kaloupek, D. G. (2002). Posttraumatic stress disorder: Diagnosis, assessment, and monitoring outcomes. In R. Yehuda (Ed.), *Treating trauma survivors with PTSD* (pp. 21–42). Washington: American Psychiatric Press.

Keane, T. M., Kaloupek, D. G., & Weathers, F. W. (1996). Ethnocultural considerations in the assessment of PTSD. In A. J. Marsella, M. J. Friedman, E. T. Gerrity, & R. M. Scurfield (Eds.), *Ethnocultural aspects of posttraumatic stress disorder: Issues, research, and clinical applications* (pp. 183–205). Washington, DC: American Psychological Association.

Keane, T. M., Kolb, L. C., Kaloupek, D. G., Orr, S. P., Blanchard, E. B., Thomas, R. G., et al. (1998). Utility of psychophysiology measurement in the diagnosis of posttraumatic stress disorder: Results from a department of Veterans Affairs cooperative study. *Journal of Consulting and Clinical Psychology, 66,* 914–923.

Keane, T. M., Malloy, P. F., & Fairbank, J. A. (1984). Empirical development of an MMPI subscale for the assessment of combat-related posttraumatic stress disorder. *Journal of Consulting and Clinical Psychology, 52,* 888–891.

Keane, T. M., Marshall, A., & Taft, C. (2006) Posttraumatic stress disorder: Epidemiology, etiology, and treatment outcome. *Annual Review of Clinical Psychology, 2,* 161–197.

Keane, T. M., Weathers, F. W., & Foa, E. B. (2000). Diagnosis and assessment. In E. B. Foa, T. M. Keane, & M. J. Friedman (Eds.), *Effective treatments for PTSD* (pp. 18–36). New York: Guilford Press.

Keane, T. M., Wolfe, J., & Taylor, K. L. (1987). Post-traumatic stress disorder: Evidence for diagnostic validity and methods for psychological assessment. *Journal of Clinical Psychology, 43,* 32–43.

Keen, S. M., Kutter, C. J., Niles, B. L. & Krinsley, K. E. (2004, November). *Psychometric properties of the PTSD Checklist.* Poster presented at the annual meeting of the International Society for Traumatic Stress Studies, New Orleans, LA.

Kendler, K. S., Prescott, C. A., Myers, J., & Neale, M. C. (2003). The structure of genetic and environmental risk factors for common psychiatric and substance use disorders in men and women. *Archives of General Psychiatry, 60,* 929–937.

Kessler, R. C., Sonnega, A., Bromet, E., Hughes, M., & Nelson, C. B. (1995). Posttraumatic stress disorder in the National Comorbidity Survey. *Archives of General Psychiatry, 52,* 1048–1060.

Kilpatrick, D. G. (1988). Rape aftermath symptom test. In M. Hersen, & A. S. Bellack, (Eds.), *Dictionary of behavioral assessment techniques* (pp. 658–669). Oxford, UK: Pergamon Press.

Kilpatrick, D., Edmunds, C., & Seymour, A. (1992). *Rape in America: A report to the nation.* Arlington, VA: National Victim Center.

Koretzky, M. B., & Peck, A. H. (1990). Validation and cross-validation of the PTSD Subscale of the MMPI with civilian trauma victims. *Journal of Clinical Psychology, 46,* 296–300.

Kraemer, H. C. (1992). *Evaluating medical tests: Objective and quantitative guidelines.* Newbury Park, CA: Sage.

Krueger, R. F. (1999). The structure of common mental disorders. *Archives of General Psychiatry, 56,* 921–926.

Krueger, R. F., Caspi, A., Moffitt, T. E., & Silva, P. A. (1998). The structure and stability of common mental disorders (DSM-III-R): A longitudinal–epidemiological study. *Journal of Abnormal Psychology, 107,* 216–227.

Krueger, R. F., Hicks, B. M., Patrick, C. J., Carlson, S. R., Iacono, W. G., & McGue, M. (2002). Etiologic connections among substance dependence, antisocial behavior, and personality: Modeling the externalizing spectrum. *Journal of Abnormal Psychology, 111,* 411–424.

Krueger, R. F., McGue, M., & Iacono, W. G. (2001). The higher-order structure of common DSM mental disorders: Internalization, externalization, and their connections to personality. *Personality and Individual Differences, 30,* 1245–1259.

Kubany, E. S., Leisen, M. B., Kaplan, A. S., & Kelly, M. P. (2000). Validation of a brief measure of post-traumatic stress disorder: The Distressing Event Questionnaire (DEQ). *Psychological Assessment, 12,* 197–209.

Kulka, R. A., Schlenger, W. E., Fairbank, J. A., Jordan, B. K., Hough, R. L., Marmar, C. R., et al. (1990). *Trauma and the Vietnam war generation: Report of findings from the National Vietnam Veterans Readjustment Study.* New York: Brunner/Mazel.

Lang, P. (1979). A bio-informational theory of emotional imagery. *Psychophysiology, 16,* 495–512.

Lang, P. J., Melamed, B. G., & Hart J. (1970). A psychophysiological analysis of fear modification using an automated desensitization procedure. *Journal of Abnormal Psychology, 76*(2), 220–234.

Levin, D. N., Cook, E. W., & Lang, P. J. (1982). Fear imagery and fear behavior: Psychophysiological analysis of clients receiving treatment for anxiety disorder. *Psychophysiology, 19,* 571–572.

Litz, B. T., & Weathers, F. (1994). The diagnosis and assessment of post-traumatic stress disorder in adults. In M. B. Williams & J. F. Sommer (Eds.), *The Handbook of Post-Traumatic Therapy* (pp. 20–37). Westport, CT: Greenwood Press.

Lyons, J. A., & Keane, T. M. (1992). Keane PTSD scale: MMPI and MMPI-2 update. *Journal of Traumatic Stress, 5,* 111–117.

Manne, S. L., DuHamel, K., Gallelli, K., Sorgen, K., & Redd, W. H. (1998). Posttraumatic stress disorder among mothers of pediatric cancer survivors: Diagnosis, comorbidity, and utility of the PTSD Checklist as a screening instrument. *Journal of Pediatric Psychology, 23,* 357–366.

Marsella, A. J., Friedman, M. J., Gerrity, E. T., & Scurfield, R. M. (Eds.). (1996). *Ethnocultural aspects of posttraumatic stress disorder.* Washington, DC: American Psychological Association.

McFall, M. E., Smith, D. E., Mackay, P. W., & Tarver, D. J. (1990). Reliability and validity of Mississippi Scale for Combat-Related Posttraumatic Stress Disorder. *Journal of Consulting and Clinical Psychology, 2,* 114–121.

McFall, M. E., Smith, D., Roszell, D. K., Tarver, D. J., & Malas, K. L. (1990). Convergent validity of measures of PTSD in Vietnam combat veterans. *American Journal of Psychiatry, 147,* 645–648.

Miller, M. W., Greif, J. L., & Smith, A. A. (2003). Multidimensional Personality Questionnaire profiles of veterans with traumatic combat exposure: Internalizing and externalizing subtypes. *Psychological Assessment, 15,* 205–215.

Miller, M. W., Kaloupek, D. G., Dillon, A. L., & Keane, T. M. (2004). Externalizing and internalizing subtypes of combat-related PTSD: A replication and extension using the PSY-5 scales. *Journal of Abnormal Psychology, 112,* 636–645.

Miller, M. W., Kaloupek, D. G., & Keane, T. M. (1999). Antisociality and physiological hyporesponsivity during exposure to trauma-related stimuli in patients with PTSD. *Psychophysiology, 36,* S81.

Miller, M. W., & Resick, P. A. (2007). Internalizing and externalizing subtypes in female sexual assault survivors: Implications for the understanding of complex PTSD. *Behavior Therapy, 38,* 58–71.

Mollica, R. F., Caspi-Yavin, Y., Bollini, P., Truong, T., Tor, S., & Lavelle, J. (1992). The Harvard Trauma

Questionnaire: Validating a cross-cultural instrument for measuring torture, trauma, and post-traumatic stress disorder in Indochinese refugees. *Journal of Nervous and Mental Disease, 180,* 111–116.

Murray, C. J., & Lopes, A. D. (1996). Evidence-based health policy: Lessons from the Global Burden of Disease Study. *Science, 274,* 740–743.

Neal, L. A., Busuttil, W., Rollins, J., Herepath, R., Strike, P., & Turnbull, G. (1994). Convergent validity of measures of post-traumatic stress disorder in a mixed military and civilian population. *Journal of Traumatic Stress, 7,* 477–455.

Newman, E., Kaloupek, D. G., & Keane, T. M. (1996). Assessment of PTSD in clinical and research settings. In B. A. van der Kolk, A. C. McFarlane, & L. Weisæth (Eds.), *Traumatic stress: The effects of overwhelming experience on mind, body, and society* (pp. 242– 275). New York: Guilford Press.

Norris, F. H. (1992). Epidemiology of trauma: Frequency and impact of different potentially traumatic events on different demographic groups. *Journal of Consulting and Clinical Psychology, 60,* 409–418.

Norris, F. H., & Riad, J. K. (1997). Standardized self-report measures of civilian trauma and posttraumatic stress disorder. In J. P. Wilson & T. M. Keane (Eds.), *Assessing psychological trauma and PTSD* (pp. 7–42). New York: Guilford Press.

Orr, S. P. (1997). Psychophysiologic reactivity to trauma-related imagery in PTSD: Diagnostic and theoretical implications of recent findings. *Annals of the New York Academy of Sciences, 821,* 114–124.

Orr, S. P., Metzger, L. J., Miller, M. W., & Kaloupek, D. G. (2004). Psychophysiological assessment of PTSD. In J. P. Wilson & T. M. Keane (Eds.), *Assessing psychological trauma and PTSD* (2nd ed., pp. 289–343). New York: Guilford Press.

Orsillo, S. M., Weathers, F. W., Litz, B. T., Steinberg, H. R., Huska, J. A., & Keane, T. M. (1996). Current and lifetime psychiatric disorders among veterans with warzone-related posttraumatic stress disorder. *Journal of Nervous and Mental Disease, 184,* 307–313.

Pitman, R. K., Orr, S. P., Altman, B., Longpre, R. E., Poire, R. E., Macklin, M. L., et al. (1996). Emotional-processing and outcome of imaginal flooding therapy in Vietnam veterans with chronic posttraumatic stress disorder. *Comprehensive Psychiatry, 37,* 409–418.

Robins, R. W., John, O. P., Caspi, A., Moffitt, T. E., & Stouthamer-Loeber, M. (1996). Resilient, overcontrolled, and undercontrolled boys: Three replicable personality types. *Journal of Personality and Social Psychology, 70,* 157–171.

Ruggiero, K. J., Del Ben, K., Scotti, J. R., & Rabalais, A. E. (2003). Psychometric properties of the PTSD Checklist—Civilian Version. *Journal of Traumatic Stress, 16,* 495–502.

Ruscio, A. M., Ruscio, J., & Keane, T. M. (2002). The latent structure of posttraumatic stress disorder: A taxonometric investigation of reactions to extreme stress. *Journal of Abnormal Psychology, 111,* 290–301.

Shalev, A. Y., Orr, S. P., & Pitman, R. K. (1992). Psychophysiologic response during script-driven imagery as an outcome measure in posttraumatic stress disorder. *Journal of Clinical Psychiatry, 53*(9), 324–326.

Sundin, E. C., & Horowitz, M. J. (2002). Impact of Event Scale: Psychometric properties. *British Journal of Psychiatry, 180,* 205–209.

Sutker, P. B., Uddo-Crane, M., & Allain, A. N. (1991). Clinical and research assessment of posttraumatic stress disorder: A conceptual overview. *Psychological Assessment, 3,* 520–530.

Weathers, F. W., Keane, T. M., & Davidson, J. R. (2001). Clinician Administered PTSD Scale (CAPS): A review of the first ten years of research. *Depression and Anxiety, 13,* 132–156.

Weathers, F. W., Keane, T. M., King, L. A., & King, D. W. (1996). Psychometric theory in the development of posttraumatic stress disorder assessment tools. In J. P. Wilson & T. M. Keane (Eds.), *Assessing psychological trauma and PTSD* (pp. 98–135). New York: Guilford Press.

Weathers, F. W., Litz, B. T., Herman, D. S., Huska, J. A., & Keane, T. M. (1993, October). *The PTSD Checklist (PCL): Reliability, validity, and diagnostic utility.* Poster presented at the 9th annual meeting of the International Society for Traumatic Stress Studies, San Antonio, TX.

Weathers, F. W., Ruscio, A. M., & Keane, T. M. (1999). Psychometric properties of nine scoring rules for the Clinician-Administered PTSD Scale (CAPS). *Psychological Assessment, 11,* 124–133.

Weiss, D., & Marmar, C. (1997). The Impact of Event Scale—Revised. In J. P. Wilson & T. M. Keane (Eds.), *Assessing psychological trauma and PTSD* (pp. 399–411). New York: Guilford Press.

Wilson, J. P., & Keane, T. M. (Eds.). (1997). *Assessing psychological trauma and PTSD.* New York: Guilford Press.

Wilson, J., & Keane, T. M. (Eds.). (2004). *Assessing psychological trauma and PTSD* (2nd ed.). New York: Guilford Press.

Wilson, J. P., & Krauss, G. E. (1984, September). *The Vietnam Era Stress Inventory: A scale to measure war stress and post-traumatic stress disorder among Vietnam veterans.* Paper presented at the Third National Conference on Post-Traumatic Stress Disorder, Baltimore.

Zayfert, C., Becker, C. B., Unger, D. L., & Shearer, D. K. (2002). Comorbid anxiety disorders in civilians seeking treatment for posttraumatic stress disorder. *Journal of Traumatic Stress, 15,* 31–38.

Zlotnick, C., Warshaw, M., Shea, T. M., Allsworth, J., Pearlstein, T., & Keller, M. B. (1999). Chronicity in posttraumatic stress disorder (PTSD) and predictors of course of comorbid PTSD in patients with anxiety disorders. *Journal of Traumatic Stress, 12,* 89–100.

Chapter 16

Early Intervention for Trauma

Brett T. Litz and Shira Maguen

Any clinician with experience treating chronic posttraumatic stress disorder (PTSD) or other psychological and functional disturbances implicated by exposure to severe trauma can speak passionately and poignantly about people whose life courses have been altered terribly and seemingly permanently by trauma. Some individuals appear to acquire habitual, difficult to change, trauma-linked ways of construing themselves and the world around them, and maladaptive ways of managing emotional demands (in addition to the burden of chronic painful recall of trauma and other PTSD symptoms). Inevitably, therapists wonder what might have happened differently had formal, careful, and caring intervention been provided soon after a given trauma. Instead, one hears stories of neglect, shaming, secrecy, and abandonment in the immediate aftermath of trauma and in the initial recovery context.

CURRENT STATE OF THE ART

Knowledge of the long-term impact of trauma and an understandable human need to help people who are in anguish drive the motivation to provide services in the immediate or acute aftermath of trauma. More recently, terrorism and the possibility of mass violence has increased awareness of the need for early intervention in the medical, mental health, public health, government, and corporate communities, and a sober examination of the evidence base (e.g., National Institute of Mental Health [NIMH], 2002). Can the course of posttraumatic adaptation be altered positively by early therapeutic intervention? Is there sufficient evidence to recommend specific strategies? In this chapter, we critically review these and related issues pertaining to early interventions for trauma in children and adults. Throughout the chapter, we critically appraise the evidence for various practices and underscore what remains to be studied. It is imperative that the community of consumers of early intervention services appreciate the importance and benefits of evidence-based interventions for trauma, the extant epidemiological evidence, and the state-of-the-art research on risk and resilience factors in posttraumatic adjustment. Because most people exposed to trauma are resilient, or recover on their own after an initial period of understandable distress and disruption, formal early intervention for all is inappropriate and a waste of resources. Inert interventions that are provided indiscriminately in the immediate aftermath of trauma can be destructive for those at risk for posttraumatic complications, because they impart the message that something useful was provided (Litz, Gray, Bryant, & Adler, 2002; McNally, Bryant, & Ehlers, 2003).

The most significant challenge in early intervention is to understand better the course and predictors of maladaptive and chronic outcomes associated with exposure to trauma. Once mechanisms of risk are understood, these can be addressed directly in novel early intervention frameworks or, at the very least, can help us identify individuals who are not likely to recover on their own over time.

The Case for Early Intervention

The risk for exposure to traumatic events across the lifespan is extraordinarily high (~50%; Breslau, Davis, & Andreski, 1998; Kessler, Sonnega, Bromet, Hughes, & Nelson, 1995). Generally, the events with the highest base rate (witnessing someone else's suffering, being the victim of an accident, being threatened) are associated with the lowest risk for PTSD, whereas events with the lowest likelihood are associated with the greatest risk for PTSD (e.g., combat, sexual abuse and sexual assault; e.g., Kessler, 2000; Kessler et al., 1995). Overall, approximately 8% of individuals in the United States have had PTSD at some point in their lives; generally the rates are higher in women than men (Kessler et al., 1995).

PTSD is associated with lower quality of life (e.g., Malik, Connor, & Sutherland, 1999), high medical care utilization (e.g., Rosenheck & Fontana, 1995), lower work productivity (e.g., Savoca & Rosenheck, 2000), and extensive functional impairments and comorbid problems (e.g., Kulka, Schlenger, & Fairbank, 1990). The most troubling aspect of PTSD is its chronic course. There is evidence that once posttraumatic adaptation difficulties are manifest, they remain chronic across the lifespan (e.g., Kessler et al., 1995; Prigerson, Maciejewski, & Rosenheck, 2001) and are resistant to treatment efforts (e.g., Kessler et al., 1995; Schnurr, Lunney, & Sengupta, 2003). Many people refuse or drop out of formal structured evidence-based psychological treatments, such as prolonged exposure therapy (e.g., Tarrier, Pilgrim, & Sommerfield, 1999; van Minnen, Arntz, & Keijsers, 2002). For some individuals with chronic PTSD, access to evidence-based psychological treatment services is either unavailable or an insufficient number of clinicians acknowledge the need for evidence-based interventions (e.g., Becker, Zayfert, & Anderson, 2004). Thus, it is important to provide effective early intervention to reduce the risk of chronic impairment. Therefore, one of the primary ways of defining the goals of early intervention for trauma is *secondary prevention* of chronic PTSD and associated impairment in individuals who are most at risk for having difficulty recovering on their own over time (e.g., Litz & Gray, 2004).

What causes trauma to be a lifelong burden for some, whereas the majority of individuals recover functioning effectively? Is there a predictable pattern of recovery over time for most people? At what point is it abnormal to continue to have adjustment problems? What is the optimal time frame to consider formal intervention? Unfortunately, there are few answers to these questions. On the other hand, a handful of studies help to provide a working conceptual framework to guide early interventions for trauma.

The Course of Recovery from Trauma

In the early intervention area, timing of treatment is a critical factor in decision making. Interventions provided too early are intrusive and inappropriate. Most of the individuals who are distressed and impaired very early on recover on their own; thus, intervention would be a waste of precious resources. If interventions are provided too late, then the treatment enters into the realm of tertiary care. If adjustment to trauma followed a predictable course for all individuals and types of traumas, then decisions about whom to target and the best time to provide services would be straightforward. Unfortunately, this is not the case.

The most cited longitudinal study bearing on the course of posttraumatic adaptation is the Rothbaum, Foa, and Riggs (1992) study of sexual assault survivors. Ninety-five female rape victims, highly variable in age, were assessed weekly for 12 weeks and ranged in age from 18 to 65 years (rape cases perpetrated by a spouse or family member were excluded). The majority of women were African American (65%), unmarried (76%), and of low socioeconomic status (60%). Of the 95 women who entered the study, 64 completed the study. Ninety-four percent of the women studied met criteria for PTSD at 1 week postrape. A sizable percentage of women improved steadily in the first few months. At the 3-month stage, those who did not improve continued to have PTSD. The victims who experienced PTSD after the 3-month period had an increased severity of PTSD symptoms at the first assessment compared to those who did not have PTSD symptoms after 3 months. This finding is consistent with other studies, which have shown that a large percentage of women improve drastically during the first several months after a rape experience (Riggs, Rothbaum, & Foa, 1995; Valentiner, Foa, Riggs, & Gershuny, 1996).

Although the Rothbaum and colleagues (1992) study is not generalizable to the general population of rape victims, it has generally been inferred from this study that most people exposed to severe trauma experience some degree of distress and impairment but recover effectively within approximately 3 months without formal intervention. It is also assumed that those who do not recover will experience chronic, unwavering PTSD, and that secondary prevention is necessary in each case. Nevertheless, there are enough small-scale, multiwave studies, such as that by Rothbaum and colleagues, with various traumatized groups to piece together a working model of the trajectory of adaptation to trauma that can inform early interventions (e.g., Blanchard, Hickling, & Forneris, 1997; Bryant & Harvey, 2000).

Figure 16.1 depicts a heuristic framework that summarizes this model of posttraumatic adaptation. It distinguishes three posttraumatic time frames that have different phenomenologies and associated needs, demands, and risks. In the *immediate impact interval* a traumatic stress reaction is still manifest. The time frame for immediate adjustment is relatively arbitrary. We recommend 0–48 hours, which corresponds with the interval for an *acute stress reaction* described in the *International Classification of Dis-*

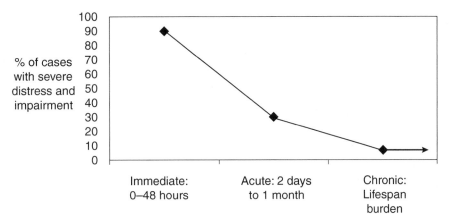

FIGURE 16.1. The "typical" course of adjustment to severe trauma.

eases (World Health Organization, 1992). By *immediate*, we mean that the event just happened, and the person or group of similarly affected individuals is contending with the most pressing and urgent initial demands. In this context, the emotional disruption caused by the trauma is fresh and intense. Individuals are likely to be disorganized, dazed, filled with anguish, and trying to cope with resource losses produced by the trauma (e.g., loss of functioning, loss of material supports). We have argued that secondary prevention interventions should be avoided in this interval, because the person is not capable of sustaining attention and the logistical resources required to benefit from psychological treatments (Litz & Gray, 2004). It is important to note that in this model, the clock starts ticking when the person is safe and secure. If the threat or insult is ongoing (e.g., a battle in a war zone, an ongoing battering relationship, or a natural disaster in which everything is destroyed and must be rebuilt), then a different approach is required, taking into account the exigencies that limit care and service delivery or underscoring other pressing needs that need to be addressed.

By contrast, the DSM-IV *acute interval* (2 days to 1 month, which corresponds to the interval for acute stress disorder; American Psychiatric Association, 1994) is a time when individuals are better prepared to receive secondary prevention interventions. For secondary prevention interventions to be effective, the person needs to be actively engaged in processing his or her trauma and acquiring new corrective knowledge, which takes sustained effort, especially as occupational, interpersonal, and self-care demands emerge over time (Litz & Gray, 2004; Shalev, 2002). In the *chronic interval*, individuals may have already developed habitual, maladaptive coping strategies, entrenched identity and role changes, and enduring posttraumatic mental health problems.

In this model's immediate phase, most people suffer considerably and have impaired functioning, as would be expected when the trauma is very severe. In this context, a goal might be to reduce this distress or assist people who are particularly impaired. The set of interventions that may be appropriate in the immediate interval has been labeled *psychological first aid* (e.g., Litz & Gray, 2004; Raphael, 1977). Psychological first aid is supportive and noninterventionist; it is not "treatment." Advice giving or other directive interventions should not be attempted. The goal is not to foster disclosure of traumatic events but to respond to the acute need that arises in many people to share their experience, while respecting those who do not wish to discuss what happened (Litz et al., 2002). In the immediate interval, individuals should also be provided information about supports and the safety of loved ones, and what they can expect in the days and weeks ahead. Intervention strategies that can be administered during the acute intervals are discussed below.

The heuristic model described earlier has guided most thinking about early intervention for extreme trauma such as direct threat to life and traumatic loss, but it fails to take into account other, known trajectories of adaptation to trauma. Another way of conceptualizing the temporal parameters of posttraumatic adaptation is reflected in Figure 16.2. In this model, some individuals are relatively *resilient*, which means they do not have noteworthy distress or impairment in functioning in the immediate and acute aftermath of trauma. Other individuals martial their coping resources effectively in the immediate impact phase but experience delayed PTSD (e.g., Gray, Bolton, & Litz, 2004). Still others may have considerable distress and impairment but grow from the experience after a period of time. Such posttraumatic growth is manifest in greater maturity, wisdom, empathy, and acceptance (Tedeschi & Calhoun, 1996), although it is important to note the mixed findings relative to the connection between PTSD symptoms and posttraumatic growth (e.g., Frazier, Conlon, & Glaser, 2001).

Ultimately, decisions about early intervention are dependent on resources. If financial, professional, public health, volunteer, and familial resources were unlimited in every trauma context, then supportive psychological first aid and evidence-based early interventions could be provided to every person to reduce suffering, hasten natural recovery, and minimize functional impacts. Unfortunately, this is impossible.

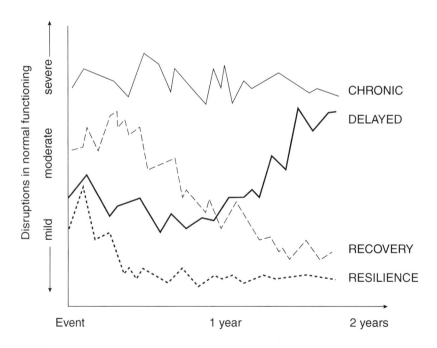

FIGURE 16.2. A heuristic framework for traumatic adjustment. From Bonanno (2004). Copyright 2004 by the American Psychological Association. Reprinted by permission.

Risk Factor Research

Given that most individuals exposed to trauma are resilient or recover over time, it is important to employ early interventions with individuals or groups that are most at risk for chronic impairments. Conceptually, three global categories of risk variables may increase an individual's probability of developing chronic PTSD, namely, *trauma impact variables, person–history variables,* and *culture and environment variables.*

Trauma impact variables include the severity of exposure to life threat; the immediate reaction to trauma; and resource losses incurred, including loss of loved ones. For example, the extent of direct threat to life or receipt of injury is strongly associated with PTSD in women exposed to violence (Resnick, Kilpatrick, & Dansky, 1993). Generally, the association between severity of exposure and PTSD is very strong across all studies, especially in the context of war (e.g., Foy, Sipprelle, & Rueger, 1984). This also holds true for traumas such as terrorist attacks; Schlenger, Caddell, and Ebert (2002) found that New York City residents were three times more likely to develop PTSD in the aftermath of September, 11, 2001, relative to those outside the city. In a meta-analysis of both military and civilian samples, Brewin, Andrews, and Valentine (2000) found that trauma severity is one of the strongest predictors of PTSD. Research has also shown that the impact of trauma is mediated by subjective appraisals of the degree of life threat (e.g., King, King, & Gudanowski, 1995).

In the aftermath of trauma, individuals are at risk for losing resources that make it possible to recover or to reduce severe stress and strain. For example, following the tragedies of September 11, 2001, Silver, Holman, and McIntosh (2002) found that individuals who lost a loved one demonstrated poorer psychological functioning. In an epidemiological study following September 11, 2001, Galea, Ahern, and Resnick (2002) found that people who lost possessions were more at risk for PTSD, whereas those who lost a loved one and/or employment were more at risk for depression. Norris and colleagues (2002) concluded that the most severe long-term psychological effects result from human-caused disasters involving one of the following resource losses: extreme and widespread damage to property; serious and ongoing financial problems for the community; and high prevalence of trauma in the form of injuries, threat to life, and loss of life.

Person and history variables include demographic variables, prior psychiatric history, trauma history, personality, and individual differences in the immediate impact of trauma. Some studies have demonstrated that women may be at greater risk for PTSD than men (e.g., Breslau et al., 1998), although this finding may be confounded by trauma type and trauma history. Following terrorism on September 11, 2001, Galea and colleagues (2002) found that Latino ethnicity was a risk factor for the development of PTSD, and Schlenger and colleagues (2002) found that younger age and female gender exacerbated the risk for PTSD. Brewin and colleagues (2000) demonstrated that lower socioeconomic status and education were also risk factors for PTSD. Lower intelligence has also been found to be uniquely associated with chronic PTSD in several studies (e.g., Macklin, Metzger, & Litz, 1998).

Preexisting psychiatric history also significantly affects risk for posttraumatic mental health problems following traumas such as terrorism (North, Nixon, & Shariat, 1999) and combat (Schnurr, Friedman, & Rosenberg, 1993). Depression in particular increases risk for development of PTSD (Freedman, Brandes, & Peri, 1999), and preexisting anxiety also places individuals at greater risk (Shalev, Peri, & Canetti, 1996). Prior exposure to trauma is a consistently robust and unique predictor of chronic PTSD stemming from a subsequent trauma (Dougall, Herberman, Delahanty, Inslicht, & Baum, 2000; King, King, Foy, Keane, & Fairbank, 1999; Stretch, Knudson, & Durand, 1998). Furthermore, risk for PTSD may increase with cumulative trauma (Martin, Rosen, Durand, Knudson, & Stretch, 2000). Personality dispositions, such as negative affectivity, have been shown to influence individual cognitions, self-concept, and worldviews, all of which may relate directly to the development of PTSD (Miller, 2004). Following exposure to trauma, individuals who demonstrate severe acute symptoms, such as intense hyperarousal, are at greater risk for chronic PTSD (Harvey & Bryant, 1999; Shalev, Freedman, & Peri, 1997).

Culture and environment variables include factors such as posttrauma social support and life adversities, and stressors that exacerbate the risk for PTSD. Individuals with taxed resources due to negative life stressors may have less energy to cope with and/or process their trauma, which may increase risk for PTSD. In a national sample of Vietnam War veterans, King, King, and Fairbank (1998) demonstrated strong mediation effects for postwar negative life events and postwar social support. Similarly, Norris and colleagues (2002) reported that a well-functioning postdisaster social support system is one of the key ingredients that may serve as a protective factor against long-term mental health consequences. In a military sample, there is evidence that social support around the time of the trauma plays a role. More specifically, unit cohesion was directly associated with fewer mental health problems, and social support from military unit leaders moderated the relationship between accumulated exposure to traumas and psychological distress (Martin et al., 2000). Overall, Brewin and colleagues (2000) found that lack of social support and the existence of additional life stressors are important predictors of posttrauma psychopathology.

What should we glean from the existing literature on risk factors for PTSD? First, researchers and clinicians need to appreciate that mental health adaptation to trauma is an unfolding, dynamic process, the understanding of which requires a longitudinal perspective. Most of the risk and resilience studies have been cross-sectional. To date, there have been very few longitudinal studies of the pretraumatic, peritraumatic, and subsequent life course influences on mental health, treatment needs, or impediments to service seeking in victims of trauma. Because posttraumatic adjustment unfolds variably over time, based on needs, demands, context changes, and so on, predictors of adjustment change dynamically over time in ways that cross-sectional studies with select small samples fail to capture. Second, very few studies have explored the causal mechanisms that place individuals at risk. The studies typically reveal that some variable of interest is correlated with chronic PTSD or predicts enduring PTSD, but we can neither discern the direction of the relationship nor glean the mechanism of action. The net effect is that risk research is unable to provide prescriptions for screening those most at risk in every traumatic context (Litz et al., 2002).

Nevertheless, it is clear that trauma exposure is a necessary but not sufficient cause of PTSD and lasting functional impairments associated with trauma exposure. People bring preexisting personal strengths, liabilities, and vulnerabilities to the trauma context and beyond. Various qualitative and quantitative features that make *every* traumatic event unique differentially affect people's coping capacities and long-term psychological health. The culture, and the immediate social and familial context, also contribute to recovery and adaptation over the long run.

From the extant risk literature, we can glean the following "rules of thumb" to guide early intervention efforts: (1) Some traumas are so heinous and severe (e.g., torture, captivity) that everyone exposed may benefit from special monitoring and follow-up, so that needs can be assessed at various intervals; and (2) depending on the context and the scope of the trauma, screening individuals with severe hyperarousal or severe depressed mood in the acute interval, if logistically feasible, is recommended. Identifying vulnerable individuals with preexisting psychiatric diagnoses (including PTSD), past traumas, severe isolation, concurrent adversity, and resource losses is also a good idea.

Goals for Early Intervention

Decisions concerning early intervention necessarily are based on the scope of the disaster or event and available resources, which, in turn, are moderated by the amount of time that has passed since the trauma. For example, following the events of September 11, 2001, the large number of victims and loved ones affected by the horrific acts of terror meant that resources needed to be allocated to those who were most in need. Even though within days an outpouring of helpers and professionals were on hand to address the putative mental health needs of victims, there were not enough trained personnel and resources to meet the mental health needs of all victims who actually required care. In addition, most people could not be reached or did not care for various services, which naturally provided time for them to recover on their own (e.g., to mobilize existing resources such as social support networks), and most individuals were able to do so. In the context of a disaster such as mass violence, terrorism, and natural disasters such as the recent tsunami that devastated entire communities in Asia and parts of Northern Africa, mental health needs are trumped by the need for food, shelter, and information about the whereabouts of loved ones and support systems. For those who had unrelenting mental health symptoms, available mental health care steadily increased over time, allowing distribution of these resources to persons in need.

As we mentioned at the start, in the trauma field, we tend to think of early intervention as secondary prevention of chronic PTSD, which in medical and public health terms means we are attempting to prevent chronic disease. If well-trained and savvy personnel are scarce, this goal is typically sufficient. On the other hand, in theory, depending on resources, many additional and no less valid or useful goals for early intervention include (1) helping people to decrease, manage, or eliminate functional incapacities caused by trauma; (2) promoting and training individuals or groups to use positive coping strategies and healthy behaviors; (3) encouraging and assisting individuals or groups to develop, nurture, and take advantage of comforting, positive, and caring social supports; (4) targeting complicated bereavement or traumatic grief (Neria & Litz, 2004); and (5) helping individuals cope with subsequent threat (e.g., Marx, Calhoun, & Wilson, 2001; Somer, Buchbinder, Peled-Avram, & Ben-Yizhack, 2004).

Increasing Functional Capacities

Disruption in functional capacities arguably varies with severity of the trauma, and goals to increase functional capacities depend on the social and occupational context of the individual. For most individuals, return to baseline likely occurs following the passage of time. After September 11, 2001, 27% of workers in New York City missed work in the days following the attacks (Melnik et al., 2002); however, as time passed, safety was ensured, and social support networks were garnered, most were able to return to work. Following September 11, 2001, some individuals living and/or working around the World Trade Center did not feel safe for fear of another attack; even the general population feared another terror attack, with two-thirds of Americans reporting fears of future terrorism (Silver et al., 2002). Furthermore, many had lost residences or were displaced, which impedes returning to daily living. Needless to say, emergency stabilization and securing basic needs such as food and housing are primary (American Red Cross, 1998). The ability to obtain these resources depends on the ecological context, where sharing and exchange function between the individual and the larger environmental/social context in which the individual operates (Hobfoll & Jackson, 1991). For those who continue to struggle and suffer significant functional impairment after these needs have been met, mental health professionals can target barriers to returning to routines of daily living and functioning.

Encouraging Positive Coping Strategies and Healthy Behaviors

For some people, trauma can set in motion an insidious downward spiral of maladaptive behavior that may be independent of PTSD status. For example, Galea and colleagues (2002) found that following September 11, 2001, 25% of individuals reported an increase in alcohol consumption, and about 10% reported an increase in smoking. From a public health standpoint, early intervention targeting early signs of unhealthy coping and maladaptive behavior serve to reduce the long-term impact and costs associated with trauma. The goal of preventing the formation or exacerbation of unhealthy behaviors (smoking, drinking, etc.) should be coupled with promoting healthy alternatives, such as appropriate diet and exercise, and providing information that healthy behaviors can fall by the wayside during exacerbated stress.

The importance of intervening to increase healthy behaviors before habits become entrenched and resistant to change was highlighted in a study of women veterans, in which those with PTSD reported greater substance abuse and smoking (Dobie et al., 2004). These veterans also were significantly more likely to endorse physical health problems, including obesity, irritable bowel syndrome, fibromyalgia, and stroke. Following the Oklahoma City bombing, Pfefferbaum, Vinekar, and Trautman (2002) reported that alcohol and cigarette use were independently associated with peritraumatic reaction, grief, posttraumatic stress, worry about safety, and functional impairment. Moreover, Buckley, Mozley, Bedard, Dewulf, and Greif (2004) found that in addition to being more likely to smoke, veterans with chronic PTSD did not engage in preventive health behaviors, such as exercise and medical screening, at levels consistent with health care guidelines. Early intervention may include educational components that teach survivors about the associations between poor coping, unhealthy behaviors, decreased physical health, and PTSD.

Increasing Opportunities for Social Support

Social support following trauma comes in many different forms, ranging from support on an individual level from loved ones to community support in larger-scale traumas. The effects of social support cannot be underestimated, although the level of effectiveness may vary with type and magnitude of the trauma. Following traumas, it is most prudent to allow individuals to activate and take advantage of existing social support networks, which in turn may increase empowerment and feelings of effectiveness. Relatedly, preventing individuals from activating these networks can be harmful in the long run and impede the natural course of recovery. Kaspersen, Matthiesen, and Götestam (2003) found that a social support network moderates between trauma exposure and posttrauma symptoms; however, there were some difference with respect to the importance of social network as a moderator, based of different types and magnitudes of exposure. More specifically, social support as a protective factor was more consistently effective for relief workers than for UN soldiers. Koenen, Stellman, and Stellman (2003) found that community involvement protected against PTSD and that discomfort in disclosing Vietnam experiences was associated with an increased risk for developing PTSD. Relatedly, Maercker and Müller (2004) found that not only social support but also social acknowledgment as a victim decreases the risk for PTSD. Overall, isolated individuals may be most sorely in need of early intervention, may have no one with whom to share or disclose their experiences, and may not be validated or supported following their trauma (e.g., Foy et al., 1984; Martin et al., 2000). Social support not only seems to be an important protective factor in the development of PTSD but may also be correlated with traumatic grief (e.g., Spooren, Henderick, & Jannes, 2000).

Managing Complicated Bereavement or Traumatic Grief

Understandably, the majority of early interventions for trauma have focused on prevention or amelioration of PTSD. Although PTSD may be the most obvious problem that develops following traumatic exposure, the danger is that exclusive emphasis on PTSD may result in other forms of posttraumatic distress not being adequately addressed. For example, survivors of disasters and other large-scale traumatic events may be mourning the death of a close friend or relative, on top of experiencing anxiety symptoms associated with the immediate trauma. For example, following September 11, 2001, an estimated 10 million individuals lost loved ones, including family members, friends, and colleagues, as a result of the terrorism tragedies (Schlenger et al., 2002). Individuals who lose loved ones due to unexpected circumstances (e.g., suicide, homicide, or accident) may also experience greater cognitive disturbances and upheaval compared to individuals with anticipated losses (Schwartzberg & Halgin, 1991). Furthermore, early intervention is important in the absence of existing social support, especially when those suffering from traumatic losses also have unstable support networks. In a study of families of homicide victims, individuals often expressed feelings indicating a lack of social support and even of betrayal from existing networks (Armour, 2002). As a result, early intervention should necessarily include attempts at both assessing and preventing complicated bereavement from traumatic loss.

Helping Individuals Cope with the Subsequent Threat

Helping individuals cope with subsequent threat is especially important in the context of terrorism, in which threats may become exacerbated and renewed over time. Most individuals cope well with the threat of terrorism and do not need professional mental health intervention. Terrorism is a low probability but horrifying possibility that may exacerbate functional impairment. For example, in the hours and days following the tragedies of September 11, 2001, many individuals feared a subsequent attack, yet for most, this fear significantly dissipated within a few months (Silver et al., 2002); however, for others, the fear perseverated.

Research in Israel, where the threat of future terrorist events has been more common over the last few years, has shown that most individuals proceed with their activities of daily living and do not become immobilized by fear of a subsequent terrorist attack. In the context of coping with future threat, there is evidence that early interventions can assist individuals in reducing anxiety. Following possible terrorist threats in Israel, promoting relaxation and positive emotionality, and challenging maladaptive self-talk were found to be helpful (e.g., Somer, Tamir, Maguen, & Litz, 2005).

We now describe the specific strategies used to prevent chronic PTSD and other mental health problems following exposure to trauma. This remains the primary goal for the early intervention because of the pervasive and extensive personal and societal costs of PTSD.

Early Intervention for Trauma in Adults

Psychological Debriefing

"Psychological debriefing" (PD) is an umbrella term describing a single-session intervention that lasts anywhere from a few hours to a few days following a traumatic event. The goal of PD is to allow victims exposed to trauma to express their emotional reactions to the trauma in the presence of other survivors, with a mental health professional present. In some cases, survivors' emotional reactions are normalized, and adaptive coping is discussed (Bisson, McFarlane, & Rose, 2000; Shalev, 2000).

The most widely used form of PD, *critical incident stress debriefing* (CISD; Mitchell & Everly, 1996), was originally intended for individuals "indirectly exposed" to a potentially traumatic event because of their roles and responsibilities as professional or first responders, rather than for "direct" victims of trauma. CISD includes a psychoeducational component and an emotional processing component, typically lasting about 3–4 hours and occurring within a few days of a "critical incident" (i.e., the potentially traumatic event). First, individuals are taught about the symptoms of stress reactions that are to be normalized. Second, they are encouraged to share and to process their recent trauma emotionally. After recounting the event, participants are invited to share cognitions and emotions that occurred during the event, which most typically are normalized by the CISD team member.

Although CISD was originally designed as an opportunity for individuals exposed to potentially traumatic events to "debrief" and share their experiences, over time it has been mistakenly conceptualized as a secondary prevention intervention broadly applied (e.g., Mitchell & Everly, 1995), despite a lack of evidence that it is beneficial or "inoculates" victims against PTSD (Deahl, Srinivasan, & Jones, 2000; Litz et al., 2002; McNally et al., 2003). A recent controlled trial comparing a single session of debriefing to stress management and no intervention among troops deployed to Kosovo revealed no differences between groups on subsequent measures of mental health and functional impairment (Litz, Williams, Wang, Bryant, & Engel, 2004).

Despite evidence for its lack of efficacy, CISD is commonly administered to professionals such as police officers, military personnel, and disaster workers who are regularly exposed to traumatic events. CISD is typically administered to all "indirectly exposed" workers who wish to participate, regardless of the degree of acute symptoms or functional impairment endorsed (Hokanson & Wirth, 2000). Consequently, individuals in one group can range from extremely distressed to resilient. Grouping individuals in this way may foster stigma among distressed individuals. Another drawback is that individuals who do not wish to disclose personal or traumatic information may feel pressure to do so in the context of CISD. Others may feel badly that their reactions differ from those around them. Thus, it can be argued that sharing within the CISD context may have detrimental consequences (Young & Gerrity, 1994). Another danger is that although participation is supposed to be voluntary, some employers may strongly suggest and subtly pressure employees to attend a debriefing session.

CISD is very attractive in that it deemphasizes pathology, and respects and encourages organizational supports. Indeed, studies have shown that participants appreciate CISD (Litz et al., 2004). However, there is no evidence whatsoever that CISD serves a secondary prevention function; it does not decrease the likelihood of developing PTSD. Although numerous uncontrolled studies have suggested that CISD and PD are generally effective in reducing posttraumatic distress (see Everly, Flannery, & Mitchell, 2000), until relatively recently there have been few randomized, controlled trials (Litz et al., 2004; Rose, Bisson, & Wessely, 2001). Past studies have had several methodological problems: (1) nonrandom assignment (especially a problem because PTSD symptoms decrease naturally over time); (2) lack of control groups; (3) self-selected samples; (4) no preintervention assessment (i.e., participants may have had few symptoms following the critical incident and did not need any intervention); and (5) lack of valid and reliable outcome measures. As a result, all of the uncontrolled studies that claim to support the efficacy of PD are invalid (Litz et al., 2002; Rose et al., 2001).

Unfortunately, existing randomized controlled trials (RCTs) of PD have been conducted on individuals rather than groups of similarly affected individuals (the putatively modal context in which CISD is employed) (Bisson, Jenkins, & Alexander, 1997; Conlon, Fahy, & Conroy, 1999; Deahl et al., 2000; Hobbs, Mayou, Harrison, & Warlock, 1996; Litz et al., 2004; Mayou, Ehlers, & Hobbs, 2000; Rose, Brewin, Andrews, & Kirk, 1999). In addition, no trial has assured the use of CISD, specifically. Nevertheless, each trial has shown that individuals receiving PD do not benefit psychologically when compared to controls, and that in fact, mean symptom improvement across both groups is virtually equal (Litz et al., 2002). Attention has focused on two studies in which subjects randomized to PD actually fare worse than controls; however, these results should be interpreted with caution. First, Mayou and colleagues (2000) reported high dropout rates, which greatly limit generalizability. Second, Bisson and colleagues (1997) studied burn victims, who may have unique demands. Third, despite randomization, individuals in Bisson and colleagues' PD group had more preintervention symptoms.

Recently, the described CISD framework has been altered to be part of a more encompassing *critical incident stress management* (CISM) program (Everly & Mitchell, 2000). The following are the articulated goals of the new program: (1) to prepare individuals mentally prior to dangerous work and its consequences; (2) to provide support to impacted individuals during critical incidents; (3) to provide CISD and additional delayed interventions; (4) to assist the families of direct victims; (5) to provide consultation to organizations and leaders; and (6) to provide referrals and follow-up interventions to those in need. Despite this expanded framework and program, there have been no controlled empirical studies of the approach or its components; as a result, it is unclear whether CISM is more effective than CISD alone.

Psychological First Aid

Although it is clear that a single-session of any formalized intervention will not prevent chronic mental health difficulties for those most at risk after trauma, there is an emerging consensus that a helpful, empathic, nonintrusive, and informative human presence in the immediate aftermath of trauma is useful when resources can sustain such activities. This approach has been labeled *psychological first aid* (e.g., Litz & Gray, 2004).

The primary aim of psychological first aid is to provide support and information that may be useful to someone who has lost personal and external resources. A supportive and empathic interaction minutes, hours, or days after a trauma may also help a victim, or a group of victims, start the process of gaining a sense of connectedness to others and regaining hopefulness and control. Psychological first aid may also mollify severe and overwhelming emotions in some. Finally, first aid after trauma may also instill positive expectancies for coping, and reduce stigma and psychological barriers to care seeking on down the line. Each of these goals for psychological first aid has yet to be studied empirically.

A consensus-based set of best practice recommendations for psychological first aid for children of all ages and adults was recently developed jointly by the National Center for PTSD and the National Child Traumatic Stress Network (*www.ncptsd.va.gov/pfa/ pfa.html*). The authors culled expert opinion and the available evidence about the psychological aftermath of disasters and terrorism. The intended goals are to assist victims and first responders soon after exposure to mass trauma "to reduce initial distress, and to foster short- and long-term adaptive functioning." This specific approach is the first consensus-based, manualized, systematic procedure for helping victims get the information, support, and empathy that they may need in the wake of disasters.

Cognitive-Behavioral Intervention

Several randomized, controlled trials have shown that the cognitive-behavioral therapy (CBT) techniques used to help treat chronic PTSD are useful to prevent chronic PTSD in acute victims (e.g., Bryant, Harvey, & Dang, 1998). These interventions are administered across multiple sessions, require therapist expertise, are more demanding of participants, are employed in the acute interval with individuals who have significant and enduring posttraumatic distress, and target specific posttraumatic symptoms and impairments.

CBT packages typically include psychoeducation, anxiety management techniques, *in vivo* and imaginal exposure, cognitive reappraisal, and homework assignments (Bryant et al., 1998; Foa, Hearst-Ikeda, & Perry, 1995). Psychoeducation provides information on PTSD and on maladaptive strategies, such as avoidance of trauma cues, that are often unsuccessful in managing distress. Anxiety management techniques include skills such as diaphragmatic breathing and other relaxation techniques. Exposure therapy occurs both in session and at home to decrease negative affect and response to trauma triggers, and to reduce avoidance that impairs function. Cognitive reappraisal includes understanding the association between thoughts and emotions, and learning to recognize and challenge maladaptive cognitions. Weekly homework assignments include monitoring symptoms and triggers, practicing anxiety management and cognitive restructuring skills, and conducting exposure exercises.

There is sufficient evidence to recommend CBT as an early intervention for trauma (e.g., Litz & Gray, 2004). However, at present, we do not know the specific, necessary elements of the approach or the most propitious time to implement the intervention. CBT has been used in controlled trials with motor vehicle accident and sexual assault survivors, but not for a wide variety of other types of trauma, particularly disasters. Also, it has never been deployed with first responders and in other organizational contexts. In addition, we do not know how much direct therapist contact is necessary. Given that homework is a necessary ingredient for behavior change and recovery from trauma, it is possible that self-management approaches may prove as effective (e.g., Litz et al., 2004). Finally, CBT requires considerable knowledge, supervised experiences, and professional expertise. As a result, in many contexts, and especially in disasters, the necessary therapist resources are not in place. Telehealth and self-management approaches may prove to be the most efficient way to provide evidence-based CBT as an early intervention (e.g., Litz et al., 2004).

Targeting Traumatic Loss

To date, early interventions for trauma have not targeted traumatic loss. Given the high base rates of loss in many traumatic contexts, and the unique phenomenology and symptoms that arise from loss due to violence and accidents, this is a glaring omission. In fact, there is rarely consensus on the specific conditions under which treatment is needed and the optimal timing to administer these preliminary treatments; furthermore, what constitutes state-of-the-art treatment for individuals who have lost loved ones due to traumatic means is still being debated. Only a few small-scale studies of individuals have targeted traumatic loss, and the evidence concerning the efficacy of these treatments is mixed.

In a randomized, controlled trial, Murphy, Johnson, and Cain (1998) conducted a 10-week group treatment attended by parents who lost their child to homicide, suicide, or accident 2 to 7 months preceding entrance into the study. The first hour of each group included teaching parents specific skills (e.g., ways to release anger and achieve closure by writing down thoughts and feelings), and the second hour focused on group emotional support, assisting parents in sharing experiences related to the death, and helping individuals reframe aspects of the death and its consequences. Whereas mothers improved on most measures of mental distress, including depression, anxiety, and fear, fathers improved on less than half of the measures. Although these results seem encouraging when compared to the control group, there were no differences on any of the outcome measures (e.g., mental distress, trauma, loss accommodation). When examined by level of distress, the intervention was beneficial for mothers with higher mental distress and grief at baseline. However, fathers with higher levels of PTSD at baseline did worse than control group fathers.

In two randomized trials of guided mourning for complicated bereavement, individuals were assigned to six sessions of either a guided mourning or an antiexposure condition (Mawson, Marks, & Ramm, 1981; Sireling, Cohen, & Marks, 1988). All participants were assigned tasks between sessions and encouraged to engage in new activities. Individuals in the guided mourning condition were exposed to avoided cognitive, affective, and behavioral cues (e.g., writing letters to the decreased, viewing pictures), whereas those in the antiexposure condition were encouraged to avoid reminders of the deceased and to focus on the future rather than thinking about the past. Interestingly, Sireling and colleagues (1988) found that both groups demonstrated improvement on a number of variables at follow-up. Arguably, support and encouragement to engage in new activities daily may have been the ingredients that facilitated improvement.

Only recently have there been efforts to develop and test interventions specifically designed to target complicated grief's unique separation distress and ruminative symptoms. Almost all of these studies use a modified exposure approach, in which painful memories of the deceased are reviewed and the negative aspects are challenged with cognitive therapy techniques. In one of the most comprehensive interventions to date, Shear, Frank, Houck, and Reynolds (2005) conducted a randomized, controlled clinical trail of a tertiary intervention for complicated grief in 95 participants with complicated grief ~2 years after their loss. The treatment package focused mainly on PTSD treatment–derived exposure therapy techniques (imaginal and *in vivo* exposure to the memory of the death and avoided people and places) and cognitive restructuring. However, Shear and colleagues secondarily employed techniques from interpersonal therapy for depression (focusing on the illness role and on interpersonal role transitions and disputes), and reviewed participants' positive memories of the deceased to help promote accommodation to losses that are more specific to complicated grief. This impressive trial showed that complicated grief therapy led to significantly faster and greater reductions in complicated grief symptoms and improved general functioning than did interpersonal therapy.

Although some promising treatments that have been described target chronic complicated grief, more methodologically rigorous, longitudinal tests of these approaches in the early intervention context are required before specific recommendations can be made. Interventions that attempt to mobilize existing social supports or to create new mechanisms of support (e.g., support groups) may be especially helpful given that social support is inversely related to symptoms of traumatic grief (Spooren et al., 2000). For traumatically bereaved individuals who avoid encountering reminders of the death or thinking about the deceased, exposure-based interventions may help by promoting individuals' acceptance of the loss and recovery from complicated bereavement despair. Finally, an important component of treatment may be assisting traumatically bereaved people to reengage in activities they enjoy and that are associated with daily living. The applicability of the traumatic grief treatments described earlier to an early intervention framework remains ambiguous at best, and the likelihood of reducing traumatic or complicated grief following such early interventions should be empirically tested.

Early Intervention with Children

Prevalence of PTSD in Children and Adolescents

Few epidemiological studies have examined mental health disorder prevalence in children and adolescents, and of the few that exist, almost none have looked at PTSD as a distinct outcome. In a multistage, stratified, area probability random-digit dialing procedure, the National Survey of Adolescents found that the prevalence rate of PTSD was 4% in boys and 6% in girls (Kilpatrick et al., 2003). Rates of PTSD in children exposed to trauma vary tremendously, due partly to variability in the type and magnitude of exposure, and partly to the complications inherent in reliably assessing PTSD in children. Kassam-Adams and Winston (2004) examined the prevalence of acute stress disorder (ASD) and PTSD, and the hypothesis that ASD is a predictor of PTSD in injured children. The authors concluded that 8% of children met symptom criteria for ASD, 6% met symptom criteria for PTSD, and that whereas symptom severity of ASD and PTSD were associated, the sensitivity for ASD as a predictor of child PTSD was low (which parallels the adult literature).

Although acute reactions to serious stressors may be severe, most children, like adults, recover naturally over time. In most studies, depending on the timing of assessment, rates of PTSD are below 25%. For example, the Laor, Wolmer, and Cohen (2001) 5-year follow-up study of Israeli children (*N* = 81) whose homes were destroyed by SCUD missiles during the 1991 Gulf War found that 8% of the children had severe PTSD symptoms. Severity of symptoms was best predicted by displacement, inadequate family cohesion, and poorer maternal mental health functioning, with younger children's symptoms strongly associated with maternal impairment. Furthermore, Laor and colleagues found that during a period of 30 months between assessments, PTSD symptoms increased in one-third of the children and decreased in another one-third, demonstrating that symptoms should be assessed longitudinally and may fluctuate over time.

Early Intervention Trials

Unfortunately, few studies of traumatized children have controlled for or reported how much time had passed since the occurrence of the trauma. Of those that have reported elapsed time since the trauma, only two studies administered treatment prior to 6 months following the traumatic incident. Both studies, which were conducted by Cohen and Mannarino (1996, 1998) on sexually abused children, used trauma-focused cognitive-behavioral therapy (TF-CBT), which included (1) psychoeducation, (2) stress inoculation training, (3) trauma narrative processing, (4) cognitive processing, and (5) a parental treatment component. Cohen and Mannarino (1996) compared TF-CBT with play therapy for 3- to 7-year-old sexually abused children and their parents. Results showed decreases in PTSD, internalizing, and externalizing symptoms in children. Cohen and Mannarino (1998) compared TF-CBT and supportive therapy for 8- to 14-year-old children who had been sexually abused. Children in the study showed a decrease in depression and an increase in social competence. In addition, children who received TF-CBT reported greater improvement in PTSD symptoms at a 1-year follow-up compared to those who received only supportive therapy.

Treatment trials for PTSD that involve children are scarce, due partially to the ethical and procedural complications of conducting methodologically sound trials with children and adolescents. Overall, there are eight known RCTs for PTSD with children: seven that target sexual abuse, and only one that specifically targets exposure to violence. The largest known study ($N = 229$) was by Stein and colleagues (2003), who compared TF-CBT with child-centered therapy (CCT) in 8- to 14-year-old children and their parents. In the final analysis, children assigned to TF-CBT improved significantly more than the CCT group on measures of PTSD, depression, shame, behavior problems, and abuse-related attributions. The six other RCTs of child sexual abuse had similar findings, with children improving significantly on at least one clinical measure, and often more (Berliner & Saunders, 1996; Celano, Hazzard, & Webb, 1996; Cohen & Mannarino, 1996, 1998; Deblinger & Heflin, 1996; Deblinger, Steer, & Lippman, 1999; King, Tonge, & Mullen, 2000). Taken together these studies provide support for the efficacy of CBT treatment for child sexual abuse–related PTSD. However, these results are not necessarily generalizable to other forms of trauma. The optimal timing and necessary components of intervention remain unclear.

In the only RCT examining child treatment of violence-related trauma, Stein and colleagues (2004) conducted a school-based intervention aimed at reducing PTSD and depression symptoms following exposure to violence. Participating sixth-grade students were randomly assigned to either a 10-week, standardized CBT early intervention group treatment (i.e., Cognitive-Behavioral Intervention for Trauma in Schools) or a wait-list delayed-intervention comparison group. The treatment, designed for an inner-city, multicultural population, comprised a combination of educational didactics, age-appropriate practices, and completion of worksheets for homework administered to groups of five to eight students. Stein and colleagues found that after three months, the intervention group reported significantly lower PTSD scores compared to the wait-list group. After 6 months, once all children had received the CBT intervention, differences between the groups disappeared.

Early Intervention for Traumatic Grief

Similar to adults, traumatic grief in children appears to be related to, yet separate from, depression and PTSD. For example, following the suicide of a friend or acquaintance, 146 adolescents were interviewed at four time intervals for up to 3 years (Melhem et al., 2004). The authors found that traumatic grief was different and independent from depression or PTSD, although traumatic grief after half a year predicted future depression and PTSD.

In childhood traumatic grief, children often perseverate on how the loved one died rather than the meaning of the death or the consequent emotional reactions. In this context, the treatment goals are to help children accept the permanence of death, and experience and cope with their emotional reactions (Goodman, 2004).

To date, there have been only two preliminary trials of traumatic grief therapy in children, both of which are ongoing (Brown, 2004; Cohen, 2004; Cohen, Mannarino, & Knudsen, 2004). These trials comprise 16 weeks of individual TF-CBT. Of these 16 sessions, half are trauma-focused and half are grief-focused. Additionally, one-fourth of the sessions include joint parent–child treatment components, and parents receive their own individual treatment in conjunction with the child's treatment. Children, ages 4 to 18 years, completed the Expanded Grief Inventory (EGI; Layne, 2004), a measure of child traumatic grief (CTG), to qualify for the study (Brown, 2004; Cohen, 2004). A battery of assessments was administered every 4 weeks to assess improvements over time. In the open pilot trial, 22 children completed the treatment (Cohen, 2004). Preliminary findings demonstrate that from pre- to posttreatment, whereas children improved on measures of traumatic grief, depression, PTSD, anxiety, and externalizing symptoms, parents improved only on measures of depression and PTSD. Also, interestingly, PTSD symptoms improved only during the trauma-focused sessions; there was no additional improvement for the grief-focused sessions. Conversely, CTG improved during both the trauma- and the grief-focused sessions.

The second study, an RCT of CTG in children whose fathers were emergency service workers (firefighters, police officers, etc.) killed on September 11, 2001 (Brown, 2004), compares CBT and client-centered therapy across three time periods. To date, 33 children from 22 families have completed the study. Although results are preliminary, children in the CBT group demonstrated improvements on measures of depression but not traumatic grief. Future studies should continue to examine traumatized children with modified versions of existing adult treatments that incorporate developmentally appropriate language and intervention techniques.

METHODOLOGICAL CONSIDERATIONS

Adults

There are a variety of challenges and limitations to planning and carrying out early intervention research. Individuals may be unable or unwilling to participate because of competing and pressing priorities and needs, as well as a confused and disorganized mental status. Even if individuals are willing to participate, access may be limited because of the reluctance of responsible and assisting agencies. Needless to say, participant burden should be minimized, researchers should do no harm, and they should provide information and intervention resources routinely (or conduct intervention research in the most ethically responsive manner). We argue that it is not appropriate or tenable to attempt to explain research and garner informed consent in the *immediate* aftermath of trauma (minutes to a day or so after an event). However, if handled delicately, sensitively, and ethically, research is possible in the acute interval (a day or so to about a month). Assessments and other subject burdens should be kept at a minimum.

In most trauma contexts, institutional review board (IRB) approvals may pose a prohibitive barrier to the speedy data collection required for an examination of adaptation and early intervention. Ideally, IRBs and granting agencies would work together to strategize about the best methods of fielding research studies rapidly and ethically (e.g., having approved protocols on the shelf).

Other exigencies pose serious challenges to research in traumatic contexts, especially surveillance and epidemiological efforts. In disasters and wars, many people become refugees, dispersing geographically. For example, we recently attempted to study the impact of Lebanon's rocket bombardment of the Northern Israeli city of Kiryat Shmona. We found that many inhabitants of the city that had fled to southern parts of Israel still had not returned to their homes several months after the attacks. When researchers try to randomize individuals into groups or merely employ typical randomized selection processes (e.g., random-digit dialing), transience foils state-of-the-art methodology and necessitates flexible solutions. A similar situation was observed post-Katrina, because many of the inhabitants who fled New Orleans have not yet returned to the city. Generally, when collecting descriptive data and implementing early intervention studies, researchers are required to consider these contextual elements that may bias the research.

Children

We know little about how best to identify and treat children in an early intervention framework. Identifying those most at risk should be a high priority given that many children and adolescents in need do not get help. For example, in a cross-sectional, random-digit dialing survey of parents with children in New York City 4 months after September 11, 2001, Fairbrother, Stuber, Galea, Pfefferbaum, and Fleischman (2004) found that whereas 10% of children received some type of counseling following the attacks, only 27% of the children who had severe posttraumatic stress reactions after the attacks received counseling services. Even if children at risk are identified, we do not know the optimal time to treat them, a problem compounded by most existing treatment studies' omission in reporting childhood PTSD and CTG of the time lag between the occurrence of trauma and the start of treatment.

Assessing PTSD symptoms in children relies on their ability to report accurately the complex aspects of their internal experience and various forms of functional impairment (e.g., Scheeringa & Zeanah, 1995). A number of brief self-report instruments are geared toward assessing PTSD in children, each with its own strengths (see Cohen, 2004). Parental reports are invaluable and indispensable in this regard. Early intervention efforts should also be informed by the assessment of symptoms related to PTSD, including depression, anxiety, and traumatic grief.

CHALLENGES FOR THE FUTURE

In this chapter, we have reviewed the current state of knowledge about early intervention for trauma and traumatic loss in children and adults. Early intervention holds promise for preventing chronic PTSD and other maladaptive posttraumatic sequelae. We have offered some bold and unprecedented goals and recommendations for early intervention, and we recognized explicitly that clinical- and organizational-level decision making about the scope of early intervention services are highly dependent on professional and financial resources. In terms of secondary prevention of PTSD, at present, multisession CBT is the early intervention of choice, if it is provided to trauma survivors manifesting significant and impairing symptoms in the acute posttraumatic interval. However, the promise of CBT has yet to be realized in the majority of clinical settings, organizations, disaster contexts, emergency services, and the military.

One of the central challenges for future research is to discover mechanisms of risk and resilience in response to trauma. These data could be used to devote resources to those most likely not to recover on their own. Those survivors who have high-risk characteristics, or live in contexts that thwart recovery, should be provided special non-stigmatizing and sensitive outreach, monitoring, and screening efforts. In addition, once the field has advanced knowledge about the causes of chronic PTSD, these variables should be targets of novel primary and secondary intervention.

In an earlier time, CISD was a promising, egalitarian, and well-designed plan of attack for early intervention in a variety of trauma contexts. CISD developed into a very well organized clinical culture that to this day is well liked, appreciated, promoted, and widely used around the world. As a result of a lack of evidence to support its efficacy and failed clinical trials on individuals, CISD has been recently rejected in the scientific community. We have argued in this chapter that any single-session intervention is bound to ineffectively address the long-term challenges of individuals severely impacted by trauma. However, it is unclear at present whether CISD is used less frequently as a result of the disenchantment with the approach. On the other hand, if every government and private-sector agency and organization rejected the CISD approach, the vacuum left behind would not be readily filled by evidence-based CBT.

CISD has appeal because it deemphasizes psychopathology and is nonstigmatizing, it is well-integrated into work cultures and organizations, it is brief, and it does not require extensive professional training and expertise. In contrast, CBT requires considerable training and supervised experience, is much more rigorous, and demands much more of trauma survivors. In addition, CBT has rarely been melded into work and organizational cultures, as has CISD. CBT will need to be overhauled, if it is to be widely used and available in most trauma contexts, and especially in disaster settings.

We recommend that the principles of behavior change and the theoretical underpinnings of posttraumatic recovery promoted by CBT be used in novel ways to accommodate various trauma contexts. The recently developed guidelines for the provision of psychological first aid are an excellent example of an appeal to the empirical literature pertaining to adjustment to disasters and expert consensus about what people need in these contexts to generate useful instructions for what to do (and perhaps more importantly, what *not* to do) that are reliable, measurable, and replicable. Another method of ensuring wider application is to use public service announcements that provide accurate information and helpful suggestions to promote adaptation, understanding, and recovery in the wake of disaster. Public service announcements and resources (e.g., pamphlets, videos, CDs) may be the only way to reach people who suffer trauma in relative isolation or without sufficient supports or resources. Another is to promote self-management and self-help using CBT principles, which lessens the demand for scarce, highly trained professionals (e.g., Litz et al., 2004). Various technologies, such as the Internet, may also be used to provide CBT-based interventions (e.g., Lange, Rietdijk, & Hudcovicova, 2003).

REFERENCES

American Psychiatric Association. (1994). *Diagnostic and statistical manual of mental disorders* (4th ed.). Washington, DC: Author.

American Red Cross. (1998). *Disaster mental health services.* Washington, DC: Author.

Armour, M. P. (2002). Journal of family members of homicide victims: A qualitative study of their post-homicide experience. *American Journal of Orthopsychiatry, 72*(3), 372–382.

Becker, C., Zayfert, C., & Anderson, E. (2004). A survey of psychologists' attitudes towards and utilization of exposure therapy for PTSD. *Behaviour Research and Therapy, 42*(3), 277–292.

Berliner, L., & Saunders, B. (1996). Treating fear and anxiety in sexually abused children: Results of a two-year follow up study to child maltreatment. *Child Maltreatment, 1*(4), 294–309.

Bisson, J. I., Jenkins, P. L., Alexander, J., & Bannister, C. (1997). Randomised controlled trial of psychological debriefing for victims of acute burn trauma. *British Journal of Psychiatry, 171*, 78–81.

Bisson, J. I., McFarlane, A. C., & Rose, S. (2000). Psychological debriefing. In E. B. Foa & T. M. Keane (Eds.), *Effective treatments for PTSD: Practice guidelines from the International Society for Traumatic Stress Studies* (pp. 39–59). New York: Guilford Press.

Blanchard, E., Hickling, E., & Forneris, C. (1997). Prediction of remission of acute posttraumatic stress disorder in motor vehicle accident victims. *Journal of Traumatic Stress, 10*(2), 215–234.

Bonanno, G. (2004). Loss, trauma, and human resilience: Have we underestimated the human capacity to thrive after extremely aversive events? *American Psychologist, 59*(1), 20–28.

Breslau, N., Davis, G., & Andreski, P. (1998). Epidemiological findings on posttraumatic stress disorder and co-morbid disorders in the general population. In B. Dohrenwend (Ed.), *Adversity, stress, and psychopathology* (pp. 319–330). London: Cambridge University Press.

Brewin, C., Andrews, B., & Valentine, J. (2000). Meta-analysis of risk factors for posttraumatic stress disorder in trauma-exposed adults. *Journal of Consulting and Clinical Psychology, 68*(5), 748–766.

Brown, E. (2004, November). *Results of a controlled randomized trial of a treatment protocol of CTG.* Symposium paper presented at the International Society for Traumatic Stress 20th Annual Meeting, New Orleans, LA.

Bryant, R., & Harvey, A. (2000). New DSM-IV diagnosis of acute stress disorder [Letter to the editor]. *American Journal of Psychiatry, 157*, 1889–1890.

Bryant, R. A., Harvey, A. G., & Dang, S. T. (1998). Treatment of acute stress disorder: A comparison of cognitive-behavioral therapy and supportive counseling. *Journal of Consulting and Clinical Psychology, 66*(5), 862–866.

Buckley, T. C., Mozley, S. L., Bedard, M. A., Dewulf, A. C., & Greif, J. (2004). Preventive health behaviors, health-risk behaviors, physical morbidity, and health-related role functioning impairment in veterans with post-traumatic stress disorder. *Military Medicine, 169*(7), 536–40.

Celano, M., Hazzard, A., & Webb, C. (1996). Treatment of traumagenic beliefs among sexually abused girls and their mothers: An evaluation study. *Journal of Abnormal Child Psychology, 24*(1), 1–17.

Cohen, J., & Mannarino, A. (1996). A treatment outcome study for sexually abused preschool children: Initial findings. *Journal of the American Academy of Child and Adolescent Psychiatry, 35*(1), 42–50.

Cohen, J. A. (2004). Early mental health interventions for trauma and traumatic loss in children and adolescents. In B. T. Litz (Ed.), *Early intervention for trauma and traumatic loss* (pp. 131–146). New York: Guilford Press.

Cohen, J. A., & Mannarino, A. P. (1998). Interventions for sexually abused children: Initial treatment outcome findings. *Child Maltreatment, 3*(1), 17–26.

Cohen, J. A., Mannarino, A. P., & Knudsen, K. (2004). Treating childhood traumatic grief: A pilot study. *Journal of the American Academy of Child and Adolescent Psychiatry, 43*, 1225–33.

Conlon, L., Fahy, T. J., & Conroy, R. (1999). PTSD in ambulant RTA victims: A randomized controlled trial of debriefing. *Journal of Psychosomatic Research, 46*(1), 37–44.

Deahl, M., Srinivasan, M., & Jones, N. (2000). Preventing psychological trauma in soldiers: The role of operational stress training and psychological debriefing. *British Journal of Medical Psychology, 73*(1), 77–85.

Deblinger, E., & Heflin, A. H. (1996). *Treating sexually abused children and their nonoffending parents: A cognitive behavioral approach.* Thousand Oaks, CA: Sage.

Deblinger, E., Steer, R. A., & Lippmann, J. (1999). Two-year follow-up study of cognitive behavioral therapy for sexually abused children suffering post-traumatic stress symptoms. *Child Abuse and Neglect, 23*(12), 1371–1378.

Dobie, D. J., Kivlahan, D. R., Maynard, C., Bush, K. R., Davis, T. M., & Bradley, K. A. (2004). Posttraumatic stress disorder in female veterans: association with self-reported health problems and functional impairment. *Archives of Internal Medicine, 164*(4), 394–400.

Dougall, A., Herberman, H., Delahanty, D., Inslicht, S. S., & Baum, A. (2000). Similarity of prior trauma exposure as a determinant of chronic stress responding to an airline disaster. *Journal of Consulting and Clinical Psychology, 68*(2), 290–295.

Everly, G. S., Flannery, R. B., & Mitchell, J. T. (2000). Critical incident stress management (CISM): A review of the literature. *Aggression and Violent Behavior, 5*(1), 23–40.

Everly, G. S., & Mitchell, J. T. (2000). The debriefing "controversy" and crisis intervention: A review of lexical and substantive issues. *International Journal of Emergency Mental Health, 2*(4), 211–225.

Fairbrother, G., Stuber, J., Galea, S., Pfefferbaum, B., & Fleischman, A. R. (2004). Unmet need for counseling services by children in New York City after the September 11th attacks on the World Trade Center: Implications for pediatricians. *Pediatrics, 113*(5), 1367–1374.

Foa, E. B., Hearst-Ikeda, D., & Perry, K. J. (1995). Evaluation of a brief cognitive-behavioral program for the prevention of chronic PTSD in recent assault victims. *Journal of Consulting and Clinical Psychology, 63*(6), 948–955.

Foy, D., Sipprelle, R., & Rueger, D. (1984). Etiology of posttraumatic stress disorder in Vietnam veterans: Analysis of premilitary, military, and combat exposure influences. *Journal of Consulting and Clinical Psychology, 52*(1), 79–87.

Frazier, P., Conlon, A., & Glaser, T. (2001). Positive and negative life changes following sexual assault. *Journal of Consulting and Clinical Psychology, 69*, 1048–1055.

Freedman, S., Brandes, D., & Peri, T. (1999). Predictors of chronic post-traumatic stress disorder: A prospective study. *British Journal of Psychiatry, 174*, 353–359.

Galea, S., Ahern, J., & Resnick, H. (2002). Psychological sequelae of the September 11 terrorist attacks in New York City. *New England Journal of Medicine, 346*(13), 982–987.

Goodman, R. (2004, November). *Clinical case conceptualization of CTG: What to treat and how.* Symposium paper presented at the International Society for Traumatic Stress 20th Annual Meeting, New Orleans, LA.

Gray, M. J., Bolton, E. E., & Litz, B. T. (2004). A longitudinal analysis of PTSD symptom course: Delayed-onset PTSD in Somalia peacekeepers. *Journal of Consulting and Clinical Psychology, 72*, 909–913.

Harvey, A., & Bryant, R. (1999). The relationship between acute stress disorder and posttraumatic stress disorder: A 2-year prospective evaluation. *Journal of Consulting and Clinical Psychology, 67*(6), 985–988.

Hobbs, M., Mayou, R., Harrison, B., & Warlock, P. (1996). A randomized trial of psychological debriefing for victims of road traffic accidents. *British Medical Journal, 313*, 1438–1439.

Hobfoll, S. E., & Jackson, A. P. (1991). Conservation of resources in community intervention. *American Journal of Community Psychology, 19*, 111–121.

Hokanson, M., & Wirth, B. (2000). The critical incident stress debriefing process for the Los Angeles County Fire Department: Automatic and effective. *International Journal of Emergency Mental Health, 2*(4), 249–257.

Kaspersen, M., Matthiesen, S. B., & Götestam, K. G. (2003). Social network as a moderator in the relation between trauma exposure and trauma reaction: A survey among UN soldiers and relief workers. *Scandinavian Journal of Psychology, 44*(5), 415–423.

Kassam-Adams, N., & Winston, F. K. (2004). Predicting child PTSD: The relationship between acute stress disorder and PTSD in injured children. *Journal of the American Academy of Child and Adolescent Psychiatry, 43*(4), 403–411.

Kessler, R. (2000). Posttraumatic stress disorder: The burden to the individual and to society. *Journal of Clinical Psychiatry, 61*(Suppl. 5), 4–14.

Kessler, R. C., Sonnega, A., Bromet, E., Hughes, M., & Nelson, C. B. (1995). Posttraumatic stress disorder in the National Comorbidity Survey. *Archives of General Psychiatry, 52*(12), 1048–1060.

Kilpatrick, D. G., Ruggiero, K. J., Acierno, R., Saunders, B. E., Resnick, H. S., & Best, C. L. (2003). Violence and risk of PTSD, major depression, substance abuse/dependence, and comorbidity: Results from the National Survey of Adolescents. *Journal of Consulting and Clinical Psychology, 71,* 692–700.

King, D., King, L., Foy, D., Keane, T. M., & Fairbank, J. A. (1999). Posttraumatic stress disorder in a national sample of female and male Vietnam veterans: Risk factors, war-zone stressors, and resilience–recovery variables. *Journal of Abnormal Psychology, 108*(1), 164–170.

King, D., King, L., & Gudanowski, D. (1995). Alternative representations of war zone stressors: Relationships to posttraumatic stress disorder in male and female Vietnam veterans. *Journal of Abnormal Psychology, 104*(1), 184–196.

King, L., King, D., & Fairbank, J. (1998). Resilience-recovery factors in post-traumatic stress disorder among female and male Vietnam veterans: Hardiness, postwar social support, and additional stressful life events. *Journal of Personality and Social Psychology, 74*(2), 420–434.

King, N. J., Tonge, B. J., & Mullen, P. (2000). Treating sexually abused children with posttraumatic stress symptoms: A randomized clinical trial. *Journal of the American Academy of Child and Adolescent Psychiatry, 39*(11), 1347–1355.

Koenen, K. C., Stellman, J. M., & Stellman, S. D. (2003). Risk factors for course of posttraumatic stress disorder among Vietnam veterans: A 14-year follow-up of American Legionnaires. *Journal of Consulting and Clinical Psychology, 71*(6), 980–986.

Kulka, R., Schlenger, W., & Fairbank, J. (1990). *Trauma and the Vietnam war generation: Report of findings from the National Vietnam Veterans Readjustment Study.* Philadelphia: Brunner/Mazel.

Lange, A., Rietdijk, D., & Hudcovicova, M. (2003). Interapy: A controlled randomized trial of the standardized treatment of posttraumatic stress through the Internet. *Journal of Consulting and Clinical Psychology, 71*(5), 901–909.

Laor, N., Wolmer, L., & Cohen, D. J. (2001). Mothers' functioning and children's symptoms 5 years after a SCUD missile attack. *American Journal of Psychiatry, 158,* 1020–1026.

Layne, C. (2004, November). *Conceptualization and measurement of child traumatic grief.* Symposium paper presented at the International Society for Traumatic Stress 20th Annual Meeting, New Orleans, LA.

Litz, B. T., & Gray, M. J. (2004). Early intervention for trauma in adults: A framework for first aid and secondary prevention. In B. Litz (Ed.), *Early intervention for trauma and traumatic loss* (pp. 87–111). New York: Guilford Press.

Litz, B. T., Gray, M. J., Bryant, R., & Adler, A. B. (2002). Early intervention for trauma: Current status and future directions. *Clinical Psychology: Science and Practice, 9,* 112–134.

Litz, B. T., Williams, L., Wang, J., Bryant, R., & Engel, C. C. (2004). A therapist-assisted Internet self-help program for traumatic stress. *Professional Psychology: Research and Practice, 35*(6), 628–634.

Macklin, M., Metzger, L., & Litz, B. (1998). Lower precombat intelligence is a risk factor for posttraumatic stress disorder. *Journal of Consulting and Clinical Psychology, 66*(2), 323–326.

Maercker, A., & Müller, J. (2004). Social acknowledgment as a victim or survivor: A scale to measure a recovery factor of PTSD. *Journal of Traumatic Stress, 17*(4), 345–351.

Malik, M. L., Connor, K. M., & Sutherland, S. M. (1999). Quality of life and posttraumatic stress disorder: A pilot study assessing changes in SF-36 scores before and after treatment in a placebo-controlled trial of fluoxetine. *Journal of Traumatic Stress, 12*(2), 387–393.

Martin, L., Rosen, L. N., Durand, D. B., Knudson, K. H., & Stretch, R. H. (2000). Psychological and physical health effects of sexual assaults and nonsexual traumas among male and female United States Army soldiers. *Behavioral Medicine, 26,* 23–33.

Marx, B. P., Calhoun, K. S., & Wilson, A. E. (2001). Sexual revictimization prevention: An outcome evaluation. *Journal of Consulting and Clinical Psychology, 69*(1), 25–32.

Mawson, D., Marks, I. M., & Ramm, E. (1981). Guided mourning for morbid grief: A controlled study. *British Journal of Psychiatry, 138,* 185–193.

Mayou, R. A., Ehlers, A., & Hobbs, M. (2000). Psychological debriefing for road traffic accident victims: Three-year follow-up of a randomised controlled trial. *British Journal of Psychiatry, 176,* 589–593.

McNally, R. J., Bryant, R. A., & Ehlers, A. (2003). Does early psychological intervention promote recovery from posttraumatic stress? *Psychological Science in the Public Interest, 4*(2), 45–79.

Melhem, N. M., Day, N., Shear, M. K., Day, R., Reynolds, C. F., III, & Brent, D. (2004). Traumatic grief among adolescents exposed to a peer's suicide. *American Journal of Psychiatry, 161,* 1411–1416.

Melnik, T. A., Baker, C. T., Adams, M. L., O'Dowd, K., Mokdad, A. H., Brown, D. W., et al. (2002). Psychological and emotional effects of the September 11 attacks on the World Trade Center—Connecticut, New Jersey, and New York, 2001. *Morbidity and Mortality Weekly Report, 51,* 784–786.

Miller, M. W. (2004). Personality and the development and expression of PTSD. *National Center for PTSD Clinical Quarterly, 15*(3), 1–3.

Mitchell, J. T., & Everly, G. S. (1995). The critical incident stress debriefing (CISD) and the prevention of work-related traumatic stress among high risk occupational groups. In G. S. Everly & J. M. Lating (Eds.), *Psychotraumatology: Key papers and core concepts in post-traumatic stress* (pp. 267–280). New York: Plenum Press.

Mitchell, J. T., & Everly, G. S. (1996). *Critical incident stress debriefing: An operations manual for the prevention of traumatic stress among emergency services and disaster workers* (2nd ed.). Ellicott City, MD: Chevron.

Murphy, S. A., Johnson, C., & Cain, K. C. (1998). Broad-spectrum group treatment for parents bereaved by the violent deaths of their 12- to 28-year-old children: A randomized controlled trial. *Death Studies, 22*(3), 209–235.

National Institute of Mental Health. (2002). *Mental health and mass violence: Evidence-based early psychological intervention for victims/survivors of mass violence: A workshop to reach consensus on best practices.* Washington, DC: U.S. Government Printing Office.

Neria, Y., & Litz, B. T. (2004). Bereavement by traumatic means: The complex synergy of trauma and grief. *Journal of Loss and Trauma, 9*(1), 73–87.

Norris, F. H., Friedman, M. J., Watson, P. J., Byrne, C. M., Diaz, E., & Kaniasty, K. (2002). 60,000 disaster victims speak: Part I. An empirical review of the empirical literature, 1981–2001. *Psychiatry, 65,* 207–239.

Pfefferbaum, B., Vinekar, S. S., & Trautman, R. P. (2002). The effect of loss and trauma on substance use behavior in individuals seeking support services after the 1995 Oklahoma City bombing. *Annals of Clinical Psychiatry, 14*(2), 89–95.

Prigerson, H., Maciejewski, P., & Rosenheck, R. (2001). Combat trauma: Trauma with highest risk of delayed onset and unresolved posttraumatic stress disorder symptoms, unemployment, and abuse among men. *Journal of Nervous and Mental Disease, 189*(2), 99–108.

Raphael, B. (1977). Preventive intervention with the recently bereaved. *Archives of General Psychiatry, 34*(12), 1450–1454.

Resnick, H., Kilpatrick, D., & Dansky, B. (1993). Prevalence of civilian trauma and posttraumatic stress disorder in a representative national sample of women. *Journal of Consulting and Clinical Psychology, 61(6), 984–991.*

Riggs, D., Rothbaum, B., & Foa, E. (1995). A prospective examination of symptoms of posttraumatic stress disorder in victims of nonsexual assault. *Journal of Interpersonal Violence, 10*(2), 201–214.

Rose, S., Bisson, J., & Wessely, S. (2004). Psychological debriefing for preventing posttraumatic stress disorder (Cochrane Review). In *The Cochrane Library,* Issue 2. Chichester, UK: Wiley. Available at *www.update-software.com*

Rose, S., Brewin, C. R., Andrews, B., & Kirk, M. (1999). A randomized controlled trial of individual psychological debriefing for victims of violent crime. *Psychological Medicine, 29*(4), 793–799.

Rosenheck, R., & Fontana, A. (1995). Do Vietnam-era veterans who suffer from posttraumatic stress disorder avoid VA mental health services? *Military Medicine, 160*(3), 136–42.

Rothbaum, B., Foa, E., & Riggs, D. (1992). A prospective examination of post-traumatic stress disorder in rape victims. *Journal of Traumatic Stress, 5*(3), 455–475.

Savoca, E., & Rosenheck, R. (2000). The civilian labor market experiences of Vietnam-era veterans: The influence of psychiatric disorders. *Journal of Mental Health Policy and Economics, 3,* 199–207.

Scheeringa, M. S., & Zeanah, C. H. (1995). Symptom expression and trauma variables in children under 48 months of age. *Infant Mental Health Journal, 16*(4), 259–270.

Schlenger, W., Caddell, J., & Ebert, L. (2002). Psychological reactions to terrorist attacks: Findings from the National Study of Americans' reactions to September 11: Reply. *Journal of the American Medical Association, 288*(21), 2684–2685.

Schnurr, P., Friedman, M., & Rosenberg, S. (1993). Preliminary MMPI scores as predictors of combat-related PTSD symptoms. *American Journal of Psychiatry, 150*(3), 479–483.

Schnurr, P., Lunney, C., & Sengupta, A. (2003). A descriptive analysis of PTSD chronicity in Vietnam veterans. *Journal of Traumatic Stress, 16*(6), 545–553.

Schwartzberg, S. S., & Halgin, R. P. (1991). Treating grieving clients: The importance of cognitive change. *Professional Psychology: Research and Practice, 22*(3), 240–246.

Shalev, A. Y. (2000). Stress management and debriefing: Historical concepts and present patterns. In B. Raphael & J. P. Wilson (Eds.), *Psychological debriefing: Theory, practice and evidence* (pp. 17–31). New York: Cambridge University Press.

Shalev, A. Y. (2002). Acute stress reactions in adults. *Biological Psychiatry, 51*(7), 532–543.

Shalev, A. Y., Freedman, S., & Peri, T. (1997). Predicting PTSD in trauma survivors: Prospective evaluation of self-report and clinician-administered instruments. *British Journal of Psychiatry, 170*, 558–564.

Shalev, A.Y., Peri, T., & Canetti, L. (1996). Predictors of PTSD in injured trauma survivors: A prospective study. *American Journal of Psychiatry, 153*(2), 219–225.

Shear, K., Frank, E., Houck, P. R., & Reynolds, C. F., III. (2005). Treatment of complicated grief: A randomized controlled trial. *Journal of the American Medical Association, 293*, 2601–2608.

Silver, R., Holman, E., & McIntosh, D. (2002). Nationwide longitudinal study of psychological responses to September 11. *Journal of the American Medical Association, 288*(10), 1235–1244.

Sireling, L., Cohen, D., & Marks, I. (1988). Guided mourning for morbid grief: A controlled replication. *Behavior Therapy, 19*(2), 121–132.

Somer, E., Buchbinder, E., Peled-Avram, M., & Ben-Yizhack, Y. (2004). The stress and coping of Israeli emergency room social workers following terrorist attacks. *Qualitative Health Research, 14*(8), 1077–1093.

Somer, E., Tamir, E., Maguen, S., & Litz, B. (2005). Brief cognitive-behavioral phone-based intervention targeting anxiety about the threat of attack: A pilot study. *Behaviour Research and Therapy, 43*, 669–679.

Spooren, D. J., Henderick, H., & Jannes, C. (2000). Survey description of stress of parents bereaved from a child killed in a traffic accident: A retrospective study of a victim support group. *Journal of Death and Dying, 42*(2), 171–185.

Stein, B. D., Jaycox, L. H., Kataoka, S. H., Wong, M., Tu, W., Elliott, M. N., et al. (2004). A mental health intervention for schoolchildren exposed to violence: A randomized controlled trial. *Journal of the American Medical Association, 290*, 603–611.

Stretch, R., Knudson, K., & Durand, D. (1998). Effects of premilitary and military trauma on the development of post-traumatic stress disorder symptoms in female and male active duty soldiers. *Military Medicine, 163*(7), 466–470.

Tarrier, N., Pilgrim, H., & Sommerfield, C. (1999). A randomized trial of cognitive therapy and imaginal exposure in the treatment of chronic posttraumatic stress disorder. *Journal of Consulting and Clinical Psychology, 67*, 13–18.

Tedeschi, R., & Calhoun, L. (1996). The Posttraumatic Growth Inventory: Measuring the positive legacy of trauma. *Journal of Traumatic Stress, 9*(3), 455–472.

van Minnen, A., Arntz, A., & Keijsers, G. (2002). Prolonged exposure in patients with chronic PTSD: Predictors of treatment outcome and dropout. *Behaviour Research and Therapy, 40*(4), 439–457.

Valentiner, D., Foa, E., Riggs, D., & Gershuny, B. S. (1996). Coping strategies and posttraumatic stress disorder in female victims of sexual and nonsexual assault. *Journal of Abnormal Psychology, 105*(3), 455–458.

World Health Organization. (1992). *International classification of diseases.* Geneva: Author.

Young, B. H., & Gerrity, E. (1994). Critical incident stress debriefing (CISD): Value and limitations in disaster response. *National Center for PTSD Clinical Quarterly, 4*, 17–19.

Chapter 17

Psychosocial Treatments for PTSD

Patricia A. Resick, Candice M. Monson, and Cassidy Gutner

Our task in this chapter is to review methodology and psychosocial treatment research for posttraumatic stress disorder (PTSD). The task is a bit daunting, because a wealth of information has been accumulated over the past two decades on the treatment of PTSD, and we are moving away from questions about whether PTSD is a treatable condition, and toward questions about who benefits from such treatment and under what circumstances. This chapter begins with a brief review of the typical progression of treatment outcome research, and the methodological rigor associated with these advances relative to PTSD. The current state of treatment outcome research on PTSD is reviewed, followed by our predictions and suggestions for future research directions.

THE PROGRESSION OF TREATMENT OUTCOME RESEARCH

When little is known about how to treat a given disorder, therapists try different approaches based usually on what they know about the treatment of other disorders. In the 1980s, this was the case with PTSD. Researchers began to apply what they knew about the treatment of anxiety and depression to the study of PTSD, and related case studies began to emerge in the literature (e.g., Frank & Stewart, 1984; Keane & Kaloupek, 1982; Kilpatrick, Veronen, & Resick, 1982). Following the usual progression from case studies to wait-list controlled trials to control for the passage of time, small, controlled PTSD treatment outcome studies were based on promising practices (Foa, Rothbaum, Riggs, & Murdock, 1991; Keane, Fairbank, Caddell, & Zimering, 1989; Resick, Jordan, Girelli, & Hutter, 1988). These early studies did not meet all of the rigorous standards and components of treatment studies that we consider optimal today (Foa & Meadows, 1997), such as blind assessment and adherence ratings (see next section). The next step in psychotherapy development is to compare the treatment with nonspecific interventions and other therapies thought or known to be effective. Examples from this stage of development are the Foa et al. (1991) trial comparing prolonged exposure to stress inoculation training, the Resick, Nishith, Weaver, Astin, and Feuer (2002) study of cognitive processing therapy and prolonged exposure, or the Taylor and colleagues (2003) study comparing eye movement desensitization and reprocessing (EMDR) and prolonged exposure.

Once a therapy is established as efficacious, then research turns to questions of external validity. One question is whether the therapy is effective in nonresearch settings, with a heterogeneous population presenting for treatment. Along these lines, studies are mounted to target people who are thought to have particularly difficult presentations and comorbidities. In these studies, there are often two treatment targets, amelioration of PTSD and another disorder, such as panic disorder or substance dependence. Subsequent studies are also conducted to determine predictors of treatment response to ascertain further who benefits from treatment. Additive studies are designed to determine whether treatment outcomes can be enhanced with the addition of other components. Dismantling studies are also conducted to determine whether all of the components of a treatment protocol are necessary, and which, if any, are sufficient alone (i.e., essential mechanisms of change). Finally, because a therapy may be determined to be both efficacious in research settings and effective in clinical settings does not mean the therapy will be adopted on a widespread basis. The final frontier in treatment outcome research is the question of how to disseminate treatments more widely into practice. Dissemination is more than simply training therapists; it includes working with systems of care to accommodate training and implementation of new therapies.

This progression of treatment research has been reified as a staged model of behavioral therapy research (Rounsaville, Carroll, & Onken, 2001) and translated into different types of grant funding through the National Institutes of Health. Stage 1 research is the most basic treatment development research in which pilot testing is conducted; treatment manuals, with accompanying adherence and competence measurement, are developed in this stage. Stage 1 usually ends with the treatment tested against a wait list or treatment as usual. Stage 1 can also include translational research from basic science to intervention development. Stage 2 studies are randomized clinical trials (RCT), often comparing a new treatment to the best available treatment, or a new treatment to a psychotherapy containing nonspecific elements. Some designs at this stage test specific mechanisms of change. When at least two successful RCTs have been completed, Stage 3 studies embark on questions of generalizability into the community or across populations, and consider other implementation issues. Because of the rapidly evolving state of psychotherapy research on PTSD, this chapter is devoted to reviewing only psychotherapeutic procedures that have evidence at Stage 2. Treatment innovations at Stage 1 are reviewed by Welch and Rothbaum, Chapter 23, this volume.

METHODOLOGICAL CONSIDERATIONS IN CLINICAL TRIALS

In critiquing the current state of psychotherapy research in PTSD it is important to appreciate state-of-the-art methodology in treatment outcome research.

Gold Standards

In the most widely cited article about methodological considerations in PTSD treatment outcome research, Foa and Meadows (1997) outline seven "gold standards" to evaluate the methodological rigor of PTSD treatment studies. The standards include (1) clearly defined target symptoms; (2) reliable and valid measures; (3) use of blind evaluators; (4) assessor training; (5) manualized, replicable, specific treatment programs; (6) unbiased assignment to treatment; and (7) treatment adherence. The standard of "clearly defined symptoms" addresses the problem of vague definitions of the purpose and outcomes of the therapy. Without a clearly defined set of symptoms, one cannot compare one treatment to another or even know what goal is agreed upon for treatment. For example, occasionally one hears therapists and researchers confuse the traumatic event with the symptoms that result. A therapist cannot treat the experience of rape, child sexual abuse, or combat. These are events that cannot change. However, a therapist can treat symptoms of PTSD, depression, or substance abuse that may result from trauma exposure. These symptoms can and must be assessed with valid and reliable measures, so that outcomes can be compared and replicated.

Clinical trials of any mental health intervention must use independent evaluations of the outcomes, because of the inherent bias of assessments conducted by therapists or investigators involved in the trial. In fact, "independence" usually means that the person conducting the assessments is blind to at least the therapy administered, if not the period of assessment. Depending on the study, the assessors may need to be blind to the identity of the therapist (if the therapist provides only one type of treatment). Those doing the assessments must be clinically skilled, accurate, and reliable at administering diagnostic interviews. This is usually demonstrated by having a portion of the interviews scored independently for reliability by a second expert rater, either live or from tape recordings.

In clinical settings, many therapists describe themselves as being eclectic in their therapeutic orientation, and may change their interventions within and across clients as their clinical judgment dictates. Treatment goals may change over the course of time and even from session to session. While providing maximum flexibility for therapist and client, this type of practice is unacceptable for treatment outcome studies. As pointed out in the first standard related to clearly defined targets, treatment goals in a trial do not change over the course of treatment, even if other problems come to light. In addition to the clear and consistent treatment goals, the therapy must also be clearly defined, so that other people can replicate the treatment in other settings and with other therapists. Therefore, treatment manuals that describe the exact course of therapy conducted are essential. Treatment results that cannot be replicated by other therapists in other settings should be considered suspect.

"Unbiased" assignment to treatment typically means that research participants are randomly assigned to treatment conditions. These procedures are outlined in the informed consent process. Unless treatment choice is a question to be addressed in the study, participants cannot choose which treatment they prefer, be referred for a particular treatment, or have the investigators assign treatments based on which treatment they think would be most beneficial to the client. True random assignment may be modified if the sample needs to be stratified based some important variable (e.g., gender, type of trauma) but must remain unbiased with regard to particular individuals. Random assignment to conditions is not difficult to accomplish in large studies, in which participants are treated individually, but it can be quite problematic for group treatments. There are ethical and pragmatic realities to having traumatized people wait for treatment until there are enough participants to form two groups for random assignment. This is not feasible in many locations and has probably dissuaded many researchers from conducting studies of group treatments.

Finally, good treatment outcome studies include some type of rating of adherence to the protocol. This is done to ensure that the therapy was not only implemented as proposed by all of the therapists in each condition, but also that proscribed components are not included. For example, if one were to compare an exposure therapy and an interpersonal therapy, it would be important to ensure that the therapist providing exposure therapy neither engages in any type of interpersonal therapy before or after the exposures nor assigns homework that might be construed as a product of interpersonal therapy. Likewise, the interpersonal therapist should not encourage clients to engage in systematic exposures. In rigorous treatment outcome studies, therapy sessions are video- or audiotaped, and independent experts on the therapies are provided checklists and evaluation forms to rate a portion of the sessions for treatment fidelity. The evaluations are specific as to which interventions are supposed to be delivered or not delivered during a particular session.

Since the Foa and Meadow (1997) article, Harvey, Bryant, and Tarrier (2003) have written about increasing the methodological rigor of PTSD outcome studies. They also drew upon the Consolidated Standards of Reporting Trials (CONSORT; Begg et al., 1996), published in the *Journal of the American Medical Association*, for conducting and reporting on clinical trials in medicine. Aside from the gold standards listed earlier, they recommend that the independent evaluators be trained and supervised throughout the course of the study to minimize observer drift, and that assessors provide some type of rating or estimate regarding the condition to which the participant was randomized, to demonstrate blindness to condition assignment. They also suggest that the random assignment be unpredictable and hidden from the investigator who is enrolling the participants. They recommend that three levels of treatment adherence, in addition to treatment fidelity, be assessed: that the treatment was administered in its pure form, without the influence of other treatments; that the treatment was received by the client; and that the enactment of the treatment was assessed to ensure that the client applied the treatment outside of the session (i.e., homework compliance).

Intention to Treat

Research designs have also improved over the past decade to include intention-to-treat principles. An intention-to-treat design includes data from all recruited participants, whether or not they complete a given treatment. Thus, even if a participant drops out of treatment, he or she is solicited and (ideally) assessed when researchers employ intention-to-treat principles. Although results from those who complete treatment are important to determine how well a therapy works if clients receive the full dose, there is an inherent bias in examining only the data from this subsample. There might be systematic differences between types of treatment with regard to who drops out. For example, if people receiving one type of therapy tend to drop out when they are not doing well, whereas people in the comparison therapy stay in treatment when not doing well, completer analyses would indicate that the former therapy was more efficacious, because the treatment nonresponders dropped out at a higher rate. The dropout bias can also work against a very effective therapy. In completer analyses, if one type of therapy leads to rapid gains, people who stop therapy before the standardized number of sessions would be counted as dropouts rather than early successes in clinical trials. Harvey and colleagues (2003) contend that intention-to-treat analyses should be the primary analyses in treatment outcome studies on PTSD. Consistent with CONSORT, we also encourage careful description of the sample, including the number of people screened, considered ineligible, or who dropped out of treatment and/or assessment, and the operational definition of treatment completion. These are important considerations regarding the external validity of the study.

The Problem of Power

One of the important tasks of a Stage 1 study is to estimate the treatment's effect size. Furthermore, Stage 2 studies increase the precision of effect size estimates, and are used to predict the number of participants necessary in a particular study to demonstrate the statistically significant effect. If a particular treatment were compared to a wait-list control group or no treatment, then one would expect from meta-analysis (Bradley, Greene, Russ, Dutra, & Westen, 2005) that an efficacious PTSD treatment would have a large effect. With .80 power to find a significant effect, and alpha set at .05, only about 17 participants per group are necessary to demonstrate the average, evidence-based PTSD treatment effect of Cohen's $d \geq 1.0$. If a treatment is compared to a condition expected to control for the nonspecific but essential effects of psychotherapy, such as time with the therapist, and support and warmth of the therapist, a more moderate effect size might be expected. Presuming such an effect size ($d = 0.50$), and the aforementioned power and alpha-level assumptions, 64 participants for each group would be needed to detect the difference between the nonspecific treatment effects and the active ingredients of the therapy.

　　If one conducts a treatment outcome study between two active treatments with known effectiveness, any difference between them would be presumed to be small; therefore, a very large sample size would be needed to detect that difference. Large sample sizes are also needed to determine which client type might benefit from different types of treatment, or to examine effects across different types of trauma or demographic groups. Given the previous assumptions, power calculations for a small effect size difference of $d = 0.20$ would require nearly 400 participants per cell.

　　The need for a large sample size is a serious dilemma for treatment researchers, especially those who conduct research with trauma victims. Treatment research, and especially psychotherapy research, is very expensive, if the researchers are attempting to achieve the research standards we reviewed earlier. To recruit appropriate participants, to assess and to have those assessments evaluated, and to treat and to have the therapy validated for fidelity are all expensive in terms of not only dollars but also time and effort. It is difficult to obtain enough treatment-seeking participants in most locations to obtain the sample sizes needed for the small effect comparisons of head-to-head trials. Multisite studies are even more difficult to mount because of the need to provide both uniform training across sites and the extra infrastructure needed to coordinate a study across sites. Some researchers argue that with a small sample, if there are no significant differences between two active treatments, then for all practical purposes there are no meaningful clinical differences. This may be true except when trends emerge or the small sample is skewed in some way. Small samples also do not answer questions regarding differential treatment effects based on race, ethnicity, gender, type of trauma, or other potentially important variables. Multiple studies and meta-analyses can be helpful in answering these questions (e.g., Bradley et al., 2005). If similar measures were used across studies, pooling data might also be a possibility. This was done recently for a conference presentation in which Cahill, Foa, Rothbaum, and Resick (2004) pooled data from four studies to determine what proportion of clients worsened as a result of exposure-based treatments.

CURRENT STATE OF THE ART

The International Society for Traumatic Stress Studies (ISTSS) established a PTSD Treatment Guidelines Task Force in 1997 to develop practice guidelines. The results were published in a 27-chapter book (Foa, Keane, & Friedman, 2000) that reviewed and developed guidelines for a wide range of therapeutic approaches. Specific topics such as early intervention (Litz & Maguen, Chapter 16) and pharmacotherapy (Friedman & Davidson, Chapter 19) are covered elsewhere in this book. With regard to psychosocial treatments, Table 17.1 is a compilation of the RCTs of adult studies reviewed in that book. It can be seen that the majority of research was conducted on individual cognitive-behavioral therapy (CBT) and EMDR. There were scattered studies on psychodynamic, group, and stress management treatments. Most of them included a wait-list comparison, but a few included head-to-head comparisons of treatments. These studies provide a solid evidence base for CBT, and especially exposure treatments, compared to wait-list and other active treatment conditions. There is also evidence for EMDR versus wait-list control. Few of these studies provided the necessary information to calculate effect sizes or used intention-to-treat principles, which is indicative of the state of psychotherapy for PTSD research at the time. A comparison of Table 17.1 and the following review of studies published subsequent to the ISTSS treatment guidelines book underscores the methodological advances made in a relatively short period of time.

Given the state of research on PTSD treatment, we have focused primarily on the controlled PTSD trials published since publication of the ISTSS treatment guidelines, as well as some treatment studies that have examined other, related questions. These studies are presented in Table 17.2. Hedge's g, including a correction for sample size, was used as the effect size measure; clinician assessment of PTSD symptoms at posttreatment was the outcome of interest. The pooled standard deviation of the compared conditions was used. Where the information required for calculating effect sizes was not available, we have provided alternative indications of the treatments' effects.

Wait-List Controlled Trials

There have been a few wait-list controlled trials published since the guidelines were released. Fecteau and Nicki (1999) compared a CBT including psychoeducation, relaxation, imaginal exposure, and cognitive restructuring to waiting list in a sample of motor vehicle accident survivors. CBT was statistically superior to waiting list, with an effect size of 1.34 for those assessed. A Brief Eclectic Psychotherapy, consisting of psychoeducation, imagery guidance, writing about the trauma, cognitive exploration, and

TABLE 17.1. Randomized Controlled Trials of PTSD Treatment of Adults Reviewed for the International Society for Traumatic Stress Studies Guidelines

Author(s)	Treatments	Population	Treatment duration	Results
Alexander, Neimeyer, Follette, Moore, & Harter (1989)	1. Interpersonal transaction 2. Process group 3. WL	Female CSA survivors (57)	10 weekly sessions	Both treatment groups produced better results than WL.
Boudweyns & Hyer (1990)	1. ST + EX 2. ST + TC	Male Vietnam veterans (51)	12–14, 50-minute sessions	ST + EX better than ST + TC at posttreatment, and results maintained at follow-up assessments.
Boudewyns & Hyer (1996)	1. EMDR + TAU 2. EMDR-EM + TAU 3. TAU	Male combat veterans (61)	5–7 sessions	All groups significantly improved, and no differences among them.
Brom, Kleber, & Defares (1989)	1. SD 2. Hypnotherapy 3. Psychodynamic 4. WL	Mixed (112)	14–19 sessions	All treatments yielded improvements relative to WL, but no differences among the treatments.
Carlson, Chemtob, Rusnak, Hedlund, & Muraoka (1998)	1. EMDR 2. BF-REL 3. TAU	Male combat veterans (35)	12 sessions	EMDR better than BF-REL and TAU. No differences between BF-REL and TAU. Results maintained at 9-month follow-up.
Cooper & Clum (1989)	1. ST 2. ST + EX	Male Vietnam veterans (26)	6–14, 90-minute sessions	ST + EX produced significant improvements over ST.
Devilly, Spence, & Rapee (1998)	1. EMDR + TAU 2. EMDR-EM + TAU 3. TAU	Male combat veterans (51)	2, 90-minute sessions	All three groups improved on all outcome measures, but EMDR and EMDR-EM showed more improvement. No differences between EMDR and EMDR-EM.
Echeburua, de Corral, Zubizarreta, & Sarasua (1996)	1. Relax 2. Coping skills training	Female sexual assault (20)	5, 60-minute sessions	Both treatment groups improved at posttreatment; Coping Skills Training group outperformed the Relax group at 12-month follow-up.
Echeburua, de Corral, Zubizarreta, & Sarasua (1997)	1. Relax 2. EX + CT	Female CSA or nonrecent adult trauma (20)	6 sessions; 7 hours for EX + CT and 4.15 hours for Relax	Both groups improved; EX + CT improved more than Relax and continued to improve at follow-up.
Foa et al. (1999)	1. EX 2. SIT 3. SIT + EX 4. WL	Female sexual and physical assault (96)	9, 90-minute sessions	All treatments produced significantly better results than WL; EX alone better than other treatment groups.

(continued)

TABLE 17.1. *(continued)*

Author(s)	Treatments	Population	Treatment duration	Results
Foa, Rothbaum, Riggs, & Murdock (1991)	1. EX 2. SIT 3. Supportive counseling 4. WL	Female sexual assault (45)	9, 90-minute sessions	At posttreatment, SIT produced better results than both supportive counseling and WL, and produced slightly better results than EX. At 3-month follow-up, EX slightly better than SIT.
Jensen (1994)	1. EMDR 2. TAU	Male combat veterans (25)	2, 50-minute sessions	EMDR did not produce better results than TAU.
Keane, Fairbank, Caddell, & Zimering (1989)	1. EX 2. WL	Male Vietnam veterans (24)	14-16 sessions; 45- to 90-minute sessions	EX better than WL.
Marcus, Marquis, & Sakai (1997)	1. EMDR 2. TAU	Mixed (67)	6, 50-minute sessions	EMDR produced significantly faster and greater improvements compared with TAU. At posttreatment, 77% of EMDR no longer met criteria for PTSD compared with 50% of TAU.
Marks, Lovell, Noshirvani, Livanou, & Thrasher (1998)	1. EX 2. CT 3. EX + CT 4. Relax	Mixed (87)	10 sessions	EX, CT, and EX + CT produced better results than Relax at posttreatment and 6-month follow-up. No differences among EX, CT, and EX + CT.
Peniston (1986)	1. SD + BIO 2. No treatment	Veterans (16)	Large number over long period of time	SD + BIO produced significant improvement at posttreatment relative to no treatment.
Peniston & Kulkosky (1991)	1. SD + BIO 2. Traditional medical Tx	Male Vietnam veterans (29)	30 sessions	SD + BIO produced better results than traditional medical Tx.
Pitman, Orr, Altman, & Longpre (1996)	1. EMDR 2. EMDR-EM	Male combat veterans (17)	6, 70- to 110-minute sessions	Both groups showed modest improvements at posttreatment; results did not maintain at 5-year follow-up.
Renfrey & Spates (1994)	1. EMDR 2. Automated EMDR 3. EMDR-EM	Mixed (23)	2–6 sessions	All groups showed improvement, with no differences among groups.
Resick, Jordan, Girelli, Hutter, & Marhoefer-Dvorak (1988)	1. SIT 2. AT 3. Supportive therapy	Female sexual assault (37)	6, 120-minute sessions	No differences among the groups.

(continued)

TABLE 17.1. *(continued)*

Author(s)	Treatments	Population	Treatment duration	Results
Resick & Schnicke (1992)	1. CPT 2. WL (natural)	Female sexual assault (19)	12 sessions	CPT produced significant improvements as compared with the natural WL. All CPT lost diagnosis at 6-month follow-up.
Rothbaum (1997)	1. EMDR 2. WL	Female sexual assault (18)	3, 90-minute sessions	EMDR better than WL.
Scheck, Schaeffer, & Gillette (1998)	1. EMDR 2. AL	Mixed (60)	2, 90-minute sessions	Both groups improved; EMDR better on most measures.
Silver, Brooks, & Obenchain (1995)	1. Milieu + BIO 2. Milieu + EMDR 3. Milieu + Relax 4. Milieu	Male Vietnam veterans (100)	Various number of sessions + Milieu, over the course of year	EMDR better than Milieu alone. No differences among the other groups.
Tarrier, Pilgrim, et al. (1999)	1. CT 2. EX	Mixed (72)	16, 60-minute sessions	CT and EX both resulted in significant improvements; no differences between them.
Vaughan, Armstrong, Gold, & O'Connor (1994)	1. EMDR 2. IHT 3. AMT	Mixed (36)	3–5 sessions, over 2–3 weeks	No differences among the treatments, and both better than WL; Results maintained at 3-month follow-up.
Watson, Tuorila, Vickers, Gearhart, & Mendez (1997)	1. Relax 2. Relax + Breathing 3. Relax + Breathing + BIO	Male Vietnam veterans (90)	10, 30-minute sessions	All groups produced only mild improvements, and no differences among the groups.
Wilson, Becker, & Tinker (1995)	1. EMDR 2. WL	Mixed (80)	3, 90-minute sessions	EMDR better than WL; results maintained at 15-month follow-up.
Zlotnick et al. (1997)	1. Affect management 2. WL	Female CSA (33)	15 sessions, over 6–8 weeks	Affect management produced better results than WL.

Note. AC, assesment control; AL, active listening; AMT, applied muscle relaxation training; AT, assertiveness training; automated EMDR, a protocol that induces left–right–left eye movements with a flashing light; BF-REL, biofeedback-assisted relaxation; BIO, biofeedback; BP, brief prevention; CSA, childhood sexual abuse; CBT, cognitive-behavioral therapy; CPT, cognitive processing therapy; CT, cognitive therapy; EMDR, eye movement desensitization and reprocessing; EMDR-EM, eye movement desensitization and reprocessing minus eye movements; EX, exposure therapy; IHT, imaginary habituation training; IMF, imaginal flooding; IVF, *in vivo* flooding; Relax, relaxation; SD, systematic desensitization; SIT, stress inoculation training; ST, standard treatment; TAU, treatment as usual; TC, traditional counseling; Tx, treatment; WL, wait-list. Data from Foa, Keane, and Friedman (2000).

TABLE 17.2. An Update of the Controlled Treatment Outcome Studies Published Since the International Society for Traumatic Stress Studies Guidelines

Author(s)	Treatment/control conditions	Sample (size)	Treatment duration	Results	Effect sizes at posttreatment based on clinician assessment for intention to treat (completers)	
					Wait-list or minimal-attention comparison	Active treatment comparison
				Wait-list control comparison		
Fecteau & Nicki (1999)	CBT WL	MVA survivors (28)	8–10 hours	CBT > WL; results maintained at 6-month follow-up	1.28[a]	
Gersons, Carlier, Lamberts, & van der Kolk (2000)	BEP WL	Police officers (42)	16 sessions	BEP > WL; results maintained at 3-month follow-up	End of treatment: 91% of BEP and 50% of WL no longer met PTSD criteria[a]	
Kubany et al. (2004)	CTT-BW WL	Formerly battered women (125)	8–11, 1.5-hour biweekly sessions	PTSD remitted in 87% of the CTT-BW group; results maintained at 6-month follow-up	1.45 (2.87)	
Monson et al. (2006)	CPT WL	Male and female veterans (60)	12 sessions	CPT > WL; results maintained at 1-month follow-up	1.12 (1.14)	
				Nonspecific treatment comparison		
Blanchard et al. (2003)	CBT ST WL	MVA survivors (78)	8–12 weekly sessions	CBT > ST > WL; results maintained at 3-month follow-up	CBT: 1.14[a] ST: 0.53[a]	CBT > ST: 0.66[a]
Neuner, Schauer, Klaschik, Karunakara, & Elbert (2004)	NET ST PsyEd	Sudanese refugees (43)	4 sessions for NET and ST, 1 session for PsyEd	NET = ST = PsyEd at posttreatment; NET > ST = PsyEd at 1-year follow-up		NET = ST: 0.06[a] NET > PsyEd: 0.19[a] ST > PsyEd: 0.13[a]
Schnurr et al. (2003)	TFGT PCGT	Vietnam veterans (360)	30, 1-hour weekly sessions	TFGT = PCGT		TFGT > PCGT: 0.12 (0.25[b])

(continued)

339

TABLE 17.2. (continued)

Author(s)	Treatment/control conditions	Sample (size)	Treatment duration	Results	Effect sizes at posttreatment based on clinician assessment for intention to treat (completers)	
				Head-to-head comparison	Wait-list or minimal-attention comparison	Active treatment comparison
Devilly & Spence (1999)	CBT EMDR	Mixed (23)	8–9 sessions	CBT > EMDR; results maintained at 3-month follow-up		CBT > EMDR: $0.72^{a,f}$
Ehlers et al. (2003)	CT SHB WL	MVA survivors (97)	CT: 12, 60–90 minute weekly sessions	CT > SHB = WL	CT > WL: $1.12–1.22^{a,e}$ SHB > WL: $.21–.26^{a,e}$	CT > SHB: $0.96–1.00^{a,e}$
Ironson, Freund, Strauss, & Williams (2002)	EMDR PE	Mixed (22)	6, 90-minute weekly sessions	EMDR = PE; EMDR treatment was faster but not more effective; results maintained at 3-month follow-up		EMDR > PE: $0.62^{a,d}$
Lee, Gavriel, Drummond, Richards, & Greenwald (2002)	SIT + PE EMDR	Mixed (24)	7, 90-minute weekly sessions	SIT + PE = EMDR; EMDR produced more improvements in intrusions		SIT + PE > EMDR: 0.59^{a}
Power et al. (2002)	EMDR E + CR WL	Mixed (105)	Up to 10, 90-minute weekly sessions	EMDR = E + CR > WL; EMDR produced better depression and social functioning outcomes than E + CR	Range of PTSD symptom change: EMDR: 2.17–2.85 E + CR: 1.58–2.55	EMDR > E + CR: $0.29–0.63^{e}$
Resick, Nishith, Weaver, Astin, & Feuer (2002)	PE CPT MA	Sexual assault victims (171)	13 hours of biweekly sessions	PE = CPT > MA; CPT produced better outcomes than PE on some aspects of guilt	PE: 0.86 (2.04) CPT: 1.13 (2.78)	CPT > PE: 0.18 (0.24)

Study	Treatment conditions	Population (N)	Treatment length	Findings	Within-group effect size	Between-group effect size
Rothbaum, Astin, & Marsteller (2005)	PE, EMDR	Sexual assault victims (74)	9, 90-minute biweekly sessions	CBT = EMDR; results maintained at 6-month follow-up		CBT = EMDR: 0.01[a]
Tarrier & Sommerfield (2004)	IE, CR	Mixed (32)	16, 1-hour sessions	At 5-year follow-up CR > IE		CR > IE: 1.12[a]
Taylor et al. (2003)	ET, RELAX, EMDR	Mixed (60)	8, 90-minute sessions	ET > RELAX = EMDR; ET produced better results on reexperiencing and avoidance symptoms		% with ≥ 2 SD reduction in PTSD symptoms: ET: 47–80% RELAX: 33–53% EMDR: 33–53%

Dismantling/additive comparison

Study	Treatment conditions	Population (N)	Treatment length	Findings	Within-group effect size	Between-group effect size
Bryant, Moulds, Guthrie, Dang, & Nixon (2003)	IE + CR, IE, ST	Mixed (58)	8, 90-minute weekly sessions	IE + CR = IE > ST; results maintained at 6-month follow-up		Symptom intensity IE + CR > IE: 0.26 (0.47) IE + CR > ST: 0.83 (1.14) IE > ST: 0.65 (0.83) Symptom frequency IE + CR > IE: 0.23 (0.44) IE + CR > ST: 0.72 (1.10) IE > ST: 0.63 (0.76)
Foa et al. (2005)	PE, PE + CR, WL	Female assault survivors (171)	9–12 sessions	PE = PE + CR > WL; results maintained at 12-month follow-up	PE: 0.65 (1.92)[c] PE + CR: 0.80 (1.80)[c]	PE + CR = PE: 0.08 (0.00)[c]
Paunovic & Öst (2001)	CBT, ET	Refugees (16)	16–20, 60- to 120-minute weekly sessions	CBT = ET; results maintained at 6-month follow-up		CBT > ET: 0.12[a]

(continued)

TABLE 17.2. (continued)

| | | | | | Effect sizes at posttreatment based on clinician assessment for intention to treat (completers) | |
Author(s)	Treatment/ control conditions	Sample (size)	Treatment duration	Results	Wait-list or minimal-attention comparison	Active treatment comparison
		Adaptations for different traumatized populations and comorbidities				
Chard (2005)	CPT-SA WL	Female sexual abuse survivors (71)	17, 90-minute weekly group sessions and 10, 60-minute individual sessions	CPT-SA > WL; results maintained at 1 year follow-up	2.32[a]	
Chemtob, Novaco, Hamada, & Gross (1997)	AT MA	Vietnam veterans (15)	12 weeks	Less frequent reexperiencing symptoms and trend for lower intensity of reexperiencing symptoms	[g]	
Cloitre, Cohen, Koenen, & Han (2002)	STAIRM-PE WL	Women with child-hood sexual abuse (58)	16, 60- to 90-minute sessions	STAIRM-PE > WL; results maintained at 9-month follow-up	1.27[a]	
Falsetti, Resnick, Davis, & Gallagher (2001)	M-CET WL	Women with mixed trauma with comorbid panic attacks (22)	12, 90-minute weekly sessions	M-CET > WL	92% of M-CET no longer met criteria for PTSD at posttreatment compared to 33% in WL[a]	
Hein, Cohen, Miele, Litt, & Capstick (2004)	SS RP WL	Women with mixed trauma with comorbid substance abuse (107)	24, 1-hour biweekly sessions	SS = RP > WL; results maintained at 9-month follow-up	SS: 0.46[h] RP: 0.40[h]	RP > SS: 0.25[h]

Note. AT, anger treatment; BEP, brief eclectic psychotherapy; CAPS, Clinician Administered PTSD Scale; CBT, cognitive-behavioral therapy; CPT, cognitive processing therapy; CPT-SA, cognitive processing therapy for sexual abuse survivors; CR, cognitive restructuring; CT, cognitive therapy; CTT-BW, cognitive trauma therapy for battered women; ET, exposure therapy; EMDR, eye movement desensitization and reprocessing therapy; E + CR, exposure and cognitive restructuring; IE, imaginal exposure; MA, minimal exposure; M-CET, multiple-channel exposure therapy; NET, narrative exposure therapy; PCGT, present-centered group therapy; PE, prolonged exposure; PsyEd, psychoeducation; RELAX, progressive relaxation; RP, relapse prevention; SHB, self-help book; SIT, stress inoculation training; SS, seeking safety; STAIRM-PE, skills training in affective and interpersonal regulation and modified prolonged exposure; ST, supportive psychotherapy/counseling; TFGT, trauma-focused group therapy; WL, wait-list.

[a] Results based on available participants.

[b] Those receiving "adequate dose" of treatment (i.e., at least 24 sessions).

[c] Effect size was based on the total sample, because there were no differences between study sites.

[d] No M (SD) for clinician assessment of PTSD available. From the Bradley et al. (2005) meta-analysis.

[e] Range across frequency and intensity of CAPS subscales.

[f] Calculated from PTSD Interview (DSM-III-R).

[g] No M (SD) for assessment of PTSD available.

[h] Reported only intent-to-treat findings.

342

"farewell rituals," was compared to waiting list in a sample of police officers (Gersons, Carlier, Lamberts, & van der Kolk, 2000). This therapy was superior to waiting list, with 91% no longer meeting PTSD criteria at the end of treatment. Interestingly, 50% of the waiting list group no longer met criteria at the end of the study, also.

Kubany, Hill, and Owens (2003; Kubany et al., 2004) conducted two studies to compare cognitive trauma therapy for battered women with PTSD and guilt cognitions with a delayed-treatment control group. The treatment comprises psychoeducation, stress management, exposure homework, cognitive restructuring for guilt-related beliefs, self-advocacy training, and management of unwanted contacts with the abuser and other, potential abusers. The findings of this treatment were very strong, with only 9% of the women retaining a PTSD diagnosis at posttreatment assessment. Furthermore, at pretreatment assessment, only 4% of the women scored in the normal range on the Beck Depression Inventory (BDI score < 10; Beck, Ward, Mendelson, Mock, & Erbaugh, 1961), but 83% scored in the normal range posttreatment. The delayed-treatment group replicated the findings of the immediately treated group, and the results for both groups were maintained at the 6-month follow-up assessment.

Monson and colleagues (2006) conducted a wait-list controlled study of CPT in male and female veterans with chronic, military-related PTSD. CPT was superior to wait-list in reducing PTSD and comorbid symptoms; 40% of the intention-to-treat sample receiving CPT no longer met criteria for a PTSD diagnosis. They also found that PTSD-related disability status was not associated with the outcomes. This trial provides some of the most encouraging results to date in the treatment of veterans with military-related PTSD.

The balance of wait-list controlled studies have focused on applying evidence-based treatments, or adaptations of them, to different trauma populations, or on addressing important comorbidities. These studies are reviewed later in the section regarding adaptations and comorbidity.

Comparison to Nonspecific Treatments

Several studies have compared active treatments and nonspecific treatments. For example, Blanchard and colleagues (2003) compared CBT to supportive counseling and a wait list in a sample of motor vehicle accident (MVA) survivors. They found that CBT was significantly better than supportive counseling, which was significantly better than wait-list. There was a 0.63 effect size advantage for CBT compared to supportive counseling in those who completed the treatment. Neuner, Schauer, Klaschik, Karunakara, and Elbert (2004) also used supportive counseling, as well as a psychoeducation intervention for control groups, and compared them to their narrative exposure therapy. In their sample of Sudanese refugees, there were no differences among the three conditions at posttreatment. However, narrative exposure therapy was significantly better than both other conditions, which were equal to one another at 1-year follow-up.

In perhaps the largest and methodologically strongest study completed thus far on PTSD treatment, Schnurr and colleagues (2003) conducted a 10-site clinical trial comparing trauma-focused and present-centered group therapy for military veterans with PTSD. There were no statistically significant differences between the two forms of group treatment in the intention-to-treat analyses; however, there was a trend toward better outcomes for the trauma-focused intervention in the sample of patients who received an adequate dose of treatment (i.e., 24 of 30 sessions). There was a slight effect size advantage for trauma-focused group therapy relative to the present-centered approach. Although the authors speculated that the failure of this study to show clini-cally significant changes, unlike most other studies, might be due to the severe and refractory Veterans Administration (VA) population, other explanations cannot be ruled out. Although the treatment included an exposure component, the exposures were not conducted with the intensity usually found in exposure treatments and were implemented in a group setting with therapists who may not have had formal training in CBT.

Treatment Comparison Trials

Of the several head-to-head trials comparing evidence-based treatments for PTSD, and the best of these have used intention-to-treat principles, few have found statistical differ-ences between the treatments. In a large sample of sexual assault victims, Resick and colleagues (2002) found no statistical difference between prolonged exposure (PE) and cognitive processing therapy (CPT), but both showed large improvement compared to the minimal attention control group. CPT was statistically better than PE for some aspects of guilt (intention-to-treat effect size advantages ranging from 0.28 to 0.44 on the global guilt and guilt cognitions subscales). At 9-month follow-up, there were intention-to-treat effect size advantages of CPT compared to PE in clinician ratings of PTSD severity (0.10) and self-reported depression (0.16).

As a follow-up to a study conducted earlier (Tarrier, Pilgrim, et al., 1999) Tarrier and Sommerfield (2004) examined imaginal exposure with cognitive restructuring 5 years posttreatment. The initial study of a mixed-trauma sample found no statistical or effect size differences between the two conditions at posttreatment. Neither group assessed at that period had worsened. However, at 5-year follow-up, there was a 1.12 effect size advantage of cognitive restructuring over imaginal exposure.

Several studies have compared EMDR to various combinations of CBT. Three of these studies report effect size advantages for EMDR in those who complete treatment. Ironson, Freund, Strauss, and Williams (2002) compared EMDR to PE in a relatively small mixed sample and found no statistical differences between the two treatments. However, there was an effect size advantage of 0.65 for EMDR compared to PE in com-pleter analyses. Lee, Gavriel, Drummond, Richards, and Greenwald (2002) compared EMDR to a combination of stress inoculation therapy and PE in another small sample and found no statistical differences between the treatments, with exception of a statisti-cal advantage of EMDR in intrusive symptoms and an overall effect size advantage of 0.62 in completer analyses.

In a large, mixed-trauma sample, Power and colleagues (2002) compared EMDR to a combination of exposure therapy and cognitive restructuring and found no signifi-cant differences. They reported effect size advantages for EMDR in the frequency and intensity of PTSD symptoms clusters. Rothbaum, Astin, and Marsteller (2005), com-pared EMDR, PE, and a wait-list control group in a very well-controlled study of 74 women with rape-related PTSD. They found no differences between the two active con-ditions on PTSD at posttreatment or the 6-month follow-up; both groups improved sub-stantially. They only found differences among the subsample with multiple comor-bidities, in which the EMDR sample with multiple comorbidities did not fare as well as the PE participants with such comorbidity.

However, Devilly and Spence (1999) found a statistical and effect size advantage (*d* = 0.67) for a CBT intervention including exposure and cognitive elements compared to EMDR. These results were maintained at 3-month follow-up assessment. Taylor and colleagues (2003) also found that exposure therapy was statistically superior to EMDR, which was not significantly different from relaxation therapy in a mixed sample. The proportion of individuals achieving clinically significant reductions in their PTSD symptoms (i.e., improvement greater than two standard deviations on clinician assessment of PTSD) was greater in the exposure group.

Ehlers and colleagues (2003) conducted a randomized control trial comparing cognitive therapy, a self-help booklet, or repeated assessments with victims of MVAs after a period of self-monitoring. They found that a small percentage of patients (12%) improved by self-monitoring alone. Those remaining patients with PTSD were randomized into one of the three conditions approximately 3 months after the MVA. Although the 64-page booklet included cognitive-behavioral principles and education about PTSD, it was found to be less effective (as were repeated assessments only) compared to cognitive therapy, which was highly effective (effect sizes > 2.0) and had no dropouts. In contrast, the educational self-help approach did not differ from repeated assessment, although both groups did improve. It should be pointed out, though, that this sample of patients with PTSD could still have been in the natural recovery stage, so the important finding is the clear superiority of the cognitive therapy.

Dismantling and Additive Studies

A few studies have been conducted to determine the active ingredients responsible for treatment gains, or the addition of interventions that can enhance treatment gains. Representative of this type of study, Bryant, Moulds, Guthrie, Dang, and Nixon (2003) compared imaginal exposure, imaginal exposure plus cognitive restructuring, and supportive counseling. They found no significant differences between imaginal exposure only and the combination of imaginal exposure and cognitive restructuring, but both were better than supportive counseling. However, there was an intention-to-treat effect size advantage for the combination over imaginal exposure and supportive counseling ($d = 0.25$ and 0.82, respectively).

Paunovic and Öst (2001) compared a package of cognitive-behavioral interventions, including exposure therapy, cognitive restructuring, and controlled breathing, to exposure therapy only in a relatively small sample of refugees. No statistical differences were found between the two conditions, but there was a small effect size advantage of 0.13 for the combination of interventions.

Most recently, in a sample of female sexual assault survivors treated in either a community rape crisis center or an academic treatment center, Foa and colleagues (2005) found no differences between PE and PE plus cognitive restructuring groups, although both showed marked improvement compared to the wait-list control. They also found no differences in the effects of treatment based on the setting in which treatment was provided.

Adaptations for Different Traumatized Populations and Comorbidities

Since the efficacy of CBT has been established, studies have emerged that examine the addition of treatment components for specific traumatized populations and comorbid disorders. For example, Chard (2005) has developed an adaptation of CPT for victims of child sexual abuse. She and others contend that victims of child sexual abuse have a range of complex posttraumatic sequelae, as well as PTSD symptoms, that need to be addressed to more fully profit from evidence-based PTSD treatment. This adaptation of

CPT includes a combination of group and individual treatment, with the processing of written exposures in the individual treatment and the cognitive interventions occurring primarily in the group context. The treatment protocol also adds modules that focus on developmental issues, communication skills, and social support seeking. In a trial comparing this 17-week treatment to a wait-list control, the treatment was highly efficacious, with a posttreatment effect size of 1.52. There was also evidence that participants continued to improve from posttreatment to 3-month assessment.

In a similar vein, Cloitre, Koenen, Cohen, and Han (2002) proposed that victims of child sexual abuse have problems of affect regulation and interpersonal effectiveness, in addition to their PTSD, that compromise their ability to profit from trauma-focused interventions. Thus, they developed a protocol that included treatment for these problems called STAIR (i.e., skills training in affective and interpersonal regulation) that included treatment for these problems prior to implementing a modification of PE. Compared to a wait-list control condition, this combination of treatment was found to be efficacious, with an effect size of 1.3 at posttreatment. However, as Cahill, Zoellner, Feeny, and Riggs (2004) have pointed out, to establish the need for these additional interventions, STAIR, with a modified version of PE and modified CPT, should now be compared to their original treatments.

Speaking to this issue, Resick, Nishith, and Griffin (2003) conducted a secondary analysis of their treatment outcome data from the comparison of CPT and PE to examine the effects of these short-term treatments on symptoms of complex PTSD. There were no overall differences between the two therapies on a measure that assessed various aspects of complex PTSD, the Trauma Symptom Inventory (TSI; Briere, 1995). The sample was divided into rape victims with (41%) and without (59%) a history of child sexual abuse. With the two forms of treatment collapsed, the authors found no differences between the two groups in PTSD and depression from pre- to posttreatment and 9-month follow-up. On the TSI subscales, participants with a child sexual abuse history did indeed score higher on many of the subscales. They improved markedly and equally with treatment compared to those without such histories, but because they started with higher scores, they ended with higher scores. When the pretreatment scores were covaried out, there were no significant differences between the two groups at the follow-up periods, indicating that even though the participants who had experienced child sexual abuse had more complicated presentations, they too benefited from brief CBT.

Another important treatment target and consideration in delivering PTSD treatment is anger. Attending to anger is important, because many people with PTSD have problems with anger (Kulka et al., 1990), and there is at least some evidence that anger expression, assessed through facial coding, may interfere with engagement in PTSD treatment (Foa, Riggs, Massie, & Yarczower, 1995). However, in a secondary analysis of their study comparing PE, stress inoculation, training, or the combination, Cahill, Rauch, Hembree, and Foa (2003) reported that with self-report measures of anger, pretreatment anger was associated with higher posttreatment PTSD scores, but that anger significantly decreased from pre- to posttreatment. In a sample of Vietnam War veterans with combat-related PTSD and anger problems, Chemtob and colleagues (1997) compared Novaco's (1983) anger treatment to a routine clinical care control. They found significant improvements in anger reaction and anger control in the anger treatment group, and the gains in anger control were sustained at 18-month follow-up. In addition, they found statistically significant improvements in the frequency, and a trend for the intensity, of PTSD reexperiencing symptoms.

Panic disorder and substance use disorders are frequently occurring, problematic conditions associated with PTSD. To address both panic disorder and PTSD, Falsetti, Resnick, Davis, and Gallagher (2001) developed multiple-channel exposure therapy (M-CET), a treatment based on the recognition of the high levels of comorbidity of panic and PTSD symptoms, as well as the potential for panic symptomatology to interfere with tolerating traditional exposure techniques (Falsetti & Resick, 1995, 1997). M-CET draws from CPT, panic control treatment developed by Barlow and Craske (1994), and stress inoculation therapy (SIT; Kilpatrick et al., 1982). In their wait-list controlled trial, the developers found that M-CET was efficacious in treating PTSD and panic disorder. At the end of treatment, 92% of the sample did not meet criteria for PTSD compared to 34% of the control group, and 50% of the treatment group compared to 93% of the control group had not had a panic attack in the past month.

The high rates of comorbidity between PTSD and substance abuse are well documented in epidemiological and treatment-seeking samples (Dansky, Saladin, Brady, Kilpatrick, & Resnick, 1995; Kessler, Sonnega, Bromet, Hughes, & Nelson, 1995). Most PTSD trials to date have excluded participants with substance dependence, and sometimes substance abuse, limiting the generalizability of these findings to this substantial proportion of PTSD sufferers. Several researchers have developed treatments to address these co-occurring disorders (Brady, Danksy, Back, Foa, & Carroll, 2001; Triffleman, Carroll, & Kellogg, 1999); however, Najavits's *Seeking Safety* (2002) is the only treatment that has been subjected to a controlled trial. Hein, Cohen, Miele, Litt, and Capstick (2004) randomized women with both substance use disorders and PTSD diagnoses to seeking safety or to relapse prevention (Carroll, Rounsaville, Gordon, & Nich, 1994), and compared them to a nonrandomized standard community care sample. No statistically significant differences were found between the two active treatments in PTSD and substance use severity, but both were superior to community care at posttreatment assessment. There was an intention-to-treat effect size advantage for relapse prevention (0.25) compared to seeking safety at posttreatment. Both treatments maintained gains in PTSD symptoms at 6- and 9-month follow-up. Relapse prevention maintained gains in substance use at both follow-up periods compared to the community condition. The maintenance of gains for substance use in the seeking safety group were significant and marginally significant at 6- and 9-month follow-up assessments, respectively, compared to the community condition.

Two secondary analyses from clinical trials by Foa and her colleagues have examined the effect of brief PTSD treatments on Axis II disorders (Feeny, Zoellner, & Foa, 2002; Hembree, Cahill, & Foa, 2004). The first study examined the trial data comparing PE, SIT, and SIT + PE, using a standardized clinical interview for personality disorders at the posttreatment assessment. Feeny and colleagues (2002) treated female assault victims who met partial or full criteria for borderline personality disorder (BPD) at posttreatment. Data were collapsed across the three treatment conditions. They found that whereas those with BPD characteristics evidenced poorer end state functioning than those without such characteristics, both groups improved significantly as a result of treatment. Two caveats are the small sample size ($n = 12$) for the BPD group and the fact that the study only assessed for personality disorder with treatment completers at posttreatment. It is unknown whether those with such traits failed to start or complete treatment at a higher rate.

In a second study that examined a larger range of Axis II disorders, Hembree, Cahill, and Foa (2004) examined data from a study that examined PE, cognitive restructuring, or the combination. A second purpose of the study was to compare expert research therapists and community counselors. Hembree and colleagues found that 39% of the 75 participants met criteria for at least one personality disorder, that there were no significant differences in PTSD outcome at posttreatment, and that those without personality disorders tended to have better end state functioning, but that community counselors tended to have better results than the research therapists with comorbid clients. Again, the personality disorder data were collected at posttreatment without regard to dropouts from treatment.

This review of outcome studies, conducted since the guidelines were published in 2000, illustrates the evidence base that has rapidly accumulated, further supporting the use of short-term, goal-directed psychotherapies for PTSD. These studies corroborate that CBT is better than no treatment and nonspecific psychotherapies. The head-to-head trials that have been conducted reveal these active treatments to be generally equivalent to each other, with some possible exceptions in the maintenance of gains and treatment of comorbidities. Dismantling and additive studies during this time suggest a possible advantage of the combination of exposure and cognitive interventions. Traumatized populations and patients with comorbidities that have previously been considered difficult to treat have also been shown to respond to these treatments. These treatments have been shown to produce improvements in childhood sexual abuse victims and multiply traumatized victims, as well as in patients with anger problems, panic disorder, substance use disorders, and personality disorders.

Predictors of Treatment and Outcome

A growing body of studies has examined predictors of PTSD treatment outcome. However, sample sizes in many studies have been too small to examine the variability among participants within groups. As a consequence, the active treatments have frequently been collapsed. Nevertheless, several studies have produced some intriguing findings. Tarrier, Sommerfield, and Pilgrim (1999) found that treatment participants receiving either cognitive therapy or imaginal exposure were affected by the emotional climate within their families (i.e., expressed emotion). Those participants with key relatives who were critical and hostile did not benefit as much from treatment as those with more supportive family members. In a second examination of this trial, Tarrier, Sommerfield, Pilgrim, and Faragher (2000) examined a range of patient, trauma, clinical symptom, and treatment variables. They found that regular therapy attendance, female gender, and low suicide risk were all predictors of better outcome at posttreatment. From pretreatment to 6-month follow-up, the variables most associated with poorer outcome were the number of missed sessions, living alone, and a comorbid generalized anxiety disorder diagnosis. Unlike some other studies, they found no relationship between initial severity scores and treatment outcome. Also interesting was that having an ongoing compensation claim had no effect on treatment.

Hembree, Street, Riggs, and Foa (2004) investigated 73 women who completed Foa and colleagues' (1991) trial comparing PE, SIT, and PE + SIT. Again, the data were combined across treatments to examine predictors. Pretreatment PTSD severity was entered first into a hierarchical regression to see whether other variables would be predictive after accounting for the significant direct relationship between pretreatment and posttreatment severity. A history of childhood trauma and physical injuries from the index assault were both significant predictors of a poorer treatment response.

Using data from their trial comparing various forms of CBT with MVA survivors, Taylor and colleagues (2001) conducted a cluster analysis to examine patterns of response over time. They found two groups: responders and partial responders. In examining these two groups, there were no differences based on a range of demographics, pending litigation, or disability payments. There were, however, some significant pretreatment differences. The partial responders had lower global functioning, greater pain severity, and pain-related interference in daily activities; greater depression and anger; and they were more likely to be using psychotropic medications. The authors suggested that outcomes might be improved by the addition of treatments for pain and depression for those with such preexisting comorbidities.

Van Minnen and her colleagues (van Minnen, Arntz, & Keijsers, 2002; van Minnen & Hagenaars, 2002) conducted two studies of predictors of treatment outcome with exposure treatment. The first study examined a range of predictors with two mixed clinical samples that had somewhat different compositions, possibly explaining why some variables correlated to treatment outcome in one study or the other, but not both. The only stable predictor of posttreatment scores across both samples was severity of PTSD at pretreatment. The second sample included a variable that the first did not—the use of benzodiazepines during treatment, which was related to poorer outcome. They suggested that the use of benzodiazepines might interfere with fear activation during exposures, an element considered to be crucial in exposure treatment.

In the second study, van Minnen and Hagenaars (2002) examined the question of fear activation during exposure treatment. They compared 21 treatment completers who improved and 13 who did not improve on indicators of within-session habituation and between-session habituation at Sessions 1 and 2. They found that those who did not improve had greater anxiety at the start of Session 1, which was assumed to be anticipatory in nature and might have interfered with the clients' accessing their trauma memory. The improvers reported no differences in within-session habituation but more habituation at home, even though there were no differences in the number of times the two groups listened to their trauma accounts. The improvers reported greater decreases in distress between Session 1 and Session 2, consistent with earlier work by Jaycox, Foa, and Morral (1998).

Cloitre, Stovall-McClough, Miranda, and Chemtob (2004) conducted an interesting analysis of therapeutic relationship factors in predicting PTSD outcomes with STAIR/ modified PE in childhood sexual abuse survivors. They found that the quality of the therapeutic relationship in the first phase of treatment predicted reduction in PTSD symptoms after the second phase. This relationship was mediated by the clients' improved capacity to regulate affect.

Therapist effects were also examined in Kubany and colleagues' (2004) study of formerly battered women. They compared the outcome of cases of the four therapists, and found that therapists with and without advanced mental health training produced equivalent outcomes. There was also no difference between male and female therapist outcomes. Of course, the ability to generalize these findings is limited by the small number of therapists involved and lack of therapist random assignment, but it is an important step that such variables were examined.

In summary, efforts to elucidate prognostic factors associated with PTSD treatment outcomes have yielded some promising results. Trauma characteristics, such as a history of childhood sexual abuse and physical violence within the traumatic event, have been associated with diminished outcomes. With regard to patient characteristics, at least one study suggests that women may respond better than men to existing PTSD treatments, and comorbidity (e.g., depression, pain, general anxiety) appears predictive of poorer treatment response. The data regarding the predictive utility of pretreatment severity of PTSD symptoms are mixed, but high levels of anticipatory anxiety at the outset of therapy may impede treatment gains. Like the prediction of nonrecovery from trauma exposure (Brewin, Andrews, & Valentine, 2000), social support appears to facilitate treatment gains. Better treatment attendance and no benzodiazepine use also appear to be associated with greater improvements. Finally, though the amount of advanced therapist training may not be associated with outcomes, the quality of the therapeutic relationship seems to be an important prognostic indicator.

Generalization of Findings

This review of extant research includes some examples of generalization that did not exist when the Foa and colleagues (2000) treatment guidelines book was published. The earliest studies focused primarily on Vietnam War veterans and crime victims, particularly rape victims. The other populations reviewed in this chapter are quite varied and include battered women (Kubany et al., 2004), victims of MVAs (Blanchard et al., 2003) police officers (Gersons et al., 2000), and mixed samples (Bryant, Moulds, Guthrie, Dang, & Nixon, 2003). Paunovic and Öst (2001), who compared their cognitive-behavioral package to exposure therapy in a highly traumatized refugee population, found that although they needed to extend the therapy over more sessions than required for most studies, these therapies could be quite effective.

CHALLENGES FOR THE FUTURE

Three and one-half decades ago, Gordon Paul (1969) proffered the ultimate clinical question: "What treatment, by whom, is most effective for this individual with that specific problem, under which set of circumstances, and how does it come about?" (p. 44). Are we moving closer to this goal? Yes and no. The newer generation of studies has controlled for a greater number of variables that might diminish confidence in results (use of blind assessments, reliability checks on assessments and treatment fidelity, intention-to-treat designs, etc.), and there is good evidence that cognitive-behavioral approaches to treatment are efficacious in remediating PTSD symptoms in many people who have suffered for years. Longer follow-ups indicate that, once treated, clients with PTSD tend not to relapse. Newer studies are beginning to expand the boundaries of PTSD cases to participants who might have been eliminated from clinical trials at one time (e.g., those with substance use disorders, serious mental illness, or Axis II disorders).

In spite of having several very efficacious therapies for PTSD described in this chapter, more work needs to be done to increase the number of people who can profit from these treatments. There is also a need to expand the range of posttraumatic sequelae that are considered treatment targets and outcomes, and to disseminate these treatments better to the patients in need of them. Those who respond to existing evidence-based therapies for PTSD tend to maintain their gains at lengthy follow-up periods. However, as many as 50% of patients in efficacy studies still meet diagnostic criteria for PTSD at the end of treatment and at follow-up periods (Bradley et al., 2005), and some PTSD symptoms (e.g., emotional numbing) have been less responsive to the existing treatments. Also, a range of posttraumatic symptoms beyond the current conceptualization of PTSD may or may not be addressed with these treatments, as well as attention to other functional outcomes such as health, work, and social functioning.

Improving PTSD Outcomes

As revealed in this chapter, the evolution of evidence-based PTSD treatments has gone from establishing several efficacious treatments to comparing them in a series of head-to-head trials designed to determine the best treatment. Keeping in mind that no existing active treatment comparison trials have had sufficient sample sizes to explore subtle differences, these trials have resulted in few differences among the treatments. Aside from needing larger trials, perhaps at multiple sites, there is a need to address the large number of patients who are not responding to the available therapies. The next generation of dismantling studies should reveal important findings about the key ingredients of efficacious treatment. In addition, therapeutic rapport is essential to all forms of psychotherapy. However, these essential but nonspecific factors are just now being studied (e.g., Cloitre et al., 2004). Treatment process studies that pinpoint specific dimensions of the therapeutic relationship that are detrimental or facilitative of trauma recovery may also be helpful in improving PTSD outcomes.

As alluded to previously, PTSD outcomes may be improved by the combination of treatments. Further studies that determine how best to time or integrate treatments will be especially valuable in lieu of mixed evidence that combined treatments do not necessarily lead to better outcomes compared to stand-alone interventions (e.g., Bryant et al., 2003; Foa et al., 1991). The various ways that components can be combined might have varying effectiveness, but again, without sufficient power, no studies will find differences.

These combination studies should also investigate the use and timing of various psychopharmacological agents for greater treatment efficacy. To date, there has not been a head-to-head comparison of drugs and evidence-based treatment, or a combination of the two. Rothbaum and colleagues (2004) have explored serotonergic agents as a first-line treatment, with a course of prolonged exposure for treatment nonresponders. Meta-analyses suggest that evidence-based treatment might be the first-line treatment, with the addition of medications for nonresponders. Psychopharmacological interventions might even be tested as adjunctive potentiators of treatment.

In addition to improving the existing treatments for PTSD, there is room for new and innovative stand-alone and adjunctive treatments for PTSD. As Welch and Rothbaum discuss in Chapter 23, this volume, some of these possibilities include technology-assisted interventions (e.g., virtual reality, Web-based), trauma-specific couple/family treatments, and adjunctive or stand-alone group treatments, that may ultimately be more cost-effective. Continued efforts to elucidate factors associated with treatment response may facilitate treatment innovation and patient matching to improve outcomes. These studies should have an eye toward identifying predictors beyond those that have been traditionally investigated, and potentially examining subtypes of PTSD (Miller, 2003).

Broadening Treatment Targets and Outcomes

PTSD treatment efforts are beginning to benefit from a broader conceptualization of the range of traumatization aftereffects. As reviewed, a few studies have examined this broader range of outcomes, such as affective regulation, sexual dysfunction, personality functioning, guilt, anger, and so forth. Although specific protocols are being tested to address comorbid conditions such as substance abuse, depression, anger, and panic, it would also be useful for studies to better determine whether existing treatments improve these comorbidities.

An additional, important facet of assessing outcomes of psychotherapy for PTSD is the realm of cost-effectiveness and utilization. Especially relative to psychopharmacological intervention, it is important to show that psychotherapy can be quite cost-effective, especially in the long term (Jaycox & Foa, 1999). This avenue of investigation is also ripe in lieu of evidence that PTSD is one of the most costly mental disorders in the United States in terms of health care expenses and job productivity loss (Greenberg et al., 1999; Kessler, 2000).

Biological outcomes are another possible area of outcomes exploration. There are at least case study data showing neurohormonal changes associated with successful PTSD treatment (Griffin, Nishith, Resick, & Yehuda, 1997; Heber, Kellner, & Yehuda, 2002). Neurohormonal assessment may be especially attractive because of the ability to use less invasive and expensive salivary collections. Griffin and Resick (2003) have also noted psychophysiological improvements in people after PTSD treatment. Brain structure and functional assessment, although expensive, would provide important information regarding the etiology of PTSD and possible mechanisms of action involved in PTSD treatment. Similar research on obsessive–compulsive disorder has revealed changes in brain functioning associated with behavioral treatment (e.g., Nakatani et al., 2003).

Disseminating Evidence-Based Treatment

Despite a substantial evidence base for several PTSD treatments, relatively few clinicians use these treatments in their clinical practice. In a recent survey of practicing clinicians, as few as 10% of clinicians used these treatments in their clinical practice (Becker, Zayfert, & Anderson, 2004). Similarly, within the VA, the world's largest provider of PTSD services, fewer than 10% of specialized PTSD providers reported routinely using manualized psychotherapies in their practice (Rosen et al., 2004). In general, more effectiveness studies are needed to determine the robustness and limits of the existing treatments. These studies can also inform training and supervision efforts.

Models of dissemination of evidence-based therapies for PTSD should be further developed and tested, with special attention to the needs of practitioners and the feedback they provide. For example, we have worked with clinicians who are committed to learning and using CBT for PTSD, but are faced with the institutional demand that they schedule patients at 30-minute intervals or provide only group treatment. Thus, dissemination efforts should look beyond the individual clinician to address larger systemic issues that can encourage or dissuade use of evidence-based treatments for PTSD, including logistical issues (e.g., time allowances for training and supervision in the treatment), attitudes toward manualized treatments, or CBT more specifically, and beliefs about the appropriate target of PTSD treatment (e.g., managing symptoms vs. addressing etiology of symptoms). Thinking systemically, graduate school training programs are an important target of dissemination efforts in this critical period of shaping clinicians' practice patterns.

REFERENCES

Alexander, P. C., Neimeyer, R. A., Follette, V. M., Moore, M. K., & Harter, S. (1989). A comparison of group treatments of women sexually abused as children. *Journal of Consulting and Clinical Psychology, 57,* 479–483.

Barlow, D. H., & Craske, M. G. (1994). *Mastery of your anxiety and panic II* (Treatment manual). Albany, NY: Graywind.

Beck, A. T., Ward, C. H., Mendelson, M., Mock, J., & Erbaugh, J. (1961). An inventory for measuring depression. *Archives of General Psychiatry, 4,* 561–571.

Becker, C. B., Zayfert, C., & Anderson, E. (2004). A survey of psychologists' attitudes towards and utilization of exposure therapy for PTSD. *Behaviour Research and Therapy, 42,* 277–292.

Begg, C., Cho, M., Eastwood, S., Horton, R., Moher, D., Olkin, I., et al. (1996). Improving the quality of reporting of randomized controlled trials: The consort statement. *Journal of the American Medical Association, 276*(8), 637–639.

Blanchard, E. B., Hickling, E. J., Devinei, T., Veazey, C. H., Galovski, T. E., Mundy, E., et al. (2003). A controlled evaluation of cognitive behavioral therapy for posttraumatic stress in motor vehicle accident survivors. *Behaviour Research and Therapy, 41,* 79–96.

Boudewyns, P. A., & Hyer, L. (1990). Physiological response to combat memories and preliminary treatment outcome in Vietnam veteran PTSD patients treated with direct therapeutic exposure. *Behavior Therapy, 21,* 63–87.

Boudewyns, P. A., & Hyer, L. A. (1996). Eye movement desensitization and reprocessing (EMDR) as treatment for post-traumatic stress disorder (PTSD). *Clinical Psychology and Psychotherapy, 3,* 185–195.

Bradley, R., Greene, J., Russ, E., Dutra, L., & Westen, D. (2005). A multidimensional meta-analysis of psychotherapy for PTSD. *American Journal of Psychiatry, 162*(2), 214–227.

Brady, K. T., Dansky, B. S., Back, S. E., Foa, E. B., & Carroll, K. M. (2001). Exposure therapy in the treatment of PTSD among cocaine-dependent individuals: Preliminary findings. *Journal of Substance Abuse Treatment, 21,* 47–54.

Brewin, C., Andrews, B., & Valentine, J. D. (2000). Meta-analysis of risk factors for posttraumatic stress disorder in trauma-exposed adults. *Journal of Consulting and Clinical Psychology, 68*(5), 748–766.

Briere, J. (1995). *The Trauma Symptom Inventory (TSI): Professional manual.* Odessa, FL: Psychological Assessment Resources.

Brom, D., Kleber, R. J., & Defares, P. B. (1989). Brief psychotherapy for PTSD. *Journal of Consulting and Clinical Psychology, 57,* 607–612.

Bryant, R. A., Moulds, M. L., Guthrie, R. M., Dang, S. T., & Nixon, R. D. V. (2003). Imaginal exposure alone and imaginal exposure with cognitive restructuring in treatment of posttraumatic stress disorder. *Journal of Consulting and Clinical Psychology, 71*(4), 706–712.

Cahill, S. P., Foa, E., Rothbaum, B., & Resick, P. A. (2004, November). *First do no harm: Worsening or improvement after prolonged exposure.* In A. Maercker & B. Gersons (Chairs), Beyond RCT Research: Evaluation of Common and New Treatment Components. Paper presented at the 20th Annual Meeting of the International Society for Traumatic Stress Studies, New Orleans, LA.

Cahill, S. P., Rauch, S. A., Hembree, E. A., & Foa, E. B. (2003). Effect of cognitive-behavioral treatments for PTSD on anger. *Journal of Cognitive Psychotherapy, 17*(2), 113–131.

Cahill, S. P., Zoellner, L. A., Feeny, N. C., & Riggs, D. S. (2004). Sequential treatment for child abuse-related posttraumatic stress disorder: Methodological comment on Cloitre, Koenen, Cohen, and Han (2002). *Journal of Consulting and Clinical Psychology, 72*(3), 543–548.

Carlson, J. G., Chemtob, C. M., Rusnak, K., Hedlund, N. L., & Muraoka, M. Y. (1998). Eye movement desensitization and reprocessing (EDMR) treatment for combat-related posttraumatic stress disorder. *Journal of Traumatic Stress, 11,* 3–24.

Carroll, K. M., Rounsaville, B. J., Gordon, L. T., & Nich, C. (1994). Psychotherapy and pharmacotherapy for ambulatory cocaine abusers. *Archives of General Psychiatry, 51*(3), 177–187.

Chard, K. M. (2005). An evaluation of cognitive processing therapy for the treatment of posttraumatic stress disorder related to childhood sexual abuse. *Journal of Consulting and Clinical Psychology, 73*(5), 965–971.

Chemtob, C. M., Novaco, R. W., Hamada, R. S., & Gross, D. M. (1997). Cognitive-behavioral treatment for severe anger in posttraumatic stress disorder. *Journal of Consulting and Clinical Psychology, 65*(1), 184–189.

Cloitre, M., Koenen, K. C., Cohen, L. R., & Han, H. (2002). Skills training in affective and interpersonal regulation followed by exposure: A phase-based treatment for PTSD related to childhood abuse. *Journal of Consulting and Clinical Psychology, 70*(5), 1067–1074.

Cloitre, M., Stovall-McClough, K. C., Miranda, R., & Chemtob, C. M. (2004). Therapeutic alliance, negative mood regulation, and treatment outcome in child abuse-related posttraumatic stress disorder. *Journal of Consulting and Clinical Psychology, 72*(3), 411–416.

Cooper, N. A., & Clum, G. A. (1989). Imaginal flooding as a supplementary treatment for PTSD in combat veterans: A controlled study. *Behavior Therapy, 20*, 381–391.

Dansky, B. S., Saladin, M. E., Brady, K. T., Kilpatrick, D. G., & Resnick, H. S. (1995). Prevalence of victimization and PTSD among women with substance use disorders: Comparison of telephone and in-person assessment samples. *International Journal of the Addictions, 30*, 1079–1100.

Devilly, G. J., & Spence, S. H. (1999). The relative efficacy and treatment distress of EMDR and a cognitive-behavior trauma treatment protocol in the amelioration of posttraumatic stress disorder. *Journal of Anxiety Disorders, 13*(1–2), 131–157.

Devilly, G. J., Spence, S. H., & Rapee, R. M. (1998). Statistical and reliable change with eye movement desensitization and reprocessing: Treating trauma within a veteran population. *Behavior Therapy, 29*, 435–455.

Echeburua, E., de Corral, P., Zubizarreta, I., & Sarasua, B. (1996). Treatment of acute posttraumatic stress disorder in rape victims: An experimental study. *Journal of Anxiety Disorders, 10*, 185–199.

Echeburua, E., de Corral, P., Zubizarreta, I., & Sarasua, B. (1997). Psychological treatment of chronic posttraumatic stress disorder in victims of sexual aggression. *Behavior Modification, 21*, 433–456.

Ehlers, A., Clark, D. M., Hackmann, A., McManus, F., Fennell, M., Herbert, C. P., et al. (2003). A randomized controlled trial of cognitive therapy, a self-help booklet, and repeated assessments as early interventions for posttraumatic stress disorder. *Archives of General Psychiatry, 60*, 1024–1032.

Falsetti, S. A., & Resick, P. A. (1995). Causal attributions, depression, and post-traumatic stress disorder in victims of crime. *Journal of Applied Social Psychology, 25*(12), 1027–1042.

Falsetti, S. A., & Resnick, H. (1997). Frequency and severity of panic attack symptoms in a treatment seeking sample of trauma victims. *Journal of Traumatic Stress, 10*(4), 683–689.

Falsetti, S. A., Resnick, H. S., Davis, J., & Gallagher, N. G. (2001). Treatment of posttraumatic stress disorder with comorbid panic attacks: Combining cognitive processing therapy with panic control treatment techniques. *Group Dynamics, 5*(4), 252–260.

Fecteau, G., & Nicki, R. (1999). Cognitive behavioural treatment of post traumatic stress disorder after a motor vehicle accident. *Behavioural and Cognitive Psychotherapy, 27*, 201–214.

Feeny, N. C., Zoellner, L. A., & Foa, E. B. (2002). Treatment outcome for chronic PTSD among female assault victims with borderline personality characteristics: A preliminary examination. *Journal of Personality Disorders, 16*(1), 30–40.

Foa, E. B., Dancu, C. V., Hembree, E. A., Jaycox, L. H., Meadows, E. A., & Street, G. P. (1999). A comparison of exposure therapy, stress inoculation training, and their combination for reducing posttraumatic stress disorder in female assault victims. *Journal of Consulting and Clinical Psychology, 67*, 194–200.

Foa, E. B., Hembree, E. A., Cahill, S. E., Rauch, S. A. M., Riggs, D. S., Feeny, N. C., et al. (2005). Randomized trial of prolonged exposure for posttraumatic stress disorder with and without cognitive restructuring: Outcome at academic and community clinics. *Journal of Consulting and Clinical Psychology, 73*(5), 953–964.

Foa, E. B., Keane, T. M., & Friedman, M. J. (Eds.). (2000). *Effective treatments for PTSD: Practice guidelines from the International Society for Traumatic Stress Studies.* New York: Guilford Press.

Foa, E. B., & Meadows, E. A. (1997). Psychosocial treatments for posttraumatic stress disorder: A critical review. *Annual Review of Psychology, 48*, 449–480.

Foa, E. B., Riggs, D. S., Massie, E. D., & Yarczower, M. (1995). The impact of fear activation and anger on the efficacy of exposure treatment for posttraumatic stress disorder. *Behavior Therapy, 26*(3), 487–499.

Foa, E. B., Rothbaum, B., Riggs, D., & Murdock, T. (1991). Treatment of posttraumatic stress disorder in rape victims: A comparison between cognitive-behavioral procedures and counseling. *Journal of Consulting and Clinical Psychology, 59,* 715–723.

Frank, E., & Stewart, B. D. (1984). Depressive symptoms in rape victims: A revisit. *Journal of Affective Disorders, 1,* 269–277.

Gersons, B. P. R., Carlier, I. V. E., Lamberts, R. D., & van der Kolk, B. A. (2000). Randomized clinical trial of brief eclectic psychotherapy for police officers with posttraumatic stress disorder. *Journal of Traumatic Stress, 13*(2), 333–347.

Greenberg, P. E., Sisitsky, T., Kessler, R. C., Finkelstein, S. N., Berndt, E. R., Davidson, J. R. T., et al. (1999). The economic burden of anxiety disorders in the 1990's. *Journal of Clinical Psychiatry, 60,* 427–435.

Griffin, M. G., Nishith, P., Resick, P. A., & Yehuda, R. (1997). Integrating objective indicators of treatment outcome in posttraumatic stress disorder. In I. R. Yehuda & A. C. McFarlane (Eds.), *Psychobiology of posttraumatic stress disorder* (pp. 388–409). New York: New York Academy of Sciences.

Griffin, M. G., & Resick, P. A. (2003, October). *Psychophysiological responses as treatment outcome indicators in PTSD.* Paper presented at the 19th annual meeting of the International Society for Traumatic Stress Studies, Chicago.

Harvey, A. G., Bryant, R. A., & Tarrier, N. (2003). Cognitive behaviour therapy for posttraumatic stress disorder. *Clinical Psychology Review, 23,* 501–522.

Heber, R., Kellner, M., & Yehuda, R. (2002). Salivary cortisol levels and the cortisol response to dexamethasone before and after EMDR: A case report. *Journal of Clinical Psychology, 58*(12), 1521–1530.

Hein, D. A., Cohen, L. R., Miele, G. M., Litt, L. C., & Capstick, C. (2004). Promising treatments for women with comorbid PTSD and substance use disorder. *American Journal of Psychiatry, 161,* 1426–1432.

Hembree, E. A., Cahill, S. E., & Foa, E. B. (2004). Impact of personality disorders on treatment outcome for female assault survivors with chronic posttraumatic stress disorder. *Journal of Personality Disorders, 18*(1), 117–127.

Hembree, E. A., Street, G. P., Riggs, D. S., & Foa, E. B. (2004). Do assault-related variables predict response to cognitive behavioral treatments for PTSD? *Journal of Consulting and Clinical Psychology, 72*(3), 531–534.

Ironson, G., Freund, B., Strauss, J. L., & Williams, J. (2002). Comparison of two treatments for traumatic stress: A community-based study of EMDR and prolonged exposure. *Journal of Clinical Psychology, 58*(1), 113–128.

Jaycox, L. H., & Foa, E. B. (1999). Cost-effectiveness issues in the treatment of posttraumatic stress disorder. In N. E. Miller & K. M. Magruder (Eds.), *Cost-effectiveness of psychotherapy: A guide for practitioners, researchers, and policymakers* (pp. 259–269). London: Oxford University Press.

Jaycox, L. H., Foa, E. B., & Morral, A. R. (1998). Influence of emotional engagement and habituation on exposure therapy for PTSD. *Journal of Consulting and Clinical Psychology, 66*(1), 185–192.

Jensen, J. A. (1994). An investigation of eye movement desensitization and reprocessing (EMD/R) as a treatment for posttraumatic stress disorder (PTSD) symptoms of Vietnam combat veterans. *Behavior Therapy, 25,* 311–325.

Keane, T. M., Fairbank, J. A., Caddell, J. M., & Zimering, R. T. (1989). Implosive (flooding) therapy reduces symptoms of PTSD in Vietnam combat veterans. *Behavior Therapy, 20,* 245–260.

Keane, T. M., & Kaloupek, D. G. (1982). Imaginal flooding in the treatment of a posttraumatic stress disorder. *Journal of Consulting and Clinical Psychology, 50,* 138–140.

Kessler, R. C. (2000). Post-traumatic stress disorder: The burden to the individual and to society. *Journal of Clinical Psychiatry, 61,* 4–14.

Kessler, R. C., Sonnega, A., Bromet, E., Hughes, M., & Nelson, C. B. (1995). Posttraumatic stress disorder in the National Comorbidity Survey. *Archives of General Psychiatry, 52*(12), 1048–1060.

Kilpatrick, D. G., Veronen, L. J., & Resick, P. A. (1982). Psychological sequelae to rape: Assessment and treatment strategies. In D. M. Doleys, R. L. Meredith, & A. R. Ciminero (Eds.), *Behavioral medicine: Assessment and treatment strategies* (pp. 473–497): Plenum Press.

Kubany, E. S., Hill, E. E., & Owens, J. A. (2003). Cognitive trauma therapy for battered women with PTSD: Preliminary findings. *Journal of Traumatic Stress, 16*(1), 81–91.

Kubany, E. S., Hill, E. E., Owens, J. A., Iannce-Spencer, C., McCaig, M. A., & Tremayne, K. J. (2004). Cognitive trauma therapy for battered women with PTSD (CTT-BW). *Journal of Consulting and Clinical Psychology, 72*(1), 3–18.

Kulka, R. A., Schlenger, W. E., Fairbank, J. A., Hough, R. L., Jordan, B. K., Marmar, C. R., et al. (1990). *Trauma and the vietnam war generation: Report of findings from the national vietnam veterans readjustment study.* New York: Brunner/Mazel.

Lee, C., Gavriel, H., Drummond, P., Richards, J., & Greenwald, R. (2002). Treatment of PTSD: Stress inoculation training with prolonged exposure compared to EMDR. *Journal of Clinical Psychology, 58*(9), 1071–1089.

Marcus, S. V., Marquis, P., & Sakai, C. (1997). Controlled study of treatment of PTSD using EMDR in an HMO setting. *Psychotherapy: Theory, Research, Practice, Training, 34*, 307–315.

Marks, I., Lovell, K., Noshirvani, H., Livanou, M., & Thrasher, S. (1998). Treatment of posttraumatic stress disorder by exposure and/or cognitive restructuring: A controlled study. *Archives of General Psychiatry, 55*, 317–325.

Miller, M. W. (2003). Personality and the etiology and expression of PTSD: A three-factor model perspective. *Clinical Psychology: Science and Practice, 10*, 379–393.

Monson, C. M., Schnurr, P. P., Resick, P. A., Friedman, M. J., Young-Xu, Y., & Stevens, S. P. (2006). Cognitive processing therapy for veterans with military-related posttraumatic stress disorder. *Journal of Consulting and Clinical Psychology, 74*, 898–907.

Najavits, L. M. (2002). *Seeking safety: A treatment manual for PTSD and substance abuse.* New York: Guilford Press.

Nakatani, E., Nakgawa, A., Ohara, Y., Goto, S., Uozumi, N., Iwakiri, M., et al. (2003). Effects of behavior therapy on regional cerebral blood flow in obsessive–compulsive disorder. *Psychiatry Research: Neuroimaging, 124*(2), 113–120.

Neuner, F., Schauer, M., Klaschik, C., Karunakara, U., & Elbert, T. (2004). A comparison of narrative exposure therapy, supportive counseling, and psychoeducation for treating posttraumatic stress disorder in an African refugee settlement. *Journal of Consulting and Clinical Psychology, 72*(4), 579–587.

Novaco, R. W. (1983). *Stress inoculation therapy for anger control: A manual for therapists.* Unpublished manuscript, University of California, Irvine.

Paul, G. L. (1969). Behavior modification research: Design and tactics. In C. M. Franks (Ed.), *Behavior therapy: Appraisal and status* (pp. 29–62). New York: McGraw-Hill.

Paunovic, N., & Öst, L. G. (2001). Cognitive-behavior therapy versus exposure therapy in the treatment of PTSD in refugees. *Behaviour Research and Therapy, 39*, 1183–1197.

Peniston, E. G. (1986). EMG biofeedback-assisted desensitization treatment for Vietnam combat veterans post-traumatic stress disorder. *Clinical Biofeedback and Health, 9*, 35–41.

Peniston, E. G., & Kulkosky, P. J. (1991). Alpha-theta brainwave neuro-feedback therapy for Vietnam veterans with combat-related post-traumatic stress disorder. *Medical Psychotherapy: An International Journal, 4*, 47–60.

Pitman, R. K., Orr, S. P., Altman, B., & Longpre, R. E. (1996). Emotional processing during eye movement desensitization and reprocessing therapy of Vietnam veterans with chronic posttraumatic stress disorder. *Comprehensive Psychiatry, 37*, 419–429.

Power, K., McGoldrick, T., Brown, K., Buchanan, R., Sharp, D., Swanson, V., et al. (2002). A controlled comparison of eye movement desensitization and reprocessing versus exposure plus cognitive restructuring versus wait list in the treatment of post-traumatic stress disorder. *Clinical Psychology and Psychotherapy, 9*, 299–318.

Renfrey, G., & Spates, C. R. (1994). Eye movement desensitization: A partial dismantling study. *Journal of Behavior Therapy and Experimental Psychiatry, 25*, 231–239.

Resick, P. A., Jordan, C. G., Girelli, S. A., & Hutter, C. K. (1988). A comparative outcome study of behavioral group therapy for sexual assault victims. *Behavior Therapy, 19*, 385–401.

Resick, P. A., Jordan, C. G., Girelli, S. A., Hutter, C. K., & Marhoeder-Dvorak, S. (1988). A comparative outcome study of behavioral group therapy for sexual assault victims. *Behavior Therapy, 19*, 385–401.

Resick, P. A., Nishith, P., & Griffin, M. G. (2003). How well does cognitive-behavioral therapy treat symptoms of complex PTSD?: An examination of child sexual abuse survivors within a clinical trial. *CNS Spectrums, 8*, 340–355.

Resick, P. A., Nishith, P., Weaver, T. L., Astin, M. C., & Feuer, C. A. (2002). A comparison of cognitive processing therapy, prolonged exposure and a waiting condition for the treatment of posttraumatic stress disorder in female rape victims. *Journal of Consulting and Clinical Psychology, 70,* 867–879.

Resick, P. A., & Schnicke, M. K. (1992). Cognitive processing therapy for sexual assault victims. *Journal of Consulting and Clinical Psychology, 60,* 748–756.

Rosen, C. S., Chow, H. C., Finney, J. F., Greenbaum, M. A., Moos, R. H., Sheikh, J. I., et al. (2004). Practice guidelines and VA practice patterns for treating posttraumatic stress disorder. *Journal of Traumatic Stress, 17,* 213–222.

Rothbaum, B. O. (1997). A controlled study of eye movement desensitization and reprocessing in the treatment of posttraumatic stress disordered sexual assault victims. *Bulletin of the Menninger Clinic, 61,* 317–334.

Rothbaum, B. O., Astin, M. C., & Marsteller, F. (2005). Prolonged exposure versus eye movement desensitization and reprocessing (EMDR) for PTSD rape victims. *Journal of Traumatic Stress, 18*(6), 607–616.

Rothbaum, B. O., Foa, E. B., Davidson, J. R. T., Cahill, S. P., Compton, J., Connor, K., et al. (2004, May). *Augmentation of sertraline with prolonged exposure in the treatment of PTSD.* Poster presented at the annual meeting of the American Psychiatric Association, New York.

Rounsaville, B. J., Carroll, K. M., & Onken, L. S. (2001). A stage model of behavioral therapies research: Getting started and moving from stage I. *Clinical Psychology: Science and Practice, 8*(2), 133–142.

Scheck, M. M., Schaeffer, J. A., & Gillette, C. (1998). Brief psychological intervention with traumatized young women: The efficacy of eye movement desensitization and reprocessing. *Journal of Traumatic Stress, 11,* 25–44.

Schnurr, P. P., Friedman, M. J., Foy, D. W., Shea, M. T., Hsieh, F. Y., Lavori, P. W., et al. (2003). Randomized trial of trauma-focused group therapy for posttraumatic stress disorder: Results from a Department of Veterans Affairs cooperative study. *Archives of General Psychiatry, 60*(5), 481–489.

Silver, S. M., Brooks, A., & Obenchain, J. (1995). Treatment of Vietnam War veterans with PTSD: A comparison of eye movement desensitization and reprocessing, biofeedback, and relaxation training. *Journal of Traumatic Stress, 8,* 337–342.

Tarrier, N., Pilgrim, H., Sommerfield, C., Faragher, B., Reynolds, M., Graham, E., et al. (1999). A randomized trial of cognitive therapy and imaginal exposure in the treatment of chronic posttraumatic stress disorder. *Journal of Consulting and Clinical Psychology, 67*(1), 13–18.

Tarrier, N., & Sommerfield, C. (2004). Treatment of chronic PTSD by cognitive therapy and exposure: 5-year follow-up. *Behavior Therapy, 35,* 231–246.

Tarrier, N., Sommerfield, C., & Pilgrim, H. (1999). Relatives' expressed emotion (EE) and PTSD treatment outcome. *Psychological Medicine, 29*(4), 801–811.

Tarrier, N., Sommerfield, C., Pilgrim, H., & Faragher, B. (2000). Factors associated with outcome of cognitive-behavioural treatment of chronic post-traumatic stress disorder. *Behaviour Research and Therapy, 38,* 191–202.

Taylor, S., Fedoroff, I. C., Koch, W. J., Thordarson, D. S., Feactau, G., & Nicki, R. M. (2001). Posttraumatic stress disorder arising after road traffic collisions: Patterns of response to cognitive-behavior therapy. *Journal of Consulting and Clinical Psychology, 69*(3), 541–551.

Taylor, S., Thordarson, D. S., Maxfield, L., Fedoroff, I. C., Lovell, K., & Orgodniczuk, J. (2003). Comparative efficacy, speed, and adverse effects of three PTSD treatments: Exposure therapy, EMDR, and relaxation training. *Journal of Consulting and Clinical Psychology, 71*(2), 330–338.

Triffleman, E., Carroll, K., & Kellogg, S. (1999). Substance dependence posttraumatic stress disorder therapy: An integrated cognitive-behavioral approach. *Journal of Substance Abuse Treatment, 17*(1–2), 3–14.

van Minnen, A., Arntz, A., & Keijsers, G. P. J. (2002). Prolonged exposure in patients with chronic PTSD: Predictors of treatment outcome and dropout. *Behaviour Research and Therapy, 40,* 439–457.

van Minnen, A., & Hagenaars, M. (2002). Fear activation and habituation patterns as early process predictors of response to prolonged exposure treatment in PTSD. *Journal of Traumatic Stress, 15*(5), 359–367.

Vaughan, K., Armstrong, M. S., Gold, R., & O'Connor, N. (1994). A trial of eye movement desensitization compared to image habituation training and applied muscle relaxation in post-traumatic stress disorder. *Journal of Behavior Therapy and Experimental Psychiatry, 25,* 283–291.

Watson, C. G., Tuorila, J. R., Vickers, K. S., Gearhart, L. P., & Mendez, C. M. (1997). The efficacies of three relaxation regimens in the treatment of PTSD in Vietnam War veterans. *Journal of Clinical Psychology, 53,* 917–923.

Wilson, S. A., Becker, L. A., & Tinker, R. H. (1995). Eye movement desensitization and reprocessing (EMDR) treatment for psychologically traumatized individuals. *Journal of Consulting and Clinical Psychology, 63,* 928–937.

Zlotnick, C., Shea, T. M., Rosen, K., Simpson, E., Mulrenin, K., Begin, A., et al. (1997). An affect-management group for women with posttraumatic stress disorder and histories of childhood sexual abuse. *Journal of Traumatic Stress, 10,* 425–436.

Chapter 18

Psychosocial Approaches for Children with PTSD

Glenn N. Saxe, Helen Z. MacDonald, and B. Heidi Ellis

Childhood trauma and posttraumatic stress disorder (PTSD) are significant public health problems. Among 500 elementary and middle school children in an inner-city community, 30% witnessed a stabbing and 26% witnessed a shooting (Bell & Jenkins, 1993). Almost 4,000 students were surveyed in six schools in two states. Among males, 3–33% reported being shot or shot at, and 6–16% reported being attacked with a knife. Among females, there were lower reported rates of victimization except for sexual abuse or assault (Singer, Anglin, & Song, 1995). The prevalence of PTSD in children exposed to trauma ranges from 20 to 30% depending on the study or trauma type. Accordingly, there is a great need to develop effective interventions for children with PTSD. A growing scientific literature has identified a number of promising approaches. This chapter offers a critical review of this scientific literature. One aspect of this literature that bears highlighting is the limited way that empirically supported interventions have penetrated frontline community and clinic practice (Hoagwood, Burns, Kiser, Ringeisen, & Schoenwald, 2001; Hoagwood & Olin, 2002).

CURRENT STATE OF THE ART

Cognitive-Behavioral Therapy

Cognitive-behavioral therapy (CBT) is among the most widely investigated treatments for childhood traumatic stress and has been recommended as the first-line treatment for childhood PTSD (American Academy of Child and Adolescent Psychiatry, 1998).

Furthermore, compared with other psychosocial interventions, CBT has thus far the strongest empirical support for its efficacy in treating children and adolescents with PTSD (Cohen, Mannarino, Berliner, & Deblinger, 2000).

Individual Cognitive-Behavioral Therapy

Individual CBT for childhood PTSD involves several distinct components, including skills training, psychoeducation, cognitive coping, stress management, muscle relaxation, thought stopping, exposure-based exercises, and relapse prevention. Additionally, most efficacious child treatments include cognitive and behavioral parental components (Cohen & Mannarino, 1993; Cohen et al., 2000; Saigh, Yule, & Inamdar, 1996). Typically, treatments begin with psychoeducation, stress management, and skills training modules, which equip children with strategies to cope with the anxiety associated with the behavioral exposures to follow in treatment. Individual CBT is flexible, because it can be catered to the child's specific traumatic history and can address issues related to the child's particular home environment.

PARENT TRAINING

Parent training is a critical component of psychotherapeutic treatment for younger traumatized children, and a secondary component of treatment for older children and adolescents with PTSD. Parent training typically consists of psychoeducation, behavior modification, and exposure-based interventions. Through psychoeducation, parents are taught about the impact of traumatic experiences on children, as well as the symptoms that traumatized children are likely to experience. Behavior modification helps parents to learn techniques to manage their child's behavior better. This component allows parents to work in conjunction with the therapist, because parents learn how to reinforce lessons the child learns in therapy. The exposure-based module of parent training also gives parents an opportunity to express their own feelings regarding their child's traumatic experience, and related behaviors and emotions. Parents may learn stress management skills to facilitate their own coping (Cohen et al., 2000).

EXPOSURE-BASED INTERVENTIONS

Exposure therapy techniques, which have long been empirically supported for the treatment of PTSD symptoms in adults (Foa, Dancu, & Hembree, 1999; Foa, Rothbaum, Riggs, & Murdock, 1991), have been extended to treat children with histories of trauma exposure (Deblinger & Heflin, 1996; Deblinger, Lippmann, & Steer, 1996). Exposure-based interventions are focused on safe, structured exposure of the patient to stimuli reminiscent of the trauma to extinguish the traumatic stress response. Several models of therapeutic exposure have been implemented with children. In gradual exposure, the child is encouraged verbally to recount aspects of the trauma to which he or she was exposed, while the therapist assists the child in processing general or minor elements of the traumatic event (Cohen & Mannarino, 1993; Deblinger & Heflin, 1996). As gradual exposure therapy progresses, the treatment works to reduce the child's distress surrounding the trauma and its reminders through various media, including writing stories, drawing pictures, talking into a tape recorder, and playing out aspects of the traumatic experience. As memories of the trauma become less activating, the therapist encourages the child to access, discuss, and process increasingly anxiety-provoking traumatic memories in the context of the safe therapeutic environment.

In contrast to gradual exposure therapy, in imaginal flooding therapy, children imagine specific details of the traumatic event, while the therapist continually monitors the child's subjective ratings of emotional distress (Saigh et al., 1996). The theory underlying exposure therapy techniques suggests that if children reexperience aspects of the traumatic event in a controlled and safe environment, the memories associated with the trauma will no longer be connected to distressing physiological and emotional reactions. Unpairing traumatic memories and disturbing affect can result in increased habituation and decreased avoidance symptoms.

COGNITIVE INTERVENTIONS

Generally, cognitive interventions can serve to challenge thoughts that maintain traumatic stress responses (Resick & Schnicke, 1992). These techniques are useful when employed to correct cognitive errors and to equip children with more adaptive coping skills. Children with traumatic stress histories often demonstrate cognitive distortions, including thoughts of a foreshortened future, self-blame, survivor guilt, and overgeneralization of the traumatic event.

Cognitive coping, a specific cognitive intervention that has been modified by Deblinger and Heflin (1996) from Beck's (1976) cognitive treatment for depression, has been adapted specifically for the treatment of traumatized children and teaches patients about the relationship between maladaptive automatic thoughts, negative emotional states, and dysfunctional behaviors. Through cognitive coping, children learn emotion regulation skills by simultaneously altering negative cognitions and assessing the impact on behavior.

Empirical treatment studies examining the efficacy of CBT for the treatment of childhood PTSD have greatly increased over the last decade. The findings of this research generally support treatments that employ a combination of the cognitive and behavioral strategies described earlier in the treatment of traumatized children. These studies are described below.

Saigh (1987a, 1987b, 1987c, 1989) made significant contributions to the childhood PTSD treatment literature with his innovative, single-case study series of children and adolescents with PTSD resulting from war trauma exposure in Lebanon. Saigh successfully employed imaginal or *in vivo* flooding therapy in the treatments of five female and three male Lebanese participants, ranging from 10 to 14 years of age. In these treatments, participants identified and described their traumatic experience over a multiple-baseline design. Saigh employed self-reported PTSD, anxiety, and depression outcome measures. These cases were associated with a reduction in PTSD and other trauma-related symptoms, including exaggerated startle response, nightmares, intrusive thoughts, avoidance, impaired concentration and memory, anxiety, depression, and guilt (Saigh et al., 1996). This important series of studies represents the only empirical literature to date demonstrating the efficacy of flooding treatments for traumatized children and adolescents. The results of these studies indicate that exposure treatments for childhood PTSD must be further examined with larger, more heterogeneous samples exposed to various types of trauma.

Celano, Hazzard, Webb, and McCall (1996) compared individual CBT and treatment-as-usual (i.e., supportive, unstructured psychotherapy) conditions. Thirty-two sexually abused girls were randomized to one of the two interventions. Participants were between ages 8 and 13 years. Seventy-five percent were African American, 22% were European American, and 3% were Hispanic. Results demonstrated that children in both groups showed significant decreases in both PTSD symptoms, and internalizing and externalizing symptoms. The experimental group did not show differential treatment effects, however.

The Celano and colleagues (1996) study has methodological shortcomings that somewhat limit the conclusions to be drawn from this study. Although the study employed random assignment to treatment condition, implemented a manualized treatment, and used standardized outcome instruments, the sample was small, the dropout rate was 35%, and there were no follow-up assessments. Additionally, PTSD diagnostic information was not collected, and the psychometric strength of the PTSD symptom instrument was questionable.

Deblinger and colleagues (Deblinger, McLeer, & Henry, 1990; Deblinger et al., 1996; Deblinger, Steer, & Lippmann, 1999) have developed a trauma-focused CBT for sexually abused children and their nonoffending caretakers. Deblinger's treatment employs gradual exposure therapy in combination with other CBT techniques, including modeling, psychoeducation, coping, and prevention skills training. The caretaker treatment focuses on behavioral management training. In their first investigation of the efficacy of their treatment, Deblinger and colleagues (1990) treated children and their caretakers individually for 12 structured treatment sessions, and 19 female children between ages 3 and 16 years who met full criteria for PTSD based on an interview they had developed. After 12 weeks of treatment, children's symptoms of PTSD, depression, and anxiety were significantly diminished. None of the children met criteria for PTSD, and parents reported that behavioral problems had greatly diminished. This study suggests that individual child–caregiver CBT holds promise in treating sexually abused children. The study used psychometrically strong self-report instruments and a standardized treatment protocol. Weaknesses of this study, however, include the lack of follow-up assessment.

In a larger, four-condition treatment study Deblinger and colleagues (1996, 1999) randomly assigned 100 sexually abused children to one of the following conditions: standard community care (no treatment provided; families provided with referral information), child CBT, nonoffending parent CBT, or child and parent CBT. The sample comprised children between ages 7 and 13, the majority of whom were female (83%). Seventy percent of the participants were white, 21% were black, 7% were Hispanic, and 2% represented other ethnic backgrounds. Results demonstrated that children who received treatment in either the child CBT or child and parent CBT conditions improved significantly across all PTSD symptom clusters from pre- to posttreatment, regardless of parent involvement, compared to children in the parent CBT or standard community care conditions. When parents received treatment, regardless of the child's involvement in treatment, children reported less severe depression symptoms, and parents reported greater parenting skills and fewer child behavior problems. Treatment gains were maintained at a 2-year follow-up (Deblinger et al., 1999).

The use of manualized treatments, independent evaluators to assess outcome, and standardized instruments, including structured clinical interviews assessing PTSD, together represent important strengths of this study. Weaknesses include the inclusion of children below threshold for a PTSD diagnosis; the heterogeneity of the community treatment condition, which included both families that sought treatment and those who did not; the limited sensitivity of some of the instruments used; and significant missing data in the follow-up study.

Cohen and Mannarino (1996) also developed a highly influential CBT treatment model for sexually abused children. Although the components of the Cohen and Mannarino and the Deblinger and colleagues (1996, 1999) treatments overlap considerably, the Deblinger and colleagues treatment lends greater emphasis to exposure-based interventions, whereas Cohen and Mannarino (1996) focus on cognitive interventions. Cohen and Mannarino's trauma-focused CBT incorporates modules of relationship–social skills building, cognitive restructuring, thought stopping, positive imagery, contingency reinforcement, and problem-solving, self-monitoring of behaviors. In their study of sexually abused children, Cohen and Mannarino (1998) randomized forty-nine 7- to 14-year-old children to 12 sessions of a trauma-focused CBT intervention or to a nondirective supportive therapy. Sixty-nine percent of treatment completers were female; 59% of the participants were European American, 37% were African American, 2% were Hispanic, and 2% were biracial. Findings revealed that children in the sexual abuse–specific CBT condition evidenced lower depression and higher social competency scores compared with children in the nondirective control group following treatment. A 1-year follow-up study indicated that the children who completed the CBT treatment demonstrated significantly lower PTSD and dissociative symptoms (Cohen, Mannarino, & Knudsen, 2005) at follow-up. Strengths of these studies include their use of manualized treatments, random assignment, standardized instruments, and blinded treatment evaluation. Weaknesses of these studies include a differentially higher dropout rate in the nondirective control group and the lack of a structured diagnostic PTSD instrument.

In their most comprehensive efficacy study to date, Cohen, Deblinger, Mannarino, and Steer (2004) compared trauma-focused CBT with child-centered therapy in a two-site treatment of 203 sexually abused 8- to 14-year-old children with high levels of PTSD symptoms. Seventy-nine percent of the sample was female, and 60% of the participants were European American, 28% were African American, 4% were Hispanic American, 7% were biracial, and 1% represented another racial category. One treatment site was a large metropolitan area, whereas the other was a suburban community. The control treatment, child-centered therapy, is a manualized treatment that aims to establish a trusting therapeutic relationship focused on self-affirmation, empowerment, and validation for parent and child. This treatment is child-directed, in that the therapist allows the child to initiate discussion of trauma. Therapeutic strategies employed in child-centered therapy included active listening, reflection, empathy, and discussion of feelings.

Results of this study revealed that, compared to the child-centered therapy condition, the trauma-focused CBT condition was associated with greater improvement in PTSD, depression, behavior problems, shame, and abuse-related attributions. Furthermore, compared to parents in the child-centered therapy condition, parents in the trauma-focused CBT condition evidenced improvement in the areas of depression, abuse-specific distress, support of the child, and effective parenting strategies. This study lends further support to the efficacy of trauma-focused CBT for the treatment of sexually abused children. This study has many strengths, including the large, diverse, and multiply traumatized sample; the use of structured diagnostic interviews and manualized treatments and independent evaluators blinded to treatment condition; and random assignment to treatment.

An additional study examining CBT treatment for childhood PTSD has also demonstrated positive results. In their treatment outcome study targeting sexually abused children, King and colleagues (2000) assigned 36 sexually abused children to either a child CBT condition, a family CBT condition, or a wait-list condition. They were particularly interested in determining the specific effects of parental involvement in treatment. Children in the study ranged from ages 5 to 17 years; 69% of the sample was female. Racial breakdown of the sample was not reported. The child CBT condition comprised graded exposures, psychoeducation, coping skills training, relaxation training, cognitive therapy, behavioral rehearsal, and relapse prevention. This treatment condition relied heavily on Deblinger and colleagues' (Deblinger et al., 1990; Deblinger & Heflin, 1996) child trauma–focused CBT model. In the family CBT condition, children and their mothers each received individual psychotherapy sessions. The child component of the family treatment condition mirrored that of the individual child condition, and the parental component was based on the models by Cohen and Mannarino (1996) and Deblinger and Heflin (1996). Mothers in the family treatment condition worked on improving communication skills, reducing avoidance, increasing behavioral management skills, and improving self-monitoring. Children in the two active treatment conditions improved significantly over children in the wait-list condition on all PTSD symptom clusters and on self-reported fear and anxiety posttreatment and at follow-up. The child CBT and family CBT conditions exhibited no differences in terms of children's PTSD symptom presentation. Although parent involvement in treatment was not found to impact treatment outcome, no conclusions can be drawn; the samples were too small to detect small effects.

This study provides further supports for the efficacy of CBT in the treatment of childhood PTSD secondary to sexual abuse. Strengths of this study include the administration of standardized self-report instruments, and use of manualized treatments and treatment integrity checks by independent evaluators. The relatively small sample size, the inclusion of children not meeting PTSD diagnostic criteria, the short follow-up period, as well as the use of study therapists to fill out clinician-based outcome assessments, possibly leading to examiner bias, represent significant weaknesses of this study.

Group Cognitive-Behavioral Therapy

In addition to individual CBT, the efficacy of group CBT interventions for childhood traumatic stress have demonstrated promising results (Chemtob, Nakashima, & Carlson, 2002; Goenjian et al., 1997; March, Amaya-Jackson, Murray, & Schulte, 1998; Stein et al., 2003) and are reviewed below. Group CBT interventions also include cognitive and behavioral exercises, including psychoeducation, cognitive therapy, exposure-based behavior therapy, and relapse prevention (March et al., 1998). The specific goal of group trauma-focused CBT is for children to integrate traumatic memories within their self-concept through group-based exposures and cognitive exercises. The unique strength of group treatments is in their ability to provide treatment to greater numbers of children representing underserved populations (e.g., many group CBT interventions occur in school settings).

One such treatment is the multimodality trauma treatment (MMTT; March et al., 1998), a manualized, trauma-focused, cognitive-behavioral group treatment for children who develop PTSD following a single-incident stressor. This 18-week treatment is delivered in a peer group format in the school setting, with individual "pullout" sessions aimed at helping children with issues particular to their specific traumatic history. March and colleagues implemented a single-case, across-setting experimental design to examine the efficacy of this treatment. The sample comprised 14 subjects between 10 and 15 years of age. Approximately 67% of the sample was female. Forty-seven percent of the participants were European American, 41% were African American, and fewer than 1% were Asian American and Native American. Children reported experiencing a range of traumatic stressors, including car accidents, severe storms, accidental injury, severe illness, accidental and criminal gunshot injury, and fires. Results indicated significant improvement across all symptom clusters in children treated with MMTT, including PTSD, anxiety, and depression. These results were maintained at 6-month follow-up. The application of this treatment has been extended to community mental health centers (Amaya-Jackson, Reynolds, & Murray, 2003).

This study has important strengths, including the use of a diagnostic interview to establish PTSD caseness; the employment of psychometrically sound outcomes measures both pretreatment, posttreatment, and at follow-up; and the implementation of a manualized treatment. The small sample size, lack of a control group, and exclusion of children with clinically significant behavior problems, however, significantly limit the generalizability of this study.

Whereas the March and colleagues (1998) treatment is designed for children exposed to a single, specific traumatic event, other group treatments have been developed for inner-city children exposed to high levels of community and domestic violence. Stein and colleagues (2003) conducted a randomized, controlled trial of their Cognitive-Behavioral Intervention for Trauma in Schools group treatment program. Their investigation of the efficacy of this treatment included 116 sixth-grade, inner-city middle school children from socioeconomically disadvantaged communities in Los Angeles. At baseline, 58% of the sample was female, and the authors reported that the participants were drawn from largely Latino communities. Children with extreme histories of exposure to violence and significant levels of PTSD symptoms were included in the study; 76% of the sample reported experiencing or witnessing violence involving a knife or gun. Participants were randomized to either the 10-session CBT early intervention group, or to a wait-list, delayed-intervention comparison group. Participants in the treatment learned CBT skills through age-appropriate activities, didactic instruction, examples, and homework. Results showed that at 3 months posttreatment, students assigned to the early intervention group demonstrated lower scores on symptoms of PTSD, depression, and psychosocial dysfunction. Overall, this study suggests that a brief, standardized CBT group intervention may significantly decrease symptoms of PTSD and depression in traumatized children within a school setting.

This study contributes importantly to the field, because it is the only randomized, controlled trial of a PTSD treatment for children exposed to community violence (Stein et al., 2003). Furthermore, this study is unique because of its relatively heterogeneous traumatized sample, including children who had been personally exposed to a range of types of violent acts, as well as children who had witnessed violence. Stein and colleagues (2003) implemented a manualized treatment protocol, standardized outcome instruments, and treatment fidelity measures, all of which represent important strengths of the study. Furthermore, the treatment was tested with a largely disadvantaged, minority sample, which is unique in this literature. In addition, the study provides a model for conducting treatment outcome research within a community setting. However, the lack of blinded evaluators or diagnostic PTSD interviews, the inability to compare the study treatment to an alternative intervention, and the short follow-up period are weaknesses of this study. Nonetheless, this study has important implications for the effectiveness and feasibility of school-based interventions targeting PTSD and depression symptoms secondary to violence exposure in children.

In a recently developed brief CBT treatment targeting postdisaster trauma symptoms in children, Chemtob and colleagues (2002) evaluated the efficacy of a school-based screening and psychosocial intervention for school-age children evidencing trauma symptoms 2 years after exposure to a hurricane. Two hundred forty-eight children ranging in age from 6 to 12 years were randomly assigned to four sessions of manualized group or individual treatment. Thirty percent of the participants were Hawaiian or part-Hawaiian, 25% were white, 20% were Filipino, and 9% were Japanese. Treatment consisted of increasing children's sense of safety, grieving over losses, improving attachments, expressing anger, and working toward closure. Following treatment, children in both groups reported lower levels of trauma-related symptoms, a finding that was maintained at a 1-year follow-up. There were no differences in symptom levels between group and individual treatment, although children in the group treatment condition were more likely than those in the individual treatment condition to complete treatment.

This study is innovative in its focus on postdisaster trauma symptoms in children. The study's large sample size, randomization of participants to treatment waves, manualization of treatments, and population-based selection of participants all represent impressive methodological strengths. A caveat, however, is for the inclusion of a less symptomatic control group against which to compare the efficacy of treated groups.

Goenjian and colleagues (1997) examined the efficacy of a school-based CBT intervention for 64 sixth and seventh graders who exhibited PTSD symptoms 1.5 years after living through the 1988 Armenian earthquake. The treatment group received a combination of group and individual therapy, components of which included processing traumatic material, correcting cognitive distortions, developing coping techniques, and managing affect. Symptoms of PTSD and depression were examined 1.5 and 3 years after the earthquake in a group of treated and untreated children. Results indicated that children who received treatment demonstrated reduced PTSD symptoms, whereas children who did not demonstrated increased PTSD and depression symptoms.

Although these results are compelling, methodological shortcomings limit the generalizability of this study. Specifically, though the study used psychometrically sound outcome instruments, no diagnostic measures were used. Furthermore, children were not randomly assigned to treatment, and there were no blind evaluators to assess treatment outcome.

The four studies we have reviewed provide initial support for the use of group CBT in the treatment of childhood PTSD (Chemtob et al., 2002; Goenjian et al., 1997; March et al., 1998; Stein et al., 2003). More research in this area is needed, however. Specifically, a large-scale randomized, controlled trial examining the efficacy of group CBT for childhood PTSD would contribute importantly to the literature. Although the study by Stein and colleagues provides important support for the effectiveness of group CBT for Latino children, there remains a need for further studies evaluating the effectiveness of group interventions with heterogeneous populations of multiply traumatized children with comorbid mental health concerns.

Eye Movement Desensitization and Reprocessing

Recent empirical studies have begun to explore the efficacy of eye movement desensitization and reprocessing (EMDR) for the treatment of PTSD in children. The theoretical underpinnings of EMDR suggest that traumatic memories can be processed neurophysiologically with dual attention tasks (Smith & Yule, 1999). In EMDR, the therapist induces rapid eye movements in the patient, who is guided through imaginal exposure of a traumatic memory. The patient reports images, thoughts, and feelings that arise as a result of each set of eye movements. The therapist continues guiding eye movements in conjunction with imaginal exposure and cognitive restructuring, until the patient's negative affect diminishes (Muris & Merckelbach, 1999). Although the efficacy of EMDR is far less well-established than CBT in the treatment of traumatized children, recent research indicates that EMDR may have utility as a treatment for childhood PTSD (Greenwald, 1998)

Chemtob and colleagues (2002) reported on the use of EMDR to treat disaster-related PTSD symptoms in 32 children whose symptoms had not improved 3.5 years after exposure to a hurricane and 1 year after receiving a school counselor–administered treatment. Children were randomly assigned to a three-session early or delayed EMDR intervention. Sixty-nine percent of the participants were female. The children ranged in age from 6 to 12 years. Thirty percent of the participants were Hawaiian and part-Hawaiian, 28% were Filipino, 12% were Japanese, and 19% were white. Participants reported significantly reduced symptoms of PTSD, depression, and anxiety. These improvements were maintained at 6-month follow-up. This is the only empirical study to date examining EMDR in treating traumatized children. Strengths of this study include random assignment, manualized treatment, and independent evaluators measuring outcomes. Weaknesses of this study include the absence of treatment fidelity ratings, a diagnostic interview for PTSD, and a control group.

In another recent study, Jaberghaderi, Greenwald, Rubin, Zand, and Dolatabadi (2004) compared CBT and EMDR interventions in 12- and 13-year-old sexually abused Iranian girls. This treatment randomized 14 girls to 12 sessions of either CBT or EMDR. The CBT intervention, based on Deblinger and Heflin's treatment (1996), focused on skills development and exposure to traumatic memories. The EMDR intervention was based on Shapiro's treatment (1995), but had a more limited focus on skills development, and processing of traumatic memories was not as stringent. Results suggested that both treatments significantly decreased levels of PTSD symptoms, with no differences between the groups. EMDR was, however, more efficient, with patients in the EMDR group completing the treatment in fewer sessions than those in the CBT group. This study has significant shortcomings limiting generalizability. First, the study's small sample significantly curtailed statistical power to detect differences between groups. Further methodological weaknesses include the lack of diagnostic interviews, treatment fidelity measures, and standardized outcome instruments.

Although the literature examining the efficacy of EMDR for children is sparse, research suggests that it may hold promise for the treatment of PTSD in children. More research is needed to evaluate these questions in larger, more representative samples. Furthermore, future studies should attempt to tease apart the unique effects of EMDR versus CBT interventions in the treatment of traumatized children.

Psychodynamic Approaches

The aim of psychodynamic treatments addressing childhood PTSD is to help diminish traumatized children's overwhelming affect through the conscious identification of the child's unconscious expression of conflicts and emotions related to the trauma. Psychodynamic theory explicitly uses the therapeutic relationship as a means of understanding the unconscious expression of trauma-related emotions, then uses this symbolic expression of traumatic relationships within the therapeutic relationship to help the child to contain these emotions. Research has shown that a significant number of practitioners use psychodynamic and nondirective play and art therapies in treating children with PTSD (Cohen, Mannarino, & Rogal, 2001). Although support for the use of psychodynamic therapies in the treatment of childhood PTSD comes primarily from theoretical and case study material (Byers, 1996; Gil, 1991; Mallay, 2002; Osofsky, Cohen, & Drell, 1995; Peri, 2004), two empirical studies to date have evaluated the efficacy of psychodynamic treatment for PTSD in children.

In one such study, Downing, Jenkins, and Fisher (1988) compared psychodynamic and reinforcement interventions in the treatment of sexually abused children and their parents. The sample comprised 22 children between 6 and 12 years of age. Both treatments were associated with positive outcomes: Reinforcement treatment was more efficacious in causing and maintaining behavior change, whereas psychodynamic treatment demonstrated more gradual improvements. The conclusions to be drawn from this study are limited by methodological problems such as nonrandom treatment assignment and lack of standardized assessment tools.

In the only randomized, controlled trial evaluating psychodynamic treatment for childhood PTSD, Trowell and colleagues (2002) compared once-weekly psychoanalytic psychotherapy with group psychoeducational psychotherapy in treating children with histories of sexual abuse. The sample comprised 58 sexually abused girls between ages 6 and 14 years. Sixty-three percent of the participants were white, 11% were black Caribbean, 10% were biracial, 7% were Chinese, 6% were of Mediterranean origin, and 3% were of unknown origin. Both treatments were manualized, and in each, caregivers were offered supportive psychotherapy. Assessments were conducted 1 and 2 years after the start of treatment. Both treatments resulted in significantly higher functioning, revealing no differences between individual and group treatments. Children in the individual therapy condition, however, showed greater improvement in PTSD symptoms compared to children in the group therapy condition.

Whereas the Trowell and colleagues (2002) study demonstrated some methodological strengths, including random assignment to treatment condition, implementation of manualized treatments, and use of diagnostic interviews, significant weaknesses in study design render the study difficult to interpret. These limitations include small sample sizes and attrition of 24% by the end of the second year. Additionally, follow-up assessments were conducted 1 year after the start of treatment rather than at the conclusion of therapy. Furthermore, whereas face-to-face treatment time was approximately equal, children in the individual psychotherapy condition received up to 30 weekly treatment sessions, but children in the group therapy condition received only up to 18 sessions. This difference may have impacted treatment efficacy. Finally, this study varies both treatment modality (individual vs. group) and theoretical orientation (psychoanalytic vs. nonspecific), making it difficult to draw conclusions about cause of therapeutic change.

Systems Approaches

There have recently been important efforts to design child mental health interventions that address ongoing stresses in children's social environments, and the complex and confusing service system in which traumatized children and families find themselves.

One of the main challenges to the development of effective interventions for traumatized children is that the same factors that place a child at risk for exposure to traumatic events also contribute to environments that impede recovery, that is, environments fraught with instability and traumatic reminders. Traumatized children frequently live in environments characterized by domestic violence, child maltreatment, parental mental illness, and substance abuse. These conditions have been demonstrated to be detrimental to child development. For some children, the experience of an unstable home environment is compounded by poverty, racism, inadequate schools, and community violence (Cicchetti & Lynch, 1993; Duncan, 1996; Garbarino & Kostelny, 1997; Gelles, 1992; Groves, 2002; Osofsky & Scheeringa, 1997; Pelton, 1978).

There have been a number of important mental health approaches designed to address these problems. For example, the Child and Adolescent Service System Program (CASSP) (Pumariega & Winters, 2003; Stroul & Friedman, 1986), developed to guide states and communities in the development of community-based systems of care for vulnerable children, outlines a number of important "guiding principles" of effective community-based intervention (Stroul & Friedman, 1986). These guiding principles concern the need to create individualized, family-oriented services that address children's physical, emotional, social, and educational needs. These services are "integrated, with linkages between child-care agencies and the programs and mechanisms for planning, developing, and coordinating services" (Stroul & Friedman, 1986, p. 17), and involve case management to coordinate the broad array of services that children might receive.

Strongly influenced by CASSP principles, Henggeler, Schoenwald, Borduin, Rowland, and Cunningham (1998) developed multisystemic therapy (MST) for children with conduct disorder. MST implements community-based interventions to target specific areas of a child's environment that are theoretically related to the development and maintenance of conduct problems. MST has demonstrated effectiveness for aggressive children by successfully targeting for intervention many fields in which the child interacts: "The child and family, school, work, peer, community, and cultural institutions are viewed as interconnected systems with dynamic and reciprocal influences on the behavior of family members" and are, thus, engaged in the treatment process (Henggeler, Schoenwald, & Pickrel, 1995, p. 70). MST targets child and family problems in the multiple systems in which families are embedded and delivers treatments in the settings that are likely to have the highest impact. Services are delivered in a variety of settings, such as the home, the school, and the community. An important limitation is that MST does not specifically target traumatic stress symptoms, though it certainly is potent in remediating social–environmental problems.

Our group designed an intervention called trauma systems therapy (TST) to be compatible with CASSP principles, and to provide an integrated and highly coordinated system of services guided by the specific understandings of the nature of child traumatic stress. TST is an intervention approach that specifically targets the interface between a traumatized child's emotional regulation capacities, and the social environment and system of care in which the child lives (Saxe, Ellis, & Kaplow, 2007). TST seeks to address what we call a "trauma system"—a child who is emotionally dysregulated and a surrounding social environment that is unable to help the child to regulate emotion. Under this model, both the child's emotion regulation abilities and stressors in the environment that directly contribute to dysregulation become targets of treatment. TST uses a phase-oriented approach, seeking to provide the most acutely affected children with immediate coping skills and to reduce or eliminate direct, ongoing threats or stresses in their social environment. Intensive home-based care, psychopharmacology, and legal advocacy may all be used in an effort to stabilize the social environment and to ameliorate the stressors causing or exacerbating a child's symptoms. Over time, treatment progresses to more advanced phases that incorporate emotion regulation skills building, cognitive processing, and ultimately, meaning making of the trauma. TST has shown promising results in an open trial of 110 children (Saxe, Ellis, Fogler, Hansen, & Sorkin, 2005), and is currently being evaluated in a randomized, controlled trial.

METHODOLOGICAL CONSIDERATIONS

The development and implementation of empirically supported interventions for child traumatic stress hold great promise. Nevertheless, a series of fundamental methodological problems must be addressed before these interventions will have widespread utility.

Hoagwood, Jensen, Petti, and Burns (1996) suggest that researchers need to assess therapeutic outcomes in child-specific treatments that go beyond the types of symptom reduction examined in research settings. They propose a model comprising five domains that should be measured in child clinical outcome research. These domains, which represent interacting and nested spheres of influence, are as follows:

1. *Symptoms*: Emotional or behavioral symptoms exhibited in any setting (e.g., impulsivity, avoidance).
2. *Functioning*: Competencies and impairments reflecting children's ability to adapt to home, school, or community demands (e.g., emotion regulation, participating in family chores).
3. *Consumer perspectives*: Children and families' subjective experience (e.g., family's beliefs about therapeutic care, ratings of quality of life).
4. *Environments*: Modifiable features of the settings in which children spend time (e.g., classroom characteristics, family constellation, community environment).
5. *Systems*: Aspects of community care provided to children and families (e.g., children's service use, foster care placements, justice system involvement).

Hoagwood and colleagues suggest that therapeutic change should be assessed on each of these levels. This model necessarily requires that attention be paid to children's functioning at home, in school, and in the community. Furthermore, this model highlights some of the limitations in the childhood traumatic stress treatment literature. Whereas many of the studies we have described capture important aspects of children's traumatic stress symptoms, other, more global measures of children's functioning, although important, have been largely neglected in this literature.

Several special characteristics associated with conducting research with children present unique challenges. As described by Forrest, Simpson, and Clancy (1997), children demonstrate developmental change, dependency on adults, different disease epidemiology, and distinct demographic characteristics. Furthermore, there are developmental difficulties associated with gathering self-report information from children that limit the capabilities of treatment outcome research. There is debate within the health care field as to whether children are able to provide reliable and valid information about their health (Forrest, 2004). There are also several, important practical issues regarding intervention research with children: (1) informed consent and obtaining the child's assent to clinical research; (2) the limits of confidentiality, particularly when working with adolescents; and (3) working with child welfare agencies when the mandated reporting of child maltreatment becomes a concern.

CHALLENGES FOR THE FUTURE

Despite gains in the development of specific treatment models for childhood PTSD, there remains a significant gap between the treatments studied and their implementation within "real-world" settings. One of the key areas for future research, identified by the 2001 National Institute of Mental Health Blueprint for Change child and adolescent mental health report, is the assessment of the transportability, sustainability, and usability of the treatments within different practice settings (Hoagwood & Olin, 2002); this challenge remains critical within child trauma intervention research. Effectiveness studies are needed to understand better which treatments work for whom, and in what settings.

One important area for future study is to understand how interventions work with different traumatized groups. Little is known about the effectiveness of trauma treatment approaches with different types of trauma, and with children of different ethnicities. The treatment literature typically involves samples of children with homogeneous trauma exposure histories, such as sexual abuse or disaster exposure. This limits our current understanding of how different treatments might generalize to children who have experienced other types of trauma, or multiple traumas. For instance, child refugees represent an increasingly significant group of traumatized children, but no published intervention studies to date clearly describe the treatment provided and the effectiveness of the treatment (Birman et al., 2005). Similarly, little is known about which treatments are effective for medically traumatized children or children with co-occurring substance abuse and PTSD. Future work will need to determine whether existing treatments can be adapted to fit the needs of these specific groups, or whether new approaches need to be developed around the unique needs of different traumatized groups.

Another area for future study involves understanding the transportability of evidence-based treatments to different service settings. Child mental health services are frequently delivered in settings other than specialty mental health clinics, such as schools (Hoagwood et al., 2001). Treatments delivered in different service settings may differ in terms of the level of training of the providers, the support for the treatment within the service system, and the types of children seen within those settings. Treatments developed within a given context, such as a specialty mental health clinic, may not fit well with the needs and constraints of different service systems. Studies establishing the effectiveness of treatments across settings are rare, although research by Stein and colleagues (2002, 2003) has begun to provide an important model for how to integrate treatment outcome research into community settings, such as schools.

Possible future directions for increasing the effectiveness of treatments includes considering adaptations to treatments that better fit different service settings, adaptations of service settings themselves that are better able to provide evidence-based treatments, or development of interventions within "real-world" settings of different service systems (Hoagwood et al., 2001). Each of these approaches requires re-conceptualizing not only the treatment approach but also treatment outcome evaluation methods. In addition to assessing outcomes typically associated with intervention research, such as symptom reduction and functioning, effectiveness research will need to consider assessment of broader factors, such as systemic change that leads to successful adoption and implementation of a specified treatment (Jensen, Hoagwood, & Petti, 1996).

Finally, the studies reviewed in this chapter focus on treatments delivered to families seeking help for their children. However, children and adolescents in general grossly underutilize mental health services (Kataoka, Zhang, & Wells, 2002). McKay, Lynn, and Bannon (2005) found that inner-city youth who have been exposed to significant levels of trauma often fail to seek or to remain engaged in services. Of 95 inner-city youth referred for care in an urban mental health clinic, 28% never attended an initial appointment, and only 9% remained in treatment after 12 weeks, despite high rates of reported trauma and mental health needs. This suggests that future work in the development of interventions for traumatized children needs to consider treatment engagement as a critical element of delivering effective care, particularly in working with youth who contend with the ongoing stressors associated with urban poverty.

In their review of childhood traumatic stress treatments, Saunders, Berliner, and Hanson (2003) reported that the only child trauma treatment meeting scientific criteria for proven effective interventions is Cohen and colleagues' (2000) trauma-focused CBT. This treatment holds great promise because it has now been tested across multiple sites and in several populations. The next step for trauma-focused CBT, and any evidence-based treatment of childhood PTSD, is to assess fully its diffusion into frontline clinical sites and services systems. This type of assessment and adaptation of treatment models is extremely important to advance the standard of care for traumatized children. At the end of the day, any treatment proposed as evidence-based will need to show that it is used by an appreciable number of children and families. This type of diffusion assessment is currently under way across community mental health sites within the federally funded National Child Traumatic Stress Network (Berliner, 2005). The process by which treatments for childhood PTSD are disseminated, adapted, and evaluated in frontline community and clinic sites across the nation will yield a great deal of information to guide the traumatic stress field in understanding principles of intervention development and replication. Together, these data will prove critical for the creation of highly useful intervention products for the significant problem of traumatic stress in children.

ACKNOWLEDGMENT

Work on this chapter was supported by Grant No. U79 SM54305 from the Substance Abuse and Mental Health Services Administration to Glenn N. Saxe.

REFERENCES

Amaya-Jackson, L., Reynolds, V., & Murray, M. (2003). Cognitive-behavioral treatment for pediatric posttraumatic stress disorder: Protocol and application in school and community settings. *Cognitive and Behavioral Practice, 10*(3), 204–213.

American Academy of Child and Adolescent Psychiatry. (1998). Practice parameters for the diagnosis and treatment of posttraumatic stress. *Journal of the American Academy of Child and Adolescent Psychiatry, 37*(10), 4S–26S.

Beck, A. T. (1976). *Cognitive therapy and the emotional disorders.* Oxford, UK: International Universities Press.

Bell, C. C., & Jenkins, E. J. (1993). Community violence and children on Chicago's southside. *Psychiatry: Interpersonal and Biological Processes, 56*(1), 46–54.

Berliner, L. (2005). The results of randomized clinical trials move the field forward. *Child Abuse and Neglect, 29,* 103–105.

Birman, D., Ho, J., Pulley, E., Batia, K., Everson, M. L., Ellis, B. H., et al. (2005). *Mental health interventions for refugee children in resettlement: White Paper II.* Los Angeles: National Child Traumatic Stress Network.

Byers, J. G. (1996). Children of the stones: Art therapy interventions in the West Bank. *Art Therapy, 13*(4), 238–243.

Celano, M., Hazzard, A., Webb, C., & McCall, C. (1996). Treatment of traumagenic beliefs among sexually abused girls and their mothers: An evaluation study. *Journal of Abnormal Child Psychology, 24,* 1–17.

Chemtob, C. M., Nakashima, J., & Carlson, J. G. (2002). Brief treatment for elementary school children with disaster-related posttraumatic stress disorder: A field study. *Journal of Clinical Psychology, 58*(1), 99–112.

Cicchetti, D., & Lynch, M. (1993). Toward an ecological/transactional model of community violence and child maltreatment: Consequences for children's development. *Psychiatry, 56*(1), 96–118.

Cohen, J. A., Deblinger, E., Mannarino, A. P., & Steer, R. A. (2004). A multisite, randomized controlled trial for children with sexual abuse-related PTSD symptoms. *Journal of the American Academy of Child and Adolescent Psychiatry, 43*(4), 393–402.

Cohen, J. A., & Mannarino, A. P. (1993). A treatment model for sexually abused preschoolers. *Journal of Interpersonal Violence, 8*(1), 115–131.

Cohen, J. A., & Mannarino, A. P. (1996). A treatment outcome study for sexually abused preschool children: Initial findings. *Journal of the American Academy of Child and Adolescent Psychiatry, 35*(1), 42–50.

Cohen, J. A., & Mannarino, A. P. (1998). Interventions for sexually abused children: Initial treatment outcome findings. *Journal of the American Professional Society on the Abuse of Children, 3*(1), 17–26.

Cohen, J. A., Mannarino, A. P., Berliner, L., & Deblinger, E. (2000). Trauma-focused cognitive behavioral therapy for children and adolescents: An empirical update. *Cognitive Behavioral Therapy, 15*(11), 1202–1223.

Cohen, J. A., Mannarino, A. P., & Knudsen, K. (2005). Treating sexually abused children: 1 year follow-up of a randomized controlled trial. *Child Abuse and Neglect, 29*(2), 135–145.

Cohen, J. A., Mannarino, A. P., & Rogal, S. (2001). Treatment practices for childhood posttraumatic stress disorder. *Child Abuse and Neglect, 25*(1), 123–135.

Deblinger, E., & Heflin, A. H. (1996). *Treating sexually abused children and their nonoffending parents: A cognitive behavioral approach.* Thousand Oaks, CA: Sage.

Deblinger, E., Lippmann, J., & Steer, R. (1996). Sexually abused children suffering posttraumatic stress symptoms: Initial treatment outcome findings. *Child Maltreatment, 1,* 310–321.

Deblinger, E., McLeer, S. V., & Henry, D. E. (1990). Cognitive behavioral treatment for sexually abused children suffering post-traumatic stress: preliminary findings. *Journal of American Academy of Child and Adolescent Psychiatry, 29,* 747–752.

Deblinger, E., Steer, R. A., & Lippmann, J. (1999). Two-year follow-up study of cognitive behavioral therapy for sexually abused children suffering post-traumatic stress symptoms. *Child Abuse and Neglect, 23*(12), 1371–1378.

Downing, J., Jenkins, S. J., & Fisher, G. L. (1988). A comparison of psychodynamic and reinforcement treatment with sexually abused children. *Elementary School Guidance and Counseling, 22*(4), 291–298.

Duncan, D. F. (1996). Growing up under the gun: Children and adolescents coping with violent neighborhoods. *Journal of Primary Prevention, 16*(4), 343–356.

Foa, E. B., Dancu, C. V., & Hembree, E. A. (1999). A comparison of exposure therapy, stress inoculation training, and their combination for reducing posttraumatic stress disorder in female assault victims. *Journal of Consulting and Clinical Psychology, 67*(2), 194–200.

Foa, E. B., Rothbaum, B. O., Riggs, D. S., & Murdock, T. B. (1991). Treatment of posttraumatic stress disorder in rape victims: A comparison between cognitive-behavioral procedures and counseling. *Journal of Consulting and Clinical Psychology, 59*(5), 715–723.

Forrest, C. B. (2004). Outcomes research on children, adolescents, and their families. *Medical Care, 42*(4), III-19–III-23.

Forrest, C. B., Simpson, L., & Clancy, C. (1997). Child health services research: Challenges and opportunities. *Journal of the American Medical Association, 277,* 1787–1793.

Garbarino, J., & Kostelny, K. (1997). What children can tell us about living in a war zone. In J. D. Osofsky (Ed.), *Children in a violent society* (pp. 32–41). New York: Guilford Press.

Gelles, R. J. (1992). Poverty and violence towards children. *American Behavioral Scientist, 35*(3), 258–274.

Gil, E. (1991). *The healing power of play: Working with abused children.* New York: Guilford Press.

Goenjian, A. K., Karayan, E., Pynoos, R. S., Minassian, D., Najarian, L. M., Steinberg, A. M., et al. (1997). Outcome of psychotherapy among early adolescents after trauma. *American Journal of Psychiatry, 154*(4), 536–542.

Greenwald, R. (1998). Eye movement desensitization and reprocessing (EMDR): New hope for children suffering from trauma and loss. *Clinical Child Psychology and Psychiatry, 32*(2), 279–287.

Groves, B. M. (2002). *Children who see too much: Lessons from the child witness to violence project.* Boston: Beacon Press.

Henggeler, S. W., Schoenwald, S. K., Borduin, C. M., Rowland, M. D., & Cunningham, P. B. (1998). *Multisystemic treatment of antisocial behavior in children and adolescents.* New York: Guilford Press.

Henggeler, S. W., Schoenwald, S. K., & Pickrel, S. G. (1995). Multisystemic therapy: Bridging the gap between university- and community-based treatment. *Journal of Consulting and Clinical Psychology, 63*(5), 709–717.

Hoagwood, K., Burns, B., Kiser, L., Ringeisen, H., & Schoenwald, S. (2001). Evidence-based practice in child and adolescent mental health services. *Psychiatric Services, 52*(9), 1179–1189.

Hoagwood, K., Jensen, P. S., Petti, T., & Burns, B. J. (1996). Outcomes of mental health care for children and adolescents: I. A comprehensive conceptual model. *Journal of the American Academy of Child and Adolescent Psychiatry, 35*(8), 1055–1063.

Hoagwood, K., & Olin, S. (2002). The NIMH Blueprint for Change report: Research priorities in child and adolescent mental health. *Journal of the American Academy of Child and Adolescent Psychiatry, 41*(7), 760–767.

Jaberghaderi, N., Greenwald, R., Rubin, A., Zand, S. O., & Dolatabadi, S. (2004). A comparison of CBT and EMDR for sexually-abused Iranian girls. *Clinical Psychology and Psychotherapy, 11,* 358–368.

Jensen, P., Hoagwood, K., & Petti, T. (1996). Outcomes of mental health care for children and adolescents: II. Literature review and application of a comprehensive model. *Journal of the American Academy of Child and Adolescent Psychiatry, 35*(8), 1064–1077.

Kataoka, S., Zhang, L., & Wells, K. (2002). Unmet need for mental health care among U.S. children: Variation by ethnicity and insurance status. *American Journal of Psychiatry, 159*(9), 1548–1555.

King, N. J., Tonge, B. J., Mullen, P., Myerson, N., Heyne, D., Rollings, S., et al. (2000). Treating sexually abused children with posttraumatic stress symptoms: A randomized clinical trial. *Journal of the American of Child and Adolescent Psychiatry, 39*(11), 1347–1355.

Mallay, J. N. (2002). Art therapy, an effective outreach intervention with traumatized children with suspected acquired brain injury. *The Arts in Psychotherapy, 29,* 159–172.

March, J. S., Amaya-Jackson, L., Murray, M. C., & Schulte, A. (1998). Cognitive-behavioral psychotherapy for children and adolescents with posttraumatic stress disorder after a single-incident stressor. *Journal of the American Academy of Child and Adolescent Psychiatry, 37*(6), 585–593.

McKay, M., Lynn, C., & Bannon, W. (2005). Understanding inner city child mental health need and trauma exposure: Implications for preparing urban service providers. *American Journal of Orthopsychiatry, 75*(2), 201–210.

Muris, P., & Merckelbach, H. (1999). Eye movement desensitization and reprocessing. *Journal of the American Academy of Child and Adolescent Psychiatry, 38*(1), 7–8.

Osofsky, J. D., Cohen, G., & Drell, M. (1995). The effects of trauma on young children: A case of 2-year-old twins. *International Journal of Psycho-Analysis, 76,* 595–607.

Osofsky, J. D., & Scheeringa, M. S. (1997). Community and domestic violence exposure: Effects of development and psychopathology. In D. Cicchetti & S. L. Toth (Eds.), *Rochester Symposium in Developmental Psychopathology: Developmental perspectives on trauma* (pp. 155–180). Rochester, NY: University of Rochester Press.

Pelton, L. H. (1978). Child abuse and neglect: The myth of classlessness. *American Journal of Orthopsychiatry, 48*(4), 608–617.

Peri, T. (2004). "It was like in the cartoons": From memory to traumatic memory and back. *Psychoanalysis and Psychotherapy, 21*(1), 63–79.

Pumariega, A. J., & Winters, N. C. (2003). *The handbook of child and adolescent systems of care: The new community psychiatry.* San Francisco: Jossey-Bass.

Resick, P. A., & Schnicke, M. K. (1992). Cognitive processing therapy for sexual assault victims. *Journal of Consulting and Clinical Psychology, 60*(5), 748–756.

Saigh, P. A. (1987a). *In vitro* flooding of an adolescent's posttraumatic stress disorder. *Journal of Clinical Child Psychology, 16,* 147–150.

Saigh, P. A. (1987b). *In vitro* flooding of a childhood posttraumatic stress disorder. *School Psychology Review, 16,* 203–211.

Saigh, P. A. (1987c). *In vitro* flooding of childhood posttraumatic stress disorders: A systematic replication. *Professional School Psychology, 2,* 133–145.

Saigh, P. A. (1989). The use of *in vitro* flooding in the treatment of traumatized adolescents. *Journal of Behavioral and Developmental Pediatrics, 10,* 17–21.

Saigh, P. A., Yule, W., & Inamdar, S. C. (1996). Imaginal flooding of traumatized children and adolescents. *Journal of School Psychology, 34*(2), 163–183.

Saunders, B., Berliner, L., & Hanson, R. (2003). *Child physical and sexual abuse guidelines for treatment.* Charleston, SC: National Crimes Victims Research and Treatment Center.

Saxe, G. N., Ellis, B. H., Fogler, J., Hansen, S., & Sorkin, B. (2005). Comprehensive care for traumatized children: An open trial examines treatment using trauma systems therapy. *Psychiatric Annals, 35*(5), 443–448.

Saxe, G. N., Ellis, B. H., & Kaplow, J. (2007). *Collaborative treatment of traumatized children and teens: The trauma systems therapy approach.* New York: Guilford Press.

Shapiro, F. (1995). *Eye movement desensitization and reprocessing: Basic principles, protocols, and procedures.* New York: Guilford Press.

Singer, M. I., Anglin, T. M., & Song, L. Y. (1995). Adolescents' exposure to violence and associated symptoms of psychological trauma. *Journal of the American Medical Association, 273*(6), 477–482.

Smith, P., & Yule, W. (1999). Eye movement desensitisation and reprocessing. In W. Yule (Ed.), *Posttraumatic stress disorders: Concepts and therapy* (pp. 267–284). New York: Wiley.

Stein, B. D., Jaycox, L. H., Kataoka, S. H., Wong, M., Tu, W., Elliot, M. N., et al. (2003). A mental health intervention for schoolchildren exposed to violence: A randomized controlled trial. *Journal of the American Medical Association, 290*(5), 603–611.

Stein, B. D., Kataoka, S. H., Jaycox, L. H., Wong, M., Fink, A., Escudero, P., et al. (2002). Theoretical basis and program design of a school-based mental health intervention for traumatized immigrant children: A collaborative research partnership. *Journal of Behavioral Health Services and Research, 29*(3), 318–326.

Stroul, B. A., & Friedman, R. M. (1986). *A system of care for children with severe emotional disturbances.* Washington, DC: Georgetown University Child Development Center, National Technical Center for Children's Mental Health, Center for Child Health and Mental Health Policy.

Trowell, J., Kolvin, I., Weeramanthri, T., Sadowski, H., Berelowitz, M., Glasser, D., et al. (2002). Psychotherapy for sexually abused girls: Psychological outcome findings and patterns of change. *British Journal of Psychiatry, 180,* 234–247.

Chapter 19

Pharmacotherapy for PTSD

Matthew J. Friedman and Jonathan R. T. Davidson

There have been significant advances in the clinical psychopharmacology of posttraumatic stress disorder (PTSD) during the past 10 years. Growing understanding of the unique pathophysiology of this disorder has laid the groundwork for rational pharmacotherapy. In addition, multisite, randomized clinical trials have established an empirical database for evidence-based treatment. The most notable benchmark is approval by U.S. Food and Drug Administration (FDA) for two medications, sertraline and paroxetine, both selective serotonin reuptake inhibitors (SSRIs), as indicated treatment for PTSD. Before reviewing the findings from clinical trials, it is important to consider methodological issues pertinent to the clinical psychopharmacology of PTSD.

To orient the reader, there is an apparent discrepancy between the following section on the pathophysiology of PTSD and a later section on pharmacotherapy. The sequence of our review of key psychobiological systems that are altered in PTSD begins with the adrenergic system and moves on to the hypothalamic–pituitary–adrenocortical (HPA), glutamate, gamma-aminobutyric acid (GABA), serotonin (5-HT), and dopamine systems, respectively. In contrast, our review of pharmacological agents progresses in a very different sequence, beginning with serotonergic medications and reserving until later consideration of adrenergic, glutamatergic, GABA-ergic and dopaminergic agents, because scientific and clinical research have moved in different directions and have not been well coordinated. Scientific research has focused on the human stress response and fear circuitry in the brain, areas that have primarily emphasized research on adrenergic and HPA mechanisms. Our understanding of other mechanisms is at a much more preliminary state. In contrast, clinical trials have predominantly tested antidepressant and anxiolytic agents that act on the serotonin system, medications that have already received FDA approval for treatment of depression and anxiety disorders other than PTSD. None of these agents was designed with the unique pathophysiology of PTSD in mind. As a result, medications that have been tested most extensively in PTSD act on systems that have been investigated much less thoroughly. This is reflected in the different organization of this chapter's sections on pathophysiology and pharmacotherapy, respectively.

METHODOLOGICAL CONSIDERATIONS

There are two methodological areas to discuss regarding the clinical psychopharmacology for PTSD. First, it is necessary to provide a conceptual context by reviewing our current state of knowledge about the pathophysiology of PTSD. Second, we must consider the methodological challenges associated with conducting randomized clinical trials.

Pathophysiology of PTSD

The presumed circuitry underlying PTSD-related biobehaviorial abnormalities focuses on excessive activation of the amygdala by stimuli that are perceived to be threatening. Such activation produces outputs to a number of brain areas that mediate memory consolidation of emotional events and spatial learning (hippocampus), memory of emotional events and choice behaviors (orbitofrontal cortex), autonomic and fear reactions (locus coeruleus, thalamus, and hypothalamus) and instrumental approach or avoidance behavior (dorsal and ventral striatum) (Davis & Whalen, 2001). In PTSD, the normal checks and balances on amygdala activation have been impaired, so that the restraining influence of the medial prefrontal cortex (PFC, especially the anterior cingulate gyrus and orbitofrontal cortex) is severely disrupted (Charney, 2004; Vermetten & Bremner, 2002). Disinhibition of the amygdala produces a vicious spiral of recurrent fear conditioning, in which ambiguous stimuli are more likely to be appraised as threatening; mechanisms for extinguishing such responses are nullified; and key limbic nuclei are sensitized, thereby lowering the threshold for fearful reactivity (Charney, 2004; Charney, Deutch, Krystal, Southwick, & Davis, 1993; Friedman, 1994; see Southwick et al., Chapter 10, this volume).

The pharmacological challenge, therefore, is to identify where and how to intervene to rein in the amygdala, and the cortical and subcortical effects it has set in motion. We consider the adrenergic, HPA, glutametergic, serotonergic, and dopaminergic systems. The mechanisms of action of medications that have been tested are shown in Table 19.1.

Adrenergic System

NOREPINEPHRINE

Animal research indicates that central noradrenergic neurons play an important role in mobilizing the human stress response. All three principal adrenergic receptor systems are involved in the fear-conditioning circuitry described previously. (A more thorough review is provided by Southwick et al., Chapter 10, this volume.) Beta- and alpha$_1$-adrenergic activity may be related to the intrusive recollections, dissociative flashbacks,

TABLE 19.1. Pharmacological Actions Affecting Stress/Fear Responses and PTSD Symptoms

Pharmacological category	Specific medication	Mechanism of action	Effect on stress/fear response
Adrenergic system	Propranolol	Beta-receptor antagonist.	All antiadrenergic agents: • Reduce amygdala activation • Enhance PFC function • Inhibit locus coeruleus activation
	Prazosin	Alpha$_1$ receptor antagonist.	
	Clonidine	Alpha$_2$ receptor agonist.	
	Guanfacine		
	Theoretical	NPY enhancer.	Antagonizes both adrenergic and CRF activation of fear/stress response.
HPA system	Antalarmin[a]	CRF antagonist (experimental).	• Suppresses adrenergic and HPA responses to stress. • Reduces CRF release. • Reduces locus coeruleus activation. • Reduces ACTH secretion with secondary GC elevation.
	Hydrocortisone and other GCs	Rectify hypocortisolism and down-regulate GC receptors.	• Reduce increased HPA activation, thereby reducing potentiation of excessive adrenergic activity. • Reduce cortisol's neurotoxic enhancement of glutamate and calcium influx into neurons.
	Ketoconazole	Blocks cortisol synthesis.	
	Mifepristone (RU-486)	GC receptor antagonist.	
Glumatergic system	D-Cycloserine	Partial NMDA receptor agonist	Enhances learning, extinction, memory function, and neurogenesis.
GABA-ergic agents	Benzodiazepines	GABA$_A$ receptor agonist.	Suppresses stress-induced amygdala activation by inhibition of NMDA receptors.
	Baclofen	GABA$_B$ receptor agonist.	Unclear • Might reduce stress-induced adrenergic/HPA activation. • Has been effective clinically in mood and anxiety disorders.

Anticonvulsants/antikindling agents	Carbamazepine	• AMPA antagonist. • Elevates GABA. • Blocks sodium channels.	• Blocks sensitization/kindling. • Suppresses adrenergic arousal.
	Valproate	• Increases brain GABA levels. • Enhances GABA-receptor sensitivity.	• Blocks sensitization/kindling. • May suppress NMDA receptors.
	Lamotrigine	• Inhibits glutamate release. • Blocks voltage-dependent sodium and calcium channels.	• Blocks sensitization/kindling. • Blocks NMDA activation of amygdala.
	Topiramate	• Suppresses glutamate function. • Enhances GABA activity.	• Blocks sensitization/kindling. • Blocks NMDA amygdala activation.
	Gabapentin	• Increases GABA turnover.	• Blocks sensitization/kindling.
	Tiagabine	• Increases GABA levels by inhibiting glial uptake.	• Blocks sensitization/kindling.
	Vigabatrin	• Increases GABA by inhibiting GABA transaminase.	• Blocks sensitization/kindling. • Blocks startle response.
Selective serotonin reuptake inhibitors	Paroxetine Sertraline Fluoxetine Fluvotamine Citalopram	SSRI	• $5HT_{1A}$ neurons potentiate GABA antagonism of amygdala NMDA activity. • Promote neurogenesis in hippocampus.
Other serotonergic antidepressants	Nefazodone[b] Trazodone	SSRI plus postsynaptic $5-HT_2$ blockade.	• In addition to potentiation of $5-HT_{1A}$ action, blockade of $5-HT_2$ receptors is anxiolytic. • Promote neurogenesis.

(continued)

379

TABLE 19.1. (*continued*)

Pharmacological category	Specific medication	Mechanism of action	Effect on stress/fear response
Tricycline antidepressants	Imipramine Amitriptyline Desipramine	Block presynaptic reuptake of norepinephrine and 5-HT.	• Enhance serotonergic actions at 5-HT$_{1A}$ receptors. • Reduce adrenergic actions by down-regulation of postsynaptic beta receptors. • Promote neurogenesis.
Monoamine oxidase inhibitors	Phenelzine	Blocks enzymatic (MAO) degradation of norepinephrine, 5-HT (and dopamine).	• Enhance serotonergic action at 5-HT$_{1A}$ receptors. • Downregulate postsynaptic beta receptors (and reduce locus coeruleus activity). • Promote neurogenesis.
	Moclobemide	Selective MAO-A inhibitor.	
Other antidepressants	Mirtazapine	• Blocks postsynaptic 5-HT$_2$ and 5-HT$_3$ receptors. • Agonist action at presynaptic adrenergic alpha$_2$ receptors.	• Anxiolytic 5-HT$_2$/5-HT$_3$ blockade. • Reduces adrenergic activity. • Promotes neurogenesis.
	Venlafaxine	Blocks presynaptic reuptake of both 5-HT and norepinephrine.	• Potentiates 5-HT and reduces adrenergic activity. • Promotes neurogenesis.
	Bupropion	Blocks presynaptic reuptake of norepinephrine and dopamine.	• Reduces adrenergic and dopaminergic activity. • Promotes neurogenesis.
Atypical antipsychotic agents	Risperidone Quetiapine Olanzapine	Dopamine (D$_2$) and serotonin (5-HT$_2$) blockade.	Promote enhanced PFC restraint of amygdala, reduce hyperarousal/hypervigilance, and block anxiogenic 5-HT$_2$ receptor actions.

Note. ACTH, corticotropin; AMPA, alpha-amino-3-hydroxy-5-methyl-4-isoxyazoleproprionic acid; CRF, corticotropin-releasing factor; FDA, U.S. Food and Drug Administration; GABA, gamma-aminobutyric acid; GC, glucocorticoid; 5-HT, serotonin; MAO, monoamine oxidase; NMDA, N-methyl-D-aspartate; NPY, neuropeptide Y; PFC, prefrontal cortex; RCT, randomized clinical trial; SSRI, selective serotonin reuptake inhibitor; TCA, tricyclic antidepressant.

a Experimental medication.

b Withdrawn from U.S. market because of liver toxicity.

380

and psychological–physiological reactivity provoked by exposure to traumatic stimuli that are usually seen among individuals with PTSD. This postsynaptic noradrenergic input promotes activation of the amygdala. In addition the amygdala's projections to the locus coeruleus generate additional adrenergic input, thereby resulting in an upward spiral of adrenergic stimulation.

Alpha$_2$-adrenergic receptors, which provide presynaptic inhibition of amygdala catecholamine release, suppress fear conditioning and reduce consolidation of emotional memories (Davies et al., 2004). They may also play a role in dissociation since the alpha$_2$ antagonist yohimbine (which disinhibits adrenergic activity) provoked dissociative flashbacks among Vietnam War veterans with PTSD (Southwick et al., 1997, 1999). Thus, from the perspective of the amygdala alone, agents that antagonize alpha$_1$- and beta-adrenergic receptors or enhance inhibitory alpha$_2$-adrenergic activity might be expected to reduce PTSD symptoms.

Although the amygdala thrives in a climate of elevated adrenergic stimulation, the opposite is true for the PFC. High levels of catecholamines impair PFC function, as well as its capacity to inhibit amygdala hyperactivity (Arnsten, 2000). Both alpha$_1$- and beta-receptor activation appear responsible for nullifying PFC activity during uncontrollable stress, and these effects can be prevented with alpha$_1$-adrenergic antagonists such as prazosin (Arnsten & Jenstsch, 1997) as well as the beta-adrenergic antagonist propranolol (Li & Mei, 1994).

To summarize, the therapeutic goal of targeting the adrenergic system is to inhibit excessive alpha$_1$- and beta-receptor activation and to augment the inhibitory influence of alpha$_2$-adrenergic receptors. The result of such treatment would be expected to reduce amygdala activation, enhance PFC function, and inhibit stimulation of the locus coeruleus and its secondary activation of other cortical and subcortical structures.

NEUROPEPTIDE Y

Neuropeptide Y (NPY), an amino acid neurotransmitter colocalized in noradrenergic neurons, inhibits the release of both norepinephrine and corticotropin-releasing factor (CRF; see below). By virtue of its endogenous antiadrenergic actions, NPY would be expected to produce the antistress/anxiolytic benefits postulated earlier for anti-adrenergic agents, thereby improving cognitive function. Indirect evidence for this assertion has been obtained in studies of military personnel exposed to extreme stress, in which there was an inverse relationship between NPY release and stress-induced performance decrements due to dissociation (Morgan et al., 2000, 2001). Clinically, it has been shown that in comparison with healthy controls, patients with PTSD exhibit both reduced baseline NPY levels and blunted release of NPY in response to yohimbine stimulation (Rasmusson et al., 2000). Based on such findings, Friedman (2002) suggested that medications that enhance NPY function might ameliorate acute stress reactions, PTSD, and other stress-induced problems. No pharmacological agents of this nature are currently available.

CRF and the HPA System

CORTICOTROPIN-RELEASING FACTOR

CRF has a dual role in the human stress response. As a neurotransmitter it promotes release of norepinephrine from the locus coeruleus, thereby enhancing amygdala activity and reducing PFC activity, as described previously. As a hypothalamic hormone activated by stressful stimuli and threat appraisal, it releases corticotropin (ACTH) from the pituitary gland, which then promotes release of cortisol and other glucocorticoids from the adrenal cortex. Vietnam War veterans with PTSD have been shown to have elevated resting levels of cerebrospinal fluid CRF (Baker et al., 1999; Bremner, Licinio, et al., 1997) and enhanced hypothalamic release of CRF (Yehuda, 2002).

GLUCOCORTICOIDS

Glucocorticoids (GCs), such as cortisol, appear to impair PFC function by enhancing catecholamine levels during activation of the stress response (Arnsten, 2000; Roozendaal, McReynolds, & McGaugh, 2004). Although excessive HPA system activity does appear to be associated with trauma exposure and PTSD, how this may be manifested is a source of controversy. On the one hand, it may be expressed by elevated cortisol levels, as has been found in some patients with PTSD and in children exposed to sexual trauma. On the other hand, it may be expressed by reduced cortisol levels associated with supersensitivity of GC receptors (DeBellis et al., 1994; Friedman et al., 2001; Heim, Newport, Bonsall, Miller, & Nemeroff, 2001; Lemieux & Coe, 1995; Rasmusson et al., 2001; Rasmusson & Friedman, 2002; Yehuda, 2002; Yehuda, Giller, Southwick, Kahana, & Boisneau, 1994).

It has been proposed that abnormal HPA activity may have neurotoxic effects through activation of excitatory amino acids, resulting in calcium influx into susceptible neurons (McEwen et al., 1992; Sapolsky, 2000). From a PTSD perspective, the theory that acute (or chronic) cortisol elevation and/or GC receptor supersensitivity is neurotoxic has been invoked to explain reduced corpus callosum and intracranial volumes observed among traumatized children (DeBellis et al., 2002) and reduced hippocampal volumes among adults with PTSD (Bremner, Randall, et al., 1997; Bremner et al., 2003; Yehuda, 1999). A demonstrated association between reduced hippocampal volume and cognitive impairment among patients with PTSD (Vermetten et al., 2003) improved after an increase in hippocampal volume following antidepressant treatment (see below).

Prevention of neurotoxicity might also be achieved with glutamate antagonists, such as certain anticonvulsants, that through blockade of excitatory amino acid actions protect neurons by preventing toxic calcium influx. Reversal of neurotoxicity might be achieved with treatments that promote neurogenesis. For example the SSRI paroxetine has been shown to increase hippocampal volume in patients with PTSD.

The Glutamate and GABA Systems

Whereas glutamate is the major excitatory neurotransmitter, GABA is the primary inhibitory neurotransmitter in the brain. Monoamines (e.g., norepinephrine, 5-HT, and dopamine) have received the most attention in the past, because effective clinical agents (e.g., antidepressants and antipsychotic medications) are known to alter monoaminergic function. An important shift in focus has begun to occur, because our growing understanding of glutamatergic and GABA-ergic mechanisms indicates their crucial function in the human stress response and their probable role in the pathophysiology of PTSD. Anticonvulsant agents, also known as mood stabilizers, exert their primary actions on glutamate and/or GABA activity. Such actions also have potential importance in ameliorating PTSD symptoms.

GLUTAMATE

There are two families of glutamate receptors: inotropic, which exert their actions through neuronal receptor ion channels, and metabotropic, which act by coupling with receptor-bound G proteins. We focus the following discussion on inotropic receptors. The three types of inotropic glutamate receptors are named after the agonists to which they are differentially sensitive: N-methyl-D-aspartate (NMDA), alpha-amino-3-hydroxy-5-methyl-4-isoxyazolepropionic acid (AMPA), and kainate.

During the fear response, NMDA receptors in the amygdala activate the fear circuit described previously. NMDA antagonists, such as certain anticonvulsants, inhibit such actions (Berlant, 2003; Davis & Whalen, 2001; Paul, Nowak, Layer, Popik, & Skolnick, 1994). In addition to enhancing the startle response and anxious behavior, AMPA receptors mediate long-term potentiation, sensitization, and kindling of brain neurons, which is an important neurobiological model of PTSD (Post, Weiss, Li, Leverich, & Pert, 1999; Post, Weiss, & Smith, 1995; Walker & Davis, 2002). Kainate receptors appear to promote fear and anxiety through actions in the periaqueductal gray and frontal cortex, where they promote reduction of benzodiazepine (e.g., GABA-ergic) sites.

NMDA receptors are crucial for all forms of learning, including fear conditioning, (Bardgett et al., 2003; Liang, Hon, & Davis, 1994; Nakazawa et al., 2002) and extinction (Davis, 2002; Falls, Miserendino, & Davis 1992; van der Meulen, Bilbija, Joosten, de Bruin, & Feenstra, 2003). They also play a major role in neurogenesis, the production of new neurons (Gould, McEwen, Tanapat, Galea, & Fuchs, 1997; Nacher, Alonso-Llosa, Rosell, & McEwen, 2003; Okuyama, Takagi, Kawai, Miyake-Takagi, & Takeo, 2004). AMPA receptors may also promote neurogenesis through activation of brain-derived neurotrophic factor (BDNF; Mackowiak, O'Neill, Hicks, Bleakman, & Skolnick, 2002).

An important model of dissociation involves the interplay of NMDA and AMPA receptors, based on the observation that NMDA receptor antagonists, such as ketamine or phencyclidine, can produce dissociative symptoms such as slowed time perception, alterations in body perceptions, and derealization. The model proposes that NMDA blockade intensifies glutamate stimulation of AMPA receptors (Chambers, Bremner, Moghaddam, Southwick, & Charney, 1999; Krystal, Bennett, Bremner, Southwick, & Charney, 1995). It is noteworthy that the dissociative effects of ketamine are blocked by lamotrigine, an anticonvulsant that inhibits glutamate release (Anand et al., 2000; Goa, Ross, & Chrisp, 1993; Xie & Hagan, 1998).

Thus, the centrality of glutamatergic actions, both in amygdala activation and with regard to cognitive deficits associated with PTSD, has strong theoretical support from laboratory research.

GAMMA-AMINOBUTYRIC ACID

GABA, the brain's major inhibitory neurotransmitter, suppresses stress-induced actions of the amygdala. GABA receptors within the basolateral amygdala inhibit glutamatergic excitation. Furthermore, serotonin enhances this GABA-ergic suppression of the amygdala (Berlant, 2003; Stutzman & LeDoux, 1999), which is a major mechanism through which serotonergic agents ameliorate both the acute stress response and PTSD symptomatology.

PTSD patients exhibit both reduced GABA plasma levels (Vaiva et al., 2000) and reduced benzodiazepine receptor activity in the amygdala, PFC, and other brain areas (Bremner et al., 2000). Since benzodiazepine receptors are a part of the $GABA_A$ receptor complex, these findings suggest that deficiencies in GABA-ergic mechanisms in the amygdala, PFC, and elsewhere result in insufficient protection against the activating effects of norepinephrine and glutamate. It is possible that intrusive recollections, hyperarousal symptoms, and disinhibited social and emotional behavior observed among patients with PTSD may be due to such deficient GABA-ergic function (Morgan, Krystal, & Southwick, 2003). It should be noted in this regard that pretreating animals later exposed to inescapable shock with benzodiazepines blocks stress-induced increases in norepinephrine in the amygdala, cortex, locus coeruleus, hypothalamus, and hippocampus (Drugan, Ryan, Minor, & Maier, 1984; Grant, Huang, & Redmond, 1980).

The Serotonin System

The serotonergic system has important interactions with the adrenergic, HPA, glutamate, GABA, and dopamine systems. Most 5-HT neurons have their origin in two midbrain loci, the dorsal and median raphe nuclei, which have extensive connections with key limbic structures mediating stressful or threatening stimuli. Excessive stress, HPA activity, or the presence of PTSD symptoms are associated with down-regulation of anxiolytic, 5-HT_{1A} and up-regulation of anxiogenic, 5-HT_{2A} receptors.

There also appear to be synergistic interactions between 5-HT_{1A} and GABA receptors with regard to acute stress and PTSD. It is thought that stimulation of 5-HT_{1A} receptors in the amygdala potentiates GABA neurons, which in turn antagonize the excitatory glutamate neurotransmission that mediates stress-related amygdala activation (Charney, 2004; Vermetten & Bremner, 2002). This model suggests three potential amygdala-based target sites for pharmacological intervention: antagonism of glutamate, potentiation of GABA, and enhancement of 5-HT neurotransmission.

As with NMDA receptors, 5-HT_{1A} receptors also promote neurogenesis in the hippocampus. It has been shown that selective serotonin reuptake inhibitors (SSRIs), as well as all clinically effective antidepressants, promote neurogenesis through activation of brain-derived neurotophic factor (BDNF) and cyclic adenosine monophosphate (cAMP) (Duman, Nakagawa, & Malberg, 2001). This obviously has important implications for prevention of neuronal degeneration and reversal of neurotoxicity, as discussed previously.

Clinical studies have long indicated that many symptoms observed among patients with PTSD, such as impulsivity, suicidal behavior, rage, aggression, depression, panic, obsessional thoughts, and chemical dependency, are associated with 5-HT deficiency (Friedman, 1990). Furthermore, because the serotonergic agonist m-chorophenyl-piperazine (MCPP) can provoke panic reactions and dissociative flashbacks in subjects with PTSD but not control subjects (Southwick et al., 1997), there is reason to presume that serotonin 5-HT_2 antagonists might be clinically useful in this regard.

The Dopaminergic System

During uncontrollable stress, amygdala activation produces PFC dopamine release (Charney, 2004). There is evidence that dopamine, D_1, receptor agonists can produce stress-induced PFC impairments in working memory (Zahrt, Taylor, Mathew, & Arnsten, 1997), and that both D_1 and D_2 receptor antagonists can prevent such cognitive deficits (Arnsten, 2000; Druzin, Kurzina, Malinina, & Kozlov, 2000).

Excessive dopamine release may have a role in PTSD hyperarousal, hypervigilance, and possibly in provoking the brief paranoid/psychotic states sometimes observed in patients with PTSD. It is surprising how little PTSD research has focused on dopamine in comparison with the neurotransmitters we discussed previously. Elevated urinary and plasma dopamine concentrations have been found in subjects with PTSD (Hamner & Diamond, 1993; Lemieux & Coe, 1995; Yehuda et al., 1995).

Randomized Clinical Trials

Randomized clinical trials (RCTs) are designed to demonstrate scientifically that beneficial outcomes following a treatment can be attributed to that specific treatment. Basic components of an RCT include (Kraemer, 2004) (1) a control or comparison group; (2) a well-defined population from which a representative sample is drawn; (3) random assignment of subjects to either the experimental or comparison treatment; and (4) "blind" assessment of clinical outcomes by study personnel who do not know whether the subject has received the experimental or comparison treatment. Successful RCTs must demonstrate that observed clinical outcomes are of sufficient magnitude to have clinical and/or policy significance. In other words, an RCT that indicates an experimental treatment is statistically better than a comparison treatment is not considered a successful trial if the magnitude of this difference is too small to influence clinical practice.

When designing RCTs to test the relative efficacies of different treatments, a number of factors must be considered (Nies, 2001):

1. *Identified specific treatment outcomes* must be quantified accurately. In the case of PTSD, a number of reliable and valid instruments, such as the Clinician Administered PTSD Scale (CAPS; Weathers, Keane, & Davidson, 2001), have become gold standards for treatment research.

2. *Reliable diagnostic, inclusion, and exclusion criteria* must be established at the outset to ensure relative homogeneity of the subjects admitted to the study. Reliable methods of assessing symptom severity are also necessary to ensure that the experimental and comparison groups are appropriately balanced in that regard.

3. *Medication dosage* must be chosen to optimize efficacy and to minimize toxicity. Both fixed and flexible dosage strategies have been utilized in RCTs; fixed-dosage protocols constrain the clinical investigator to treat all subjects with the same dose of medication, whereas flexible-dosage protocols are more naturalistic and permit dosage titration within a specified range to adjust the administered dose according to clinical results and the subject's capacity to tolerate the medication.

4. *Nonspecific benefits* from receiving any treatment must be controlled, so that any positive outcomes are not erroneously attributed to the experimental medication. In pharmacological RCTs this is often accomplished by administering a pill placebo to comparison subjects. Although this strategy satisfies the scientific requirements of RCTs, it does not always satisfy ethical requirements (see below).

5. It is always desirable to assess *subject compliance* with a treatment regimen, because failure to take medication as prescribed may result in misleading underestimates of treatment efficacy. Techniques to determine compliance include plasma steady-state drug levels, biological markers attributable to medication effects (e.g., inhibition of a specific enzyme), and pill counts to determine how much prescribed medication was actually taken during the previous week.

6. The *sample size* should be estimated before the trial and have sufficient statistical power to detect clinically significant effects. A number of techniques are currently uti-

lized to analyze data from RCTs. At present, the most common measure is the effect size (Cohen, 1988), which is the standardized mean difference between the experimental and comparison groups. Kraemer (2004) criticized this approach and suggested that alternate analytic strategies, such as area under the curve (AUC) or number needed to treat (NNT), are better statistical methods.

7. *Ethical considerations* may affect recruitment and design issues. Research with children or subjects with certain psychiatric impairments inevitably raises questions about a potential subject's capacity to give informed consent. A major ethical concern is the appropriateness of a placebo comparison group. A source of current controversy is whether assignment to a placebo group is ethical when effective treatments are available. Strong opinions have been expressed on both sides (Charney et al., 2002). From the perspective of an RCT, this means that a new antidepressant would have to be compared with an established antidepressant (e.g., an SSRI) rather than with a placebo. The matter is even murkier in PTSD research, where only two FDA-approved medications, sertraline and paroxetine, contrast the wide array of available, FDA-approved antidepressants.

Many other issues concerning RCTs are beyond the scope of this brief overview, such as age, gender, drug–drug interactions, and genetic factors (Nies, 2001).

An important development exemplified in the sequenced treatment alternatives to relieve depression (STAR*D) study (Rush, 2001), the concept of "equipoise," expands traditional RCTs, in which two (or, at most, three) treatment groups are compared to an experimental design, and can accommodate multiple treatment options. Equipoise extends the accessibility of a clinical trial, because most subjects are permitted to select in advance from a wide variety of treatment options to which they may either accept or refuse randomization (Lavori et al., 2001).

CURRENT STATE OF THE ART

Before 2000, very few RCTs had been published on the efficacy of different medications for PTSD. Prior to that, a handful of single-site RCTs had been conducted with tricyclic antidepressants (TCAs), monoamine oxidase inhibitors (MAOIs), SSRIs, and benzodiazepines. The impetus for new research activity came from the discovery that SSRI antidepressants are also effective treatments for several different anxiety disorders, such as panic disorder, social phobia, and obsessive–compulsive disorder. Therefore, a number of multisite RCTs led to FDA approval of the SSRIs sertraline and paroxetine as indicated treatments for PTSD (Brady et al., 2000; Davidson, Rothbaum, van der Kolk, Sikes, & Farfel, 2001; Marshall, Beebe, Oldham, & Zaninelli, 2001; Tucker et al., 2001). RCTs with other SSRIs and other, newer antidepressants are ongoing at this time. Other classes of medications that have recently become the focus for open-label trials and RCTs are new anticonvulsants, which have shown efficacy in recurrent bipolar and unipolar affective disorders, and atypical antipsychotic agents developed for treatment of schizophrenia.

At this point in time, treatment research has therefore focused primarily on clinical trials with medications initially developed for treating depression, seizure disorders, mood fluctuations, and schizophrenia. Serotonergic mechanisms have clearly received the most attention, whereas medications that target CRF, norepinephrine, glutamate, and other neurotransmitters or neuromodulator systems have been tested sparingly.

This is reflected in the organization of Table 19.2, which summarizes findings on classes of medications that have been tested most frequently and for which evidence of efficacy is strongest.

Antidepressants

Selective Serotonin Reuptake Inhibitors

SSRIs are the treatment of choice for patients with PTSD, as attested by four independent clinical practice guidelines (American Psychiatric Association, 2004; Davidson et al., 2005; Friedman, Davidson, Mellman, & Southwick, 2000; Veterans Administration/ Department of Defense [VA/DoD], 2004). As stated previously, two SSRIs, sertraline and paroxetine, have received FDA approval for PTSD treatment. Multisite RCTs with *sertraline* (Brady et al., 2000; Davidson, Rothbaum, et al., 2001) and *paroxetine* (Marshall et al., 2001; Tucker et al., 2001) demonstrated that both agents significantly reduced PTSD symptoms in contrast to placebo. It was also shown that when sertraline treatment was extended from 12 to 36 weeks, 55% of nonresponding patients converted to medication responders (Londborg et al, 2001). Finally, discontinuation of SSRI treatment is associated with clinical relapse and a return of PTSD symptoms (Davidson, Pearlstein, et al., 2001; Martenyi, Brown, Zhang, Koke, & Prakash, 2002; Rapaport, Endicott, & Clary, 2002).

RCTs with *fluoxetine* (Martenyi, Brown, Zhang, Prakash, & Koke, 2002; van der Kolk et al., 1994), and open-label studies with *fluvoxamine* (De Boer et al., 1992; Escalona, Canive, Calais, & Davidson, 2002; Marmar et al., 1996) and *citalopram* (Seedat, Lockhart, Kaminer, Zungu-Dirwayi, & Stein, 2001) indicate that these SSRIs are also effective agents. SSRIs have a broad spectrum of action and are effective for PTSD reexperiencing, avoidance/numbing, and hyperarousal symptoms. They also appear to promote rapid improvement in quality of life, which is sustained during treatment (Rapaport et al., 2002).

A very exciting study with the SSRI paroxetine addresses neurogenesis and its possible role in ameliorating cognitive deficits among patients with PTSD. Vermetten and associates (2003) assessed declarative memory and hippocampal volume among 20 patients with PTSD who completed 9–12 months of treatment with paroxetine. They observed a significant improvement in logical, figural, and visual memory after treatment. Most remarkably, these investigators also observed a 4.6% increase in patients' mean hippocampal volume, as measured by magnetic resonance imaging (MRI).

Other Serotonergic Antidepressants

Nefazodone and *trazodone* are antidepressants that enhance serotonergic activity through a dual mechanism that combines an SSRI action with postsynaptic 5-HT$_2$ blockade. An RCT showed that nefazodone is as effective as sertraline (Saygin, Sungur, Sabol, & Cetinkaya, 2002). Similar positive results have been obtained in open-label nefazodone trials (Davis, Nugent, Murray, Kramer, & Petty, 2000; Hertzberg, Feldman, Beckham, Moore, & Davidson, 2002; Hidalgo et al., 1999). Despite these promising results, nefazodone was withdrawn from the American market because of liver toxicity. The reason for including nefazodone in this review is to demonstrate the utility of non-SSRI medications that enhance serotonergic actions in patients with PTSD.

Trazodone has limited efficacy in monotherapy (e.g., in trials that test only one medication) for PTSD. Due to its sedating effects and serotonergic action, it is often

TABLE 19.2. Medications for PTSD: Indications and Contraindications

Class	Medication	Daily dose	RCTs	Indications	Contraindications
Selective serotonin reuptake inhibitors (SSRIs)	Paroxetine[a]	10–60 mg	3	• Reduce B, C, and D symptoms. • Produce clinical global improvement. • Effective treatment for depression, panic disorder, social phobia, and obsessive–compulsive disorder. • Reduce associated symptoms (rage, aggression, impulsivity, suicidal thoughts).	• May produce insomnia, restlessness, nausea, decreased appetite, daytime sedation, nervousness, and anxiety. • May produce sexual dysfunction, decreased libido, delayed orgasm or anorgasmia. • Clinically significant interactions for people prescribed MAOIs. • Significant interactions with hepatic enzymes produce other drug interactions.
	Sertraline[a]	50–200 mg	2		
	Fluoxetine	20–80 mg	2		
	Citalopram	20–60 mg	—		
	Fluvoxamine	50–300 mg	—		
Other serotonergic antidepressants	Nefazodone	200–600 mg	1	• May reduce B, C, and D symptoms. • Effective antidepressants. • Trazodone has limited efficacy by itself but is synergistic with SSRIs and may reduce SSRI-induced insomnia.	• Reports of hepatotoxicity associated with nefazodone treatment. • May be too sedating, rare priapism with trazodone.
	Trazodone	150–600 mg	—		
Other second-generation antidepressants	Mirtazapine	15–45 mg	1	• Efficacy in PTSD has been demonstrated. • Effective antidepressants.	• Mirtazapine may produce somnolence, increased appetite, and weight gain. • Venlafaxine may exacerbate hypertension. • Bupropion may exacerbate seizure disorder.
	Venlafaxine	75–225 mg	2		
	Bupropion[b]	200–450 mg	—		
Monoamine oxidase inhibitors (MAOIs)	Phenelzine	15–90 mg	1	• Reduce B symptoms. • Produces global improvement. • Effective agents for depression, panic, and social phobia. • Efficacy in PTSD has not been demonstrated for other MAOIs.	• Risk of hypertensive crisis makes it necessary for patients to follow a strict dietary regimen. • Contraindicated in combination with most other antidepressants, central nervous system stimulants, and decongestants. • Contraindicated in patients with alcohol/substance abuse dependency. • May produce insomnia, hypotension, and anticholinergic and severe liver toxicity.
Tricyclic antidepressants	Imipramine	150–300 mg	1,1[c]	• Reduce B symptoms. • Produce global improvement.	• Anticholinergic side effects (dry mouth, rapid pulse, blurred vision, constipation).
	Amitriptyline	150–300 mg	1		

388

					Side effects
(TCAs)	Desipramine	100–300 mg	1	• Effective antidepressant and antipanic agents. • Imipramine reduced acute stress disorder in pediatric burn patients. Desipramine ineffective in one RCT. Other TCAs have not been tested in PTSD.	• May produce ventricular arrhythmias. • May produce orthostatic hypotension, sedation, and arousal.
Antiadrenergic agents	Propranolol	40–160 mg	1[c]	• Reduce B and D symptoms.	• May produce hypotension or bradycardia. Use cautiously with hypotensive patients. Titrate prazosin starting at 1 mg at bedtime and monitor blood pressure. • Propranolol may produce depressive symptoms, psychomotor slowing, or bronchospasm.
	Prazosin	6–10 mg	1	• Produce global improvement.	
	Clonidine[b]	0.2–0.6 mg	—	• Prazosin shown to have marked efficacy for PTSD nightmares and insomnia.	
	Guanfacine[b]	1–3 mg	—	• Propranolol reduced physiological hyperreactivity in acutely traumatized individuals.	
Glucocorticoids	Hydrocortisone		1[c]	• Prevents later development of PTSD in septic shock and cardiac surgery patients.	
Anticonvulsants	Carbamazepine[b]	400–1,600 mg	—	• Effective on B and D symptoms. • Effective in bipolar affective disorder. • Possibly effective in reducing impulsive, aggressive, and violent behavior.	• Neurological symptoms, ataxia, drowsiness, low sodium, leukopenia.
	Valproate[b]	750–1,750 mg	—	• Effective on C and D symptoms. • Effective in bipolar affective disorder.	• Gastrointestinal problems, sedation, tremor, and thrombocytopenia. • Valproate is teratogenic and should not be used in pregnancy.
	Gabapentin[b]	300–3,600 mg	—	• Small trials suggesting favorable effects.	Sedation and ataxia.
	Lamotrigine[b]	50–400 mg	1	• Efficacy of gabapentin, lamotrigine, and topiramate has not been demonstrated in PTSD.	Stevens–Johnson syndrome, skin rash, and fatigue.
	Topiramate[b]	200–400 mg	—		Glaucoma, sedation, dizziness, and ataxia.

(continued)

389

TABLE 19.2. (continued)

Class	Medication	Daily dose	RCTs	Indications	Contraindications
Anticonvulsants (cont.)	Tiagabine[b]	4–12 mg	—		Dizziness, somnolence, and tremor.
	Vigabatrin[b]	250–500 mg	—		Construction of visual fields.
Glutamatergic agent	D-Cycloserine	50–500 mg	1[d]	• Reduction in PTSD severity. • Improved cognition.	• Somnolence, headache, tremor, dysarthria, vertigo, and confusion.
GABA$_B$ agent	Baclofen[b]	30–80 mg	—	• Improvement in PTSD severity.	
Benzodiazepines	Alprazolam	0.5–6.0 mg	1	• Not recommended. • Do not reduce core B and C symptoms. • Effective only for general anxiety and insomnia. • Other benzodiazepines have not been tested in PTSD.	• Sedation, memory impairment, ataxia. • Not recommended for patients with past or present alcohol/drug abuse/dependency because of risk for dependence. • May exacerbate depressive symptoms. • Alprazolam may produce rebound anxiety.
	Clonazepam	1–8 mg			
Conventional antipsychotics	Thioridazine	20–800 mg	—	Not recommended.	Sedation, orthostatic hypotension, anticholinergic extrapyramidal effects, tardive dyskinesia, neuroleptic malignant syndrome, endocrinopathies, EKG abnormalities, blood dyscrasias, and hepatotoxicity.
	Chlorpromazine	30–800 mg	—		
	Haloperidol	1–100 mg	—		
Atypical antipsychotics	Risperidone	4–16 mg	2[d]	• Preliminary data suggests effectiveness against PTSD symptom clusters and aggression. • May have a role as augmentation treatment for partial responders to other agents.	• Weight gain with all agents. • Risk of Type II diabetes with olanzapine.
	Olanzapine	5–20 mg	2[d]		
	Quetiapine	50–750 mg	—		

Note. Data from Friedman (2003). RCT, randomized clinical trial; B symptoms, intrusive recollections; C symptoms, avoidant/numbing; D symptoms, hyperarousal.
[a] FDA approval as indicated treatment for PTSD.
[b] The only data are from small trials and case reports.
[c] RCT to prevent PTSD.
[d] Utilized as adjunctive agent.

390

used in conjunction with SSRIs to counter medication-induced insomnia (Friedman, 2003).

Tricyclic Antidepressants

TCAs block presynaptic reuptake of both 5-HT and norepinephrine. Whereas some TCAs exert their actions primarily on 5-HT reuptake (e.g., amitriptyline), others act primarily on norepinephrine reuptake (e.g., desipramine), and still others act on both neurotransmitter systems (e.g., imipramine). From the previous discussion, it is clear why 5-HT enhancement might be beneficial for patients with PTSD. Blockade of adrenergic reuptake (which is also effective for panic disorder) probably exerts its therapeutic action through either enhancement of (presynaptic inhibitory) alpha$_2$-receptors and/or down-regulation of postsynaptic beta-receptors. In either case, the end result is a reduction in adrenergic activity in the amygdala, PFC, and locus coeruleus. RCTs with *imipramine* (Kosten, Frank, Dan, McDougle, & Giller, 1991) and *amitriptyline* (Davidson et al., 1990), but not *desipramine* (Reist et al., 1989), have demonstrated symptom reduction in patients with PTSD.

One consequence of the fact that SSRIs and other, new antidepressants have more benign side effect profiles than older agents has been a loss of investigator interest in TCAs and MAOIs, despite the fact that these medications are effective. Another reason why TCAs and MAOIs have not been tested recently is that pharmaceutical companies are not motivated to fund such studies. A remarkable exception to this is a prospective RCT comparing imipramine with the hypnotic, chloral hydrate, among pediatric patients; imipramine treatment was effective in treating young burn victims with acute stress disorder (Robert, Blakeney, Villarreal, Rosenberg, & Meyer, 1999).

Monoamine Oxidase Inhibitors

MAOIs block the intraneuronal metabolic breakdown of 5-HT, norepinephrine, dopamine, and other monoamines. By preventing enzymatic destruction, more of these neurotransmitters are available for presynaptic release. Thus their therapeutic action may result from down-regulation of postsynaptic receptors and, possibly, by down-regulating adrenergic activity in the locus coeruleus (Davidson, Walker, & Kilts, 1987). An RCT with the MAOI *phenelzine* with Vietnam War combat veterans was extremely successful in reducing reexperiencing and arousal PTSD symptoms (Kosten et al., 1991). Results have been mixed in open-label trials (Davidson et al., 1987; Lerer, Ebstein, Shestatzky, Shemesh, & Greenberg, 1987; Milanes, Mack, Dennison, & Slater, 1984) and a small, methodologically flawed 5-week crossover study had negative results (Shestatzky, Greenberg, & Lerer, 1988). Finally, in an open trial with the reversible MAO$_A$ inhibitor *moclobemide*, Neal, Shapland, and Fox (1997) reported improvement in all three PTSD symptom clusters.

Newer Antidepressants

MIRTAZAPINE

Mirtazapine has both serotonergic actions (blockade of postsynaptic 5HT$_2$ and 5HT$_3$ receptors) as well as action at presynaptic alpha$_2$-adrenergic receptors. In one randomized trial (vs. placebo; Davidson et al., 2003) and one open-label 8-week trial in Korea (Bahk et al., 2002) mirtazapine effectively reduced PTSD symptom severity. An interesting case report describes the usefulness of mirtazapine for traumatic nightmares (or for postawaking memory of such nightmares) among 300 refugees who had previously failed to benefit in this regard from other medications (Lewis, 2002).

Venlafaxine blocks presynaptic reuptake of norepinephrine and 5-HT. It also has a much less potent effect on blocking dopamine reuptake. Two large, multicenter trials of venlafaxine-XR (extended release), a dual reuptake inhibitor of 5-HT and norepinephrine, have shown its superiority relative to placebo; one of these trials was of 12 weeks' duration (Davidson et al., 2006), and the other, 6 months' duration (Davidson et al., 2006). Of interest is the finding that the drug differed significantly from placebo in both studies on measures of resilience, or the ability to deal with daily stress. The long-term trial also showed that a substantial percentage of patients eventually achieve remission, but it takes several months to occur.

Bupropion blocks presynaptic reuptake of norepinephrine and dopamine, but not 5-HT. Anecdotal evidence and open trials suggest that it may be effective in treating PTSD (Canive, Clark, Calais, Qualls, & Tuason, 1998).

Antiadrenergic Agents

Compared to research with antidepressants, there are few studies with antiadrenergic agents. There is one RCT with this class of medications. Research with the alpha$_1$ antagonist *prazosin* has indicated that nightmares and other PTSD symptoms are reduced by treatment (Raskind et al., 2003). Prazosin would be expected to increase PFC activation and reduce amygdala activation.

Cahill and McGaugh (1996) have shown that the beta-adrenergic antagonist *propranolol* reduced enhancement of emotional memories among volunteers. In small studies with clinical populations, propranolol has had beneficial effects on PTSD symptoms (including intrusive recollections and reactivity to traumatic stimuli) (Famularo, Kinscherff, & Fenton, 1988; Kolb, Burris, & Griffiths, 1984). Propranolol has also showed promise as a prophylactic agent to prevent the later development of PTSD in acutely traumatized individuals (Pitman et al., 2002; Taylor & Cahill, 2002; Vaiva et al., 2003).

Alpha$_2$ adrenergic agonists such as *clonidine* and *guanfacine* would also be expected to improve PFC function, in addition to directly reducing amygdala activity. Animal research has shown that alpha$_2$ agonists enhance PFC working memory function (Franowicz et al., 2002; Mao, Arnsten, & Li, 1999). The sparse literature on the clinical efficacy of these agents in PTSD is generally favorable (Kinzie & Friedman, 2004; Kolb et al., 1984).

Anticonvulsant/Antikindling Agents

Anticonvulsant agents have been sporadically tested in small, single-site studies for almost 20 years. Only one, small RCT has been carried out. Interest in this class of medications was initially prompted by their antikindling actions, because there has been great interest in sensitization/kindling hypotheses for a long time (Friedman, 1994; Post et al., 1995, 1999). More recently, appreciation of glutamatergic and GABA-ergic actions of anticonvulsants, as well as detection of abnormalities in these two systems among patients with PTSD, has raised the level of interest in this class of medications (Zarate, Quirox, Payne, & Manji, 2002; Zullino, Krenz, & Besson, 2003). Finally, the development of several new anticonvulsant/mood stabilizers in recent years has motivated the pharmaceutical industry to support clinical trials utilizing these agents with patients with PTSD.

All anticonvulsants block sensitization/kindling, although their specific mechanisms of action differ, as shown in Table 19.1.

1. *Carbamazepine*: Three open-label studies with veterans and adolescents reported improvement in PTSD symptom severity and in impulse control, anger, and violent behavior (Lipper et al., 1986; Loof, Grimley, Kuller, Martin, & Schonfield, 1995; Wolfe, Alavi, & Mosnaim, 1988). A large retrospective study with military personnel indicated the effectiveness of carbamazepine in PTSD (Viola et al., 1997). Case reports with carbamazepine (Steward & Bartucci, 1986) and its close relative oxcarbamazepine (Berigan, 2002b) have also been positive.

2. *Valproate*: Four open-label trials and two case reports indicate the effectiveness of valproate for PTSD (Berigan & Holzgang, 1995; Clark, Canive, Calais, Qualls, & Tuason, 1999; Fesler, 1991; Goldberg, Cloitre, Whiteside, & Han, 2003; Petty et al., 2002; Szymanski & Olympia, 1991).

3. *Lamotrigine*: The one study on lamotrigine in PTSD treatment is the only RCT with any anticonvulsant. In this 10-week trial, 10 patients were randomized to lamotrigine and five to placebo monotherapy. Although the investigators reported amelioration of PTSD in 50% (5/10) of the lamotrigine group in contrast to 25% (1/4) in the placebo group (Hertzberg et al., 1999), their interpretation was challenged (Berlant, 2003) based on a reanalysis of this data.

4. *Topiramate*: An open-label trial with 35 patients with PTSD focused exclusively on reexperiencing symptoms, such as nightmares, intrusive recollections, and flashbacks. Overall, 71% patients had complete remission of those symptoms, and 21% reported a partial response (Berlant & van Kammen, 2002). Further trials that monitor the full spectrum of PTSD symptoms are warranted.

5. *Gabapentin*: Three case reports describe amelioration of PTSD symptoms (Berigan, 2002a; Brannon, Labbate, & Huber, 2000; Malek-Ahmadi, 2003). The most extensive report is a retrospective chart review of 30 patients, most of whom (90%) received another medication in addition to gabapentin (Hamner, Brodrick, & Labbate, 2001).

6. *Tiagabine*: Three case reports describe its effectiveness in PTSD (Berigan, 2002c; Schwartz, 2002; Taylor, 2003).

7. *Vigabatrin*: A report on five patients with PTSD treated with this agent emphasized its reduction of the startle response, along with improvement in anxiety and insomnia (Macleod, 1996).

To summarize, anticonvulsant agents have diverse actions on glutamate, GABA, and other neurotransmitters. Findings are generally favorable regarding amelioration of PTSD symptoms, but all but one of these reports are either open-label trials or case reports. The single RCT (with lamotrigine) included only 15 patients and lacked sufficient statistical power.

Partial NMDA Agonist

D-*Cycloserine* is a partial NMDA receptor agonist that has positive effects on memory deficits in animals (Monahan, Handelman, Hood, & Cordi, 1989; Thompson, Moskal, & Disterhoft, 1992), in elderly volunteers (Jones, Wesnes, & Kirby, 1991) and in patients with Alzheimer's disease (Schwartz, Hashtroudi, Herting, Schwartz, & Deutsch, 1996). In a 12-week double-blind, placebo-controlled crossover design, patients with PTSD currently treated with other medications were randomized to augmentation treatment with either D-cycloserine or placebo. Significant within-treatment pre- to posttreatment reductions in PTSD and anxiety (but not depression) symptom severity were observed. The only difference between D-cycloserine and placebo treatment was that the medicated group performed significantly better on the Wisconsin Card Sorting Test (Heresco-Levy et al., 2002).

GABA-Ergic Agonists

One might expect that treatment with benzodiazepines, which act at $GABA_A$ receptors, might ameliorate PTSD symptoms. This has not been the case. An RCT with *alprazolam* did not reduce core reexperiencing or avoidant/numbing symptoms, although it did lead to improvement in insomnia and generalized anxiety (Braun, Greenberg, Dasberg, & Lerer, 1990). Treatment of recently traumatized emergency room patients with *clonazepam* (Gelpin, Bonne, Peri, Brandes, & Shalev, 1996) or the hypnotic benzodiazepine *temazepam* (Mellman, Bustamante, David, & Fins, 2002) did not prevent the later development of PTSD. Other open trials with benzodiazepines have also been unsuccessful (Friedman et al., 2000).

Baclofen is a medication that activates $GABA_B$ receptors. Previous studies have shown that $GABA_B$ agonists have been effective in treating mood and anxiety disorders (Breslow et al., 1989; Krupitsky et al., 1993). In an open-label study with baclofen, 9 of 11 veterans with PTSD experienced improvement in overall PTSD symptom severity, although there was no improvement in reexperiencing symptoms (Drake et al., 2003). It does appear, however, that further trials with baclofen are warranted.

Atypical Antipsychotic Medications

There is a small but growing literature on favorable results with atypical antipsychotic agents. These findings are particularly impressive, because they have been generated among the more treatment-resistant patient groups (e.g., American military veterans receiving treatment at Department of Veterans Affairs Medical Centers) in which it has traditionally been difficult to demonstrate drug–placebo differences. This contrasts with a general consensus that conventional antipsychotic agents (e.g., *chlorpromazine* or *haloperidol*) have no place in PTSD treatment because of a very unfavorable risk:benefit ratio due to questionable clinical usefulness, plus serious side effects, especially tardive dyskinesia (Friedman et al., 2000). One the other hand, atypical antipsychotics have two actions, D_2 receptor blockade (which they share with conventional antipsychotics) and a unique 5-HT_2 receptor antagonism. As a result, they have not only a much more benign side effect profile (e.g., rare extrapyramidal complications) but also unique therapeutic actions, such as efficacy against negative symptoms of schizophrenia. In PTSD treatment, atypicals have usually been utilized as adjunctive agents for refractory patients who have failed to respond to SSRIs or other antidepressants. Although there is little empirical evidence to guide general practice, these medications are usually prescribed to ameliorate dissociation, hypervigilance/paranoia, psychosis, hyperarousal, irritability, and aggression.

There are published reports on three antipsychotic medications—risperidone, quetiapine, and olanzapine—that have been tested as adjuncts to ongoing pharmacotherapy. Results from two randomized trials (Bartzokis, Lu, Turner, Mintz, & Saunders, 2005; Hamner et al., 2003), an open-label trial (Monnelly, Ciraulo, Knapp, & Keane, 2003), and several case reports with *risperidone* as adjunctive therapy suggest that it reduces overall PTSD symptom severity, and dissociative flashbacks and aggressive behavior. Similar findings have been obtained with *quetiapine* as an adjunctive agent in which an open-label trial (Hammer, Deitsch, Ulmer, Brodrick, & Loberbaum, 2001), a retrospective client review (Sokolski, Densen, Lee, & Reist, 2003), and several case reports indicate beneficial effects in reducing PTSD symptoms among refractory patients who had failed to respond to SSRIs and other medications. Finally, although a randomized trial with *olanzapine* (Stein, Kline, & Matloff, 2002) indicated its effectiveness as an adjunctive agent in reducing PTSD symptoms among chronic patients who had failed to respond to other agents, there is also a negative randomized trial with olanzapine as adjunctive treatment for PTSD (Butterfield et al., 2001).

To summarize, the scant research conducted on dopamine mechanisms in patients with PTSD suggests that dopamine blockade might be a beneficial approach. Small, open-label and randomized trials with atypical antipsychotics as adjunctive agents for nonrespondent patients with chronic PTSD have been encouraging.

HPA System

Preclinical studies with the CRF receptor antagonist *antalarmin* have demonstrated reductions in cerebrospinal fluid CRF, reduced stress-induced fearful behavior, and suppression of both adrenergic and HPA responses to stress (Habib et al., 2000). Given its key role in mobilizing the human stress response, as well as its increased expression among patients with PTSD, there is good reason to predict that CRF antagonists might have beneficial clinical effects on PTSD-related symptoms. Although CRF antagonists are utilized in animal research and under development by pharmaceutical companies, none are available for clinical use.

It is also useful to consider pharmacological strategies that might either prevent or ameliorate PTSD-related neurotoxic effects hypothesized to result from excessive HPA activity discussed previously. With regard to early intervention, potential treatments might include CRF antagonists or NPY enhancers that would reduce the intensity of the acute stress response (Friedman, 2002). If the problem is excessive cortisol levels, a medication that inhibits cortisol synthesis (e.g., *ketoconazole*) or that blocks GC receptors (e.g., *mifepristone, RU-486*) might be considered. If the problem is reduced cortisol and supersensitive GC receptors, the opposite approach might be indicated, in which GCs would be administered to down-regulate supersensitive GC receptors. Indeed, it has been shown that acute *hydrocortisone* treatment for septic shock effectively prevents the later development of PTSD (Schelling et al., 2001).

GENERALIZABILITY OF CURRENT FINDINGS

The major focus for clinical trials in PTSD has been on SSRIs. Thus, the discussion about generalizability of findings applies primarily to data from RCTs with the SSRIs we reviewed previously. Most clinical trials with medication have recruited middle-aged white females traumatized sexually as children, or adults or Vietnam War veterans receiving treatment in Veterans Affairs (VA) hospitals. The trials with women have generally had more favorable results than RCTs with veterans. In the two RCTs with sertraline (Brady et al., 2000; Davidson, Rothbaum, et al., 2001) there were too few male subjects to generate sufficient statistical power to demonstrate the efficacy of sertraline in males. Indeed, FDA approval for sertraline is limited to treatment for women with PTSD. The two trials with paroxetine (Marshall et al., 2001; Tucker et al., 2001), however, had many more total participants and sufficient statistical power to demonstrate efficacy in both males and females.

Negative findings with Vietnam War veterans in VA hospital treatment settings had initially caused speculation that PTSD due to combat trauma is less responsive to pharmacotherapy than PTSD due to other causes. This erroneous conclusion was corrected in both the paroxetine and fluoxetine RCTs. In the paroxetine studies (Marshall et al., 2001; Tucker et al., 2001) veterans recruited from the general population (rather than from VA hospital treatment settings) exhibited as much benefit from SSRI treatment as did male and female nonveterans. The RCT with fluoxetine (Martenyi et al., 2002b), which recruited mostly male veterans of recent (United Nations and NATO) deployments (rather than Vietnam War veterans) had positive results. Indeed, exposure to combat trauma actually predicted a successful response to fluoxetine pharmacotherapy. These recent findings illustrate three things. First, people with PTSD due to combat trauma are as likely to respond to SSRI treatment as those with PTSD due to other traumatic events. Second, both men and women can benefit from SSRI treatment. Third, male Vietnam War veterans in VA settings are a particularly chronic and treatment-refractory cohort that appears unlikely to benefit from pharmacotherapy or from psychosocial treatments (Friedman, 1997; Schnurr et al., 2003). These findings are a strong argument for early detection and treatment of PTSD, because decades of chronicity appear to reduce the prognosis for a favorable outcome.

The previous discussion has focused on sexual abuse and combat trauma, because most PTSD subjects recruited for RCTs fall into one of these two categories. There is much less evidence on PTSD due to motor vehicle accidents, urban violence, natural disasters, terrorist attacks, or other traumatic events. Although there is no evidence that PTSD due to nonsexual or noncombat trauma is any different, these areas require further research before we can generalize confidently based on the data at hand.

Several important gaps in knowledge limit generalizability of current findings. First, there is insufficient information on nonwhite subjects from both the United States and other nations. This applies especially to refugees or internally displaced people, who are at very high risk for PTSD (Green et al., 2003) and are predominantly nonwhites. Within the United States, it is known (Kessler, Sonnega, Bromet, Hughes, & Nelson, 1995; Kulka et al., 1990) that being nonwhite is a risk factor for trauma exposure and PTSD.

Another serious gap in knowledge concerns children and adolescents. Childhood sexual, physical, or emotional abuse is a major cause of PTSD. Accidents are another potential precipitant of PTSD among this age group. Prompt treatment is desirable to prevent chronic PTSD (and associated comorbidities and functional impairment). Concerns about safety, however, have slowed the pace of launching RCTs for children with PTSD. In addition, mounting concerns about increased suicides among children and adolescents treated with SSRIs for depression (U.S. Food and Drug Administration, 2004) will undoubtedly generate even more caution and resistance to conducting clinical trials with medications for younger people with PTSD.

There is also a paucity of information on the efficacy of medications for older people with PTSD. Concerns about safety, age-related pharmacokinetic capacity, drug–drug interactions, and comorbid medical conditions must always be factored into decisions for treating older individuals (Cook, Cassidy, & Ruzek, 2001). Given that PTSD is a risk factor for medical illness (Schnurr & Green, 2003) and primary care practitioners have just begun to identify PTSD among their patients, treatment of older individuals with PTSD is emerging as an important clinical challenge, with very little evidence to guide practice.

CHALLENGES FOR THE FUTURE

The major challenge for the future is to develop and test pharmacological agents that target specific pathophysiological abnormalities associated with PTSD. As mentioned earlier, the lion's share of medication trials so far have utilized agents with established efficacy for other disorders, such as antidepressants, antiadrenergics, anticonvulsants, and (atypical) antipsychotics.

Medications designed primarily for PTSD might include CRF antagonists, NPY enhancers, or more specific serotonergic, glutamatergic, or GABA-ergic agents (Friedman, 2002). Agents that promote neurogenesis should be a focus for future research. The potential importance of fear conditioning, resistance to extinction, and sensitization/kindling (Charney, 2004) should direct our attention to glutamatergic agents and medications that can modulate these mechanisms, such as D-cycloserine, lamotrigine, and other anticonvulsants. Emerging knowledge on the psychobiology of dissociation points to possible roles for medications acting on NMDA, AMPA, alpha$_2$-adrenergic, and 5HT-$_2$ receptors. Further research with promising but inadequately tested classes of medication would include RCTs with antiadrenergic and anticonvulsant agents in particular.

Two important areas for future research are pharmacotherapy for acute posttraumatic reactions and prevention of PTSD. Thus far, a handful of studies with propranolol, hydrocortisone, and imipramine (Pitman et al., 2002; Robert et al., 1999; Schelling et al., 2001) suggest that early intervention might have a favorable influence on ameliorating posttraumatic symptomatology. Designing such a "morning after pill" (Friedman, 2002) is a major priority that should focus on CRF, NPY, adrenergic, glutamatergic, and possibly anti-inflammatory agents.

Primary prevention against PTSD would have to be predicated on a better understanding of the difference between resilient and vulnerable individuals. Because more than half of all American adults are likely to encounter at least one traumatic event in the course of their lives (Kessler et al., 1995), it would be useful for individuals to have some information in advance about their capacity to cope with traumatic events. Methodologies can and should be developed as part of normal health maintenance to provide individuals with a psychobiological stress profile (analogous to a serum cholesterol and lipid profile for vulnerability to heart disease) that would help them fortify their resilience to traumatic stress, if indicated (Friedman, 2002). For example, individuals with high cholesterol and low high-density lipoprotein levels are encouraged to take statin medications. Analogous prophylactic measures need to be developed to bolster resilience against traumatic stress.

A final challenge for future researchers might involve a paradigm shift in which PTSD is reconceptualized as a spectrum disorder, in which several distinct, pathological posttraumatic disorders are operationalized symptomatically and psychobiologically. Different posttraumatic disorders might be characterized by different patterns of alterations in adrenergic, HPA, glutamatergic, and so forth, mechanisms. Distinct posttraumatic disorders might be acute or chronic, psychiatric or medical. Under such a scenario, optimal pharmacotherapy for one disorder might not necessarily be the best treatment for another. In short, there is no lack of challenges for the future.

REFERENCES

American Psychiatric Association. (2004, November). Practice guideline for the treatment of patients with acute stress disorder and posttraumatic stress disorder. *American Journal of Psychiatry, 161*, 1–57.

Anand, A., Charney, D. S., Oren, D. A., Berman, R. M., Hu, X. S., Capiello, A., et al. (2000). Attenuation of the neuropsychiatric effects of ketamine with lamotrigine: Support for hyperglutamatergic effects of N-methyl-D-aspartate receptor antagonists. *Archives of General Psychiatry, 57*, 270–276.

Arnsten, A. F. T. (2000). Stress impairs PFC function in rats and monkeys: Role of dopamine D1 and norepinephrine alpha-1 receptor mechanisms. *Progress in Brain Research, 126*, 183–192.

Arnsten, A. F. T., & Jentsch, J. D. (1997). The alpha-1 adrenergic agonist, cirazoline, impairs spatial working memory performance in aged monkeys. *Pharmacology, Biochemistry, and Behavior, 58*, 55–59.

Bahk, W.-M., Pae, C.-U., Tsoh, J., Chae, J.-H., Jun, T.-Y., Kim, C.-L., et al. (2002). Effects of mirtazapine in patients with post-traumatic stress disorder in Korea: A pilot study. *Human Psychopharmacology, 17*, 341–344.

Baker, D. G., West, S. A., Nicholson, W. E., Ekhator, N. N., Kasckow, J. W., Hill, K. K., et al. (1999). Serial CSF corticotropin-releasing hormone levels and adrenocortical activity in combat veterans with posttraumatic stress disorder. *American Journal of Psychiatry, 156*, 585–588.

Bardgett, M. E., Boeckman, R., Krochmal, D., Fernando, H., Ahrens, R., & Csernansky, J. G. (2003). NMDA receptor blockade and hippocampal neuronal loss impair fear conditioning and position habit reversal in C 57Bl/6 mice. *Brain Research Bulletin, 60*, 131–142.

Bartzokis, G., Lu, P. H., Turner, J., Mintz, J., & Saunders, C. S. (2005). Adjunctive resperidone in the treatment of chronic combat-related posttraumatic stress disorder. *Biological Psychiatry, 57*, 474–479.

Berigan, T. (2002a). Gabapentin and PTSD [Letter]. *Journal of Clinical Psychiatry, 63*, 744.

Berigan, T. (2002b). Oxcarbamazepine treatment of posttraumatic stress disorder. *Canadian Journal of Psychiatry, 10*, 973–974.

Berigan, T. (2002c). Treatment of posttraumatic stress disorder with tiagabine [Letter]. *Canadian Journal of Psychiatry, 8*, 788.

Berigan, T. R., & Holzgang, A. (1995). Valproate as an alternative in post-traumatic stress disorder: A cast report. *Military Medicine, 6*, 318.

Berlant, J. L. (2003). Antiepileptic treatment of posttraumatic stress disorder. *Primary Psychiatry, 10*, 41–49.

Berlant, J. L., & van Kammen, D. P. (2002). Open-label topiramate as primary or adjunctive therapy in chronic civilian post-traumatic stress disorder: A preliminary report. *Journal of Clinical Psychiatry, 63*, 15–20.

Brady, K., Pearlstein, T., Asnis, G. M., Baker, D., Rothbaum, B., Sikes, C. R., et al. (2000). Efficacy and safety of sertraline treatment of posttraumatic stress disorder. *Journal of the American Medical Association, 283*, 1837–1844.

Brannon, N., Labbate, L., & Huber, M. (2000). Gabapentin treatment for posttraumatic stress disorder [Letter]. *Canadian Journal of Psychiatry, 45*, 84.

Braun, P., Greenberg, D., Dasberg, H., & Lerer, B. (1990). Core symptoms of posttraumatic stress disorder unimproved by alprazolam treatment. *Journal of Clinical Psychiatry, 51*, 236–238.

Bremner, J. D., Innis, R. B., Southwick, S. M., Staib, L., Zoghbi, S., & Charney, D. S. (2000). Decreased benzodiazepine receptor binding in prefrontal cortex in combat related posttraumatic stress disorder. *American Journal of Psychiatry, 157*, 1120–1126.

Bremner, J. D., Licinio, J., Darnell, A., Krystal, J. H., Owens, M. J., Southwick, S. M., et al. (1997). Elevated CSF corticotropin-releasing factor concentrations in posttraumatic stress disorder. *American Journal of Psychiatry, 154*, 624–629.

Bremner, J. D., Randall, P. K., Vermetten, E., Staib, L. H., Bronen, R. A., Mazure, C., et al. (1997). Magnetic resonance imaging-based measurement of hippocampal volume in posttraumatic stress disorder related to childhood physical and sexual abuse: A preliminary report. *Biological Psychiatry, 41*, 23–32.

Bremner, J. D., Vythilingam, M., Vermetten, E., Southwick, S. M., McGlashan, T., Staib, L. H., et al. (2003). MRI and PET study of deficits in hippocampal structure and function in women with childhood sexual abuse and posttraumatic stress disorder. *American Journal of Psychiatry, 160*, 924–932.

Breslow, M. F., Fankhauser, M. P., Potter, R. L., Meredith, K. E., Misiaszek, J., & Hope, D. G., Jr. (1989). Role of gamma-aminobutyric acid in antipanic drug efficacy. *American Journal of Psychiatry, 146*, 353–356.

Butterfield, M. I., Becker, M. E., Connor, K. M., Sutherland, S., Churchill, L. E., & Davidson, J. R. (2001). Olanzapine in the treatment of post-traumatic stress disorder: a pilot study. *International Clinical Psychopharmacology, 16*, 197–203.

Cahill, L., & McGaugh, J. L. (1996). Modulation of memory storage. *Current Opinion in Neurobiology, 6*, 237–242.

Canive, J. M., Clark, R. D., Calais, L. A., Qualls, C., & Tuason, V. B. (1998). Bupropion treatment in veterans with posttraumatic stress disorder: An open study. *Journal of Clinical Psychopharmacology, 18*, 379–383.

Chambers, R. A., Bremner, J. D., Moghaddam, B., Southwick, S., Charney, D. S., & Krystal, J. H. (1999). Glutamate and PTSD: Toward a psychobiology of dissociation. *Seminars in Clinical Neuropsychiatry, 4*, 274–281.

Charney, D. S. (2004). Psychobiological mechanisms of resilience and vulnerability: Implications for the successful adaptation to extreme stress. *American Journal of Psychiatry, 161*, 195–216.

Charney, D. S., Deutch, A. Y., Krystal, J. H., Southwick, S. M., & Davis, M. (1993). Psychobiologic mechanisms of posttraumatic stress disorder. *Archives of General Psychiatry, 50*, 295–305.

Charney, D. S., Nemeroff, C. B., Lewis, L., Laden, S. K., Gorman, J. M., Laska, E. M., et al. (2002). National Depressive and Manic–Depressive Association consensus statement on the use of placebo in clinical trials of mood disorders. *Archives of General Psychiatry, 59*, 262–270.

Clark, R. D., Canive, J. M., Calais, L. A., Qualls, C. R., & Tuason, V. B. (1999). Divalproex in posttraumatic stress disorder: An open-label clinical trial. *Journal of Traumatic Stress, 12*, 395–401.

Cohen, J. (1988). *Statistical power analysis for the behavioral sciences.* Hillsdale, NJ: Erlbaum.

Cook, J. M., Cassidy, E. L., & Ruzek, J. I. (2001). Aging combat veterans in long-term care. *National Center for PTSD Clinical Quarterly, 10*, 2–30.

Davidson, J. R. T., Baldwin, D. S., Stein, D. J., Kuper, E., Benattia, I., Ahmed, S., et al. (2006). Treatment of posttraumatic stress disorder with venlafaxine extended release: A 6-month randomized, controlled trial. *Archives of General Psychiatry, 63*, 1158–1165.

Davidson, J. R. T., Bernick, M., Connor, K. M., Friedman, M. J., Jobson, K., Kim, Y., et al. (2005). A psychopharmacology algorithm for treating posttraumatic stress disorder. *Psychiatric Annals, 35*, 887–898.

Davidson, J. R. T., Kudler, H., Smith, R., Mahorney, S. L., Lipper, S., Hammett, E. B., et al. (1990). Treatment of post-traumatic stress disorder with amitriptyline and placebo. *Archives of General Psychiatry, 47*, 259–266.

Davidson, J., Pearlstein, T., Londborg, P., Brady, K. T., Rothbaum, B., Bell, J., et al. (2001). Efficacy of sertraline in preventing relapse of posttraumatic stress disorder: Results of a 28-week double-blind, placebo-controlled study. *American Journal of Psychiatry, 158*, 1974–1981.

Davidson, J. R. T., Rothbaum, B. O., Tucker, P., Asnis, G., Benattia, I., & Musgnung, J. (2006). Venlafaxine extended release in posttraumatic stress disorder: A sertraline- and placebo-controlled study. *Journal of Clinical Psychopharmacology, 26*, 259–267.

Davidson, J. R. T., Rothbaum, B. O., van der Kolk, B. A., Sikes, C. R., & Farfel, G. M. (2001). Multicenter, double-blind comparison of sertraline and placebo in the treatment of posttraumatic stress disorder. *Archives of General Psychiatry, 58*, 485–492.

Davidson, J. R. T., Walker, J. U., & Kilts, C. (1987). A pilot study of phenelzine in posttraumatic stress disorder. *British Journal of Psychiatry, 150*, 252–255.

Davidson, J. R. T., Weisler, R. H., Butterfield, M. I., Casat, C. D., Connor, K. M., Barnett, S., et al. (2003). Mirtazapine vs. placebo in posttraumatic stress disorder: A pilot trial. *Biological Psychiatry, 53*, 188–191.

Davies, M. F., Tsui, J., Flannery, J. A., Li, X. C., DeLorey, T. M., & Hoffman, B. B. (2004). Activation of alpha$_2$ adrenergic receptors suppresses fear conditioning: Expression of c-Fos and phosphorylated CREB in mouse amygdala. *Neuropsychopharmacology, 29*, 229–239.

Davis, L. L., Nugent, A. L., Murray, J., Kramer, G. L., & Petty, F. (2000). Nefazodone treatment for chronic posttraumatic stress disorder: An open trial. *Journal of Clinical Psychopharmacology, 20*, 159–164.

Davis, M. (2002). Role of NMDA receptors and MAP kinase in the amygdala in extinction of fear: Clinical implications for exposure therapy. *European Journal of Neuroscience, 16*, 395–398.

Davis, M., & Whalen, P. J. (2001). The amygdala: Vigilance and emotion. *Molecular Psychiatry, 1,* 13–34.

DeBellis, M. D., Chrousos, G. P., Dorn, L. D., Burke, L., Helmers, K., Kling, M. A., et al. (1994). Hypothalamic–pituitary–adrenal axis dysregulation in sexually abused girls. *Journal of Clinical Endocrinology and Metabolism, 78,* 249–255.

DeBellis, M. D., Keshaven, M. S., Shiflett, H., Iyengar, S., Beers, S. R., Hall, J., et al. (2002). Brain structure in pediatric maltreatment-related posttraumatic stress disorder: A sociodemographically matched study. *Biological Psychiatry, 52,* 1066–1078.

De Boer, M., Op den Velde, W., Falger, P. J., Hovens, J. E., De Groen, J. H., & Van Duijn, H. (1992). Fluvoxamine treatment for chronic PTSD: A pilot study. *Psychotherapy and Psychosomatics, 57,* 158–163.

Drake, R. G., Davis, L. L., Cates, M. E., Jewell, M. E., Ambrose, S. M., & Lowe, J. S. (2003). Baclofen treatment for chronic posttraumatic stress disorder. *Annals of Pharmacotherapy, 37,* 1177–1181.

Drugan, R. C., Ryan, S. M., Minor, T. R., & Maier, S. F. (1984). Librium prevents the analgesia and shuttlebox escape deficit typically observed following inescapable shock. *Pharmacology, Biochemistry, and Behavior, 21,* 749–754.

Druzin, M. Y., Kurzina, N. P., Malinina, E. P., & Kozlov, A. P. (2000). The effects of local application of D_2 selective dopaminergic druges into the medial prefrontal cortex of rats in a delayed spatial choice task. *Behavioural Brain Research, 109,* 99–111.

Duman, R. S., Nakagawa, S., & Malberg, J. (2001). Regulation of adult neurogenesis by antidepressant treatment. *Neuropsychopharmacology, 25,* 836–844.

Escalona, R., Canive, J. M., Calais, L. A., & Davidson, J. R. (2002). Fluvoxamine treatment in veterans with combat-related post-traumatic stress disorder. *Depression and Anxiety, 15,* 29–33.

Falls, W. A., Miserendino, M. J., & Davis, M. (1992). Extinction of fear-potentiated startle: Blockage by infusion of an NMDA antagonist into the amygdala. *Journal of Neuroscience, 12,* 854–863.

Famularo, R., Kinscherff, R., & Fenton, T. (1988). Propranolol treatment for childhood posttraumatic stress disorder, acute type. *American Journal of Diseases of Children, 142,* 1244–1247.

Fesler, F. A. (1991). Valproate in combat-related posttraumatic stress disorder. *Journal of Clinical Psychiatry, 52,* 361–364.

Franowicz, J. S., Kessler, L., Dailey-Borja, C. M., Kobilka, B. K., Limbird, L. E., & Arnsten, A. F. T. (2002). Mutation of the $alpha_{2A}$-adrenoceptor impairs working memory performance and annuls cognitive enhancement by guanfacine. *Journal of Neuroscience, 22,* 8771–8777.

Friedman, M. J. (1990). Interrelationships between biological mechanisms and pharmacotherapy of posttraumatic stress disorder. In M. E. Wolfe & A. D. Mosnaim (Eds.), *Posttraumatic stress disorder: Etiology, phenomenology, and treatment* (pp. 204–225). Washington, DC: American Psychiatry Press.

Friedman, M. J. (1994). Neurobiological sensitization models of post-traumatic stress disorder: Their possible relevance to multiple chemical sensitivity syndrome. *Toxicology and Industrial Health, 10,* 449–462.

Friedman, M. J. (1997). Drug treatment for PTSD: Answers and questions. *Annals of the New York Academy of Sciences, 821,* 359–371.

Friedman, M. J. (2002). Future pharmacotherapy for PTSD: Prevention and treatment. *Psychiatric Clinics of North America, 25,* 1–15.

Friedman, M. J. (2003). Pharmacologic management of posttraumatic stress disorder. *Primary Psychiatry, 10,* 66–68, 71–73.

Friedman, M. J., Davidson, J. R. T., Mellman, T. A., & Southwick, S. M. (2000). Pharmacotherapy. In E. B. Foa, T. M. Keane, & M. J. Friedman (Eds.), *Effective treatments for PTSD: Practice guidelines from the International Society for Traumatic Stress Studies* (pp. 84–105). New York: Guilford Press.

Friedman, M. J., McDonagh-Coyle, A. S., Jalowiec, J. E., Wang, S., Fournier, D. A., & McHugo, G. J. (2001). *Neurohormonal findings during treatment of women with PTSD due to childhood sexual abuse.* Presented at 17th annual meeting of the International Society for Traumatic Stress Studies, New Orleans, LA.

Gelpin, E., Bonne, O., Peri, T., Brandes, D., & Shalev, A. Y. (1996). Treatment of recent trauma survivors with benzodiazepines: A prospective study. *Journal of Clinical Psychiatry, 57,* 390–394.

Goa, K. L., Ross, S. P., & Chrisp, P. (1993). Lamotrigine: A review of pharmacological properties and clinical efficacy in epilepsy. *Drugs, 46,* 152–176.

Goldberg, J. F., Cloitre, M., Whiteside, J. E., & Han, H. (2003). An open-label pilot study of divalproex sodium for posttraumatic stress disorder related to childhood abuse. *Current Therapeutic Research, 64,* 45–54.

Gould E., McEwen, B. S., Tanapat, P., Galea, L. A. M., & Fuchs, E. (1997). Neurogenesis in the dentate gyrus of the adult tree shrew is regulated by psychosocial stress and NMDA receptor activation. *Journal of Neuroscience, 17,* 2492–2498.

Grant, S. J., Huang, Y. H., & Redmond, D. E. (1980). Benzodiazepines attenuate single unit activity in the locus coeruleus. *Life Sciences, 27,* 2231.

Green, B. L., Friedman, M. J., de Jong, J. T. V. M., Solomon, S. D., Keane, T. M., Fairbank, J. A., et al. (Eds.). (2004). *Trauma interventions in war and peace: Prevention, practice, and policy.* New York: Kluwer Academic.

Habib, K. E., Weld, K. P., Rice, K. C., Pushkas, J., Champoux, M., Listwak, S., et al. (2000). Oral administration of a corticotropin-releasing hormone receptor antagonist significantly attenuates behavioral, neuroendocrine, and autonomic responses to stress in primates. *Proceedings of the National Academy of Sciences USA, 97,* 6079–6084.

Hamner, M. B., Brodrick, P. S., & Labbate, L. A. (2001). Gabapentin in PTSD: A retrospective, clinical series of adjunctive therapy. *Annals of Clinical Psychiatry, 3,* 141–146.

Hamner, M. B., Deitsch, S. E., Ulmer, H. G., Brodrick, P. S., & Lorberbaum, J. P. (2001). *Quetiapine treatment in posttraumatic stress disorder: A preliminary trial of add-on therapy.* Presented at the 41st annual meeting of the New Clinical Drug Evaluation Unit, Phoenix, AZ.

Hamner, M. B., & Diamond, B. I. (1993). Elevated plasma dopamine in posttraumatic stress disorder: A preliminary report. *Biological Psychiatry, 33,* 304–306.

Hamner, M. B., Faldowski, R. A., Ulmer, H. G., Frueh, B. C., Huber, M. G., & Arana, G. W. (2003). Adjunctive risperidone treatment in post-traumatic stress disorder: A preliminary controlled trial of effects on comorbid psychotic symptoms. *International Clinical Psychopharmacology, 18,* 1–8.

Heim, C., Newport, J. D., Bonsall, R., Miller, A. H., & Nemeroff, C. B. (2001). Altered pituitary-adrenal axis responses to provocative challenge tests in adult survivors of childhood abuse. *American Journal of Psychiatry, 158,* 575–581.

Heresco-Levy, U., Kremer, I., Javitt, D. C., Goichman, R., Reshef, A., Blanaru, M., et al. (2002). Pilot-controlled trial of D-cycloserine for the treatment of post-traumatic stress disorder. *International Journal of Neuropsychopharmacology, 5,* 301–307.

Hertzberg, M. A., Butterfield, M. I., Feldman, M. E., Beckham, J. C., Sutherland, S. M., Connor, K. M., et al. (1999). A preliminary study of lamotrigine for the treatment of posttraumatic stress disorder. *Biological Psychiatry, 45,* 1226–1229.

Hertzberg, M. A., Feldman, M. E., Beckham, J. C., Moore, S. D., & Davidson, J. R. (1998). Open trial of nefazodone for combat-related posttraumatic stress disorder. *Journal of Clinical Psychiatry, 59,* 460–464.

Hidalgo, R., Hertzberg, M. A., Mellman, T., Petty, F., Tucker, P., Weisler, R., et al. (1999). Nefazodone in posttraumatic stress disorder: Results from six open-label trials. *International Clinical Psychopharmacology, 14,* 61–68.

Jones, R. W., Wesnes, K. A., & Kirby, J. (1991). Effects of NMDA modulation in scopolamine dementia. *Annals of the New York Academy of Science, 640,* 241–244.

Kessler, R. C., Sonnega, A., Bromet, E., Hughes, M., & Nelson, C. B. (1995). Posttraumatic stress disorder in the National Comorbidity Survey. *Archives of General Psychiatry, 52,* 1048–1060.

Kinzie, J. D., & Friedman, M. J. (2004). Psychopharmacology for refugee and asylum seeker patients. In J. P. Wilson & B. Drozdek (Eds.), *Broken spirits: The treatment of asylum seekers and refugees with PTSD* (pp. 579–600). New York: Brunner–Routledge Press.

Kolb, L. C., Burris, B. C., & Griffiths, S. (1984). Propranolol and clonidine in the treatment of the chronic post-traumatic stress disorders of war. In B. A. van der Kolk (Ed.), *Post-traumatic stress disorder: Psychological and biological sequelae* (pp. 97–107). Washington, DC: American Psychiatric Press.

Kosten, T. R., Frank, J. B., Dan, E., McDougle, C. J., & Giller, E. L. (1991). Pharmacotherapy for post-traumatic stress disorder using phenelzine or imipramine. *Journal of Nervous and Mental Disease, 179,* 366–370.

Kraemer, H. C. (2004) Statistics and clinical trial design in psychopharmacology. In A. Schatzberg & C. Nemeroff (Eds.), *Textbook of psychopharmacology* (pp. 173–183). Washington, DC: American Psychiatric Press.

Krupitsky, E. M., Burakov, A. M., Ivanov, G. F., Krandashova, G. F., Lapin, I. P., Grinenko, A. J., et al. (1993). Baclofen administration for the treatment of affective disorders in alcoholic patients. *Drug and Alcohol Dependence, 33,* 157–163.

Krystal, J., Bennett, A. L., Bremner, J. D., Southwick, S. M., & Charney, D. S. (1995). Toward a cognitive neuroscience of dissociation and altered memory functions in post-traumatic stress disorder. In M. J. Friedman, D. S. Charney, & A. Y. Deutch (Eds.), *Neurobiological and clinical consequences of stress: From normal adaptation to post-traumatic stress disorder* (pp. 239–269). Philadelphia: Lippincott–Raven.

Kulka, R. A., Schlenger, W. E., Fairbank, J. A., Hough, R. L., Jordan, K. B., Marmar, C. R., et al. (1990). *Trauma and the Vietnam War generation: Report of findings from the National Vietnam Veterans Readjustment Study.* New York: Brunner/Mazel.

Lavori, P. W., Rush, J. A., Wisniewski, S. R., Alpert, J., Fava, M., Kupfer, D. J., et al. (2001). Strengthening clinical effectiveness trials: Equipoise stratified randomization. *Biological Psychiatry, 50,* 792–801.

Lemieux, A. M., & Coe, C. L. (1995). Abuse-related posttraumatic stress disorder: Evidence for chronic neuroendocrine activation in women. *Psychosomatic Medicine, 57,* 105–115.

Lerer, B., Ebstein, R. P., Shestatzky, M., Shemesh, Z., & Greenberg, D. (1987). Cyclic AMP signal transduction in posttraumatic stress disorder. *American Journal of Psychiatry, 144,* 1324–1327.

Lewis, J. D. (2002). Mirtazapine for PTSD nightmares [Letter]. *American Journal of Psychiatry, 159,* 1948–1949.

Li, B.-M., & Mei, Z.-T. (1994). Delayed response deficit induced by local injection of the alpha-2 adrenergic antagonist yohimbine into the dorsolateral prefrontal cortex in young adult monkeys. *Behavioral and Neural Biology, 62,* 134–139.

Liang, K. C., Hon, W., & Davis, M. (1994). Pre- and posttraining infusion of N-methyl-D-aspartate receptor antagonists into the amygdala impair memory in an inhibitory avoidance task. *Behavioral Neuroscience, 108,* 241–253.

Lipper, S., Davidson, J. R. T., Grady, T. A., Edinger, J. D., Hammett, E. B., Mahorney, S. L., et al. (1986). Preliminary study of carbamazepine in posttraumatic stress disorder. *Psychosomatics, 27,* 849–854.

Londborg, P. D., Hegel, M. T., Goldstein, S., Goldstein, D., Himmelhoch, J. M., Maddock, R., et al. (2001). Sertraline treatment of posttraumatic stress disorder: Results of weeks of open-label continuation treatment. *Journal of Clinical Psychiatry, 62,* 325–331.

Loof, D., Grimley, P., Kuller, F., Martin, A., & Shonfield, L. (1995). Carbamazepine for PTSD. *Journal of the American Academy of Child and Adolescent Psychiatry, 6,* 703–704.

Mackowiak, M., O'Neill, M. J., Hicks, C. A., Bleakman, D., & Skolnick, P. (2002). An AMPA receptor potentiator modulates hippocampal expression of BDNF: An *in vivo* study. *Neuropharmacology, 43,* 1–10.

Macleod, A. D. (1996). Vigabatrin and posttraumatic stress disorder. *Journal of Clinical Psychopharmacology, 2,* 190–191.

Malek-Ahmadi, P. (2003). Gabapentin and posttraumatic stress disorder. *Annals of Pharmacotherapy, 37,* 664–666.

Mao, Z.-M., Arnsten, A. F. T., & Li, B.-M. (1999). Local infusion of alpha-1 adrenergic agonist into the prefrontal cortex impairs spatial working memory performance in monkeys. *Biological Psychiatry, 46,* 1259–1265.

Marmar, C. R., Schoenfeld, F., Weiss, D. S., Metzler, T., Zatzick, D., Wu, R., et al. (1996). Open trial of fluvoxamine treatment for combat-related posttraumatic stress disorder. *Journal of Clinical Psychiatry, 57,* 66–70.

Marshall, R. D., Beebe, K. L., Oldham, M., & Zaninelli, R. (2001). Efficacy and safety of paroxetine treatment for chronic PTSD: A fixed-dose–placebo-controlled study. *American Journal of Psychiatry, 158,* 1982–1988.

Martenyi, F., Brown, E. B., Zhang, H., Koke, S. C., & Prakash, A. (2002). Fluoxetine v. placebo in prevention of relapse in post-traumatic stress disorder. *British Journal of Psychiatry, 181,* 315–320.

Martenyi, F., Brown, E. B., Zhang, H., Prakash, A., & Koke, S. C. (2002). Fluoxetine versus placebo in posttraumatic stress disorder. *Journal of Clinical Psychiatry, 63,* 199–206.

McEwen, B. S., Angulo, J., Cameron, H., Chao, H. M., Daniels, D., Gannon, M. N., et al. (1992). Paradoxical effects of adrenal steroids on the brain: Protection versus degeneration. *Biological Psychiatry, 31,* 177–199.

Mellman, T. A., Bustamante, V., David, D., & Fins, A. I. (2002). Hypnotic medication in the aftermath of trauma (Letter). *Journal of Clinical Psychiatry, 63,* 1183–1184.

Milanes, F. J., Mack, C. N., Dennison, J., & Slater, V. L. (1984). Phenelzine treatment of post-Vietnam stress syndrome. *VA Practitioner, 1,* 40–49.

Monahan, J. B., Handelman, G. E., Hood, W. F., & Cordi, A. A. (1989). DCS, a positive modulator of N-methyl-D-aspartate receptor, enhances performance of learning tasks in rats. *Pharmacology, Biochemistry, and Behavior, 34,* 649–653.

Monnelly, E. P., Ciraulo, D. A., Knapp, C., & Keane, T. (2003). Low dose risperidone as adjunctive therapy for irritable aggression in posttraumatic stress disorder. *Journal of Clinical Psychopharmacology, 19,* 377–378.

Morgan, C. A., Krystal, J. H., & Southwick, S. M. (2003). Toward early pharmacologic post-traumatic stress intervention. *Biological Psychiatry, 53,* 834–843.

Morgan, C. A., Wang, S., Mason, J., Southwick, S. M., Fox, P., Hazlett, G., et al. (2000). Hormone profiles in humans experiencing military survival training. *Biological Psychiatry, 47,* 891–901.

Morgan, C. A., III, Wang, S., Rasmusson, A., Hazlett, G., Anderson, G., & Charney, D. S. (2001). Relationship among cortisol, catecholamines, neuropeptide-Y and human performance during exposure to uncontrollable stress. *Psychosomatic Medicine, 63,* 412–422.

Nacher, J., Alonso-Llosa, G., Rosell, D. R., & McEwen, B. S. (2003). NMDA receptor antagonist treatment increases the production of new neurons in the aged rat hippocampus. *Neurobiology of Aging, 24,* 273–284.

Nakazawa, K., Quirk, M. C., Chitwood, R. A., Watanabe, M., Yeckel, M. F., Sun, L. D., et al. (2002). Requirement for hippocampal CA3 NMDA receptors in associative memory recall. *Science, 297,* 211–218.

Neal, L. A., Shapland, W., & Fox, C. (1997). An open trial of moclobemide in the treatment of posttraumatic stress disorder. *International Journal of Clinical Psychopharmacology, 12,* 231–232.

Nies, A. S. (2001). Principles of therapeutics. In J. G. Hardman, L. L. Limbird, & A. G. Gilman (Eds.), *Goodman and Gilman's the pharmacological basis of therapeutics* (10th ed., pp. 45–66). New York: McGraw-Hill.

Okuyama, N., Takagi, N., Kawai, T., Miyake-Takagi, K., & Takeo, S. (2004). Phosphorylation of extracellular-regulating kinase in NMDA receptor antagonist-induced newly generated neurons in the adult rate dentate gyrus. *Journal of Neurochemistry, 88,* 717–724.

Paul, I. A., Nowak, G., Layer, R. T., Popik, P., & Skolnick, P. (1994). Adaptation of the N-methyl-D-aspartate receptor complex following chronic antidepressant treatments. *Journal of Pharmacology and Experimental Therapeutics, 1,* 95–102.

Petty, F., Davis, L. L., Nugent, A. L., Kramer, G. L., Teten, A., Schmitt, A., et al. (2002). Valproate therapy for chronic, combat-induced posttraumatic stress disorder. *Journal of Clinical Psychopharmacology, 1,* 100–101.

Pitman, R. K., Sanders, K. M., Zusman, R. M., Healy, A. R., Cheema, F., Lasko, N. B., et al. (2002). Pilot study of secondary prevention of posttraumatic stress disorder with propranolol. *Biological Psychiatry, 51,* 189–192.

Post, R. M., Weiss, S. R. B., Li, H., Leverich, G. S., & Pert, A. (1999). Sensitization components of posttraumatic stress disorder: Implications for therapeutics. *Semininars in Clinical Neuropsychiatry, 4,* 282–294.

Post, R. M., Weiss, S. R. B., & Smith, M. A. (1995). Sensitization and kindling: Implications for the evolving neural substrate of PTSD. In M. J. Friedman, D. S. Charney, & A. Y. Deutch (Eds.), *Neurobiological and clinical consequences of stress: From normal adaptation to post-traumatic stress disorder* (pp. 135–147). Philadelphia: Lippincott–Raven.

Rapaport, M. H., Endicott, J., & Clary, C. M. (2002). Posttraumatic stress disorder and quality of life: Results across 64 weeks of sertraline treatment. *Journal of Clinical Psychiatry, 63,* 59–65.

Raskind, M. A., Peskind, E. R., Kanter, E. D., Petrie, E. C., Radont, A., Thompson, C. E., et al. (2003). Reduction of nightmares and other PTSD symptoms in combat veterans by prazosin: A placebo-controlled study. *American Journal of Psychiatry, 160,* 371–373.

Rasmusson, A. M., & Friedman, M. J. (2002). Gender issues in the neurobiology of PTSD. In R. Kimerling, P. Ouimette, & J. Wolfe (Eds.), *Gender and PTSD* (pp. 43–75). New York: Guilford Press.

Rasmusson, A. M., Hauger, R. L., Morgan, C. A., III, Bremner, J. D., Southwick, S. M., & Charney, D. S. (2000). Low baseline and yohimbine stimulated plasma neuropeptide Y (NPY) levels in combat-related PTSD. *Biological Psychiatry, 47,* 526–539.

Rasmusson, A. M., Lipschitz, D. S., Wang, S., Hu, S., Vojvoda, D., Bremner, J. D., et al. (2001). Increased pituitary and adrenal reactivity in premenopausal women with PTSD. *Biological Psychiatry, 50,* 965–977.

Reist, C., Kauffman, C. D., Haier, R. J., Sangdahl, C., DeMet, E. M., Chicz-DeMet, A., et al. (1989). A controlled trial of desipramine in 18 men with post-traumatic stress disorder. *American Journal of Psychiatry, 146,* 513–516.

Robert, R., Blakeney, P. E., Villarreal, C., Rosenberg, L., & Meyer, W. J., III. (1999). Imipramine treatment in pediatric burn patients with symptoms of acute stress disorder: A pilot study. *Journal of the American Academy of Child and Adolescent Psychiatry, 38,* 873–882.

Roozendaal, B., McReynolds, J. R., & McGaugh, J. L. (2004). The basolateral amygdala interacts with the medial prefrontal cortex in regulating glucocorticoid effects on working memory impairment. *Journal of Neuroscience, 24,* 1385–1392.

Rush, A. J. (2001, May 5–10). *Sequence treatment alternatives to relieve depression (STAR*D).* Presented in Syllabus and Proceedings Summary, American Psychiatric Association 154th Annual Meeting, New Orleans, LA.

Sapolsky, R. M. (2000). Glucocorticoids and hippocampal atrophy in neuropsychiatric disorders. *Archives of General Psychiatry, 57,* 925–935.

Saygin, M. Z., Sungur, M. Z., Sabol, E. U., & Cetinkaya, P. (2002). Nefazodone versus sertraline in treatment of posttraumatic stress disorder. *Bulletin of Clinical Psychopharmacology, 12,* 1–5.

Schelling, G., Briegel, J., Roozendaal, B., Stoll, C., Rothenhäusler, H.-B., & Kapfhammer, H.-P. (2001). The effect of stress doses of hydrocortisone during septic shock on posttraumatic stress disorder in survivors. *Biological Psychiatry, 50,* 978–985.

Schnurr, P. P., Friedman, M. J., Foy, D. W., Shea, M. T., Hsieh, F. Y., Lavori, P. W., et al. (2003). Randomized trial of trauma-focused group therapy for posttraumatic stress disorder. *Archives of General Psychiatry, 60,* 481–489.

Schnurr, P. S., & Green, B. L. (Eds.). (2004). *Trauma and health: Physical health consequences of exposure to extreme stress.* Washington, DC: American Psychological Association.

Schwartz, B. L., Hashtroudi, S., Herting, R. L., Schwartz, P., & Deutsch, S. I. (1996). D-cycloserine enhances implicit memory in Alzheimer patients. *Neurology, 46,* 420–424.

Schwartz, T. L. (2002). The use of tiagabine augmentation for treatment-resistant anxiety disorders: A case series. *Psychopharmacology Bulletin, 2,* 53–57.

Seedat, S., Lockhart, R., Kaminer, D., Zungu-Dirwayi, N., & Stein, D. J. (2001). An open trial of citalopram in adolescents with post-traumatic stress disorder. *International Clinical Psychopharmacology, 16,* 21–25.

Shestatzky, M., Greenberg, D., & Lerer, B. (1988). A controlled trial of phenelzine in posttraumatic stress disorder. *Psychiatry Research, 24,* 149–155.

Sokolski, K. N., Densen, T. F., Lee, R. T., & Reist, C. (2003). Quetiapine for treatment of refractory symptoms of combat-related post-traumatic stress disorder. *Military Medicine, 168,* 486–489.

Southwick, S. M., Krystal, J. H., Bremner, J. D., Morgan, C. A., Nicolaou, A. L., Nagy, L. M., et al. (1997). Noradrenergic and serotonergic function in posttraumatic stress disorder. *Archives of General Psychiatry, 54,* 749–758.

Southwick, S. M., Paige, S. R., Morgan, C. A., Bremner, J. D., Krystal, J. H., & Charney, D. S. (1999). Adrenergic and serotonergic abnormalities in PTSD: Catecholamines and serotonin. *Seminars in Clinical Neuropsychiatry, 4,* 242–248.

Stein, M. B., Kline, N. A., & Matloff, J. L. (2002). Adjunctive olanzapine for SSRI-resistant combat-related PTSD: A double-blind, placebo-controlled study. *American Journal of Psychiatry, 159,* 1777–1779.

Steward, J. T., & Bartucci, R. J. (1986). Posttraumatic stress disorder and partial complex seizures. *American Journal of Psychiatry, 1,* 113–114.

Stutzmann, G. E., & LeDoux, J. E. (1999). GABAergic antagonists block the inhibitory effects of serotonin in the lateral amygdala: A mechanism for modulation of sensory inputs related to fear conditioning. *Journal of Neuroscience, 11,* RC8.

Szymanski, H. V., & Olympia, J. (1991). Divalproex in post-traumatic stress disorder. *American Journal of Psychiatry, 8,* 1086–1087.

Taylor, F. B. (2003). Tiagabine for posttraumatic stress disorder: A case series of 7 women. *Journal of Clinical Psychiatry, 64,* 1421–1425.

Taylor, F. B., & Cahill, L. (2002). Propranolol for reemergent posttraumatic stress disorder following an event of retraumatization: A case study. *Journal of Traumatic Stress, 15,* 433–437.

Thompson, L. T., Moskal, J. R., & Disterhoft, J. F. (1992). Hippocampus-dependent learning facilitated by a monoclonal antibody or D-cycloserine. *Nature, 356,* 638–641.

Tucker, P., Zaninelli, R., Yehuda, R., Ruggiero, L., Dillingham, K., & Pitts, C. D. (2001). Paroxetine in the treatment of chronic posttraumatic stress disorder: Results of a placebo-controlled, flexible-dosage trial. *Journal of Clinical Psychiatry, 62*, 860–868.

U.S. Food and Drug Administration. (2004, October). FDA launches a multi-pronged strategy to strengthen safeguards for children treated with antidepressant medications. *FDA News.* Available at *www.fda.gov/bba/topics/news/2004/NEW01124.html*

VA/DoD Clinical Practice Guideline for Management of Post-Traumatic Stress, Veterans Health Administration. (2004). Available at *www.oqp.med.va.gov/cpg/pts/pts_base.htm*

Vaiva, G., Boss, V., Addesa, G., Cottencin, O., Fontaine, C., Fontaine, M., et al. (2000, November 17). *Low GABA levels and posttraumatic stress disorder.* Paper presented at the annual meeting of the International Society for Traumatic Stress Studies, San Antonio, TX.

Vaiva, G., Ducrocq, F., Jezequel, K., Averland, B., Lestavel, P., Brunet, A., et al. (2003). Immediate treatment with propranolol decreases posttraumatic stress disorder two months after trauma. *Biological Psychiatry, 54*, 947–949.

van der Kolk, B. A., Dreyfuss, D., Michaels, M., Shera, D., Berkowitz, R., Fisler, R., et al. (1994). Fluoxetine versus placebo in posttraumatic stress disorder. *Journal of Clinical Psychiatry, 55*, 517–522.

van der Meulen, J. A., Bilbija, L., Joosten, R. N., de Bruin, J. P., & Feenstra, M. G. (2003). The NMDA-receptor antagonist MK-801 selectively disrupts reversal learning in rats. *NeuroReport, 14*, 2225–2228.

Vermetten, E., & Bremner, J. D. (2002). Circuits and systems in stress: II. Applications to neurobiology and treatment in posttraumatic stress disorder. *Depression and Anxiety, 16*, 14–38.

Vermetten, E., Vythilingam, M., Southwick, S. M., Charney, D. S., & Bremner, J. D. (2003). Long-term treatment with paroxetine increases verbal declarative memory and hippocampal volume in posttraumatic stress disorder. *Biological Psychiatry, 54*, 693–702.

Viola, J., Ditzler, T., Batzer, W., Harazin, J., Adams, D., Lettich, L., et al. (1997). Pharmacological management of posttraumatic stress disorder: Clinical summary of a five-year retrospective study, 1990–1995. *Military Medicine, 162*, 616–619.

Walker, D. L., & Davis, M. (2002). The role of amygdala glutamate receptors in fear learning, fear-potentiated startle, and extinction. *Pharmacology, Biochemistry, and Behavior, 3*, 379–392.

Weathers, F. W., Keane, T. M., & Davidson, J. R. T. (2001). Clinician-Administered PTSD Scale: A review of the first ten years of research. *Depression and Anxiety, 13*, 132–156.

Wolfe, M. E., Alavi, A., & Mosnaim, A. D. (1988). Posttraumatic stress disorder in Vietnam veterans: Clinical and EEG findings: possible therapeutic effects of carbamazepine. *Biological Psychiatry, 23*, 642–644.

Xie, X., & Hagan, R. M. (1998). Cellular and molecular actions of lamotrigine: Possible mechanisms of efficacy in bipolar disorder. *Neuropsychobiology, 38*, 119–130.

Yehuda, R. (1999). Linking the neuroendocrinology of post-traumatic stress disorder with recent neuroanatomic findings. *Seminars in Clinical Neuropsychiatry, 4*, 256–265.

Yehuda, R. (2002). Current status of cortisol findings in post-traumatic stress disorder. *Psychiatric Clinics of North America, 2*, 341–368.

Yehuda, R., Giller, E. L., Southwick, S. M., Kahana, B., Boisneau, D., Ma, X., et al. (1994). Relationship between catecholamine excretion and PTSD symptoms in Vietnam combat veterans and holocaust survivors. In M. M. Murburg (Ed.), *Catecholamine function in post-traumatic stress disorder: Emerging concepts* (pp. 203–220). Washington, DC: American Psychiatric Press.

Zahrt, J., Taylor, T. R., Mathew, R. G., & Arnsten, A. F. T. (1997). Supranormal stimulation of dopamine D1 receptors in the rodent prefrontal cortex impairs spatial working memory performance. *Journal of Neuroscience, 17*, 8525–8535.

Zarate, C. A., Quirox, J., Payne, J., & Manji, H. K. (2002). Modulators of the glutamatergic system: Implications for the development of improved therapeutics in mood disorders. *Psychopharmacology Bulletin, 4*, 35–83.

Zullino, D. F., Krenz, S., & Besson, J. (2003). AMPA blockade may be the mechanism underlying the efficacy of topiramate in PTSD. *Journal of Clinical Psychiatry, 4*, 219–220.

Chapter 20

Trauma Exposure and Physical Health

Paula P. Schnurr, Bonnie L. Green, and Stacey Kaltman

Exposure to traumatic events is associated with poor physical health (Friedman & Schnurr, 1995; Green & Kimerling, 2004; Resnick, Acierno, & Kilpatrick, 1997; Schnurr & Jankowski, 1999). This chapter presents a model that explains the association through psychological, biological, attentional, and behavioral mechanisms (Schnurr & Green, 2004). First we describe the model and methodological issues, review supporting evidence, then discuss clinical, systems, and policy implications. The chapter ends with a proposed agenda for basic and applied research.

A MODEL OF HOW TRAUMATIC EXPOSURE AFFECTS PHYSICAL HEALTH

Think of a man who fractures his back in a motor vehicle accident or a woman who is infected with a sexually transmitted disease while being raped. It is easy to understand how they might have serious health problems that require treatment and impair their quality of life. Yet most individuals are not seriously injured or exposed to disease during a traumatic event (e.g., Kulka et al., 1990; Resnick, Kilpatrick, Dansky, Saunders, & Best, 1993). Furthermore, the types of health problems typically reported by trauma survivors are not directly related to the types of events experienced. For example, in a classic study, Felitti and colleagues (1998) found that childhood trauma was associated with increased likelihood of adult cancer, ischemic heart disease, chronic lung problems, and other conditions that had no known or direct etiological basis in the childhood events. How could traumatic exposure lead to such seemingly unrelated health problems?

Building upon prior work (Friedman & Schnurr, 1995; Schnurr & Jankowski, 1999), we have proposed a model (Schnurr & Green, 2004) to explain how a traumatic event could affect physical health. The model, depicted in Figure 20.1, is based on two key assumptions. The first assumption is that following trauma exposure, distress, manifested as either posttraumatic stress disorder (PTSD) or other serious psychological conditions, is necessary for adverse health outcomes. Even when an individual suffers direct physical consequences of exposure, posttraumatic distress is likely (e.g., trauma-related injury is associated with increased risk of developing PTSD) (Green, Grace, Lindy, Gleser, & Leonard, 1990; Resnick et al., 1993; Schnurr, Ford, et al., 2000). Thus, the model applies broadly to all trauma survivors.

The second assumption is that the effects of PTSD and other posttraumatic distress reactions are mediated though interacting biological, psychological, attentional, and behavioral mechanisms. *Biological mechanisms* include alterations of the two primary systems of the stress response: the locus coeruleus/norepinephrine (LC/NE) sympathetic system and the hypothalamic–pituitary–adrenal (HPA) system. Friedman and McEwen (2004) summarized the literature on these systems, as well as other neurobiological changes associated with PTSD, and discussed the possible implications of the changes for physical health. *Psychological mechanisms* include depression, hostility, and poor coping, all of which have been linked to adverse health effects. For example, depression is associated with greater likelihood of cardiovascular disease and the mechanisms that could explain this association, including greater platelet activation, de-

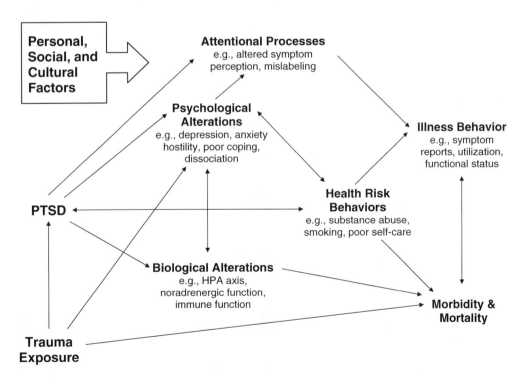

FIGURE 20.1. A model relating traumatic exposure and PTSD to physical health outcomes. From Schnurr and Green (2004, p. 248). In the public domain.

creased heart rate variability, and greater likelihood of hypertension (Ford, 2004). With respect to *attentional factors*, many explanations have been offered for why trauma may increase negative health perceptions and illness behavior. For example, Pennebaker (2000) has suggested avoidance of thinking about a trauma, and mislabeling of the autonomic and emotional consequences of such avoidance (in addition to actual biological changes and secondary gain), both of which are associated with poor health. *Behavioral mechanisms* associated with trauma and PTSD include both substance use or abuse (smoking, alcohol, drugs, and food) and failure to engage in preventive strategies (exercise, diet, safe sex, regular health care) (Rheingold, Acierno, & Resnick, 2004). Failure to adhere to medical regimens also falls into this category (Buckley, Green, & Schnurr, 2004). Although behavioral mechanisms are not likely to explain fully the relationship between trauma and poor health, they appear to play an important role.

Allostatic load is defined as "the strain on the body produced by repeated ups and downs of physiologic response, as well as the elevated activity of physiologic systems under challenge, and the changes in metabolism and wear and tear on a number of organs and tissues" (McEwen & Stellar, 1993, p. 2094). It has been proposed as a unifying mechanism to explain how the numerous and sometimes subtle neurobiological, psychological, and behavioral changes associated with PTSD may jointly affect health (Friedman & McEwen, 2004; Schnurr & Green, 2004; Schnurr & Jankowski, 1999). Because the construct emphasizes cumulative and interactive effects across multiple systems, it is useful for understanding how changes that by themselves are clinically insignificant could combine to produce disease. Schnurr and Jankowski (1999) gave the example of elevated levels of arousal and hyperreactivity in PTSD. These changes alone may not lead to cardiovascular disease, even though they may strain the system (i.e., increase allostatic load). However, in combination with behavioral risk factors that are also associated with PTSD, such as substance abuse and smoking, allostatic load might be sufficiently increased to cause disease (see Figure 20.2). Schnurr and Jankowski proposed that allostatic load might be greater in PTSD relative to other disorders, but this hypothesis has not yet been tested.

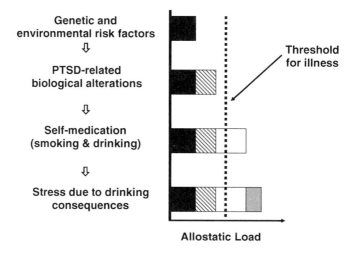

Allostatic Load

FIGURE 20.2. A hypothetical example of how biological and behavioral factors could combine to increase allostatic load in an individual with PTSD.

Psychiatric disorders such as depression and substance abuse, which have known effects on physical health, are frequently comorbid with PTSD (Kessler, Sonnega, Bromet, Hughes, & Nelson, 1995). According to the model, PTSD is the key mechanism through which trauma leads to poor health, although other types of distress, particularly depression, may lead to poor health in the absence of PTSD. Furthermore, depression and other types of comorbid distress may mediate the effects of PTSD. Yet PTSD appears to have a distinctive effect on health beyond that associated with these comorbid conditions. Below we present evidence in support of these points after reviewing evidence on how trauma and PTSD are associated with physical health. First, we describe key methodological issues that are important for understanding the evidence.

METHODOLOGICAL CONSIDERATIONS

Physical health consists of interacting components manifested in both objective and subjective domains. According to Wilson and Cleary (1995), these domains exist on a continuum of increasing complexity. Biological and physiological variables—the underlying changes that represent disease or alterations of the physical system—comprise the most basic level. Next are the symptoms the individual experiences, which are imperfectly correlated with biological and physiological variables. Functional status—how the individual is affected by symptoms—is the next level, followed by health perceptions (e.g., "My health is poor [fair, good, excellent]"). Health-related quality of life is at the most complex level. Personal and environmental factors influence all levels of this continuum. Schnurr and Green (2004) gave the example of functional status, which is affected by not only experienced symptoms but also a person's motivation and socioeconomic supports. An example at a more basic level is that diet can affect a person's cholesterol level, which also is affected by genetic factors.

Measurement Issues

Given its multidimensional character, physical health may be measured in numerous ways. Self-reports can be used to assess overall health, individual symptoms, functional status, physical conditions, and utilization. Laboratory tests and physicians' exams are useful for assessing biological changes and disease. Archival records may be used to assess utilization and mortality.

A particular implication of Wilson and Cleary's (1995) model, or any other model that includes both objective and subjective components, is that self-reports are valid indicators of health. Self-reports can be used to assess outcomes across almost the entire range of Wilson and Cleary's continuum, from the biological (e.g., "What is your weight?") through health-related quality of life. However, self-reports may not always map onto other levels. Studies that have compared self-reports with information in medical charts or databases typically find some level of disagreement between the two sources of information. Beckham and colleagues (1998) found kappas in the low-to-moderate range for self-reported medical conditions among Vietnam veterans; accuracy did not differ between PTSD and no PTSD groups. Edwards and colleagues (1994), in a study by the National Center for Health Statistics, found that self-reported medical conditions were both overreported and underreported relative to information in medical charts, depending on the type of condition and whether an individual was receiving ongoing treatment for that condition. One study found that 24% of patients who had been hospitalized in the prior year failed to report the hospitalization (Wallihan, Stump, & Callahan, 1999), and another found that approximately 20% of patients reported their current prescriptions inaccurately (Sjahid, van der Linden, & Stricker, 1998). When interpreting these findings, it is essential to remember that archival sources may not be complete. Individuals may seek care from more than one source (e.g., from both a health maintenance organization [HMO] and providers who are outside of the plan). Using utilization records or medical charts from only the HMO to verify self-reports could cause accurately reported information to appear inaccurate.

Moreover, there is a history of controversy about the use of self-report methods for studying physical health, because self-reports are affected by psychological and emotional factors such as negative affectivity (Watson & Pennebaker, 1989). Symptom reports are most affected, although all reports of health status and functional health are affected to some extent (see Schnurr & Jankowski, 1999). The concern that self-reports do not actually reflect physical status so much as emotional status is only a problem if one assumes that health is strictly biological. Rather than dismissing self-reports as invalid, readers should view them as one of a range of necessary approaches for thoroughly capturing the multidimensional complexity of physical health.

Design Issues

Physical health is affected by many factors. It is important to identify the way these factors interact with trauma. For example, depression and substance abuse, which have known adverse effects on physical health, are more common in individuals with PTSD than in those without PTSD (Kessler et al., 1995). In the majority of cases, the onset of the comorbid problems is secondary to the onset of PTSD (Kessler et al., 1995). Study designs that statistically control for the effects of these problems—to eliminate the potential "confounding" they introduce—reduce the observed effect of PTSD if the problems are actually mechanisms through which PTSD influences health. This control approach is appropriate only if the goal is to determine the unique effect of PTSD, beyond that mediated by its consequences (e.g., to know whether the effects of PTSD are fully explained by other behavioral mechanisms). The approach is inappropriate if one's goal is to determine the total effect of PTSD or to understand how PTSD affects health. In the former case, either simultaneous multiple regression analysis or analysis of covariance is an acceptable analytic technique. In the latter, hierarchical regression, path analysis, or structural equation modeling should be used to test for mediation.

As we have indicated, most trauma survivors are not injured or made ill as a direct consequence of their exposure (e.g., Resnick et al., 1993). Nevertheless, physical harm can result from almost any directly experienced traumatic event, and some types of traumatic exposure, such as torture, accidents, and physical assault, are likely to involve injury or illness. For example, take the case of chronic pain and PTSD in motor vehicle accident (MVA) survivors. PTSD and chronic pain often co-occur, even in individuals who do not have trauma-related injury (Otis, Keane, & Kerns, 2003). This may be because pain and PTSD mutually reinforce one another (Sharp & Harvey, 2001), or because anxiety sensitivity predisposes individuals to both conditions (Asmundson, Coons, Taylor, & Katz, 2002). In such cases, designs and analytic strategies need to distinguish between health effects that are a direct result of trauma and those that are a result of mechanisms stemming from posttraumatic reactions.

Schnurr and Green (2004) noted that health is influenced by many factors in addition to trauma: personal characteristics, including genetics, social factors, and ethnic and cultural background (Wilson & Cleary, 1995). Thus, it may seem necessary to isolate the effect of exposure and its consequences as distinct from these factors. However, this is not always possible or even desirable. What is necessary is to ensure that these factors are adequately controlled insofar as they may provide an alternative explanation for a given finding.

CURRENT STATE OF THE ART

Literature reviews have summarized the evidence showing that trauma exposure and PTSD are associated with poor health (Friedman & Schnurr, 1995; Green & Kimerling, 2004; Resnick et al., 1997; Schnurr & Jankowski, 1999). Rather than review the existing literature in detail, we present selected studies as examples of more general findings or to emphasize key points.

Is Trauma Associated with Poor Health?

Evidence of the relationship between trauma exposure and self-reported health problems comes from large samples of civilians, veterans, and military personnel (e.g., Flett, Kazantzis, Long, MacDonald, & Millar, 2002; Martin, Rosen, Durand, Knudson, & Stretch, 2000; Schnurr, Spiro, Aldwin, & Stukel, 1998; Ullman & Siegel, 1996). Additional evidence comes from studies of special populations, including sexual assault victims (Golding, 1996), adults who experienced childhood trauma (Felitti et al., 1998), and older adults (Higgins & Follette, 2002).

Other studies have demonstrated that trauma exposure is associated with objective indicators of poor health. In a sample of 1,225 randomly selected female subscribers of a large HMO, Walker and colleagues (1999) found that women who experienced childhood maltreatment had more charted medical diagnoses, including minor infectious disease, pain disorders, and other conditions (hypertension, diabetes, asthma, allergy, and abnormal uterine bleeding) in the past year, than women without maltreatment, consistent with their self reports. Sibai, Armenian, and Alam (1989) observed a relationship between wartime trauma exposure and coronary artery disease (as measured by coronary angiography) that remained when well-known risk factors for coronary artery disease were controlled.

Individuals who are exposed to trauma also use more medical services than unexposed individuals. Although a number of studies have demonstrated a link between trauma exposure and self-reported utilization, more robust evidence is supplied by studies that use objective indicators of utilization (see Walker, Newman, & Koss, 2004). Some research has assessed the costs associated with increased utilization. For example, Walker and colleagues (1999) studied utilization and costs in HMO participants, using the automated cost accounting system of an HMO. Women with any childhood abuse or neglect had significantly higher median annual health care costs than women who reported no childhood maltreatment. Women who reported childhood sexual abuse had particularly high median annual health care costs, as well as higher primary care costs, higher outpatient costs, and more emergency department visits.

Most investigations of the relationship between trauma exposure and mortality have found that exposure is related to increased mortality. Studies of veterans have shown that the increase is primarily due to external causes (e.g., accidents and suicide) and not disease, although there are a few exceptions (e.g., Visintainer, Barone, McGee,

& Peterson, 1995). For example, Vietnam War veterans in Vietnam Experience Study had a 7% higher mortality risk than Vietnam Era veterans over a 30-year period (Catlin Boehmer, Flanders, McGeehin, Boyle, & Barrett, 2004). The excess mortality was attributable to increased deaths due to external causes during the first 5 years after discharge. Less is known about mortality in nonveteran populations. White and Widom (2003) found no relationship between mortality and abuse in a study of young adults. In contrast, Sibai, Fletcher, and Armenian (2001) found that exposure was associated with increased risk of both cardiovascular-related death and all-cause mortality among men and women exposed to war-related stressors in Lebanon. Women who experienced loss-related trauma and individuals displaced by war-related events had the greatest risk.

Is PTSD Associated with Poor Health?

As described in our model, a primary pathway from the experience of trauma exposure and adverse health outcomes is the reaction to the exposure, specifically, PTSD. There is now solid empirical evidence of the link between PTSD and poor health. Studies of veterans have revealed that PTSD is associated with poorer perceived health, more chronic health conditions, and poorer self-rated functional status (e.g., Barrett et al., 2002; Kulka et al., 1990; Neria & Koenen, 2003; Schnurr, Ford, et al., 2000). For example, among veterans in the Vietnam Experience Study (Boscarino, 1997), a lifetime diagnosis of PTSD was associated with increased risk for self-reported chronic disorders, including circulatory, digestive, musculoskeletal, endocrine, and respiratory disorders, as well as non-sexually-transmitted infectious disease. Importantly, these relationships were observed in analyses that also controlled for numerous factors thought to affect onset of illness, including demographic characteristics, hypochondriasis, smoking, and substance abuse.

PTSD also is associated with poor health in nonveterans. Sareen, Cox, Clara, and Asmundson (2005) recently reported findings from a large national probability sample of U.S. adults (Kessler et al., 1995) in which medical disorders had been assessed with the National Health Interview Survey, a measure with acceptable validity (Edwards et al., 1994). PTSD was associated with increased odds of neurological, vascular, gastrointestinal, metabolic or autoimmune, and bone or joint conditions. Zatzick, Jurkovich, Gentillelo, Wisner, and Rivara (2002) found an association between self-reported health and PTSD in a civilian sample of 101 acutely injured patients who were hospitalized. At 1-year follow-up, individuals with PTSD reported poorer perceived health than those without PTSD. PTSD is also associated with accelerated disease progression, worse functional outcomes, and an increased likelihood of being seen in medical settings (Kimerling et al., 1999; Stein, McQuaid, Pedrelli, Lenox, & McCahill, 2000).

The studies mentioned previously have used self-report methods to assess health. PTSD is associated with poor outcomes when physical health is measured by objective indicators as well. Most of the evidence to date comes from studies of veteran populations. In a longitudinal study of 605 male veterans of the Korean Conflict and World War II, PTSD symptoms were associated with risk of onset for several categories of physician-diagnosed medical problems: arterial, gastrointestinal, and musculoskeletal disorders, and dermatological problems even when other factors predictive of health status (age, smoking, body mass index [BMI], and alcohol use) were statistically controlled (Schnurr, Spiro, & Paris, 2000). In a sample of individuals seeking treatment within the Department of Veterans Affairs health care system, PTSD diagnosis and symptoms were associated with an increased prevalence of circulatory and musculoskeletal disorders, as well as an increased number of medical conditions (Ouimette et al., 2004).

Several studies have examined the specific association between PTSD and cardio-vascular health (e.g., Beckham et al., 2002). In the Centers for Disease Control and Prevention (CDC) Vietnam study sample, PTSD was associated with electrocardiograph (ECG) abnormalities, atrioventricular defects, and infarctions, even when other factors related to coronary heart disease, such as demographic characteristics, drug, alcohol, and tobacco use, and BMI were controlled (Boscarino & Chang, 1999). In a study of Israeli veterans, individuals with PTSD had worse performance than noncombat veterans on laboratory stress tests, although no differences were found on other indices, including heart rate, blood pressure, or physical exam findings (Shalev, Bleich, & Ursano, 1990). In a recent study of almost 20,000 former prisoners of war (POWs) from World War II, odds of circulatory disease, hypertension, and chronic heart disorder were elevated in former POWs with PTSD relative to those who did not have PTSD (Kang, Bullman, & Taylor, 2006).

Evidence that PTSD is associated with higher utilization of the health care system is also accumulating. In a sample of 156 veterans, Deykin and colleagues (2001) found that high users of health care services were more likely to meet diagnostic criteria for PTSD than low users of health care services. Participants with current PTSD made on average 30% more health care visits than participants with partial or no PTSD. In their study of 1,225 women HMO participants, Walker and colleagues (2003) used data from an automated cost accounting system and found that PTSD was associated with increased health care costs even after controlling for depression, chronic medical illness, and psychological distress.

The few studies examining mortality among individuals with PTSD have found that PTSD is associated with excess mortality. Boscarino (2006) analyzed data from a sample of over 15,000 Vietnam Era and Vietnam Theater veterans who were originally assessed for PTSD in the 1980s. PTSD was associated with all-cause mortality, and mortality due to cardiovascular disease and external causes over a 16-year interval. Among Vietnam Era veterans, PTSD was associated with all-cause mortality only. Among Vietnam Theater veterans, PTSD was associated with all-cause mortality and mortality due to cardiovascular disease, cancer, and external causes. Other studies have found that the excess mortality associated with PTSD is primarily due to external causes. Bullman and Kang (1994) studied mortality rates of 16,257 Vietnam War veterans from the Agent Orange Registry. Veterans with PTSD had a 71% higher mortality rate than the veterans without PTSD, as well as higher mortality rates due to accidents and suicide. The veterans with PTSD were also more likely to die from digestive system diseases, including cirrhosis of the liver, than were age and ethnicity-matched general population controls. Kasprow and Rosenheck (2000) found that male veterans who received PTSD outpatient treatment in 1989 or 1990 had higher mortality (through 1999) than U.S. men in the general population. Drescher, Rosen, Burling, and Foy (2003) examined causes of death in 1,866 male veterans who received residential treatment for PTSD. Approximately 6% of the veterans died during the study period, which is higher than expected for a general population sample with similar age and racial distributions. Behavioral causes of death—accidents, suicide, effects of chronic substance use, and HIV/hepatitis—were almost six times more than expected. Deaths due to other diseases were not significantly more than expected.

Does PTSD Mediate the Relationship between Trauma Exposure and Poor Health?

Evidence consistent with the hypothesis that PTSD mediates the relationship between trauma exposure and self-reported health has been observed in a diverse range of samples, including female Vietnam War veterans (Wolfe, Schnurr, Brown, & Furey, 1994), bus drivers who experienced work-related accidents (Vedantham et al., 2001), and primary care patients (Weisberg et al., 2002). In path-analytic data from 1,632 male and female Vietnam veterans, Taft, Stern, King, and King (1999) found that PTSD mediated 58% of the effect of combat exposure on self-reported health in men and 35% of the effect in women. Similarly, in path-analytic data from over 900 older male veterans, Schnurr and Spiro (1999) found that 90% of the effect of combat exposure on health was mediated through PTSD. In a prospective study of 2,301 Gulf War veterans, Wagner, Wolfe, Rotnitsky, Procter, and Erickson (2000) found that combat exposure was a significant predictor of health status 18–24 months following return from the Gulf. This relationship was substantially reduced when PTSD was included in the regression model, which is consistent with the idea that PTSD is mediating the relationship. Using structural equation modeling, Ford and colleagues (2004) found that the best model to explain health problems in a sample of World War II veterans exposed to mustard gas included PTSD as a mediator of the relationship between exposure and health.

One study used objective indicators of health status to examine the mediating role of PTSD in a sample of World War II and Korean Conflict veterans. Schnurr, Spiro, and Paris (2000) found that combat exposure predicted onset of physician-diagnosed arterial, pulmonary, upper gastrointestinal, and heart disorders over a 30-year interval. However, PTSD appeared to mediate only the effect of exposure on arterial disorders.

Are the Effects of PTSD Distinct from the Effects of Other Mental Disorders?

Because PTSD is often comorbid with other disorders (Kessler et al., 1995), an important question is whether PTSD has a specific impact beyond that caused by these disorders. Some researchers have examined the unique contribution of PTSD in explaining physical health problems by controlling for other mental disorders (e.g., Beckham et al., 2002). In a veteran sample, for instance, Beckham and colleagues (1998) found that PTSD predicted several health outcomes even in analyses that were statistically controlled for demographics, combat exposure, smoking, alcohol use, depression, and hypochondriasis. In a study of female veterans, Frayne and colleagues (2004) found that veterans with PTSD reported more medical conditions and worse physical health status than veterans with depression (without PTSD) and those with neither condition. Other studies of veterans have shown a specific impact of PTSD when researchers controlled for substance abuse, anxiety disorders, and depression (e.g., Boscarino, 1997; Boscarino & Chang, 1999; Schnurr, Friedman, Sengupta, Jankowski, & Holmes, 2000).

Further evidence for a specific impact of PTSD has been found in civilian samples (e.g., Walker et al., 2003; Zatzick et al., 2002; Zayfert, Dums, Ferguson, & Hegel, 2002; Zoellner, Goodwin, & Foa, 2000). For example, Clum, Calhoun, and Kimerling (2000) found that both depression and PTSD in rape victims contributed unique variance to predicting self-reported symptoms and global health perceptions. Among 1,007 HMO patients, PTSD significantly increased the risk of somatic symptoms, over and above the total number of psychiatric disorders (Andreski, Chilcoat, & Breslau, 1998). The variance explained by PTSD was substantially reduced when the other disorders were included in the model, which is consistent with the hypothesis that the effects of PTSD are partially mediated through comorbid psychological alterations.

Does Treating PTSD Affect Physical Health?

It follows that if PTSD is a pathway through which individuals exposed to trauma develop physical health impairment, then treating PTSD should also improve physical health outcomes. However, there is insufficient evidence to determine whether successfully treating PTSD improves health. In an uncontrolled case report of six women with chronic pain following MVAs, Shipherd, Beck, Hamblen, Lackner, and Freeman (2003) examined the impact of a manualized PTSD intervention. Five of the six women no longer met criteria for PTSD at the end of treatment. The sample also had pain-related functional improvements, although there was no change in subjective pain. Results from two randomized clinical trials are similarly discouraging. Schnurr and colleagues (2003), in a large study of two types of group psychotherapy for PTSD in Vietnam War veterans, found no improvement in self-reported health on the Short Form–36 Health Survey (SF-36), despite improvement in PTSD and other symptoms. Malik and colleagues (1999) reported similar negative findings in a small, randomized, placebo-controlled pilot study of fluoxetine. It is clear that more evidence is needed before conclusions can be drawn about the effects of PTSD treatment on physical health.

CHALLENGES FOR THE FUTURE

We have discussed practice and policy issues that arise from the evidence on the adverse physical health consequences of trauma and PTSD (Schnurr & Green, 2004). Below we summarize these issues and suggest a research agenda for the future.

Practice

The relationship between trauma and poor health has significant implications for the provision of health care. If trauma increases the likelihood of disease, then attention should be paid to the physical health needs of trauma survivors in mental health care settings. Yet the majority of individuals with PTSD do not seek mental health care. Instead, they are likely to seek care in a medical setting (Samson, Benson, Beck, Price, & Nimmer, 1999), where their PTSD may go unrecognized (Taubman-Ben-Ari, Rabinowitz, Feldman, & Vaturi, 2001). Thus, attention also should be paid to the mental health needs of trauma survivors in medical settings.

Mental Health Care Settings

Mental health clinicians need to attend to the physical health problems of traumatized patients. Kilpatrick, Resnick, and Acierno (1997) suggested that a patient's physical health problems and visits to providers outside the mental health system should be assessed to assess the patient comprehensively and to increase his or her awareness of the interdependency between mental and physical health. Psychoeducation is a critical aspect of treating a trauma patient. Helping individuals who have PTSD or other trauma-related disorders to understand the links between their distress and their physical health can facilitate management of both physical and mental health problems (Kilpatrick et al., 1997).

Mental health clinicians also should attend to health risk behaviors such as smoking and substance abuse. Although these behaviors are sometimes the focus of treatment, they may need to be addressed differently if they are being used as coping strategies to manage trauma-related distress. Helping patients understand linkages between their symptoms and health risk behaviors may be necessary to achieve behavior change. It may also be important to help patients find alternatives for the risk behaviors used as coping strategies (Rheingold et al., 2004).

In some settings, offering physical health care in a mental health clinic can be another useful strategy. Druss, Rohrbaugh, Levinson, and Rosenheck (2001) compared two models of providing primary care for patients with serious psychiatric disorder (including PTSD): care integrated in a mental health clinic versus usual care in a general medical clinic. The integrated care group had more favorable medical outcomes, was more likely to receive preventive care, and had higher satisfaction—all without increased costs. Further studies like this are needed to generate an empirical basis for determining how best to meet the medical needs of traumatized patients.

Medical Care Settings

Identifying individuals who have PTSD or other types of posttraumatic distress is the first step in treating trauma-related problems. Green and Kimerling (2004) noted that studies in this area have tended to recommend universal screening procedures, even though screening is not always possible or desirable. The ideal screening procedure has optimal efficiency and adds minimal burden in terms of cost or other resource demands. In addition, there need to be effective interventions available for individuals who screen positive. It can be difficult to satisfy these conditions, and to date, no research with trauma survivors indicates whether such "screen-and-treat" models are useful. However, depression screening that is accompanied by integrated care has proven beneficial to patients (e.g., Oxman, Dietrich, Williams, & Kroenke, 2002) and may be cost-effective (e.g., Schoenbaum et al., 2001). It is reasonable to suggest that similar findings may be achieved in traumatized patients.

A brief self-report screen collected as part of a medical history is an easy and relatively inexpensive way to obtain information about trauma. But what, exactly, should be measured? Screening for depression alone is inadequate, unless patients are followed with a more thorough exam prior to the initiation of treatment, because different treatments may be indicated for PTSD than for depression. This is particularly important if an individual fails to respond to a selective serotonin reuptake inhibitor (SSRI), which may be a firstline strategy for both conditions (American Psychiatric Association, 2000; Foa, Keane, & Friedman, 2000). Thus, depression screening should be supplemented with a brief PTSD screen even if trauma history is not assessed.

There are several good PTSD screens available. The PTSD Checklist (PCL) is a questionnaire in which each of the 17 DSM-IV symptoms of PTSD are rated on a 5-point (1–5) Likert scale. The cutpoint needed for diagnosis may vary as a function of the population being treated. In two studies of primary care patients, one study found that a cutoff of 30 was optimal in women seen in an HMO (Walker, Newman, Dobie, Ciechanowski, & Katon, 2002), whereas the other study found that 48 was the optimal cutoff in male and female military veterans seen at a VA hospital (A. Prins, personal communication, October 20, 2004). Brief versions of the PCL have been validated for use in primary care screening (Lang & Stein, 2004); a six-item version has the best overall trade-off between length and psychometric properties. Another brief screen is the four-item Primary Care PTSD Screen (PC-PTSD), in which each of four PTSD symptom clusters (intrusion, avoidance, numbing, and hyperarousal) is assessed in dichotomous format (yes–no). A score of 3 or higher (endorsing 3 or more clusters) has optimal sensitivity and specificity for detecting PTSD (Prins et al., 2004).

Although all of these instruments link intrusion and avoidance symptoms to the occurrence of a traumatic event, and the PC-PTSD anchors questions about symptoms to a lead-in that references the symptoms to traumatic exposure, none contain specific questions about exposure. Primary care providers may feel more comfortable asking only about symptoms rather than the details of an event such as child abuse. Yet without questions to distinguish between nontraumatic and traumatic stressors, it is possible that individuals experiencing distress due to a nontraumatic life event might be categorized as having PTSD when in fact they do not. The specificity of the existing screens is very good (e.g., .72 for Lang & Stein's [2005] six-item PCL and .87 for Prins et al.'s [2004] PC-PTSD), but there have been no investigations of whether diagnostic performance could be enhanced if trauma history information were additionally used for making a screening diagnosis.

Education is essential. Medical providers need information to enable them to address trauma-related issues. They need to be aware of trauma reactions and how to discuss them with patients, especially because patients with trauma-related disorders can be challenging to treat. For example, in a study of economically disadvantaged women, those who had PTSD, compared to women exposed to violence who did not have PTSD, were more concerned that they would not get good care, did not trust their doctors, found medical staff rude, and felt that staff did not understand their problems (Bassuk, Dawson, Perloff, & Weinreb, 2001). Patients may need basic information about trauma and PTSD. They also may need to understand how their symptoms may relate to their physical problems and self-care behaviors. Depending on a practice's resources, this information can be delivered by the provider, other staff, written materials, or the Internet.

Blount (1998) described nine reasons to integrate primary and mental health care, four of which are especially relevant here: (1) Integrated care is consistent with the way patients typically present their distress (i.e., they do not make a sharp distinction between physical and mental health); (2) primary care is still the main setting in which patients prefer to receive care for problems that have a psychological basis; (3) adherence to treatment regimens and outcomes is likely to be enhanced, because a setting in which mental and physical health care are integrated reflects how patients present (i.e., with mixed physical and mental health symptoms); and (4) primary care providers, even those trained in mental health, cannot be expected to treat the range of psychological problems they encounter, and referrals outside the practice often fail. Blount also noted that integrated care leads to higher satisfaction among providers and patients, usually without additional cost.

Integrating medical care and mental health care can be accomplished in many ways. Involvement may range from a courtesy report of a patient contact to active collaboration between mental health and medical providers (Blount, 1998). The form of active collaboration may vary as well. Mental health services may be directly provided in the medical setting (e.g., Katon et al., 1996); conversely, medical care services may be integrated in a mental health setting (e.g., Druss et al., 2001). Also, mental health providers may help to enhance the skills of medical care providers to treat mental health issues (Dietrich et al., 2004) through education, consultation, and supervision. For example, a psychiatrist supervising the administration of psychiatric medication by primary care nurse practitioners worked well in a study of low-income women with depression who were treated in their primary care setting (Green et al., 2006; Miranda et al., 2003). One-third to one-half of the women had PTSD. Compared with women referred to community treatment, women with and without comorbid PTSD improved on paroxetine (Paxil) administered by nurse practitioners at 3 and 6 months, although the women with PTSD continued to have more symptoms and poorer functioning throughout the 12-month follow-up period (Green et al., 2006).

To date, there is no evidence about which of these approaches works best for addressing the needs of traumatized patients. We suggest that the way these needs are met should depend on the practice setting, as well as patient characteristics. Providing primary care in a mental health clinic may be best for patients with extremely chronic PTSD, who otherwise might fail to attend other medical appointments. Likewise, providing mental health care in a primary care clinic may be best for patients who find the concept of mental illness stigmatizing (e.g., Hoge et al., 2004) and who might reject or not follow through with a referral to a mental health clinic.

Policy and Systems Issues

The relationship between trauma and poor health has important implications for health policy. Trauma may be seen as one of the root causes of serious public health concerns—both the behavioral risk factors, such as smoking and lack of exercise, that may lead to disease and the diseases themselves. Furthermore, trauma and PTSD increase costs for individuals, health care systems, and society as a whole (Walker et al., 2004). For example, in one HMO, women with severe PTSD had annual median costs that were 104% higher than costs among women who did not have PTSD (Walker et al., 2003). We suggest that trauma be included in public health efforts to enhance both the recognition and treatment of mental disorder, and the prevention of exposure to accidents, violence, and other (possibly) avoidable events.

Integration of care is a key issue. Although individual providers may develop locally integrated services, it is also necessary to make changes at the systems level. Walker and colleagues (2004) offered an approach to treating patients with trauma-related medical and psychological problems in the context of an HMO or other large medical system. The approach focuses on stabilization of chronic problems. It involves case management and special tracks in primary care to increase recognition of trauma-related distress. The premise is that not only will outcomes be enhanced for patients but also initial costs will be offset by lower costs once the chronic problems are stabilized and appropriate maintenance care is initiated. However, the authors note that despite the theoretical logic—investing up front to save later—the proposed approach would be unlikely to yield cost savings because of the high turnover rates among health plan enrollees. There is no benefit to a system that invests in screening and treatment, unless participants remain in the system long enough for the costs of additional care to be offset by the reduction of inappropriate or unnecessarily expensive care. Walker and colleagues (2004) suggest that the provision of national health insurance is one strategy that could allow the offset to be realized.

Research

Research is needed to help us understand and treat the physical health consequences of traumatic exposure. We have proposed the following agenda to help generate the necessary empirical base (Schnurr & Green, 2004).

First are methodological issues. Future research should be based on large, representative samples to enhance the generalizability of findings. There is a particular need for studies of populations outside of North America, including those in developing countries. This research should include measures of PTSD and other significant posttraumatic reactions either in addition to or even instead of measures of traumatic exposure. Measures of morbidity based on physical exams or laboratory tests (and not just self-reports) are essential.

Next are the content issues that need to be addressed. We need to know which physical health problems are associated with PTSD. Although the range of behavioral and neurobiological correlates of PTSD could affect multiple body systems, some types of problems (e.g., cardiovascular) may be more likely than others. By knowing the specific outcomes associated with PTSD, we could begin to examine the mechanisms through which PTSD leads to poor health, particularly the biological mechanisms. We have recommended that studies of biological factors in PTSD include measures of health status to permit tests of how these factors relate to health (Schnurr & Green, 2004).

We also have recommended that future research include measures of both PTSD and other types of distress to address questions about the unique effects of PTSD on physical health. One question is about the extent to which PTSD affects health independent of comorbid disorders. A related question concerns the extent to which posttraumatic reactions other than PTSD affect physical health. Depression is a particularly important construct to include in studies of PTSD and health given its comorbidity with PTSD (Kessler et al., 1995) and diverse effects on physical health (Leon et al., 2003).

Other research needs to address treatment-related changes. We need to evaluate whether preventive, educational, and supportive interventions for PTSD and other outcomes in trauma survivors improve physical health. Although the few existing treatment studies have not found that improvements in PTSD lead to improvements in health, more research is needed, especially studies with sufficiently long follow-up periods to observe changes in health. Another important question regarding treatment is whether interventions designed to improve physical health affect PTSD and other clinically significant distress reactions. Furthermore, trauma survivors may need targeted health promotion interventions to address the ways their symptoms prevent them from engaging in positive health practices, and such interventions should be evaluated.

Interventions that target providers and systems of care also need to be evaluated, especially outside of mental health care settings (e.g., to address questions involving screening, integrated vs. segregated care, etc.). Recent research on the treatment of depression in primary care (e.g., Dietrich et al., 2004; Oxman et al., 2002) offers a useful example of how this work might proceed. The cost-effectiveness of clinical interventions should be evaluated as well (Schoenbaum et al., 2001; Simon et al., 2001).

SUMMARY AND CONCLUSIONS

Experiencing a traumatic event may lead to physical health problems in a person who develops a clinically significant distress reaction, especially PTSD. Although most of the evidence linking PTSD and poor health is based on self-report, there is enough evidence to firmly conclude that PTSD is associated with actual morbidity. There is not enough evidence to conclude definitively which physical problems and disorders are associated with PTSD. According to our proposed model of potential mechanisms, multiple body systems may be affected.

The observation of physical health consequences following traumatic exposure has important public health implications. Trauma may substantially contribute to many of the behavioral factors that are the target of current public health programs, such as smoking, exercise, diet, and risky sexual behavior. Prevention is a key issue. It may be possible to reduce the likelihood exposure to trauma through existing public health campaigns focused on high-risk behaviors that lead to accidents, disasters, child abuse, and sexual assault. But it is not likely that all trauma can be eliminated. Secondary prevention is also important, because it may be possible to prevent the physical health consequences of traumatic exposure. These consequences occur primarily in individuals who develop trauma-related distress. Therefore, strategies aimed at enhancing the detection and treatment of PTSD could result in improved physical and mental health outcomes among individuals who have experienced a traumatic event.

REFERENCES

American Psychiatric Association. (2000). Practice guideline for the treatment of patients with major depressive disorder. *American Journal of Psychiatry, 157*(Suppl. 4), 1–45.

Andreski, P., Chilcoat, H. D., & Breslau, N. (1998). Post-traumatic stress disorder and somatization symptoms: A prospective study. *Psychiatry Research, 79*, 131–138.

Asmundson, G. J. G., Coons, M. J., Taylor, S., & Katz, J. (2002). PTSD and the experience of pain: Research and clinical implications of shared vulnerability and mutual maintenance models. *Canadian Journal of Psychiatry, 47*, 930–937.

Barrett, D. H., Doebbeling, C. C., Schwartz, D. A., Voelker, M. D., Falter, K. H., Woolson, R. F., et al. (2002). Posttraumatic stress disorder and self-reported physical health status among U.S. military personnel serving during the Gulf War period: Population-based study. *Psychosomatics, 43*, 195–205.

Bassuk, E. L., Dawson, R., Perloff, J., & Weinreb, L. (2001). Post-traumatic stress disorder in extremely poor women: Implications for health care clinicians. *Journal of the American Medical Women's Association, 56*, 79–85.

Beckham, J. C., Moore, S. D., Feldman, M. E., Hertzberg, M. A., Kirby A. C., & Fairbank, J. A. (1998). Health status, somatization, and severity of posttraumatic stress disorder in Vietnam combat veterans with posttraumatic stress disorder. *American Journal of Psychiatry, 155*, 1565–1569.

Beckham, J. C., Vrana, S. R., Barefoot, J. C., Feldman, M. E., Fairbank, J. A., & Moore, S. D. (2002). Magnitude and duration of cardiovascular responses to anger in Vietnam veterans with and without posttraumatic stress disorder. *Journal of Consulting and Clinical Psychology, 70*, 228–234.

Blount, A. (1998). Introduction to integrated primary care. In A. Blount (Ed.), *Integrated primary care: The future of medical and mental health collaboration* (pp. 1–43). New York: Norton.

Boscarino, J. (2006). Posttraumatic stress disorder among U.S. Army veterans 30 years after military service. *Annals of Epidemiology, 16*, 248–256.

Boscarino, J. A. (1997). Diseases among men 20 years after exposure to severe stress: Implications for clinical research and medical care. *Psychosomatic Medicine, 59*, 605–614.

Boscarino, J. A., & Chang, J. (1999). Electrocardiogram abnormalities among men with stress-related psychiatric disorders: Implications for coronary heart disease and clinical research. *Annals of Behavioral Medicine, 21*, 227–234.

Buckley, T. C., Green, B. L., & Schnurr, P. P. (2004). Trauma, PTSD, and physical health: Clinical issues. In J. P. Wilson & T. M. Keane (Eds.), *Assessing psychological trauma and PTSD* (2nd ed., pp. 441–465). New York: Guilford Press.

Bullman, T. A., & Kang, H. K. (1994). Posttraumatic stress disorder and the risk of traumatic deaths among Vietnam veterans. *Journal of Nervous and Mental Disease, 182*, 604–610.

Catlin Boehmer, T. K., Flanders, D., McGeehin, M. A., Boyle, C., & Barrett, D. H. (2004). Postservice mortality in Vietnam veterans: 30-year follow-up. *Archives of Internal Medicine, 164*, 1908–1916.

Clum, G. A., Calhoun, K. S., & Kimerling, R. (2000). Associations among symptoms of posttraumatic stress disorder and self-reported health in sexually assaulted women. *Journal of Nervous and Mental Disease, 188*, 671–678.

Deykin, E. Y., Keane, T. M., Kaloupek, D., Fincke, G., Rothendler, J., Siegreid, M., et al. (2001). Posttraumatic stress disorder and the use of health services. *Psychosomatic Medicine, 63*, 835–841.

Dietrich, A. J., Oxman, T. E., Williams, J. W., Jr., Schulberg, H. C., Bruce, M. L., Lee, P. W., et al. (2004). Re-engineering systems for the primary care treatment of depression: A randomised controlled trial. *British Medical Journal, 329*, 602–608.

Drescher, K. D., Rosen, C. S., Burling, T. A., & Foy, D. W. (2003). Causes of death among male veterans who received residential treatment for PTSD. *Journal of Traumatic Stress, 16*, 535–543.

Druss, B. G., Rohrbaugh, R. M., Levinson, C. M., & Rosenheck, R. A. (2001). Integrated medical care for patients with serious psychiatric illness: A randomized trial. *Archives of General Psychiatry, 58*, 861–868.

Edwards, W. S., Winn, D. M., Kurlantzick, V., Sheridan, S., Berk, M. L., Retchin, S., et al. (1994). *Evaluation of National Health Interview Survey diagnostic reporting.* Hyattsville, MD: National Center for Health Statistics.

Felitti, V. J., Anda, R. F., Norenberg, D., Williamson, D. F., Spitz, A. M., Edwards, V., et al. (1998). Relationship of childhood abuse and household dysfunction to many of the leading causes of death in adults. *American Journal of Preventative Medicine, 14*, 245–258.

Flett, R. A., Kazantzis, N., Long, N. R., MacDonald, C., & Millar, M. (2002). Traumatic events and physical health in a New Zealand community sample. *Journal of Traumatic Stress, 15*, 303–312.

Foa, E. B., Keane, T. M., & Friedman, M. J. (Eds.). (2000). *Effective treatments for PTSD: Practice guidelines from the International Society for Traumatic Stress Studies.* New York: Guilford Press.

Ford, D. (2004). Depression, trauma, and cardiovascular health. In P. P. Schnurr & B. L. Green (Eds.), *Trauma and health: Physical health consequences of exposure to extreme stress* (pp. 73–97). Washington, DC: American Psychological Association.

Ford, J. D., Schnurr, P. P., Friedman, M. J., Green, B. L., Adams, G., & Jex, S. (2004). Posttraumatic stress disorder symptoms, physical health, and health care utilization 50 years after repeated exposure to a toxic gas. *Journal of Traumatic Stress, 17*, 185–194.

Frayne, S. M., Seaver, M. R., Loveland, S., Christiansen, C. L., Spiro, A., III, Parker, V. A., et al. (2004). Burden of medical illness in women with depression and posttraumatic stress disorder. *Archives of Internal Medicine, 164*, 1306–1311.

Friedman, M. J., & McEwen, B. S. (2004). Posttraumatic stress disorder, allostatic load, and medical illness. In P. P. Schnurr & B. L. Green (Eds.), *Trauma and health: Physical health consequences of exposure to extreme stress* (pp. 157–188). Washington, DC: American Psychological Association.

Friedman, M. J., & Schnurr, P. P. (1995). The relationship between PTSD, trauma, and physical health. In M. J. Friedman, D. S. Charney, & A. Y. Deutch (Eds.), *Neurobiological and clinical consequences of stress: From normal adaptation to PTSD* (pp. 507–527). Philadelphia: Lippincott–Raven.

Golding, J. M. (1996). Sexual assault history and women's reproductive and sexual health. *Psychology of Women Quarterly, 20*, 101–121.

Green, B. L., Grace, M. C., Lindy, J. D., Gleser, G. C., & Leonard, A. (1990). Risk factors for PTSD and other diagnoses in a general sample of Vietnam veterans. *American Journal of Psychiatry, 147*, 729–733.

Green, B. L., & Kimerling, R. (2004). Trauma, posttraumatic stress disorder, and health status. In P. P. Schnurr & B. L. Green (Eds.), *Trauma and health: Physical health consequences of exposure to extreme stress* (pp. 13–42). Washington, DC: American Psychological Association.

Green, B. L., Krupnick, J. L., Chung, J., Siddique, J., Krause, E., Revicki, D., et al. (2006). Impact of PTSD comorbidity on one-year outcomes in a depression trial. *Journal of Clinical Psychology, 62*, 815–835.

Higgins, A. B., & Follette, V. M. (2002). Frequency and impact of interpersonal trauma in older women. *Journal of Clinical Geropsychology, 8*, 215–226.

Hoge, C., Castro, C. A., Messer, S.C., McGurk, D., Cotting, D. I., & Koffman, R. L. (2004). Combat duty in Iraq and Afghanistan, mental health problems, and barriers to care. *New England Journal of Medicine, 351*, 13–22.

Kang, H. K., Bullman, T. A., & Taylor, J. T. (2006). Risk of selected cardiovascular diseases and posttraumatic stress disorder among former World War II prisoners of war. *Annals of Epidemiology, 16*, 381–386.

Kasprow, W. J., & Rosenheck, R. (2000). Mortality among homeless and nonhomeless mentally ill veterans. *Journal of Nervous and Mental Disease, 188*, 141–147.

Katon, W., Robinson, P., Von Korff, M., Lin, E., Bush, T., Ludman, E., et al. (1996). Multi-faceted intervention to improve treatment of depression in primary care. *Archives of General Psychiatry, 53*, 924–932.

Kessler, R. C., Sonnega, A., Bromet, E., Hughes, M., & Nelson, C. B. (1995). Posttraumatic stress disorder in the National Comorbidity Survey. *Archives of General Psychiatry, 52*, 1048–1060.

Kilpatrick, D. G., Resnick, H., & Acierno, R. (1997). Health impact of interpersonal violence: 3. Implications for clinical practice and public policy. *Behavioral Medicine, 23*, 79–85.

Kimerling, R., Calhoun, K. S., Forehand, R., Armistead, L., Morse, E., Morse, P., et al. (1999). Traumatic stress in HIV-infected women. *AIDS Education and Prevention, 11*, 321–330.

Kulka, R. A., Schlenger, W. E., Fairbank, J. A., Hough, R. L., Jordan, B. K., Marmar, C. R., et al. (1990). *Trauma and the Vietnam War generation: Report of findings from the National Vietnam Veterans Readjustment Study.* New York: Brunner/Mazel.

Lang, A. J., & Stein, M. B. (2004). An abbreviated PTSD checklist for use as a screening instrument in primary care. *Behaviour Research and Therapy, 43*, 585–594.

Leon, F. G., Keller Ashton, A., D'Mello, D. A., Dantz, B., Hefner, J., Matson, G. A., et al. (2003). Depression and comorbid medical illness: Therapeutic and diagnostic challenges. *Journal of Family Practice*, (Suppl. 52), S19–S33.

Malik, M. L., Connor, K. M., Sutherland, S. M., Smith, R. D., Davison, R. M., & Davidson, J. R. T. (1999). Quality of life and posttraumatic stress disorder: A pilot study assessing changes in SF-36 scores before and after treatment in a placebo-controlled trial of fluoxetine. *Journal of Traumatic Stress, 12*, 387–393.

Martin, L., Rosen, L. N., Durand, D. B., Knudson, K. H., & Stretch, R. H. (2000). Psychological and physical health effects of sexual assaults and nonsexual traumas among male and female United States Army soldiers. *Behavioral Medicine, 26*, 23–33.

McEwen, B. S., & Stellar, E. (1993). Stress and the individual: Mechanisms leading to disease. *Archives of Internal Medicine, 153*, 2093–2101.

Miranda, J., Chung, J. Y., Green, B. L., Krupnick, J., Siddique, J., Revicki, D. A., et al. (2003). Treating depression in predominantly low-income young minority women: A randomized controlled trial. *Journal of the American Medical Association, 290*, 57–65.

Neria, Y., & Koenen, K. C. (2003). Do combat stress reaction and posttraumatic stress disorder relate to physical health and adverse health practices?: An 18-year follow-up of Israeli war veterans. *Anxiety, Stress, and Coping, 16*, 227–239.

Otis, J. D., Keane, T. M., & Kerns, R. D. (2003). An examination of the relationship between chronic pain and posttraumatic stress disorder. *Journal of Rehabilitation Research and Development, 40*, 397–406.

Ouimette, P., Cronkite, R., Henson, B. R., Prins, A., Gima, K., & Moos, R. H. (2004). Posttraumatic stress disorder and health status among female and male medical patients. *Journal of Traumatic Stress, 17*, 1–9.

Oxman, T. E., Dietrich, A. J., Williams, J. W., & Kroenke, K. (2002). A three-component model for reengineering systems for the treatment of depression in primary care. *Psychosomatics, 43*, 441–450.

Pennebaker, J. (2000). Psychological factors influencing the reporting of physical symptoms. In A. A. Stone, J. S. Turkkan, C. A. Bachrach, J. B. Jobe, H. S. Kurtzman, & V. S. Cain (Eds.), *The science of self-report: Implications for research and practice* (pp. 299–315). Mahwah, NJ: Erlbaum.

Prins, A., Ouimette, P., Kimerling, R., Cameron, R. P., Hugelshofer, D., Shaw-Hegwer, J., et al. (2004). The Primary Care PTSD Screen (PC-PTSD): Development and operating characteristics. *Primary Care Psychiatry, 9*, 9–14.

Resnick, H. S., Acierno, R., & Kilpatrick, D. G. (1997). Health impact of interpersonal violence 2: Medical and mental health outcomes. *Behavioral Medicine, 23*, 65–78.

Resnick, H. S., Kilpatrick, D. G., Dansky, B. S., Saunders, B. E., & Best, C. L. (1993). Prevalence of civilian trauma and posttraumatic stress disorder in a representative national sample of women. *Journal of Consulting and Clinical Psychology, 61*, 984–991.

Rheingold, A. A., Acierno, R., & Resnick, H. S. (2004). Trauma, posttraumatic stress disorder, and health risk behaviors. In P. P. Schnurr & B. L. Green (Eds.), *Trauma and health: Physical health consequences of exposure to extreme stress* (pp. 217–243). Washington, DC: American Psychological Association.

Samson, A. Y., Benson, S., Beck, A., Price, D., & Nimmer, C. (1999). Posttraumatic stress disorder in primary care. *Journal of Family Practice, 48*, 222–227.

Sareen, J., Cox, B. J., Clara, I. P., & Asmundson, G. J. G. (2005). The relationship between anxiety disorders and physical disorders in the U.S. National Comorbidity Survey. *Depression and Anxiety, 21*, 193–202.

Schnurr, P. P., Ford, J. D., Friedman, M. J., Green, B. L., Dain, B. J., & Sengupta, A. (2000). Predictors and outcomes of PTSD in World War II veterans exposed to mustard gas. *Journal of Consulting and Clinical Psychology, 68*, 258–268.

Schnurr, P. P., Friedman, M. J., Foy, D. W., Shea, M. T., Hsieh, F. Y., Lavori, P. W., et al. (2003). Randomized trial of trauma-focused group therapy for posttraumatic stress disorder. *Archives of General Psychiatry, 60*, 481–489.

Schnurr, P. P., Friedman, M. J., Sengupta, A., Jankowski, M. K., & Holmes, T. (2000). PTSD and utilization of medial treatment services among male Vietnam veterans. *Journal of Nervous and Mental Disease, 188*, 496–504.

Schnurr, P. P., & Green, B. L. (2004). Understanding relationships among trauma, posttraumatic stress disorder, and health outcomes. In P. P. Schnurr & B. L. Green (Eds.), *Trauma and health: Physical health consequences of exposure to extreme stress* (pp. 247–275). Washington, DC: American Psychological Association.

Schnurr, P. P., & Jankowski, M. K. (1999). Physical health and post-traumatic stress disorder: Review and synthesis. *Seminars in Clinical Neuropsychiatry, 4*, 295–304.

Schnurr, P. P., & Spiro, A. (1999). Combat exposure, posttraumatic stress disorder symptoms, and health behaviors as predictors of self-reported physical health in older veterans. *Journal of Nervous and Mental Disease, 187*, 353–359.

Schnurr, P. P., Spiro, A., Aldwin, C. M., & Stukel, T. A. (1998). Physical symptom trajectories following trauma exposure: Longitudinal findings from the Normative Aging Study. *Journal of Nervous and Mental Disease, 186*, 522–528.

Schnurr, P. P., Spiro, A., & Paris, A. H. (2000). Physician-diagnosed medical disorders in relation to PTSD symptoms in older male military veterans. *Health Psychology, 19*, 91–97.

Schoenbaum, M., Unutzer, J., Sherbourne, C., Duan, N., Rubenstein, L. V., Miranda, J., et al. (2001). Cost-effectiveness of practice-initiated quality improvement for depression: Results of a randomized controlled trial. *Journal of the American Medical Association, 286*, 1325–1330.

Shalev, A., Bleich, A., & Ursano, R. J. (1990). Posttraumatic stress disorder: Somatic comorbidity and effort tolerance. *Psychosomatics, 31*, 197–203.

Sharp, T. J., & Harvey, A. G. (2001). Chronic pain and PTSD: Mutual maintenance? *Clinical Psychology Review, 21*, 857–877.

Shipherd, J. C., Beck, J. G., Hamblen, J. L., Lackner, J. M., & Freeman, J. B. (2003). A preliminary examination of treatment for posttraumatic stress disorder in chronic pain patients: A case study. *Journal of Traumatic Stress, 16*, 451–457.

Sibai, A. M., Armenian, H. K., & Alam, S. (1989). Wartime determinants of arteriographically confirmed coronary artery disease in Beirut. *American Journal of Epidemiology, 130*, 623–631.

Sibai, A. M., Fletcher, A., & Armenian, H. K. (2001). Variations in the impact of long-term wartime stressors on mortality among the middle-aged and older population in Beirut, Lebanon, 1983–1993. *American Journal of Epidemiology, 154*, 128–137.

Simon, G. E., Katon, W. J., Von Korff, M., Unutzer, J., Lin, E. H. B., Walker, E. A., et al. (2001). Cost-effectiveness of a collaborative care program for primary care patients with persistent depression. *American Journal of Psychiatry, 158*, 1638–1644.

Sjahid, S. I., van der Linden, P. D., & Stricker, B. H. C. (1998). Agreement between the pharmacy medication history and patient interview for cardiovascular drugs: The Rotterdam Elderly Study. *British Journal of Clinical Pharmacology, 45*, 591–595.

Stein, M. B., McQuaid, J. R., Pedrelli, P., Lenox, R., & McCahill, M. E. (2000). Posttraumatic stress disorder in the primary care medical setting. *General Hospital Psychiatry, 22*, 261–269.

Taft, C. T., Stern, A. S., King, L. A., & King, D. W. (1999). Modeling physical health and functional health status: The role of combat exposure, posttraumatic stress disorder, and personal resource attributes. *Journal of Traumatic Stress, 12*, 3–23.

Taubman-Ben-Ari, O., Rabinowitz, J., Feldman, D., & Vaturi, R. (2001). Post-traumatic stress disorder in primary-care settings: Prevalence and physicians' detection. *Psychological Medicine, 31*, 555–560.

Ullman, S. E., & Siegel, J. M. (1996). Traumatic events and physical health in a community sample. *Journal of Traumatic Stress, 9*, 703–720.

Vedantham, K., Brunet, A., Boyer, R., Weiss, D. S., Metzler, T. J., & Marmar, C. R. (2001). Posttraumatic stress disorder, trauma exposure, and the current health of Canadian bus drivers. *Canadian Journal of Psychiatry, 46*, 149–155.

Visintainer, P. F., Barone, M., McGee, H., & Peterson, E. L. (1995). Proportionate mortality study of Vietnam-era veterans of Michigan. *Journal of Occupational and Environmental Medicine, 37*, 423–428.

Wagner, A. W., Wolfe, J., Rotnitsky, A., Proctor, S. P., & Erickson, D. J. (2000). An investigation of the impact of posttraumatic stress disorder on physical health. *Journal of Traumatic Stress, 13*, 41–55.

Walker, E. A., Gelfand, A. N., Katon, W. J., Koss, M. P., Von Korff, M., Bernstein, D. E., et al. (1999). Adult health status of women with histories of childhood abuse and neglect. *American Journal of Medicine, 107*, 332–339.

Walker, E. A., Katon, W., Russo, J., Ciechanowski, P., Newman, E., & Wagner, A. (2003). Health care costs associated with posttraumatic stress disorder symptoms in women. *Archives of General Psychiatry, 60*, 369–374.

Walker, E. A., Newman, E., Dobie, D. J., Ciechanowski, P., & Katon, W. J. (2002). Validation of the PTSD Checklist in an HMO sample of women. *General Hospital Psychiatry, 24*, 375–380.

Walker, E. A., Newman, E., & Koss, M. P. (2004). Costs and health care utilization associated with traumatic experiences. In P. P. Schnurr & B. L. Green (Eds.), *Trauma and health: Physical health consequences of exposure to extreme stress* (pp. 43–69). Washington, DC: American Psychological Association.

Wallihan, D. B., Stump, T. E., & Callahan, C. M. (1999). Accuracy of self-reported health service use and patterns of self-care among urban older adults. *Medical Care, 37*, 662–670.

Watson, D., & Pennebaker, J. W. (1989). Health complaints, stress, and distress: Exploring the central role of negative affectivity. *Psychological Review, 96*, 234–254.

Weisberg, R. B., Bruce, S. E., Machan, J. T., Kessler, R. C., Culpepper, L., & Keller, M. B. (2002). Nonpsychiatric illness among primary care patients with trauma histories and posttraumatic stress disorder. *Psychiatric Services, 53*, 848–854.

White, H. R., & Widom, C. S. (2003). Does childhood victimization increase the risk of early death?: A 25-year prospective study. *Child Abuse and Neglect, 27*, 841–853.

Wilson, I. B., & Cleary, P. D. (1995). Linking clinical variables with health-related quality of life. *Journal of the American Medical Association, 273*, 59–65.

Wolfe, J., Schnurr, P. P., Brown, P. J., & Furey, J. (1994). Posttraumatic stress disorder and war-zone exposure as correlates of perceived health in female Vietnam War veterans. *Journal of Consulting and Clinical Psychology, 62*, 1235–1240.

Zatzick, D. F., Jurkovich, G. J., Gentilello, L., Wisner, D., & Rivara, F. P. (2002). Posttraumatic stress, problem drinking, and functional outcomes after injury. *Archives of Surgery, 137*, 200–205.

Zayfert, C., Dums, A. R., Ferguson, R. J., & Hegel, M. T. (2002). Health functioning impairments associated with posttraumatic stress disorder, anxiety disorders, and depression. *Journal of Nervous and Mental Disease, 190*, 233–240.

Zoellner, L. A., Goodwin, M. L., & Foa, E. B. (2000). PTSD severity and health perceptions in female victims of sexual assault. *Journal of Traumatic Stress, 13*, 635–649.

Chapter 21

Cultural Issues and Trauma

Janet E. Osterman and Joop T. V. M. de Jong

Traumatic events are, unfortunately, ubiquitous human experiences that affect people from diverse cultural groups across the globe. Clinicians and researchers typically work with individuals, families, and communities that differ from their own cultural groups. Cultural competency skills are essential to either effectively treat or research any medical condition, and especially psychiatric disorders, because of the interplay between culture and concepts of health or illness, expressions of distress, and healing beliefs and practices. These broad concepts cross the fields of anthropology, psychiatry, and psychology. Across all disease and illnesses (medical and psychiatric), culture determines the local expression of symptoms (idioms of distress, illness behavior), illness attributions (explanatory models), coping, locally sanctioned treatments, as well as, acceptance of treatments, both Western and non-Western (Becker & Kleinman, 2000; Kleinman, 1980; Kleinman & Good, 1985).

The vast diversity of cultural and subcultural groups creates barriers to knowledge of all cultures and subcultures. However, a clinician or researcher can develop cultural competency by applying core concepts. This chapter presents several core concepts in cultural medicine and current methodology in mental health treatment and research to provide the reader with an approach to cultural issues. Brief examples from a variety of cultures around the world are used to illustrate some of these core concepts. Current literature pertaining to psychological trauma is reviewed, and its generalizability is discussed. Future directions for research to increase our knowledge of the interplay between culture and the impact of traumatic events are presented.

METHODOLOGICAL CONSIDERATIONS

Core Concepts

Several core concepts from the field of medical anthropology that researchers and clinicians must consider in working with patients or populations from other cultures and subcultures are summarized here, but there is an extensive literature available to the student of culture (cf. Kleinman & Good, 1985; Pedersen, Dragus, Lonner, & Trimble, 1996; Ponterotto, Casas, Suzuki, & Alexander, 1995).

The first concept is the etic–emic dichotomy. An etic framework is culture-general and as such does not make adaptations for a particular culture. An emic approach is culture-specific; the clinician or researcher adapts assessment and/or therapeutic interventions to fit the culture of the patient or group, typically working within the culture to learn these concepts. An etic approach risks ethnocentrism when the therapist's or researcher's concepts, assumptions, and biases develop only from his or her own culture and fail to consider the patient's cultural context (Marsella, Friedman, & Spain, 1996). Wrenn (1962) used the term "encapsulation" to describe this bias. For example, using a Western diagnostic tool, such as the Diagnostic Interview Schedule (DIS) or Structured Clinical Intervention for DSM-IV (SCID) in a non-Western culture is an example of an etic approach that might risk diagnostic accuracy. Kleinman (1977) used the term "category fallacy" to identify this type of diagnostic error, based on the presumption that a particular mental disorder will be similar across cultures. These diagnostic errors occur when researchers look for and find evidence of a "disorder" in the culture under study that supports their beliefs and knowledge about a disorder diagnosed in their own cultural group. This issue of validity of a diagnosis across cultures is a critical concept both in clinical work and research.

A second core concept, "idiom of distress," is described by Kleinman (1982) as the cultural expression of a person's illness. It is imperative that the clinician or researcher understands the terms used to define distress and illness as they are typically understood and sanctioned by a particular culture. This term is closely linked to an additional core concept in medical anthropology, the disease–illness dichotomy. "Illness" is the subjective, culturally influenced nature of a patient's experience of his or her sickness, whereas "disease" is the domain of the medical professional (Kleinman, Eisenberg, & Good, 1978). These concepts of illness and disease may be divergent due to conflicting influences from both the local culture and the culture of medicine.

A third widely used and related concept is that of explanatory model (Kleinman, 1980), which includes cultural attribution of illness or the patient's understanding of the illness and the interactive experiences of patients, their families and health professionals, including culturally sanctioned healers. Differences in attribution frequently exist between patients and medical health professionals.

Research Methods in Cultural Medicine

To guard against ethnocentrism, an emic approach is warranted and requires detailed knowledge of the culture, the cultural language of illness and health, explanatory models, and cultural healing practices and beliefs to ensure cultural competence. Ethnography is a branch of anthropology that develops a scientific description of human cultures based on participant observations (Bernard, 1994). Ethnography is critical to an emic approach. Mental health researchers often conduct ethnographic research in collaboration with anthropologists and, if possible, with bicultural and bilingual mental health professionals. An ethnographic researcher typically lives with the cultural community for some time period, conducting interviews with individuals within the community. Psychiatric ethnographic research has been conducted in a large number of non-Western populations: for example, Chinese (Kleinman, 1975), West African (de Jong, 1987), Cambodian (van de Put & Eisenbruch, 2002), Sri Lankan (Somasundaram & Jamunanatha, 2002), Congolese (Roy, 2002), and Somalian (Zarowsky, 2004). Ethnographic research provides the foundation for developing clinical approaches and organizing systems of care, and for research studies. Bolton and Tang (2004) describe an ethnographic approach that can assist in developing and implementing postdisaster mental health programs in non-Western settings by incorporating knowledge about local views of mental health and mental health problems.

Methodologies in ethnographic studies include the use of focus groups, key informants, snowball sampling, and illness narratives. We will begin with snowballing technique, because this approach may need to be used to find members for focus groups, key informants, and individuals and family members to interview for illness narratives. Snowball sampling was developed to assist researchers in finding participants in a study population that may be difficult to locate (Biernacki & Waldorf, 1981; Goodman, 1961). In this methodological approach, an individual referred by knowledgeable people within the population under study is asked to identify others who may share the condition that is being studied. A person is then randomly selected for an interview and is asked to provide a list of others; the process is repeated several times (Ding et al., 2005; Gernaat, Malwand, Laban, Komproe, & de Jong, 2002; Momartin, Silove, Manicavasagar, & Steel, 2004). Snowball sampling may identify key informants within the culture—such as community leaders, local healers, health care providers, or religious leaders—who may want to collaborate and can provide insight and information about the proposed research problem or help recruit members for focus groups or in-depth interviews. In some cultures, community leaders may need to approve the study or give permission for members of the community to be interviewed. People recruited in this way can assist researchers to develop knowledge of the culture and to refine the problem(s) to be studied (de Jong & Van Ommeren, 2002). This methodology does not provide a representative sample of the population, which limits the generalizability of the sample to the population as a whole. Momartin and colleagues (2004) noted that the use of snowball sampling results in a study population with higher levels of psychopathology.

A focus group is a small group, typically 10 or fewer members from the population under study (Krueger, 1994; Morgan, 1997). The purpose of focus groups is to gather qualitative information about the culture, such as idioms of distress, explanatory models, and health care systems. Focus groups also provide knowledge and insight into the problems, risk and protective factors at the population level, as well as the sociocultural, socioeconomic, and political context of the population under study. Qualitative information from focus groups may help the researcher to define or refine the research topic or questions and to identify which variables to measure. Focus group members and a group facilitator discuss one or two research questions that are critical to the study (cf. de Jong & Van Ommeren, 2002). For example, focus groups might concentrate on current health problems in the community, coping, illness behaviors, or exposure to traumatic events (Aheto & Gbesemete, 2005; Ayuku, Odero, Kaplan, De Bruyn, & De Vries, 2003; Ding et al., 2005; Hollifield et al., 2005). Depending on the study and culture, focus groups may be organized on several levels: community leaders, healers, health care professionals, family members, male only, female only, educational level (de Jong & Van Ommeren, 2002). Several focus groups may be needed to answer the same questions to avoid informant biases or reluctance to discuss topics that may cause conflict with other members of the community, result in shame or embarrassment, or be cultural taboos within mixed genders or ages. Information gathered from in-depth interviews of key informants help guide the composition of focus groups to avoid these pitfalls and result in better data collection.

In-depth interviews may be conducted with identified persons in the group under study, their family members, or key informants. Anthropologists label these in-depth interviews "illness narratives," "thick description," or "person-centered ethnographies" (Holman, 1997; Kleinman, 1988), whereas personality psychologists use the term "psychobiography" (McAdams & Ochberg, 1988). Illness narratives elicit detailed, focused information about an individual's illnesses, idioms of distress, explanatory models, cultural significance of the illness, symptoms, and coping mechanisms, as well as family and cultural responses. Information from these interviews can identify concerns of the local population to ensure relevance of the proposed research (de Jong & Van Ommeren, 2002, 2005) and facilitate the successful implementation of treatment programs (Bolton & Tang, 2004; de Jong, 2002a, 2002b).

An ethnographic study using in-depth interviews was conducted with 1,100 healers in Cambodia (van de Put & Eisenbruch, 2002). Through this research, the extensive Cambodian system of healing and social support through local healers and monks was detailed; idioms of distress and explanatory models were defined. In this system of care, each type of healer has a sanctioned healing role within the culture. Buddhist monks (*preah sang*) give advice to decrease anxiety. *Kruu* (trained healers) offer medication remedies and magical rituals to purge invading spells and spirits causing the problem. Public rituals by trained healers allow for reintegration into the community. Mediums (*kruu chool ruub*), primarily women, communicate and intercede with the ancestors to help women who are distressed about their future. Traditional birth attendants (*cmap*) assist with childbirth and throughout the puerperium, providing support to new mothers and families. "Madness" in this culture has many causes: offended ancestors cause ancestral madness; offended people can invoke sorcery or magic to cause illness, and offended spirits cause spirit madness. These forms of "madness" tend to cause acute illnesses that respond well to indigenous healing methods. Some indigenous healers are quite famed for treating specific problems, and people may travel quite a distance to seek healing. However, most healers accept the limits of their capacity and do not treat chronic psychotic disorders (Eisenbruch, 1994). Healers may help the family to understand why a person may have developed this kind of "madness," or *chuet*, and facilitate the reintegration of the ill person into the community. This ethnographic research was used to develop clinical programs to reestablish and reintegrate indigenous healing methods that had been disrupted during the Pol Pot era (van de Put & Eisenbruch, 2002) and to promote epidemiological research (de Jong et al., 2001).

Adaptation of Assessment Instruments

One diagnostic assessment method utilized to avoid ethnocentric biases that lead to category fallacy is a combined etic–emic approach where an etic construct is modified using emic concepts. Several research methodologies have been developed to avoid the pitfalls of such biases and increase cross-cultural validity of assessment instruments (Brislin, 1986; Brislin, Lonner, & Thorndike, 1973; Flaherty et al., 1988; Marsella, Friedman, et al., 1996; Marsella & Kameoka, 1989; Sartorius & Janca, 1996). Marsella and Kameoka (1989; Marsella, Friedman, et al., 1996) have recommended the use of ethnosemantic methods rooted in the culture under study to reduce Western ethno-centricity and bias. The tendency to define personal cultural views, patterns of thinking and behavior as the correct view risks discounting or invalidating those of people from other cultural groups. An ethnosemantic approach uses subjective experiences of the cultural group to construct a culturally appropriate translation and to reduce cultural insensitivity or inappropriate materials and procedures. For example, focus group members from a specific cultural group are asked whether an assessment instrument has been appropriately translated, both linguistically and culturally, to reduce cultural insensitivity or inappropriate materials and procedures (Brislin, 1986; Brislin et al., 1973; de Jong & Van Ommeren, 2002; Flaherty et al., 1988; Keane, Kalopeuk, & Weathers, 1996; Marsella, Friedman, & Spain, 1996; Van Ommeren et al., 1999; Westermeyer & Sines, 1979).

Manson (1997) has proposed that all utilized terms satisfy criteria for compre-hensibility, acceptability, relevance, and completeness. A translated term that is not understood or that differs from the original meaning is incomprehensible. For exam-ple, in attempting to study posttraumatic stress disorder (PTSD) among the Kalahari Bushmen, McCall and Resick (2003) found that the concept of numbing could not be translated or understood in that culture. It was not a relevant concept. An unacceptable term may be insulting or fail to recognize a cultural taboo, such as asking about premar-ital sexual activity among Islamic woman. Irrelevant terms describe experiences that do not occur in the culture or are seen as unrelated to the underlying concept. For exam-ple, asking about loss of appetite is not relevant in a setting of famine. Incomplete ques-tions are etic and do not assess for cultural differences.

Flaherty and colleagues (1988) identified five forms of equivalence (content, semantic, technical, criterion, and conceptual) that must be met between the original and translated instruments. Content equivalence is met if the content of each assess-ment item is relevant to the experience of the culture. For example, in populations undergoing famine, answers to questions about appetite to assess depression often reflect preoccupation with availability of food (e.g., "If I have enough food, I like to eat") (de Jong, 1987). Semantic equivalence is achieved if the item has been translated into the local language and idioms of distress to assess the initial question. For example, in Nepal, Van Ommeren and colleagues (1999) found that there was no locally used word for suicide. The formal Nepali and Sanskrit word *aatmahatyaa* was not known or used by the villagers. To achieve semantic equivalence, *suicide* was translated in the instruments as "death by hanging, taking poison, or jumping off a cliff," concepts well known in the local community. Technical equivalence, which assesses comparability of assessment methods and data acquisition across cultures, addresses the question of the most sensible way to administer a research instrument or procedure in another culture, so that it does not create systematic biases (de Jong & van Ommeren, 2002). For exam-ple, technical equivalence assesses whether the interviewer is associated with an emo-tionally loaded (political, religious, tribal, etc.) institution; whether questions need to be presented in the positive, as well as negative, form to obtain accurate information; whether there is enough privacy in the research setting; whether optimal interpersonal distance in that culture has been taken into account; and whether communication between interviewer and respondent may be disturbed by differences in sex, ethnicity, socioeconomic status, or rural background (de Jong, 1987). Criterion equivalence is achieved if the interpretation of the measurement of the variable remains the same when compared to the norm of each culture under study, or if the outcome of measure-ment of the variable is in agreement with another criterion, such as an independent assessment by a psychiatrist with knowledge about the culture. For example, many well-adjusted respondents in African countries may endorse the statement that someone wants to harm them. A culturally informed psychiatrist would not consider this as evidence for a paranoid delusion, because fear of witchcraft or sorcery is a common cultural belief (de Jong, 1987). Conceptual equivalence is attained when the instrument measures the same theoretical construct across cultures. For example, in assessing PTSD, an instrument may include questions about nightmares. Nightmares may be as normal and accepted as visits from an ancestor in an animistic or Buddhist culture. Problems in both semantic and conceptual equivalence may be encountered in translating symptoms of PTSD. This may require several focus groups in a reiterative process to find the most appropriate translation.

Brislin (1986; Brislin et al., 1973) proposed five steps to achieve proper translation and cultural adaptation of assessment instruments: translation; blind back-translation; a comparison of the original instrument, the translated and back-translated instruments; a pilot study; and examination of the pilot study data and subjects. The World Health Organization (WHO; Sartorius & Janca, 1996) expanded this to include seven stages: (1) establishment of a bilingual group of experts; (2) examination of the conceptual structure of the instruments by the experts; (3) translation; (4) examination of the translation by the experts; (5) examination of the instruments by a monolingual focus group; (6) blind back-translation; and (7) examination of the blind back-translations by the experts. Van Ommeren and colleagues (1999) developed a translation monitoring form to ensure completion of these complicated steps necessary to translate and culturally adapt an instrument to meet content, semantic, technical, criterion, and conceptual equivalence. Once the instrument is prepared for the study population, it must be tested and validated to see whether it accurately measures the disorder. For example, a translated form that assesses depression must find depression in a person who has been clinically determined to suffer from depression. When analyzing the data, the researchers must again seek to uncover any potential biases. This methodology has been used to translate instruments for epidemiological and cost-effectiveness studies in several non-Western populations (de Jong et al., 2001; de Jong, Komproe, & Van Ommeren, 2003; Van Ommeren et al., 1999; Van Ommeren, de Jong, et al., 2001).

A recent study of the prevalence of Disorders of Extreme Stress (DES) across three cultures found that this Western diagnostic construct was difficult to assess, because of problems in conceptual, content, semantic, and/or technical equivalence (de Jong, Komproe, Spinazzola, van der Kolk, & Van Ommeren, 2005). For example, questions about dissociation created problems in both semantic and conceptual equivalence. The cultures of Algeria, Palestine, and Ethiopia had difficulty with concepts about feeling "like there are two people living inside you who control how you behave at different times." This concept asks a person to identify different emotional states and ascribe them to subpersonalities or "part of" him- or herself. In cultures that attribute dissociation or trance to a visit from an ancestor, spirit possession, or part of his or her soul leaving the body, this question typically results in a negative endorsement despite the presence of dissociation. Shamans and healers who go into trance states regularly would also not endorse dissociation with this question or construct. Similarly, "spacing out when you feel frightened or are under stress" is not a concept known in many cultures, and it may be attributed to spirit possession.

International standardization of diagnoses may mask sociocultural influences and challenge diagnostic validity in non-Western cultures (de Jong & Van Ommeren, 2002). In addition to careful cultural adaptation and translation, researchers and clinicians must examine skip rules in diagnostic instruments (e.g., Composite International Diagnostic Interview [CIDI], SCID, DIS). In usual instrument administration, nonendorsement of a screening question allows the interviewer to skip a full set of diagnostic questions. There is a risk that meaningless prevalence rates may be obtained across cultures if the concept in the screening question is neither culturally relevant nor culturally equivalent (de Jong & Van Ommeren, 2002). One particularly challenging area is with screening questions for psychosis. In animistic cultures across Africa, visions are common and attributed to a visit from an ancestor who gives support and advice. To deal with this, the CIDI asks about seeing visions, then uses open-ended questions that require the interviewer to assess the context of the visions. A researcher or clinician would then interpret and score this information as pathological or normative, thus allowing for cultural sensitivity in the instrument (Kessler et al., 2005). We recommend using focus groups to determine the likelihood that using skip rules would result in diagnostic inaccuracies. If there is any question of diagnostic accuracy, then all questions within a diagnostic set should be asked.

Much of this methodology is not practical for the clinical setting; thus, the clinician needs to rely on the literature, both in mental health and anthropology, to develop cultural competency. Key informants for the clinician are medical interpreters, who are often members of the patient's cultural group and can act as cultural brokers, informing the clinician about accepted or deviant behaviors, idioms of distress, explanatory models, and cultural supports and healing practices.

CURRENT STATE OF THE ART: PTSD AND CULTURE

The epidemiological literature supports the diagnosis of PTSD in survivors of war and violence (for reviews, see de Jong, 2002b; Green et al., 2003) and natural disasters across a number of cultures (for reviews, see Katz, Pelligrino, Pandya, Ng, & DeLisi, 2002; Norris, Friedman, & Watson, 2002; Norris, Friedman, Watson, Byrne, et al., 2002). The vast majority of war-related PTSD surveys have been conducted in the West with resettled refugees, and have primarily included treatment- or help-seeking refugees and asylum seekers. Fewer surveys have recruited community samples of refugees living in the West, and far fewer have surveyed populations in non-Western regions. Disasters have been more extensively researched than other types of trauma in both Western and non-Western settings. Disaster studies include those of treatment seekers as well as population studies that use random, typically large samples from the general population. Kokai, Fujii, Shinfuku, and Edwards (2004), in a review of disaster and mental health, noted that of the approximately 3 billion people in the world affected by disasters between 1967 and 1991, 85% were living in Asia.

Although an extensive body of literature supports PTSD treatment efficacy in Western samples (for reviews, see Ballenger et al., 2000; Foa, Keane, & Friedman, 2000; Keane & Barlow, 2002) current treatment research does not provide sufficient knowledge to guide evidence-based practice in non-Western cultural groups. Traumatized people in low-income and non-Western countries have limited access to [Western] medical or mental health care. Bracken, Giller, and Summerfield (1995) propose that implementation of Western mental health programs to other cultures may do harm. Treatments that are efficacious in Western laboratory settings should not be exported to non-Western patients without empirical data as a guide. This area of research is in its early stages, but it is an actively growing field.

This section provides an overview of epidemiological studies of refugees in the West (both in community and treatment-seeking cohorts), survivors of war and natural disasters in non-Western countries, and the current data on PTSD treatment studies of individuals from other cultures. Several publications are available to the clinician that may provide insight and guidance about culture in clinical practice (see Gaw, 1993) and about PTSD in American subcultures and other cultures (see Green et al., 2003; Marsella, Friedman, Gerrity, & Scurfield, 1996).

Epidemiological Surveys of War and Violence

Treatment or Service-Seeking Populations in Western Settings

Reported prevalence rates of PTSD in treatment-seeking populations range from 14 to 95% (Favaro, Maiorani, Colombo, & Santonastaso, 1999; Gorst-Unsworth & Goldenberg, 1998; Hinton et al., 1993; Kinzie et al., 1990; Kinzie, Sack, Angell, Clark, & Ben, 1989; Kroll et al., 1989; Lavik, Hauff, Skronda, & Solberg, 1996; Mollica et al., 1998; Solvig & Göran, 1997; Weine et al., 1995; Weine, Razzano, et al., 2000; Weine, Vojvoda, et al., 1998). Depressive disorders are a common comorbid finding. For example, Kinzie and colleagues (1989) reported that 48% of 27 young Cambodians had PTSD and 41% had depression. PTSD prevalence was 93% in the Mein, 65% in Laotians, and 53% in Southeast Asian patients in an Indochinese clinic in the United States (Kinzie et al., 1990). Kroll and colleagues (1989), in a study of 404 Southeast Asian refugees (Laotians, Cambodians, and Vietnamese) in a U.S. community mental health clinic, reported that 14% of refugees were diagnosed with PTSD and nearly 50% had major depression. In Oslo, 46.6% of treatment-seeking refugees met diagnostic criteria for PTSD (Lavik et al., 1996). Treatment-seeking Bosnian refugees had high endorsement of PTSD (65%) (Weine et al., 1995). In Australia, 37% of asylum seekers attending a welfare center were diagnosed with PTSD (Silove, Sinnerbrink, Field, Manicavasagar, & Steel, 1997). The rates of illness vary substantially across settings. One of the problems with this body of knowledge is that, for the most part, these are samples of convenience (e.g., treatment- or asylum-seeking refugees). Therefore, it is difficult to generalize the findings to the population at large. Treatment-seeking refugees may have a higher incidence of morbidity than those living in the community. Asylum seekers may have experienced more hardships, such as torture, or may feel compelled to endorse more complaints to enhance the likelihood of achieving asylum during the claimant process. In addition, sample sizes varied, with some having a very small sample size (e.g., Kinzie et al., 1989), as did assessment measures, limiting comparisons across groups.

Community Samples in Western Settings

Community samples of refugees living in Western countries support diagnoses of PTSD, other anxiety disorders, and major depressive disorder. A recent meta-analysis of 20 surveys assessing community samples of refugees ($n = 6,743$) in seven Western countries found a prevalence of 9% for PTSD and 5% for major depression (Fazel, Wheeler, & Danesh, 2005). A community survey of Vietnamese refugees in the United States found that Vietnamese political prisoners suffered higher rates of both PTSD (88.2%) and depression (56.9%) than nondetained Vietnamese refugees, with PTSD rates of 77.3% and depression rates of 36.4% (Mollica et al., 1998). In an Italian camp for Yugoslavian refugees, 50% met diagnostic criteria for PTSD, whereas 35% had major depressive disorder (Favaro et al., 1999). A community sample in the Netherlands reported that 31.5% of Somali refugees met criteria for PTSD, as assessed with the Harvard Trauma Questionnaire; 36% met criteria for an anxiety disorder and 63% for a depressive disorder (Roodenrijs, Scherpenzeel, & de Jong, 1998). Gernaat and colleagues (2002) used snowball sampling to determine 12-month prevalence rates of disorders among 51 Afghan refugees in the Netherlands and found that 65% had a psychiatric disorder, 57% had a depressive disorder, and 35% had PTSD. Psychopathology in this community population was related to poor language skills, a lower level of education, and current unemployment. Nearly 10% of a non-treatment-seeking group of male Iraqi refugees in the United States (45% Kurdish, 55% Shia) had PTSD (Gorst-Unsworth & Goldenberg, 1998). In a recent random sample of 294 Iraqi refugees in the Netherlands, using the CIDI 2.1, Laban, Gernaat, Komproe, Van der Tweel, and de Jong (2005) found a lifetime PTSD prevalence of 37%; other anxiety disorders, 22%; and depressive disorder, 35%.

In the child literature, five surveys of 260 refugee children from three Western countries reported that 11% had PTSD (Fazel et al., 2005). Of 241 surveyed Sudanese minors in refugee foster care programs, 20% had PTSD, with social isolation and a history of injury associated with PTSD risk (Geltman et al., 2005).

Community samples offer greater generalizability than treatment- or asylum-seeking samples; however, the generalizability may be limited by variations in prior experiences and trauma exposure (e.g., political detainees or nondetainees), flight experiences, current living conditions (e.g., a refugee camp), or the level of acculturation stress within the new community. In addition, the length of time awaiting asylum may contribute to psychopathology (Laban et al., 2005).

Community Surveys in Non-Western Settings

Few studies of survivors or refugees have taken place in low-income, postconflict countries. A large epidemiological study (de Jong et al., 2001) in four postconflict countries [Cambodia (n = 610), Algeria (n = 653), Ethiopia (n = 1,200), and Gaza (n = 585)] assessed prevalence of PTSD, depression, anxiety, risk factors, and nature of ongoing stressors. PTSD prevalence rates were 28.4% in Cambodia, 37.4% in Algeria, 15.8% in Ethiopia, and 17.8% in Gaza. The only risk factor for PTSD present in all groups was conflict-related trauma after age 12. Torture was a PTSD risk factor in Ethiopia, Gaza, and Algeria. Youth domestic stress, death or separation in the family, and alcohol abuse in parents all were associated with PTSD in Cambodians. A prior psychiatric history and current illness were PTSD risk factors in Cambodia and Ethiopia. Poor refugee camp conditions in Gaza and Algeria, and daily life hassles in Algeria, were associated with PTSD. Risk ratio for PTSD [JC3]ranged from 10.03 in Gaza to 3.14 in Algeria; risk ratios for mood disorders were 6.06 in Ethiopia and 4.53 in Gaza. For other anxiety disorders, the risk ratios ranged from 2.10 to 3.16 in Ethiopia, Algeria, and Gaza. Moreover, disability was more associated with mood and anxiety disorders than with PTSD. The advantage of this study was its rigorous adherence to translation of study instruments and large sample sizes for each population.

A population-based survey of Bhutanese refugees displaced in Nepal found that PTSD was more commonly diagnosed in survivors of torture (14%) compared to nontortured refugees (3%). In addition, the torture survivors had higher anxiety and depression symptoms (Shrestha et al., 1998). A survey of 993 Cambodian refugees in a camp along the Thailand–Cambodia border reported that 15% of refugees had symptoms of PTSD and 55% had depressive symptoms (Mollica et al., 1993). Dahl, Mutapcic, and Schei (1998) studied 209 displaced Bosnian women in a war zone in Bosnia–Herzegovina in 1994. Women with high levels of exposure to traumatic events, such as concentration camp interment or other detention, were identified as having posttraumatic distress, endorsing six or more of the 10 symptoms in a survey. Other factors associated with having symptoms of posttraumatic distress were over age 25, motherhood, and an absent husband. In a study of 80 Senegalese refugees in two camps in The Gambia (West Africa), Tan and Fox (2001) found that 10% had symptoms of PTSD, 46.3% had anxiety symptoms, and 58.8% had depressive symptoms. PTSD was found in 27% of 101 civilians living in a conflict area in Sri Lanka, and 41% of these civilians had symptoms of somatization (Somasundaram & Sivayokan, 1994). Allden and colleagues (1996), in a study of 104 Burmese refugees in Thailand, found that 38% had depressive symptoms and 23% met criteria for PTSD. As with other studies of postconflict mental health, there was a range of population sizes and methodologies. Multiple stressors, as noted in the de Jong and colleagues (2001) study, may affect the level of distress within the population.

Epidemiological Surveys of Disaster

The disaster research, primarily conducted through population surveys, has also reported PTSD, depression, and other anxiety disorders following trauma exposure. In a review of 160 postdisaster studies across five continents, Norris, Friedman, Watson, Byrne, and colleagues (2002) reported that suffering from a mental disorder with severe or very severe impairment was highest within developing countries in Asia, Africa, eastern Europe, and the Americas. Compared to U.S. samples with 25% reporting severe or very severe impairment, 78% of the population samples from developing countries met this critical level of impairment. Norris and colleagues felt that some factors accounting for the higher percentage of severe or very severe impairment in developing populations was due to greater severity of traumatic exposure and limited access to resources (food, shelter, health services) in the postdisaster period.

Some studies of the impact of disaster explored ethnicity as a potential risk factor in Western populations, in non-Western settings, and in treatment-seeking samples. Ethnicity was assessed in a population survey following the September 11, 2001, terrorist attacks (Galea et al., 2004). Dominican and Puerto Rican Hispanics reported more symptoms of PTSD than Hispanics from other origins and non-Hispanics. The higher incidence for Dominican (14.3%) and Puerto Rican (13.2%) compared to other Hispanics (6.1%) and non-Hispanics (5.2%) was thought to be mediated by greater exposure, lower incomes, lower social support, and younger age. Webster, McDonald, Lewin, Lewin, and Carr (1995) reported that non-English-speaking immigrants suffered more psychological distress following an earthquake than did native Australians. Again, level of exposure contributed more to psychological distress than ethnicity.

Several studies assessed symptoms of PTSD, depression, and anxiety following an earthquake in Japan that affected an estimated 3.5 million people (Kato, Asukai, Miyake, Minakawa, & Nishiyama, 1996; Kokai et al., 2004). Kato and colleagues (1996) assessed PTSD and depressive symptoms in 67 people younger than age 60 and 75 elderly evacuees at two time points: 3 and 8 weeks after the earthquake. Complaints of poor sleep, depression, irritability, and high startle were common. Younger evacuees continued to show symptoms at 8 weeks, whereas the elderly showed a reduction in the number of symptoms. This decrease in symptoms in the elderly was attributed to their access to more extensive social networks and prior experience with trauma, although

Kokai and colleagues (2004) advised caution in accepting prior trauma as a protective factor. Kokai and colleagues reviewed three published surveys conducted in a clinical population in Japan. Initial anxiety and depression detected 1 month after the earthquake had decreased at 1 year. They note that different estimates of the number of PTSD cases were mediated by the survey instrument, with 21.1% meeting ICD-10 PTSD criteria and only 5% meeting DSM-IV PTSD criteria. A population survey of 252 randomly selected survivors 10 months after an earthquake in Taiwan reported a PTSD prevalence rate of 10.3% and a partial PTSD rate of 19%. Current major depression was reported to be 17.5%. Neurasthenia, a disorder well known in China, characterized by fatigue, weakness, and somatic complaints, was reported by 38.9% of survivors (Lai, Chang, Connor, Lee, & Davidson, 2004).

A survey of Nicaraguan adolescents from three cities was conducted after Hurricane Mitch (Goenjian et al., 2001). The rates of PTSD and depression were related to the level of exposure; PTSD and depression rates in the most exposed city were 90% and 81%, respectively, compared to 55% and 51%, respectively, in the city with midlevel exposure, and 14% and 29%, respectively, in the least exposed city. A survey of both treatment-seeking and non-treatment-seeking samples ($n = 2,627$) after a terrorist bombing of the American embassy in Nairobi, Kenya, revealed that 35% of Kenyans exposed to the trauma endorsed sufficient symptoms of PTSD to meet diagnostic criteria (Njenga, Nicholls, Naymai, Kigamwa, & Davidson, 2004). Among risk factors, closer proximity and injury predicted symptoms of PTSD. Of note, this finding that higher rates of PTSD and depression were found in survivors with greater levels of exposure has been seen in other, non-treatment-seeking populations and in other types of disasters in Western settings, for example, after earthquakes in Turkey (Kilic & Ulusoy, 2003; Salcioglu, Basoglu, & Livanou, 2003), an industrial disaster in Toulouse (Godeau et al., 2005), and terrorists attacks in New York City (Galea et al., 2002). Similar to the war-related literature, the disaster literature is based on varying methodology, including study instruments and populations.

Summary of Epidemiological Surveys

Epidemiological surveys suggest that PTSD is an acceptable classification for the psychological reactions in survivors across cultural settings and appears to be a universal reaction to severe stressors that has transcultural diagnostic validity. The studies differ as to whether PTSD or major depressive disorder is the most prevalent sequela to psychological trauma. The rates of PTSD in non-Western settings are similar to findings in Western areas, but significant variations in methodology in both settings may account for the significant variations in prevalence rates of PTSD, other anxiety disorders, and major depression. Studies indicate that greater exposure (e.g., closer proximity) and more severe trauma (e.g., sustaining injuries) are significant risk factors across the world. For example, a recent community survey that compared prevalence of PTSD in African Americans in Oklahoma City and Kenyans after terrorist attacks found that both groups had similar rates of PTSD (North et al., 2005).

Treatment

Culturally informed therapy in the treatment of PTSD has been advocated (for reviews, see de Jong, 2002b; de Jong & Clarke, 1996; De Vries, 1996; Green et al., 2003; Marsella, Friedman, et al., 1996; Westermeyer, 1989); however, few empirical studies have been conducted. The majority of empirical studies have been of therapy. Despite knowledge of genetic variations in drug metabolism, there are no currently published PTSD medication trials in non-Western populations. Ethnopharmacology has identified genetic variants in the cytochrome P450 system that affect drug metabolism (see Gaw, 1993, for a review). Psychotherapy and psychopharmacology treatment studies in non-Western populations are currently in early stages and are for the most part case studies, open studies, or small pilot studies. Although the effect size in some studies are robust, these preliminary findings warrant further study. Thus, there are few data to guide the clinician and many avenues for researchers to explore.

General Concepts

Pedersen (1997) noted that there is basic agreement among theoreticians, researchers, and clinicians in cross-cultural counseling that techniques must be culturally adapted when applied to a different cultural milieu, because therapeutic approaches cannot be transposed automatically across cultures. Problems in providing culturally sensitive treatment increase in relation to the cultural differences between the clinician and client, because (1) expressions of distress differ across cultures, (2) these modes of self-presentation and communication of distress impact the therapy, and (3) norms and expectations are culturally determined. Cultural constructs and clinical approaches that broadly guide the clinician have been described for PTSD treatment in African Americans (Allen, 1996), Hispanic Americans (Hough, Canino, Abueg, & Gusman, 1996), Native Americans (Manson, 1996; Manson et al., 1996; Robin, Chester, & Goldman, 1996), and Asian Americans (Abueg & Chun, 1996).

De Jong (2002a) has addressed the need to work closely with each cultural group to determine culturally appropriate programs for the delivery of mental health services. The needs of the members of the local culture, their leaders, and local healers must be a core component of program development. For refugee or immigrant programs, recruitment of immigrant staff is a good way to enhance treatment services, both to inform nonimmigrant staff about ethnic and cultural differences, and to provide therapists or counselors for the same ethnic and linguistic group (de Jong & van Ommeren, 2005). Sue, Fujino, Hu, Takeuchi, and Zane (1991) noted that ethnic and linguistic matches decreased dropout rates and had better clinical outcomes compared to nonmatched patients and providers. However, Beutler, Machado, and Neufeldt (1994) disputed this finding, reporting no differences in treatment outcome when clinicians and clients were mismatched for age, sex, or ethnicity. On the other hand, they also found that patients were likely to remain in treatment longer if matched on these dimensions (Beutler et al., 1994; Kirmayer, Groleau, Guzder, Blake, & Jarvis, 2003).

Immigrant staff can inform others about normal or deviant behaviors and assess what social supports are normal for the culture and are available to the patient and family. Cultural healing traditions, as well as the use of culturally related vocabulary and metaphors, may be helpful in treatment (Hinton et al., 2004; Otto et al., 2003). For example, a Native American ritual, the sweat lodge, has been incorporated into the treatment of Native American Vietnam veterans (Silver & Wilson, 1988). Cognitive-behavioral therapy may be an acceptable and effective therapeutic modality for Southeast Asian populations due to similarity between its core aspects and Buddhist principles (Bemak, Chung, & Borneman, 1996; Boehnlein, 1987; de Jong & Van Ommeren, 2005; Hinton et al., 2004; Otto et al., 2003).

Treatment Outcome Studies

A recently reported 10-year follow-up study of 23 Cambodian refugees in continuous treatment that included medication, supportive psychotherapy, socialization, group therapy, and social support reported that 60% of patients with PTSD improved and 83% with depression improved with treatment. However, the authors found that relapses were common, with 14 patients having at least one severe relapse, and many having several severe relapses. They commented on the difficulty in assessing recovery given the nature of PTSD and that a patient can move in or out of a disorder status depending on current stressors. The authors suggested a relapse prevention model (Boehnlein et al., 2004).

An early treatment intervention study of Bosnian refugees in the Netherlands compared a 24-week phase-oriented group psychotherapy, medication, and combined group therapy and medication treatment (Drozdek, 1997). Medications were uncontrolled and consisted of variable doses of benzodiazepines and tricyclic antidepressants. All treatments were found to be equally effective, with 73% of refugees in the treatment groups no longer meeting diagnostic criteria for PTSD, whereas 90% of the control group that refused treatment continued to have PTSD.

Weine, Kulenovic, Pavkovic, and Gibbons (1998) conducted a pilot study of testimony psychotherapy, a brief (four to eight sessions) intervention, in 20 Bosnian refugees in the United States, assessing pre- and posttreatment effects. The authors reported a reduction in PTSD diagnoses and depressive symptomatology, and an improvement in functioning. All parameters continued to show gains at the 2- and 6-month follow-up assessments. A randomized study compared 16–20 sessions of CBT (exposure, cognitive therapy, and controlled breathing) and prolonged exposure for the treatment of PTSD in 16 refugees living in Sweden (Paunovic & Öst, 2001). Pre- and posttreatment and 6-month follow-up assessments revealed that both treatment interventions resulted in nearly 50% improvement in symptoms of PTSD, generalized anxiety, and depression, with improvements also in quality of life and cognitive schemas.

A recent study by Otto and colleagues (2003; Otto & Hinton, 2004) modified a cognitive-behavioral therapy (CBT) treatment of PTSD, cognitive processing therapy (CPT; Resick & Schnicke, 1992), for Cambodian women refugees in the United States, with an emic approach based on the previously described ethnographic studies using the local idioms of distress *kyol goeu*, or wind overload (Hinton, Um, & Ba, 2001a, 2001b). Ten treatment-resistant Cambodian women with PTSD were randomized to 10-session group CPT and sertraline, or sertraline alone. The combined treatment had greater efficacy for PTSD, somatization, anxiety, and in particular anxiety related to culture-specific fears. The effect sizes for reexperiencing and avoidance/numbing symptoms, 0.82 and 0.85, respectively, were more robust than that for hyperarousal symptoms (0.45). The effect size for culture-specific symptoms of anxiety was 1.77, whereas both somatization and general anxiety symptoms had effect sizes of about 0.60.

An open study of a 12-week manualized treatment that included individual, family, and group therapy focused on psychoeducation, trauma and grief activities, and creative and relaxation techniques, was conducted in 10 Kosovan child and adolescent refugees in Germany (Mohlen, Parzer, Resch, & Brunner, 2005). Three of the six children who had a pretreatment diagnosis of PTSD no longer met criteria for PTSD, and 9 of the 10 children showed gains in psychosocial functioning.

In an open study of a 6-week program of meditation, biofeedback, drawing, autogenic training, guided imagery, movement, and relaxation in Kosovo, Gordon, Staples, Blyta, and Bytyqi (2004) reported that adolescents benefited from this treatment. Chemtob, Nakashima, and Carlson (2001) conducted a controlled study of a brief postdisaster intervention in 32 school-age Hawaiian children. Three sessions of eye movement desensitization and reprocessing resulted in fewer PTSD symptoms, with gains sustained at 6-month follow-up.

A treatment efficacy study of narrative exposure therapy (NET), a brief therapy based on cognitive-behavioral theory and testimony therapy, was conducted in a Ugandan refugee settlement. Forty-three Sudanese refugees diagnosed with PTSD were randomly assigned to four sessions of NET, four sessions of supportive counseling, or one session of psychoeducation. Outcome measures 1 year after treatment supported efficacy of NET (29% having PTSD) compared to supportive counseling (79%) and psychoeducation groups (80%) (Neuner, Schauer, Klaschik, Karunakara, & Elbert, 2004).

To date, there are no studies of psychopharmacology treatment in non-western populations. One study assessed treatment of post-conflict PTSD. Martenyi, Brown, Zhang, Prakash, and Koke (2002) conducted the only double-blind, placebo-controlled trial of fluoxetine in patients from several war-torn areas in Europe, Israel, and South Africa. Eighteen study centers in Belgium, Bosnia, Croatia, Israel, and Yugoslavia recruited 301 patients with postconflict PTSD. The patients were predominately male (81%), white (91%), had experienced multiple combat-related events (48%) and/or were survivors of war or witnesses to war (47%), and/or had witnessed another person's death (33%). Random assignment resulted in 226 subjects being treated with fluoxetine in dosages titrated from 20 to 80 mg (mean dose, 57 mg) and 75 subjects receiving placebo. In clinician-rated measures, fluoxetine resulted in significant improvement in total PTSD scores, as well as the intrusive and hyperarousal subscales, but not the avoidance and numbing subscales. These findings were statistically significant by 6 weeks and maintained significance through the 12-week study. In addition, clinician-administered measures of depression indicated significant improvement in depression symptoms in the fluoxetine-treated sample. Overall, the effect size was 0.40 in favor of fluoxetine. An analysis of groups found that for post-combat-related trauma, the effect size was 0.78, and in persons without dissociation, it was 1.20. Of note, the patient-rated measure for both PTSD and depression failed to show a difference between placebo and fluoxetine.

GENERALIZABILITY OF CURRENT FINDINGS

The most urgent question facing the mental health field at present is the generalizability of our current knowledge of PTSD and its treatment. How to adapt treatment to meet the needs of people across the globe is a significant challenge. Based on the literature, PTSD is a disorder that does not spare people exposed to severe trauma, whether in Western or non-Western cultures. As noted, the issue of category fallacy must be considered in any assessment of PTSD across cultures, and we should question whether some symptoms are modified by culture, and whether the epidemiology reflects accurate prevalence rates. Cluster B and C criteria appear to have symptoms that are potentially subject to cultural beliefs (e.g., a dream may be seen as a visit from an ancestor), practices, or life circumstances (e.g., sense of foreshortened future may not be endorsed in a country with an average lifespan of 45 years or in areas where AIDS is epidemic). Cluster D symptoms may be more solidly based in neurobiological changes that we might expect across cultures. Further investigation is necessary to clarify which symptoms and symptom clusters are affected by culture, and which may be ubiquitous neurobiological responses to overwhelming stress.

In 1994, Sue, Zane, and Young noted that the generalizability of treatment outcome research findings from the West to populations outside the West is largely unknown. This is true today of PTSD. We are currently in the early stages of exploring these questions. As noted, most studies are open studies or small pilot studies of refugees, with few in the local setting using available resources. There is very little effectiveness research conducted in the naturalistic setting with people living in the local or non-Western community to increase our knowledge about the impact of treatment interventions in less sterile environments than those typically designed in Western efficacy studies. This type of knowledge will be critical to developing programs to address widespread calls reaffirming good health as a human right and to close the gap between those with and those without services (Flanagin & Winker, 2003).

CHALLENGES FOR THE FUTURE

Collaboration with other fields in science (e.g., neuroscience) and the social sciences (e.g., anthropology, sociology) is needed to address future challenges in this emerging cultural field of traumatic stress. It is time for the fields of anthropology and mental health to end the debate about the validity of the diagnosis of PTSD, now part of our diagnostic nomenclature for nearly 25 years (de Jong, in press). The growing knowledge in neuroscience provides insights into the basis of mental illness. The role of brain structures, such as the amygdala in fear response, seen in both animal models (LeDoux, 1996) and neuroimaging of patients with PTSD (Bremner, 2002; Shin et al., 1997), and of changes in neurotransmitters such as norepinephrine (Southwick et al., 1993) and neurohumoral responses as seen with cortisol (Yehuda, 2002), are critical to developing a comprehensive model of the neurobiology of PTSD. Such knowledge should enable us to parse out the unique and interactive contributions of biology and culture to the PTSD syndrome. With that advance, we will also be in a position to understand how PTSD or other posttraumatic idioms of distress are modified by cultural beliefs and meaning systems.

Such a new collaborative effort could help to answer the challenging questions that are unanswered or only partially answered by our current knowledge. Given the number of cultures and subcultures, and the level of trauma exposure around the globe, the task of understanding cultural influences in traumatic stress seems daunting. It is therefore crucial to develop models of traumatic stress that can incorporate the cultural knowledge that will emerge from this new collaboration. A cultural traumatic stress model will provide a context within which researchers and clinicians can develop and adapt culturally competent scientific hypotheses to consider and adapt their assessments techniques, and treatment approaches.

A culturally competent model of traumatic stress should explore how culture is implicated in the process of traumatization, including appraisal, vulnerability, and protective factors. Critical to this field is learning the language—the idioms of distress—that individuals within a cultural group use as acceptable means of communicating distress. Equally important is for researchers and therapists to understand the explanatory models, the attribution of illness, and the culturally sanctioned local healing methods. These factors are critical to the success of research and clinical care, and effective and cost-effective health care delivery systems that are not bound by Western attitudes and knowledge, but are firmly based in the reality of the culture, the available resources, and the current individual or sociocultural situation.

Mental health clinicians by themselves cannot meet the mental health needs of a world that faces daily traumatic stress. Knowledge of how culture intertwines with negative and traumatic life events and responses at both the individual and community level must grow from a broad collaboration to meet the mounting need for interventions to reduce both the outcomes and cycles of violence rooted in both culture and traumatic stress disorders.

REFERENCES

Abueg, F., & Chun, K. M. (1996). Traumatization stress among Asians and Asian Americans. In A. J. Marsella, M. J. Friedman, E. T. Gerrity, & R. M. Scurfield (Eds.), *Ethnocultural aspects of posttraumatic stress disorder: Issues, research, and clinical applications* (pp. 285–300). Washington, DC: American Psychological Association.

Aheto, D. W., & Gbesemete, K. P. (2005). Rural perspectives on HIV/AIDS prevention: A comparative study of Thailand and Ghana. *Health Policy, 72*, 25–40.

Allden, K., Poole, C., Chantavanich, S., Ohmar, K., Aung, N. N., & Mollica, R. F. (1996). Burmese political dissidents in Thailand: Trauma and survival among young adults in exile. *American Journal of Public Health, 86*, 1561–1569.

Allen, I. M. (1996). PTSD among African Americans. In A. J. Marsella, M. J. Friedman, E. T. Gerrity, & R. M. Scurfield (Eds.), *Ethnocultural aspects of posttraumatic stress disorder: Issues, research, and clinical applications* (pp. 209–238). Washington, DC: American Psychological Association.

Ayuku, D., Odero, W., Kaplan, C., De Bruyn, R., & De Vries, M. (2003). Social network analysis for health and social interventions among Kenyan scavenging street children. *Health Policy and Planning, 18*, 109–118.

Ballenger, J. C., Davidson, J. R. T., Lecrubier, Y., Nutt, D., Foa, E. B., Kessler, R. C., et al. (2000). Consensus statement on posttraumatic stress disorder from the International Consensus Group on Anxiety and Depression. *Journal of Clinical Psychiatry, 61*(Suppl. 5), 60–66.

Becker, A. E., & Kleinman, A. (2000). Anthropology and psychiatry. In B. I. Kaplan & V. A. Sadock (Eds.), *Comprehensive textbook of psychiatry* (7th ed., Chapter 4.1). Baltimore: Williams & Wilkins.

Bemak, F., Chung, R.C.-Y., & Bornemann, T. (1996). Counseling and psychotherapy with refugees. In P. Pedersen, J. Draguns, W. Lonner, & J. Trimble (Eds.), *Counseling across cultures* (4th ed., pp. 243–265). Thousand Oaks, CA: Sage.

Bernard, R. H. (1994). *Research methodology in anthropology: Qualitative and quantitative approaches.* London: Sage.

Beutler, L. E., Machado, P. P., & Neufeldt, S. A. (1994). Therapist variables. In A. E. Bergin & S. L. Garfield (Eds.), *Handbook of psychotherapy and behavior change* (4th ed., pp. 229–269). New York: Wiley.

Biernacki, P., & Waldorf, D. (1981). Snowball sampling: Problems and techniques of chain referral sampling. *Social Methods Research, 2*, 141–163.

Boehnlein, J. K. (1987). Clinical relevance of grief and mourning among Cambodian refugees. *Social Science and Medicine, 25*, 765–772.

Boehnlein, J. K., Kinzie, J. D., Sekiya, U., Riley, C., Pou, K., & Rosborough, B. (2004). A ten-year treatment outcome study of traumatized Cambodian refugees. *Journal of Nervous and Mental Disease, 192*, 658–663.

Bolton, P., & Tang, A. M. (2004). Using ethnographic methods in the selection of post disaster, mental health interventions. *Prehospital and Disaster Medicine, 19*, 97–101.

Bracken, P. J., Giller, J. E., & Summerfield, D. (1995). Psychological responses to war and atrocity: The limitations of current concepts. *Social Science and Medicine, 40*, 1073–1082.

Bremner, J. D. (2002). *Does stress damage the brain? Understanding trauma related disorders from a mind–body perspective.* New York: Norton.

Brislin, R. W. (1986). The wording and translation of research instruments. In J. W. Lonner & J. W. Berry (Eds.), *Field methods in cross-cultural research* (pp. 137–164). Newbury Park, CA: Sage.

Brislin, R. W., Lonner, W. J., & Thorndike, R. M. (1973). *Cross-cultural research methods.* New York: Wiley.

Chemtob, C. M., Nakashima, J., & Carlson, J. G. (2001). Brief treatment for elementary school children with disaster-related posttraumatic stress disorder: A field study. *Journal of Clinical Psychology, 58,* 99–112.

Dahl, S., Mutapcic, A., & Schei, B. (1998). Traumatic events and predictive factors for posttraumatic symptoms in displaced Bosnian women in a war zone. *Journal of Traumatic Stress, 11,* 137–145.

de Jong, J. (2000). Traumatic stress among ex-combatants. In N. Pauwels (Ed.), *Work force to work force: Global perspectives on demobilization and reintegration.* Baden-Bande: Nomos Verlag.

de Jong, J. (2002a). Public mental health, traumatic stress and human rights violations in low-income countries: A culturally appropriate model in times of conflict, disaster and peace. In J. de Jong (Ed.), *Trauma, war and violence: Public mental health in sociocultural context* (pp. 1–91). New York: Kluwer Academic/Plenum Press.

de Jong, J. (2002b). *War, trauma, and violence: Public mental health in the sociocultural context.* New York: Kluwer Academic/Plenum Press.

de Jong, J. T. V. M. (1987). *A descent into African psychiatry.* Amsterdam: Royal Tropical Institute.

de Jong, J. T. V. M. (2004). Public mental health and culture: Disasters as a challenge to western mental health care models, the self, and PTSD. In J. P. Wilson & B. Drozdek (Eds.), *Broken spirits: The treatment of asylum seekers and refugees with PTSD* (pp. 157–176). New York: Brunner/Routledge.

de Jong, J. T. V. M. (in press). Deconstructing critiques on the internationalization of PTSD. *Culture, Medicine and Psychiatry.*

de Jong, J. T. V. M., & Clarke, L. (Eds.). (1996). *Mental health of refugees.* Geneva: World Health Organization. Available at whqlibdoc.who.int/hq/1996/a49374.pdf

de Jong, J. T. V. M., Komproe, I. H., Spinazzola, J., van der Kolk, B. A., & Van Ommeren, M. H. (2005). DESNOS in three postconflict settings: Assessing cross-cultural construct equivalence. *Journal of Traumatic Stress, 18,* 13–21.

de Jong, J. T. V. M., Komproe, I., & Van Ommeren, M. (2003). Terrorism, human-made and natural disasters as a professional and ethical challenge to psychiatry. *International Psychiatry, 1*(7), 8–9.

de Jong, J. T. V. M., Komproe, I. H., Van Ommeren, M., El Masri, M., Araya, M., Khaled, N., et al. (2001). Lifetime events and posttraumatic stress disorder in four postconflict settings. *Journal of the American Medical Association, 286,* 555–562.

de Jong, J. T. V. M., & Van Ommeren, M. H. (2002). Toward a culture informed epidemiology: Combining qualitative and quantitative psychiatric research in transcultural contexts. *Transcultural Psychiatry, 39,* 422–433.

de Jong, J. T. V. M., & Van Ommeren, M. (2005). Mental health services in a multicultural society: Interculturalization and its quality surveillance. *Transcultural Psychiatry, 42,* 437–456.

de Jong, J. T. V. M., & van Schaik, M. M. (1994). Culturele en religieuze aspecten vantraumaverwerking naar aanleiding van de Bijlmerramp [Cultural and religious aspects of coping with trauma after the Bijlmer disaster]. *Tijdschrift voor Psychiatrie, 36,* 291–304.

De Vries, M. W. (1996). Trauma in cultural perspective. In B. A. van der Kolk, A. C. McFarlane, & L. Weisæth (Eds.), *Traumatic stress: The effects of overwhelming experience on mind, body, and society.* (pp. 398–416). New York: Guilford Press.

Ding, Y., Detels, R., Zhao, Z., Zhu, Y., Zhu, G., Zhang, B., et al. (2005). HIV infection and sexually transmitted diseases in female commercial sex workers in China. *Journal of Acquired Immune Deficiency Syndromes, 38,* 314–319.

Drozdek, B. (1997). Follow-up study of concentration camp survivors from Bosnia Herzegovina: Three years later. *Journal of Nervous and Mental Disease, 185,* 690–694.

Eisenbruch, M. (1994). Mental health and the Cambodian traditional healer for refugees who resettled, were repatriated or internally displaced, and for those who stayed at home. *Collegium Antropologicum, 18,* 219–230.

Favaro, A., Maiorani, M., Colombo, G., & Santonastaso, P. (1999). Traumatic experiences, posttraumatic stress disorder, and dissociative symptoms in a group of refugees from former Yugoslavia. *Journal of Nervous and Mental Disease, 187,* 306–308.

Fazel, M., Wheeler, J., & Danesh, J. (2005). Prevalence of serious mental disorder in 7000 refugees resettled in western countries: A systematic review. *Lancet, 365,* 1309–1314.

Flaherty, J. A., Gaviria, F. M., Pathak, D., Mitchell, T., Wintrob, R., Richman, J. A., et al. (1988).

Developing instruments for cross-cultural psychiatric research. *Journal of Nervous and Mental Disease, 76*, 257–263.

Flanagin, A., & Winker, M. A. (2003). Global health-targeting problems and achieving solutions. *Journal of the American Medical Association, 10*, 1382–1384.

Foa, E. B., Keane, T. M., & Friedman, M. J. (2000). *Effective treatments for PTSD: Practice guidelines from the International Society for Traumatic Stress Studies.* New York: Guilford Press.

Galea, S., Resnick, H., Ahern, J., Gold, J., Bucuvalas, M., Kilpatrick, D., et al. (2002). Posttraumatic stress disorder in Manhattan, New York City, after the September 11th terrorist attacks. *Journal of Urban Health, 79*, 340–353.

Galea, S., Vlahov, D., Tracy, M., Hoover, D. R., Resnick, H., & Kilpatrick, D. (2004). Hispanic ethnicity and post-traumatic stress disorder after a disaster: Evidence from a general population survey after September 11, 2001. *Annals of Epidemiology, 14*, 520–531.

Gaw, A. (Ed.). (1993). *Culture, ethnicity, and mental illness.* Washington, DC: American Psychiatric Press.

Geltman, P. L., Grant-Knight, W., Mehta Supriya, D., Lloyd-Travaglini, C., Lustig, S., Landgraf, J. M., et al. (2005). The "lost boys of Sudan": Functional and behavioral health of unaccompanied refugee minors resettled in the United States. *Archives of Pediatrics and Adolescent Medicine, 159*, 585–591.

Gernaat, H. B. P. E., Malwand, A. D., Laban, C. J., Komproe, I., & de Jong, J. T. V. M. (2002). [Many psychiatric disorders among Afghan refugees in Drenthe, the Netherlands, with a residence status, in particular depressive and posttraumatic stress disorders: Community based study.] *Nederlands Tijdschrift voor Geneeskunde, 146*, 1127–1131.

Godeau, E., Vignes, C., Navarro, F., Iachan, R., Ross, J., Pasquier, C., et al. (2005). Effects of a large-scale industrial disaster on rates of symptoms consistent with posttraumatic stress disorders among schoolchildren in Toulouse. *Archives of Pediatrics and Adolescent Medicine, 159*, 579–584.

Goenjian, A. K., Molina, L., Steinberg, A. M., Fairbanks, L. A., Alvarez, M. L., Goenjian, H. A., et al. (2001). Posttraumatic stress and depressive reactions among Nicaraguan adolescents after hurricane Mitch. *American Journal of Psychiatry, 158*, 788–794.

Goodman, L. A. (1961). Snowball sampling. *Annals of Mathematics and Statistics, 32*, 148–170.

Gordon, J. S., Staples, J. K., Blyta, A., & Bytyqi, M. (2004). Treatment of posttraumatic stress disorder in postwar Kosovo high school students using mind–body skills groups: A pilot study. *Journal of Traumatic Stress, 17*, 143–147.

Gorst-Unsworth, C., & Goldenberg, E. (1998). Psychological sequelae of torture and organized violence suffered by refugees from Iraq. *British Journal of Psychiatry, 172*, 90–94.

Green, B. L., Friedman, M. J., de Jong, J. T. V. M., Solomon, S. D., Keane, T. M., Fairbank, J. A., et al. (Eds.). (2003). *Trauma interventions in war and peace.* New York: Kluwer Academic/Plenum Press.

Hinton, D., Pham, T., Tran, M., Safren, S. A., Otto, M. W., & Pollack, M. H. (2004). CBT for Vietnamese refugees with treatment-resistant PTSD and panic attacks: A pilot study. *Journal of Traumatic Stress, 17*, 429–433.

Hinton, D., Um, K., & Ba, P. (2001a). *Kyol goeu* ("wind overload") part I: A cultural syndrome of orthostatic panic among Cambodian refugees. *Transcultural Psychiatry, 38*, 403–432.

Hinton, D., Um, K., & Ba, P. (2001b). *Kyol goeu* ("wind overload") part II: Prevalence, characteristics and mechanisms of *kyol goeu* and *near-kyol goeu* episodes of Cambodian patients attending a psychiatric clinic. *Transcultural Psychiatry, 38*, 433–460.

Hinton, W. L., Chen, Y. C., Du, N., Tran, C. G., Lu, F. G., Miranda, J., et al. (1993). DSM III-R disorders in Vietnamese refugees: Prevalence and correlates. *Journal of Nervous and Mental Disease, 181*, 113–122.

Hollifield, M., Eckert, V., Warner, T. D., Jenkins, J., Krakow, B., Ruiz, J., et al. (2005). Development of an inventory for measuring war-related events in refugees. *Comprehensive Psychiatry, 46*, 67–80.

Holman, D. (1997). The relevance of person centered ethography to cross-cultural psychiatry. *Transcultural Psychiatry, 34*, 219–234.

Hough, R. L., Canino, G. J., Abueg, F. R., & Gusman, F. D. (1996). PTSD and related disorders among Hispanics. In A. J. Marsella, M. J. Friedman, E. T. Gerrity, & R. M. Scurfield (Eds.), *Ethnocultural aspects of posttraumatic stress disorder: Issues, research, and clinical applications* (pp. 301–340). Washington, DC: American Psychological Association.

Kato, H., Asukai, N., Miyake, Y., Minakawa, K., & Nishiyama, A. (1996). Post-traumatic symptoms among younger and elderly evacuees in the early stages following the 1995 Hanshin-Awaji earthquake in Japan. *Acta Pscyhiatrica Scandinavica, 93*, 441–447.

Katz, C. L., Pellegrino, L., Pandya, A., Ng, A., & DeLisi, L. E. (2002). Research on psychiatric outcomes and interventions subsequent to disasters: A review of the literature. *Psychiatry Research, 110,* 201–217.

Keane, T. M., & Barlow, D. H. (2002). Posttraumatic stress disorder. In D. H. Barlow (Ed.), *Anxiety and its disorders: The nature and treatment of anxiety and panic* (2nd ed., pp. 418–453). New York: Guilford Press.

Keane, T. M., Kaloupek, D. G., & Weathers, F. W. (1996). Ethnocultural considerations in the assessment of PTSD. In A. J. Marsella, M. J. Friedman, E. T. Gerrity, & R. M. Scurfield (Eds.), *Ethnocultural aspects of posttraumatic stress disorder: Issues, research, and clinical applications* (pp. 183–205). Washington, DC: American Psychological Association.

Kessler, R. C., Birnbaum, H., Demler, O., Falloon, I. R. H., Gagnon, E., Guyer, M., et al. (2005). The prevalence and correlates of nonaffective psychosis in the national comorbity survey replication (NCS-R). *Biological Psychiatry, 58,* 668–676.

Kilic, C., & Ulusoy, M. (2003). Psychological effects of the November 1999 earthquake in Turkey: An epidemiological study. *Acta Psychiatrica Scandinavica, 108,* 232–238.

Kinzie, J. D., Boehnlein, J. K., Leung, P. K., Moore, L. J., Riley, C., & Smith, D. (1990). The prevalence of posttraumatic stress disorder and its clinical significance among Southeast Asian refugees. *American Journal of Psychiatry, 147,* 913–917.

Kinzie, J. D., Sack, W. H., Angell, R. H., Clarke, G., & Ben, R. (1989). A three-year follow up of Cambodian young people traumatized as children. *Journal of the American Academy of Child and Adolescent Psychiatry, 28,* 501–504.

Kirmayer, L. J., Groleau, D., Guzder, J., Blake, C., & Jarvis, E. (2003). Cultural consultation: A model of mental health service for multicultural societies. *Canadian Journal of Psychiatry, 48,* 145–153.

Kleinman, A. (1977). Depression, somatization, and the "new" cross-cultural psychiatry. *Social Science in Medicine, 11,* 3–10.

Kleinman, A. (1982). Neurasthenia and depression: A study of somatization and culture in China. *Culture, Medicine, and Psychiatry, 6,* 117–190.

Kleinman, A. (1980). *Patients and healers in the context of culture: An exploration of the borderland between anthropology, medicine, and psychiatry.* Berkeley: University of California Press.

Kleinman, A. (1988). *The illness narrative.* New York: Basic Books.

Kleinman, A., & Good, B. (Eds.). (1985). *Culture and depression.* Los Angeles: University of California Press.

Kleinman, A., Eisenberg, L., & Good, B. (1978). Culture, illness, and care: Clinical lessons from anthropologic and cross-cultural research. *Annals of Internal Medicine, 88,* 251–258.

Kleinman, A. M. (1975). Medical and psychiatric anthropology and the study of traditional forms of medicine in modern Chinese culture. *Bulletin of the Institute of Ethnology Academy Sinica, 39,* 107–123.

Kokai, M., Fujii, S., Shinfuku, N., & Edwards, G. (2004). Natural disaster and mental health in Asia. *Psychiatry and Clinical Neurosciences, 58,* 110–116.

Kroll, J., Habenicht, M., Mackenzie, T., Yang, M., Chan, S., Vang, T., et al. (1989). Depression and posttraumatic stress disorder in Southeast Asian refugees. *American Journal of Psychiatry, 146,* 1592–1597.

Krueger, R. A. (1994). *Focus groups: A practical guide for applied research* (2nd ed.). Thousand Oaks, CA: Sage.

Lai, T. J., Chang, C. M., Connor, K. M., Lee, L. C., & Davidson, J. R. (2004). Full and partial PTSD among earthquake survivors in rural Taiwan. *Journal of Psychiatric Research, 38,* 313–322.

Laban, C. J., Gernaat, H. B. P. E., Komproe, I. H., Schreuders, B. A., & de Jong, J. T. V. M. (2004). Impact of a long asylum procedure on the prevalence of psychiatric disorders in Iraqi asylum seekers in the Netherlands. *Journal of Nervous and Mental Disease, 192,* 843–851.

Laban, C. J., Gernaat, H. B. P. E., Komproe, I. H., Van der Tweel, I., & de Jong, J. T. V. M. (2005). Post migration living problems and common psychiatric disorders in Iraqi asylum seekers in the Netherlands. *Journal of Nervous and Mental Disease, 193,* 825–832.

Lavik, N. J., Hauff, E., Skrondal, A., & Solberg, O. (1996). Mental disorder among refugees and the impact of persecution and exile: Some findings from an outpatient population. *British Journal of Psychiatry, 169,* 726–732.

LeDoux, J. (1996). *The emotional brain.* New York: Simon & Schuster.

Manson, S., Beals, J., O'Nell, T., Piasecki, J., Bechtold, D., Keane, E., et al. (1996). Wounded spirits, ailing hearts: PTSD and related disorders among American Indians. In A. J. Marsella, M. J. Friedman, E. T. Gerrity, & R. M. Scurfield (Eds.), *Ethnocultural aspects of posttraumatic stress disorder: Issues, research, and clinical applications* (pp. 255–284). Washington, DC: American Psychological Association.

Manson, S. M. (1996). The wounded spirit: A cultural formulation of post-traumatic stress disorder. *Culture, Medicine and Psychiatry, 20*, 489–498.

Manson, S. M. (1997). Cross-cultural and multi-ethnic assessment of trauma. In J. P. Wilson & T. M. Keane (Eds.), *Assessing psychological trauma and PTSD: A handbook for practitioners* (pp. 239–266). New York: Guilford Press.

Marsella, A. J., Friedman, M. J., Gerrity, E. T., & Scurfield, R. M. (Eds.). (1996). *Ethnocultural aspects of posttraumatic stress disorder: Issues, research, and clinical applications*. Washington, DC: American Psychological Association.

Marsella, A. J., Friedman, M. J., & Spain, E. H. (1996). Ethnocultural aspects of PTSD: An overview of issues and research directions. In A. J. Marsella, M. J. Friedman, E. T. Gerrity, & R. M. Scurfield (Eds.), *Ethnocultural aspects of posttraumatic stress disorder: Issues, research, and clinical applications* (pp. 105–130). Washington, DC: American Psychological Association.

Marsella, A. J., & Kameoka, V. (1989). Ethnocultural issues in the assessment of psychopathology. In S. Wetzler (Ed.), *Measuring mental illness: Psychometric assessment for clinicians* (pp. 229–256). Washington, DC: American Psychiatric Press.

Martenyi, F., Brown, E. B., Zhang, H., Prakash, A., & Koke, S. C. (2002). Fluoxetine versus placebo in posttraumatic stress disorder. *Journal of Clinical Psychiatry, 63*, 199–206.

McAdams, D. P., & Ochberg, R. L. (1988). *Psychobiology and life narratives*. Durham, NC: Duke University.

McCall, G. J., & Resick, P. A. (2003). A pilot study of PTSD symptoms among Kalahari Bushmen. *Journal of Traumatic Stress, 16*, 445–450.

Mohlen, H., Parzer, P., Resch, F., & Brunner, R. (2005). Psychosocial support for war traumatized child and adolescent refugees: Evaluation of a short-term treatment program. *Australian and New Zealand Journal of Psychiatry, 39*, 81–87.

Mollica, R. F., Donelan, K., Tor, S., Lavelle, J., Elias, C., Frankel, M., et al. (1993). The effect of trauma and confinement on functional and mental health status of Cambodians living in Thailand–Cambodia border camps. *Journal of the American Medical Association, 270*, 581–586.

Mollica, R. F., McInnes, K., Pham, T., Smith Fawzi, M. C., Murphy, E., & Lin, L. (1998). The dose–effect relationships between torture and psychiatric symptoms in Vietnamese ex-political detainees and a comparison group. *Journal of Nervous and Mental Disease, 186*, 543–553.

Momartin, S., Silove, D., Manicavasagar, V., & Steel, Z. (2004). Complicated grief in Bosnian refugees: Associations with posttraumatic stress disorder and depression. *Comprehensive Psychiatry, 45*, 475–482.

Morgan, D. L. (1997). *Focus groups as qualitative research* (2nd ed.). Thousand Oaks, CA: Sage.

Neuner, F., Schauer, M., Klaschik, C., Karunakara, U., & Elbert, T. (2004). A comparison of narrative exposure therapy, supportive counseling, and psychoeducation for treating posttraumatic stress disorder in an African refugee settlement. *Journal of Consulting and Clinical Psychology, 72*, 579–587.

Njenga, F. G., Nicholls, P. J., Nyamai, C., Kigamwa, P., & Davidson, J. R. (2004). Post traumatic stress after terrorist attack: Psychological reactions following the US embassy bombing in Nairobi: Naturalistic study. *British Journal of Psychiatry, 185*, 328–333.

Norris, F. H., Friedman, M. J., Watson, P. J., Byrne, C. M., Diaz, E., & Kaniasty, K. (2002). 60,000 disaster victims speak: Part I. An empirical review of the empirical literature, 1981–2001. *Psychiatry, 65*, 207–239.

Norris, F. H., Friedman, M. J., & Watson, P. J. (2002). 60,000 disaster victims speak: Part II. Summary and implications of the disaster mental health research. *Psychiatry, 65*, 240–260.

North, C. S., Pfefferbaum, B., Narayanan, P., Thielman, S., McCoy, G., Dumont, C., et al. (2005). Comparison of post-disaster psychiatric disorders after terrorist bombings in Nairobi and Oklahoma City. *British Journal of Psychiatry, 186*, 487–493.

Otto, M. W., Hinton, D., Korbly, N. B., Chea, A., Ba, P., Gershuny, B. S., et al. (2003). Treatment of pharmacotherapy-refractory posttraumatic stress disorder among Cambodian refugees: A pilot study of combination treatment with cognitive-behavior therapy vs. sertraline alone. *Behaviour Research and Therapy, 41*, 1271–1276.

Paunovic, N., & Öst, L. (2001). Cognitive-behavior vs. exposure therapy in the treatment of PTSD in refugees. *Behavior Research and Therapy, 39*, 1183–1187.

Pedersen, P., Dragus, J. G., Lonner, W. J., & Trimble, J. E. (Eds.). (1996). *Counseling across cultures* (4th ed.). Thousand Oaks, CA: Sage.

Pedersen, P. B. (1997). *Culture-centered counseling interventions: Striving for accuracy.* Thousand Oaks, CA: Sage.

Ponterotto, J. G., Casas, J. M., Suzuki, L. A., & Alexander, C. M. (Eds.). (1995). *Handbook of multicultural counseling.* Thousand Oaks, CA: Sage.

Resick, P. A., & Schnicke, M. K. (1992). Cognitive processing therapy for sexual assault victims. *Journal of Consulting and Clinical Psychology, 60*, 748–756.

Robin, R. W., Chester, B., & Goldman, D. (1996). Cumulative trauma and PTSD in American Indian communities. In A. J. Marsella, M. J. Friedman, E. T. Gerrity, & R. M. Scurfield (Eds.), *Ethnocultural aspects of posttraumatic stress disorder: Issues, Research, and clinical applications* (pp. 239–254). Washington, DC: American Psychological Association.

Roodenrijs, T. C., Scherpenzeel, R. P., & de Jong, J. T. V. M. (1997). [Traumatic experiences and psychopathology among Somalian refugees in the Netherlands]. *Tijdschrift voor Psychiatrie, 98*, 132–143.

Roy, J. L. (2002). How can participation of the community and traditional healers improve primary health care in Kinshasa, Congo. In J. de Jong (Ed.), *War, trauma, and violence: Public mental health in the sociocultural context* (pp. 405–440). New York: Kluwer Academic/Plenum Press.

Salcioglu, E., Basoglu, M., & Livanou, M. (2003). Long-term psychological outcome for non-treatment-seeking earthquake survivors in Turkey. *Journal of Nervous and Mental Disease, 191*, 154–160.

Sartorius, N., & Janca, A. (1996). Psychiatric assessment instruments developed by the World Health Organization. *Social Psychiatry and Psychiatric Epidemiology, 31*, 55–69.

Shin, L. M., Kosslyn, S. M., McNally, R. J., Alpert, N. M., Thompson, W. L., Rauch, S. L., et al. (1997). Visual imagery and perception in posttraumatic stress disorder: A positron emission tomographic investigation. *Archives of General Psychiatry, 54*, 233–241.

Shrestha, N. M., Sharma, B., Van Ommeren, M., Regmi, S., Makaju, R., Komproe, I., et al. (1998). Impact of torture of refugees displaced within the developing world: Symptomotology among Bhutanese refugees in Nepal. *Journal of the American Medical Association, 280*, 443–448.

Silove, D., Sinnerbrink, I., Field, A., Manicavasagar, V., & Steel, Z. (1997). Anxiety, depression, and PTSD in asylum-seekers: Association with pre-migration trauma and post-migration stressors. *British Journal of Psychiatry, 170*, 351–357.

Silver, S., & Wilson, J. P. (1988). Native American healing and purification rituals for war stress. In J. P. Wislon, Z. Harel, & B. Kahana (Eds.), *Human adaptation to extreme stress: From the Holocaust to Vietnam* (pp. 221–228). New York: Plenum Press.

Solvig, E., & Göran, R. (1997). Diagnosing posttraumatic stress disorder in multicultural patients in a Stockholm psychiatric clinic. *Journal of Nervous and Mental Disease, 185*, 102–107.

Somasundaram, D., & Jamunanantha, C. S. (2002). Psychosocial consequences of war. In J. de Jong (Ed.), *War, trauma, and violence: Public mental health in the sociocultural context* (pp. 205–258). New York: Kluwer Academic/Plenum Press.

Somasundaram, D. J., & Sivayokan, S. (1994). War trauma in a civilian populations. *British Journal of Psychiatry, 165*, 524–527.

Southwick, S. M., Krystal, J. H., Morgan, C. A., Johnson, D., Nagy, L. M., Nicolaou, A., et al. (1993). Abnormal noradrenergic function in posttraumatic stress disorder. *Archives of General Psychiatry, 50*, 266–274.

Sue, S., Fujino, D. C., Hu, L., Takeuchi, D. T., & Zane, N. (1991). Community mental health services for ethnic minority groups: A test of the cultural responsiveness hypothesis. *Journal of Consulting and Clinical Psychology, 59*, 533–540.

Sue, S., Zane, N., & Young, K. (1994). Research on psychotherapy with culturally diverse populations. In A. E. Bergin & S. L. Garfield (Eds.), *Handbook of psychotherapy and behavior change* (4th ed., pp. 783–820). New York: Wiley.

Tang, S., & Fox, S. H. (2001). Traumatic experiences and mental health of Senegalese refugees. *Journal of Nervous and Mental Disease, 189*, 507–512.

van de Put, W. A. C., & Eisenbruch, M. (2002). The Cambodian experience. In J. de Jong (Ed.), *Trauma, war, and violence: Public mental health in sociocultural context* (pp. 93–156). New York: Kluwer Academic/Plenum Press.

Van Ommeren, M., de Jong, J. T. V. M., Sharma, B., Komproe, I., Thapa, S., & Cardeña, E. (2001). Prevalence of psychiatric disorders among tortured Bhutanese refugees in Nepal. *Archives of General Psychiatry, 58,* 475–482.

Van Ommeren, M., Sharma, B., Komproe, P. B. N., Sharma, G. K., Carena, E., & de Jong, J. T. M. V. (2001). Trauma and loss as determinants of medically unexplained illness in a Bhutanese refugee camp. *Psychological Medicine, 31,* 1259–1267.

Van Ommeren, M., Sharma, B., Thapa, S., Makaju, R., Prasain, D., Bhattarai, R., et al. (1999). Preparing instruments for transcultural research: Use of a translation monitoring form with Nepalispeaking Bhutanese refugees. *Transcultural Psychiatry, 36,* 285–301.

Webster, R. A., McDonald, R., Lewin, T. J., & Carr, V. J. (1995). Effects of a natural disaster on immigrants and host population. *Journal of Nervous and Mental Disease, 183,* 390–397.

Weine, S. M., Becker, D. F., McGlashan, T. H., Laub, D., Lazrove, S., Vojvoda, D., et al. (1995). Psychiatric consequences of "ethnic cleansing": Clinical assessments and trauma testimonies of newly resettled Bosnian refugees. *American Journal of Psychiatry, 152,* 536–542.

Weine, S. M., Kulenovic, A. D., Pavkovic, I., & Gibbons, R. (1998). Testimony psychotherapy in Bosnian refugees: A pilot study. *American Journal of Psychiatry, 155,* 1720–1726.

Weine, S. M., Razzano, L., Nenad, B., Ramic, A., Miller, K., Smajkic, A., et al. (2000). Profiling the trauma related symptoms of Bosnian refugees who have not sought mental health services. *Journal of Nervous and Mental Disease, 188,* 416–421.

Weine, S. M., Vojvoda, D., Becker, D. F., McGlashan, T. H., Hodzic, E., Laub, D., et al. (1998). PTSD symptoms in Bosnian refugees 1 year after resettlement in the United States. *American Journal of Psychiatry, 155,* 562–567.

Westermeyer, J. (1989). *Psychiatric care of migrants: A clinical guide.* Washington, DC: American Psychiatric Press.

Westermeyer, J., & Sines, L. (1979). Reliability of cross-cultural psychiatric diagnosis with assessment of two rating contexts. *Journal of Psychiatric Research, 15,* 199–213.

Wrenn, G. (1962). The culturally encapsulated counselor. *Harvard Educational Review, 32,* 444–449.

Yehuda, R. (2002). Current concepts: Post-traumatic stress disorder. *New England Journal of Medicine, 346,* 108–114.

Zarowsky, C. (2004). Writing trauma: Emotion, ethnography, and the politics of suffering among Somali returnees in Ethiopia. *Culture, Medicine, and Psychiatry, 28,* 189–209; discussion, 211–220.

Part IV

UNCHARTED TERRITORY

Chapter 22

PTSD and the Law

Landy F. Sparr and Roger K. Pitman

> Marriage is like life—it is a field of battle, not a bed of roses.
> —Robert Louis Stevenson

The sometimes rocky relationship between psychiatry and the law has been described in colorful ways—for example, "unhappy marriage" (Walters, 1984) "and theatre of the absurd" (Lesse, 1982)—but most see the union as necessary and beneficial. The diagnosis of posttraumatic stress disorder (PTSD), first introduced in 1980, has solidified the bond. A forensic psychologist once remarked that if mental illnesses were rated on the New York Stock Exchange, PTSD would be a growth stock worth watching (Lees-Haley, 1986). Professor Allan Stone (1993) observed, "No diagnosis in the history of American psychiatry has had a more dramatic and pervasive impact on law and social justice than PTSD. . . . The diagnosis of PTSD has also given a new credibility to a variety of victims who come before the courts either as defendants or plaintiffs" (pp. 23–24). Whether the legal issue is criminal or civil, or involves culpability or compensation, PTSD has become relevant to law in general and to litigation in particular. PTSD is now as firmly entrenched in the legal landscape as it is in contemporary psychiatric textbooks. Stone's observations were prophetic. Despite skepticism from mental health providers and lawyers alike, PTSD-based legal claims have burgeoned to the point that some have come to see PTSD threatening to overwhelm the personal injury, workers' compensation, and disability insurance litigation systems (Pitman & Sparr, 1998).

Slovenko (1994) has noted that PTSD is a favored diagnosis in tort law, because it is incident-specific, easy to understand, and tends to rule out other factors potentially involved in causation. Through PTSD, plaintiffs attempt to establish that the psychological problems they are claiming issue from an alleged traumatic event rather than from myriad other possible sources. A diagnosis of depression, in contrast, may expose the causation issue to many etiological considerations. Spaulding (1988) has observed, "The further from the diagnosis of PTSD that the evaluator strays, the more speculative the opinion on causation will become" (p. 13). Special features of PTSD also help it overcome legal barriers under workers' compensation. Whereas affective and other anxiety disorders may be argued to represent "ordinary diseases of life," the recognition that PTSD is caused by a discrete external event (e.g., a workplace accident) removes it from this exclusionary category.

In previous articles we have discussed the relationship between PTSD and the law, particularly in regard to assessment of criminal intent (Sparr, 1996), criminal behavior (Sparr, Reaves, & Atkinson, 1987), the insanity defense (Sparr & Atkinson, 1986), civil issues (Pitman & Sparr, 1998; Pitman, Sparr, Saunders, & McFarlane, 1996; Sparr, 1990), tort actions (Sparr & Boehnlein, 1990), and factitious behavior (Sparr & Pankratz, 1983). This chapter is intended to supplement, not to replace, these earlier contributions. As PTSD has aged as an official psychiatric diagnosis, its forensic face has changed as well. Initial enthusiasm for PTSD as a criminal defense has waned, and initial fears about misuse have not materialized. Appelbaum and colleagues (1993) have shown that despite early concerns, the PTSD insanity defense is raised infrequently and, like other insanity pleas, is usually unsuccessful. Instead, the primary thrust of PTSD criminal defenses has been as an occasional factor in diminished capacity considerations, pretrial plea bargaining, or sentencing (Pitman & Sparr, 1998). The more significant growth of PTSD in forensic deliberations has been in the civil area. In this chapter we suggest that this is due in part to changes in the PTSD stressor criteria. As a result, civil issues now dominate the relationship between PTSD and the law. PTSD's broad and vast influence applies to workers' compensation, social security disability, tort litigation, and Veterans' Affairs disability compensation. On the criminal side we focus on two areas: battered woman syndrome and the nascent relationship between PTSD and deficits in explicit memory function.

CRIMINAL ISSUES

Few topics in American jurisprudence have been written about more passionately than mental incapacity defenses. Despite the great public controversies that insanity and other mental status defenses have engendered, empirical research reveals that these defenses are seldom used and, when used, are seldom successful. Reportedly, the insanity defense is raised in less than 4% of criminal cases, and a successful insanity defense occurs only 1% of the time (Gutheil, 1999). Moreover, insanity acquittees "often spend more time behind the walls than if they had been tried, found guilty and given a standard sentence with a length predetermined by statute" (Steadman, 1985). Although PTSD has been used as an insanity defense, it is no more likely to succeed than pleas based on any other diagnosis (Appelbaum et al., 1993). One reason is that the behavior of an individual with PTSD rarely fits insanity defense standards. Some effort has been made to correlate so-called PTSD dissociative states with an inability to distinguish right from wrong or to control conduct, but these cases are infrequent. A lesser known use of mental incapacity defenses is in *mens rea*/diminished capacity (United States) or diminished responsibility (Europe) formulations. PTSD as a successful basis for these "failure of proof" and "excuse" defenses, which are complex and discussed extensively elsewhere, is also rare (Sparr, 2005). However, a novel use of PTSD as a criminal defense that has survived somewhat unscathed over the years is battered woman syndrome.

Battered Woman Syndrome

A number of controversial "syndromes" of varying validity, purportedly related to PTSD in one way or another, have been proposed from time to time in the context of criminal defenses (e.g., "urban response syndrome" [Parson, 1994], "child abuse syndrome" [Summit, 1983], "television intoxication" [Falk, 1996], and "battered woman syndrome" [Walker, 1980]). In addition, infrequently used mental state defenses, such as extreme emotional disturbance and automatism, are controversial no matter what mental condition is alleged (Parry & Drogin, 2000). The key question regarding the forensic use of these conditions is whether the court will allow evidence and testimony to support them as a defense. In addition to determining the admissibility of scientific evidence (discussed below), the court must decide whether the testimony will assist the trier-of-fact in understanding the ultimate issue.

Battered woman syndrome draws on the social science research of Lenore Walker (1980), among others. It first entered the criminal law landscape as a self-defense defense in homicide cases, where it has found some degree of success (Lustberg & Jacoby, 1992). Although it is often seen as a variation of PTSD, it cuts across a spectrum of underlying diagnostic categories rather than being an official diagnosis (Goodstein & Page, 1981). Although the scientific rigor and generalizability of the underlying research have been sharply debated, a majority of states have accepted the defense. Collectively, 31 states have allowed use of expert testimony on battered woman syndrome (Shumam, 2003a). Expert testimony may assist the jury to understand a woman's emotions and behavior in the years preceding the homicide, corroborating the history of violence in the relationship. Expert testimony may also be used to explain that women may remain in abusive relationships because they may be terrified of leaving and believe they cannot survive on their own (Shuman, 2003a). Increasing acceptance of battered woman syndrome has led some courts to hold that indigent defendants must be provided funds to retain a psychiatric or psychological expert to aid and assist in this defense (*Dunn v. Roberts*, 1992; *Lewis v. State*, 1995). The consequence of invoking a battered woman syndrome defense varies by jurisdiction. In some, evidence of the syndrome may be used to support a claim of self-defense, that is, a normal response to a threat (*State v. Grubbs*, 2003; *State v. Kelly*, 1984), whereas in others it may be used to support an insanity defense, that is, a disordered response to a threat (*State v. Necaise*, 1985).

To mount a successful self-defense claim, a woman must prove that she was operating under a reasonable belief of imminent danger. Furthermore, the defendant must prove that the force she used was reasonable and necessary, that she was not the aggressor, and, depending on the jurisdiction, that she did not have an opportunity to retreat safely from confrontation. In particular, battered women have had a difficult time proving the two elements of reasonable belief and imminent harm. Most courts rely on an objective standard when assessing reasonableness, which requires a judge or jury to assess the defendant's beliefs from the standpoint of a "reasonable man." The second element, imminent harm, presents difficulty when a woman kills in a nonconfrontational situation, for example, when the victim is sleeping or his back it turned (Blowers & Bjerregaard, 1994).

Expert testimony has been used to show that abused women often perceive their situation differently than would a usual "reasonable (wo)man," because such abuse typically increases in both frequency and severity over time. Thus, it may be reasonable for an abused woman to believe that subsequent encounters may prove more deadly than preceding confrontations. The second characteristic of battered woman syndrome often involves learned helplessness. This theory posits that the woman's ability to control her situation is significantly impaired when she realizes that the abuse she receives is not contingent on her behavior or actions. Thus, an abused woman may develop the belief that she is helpless to control her situation and not perceive escape options (Walker, 2000).

In *State v. Kelly* (1984), the court held that in asserting self-defense, the defendant may offer expert testimony on battered woman syndrome to aid a jury in understanding how a history of abuse may support a woman's claim that she believes she was in imminent danger, and that her belief was reasonable. In particular, the court declared that the subject was "beyond the ken of the average juror and thus suitable for explanation through expert testimony." The court also held that the syndrome had sufficient scientific basis to fulfill an expert witness reliability requirement. In *State v. McClain* (1991), however, a court limited the applicability of battered woman syndrome evidence by holding that because the defendant was not in imminent danger, such evidence was irrelevant to the question of reasonableness.

Assuming that the legitimate purposes of expert testimony justify the use of battered woman syndrome as a defense, a significant question is whether the expert testimony is sufficiently reliable to be regarded as admissible. The applicable tests are the venerable *Frye* Rule (*Frye v. United States*, 1923), or in states that have modeled their laws after the federal example, the Federal Rules of Evidence, as interpreted in *Daubert v. Merrill Dow Pharmaceuticals* (1993). The former requires general acceptance of the principle underlying the testimony by the relevant scientific community, whereas the latter requires a determination by the judge that the testimony is scientifically reliable. Whether the standard is *Frye* or *Daubert*, it has been insufficient to argue that the experts had examined hundreds of battered women, that their research was supported by the federal government, and/or that they had written articles and books about the syndrome. None of this, per se, establishes reliability. The syndrome does not appear in the *Diagnostic and Statistical Manual of Mental Disorders* (DSM), and there are some who say the whole concept is another instance of a creative criminal defense masquerading as science. Although *Frye*'s general acceptance test does not require the DSM's blessing, having the latter usually equates with general acceptance and thus admissibility. Despite being viewed generally as more liberal than *Frye*, *Daubert*'s five-prong assessment of testability, established standards of measurement, known error rate, peer review/publication, and general acceptance may in the case of battered woman syndrome be more difficult to satisfy, particularly if the first three criteria are taken seriously.

Ten state legislatures have passed statutes dealing with the battered woman syndrome conundrum. Several of them begin by resolving the admissibility issue explicitly as a matter of law; most statutes allow the concept as part of a self-defense. Battered woman syndrome may also be used for sentence mitigation, especially when the facts fall short of establishing a full legal defense (Brakel & Brooks, 2001). *State v. Pascal* (1987) is such a "downward departure" case. Furthermore, the concept has been, and continues to be, widely used to obtain executive clemency.

Unlike many novel defenses associated with PTSD, defenses based on battered woman syndrome continue to flourish both in number and success of outcome. One reason is that the concept dovetails with an increased societal awareness of domestic battery as a serious and pervasive problem. The law has moved toward greater protection of women in other ways as well, including antistalking laws and restraining orders. These developments are intended to combat a pattern of inadequate law enforcement response to domestic violence, which has been rationalized by pointing to the frequent ambiguity of such situations, including the unwillingness of some victims to pursue charges against their abusers (Brakel & Brooks, 2001).

PTSD and Memory

The nature of remembrance of traumatic events has been particularly controversial during the past decade as vigorous new research has reshaped thinking about trauma and memory. There are different types of memory and empirical studies have associated PTSD with both a strengthening of some and a weakening of others. Persons with PTSD form stronger conditioned fear responses to both traumatic (Pitman, Orr, & Forgue, 1987) and de novo (Orr, Metzger, & Lasko, 2000) events. In contrast, deficiencies have been reported in their explicit, or declarative, memory (i.e., the ability to report accurately newly learned information [Bremner, Vermetten, & Afzal, 2004; Gilbertson, Gurvits, & Lasko, 2001]). However, patients with PTSD may have better explicit memory for trauma-related material (McNally, 1998). Accuracy of memory has received particular scrutiny, because considerable importance may be attached to victims' declarative recollections (Brown, Scheflin, & Hammond, 1998). In 1998, at the International War Crimes Tribunal in The Hague, a Bosnian–Croatian soldier was tried for aiding and abetting the rape of a Muslim woman (*Prosecutor v. Anto Furundzija*, 1998). The defendant's lawyers suggested that the woman's memory was inaccurate, because it had been adversely affected by her traumatic experiences, and that the defendant she identified was not actually present during her interrogation and abuse. The prosecution disagreed and argued that memories of traumatic experiences in individuals with PTSD are characteristically hyperaccessible. Expert witnesses on both sides were brought in to provide medicolegal testimony about the scientific parameters of stress and its long-term effects on brain regions associated with memory (Sparr & Bremner, 2005).

This case highlights on multiple levels the politics of psychological trauma and the law. PTSD researchers have identified psychological and biological markers that are characteristic of the disorder. This research has been used to champion PTSD as a real disorder, implying that trauma victims deserve financial compensation under the law and treatment from medical providers. The case of *Anto Furundzija*, however, turns the heretofore peaceful alignment of forces between trauma victims and clinicians/researchers on its head. The disruption this entails can be seen in the controversy related to this case in general, and the presentation and discussion of the scientific evidence in particular. The case taps into a recently evolving area of research that suggests PTSD victims have more memory fallibility of a certain type than persons without the disorder. The same framework of physiological disturbances that was previously used in support of victims has now acquired the potential to undermine them in a fundamental way.

A controversial forensic question that emerges from the *Furundzija* case is whether testimony from patients with PTSD should be thrown out of a court of law. Assuming for the moment that patients with PTSD do have more fallible memories, at least in some areas, does this imply that all their memories are inherently suspect? There is now a wealth of evidence that memories in all individuals are subject to a range of inaccuracies, and much of this literature has been related to research on witness testimony in court. Does that mean that all witness testimony should be discounted? Perhaps this merely highlights the importance of obtaining collateral information for cases involving victims with and without PTSD. The model of children's testimony may be used as an example. Although memory in children may not be as accurate or take the same form as memory in adults, children's testimony is allowed in court.

CIVIL ISSUES

The following discussion features the role played by PTSD in plaintiffs' efforts to assert, substantiate, and quantify mental impairment and/or mental harm to preserve entitlements or rights in civil litigation, and defense's effort to rebut such assertions. In this instance, Stone (1993) notes that "by giving diagnostic credence and specificity to the concept of psychic harm, PTSD has become the lightning rod for a wide variety of claims of stress-related psychopathology in the civil arena" (p. 29). Unlike the diagnostic concept of neurosis, which emphasizes a complex etiology, PTSD posits a straightforward causal relationship that plaintiffs' lawyers welcome. Beyond its significance as an apparent solution to the legal problem of causation, PTSD's greatest importance is that it seems to make scientific and objective matters that the court once considered too subjective for legal resolution (Lesse, 1982). Along with this legitimization has come a geometric increase in claims, particularly since the introduction of DSM-IV (American Psychiatric Association, 1994) with its change in the PTSD stressor criteria. The previous edition, DSM-III-R (American Psychiatric Association, 1987), required that the traumatic event be "outside the range of usual human experience." With the removal of this limiting criterion in DSM-IV, the prevalence of PTSD as a factor in civil litigation has increased. DSM-III-R made it difficult for individuals involved in, for example, motor vehicle accidents or other common occurrences to claim PTSD. Resnick (1998) believes that this change has also made it easier for trauma victims to fake the PTSD stressor criteria, and Melton, Petrila, Poythress, and Slobogin (1997) predicted that the use of PTSD in litigation would increase.

Still, the DSM-IV requirement that the event be one that involves death, serious injury, or a threat to physical integrity may be difficult to establish in some workers' compensation cases when the claimed stressor is, say, a tedious job that gradually affects the individual over time (Melton et al., 1997). In civil law, particular entitlements and rights require proof of either mental impairment or mental harm. The definitions of these disabilities are frequently dictated by statute. Special features of PTSD have helped it overcome legal barriers under workers' compensation. Whereas affective disorders and other anxiety disorders may be argued to represent "ordinary diseases of life," the recognition that PTSD is caused by a discrete external event (e.g., a workplace accident) removes it from this exclusionary category. Where mental impairment due to PTSD is at issue, the major sources of litigation are (1) Social Security disability; (2) private disability and health insurance; and (3) veterans disability benefits. Where mental harm secondary to PTSD is at issue, the two main sources of litigation involve (1) workers' compensation and (2) personal injury (Parry & Drogin, 2001). Mental harm claims, in particular, have increased dramatically in the past two decades.

By the early 1990s, the costs of Workers' Compensation cases had escalated at a rate 50% greater than the inflation in total health care spending; in absolute terms, the total cost to the system was $70 billion in 1992, a tripling in costs since 1982 ("Sticking It to Business," 1993). Commentators cited three reasons for this escalation: the rising costs of medical care, increased litigiousness associated with the Workers' Compensation system, and, most important for our purposes, the expanding definition of "compensable injury," particularly with respect to job-related stress and emotional or mental injury (Melton et al., 1997). U.S. Chamber of Commerce statistics show that the number of mental stress claims recorded by employees under Workers' Compensation jumped nearly 800% between 1979 and 1990, making stress-related disorders the fastest growing disease category (deCarteret, 1994). In a study of more than 700,000 claims filed in 11 states, the National Council on Compensation Insurance reported that costs of stress claims averaged about 52% more than physical injury claims (Calise, 1993). This has led to many Workers' Compensation reforms proposed or adopted by the states. An important contributor to this rise in mental injury claims has been an increase in the use of the PTSD diagnosis. As Slovenko (1994) has noted, "A lot of distressed people are feeling better these days—thanks to the courts" (p. 439).

Mental Impairment

Social Security Disability Benefits

A number of federal disability programs provide benefits to disabled persons and their dependents. Prominent among these programs are Social Security disability insurance (SSDI; 42 U.S.C. § 423) and Supplemental Security Income (SSI; 42 U.S.C. § 1381). The SSDI program provides benefits to disabled wage earners on whose behalf Social Security taxes have been paid for requisite quarters of coverage. Specifically it provides benefits for disabled people who have worked and paid into the Social Security trust fund in 20 of the 40 calendar quarters prior to the beginning of disability. The SSI program is federalized public assistance to disabled persons who meet certain income and asset limitations but lack the required quarters of coverage. Both programs are governed by the same disability standard that requires the claimant to demonstrate an "inability to engage in any substantial gainful activity by reason of a medically determinable physical or mental impairment which can be expected to result in death or which has lasted or can be expected to last for a continuous period of not less than twelve months" [42 U.S.C. § 423 (d) (1) (A)]. Consideration of the nature of the disability entails review of the administration's listing of impairments that presumptively satisfy the requirement for disability with the requisite degree of severity to preclude the claimant's capacity to work. The list of mental impairments does not specifically correspond to DSM-IV criteria; however, it is designed to measure severity even in the absence of diagnostic agreement (Shuman, 2003b). On the list of impairments under the anxiety disorders category appears "recurrent and intrusive recollections of a traumatic experience, including dreams which are a source of marked distress (20 C.F.R. Part 404, Subpart P, Appendix 1, § 12.06). Most readers will recognize these as two of the 17 DSM-IV PTSD criteria. In 2002, 6 out of 10 SSI recipients were diagnosed with a mental disorder (Social Security Administration Office of Policy, 2003). These beneficiaries received monthly payments that, in many cases, represented their major source of income. The benefits also provide access to health care through Medicare and Medicaid (Okpaku, Sibulkin, & Schenzler, 1994).

In evaluating functional loss from mental impairment, the Social Security Administration (SSA) divides inquiries into four areas: restriction of activities of daily living, difficulty in maintaining social functioning, deficiencies of concentration and persistence resulting in failure to complete tasks, and episodes of deterioration or decompensation in work or work-like situations that cause the individual to withdraw from that situation (Melton et al., 1997). All types of mental disorders that affect the ability to work are subject to SSDI or SSI claims. Although no specific data are available, it is not unusual for PTSD to be the basis for a claim. The SSA employs independent examiners to make diagnostic assessments. Overall, SSI claims for mental impairment have risen steadily in the past several decades (Social Security Administration Office of Policy, 2003).

Private Disability and Health Insurance

The essential difference between private disability insurance and health insurance is that disability insurance generally focuses on the insured person's ability to work, whereas health insurance is concerned with whether the insured has a medical condition, the treatment of which is covered within the terms of a given policy (Parry & Drogin, 2001). Private disability and health insurance policies tend to be divided into two general categories. The first, "indemnity" plans, involve a direct contractual relationship between the insured and the insurer. The second involves policies pertaining to health maintenance organizations (HMOs) with rates set (and with care more tightly managed) by a highly specific contract between the insured and the employer (Edelman, 1990; Furrow, 1998). Definitions of "mental illness" and/or "mental disability" in health insurance policies vary considerably, limited only by the regulatory restrictions of jurisdictions in which the plans are offered.

Morrison (2000) states, "The growing recognition that mental illnesses are biologically based and are sometimes related to other health problems may increase the number of legal challenges against health plans providing unequal benefits for mental and physical illnesses" (p. 31). The essential difference found in private disability policies is that any mental condition must be linked to an impairment level that prevents or limits the insured person from working. Often, inability to work in one's own field (as opposed to any field) is sufficient for recovery under the terms of the policy. Typically, such policies avoid specific diagnoses in their definition of what constitutes mental illness. More commonly, one finds definitions that refer to whether an affliction is "biologically based," involves "psychotic" symptoms, or may be treated by "psychiatric methods" (Parry & Drogin, 2001). Another approach is to be more inclusive about what constitutes "mental illness" in general but to limit reimbursement to those conditions that are sufficiently "severe" to cause unemployment (Tommasini, 1994).

Litigation at the state level concerning mental disability policies is often focused upon barriers to recovery based on a specific exclusions or exemptions. An important issue regarding private health insurance is the availability of adequate coverage for the full range of diagnosable mental disorders. Over the past decade, in particular, several state legislatures have witnessed the filing of "comprehensive parity" bills designed to require insurance companies or "third-party providers" to offer the same degree of coverage for mental illness as for physical illness (Sing & Hill, 2001). Although Congress did enact the Federal Mental Health Parity Act (1996) equating aggregate lifetime limits and annual limits for mental health benefits with similar existing limits concerning medical and surgical benefits, the act "lacks regulatory teeth to have much real effect" (Gould, 2001, p. A1).

Private insurers often hire so-called "independent" medical examiners to conduct an assessment, make a diagnosis, review the propriety of a specific treatment, and/or render an opinion as to whether the treatment is reasonable and necessary. Usually an independent medical exam (IME) includes a determination of the permanence of the impairment (e.g., "medically stationary": so-called "medical end point"), and the residual functional capacities (RFCs) of the examinee. Unlike the normal clinical situation, the IME may take place in the context of an adversarial medicolegal setting. In most cases, the patient is coming to the IME for a scrutinizing exam rather than help. He or she is concerned about the written report that will be sent to the insurance company and, perhaps inevitably, to the employer. Possible negative outcomes include a denial of coverage for psychological services that have already been initiated and/or no compensation for alleged injuries. IME examiners frequently encounter claimants who allege PTSD from exposure to a stress that does not satisfy DSM-IV stressor criteria (Sparr, 2003). Common stressors are motor vehicle accidents (MVAs), workplace falls, and interpersonal conflicts with coworkers. Some stressors that would not have qualified under DSM-III-R criteria may now qualify under DSM-IV (e.g., motor vehicle accidents).

The reported PTSD rates in victims of serious MVAs have ranged from 8 to 46% (Malt & Blikra, 1993; Mayou, Bryant, & Duthie, 1993). In a cross-sectional study, Blanchard, Hickling, Taylor, and Loos (1995) found that a history of trauma or major depression was a significant risk factor for developing PTSD after a serious MVA (Blanchard et al., 1995). Ursano, Fullerton, and Epstein (1999) found a PTSD rate of 17.6% nine months postaccident. These data suggest that although PTSD may exist in the aftermath of serious MVAs, the incidence is moderate. IME examinations frequently reveal preexisting psychiatric problems among those who have been traumatized and who have PTSD. As a result, a struggle often occurs about the relative weight of preexistent mental health conditions, possibly related to previous trauma, versus the causal primacy of the claimed traumatic event. Only a few long-term follow-up studies have been carried out, one reporting psychiatric morbidity (mostly depressive disorders) in 22% of accident victims over an observation period of 28 months (Malt, 1988), and another reporting PTSD in 8% of victims 5 years after the accident (Mayou, Tyndell, & Bryant, 1997). In a study in Switzerland by Schnyder, Moergeli, Klaghofer, and Buddeberg (2001) in a cohort of severely injured accident victims who were healthy before experiencing trauma, the incidence of full and "subsyndromal" PTSD was low (2 and 12%, respectively). One third of the variance of PTSD symptoms at 1-year follow-up could be predicted by psychosocial variables (e.g., biographical risk factors, sense of impending death).

Veterans' Disability Benefits

In addition to providing medical care for veterans, the U.S. Department of Veterans Affairs (VA) administers a system that provides benefits, including monetary compensation, treatment, and rehabilitation services for veterans with a "service-connected disability." In VA parlance, service-connected disabilities are disorders that develop during military service, including, but not limited to, those directly related to combat. Veterans may apply for disability payments for medical or psychiatric conditions that had their onset during or within 1 year following their military service (Sparr, White, Friedman, & Wiles, 1994). Congress determines the list of compensable disorders. PTSD did not become separately compensable until its appearance in DSM-III (American Psychiatric

Association, 1980). When it did, however, in recognition of the DSM-III provision that PTSD could appear in delayed form, the requirement of onset within 1 year following military service was waived, opening the floodgates to claims of PTSD from wars that had occurred many years earlier. As a result, PTSD is now the most common psychiatric condition for which veterans seek service connection. In 2003, the Veterans Benefits Administration paid approximately $121 million each month in compensation to 193,859 veterans with service-related PTSD (U.S. Department of Veterans Affairs, 2003).

In the years following 1980, the process for establishing veterans' eligibility for PTSD benefits was often erratic. Clinicians disputed the diagnostic validity of PTSD, and claims specialists, charged with developing veterans' claims, lacked sufficient standardized protocols to follow. Frequently, they could not find sufficient corroborating evidence to support favorable awards (Henderson & Sparr, 1994). Not surprisingly, heterogeneity entered the claims development process. Depending on the region in which claims were filed, approval rates ranged from 36 to 74% (Sparr et al., 1994). There is a growing trend in the VA system for veterans to seek such payments. Data show that Gulf War veterans draw disability compensation at a much higher rate than veterans of any previous conflicts, and at almost twice the current rate of World War II veterans (16% compared with 8.6%) ("Gulf War Veterans," 1999). Furthermore, 69–94% of veterans who seek treatment for PTSD in the VA system apply for psychiatric disability (Frueh et al., 2003).

Briefly, the adjudication process begins with the receipt of the veteran's disability claim. The gatekeepers in this system, the VA regional offices, are staffed by a small army of claims examiners that includes supervisors, clerks, rating specialists, adjudicators, quality assurance reviewers, and hearing officers. Initially, the authorization section reviews the claim for basic eligibility criteria, such as dates of service, combat or other potentially traumatic military service (e.g., graves registration), if PTSD is claimed, and character of discharge. If the claim survives authorization review, it is referred to one of the local regional office rating boards. The boards currently comprise three members (called rating specialists), one of whom is a medical specialist. The board weighs the evidence and makes a rating decision, and the adjudication section notifies the applicant. If the applicant disagrees with the decision, he or she may initiate an appeal at the local regional office by filing a notice of disagreement within 1 year of the VA letter of denial (Sparr et al., 1994).

In addition to the claimant's military records, the rating board schedules a psychological examination. In some facilities, VA mental health clinicians who work in direct patient care also serve as VA disability examiners, with the consequent possibility of conflict of interest between the patient's clinical needs and remunerative wishes. In other facilities, outside (non-VA) clinicians perform the examinations on a fee-for-service basis. Once the symptoms have been described and the diagnosis(es) established, the board sets a disability percentage by determining where the veteran's disability falls on the rating schedule. General descriptions of psychiatric disability level are found in the mental disorders section of the VA schedule for rating disabilities. The schedule assigns percentages of disability to reflect average impairment of earning capacity based on general formulas. Personality disorder diagnoses cannot be service-connected. Disability percentages are set in increments of 10 but also may be zero (0% disability confers eligibility for treatment only). As of December 2004, for example, 10% disability for a veteran without dependents conferred $1,296 per year, whereas 100% disability conferred $27,586 per year (the equation is not necessarily linear).

Once a veteran's service connected disability is established, the rating is periodically reevaluated by examiners who review the veteran's recent medical record and conduct face-to-face interviews. Knowledge of this step may discourage veterans from reporting symptomatic improvement that may be documented in their record and available for review by disability examiners. Disability levels may be adjusted by the veteran's local rating board in accordance with the examiner's findings. Disability that is nonstatic (e.g., that may improve with treatment) is reviewed at approximately 2-year intervals. Disability that has been present for more than 20 years is protected and not subject to review. Disability monetary compensation is not taxable. SSDI or retirement benefits are not reduced upon receipt of VA service-connected disability; however, SSI is reduced (Sparr et al., 1994).

In 1996, total disability expenditures for the VA were estimated to be just over $18 billion, with 2.2 million of the surviving 25.4 million veterans (8.9%) receiving some level of service-connected disability benefits (Oboler, 2000). When disability claims are denied or only partially granted, veterans may appeal the decision an indefinite number of times. These repeat claims outnumber original claims almost three to one and dominate the VA adjudication and appeals system. For many veterans who have chronic illnesses the process of obtaining and maintaining disability payments is a protracted struggle; however, the financial incentives are significant (Frueh et al., 2003).

Claim approval rates and the mean degree of service connection granted have gradually increased since 1993 and showed a dramatic increase after 1997. Murdoch, Nelson, and Fortier (2003) assessed the interaction between veterans' period of service and branch of service with the odds of being service-connected for PTSD between 1980 and 1998. Once service period and branch of service were considered, observed rates of service connection ranged from 80 to 100% for combat-injured men. In contrast, observed rates of service connection for PTSD ranged from 18 to 63% for men without combat injuries. Their study also showed that regional variability in estimated claim approval rates remains prevalent. Only 4.3% of men appealed denied claims for PTSD disability, which is in marked contrast to the 11% appeal rate reported for veterans claiming other psychiatric disabilities, or the 20% appeal rate for veterans claiming disability for nonpsychiatric disorders (Murdoch et al., 2003). Battista (1985) has observed that the VA system has less of the adversarial element characteristic of other disability programs.

In a system that has been described as "countertherapeutic jurisprudence" (Mossman, 1996), studies consistently demonstrate that combat veterans who are evaluated for PTSD exhibit both diffuse levels of psychopathology as assessed by standardized interviews and show extreme elevations on the validity scales of the Minnesota Multiphasic Personality Inventory (MMPI). In a recent study of 320 adult male combat veterans, Frueh and colleagues (2003) showed that compensation-seeking veterans reported significantly more psychopathology, even after the effects of income had been controlled for, despite an absence of difference in PTSD diagnosis between compensation-seeking and non-compensation-seeking groups. Compensation-seeking veterans were much more likely to overreport or exaggerate their symptoms. These data suggest that VA disability policies have problematic implications for the delivery of clinical care, evaluation of treatment outcome, and rehabilitation efforts. Disincentives for reporting improvement, and even possibly for improving, are present at virtually every step of the VA claims adjudication process.

MENTAL HARM

Tort Actions

In the past 30 years, courts have been tracing a somewhat irregular line between compensable and noncompensable psychic impairment in personal injury cases. Many years ago, liability for psychic impairment was contingent on physical injury or impact. Other than that, there was no tort liability for a "broken mind." Once the concept of psychological injury became more accepted, the courts became more willing to compensate for emotional distress in the absence of physical impact or injury (Lambert, 1978). Specific examples include individuals who are within the "zone of danger" (radius of risk) from negligent physical contact and, as a result, suffer an emotional disturbance; individuals who suffer emotional distress after witnessing severe harm to a third person, such as a spouse or child; or individuals who do not actually see the physical injury of another but suffer a severe shock when hearing of it or seeing the results (Sparr et al., 1987).

Nevertheless, the legal system has traditionally manifested hostility toward claims for mental distress damages for as long as litigants have presented the claims. Its antagonism has been mostly aimed at claims of intentionally and negligently inflicted mental distress. Numerous policy concerns have been expressed, including the impossibility of measuring mental disturbance in terms of money, lack of "proximate cause," lack of precedent, the possibility of recovery for fraudulent or trivial claims, significant increases in liability of defendants in amounts disproportionate to culpability level, and the inability of courts to set appropriate limits on claims (Davies, 1992). In England, fear of unfettered tort claims due to mental distress has led to a series of recent decisions that have placed limitations on eligibility (Adamou & Hale, 2003; McCulloch, Jones, & Bailey, 1995).

Still, despite the aforementioned concerns, the trend of the law has been to give increasing protection to feelings and emotions of injured parties, and to enlarge redress and reparation for psychic injury. One estimate is that approximately 2–3% of all torts are associated with psychiatric disability (Slovenko, 1973). Whatever the source of the plaintiff's psychological harm, his or her right to damages depend, in most cases, on proof that such harm resulted from the defendant's conduct. Thus, not surprisingly, psychiatrists and psychologists with expertise in assessing psychological injury and its causation have come to play a significant role in personal injury litigation. The Modlin and Felthous (1989) 12-year survey of their forensic psychiatry practices yielded 403 civil cases, with 55% involving personal injury lawsuits or workers' compensation claims.

The primary purpose of tort law is to provide compensation for private wrongs. Thus, a tort is not the same thing as a crime and does not normally contemplate an evil intent or motive. Nor is it an action for breach of contract, which is based on violation of an explicit understanding about duties between two or more parties. Conversely, an action that might be a moral wrong is not necessarily tortuous conduct, if the actor's conduct is "within the rules." Thus, for instance, failure to save a drowning child would not be considered a tort unless one has an affirmative obligation (duty) to act, as would be true of parents of a child (Lee, Lindahl, & Dooley, 2002). The most common modern examples of torts are motor vehicle negligence, product liability, and professional malpractice. Less common examples include invasion of privacy, defamation, misrepresentation, nuisance, assault and battery, and false imprisonment (Hoffman & Spiegel, 1989). Legal claims involving physical injury may or may not be accompanied by claims of psychic or emotional trauma. Although definitional criteria differ for individual torts, certain core concepts define whether or not an actionable wrong has been committed. These are commonly known under the mnemonic of the "four D's": (1) The defendant must have a *duty* of care to the Plaintiff; (2) there must be a *dereliction*, or breach, of that duty; and (3) the breach must *directly* (proximately) cause (4) *damages* that are recognized as compensable (Melton et al., 1997).

PTSD is often asserted in personal injury cases and, as in other civil jurisprudence involving PTSD, what is frequently overlooked is the dire nature of the "stressor" required for the diagnosis (Perr, 1992; Rosen, 1995). In addition to the "objective" requirement that the experienced, witnessed, or confronted traumatic event must involve death, serious injury, or a threat to physical integrity, there is the added subjective DSM-IV requirement that the event must produce a "subjective" response of intense fear, helplessness or horror. Finally, the resulting, long-term symptoms must be of documented type and severity. These requirements substantially limit, or should limit, PTSD's applicability in mental injury cases.

Workers' Compensation

Workers' Compensation stress claims may be viewed as part of an extensive societal, medicolegal inquiry into the emotional underlife of work organizations (Bale, 1990). Each state has its own Workers' Compensation statutes. Most were passed in the early 1900s, around the time of World War I. In some basic respects, the various state statutes are similar and provide compensation to injured workers for certain consequences of their work injury. Compensation includes medical expenses, lost wages during recuperation, and any permanent loss of earning capacity (Sersland, 1984).

Before the first Workers' Compensation law in 1911, employers were liable only for injuries resulting from negligence; hence, employees had to prove fault to receive an award for an injury arising in the workplace. To facilitate recovery and a quick return to work, Workers' Compensation laws relieved injured workers of this legal burden of proof, and, without fault, employers were responsible for all injury costs. In exchange, employees relinquished right to sue their employers (but not third parties, e.g., in the case of the faulty manufacture of equipment used on the job). Borrowing from the doctrine of proximate cause in tort law, Workers' Compensation law creates a two-part requirement for workplace causation: The injury must arise out of, and occur in the course of, employment (London, Zonana, & Loeb, 1988). In common jargon, four basic terms are used to describe Workers' Compensation claims based on inferred cause and effect: physical–physical, mental–physical, physical–mental, and mental–mental.

In a mental–physical claim, psychological stress or trauma leads to a documented physical disease or disorder. In such a claim, the sticking point is typically causation; however, if the adjudicator can be persuaded that PTSD is a "physical" (e.g., biological) disorder, it may be useful to establish a mental–physical claim, assuming that the psychologically stressful event meets the required criteria described earlier. This is another example of the power of PTSD to boost the chances of recovery in civil actions. In a physical–mental claim, a physical injury leads to some sort of mental distress or impairment. An example would be PTSD following a life-threatening injury. A mental–mental claim, of course, means that mental stress has resulted in a mental problem. An example would be PTSD resulting from the threat, but not actual occurrence, of severe physical injury (Sparr & Boehnlein, 1990). The 1960 landmark Michigan case of *Carter v. General Motors* (1961) was the first to compensate for a mental disorder precipitated solely by a mental stimulus.

Acceptance of stress claims varies considerably by state. Whereas a worker may be compensated for a certain job-related mental disability in one state, a worker with a similar disability may be denied compensation in another. More than half the state's Workers' Compensation systems recognize mental disabilities in one form or another as compensable. Some states require evidence that some type of physical contact or physical disability occurred first. Several states compensate for disabling stress claims only if the disability resulted from a sudden, unexpected, shocking event (e.g., one that produced intense fear, helplessness, or horror—yet another example of the forensic power of the PTSD diagnosis). Others compensate stress claims if the source of stress is more than normally expected in the course of everyday life or employment, and the event(s) was a substantial contributing cause of the mental injury. This type of stress would result from events outside the worker's usual work experience that would evoke significant symptoms of distress in another worker in similar circumstances. Finally, some states compensate workers who file stress claims even if the source is not unusual or in excess of the stress of daily living. These states make no attempt to differentiate between stress claims and any other type of Workers' Compensation claim (deCarteret, 1994).

Because of the rapid expansion of mental stress claims, particularly in the past several decades, some states have redefined occupational diseases and/or excluded some mental disorders from the Workers' Compensation system. In 1988, the Oregon legislature passed new Workers' Compensation statutes specifying that employment conditions producing the mental stress disorder exist in "real and objective sense," and that the conditions must not be "generally inherent in every working situation" or be "reasonable disciplinary, corrective, or job performance evaluation actions by the employer, or [be] cessation of employment." In addition, there must be "clear and convincing evidence that the mental disorder arose out of and in the course of employment" (Helmer, 1996). These changes were made in Oregon, and in other states, because of the rapid rise in employer expenditures for Workers' Compensation benefits, which had increased from $2 billion in 1960 to $5 billion in 1970, to $21 billion in 1980, to an estimated $62 billion in 1992 (Skoppek, 1995). Many Workers' Compensation claims are advanced secondary to a diagnosis of PTSD by the worker's care provider. Such claims may or may not be accepted by the employer's fiduciary insurance program and may result in an IME. PTSD is generally compensable, if sufficiently related to a requisite accident or injury. Although many workers' PTSD claims are well substantiated, some are dubious. In England, stress claims have encountered the same level of controversy; however, injured workers are still limited to tort remedies (Adamou & Hale, 2003; McCulloch et al., 1995; Wheat, 2002).

FORENSIC PTSD ASSESSMENT

Forensic assessment for PTSD has been covered extensively in our previous publications (Pitman & Orr, 2003; Pitman et al., 1996; Sparr, 1990; Sparr & Boehnlein, 1990; Sparr & Pitman, 1999). Many persons who seek redress in the legal system after a traumatic event have genuine claims. Others, however, come with the purpose of exaggerating a claim for compensation. In a forensic context in particular, clinicians who evaluate patients after a major stressor must consider malingering in their differential diagnosis. As mentioned earlier, plaintiffs' attorneys strongly favor the diagnosis of PTSD, because the diagnosis itself constitutes evidence that the symptoms are due to the traumatic event in question. Resnick (1998) observes that PTSD has been described by various names, many of which are pejorative and suggestive of malingering (e.g., litigation neurosis, compensation neurosis). Lees-Haley and Dunn (1994) have demonstrated that a significant majority of untrained college students (86%) were able to endorse symptoms to meet criteria for a PTSD diagnosis from examiners.

Resnick (1998) has developed a list of clues to malingered PTSD:

1. Malingerers are more likely to be marginal members of society, with few binding ties or committed, long-standing financial responsibilities, such as home ownership.
2. The malingerer may have a history of spotty employment, previous incapacitating injuries, and extensive absences from work.
3. Malingerers frequently depict themselves and their prior functioning in exclusively complimentary terms.
4. The malingerer may incongruously assert an inability to work but retain the capacity for recreation. In contrast, the patient with genuine PTSD is more likely to withdraw from recreational activities, as well as work.
5. The malingerer may pursue a legal claim with impressive tenacity, while alleging depression or incapacitation in other pursuits.
6. Malingerers are unlikely to volunteer information about sexual dysfunction, although they are generally eager to emphasize their physical complaints.
7. Malingerers are also unlikely to volunteer information about nightmares, unless they have read the diagnostic criteria for PTSD. When they occur in PTSD, genuine nightmares typically show variations on the theme of the traumatic event. In contrast, the malingerer may claim repetitive dreams that always reenact the traumatic event in exactly the same way.

The MMPI-II has two scales designed to assess combat-related PTSD (PK and PS scales). The more commonly used PK scale was developed by Keane, Malloy, and Fairbank (1984) to determine the difference between individuals with a genuine PTSD diagnosis and those with other diagnoses. The content of the PK scale is suggestive of emotional turmoil. The authors have indicated that caution should be used with the PK scale, because its results may be susceptible to faked answers by veterans who are motivated to appear to have PTSD to gain monetary compensation.

Raifman (1993) has proposed that expert witness testimony regarding PTSD should be "increasingly supported by empirically based research data" (p. 115). Data obtained through laboratory testing has the potential to enhance expert testimony in the area of PTSD. Although this effort is in its infancy, objective measurement of psychophysiological responses during the structured (script-driven) recollection of the traumatic event has reliably distinguished between trauma victims with and without PTSD in clinical research (Pitman et al., 1987), and it has been used successfully in the forensic setting (Pitman & Orr, 2003).

CONCLUSION

In the intervening years since we first addressed this issue (Pitman & Sparr, 1998; Pitman et al., 1996; Sparr, 1990, 1996; Sparr & Atkinson, 1986; Sparr & Boehnlein, 1990; Sparr & Pankratz, 1983; Sparr et al., 1987), PTSD has continued to influence, and be influenced by, the law. The most dramatic change is the geometric rise in PTSD claims in civil litigation. Much of this can be attributed generally to society's increasing concern with, and acceptance of, psychological trauma, and specifically to liberalization of the stressor criteria in DSM-IV. A continued concern is laws and regulations that provide financial incentives for plaintiffs or claimants to remain ill and to disavow responsibility for their emotional problems. In particular, these have become key issues in the conduct of Workers' Compensation and VA disability claims. In this context, evaluating professionals are obligated not only to educate themselves in the way the diagnosis can be used and abused in the legal setting but also to remain cognizant of the vulnerability of PTSD patients and their capacity for retraumatization by the legal process. Healthy skepticism must be tempered with an ethical obligation to deal with PTSD claimants in an honest and empathic manner.

REFERENCES

Adamou, M.C., & Hale, A. S. (2003). PTSD and the law of psychiatric injury in England and Wales: Finally coming closer? *Journal of the American Academy of Psychiatry and the Law, 31*, 327–332.

American Psychiatric Association. (1980). *Diagnostic and statistical manual of mental disorders* (3rd ed.). Washington, DC: Author.

American Psychiatric Association. (1987). *Diagnostic and statistical manual of mental disorders* (3rd ed., rev.). Washington, DC: Author.

American Psychiatric Association. (1994). *Diagnostic and statistical manual of mental disorders* (4th ed.). Washington, DC: Author.

Appelbaum, P. S., Jick, R. Z., Grisso, T., Givelber, D., Silver, E., & Steadman, H. J. (1993). Use of post-traumatic stress disorder to support an insanity defense. *American Journal of Psychiatry, 150*, 229–234.

Bale, A. (1990). Medicolegal stress at work. *Behavioral Sciences and the Law, 8*, 399–420.

Battista, M. E. (1985). The disability benefits matrix: Medical legal issues of physician participation. In C. H. Wecht (Ed.), *Legal medicine* (pp. 367–393). New York: Praeger Scientific.

Blanchard, E. B., Hickling, E. J., Taylor, A. E., & Loos, W. (1995). Psychiatric morbidity associated with motor vehicle accidents. *Journal of Nervous and Mental Disease, 183*, 495–504.

Blowers, A. N., & Bjerregaard, B. (1994). The admissibility of expert testimony on the battered woman syndrome in homicide cases. *Journal of Psychiatry and the Law, 22*, 527–560.

Brakel, S. J., & Brooks, A. D. (2001). *Law and psychiatry in the criminal justice system.* Littleton, CO: Rothman.

Bremner, J. D., Vermetten, E., & Afzal, N. (2004). Deficits in verbal declarative memory function in women with childhood sexual abuse-related posttraumatic stress disorder. *Journal of Nervous and Mental Disease, 192*, 643–649.

Brown, D., Scheflin, A. W., & Hammond, D. C. (1998). *Memory, trauma treatment, and the law.* New York: Norton.

Calise, A. (1993, August 30). Workers compensation mental stress claims in decline. *National Underwriter*, pp. 3, 8, 31.

Carter v. General Motors, 106 NW 2d (361 Mich. 1961).

Daubert v. Merrill Dow Pharmaceuticals, Inc., 509 U.S. 579, 595 (1993).

Davies, J. A. (1992). Direct actions for emotional harm: Is compromise possible? *Washington Law Review, 67*, 1–53.

deCarteret, J. C. (1994). Occupational stress claims: Effects on workers compensation. *American Association of Occupational Health Nurses Journal, 42*, 294–498.

Dunn v. Roberts, 963 F2d 308, 1992 U.S. App. LEXIS 8783 (1992).

Edelman, P. S. (1990, June 1). Indemnity insurance policies. *New York Law Journal, 203*, 3.

Falk, P. J. (1996). Novel theories of criminal defense based upon the toxicity of the social environment: Urban psychosis, television intoxication, and black rage. *North Carolina Law Review, 74*, 731–811.

Federal Mental Health Parity Act, 25 U.S.C. § 1185a (1996).

Frueh, B. C., Elhai, J. D., Gold, P. B., Monnier, J., Magruder, K. M., Keane, T. M., et al. (2003). Disability compensation seeking among veterans evaluated for posttraumatic stress disorder. *Psychiatric Services, 54*, 84–91.

Frye v. United States, 293 F. 1013 (D.C. Cir. 1923).

Furrow, P. R. (1998). Regulating the managed care revolution: Private accreditation and a new system ethos. *Villanova Law Review, 43*, 361–407.

Gilbertson, M. W., Gurvits, T. V., & Lasko, N. B. (2001). Multivariate assessment of explicit memory function in combat veterans with PTSD. *Journal of Traumatic Stress, 14*, 437–456.

Goodstein, R. K., & Page, A. W. (1981). Battered wife syndrome: Overview of dynamics and treatment. *American Journal of Psychiatry, 138*, 1036–1044.

Gould, E. (2001, January 1). Nine million gaining upgraded benefit for mental care. *New York Times*, p. A1.

Gulf War Veterans draw disability compensation at a higher rate than those of any other conflict. (1999, October 27). *Wall Street Journal*, p. A1.

Gutheil, T. G. (1999). A confusion of tongues: Competence, insanity, psychiatry, and the law. *Psychiatric Services, 50*, 767–773.

Helmer, G. (1996, November). *Mental health stress claims, Oregon, 1991–1995*. Salem: Research and Analysis Section, Oregon Department of Consumer and Business Services.

Henderson, R., & Sparr, L. (1994). Psychiatric file reviews in the compensation and pension assessment process. *Federal Practitioner, 11*, 92–96.

Hoffman, B. F., & Spiegel, H. (1989). Legal principles in the psychiatric assessment of personal injury litigants. *American Journal of Psychiatry, 146*, 304–310.

Keane, T. M., Malloy, P. F., & Fairbank, J. A. (1984). Empirical development of an MMPI subscale for the assessment of combat-related posttraumatic stress disorder. *Journal of Consulting and Clinical Psychology, 52*, 888–891.

Lambert, T. F. (1978). Tort liability for psychic injuries: Overview and update. *Journal of the Association of Trial Lawyers of America, 37*, 1–31.

Lee, J. D., Lindahl, B. A., & Dooley, J. A. (2002). *Modern tort law: Liability and litigation*. St. Paul, MN: West Group.

Lees-Haley, P. R. (1986). Pseudo post-traumatic stress disorder. *Trial Diplomacy Journal, 9*, 17–20.

Lees-Haley, P. R., & Dunn, J. T. (1994). The ability of naive subjects to report symptoms of mild brain injury, posttraumatic stress disorder, major depression, and generalized anxiety disorder. *Journal of Clinical Psychology, 50*, 252–256.

Lesse, S. (1982). The psychiatrist in court: Theatre of the absurd [Editorial]. *American Journal of Psychotherapy, 36*, 287–291.

Lewis v. State, S95A0250 265 Ga. 451, 457 S.E.2d 173 (1995).

London, D. B., Zonana, H. V., & Loeb, R. (1988). Workers' compensation and psychiatric disability. In R. C. Larson & J. S. Felton (Eds.), *Occupational medicine: Psychiatric injury in the workplace* (pp. 595–609). Philadelphia: Hanley & Belfus.

Lustberg, L. S., & Jacoby, J. V. (1992). The battered woman as reasonable person: A critique of the appellate division decision in State v. McClain. *Seton Hall Law Review, 22*, 365–388.

Malt, U. (1988). The long-term psychiatric consequences of accidental injuries: A longitudinal study of 107 adults. *British Journal of Psychiatry, 153*, 810–818.

Malt, U. F., & Blikra, G. (1993). Psychosocial consequences of road accidents. *European Psychiatry, 8*, 227–228.

Mayou, R., Bryant, B., & Duthie, R. (1993). Psychiatric consequences of road traffic accidents. *British Medical Journal, 307*, 647–651.

Mayou, R., Tyndell, S., & Bryant, B. (1997). Long-term outcome of motor vehicle accident injury. *Psychosomatic Medicine, 59*, 578–584.

McCulloch, M., Jones, C., & Bailey, J. (1995). Posttraumatic stress disorder: Turning the tide without opening the floodgates. *Medical Science Law, 35*, 287–293.

McNally, R. J. (1998). Experimental approaches to cognitive abnormality in posttraumatic stress disorder. *Clinical Psychology Review, 18*, 971–982.

Melton, G. B., Petrila, J., Poythress, N. G., & Slobogin, C. (Eds.). (1997). *Psychological evaluations for the courts: A handbook for mental health professionals and lawyers* (2nd ed.). New York: Guilford Press.

Modlin, H. C., & Felthous, A. (1989). Forensic psychiatry and private practice. *Bulletin of the American Academy of Psychiatry and the Law, 17,* 69–82.

Morrison, M. A. (2000). Changing perceptions of mental illness and the emergence of expansive mental health parity legislation. *South Dakota Law Review, 45,* 8–32.

Mossman, D. (1996). Veterans Affairs disability compensation: A case study in counter therapeutic jurisprudence. *Bulletin of the American Academy of Psychiatry and the Law, 24,* 27–44.

Murdoch, M., Nelson, D. B., & Fortier, L. (2003). Time, gender, and regional trends in the application for service-related posttraumatic stress disorder disability benefits, 1990–1998. *Military Medicine, 168,* 662–670.

Oboler, S. (2000). Disability evaluations under the Department of Veterans Affairs. In R. D. Rondinelli & R. T. Katz (Eds.), *Impairment ratings and disability evaluations* (pp. 187–217). Philadelphia: Saunders.

Okpaku, S. O., Sibulkin, A. E., & Schenzler, C. (1994). Disability determinations for adults with mental disorders: Social Security Administration v. independent judgments. *American Journal of Public Health, 84,* 1791–1795.

Oregon Revised Statues (2005 ed.). Volume 14, Chapter 656 (Workers' Compensation), Section 656.802 (Occupational disease; mental disorder; proof). Available at *www.leg.state.or.us/ors*

Orr, S. P., Metzger, L. J., & Lasko, N. B. (2000). De novo conditioning in trauma-exposed individuals with and without post-traumatic stress disorder. *Journal of Abnormal Psychology, 109,* 290–298.

Parry, J., & Drogin, E. Y. (2000). *Criminal law handbook on psychiatric and psychological evidence.* Washington, DC: American Bar Association.

Parry, J., & Drogin, E. Y. (2001). *Civil law handbook on psychiatric and psychological evidence and testimony.* Washington, DC: American Bar Association.

Parson, E. A. (1994). Inner city children of trauma: Urban violence traumatic stress response syndrome and therapists responses. In J. P. Wilson & J. D. Wilson (Eds.), *Countertransference and the treatment of PTSD* (pp. 151–178). New York: Guilford Press.

Perr, I. N. (1992). Asbestos exposure and psychic injury—a review of 48 claims. *Bulletin of the American Academy of Psychiatry and the Law, 20,* 383–393.

Pitman, R. K., & Orr, S. P. (2003). Forensic laboratory testing for post-traumatic stress disorder. In R. I. Simon (Ed.), *Posttraumatic stress disorder in litigation: Guidelines for forensic assessment* (2nd ed., pp. 207–223). Washington, DC: American Psychiatric Press.

Pitman, R. K., Orr, S. P., & Forgue, D. F. (1987). Psychophysiologic assessment of post-traumatic stress disorder imagery in Vietnam combat veterans. *Archives of General Psychiatry, 44,* 970–975.

Pitman, R. K., & Sparr L. F. (1998). PTSD and the law. *PTSD Research Quarterly, 9,* 1–6.

Pitman, R. K., Sparr, L. F., Saunders, L. S., & McFarlane, A. C. (1996). Legal issues in posttraumatic stress disorder. In B. A. van der Kolk, A. C. McFarlane, & L. Weisæth (Eds.), *Traumatic stress: The effects of overwhelming experience on mind, body, and society* (pp. 378–397). New York: Guilford Press.

Prosecutor v. Anto Furundzija. (1998). International Criminal Tribunal for Former Yugoslavia Case No. IT-95-17/1-T.

Raifman, L. J. (1993). Problems of diagnosis and legal causation in courtroom use of post-traumatic stress disorder. *Behavioral Sciences and the Law, 1,* 115–131.

Resnick, P. J. (1998). Malingering of posttraumatic stress disorders. *Journal of Practical Psychiatry and Behavioral Health, 4,* 329–339.

Rosen, G. M. (1995). The Aleutian Enterprise sinking and posttraumatic stress disorder: Misdiagnosis in clinical and forensic settings. *Professional Psychology: Research and Practice, 26,* 82–87.

Schnyder, U., Moergeli, H., Klaghofer, R., & Buddeberg, C. (2001). Incidence and prediction of posttraumatic stress disorder symptoms in severely injured accident victims. *American Journal of Psychiatry, 158,* 594–599.

Sersland, S. J. (1984). Mental disability caused by mental stress: Standards of proof in Workers Compensation cases. *Drake Law Review, 33,* 751–816.

Shuman, D. W. (2003a). Criminal proceedings: trial. In *Psychiatric and psychological evidence* (2nd ed., § 12.01–§ 12.15). St. Paul, MN: West Group.

Shuman, D. W. (2003b). Personal injury litigation. In *Psychiatric and psychological evidence* (2nd ed., § 14.01–§ 14.20). St. Paul, MN: West Group.

Sing, M., & Hill, S. C. (2001). Economic grand rounds: The costs of parity mandates for mental health and substance abuse insurance benefits. *Psychiatric Services, 52*, 437–440.

Skoppek, J. (1995). *Stress claims in Michigan: Workers Compensation entitlement for mental disability*. Midland, MI: Mackinac Center for Public Policy.

Slovenko, R. (1973). *Tort liability and claims of the mentally incompetent in psychiatry and law*. Boston: Little, Brown.

Slovenko, R. (1994). Legal aspects of posttraumatic stress disorder. *Psychiatric Clinics of North America, 17*, 436–439.

Social Security Administration Office of Policy. (2003, August). *SSI Annual Statistical Report for 2002* (SSA Publication No. 13-11827). Washington, DC: Author.

Sparr, L. F. (1990). Legal aspects of posttraumatic stress disorder: Uses and abuses. In M. E. Wolf & A. D. Mosnaim (Eds.), *Posttraumatic stress disorder: Ideology, phenomenality, and treatment* (pp. 239–264). Washington, DC: American Psychiatric Press.

Sparr, L. F. (1996). Mental defenses and posttraumatic stress disorder: Assessment of criminal intent. *Journal of Traumatic Stress, 9*, 405–425.

Sparr, L. F. (2003, May). *The uses and abuses of psychiatric independent medical examinations: An ethical dilemma?* Paper presented at the annual meeting of the American Psychiatric Association, San Francisco.

Sparr, L. F. (2005). Mental incapacity defenses at the War Crimes Tribunal: Questions and controversy. *Journal of the American Academy of Psychiatry and the Law, 33*, 59–70.

Sparr, L. F., & Atkinson, R. M. (1986). Posttraumatic stress disorder as an insanity defense: Medicolegal quicksand. *American Journal of Psychiatry, 143*, 608–613.

Sparr, L. F., & Bremner, J. D. (2005). Posttraumatic stress disorder and memory: Prescient medicolegal testimony at the International War Crimes Tribunal. *Journal of the American Academy of Psychiatry and the Law, 33*, 71–78.

Sparr, L. F., & Boehnlein, J. K. (1990). Posttraumatic stress disorder and tort actions: Forensic minefield. *Bulletin of the American Academy of Psychiatry and the Law, 18*, 283–302.

Sparr, L. F., & Pankratz, L. D. (1983). Factitious posttraumatic stress disorder. *American Journal of Psychiatry, 140*, 1016–1019.

Sparr, L. F., & Pitman, R. K. (1999). Forensic assessment of traumatized adults. In J. D. Bremner & P. Saigh (Eds.), *Posttraumatic stress disorder: A comprehensive text* (pp. 284–308). Boston: Allyn & Bacon.

Sparr, L. F., Reaves, M. E., & Atkinson, R. M. (1987). Military combat, posttraumatic stress disorder, and criminal behavior in Vietnam veterans. *Bulletin of the American Academy of Psychiatry and the Law, 15*, 141–162.

Sparr, L. F., White, R., Friedman, M. J., & Wiles, D. B. (1994). Veterans psychiatric benefits: Enter courts and attorneys. *Bulletin of the American Academy of Psychiatry and the Law, 22*, 205–222.

Spaulding, W. J. (1988). Compensation for mental disability. In J. O. Cavenar (Ed.), *Psychiatry* (Vol. 3, pp. 1–27). Philadelphia: Lippincott.

State v. Grubbs, 353 S.C. 374, 381, 577 S.E.2d 493, 497 (Ct. App. 2003).

State v. Kelly, 97 N.J. 178, 478, A.2d 364 (1984).

State v. McClain, 248 N.J. Super. 409, 591 A. 2d 652 (N.J. Super. A.D. 1991).

State v. Necaise, 466 So. 2d 660 (La. Ct. App. 1985).

State v. Pascal, 736 P. 2d 1065 (Wash. 1987).

Steadman, H. J. (1985). Empirical research on the insanity defense. *Annals of the American Academy of Policy and Social Science, 477*, 58–64.

Sticking it to business: A company's struggle with an out-of-control workers compensation system. (1993, March 8). *U.S. News & World Report*, p. 59.

Stone, A. A. (1993). Posttraumatic stress disorder and the law: Critical review of the new frontier. *Bulletin of the American Academy of Psychiatry and the Law, 21*, 23–36.

Summit, R. (1983). The child sexual abuse accommodation syndrome. *Child Abuse and Neglect, 7*, 177–193.

Tommasini, N. R. (1994). Private insurance coverage for the treatment of mental illness versus general medical care. *Archives of Psychiatric Nursing, 8*, 9–13.

Ursano, R. J., Fullerton, C. S., & Epstein, R. S. (1999). Acute and chronic posttraumatic stress disorder in motor vehicle accident victims. *American Journal of Psychiatry, 156,* 489–595.

U.S. Department of Veterans Affairs. (2003). *Annual Benefits Report of the Secretary of Veterans Affairs Fiscal Year 2003.* Washington, DC: Author.

Walker, L. E. (1980). *Battered woman.* New York: HarperCollins.

Walker, L. E. (2000). *The battered woman syndrome* (2nd ed.). New York: Springer.

Walters, K. S. (1984). The unhappy marriage of psychiatry and the law: Willard Gaylin's *The Killing of Bonnie Garland. Academy Forum, 28,* 15–17.

Wheat, K. (2002). Psychiatric injury and employment. In *Napier & Wheat's recovering damages for psychiatric injuries* (2nd ed., pp. 143–173). London: Oxford University Press.

Chapter 23

Emerging Treatments for PTSD

Stacy Shaw Welch and Barbara Olasov Rothbaum

Extensive research and clinical efforts over the past decade have resulted in a number of efficacious treatments for posttraumatic stress disorder (PTSD); both pharmacological and psychosocial treatments, with strong evidence of their efficacy, are detailed in other chapters in this volume. Despite the impressive clinical gains demonstrated by these treatments, however, there remain patients that they do not reach because of contraindications, availability, attrition/dropout, refusal to participate, noncompliance, and failure or incomplete response. In this chapter, we review other psychosocial treatments that have been tested, some of which show promise but thus far have limited research to back them. Although not an exhaustive list, these include (1) imagery-based treatments, including imagery rescripting and dream/imagery rehearsal for nightmares, as well as the so-called "power therapies"; (2) treatments using novel technology, such as Internet-delivered approaches and virtual reality exposure therapy; (3) treatments that emphasize social support, including group therapy, family or couple therapy; and (4) treatments that may be useful for patients who refuse or have problems tolerating exposure therapy, such as dialectical behavior therapy and acceptance and commitment therapy. Representative research from these areas is summarized in Table 23.1, and Table 23.2 presents a summary of conclusions from the research.

Several years ago, Foa and Meadows (1997) outlined several "gold standards" to evaluate the methodological rigor of research studies in the PTSD treatment literature. The seven standards include (1) clearly defined symptoms; (2) reliable and valid measures; (3) use of independent evaluators; (4) trained assessors; (5) manualized, replicable, specific treatment programs; (6) treatment adherence; and (7) unbiased assignment to treatment. We evaluate each of the new treatments presented in this chapter, using these methodological standards (see Table 23.1). Of course, when developing new

TABLE 23.1. Emerging Treatments and Associated Evidence

Study	N	Clearly defined symptoms	Reliable/valid measures	Independent evaluators	Trained assessors	Manualized, replicable treatment	Treatment adherence	Unbiased assignment to treatment	Generalizability
Imagery rescripting (IR)									
Rusch, Grunert, Mendelsohn, & Smucker (2000)	11		1			✓		2	Individuals with disturbing, repetitive images related to life experience but not actual memories (most related to industrial accidents). 1. SUDS only 2. Uncontrolled study
Imagery rehearsal thearpy (IRT)									
Forbes et al. (2003)	12	✓	✓	1	✓	✓		2	Veterans with nightmares related to chronic, combat-related PTSD; excluded for organic mental disorder or psychosis, severe depression, alcohol abuse, and illiteracy. 1. Self-report measures 2. No control group
Krakow et al. (2001d)	19	✓	✓	1	1	✓		2	Involuntarily adjudicated adolescent girls with nightmares at least once per week, high rates of victimization and PTSD. 1. Not reported/self-report measures used 2. Nonrandomized control
Krakow et al. (2001b)	62	✓	✓	1	1	✓		2	Adult victims of various crimes reporting weekly episodes of insomnia and nightmares. All met PTSD criteria and most had prior abuse. Excluded for psychosis, alcohol/drug relapse/withdrawal, or trauma < 6 months before study intake. 1. Not reported/self-report measures used 2. No control group

Power therapies

Study	n						Notes
TIR (trauma incident reduction), VK/D (visual kinesthetic dissassociation), TFT (thought field therapy): Carbonell & Figley (1999)	39	✓				✓	Participants were required to articulate only a trauma or phobia that was interfering with their daily functioning; traumas were varied and included childhood abuse, combat exposure, criminal victimization, motor vehicle accidents, and accidental shooting.
TIR (trauma incident reduction): Valentine & Smith (2001)	123	✓	1	1	✓	√²	Participants were inmates with self-reported trauma histories and some symptoms of anxiety, depression, or PTSD. Excluded for antipsychotic medications, hospitalization within the past 3 years with a bipolar or schizophrenia diagnosis; experience of hallucinations, delusions, or bizarre behavior, drug/alcohol disorders, or victimization within 3 months of the study. 1. Not reported 2. No scale, but sections of tapes were reviewed by a TIR expert
TFT (thought field therapy): Sakai et al. (2001)	714	✓					714 patients at Kaiser Behavioral Medicine Services, a very large HMO. Patients were assessed by a SUDS rating and included "acute stress, adjustment disorder with anxiety and depression, alcohol cravings, anger, anxiety, anxiety due to medical condition, bereavement, chronic pain, depression, fatigue, major depressive disorder, maladaptive food cravings, nausea, neurodermatitis, nicotine cravings, obsessive traits, obsessive–compulsive disorder, obsessive–compulsive personality disorder, panic isorder without agoraphobia, parent–child stress, partner relational stress, PTSD, relationship stress, social phobia, specific phobia, tremor, trichotillomania, Type A personality traits or histrionic traits, and work stress." Since no standardized assessment of any kind was made, the generalizability of these results is questionable.
TFT: Johnson, Shala, Sejdijaj, Odell, & Dabishevci (2001)	105	✓					Ethnic Albanian residents of Kosovo, ages 4–78, with varying traumas. No standard assessment of any kind was made.

(continued)

TABLE 23.1. (continued)

Study	N	Clearly defined symptoms	Reliable/valid measures	Independent evaluators	Trained assessors	Manualized, replicable treatment	Treatment adherence	Unbiased assignment to treatment	Generalizability
Interapy									
Lange et al. (2003)	101	✓	✓	1	1	✓	2	✓	Excluded participants with severely depressed mood, tendency to psychological dissociation, risk of psychosis, substance abuse, trauma occurrence within past 3 months, incest, age younger than 18 years. Included a community sample of posttraumatic stress and grief in a group with mild to relatively severe trauma symptoms (PTSD not diagnosed) 1. N/A (all assessments were computerized) 2. Therapists used standardized instructions/feedback, tailored them to clients, and had weekly supervision
Lange, van de Ven, Schrieken, & Emmelkamp (2001)	25	✓	✓	1	1	✓	2	✓	Students who had experienced a traumatic event; excluded for substance abuse, severe major depression, psychological dissociation, psychotic disorder or the use of antipsychotic medication, extremely high scores in general psychopathology, or involvement in other psychological treatment. 1. N/A (all assessments were computerized) 2. Therapists used standardized instructions/feedback, tailored them to clients, and had ongoing supervision
Virtual reality									
Rothbaum, Hodges, Ready, Graap, & Alarcon (2001)	10	✓	✓	✓	✓	✓		1	Vietnam veterans with chronic, treatment-refractory PTSD. Excluded for active addiction, serious heart conditions, psychosis, bipolar disorder, unstable medication regimens, uncontrolled suicidal intention, and/or lack of approval from treating physicians/teams. 1. Uncontrolled study
Family/couple therapy									
Behavioral family therapy: Glynn et al. (1999)	42	✓	✓	✓	✓	✓	✓	✓	Vietnam veterans with combat-related PTSD, stabilized on meds, plus family member willing to participate; excluded for severe cardiovascular

Study	N								Inclusion/exclusion criteria and generalizability
				✓	✓	✓	✓	2	disease, organic brain, dissociative, or psychotic disorder; current substance dependence, and evidence of overt physical aggression to self or others during the past year.
CBCT (cognitive-behavioral couple treatment): Monson, Schnurr, Stevens, & Guthrie (in press)	7	✓	✓	✓	✓	✓	✓¹	2	Veterans with military-related PTSD and partners; excluded for substance abuse/dependence not in remission for at least 3 months, current uncontrolled bipolar or psychotic disorder, severe cognitive impairment, and couples who were experiencing severe intimate aggression or a desire to separate or end their intimate relationship. 1. No scale; authors observed each other's treatment 2. Uncontrolled study
Group therapy									
Trauma-focused and supportive groups: Schnurr et al. (2003)	325	✓	✓	✓	✓	✓	✓	✓	Male veterans with combat-related PTSD, stable med regimen for 2 weeks prior to the study; excluded for psychotic disorder, mania, bipolar disorder, major depressive disorder with psychotic features, current alcohol or drug dependence, substance abuse at treatment or work, cognitive impairment, and severe cardiovascular disorder.
DBT-augmented exposure									
Cloitre, Koenen, Cohen, & Han (2002)	46	✓	✓¹	✓	2	✓	✓	✓	DSM-IV diagnosis of PTSD related to CSA, physical abuse, or both; excluded for current diagnosis of organic or psychotic mental disorders, substance dependence, eating disorder, dissociative disorder, bipolar I disorder or BPD and the presence of suicide attempt or psychiatric hospitalization within the last 3 months. 1. At pre- and posttreatment, not at follow-up 2. Not reported
Bradley & Follingstad (2003)	31	✓	1	✓	1	✓	✓²	✓	Generalizable to incarcerated women in a medium-security prison, histories of CSA and high rates of experiences with interpersonal victimization. 1. Used self-report measures 2. Used therapist's notes/chart records

ACT—no data available

Note. For explanation of superscript numerals 1 and 2, see "Generalizability" column.

TABLE 23.2. Evidence Base and Recommendations

Treatment	Number of RCTs	Typical number of sessions	Includes evidence-based components?	Recommendations
Imagery rescripting (IR)	0	1 to 9	Yes (imaginal exposure, cognitive restructuring)	Promising treatment for nonresponders to PE, CPT, or other CBT treatments, but more research is needed. Best known treatment for disturbing images loosely related to life events.
Imagery rehearsal therapy (IRT)	1	2 3-hour sessions	Yes (cognitive restructuring, CBT skills, some [although deemphasized] imaginal exposure)	Effective treatment for nightmares. Has good potential for a treatment for PTSD generally but more research is needed.
Trauma incident reduction	1	1 to 10	Yes (imaginal exposure, cognitive restructuring)	Not recommended.
Thought field therapy	0	1	Yes (imaginal exposure)	Not recommended.
Visual–kinesthetic dissassociation	0	1 to 3	Yes (imaginal exposure)	Not recommended.
Interapy	2	10	Yes (cognitive restructuring, exposure through writing about trauma)	Very promising Internet-delivered treatment for PTSD; probably efficacious for low severity, more research needed for higher severity.
Virtual reality	0	13	Yes (imaginal exposure)	Promising treatment for PTSD; needs more research.
Behavioral family therapy	1 (for PTSD)	18 exposure only/34 exposure + BFT	Yes (imaginal exposure, cognitive restructuring)	Does not appear to have additional impact on PTSD symptoms; appears that standard prolonged exposure suffices.
Cognitive-behavioral couple treatment	0	15	Yes (cognitive restructuring)	Some promise; clinicians and partners rated PTSD symptoms improved, but veterans did not.
Trauma-focused and supportive groups	1	30 (plus 5 booster sessions)	Yes (imaginal exposure, cognitive restructuring)	No strong evidence that it is more effective than supportive therapy.
Dialectical behavior therapy (as an adjunct to exposure treatment)	2	16 to 18	Yes (imaginal exposure, cognitive restructuring)	May help very emotionally dysregulated patients and therapists hesitant to use exposure techniques; more research is needed.
Acceptance and commitment therapy	0	8 to 32	Yes (cognitive restructuring, some expoure)	May help avoidant patients unwilling to do exposure therapy; more research is needed.

treatments, it is not always feasible or prudent to expend the costs necessary to conduct such rigorous trials early in the testing. We therefore include new treatments that are in the development stages of small pilot and case studies, but evaluate their promise based on the soundness of their theoretical rationale and associated evidence. It is widely thought that two main factors are necessary to treat PTSD successfully through psychosocial therapies: habituation to aversive stimuli, achieved by some kind of exposure to the traumatic or avoided stimuli (Jaycox, Foa, & Morral, 1998), and cognitive reappraisal of the traumatic experiences (Ehlers & Clark, 2000). Typically, this is achieved through having patients engage the traumatic memories by talking about them in detail or writing about them, and/or explicit cognitive restructuring. These techniques are included to various degrees in the established treatments for PTSD reviewed earlier in this volume. We note their inclusion in the less established treatments reviewed in this chapter.

IMAGERY-BASED TREATMENTS

Repetitive and prolonged imagery of the traumatic event is the core of prolonged exposure (PE) therapy, one of the best available treatments for PTSD (Rothbaum, Meadows, Resick, & Foy, 2000). *Imagery rescripting* (IR), developed by Smucker and colleagues (Smucker & Dancu, 1999; Smucker, Dancu, Foa, & Niederee, 1995; Smucker & Niederee, 1995) departs slightly from PE by asking clients to manipulate recurrent, distressing images mentally through "rescripting," which is intended to modify the images and challenge trauma-based beliefs, particularly those containing themes of powerlessness and helplessness. New images are patient-generated; the therapist assists the process through Socratic questioning (i.e., "If you could now change that part of your image in any way to make it less threatening or frightening, what change or changes would you introduce?"). Patients are encouraged to use mastery imagery to replace victimization imagery; for instance, kicking a rapist in the crotch and rendering him helpless. The treatment was originally developed for PTSD related to childhood sexual abuse (CSA). Smucker, Grunert, and Weiss (2003) have more recently proposed an algorithm for the use of IR, in which they recommend beginning with prolonged imaginal exposure and adding imagery-based cognitive restructuring such as IR only when non-fear-based emotions such as guilt, helplessness, anger, or shame appear primary.

IR has been adapted to treat industrial accident victims who fail to benefit from PE alone (Grunert, Smucker, Weis, & Rusch, 2003), as well as repetitive, disturbing images that are loosely related to life experiences but do not correspond with actual experience or memory (Rusch, Grunert, Mendelsohn, & Smucker, 2000). An example is a patient who begins experiencing highly distressing images of self-inflicted injury (with no history, urges, or intent to engage in the behavior) following a work-related accident. One estimate rates the occurrence of this phenomenon in trauma samples at 11% (Reynolds & Brewin, 1998). The treatment is typically conducted over nine 1.5- to 2-hour sessions, although the some patients (Rusch et al., 2000) have been treated successfully in one session.

Data for IR thus far are limited to small pilot studies and case illustrations. Still, the limited available evidence to date is encouraging given the combination of a strong, data-driven rationale for the treatment (Smucker et al., 1995) and data profiles characterized by unremitting SUDs (subjective units of distress) ratings during prolonged exposure, followed by dramatic drops after the IR. Smucker and colleagues hypothesize that the mechanisms at play are primarily cognitive, such as changing patients' schemas about abuse and its meaning, as well as increasing their sense of mastery and control. It is possible that these changes lead to decreased avoidance of thoughts, images, and other stimuli, or that a feeling of mastery might facilitate more therapeutic engagement with the traumatic material and lead to better outcomes. The idea of IR for clients who do not respond well to PE is promising, and it may also turn out to be a good alternative for clients who do not respond well to explicit verbal cognitive interventions, such as those found in cognitive therapy (CT) or cognitive processing therapy (CPT) (see Kimerling, Ouimette, & Weitlauf, Chapter 12, this volume). Although we need more data to make conclusions about IR's efficacy, the time seems ripe for a randomized, controlled trial (RCT). However, such an RCT will not be easy to conduct. In general, well-delivered PE leads to significant improvement in about 60–80% of treated patients (Foa, Rothbaum, & Furr, 2003). Therefore, if IR is used to treat PE nonresponders, a very large number of patients would be needed to have a large enough sample to be randomized. Conversely, if patients are randomized to receive either PE or IR, an enormous number of participants would still be required to detect differences in treatment outcome/predictors of outcome, because the two treatments share so many of the presumably active ingredients of therapy. IR also appears quite promising to treat the kinds of distressing images loosely linked to traumatic life events described earlier, and in the absence of other treatments for the phenomenon may be the best course of action currently. Of course, more methodologically rigorous study is crucial to confirm our conclusion that IR holds promise.

The second imagery-based therapy is *imagery rehearsal therapy*, or IRT, developed by Krakow and colleagues (Krakow, Hollifield, et al., 2000; Krakow, Hollifield, et al., 2001). IRT is a brief, group-based treatment developed specifically to treat nightmares, but the results of several clinical trials suggest that it may also be helpful for general sleep problems and PTSD symptomatology. In the largest study to date (N = 168), a randomized trial of sexual assault survivors with PTSD, IRT showed moderate to large effect sizes compared to a wait-list control. Treatment gains were maintained at 3- and 6-month posttreatment, and PTSD symptoms decreased significantly in 65% of the treated group compared with symptom exacerbation or no change in 69% of controls (Krakow, Hollifield, et al., 2000, 2001). Similar improvements, maintained and even enhanced at a 12-month follow-up, were found in two uncontrolled studies: a small study of male combat veterans (N = 12; Forbes et al., 2003; Forbes, Phelps, & McHugh, 2001) and a study of crime victims with PTSD (N = 62; Krakow, Johnston, et al., 2001). IRT reduced nightmares but did not improve overall sleep quality or PTSD symptoms compared to wait-list controls in a residential facility group of 19 adjudicated adolescent girls with high rates of sexual abuse and PTSD (Krakow, Sandoval, et al., 2001). A more comprehensive sleep treatment, "sleep dynamic therapy," which included IRT, significantly improved sleep and PTSD symptoms in an uncontrolled study of 69 natural disaster victims (Krakow, Melendrez, et al., 2002).

IRT is typically delivered to small groups of four to eight participants via two 3-hour sessions spaced 1 week apart, with a 1-hour follow-up 3 weeks later. The first session is primarily psychoeducational; the major concept is that although nightmares are trauma-induced, they can be viewed as learned, habit-sustained behavior, and as such, can be controlled. Session 1 also includes teaching/practice of pleasant imagery, as well as cognitive-behavioral strategies for coping with the emergence of unpleasant images, including thought stopping, breathing techniques, grounding, talking, writing, acknowledging, and choosing. Participants are asked to practice the pleasant imagery exercises as homework. Session 2 reviews the imagery practice and any difficulties that arise. Imagery is then practiced on a single nightmare. Participants are asked to write down their nightmare, then are instructed to "change the nightmare any way (they) wish," write down the changes, then imagine the "new dream" for 10–15 minutes. They then describe the old and new nightmares to the group. After this, they are told to repeat the process mentally (not in writing), rehearse the new dream for 5–20 minutes a day, and are warned not to work on more than two new dreams in a given week. Krakow and colleagues state that they deemphasize exposure by discouraging descriptions of traumatic content in nightmares by telling participants to begin with a nightmare of lesser intensity, or one that does not seem like a replay of the trauma. The third and final session is devoted to discussing progress, sharing, and asking questions. This has been adapted to one 6-hour session (Krakow, Sandoval, et al., 2001) and six weekly sessions in which the nightmares that patients select appear to be fairly accurate replays of the actual traumas (Forbes et al., 2003).

Like IR, IRT employs the two major elements found in other successful treatments: exposure to traumatic material, although deemphasized, and a variant of cognitive restructuring. However, IRT is novel in its emphasis on nightmares as learned behaviors that can be altered, and its emphasis on sleep quality may also be important. It is possible that as participants' nightmares decrease and sleep quality increases, their daytime energy, ability to cope with distress, and approach of previously avoided trauma cues might increase. This theory would explain why in one study, when patients' sleep quality did not improve, neither did their PTSD symptoms. Very interestingly, emerging data suggest that treatments for sleep-disordered breathing (prevalent in PTSD samples) with continuous positive airway pressure breathing masks have been associated with decreases in nightmares, insomnia, and PTSD symptoms (Krakow, Artar, et al., 2000; Krakow, Germain, et al., 2000, 2002; Krakow, Haynes, et al., 2004; Krakow, Lowry, et al., 2000; Krakow, Melendrez, et al., 2001).

Thus far, IRT should be considered an effective treatment for nightmares compared to no treatment. Although it holds some potential as a brief treatment for PTSD more generally, much more caution is warranted here. First, not all studies have demonstrated reductions in PTSD symptoms. Second, the pre- to posttreatment reductions in PTSD symptoms, although statistically significant, are not as robust as those found in more established PTSD treatments (Foa et al., 1999; Resick, Nishith, & Griffin, 2003). Third, most consider sleep disturbance as part of the PTSD symptom picture; in fact, most effective treatments for PTSD also reduce nightmares and sleep disturbance. It would be very useful to compare the more established treatments for PTSD with IRT to help determine the most efficient and effective means of treatment and IRT's contribution.

POWER THERAPIES

The next group of newer treatments for PTSD include the so-called "power therapies" (Commons, 2000). These treatments are purported to be so much more powerful than traditional CBT that they require fewer sessions (in some cases, as few as one) to ameliorate symptoms dramatically. The first, "trauma incident reduction," or TIR (Gerbode, 1985), is purported to blend Rogerian and Freudian concepts to create a treatment that is superior to more direct exposure treatments such as PE. However, the actual treatment involves having clients repeatedly imagine their trauma. TIR proponents claim that the treatment differs from other exposure treatments by being more client-focused, restricting the therapist's role to basic imagery instructions and unconditional positive regard, increasing clients' positive emotions (rather than amelioration of negative emotions), and helping them develop more insight about the trauma. Claims are lofty, such as the following:

> In the great majority of cases, TIR correctly applied results in the complete and permanent elimination of PTSD symptomatology. It also provides valuable insights, which the viewer arrives at quite spontaneously, without any prompting from the facilitator and hence can "own" entirely as his own. By providing a means for completely confronting a painful incident, TIR can and does deliver the positive gain a person would have had if he had been able to fully confront the trauma at the time it occurred. (*healing-arts.org/tir/frametirfaq.htm*, September 2004)

Data for TIR are limited to uncontrolled case studies/case series (Carbonell & Figley, 1999; Figley & Carbonell, 1999), an unpublished dissertation (Bisbey, 1995), and a controlled trial (Valentine & Smith, 2001) that compared TIR to a wait-list control in a sample of inmates who reported trauma histories and experienced one or more symptoms associated with PTSD, depression, or anxiety. Compared to a no-contact wait-list control, TIR was significantly better at increasing self-efficacy and reducing depression, anxiety, and some PTSD symptoms at posttest and at a 3-month follow-up. It is highly doubtful, however, that TIR's efficacy results from anything unique about the treatment. It is much more likely that results simply stem from imaginal exposure to the trauma and cognitive restructuring, already well-established ingredients of PTSD treatments. At first glance, it appears noteworthy that TIR was apparently delivered in one session, albeit one 3–4 hours long. However, this becomes less impressive in light of the relatively mild pretreatment PTSD symptom severity, indicating that many in the sample likely did not meet full diagnostic criteria for PTSD or were only minimally symptomatic.

The second "power therapy" is visual kinesthetic dissassociation, or VK/D (Bandler & Grinder, 1979, Cameron-Bandler, 1978). Proponents of the treatment claim that it temporarily induces disassociation from negative emotions associated with traumatic memories through a "visual review of the events from a different perspective" (Commons, 2000, p. 1). This includes "directed meta-self-visualization" (Commons, 2000, p. 1), which instructs clients to imagine themselves in a traumatic scene, altering elements of the memory (movement, proximity, etc.) to promote the resolution and reduction of negative emotions. Afterwards, clients are "reassociated" and helped to maintain the learning that occurred during the dissociation phase (Gallo, 1996). Proponents of V/KD claim that the therapy differs from other imagery-based treatments by directly promoting this "outside-observer" perspective. Research is limited to a case series (Carbonell & Figley, 1999) and one small, multiple baseline study (Hossack & Bentall, 1996) that also included guided imagery and relaxation and did not provide strong support for VK/D. Despite these claims, it is unclear how VK/D differs in any substantial way from the processes involved in prolonged exposure. It is difficult to understand how the slight variation in instruction to the client and significant differences in underlying rationale/proposed mechanisms are meaningful. Thus, its unique contribution to the PTSD treatment literature is questionable.

Eye movement desensitization and reprocessing (EMDR), which has also been classified as a power therapy, was reviewed by Resick, Monson, and Gutner in Chapter 17, this volume. Finally, there is thought field therapy, or TFT, which has been developed and aggressively marketed for the last two decades by Callahan and Callahan (1997). The treatment involves imagining an anxiety-producing or traumatic experience, rating subjective discomfort, then tapping oneself on various body parts as directed by a therapist. These techniques are repeated if the patient does not experience a decrease in subjective anxiety. Callahan claims that TFT's efficacy lies in the use of circulatory fields, or "meridians," within the body. The procedure supposedly directs various "thought fields" in a way that eliminates the patient's symptoms ("perturbations") permanently. Unlike the aforementioned approaches, TFT can be used over the telephone, through audio- and videotapes, or to treat groups of people simultaneously (Callahan, 1985; Callahan & Callahan, 1997). Callahan also advertises his discovery of "codes" carried in the human voice that correspond to different perturbations in the thought field; practitioners who "wish to take this work to its highest level of effectiveness and have a sincere desire to decrease human suffering," and who are willing to pay $100,000 U.S. for the 3-day training and materials, can acquire his "voice technology training" (*www.tftrx.com/training/6advance.html#vt_list*, September 2004). The claims of efficacy go beyond treatment for PTSD, to addictions, weight problems, and even health conditions; for instance, one voice technology practitioner claims that

> during teaching in Dublin Ireland he [the practitioner] had his life saved by a TFT diagnostic practitioner, whilst suffering from an anaphylatic (extreme allergy attack) reaction to peanuts. The practitioner immediately brought back his breathing and his rapidly lowering blood pressure and all symptoms disappeared within minutes. ... (*www.tftrx.com/vt_milbank.html*, September 2004)

There is no controlled, scientific evidence to back these claims. Although there are books and manuals in multiple languages on TFT, we are aware of only uncontrolled case series/studies of the treatment, most of which have not been submitted to standard peer-review procedures. These have included an uncontrolled report of the use of TFT in a large outpatient service (Sakai et al., 2001), an uncontrolled report of the use of TFT in Kosovo, including teams traveling to remote villages (Johnson, Shala, Sejdijaj, Odell, & Dabishevci, 2001), an uncontrolled case series at a university-based demonstration (Figley & Carbonell, 1999), an unpublished dissertation (Wade, 1990), two studies involving callers to radio talk shows (Callahan, 1987), and some heavily touted but unsophisticated case reports using heart rate variability as an outcome measure (Callahan & Callahan, n.d.). All of the research fails to conform to minimal standards for clinical research and has been heavily criticized (Herbert & Gaudiano, 2001; Lohr, 2001; Rosen & Davidson, 2001; Rosner, 2001). The response from Callahan to the criticism is a sweeping rejection of control groups, tests of statistical significance, questionnaires, and concern about placebo effects (Callahan, 2001). He argues that control groups and statistical tests are not needed with a treatment as clearly powerful as TFT. Furthermore, he reasons that since "TFT is successful with horses, dogs, cats, infants, and very young children" (Callahan, 2001, p. 1255; he does not provide a basis for these claims) there is no validity to his critics' concern about demand characteristics and common factors that must be considered in any psychotherapy research. In summary, there is no convincing evidence for the theory of TFT. Indeed, based on some of its proponents' claims, it is scientifically untestable (Herbert & Gaudiano, 2001), and the existing data are too limited and methodologically flawed to interpret. TFT, V/KD and TIR do include a few scientifically valid components that have been rigorously tested in established treatments for PTSD, such as imaginal exposure to the trauma memory, common factors of psychotherapy, and cognitive restructuring. In the absence of any data supporting the hypothesized mechanism of action for any of these power therapies, it seems likely that any efficacy is due to the proven CBT components of each of these novel treatment approaches. Given the lack of empirical evidence to support the glowing testimonials of success put forth by power therapy proponents, we strongly discourage their use. It is of grave concern to us that in the presence of well-established treatments for PTSD, a treatment with no empirical support (e.g., TFT) has been used by a major health maintenance organization (HMO) such as Kaiser.

NEW TECHNOLOGIES

A few new treatments utilize technology to deliver PTSD treatments: The first one we review delivers treatment through the Internet, which is changing society in striking ways, including modes of psychological interventions. Providing treatment through the Internet has many pros and cons; the reader is referred to Tate and Zabinski (2004) for an excellent discussion of the relative advantages and challenges of different types of interventions. At least one Web-based treatment has been developed for PTSD and bereavement by Lange and colleagues (2000, 2003; Lange, van de Ven, Schrieken, & Emmelkamp, 2001) at the University of Amsterdam. The treatment, called "Interapy," has modified traditional CBT approaches and has four major foci: self-confrontation of traumatic memories (e.g., exposure), cognitive reappraisal, social sharing, and social support/empathy, all with evidence for their helpfulness in treating trauma (Lange, 1996; Rime, 1995).

As studied, Interapy extended for 5 weeks. Participants completed ten, 45-minute writing sessions twice weekly that were submitted to a website. Therapist reaction to these submissions was also provided through a protected website. Treatment was divided into three phases. After psychoeducation/rationale, Phase I included the first four writing sessions and "self-confrontation," in which participants were instructed to describe their traumatic event in detail, including their intimate fears and thoughts concerning the event. Therapists provided feedback on written submissions. For instance, a therapist might ask the participant to add more details about the trauma, more sensory detail, feeling, and so forth. Phase II focused on cognitive restructuring. Participants were asked to imagine that a hypothetical friend had gone through a situation similar to their own and to formulate advice, addressing issues such as what the friend might have learned from the trauma. Again, therapist feedback was given halfway through the phase, focusing on helping the participant to adopt a new view of the event and regain a sense of control. In this sense, then, the first two phases are similar to the exposure and cognitive restructuring components of CPT (Resick & Schnicke, 1992). Phase III, the "sharing and farewell ritual," began with psychoeducation about the positive effects of sharing. Participants were asked to take "symbolic leave" of their traumatic experience by writing a letter, either to themselves or to a significant other who had somehow been involved in the traumatic event. Feedback given in this phase focused on encouraging participants to write about the way the traumatic event changed them, and how they might cope with it now and in the future. These letters were not necessarily sent.

The largest trial of Interapy was conducted in a community sample with 69 participants, including 32 wait-list controls (Lange et al., 2003). Screening, informed consent, and data collection were all conducted online. The treated participants improved significantly more than those on the wait list (although there was some improvement in that group), with large effect sizes for both trauma-related symptoms and general psychopathology. On most outcome subscales, approximately 50% of the Interapy-treated group showed both reliable change and clinically significant improvement (as proposed by Jacobson & Truax, 1991). Investigation of mediating variables suggested that the treatment was most beneficial for those who had suffered an intentional trauma (e.g., abuse) and those who had not previously disclosed their traumatic experiences to significant others. Treatment gains appeared to be maintained at a 6-week follow-up.

The Interapy data are quite encouraging. It is exciting to consider the possibilities for reaching a much larger population needing treatment than previously feasible. Multiple studies have indicated that computerized assessments result in more accurate prediction and increased client disclosure (see Newman, 2004). As such, they may attract clients with high levels of stigma or shame, which prevents many individuals with PTSD from seeking treatment (Hoge et al., 2004). They may also increase access for people in underserved or rural areas or for those who have limited mobility, or they may simply be more convenient in terms of time. Given these data and the results of the Interapy trial, the idea of adapting PTSD treatments to the Web seems quite promising. Still, several areas require critiques and further study. First, generalizability is questionable. The exclusion criteria ruled out individuals with severe levels of psychopathology, and many of the "traumas" would not have met DSM criterion A (i.e., sudden loss of a loved one, loss of health/house/job, divorce, or other traumatic events within the family). On the other hand, as the authors point out, Impact of Event scale (IES) scores indicated that the sample was quite symptomatic, with means that Intrusion and Avoidance subscale scores were within the upper regions of the norm table for the Dutch PTSD patients, and 90% of the final participants scored above Dutch cutoff scores for PTSD. Still, conducting treatment over the Internet raises some interesting ethical dilemmas about treating patients with PTSD. Should the treatment be reserved for less severely traumatized populations? Do concerns inherent in any Internet treatment (concerns about confidentiality, urgent issues, and crises) outweigh the benefit of providing some treatment to greater percentages of people? Can computer-based treatments ever be as effective as traditional approaches? These are questions that the field will increasingly have to address. Encouragingly, recent reports suggest that despite fears about compromises to therapeutic alliance with Internet therapies, many studies find equal or higher rates of treatment initiation, satisfaction, and retention when traditional and technologically driven treatments are compared (Ghosh, Marks, & Carr, 1988; Newman, 2004).

A second technology-based approach that provides a new method for delivering exposure therapy, virtual reality exposure (VRE), has been utilized with Vietnam veterans with PTSD. Veterans are exposed to a computer-generated view of a "virtual Vietnam" that changes in real time with head motion. During VRE therapy sessions, patients wear a head-mounted display device that contains two television screens and stereo speakers that expose patients to both sights and sounds consistent with the Vietnam experience. Two scenarios have been created for VRE: a virtual Huey helicopter that flies over various Vietnam terrain scenarios (jungles, river, and rice paddies), and a virtual clearing (helicopter landing zone) surrounded by jungle. In all the environments, the patient experiences only computer-generated audio and visual stimuli, while "real-world" stimuli are shut out. The therapist communicates with the patient through a microphone connected to the computer and then to the headphones. While immersed in the virtual stimuli, patients are instructed to recall the details of their traumatic Vietnam memories, as in standard prolonged imaginal exposure. For the helicopter ride environment, the patient sits in a special chair that provides tactile stimuli via a bass speaker integrated in the chair. For the clearing environment, the patient stands on a raised platform surrounded by handrails on all sides. The patient "walks" in the environment by pushing a button on a hand-held joystick. Audio, head-tracking, and real-time graphics are computed on a PC. The VRE allows the addition of various stimuli under the therapist's control, such as audio effects (helicopter blades, gunfire, male voices shouting, "Move out! Move out!," and radio chatter and static), including directional sound and visual effects (e.g., night or daylight, fog, landing–taking off in the helicopter, helicopters flying nearby, flying over various terrain). The therapist is able to control the apparent closeness of the stimuli with the audio effects and volume control, and attempts to match in virtual reality the traumatic memory that the patient is describing.

Initial evaluation of this procedure indicates that it is a successful mode of treatment for PTSD (Rothbaum et al., 1999; Rothbaum, Hodges, Ready, Graap, & Alarcon, 2001). In the Rothbaum and colleagues (1999) study, 10 Vietnam veterans completed a course of VRE therapy using the two virtual reality war environments. After an average of thirteen 90-minute exposure therapy sessions delivered over 5 to 7 weeks, there was a significant reduction in PTSD and related symptoms. Clinician-rated PTSD symptoms as measured by the Clinician Administered PTSD Scale (CAPS), the primary outcome measure at 6-month follow-up indicated an overall statistically significant reduction from baseline in symptoms associated with specific reported traumatic experiences. Eight of the 10 participants at the 6-month follow-up reported reductions in PTSD symptoms ranging from 15 to 67%. Significant decreases were seen in all three symptom clusters. Patients' self-reported intrusion and avoidance symptoms, as measured by the IES, were significantly lower at 3 months than at baseline but not at 6 months, although there was a clear trend toward fewer intrusive thoughts and somewhat less avoidance. The main drawbacks to VRE therapy include (1) the limitations of the technology, (2) cost, and (3) acceptance. Regarding the technology, the virtual environment is limited to what it is programmed to include. Although it can be used differently based on individual needs, it may not be appropriate for most patients, because the elements may not match their trauma well enough. At present, the only virtual environment created specifically for PTSD is the virtual Vietnam, although the virtual airplane has been used clinically for patients with PTSD as a result of airplane accidents, and a virtual Iraq is under construction. Regarding cost, both the cost of creating the virtual environments and the costs for the practitioner limit wide availability. Finally, acceptance of the technology on the part of patients, practitioners, and institutions, such as VA hospitals, remains limited (this appears in large part related to the widespread lack of acceptance of exposure-based therapies, discussed by Resick et al., Chapter 17, this volume). Obviously, more research and controlled studies are needed to provide an acceptable evidence-base for VRE therapy for PTSD, because evidence so far is based on open clinical trials.

TREATMENTS CAPITALIZING ON SOCIAL SUPPORT

There are several reasons why marital and family and group approaches are of interest to clinicians and researchers seeking better treatments for PTSD. In a broad sense, the role of social support has a strong and clear influence on both the development of PTSD following trauma (for reviews, see Bailham & Joseph, 2003; Tedstone & Tarrier, 2003; Resick, 1993; Steketee & Foa, 1988) and recovery from PTSD (Koenen, Stellman,

Stellman, & Sommer, 2003; Mertin & Mohr, 2001), leading some clinicians to call for PTSD treatments focused more specifically on social support, such as family treatments, at least as adjuncts to CBT (Tarrier & Humphreys, 2003).

Family and Couple Therapy

Families and spouses, who clearly constitute a major source of social support (or lack thereof), may play a very important role in the development, maintenance, and impact of PTSD (Riggs, 2000; Solomon, Waysman, & Mikulincer, 1990). The features associated with PTSD also impact families (Figley, 1985; Riggs, 2000). Riggs (2000) has reviewed the two major philosophies and associated couple/family treatment for PTSD, which loosely adhere to the two following lines of reasoning: The first, which he calls "systemic treatments," try to remedy the systemic disruption to marriages and families caused by both PTSD symptoms and the trauma itself. Emphasis is on reducing stress and disruption to the family system as opposed to the individual's PTSD symptoms, and interventions are generally drawn from the family/couple therapy literature. The second group of support treatments attempt to help maximize the social support of the family to help the traumatized individual recover and typically emphasize psychoeducation and skills training. Unfortunately, there is a serious lack of evidence regarding these interventions, so no firm conclusions can be made. While several treatment approaches have been described (Erickson, 1989; Figley, 1985) such as systemic family therapy (Harris, 1991) and emotion-focused marital therapy (Johnson, 1989; Johnson & Williams-Keeler, 1998), there are no data for any of these treatments as they pertain to trauma survivors. In fact, only a few empirical examinations of couple's therapy for PTSD exist, two of them unpublished dissertation studies. Still, the controlled (Glynn et al., 1999[1]; Sweany, 1987) and uncontrolled (Cahoon, 1984) studies suggest that there may be some promise for couple's treatments in reducing marital distress and PTSD symptoms, but results have been far from strong. The most methodologically rigorous study (Glynn et al., 1999) found that there was no additional benefit of the family treatment following an effective treatment of direct therapeutic exposure to traumatic material.

A more recent systemic couple treatment designed specifically for PTSD is cognitive-behavioral couple therapy (CBCT) for PTSD (Monson, Schnurr, Stevens, & Guthrie, 2004). Drawing from previous CBCT trials that have successfully treated other Axis I disorders (Daiuto, Baucom, Epstein, & Dutton, 1998; Jacobson, Dobson, Fruzzetti, Schmaling, & Salusky, 1991; O'Farrell & Fals-Stewart, 2000), Monson and colleagues (2004) conducted a clinical trial of the treatment that met six out of seven of the Foa and Meadows (1997) gold standard criteria. Seven couples, at least one member of which met criteria for military-related PTSD, participated in the treatment. The study was uncontrolled, and indicated significant changes pre- to posttreatment on both CAPS and partner ratings of the veteran's PTSD symptoms, and veteran's self-reported ratings of depression and anxiety, as well as large effect sizes. However, veteran's self-reported PTSD symptoms and relationship satisfaction ratings did not change significantly.

Although the theoretical rationale for couple therapy for PTSD is quite strong, the data thus far are less strong. Whereas the Monson and colleagues (2004) trial showed

---

[1] This study examined behavioral family therapy, but 90% of the sample comprised domestic partners.

some promise, results were still mixed, in that veterans did not report improvement in their PTSD symptoms, but there is no way to rule out the potential role of secondary gains (some veterans may have felt they needed to report no improvement because of fear of losing their benefits). Furthermore, given the results of the Glynn and colleagues (1999) study any optimism must be tempered with caution. In the absence of further data, we agree with the conclusions of Riggs (2000), who recommended that whereas therapists should evaluate levels of family distress and need for treatment, couple treatment should be considered an adjunct to more established therapy for PTSD at this time.

Group Therapy

Group treatments for PTSD are not new; on the contrary, they are one of the most widespread modalities of trauma treatment, particularly in VA settings. Group therapy is also common in community mental health settings, where traumatized populations abound and resources are limited. There are dozens upon dozens of different group treatments; even a cursory literature search reveals a large number of group treatment descriptions for PTSD, ranging from trauma-focused groups for incarcerated juvenile offenders (McMackin, Leisen, Sattler, Krinsley, & Riggs, 2003) to supportive group therapy for adolescents with dissociative disorders (Brand, 1996), and a wide range in between. The problem is that group therapies for PTSD remain a woefully understudied phenomenon. In an excellent chapter, Foy and colleagues (2000) reviewed the three major applications of group therapy for PTSD—supportive group therapy, psychodynamic group therapy, and cognitive-behavioral group therapy—and the data to support each treatment. Although the data for group treatments initially appear promising, no single treatment emerges as the obvious choice. Very few studies (to date, two) have compared two or more active treatments, and both failed to find compelling differences. For instance, Resick, Jordan, Girelli, Hutter, & Marhoefer-Dvorak, (1988) found similarly positive effects when comparing stress inoculation training, assertion training, and supportive group therapy compared to each other and to a wait-list control for female sexual assault survivors; this may, however, been due to lack of sufficient power to detect treatment differences.

One of the more well-known group treatments for trauma, and especially for Vietnam veterans, is trauma-focused group therapy (TFGT), developed by Foy, Ruzek, Glynn, Riney, and Gusman (2003) for Vietnam veterans. The format is manualized and well organized, a step forward in the study of group treatments, and incorporates hypothesized active treatment components, including psychoeducation about PTSD, coping and relapse prevention skills, personal autobiography, prolonged exposure, cognitive restructuring, and group cohesion. Disappointingly, TFGT did not perform significantly better than a present-centered group treatment that avoided trauma focus, cognitive restructuring, and other TFGT components in a multisite VA cooperative study (Schnurr et al., 2003), although there was some evidence that TFGT, which had higher dropouts, performed slightly better than the control treatment in comparisons of adequate dose groups. The good news is that both treatments helped to decrease PTSD symptoms; the bad news is that we still lack substantive data for a group treatment of choice. It may be, as Schnurr and colleagues suggest, that adding a motivational enhancement component to TFGT may boost effects. Clearly, more research is needed on (1) determining what the active elements of group treatments are and (2) how (and indeed, whether it is possible) to adapt individual treatments with proven efficacy, such as prolonged exposure, to a group model. Cognitive processing therapy, for instance, has been applied successfully in both group and individual settings; more research on this topic would be extremely useful. For now, group treatment should be considered potentially effective based on the strength of the evidence. Future studies should evaluate whether the gains reported in studies of group treatments are equivalent to the more substantiated gains made in individual treatments for PTSD.

TREATMENTS TO ENHANCE TOLERANCE AND ACCEPTABILITY OF EXPOSURE TO TRAUMATIC MATERIAL

Although exposure therapy has demonstrated efficacy and is one of the best treatments for PTSD, there has been some concern that clients with poor emotion regulation skills may have trouble with the treatment. Concerns about these clients include the following: (1) Overwhelming anxiety may be detrimental to exposure treatment (Jaycox, Foa, & Morral, 1998); (2) although this is by no means established and remains quite controversial, some authors have suggested that certain clients may find it difficult to tolerate exposure, which may diminish treatment compliance and outcomes (Scott & Stradling, 1997; Tarrier & Humphreys, 2000); and (3) factors including difficulty with distress tolerance (particularly with anger and anxiety), a tendency to dissociate under stress, and problems in the therapeutic relationship have predicted less favorable outcomes in exposure therapy (Chemtob, Novaco, Hamada, Gross, & Smith, 1997; Cloitre & Koenen, 2001; Jaycox et al., 1998). Two treatments, dialectical behavior therapy and acceptance and commitment therapy, have been proposed to address some of these problems.

Dialectical Behavior Therapy

Dialectical behavior therapy (DBT), a behavioral treatment developed by Linehan (1993) for chronically suicidal individuals with borderline personality disorder (BPD), is based on the theory that the primary source of dysfunction in these individuals is one of emotion regulation (for more on the theory of DBT, see Linehan, 1993). The treatment adds a heavy emphasis on validation to traditional tenets such as behavioral analysis and skills building in the push toward change; thus, the primary *dialectic* in DBT is one of acceptance and change, which are balanced throughout the treatment. Traditional DBT combines several elements, among them weekly individual psychotherapy, in which the therapist uses a variety of acceptance and change-based strategies to increase motivation and commitment on the part of the client. Weekly skills groups teach clients the skills thought to be necessary to regulate strong affect, including mindfulness, emotion regulation, distress tolerance, and interpersonal effectiveness. DBT has demonstrated efficacy in treating BPD (for reviews, see Koerner & Dimeff, 2000; Koerner & Linehan, 2000) and is increasingly being applied to a broader range of individuals in whom emotion regulation is hypothesized to be a central area of dysfunction, such as people with eating disorders (Palmer et al., 2003; Telch, Agras, & Linehan, 2001; Wisniewski & Kelly, 2003) incarcerated men (McCann, Ball, & Ivanoff, 2000), suicidal adolescents (Katz, Cox, Gunasekara, & Miller, 2004), female juvenile offenders (Trupin, Steward, Beach, & Boesky, 2002), older adults with depression (Lynch, Morse, Mendelson, & Robins, 2003), and clients with BPD substance use disorder (Linehan et al., 1999, 2002; van den Bosch, Verheul, Schippers, & van den Brink, 2002). Because of the pattern of emotion dysregulation, treatment noncompliance, dropout, and avoidance in patients with PTSD, areas specifically targeted in DBT, there is growing interest in the application of the treatment for these patients. The approaches can generally be divided into two camps. First is DBT to treat "Stage II" clients with BPD, who have successfully undergone treatment for severe behavior dyscontrol (Stage I) but still experience significant problems with emotion regulation and experience. Linehan termed this the stage of "quiet desperation" (Linehan, 1993; Wagner & Linehan, 2006). Because many individuals with BPD have histories of trauma and meet criteria for PTSD, trauma-focused exposure therapy would take place once the client had enough skills to tolerate it.

Wagner and Linehan (2006) have recently written on the topic of when a client would be recommended for Stage II DBT as opposed to another efficacious PTSD treatment. They suggest that individuals with (1) significant self-invalidation (see Linehan, 1993, for a full discussion of the role of invalidation in BPD), (2) severe emotion dysregulation, and (3) recent histories of dysfunctional behaviors related to emotion dysregulation would be recommended for Stage II DBT. They note that because Stage II DBT is designed to treat a full range of factors maintaining emotional reactivity to trauma and other topics, it may also be indicated for individuals whose problems with emotional experiencing cannot adequately be conceptualized with current theories of acute PTSD (e.g., classically conditioned fear responses maintained by avoidance and faulty cognitions). Stage II DBT would incorporate the same factors as Stage I DBT but might increase use of formal exposure, informal exposure, and response prevention. For instance, a client who dissociates when presented with trauma-related cues, such as an angry voice, might practice staying alert and aware of the present moment, while the therapist speaks in an increasingly angry voice. The goal is to prevent or block dissociation should it occur (see Wagner & Linehan, 2006). To our knowledge, no data exist on using Stage II DBT for PTSD.

The second type of application of DBT to PTSD is more common and somewhat more studied. Here, DBT is conceptualized as a helpful adjunct or pre-PTSD treatment therapy to help clients increase their ability to regulate strong affect before or during more traditional exposure (see Becker & Zayfert, 2001, for an excellent discussion). Becker and Zayfert (2001) have also made the point that the skills may be equally useful in giving *therapists* the skills and confidence they need to deliver exposure treatment effectively, because helping the client find a balance between over- and underengagement is not a simple task.

Two uncontrolled treatment studies, one at a VA hospital and the other in a partial hospitalization program, have reported anecdotally promising results using DBT with their patient populations with severe PTSD (Simpson et al., 1998; Spoont, Sayer, Thuras, Erbes, & Winston, 2003), and two controlled studies have been published. Cloitre, Koenen, Cohen, and Han (2002) used DBT emotion regulation and interpersonal skills to inform a two-phase treatment for patients with PTSD and childhood sexual abuse histories (this approach also included therapeutic work on cognitive schemas, which is not generally part of DBT). Called STAIR (skills training in affective and interpersonal regulation), the Phase I treatment included eight sessions that comprise psychoeducation, skills acquisition, application/practice, and homework between sessions. Phase II was modified prolonged exposure (PE) in which the *in vivo* element was eliminated, and several components were added that focus on postexposure coping, emotion regulation, cognitive restructuring, and skills review. Compared to a control group that had 12 weeks of weekly, 15-minute phone contact with an assessor, patients in the two-phase treatment improved significantly on affect regulation problems and interpersonal problems in Phase I and PTSD symptoms in the PE phase. Gains were maintained at both 3- and 9-month follow-up. A somewhat similar approach was taken by Bradley and Follingstad (2003), who also used some DBT skills in a two-phase treatment. Nine sessions combined psychoeducation about interpersonal victimization and affect regulation, primarily identifying and naming emotions and environmental cues, and breathing exercises. This was followed by nine sessions of structured writing assignments about life experiences, including interpersonal victimization. Comparison to a no-contact control group revealed significant decreases in PTSD symptoms, and mood and interpersonal problems. A high dropout rate during the writing phase led the authors to hypothesize that more skills training prior to the writing component might have been helpful.

Studies like these have been criticized on several grounds. For instance, Cahill, Zoellner, Feeny, and Riggs (2004) point out that it is premature to conclude that STAIR facilitated gains made with the exposure using a wait-list control, because a STAIR + exposure and exposure only comparison would be necessary to make this determination. They also note that the data preclude the conclusion that exposure exacerbates symptoms or dropout more than other treatments, or that victims of CSA are more likely to experience problems with exposure than other people with PTSD. Cloitre and colleagues are currently conducting a study comparing STAIR + modified PE with two groups: supportive counseling + modified PE and supportive counseling + STAIR. The results of this study will be a most welcome contribution to the literature. Until then, the treatment should be considered safe and efficacious; its utility over and above PE is unknown and needs to be established before widespread use of the treatment is recommended. Furthermore, to learn who might benefit from an amalgamated treatment, we need more studies to determine what characteristics predict an overengaged or dysregulated response to exposure treatment.

DBT may provide clinicians treating PTSD with a set of well-manualized and tested skills to help their clients tolerate exposure treatments. It may, as Becker and Zayfert (2001) put it, keep them from having to reinvent the wheel. Given DBT's efficacy in reducing dropout in highly avoidant BPD populations, some of the techniques used in the individual therapy component may prove to be very helpful. The Cloitre study reported a dropout rate of 29% in the STAIR group. Whether this offers protection against dropout is debatable; it is better than the dropout rate in the only other PTSD treatment study composed exclusively of CSA survivors (41%, McDonagh-Coyle et al., 2000) but not an improvement over averages of 20.6% in 12 trials of exposure therapy, 22.1% for either cognitive therapy or stress inoculation training alone, and 26% over 12 trials of exposure therapy combined with cognitive therapy or stress inoculation training (Hembree et al., 2003). Based on these data, Hembree and colleagues point out that adding complexity to treatment may be worse in terms of dropout, not better, and this may apply to additive DBT skills as well. Foa and colleagues (2003) concluded that adding other components did not increase the efficacy of well-delivered PE, although most of the studies were underpowered to detect differences between two or more active treatments. However, the research is clearly in its infancy. Which components of the treatment are crucial, at what level of intensity, at what points in treatment, and for which clients, are questions that do not yet have answers. Neither of the aforementioned studies used the distress tolerance or mindfulness skills that might be expected to greatly assist patients with PTSD before or during PE treatments. The question of whether the skills might help therapists feel that, in addition to intentionally inducing distress in their clients with exposure, they are also giving them the skills to cope with it is an excellent one, because it often seems that the biggest impediment to the use of exposure is the therapist's apprehension about this technique.

Acceptance and Commitment Therapy

Acceptance and commitment therapy, or ACT, another treatment that may hold promise for treating PTSD, is a behavior therapy based in "functional contextualism" and relational frame theory. ACT therapists view the attempt to avoid or change private experience (e.g., painful thoughts, memories, emotions, etc.) as the chief determinant of psychopathology, as opposed to the experiences themselves (the interested reader is referred to Hayes, Strosahl, and Wilson [1999] for more on the theoretical background). The treatment aims to help clients become more accepting and willing to experience painful private experiences. It also advocates a systematic formulation of clients' goals based on their individual value systems, as opposed to symptom reduction per se. Clients are encouraged to commit to useful behavioral change in the service of these goals, even when the behavior is emotionally difficult and results in previously avoided experiences.

Given the hesitancy of some therapists to use exposure therapy, Orsillo and Batten (2005) have suggested that ACT may be particularly useful in PTSD treatment for clients who refuse exposure therapy. They propose that ACT may be helpful based on several grounds. First, there is ample evidence that avoidance and emotional numbing are key features of PTSD. These phenomena may represent another means of avoidance through control/escape from intense negative emotions when effortful avoidance is not successful (Foa, Riggs, Massie, & Yarczower, 1995). Furthermore, the kinds of avoidance strategies often employed by patients with PTSD, such as thought or emotional suppression and wishful thinking (Amir et al., 1997), actually increase the frequency of the avoided events, distress associated with them, and memory impairment (Gross, 2002; Gross & John, 2003; Roemer & Borkovec, 1994; Shipherd & Beck, 1999), and have been specifically linked to PTSD symptomatology (Clohessy & Ehlers, 1999; Valentiner, Foa, Riggs, & Gershuny, 1996).

Avoidance of painful internal stimuli, such as thoughts of trauma, is directly targeted by ACT. However, the approach might be more gradual than that in PE. ACT prescribes beginning treatment with techniques designed to instill a sense of "creative hopelessness" in a client. This refers to the sense that his or her previous and seemingly rational attempts to decrease suffering, for example, by avoiding painful memories, have not and will not work. Ideally, this sets the stage for working with clients to help them increase their acceptance of previously avoided stimuli. As illustrated in their case example (see Orsillo & Batten, 2005), this might begin with a client attempting to go to work, if this had been identified by the client as a core value, even when feeling anxious and depressed.

Exposure to traumatic memories might be conducted later in therapy, but the emphasis would be on (1) doing so to achieve the client's personal goal, if it becomes clear that the avoidance of traumatic thoughts is preventing the goal, and (2) demonstrating to clients that their struggle with thoughts and feelings is the primary problem, rather than the thoughts and feelings themselves. Reduction of PTSD symptoms, then, is not the explicit goal. Orsillo and Batten (2005) have suggested that this focus on the client's overall quality of life may be particularly relevant in attracting and maintaining this avoidant group of patients in treatment. They also suggest that ACT may be helpful because of its focus on other emotions besides fear and anxiety, which are the primary focus of PE, and also its more detailed description of multiple strategies to help fearful and avoidant clients engage in the difficult work of exposure and emotional processing.

Direct evidence for the efficacy of ACT for PTSD is limited, although it has been shown to be effective with several other disorders (see Batten & Hayes, 2005). There have been a few preliminary reports on the feasibility and acceptability of ACT in an inpatient PTSD treatment setting (Walser, Loew, Westrup, Gregg, & Rogers, 2002; Walser, Westrup, Rogers, Gregg, & Loew, 2003), where patients rated ACT as helpful and applicable to their lives. In these early reports, the presence of automatic thoughts did not change from pre- to posttreatment assessments; however, both the believability of the thoughts and the intensity of depressive symptoms showed significant improvement from pre- to postassessment. The efficacy of ACT as a replacement, adjunct, or precursor to exposure treatment is not established and is subject to similar criticisms such as those described earlier in the discussion of DBT. It is also unclear whether the distinction between decreasing efforts to avoid trauma-related stimuli and efforts to avoid symptoms themselves results in better outcomes. Again, controlled studies are warranted.

GENERALIZABILITY OF CURRENT FINDINGS

As can be seen in Tables 23.1 and 23.2, extreme caution must be taken when considering the generalizability of these treatments. New treatments for families, couples, and groups have been primarily studied in veteran populations. DBT-augmented exposure treatments have shown some promise in females with PTSD related to childhood abuse, but the generalizability of the STAIR treatment may be limited. Although it was designed for a more severe and more fragile PTSD population, many comorbidities that would typically appear in clinical settings were excluded. Of the imagery-based treatments, IRT is by far the best studied. It appears to be generalizable to most types of PTSD for sleep-related problems; however, as discussed earlier, its efficacy as a treatment for PTSD more generally needs further examination. The data on IR are much more limited; although it makes good conceptual sense to use IR in place of traditional prolonged exposure, the data preclude firm conclusions. VRE for PTSD can be generalized to veterans only at this time. The "power therapies" make broad claims of generalizability to nearly every psychological disorder, including PTSD. However, as reviewed earlier, the data are so fraught with methodological problems that their efficacy cannot be assumed with any population. We recommend extreme caution against the wide dissemination of any therapy that has not received strong empirical support.

CONCLUSION

Although we certainly have room for improvement in the treatment of PTSD and trauma-related disorders, we also have much to celebrate. Elsewhere in this volume, very successful psychotherapeutic and pharmacological interventions have been reviewed, with strong evidence from several well-controlled studies. In this chapter, we have reviewed less well researched although sometimes promising emerging therapies for PTSD and related disorders. In light of the review, we offer several recommendations for the future. First, we need to focus our efforts as a field. We do not need to discover an effective treatment for PTSD; this has already been done. We do need to determine more specifically where these treatments predictably fail, where they need improvement, and for whom they do not work. Some of this work has already begun, and it is within this context that these newer treatments seem to hold the most promise. Virtual realities to enhance or improve exposure treatments, imagery techniques for clients who do not do well in exposure, and Internet-based treatment to increase access are logical steps to move the field forward. Attempts to propagate new treatments without data to support their efficacy, such as the so-called "power therapies," only serve to confuse the public and divide the field. Second, we need more research to determine the need for treatments such as DBT and ACT as adjuncts to traditional exposure treatment. Data indicate that exposure-based treatments do not detrimentally exacerbate symptoms (Foa, Zoellner, Feeny, Hembree, & Alvarez-Conrad, 2002) and are well tolerated by most clients, including those with complex PTSD, multiple traumas, and features of BPD, among other complications (Feeny, Zoellner, & Foa, 2002; Hembree et al., 2003; Resick et al., 2003; Rothbaum & Schwartz, 2002). Still, widespread concerns about exposure techniques remain in the clinical community, and most clinicians are simply not using them. This is probably the greatest challenge to the field—disseminating effective treatments and getting community clinicians to deliver the treatments that have been shown most effective. There are some holes in the data that do need to be filled; for instance, the study on patients with BPD (Feeny et al., 2002) excluded those who were acutely parasuicidal, so whether such patients can tolerate exposure remains an empirical question. There are certainly clients who refuse exposure treatments, and the data indicate that there is an optimal level of arousal that many patients have difficulty maintaining. We need to find out whether it is useful to augment therapist skills with techniques derived from other treatments. More data will certainly help focus our efforts. Third, we need more studies of group therapy. It is likely that even without a great deal of data, economic realities will result in the widespread use of group treatments, especially in the VA system. Given the widespread use and paucity of data, this needs to be a priority for future research. Studies randomizing patients to group versus individual treatments would also be extremely useful. More research and development of other ways to capitalize on the role of social support, a clear predictor for the course of PTSD, would also be very useful. Finally, a major focus for the future is to discover interventions immediately following exposure to trauma to prevent the development of chronic disturbances.

REFERENCES

Amir, M., Kaplan, Z., Efroni, R., Levine, Y., Benjamin, J., & Kotler, M. (1997). Coping styles in post-traumatic stress disorder (PTSD) patients. *Personality and Individual Differences, 23,* 399–405.

Bailham, D., & Joseph, S. (2003). Post-traumatic stress following childbirth: A review of the emerging literature and directions for research and practice. *Psychology Health and Medicine, 8*(2), 159–168.

Bandler, R., & Grinder, J. (1979). Frogs into princes: Neuro-linguistic programming. Moab, UT: Real People Press.

Batten, S. V., & Hayes, S. C. (2005). Acceptance and commitment therapy in the treatment of comorbid substance abuse and posttraumatic stress disorder: A case study. *Clinical Case Studies.*

Becker, C., & Zayfert, C. (2001). Integrating DBT-based techniques and concepts to facilitate exposure treatment for PTSD. *Cognitive and Behavioral Practice, 8,* 107–122.

Bisbey, L. B. (1995). No longer a victim: A treatment outcome study for crime victims with post-traumatic stress disorder. *Dissertation Abstracts International, Section B: The Sciences and Engineering, 56*(3-B), 1692.

Bradley, R. G., & Follingstad, D. R. (2003). Group therapy for incarcerated women who experienced interpersonal violence: A pilot study. *Journal of Traumatic Stress, 16*(4), 337–340.

Brand, B. (1996). Supportive group psychotherapy for adolescents with dissociative disorders. In J. L. Silberg (Ed.), *The dissociative child: Diagnosis, treatment, and management* (pp. 219–234). Baltimore: Sidran Press.

Cahill, S. P., Zoellner, L. A., Feeny, N. C., & Riggs, D. S. (2004). Sequential treatment for child abuse related posttraumatic stress disorder: Methodological comment on Cloitre, Koenen, Cohen, and Han (2002). *Journal of Consulting and Clinical Psychology, 72,* 543–548.

Cahoon, E. P. (1984). *An examination of the relationship between post-traumatic stress disorder, marital distress, and response to therapy by Vietnam veterans.* Unpublished doctoral dissertation, University of Connecticut, Storrs.

Callahan, J., & Callahan, R. (n.d.). Pre and post HRV measurements: Case studies. Available at *www.tftrx.com/ref.php?art_id=102&art_catid=1*

Callahan, R. (1987). *Successful psychotherapy by telephone and radio.* Presented at the International College of Applied Kinesiology. Proprietary archive, as cited in J. D. Herbert & B. A. Gaudiano (2001). The search for the Holy Grail: Heart rate variability and thought field therapy. *Journal of Clinical Psychology, 57*(10), 1207–1214.

Callahan, R. (1995). *Five minute phobia cure.* Wilmington, DE: Enterprise.

Callahan, R. J. (2001). Thought field therapy: Response to our critics and a scrutiny of some old ideas of social science. *Journal of Clinical Psychology, 57*(10), 1251–1260.

Callahan, R. J., & Callahan, J. (1997). Thought field therapy: Aiding the bereavement process. In C. Figley, B. Bride, & N. Mazza (Eds.), *Death and trauma: The traumatology of grieving* (pp. 249–267). Washington, DC: Taylor & Francis.

Cameron-Bandler, L. (1978). *They lived happily ever after.* Cupertino, CA: Meta.

Carbonell, J. L., & Figley, C. (1999). A systematic clinical demonstration of promising PTSD treatment approaches. *Traumatology, 5*(1). Available at *www.fsu.edu/~trauma/promising.html*

Chemtob, C. M., Novaco, R. W., Hamada, R. S., Gross, D. M., & Smith, G. (1997). Anger regulation deficits in combat-related posttraumatic stress disorder. *Journal of Traumatic Stress, 10,* 17–35.

Clohessy, S., & Ehlers, A. (1999). PTSD symptoms, response to intrusive memories and coping in ambulance service workers. *British Journal of Clinical Psychology, 38,* 251–265.

Cloitre, M., & Koenen, K. (2001). Interpersonal group process treatment for CSA-related PTSD: A comparison study of the impact of borderline personality disorder on outcome. *International Journal of Group Psychotherapy, 51,* 379–398.

Cloitre, M., Koenen, K. C., Cohen, L. R., & Han, H. (2002). Skills training in affective and interpersonal regulation followed by exposure: A phase-based treatment for PTSD related to childhood abuse. *Journal of Consulting and Clinical Psychology, 70*(5), 1067–1074.

Commons, M. L. (2000). The power therapies: A proposed mechanism for their action and suggestions for future empirical validation. *Traumatology, 6*(2).

Daiuto, A. D., Baucom, D. H., Epstein, N., & Dutton, S. S. (1998). The application of behavioral couples therapy to the assessment and treatment of agoraphobia: Implications of empirical research. *Clinical Psychology Review, 18,* 663–687.

Ehlers, A., & Clark, D. M. (2000). A cognitive model of posttraumatic stress disorder. *Behaviour Research and Therapy, 38,* 319–345.

Erickson, C. A. (1989). Rape and the family. In C. R. Figley (Ed.), *Treating stress in families* (pp. 257–289). New York: Brunner/Mazel.

Feeny, N. C., Zoellner, L. A., & Foa, E. B. (2002). Treatment outcome for chronic PTSD among female assault victims with borderline personality characteristics: A preliminary examination. *Journal of Personality Disorders, 16*(1), 30–40.

Figley, C. R. (1985). From victim to survivor: Social responsibility in the wake of catastrophe. In C. R. Figley, (Ed.), *Trauma and its wake: Volume II. The study and treatment of post-traumatic disorder* (pp. 39–54). New York: Brunner/Mazel.

Figley, C. R., & Carbonell, J. (1999). Promising treatment approaches. *Electronic Journal of Traumatology, 5*(1). Available at *www.fsu.edu/~trauma/promising.html*

Foa, E. B., Dancu, C. V., Hembree, E. A., Jaycox, L. H., Meadows, E. A., & Street, G. P. (1999). The efficacy of exposure therapy, stress inoculation training and their combination in ameliorating PTSD for female victims of assault. *Journal of Consulting and Clinical Psychology, 67,* 194–200.

Foa, E. B., & Meadows, E. A. (1997). Psychosocial treatments for posttraumatic stress disorder: A critical review. *Annual Review of Psychology, 48,* 449–480.

Foa, E. B., Riggs, D. S., Massie, E. D., & Yarczower, M. (1995). The impact of fear activation and anger on the efficacy of exposure treatment for posttraumatic stress disorder. *Behavior Therapy, 26,* 487–499.

Foa, E. B., Rothbaum, B. O., & Furr, J. M.(2003). Is the efficacy of exposure therapy for posttraumatic stress disorder augmented with the addition of other cognitive behavior therapy procedures? *Psychiatric Annals, 33*(1), 47–53.

Foa, E. B., Zoellner, L. A., Feeny, N. C., Hembree, E. A., & Alvarez-Conrad, J. (2002). Does imaginal exposure exacerbate PTSD symptoms? *Journal of Consulting and Clinical Psychology, 70*(4), 1022–1028.

Forbes, D., Phelps, A. J., & McHugh, A. F. (2001). Imagery rehearsal in the treatment of posttraumatic nightmares in combat-related PTSD. *Behaviour Research and Therapy, 39,* 977–986.

Forbes, D., Phelps, A. J., McHugh, A. F., Debenham, P., Hopwood, M., & Creamer, M. (2003). Imagery rehearsal in the treatment of posttraumatic nightmares in Australian veterans with chronic combat-related PTSD: 12-month follow-up data. *Journal of Traumatic Stress, 16*(5), 509–513.

Foy, D. F., Glynn, S. M., Schnurr, P. P., Jankowski, M. K., Wattenberg, M. S., Weiss, D. S., et al. (2000). Group therapy. In E. B. Foa, T. M. Keane, & M. J. Friedman (Eds.), *Effective treatments for PTSD: Practice guidelines from the International Society for Traumatic Stress Studies* (pp. 336–338). New York: Guilford Press.

Foy, D. W., Ruzek, J. I., Glynn, S. M., Riney, S. J., & Gusman, F. D. (2002). Trauma-focused group therapy for combat-related PTSD: An update. *Journal of Clinical Psychology, 58*(8), 907–918.

Gallo, F. P. (1996). Reflections on active ingredients in efficient treatments of PTSD, part 1. *Traumatology, 2*(1).

Gerbode, F. (1985). *Beyond psychology: An introduction to meta-psychology.* Palo Alto, CA: IRM Press.

Ghosh, A., Marks, I. M., & Carr, A. C. (1988). Therapist contact and outcome of self-exposure treatment for phobias: A controlled study. *British Journal of Psychiatry, 152,* 234–238.

Glynn, S. M., Eth, S., Randolph, E. T., Foy, D. W., Urbaitis, M., Boxer, L., et al. (1999). A test of behavioral family therapy to augment exposure for combat-related posttraumatic stress disorder. *Journal of Consulting and Clinical Psychology, 67,* 243–251.

Gross, J. J. (2002). Emotion regulation: Affective, cognitive, and social consequences. *Psychophysiology, 39,* 281–291.

Gross, J. J., & John, O. P. (2003). Individual differences in two emotion regulation processes: Implications for affect, relationships, and well-being. *Journal of Personality and Social Psychology, 85,* 348–362.

Grunert, B. K., Smucker, M. R., Weis, J. M., & Rusch, M. D. (2003). When prolonged exposure fails: Adding an imagery-based cognitive restructuring component in the treatment of industrial accident victims suffering from PTSD. *Cognitive and Behavioral Practice, 10*(4), 333–346.

Hayes, S. C., Strosahl, K. D., & Wilson, K. G. (1999). *Acceptance and commitment therapy: An experiential approach to behavior change.* New York: Guilford Press.

Hembree, E. A., Foa, E. B., Dorfan, N. M., Street, G. P., Kowalski, J., & Tu, X. (2003). Do patients drop out prematurely from exposure therapy for PTSD? *Journal of Traumatic Stress, 16,* 555–562.

Herbert, J. D., & Gaudiano, B. A. (2001). The search for the Holy Grail: Heart rate variability and thought field therapy. *Journal of Clinical Psychology, 57*(10), 1207–1214.

Hoge, C. W., Castro, C. A., Messer, S. C., McGurk, D., Cotting, D. I., & Koffman, R. L. (2004). Combat duty in Iraq and Afghanistan, mental health problems, and barriers to care. *New England Journal of Medicine, 35*(1), 13–22.

Hossack, A., & Bentall, R. P. (1996). Elimination of posttraumatic symptomatology by relaxation and visual-kinesthetic dissociation. *Journal of Traumatic Stress, 9*(1), 99–111

Jacobson, N. S., Dobson, K., Fruzzetti, A. E., Schmaling, K. B., & Salusky, S. (1991). Marital therapy as a treatment for depression. *Journal of Consulting and Clinical Psychology, 59,* 547–557.

Jacobson, N. S., & Truax, P. (1991). Clinical significance: A statistical approach to meaningful change in psychotherapy research. *Journal of Consulting and Clinical Psychology, 59,* 12–19.

Jaycox, L. H., Foa, E. B., & Morral, A. T. (1998). Influence of emotional engagement and habituation on exposure therapy for PTSD. *Journal of Consulting and Clinical Psychology, 66,* 185–192.

Johnson, C., Shala, M., Sejdijaj, X., Odell, R., & Dabishevci, K. (2001). Thought field therapy—soothing the bad moments of Kosovo. *Journal of Clinical Psychology, 57*(10), 1237-1240.

Johnson, S. M. (1989). Integrating marital and individual therapy for incest survivors: A case study. *Psychotherapy, 21*(6), 96-103.

Johnson, S. M., & Williams-Keeler, L. (1998). Creating healing relationships for couples dealing with trauma: The use of emotionally focused marital therapy. *Journal of Marital and Family Therapy, 24,* 25-40.

Katz, L. Y., Cox, B. J., Gunasekara, S., & Miller, A. L. (2004). Feasibility of dialectical behavior therapy for suicidal adolescent inpatients. *Journal of the American Academy of Child Psychiatry, 43*(3), 276-282.

Koenen, K. C., Stellman, J. M., Stellman, S. D., & Sommer, J. F. (2003). Risk factors for course of posttraumatic stress disorder among Vietnam veterans: A 14-year follow-up of American Legionnaires. *Journal of Consulting and Clinical Psychology, 71*(6), 980-986.

Koerner, K., & Dimeff, L. A. (2000). Further data on dialectical behavior therapy. *Clinical Psychology: Science and Practice, 7*(1), 104-112.

Koerner, K., & Linehan, M. (2000). Research on dialectical behavior therapy for patients with borderline personality disorder. *Psychiatric Clinics of North America, 23*(1), 151-167.

Krakow, B., Artar, A., Warner, T. D., Melendrez, D., Johnston, L., Hollifield, M., et al. (2000). Sleep disorder, depression, and suicidality in female sexual assault survivors. *Crisis, 21*(4), 163-170.

Krakow, B., Germain, A., Tandberg, D., Koss, M., Schrader, R., Hollifield, M., et al. (2000). Sleep breathing and sleep movement disorders masquerading as insomnia in sexual assault survivors with PTSD. *Comprehensive Psychiatry, 41,* 49-56.

Krakow, B., Germain, A., Warner, T., Schrader, R., Koss, M., Hollifield, M., et al. (2002). The relationship of sleep quality and posttraumatic stress to potential sleep disorders in sexual assault survivors with nightmares, insomnia and PTSD. *Journal of Traumatic Stress, 14*(4), 647-665.

Krakow, B., Haynes, P. L., Warner, T. D., Santana, E., Melendrez, D., Johnston, L., et al. (2004). Nightmares, insomnia, and sleep-disordered breathing in fire evacuees seeking treatment for posttraumatic sleep disturbance. *Journal of Traumatic Stress, 17*(3), 257-268.

Krakow, B., Hollifield, M., Johnston, L., Koss, M., Schrader, R., Warner, T. D., et al. (2001). Imagery rehearsal for chronic nightmares in sexual assault survivors with posttraumatic stress disorder: A randomized trial. *Journal of the American Medical Association, 286,* 537-545.

Krakow, B., Hollifield, M., Schrader, R., Koss, M., Tandberg, D., Lauriello, J., et al. (2000). A controlled study of imagery rehearsal for chronic nightmares in sexual assault survivors with PTSD: A preliminary report. *Journal of Traumatic Stress, 13*(4), 589-609.

Krakow, B., Johnston, L., Melendrez, D., Hollifield, M., Warner, T. D., Chavez-Kennedy, D., et al. (2001). A open-label trial of evidence-based cognitive behavior therapy for nightmares and insomnia in crime victims with PTSD. *American Journal of Psychiatry, 158*(12), 2043-2047.

Krakow, B., Lowry, C., Germain, A., Gaddy, L., Hollifield, M., Koss, M., et al. (2000). A retrospective study on improvements in nightmares and posttraumatic stress disorder following treatment for co-morbid sleep-disordered breathing, *Journal of Psychosomatic Research, 49*(5), 291-298.

Krakow, B., Melendrez, D., Johnston, L., Clark, J., Santana, E., Warner, T., et al. (2002). Sleep dynamic therapy for Cerro Grande fire evacuees with posttraumatic stress symptoms: A preliminary report. *Journal of Clinical Psychiatry, 63*(8), 673-684.

Krakow, B., Melendrez, D., Pedersen, B., Johnston, L., Hollifield, M., Germain, A., et al. (2001). Complex insomnia: Insomnia and sleep-disordered breathing in a consecutive series of crime victims with nightmares and PTSD. *Biological Psychiatry, 49*(11), 948-953.

Krakow, B., Sandoval, D., Schrader, R., Keuhne, B., McBride, L., Yau, C. L., et al. (2001). Treatment of chronic nightmares in adjudicated adolescent girls in a residential facility. *Journal of Adolescent Health, 29,* 94-100.

Lange, A. (1996). Using writing assignments with families managing legacies of extreme traumas. *Journal of Family Therapy, 18,* 375-388.

Lange, A., Rietdijk, D., Hudcovicova, M., van de Ven, J. P., Schrieken, B., & Emmelkamp, P. M. G. (2003). Interapy: A controlled randomized trial of the standardized treatment of posttraumatic stress through the Internet. *Journal of Consulting and Clinical Psychology, 71*(5), 901-909.

Lange, A., Schrieken, B., Van de Ven, J.-P., Bredeweg, B., Emmelkamp, P. M. G., van der Kolk, J., et al.

(2000). "Interapy": The effects of a short protocolled treatment of posttraumatic stress and pathological grief through the Internet. *Behavioral and Cognitive Psychotherapy, 28*, 175–192.

Lange, A., van de Ven, J. P., Schrieken, B., & Emmelkamp, P. M. G. (2001). Interapy treatment of posttraumatic stress through the internet: A controlled trial. *Journal of Behavior Therapy and Experimental Psychiatry, 32*, 73–90.

Linehan, M. M. (1993). *Cognitive-behavioral treatment for borderline personality disorder.* New York: Guilford Press.

Linehan, M. M., Dimeff, L. A., Reynolds, S. K., Comtois, K. A., Welch, S. S., Heagerty, P., et al. (2002). Dialectical behavior therapy versus comprehensive validation therapy plus 12-step for the treatment of opiod dependent women meeting criteria for borderline personality disorder. *Drug and Alcohol Dependence, 67*(1), 13–26.

Linehan, M. M., Schmidt, H., Dimeff, L. A., Craft, C. J., Kanter, J., & Comtois, K. A. (1999). Dialectical behavior therapy for patients with borderline personality disorder and drug-dependence. *American Journal on Addictions, 8*(4), 279–292.

Lohr, J. M. (2001). Sakai et al. is not an adequate demonstration of TFT effectiveness. *Journal of Clinical Psychology, 57*(10), 1229–1235.

Lynch, T. R., Morse, J. Q., Mendelson, T., & Robins, C. J. (2003). Dialectical behavior therapy for depressed older adults: A randomized pilot study. *American Journal of Geriatric Psychiatry, 11*(1), 33–45.

McCann, R. A., Ball, E. M., & Ivanoff, A. (2000). DBT with an inpatient forensic population: The CMHIP forensic model. *Cognitive and Behavioral Practice, 7*(4), 447–456.

McDonagh-Coyle, A., Friedman, M. J., McHugo, G., Ford, J., Mueser, K., Demment, C. C., et al. (2000, November). Cognitive restructuring and exposure treatment for CSA survivors with PTSD. In M. Cloitre (Chair), *Empirically based treatments for childhood abuse and the multiply traumatized.* Symposium conducted at the meeting of the International Society for Traumatic Stress Studies, Miami, FL.

McMackin, R. A., Leisen, M. B., Sattler, L., Krinsley, K., & Riggs, D. S. (2002). Preliminary development of trauma-focused treatment groups for incarcerated juvenile offenders. In R. Greenwalk (Ed.), *Trauma and juvenile delinquency: Theory, research, and interventions* (pp. 175–199). Binghamton, NY: Haworth Maltreatment and Trauma Press/Haworth Press.

Mertin, P., & Mohr, P. B. (2001). A follow-up study of posttraumatic stress disorder, anxiety, and depression in Australian victims of domestic violence. *Violence and Victims, 16*(6), 645–654.

Monson, C. M., Schnurr, P. P., Stevens, S. P., & Guthrie, K. A. (2004). Cognitive-behavioral couple's treatment for posttraumatic stress disorder: Initial findings. *Journal of Traumatic Stress, 17*, 341–344.

Newman, M. G. (2004). Technology in psychology: An introduction. *Journal of Clinical Psychology, 60*(2), 141–145.

O'Farrell, T. J., & Fals-Stewart, W. (2000). Behavioral couples therapy for alcoholism and drug abuse. *Journal of Substance Abuse Treatment, 18*, 51–54.

Orsillo, S. M., & Batten, S. V. (2005). Acceptance and commitment therapy in the treatment of posttraumatic stress disorder. *Behavior Modification, 29*, 95–129.

Palmer, R. L., Birchall, H., Damani, S., Gatward, N., McGrain, L., & Parker, L. (2003). A dialectical behavior therapy program for people with an eating disorder and borderline personality disorder—description and outcome. *International Journal of Eating Disorders, 33*(3), 281–286.

Resick, P. A. (1993). The psychological impact of rape. *Journal of Interpersonal Violence, 8*(2), 223–255.

Resick, P. A., Jordan, C. G., Girelli, S. A., Hutter, C. K., & Marhoefer-Dvorak, S. (1988). A comparative outcome study of behavioral group therapy for sexual assault victims. *Behavior Therapy, 19*, 385–401.

Resick, P. A., Nishith, P., & Griffin, M. G. (2003). How well does cognitive-behavioral therapy treat symptoms of complex PTSD?: An examination of child sexual abuse survivors within a clinical trial. *CNS Spectrums, 8*(5), 351–355.

Resick, P. A., & Schnicke, M. K. (1992). Cognitive processing therapy for sexual assault victims. *Journal of Consulting and Clinical Psychology, 60*, 748–756.

Reynolds, M., & Brewin, C. R. (1998). Intrusive cognitions, coping strategies, and emotional responses in depression, post-traumatic stress disorder and a non-clinical population. *Behaviour Research and Therapy, 36*, 135–147.

Riggs, D. S. (2000). Marital and family therapy. In E. B. Foa, T. M. Keane, & M. J. Friedman (Eds.), *Effective treatments for PTSD: Practice guidelines from the International Society for Traumatic Stress Studies* (pp. 354–355). New York: Guilford Press.

Rime, B. (1995). Mental rumination, social sharing, and the recovery from emotional exposure. In J. W. Pennebaker (Ed.), *Emotion, disclosure, and health* (pp. 271–291). Washington, DC: American Psychological Association.

Roemer, L., & Borkovec, T. D. (1994). Effects of suppressing thoughts about emotional material. *Journal of Abnormal Psychology, 103,* 467–474.

Rosen, G. M., & Davison, G. C. (2001). "Echo attributions" and other risks when publishing on novel therapies without peer review. *Journal of Clinical Psychology, 57*(10), 1245–1250.

Rosner, R. (2001). Between search and research: How to find your way around?: Review of the article "Thought field therapy: Soothing the bad moments of Kosovo." *Journal of Clinical Psychology, 57*(10), 1241–1244.

Rothbaum, B. O., Hodges, L., Alarcon, R., Ready, D., Shahar, F., Graap, K., et al. (1999). Virtual reality exposure therapy for Vietnam veterans with posttraumatic stress disorder. *Journal of Traumatic Stress, 12,* 263–271.

Rothbaum, B. O., Hodges, L. F., Ready, D., Graap, K., & Alarcon, R. D. (2001). Virtual reality exposure therapy for Vietnam veterans with posttraumatic stress disorder. *Journal of Clinical Psychiatry, 62*(8), 617–622.

Rothbaum, B. O., & Schwartz, A. C. (2002). Exposure therapy for posttraumatic stress disorder. *American Journal of Psychotherapy, 56*(1), 59–75.

Rusch, M., Grunert, B., Mendelsohn, R., & Smucker, M. (2000). Imagery rescripting for recurrent, distressing images. *Cognitive and Behavioral Practice, 7,* 173–182.

Sakai, C., Paperney, D., Matthews, M., Tanida, G., Boyd, G., Simmons, A., et al. (2001). Thought field therapy clinical applications: Utilization in an HMO in behavioral medicine and behavioral health sciences. *Journal of Clinical Psychology, 57*(10), 1215–1227.

Scott, M. J., & Stradling, S. G. (1997). Client compliance with exposure treatments for posttraumatic stress disorder. *Journal of Traumatic Stress, 10,* 523–526.

Schnurr, P., Friedman, M. J., Foy, D. W., Shea, T. M., Hsieh, F. Y., Lavori, P. W., et al. (2003). Randomized trial of trauma focused group therapy for post-traumatic stress disorder: Results from a Department of Veteran's Affairs cooperative study. *Archives of General Psychiatry, 60*(5), 481–489.

Shipherd, J. C., & Beck, J. G. (1999). The effects of suppressing trauma-related thoughts on women with rape-related posttraumatic stress disorder. *Behaviour Research and Therapy, 37,* 99–112.

Simpson, E. B., Pistorello, J., Begin, A., Costello, E., Levinson, J., Mulberry, S., et al. (1998). Focus on women: Use of dialectical behavior therapy in a partial hospitalization program for women with borderline personality disorder. *Psychiatric Services, 49,* 669–673.

Smucker, M. R., & Dancu, C. V. (1999). *Cognitive behavioral treatment for adult survivors of childhood trauma: Rescripting and reprocessing.* Northvale, NJ: Aronson.

Smucker, M. R., Dancu, C., Foa, E. B., & Niederee, J. L. (1995). Imagery rescripting: A new treatment for survivors of childhood sexual abuse suffering from posttraumatic stress. *Journal of Cognitive Psychotherapy: An International Quarterly, 9,* 3–17.

Smucker, M. R., Grunert, B. K., & Weis, J. M. (2003). Posttraumatic stress disorder: A new algorithm treatment model. In R. L. Leahy (Ed.), *Roadblocks in cognitive-behavioral therapy: Transforming challenges into opportunities for change* (pp. 175–194). New York: Guilford Press.

Smucker, M. R., & Niederee, J. (1995). Treating incest-related PTSD and pathogenic schemas through imaginal exposure and rescripting. *Cognitive and Behavioral Practice, 2,* 63–93.

Solomon, Z., Waysman, M., & Mikulincer, M. (1990). Family functioning, perceived social support, and combat-related psychopathology: The moderating role of loneliness. *Journal of Social and Clinical Psychology, 9,* 456–472.

Spoont, M. R., Sayer, N. A., Thuras, P., Erbes, C., & Winston, E. (2003). Adaptation of dialectical behavior therapy by a VA medical center. *Psychiatric Services, 54*(5), 627–629.

Steketee, G., & Foa, E. B. (1988). Rape victims: Post-traumatic stress responses and their treatment: A review of the literature. *Journal of Anxiety Disorders, 1*(1), 69–86.

Sweany, S. L. (1987). *Marital and life adjustment of Vietnam combat veterans: A treatment outcome study.* Unpublished doctoral dissertation, University of Washington, Seattle.

Tarrier, N., & Humphreys, A. L. (2000). Subjective improvement in PTSD patients with treatment by imaginal exposure or cognitive therapy: Session by session changes. *British Journal of Clinical Psychology, 39,* 27–34.

Tarrier, N., & Humphreys, A. L. (2003). PTSD and the social support of the interpersonal environment: The development of social cognitive behavior therapy. *Journal of Cognitive Psychotherapy, 17*(2), 187–198.

Tate, D. F., & Zabinski, M. F. (2004). Computer and Internet applications for psychological treatment: update for clinicians. *Journal of Clinical Psychology, 60*(2), 209–220.

Tedstone, J. E., & Tarrier, N. (2003). Posttraumatic stress disorder following medical illness and treatment. *Clinical Psychology Review, 23*(3), 409–448.

Telch, C. F., Agras, S. W., & Linehan, M. M. (2001). Dialectical behavior therapy for binge eating disorder. *Journal of Consulting and Clinical Psychology, 69*(6), 1061–1065.

Trupin, E. W., Stewart, D. G., Beach, B., & Boesky, L. (2002). Effectiveness of dialectical behaviour therapy program for incarcerated female juvenile offenders. *Child and Adolescent Mental Health, 7*(3), 121–127.

Valentine, P. V., & Smith, T. E. (2001). Evaluating traumatic incident reduction therapy with female inmates: A randomized controlled clinical trial. *Research on Social Work Practice, 11*(1), 40–52.

Valentiner, D. P., Foa, E. B., Riggs, D. S., & Gershuny, B. S. (1996). Coping strategies and posttraumatic stress disorder in female victims of sexual and nonsexual assault. *Journal of Abnormal Psychology, 105,* 455–458.

Van den Bosch, L. M., Verheul, R., Schippers, G. M., & van den Brink, W. (2002). Dialectical behavior therapy of borderline patients with and without substance use problems: Implementation and long-term effects. *Addictive Behaviors, 27*(6), 911–923.

Wade, J. F. (1990). *The effects of the Callahan phobia treatment technique on self concept.* Unpublished doctoral dissertation, Professional School of Psychological Studies.

Wagner, A. W., & Linehan, M. M. (2006). Applications of dialectical behavior therapy to PTSD and related problems. In V. M. Follette & J. I. Ruzek (Eds.), *Cognitive-behavioral therapies for trauma* (2nd ed., pp. 117–146). New York: Guilford Press.

Walser, R. D., Loew, D., Westrup, D., Gregg, J., & Rogers, D. (2002). *Acceptance and commitment therapy: Theory and treatment of complex PTSD.* Paper presented at the annual meeting of the International Society of Traumatic Stress Studies, Baltimore.

Walser, R. D., Westrup, D., Rogers, D., Gregg, J., & Loew, D. (2003, November). *Acceptance and commitment therapy for PTSD.* Presented at the annual meeting of the International Society of Traumatic Stress Studies, Chicago.

Wisniewski, L., & Kelly, E. (2003). The application of dialectical behavior therapy to the treatment of eating disorders. *Cognitive and Behavioral Practice, 10*(2), 131–138.

Chapter 24

Risk, Vulnerability, Resistance, and Resilience

TOWARD AN INTEGRATIVE CONCEPTUALIZATION OF POSTTRAUMATIC ADAPTATION

**Christopher M. Layne, Jared S. Warren,
Patricia J. Watson, and Arieh Y. Shalev**

In a keynote address delivered at the International Society for Traumatic Stress Studies (ISTSS) annual meeting some 10 years ago, Dr. Norman Garmezy, a pioneer in the field of developmental psychopathology, adopted a surprisingly severe tone in chastising Society members for their simplistic approach to studying resilience. In taking this unsophisticated approach, he chided, traumatic stress researchers had largely failed to enrich and expedite their work by drawing upon the work of related fields. In particular, Dr. Garmezy drew attention to, and criticized, the lack of a systematic conceptual framework for defining, scientifically investigating, and applying information relating to terms such as "protective factor," "risk factor," "vulnerability," and "resilience." Instead, he noted that the traumatic stress field was mired at the level of disseminating and using lists of "pet" protective and risk factors that appeared to be drawn primarily from clinical observation and anecdotal sources without a clear understanding of how, why, or in which contexts they worked. He then strongly advised the Society to draw upon over 15 years' worth of research within the field of developmental psychopathology, as well as to learn and borrow from other fields. Given the stern nature of Garmezy's admonition and the passing of a very eventful and productive decade since its delivery, it seems appropriate to conduct a "10-year checkup" by reviewing portions of the traumatic stress research literature that focus on resilience-related terms, concepts, findings, and applications.

HOW TIMELY WAS GARMEZY'S COUNSEL?

The last decade has seen a remarkable explosion in traumatic stress studies utilizing the key word *resilience* (Layne et al., 2004). This trend was most clearly manifest in an exponential increase in published studies beginning in the mid-1990s. Given the tremendous investment and sacrifice that such a collective research focus entails, not the least of which is its tendency to distract investigators from other lines of inquiry, a detailed study of the literature poses a number of basic questions:

1. How do we, as social scientists, define and communicate about resilience-related concepts, and what have we learned so far about them?
2. Is this recent explosion in resilience-related studies justified in the sense that it holds broad relevance for scientists and consumers alike; or does it instead have the markings of a "fad" topic?
3. How well have we followed Dr. Garmezy's counsel during the past decade? That is, are there still problems with the ways we study resilience-related phenomena that impede our efforts? And if such problems are found, how can we correct weaknesses in our theoretical and experimental approaches, so that we can generate the knowledge base needed to intervene with at-risk groups in as effective, efficient, and sustainable a manner as possible? (see Kazdin, 1999; Kazdin & Nock, 2003).

In this chapter, we review the traumatic stress literature as it pertains to resilience-related concepts. Our hope is to provide a broad overview of this area of investigation and application as it pertains to trauma-exposed populations, and to provide a guiding framework by which to criticize it constructively, enrich it, and improve its theoretical potency, methodological sophistication, and relevance for intervention. We commence with an overview of what is currently known about resilience, including a review of ways in which resilience-related phenomena are defined and conceptualized in the developmental psychopathology and traumatic stress literatures. We follow this with an examination of the degree to which the traumatic stress field, collectively, has followed Dr. Garmezy's admonition, with an eye toward identifying significant weaknesses or problems that may impede our collective ability to learn efficiently about and apply resilience-related phenomena. We expand this section of the chapter to address the clinical implications of this area of investigation. We conclude with recommendations for future traumatic stress research studies that show promise for strengthening weaknesses in the literature, particularly with respect to improving its conceptual clarity and informational yield.

A REVIEW OF RESILIENCE-RELATED CONCEPTS AND RESEARCH FINDINGS

How Do the Developmental Psychopathology and Traumatic Stress Fields Conceptualize and Communicate about Resilience?

Our review of the existing traumatic stress and developmental psychopathology literatures uncovered a proliferation of meanings imputed to terms such as *resilience, risk factor, vulnerability factor*, and *stress resistance* (Layne et al., 2004). This great variety, as a natural consequence, introduces many terminological inconsistencies in the meanings of concepts identified by the same names. For example, our review uncovered at least eight distinct meanings conferred on the term *resilience*. These include "the individual's capacity for adapting successfully and functioning competently despite experiencing chronic stress or adversity, or following exposure to prolonged or severe trauma" (Cicchetti & Rogosch, 1997, p. 797) and "the possession and sustaining of key resources that prevent or interrupt loss cycles" (Hobfoll, Ennis, & Kay, 2000, p. 277). Other examples include "under adversity, an individual can bend, lose some of his or her power and capability, yet subsequently recover and return to the prior level of adaptation as stress is reduced or compromised" (Garmezy, 1993, p. 132) and "pathways to competent adaptation despite exposure to conditions of adversity" (Cicchetti, 1996, p. 255). Unfortunately, many of these meanings, albeit widely used, lack the precision needed for specific research and clinical applications; others are technically inaccurate and refer to other terms (e.g., stress *resistance*). These many meanings imputed to the term *resilience* alone—which range between the extremes of an "absence of pathology" on the one hand and "heroism" on the other, to "differences in developmental pathways"—strongly underscore the need for a more precise professional terminology with which to conceptualize, define, measure, and apply resilience-related phenomena. Indeed, this diversity of meanings raises concerns regarding the lack of consensus and precision—and the consequent ambiguity—now found in our current professional terminology and conceptual frameworks.

Given these ambiguities in the meanings and implications of resilience-related concepts, we next provide working definitions of the relevant terms to which we adhere for the remainder of the chapter. We follow these definitions with a review of important findings relating to resilience that have emerged within the developmental psychopathology literature during the past 25 years.

Working Definitions of Resilience-Related Concepts

Drawing upon the work of seminal theorists in the field (see Kraemer et al., 1997; Kraemer, Stice, Kazdin, Offord, & Kupfer, 2001; Masten & Gewirtz, 2006; Pine, Costello, & Masten, 2005; Steinberg & Ritzmann, 1990), we define "risk marker" as a measurable attribute of individuals, their interpersonal relationships, or their social and physical environments, whose presence or magnitude signals (1) a significant increase in the statistical likelihood of attaining a negative or undesirable current or future outcome (e.g., dropping out of school) and/or (2) a significant decrease in the likelihood of attaining a positive or adaptive current or future outcome (e.g., attending college). Risk markers may constitute causal risk factors, but may also consist only of proxies (e.g., correlates, outcomes, or by-products) for underlying, often unmeasured, causal variables. "Causal risk factor" refers to a measurable attribute of individuals, their interpersonal relationships, or their social and physical environments that demonstrably increase the likelihood of negative or undesirable current or future outcomes. "Vulnerability factor" refers to a measurable attribute of individuals, their interpersonal relationships, or their social and physical environments whose operations decrease stress resistance, thereby increasing the susceptibility of the organism to the adverse effects of exposure to stress. Similarly, "protective factor" refers to a measurable attribute of individuals, their interpersonal relationships, or their social and physical environments whose operations increase stress resistance, thereby decreasing the susceptibility of the organism to the adverse effects of stress. By definition, vulnerability and protective factors are moderator variables that interact with specific risk factors, such that they augment–exacerbate or attenuate–mitigate, respectively, the adverse effects of that risk factor on specific outcome variables. "Promotive factors" (also called "protective assets") comprise measurable attributes of individuals, their interpersonal relationships, or their social and physical environments that regardless of adversity or risk level directly promote (1) movement *away* from pathways leading to adverse outcomes, or (2) movement *toward or along* pathways leading to positive or adaptive outcomes.

"Intervening variables" comprise a general class of variables that "come between" the trauma exposure → posttraumatic adjustment cause–effect link, including moderator variables (e.g., prior history of significant loss) and mediator variables (e.g., persisting secondary adversities generated by the traumatic event, which in turn causally influence one or more focal outcome variables). "Developmental tasks" are standards of performance, achievement, or competence in one or more psychosocial domains of adaptation that are expected by a given culture or other social group for individuals within a given developmental period. Developmental tasks are theorized to vary by culture, gender, physical circumstance, and historical time period. "Mechanisms" consist of variables that exert a causal influence, whether beneficial, neutral, or detrimental, on one or more other variables, thereby generating cause–effect linkages. "Processes" are the dynamic sequences of events by which mechanisms operate over time, thereby exerting their influences on other variables. Processes may be conceptualized and measured at the level of a given mechanism, such as by explicating how socially supportive transactions take place between family members. Processes may also be conceptualized and measured at the intermechanism level, such as by explicating how social support (one mechanism) decreases depressogenic thoughts (a second mechanism), thereby alleviating depressive affect (an outcome).

"Pathways of influence" comprise a series of one or more mechanisms, their associated processes, and focal outcome variables that interlink to form chains of causal influence between two or more variables. Pathways of influence can be unidirectional (denoted by one-way "cause → effect" arrows between two or more variables; or, in more complex systems, they may be bidirectional (denoted by two arrows pointing in opposite directions between variables). "Stress resistance" refers to the capacity of a system to use effective adjustment processes to *maintain* homeostatic balance, thus maintaining adaptive functioning within a given psychosocial domain, during and after exposure to stress. Last, "resilience" refers to the capacity of a given system to implement early, effective adjustment processes to alleviate strain imposed by exposure to stress, thus efficiently restoring homeostatic balance or adaptive functioning within a given psychosocial domain following a temporary perturbation therein. We conceptualize both stress resistance and resilience as multidimensional processes, in that individuals may potentially manifest stress resistance in one life domain (e.g., maintaining stable and healthy peer relationships following a school shooting) and resilience in another (e.g., manifesting a temporary decrement in academic functioning followed by expeditious recovery) (see Kraemer et al., 1997, 2001; Masten & Gewirtz, 2006; Pine et al., 2005; Steinberg & Ritzmann, 1990).

What Is Currently Known about Resilience-Related Concepts?

The study of positive adaptation in the face of significant adversity has held the interest of researchers for over three decades (Garmezy, 1971, 1974; Luthar, 2006; Masten, 2001; Werner, Bierman, & French, 1971; Werner & Smith, 1982). Historically, research on resilience focused primarily on children and youth, beginning with the pioneering work of investigators such as Norman Garmezy and Emmy Werner. Early work by Garmezy (1974) focused on the children of mothers with schizophrenia, noting the positive adaptation demonstrated by many of these children despite their increased risk for poor outcomes (see also Garmezy, 1987). Furthermore, seminal longitudinal studies of at-risk children in Hawaii (Werner et al., 1971; Werner & Smith, 1982) stimulated interest in the identification of multiple factors that appeared to moderate the effects of stressful or high-risk conditions. These lines of research stimulated a burgeoning interest throughout multiple disciplines in identifying the mechanisms that promote the abilities of individuals, families, and communities to maintain, or to quickly regain, adaptive functioning in highly stressful contexts.

In the decades that followed, researchers sought primarily to identify "protective attributes" linked to stress resistance and resilience, including child-intrinsic traits, family attributes, and characteristics of their broader social environments. "Protective" attributes most frequently reported in the child/adolescent literature include an easygoing temperament, high intellectual ability, positive family environment, internal locus of control, socioeconomic advantage, and supportive relationships with peers, family members, and other adults (Cowen, Wyman, & Work, 1996; Garmezy, 1985; Grossman et al., 1992; Luthar, 1991; Luthar & Zigler, 1991; Masten, Best, & Garmezy, 1990; Masten et al., 1988; Parker, Cowen, Work, & Wyman, 1990; Rutter, 1987; Werner, 1989). Importantly, these studies have typically utilized correlational or quasi-experimental methods (i.e., based on preexisting attributes such as good academic performance under conditions of high stress) designed to identify attributes that reliably discriminate between "resilient" and "nonresilient" groups.

A distinct disadvantage associated with such studies is that the knowledge base available to interventionists is inevitably restricted to comparatively static lists of variables that reliably discriminate between "resilient" and "nonresilient" groups. Importantly, studies that yield lists of such distinguishing characteristics—even though they do not shed light on the mechanisms responsible and the processes through which they operate—are nevertheless very helpful. Such studies facilitate efforts to identify individuals at risk for marked persisting distress, functional impairment, and developmental disturbance on one hand, and strengths within individuals and their surrounding contexts that show potential for being fostered and strengthened on the other hand. Both these activities promote effective and efficient risk identification and, potentially, intervention. However, these advantages are offset by the caveat that the very methodologies of such studies lack the explanatory power needed to elucidate the underlying mechanisms, processes, and pathways of influence that interlink key risk, vulnerability, protective, and promotive factors with outcome variables. This is a critical weakness whose potential consequences we discuss in a later section.

In light of these disadvantages, researchers have increasingly urged the utilization of more sophisticated designs that permit closer examination of the underlying protective mechanisms and associated processes involved in the achievement of resistance- and resilience-related outcomes (Cowen et al., 1997; Luthar, Cicchetti, & Becker, 2000). As a result, research attention has largely turned from a focus on comparatively static "resilience"-related *attributes* in individuals and their social and physical environments (e.g., engaging personality, intelligence), to a focus on more dynamic resilience-related *mechanisms*, the *processes* through which they operate over time, and the *pathways of influence* that are formed as they interlink (e.g., how individuals recruit social support to cope with specific stressors). The rationale underlying this movement is that by examining precisely how risk, vulnerability, and protective mechanisms and associated processes intersect to influence ongoing adaptive and maladaptive psychosocial adjustment, multilevel prevention and intervention strategies can be developed (Masten & Coatsworth, 1998; Rolf & Johnson, 1999).

Another notable trend is that stress resistance and resilience are being conceptualized as more widespread phenomena than previously assumed. Although early researchers labeled individuals who demonstrated adaptive outcomes under stress as "remarkable" and "extraordinary" (sometimes using such misleading and globally heroizing labels as "invulnerable" or "invincible"), more recent literature suggests that positive adaptation under stressful conditions is a relatively common phenomenon that stems from normal developmental processes (Masten, 2001). Specifically, stress resistance and resilience are now largely regarded as the outcome of common, developmentally linked regulatory processes (e.g., coping skills, social support, self-concept, self-efficacy, self-regulation) that promote positive adaptation not only in conditions of heightened or extreme adversity but also in conditions of normal, everyday stress and strain (see Masten, 2001). The study of resistance and resilience under conditions of heightened adversity, then, consists of an *extension* of the study of normal adaptation into the realms of heightened stress and adversity rather than constituting a qualitatively distinct and separate field of research. This linkage between resilience under extremely stressful circumstances and basic processes of human adaptation under conditions of everyday stress has generated two outcomes. First, it has challenged the adequacy of psychopathology-oriented approaches that utilize deficit-based models (e.g., low intelligence, lack of an engaging personality) to explain how maladaptation develops, and that emphasize traditional psychotherapeutic treatment approaches to remediate dysfunction. Second, it has stimulated the development of prevention and intervention approaches that seek to strengthen basic protective systems within contexts of both normative and high-magnitude stress. Recent foci of such programs often center on developing an adaptive coping repertoire, including problem-solving and support-seeking skills, improving parenting skills, and buttressing extended social support networks (e.g., Luthar et al., 2000; Masten, 2001; Masten & Coatsworth, 1998). A distinct advantage is that these efforts seek to promote positive adaptation within multiple contexts, including everyday stresses and normal developmental challenges, as well as extreme stress and adversity. This suggests that resilience-enhancing factors may be essentially promotive rather than protective in nature given that they do not produce their effects only under conditions of extreme stress. Such efforts show promise for yielding a more efficient, effective, and sustainable return on resource investments given that these skills are more likely to be employed and to induce beneficial effects on a regular basis, rather than being "dusted off" and used only under extraordinarily stressful circumstances.

Developmental Psychopathology Perspectives on Posttraumatic Adjustment

The popularity of stress resistance and resilience as concepts has stimulated investigations of the correlates of adaptive outcomes for a broad range of populations including individuals exposed to adverse life events (Parker et al., 1990), children of divorced parents (Wolchik, Ruehlman, Braver, & Sandler, 1989), children of Holocaust victims (Baron, Eisman, Scuello, Veyzer, & Lieberman, 1996), victims of sexual abuse (Valentine & Feinauer, 1993), and children whose parents abuse alcohol or drugs (Johnson, Glassman, Fiks, & Rosen, 1990; Roosa, Beals, Sandler, & Pillow, 1990), among others. Nevertheless, despite such broad attention to resilient outcomes, relatively little is known concerning how theorized protective mechanisms and processes precisely contribute, as articulated in a common terminology of risk, vulnerability, protective, and promotive factors (Kraemer et al., 1997, 2001), to posttraumatic adjustment. Clearly, the robustness and generalizability of at least several protective factors identified within the developmental psychopathology literature (e.g., social support, positive parenting behaviors) suggest that the same factors may potentially manifest protective effects in trauma-exposed populations (see Brewin, Andrews, & Valentine, 2000).

The relevance of resilience-related constructs for understanding trauma-exposed populations is further highlighted by conclusions in recent literature reviews that identified intervening variables as among the best predictors of posttraumatic adaptation (Brewin et al., 2000; Silverman & La Greca, 2002). Of particular note is Brewin and colleagues' (2000) meta-analysis of risk factors for the development of posttraumatic stress disorder (PTSD) in adults: The absence of social support and the presence of contextual life stress were two of the top three "risk factors," carrying larger effect sizes than traditionally emphasized risk or moderator variables such as child abuse history, intelligence, socioeconomic status, lack of education, and gender. Importantly, the effect sizes of these two intervening variables were comparable in magnitude or greater than the "gold standard" risk variable of magnitude of trauma exposure.

Despite the scarcity of well-conducted, resilience-focused studies within the traumatic stress literature, researchers are increasingly focusing on intervening variables in an effort to explain the often high degree of variability in outcomes observed among trauma-exposed individuals. For example, in addressing the need to understand the effects of violence on children, Garbarino (2001) has argued that assessment of *opportunity* (e.g., social supports, positive family environment, financial resources, ease of access to higher education) must accompany the assessment of risk and vulnerability factors in order to better predict and understand long-term outcomes. The author asserts that despite the presence of risk and vulnerability factors, at least some protective or promotive factors operate within most children's lives to at least partially offset the adverse influences of those risks and vulnerabilities. This assertion coincides with Hobfoll's (1998) proposal that the particular trajectories of posttraumatic adaptation that individuals manifest are determined by the intersection of preexisting resources, ongoing resource-depleting loss cycles, and ongoing resource-replenishing gain cycles. Other researchers, upon observing the traditionally narrow focus on psychopathology in the psychological sciences in general (and in stress and trauma research in particular), have urged that increased attention be given to adaptive phenomena, such as resilience (Masten, 2001; Shakoor & Fister, 2000; Witmer & Culver, 2001). As emphasized by McFarlane (1996), most trauma-exposed individuals do *not* develop clinically significant distress and functional impairment, suggesting that chronic, severe maladjustment following trauma is the exception rather than the rule. Clearly, an increased focus on the risk, vulnerability, protective, and promotive mechanisms and processes involved can shed light on how, why, and when adaptive posttraumatic adjustment and recovery does or does not occur.

As noted earlier, empirical studies of risk, vulnerability, and protective and promotive factors in trauma-exposed populations are still comparatively scarce (although the literature is growing—see Layne et al., 2004) given that most trauma-focused studies have, until recently, focused almost exclusively on the adverse sequelae of trauma (Witmer & Culver, 2001). Nevertheless, initial findings within this literature appear promising. In one study of Kuwaiti children exposed to the Gulf War of 1990, Llabre and Hadi (1997) explored the relations between social support, PTSD symptoms, and depressive symptoms. Both moderator and mediator hypotheses were tested in this sample of 9- to 13-year-olds, many of whose fathers or other relatives had been killed, arrested, or were still missing. Results indicated that for girls, perceived emotional support moderated the effects of traumatic events, such that girls experiencing significant trauma with low levels of perceived support reported the highest level of PTSD symptoms. The authors concluded that girls may be at greater risk than boys for developing trauma-induced psychological distress, and that high levels of social support may significantly mitigate this risk in girls. Similarly, Ferren (1999) speculated that social support (which he did not directly investigate) may nevertheless have accounted for his unexpected observation that perceived self-efficacy in a sample of trauma-exposed adolescent Bosnian and Croatian refugees exceeded that of nontraumatized controls.

Further evidence for the relevance of social support in mitigating the adverse influences of trauma is found in a study of postwar adjustment in a large sample of Vietnam War veterans (King, King, Fairbank, Keane, & Adams, 1998). An examination of both *structural* social support (measured via the size and complexity of veterans' social networks) and *functional* social support (measured in the forms of emotional and instrumental assistance) indicated that both types of support mediated the relationship between war exposure and PTSD. In contrast, support for a hypothesized moderating role for social support was not found; the authors suggested that this was perhaps the result of a statistical artifact related to a restricted range of war-zone stressors, as reported by participants.

Although a number of other studies have examined correlates of adaptive outcomes in trauma-exposed individuals, few have tested mediator or moderator models for the variables in question, thus rendering it difficult to draw clear conclusions regarding "stress-buffering" (i.e., interactive or moderated) effects. For example, in one review of the literature on trauma and resilience in Bosnian refugee families (Witmer & Culver, 2001), several factors were identified that predicted positive outcomes, including a strong family and cultural identity, maintenance of cultural traditions, and perceived social support. However, these factors were apparently tested using only a main-effects model, thus calling into question whether interactive (i.e., stress-buffering or protective) effects were in operation. Importantly, studies that examine simple direct effects models can be useful in identifying factors of significance to individuals in a given population, irrespective of their stress levels. However, these studies are less helpful when the goal is to identify promotive or protective factors of particular importance to individuals exposed to high-magnitude stressors.

BIOLOGICAL FACTORS RELATING TO RESILIENCE

Mounting evidence indicates that the study of stress resistance and resilience must extend beyond psychological, environmental, and social factors to include biological mechanisms and processes. Individual differences in responses to trauma are indeed determined by a complex interplay between psychological, behavioral, social, and biological factors. Thus, understanding the role that the biological correlates of resilience-related phenomena may play is integral to explaining and predicting individual responses to traumatic events. In reviewing this literature, we focus on two influential biological concepts, *allostatic load* and *predisposing genetic factors*.

The natural response of systems exposed to stress is an attempt to maintain system stability, or *homeostasis*. Sterling and Eyer (1988) originally used the term "allostasis" with reference to an organism's efforts to maintain stability through change. In accordance with this definition, allostatic stress response patterns are biological changes enacted to meet perceived external demands created by a traumatic event or circumstance. "Allostasis" is thus defined as an ongoing process that may be either adaptive or ultimately maladaptive, depending on the specific context (i.e., its intensity, duration, and repetition) and the degree of exertion by the organism. Specifically, allostasis may be considered adaptive in the short term if it promotes stability in the presence of life-threatening circumstances. Conversely, allostatic processes may induce damage over time if the levels of neurotransmitters, neuropeptides, and hormones linked to acute psychobiological response patterns do not return to pretrauma levels. Specifically, if homeostatic responses to these acute psychobiological responses, as induced via allostasis, are not sufficient to return the system to pretrauma levels, deleterious effects may occur in psychological and physiological functioning. In an extensive review of the neuropsychological effects of acute allostatic adaptation, Charney (2004) invokes the term "allostatic load" to refer specifically to the physiological and psychological burden consequent to repeatedly adapting to challenges that is imposed on organisms subjected to stress. Trauma exposure thus activates allostatic stress response patterns that are part of the sensory nervous system's neurohormonal stress response mechanism. The concepts of allostasis and allostatic load thus link, respectively, the protective and survival values of the organism's acute response to stress to the adverse consequences that result if the acute response persists.

Charney's (2004) review of the stress hormone cortisol illustrates this trade-off. Much evidence indicates that a variety of forms of psychological stress increase cortisol synthesis and secretion. In turn, cortisol secretion promotes the acute stress response, including increased arousal, vigilance, attentional focus, and memory formation; mobilization and replenishment of energy stores; inhibition of reproductive and growth systems; and containment of the immune response. Notwithstanding the survival value these effects impart, the stress-induced increases in cortisol, if not constrained through elaborate negative feedback systems involving glucocorticoid and mineral corticoid receptors, may generate serious adverse effects on physical health (Charney, 2004; Friedman & McEwen, 2004; see also Boyce & Ellis, 2005; Curtis & Cicchetti, 2003; Ellis, Essex, & Boyce, 2005). Given these observations, Charney (2004) proposed a neurochemical profile characteristic of psychobiological resilience. Specifically, he predicted that resilient individuals will score in the highest range for measures of DHEA, neuropeptide Y, galanin, testosterone, serotonin receptor (5-HT$_{1A}$), and benzodiazapine receptor function, and in the lowest range for the hypothalamic–pituitary–adrenal (HPA) axis, corticotropin-releasing hormone (CRH), and locus cereleus–norepinephrine activity. Individuals who are most vulnerable to stress are hypothesized to have opposite profiles.

Studies in genetics are also beginning to reveal the impact of certain genotypes on adaptive stress responses, which may eventually yield genetic interventions to prevent the development of PTSD. Caspi and colleagues (2003) found that individuals with one or two copies of the short allele of the 5-HT T promoter polymorphism exhibited more depressive symptoms, diagnosable depression, and suicidality in relation to stressful life events than individuals homozygous for the long allele. Twin studies are also being conducted; in one such study using the Vietnam Twin Registry, True and colleagues (1993) found that inherited factors explained up to 32% of the variance in PTSD symptoms.

Having reviewed relevant resilience-related findings, we now proceed to a 10-year checkup pertaining to how well the traumatic stress field has responded to Dr. Garmezy's admonitions.

DURING THE PAST DECADE, HOW WELL
HAVE WE FOLLOWED DR. GARMEZY'S ADMONITION?

Our review of the traumatic stress and developmental psychopathology literatures (Layne et al., 2004) suggests that Dr. Garmezy's counsel was prescient given the recent marked upsurge in published studies within the mental health literature that use the terms *risk* and *resilience*. This powerful trend has been accompanied by an increased emphasis in both literatures on pretrauma (often childhood era) and posttrauma risk and intervening (i.e., mediating or moderating) variables. Less encouraging was our finding that despite this marked increase in interest, there is a concomitant lack of sophistication in both literatures in terminology, theory, methods, and description of the implications of study findings for intervention efforts. These problems appear to fall into four interrelated categories.

   • *Problem 1: In the traumatic stress and developmental psychopathology fields, we collectively tend to use only the most basic concepts, terms, and methods, and to neglect those that are more precise and (when used collectively) comprehensive.* As a consequence, our terminology, conceptual frameworks, research methods, and intervention efforts lack precision, breadth, and the power to instruct, guide, and promote recovery as well as they might. In particular, we conclude in our review that particularly within the traumatic stress literature, the term "resilience" has generally come to refer to *any* form of positive adjustment, or lack of maladjustment, manifest within the context of significant adversity. Such global, undifferentiated terminology glosses over critical considerations such as the multidimensional nature of adjustment processes (e.g., at school, at home, with peers, etc.?), ongoing development, and the particular trajectory of adaptation manifest over time following initial exposure to a focal stressor (e.g., high stable functioning, a transient dip followed by recovery, a protracted dip followed by recovery, intermittent fluctuations between positive and poor adaptation?).

This concern is well summarized by the rhetorical question, "How many words do we need to describe snow?" (S. Hobfoll, personal communication, April 2005). For cultures that inhabit the lowlands and only observe snow on distant mountaintops, one word suffices. In contrast, cultures that are involved frequently and directly with this medium require a more rich and differentiated terminology. Notably, our review of the traumatic stress literature published between 1980 and 2005 indicates that whereas 381 published articles contained the key words *trauma* and *resilience*, only 215 contained the term *trauma* and an equally relevant term *resistance*, and only 8 articles contained the terms *trauma* and *developmental trajectory*. Given the recent explosion in the number of resilience-focused studies in the traumatic stress and developmental psychopathology literatures, this comparative overuse of the term *resilience*, and corresponding underuse of other, equally useful but distinctly different terms, suggest that our increased contact with this domain of study has not been accompanied by a commensurate increase in the differentiation and sophistication of our terminology. As a consequence, and to a significant degree, our field appears to be spinning its wheels in vague, ambiguous waters of our own collective making.

   • *Problem 2: Both the traumatic stress and the developmental psychopathology literatures relating to resilience and related concepts contain many terminological inconsistencies and inaccuracies that confuse readers and obscure important information.* Based on our review, the extant literature lacks a common terminological and conceptual framework for understanding and measuring risk, vulnerability, resistance, resilience, and related concepts.

This lack is particularly evidenced by a lack of consistency and clarity in the use of basic terminology and associated meanings. This problem is primarily manifest in the form of three terminological errors: (1) using two different terms to define the same concept (e.g., some authors appear to use the terms *stress resistance* and *resilience* interchangeably); (2) using the same term in reference to two different concepts (e.g., the term *resilience* is used in reference to at least eight distinct concepts, as described earlier); and (3) using one conceptually distinct term to define another conceptually distinct term (e.g., using *stress resistance* to define *resilience*).

These difficulties are compounded by the incorrect usage of key terms. The original usages of the terms *resistance* and *resilience*, as well as *stress* and *strain*, denote important differences in their meanings and potential applications. Indeed, these terms were imported into the social sciences from the field of engineering, wherein they were used to describe important properties of metals (see Mercier, 1904, for an early but eerily similar lamentation regarding an unchecked proliferation in uses of the term *stress*). Specifically, *resistance* referred to the capacity of a given metal to resist *stress* (the imposition of which places the metal under *strain*) without flexing or bending. In contrast, *resilience* referred to the capacity of that metal, once flexed as a result of stress, to "flex back," thus returning to its former state without being bent permanently. Therefore, as applied to a variety of materials, metals such as iron are highly resistant but not resilient, whereas rubber is highly resilient but not resistant. As applied to the social sciences, *stress-resistant* individuals or other systems are characterized by the ability to maintain homeostatic (i.e., adaptive) functioning under stressful circumstances, thus manifesting consistently high adaptive trajectories before, during, and following exposure to stress. In contrast, *resilient* individuals or systems exhibit a relatively transient but significant decrement in functioning, followed by quick recovery, thus "flexing back" to an adaptive level of functioning (Steinberg & Ritzmann, 1990; note the consistency of this definition with Garmezy's [1993] early usage). In summary, the adaptational trajectories of resistant individuals and other systems are relatively flat and high within a given psychosocial domain across pre-, peri-, and posttraumatic periods, whereas those of resilient individuals have a pronounced "V" shape (see Figure 24.1).

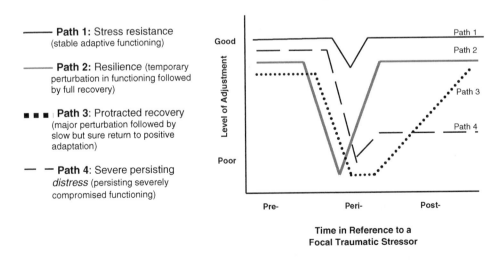

FIGURE 24.1. Approach resistance, resilience, protracted recovery, and chronic distress as distinct dynamic processes that are distinguished by different adaptational trajectories over time.

Additional subtypes, *protracted recovery* and *severe persisting distress*, are characterized by significant perturbations, followed by slow recovery curves, and a marked and persistent diminution in functioning, respectively.

• *Problem 3: There is a surplus of basic studies that identify attributes that characterize "resilient" youth, whereas there is a scarcity of sophisticated studies that elucidate the mechanisms, processes, and pathways of influence that underlie resilience-related outcomes.* Unfortunately, research findings in both the traumatic stress and developmental psychopathology literatures that identify "shopping lists" of attributes (e.g., optimism, personal hardiness, engaging personality, sense of mastery, engagement-oriented coping style, having a close friend, above-average intelligence) that distinguish between stress-resistant or resilient individuals and individuals who are less resistant or resilient are best suited to guide the development of "shopping list" or attribute-centered interventions (Rutter, 1996). Specifically, in the absence of precise information relating to *how*, *why*, and *when* the mechanisms and processes that underlie positive adaptation in stressful contexts operate, interventionists are primarily limited to two choices: either intervention at a comparatively superficial level by promoting the development or use of the "distinguishing" attributes identified earlier, such as by admonishing clients to "be optimistic," "be hardy," "be resourceful," "have a sense of mastery and self-efficacy," "be competent," "talk to a friend," or "be smart" on one hand (cf. American Psychological Association, 2004).

On the other hand, well-intentioned but inadequately informed interventionists may be constrained to develop programs based primarily on speculation regarding underlying mechanisms and processes, untested "pet" theories, clinical experience, intuition, or some combination thereof. Such approaches unfortunately carry a number of disadvantages, including an increased likelihood for developing intervention programs that lack adequate scope, effectiveness, or efficiency, or that contain therapeutically inert or potentially harmful components. Such interventions may also lack needed theoretical coherence or a solid evidence base, and may thus be difficult to disseminate with fidelity, to evaluate rigorously, or to replicate. Interventionists who follow such strategies are also disadvantaged when confronted with the need to revise their treatment plans, such as when clients are making inadequate progress, or when elements of the intervention are weak, inconsistent in their effects, inert, or harmful (see Kazdin, 1999; Kazdin & Nock, 2003).

Theoretically sophisticated studies are needed for two additional reasons. First, measurable attributes that statistically discriminate between resilient and nonresilient groups may or may not actually *cause* the individuals therein to *be* resilient or nonresilient. Instead of serving as "active ingredients," such attributes may instead function as causally inert markers (e.g., correlates, outcomes, or by-products) of other misunderstood, neglected, or unknown mechanisms. Second, even when it can be demonstrated that selected resilience-related variables *do* exert a causal influence in promoting resistance- and resilience-related outcomes upon exposure to traumatic stress, we, as a field, still do not necessarily understand the processes and pathways through which their underlying mechanisms operate. As a consequence, we may lack the know-how needed to promote effectively or harness the undergirding protective mechanisms on the one hand, or to prevent, interrupt, or mitigate the actions and consequences of risk and vulnerability-enhancing mechanisms on the other. Thus, even in the best of scenarios—when investigators discover *whether* a given intervention program is effective in a general sense—program evaluators may still be unable to explain *why* and *how* beneficial changes were produced, how the program's effectiveness and impact might be strengthened (by adding potent components), or how the program might be trimmed and streamlined (by removing redundant, weak, inert, or harmful components) to increase its efficiency, impact, sustainability, or attractiveness to consumers (Kazdin, 1999; Kazdin & Nock, 2003).

• *Problem 4: The intervention-related implications of resilience-related concepts are often not clearly and systematically explicated.* As an outgrowth of the lack of studies that elucidate the mechanisms and processes undergirding the operations of risk, vulnerability, and protective factors, the current traumatic stress literature (and to a considerable degree, the developmental psychopathology literature) provides comparatively sparse yields of intervention-friendly information (Rutter, 1996). By extension, program developers, interventionists, administrators, and policymakers are deprived of the information that they need to make good decisions regarding where, when, and precisely how to intervene; what to monitor during the course of treatment; and how to evaluate program effectiveness, efficiency, and sustainability.

In summary, the problem we collectively face as a field might be summarized by the adage, "If we keep doing what we're doing, we'll keep getting what we're getting." The traumatic stress field generally appears to be using the most basic terms that the developmental psychopathology literature proffers, and often in a manner that is inconsistent, imprecise, inaccurate, or lacking in clear implications for intervention. Thus, even though the term *resilience* is used with increasing frequency in the traumatic stress literature, we cannot be sure about its precise meaning or whether another term (e.g., *stress resistance*) might be more accurately and informatively used. Furthermore, although we have learned that factors such as social support, sense of mastery, and optimism are *associated* with positive outcomes (e.g., Brewin et al., 2000), we still lack knowledge concerning *why, how,* and *in which contexts* they work and, by extension, how to facilitate and harness them effectively. As a result, our collective findings are probably not nearly as clear, informative, and practically useful as they might be.

Taken together, these concerns suggest that the traumatic stress field has not abided by Dr. Garmezy's counsel in a number of important aspects. This is not a petty squabble between academics; it is instead a problem with broad relevance for the entire field. If continued, our current collective approach may hinder scientific inquiry in a variety of ways, including inefficient progress in scientific inquiry, characterized by a lack of clear understanding relative to what we do and do not know, and an inability to determine which directions we should collectively pursue. We may also lack theories that possess sufficient power to guide the design of creative and informative studies. Therefore, if our goal as a traumatic stress field is to create more effective, efficient, and sustainable prevention and intervention programs, we must seek to understand how risk, vulnerability, and protective and promotive mechanisms and their associated processes actually work. This requires that in parallel with the increased number of resilience-related studies we are collectively publishing, we concomitantly increase our clarity, precision, and refinement in the areas of terminology, theory, research methodology, and intervention.

This conclusion raises the broader question, "How can we conduct more meaningful research studies so that we are in a stronger position, when designing and implementing intervention programs, to know *what* we are trying to achieve, with *whom, why* we want to do so, and *how, where,* and *when* to carry it out?" In the remainder of this chapter, we explore two major avenues for future research and application. We first seek to increase our collective ability to learn efficiently from conducting research studies by addressing Problems 1 through 3, discussed earlier. We then address Problem 4 by underscoring the relevance this theoretical framework holds for interventionists, exploring its implications for designing, implementing, and evaluating intervention programs for at-risk groups.

HOW CAN WE AS A FIELD STRENGTHEN OUR COLLECTIVE ABILITY TO INTERVENE WITH AT-RISK GROUPS?

Exploring these concerns raises an obvious but important first question: "Can we effectively remedy this problem by directly importing constructs, definitions, and methodologies from the field of developmental psychopathology?" Notwithstanding the convenience of this strategy, a careful analysis of the developmental psychopathology literature suggests, for two primary reasons, that this strategy is inadvisable, at least in the absence of much modification and supplementation. First, similar to the field of traumatic stress, our review of developmental psychopathology's terminology indicates that this field similarly suffers from pervasive problems associated with inconsistencies and inaccuracies that consequently reduce the clarity, precision, and the overall utility of the findings. Second, our review of developmental psychopathology terms, conceptual frameworks, and primary research methods suggests that these are also not appropriate, in many instances, for our needs as a traumatic stress field, due to often marked differences in the nature of phenomena and populations studied (Layne et al., 2004).

In particular, our review of the developmental psychopathology literature indicates that the studies published therein typically involve populations exposed to chronic stress. Typical examples include the effects of poverty; persistent, severe marital discord; parental incarceration; or living with a parent diagnosed with a psychiatric disorder. In contrast, publications in the traumatic stress literature tend, not surprisingly, to be based on the study of trauma-exposed populations. These include acute traumas (e.g., motor vehicle accidents); chronic or ongoing traumas (e.g., prisoner of war); serial traumas (e.g., repeated childhood sexual abuse); or sequential traumas (e.g., childhood physical abuse, followed by gang violence exposure in adolescence, followed by combat exposure in early adulthood) (e.g., Casey & Nurius, 2005; see also Layne et al., 2006). These traumatic events are often studied in conjunction with acute and/or persistent secondary stressors, such as short- and long-term adversities secondary to parental homicide (e.g., Kaslow, Kingree, Price, Thompson, & Williams, 1998).

Strikingly, the two literatures may perhaps be best distinguished according to their differential relationships to *time*. In particular, the developmental psychopathology literature focuses primarily on ontogenetic development over an extended period, such as that seen in mapping out developmental trajectories across multiple psychosocial domains in the context of chronic or intermittent adversities (e.g., Cicchetti & Hinshaw, 2003). The resilience-related aspects of the developmental psychopathology literature emphasize the study of positive adaptation in the context of significant and ongoing adverse life circumstances. In contrast, the traumatic stress research literature often focuses on examining adaptation with reference to one or more focal traumatic events or circumstances. Importantly, our review of the literature indicates that although both the developmental psychopathology and the traumatic stress literature study adaptation following exposure to chronic stressors (e.g., poverty, unemployment, physical illness or disability), some of which constitute traumatic stressors (e.g., recurrent child sexual or physical abuse), the study of adaptation in reference to *acute* (i.e., circumscribed, single-incident) traumatic events appears to be quite unique to the traumatic stress literature (Layne et al., 2004). In particular, the periods *before, during,* and *following* a focal traumatic life event or circumstance are likely to be of primary interest to many traumatic stress researchers and interventionists. This finding suggests that *precisely when* one measures adaptation—pretrauma, peritrauma, acute posttrauma, or intermediate or long-term posttrauma—may be as consequential a decision as choosing what to measure and how to measure it. In contrast, precise timing with reference to a discrete event appears to be a much less pivotal concern within the developmental psychopathology literature given that the phenomena under study are generally ongoing and often do not have a discrete (hence, clearly identifiable) beginning, middle, or end.

There are clear logistical differences between the traumatic stress and developmental psychopathology literatures as well. The developmental psychopathology literature on resilience relies heavily on the use of prospective longitudinal designs in studying populations at increased risk for maladaptation in the context of significant stress (Werner, 2005). In contrast, traumatic stress researchers often face significant pragmatic challenges as they attempt to conduct field research in traumatically stressed contexts, including war-related settings, sites of terrorist attacks, or in the aftermath of natural disasters. Given the unpredictable nature of many catastrophic events, systematically gathering preexposure data is often very difficult or impossible, and researchers must instead implement studies using retrospective reports of pretrauma and trauma-related exposure, often with convenience samples (e.g., Pfefferbaum, Call, & Lensgraf, 2001).

In summary, it is apparent that the inherent complexity and distinctness of trauma and its surrounding contexts require an approach that is both broader and more precise—requiring substantially different theory, terminology, methodology, and interventions—compared to fields that focus primarily on the sequelae of chronic hardships alone. This suggests that, as a traumatic stress field, we must adapt our terms, concepts, methods, and interventions from other fields for our own populations, contexts, questions, and problems, and develop our own as needed.

We now provide four recommendations regarding how we may increase the efficiency, informational yield, and utility of our published studies.

CLARIFYING AND ENRICHING THE TRAUMATIC STRESS LITERATURE: FOUR RECOMMENDATIONS

• *Recommendation 1: Define and conceptualize resistance, resilience, prolonged recovery, and severe persisting distress as distinct processes.* As depicted in Figure 24.1, resistance, resilience, protracted recovery, and severe persistent distress can be defined as ongoing dynamic processes that are distinguishable by differences in their slopes, levels, and associated trajectories of posttraumatic adaptation in reference to a focal stressor. Such conceptual frameworks add precision, clarity, and richness to the many manifestations that exposure to trauma and its aftermath may evoke. These frameworks will assist the traumatic stress field in moving well beyond the very broad, nebulous, and only moderately useful conceptualization of resilience as comprising any and all forms of "doing acceptably well under conditions of heightened adversity."

• *Recommendation 2: Approach resistance, resilience, protracted recovery, and chronic distress as domain-specific phenomena.* In keeping with recommendations made by leading developmental psychopathology researchers (Luthar et al., 2000; Masten & Coatsworth,

1998), the traumatic stress field will benefit from a richer, more differentiated conceptual framework that portrays resilience-related phenomena as *domain specific*. Indeed, understanding that adaptation is a multidimensional construct is vital for investigating the sequelae of trauma exposure and its aftermath. This multidimensional framework permits a focus on the often differentiated, domain-specific nature of posttraumatic adaptation, in which trauma survivors are observed to function competently in one psychosocial domain while manifesting significant difficulties in others. For example, Figure 24.2 illustrates a fictional adaptational trajectory of a female rape survivor who maintains a relatively high level of functioning at work and in childrearing, while experiencing serious persisting difficulties in her marriage, relationships with her family of origin, and friends.

In conceptualizing stress resistance and resilience, it is imperative that trauma researchers remain mindful that these forms of adaptive functioning consist of more than the mere absence of psychopathology. Rather, for a given individual to be judged as stress resistant or resilient, two judgments must be made. These include verification that (1) exposure to extreme adversity has occurred, and (2) the individual is exhibiting signs of developmentally salient competence (Masten & Gerwirtz, 2006). Thus, it is important to assess positive or healthy adaptation rather than the mere absence of psychopathology in populations exposed to single-episode or recurrent trauma. This recommendation raises unanswered questions regarding how positive adaptation—that is, "acceptable competence"—should be conceptualized and measured within chronically stressful or traumatic contexts. Admittedly, one can assess children and adolescents' adaptation according to Western notions of age-salient developmental tasks, such as the formation of healthy peer relationships by certain ages (see Summerfield, 1999, for a caveat). However, whether such developmental milestones are similar between *moderately to severely stressed* (e.g., chronic poverty; incarceration of a parent) versus *trauma-exposed* populations is unknown—especially if the latter involves exposure to recurring traumas, such as ongoing sexual or physical abuse. Thus, positive adaptation should

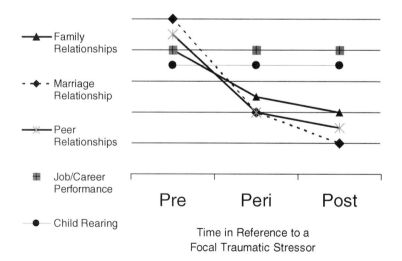

FIGURE 24.2. Approach resistance, resilience, recovery, and chronic distress as domain-specific phenomena.

perhaps be regarded as context-specific—contingent upon stressor type, culture, availability of protective resources, and other contextual factors (J. Obradovic, personal communication, July 2005).

- *Recommendation 3: Undertake sophisticated studies and analysis strategies that discover the pathways of influence leading to stress-resistant, resilient, and other outcomes.* Authoritative calls have recently been issued to increase the theoretical and methodological sophistication of resilience-focused studies to uncover the dynamic processes that underlie positive versus negative adaptation in stressful contexts (Luthar et al., 2000; Luthar & Zelazo, 2003; Masten & Coatsworth, 1998). Researchers have emphasized the need for increased sophistication in methods of measuring risk, examination of adaptation across multiple domains of functioning, and closer examination of precise mechanisms of adaptation using longitudinal designs, thereby demonstrating that change in the proposed mechanism preceded change in the outcome of interest (Cicchetti & Hinshaw, 2003; Kazdin & Nock, 2003; Luthar & Zelazo, 2003). Longitudinal designs with trauma-exposed population are increasingly being undertaken, although the logistical challenges are formidable: Study designs that comprise at least four waves of data collection over an extended period (often 2 years or more following trauma exposure) may be needed to elucidate properly the workings of mechanisms, processes, and pathways of influence that underlie positive adaptation following trauma exposure (Ferrer & McArdle, 2003; see also Brookmeyer, Henrich, & Schwab-Stone, 2005; Raudenbush, 2001).

- *Recommendation 4: Investigate and disseminate the intervention-related implications of resilience-focused theories.* The legitimate question "Does making such *distinctions* in concepts and terms actually make a real *difference* in how we intervene with at-risk groups?" has at least three answers. First, increasing the specificity of our theories will simplify and clarify our terminology by reducing the number of discrepant meanings that a given term, such as *resilience*, is forced to assume. Using different terms to describe different concepts allows interventionists collectively to communicate more precisely, systematically, and comprehensively as they conceptualize, assess, and intervene with various individuals and groups placed at differential risk for various outcomes. Second, increased precision and differentiation among the variables involved will promote greater understanding of behavior by increasing our collective capacity to map out the complex pathways of influence through which risk factors, vulnerability factors, protective factors, and promotive factors function and transmit their influences over time (Luthar et al., 2000). This knowledge in turn assists in building powerful theories that are capable of describing, explaining, predicting, and controlling behavior. Third, powerful theories enhance the accuracy with which interventionists can predict or prognosticate the likelihood of adaptive versus maladaptive outcomes. Thus, such theories facilitate the design of evidence-based triage algorithms by addressing questions such as "Which risk and vulnerability factors place youths at increased risk for which maladaptive outcomes, or at decreased risk for which positive outcomes, via which pathways?", "Which protective or promotive factors place youths at decreased risk for which maladaptive outcomes, or at increased likelihood for which positive outcomes, via which pathways?", "With which risk factor(s) and outcome variable is a given protective factor maximally protective?", and, "With which risk factor(s) and outcome variable is a given vulnerability factor maximally influential?" For example, Smith, Perrin, Yule, and Rabe-Hesketh (2001) reported that war trauma exposure was the best predictor of posttraumatic stress reactions and anxiety symptoms in Bosnian youth. Conversely, the best predictor of youths' depression was the mother's mental health, particularly depression and trauma-related intrusion symptoms.

WHAT CONCRETE IMPLICATIONS FOR INTERVENTION DO RESILIENCE-BASED THEORIES HOLD?

The relative strengths and weaknesses of trauma-, versus "resilience"-focused interventions suggest that each may compliment the other. Trauma-focused interventions undertaken in the aftermath of mass casualty events have been criticized for lacking sufficient breadth, generalizability, cultural and developmental sensitivity, and ecological validity. In contrast, "resilience"-focused interventions mounted within similar contexts have been criticized for quite different weaknesses, including a lack of convincing evidence regarding effectiveness, relevance for high-magnitude stressors, understanding of both trauma theory and posttraumatic adjustment processes, and clear prioritization of needs and objectives (Norris, Murphy, Baker, & Perilla, 2003).

These problems can be reduced through the adoption of conceptual models that facilitate better risk identification methods, triage algorithms, and interventions based on an understanding of *who* is placed at elevated risk for *what* adverse outcome(s) via *which* constellations of risk, vulnerability, protective, and promotive mechanisms operating through *which* pathways of influence. Thus guided, interventionists are better positioned to carry out intervention-related tasks such as risk identification, case conceptualization, treatment planning, treatment monitoring, and program evaluation in a theoretically grounded, evidence-informed manner. Such increases in precision enhance case conceptualization and allow treatment objectives and intervention strategies to be specified in clearer, theoretically sound, priority-driven, and often more flexible ways, as articulated according to what one wishes to accomplish and how, where, and when to do it. Interventions may systematically adopt and prioritize one or more foci, including (1) using risk markers to identify subgroups at elevated risk, (2) preventing and/or mitigating the effects of *risk factors*, (3) preventing and/or mitigating the effects of *vulnerability factors*, and (4) increasing the availability, potency, and effective utilization of naturally occurring *protective or promotive factors*. These options have a strong "ecologically-based intervention" emphasis and focus, and can be supplemented, when and where needed, with specialized therapeutic interventions. This guidance also facilitates "course corrections" with individuals who benefit too little from standard interventions. Highly distressed or treatment-refractory clients may require potent, multisystemic interventions that systematically target multiple risk, vulnerability, protective, and promotive factors over a protracted period, and that blend efforts to enhance naturally existing protective and promotive factors with specialized therapeutic interventions (Gottlieb, 1996). Thus, such approaches promote the effectiveness, efficiency, impact, and sustainability of intervention efforts, and minimize the risk for negative outcomes.

As a case example, an intervention program for youth whose fathers were killed in armed conflict may focus directly on *etiological risk variables* linked to the traumatic loss (e.g., via trauma processing and therapeutic grief work). The intervention may focus also, or instead, on *intervening* variables that form part of consequent pathways of influence, with the objectives of preventing or mitigating the effects of risk and vulnerability variables, or enhancing the effects of protective or promotive variables, respectively. Intervening variables may include disturbances in family functioning, inadequate living circumstances, and nontherapeutic exposure to distressing reminders; facilitating supportive transactions with peers and extended family members; and enhancing adaptive coping skills. The intervention may also directly focus on *outcome* variables, both maladaptive and adaptive, via symptom management strategies, the promotion of academic skills, and engagement in prosocial activities at school and in the community. *Monitoring* activities may focus on targeted outcomes, such as distress reactions and academic performance, as well as on hypothesized mechanisms of change, such as alterations in maladaptive cognitions. *Surveillance* could involve intermittently assessing for exposure to new or ongoing stressful life events and circumstances, maladaptive coping responses, and so forth. How these potential sites for intervention are prioritized and approached should be guided by theory, empirical evidence, social and cultural values, and logistical/practical considerations (Saltzman, Layne, Steinberg, & Pynoos, 2006).

As a second case example, partitioning events into pre-, peri-, and postevent time periods with reference to a focal traumatic event assists in developing clear, intervention-related priorities for each period, including (1) preventing the traumatic stressor where possible *before stress occurs*, (2) promoting stress resistance in those likely to be exposed *before stress occurs*, (3) promoting resilient recovery among those adversely affected *shortly after the stress has occurred*, and (4) promoting, *after the traumatic stressor has occurred*, protracted recovery in adversely affected subgroups that do not show resilient recovery, with the goal of preventing persistent, severe distress and functional impairment. On a broader scale, such time-based partitions assist in developing systematic intervention programs that target prevention of the stressor where possible and the buildup of materials, expert networks, and disaster plans *before* the stressor occurs; adaptive coping *while* it is occurring (e.g., earthquake, tornado, or hurricane safety drills); and multistage intervention programs *after* it has occurred (e.g., promoting ongoing stress resistance where possible, resilient recovery if needed, and protracted recovery where necessary).

When evaluating the effectiveness of such theory-based interventions, one should examine targeted outcomes in accordance with the aims of the intervention. For example, two days of rest, shelter, and food prescribed to reduce combat attrition are intended to promote stress resistance rather than to prevent disease processes. Hence, this intervention should be evaluated in reference to its ability to promote quick stabilization and retention in the armed forces. In contrast, delayed clinical intervention for trauma survivors who have already developed acute stress disorder has as its goal the enhancement of resilient recovery. Thus, such interventions can be appropriately evaluated according to the incidence of subsequent PTSD, other diagnoses, and the degree of functional impairment (Watson, Ritchie, Demer, Barton, & Pfefferbaum, 2006).

In closing, we note that this evolving conceptual framework must account for changes in learning and adaptation that follow major traumatic events. In many cases, a return to previous functioning is unachievable and, ultimately, undesirable. Trauma exposure will inevitably induce change, whether it be internal (e.g., alterations in one's worldviews) or external (e.g., relocations following a natural disaster). From this perspective, the idea of "full" recovery, as defined by one's return to a former state, is often unrealistic. Thus, future theoretical refinements should expand beyond concepts such as "returning to adaptive functioning" to include elements such as acceptance of loss, positive adaptation to enduring or ongoing change, "reasonably good" survival, and posttraumatic growth (Tedeschi, 1999). Much like the concept of resilience itself, posttraumatic adaptation will undoubtedly require a richer, multifaceted conceptual framework.

CONCLUSION

In this chapter, we have both summarized and critiqued the resilience-related traumatic stress literature. We underscored the relevance of resilience-related concepts for the field and discussed problems that impede the discovery, interpretation, dissemination, and utilization of our findings. We made recommendations for increasing the clarity, efficiency, and informational yield of our studies. Put simply, if we wish to advance from the Industrial Age to the Information Age in our ability to describe, explain, predict, and influence the course of resilience-related phenomena, we, as a field, must create, adapt, and use more sophisticated tools—particularly our terminology, concepts, research methods, and theories. The speed and degree of our evolution as a *society* for traumatic stress studies are closely linked to the technologies we collectively use (Diamond, 1997). We hope that the terminological and conceptual distinctions we have proposed *will* make a difference, in terms of both improving the chances for positive adaptation and reducing the likelihood of maladaptation among individuals exposed to extreme stress.

ACKNOWLEDGMENTS

Support for this work was provided by UNICEF Bosnia and Hercegovina, and by research grants to Christopher M. Layne from the Brigham Young University School of Family Life, the Brigham Young University Kennedy International Studies Center, and the UCLA Trauma Psychiatry Bing Fund.

We gratefully acknowledge the assistance of members of the BYU Developmental Psychopathology Research Lab, consisting of Bradley Cohn, Nicole Niebaur, Benjamin Carter, Jacob Tanner, Paul McClaren, Richard Hagen, Callie Beck, Brendan Rowlands, Sarah Turner, Benjamin Walser, Joshua Downs, John-Paul Legerski, Kristy Money, Marko Moreno, Stephanie Donnely, and Ryan Curtis. We also gratefully acknowledge the assistance of Jelena Obradovic, who provided constructive feedback on an earlier version of this chapter.

REFERENCES

American Psychological Association. (2004). *Fostering resilience in response to terrorism: For psychologists working with children.* Retrieved November 5, 2004, from *www.apa.org/psychologists/pdfs/children.pdf*

Baron, L., Eisman, H., Scuello, M., Veyzer, A., & Lieberman, M. (1996). Stress resilience, locus of control, and religion in children of Holocaust victims. *Journal of Psychology, 130,* 513–525.

Boyce, W. T., & Ellis, B. J. (2005). Biological sensitivity to context: I. An evolutionary–developmental theory of the origins and functions of stress reactivity. *Development and Psychopathology, 17,* 271–301.

Brewin, C., Andrews, B., & Valentine, J. (2000). Meta-analysis of risk factors for posttraumatic stress disorder in trauma-exposed adults. *Journal of Consulting and Clinical Psychology, 68,* 748–766.

Brookmeyer, K. A., Henrich, C. C., & Schwab-Stone, M. (2005). Adolescents who witness community violence: Can parent support and prosocial cognitions protect them from committing violence? *Child Development, 76,* 917–929.

Casey, E. A., & Nurius, P. S. (2005). Trauma exposure and sexual revictimization risk: Comparisons across single, multiple incident, and multiple perpetrator victimizations. *Violence Against Women, 11,* 505–530.

Caspi, A., Sugden, K. Moffitt, T. E., Taylor, A., Craig, I. W., Harrington, H., et al. (2003). Influence of life stress on depression: Moderation by a polymorphism in the 5-HTT gene. *Science, 301*, 386–389.

Charney, D. S. (2004). Psychobiological mechanisms of resilience and vulnerability: Implications for successful adaptation to extreme stress. *American Journal of Psychiatry, 161*, 195–216.

Cicchetti, D. (1996). Developmental theory: Lessons from the study of risk and psychopathology. In S. Matthysse, D. L. Levy, J. Kagan, & F. M. Benes, (Eds.), *Psychopathology: The evolving science of mental disorder* (pp. 253–284). New York: Cambridge University Press.

Cicchetti, D., & Hinshaw, S. (Eds.). (2003). Conceptual, methodological, and statistical issues in developmental psychopathology: A special issue in honor of Paul E. Meehl. *Development and Psychopathology, 15*.

Cicchetti, D., & Rogosch, F. A. (1997). The role of self-organization in the promotion of resilience in maltreated children. *Development and Psychopathology, 9*, 797–815.

Cowen, E. L., Wyman, P. A., & Work, W. C. (1996). Resilience in highly stressed urban children: Concepts and findings. *Bulletin of the New York Academy of Medicine, 73*, 267–284.

Cowen, E. L., Wyman, P. A., Work, W. C., Kim, J. Y., Fagen, D. B., & Magnus, B. B. (1997). Follow-up study of young stress-affected and stress-resilient urban children. *Development and Psychopathology, 9*, 565–577.

Curtis, W. J., & Cicchetti, D. (2003). Moving research on resilience into the 21st century: Theoretical and methodological considerations in examining the biological contributors to resilience. *Development and Psychopathology, 15*, 773–810.

Diamond, J. (1997). *Guns, germs, and steel: The fates of human societies*. New York: Norton.

Ellis, B. J., Essex, M. J., & Boyce, W. T. (2005). Biological sensitivity to context : II. Empirical explorations of an evolutionary–developmental theory. *Development and Psychopathology, 17*, 303–328.

Ferren, P. M. (1999). Comparing perceived self-efficacy among adolescent Bosnian and Croatian refugees with and without posttraumatic stress disorder. *Journal of Traumatic Stress, 12*, 405–420.

Ferrer, E., & McArdle, J. J. (2003). Alternative structural models for multivariate longitudinal data analysis. *Structural Equation Modeling, 10*, 493–524.

Friedman, M. J., & McEwen, B. S. (2004). Posttraumatic stress disorder, allostatic load, and medical illness. In P. P. Schnurr & B. L. Green (Eds.), *Trauma and health: Physical health consequences of exposure to extreme stress* (pp. 157–188). Washington, DC: American Psychological Association.

Garbarino, J. (2001). An ecological perspective on the effects of violence on children. *Journal of Community Psychology, 29*, 361–378.

Garmezy, N. (1971). Vulnerability research and the issue of primary prevention. *American Journal of Orthopsychiatry, 41*, 101–116.

Garmezy, N. (1974). The study of competence in children at risk for severe psychopathology. In E. J. Anthony & C. Koupernik (Eds.), *The child in his family: Vol. 3. Children at psychiatric risk* (pp. 77–97). New York: Wiley.

Garmezy, N. (1985). Stress-resistant children: The search for protective factors. In J. E. Stevenson (Ed.), *Recent research in developmental psychopathology: Journal of Child Psychology and Psychiatry* (Book Suppl. No. 4, pp. 213–233). Oxford, UK: Pergamon Press.

Garmezy, N. (1987). Stress, competence, and development: Continuities in the study of schizophrenic adults, children vulnerable to psychopathology, and the search for stress-resistant children. *American Journal of Orthopsychiatry, 57*, 159–174.

Garmezy, N. (1993). Children in poverty: Resilience despite risk. *Psychiatry: Interpersonal and Biological Processes, 56*, 127–136.

Gottlieb, B. H. (1996). Theories and practices of mobilizing support in stressful circumstances. In C. L. Cooper (Ed.), *Handbook of stress, medicine, and health* (pp. 339–356). Boca Raton, FL: CRC Press.

Grossman, F. K., Beinashowitz, J., Anderson, L., Sakurai, M., Finnin, L., & Flaherty, M. (1992). Risk and resilience in young adolescents. *Journal of Youth and Adolescence, 21*, 529–550.

Hobfoll, S. E. (1998). *Stress, culture, and community: The psychology and philosophy of stress*. New York: Plenum Press.

Hobfoll, S. E., Ennis, N., & Kay, J. (2000). Loss, resource, and resiliency in close interpersonal relationships. In H. J. Harvey & D. E. Miller (Eds.), *Loss and trauma: General and close relationship perspectives* (pp. 267–285) New York: Brunner–Routledge.

Johnson, H. L., Glassman, M. B., Fiks, K. B., & Rosen, T. S. (1990). Resilient children: Individual differences in developmental outcome of children born to drug abusers. *Journal of Genetic Psychology, 151,* 523–539.

Kaslow, N. J., Kingree, J. B., Price, A. W., Thompson, M. P., & Williams, K. (1998). Role of secondary stressors in the parental death–child distress relation. *Journal of Abnormal Child Psychology, 26,* 357–366.

Kazdin, A. E. (1999). Current (lack of) status of theory in child and adolescent psychotherapy research. *Journal of Clinical Child Psychology, 28,* 533–543.

Kazdin, A. E., & Nock, M. K. (2003). Delineating mechanisms of change in child and adolescent therapy: Methodological issues and research recommendations. *Journal of Child Psychology and Psychiatry, 44,* 1116–1129.

King, L. A., King, D. W., Fairbank, J. A., Keane, T. M., & Adams, G. A. (1998). Resilience–recovery factors in post-traumatic stress disorder among female and male Vietnam veterans: Hardiness, postwar social support, and additional stressful life events. *Journal of Personality and Social Psychology, 74,* 420–434.

Kraemer, H. C., Kazdin, A. E., Offord, D. R., Kessler, R. C., Jensen, P. S., & Kupfer, D. J. (1997). Coming to terms with the terms of risk. *Archives of General Psychiatry, 54,* 337–343.

Kraemer, H. C., Stice, E., Kazdin, A. E., Offord, D. R., & Kupfer, D. J. (2001). How do risk factors work together?: Mediators, moderators, and independent, overlapping, and proxy risk factors. *American Journal of Psychiatry, 158,* 848–856.

Layne, C. M., Steinberg, A., Warren, J., Cohn, B., Neibauer, N., Carter, B., et al. (2004, November). *Risk, resistance, and resilience following disaster.* In R. Pynoos (Chair), Risk, Resistance, and Resilience in Trauma-Exposed Populations: Emerging Concepts, Methods, and Intervention Strategies. Invited symposium at the annual meeting of the International Society for Traumatic Stress Studies, New Orleans, LA.

Layne, C. M., Warren, J. S., Saltzman, W. R., Fulton, J., Savjak, N., Popovic, T., et al. (2006). Contextual influences on post-traumatic adjustment: Retraumatization and the roles of distressing reminders, secondary adversities, and revictimization. In L. A. Schein, H. I. Spitz, G. M. Burlingame, & P. R. Muskin (Eds.), *Group approaches for the psychological effects of terrorist disasters* (pp. 235–286). New York: Haworth Press.

Llabre, M. M., & Hadi, F. (1997). Social support and psychological distress in Kuwaiti boys and girls exposed to Gulf Crisis. *Journal of Clinical Child Psychology, 26,* 247–255.

Luthar, S. S. (1991). Vulnerability and resilience: A study of high-risk adolescents. *Child Development, 62,* 600–616.

Luthar, S. S. (2006). Resilience in development: A synthesis of research across five decades. In D. Cicchetti & D. J. Cohen (Eds.), *Developmental psychopathology: Risk, disorder, and adaptation* (2nd ed., Vol. 3, pp. 739–795). New York: Wiley.

Luthar, S. S., Cicchetti, D., & Becker, B. (2000). The construct of resilience: A critical evaluation and guidelines for future work. *Child Development, 71,* 543–562.

Luthar, S. S., & Zelazo, L. B. (2003). Research on resilience: An integrative review. In S. S. Luthar (Ed.), *Resilience and vulnerability: Adaptation in the context of childhood adversities* (pp. 510–549). Cambridge, UK: Cambridge University Press.

Luthar, S. S., & Zigler, E. (1991). Vulnerability and competence: A review of research on resilience in childhood. *American Journal of Orthopsychiatry, 61,* 6–22.

Masten, A. S. (2001). Ordinary magic: Resilience processes in development. *American Psychologist, 56,* 227–238.

Masten, A. S., Best, K. M., & Garmezy, N. (1990). Resilience and development: Contributions from the study of children who overcome adversity. *Development and Psychopathology, 2,* 425–444.

Masten, A. S., & Coatsworth, J. D. (1998). The development of competence in favorable and unfavorable environments: Lessons from research on successful children. *American Psychologist, 53,* 205–220.

Masten, A. S., Garmezy, N., Tellegen, A., Pellegrini, D. S., Larkin, K., & Larsen, A. (1988). Competence and stress in school children: The moderating effects of individual and family qualities. *Journal of Child Psychology and Psychiatry, 29,* 745–764.

Masten, A. S., & Gewirtz, A. H. (2006). Vulnerability and resilience in early child development. In K.

McCartney & D. Phillips (Eds.), *Handbook of early childhood development* (pp. 22–43). New York: Blackwell.

McFarlane, A. C. (1996). Resilience, vulnerability, and the course of posttraumatic reactions. In B. A. van der Kolk, A. C. McFarlane, & L. Weisaeth (Eds.), *Traumatic stress: The effects of overwhelming experience of mind, body, and society* (pp. 155–181). New York: Guilford Press.

Mercier, C. (1904). Stress. *Journal of Mental Science, 50,* 281–283.

Norris, F. H., Murphy, A. D., Baker, C. K., & Perilla, J. L. (2003). Severity, timing, and duration of reactions to trauma in the population: An example from Mexico. *Biological Psychiatry, 53*(9), 769–778.

Parker, G. R., Cowen, E. L., Work, W. C., & Wyman, P. A. (1990). Test correlates of stress resilience among urban school children. *Journal of Primary Prevention, 11,* 19–35.

Pfefferbaum, B., Call, J. A., & Lensgraf, S. J. (2001). Traumatic grief in a convenience sample of victims seeking support services after a terrorist incident. *Annals of Clinical Psychiatry, 13,* 19–24.

Pine, D. S., Costello, J., & Masten, A. (2005). Trauma, proximity, and developmental psychopathology: The effects of war and terrorism on children. *Neuropsychopharmacology, 30,* 1781–1792.

Raudenbush, S. W. (2001). Comparing personal trajectories and drawing causal inferences from longitudinal data. *Annual Review of Psychology, 52,* 501–525.

Rolf, J. E., & Johnson, J. L. (1999). Opening doors to resilience intervention for prevention research. In M. D. Glantz & J. L. Johnson (Eds.), *Resilience and development: Positive life adaptations* (pp. 229–249). New York: Plenum Press.

Roosa, M. W., Beals, J., Sandler, I. N., & Pillow, D. R. (1990). The role of risk and protective factors in predicting symptomatology in adolescent self-identified children of alcoholic parents. *American Journal of Community Psychology, 18,* 725–741.

Rutter, M. (1987). Psychosocial resilience and protective mechanisms. *American Journal of Orthopsychiatry, 53,* 316–331.

Rutter, M. (1996). Stress research: Accomplishments and tasks ahead. In R. J. Haggerty, L. R. Sherrod, N. Garmezy, & M. Rutter, (Eds.), *Stress, risk, and resilience in children and adolescents: Processes, mechanisms, and interventions* (pp. 354–385). New York: Cambridge University Press.

Saltzman, W. R., Layne, C. M., Steinberg, A. M., & Pynoos, R. S. (2006). Trauma/grief-focused group psychotherapy with adolescents. In L. A. Schein, H. I. Spitz, G. M. Burlingame, & P. R. Muskin (Eds.), *Group approaches for the psychological effects of terrorist disasters* (pp. 731–786). New York: Haworth Press.

Shakoor, M., & Fister, D. L. (2000). Finding hope in Bosnia: Fostering resilience through group process intervention. *Journal for Specialists in Group Work, 25,* 269–287.

Silverman, W. K., & La Greca, A. M. (2002). Children experiencing disasters: Definitions, reactions, and predictors of outcomes. In A. M. La Greca, W. K. Silverman, E. M. Vernberg, & M. C. Roberts (Eds.), *Helping children cope with disasters and terrorism* (pp. 11–33). Washington, DC: American Psychological Association.

Smith, P., Perrin, S., Yule, W., & Rabe-Hesketh, S. (2001). War exposure and maternal reactions in the psychological adjustment of children from Bosnia-Hercegovina. *Journal of Child Psychology and Psychiatry and Allied Disciplines, 42*(3), 395–404.

Steinberg, A., & Ritzmann, R. F. (1990). A living systems approach to understanding the concept of stress. *Behavioral Science, 35,* 138–147.

Sterling, P., & Eyer, J. (1988). Allostasis: A new paradigm to explain arousal pathology. In S. Fisher & J. Reason (Eds.), *Handbook of life stress, cognition and health* (pp. 629–649). Oxford, UK: Wiley.

Summerfield, D. (1999). A critique of seven assumptions behind psychological trauma programmes in war-affected areas. *Social Science and Medicine, 48,* 1449–1462.

Tedeschi, R. G. (1999). Violence transformed: Posttraumatic growth in survivors and their societies. *Aggression and Violent Behavior, 4,* 319–341.

True, W. R., Rice, J., Eisen, S. A., Heath, A. C., Goldberg, J., Lyons, M. J., et al. (1993). A twin study of genetic and environmental contributions to liability for posttraumatic stress symptoms. *Archives of General Psychiatry, 50,* 257–264.

Valentine, L., & Feinauer, L. L. (1993). Resilience factors associated with female survivors of childhood sexual abuse. *American Journal of Family Therapy, 21,* 216–224.

Watson, P. J., Ritchie, E. C., Demer, J., Barton, P., & Pfefferbaum, B. J. (2006). Improving resilience trajectories following mass violence and disaster. In E. C. Ritchie, P. J. Watson, & M. J. Friedman (Eds.), *Interventions following mass violence and disasters: Strategies for mental health practice* (pp. 37–53). New York: Guilford Press.

Werner, E. E. (1989). High-risk children in young adulthood: A longitudinal study from birth to 32 years. *American Journal of Orthopsychiatry, 59,* 72–81.

Werner, E. E. (2005). What can we learn about resilience from large-scale longitudinal studies? In S. Goldstein & R. B. Brooks (Eds.), *Handbook of resilience in children* (pp. 91–105). New York: Kluwer Academic/Plenum Press.

Werner, E. E., Bierman, J. M., & French, F. E. (1971). *The children of Kauai.* Honolulu: University of Hawaii Press.

Werner, E. E., & Smith, R. S. (1982). *Vulnerable but invincible: A study of resilient children.* New York: McGraw-Hill.

Witmer, T. A. P., & Culver, S. M. (2001). Trauma and resilience among Bosnian refugee families: A critical review of the literature. *Journal of Social Work Research, 2,* 173–187.

Wolchik, S. A., Ruehlman, L. S., Braver, S. L., & Sandler, I. N. (1989). Social support of children of divorce: Direct and stress buffering effects. *American Journal of Community Psychology, 17,* 485–501.

Chapter 25

Public Mental Health Interventions Following Disasters and Mass Violence

Patricia J. Watson, Laura Gibson, and Josef I. Ruzek

This chapter focuses on public mental health interventions at the community level following disasters and mass violence. Due to the dearth of empirical studies in this area, discussion and recommendations that follow draw heavily from consensus recommendations evolving from expert panel discussions and consensus conferences. We end the chapter with a discussion of necessary next steps regarding the further development and refinement of public mental health interventions for mass violence and disaster.

CONSENSUS RECOMMENDATIONS FOR AN EFFECTIVE PUBLIC MENTAL HEALTH DISASTER RESPONSE

Research, practice, and policy experts from several consensus conferences (National Institute of Mental Health, 2002; Watson, 2004) recommend that planners of an effective disaster mental health program should make an effort to be:

1. *Proactive:* Experts concurred that many variables affect the mental health of a community in situations of mass violence (i.e., the quality of communication and coordination among responders, public confidence in leaders, and the accuracy and effectiveness of communications to the public about the risks and the appropriate actions to be taken). Mental health interventions should not be conceptualized or implemented in

isolation; rather, they should be streamlined with other components of the overall community disaster response. Public mental health disaster response should therefore be integrated into the local, state, and federal emergency preparedness response community (emergency management associations, public health offices, hospitals, faith-based community, law enforcement, etc.). Prior to an event, Department of Mental Health (DMH) planners should have provided training to mental health professionals, media, government, public agencies, and educational institutions, as well as the agencies mentioned earlier, and should provide ongoing updates. There must be a recognition among community leaders and planners that each aspect of disaster response can potentially impact community mental health. As a result, there are many possible roles for mental health providers following mass violence. Accordingly, tasks should be matched to professionals' skills, and a proportion of community professionals should be held in reserve to provide services across the entire developmental recovery period rather than flooding in to respond in the immediate phase postevent.

2. *Protective:* In the aftermath of an event, the DMH response should limit inappropriate interventions from professionals, initiate psychological first aid to those who need it, and identify the needs of at-risk individuals who may require additional surveillance and evidence-based intervention over time. Although most public mental health efforts to date have focused on the prevention of posttraumatic stress disorder (PTSD), experts have advocated inclusion of programs designed to prevent development of other problems, including alcohol abuse, drug abuse (including inappropriate use of medications), depression, interference in appropriate developmental functioning, relational and occupational functioning, and anxiety disorders, as well as the exacerbation of preexisting mental disorders.

3. *Pragmatic:* Programs should be pragmatic and culturally competent, enhancing natural resilience and providing tools that enable less-resilient individuals to increase their capacity to prepare for and withstand traumatic events. A public mental health response goes beyond the provision of individual interventions by assisting communities in building inherent strengths that enhance resilience/recovery, organizing training, and supporting natural community groups in helping themselves.

4. *Principle-driven:* Disaster mental health programs should include in all endeavors ways to increase the evidence-informed principles of safety, efficacy, hope, connectedness, and calming (Hobfall, personal communication, November 2004), as described in more detail below.

5. *Proven:* Finally, programs should periodically monitor at-risk individuals, and evaluate services to establish feedback from the community and an evidence base for future interventions.

KEY COMPONENTS
OF PUBLIC MENTAL HEALTH INTERVENTIONS

Mental health interventions can be conceptualized as including a set of key components related to the promotion of individual and community-wide recovery following episodes of mass violence. These components overlap in time; are provided by a range of individuals, organizations, and professionals; and create an overall framework within which recovery from mass violence can be maximized. An expert panel on mass violence and mental health interventions recommended a set of related key components

(National Institute of Mental Health, 2002). In a more recent expert panel those components have been modified somewhat into the following set:

Systems issues/program management process
1. Prepare/foster capacity and resilience.
2. Conduct needs assessments.
3. Monitor the rescue and recovery environment.
4. Foster recovery.
5. Evaluate outcomes.

Interventions/direct survivor care
1. Provision for basic needs.
2. Triage.
3. Psychological first aid.
4. Outreach and information dissemination.
5. Technical assistance, consultation, and training.
6. Treatment.

Systems Issues/Program Management Process

Experts have recommended a stepped care approach in designing a public mental health system following mass violence, such that care delivery helps most people in early adaptation, but as time progresses, reserves more individualized and time-consuming interventions for the minority of people who require it. This approach has had some success in situations of individual trauma from injury and assault (Zatzick et al., 2004). Attention is directed toward interlinking types and modes of services to ensure continuity of care, with appropriate triage and longer-term follow-up. The developmental perspective is to fit services across the developmental lifespan, and to distinguish between services for persons who have been stressed and those who have had higher levels of traumatic stress. People's reactions will be different depending on their experience with threat and their personal history, so planning includes a full range of services and a way to identify those at risk. Both individual and community-level interventions are planned. For instance, use of the media or the Internet to prepare the community at large may involve constructs that are different (e.g., fostering resilience) than programs that support individuals (e.g., teaching coping skills). All planned components should combine to achieve the following goals:

1. Keeping people active, involved, and informed.
2. Honoring cultural variation.
3. Promoting self-efficacy and a communal sense of efficacy.
4. Keeping the program multimodal.
5. Remaining flexible to changes in course.
6. Promoting top-down and bottom-up communication and solutions at each level.
7. Promoting a sense of safety, connectedness, calm, hope, and efficacy at every level of intervention.

It is recommended that the following components be incorporated into all systems involved in disaster mental health recovery to achieve these goals.

Prepare and Foster Capacity and Resilience

The critical work that occurs prior to disasters and mass violence includes preparation and fostering capacity and resilience. Experts recommend that ongoing surveillance be put in place to have a useful baseline from which to create programs, to identify populations most at risk for negative outcomes after an event, and to determine the assets and resources that may be employed to meet people's needs during and after events. In addition, a preexisting surveillance system can (1) establish both the baseline prevalence of psychopathology in the population and prevalence differences between groups (e.g., racial/ethnic groups), making the interpretation of estimates of the mental health burden after a disaster easier; (2) rapidly and reliably establish the change in psychopathology after a disaster and provide estimates of the cases attributable to the disaster that may benefit from public health intervention; and (3) establish a clearer understanding of psychopathology after disasters and allow public health practitioners to gauge the cost-effectiveness of postdisaster intervention (Galea & Norris, 2006).

In addition, diverse, community-wide efforts to build capacity and resilience should be established ahead of time. For instance, Adger, Hughes, Folke, Carpenter, and Rockstrom (2005) point out that social–ecological resilience is an important determinant in the recovery from disasters, particularly the ability of communities to mobilize assets, networks, and social capital, both to anticipate and to react to disasters. For instance, recovery from Hurricane Andrew was fostered by strong social institutions, such as early warning systems and federal aid programs. Expert consensus additionally recommends a multilevel communication channel as a means of conveying safety messages and receiving feedback on community needs. Community safety and skills-building initiatives may also be implemented ahead of time to help build and strengthen relationships between communities and governments/organizations that have access to the resources communities need (general practitioners, hospitals, social workers, police, schools, etc.).

Conduct Needs Assessments

A systematic assessment of the current status of groups, neighborhoods, and the overall affected community is critical in implementing an effective public mental health response. Included in the assessment should be evaluations of the degree of exposure and impact, whether survivors' needs are being adequately addressed, characteristics of the recovery environment, and what additional interventions and resources are required.

Monitor the Rescue and Recovery Environment

Ongoing monitoring and surveillance of the environment postevent should incorporate a quantitative estimate of emerging social, behavioral, and functional changes, as well as direct service provision. Qualitative attention should be paid to emerging concerns, lack of resources, ongoing stressors or toxins, media coverage and rumors, and perceptions of resource distribution that are inequitable or stigmatizing. Those most affected by the incident particularly should be observed and monitored for potential behavioral and physical health sequelae.

Foster Recovery

Fostering recovery includes both individual and community interventions designed to support the community's natural recovery from disasters. Any disaster mental health system should engage in proactive practical steps that enhance protective interventions and result in increased self-efficacy, safety behaviors, and sense of safety, and lower rates of long-term distress.

On a societal level, Adger and colleagues (2005, p. 1038) note that "the challenge is to enhance adaptive capacity to deal with disturbance, and to build preparedness for living with change and uncertainty." In their view, social resilience includes "knowledgeable, prepared, and responsive . . . institutions for collective action, robust governance systems, and a diversity of livelihood choices," as well as "promotion of self-mobilization in civil society and private corporations" and "formal and informal institutions with the capacity to respond to rapid change in environmental and social conditions." They note that this type of response led to prevention of widespread infection and disease following the 2004 Tsunami and rapid rebuilding of public infrastructure following Hurricane Ivan in 2004.

On a community level, because negative social support has been found to be a predictor of PTSD following trauma (Brewin, 2001), it has been recommended that individuals who work with survivors should actively explore how well survivors are able to access and use social support, for instance, by focusing on prosocial behaviors or ways individuals/communities can feel more helpful. Interventions, when feasible and desired by the individual, are often designed to include the families, friends, and work colleagues, as well as the more formal helping networks within the local community (e.g., primary care providers, clergy). Community recovery can additionally be fostered by acknowledging the diversity within any given population, and eliciting input from people in the community around issues that are most relevant and timely for them. The more leaders know about the values of individuals and groups, the better they are able to develop and apply metaphors, examples, and stories to help them make sense of a chaotic, overwhelming situation.

A growing body of literature indicates that recovery on an individual level seems to be facilitated by social, educational, and supportive interventions. The empirical literature shows that recovery is promoted by decreasing ongoing stressors, finding benefit in a sense of relationship with God, cognitive-behavioral therapy (CBT)–type treatment for trauma individually chosen disclosure and social support, the perception that the social milieu accepts one's reactions and welcomes disclosure, and seeing oneself as a hero or survivor rather than a victim (Bonanno, 2004). Interventions should seek to assess, support, and facilitate natural strengths, and promote those factors that contribute to recovery. Although "traumatic events" can also be defined as events exceeding the person's coping resources, individual responses themselves may not be as important as the degree to which recovery efforts foster the survivor's ability to continue task-oriented activity, regulate emotion, sustain positive self-value, and maintain and enjoy rewarding interpersonal contacts (Shalev, personal communication, April 2004).

Evaluate Outcomes

Disaster mental health services are often delivered in a chaotic, rapidly evolving environment, in which decisions need to be made quickly on the basis on limited information. As a result of this context, program activities may be implemented inefficiently, and interventions may be only partially successful or have unintended consequences (Rosen, Young, & Norris, 2006). During the crisis, there may be little time for, or interest in, collecting systematic information on how the program is working, which makes it difficult to monitor program progress or to evaluate program achievements. Without systematic evaluation, programs cannot report to others about which interventions were or were not effective. Evaluation may also identify key challenges that need to be addressed. In addition, because empirical knowledge about best practices is still very limited, ineffective counseling practices may be perpetuated, while innovations and improvements are not disseminated. By encouraging pilot testing of new innovations, such problems may be avoided in the future. Expert panelists recommend tracking program impact throughout the full intervention phase, and widely disseminating program evaluation data through multiple channels to plan for future programs.

In the process of evaluating intervention success, multiple indices can be utilized, based on previous empirical literature about the impact of disasters (Norris et al., 2002). Selected outcome indicators should be based in part on input from providers and community members. These indicators include individual indices, such as specific psychological symptoms/conditions/distress, physical health concerns, behavioral health problems (substance use, poor self-care, etc.), chronic problems in living (exacerbation of premorbid functioning or chronic stressors), psychosocial resource loss (loss of hope keyed to the disaster, and loss of personal relationships, feelings of safety, trust in perceived community resources), coping self-efficacy, skills in recruiting and receiving social support, reduced functional impairment, and religious/spiritual coping. Group indices may also be assessed, such as family functioning, communal mastery/efficacy, community cohesion/support, knowledge of community resources, and race-related stress (i.e., Islamic communities' tension in accessing services after September 11, 2001).

Interventions/Direct Survivor Care

To date, there is little empirical support for interventions offered in the first 14 days following mass violence (Watson et al., 2003). In treatment trials with individual traumas, neither CBT nor eye movement desensitization and reprogramming (EMDR) has been empirically examined in the immediate aftermath (0–14 days) of trauma. Recent work with injury and accident victims has sought to evaluate services in the acute phase postincident, but most interventions generally occur 14 or more days posttrauma (Bisson, 2003; Bisson et al., 2004; Zatzick & Roy-Byrne, 2003; Zatzick et al., 2004). As a number of reviews of the literature have concluded, there is no evidence that critical incident stress debriefing (CISD), a structured, group model designed to explore facts, thoughts, reactions, and coping strategies following trauma, prevents long-term negative outcomes. Additionally, two randomized, controlled trials (RCTs) of CISD reported a higher incidence of negative outcomes in those who received CISD compared to those who did not receive an intervention (for reviews, see Bisson, 2003; Litz, Gray, Bryant, & Adler, 2002; McNally, Bryant, & Ehlers, 2003; Watson et al., 2003). Given that many of the CISD studies, particularly those showing negative outcomes, have methodological flaws, it has been recommended that all one-session interventions requiring emotional processing be more fully researched prior to their routine practice postdisaster (Watson, 2004). It is possible that future research will demonstrate that CISD may be useful for some populations, or in particular settings, or that it has more subtle effects, such as perceived social support. In the meantime, numerous reviews of the best-controlled studies conclude that it cannot be endorsed as an intervention that prevents long-term distress or psychopathology, given the current state of the research (Gray & Litz, 2005; McNally et al., 2003; Rose, Bisson, & Wessely, 2003). There are particularly strong recommendations against its use in postdisaster settings involving mass trauma, due to the chaotic postincident environment, the need for attention to pragmatic material needs, possible cultural and bereavement issues, and multiple recovery trajectories based on complex variables (Watson, Friedman, Ruzek, & Norris, 2002; Litz & Maguen, Chapter 16, this volume). Only one RCT has been published to date on the use of psychopharmacological interventions for acute stress responses. Pitman and colleagues (2002), in a randomized, double-blind pilot study, administered the beta-adrenergic antagonist propranolol within 6 hours of a traumatic event (hypothesizing that the medication might interfere with both fear conditioning and encoding of traumatic memories). Although the propranolol group did not exhibit decreased PTSD symptom severity 3 months later, it did exhibit significantly reduced physiological reactivity. These provocative results indicate that more work is needed with a larger sample size to understand these findings better.

A few other studies have used psychopharmacological interventions in the acute stages after trauma, but they suffer from serious methodological weaknesses that limit their interpretability and generalizability. Given the lack of evidence with pharmacological agents in the acute phases posttrauma, experts recommend use of pharmacology for symptomatic relief only, particularly when individuals exhibit intense psychiatric symptoms that impair functioning (i.e., prolonged insomnia, suicidality, psychosis, intense anxiety, mania, etc.) (Simon & Gorman, 2004).

Because of the lack of empirical support for interventions in the acute phase postdisaster or after mass violence, expert consensus recommends the multitiered approach to disaster intervention in the next section. There is overlap in a number of the key intervention components listed, but a concerted effort to include each of these components in some combination, based on the community and disaster context, is recommended for an effective disaster intervention program.

Provide for Basic Needs

During the immediate response period, all responders (including mental health providers) should focus primarily on helping survivors to meet their basic needs (e.g., safety, shelter, food, rest), as well as provide soothing human contact. An incident command structure offers assistance with public messaging; gathering factual information about the incident and gaining authorization to inform those with a need to know; and identifying the role of behavioral health in the overall response, including its limitations. It is recommended that incident command structure personnel work with behavioral health staff in setting up a family assistance center (FAC) that is feasible and accessible. Although behavioral health staff do not offer formal treatment to victims or rescue workers at this stage, it is appropriate that they have a presence in the FAC. On the other hand, it may be appropriate to offer psychological first aid in the immediate aftermath of a trauma to some survivors, family members, witnesses, children, and others who have been affected. This involves offering concrete assistance with basic needs such as shelter, food, and medical attention. Traditional "treatment" is neither the appropriate intervention nor the goal at this point.

Triage

Beginning in the immediate aftermath of mass violence, there is a need to triage survivors and provide emergency hospitalization or mental health referral when indicated. A functional assessment of an individual's ability to cope is recommended at this stage. For instance:

1. Can the survivor continue task-oriented activity?
2. How well organized, goal-directed, and effective is such activity?
3. Is the survivor overwhelmed by strong emotions most of the time?
4. Can emotions be modulated when such modulation is required?
5. How isolated, alienated, or withdrawn is the survivor? (Watson & Shalev, 2005)

Another goal of assessment should be to identify individuals and groups at elevated risk for development of problems over time. The "screen and treat" model proposes that immediate intervention be restricted to providing information, support, and education, but that survivors be followed up to detect individuals with persistent symptoms, who can then be treated with empirically supported interventions (Brewin, 2005). Research indicates that levels of symptoms assessed very soon after an event do not predict well the future course of disorder (McNally et al., 2003). Therefore, it is not appropriate to screen for symptoms, in the hope of predicting future adjustment, in the immediate aftermath (days) of mass violence.

After a few weeks, however, symptoms provide an excellent guide to future risk of PTSD (McNally et al., 2003). This suggests that symptom-based instruments offer the most promising approach to detecting trauma-related symptoms, particularly if administered to the most high-risk populations: those who were directly involved in the traumatic scene, who witnessed the traumatic scene at first hand, or were physically close to the traumatic scene; those who had reason to fear death or serious injury to themselves or their loved ones; those who were injured; those who lost close friends or family members in the incident; those who reacted most extremely; those who subsequently were left with the most negative beliefs about the event, their role in it, and their reactions; and those who were subsequently criticized or found wanting by other people (Brewin, 2005).

Experts recommend the flexible use of assessment and screening for different contexts, exposure types, and needs. It is important to take into account the experience of the person who is traumatized to maximize acceptability of screening and to engage such individuals personally in case follow-up is indicated. For instance, informal, conversational questioning is more appropriate than paper-and-pencil screening tools in chaotic or informal settings. It is also best to limit needs assessment questions to the fewest essential items during the acute posttraumatic phase (see Table 25.1). Additionally, developmental and cultural issues must be addressed in setting up screening protocols. All assessment should be practical, achievable, and implementable at the local level, and informed by the resources in the entire system of care. Whenever possible, it is obviously best to put systems in place prior to traumatic incidents, with planning coordinated at federal, regional, state, and local levels.

Psychological First Aid

Early intervention, beginning immediately at the scene of an incident and continuing for several days or weeks, is increasingly being organized around a set of actions collec-

TABLE 25.1. Screening Categories across Phases

What are we screening for in acute phases?	What are we screening for in recovery/return phases?

• Basic needs (food, housing, medical, information)
 • Immediate risk to life/suicidality
 • Patient-focused self-report of what they think they need to further recovery
 • Functional capacity/impairment
 • Factors that prevent recovery
   • Continuation of adversity
   • Secondary stressors (loss of resources)
   • Uncontrolled reactions
   • Major risk factors (i.e., past trauma, bereavement, exposure level)
 • Strengths/resources (social support, coping skills, finances, etc.)
 • Information availability (TV, newspapers, Internet access, transportation)
 • Attitudes toward stigma
 • Prediction of chronic dysfunction
	• Same as acute phase, except for prediction of chronic dysfunction
	 • Add symptom scales:
	   • PTSD
	   • Bereavement
	   • Depression
	   • Anxiety disorder
	   • Psychosomatic specturm
	   • Substance abuse/dependence
	   • Alcohol abuse/dependence
	 • Focus on functional "failure to thrive" in work/school/home
	 • Add positive outcomes (i.e., posttraumatic growth, accelerated moral functioning, resilience, etc.)

tively labeled as "psychological first aid" (PFA). Many of these actions are not specifically psychological in nature but are essential for improving function and mental health response. They focus on meeting basic needs for physical safety, connectedness, security, and survival. PFA also involves orienting survivors to the disaster response site and helping them to access available services. It provides survivors who wish to share their thoughts, feelings, or experiences an opportunity to do so. On the other hand, PFA, which is not coercive, permits those who do not wish to discuss the trauma to avoid doing so. In this way, PFA is flexible with regard to encouraging survivors to disclose their specific traumatic experiences.

A recent national expert group developing PFA modules designed PFA to be consistent with research evidence, applicable in field settings, tailored to the full developmental spectrum, and culturally informed (National Child Traumatic Stress Network and National Center for PTSD, 2005). Different components of PFA can be delivered by either mental health or lay responders who provide acute assistance following trauma in a variety of settings (shelters, schools, workplace, etc.). Later phase interventions, labeled secondary psychological assistance (SPA), are delivered in the first weeks and months after critical incidents, and into the longer term recovery phase. The following goals of PFA and SPA drive the interventions listed below:

Goals of PFA
1. *Contact and engagement*: Establish a human connection in a nonintrusive, compassionate manner.
2. *Safety and comfort*: Enhance immediate and ongoing safety, and provide physical and emotional comfort.
3. *Stabilization:* Calm and orient emotionally overwhelmed/distraught survivors.
4. *Information gathering:* Help survivors to articulate immediate needs and concerns; gather additional information as appropriate.

5. *Practical assistance:* Offer practical assistance and information to help survivors address their immediate needs and concerns.
6. *Connection with social supports:* Connect survivors as soon as possible to social support networks, including family members, friends, neighbors, and community helping resources.
7. *Information on coping:* Provide information that may help survivors cope with the psychological impact of disasters.
8. *Linkage with collaborative services:* Facilitate continuity in disaster response efforts by clarifying how long the PFA provider will be available and (when appropriate) linking the survivor to another member of a disaster response team or to indigenous recovery systems, public-sector services, and organizations.

Goals of SPA
1. *Contact and engagement:* Establish a human connection in a nonintrusive, compassionate manner.
2. *Information gathering:* Help survivors to articulate their current needs and concerns; gather additional information, such as ability to cope since the event, resources, and risk factors.
3. *Comfort and support:* Help people tolerate the unknown and ever-changing situation; support positive coping styles, foster perserverence, reduce distress.
4. *Practical assistance:* Offer assistance with ongoing practical needs and help survivors access the resource network.
5. *Connection with social supports:* Empower/mobilize natural support systems and facilitate simple task groups.
6. *Information on coping:* Provide information on posttrauma reactions and reminders, stress management, positive coping, reframing of negative cognitions, coping with varied recovery trajectories in families, and anxiety management to deal with avoidance, experiencing, and intrusive thoughts.
7. *Problem solving:* Promote effective problem solving in relation to immediate needs, concerns, and goals.
8. *Risk reduction:* Promote understanding and effective use of postdisaster/terrorism risk-related information.
9. *Resilence and recovery:* Promote adaptive youth, family, and adult developmental progression.
10. *Linkage with collaborative services:* Make every effort to ensure that survivors' connect with resources and providers in the community.

Such actions are increasingly endorsed for universal application after mass violence or disaster, in part because they are considered to hold little potential for harm and do not contain elements (e.g., systematic emotional processing) hypothesized to be potentially harmful for some in the immediate aftermath of trauma. Although a small proportion of survivors may need to be triaged immediately to more formal psychiatric or psychological interventions, epidemiological studies and anecdotal evidence suggest that most individuals are capable of recovering from traumatic stress with appropriate education, information, and social and practical support in the very early phase following exposure to disaster or mass violence. Observations from the field suggest that most individuals are not interested in receiving formalized mental health interventions in this very early stage after mass violence or disaster, and because resilience is considered to be the norm following trauma exposure, compulsory procedures that impose a particular model or time line of recovery on all survivors of mass violence have been discouraged.

Although PFA has not yet been studied systematically, experience in the field suggests that it is generally acceptable to and well received by consumers. Experts generally concur that PFA practices are evidence-consistent, if not evidence-based (National Child Traumatic Stress Network and National Center for PTSD, 2005). A draft of the manual, which is currently being pilot-tested, is available at *www.ncptsd.va.gov/pfa/pfa.html*.

Outreach and Information Dissemination

Education is widely held to be a critical component of individual, group, and community interventions offered in the aftermath of disasters. As a relatively brief, non-stigmatizing, low-cost form of care, postdisaster education is generally designed to help survivors (1) better understand a range of posttrauma responses; (2) view their posttrauma reactions as expectable and understandable (not as reactions to be feared, signs of personal failure or weakness, or signs of mental illness); (3) recognize and deal with reminders that may trigger a posttrauma response; (4) understand the circumstances under which they should consider seeking further counseling; (5) know how and where to access additional help, including mental health counseling; (6) increase use of social supports and other adaptive ways of coping with the trauma and its effects; (7) decrease use of problematic forms of coping (e.g., excessive alcohol consumption, extreme social isolation); and (8) increase ability to help family members cope (e.g., information about how to talk to children about what happened). Accurate and timely information regarding the nature of the unfolding disaster situation is also an important part of education. Whereas self-help education is not as effective as more formal, cognitive-behavioral intervention for survivors of trauma, cognitive-behavioral self-help interventions have been found to be effective for treatment of non-trauma-related anxiety in a number of controlled treatment outcome studies (e.g., Gould & Clum, 1995; Lidren et al., 1994) and require more investigation with survivors of mass trauma.

Effectively disseminating educational materials is a key goal for a disaster mental health response. Many of those affected by terrorist attacks or other disasters do not seek mental health care or use available services (e.g., Smith, Kilpatrick, Falsetti, & Best, 2002). For instance, 3–6 months after the World Trade Center attacks in New York City, only 27% of those reporting severe psychiatric symptoms had obtained mental health treatment (Delisi et al., 2004). Generally, relatively little is known about how survivors make decisions about self-referral, use of services, or acceptance of referral for more intensive counseling. Because many trauma survivors are reluctant to use mental health services, it is recommended that disaster mental health programs take steps to educate survivors about effective self-recovery and peer-support strategies, as well as the availability and range of support services. Specific outreach strategies may differ depending on the preexisting mental health infrastructure, and the areas and individuals affected. Cultural sensitivity in design of services may affect rates of self-referral, engagement with care, and retention in counseling services.

Another component of information dissemination and outreach is use of the media to provided information and guidance to communities. Experts recommend providing the media with materials (e.g., interviews, releases, and programs) to help increase knowledge about trauma and recovery, for instance, information about self-protection, sleep hygiene, active coping, social connectedness, and so on. Successful delivery of such messages by a calm and honest leader of the community, who is able to ease the public's concern and provide instruction as to how to respond to the disaster, and provide timely, accurate, and sensitive communications, is felt to have the potential to decrease anxiety in the general public. Hotline numbers, resource contact information, and public education information can be disseminated through the media in an effort to reach a broad base of community members affected in the disaster area. For instance, Project Liberty (the public mental health program established in New York after the September 11, 2001, terrorist attacks) utilized well-known actors to attract the public's attention to call their lifeline hotline. Callers responded directly to the instruction to call the program (e.g., "Alan Alda told me to call"). Staff of the lifeline hotline could then direct callers to information, education, or intervention services. Additionally, peer outreach workers available on "warm lines" provided support and information to recipients who expressed concern that in calling for disaster relief they might be identified, stigmatized, or hospitalized.

Web-based services may be an effective and efficient manner in which to disseminate public education and information regarding disaster responses and available resources. The goal of the website set up in New York following September 11, 2001, was to provide information and referral, but a significant number of affected individuals e-mailed the disaster response program, communicating a need for an immediate crisis counseling response. Many individuals who accessed the website for support were concerned about confidentiality (particularly rescue workers), and felt that e-mail correspondence was a more confidential modality. Thus, offering Web-based crisis counseling may be a successful way to engage persons who otherwise would not accept such support. As noted below (and in Litz & Maguen, Chapter 16, this volume), new CBT-based protocols are currently being tested for Web-based intervention. In addition, for those seeking helpful information on trauma and recovery, websites are an excellent source of more in-depth information than can be found in news media. For example, the National Center for PTSD website (*www.ncptsd.va.gov*) saw a 10-fold increase in users seeking information following September, 11, 2001.

Finally, following disasters, dissemination of knowledge about recommended practice to community clinicians is crucial in dealing with heightened surge capacity in community, public, and mental health settings. Therefore, rapid dissemination of information to mental health responders on triage, acute intervention, and referral is necessary in the early phases following disaster, followed by more in-depth training and evidence-based practice in the long-term aftermath (see next section).

Technical Assistance, Consultation, and Training

Organizations, leaders, responders, and caregivers have to be supported at all phases postdisaster to provide what is needed to reestablish community structure, foster family recovery/resilience, and safeguard the community. This can be accomplished via the multiphase, multimodality dissemination of knowledge, consultation, and training.

It is important that different levels of training in disaster recovery be routinely delivered, from acute to long-term phases, to a broad range of professionals, including those outside the traditional mental health area. For example, trained professionals can include clergy, nurses, and teachers, especially in rural areas. Primary care physicians are also an important group to train, because many survivors with physical health complaints seek help from their doctor rather than visit a mental health setting. Hospital emergency room medical personnel are also important targets for training, because they are a first line of contact for some survivors (e.g., seriously injured survivors and their families, those concerned about exposure to biological or chemical agents). Acute-phase consultation and training in PFA, triage, and support of basic needs should be implemented as rapidly as possible to a broad range of disciplines, followed by more in-depth training in later phases.

The challenge for long-term training, often referred to as knowledge management (KM; see Ruzek, Friedman, & Murray, 2005) is to design and implement a process by which evidence-based treatments are not only disseminated but also utilized by clini-cians. A combination of workshop training and ongoing supervision has been success-ful in training community service providers in empirically-supported treatments for PTSD (Gillespie, Duffy, Hackmann, & Clark, 2002; Levitt, Davis, Martin, & Cloitre, 2003; Marshall, Amsel, Neria, & Suh, 2006). Recent evidence and experience suggest that mental health professionals can be rapidly trained in the delivery of evidence-based trauma treatments. Following September 11, 2001, several efforts to train mental health providers in evidence-based treatments were undertaken. Neria, Suh, and Mar-shall (2003) described their efforts to provide systematic training and supervision in prolonged exposure treatment for PTSD (Foa & Rothbaum, 1998) for New York City trauma therapists. Training was initiated approximately 2 months after the attacks, and more than 500 local clinicians were trained over a 12-month period. This relatively short training was found to increase knowledge and motivation, but not confidence in abilities to deliver the treatment. Effective training utilized lectures, role play, and clini-cal demonstrations by experts, as well as follow-on supervision. This training effort sig-nificantly increased the number of well-trained clinicians available to provide an effec-tive community response to a whole range of posttraumatic clinical problems.

Treatment

Survivors may present with a range of problems after experiencing mass violence or disaster (e.g., depression, sleep problems, fear, guilt, substance misuse), so that flexible or modularized, evidence-informed interventions with different components targeting individual needs have been included in current best practices recommendations (NIMH, 2002). In Western culture, CBT is the treatment with the best empirical sup-port to date for several problems that may emerge or be exacerbated after exposure to mass violence, such as PTSD, anxiety, panic, depression, guilt, and so forth (see Litz & Maguen [Chapter 16] and Resick, Monson, & Gutner [Chapter 17], this volume). RCTs with EMDR and selective serotonin reuptake inhibitors (SSRIs) have not shown the same efficacy as CBT (Brady et al., 2000; Davidson, Rothbaum, van der Kolk, Sikes, & Farfel, 2001; Rothbaum, 1997). Although there is growing support for EMDR, clinical trials comparing EMDR and CBT have shown more lasting effects from CBT. There-fore, EMDR would only be recommended if CBT were not available.

Although there is little RCT evidence for the effectiveness of CBT following mass violence, one RCT comprised a 15-session cognitive-behavioral intervention (CBI, compared to a wait-list control group) with children in the Gaza Strip. Distinctive psychosocial benefits were reported in children ages 6–11 and in adolescent girls; bene-fits included more effective communication; decreased hyperactivity, arousal, and dis-ruptive behavior; and increased sense of control, efficacy, and prosocial behavior (Khamis, Macy, & Coignez, 2004). One non-controlled study supports the effectiveness of CBT following a terrorist event in Northern Ireland (Gillespie et al., 2002). In patients who received a mean of eight sessions of CBT (focused primarily on cognitive restructuring rather than exposure), 97% showed varying degrees of improvement, most commonly in the 70–90% range, comparable to findings reported in controlled research with other traumatized populations.

Research also supports the use of SSRIs with depression and PTSD, which might be recommended for individuals who prefer a medication approach, or when psychotherapeutic approaches are not available (see Friedman & Davidson, Chapter 19, this volume). SSRIs might also serve as a supplement to CBT or other psychotherapeutic interventions. Given the state of the research literature, the chaotic and demanding recovery environment, the need to focus energy on pragmatic issues in the early phases postdisaster, and the need to reduce pathologizing of symptoms in the early phases postdisaster, it is generally recommended that these interventions be used at least 3 weeks after the trauma (Watson, 2004). Earlier pharmacological interventions should focus on targeted symptom alleviation only (e.g., insomnia).

Anecdotal reports following disasters over the last few decades suggest that numerous individuals wait many months before pursuing traditional mental health services. Sometimes, anniversary reactions or the holidays motivate those who have lost loved ones to seek treatment many months, or even years, after the loss.

As part of the Federal Emergency Management Agency (FEMA)–funded "Project Liberty" crisis counseling program in New York after September 11, 2001, more intensive "enhanced services" (10–12 session CBT-based individual intervention) were made available to some people (Hamblen et al., 2003). One specialized version of these services was tailored for traumatic grief, predicated on the recognition that intense grief reactions are not "pathological," and that continuing intense distress is common many months after bereavement (*www.projectliberty.state.ny.us/enhanced%20services/ esadult_factsheet.htm*). Because most individuals adapt over time without intervention, these enhanced services targeted persons whose grief reactions were especially long-lasting or disruptive. All enhanced services programs were federally funded for the first time as part of the New York response. As a result, it was possible to offer empirically supported services to all survivors of the disaster, rather than limiting them to those with adequate mental health insurance or financial means to pay for services.

ETHNOCULTURAL ISSUES

Because individuals are embedded in a broader familial, interpersonal, and social context, services should be tailored to meet the needs of as many community members as possible. Minorities deserve special attention for the following reasons: (1) They have more adverse postdisaster mental health outcomes and are more likely to retain the effects of disaster for a longer period of time (Norris et al, 2002); (2) the concept of pathology and health is different in every community; (3) disaster itself can create a temporary culture that highlights the discrepancies present in communities (Ursano et al., 2004); (4) victims may feel that they have been discriminated against and denied services because of their ethnicity, their exposure, and disparities in the interventions offered; and (5) because of the many types of barriers to seeking services among minorities that must be addressed, information should be gathered about minority populations' basic needs, barriers to care, and concepts of recovery, utilizing phrases and language that survivors can best understand.

To reach minorities, experts recommend working with primary care doctors, established community structures and natural leaders, and clergy. Minorities are more likely to delay seeking help until symptoms are severe and less likely to seek help from mental health specialists. Services should, as much as possible, be free and accessible, close to home, community-based, and offered in concert with other activities. It is important to destigmatize distress and help seeking, emphasize strengths, and value both interdependence and independence as appropriate goals. Whenever possible, natural support systems that are already present in the communities should be utilized, and outreach should be provided to those who do not seek services in a typical way.

PROVIDING SERVICES
UNDER CONDITIONS OF ONGOING THREAT

In many terrorism situations there is a continuing possibility of additional attacks. Such circumstances may create or maintain anxiety reactions (Silver, Holman, McIntosh, Poulin, & Gil-Rivas, 2002), and the disrupted daily routines created by these circumstances may be associated with increased symptomatology (Shalev, Tuval, Frenkiel, Hadar, & Eth, 2006). Postdisaster mental health care has sometimes been assumed to take place in conditions of relative safety, but when exposure to ongoing or subsequent attacks is a realistic possibility, mental health responders need to help survivors cope with anticipatory fears related to the threat of imminent or future attack, as well as distress from previous traumatic exposure. One study employed a cognitive-behavioral (e.g., relaxation breathing and challenging maladaptive thoughts) hotline intervention in Israel before the 2003 U.S. invasion of Iraq (see Somer, Tamir, Maguen, & Litz, 2005). Results indicated decreased anxiety on several measures. Shalev and colleagues (2003) described modifications in delivery of CBT for PTSD related to terrorist attacks in Israel, designed to reflect a terror-ridden environment. For instance, during *in vivo* exposure assignments, survivors were encouraged to expose themselves to situations that were clearly safe, not to situations widely considered dangerous and avoided by most of the populace (e.g., city centers where repeated bombings had occurred). Such pragmatic tailoring of interventions is needed to provide care in the midst of continuing danger and other challenging aspects of the postterrorism environment.

In some events, it may be difficult for mental health providers to establish face-to-face contact with survivors, due to restrictions on travel by authorities or perceptions of ongoing environmental danger. Telephone- or Internet-delivered services may be useful in these circumstances. Both cognitive-behavioral telephone (Greist, Osgood-Hynes, Baer, & Marks, 2000; Mohr, Lutz, Fantuzzo, & Perry, 2000; Somer, Buchbinder, Peled-Avram, & Ben-Yizhack, 2004) and Internet interventions (Gega, Marks, & Mataix-Cols, 2004) have proven helpful with a variety of mental health problems.

In New York after September 11, 2001, a LifeNet hotline established as part of Project Liberty received heavy use; it provided 24-hour mental health counseling, information, and referral, offering assistance in multiple languages. A study by Gidron and colleagues (2001) reported reductions in PTSD symptoms at 3- to 4-month follow-up utilizing a CBT-based telephone intervention. And as noted earlier, the telephone has also been used to reduce anxiety levels among individuals fearing future terrorist attacks (Somer, Tamir, Maguen, & Litz, 2005).

Finally, the Internet represents a delivery system with significant potential that has yet to be effectively harnessed to meet the needs of trauma survivors. A writing-based, cognitive-behavioral protocol delivered to students over the Internet, "Interapy," was associated with lower general psychopathology scores, more improvement in mood, greater reduction in PTSD symptoms, anxiety, depression, somatization, and sleep problems compared to a wait-list control (Lange, van den Ven, Schrieken, & Emmelkamp, 2001; Lange et al., 2003). Litz and colleagues (B. Litz, personal communication, November 6, 2005) have designed a cognitive-behavioral therapist-assisted, Internet-based self-help intervention to treat large numbers of traumatized individuals; it uses a form of stress inoculation training for both secondary prevention of PTSD and treatment of the chronic disorder.

CONCLUSIONS

As can be seen in this review of the empirical literature on interventions following mass violence, there are few well-controlled studies related to any particular intervention in this context. Rather, expert consensus based on both empirical literature and experiential practice endorses a multifaceted approach to the management of traumatic stress following disasters and mass violence. Experts in this field are currently attempting to address a number of mass violence situations and contexts effectively, including situations of ongoing threat, ethnocultural contexts, and situations of infectious disease. As we consider the many specific components of intervention, great need for program evaluation that evaluates the effectiveness of each component is apparent. There is a need to evaluate the optimal timing and balance of psychological intervention and community support following mass trauma. Of particular need is specifying which components are most effective in non-Western cultures and complex, multicultural environments. Research is needed that addresses a range of outcomes, including not only PTSD but also substance abuse, depression, anger and violence, interpersonal and role functioning, and physical health. In addition to such individual outcomes, research is needed that focuses on group, organizational, and community outcomes, such as behavioral, emotional, and functional consequences most likely to be expressed in the school or workplace (staff turnover, organizational cohesion, morale, absenteeism, performance deficits, or medical symptoms).

In addition to research, it is important to synthesize and disseminate the collective experiential knowledge of professional responders to develop proactive, practical strategies for disseminating evidenced-based information on intervention strategies to policymakers and practitioners in the field. Such actions facilitate the science-to-service and service-to-science communication needed to promote the best research and practice in this highly complex setting.

REFERENCES

Adger, W. N., Hughes, T. P., Folke, C., Carpenter, S. R., & Rockstrom, J. (2005). Social–ecological resilience to coastal disasters. *Science, 309*, 1036–1039.

Bisson, J. I. (2003). Single-session early psychological interventions following traumatic events. *Clinical Psychology Review, 23*, 481–499.

Bisson, J. I., Shepherd, J. P., Joy, D., Probert, R., & Newcombe, R. G. (2004). Early cognitive-behavioural therapy for post-traumatic stress symptoms after physical injury: Randomised controlled trial. *British Journal of Psychiatry, 184*, 63–69.

Bonanno, G. (2004). Loss, trauma and human resilience: Have we underestimated the human capacity to thrive after extremely aversive events? *American Psychologist, 59*(1), 20–28.

Brady, K., Pearlstein, T., Asnis, G. M., Baker, D., Rothbaum, B., Sikes, C. R., et al. (2000). Double-blind placebo-controlled study of the efficacy and safety of sertraline treatment of posttraumatic stress disorder. *Journal of the American Medical Association, 283*, 1837–1844.

Brewin, C. R. (2005). Risk factor effect sized for PTSD: What this means for intervention. *Journal of Trauma and Dissociation, 6*(2), 123–130.

Davidson, J. R. T., Rothbaum, B. O., van der Kolk, B. A., Sikes, C. R., & Farfel, G. M. (2001). Multicenter, double-blind comparison of sertraline and placebo in the treatment of posttraumatic stress disorder. *Archives of General Psychiatry, 58*, 485–492.

Delisi, L. E., Maurizio, A. Y., Yost, M., Papparozzi, C. F., Fulchino, C., Katz, C. I., et al. (2003). A survey of New Yorkers after the Sept. 11, 2001, terrorist attacks. *American Journal of Psychiatry, 160*(4), 780–783.

Foa, E. B., & Rothbaum, B. O. (1998). *Treating the trauma of rape: Cognitive-behavioral therapy for PTSD.* New York: Guilford Press.

Galea, S., & Norris, F. (2006). Public mental health surveillance and monitoring. In F. H. Norris, S. Galea, F. Friedman, & P. J. Watson (Eds.), *Methods for disaster mental health research* (pp. 177–193). New York: Guilford Press.

Gega, L., Marks, I. M., & Mataix-Cols, D. (2004). Computer-aided CBT self-help for anxiety and depressive disorders: Experience of a London clinic and future directions. *Journal of Clinical Psychology, 60*, 147–157.

Gidron, Y., Gal, R., Freedman, S. A., Twiser, I., Lauden, A., Snir, Y., et al. (2001). Translating research findings to PTSD prevention: Results of a randomized-controlled pilot study. *Journal of Traumatic Stress, 14*, 773–780.

Gillespie, K., Duffy, M., Hackmann, A., & Clark, D. M. (2002). Community based cognitive therapy in the treatment of post-traumatic stress disorder following the Omagh bomb. *Behaviour Research and Therapy, 40*, 345–357.

Gould, R. A., & Clum, A. A. (1995). Self-help plus minimal therapist contact in the treatment of panic disorder: A replication and extension. *Behavior Therapy, 26*, 533–546.

Gray, M. J., & Litz, B. T. (2005). Behavioral interventions for recent trauma: Empirically informed practice guidelines. *Behavior Modification, 29*(1), 189–215.

Greist, J. H., Osgood-Hynes, D. J., Baer, L., & Marks, I. M. (2000). Technology-based advances in the management of depression: Focus on the COPE Program. *Disease Management and Health Outcomes, 7*(4), 193–200.

Hamblen, J., Gibson, L. E., Mueser, K., Rosenberg, S., Jankowski, K., Watson, P., et al. (2003). *The National Center for PTSD's Brief Intervention for continuing postdisaster distress.* New York: Project Liberty.

Khamis, V., Macy, R., & Coignez, V. (2004). *The impact of the classroom/community/camp-based intervention (CBI) program* (U.S. Agency for International Development [USAID/WBG] technical report). Retrieved June 2005, from *www.usaid.gov/wbg/reports/Save2004_eng.pdf*

Lange, A., Rietdijk, D., Hudcovicova, M., van de Ven, J. Q. R., Schrieken, B., & Emmelkamp, P. M. G. (2003). Interapy: A controlled randomized trial of the standardized treatment of posttraumatic stress through the Internet. *Journal of Consulting and Clinical Psychology, 71*(5), 901–909.

Lange, A., van de Ven, J. Q. R., Schrieken, B., & Emmelkamp, P. M. G. (2001). Interapy: treatment of posttraumatic stress through the Internet: A controlled trial. *Journal of Behavior Therapy and Experimental Psychiatry, 32*(2), 73–90.

Levitt, J. T., Davis, L., Martin, A., & Cloitre, M. (2003, November). *Bringing a manualized treatment for PTSD to the community in the aftermath of 9/11.* Paper presented at the annual meeting of the Association for Advancement of Behavior Therapy, Boston.

Lidren, D. M., Watkins, P. L. Gould, R. A., Clum, G. A., Asterino, M., & Tulloch, H. L. (1994). A comparison of bibliotherapy and group therapy in the treatment of panic disorder. *Journal of Consulting and Clinical Psychology, 62*, 865–869.

Litz, B. T., Gray, M. J., Bryant, R. A., & Adler, A. B. (2002). Early intervention for trauma: Current status and future directions. *Clinical Psychology: Science and Practice, 9*, 112–134.

Marshall, R. D., Amsel, L., Neria, Y., & Suh, E. J. (2006). Strategies for dissemination of evidence-based treatments: Training clinicians after large-scale disasters. In F. H. Norris, S. Galea, M. J., Friedman, & P. J. Watson (Eds.), *Methods for disaster mental health research.* New York: Guilford Press.

McNally, R., Bryant, R., & Ehlers, A. (2003). Does early psychological intervention promote recovery from posttraumatic stress? *Psychological Science in the Public Interest, 4*, 45–79.

Mohr, W. K., Lutz, M. J. N., Fantuzzo, J. W., & Perry, M. A. (2000). Children exposed to family violence: A review of empirical research from a developmental-ecological perspective. *Trauma, Violence, and Abuse: A Review Journal, 1*(3), 264–283.

National Child Traumatic Stress Network and National Center for PTSD. (2005). *The psychological first aid field operations guide* (2nd edition). Retrieved August 17, 2006, from *www.ncptsd.va.gov/pfa/ pfa.html*

National Institute of Mental Health. (2002). *Mental health and mass violence: Evidence-based early psychological intervention for victims/survivors of mass violence. A workshop to reach consensus on best practices* (NIH Publication No. 02-5138). Washington, DC: U.S. Government Printing Office.

Neria, Y., Suh, E. J., & Marshall, R. D. (2003). The professional response to the aftermath of September 11, 2001, in New York City: Lessons learned from treating victims of the World Trade Center attacks. In B. T. Litz (Ed.), *Early intervention for trauma and traumatic loss* (pp. 201–215). New York: Guilford Press.

Norris, F. H., Friedman, M. J., Watson, P. J., Byrne, C. M., Diaz, E., & Kaniasty, K. (2002). 60,000 disaster victims speak: Part I. An empirical review of the empirical literature, 1981–2001. *Psychiatry, 65*, 207–239.

Pitman, R. K., Sanders, K. M., Zusman, R. M., Healy, A. R., Cheema, F., Laski, N. B., et al. (2002). Pilot study of secondary prevention of post-traumatic stress disorder with propranolol. *Biological Psychiatry, 51*, 189–192.

Rose, S., Bisson, J. I., & Wessely, S. C. (2003). A systematic review of single-session psychological interventions ("debriefing") following trauma. *Psychotherapy and Psychosomatics, 72*(4), 176–184.

Rosen, C. S., Young, H. E., & Norris, F. H. (2006). On a road paved with good intentions, you still need a compass: Monitoring and evaluating disaster mental health services. In E. C. Ritchie, P. J. Watson, & M. J. Friedman (Eds.), *Interventions following mass violence and disasters: Strategies for mental health practices* (pp. 206–223). New York: Guilford Press.

Rothbaum, B. O. (1997). A controlled study of eye movement desensitization and reprocessing in the treatment of posttraumatic stress disordered sexual assault victims. *Bulletin of the Menninger Clinic, 61*, 317–334.

Ruzek, J. I., Friedman, M. J., & Murray, S. (2005). Toward a knowledge management system for post-traumatic stress disorder. *Psychiatric Annals, 35*(11), 911–920.

Shalev, A. Y., Addesky, R., Boker, R., Bargai, N., Cooper, R., Freedman, S. A., et al. (2003). Clinical intervention for survivors of prolonged adversities. In R. J. Ursano, C. S. Fullerton, & A. E. Norwood (Eds.), *Terrorism and disaster: Individual and community mental health interventions* (pp. 162–188). Cambridge, UK: Cambridge University Press.

Shalev, A. Y., Tuval, R., Frenkiel-Fishman, S., Hadar, H., & Eth, S. (2006). Psychological responses to continuous error: A study of two communities in Israel. *American Journal of Psychiatry, 163*(4), 667–673.

Silver, R. C., Holman, E. A., McIntosh, D. N., Poulin, M., & Gil-Rivas, V. (2002). Nationwide longitudinal study of psychological responses to September 11. *Journal of the American Medical Association, 288*, 1235–1244.

Simon, A., & Gorman, J. M. (2004). Psychopharmacological possibilities in the acute disaster setting. *Psychiatric Clinics of North America, 27*(3), 425–458.

Smith, D. W., Kilpatrick, D. G., Falsetti, S. A., & Best, C. L. (2002). Post-terrorism services for victims and surviving family members: Lessons from Pan Am 103. *Cognitive and Behavioral Practice, 9*, 280–286.

Somer, E., Buchbinder, E., Peled-Avram, M., & Ben-Yizhack, Y. (2004). The stress and coping of Israeli emergency room social workers following terrorist attacks. *Qualitative Health Research, 14*(8), 1077–1093.

Somer, E., Tamir, E., Maguen, S., & Litz, B. T. (2005). Brief cognitive-behavioral phone-based intervention targeting anxiety about the threat of attack: A pilot study. *Behaviour Research and Therapy, 43*(5), 669–679.

Ursano, R. J., Bell, C. C., Eth, S., Friedman, M. J., Norwood, A. E., Pfefferbaum, B. C., et al. (2004). *American Psychiatric Association Work Group on ASD and PTSD: Practice guidelines for the treatment of acute stress and posttraumatic stress disorder* (American Psychiatric Association Steering Committee on Practice Guidelines). Washington, DC: American Psychiatric Association.

Watson, P. (2004). Mental health interventions following mass violence. *Stresspoints, 12*(2), 4–5.

Watson, P. J., Friedman, M. J., Gibson, L. E., Ruzek, J. I., Norris, F. H., & Ritchie, E. C. (2003). Early intervention for trauma-related problems. *Review of Psychiatry, 22*, 97–124.

Watson, P. J., Friedman, M. J., Ruzek, J. I., & Norris, F. H. (2002). Managing acute stress response to major trauma. *Current Psychiatry Reports, 4*(4), 247–253.

Watson, P. J., & Shalev, A. Y. (2004). Assessment and treatment of adult acute responses to traumatic stress following mass traumatic events. *CNS Spectrums, 10*(2), 123–131.

Zatzick, D. F., & Roy-Byrne, P. P. (2003). Developing high-quality interventions for posttraumatic stress disorder in the acute care medical setting. *Seminars in Clinical Neuropsychiatry, 8*(3), 158–167.

Zatzick, D. F., Roy-Byrne, P. P., Russo, J. E., Rivara, F. P., Droesch, R., Wagner, A. W., et al. (2004). A randomized effectiveness trial of stepped collaborative care for acutely injured trauma survivors. *Archives of General Psychiatry, 61*(5), 498–506.

Chapter 26

Key Questions and an Agenda for Future Research

Matthew J. Friedman, Patricia A. Resick,
and Terence M. Keane

There has been remarkable progress in advancing our conceptual and clinical understanding of posttraumatic stress disorder (PTSD) during the past 25 years. This volume attests to the depth and breadth of scientific research on psychological and psychobiological mechanisms that mediate or moderate the processing of trauma-related stimuli. It also documents the many significant advances in the development and testing of evidence-based psychosocial and pharmacological treatments for PTSD that are now available to clinicians. In this chapter, we briefly review 13 key crosscutting questions with important implications for science and practice.

Question 1. Does the scientific evidence support the APA/WHO initiative to categorize PTSD as a "stress-related fear circuitry disorder"?

The American Psychiatric Association and the World Health Organization have launched an important and exciting initiative to merge their overlapping but different nosological schemes for psychiatric diagnosis, the *Diagnostic and Statistical Manual of Mental Disorders* (DSM) and the *International Classification of Diseases* (ICD), respectively. As things currently stand, differences in PTSD diagnostic criteria between DSM-IV and ICD-10 result in a much lower threshold for ICD-10, primarily because it lacks both DSM-IV's F criterion (functional impairment) and C criterion (numbing symptoms) (see Peters, Slade, & Andrews, 1999).

The proposal, currently under consideration for DSM-V/ICD-11, would be a new category, stress-related fear circuitry disorders. Along with PTSD, panic disorder, simple phobia, and social phobia would comprise this group of diagnoses. The basis for this proposal, common neurocircuitry, cognitive alterations, and neurohormonal alterations, has been reviewed extensively in a recent book (Andrews, Charney, Sirovatka, & Regier, in press). If this proposal is adopted, these four diagnoses would be classified together as exemplars of this new category.

The major argument for PTSD emphasizes the amygdala's key role in processing threatening or fearful stimuli. Amygdala activation produces outputs to the hippocampus, medial prefrontal cortex, locus coeruleus, thalamus, hypothalamus, and striatum. This neurocircuitry mediates and moderates the afferent processing, appraisal, encoding, and retrieval of trauma-related information and coordinates the brain's reaction to such stimuli (Davis & Whalen, 2001; Friedman & Karam, in press). In PTSD, the normal restraining influence of the medial prefrontal cortex (PFC), especially the anterior cingulate gyrus and orbitofrontal cortex, has been severely disrupted (Charney 2004; Vermetten & Bremner, 2002; Woodward et al., 2006). The resulting disinhibition of the amygdala increases the likelihood of recurrent fear conditioning, because ambiguous stimuli are more likely to be misinterpreted as threatening; normal counterbalancing inhibitory PFC restraint is nullified, and key limbic nuclei may be sensitized, thereby lowering the threshold for fearful reactivity (Charney, 2004; Charney, Deutch, Krystal, Southwick, & Davis, 1993).

In addition to animal and brain imaging studies, Pavlovian fear conditioning has been repeatedly proposed as a model for PTSD in and of itself (Kolb et al., 1989), as a component of a two-factor theory model (Keane, Zimering, & Caddell, 1985) and within a cognitive context of activated fear networks (Foa & Kozak, 1986; Lang, 1977; see Chapter 3 by Monson, Friedman, & La Bash and Chapter 4 by Cahill and Foa in this volume).

The great body of data on the human stress response has also provided a major theoretical and clinical context through which to explicate the pathophysiology of PTSD with respect to both adrenergic hyperreactivity and hypothalamic–pituitary–adrenocortical (HPA) dysregulation (Charney, 2004; Friedman & McEwen, 2004; see Chapter 10 by Southwick et al., this volume). In short, PTSD may exemplify the prototypical stress-related fear circuitry disorder.

The problem with this formulation, however, is that other emotions, such as sadness, grief, anger, guilt, shame, and disgust, are also associated with PTSD. From a DSM/ICD classification perspective, this may not be a problem, because the nosological goal is to cluster diagnoses according to a common denominator. For understanding PTSD, however, it would be ill-advised to conceptualize this disorder entirely within the context of fear-based appraisals and reactions. PTSD consistently has been more comorbid with depression than with any of the anxiety disorders (Kessler, Sonnega, Bromet, Hughes, & Nelson, 1995; Kilpatrick et al., 2003; Nixon, Resick, & Nishith, 2004; Resick, Nishith, Weaver, Astin, & Feuer, 2002). In fact, when Cox, Clark, and Enns (2002) conducted a factor analysis of disorders assessed within the National Comorbidity Study (NCS), PTSD loaded, albeit weakly, on the depressive factor and not on the anxiety disorders factor. As pointed out by Keane, Brief, Pratt, and Miller (Chapter 15, this volume), there is a growing body of evidence tying most disorders to underlying externalizing and internalizing subtypes. Miller, Kaloupek, Dillon, and Keane (2004) have found three types of people with PTSD; people with simple PTSD; those with an internalizing subtype, who have a depressive and anxious response; and people with an externalizing subtype, who are more impulsive, angry, and prone to substance abuse. To conceptualize PTSD as a fear circuitry problem may mean ignoring those people who respond to their flashbacks, nightmares, shame, and anger with substance abuse or aggression. And if we are only to diagnose PTSD in people who have a fear-based disorder, then we will need to develop another diagnosis for those people who do not fear for themselves but are overcome with horror or helplessness, such as nurses, emergency medical personnel, graves registration workers, disaster first-responders, or families of homicide victims. These findings call into question the theories that imply only a biological or conditioned etiology to the disorder, or that may only apply to a portion of people who meet criteria for PTSD. At a minimum we need to consider different pathways to developing posttraumatic psychopathology or failure to recover from initial normal distress, as well as different pathways to reduction of symptoms.

Question 2. How well is the PTSD construct working?

Overall, the evidence exists that the construct of PTSD appears to be working (see Keane et al., Chapter 15, this volume, for psychometric review). Psychometric studies consistently show that measures of PTSD exhibit very high rates of internal consistency (Kilpatrick et al., 1998; Weathers, Ruscio, & Keane, 1999). Construct validity has been established, in that PTSD correlates with the outcomes that we would predict to correlate, such as work or family adjustment (Kulka et al., 1990), or health (see Schnurr, Green, & Kaltman, Chapter 20, this volume). Furthermore, PTSD does not correlate with things it should not be related to (e.g., no past traumatic events) (Kilpatrick et al., 1998; Orsillo et al., 1996; Weathers et al., 1999). Finally, PTSD has good predictive validity over time, in that it predicts long-term outcomes (Kessler et al., 1995; Kulka et al., 1990). That said, there is room for improvement.

In some ways, content validity is questionable. The symptom clusters and some of the individual items have not withstood close scrutiny. The question here is whether we have selected the optimal set(s) of symptoms to identify people with the disorder. Ideally we would want the most parsimonious set of symptoms to identify the disorder, and not an exhaustive list of all possible symptoms. Problems with concentration or sleep, for example, are common to many disorders. Trauma-related nightmares, on the other hand, would be specific to PTSD. Not remembering an important aspect of the traumatic event could reflect psychogenic amnesia, a dissociative process that might respond to treatment; however, one could also respond affirmatively to that item with a head injury from the trauma that will never resolve with treatment, because there is no memory that can be retrieved from long-term storage. Some factor-analytic studies have found that the amnesia item loads weakly or not at all (Buckley, Blanchard, & Hickling, 1998; Foa, Riggs, & Gershuny, 1995; King, Leskin, King, & Weathers, 1998; Taylor, Kuch, Koch, Crockett, & Passey, 1998).

Since the three PTSD symptom clusters with their 1, 3, 2 configuration were introduced, many factor-analytic studies have attempted to confirm the underlying structure of the disorder (Amdur & Liberzon, 2001; Asmundson et al., 2000; Buckley et al., 1998; Foa et al., 1995; King et al., 1998; Simms, Watson, & Doebbeling, 2002; Taylor et al., 1998). Thus far, no one has ever found the three DSM clusters of reexperiencing, avoidance, and arousal with the 17 current items. Most studies have found either two (intrusion/avoidance and arousal/numbing) or four factors, which most frequently separate out effortful avoidance from numbing. Someone who actively avoided but did not have numbing symptoms, or vice versa, would not meet the criteria for the disorder under the four-factor solution or might not meet current criteria with the requirement of three types of avoidance. This does not mean that a person does not have significant distress from reexperiencing or arousal symptoms, or impairment in functioning. It may just mean that our lists of avoidance and numbing symptoms may be too limited to capture the range of avoidant coping in which people engage to stop their flashbacks, thoughts, or emotions. For example, when we ask traumatized people about their avoidance, we do not specifically ask them, as part of PTSD criteria, whether they use alcohol before going to bed to suppress their nightmares or to shut down their strong affect when upset (Nishith, Resick, & Mueser, 2001; Sharkansky, Brief, Peirce, Meehan, & Mannix, 1999). Perhaps we need to move to a more functional examination of symptoms and how they interact.

Question 3. What is the evidence for subsyndromal PTSD as a distinct diagnostic entity? Should PTSD be considered a dimensional rather than a categorical disorder?

When the findings from the National Vietnam Veterans Readjustment Study (NVVRS) were first published (Kulka et al., 1990) results were reported with respect to both full and "partial" PTSD. Partial PTSD had been determined through an adjudication process by which individuals who met most full PTSD diagnostic criteria (e.g., 1B, 3C, and 2D cluster symptoms) and also exhibited severe overall symptom severity on the Mississippi and MMPI PTSD Scales were given that designation (Weiss et al., 1992). Indeed, most veterans diagnosed with partial PTSD lacked only one C cluster symptom to meet full PTSD diagnostic criteria. The rationale for this procedure was that veterans with partial PTSD exhibited significant posttraumatic distress that often required clinical attention. This impression has certainly been echoed by many clinicians who have accepted patients with partial PTSD for treatment.

Since that time, other investigators have also identified partial or subsyndromal cohorts in research studies. In most cases, the criteria for this designation have been specified more clearly than those in NVVRS, to eliminate an adjudication process that might be open to some inconsistencies (Breslau, Lucia, & Davis, 2004; Friedman, Schnurr, Sengupta, Holmes, & Ashcraft, 2004; Schnurr et al., 2000). Breslau and colleagues (2004) measured both symptom severity and functional impairment; they found that on all domains measured, patients with partial/subsyndromal PTSD were significantly more impaired than normal comparison subjects, and significantly less impaired than subjects with full PTSD.

What all of this suggests is that PTSD is a spectrum disorder in which posttraumatic stress symptoms are distributed along a mild-to-severe continuum. According to this argument, people who meet PTSD diagnostic criteria generally represent those affected most severely, but the line separating full and partial/subsyndromal PTSD is arbitrary at best.

There is precedent in DSM-IV for the addition of a subsyndromal entity as a recognized diagnosis in its own right. For example, dysthymia is a subsyndromal major depressive disorder, and cyclothymia is subsyndromal bipolar affective disorder. Therefore, the argument goes, addition of partial/subsyndromal PTSD to DSM-V would acknowledge the dimensional nature of posttraumatic distress and provide a diagnostic niche for people requiring clinical attention who do not meet full PTSD diagnostic criteria.

When considering partial/subsyndromal PTSD, it may be important to distinguish between individuals who once met full PTSD criteria and are now in partial remission and those who have never exceeded the full PTSD threshold (Zlotnick et al., 2004). It has also been suggested that the partial/subsyndromal diagnostic designation be restricted to individuals who meet the F criterion for symptom severity and functional impairment (Mylle & Maes, 2004).

Clearly, much more research is needed to address this issue. For starters, a consistent set of criteria for partial/subsyndromal PTSD must be adopted, so that all research on this putative disorder will be conducted on people who meet the same diagnostic criteria. Next, research is needed to demonstrate that partial/subsyndromal PTSD is clinically significant in terms of symptom severity and functional impairment. Finally, it would be important to determine whether partial/subsyndromal PTSD responds to treatments shown to be effective for full PTSD, or whether better results might be achieved from different therapeutic approaches.

Question 4. What new directions should be considered with respect to psychosocial treatments?

Cognitive-behavioral therapy (CBT), especially cognitive therapy, cognitive processing therapy, and prolonged exposure have proven to be very effective treatments for PTSD. They are recognized by all major practice guidelines as evidence-based approaches. Each of these CBT approaches provides a technique for focusing on traumatic material, either through the extinction of trauma-related fear networks, as in prolonged exposure, or in the correction of trauma-related erroneous cognitions, as in the cognitive therapies. The success of these various trauma-focused approaches has led to the general belief that the theoretical underpinning for effective psychosocial treatment is the processing of traumatic material (see Cahill & Foa, Chapter 4, and Resick, Monson, & Gutner, Chapter 17, this volume).

There is evidence to the contrary. Most notable is stress inoculation therapy (SIT), a CBT approach that focuses on symptom management rather than trauma processing. Although not tested recently, older studies have shown comparable results from SIT and from prolonged exposure among participants with PTSD related to sexual trauma (Foa et al., 1999; Foa, Rothbaum, Riggs, & Murdock, 1991). Other studies have shown that supportive or present-centered therapies that avoid traumatic material and focus instead on current symptoms and problem-solving techniques, produce significant pre- to posttreatment effects (which are not always equal to those achieved by CBT) (McDonagh-Coyle et al., 2005; Schnurr et al., 2007).

A possible advantage of present-centered therapies over CBT, especially prolonged exposure, is in the area of recruitment and retention. Clients offered CBT within the context of randomized trials are sometimes hesitant to participate because of reluctance to confront the traumatic material that has made their PTSD so difficult to tolerate. Furthermore, in randomized trials of CBT with present-centered or supportive therapy comparison conditions, failure to complete (usually 10–12 sessions) therapy is sometimes higher in the CBT conditions. At present, it is not known whether there are differences between individuals most likely to benefit from CBT or from a present-centered approach. This should be an area for fruitful research.

A second important focus for research concerns, eye movement desensitization and reprocessing (EMDR) is clearly an effective evidence-based treatment for PTSD (see

Monson et al., Chapter 3, and Resick et al., Chapter 17, this volume). What is unclear is how it works. Is it another variant of CBT (Lohr, Tolin, & Lilienfeld, 1998)? Is it a unique amalgam of proven client-centered approaches (Hyer & Brandsma, 1997; Lohr et al., 1998)? Or does it achieve its effects through a novel mechanism of action, as maintained by its advocates (Shapiro & Maxfield, 2002)? Although dismantling studies suggest that repetitive motor movements are not necessary for EMDR's success (Pitman et al., 1996; Renfrey & Spates, 1994), we lack a clear idea of how it works. Assuming that EMDR works differently than CBT, elucidating its active ingredient(s) might pave the way for other novel and effective psychosocial approaches.

The last decade has seen the emergence of the "third wave" of cognitive-behaviorism influenced by Eastern and "mindfulness" approaches that emphasize acceptance, although rigorous evidence supporting their utilization is currently lacking. These include dialectical behavior therapy, mindfulness-based cognitive therapy, and acceptance and commitment therapy (see Monson et al., Chapter 3, and Welch & Rothbaum, Chapter 23, this volume). Clearly, such treatments should be evaluated in randomized, controlled clinical trials (RCTs) to determine whether their current popularity is warranted.

After the first wave of studies that found efficacious treatments for PTSD, the next wave began to examine a number of questions: whether these findings translate to practice settings (e.g., Foa et al., 2005), and how best to disseminate and train practitioners in these approaches, and determine the necessary and sufficient conditions for treatment. Researchers are pushing the envelope of PTSD treatment by examining the effectiveness of PTSD treatment in different substance abuse populations (Brady, Dansky, Beck, Foa, & Carroll, 2001; Najavits, 2004) and in those with severe mental illness (Mueser, Rosenberg, Goodman, & Trambetta, 2002; Rosenberg, Mueser, Jankowski, Salyers, & Ackers, 2004), or personality disorders (Feeny, Zoellner, & Foa, 2002). People with comorbid PTSD and medical or neurological conditions represent another clinical challenge. For example, because of the war in Iraq, which has been notable for blast injuries, there are troops returning with both PTSD and traumatic brain injury. Treatment-refractory patients or those who drop out of treatment are also likely to be included in the next wave of studies. More studies of predictors of treatment outcome and, eventually, research on patient–treatment matching will be important in understanding how to modify treatments for difficult populations.

A final new direction for psychosocial treatments involves the application of new technologies and pushing the envelope with regard to the method of treatment delivery. These include Web-based treatments, virtual reality, and telehealth approaches (see Welch & Rothbaum, Chapter 23, this volume). This very exciting area will certainly attract a great deal of attention in the foreseeable future.

Question 5. What are the major questions regarding biological alterations associated with PTSD and how might they be related to psychological processes?

Despite great advances in explicating biological alterations associated with PTSD (see Neumeister, Henry, & Krystal, Chapter 9, and Southwick et al., Chapter 10, this volume), progress in developing effective pharmacotherapy has not kept pace (see Friedman & Davidson, Chapter 19, this volume). This suggests that we may need to develop and test medications that act primarily on mechanisms other than serotonergic or adrenergic receptor systems, such as corticotropin-releasing factor (CRF), neuropeptide

Y (NPY), glutamate, gamma-aminobutyric acid (GABA), or other factors (see Charney, 2004). Our capacity to target key dysregulated pharmacological mechanisms would be greatly enhanced if it was predicated on a more comprehensive and fine-grained understanding of neurobiological abnormalities associated with PTSD.

Historically, PTSD has emerged from a wide variety of antecedent syndromes, each named after the putative etiological traumatic experience, such as soldier's heart, railway spine, shell-shock, combat fatigue, KZ (concentration camp) syndrome, rape trauma syndrome, post-Vietnam syndrome, and so on (see van der Kolk, Chapter 2, and Monson et al., Chapter 3, this volume). At present, the preponderance of evidence indicates that similar alterations in neurophysiology, neurobiology, and functional brain imaging are detected among individuals with PTSD, regardless of the nature of the precipitating traumatic experience.

An interesting exception, however, was the finding by Southwick and colleagues (1997), who suggested two different biological endophenotypes for PTSD: an adrenergic subtype sensitive to intravenous infusions of the adrenergic alpha$_2$ antagonist yohimbine, and a serotonergic subtype sensitive to the serotonin 5-HT$_2$ agonist meta-chlorophenylpiperazine (MCPP). The first group exhibited panic attacks, intrusive recollections, and flashbacks to yohimbine but not MCPP, whereas the second group exhibited similar symptoms to MCPP but not yohimbine. Such results suggest that PTSD may be an end-stage syndrome that can develop through alterations in any one of several biological systems (in the same way that fever or edema might result from a number of different, independent abnormalities). Explication of such different pathophysiological pathways to PTSD would not only enhance our current understanding of this disorder but might also enable us to develop more effective treatment matching strategies. For example, the MCPP subtype might respond preferentially to medications affecting 5-HT receptors, whereas the yohimbine subtype might benefit more from adrenergic agents.

Consideration of endophenotypes leads to speculation about differences in genotype among people with PTSD (see Segman, Shalev, & Gelernter, Chapter 11, this volume). As discussed elsewhere (see Question 13), genetic research may advance our understanding of differences between vulnerable and resilient individuals following exposure to traumatic events. Genotyping may also enable us to predict which individuals are most likely to respond to which pharmacological agents or psychosocial interventions. For example, individuals homozygous for the two short alleles of the 5-HT transporter gene might be better candidates for serotonergic medications than for other agents. A final area in which genotyping should prove to be an important tool concerns information it can provide about drug metabolism. Individual differences in metabolic capacity can often determine whether a particular agent will be effective, ineffective, or toxic. The usefulness of such information for selecting the best medications for patients with PTSD is potentially no different than that for other disorders.

Question 6. What new directions should be considered with respect to pharmacotherapy?

The major challenge for the future is to develop and test pharmacological agents that can do justice to the many potential receptor intervention sites that appear to mediate or moderate psychobiological abnormalities associated with PTSD. Most current research has focused on serotonergic mechanisms and, to a lesser extent, the adrenergic system. The emergence of many new anticonvulsant/mood stabilizer agents has shifted attention to glutamatergic and GABA-ergic mechanisms. The obvious goal for future pharmacotherapy is to target specific pathophysiological abnormalities associated with PTSD. The lion's share of medication trials so far have exemplified an empirical rather than conceptually driven approach, and have utilized agents with established efficacy for other disorders, such as antidepressants, antiadrenergics, anticonvulsants, and atypical antipsychotics.

Medications designed primarily for PTSD might include CRF antagonists, NPY enhancers or more specific serotonergic, glutamatergic, or GABA-ergic agents. Agents that promote neurogenesis should also be a focus for future research (Friedman, 2002). The potential importance of fear conditioning, resistance to extinction, and sensitization/kindling (Charney, 2004) should direct attention to glutamatergic agents and medications that can modulate these mechanisms, such as D-cycloserine, lamotrigine, and other anticonvulsants. Emerging knowledge of the psychobiology of dissociation points to possible roles for medications acting on N-methyl-D-aspartate (NMDA), alpha-amino-3-hydroxy-5-methyl-4-isoxyazolepropionic acid (AMPA), $alpha_2$ adrenergic, and 5-HT_2 receptors (Chambers et al., 1999). Further research with promising but inadequately tested classes of medication would include RCTs with antiadrenergic and anticonvulsant agents in particular.

In practice, the problem for pharmacotherapy is that full remission is achieved in only a minority of cases. Indeed, the multisite trials that resulted in U.S. Food and Drug Administration (FDA) approval for sertraline and paroxetine as indicated treatments for PTSD documented remission rates of approximately 30%. Another 20% showed little or no improvement, whereas approximately 50% exhibited notable improvement but only partial remission. Finally, selective serotonin reuptake inhibitor (SSRI)–related improvement in PTSD was independent of improvement in depressive symptomatology (Brady et al., 2000; Davidson, Rothbaum, van der Kolk, Sikes, & Farfel, 2001; Marshall, Beebe, Oldham, & Zaninelli, 2001; Tucker et al., 2001). As a result, there has been renewed interest in augmentation strategies in which a partial responder to sertraline or paroxetine, for example (while remaining on the current medication), would in addition receive an antiadrenergic, anticonvulsant, atypical antipsychotic, or some other agent. As reviewed (see Friedman & Davidson, Chapter 19, this volume), a few small RCTs with atypical antipsychotic agents have been encouraging, but much more work needs to be done.

Frankly, a more successful augmentation strategy for partial responders might be with CBT, because monotherapy trials have shown greater success with this approach than with medication. In short, until we can identify more effective medication, systematic exploration of augmentation with other medications or CBT is a top priority.

Other important areas for future research are treatment for children (see Question 9), pharmacotherapy for acute posttraumatic reactions, and prevention of PTSD. Thus far, a handful of studies with propranolol, hydrocortisone, and imipramine (Pitman et al., 2002; Robert, Blakeney, Villarreal, Rosenberg, & Meyer, 1999; Schelling et al., 2001) suggest that early pharmacological intervention might have a favorable influence on ameliorating posttraumatic symptomatology. Designing such a "morning after pill" (Friedman, 2002) is a major priority that should focus on CRF, NPY, adrenergic, glutamatergic, and possibly anti-inflammatory agents.

Question 7. Can neuroimaging inform us of the mechanisms of action for CBT?

Perhaps the most intriguing and conceptually rich area for biological research concerns the explication of mechanisms underlying successful psychosocial treatment. There are several potential areas for investigation: the locus of action of effective psychosocial versus pharmacological treatment; different biological alterations produced by different CBT treatments; and biological alterations associated with EMDR.

Utilizing functional brain imaging, Mayberg and associates (Goldapple et al., 2004) identified different loci of action among successfully treated depressed patients who received either CBT or pharmacotherapy. In short, they report a "top-down" cortical, especially prefrontal cortical, target area among treatment-responsive depressed CBT patients. This contrasts with a "bottom-up" subcortical locus of action among depressed patients who responded to medications. Different target areas for these different therapeutic approaches suggest both how and why conjoint CBT–medication treatment might be more effective than either treatment alone. This is a very exciting area of investigation that, we hope, will be extended from depression to PTSD.

An additional question that might be addressed by such research concerns alterations in brain function associated with successful treatment of comorbid PTSD and depression. Because the two disorders frequently occur simultaneously, comparisons between pre- and posttreatment functional brain imaging might help us understand whether comorbid PTSD and depression represents the co-occurrence of two distinct DSM-IV Axis I disorders, or whether PTSD–depression is really a single entity, either a depressive subtype of PTSD or a posttraumatic subtype of major depressive disorder.

A number of exciting questions might be addressed by conjoint biological–CBT research. First, it would be useful to discover what PTSD-related biological alterations are normalized following successful CBT treatment. The very sparse research in this area comprises a few small studies on heart rate, skin conductance, and HPA function (Friedman, McDonagh-Coyle, Jalowiec, McHugo, & Wang, 2006; Griffin, Nishith, Resick, & Yehuda, 1997; Heber, Kellner, & Yehuda, 2002).

The most theoretically interesting focus, however, would be functional brain imaging before and after different CBT approaches. Different mechanisms of action have been postulated for different CBT approaches (Question 4, see Monson et al., Chapter 3; Cahill & Foa, Chapter 4; Resick et al., Chapter 17; and Welch & Rothbaum, Chapter 23, this volume). For example, Resick and colleagues (2002) conducted a large randomized clinical trial comparing prolonged exposure and cognitive processing therapy. Contrary to what might have been expected, both treatments appeared to be equally effective in reducing PTSD and depressive symptoms. These results suggest that both treatments may either have in common a neurocognitive locus of action or achieve the same results through different pathways. Functional brain imaging before, during, and after treatment would be an excellent way to address this issue. It would also be of great interest to use such an approach for patients with PTSD receiving psychosocial treatments that do not focus on processing traumatic material, but emphasize present-centered symptom relief and problem-solving techniques, such as SIT, given that such approaches have compared well with prolonged exposure in previous trials (see Question 4; Foa et al., 1999; Foa, Rothbaum, Riggs, & Murdock, 1991).

Finally, functional brain imaging might help to settle the question about the mechanism of action in EMDR (see Question 5). Altered functional brain imaging following successful EMDR treatment resembling that observed following successful CBT would suggest that EMDR is a variant of CBT. If, on the other hand, evidence that EMDR and

CBT appear to be mediated and moderated by different brain mechanisms would support claims by EMDR advocates that there is indeed a unique therapeutic approach for PTSD.

Question 8. What are the major challenges in research on memory and dissociation, and how might such findings influence both clinical and forensic practice?

Among other things, PTSD is a disorder of memory. On the one hand, some people with PTSD cannot escape intolerable, intrusive recollections of their traumatic experiences. On the other hand, some survivors of such experiences cannot retrieve memories of part, or all, of such events. These clinical observations have spawned in the past decade considerable research on fundamental mechanisms of cognition and memory, and how such mechanisms may be altered among individuals exposed to traumatic events and those who have developed PTSD.

It is generally accepted that different cognitive and neurobiological mechanisms underlie the acquisition, encoding, and retrieval of emotionally charged information compared to more neutral input. It also appears that such cognitive processing is altered among people with PTSD. Such abnormalities in cognition and memory appear to be implicated in expression of clinical symptoms such as re-experiencing, fragmented thoughts, amnesia, and dissociation (see Brewin, Chapter 7, this volume). Indeed, trauma-related dissociation and dissociative amnesia are topics of renewed interest, because of their prominence in PTSD and other trauma-related disorders (see DePrince & Freyd, Chapter 8, this volume).

Questions about PTSD-induced memory alterations and dissociation have prompted innovative basic and clinical research. Investigators utilizing sophisticated cognitive psychology paradigms and/or functional brain imaging protocols designed with these questions in mind have begun to enlarge our understanding of fundamental mechanisms that mediate and moderate information processing, encoding, and memory retrieval. Much more research is needed to help us understand how such mechanisms are altered in PTSD and explicate the psychopathology and the pathophysiology of this disorder. We hope this knowledge will inform therapeutic techniques that address and ameliorate such abnormalities.

Question 9. What new directions in developmental issues should be considered with respect to children, adolescents, and older adults?

In recent years there has been increased attention to the impact of traumatic exposure on younger and older individuals. We have learned not to generalize from findings with 30-year-old adults to children, adolescents, or older adults. Each age group appears to respond differently to exposure to traumatic events. Thus, a developmental perspective is needed to inform theory and practice across the lifespan.

Many of the cognitive, emotional, and behavioral challenges associated with normal development mediate or moderate the impact of trauma exposure in the young (see Fairbank, Putnam, & Harris, Chapter 13, this volume). Key trajectories influencing this process include neurobiological maturation, affect regulation, cognitive–emotional development, coping capacity, beliefs about oneself and the environment, social embeddedness, safety and security at home, and prior/ongoing exposure to severe or traumatic stress. Such developmental differences may influence not only the appraisal, cognitive processing, encoding, and retrieval of traumatic material but also affect the posttraumatic psychological, emotional, and behavioral expression of such experiences. Thus, treatments must be developmentally sensitive and appropriate. One size definitely does not fit all.

The good news is that an emerging body of clinical research shows that there are effective, evidence-based treatments for children and adolescents. These include individual therapy, conjoint child–parent approaches, school-based treatments, and interventions designed to affect the social environment (see Saxe, MacDonald, & Ellis, Chapter 18, this volume). The challenge for the future is to accelerate the pace of such research to develop a solid body of evidence-based treatments for children and adolescents at all developmental levels.

Research on pharmacotherapy for children is at an early stage, with few published RCTs. The single study showing the efficacy of imipramine for children hospitalized on a burn unit with acute stress disorder (Robert et al., 1999) is an important finding, but much more research is needed. Recent concerns about suicidal thoughts among depressed children prescribed SSRIs (U.S. Food and Drug Administration, 2004) are a barrier to research that needs to be, and will be, overcome if addressed thoughtfully.

At the other end of the age continuum are the young-old, middle-old, and old-old adults, all of whom receive much less attention either conceptually or with respect to basic or clinical research. Indeed, medication trials generally exclude older adults as participants, and what we know about the treatment of PTSD in the elderly is affected by this limitation. Studies that possess a sufficient sample of older adults might analyze treatment findings as a function of age to address systematically this important void in the literature. Some of the unique challenges regarding older adults with PTSD concern the impact of retirement, reduced physical capacity, concurrent physical illnesses, impaired cognition and memory caused either by normal aging or neurodegenerative processes, loss of social support through death and illness, and metabolic changes affecting pharmacotherapy (see Cook & Niederehe, Chapter 14, this volume). Finally, because the processing of traumatic material is often carried out within the context of life review, therapy with older adults presents exciting challenges and opportunities for the development of age-specific components in psychological treatments.

Question 10. What are the major questions about gender differences with respect to posttraumatic reactions, and how should they be addressed in research and practice?

Although gender has played an important role in epidemiological studies, comparisons of men and women in laboratory, information processing, and risk factor studies have generally been lacking. Gender comparison has also been insufficient in treatment research. Treatment studies in the United States have tended to focus on one type of trauma or another (i.e., combat or rape), resulting in gender segregation. For example, treatment for combat trauma among women and rape trauma among men has not been adequately studied. Some of the British and Australian clinical trials including mixed trauma samples have therefore studied both genders; however, gender comparison could not be studied, because sample sizes were too small. Therefore, it is generally unknown whether the two genders respond to PTSD treatment differently. Another large area with a dearth of information regards biological underpinnings of PTSD relative to gender (see Kimerling, Ouimette, & Weitlauf, Chapter 12, this volume).

***Question 11. What new directions in research and practice will advance
our understanding of PTSD within a cross-cultural context?***

PTSD has been identified in traumatized individuals around the world. Despite ethnic, cultural, and other differences between cohorts, the pattern of posttraumatic symptoms operationalized by DSM-IV has been detected in both Western industrialized settings and traditional cultures across the globe. The dose–response curve between trauma severity and PTSD prevalence has proven to be a robust relationship. Therefore, the question is no longer whether PTSD is solely a European American, culture-bound syndrome with no relevance for other people, but whether PTSD is the best posttraumatic idiom of distress for individuals from traditional cultures.

It is at present an unanswerable question given that few investigators have addressed this issue systematically. Mexican men and women exposed to a variety of traumatic events reported both PTSD and culture-specific idioms of distress (e.g., *ataques de nervios*; Norris, Murphy, Baker, & Perilla, 2003). Among Puerto Rican survivors of the 1985 floods and mudslides, 17% of those reporting *ataques de nervios* also met criteria for PTSD (Guarnaccia, Canino, Rubio-Stipec, & Bravo, 1993). Much more research is needed to investigate the degree of overlap between PTSD and a variety of culture-specific posttraumatic idioms of distress.

North and colleagues (2005), in the only study that directly compared distinct ethnocultural groups exposed to a similar trauma, compared Africans exposed to the American embassy bombing in Nairobi and Americans exposed to the Murrah Federal Building bombing in Oklahoma City. They found similar outcomes for the individuals exposed to these events. Morbidity with respect to PTSD symptoms and functional impairment was remarkably similar in Africans and Americans. Given these quantitative data, it would have been extremely interesting to have carried out qualitative research (e.g., focus groups, key informants, etc.; see Palinkas, 2006) to understand similarities and differences between African and American representatives of these two cohorts. This very important question suggests the combination of quantitative and qualitative approaches to deepen our understanding of major cross-cultural questions.

It has previously been proposed that PTSD reexperiencing and hyperarousal symptoms are universal aspects of a posttraumatic response, because they can be precipitated in a laboratory with psychological (e.g., script-driven imagery) or pharmacological (e.g., yohimbine) experimental probes. According to this suggestion, the avoidant/numbing symptoms are least hardwired and most subject to cultural factors that might influence the appraisal of a given situation as tolerable or threatening (Friedman & Marsella, 1996). Furthermore, if the avoidance/numbing criterion reflects attempts to cope with the reexperiencing and arousal symptoms, then coping would be strongly affected by cultural practices. By extension, it would be expected that the A2 criterion, the subjective experience of fear, helplessness or horror, would also be significantly affected by cultural influences. A potential problem with this formulation is the great diversity of individual thresholds for threat appraisals within a homogeneous culture. Nevertheless, this is a question that could be investigated systematically.

Such speculations lead inevitably to questions about ethnocultural differences in psychobiological reactivity associated with either PTSD or culture-specific idioms of posttraumatic distress. Two questions merit attention in this regard. First, do people diagnosed with PTSD from industrialized and traditional cultures exhibit the same pattern of biological alterations? Second, do Mexicans, for example, exposed to the same traumatic event, diagnosed either with PTSD or *ataques de nervios*, exhibit similar or different patterns of biological alterations? Designing experiments to address such questions is straightforward. The challenge is the implementation of such designs in settings that deepen our understanding of posttraumatic reactions in different ethnocultural settings.

Question 12. How should our emerging understanding of the association between PTSD and physical disorders influence research and practice?

PTSD appears to be a major risk factor for physical illness. In comparison with nonaffected people, individuals with PTSD are more likely to develop a wide spectrum of medical problems (Schnurr & Green, 2004; see Schnurr et al., Chapter 20, this volume). The mechanisms underlying this association are unclear, but a variety of psychological (e.g., depression, hostility), behavioral (e.g., risky behaviors, substance abuse), and biological alterations (e.g., adrenergic, HPA, and immunological dysregulations) have been proposed (Friedman & McEwen, 2004; Friedman & Schnurr, 1995; Schnurr & Jankowski, 1999).

A sampling of important observations in this area includes (1) distinctive electrocardiographic abnormalities among Vietnam War veterans with PTSD (Boscarino & Chang, 1999); (2) earlier onset of peripheral vascular disease among World War II and Korean Conflict veterans with PTSD (Schnurr, Spiro, Vielhauer, Findler, & Hamblen, 2002); and (3) increased mortality due to cardiovascular disease or cancer among Vietnam War veterans with PTSD (Boscarino, 2006). In addition, medical problems known to be influenced by stress, such as chronic fatigue syndrome and fibromyalgia, are exacerbated following exposure to traumatic situations (Ciccone, Elliott, Chandler, Navak, & Raphael, 2005; Eisen et al., 2005; Kang, Natelson, Mahan, Lee, & Murphy, 2003; Raphael, Janal, & Navak, 2004).

The observation that exposure to traumatic stress may be followed by medical problems is not new. Stress-related cardiovascular syndromes such as soldier's heart, DaCosta's syndrome, and neurocirculatory asthenia have been diagnosed among military veterans since the American Civil War (Pizarro, Silver, & Prause, 2006; see van der Kolk, Chapter 2, and Monson et al., Chapter 3, this volume). What is new is the discovery that exposure to traumatic stress per se is not associated with such outcomes. The association between traumatic stress and medical illness appears to be mediated by PTSD (Schnurr & Green, 2004). The research in this area, while compelling, is not conclusive at this time. Therefore, the first order of business is more research to document this relationship and to understand how pathophysiological alterations associated with PTSD increase the risk for medical illness.

Translating science into practice, these findings suggest that medical practitioners, especially primary care providers, should routinely screen their patients for PTSD symptoms. Such information is potentially as relevant for effective treatment and health management as current medical concerns about smoking, alcohol abuse, and obesity. Recognition that PTSD may affect both the etiology and management of medical illness has prompted the Department of Veterans Affairs (VA) to institute a new policy, namely, that all veterans seeking primary care treatment will be screened for PTSD symptoms on an annual basis. Other treatment implications include the integration of mental health experts within the primary care setting, where most patients with PTSD and other psychological conditions are treated.

Most patients with PTSD access the health care system through primary or specialty medical treatment settings rather than directly seeking mental health assistance. There are several possible reasons for this. Coexistence of medical illness and PTSD may drive such requests for treatment. Physical symptoms associated with PTSD may be interpreted as a medical problem needing attention. And stigma against acknowledgment of mental distress may drive the initial choice of clinical treatment. Whatever the reason, within VA treatment settings, 11.5–36.0% of patients with PTSD seek primary rather than mental health care for their symptoms (Dobie et al., 2006; Magruder et al., 2004).

Attention to the relationship between medical and psychiatric illness was initiated with the study of depression in primary care settings. This led to the emergence of integrated primary–behavioral care treatment models, in which depression is identified and treated, initially, within the primary care setting. Both pharmacological interventions and CBT are utilized under such circumstances. There are now approximately 15 successful trials of integrated primary and behavioral health care for depression within primary care settings (see Dietrich et al., 2004). To date, there are no systematic investigations of such an approach for PTSD. Recognition of the importance of such research has prompted the recent initiation of a few pilot studies by the U.S. Army and the VA.

The final translation of science into practice with respect to this association concerns treatment strategies for people with comorbid PTSD and medical illness. If pathophysiological alterations associated with PTSD also affect symptom expression of the comorbid medical illness, effective treatment of PTSD might also ameliorate comorbid medical symptoms. For example, irritable bowel syndrome (IBS) is a medical disorder known to be exacerbated by stress and to be associated with PTSD (Blanchard, Keefer, Payne, Turner, & Galovski, 2002; Irwin et al., 1996). Therefore, it might be expected that among patients with comorbid PTSD and IBS, those who receive IBS treatment-as-usual will have significantly poorer outcomes than those who receive IBS treatment in conjunction with treatment for PTSD (Weaver, Nishith, & Resick, 1998). Should this prediction be confirmed, it might be tested further with other medical disorders comorbid with PTSD. Demonstration of better outcomes from concurrent medical–PTSD treatment might then influence general medical practice for physical health problems that occur in the context of a psychological condition. We offer this as but one of any number of research designs for developing evidence-based treatment algorithms for treating PTSD along with a comorbid medical condition. Other common, experimental approaches might also be used to investigate these interactions.

Question 13. What are the major priorities for research and practice concerning prevention and public health interventions following mass casualties and disasters?

Epidemiological research indicates that the vast majority of the population is resilient and will develop neither PTSD nor some other psychiatric syndrome following exposure to a mass casualty or disaster. It is also apparent that almost everyone will be upset during the immediate posttraumatic aftermath, so that distinguishing between vulnerable and resilient individuals during the immediate postimpact phase is very difficult. Thus, a "wellness" public health approach needs to focus on resilience, prevention, identification of populations at risk, early intervention, community (societal) interventions, and traditional clinical approaches for individual patients (Friedman, 2005).

Resilience may be expressed variably in genetic, molecular, behavioral, social, and other domains (see Layne, Warren, Watson, & Shalev, Chapter 24, this volume). Research with depressed children also suggests that vulnerability in the genetic domain (e.g., homozygosity for the short allele of the 5-HT transporter gene) may be offset by resilience elsewhere (e.g., social support; Kaufmann et al., 2004). Our understanding of resilience among people exposed to traumatic stress is at an early stage. A crucial imperative of such research is to move beyond traditional approaches identifying risk and protective factors (see Vogt, King, & King, Chapter 6, this volume) to discover dynamic biopsychosocial mechanisms that mediate or moderate resilience.

Discovery of such mechanisms could be translated into a public health strategy that might involve childhood screening (as in sickle-cell anemia), repeated monitoring (as in periodic health maintenance checkups for children and adults), routine skills training on coping with stress for all school children, and more intensive monitoring for at-risk populations (as with chest X-ray or serum cholesterol level monitoring as either a preventive measure or an indication to initiate preventive treatment; Friedman, 2002).

The goal of such a wellness-oriented preventive public health approach is twofold. First, given that more than half of the American population can anticipate exposure to at least one traumatic stressor during the course of their lives (Kessler et al., 1995), and even higher anticipated rates in nations subjected to war or internal conflict (de Jong et al., 2001), it makes sense to prepare the population-at-large as much as possible, before it is exposed to such events. Such a strategy is predicated on the expectation that although most people are resilient, psychological recovery from the impact of traumatic events can be accelerated by both enhancement of people's natural resilience and promotion of new strategies for coping with traumatic stress.

The second preventive public health goal is to identify individuals who may have serious deficiencies in resilience. Such individuals might benefit from acquisition of skills that compensate for deficiencies identified in advance. For example, genetic vulnerabilities might be offset by behavioral (e.g., reduced conditionability), social (e.g., increased capacity to obtain and utilize social support) or pharmacological (e.g., NPY enhancers) interventions.

Psychoeducation for the public-at-large may be an important preventive mental health strategy for resilient and vulnerable individuals alike. As with national smoking cessation initiatives, such an approach would provide the public with key information about what to anticipate following exposure to traumatic stress, how to distinguish between normal and abnormal posttraumatic reactions within themselves and among loved ones, what to do if such events occur, and what mental health resources might be available. Such information could be made available and accessible on the Internet, in naturalistic settings (schools, churches, workplaces, etc.), through public service announcements, and so forth.

The current war in Iraq is spawning a number of predeployment stress inoculation strategies that utilize well-established findings concerning psychological toughening, fear conditioning, and trauma-induced erroneous cognitions. Such approaches need to be evaluated systematically to demonstrate their efficacy; if they are effective, then we need to know for whom they are effective and how they are working. Demonstration of effectiveness in military settings should set in motion similar tests of stress inoculation among civilian cohorts. A good place to start might be with children, who will probably be exposed to urban or domestic violence, or with people who live in geographic areas where the probability of natural disasters is high. It is a very hopeful sign of the times that the trauma field has shifted from an exclusive interest in diagnosis and treatment of chronic PTSD to an interest in resilience and prevention.

Despite the best prevention in the world, traumatic stress does occur, so that a comprehensive public mental health strategy needs to extend beyond resilience building and prevention to early detection and intervention for people at risk to develop chronic posttraumatic problems. CBT for severely affected people (e.g., those with acute stress disorder), provided several weeks after traumatic exposure, appears to be very successful (Bryant, Moulds, & Nixon, 2003; see Litz & Maguen, Chapter 16, this volume). There remain unanswered questions about timing (how soon after the traumatic event), dosage (how much treatment), developmental, cultural, and other differences that need to be addressed systematically. Future research might address the issue of how to engage people in early intervention programs when their strong inclination will be to escape or avoid any reminders of the traumatic event and hope that they can forget or "just get over it." Most early intervention studies employ very small sample sizes and have struggled to recruit participants. Other barriers to early intervention include the pragmatic realities of some traumatic event settings that render psychological care irrelevant or at least a luxury for some time. For example, after Hurricane Katrina, safety, shelter, food, and finding loved ones were far more important tasks than relieving distress through counseling. Because postdisaster relief rightly focuses on the lower rungs of Maslow's hierarchy of needs, the window of opportunity for early intervention may pass. Nonetheless, even in the aftermath of the most devastating disasters, there are opportunities to manage people's short-term and intermediate reactions to their losses. Pairing basic psychological interventions with basic needs might well be the wave of the future for emergency management strategies. Such an approach is exemplified by psychological first aid (see below).

Tests of pharmacological intervention in the aftermath of disasters lag far behind psychosocial interventions, partly because many clinicians are hesitant to "pathologize" a posttraumatic reaction during the immediate postimpact phase from which untreated people may recover completely within a matter of days or weeks. This uncertainly is not only particularly relevant to the management of acute combat stress reactions in military settings, but it also applies to civilians acutely exposed to mass casualties or disasters.

Confronting such issues within a traditional clinical setting is little preparation for public mental health interventions for the population-at-large. As noted elsewhere (Friedman, 2005; Ritchie et al. 2006; see Watson, Gibson, & Ruzek, Chapter 25, this volume), intervention strategies need to be embedded within the existing social and community infrastructure, and institutions such as neighborhoods, schools, religious communities, workplace settings and different ethnocultural enclaves. The tools for implementing such approaches include social procedures and activities such as legislation, public safety, public education, family self-help networks, community outreach, Web-based information, public service announcements, and the media. Others have emphasized that a useful context for understanding such a multilayered, multifaceted approach is the inverted psychosocial pyramid, which emphasizes four levels of intervention: societal, community, family, and individual (de Jong, 2002; Green et al., 2003; Marsella, 1998).

As with psychosocial or pharmacological interventions, such approaches need to be tested rigorously with regard to the effectiveness with which they improve posttraumatic outcomes. Elsewhere, we have suggested that key measurable public health outcomes should be available to the general population; be relatively inexpensive; have a many pronged pre- and posttraumatic public education component; ameliorate widespread distress through effective posttraumatic risk communication; accelerate the timetable for normal recovery among resilient individuals who experience transient posttraumatic distress; provide effective outreach, especially to communities at greatest risk; empower families and communities to achieve recovery; and provide screening, referral, and therapeutic services for those requiring clinical intervention (Friedman, 2005).

It is heartening that a developmentally sensitive psychological first aid manual was finalized and made available during the immediate aftermath of Hurricane Katrina (Ritchie et al., 2006; see Watson et al., Chapter 25, this volume). This approach, developed jointly by the National Center for PTSD and the National Center for Child Traumatic Stress, is predicated on consensus opinions of leaders in both civilian and military settings, and extrapolation from RCTs with psychosocial interventions (see Watson et al., 2003). Unlike psychological debriefing that promotes emotional processing of very recent traumatic events and may be less effective or potentially harmful (see Litz & Maguen, Chapter 16, this volume), psychological first aid is a very pragmatic approach that emphasizes safety, security, communication, reunification with loved ones, psychoeducation, and information about available resources should clinical evaluation seem warranted. Although this approach appears to be a reasonable one, its effectiveness must be evaluated by empirical trials. Acknowledging that rigorous research on acute disaster mental health interventions can be a daunting challenge, there are a number of methodological approaches such as dismantling studies, paired cohort comparisons, utilization of ongoing surveillance databases (for quasi-prospective studies), and other strategies that might be easier to implement than randomized, controlled designs to evaluate the effectiveness of psychological first aid (see Norris, Friedman, Reisman, & Watson, 2006).

FINAL THOUGHTS

Twenty-five years of research and clinical experience support the validity of PTSD as a unique, prevalent, and potentially disabling psychiatric diagnosis. PTSD also provides a valuable scientific heuristic within which to understand the impact of traumatic stress at molecular, neurobiological, cognitive, behavioral, and sociocultural levels. Integration among these levels of analysis would be a welcome advance to the field. As we learn more about how traumatic stress affects these basic mechanisms, the next task is to understand how such alterations influence gene expression, brain function, psychological processes, and clinical abnormalities. The ultimate goal, however, is to translate such scientific findings into effective and widely disseminated evidence-based practices for people with PTSD and, whenever possible, to intervene early or even prevent onset of the disorder.

REFERENCES

Amdur, R. L., & Liberzon, I. (2001). The structure of posttraumatic stress disorder symptoms in combat veterans: A confirmatory factor analysis of the impact of event scale. *Journal of Anxiety Disorders, 15,* 345–357.

Andrews, G., Charney, D., Sirovatka, P., & Regier, D. (Eds.). (in press). *Stress-induced fear circuitry dis-*

orders: Refining the research agenda for DSM-V. Washington, DC: American Psychiatric Association.

Asmundson, G. J. G., Frombach, I., McQuaid, J., Pedrelli, P., Lenox, R., & Stein, M. B. (2000). Dimensionality of posttraumatic stress symptoms: A confirmatory factor analysis of DSM-IV symptom clusters and other symptom models. *Behaviour Research and Therapy, 38,* 203–214.

Blanchard, E. B., Keefer, L., Payne, A., Turner, S. M., & Galovski, T. E. (2002). Early abuse, psychiatric diagnoses and irritable bowel syndrome. *Behaviour Research and Therapy, 40,* 289–298.

Boscarino, J. A. (2006). Posttraumatic stress disorder and mortality among U.S. Army veterans 30 years after military service. *Annals of Epidemiology, 16,* 248–256.

Boscarino, J. A., & Chang, J. (1999). Electrocardiogram abnormalities among men with stress-related psychiatric disorders: Implications for coronary heart disease and clinical research. *Annals of Behavioral Medicine, 21,* 227–234.

Brady, K., Pearlstein, T., Asnis, G. M., Baker, D., Rothbaum, B., Sikes, C. R., et al. (2000). Efficacy and safety of sertraline treatment of posttraumatic stress disorder. *Journal of the American Medical Association, 283,* 1837–1844.

Brady, K. T., Dansky, B. S., Back, S. E., Foa, E. B., & Carroll, K. M. (2001). Exposure therapy in the treatment of PTSD among cocaine-dependent individuals: Preliminary findings. *Journal of Substance Abuse Treatment, 21,* 47–54.

Breslau, N., Lucia, V. C., & Davis, G. C. (2004). Partial PTSD versus full PTSD: An empirical examination of associated impairment. *Psychological Medicine, 34,* 1205–1214.

Bryant, R. A., Moulds, M. L., & Nixon, R. D. V. (2003). Cognitive therapy of acute stress disorder: A four-year follow-up. *Behaviour Research and Therapy, 41,* 489–494.

Buckley, T. C., Blanchard, E. B., & Hickling, E. J. (1998). A confirmatory factor analysis of posttraumatic stress symptoms. *Behaviour Research and Therapy, 36,* 1091–1099.

Chambers, R. A., Bremner, J. D., Moghaddam, B., Southwick, S., Charney, D. S., & Krystal, J. H. (1999). Glutamate and PTSD: Toward a psychobiology of dissociation. *Seminars in Clinical Neuropsychiatry, 4,* 274–281.

Charney, D. S. (2004). Psychobiological mechanisms of resilience and vulnerability: Implications for the successful adaptation to extreme stress. *American Journal of Psychiatry, 161,* 195–216.

Charney, D. S., Deutch, A. Y., Krystal, J. H., Southwick, S. M., & Davis, M. (1993). Psychobiologic mechanisms of posttraumatic stress disorder. *Archives of General Psychiatry, 50,* 295–305.

Ciccone, D. S., Elliott, D. K., Chandler, H. K., Navak, S., & Raphael, K. G. (2005). Sexual and physical abuse in women with fibromyalgia syndrome: A test of the trauma hypothesis. *Clinical Journal of Pain, 21,* 378–386.

Cox, B. J., Clark, I. P., & Enns, M. W. (2002). Posttraumatic stress disorder and the structure of common mental disorders. *Depression and Anxiety, 15,* 168–171.

Davidson, J. R., Rothbaum, B. O., van der Kolk, B. A., Sikes, C. R., & Farfel, G. M. (2001). Multicenter, double-blind comparison of sertraline and placebo in the treatment of posttraumatic stress disorder. *Archives of General Psychiatry, 58,* 485–492.

Davis, M., & Whalen, P. J. (2001). The amygdala: Vigilance and emotion. *Molecular Psychiatry, 1,* 13–34.

de Jong, J. T. V. M. (2002). Public mental health, traumatic stress and human rights violations in low-income countries: A culturally appropriate model in times of conflict, disaster and peace. In *Trauma, war and violence: Public mental health in sociocultural context* (pp. 1–91). New York: Kluwer Academic/Plenum Press.

de Jong, J. T. V. M., Komproe, I. H., Van Ommeren, M., El Masri, M., Mesfin, A., Khaled, N., et al. (2001). Lifetime events and posttraumatic stress disorder in four postconflict settings. *Journal of the American Medical Association, 286,* 555–562.

Dietrich, A. J., Oxman, T. E., Williams, J. W., Jr., Schulberg, H. C., Bruce, M. L., Lee, P. W., et al. (2004). Re-engineering systems for the treatment of depression in primary care: Cluster randomized controlled trial. *British Medical Journal, 329,* 602.

Dobie, D. J., Maynard, C., Kivlahan, D. R., Johnson, K. M., Simpson, T. L., David, A. C., et al. (2006). Posttraumatic stress disorder screening status is associated with increased VA medical and surgical utilization in women. *Journal of Internal Medicine, 21,* s58–s64.

Eisen, S. A., Kang, H. K., Murphy, F. M., Blanchard, M. S., Reda, D. J., Henderson, W. G., et al. (2005). Gulf War veterans' health: Medical evaluation of a U.S. cohort. *Annals of Internal Medicine, 142,* 881–890.

Feeny, N. C., Zoellner, L. A., & Foa, E. B. (2002). Treatment outcome for chronic PTSD among female assault victims with borderline personality characteristics: A preliminary examination. *Journal of Personality Disorders, 16*, 30–40.

Foa, E. B., Cahill, S. P., Boscarino, J. A., Hobfoll, S. E., Lahad, M., McNally, R. J., et al. (2005). Social, psychological, and psychiatric interventions following terrorist attacks: Recommendations for practice and research. *Neuropsychopharmacology, 30*, 1806–1817.

Foa, E. B., Dancu, C. V., Hembree, E. A., Jaycox, L. H., Meadows, E. A., & Street, G. P. (1995). A comparison of exposure therapy, stress inoculation training, and their combination for reducing post-traumatic stress disorder in female assault victims. *Journal of Consulting and Clinical Psychology, 67*, 194–200.

Foa, E. B., & Kozak, M. J. (1986). Emotional processing of fear: Exposure to corrective information. *Psychological Bulletin, 99*, 20–35.

Foa, E. B., Riggs, D. S., & Gershuny, B. S. (1995). Arousal, numbing and intrusion: Symptom structure of PTSD following assault. *American Journal of Psychiatry, 152*, 116–120.

Foa, E. B., Rothbaum, B. O., Riggs, D. S., & Murdock, T. B. (1991). Treatment of posttraumatic stress disorder in rape victims: A comparison of cognitive-behavioral procedures and counseling. *Journal of Consulting and Clinical Psychology, 59*, 715–723.

Friedman, M. J. (2002). Future pharmacotherapy for post-traumatic stress disorder: Prevention and treatment. *Psychiatric Clinics of North America, 25*, 427–441.

Friedman, M. J. (2005). Toward a public mental health approach to survivors of terrorism. *Journal of Aggression, Maltreatment, and Trauma, 10*, 527–539.

Friedman, M. J., & Karam, E. G. (in press). PTSD: Looking toward DSM-V and ICD-11. In G. Andrews, D. Charney, P. Sirovatka, & D. Regier (Eds.), *Stress-induced fear circuitry disorders: Refining the research agenda for DSM-V*. Washington, DC: American Psychiatric Association.

Friedman, M. J., & Marsella, A. J. (1996). Post-traumatic stress disorder: An overview of the concept. In A. J. Marsella, M. J. Friedman, E. T. Gerrity, & R. M. Scurfield (Eds.), *Ethnocultural aspects of post-traumatic stress disorder: Issues, research and applications* (pp. 11–32). Washington, DC: American Psychological Association.

Friedman, M. J., McDonagh-Coyle, A., Jalowiec, J. J., McHugo, G., & Wang, S. (2006). Urinary neurohormone levels in women with PTSD due to childhood sexual abuse: The effect of cognitive-behavioral therapy. *Journal of Clinical Psychiatry*.

Friedman, M. J., & McEwen, B. S. (2004). PTSD, allostatic load, and medical illness. In P. P. Schnurr & B. L. Green (Eds.), *Trauma and health: Physical health consequences of exposure to extreme stress* (pp. 157–188). Washington, DC: American Psychological Association.

Friedman, M. J., & Schnurr, P. P. (1995). The relationship between trauma and physical health. In M. J. Friedman, D. S. Charney, & A. Y. Deutch (Eds.), *Neurobiological and clinical consequences of stress: From normal adaptation to post-traumatic stress disorder* (pp. 507–526). Philadelphia: Lippincott–Raven.

Friedman, M. J., Schnurr, P. P., Sengupta, A., Holmes, T., & Ashcraft, M. (2004). The Hawaii Vietnam Veterans Project: Is minority status a risk factor for posttraumatic stress disorder? *Journal of Nervous and Mental Disease, 192*, 42–50.

Goldapple, K., Zindel, S., Garson, C., Lau, M., Bieling, P., Kennedy, S., et al. (2004). Modulation of cortical-limbic pathways in major depression: Treatment-specific effects of Cognitive Behavior Therapy. *Archives of General Psychiatry, 61*, 34–41.

Green, B. L., Friedman, M. J., de Jong, J., Solomon, S., Keane, T., Fairbank, J. A., et al. (2004). *Trauma interventions in war and peace: Prevention, practice, and policy*. Amsterdam: Kluwer Academic/Plenum Press.

Griffin, M. G., Nishith, P., Resick, P. A., & Yehuda, R. (1997). Integrating objective indicators of treatment outcome in posttraumatic stress disorder. *Annals of the New York Academy of Sciences, 821*, 388–409.

Guarnaccia, P. J., Canino, G. J., Rubio-Stipec, M., & Bravo, M. (1993). The prevalence of ataques de nervios in the Puerto Rico Disaster Study: The role of culture in psychiatric epidemiology. *Journal of Nervous and Mental Disease, 181*, 157–165.

Heber, R., Kellner, M., & Yehuda, R. (2002). Salivary cortisol levels and the cortisol response to dexamethasone before and after EMDR: A case report. *Journal of Clinical Psychology, 58*, 1521–1530.

Hyer, L. A., & Brandsma, J. M. (1997). EMDR minus eye movements equals good psychotherapy. *Journal of Traumatic Stress, 10*, 515–522.

Irwin, C., Falsetti, S. A., Lydiard, R. B., Ballenger, J. C., Brock, C. D., & Brener, W. (1996). Comorbidity of posttraumatic stress disorder and irritable bowel syndrome. *Journal of Clinical Psychiatry, 57*, 576–578.

Kang, H. K., Natelson, B. H., Mahan, C. M., Lee, K. Y., & Murphy, F. M. (2003). Post-traumatic stress disorder and chronic fatigue syndrome-like illness among Gulf War veterans: A population-based survey of 30,000 veterans. *American Journal of Epidemiology, 157*, 141–148.

Kaufman, J., Yang, B.-Z., Douglas-Palumberi, H., Houshyar, S., Lipschitz, D., Krystal, J. H., et al. (2004). Social supports and serotonin transporter gene moderate depression in maltreated children. *Proceedings of the National Academy of Sciences USA, 101*, 17316–17321.

Keane, T. M., Zimering, R. T., & Caddell, J. M. (1985). A behavioral formulation of posttraumatic stress disorder in Vietnam veterans. *Behavior Therapist, 8*, 9–12.

Kessler, R. C., Sonnega, A., Bromet, E., Hughes, M., & Nelson, C. B. (1995). Posttraumatic stress disorder in the National Comorbidity Survey. *Archives of General Psychiatry, 52*(12), 1048–1060.

Kilpatrick, D. G., Resnick, H. S., Freedy, J. R., Peleovitz, D., Resick, P. A., Roth, S. H., et al. (1998). Posttraumatic stress disorder field trial: Evaluation of the PTSD construct criteria A through E. In T. A. Widiger (Ed.), *DSM-IV sourcebook* (Vol. 4, pp. 803–838). Washington, DC: American Psychiatric Association.

Kilpatrick, D. G., Ruggiero, K. J., Acierno, R., Saunders, B. E., Resnick, H. S., & Best, C. L. (2003). Violence and risk of PTSD, major depression, substance abuse/dependence, and comorbidity: Results from the National Survey of Adolescents. *Journal of Consulting and Clinical Psychology, 71*(4), 692–700.

King, D. W., Leskin, G. A., & King, L. A. (1998). Confirmatory factor analysis of the Clinician-Administered PTSD scale: Evidence for the dimensionality of posttraumatic stress disorder. *Psychological Assessment, 10*, 90–96.

Kolb, L. C. (1989). Heterogeneity of PTSD [Letter]. *American Journal of Psychiatry, 146*, 811–812.

Kulka, R. A., Schlenger, W. E., Fairbank, J. A., Hough, R. L., Jordan, K. B., Marmar, C. R., et al. (1990). *Trauma and the Vietnam War generation: Report of findings from the National Vietnam Veterans Readjustment Study*. New York: Brunner/Mazel.

Lang, P. J. (1977). Imagery in therapy: An information processing analysis of fear. *Behavior Therapy, 8*, 862–886.

Lohr, J. M., Tolin, D. F., & Lilienfeld, S. O. (1998). Efficacy of eye movement, desensitization, and reprocessing: Implications for behavior therapy. *Behavior Therapy, 29*, 123–156.

Magruder, K. M., Frueh, B. C., Knapp, R. G., Johnson, M. R., Vaughan, J. A., Carson, T. C., et al. (2004). PTSD symptoms, demographic characteristics, and functional status among veterans treated in VA primary care clinics. *Journal of Traumatic Stress, 17*, 293–301.

Marsella, A. J. (1998). Toward a "global-community psychology": Meeting the needs of a changing world. *American Psychologist, 53*, 1282–1291.

Marshall, R. D., Beebe, K. L., Oldham, M., & Zaninelli, R. (2001). Efficacy and safety of paroxetine treatment for chronic PTSD: A fixed-dose-placebo-controlled study. *American Journal of Psychiatry, 158*, 1982–1988.

McDonagh-Coyle, A. S., Friedman, M. J., McHugo, G., Ford, J., Sengupta, A., Mueser, K., et al. (2005). Randomized trial of cognitive behavioral therapy for chronic PTSD. *Journal of Clinical and Consulting Psychiatry, 73*, 515–524.

Miller, M. W., Kaloupek, D. G., Dillon, A. L., & Keane, T. M. (2004). Externalizing and internalizing subtypes of combat-related PTSD: A replication and extension using the PSY-5 scales. *Journal of Abnormal Psychology, 113*, 636–645.

Mylle, J., & Maes, M. (2004). Partial posttraumatic stress disorder revisited. *Journal of Affective Disorders, 78*, 37–48.

Najavits, L. M. (2004). Treatment of posttraumatic stress disorder and substance abuse: Clinical guidelines for implementing "Seeking Safety" therapy. *Alcoholism Treatment Quarterly, 22*, 43–62.

Nixon, R. D. V. Resick, P. A., & Nishith, P. (2004). An exploration of comorbid depression among female victims of intimate partner violence with posttraumatic stress disorder. *Journal of Affective Disorder, 82*(2), 315–320.

Norris, F. H., Friedman, M. J., Reisman, D., & Watson, P. J. (2006). *Clinical research in the wake of disasters and terrorism.* New York: Guilford Press.

Norris, F. H., Murphy, A. D., Baker, C. K., & Perilla, J. L. (2003). Severity, timing and duration of reactions to trauma in the population: An example from Mexico. *Biological Psychiatry, 53,* 769–778.

North, C. S., Pfefferbaum, B., Narayanan, P., Thielman, S. B., McCoy, G., Dumont, C. E., et al. (2005). Comparison of post-disaster psychiatric disorders after terrorist bombings in Nairobi and Oklahoma City. *British Journal of Psychiatry, 186,* 487–493.

Palinkas, L. A. (2006). Qualitative approaches to studying the effects of disasters. In F. Norris, S. Galea, M. J. Friedman, & P. Watson (Eds.), *Research methods for studying mental health and disasters* (pp. 158–173). New York: Guilford Press.

Peters, L., Slade, T., & Andrews, G. (1999). A comparison of ICD-10 and DSM-IV criteria for posttraumatic stress disorder. *Journal of Traumatic Stress, 12,* 335–343.

Pitman, R. K., Orr, S. P., Altman, B., Longpre, R. E., Poiré, R. E., & Macklin, M. L. (1996). Emotional processing during eye movement desensitization and reprocessing therapy of Vietnam veterans with chronic posttraumatic stress disorder. *Comprehensive Psychiatry, 37,* 419–429.

Pitman, R., Sanders, K. M., Zusman, R. M., Healy, A. R., Cheema, F., Lasko, N. B., et al. (2002). Pilot study of secondary prevention of posttraumatic stress disorder with propranolol. *Biological Psychiatry, 51,* 189–192.

Pizarro, J., Silver, R. C., & Prause, J. (2006). Physical and mental health costs of traumatic war experiences among Civil War veterans. *Archives of General Psychiatry, 63,* 193–200.

Raphael, K. G., Janal, M. N., & Navak, S. (2004). Comorbidity of fibromyalgia and posttraumatic stress disorder symptoms in a community sample of women. *Pain Medicine, 5,* 33–41.

Renfrey, G. S., & Spates, C. R. (1994). Eye movement desensitization: A partial dismantling study. *Journal of Behavior Therapy and Experimental Psychiatry, 25,* 231–239.

Resick, P. A., Nishith, P., Weaver, T. L., Astin, M. C., & Feuer, C. A. (2002). A comparison of cognitive processing therapy with prolonged exposure and a waiting condition for the treatment of chronic posttraumatic stress disorder in female rape victims. *Journal of Consulting and Clinical Psychology, 70,* 867–879.

Ritchie, E. C., Friedman, M. J., & Watson, P. J. (2005). *Interventions following mass violence and disasters: Strategies for mental health practice.* New York: Guilford Press.

Robert, R., Blakeney, P. E., Villarreal, C., Rosenberg, L., & Meyer, W. J., III. (1999). Imipramine treatment in pediatric burn patients with symptoms of acute stress disorder: A pilot study. *Journal of the American Academy of Child and Adolescent Psychiatry, 38,* 873–882.

Rosenberg, S. D., Mueser, K. T., Jankowski, M. K., Salyers, M. P., & Acker, K. (2004). Cognitive-Behavioral Treatment of PTSD in severe mental illness: Results of a pilot study. *American Journal of Psychiatric Rehabilitation, 7,* 171–186.

Schelling, G., Briegel, J., Roozendaal, B., Stoll, C., Rothenhäusler, H-B., & Kapfhammer, H.-P. (2001). The effect of stress doses of hydrocortisone during septic shock on posttraumatic stress disorder in survivors. *Biological Psychiatry, 50,* 978–985.

Schnurr, P. P., Ford, J. D., Friedman, M. J., Green, B. L., Dain, B. J., & Sengupta, A. (2000). Predictors and outcomes of posttraumatic stress disorder in World War II veterans exposed to mustard gas. *Journal of Consulting and Clinical Psychology, 68,* 258–268.

Schnurr, P. P., Friedman, M. J., Engel, C. C., Chow, B., Foa, E. B., Resick, P. A., et al. (2007). Randomized clinical trial of cognitive-behavioral therapy for posttraumatic stress disorder in women in VA and DoD settings. *Journal of the American Medical Association, 297,* 820–830.

Schnurr, P. P., & Green, B. L. (Eds.). (2004). *Trauma and health: Physical health consequences of exposure to extreme stress.* Washington, DC: American Psychological Association.

Schnurr, P. P., & Jankowski, M. K. (1999). Physical health and post-traumatic stress disorder: Review and synthesis. *Seminars in Clinical Neuropsychiatry, 4,* 295–304.

Schnurr, P. P., Spiro, A., Vielhauer, M. J., Findler, M. N., & Hamblen, J. L. (2002). Trauma in the lives of older men: findings from the Normative Aging Study. *Journal of Clinical Geropsychology, 8,* 175–187.

Shapiro, F., & Maxfield, L. (2002). Eye movement desensitization and reprocessing (EMDR): Information processing in the treatment of trauma. *Journal of Clinical Psychology, 58,* 933–946.

Sharkansky, E. J., Brief, D. J., Peirce, J. M., Meehan, J. C., & Mannix, L. M. (1999). Substance abuse patients with posttraumatic stress disorder (PTSD): Identifying specific triggers of substance use and their associations with PTSD symptoms. *Psychology of Addictive Behaviors, 13,* 89–97.

Simms, L. J., Watson, D., & Doebbeling, B. N. (2002). Confirmatory factor analyses of posttraumatic stress symptoms in deployed and nondeployed veterans of the Gulf War. *Journal of Abnormal Psychology, 111,* 637–647.

Southwick, S. M., Krystal, J. H., Bremner, J. D., Morgan, C. A., Nicolaou, A. L., Nagy, L. M., et al. (1997). Noradrenergic and serotonergic function in posttraumatic stress disorder. *Archives of General Psychiatry, 54,* 749–758.

Taylor, S., Kuch, K., Koch, W. J., Crockett, D. J., & Passey, G. (1998). The structure of posttraumatic stress symptoms. *Journal of Abnormal Psychology, 107,* 154–160.

Tucker, P., Zaninelli, R., Yehuda, R., Ruggiero, L., Dillingham, K., & Pitts, C. D. (2001). Paroxetine in the treatment of chronic posttraumatic stress disorder: Results of a placebo-controlled, flexible-dosage trial. *Journal of Clinical Psychiatry, 62,* 860–868.

U.S. Food and Drug Administration. (2004). *Worsening depression and suicidality in patients being treated with antidepressant medications.* Washington, DC: Author.

Vermetten, E., & Bremner, J. D. (2002). Circuits and systems in stress: II. Applications to neurobiology and treatment in posttraumatic stress disorder. *Depression and Anxiety, 16,* 14–38.

Watson, P. J., Friedman, M. J., Gibson, L., Ruzek, J. I., Norris, F., & Ritchie, E. C. (2003). Early intervention for trauma-related problems. In R. Ursano & A. E. Norwood (Eds.), *Trauma and disaster responses and management.* Washington, DC: American Psychiatric Press.

Weathers, F. W., Ruscio, A. M., & Keane, T. M. (1999). Psychometric properties of nine scoring rules for the Clinician-Administered Posttraumatic Stress Disorder scale. *Psychological Assessment, 11,* 124–133.

Weaver, T. L., Nishith, P., & Resick, P. A. (1998). Prolonged exposure therapy and irritable bowel syndrome: A case study examining the impact of a trauma-focused treatment on a physical condition. *Cognitive and Behavioral Practice, 5,* 103–122.

Weiss, D. S., Marmar, C. R., Schlenger, W. E., Fairbank, S. A., Jordan, K. B., Hough, R. L., et al. (1992). The prevalence of lifetime and partial post-traumatic stress disorder in Vietnam theater veterans. *Journal of Traumatic Stress, 5,* 365–376.

Woodward, S. H., Kaloupek, D. G., Streeter, C. C., Martinez, C., Schaer, M., & Eliez, S. (2006). Decreased anterior cingulate volume in combat-related PTSD. *Biological Psychiatry, 59,* 582–587.

Zlotnick, C., Rodriguez, B. F., Weisberg, R. B., Bruce, S. E., Spencer, M. A., Culpepper, L., et al. (2004). Chronicity in posttraumatic stress disorder and predictors of the course of posttraumatic stress disorder among primary care patients. *Journal of Nervous and Mental Disease, 192,* 153–159.

Author Index

Subject Index

DATE DUE